KUBY
Immunology

Icons Used in This Book

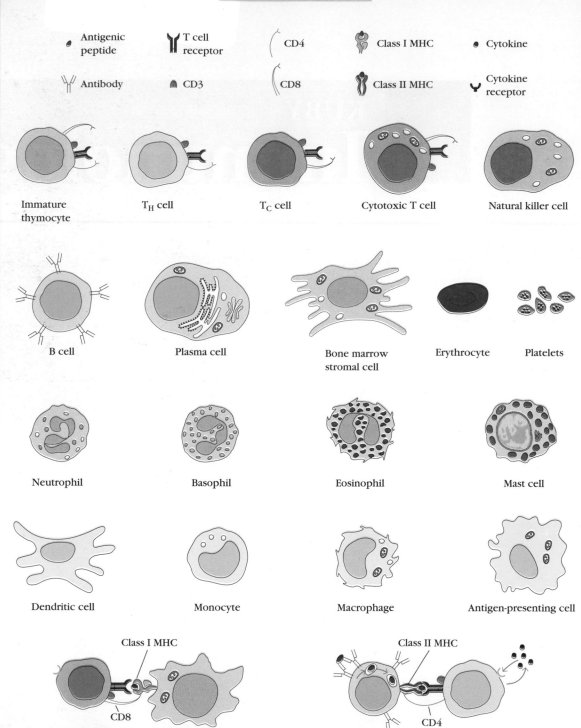

Antigenic peptide

T cell receptor

CD4

Class I MHC

Cytokine

Antibody

CD3

CD8

Class II MHC

Cytokine receptor

Immature thymocyte

T_H cell

T_C cell

Cytotoxic T cell

Natural killer cell

B cell

Plasma cell

Bone marrow stromal cell

Erythrocyte

Platelets

Neutrophil

Basophil

Eosinophil

Mast cell

Dendritic cell

Monocyte

Macrophage

Antigen-presenting cell

Class I MHC

T_C cell

CD8

Altered self cell

Class II MHC

B cell

CD4

T_H cell

KUBY
Immunology

Judith A. Owen
Haverford College

Jenni Punt
Haverford College

Sharon A. Stranford
Mount Holyoke College

with contributions by
Patricia P. Jones
Stanford University

Seventh Edition

W. H. Freeman and Company • New York

Publisher: Susan Winslow

Senior Acquisitions Editor: Lauren Schultz

Associate Director of Marketing: Debbie Clare

Marketing Assistant: Lindsay Neff

Developmental Editor: Erica Champion

Developmental Editor: Irene Pech

Developmental Coordinator: Sara Ruth Blake

Associate Media Editor: Allison Michael

Supplements Editor: Yassamine Ebadat

Senior Project Manager at Aptara: Sherrill Redd

Photo Editor: Christine Buese

Photo Researcher: Elyse Reider

Art Director: Diana Blume

Text Designer: Marsha Cohen

Illustrations: Imagineering

Illustration Coordinator: Janice Donnola

Production Coordinator: Lawrence Guerra

Composition: Aptara®, Inc.

Printing and Binding: RR Donnelley

Library of Congress Control Number: 2012950797

North American Edition
Cover image:
©2009 Pflicke and Sixt. Originally published in
The Journal of Experimental Medicine. 206:2925-2935.
doi:10.1084/jem.20091739.
Image provided by Holger Pflicke and Michael Sixt.

International Edition
Cover design: Dirk Kaufman
Cover image: Nastco/iStockphoto.com

North American Edition
ISBN-13: 978-14292-1919-8
ISBN-10: 1-4292-1919-X

International Edition
ISBN-13: 978-14641-3784-6
ISBN-10: 1-4641-3784-6

Printed in the United States of America

First printing

North American Edition
W. H. Freeman and Company
41 Madison Avenue
New York, NY 10010
www.whfreeman.com

International Edition
Macmillan Higher Education
Houndmills, Basingstoke
RG21 6XS, England
www.macmillanhighered.com/international

To all our students, fellows, and colleagues who have made our careers in immunology a source of joy and excitement, and to our families who made these careers possible. We hope that future generations of immunology students will find this subject as fascinating and rewarding as we have.

About the Authors

All four authors are active scholars and teachers who have been/are recipients of research grants from the NIH and the NSF. We have all served in various capacities as grant proposal reviewers for NSF, NIH, HHMI, and other funding bodies as well as evaluating manuscripts submitted for publication in immunological journals. In addition, we are all active members of the American Association of Immunologists and have served our national organization in a variety of ways.

Judy Owen holds B.A. and M.A. (Hons) degrees from Cambridge University. She pursued her Ph.D. at the University of Pennsylvania with the late Dr. Norman Klinman and her post-doctoral fellowship with Dr. Peter Doherty in viral immunology. She was appointed to the faculty of Haverford College, one of the first undergraduate colleges to offer a course in immunology, in 1981. She teaches numerous laboratory and lecture courses in biochemistry and immunology and has received several teaching and mentorship awards. She is a participant in the First Year Writing Program and has been involved in curriculum development across the College.

Jenni Punt received her A.B. from Bryn Mawr College (*magna cum laude*) majoring in Biology at Haverford College, She received her VMD (*summa cum laude*) and Ph.D. in immunology from the University of Pennsylvania and was a Damon Runyon-Walter Winchell Physician-Scientist fellow with Dr. Alfred Singer at the National Institutes of Health. She was appointed to the faculty of Haverford College in 1996 where she teaches cell biology and immunology and performs research in T cell development and hematopoiesis. She has received several teaching awards and has contributed to the development of college-wide curricular initiatives.

Together, Jenni Punt and Judy Owen developed and ran the first AAI Introductory Immunology course, which is now offered on an annual basis.

Sharon Stranford obtained her B.A. with Honors in Biology from Arcadia University and her Ph.D. in Microbiology and Immunology from Hahnemann (now Drexel) University, where she studied autoimmunity with funding from the Multiple Sclerosis Foundation. She pursued postdoctoral studies in transplantation immunology at Oxford University in England, followed by a fellowship at the University of California, San Francisco, working on HIV/AIDS with Dr. Jay Levy. From 1999 to 2001, Sharon was a Visiting Assistant Professor of Biology at Amherst College, and in 2001 joined the faculty of Mount Holyoke College as a Clare Boothe Luce Assistant Professor. She teaches courses in introductory biology, cell biology, immunology, and infectious disease, as well as a new interdisciplinary course called Controversies in Public Health.

Pat Jones graduated from Oberlin College in Ohio with Highest Honors in Biology and obtained her Ph.D. in Biology with Distinction from the Johns Hopkins University. She was a postdoctoral fellow of the Arthritis Foundation for two years in the Department of Biochemistry and Biophysics at the University of California, San Francisco, Medical School, followed by two years as an NSF postdoctoral fellow in the Departments of Genetics and Medicine/Immunology at Stanford University School of Medicine. In 1978 she was appointed Assistant Professor of Biology at Stanford and is now a full professor. Pat has received several undergraduate teaching awards, was the founding Director of the Ph.D. Program in Immunology, and in July, 2011, she assumed the position of Director of Stanford Immunology, a position that coordinates activities in immunology across the university.

Contents

Benjamin F. Bravo

Chapter 3

Receptors and Signaling: B and T-Cell Receptors   65

Chapter 4

Receptors and Signaling: Cytokines and Chemokines   105

Chapter 5

Innate Immunity   141

Chapter 6

The Complement System   187

Chapter 7

The Organization and Expression of Lymphocyte Receptor Genes   225

Chapter 8

The Major Histocompatibility Complex and Antigen Presentation   261

The Structure and Function of MHC Molecules   262

Chapter 9

T-Cell Development 299

Chapter 10

B-Cell Development 329

Chapter 11

T-Cell Activation, Differentiation, and Memory   357

Chapter 12

B-Cell Activation, Differentiation, and Memory Generation   385

Chapter 13

Effector Responses: Cell-and Antibody-Mediated Immunity   415

Chapter 14

The Immune Response in Space and Time   451

Chapter 15

Allergy, Hypersensitivities, and Chronic Inflammation   485

Chapter 16

Tolerance, Autoimmunity, and Transplantation 517

Chapter 17

Infectious Diseases and Vaccines 553

Chapter 18

Immunodeficiency Disorders 593

Chapter 19

Cancer and the Immune System 627

Chapter 20

Experimental Systems and Methods   653

Appendix I

CD Antigens   A-1

Appendix II

Cytokines   B-1

Appendix III

Chemokines and Chemokine Receptors   C-1

Feature Boxes in Kuby 7e

Preface

Like all of the previous authors of this book, we are dedicated to the concept that immunology is best taught and learned in an experimentally-based manner, and we have retained that emphasis with this edition. It is our goal that students should complete an immunology course not only with a firm grasp of content, but also with a clear sense of *how* key discoveries were made, what interesting questions remain, and how they might best be answered. We believe that this approach ensures that students both master fundamental immunological concepts and internalize a vision of immunology as an active and ongoing process. Guided by this vision, the new edition has been extensively updated to reflect the recent advances in all aspects of our discipline.

New Authorship

As a brand-new team of authors, we bring experience in both research and undergraduate teaching to the development of this new edition, which continues to reflect a dedication to pedagogical excellence originally modeled by Janis Kuby. We remain deeply respectful of Kuby's unique contribution to the teaching of immunology and hope and trust that this new manifestation of her creation will simply add to her considerable legacy.

Understanding Immunology As a Whole

We recognize that the immune system is an integrated network of cells, molecules, and organs, and that each component relies on the rest to function properly. This presents a pedagogical challenge because to understand the whole, we must attain working knowledge of many related pieces of information, and these do not always build upon each other in simple linear fashion. In acknowledgment of this challenge, this edition presents the "big picture" twice; first as an introductory overview to immunity, then, thirteen chapters later, as an integration of the details students have learned in the intervening text.

Specifically, Chapter 1 has been revised to make it more approachable for students who are new to immunology. The chapter provides a short historical background to the field and an introduction to some of the key players and their roles in the immune response, keeping an eye on fundamental concepts (Overview Figure 1-9). A new section directly addresses some of the biggest conceptual hurdles, but leaves the cellular and molecular details for later chapters.

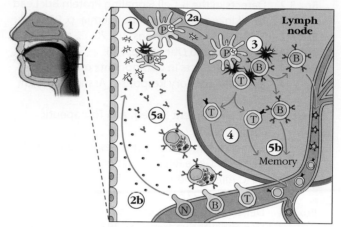

OVERVIEW FIGURE 1-9 Collaboration between innate and adaptive immunity in resolving an infection.

A new capstone chapter (Chapter 14) integrates the events of an immune response into a complete story, with particular reference to the advanced imaging techniques that have become available since the writing of the previous edition. In this way, the molecular and cellular details presented in Chapters 2-13 are portrayed in context, a moving landscape of immune response events in time and space (Figure 14-5).

FIGURE 14-5 A T cell (blue) on a fibroblastic reticular network (red and green) in the lymph node.

Focus on the Fundamentals

The order of chapters in the seventh edition has been revised to better reflect the sequence of events that occurs naturally during an immune response in vivo. This offers instructors the opportunity to lead their students through the steps of an immune response in a logical sequence, once they have learned the essential features of the tissues, cells, molecular structures, ligand-receptor binding interactions, and signaling pathways necessary for the functioning of the immune system. The placement of innate immunity at the forefront of the immune response enables it to take its rightful place as the first, and often the only, aspect of immunity that an organism needs to counter an immune insult. Similarly, the chapter on complement is located within the sequence in a place that highlights its function as a bridge between innate and adaptive immune processes. However, we recognize that a course in immunology is approached differently by each instructor. Therefore, as much as possible, we have designed each of the chapters so that it can stand alone and be offered in an alternative order.

Challenging All Levels

While this book is written as a text for students new to immunology, it is also our intent to challenge students to reach deeply into the field and to appreciate the connections with other aspects of biology. Instead of reducing difficult topics to vague and simplistic forms, we instead present them with the level of detail and clarity necessary to allow the beginning student to find and understand information they may need in the future. This offers the upper level student a foundation from which they can progress to the investigation of advances and controversies within the current immunological literature. Supplementary focus boxes have been used to add nuance or detail to discussions of particular experiments or ideas without detracting from the flow of information. These boxes, which address experimental approaches, evolutionary connections, clinical aspects, or advanced material, also allow instructors to tailor their use appropriately for individual courses. They provide excellent launching points for more intensive in class discussions relevant to the material.

Some of the most visible changes and improvements include:

- A rewritten chapter on the cells and organs of the immune system (Chapter 2) that includes up to date images reflecting our new understanding of the microenvironments where the host immune system develops and responds.

- The consolidation of signaling pathways into two chapters: Chapter 3 includes a basic introduction to ligand:receptor interactions and principles of receptor signaling, as well as to specific molecules and pathways involved in signaling through antigen receptors. Chapter 4 includes a more thorough introduction to the roles of cytokines and chemokines in the immune response.

- An expanded and updated treatment of innate immunity (Chapter 5), which now includes comprehensive coverage of the many physical, chemical, and cellular defenses that constitute the innate immune system, as well as the ways in which it activates and regulates adaptive immunity.

- Substantial rewriting of chapters concerned with complement (Chapter 6) and antigen receptor gene rearrangement (Chapter 7). These chapters have been extensively revised for clarity in both text and figures. The description of the complement system has been updated to include the involvement of complement proteins in both innate and adaptive aspects of immunity.

- A restructured presentation of the MHC, with the addition of new information relevant to cross-presentation pathways (Chapter 8) (Figure 8-22b).

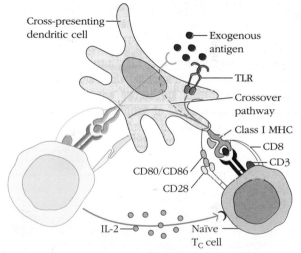

(b) DC cross-presentation and activation of CTL

FIGURE 8-22b Exogenous antigen activation of naïve T_C cells requires DC licensing and cross-presentation

- The dedication of specialized chapters concerned with T cell development and T cell activation (Chapters 9 and Chapter 11, respectively). Chapter 11 now includes current descriptions of the multiple helper T cell subsets that regulate the adaptive immune response.

- Substantially rewritten chapters on B cell development and B cell activation (Chapters 10 and 12, respectively) that address the physiological locations as well as the nature of the interacting cells implicated in these processes.

- An updated discussion of the role of effector cells and molecules in clearing infection (Chapter 13), including a more thorough treatment of NK and NKT cells.

- A new chapter that describes advances in understanding and visualizing the dynamic behavior and activities of immune cells in secondary and tertiary tissue (Chapter 14).

- Substantial revision and updating of the clinical chapters (Chapters 15-19) including the addition of several new clinically relevant focus boxes.

- Revised and updated versions of the final methods chapter (Chapter 20), and the appendices of CD antigens, chemokines, and cytokines and their receptors.

Throughout the book, we attempt to provide a "big picture" context for necessary details in a way that facilitates greater student understanding.

Recent Advances and Other Additions

Immunology is a rapidly growing field, with new discoveries, advances in techniques, and previously unappreciated connections coming to light every day. The 7th edition has been thoroughly updated throughout, and now integrates the following new material and concepts:

- New immune cell types and subtypes, as well as the phenotypic plasticity that is possible between certain subtypes of immune cells.

- A greater appreciation for the wide range of mechanisms responsible for innate immunity and the nature and roles of innate responses in sensing danger, inducing inflammation, and shaping the adaptive response (Figure 5-18).

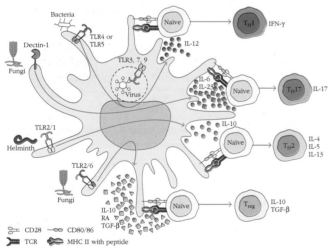

FIGURE 5-18 Differential signaling through dendritic cell PRRs influences helper T cell functions.

- Regulation of immunity, including new regulatory cell types, immunosuppressive chemical messengers and the roles these play, for example, in tolerance and in the nature of responses to different types of antigens (Figure 9-10).

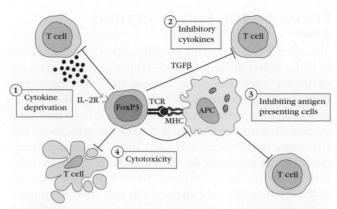

FIGURE 9-10 How regulatory T cells inactivate traditional T cells.

- The roles of the microbiome and commensal organisms in the development and function of immunity, as well as the connections between these and many chronic diseases.

- A new appreciation for the micro environmental substructures that guide immune cell interactions with antigen and with one another (Figure 14-11a).

Antigen delivery to T cells

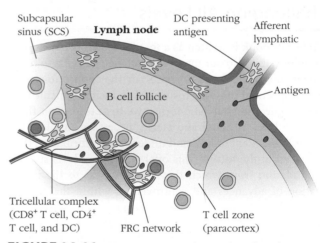

FIGURE 14-11a How antigen travels into a lymph node.

- Many technical advances, especially in the areas of imaging and sequencing, which have collectively enhanced our understanding of immune function and cellular interactions, allowing us to view the immune response in its natural anatomical context, and in real time (see Figure 14-5).

Connections to the Bench, the Clinic, and Beyond

We have made a concerted effort in the 7th edition to integrate experimental and clinical aspects of immunology into the text. In Chapter 2, illustrations of immune cells and tissues are shown alongside histological sections or, where possible, electron micrographs, so students can see what they actually look like. Throughout the text, experimental data are used to demonstrate the bases for our knowledge (Figure 3-4b), and the clinical chapters at the end of the book (Chapters 15 through 19) describe new advances, new challenges, and newly appreciated connections between the immune system and disease.

FIGURE 3-4b Targeted delivery of cytokines (pink).

Featured Boxes

Associated with each chapter are additional boxed materials that provide specialized information on historically-important studies (Classic Experiments) that changed the way immunologists viewed the field, noteworthy new breakthroughs (Advances) that have occurred since the last edition, the clinical relevance of particular topics (Clinical Focus) and the evolution of aspects of immune functioning (Evolution). Examples of such boxes are "The Prime and Pull Vaccine strategy," "Genetic defects in components of innate and inflammatory responses associated with disease," "The role of miRNAs in the control of B cell development" and an updated "Stem cells: Clinical uses and potential." We have involved our own undergraduate students in the creation of some of these boxes, which we believe have greatly benefitted from their perspective on how to present interesting material effectively to their fellow students.

Critical Thinking and Data Analysis

Integration of experimental evidence throughout the book keeps students focused on the how and why. Detailed and clear descriptions of the current state of the field provide students with the knowledge, skills, and vocabulary to read critically in the primary literature. Updated and revised study questions at the end of the chapter range from simple recall of information to analyzing original data or proposing hypotheses to explain remaining questions in the field. Classic Experiment boxes throughout the text help students to appreciate the seminal experiments in immunology and how they were conducted, providing a bridge to the primary research articles and emphasizing data analysis at every step.

Media and Supplements

NEW! ImmunoPortal (courses.bfwpub.com/immunology7e)
   This comprehensive and robust online teaching and learning tool combines a wealth of media resources, vigorous assessment, and helpful course management features into one convenient, fully customizable space.

ImmunoPortal Features:

NEW! Kuby Immunology Seventh Edition e-Book—also available as a standalone resource (ebooks.bfwpub.com/immunology7e)
   This online version of the textbook combines the contents of the printed book, electronic study tools, and a full complement of student media, including animations and videos. Students can personalize their e-Book with highlighting, bookmarking, and note-taking features. Instructors can customize the e-Book to focus on specific sections, and add their own notes and files to share with their class.

NEW! LearningCurve—A Formative Quizzing Engine

With powerful adaptive quizzing, a game-like format, and the promise of significantly better grades, *LearningCurve* gives instructors a quickly implemented, highly effective new way to get students more deeply involved in the classroom. Developed by experienced teachers and experts in educational technology, *LearningCurve* offers a series of brief, engaging activities specific to your course. These activities put the concept of "testing to learn" into action with adaptive quizzing that treats each student as an individual with specific needs:

- Students work through *LearningCurve* activities one question at a time.

- With each question, students get immediate feedback. Responses to incorrect answers include links to book sections and other resources to help students focus on what they need to learn.

- As they proceed toward completion of the activity, the level of questioning adapts to the level of performance. The questions become easier, harder, or the same depending on how the student is doing.

- And with a more confident understanding of assigned material, students will be more actively engaged during classtime.

Resources

The Resources center provides quick access to all instructor and student resources for Kuby Immunology.

For Instructors—

All instructor media are available in the ImmunoPortal and on the Instructor Resource DVD.

NEW! test bank—over 500 dynamic questions in PDF and editable Word formats include multiple-choice and short-answer problems, rated by level of difficulty and Bloom's Taxonomy level.

Fully optimized JPEG files of every figure, photo, and table in the text, featuring enhanced color, higher resolution, and enlarged fonts. Images are also offered in PowerPoint® format for each chapter.

Animations of complex text concepts and figures help students better understand key immunological processes.

Videos specially chosen by the authors to complement and supplement text concepts.

For Students—

All of these resources are also available in the ImmunoPortal.

- Student versions of the Animations and Videos, to help students understand key mechanisms and techniques at their own pace.

- Flashcards test student mastery of vocabulary and allow students to tag the terms they've already learned.
- Immunology on the Web weblinks introduce students to a world of online immunology resources and references.

Assignments

In this convenient space, ImmunoPortal provides instructors with the ability to assign any resource, as well as e-Book readings, discussion board posts, and their own materials. A gradebook tracks all student scores and can be easily exported to Excel or a campus Course Management System.

Acknowledgements

We owe special thanks to individuals who offered insightful ideas, who provided detailed reviews that led to major improvements, and who provided the support that made writing this text possible. These notable contributors include Dr. Stephen Emerson, Dr. David Allman, Dr. Susan Saidman, Dr. Nan Wang, Nicole Cunningham, and the many undergraduates who provided invaluable students' perspectives on our chapters. We hope that the final product reflects the high quality of the input from these experts and colleagues and from all those listed below who provided critical analysis and guidance.

We are also grateful to the previous authors of Kuby's Immunology, whose valiant efforts we now appreciate even more deeply. Their commitment to clarity, to providing the most current material in a fast moving discipline, and to maintaining the experimental focus of the discussions set the standard that is the basis for the best of this text.

We also acknowledge that this book represents the work not only of its authors and editors, but also of all those whose experiments and writing provided us with ideas, inspiration and information. We thank you and stress that all errors and inconsistencies of interpretation are ours alone.

We thank the following reviewers for their comments and suggestions about the manuscript during preparation of this seventh edition. Their expertise and insights have contributed greatly to the book.

Lawrence R. Aaronson, *Utica College*
Jeffrey K. Actor, *University of Texas Medical School at Houston*
Richard Adler, *University of Michigan-Dearborn*
Emily Agard, *York University, North York*

Karthik Aghoram, *Meredith College*
Rita Wearren Alisauskas, *Rutgers University*
John Allsteadt, *Virginia Intermont College*
Gaylene Altman, *University of Washington*
Angelika Antoni, Kutztown University
Jorge N. Artaza, Charles R. *Drew University of Medicine and Science*
Patricia S. Astry, *SUNY Fredonia*
Roberta Attanasio, *Georgia State University*
Elizabeth Auger, *Saint Joseph's College of Maine*
Avery August, *Penn State University*
Rajeev Aurora, *Saint Louis University Hospital*
Christine A. Bacon, *Bay Path College*
Jason C. Baker, *Missouri Western State College*
Kenneth Balazovich, *University of Michigan-Dearborn*
Jennifer L. Bankers-Fulbright, *Augsburg College*
Amorette Barber, *Longwood University*
Brianne Barker, *Hamilton College*
Scott R. Barnum, *University of Alabama at Birmingham*
Laura Baugh, *University of Dallas*
Marlee B. Marsh, *Columbia College*
Rachel Venn Beecham, *Mississippi Valley State University*
Fabian Benencia, *Ohio University Main Campus*
Charlie Garnett Benson, *Georgia State University*
Daniel Bergey, *Black Hills State University*
Carolyn A. Bergman, *Georgian Court College*
Elke Bergmann-Leitner, *WRAIR/Uniformed Services University of Health Services*

Brian P. Bergstrom, *Muskingum College*

Susan Bjerke, *Washburn University of Topeka*

Earl F. Bloch, *Howard University*

Elliott J. Blumenthal, *Indiana University–Purdue University Fort Wayne*

Kathleen Bode, *Flint Hills Technical College*

Dennis Bogyo, *Valdosta State University*

Mark Bolyard, *Union University*

Lisa Borghesi, *University of Pittsburgh*

Phyllis C. Braun, *Fairfield University*

Jay H. Bream, *Johns Hopkins University School of Medicine*

Heather A. Bruns, *Ball State University*

Walter J. Bruyninckx, *Hanover College*

Eric L. Buckles, *Dillard University*

Sandra H. Burnett, *Brigham Young University*

Peter Burrows, *University of Alabama at Birmingham*

Ralph Butkowski, *Augsburg College*

Jean A. Cardinale, *Alfred University*

Edward A. Chaperon, *Creighton University*

Stephen K. Chapes, *Kansas State University*

Christopher Chase, *South Dakota State University*

Thomas Chiles, *Boston College*

Harold Chittum, *Pikeville College*

Peter A. Chung, *Pittsburg State University*

Felicia L. Cianciarulo, *Carlow University*

Bret A. Clark, *Newberry College*

Patricia A. Compagnone-Post, *Albertus Magnus College*

Yasemin Kaya Congleton, *Bluegrass Community and Technical College*

Vincent A. Connors, *University of South Carolina-Spartanburg*

Conway-Klaassen, *University of Minnesota*

Lisa Cuchara, *Quinnipiac University*

Tanya R. Da Sylva, *York University, North York*

Kelley L. Davis, *Nova Southeastern University*

Jeffrey Dawson, *Duke University*

Joseph DeMasi, *Massachusetts College of Pharmacy & Allied Health*

Stephanie E. Dew, *Centre College*

Joyce E. S. Doan, *Bethel University*

Diane Dorsett, *Georgia Gwinnett College*

James R. Drake, *Albany Medical College*

Erastus C. Dudley, *Huntingdon College*

Jeannine M. Durdik, *University of Arkansas Fayetteville*

Karen M. Duus, *Albany Medical College*

Christina K. Eddy, *North Greenville University*

Anthony Ejiofor, *Tennessee State University*

Jennifer Ellington, *Belmont Abbey College*

Samantha L. Elliott, *Saint Mary's College of Maryland*

Lehman L. Ellis, *Our Lady of Holy Cross College*

Sherine F. Elsawa, *Northern Illinois University*

Uthayashanker Ezekiel, *Saint Louis University Medical Center*

Diana L. Fagan, *Youngstown State University*

Rebecca V. Ferrell, *Metropolitan State College of Denver*

Ken Field, *Bucknell University*

Krista Fischer-Stenger, *University of Richmond*

Howard B. Fleit, *SUNY at Stony Brook*

Sherry D. Fleming, *Kansas State University*

Marie-dominique Franco, *Regis University*

Joel Gaikwad, *Oral Roberts University*

D. L. Gibson, *University of British Columbia-Okanagan*

Laura Glasscock, *Winthrop University*

David Glick, *Kings College*

Elizabeth Godrick, *Boston University*

Karen Golemboski, *Bellarmine University*

Sandra O. Gollnick, *SUNY Buffalo*

James F. Graves, *University of Detroit-Mercy*

Demetrius Peter Gravis, *Beloit College*

Anjali D. Gray, *Lourdes University*

Valery Z. Grdzelishvili, *University of North Carolina-Charlotte*

Carla Guthridge, *Cameron University*

David J. Hall, *Lawrence University*

Sandra K. Halonen, *Montana State University*

Michael C. Hanna, *Texas A & M-Commerce*

Kristian M. Hargadon, *Hampden-Sydney College*

JL Henriksen, *Bellevue University*

Michelle L. Herdman, *University of Charleston*

Jennifer L. Hess, *Aquinas College*

Edward M. Hoffmann, *University of Florida*

Kristin Hogquist, *University of Minnesota*

Jane E. Huffman, *East Stroudsburg University of Pennsylvania*

Lisa A. Humphries, *University of California, Los Angeles*

Judith Humphries, *Lawrence University*

Mo Hunsen, *Kenyon College*

Vijaya Iragavarapu-Charyulu, *Florida Atlantic University*

Vida R. Irani, Indiana *University of Pennsylvania*

Christopher D. Jarvis, *Hampshire College*

Eleanor Jator, *Austin Peay State University*

Stephen R. Jennings, *Drexel University College of Medicine*

Robert Jonas, *Texas Lutheran University*

Vandana Kalia, *Penn State University- Main Campus*

Azad K. Kaushik, *University of Guelph*

George Keller, *Samford University*

Kevin S. Kinney, *De Pauw University*

Edward C. Kisailus, *Canisius College*

David J. Kittlesen, *University of Virginia*

Dennis J. Kitz, *Southern Illinois University-Edwardsville*

Janet Kluftinger, *University of British Columbia -Okanagan*

Rolf König, *University of Texas Medical Branch at Galveston*

Kristine Krafts, *University of Minnesota-Duluth*

Ruhul Kuddus, *Utah Valley University*

Narendra Kumar, *Texas A&M Health Science Center*

N. M. Kumbaraci, *Stevens Institute of Technology*
Jesse J. Kwiek, *The Ohio State University Main Camp*
John M. Lammert, *Gustavus Aldolphus College*
Courtney Lappas, *Lebanon Valley College*
Christopher S. Lassiter, *Roanoke College*
Jennifer Kraft Leavey, *Georgia Institute of Technology*
Melanie J. Lee-Brown, *Guilford College*
Vicky M. Lentz, SUNY *College at Oneonta*
Joseph Lin, *Sonoma State University*
Joshua Loomis, *Nova Southeastern University*
Jennifer Louten, *Southern Polytechnic State University*
Jon H. Lowrance, *Lipscomb University*
Milson J. Luce, *West Virginia University Institute of Technology*
Phillip J. Lucido, *Northwest Missouri State University*
M.E. MacKay, *Thompson Rivers University*
Andrew P. Makrigiannis, *University of Ottawa*
Greg Maniero, *Stonehill College*
David Markwardt, *Ohio Wesleyan University*
John Martinko, *Southern Illinois University*
Andrea M. Mastro, *Penn State University-Main Campus*
Ann H. McDonald, *Concordia University*
Lisa N. McKernan, *Chestnut Hill College*
Catherine S. McVay, *Auburn University*
Daniel Meer, *Cardinal Stritch University*
JoAnn Meerschaert, *Saint Cloud State University*
Brian J. Merkel, *University Wisconsin-Green Bay*
Jiri Mestecky, *University of Alabama at Birmingham*
Dennis W. Metzger, *Albany Medical College*
Jennifer A. Metzler, *Ball State University*
John A. Meyers, *Boston University Medical School*
Yuko J. Miyamoto, *Elon College*
Jody M. Modarelli, *Hiram College*
Devonna Sue Morra, *Saint Francis University*
Rita B. Moyes, *Texas A&M*
Annette Muckerheide, *College of Mount Saint Joseph*
Sue Mungre, *Northeastern Illinois University*
Kari L. Murad, *College of Saint Rose*
Karen Grandel Nakaoka, *Weber State University*
Rajkumar Nathaniel, *Nicholls State University*
David Nemazee, *University of California, San Diego*
Hamida Rahim Nusrat, *San Francisco State University*
Tracy O'Connor, *Mount Royal College*
Marcos Oliveira, *University of the Incarnate Word*
Donald Ourth, *University of Memphis*
Deborah Palliser, *Albert Einstein College of Medicine*
Shawn Phippen, *Valdosta State University*
Melinda J. Pomeroy-Black, *La Grange College*
Edith Porter, *California State University, Los Angeles*
Michael F. Princiotta, *SUNY Upstate Medical University*
Gerry A Prody, *Western Washington University*
Robyn A. Puffenbarger, *Bridgewater College*

Aimee Pugh-Bernard, *University of Colorado at Denver*
Pattle Pun, *Wheaton College*
Sheila Reilly, *Belmont Abbey College*
Karen A. Reiner, *Andrews University*
Margaret Reinhart, *University of the Sciences in Philadelphia*
Stephanie Richards, *Bates College*
Sarah M. Richart, *Azusa Pacific University*
James E. Riggs, *Rider University*
Vanessa Rivera-Amill, *Ponce School of Medicine*
Katherine Robertson, *Westminster College*
James L. Rooney, *Lincoln University*
Robin S. Salter, *Oberlin College*
Sophia Sarafova, *Davidson College*
Surojit Sarkar, *Penn State University-Main Campus*
Perry M. Scanlan, *Austin Peay State University*
Ralph Seelke, *University Wisconsin-Superior*
Diane L. Sewell, *University Wisconsin-La Crosse*
Anding Shen, Calvin College
Penny Shockett, Southeastern Louisiana University
Michael Sikes, North Carolina State University
Maryanne C. Simurda, *Washington and Lee University*
Paul K. Small, *Eureka College*
Jonathan Snow, *Williams College*
Ralph A. Sorensen, *Gettysburg College*
Andrew W. Stadnyk, *Dalhousie University Faculty of Medicine*
Douglas A. Steeber, *University Wisconsin-Milwaukee*
Viktor Steimle, *University of Sherbrooke, Sherbrooke*
Douglas J. Stemke, *University of Indianapolis*
Carolyn R. Stenbak, *Seattle University*
Jennifer Ripley Stueckle, *West Virginia University*
Kathleen Sullivan, *Louisiana Technical College Alexandria*
Susmit Suvas, *Oakland University*
Gabor Szalai, *University of South Carolina*
Seetha M Tamma, *Long Island University-C.W. Post*
Matthew J. Temple, *Nazareth College*
Kent R. Thomas, *Wichita State University*
Diane G. Tice, *SUNY Morrisville*
Sara Sybesma Tolsma, *Northwestern College*
Clara Tóth, *Saint Thomas Aquinas College*
Bebhinn Treanor, *University of Toronto Scarborough*
Allen W. Tsang, *Bowman Gray Medical School*
Amar S. Tung, *Lincoln University*
Lloyd Turtinen, *University Wisconsin-Eau Claire*
Timothy M VanWagoner, *Oklahoma Christian/ University of Oklahoma HSC*
Evros Vassiliou, *Kean University*
Vishwanath Venketaraman, *Western University of Health Sciences*
Kathleen Verville, *Washington College*
Katherine A. Wall, *University of Toledo*
Helen Walter, *Mills College*

Christopher Ward, *University of Alberta*

Benjamin S. Weeks, *Adelphi University*

Ben B. Whitlock, *University of Saint Francis*

Robert Winn, *Northern Michigan University*

Candace R. Winstead, *California Polytechnic State University-San Luis Obispo*

Dorothy M. Wrigley, *Minnesota State University*

Jodi L. Yorty, *Elizabethtown College*

Sheryl Zajdowicz, *Metropolitan State University of Denver*

Mary Katherine Zanin, *The Citadel The Military College of South Carolina*

Gary Zieve, *SUNY at Stony Brook*

Michael I. Zimmer, *Purdue Calumet*

Gilbert L. Zink, *University of the Sciences in Philadelphia*

Patty Zwollo, *College of William & Mary*

Finally, we thank our experienced and talented colleagues at W. H. Freeman and Company. Particular thanks to the production team members Philip McCaffrey, Sherrill Redd, Heath Lynn Silberfeld, Diana Blume, Lawrence Guerra, Janice Donnola, Christine Buese, and Elyse Reider. Thanks are also due to the editorial team of Lauren Schultz, Susan Winslow, Allison Michael, Yassamine Ebadat, and Irene Pech.

However, a very special thanks go to our developmental editor, Erica Champion, and our developmental coordinator, Sara Ruth Blake. Erica has guided us from the beginning with a probing vision, endless patience, and keen eye for narrative and clarity. Sara kept us organized and true to deadlines with heroic resolve. The involvement of these two extraordinarily talented team members has made this edition, and its ambitious aspirations, possible.

Overview of the Immune System

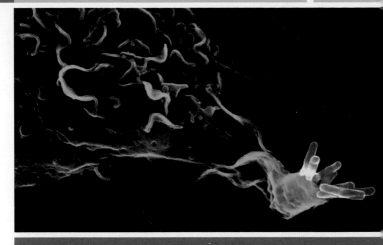

- A Historical Perspective of Immunology

- Important Concepts for Understanding the Mammalian Immune Response

- The Good, Bad, and Ugly of the Immune System

The immune system evolved to protect multicellular organisms from pathogens. Highly adaptable, it defends the body against invaders as diverse as the tiny (~30 nm), intracellular virus that causes polio and as large as the giant parasitic kidney worm *Dioctophyme renale*, which can grow to over 100 cm in length and 10 mm in width. This diversity of potential pathogens requires a range of recognition and destruction mechanisms to match the multitude of invaders. To accomplish this feat, vertebrates have evolved a complicated and dynamic network of cells, molecules, and pathways. Although elements of these networks can be found throughout the plant and animal kingdoms, the focus of this book will be on the highly evolved mammalian immune system.

The fully functional immune system involves so many organs, molecules, cells, and pathways in such an interconnected and sometimes circular process that it is often difficult to know where to start! Recent advances in cell imaging, genetics, bioinformatics, as well as cell and molecular biology, have helped us to understand many of the individual players in great molecular detail. However, a focus on the details (and there are many) can make taking a step back to see the bigger picture challenging, and it is often the bigger picture that motivates us to study immunology. Indeed, the field of immunology can be credited with the vaccine that eradicated smallpox, the ability to transplant organs between humans, and the drugs used today to treat asthma. Our goal in this chapter is therefore to present the background and concepts in immunology that will help bridge the gap between the cellular and molecular detail presented in subsequent chapters and the complete picture of an immune response. A clear understanding of each of the many players involved will help one appreciate the intricate coordination of an immune system that makes all of this possible.

The study of immunology has produced amazing and fascinating stories (some of which you will see in this book), where host and microbe engage in battles waged over both minutes and millennia. But the immune system is also much more than an isolated component of the body, merely responsible for search-and-destroy missions. In fact, it interleaves with many of the other body systems, including the endocrine, nervous, and metabolic systems, with more connections undoubtedly to be discovered in time. Finally, it has become increasingly clear that elements of immunity play key roles in regulating homeostasis in the body for a healthy balance. Information gleaned from the study of the immune system, as well as its connections with other systems, will likely have resounding repercussions across many basic science and biomedical fields, not to mention in the future of clinical medicine.

This chapter begins with a historical perspective, charting the beginnings of the study of immunology, largely driven by the human desire to survive major outbreaks of infectious disease. This is followed by presentation of a few key concepts that are important hallmarks of the mammalian immune response, many of which may not have been encountered elsewhere in basic biology. This is not meant as a comprehensive overview of the mammalian immune system but rather as a means for jumping the large conceptual hurdles frequently encountered as one begins to describe the complexity and interconnected nature of the immune response. We hope this will whet the appetite and prepare the reader for a more thorough discussion of the specific components of immunity presented in the following chapters. We conclude with a few challenging clinical situations, such as instances in which the immune system fails to act or becomes the aggressor, turning its awesome powers against the host. More in-depth coverage of these and other medical aspects of immunology can be found in the final chapters of this book.

A Historical Perspective of Immunology

The discipline of immunology grew out of the observation that individuals who had recovered from certain infectious diseases were thereafter protected from the disease. The Latin term *immunis*, meaning "exempt," is the source of the English word **immunity**, a state of protection from infectious disease. Perhaps the earliest written reference to the phenomenon of immunity can be traced back to Thucydides, the great historian of the Peloponnesian War. In describing a plague in Athens, he wrote in 430 BC that only those who had recovered from the plague could nurse the sick because they would not contract the disease a second time. Although early societies recognized the phenomenon of immunity, almost 2000 years passed before the concept was successfully converted into medically effective practice.

Early Vaccination Studies Led the Way to Immunology

The first recorded attempts to deliberately induce immunity were performed by the Chinese and Turks in the fifteenth century. They were attempting to prevent smallpox, a disease that is fatal in about 30% of cases and that leaves survivors disfigured for life (Figure 1-1). Reports suggest that the dried crusts derived from smallpox pustules were either inhaled or inserted into small cuts in the skin (a technique called *variolation*) in order to prevent this dreaded disease. In 1718, Lady Mary Wortley Montagu, the wife of the British ambassador in Constantinople, observed the positive effects of variolation on the native Turkish population and had the technique performed on her own children.

The English physician Edward Jenner later made a giant advance in the deliberate development of immunity, again targeting smallpox. In 1798, intrigued by the fact that milkmaids who had contracted the mild disease cowpox were subsequently immune to the much more severe smallpox, Jenner reasoned that introducing fluid from a cowpox pustule into people (i.e., inoculating them) might protect them from smallpox. To test this idea, he inoculated an eight-year-old boy with fluid from a cowpox pustule and later intentionally infected the child with smallpox. As predicted, the child did not develop smallpox. Although this represented a major breakthrough, as one might imagine, these sorts of human studies could not be conducted under current standards of medical ethics.

Jenner's technique of inoculating with cowpox to protect against smallpox spread quickly through Europe. However, it

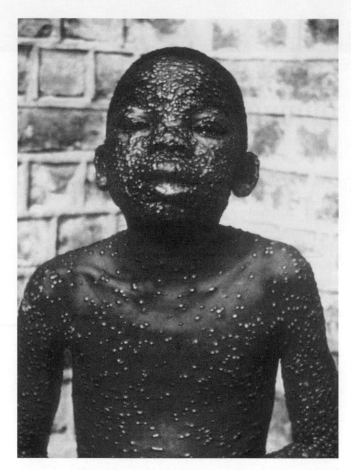

FIGURE 1-1 African child with rash typical of smallpox on face, chest, and arms. Smallpox, caused by the virus *Variola major,* has a 30% mortality rate. Survivors are often left with disfiguring scars. [Centers for Disease Control.]

was nearly a hundred years before this technique was applied to other diseases. As so often happens in science, serendipity combined with astute observation led to the next major advance in immunology: the induction of immunity to cholera. Louis Pasteur had succeeded in growing the bacterium that causes fowl cholera in culture, and confirmed this by injecting it into chickens that then developed fatal cholera. After returning from a summer vacation, he and colleagues resumed their experiments, injecting some chickens with an old bacterial culture. The chickens became ill, but to Pasteur's surprise, they recovered. Interested, Pasteur then grew a fresh culture of the bacterium with the intention of injecting this lethal brew into some fresh, unexposed chickens. But as the story is told, his supply of fresh chickens was limited, and therefore he used a mixture of previously injected chickens and unexposed birds. Unexpectedly, only the fresh chickens died, while the chickens previously exposed to the older bacterial culture were completely protected from the disease. Pasteur hypothesized and later showed that aging had weakened the virulence of the pathogen and that such a weakened or **attenuated** strain could be administered to provide immunity against the disease. He called this attenuated strain a

vaccine (from the Latin *vacca*, meaning "cow"), in honor of Jenner's work with cowpox inoculation.

Pasteur extended these findings to other diseases, demonstrating that it was possible to attenuate a pathogen and administer the attenuated strain as a vaccine. In a now classic experiment performed in the small village of Pouilly-le-Fort in 1881, Pasteur first vaccinated one group of sheep with anthrax bacteria (*Bacillus anthracis*) that were attenuated by heat treatment. He then challenged the vaccinated sheep, along with some unvaccinated sheep, with a virulent culture of the anthrax bacillus. All the vaccinated sheep lived and all the unvaccinated animals died. These experiments marked the beginnings of the discipline of immunology. In 1885, Pasteur administered his first vaccine to a human, a young boy who had been bitten repeatedly by a rabid dog (Figure 1-2). The boy, Joseph Meister, was inoculated with a series of attenuated rabies virus preparations. The rabies vaccine is one of very few that can be successful when administered shortly after exposure, as long as the virus has not yet reached the central nervous system and begun to induce neurologic symptoms. Joseph lived, and later became a caretaker at the Pasteur Institute, which was opened in 1887 to treat the many rabies victims that began to flood in when word of Pasteur's success spread; it remains to this day an institute dedicated to the prevention and treatment of infectious disease.

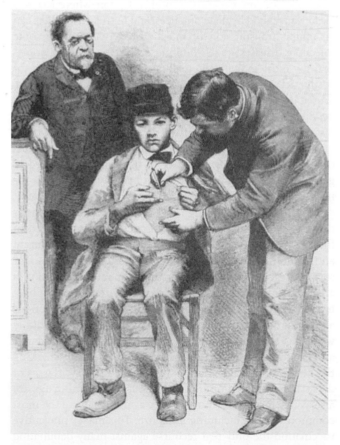

FIGURE 1-2 Wood engraving of Louis Pasteur watching Joseph Meister receive the rabies vaccine. [*Source: From* Harper's Weekly 29:836; *courtesy of the National Library of Medicine.*]

Vaccination Is an Ongoing, Worldwide Enterprise

The emergence of the study of immunology and the discovery of vaccines are tightly linked. The development of effective vaccines for some pathogens is still a major challenge, discussed in greater detail in Chapter 17. However, despite many biological and social hurdles, vaccination has yielded some of the most profound success stories in terms of improving mortality rates worldwide, especially in very young children.

In 1977, the last known case of naturally acquired smallpox was seen in Somalia. This dreaded disease was eradicated by universal application of a vaccine similar to that used by Jenner in the 1790s. One consequence of eradication is that universal vaccination becomes unnecessary. This is a tremendous benefit, as most vaccines carry at least a slight risk to persons vaccinated. And yet in many cases every individual does not need to be immune in order to protect most of the population. As a critical mass of people acquire protective immunity, either through vaccination or infection, they can serve as a buffer for the rest. This principle, called **herd immunity**, works by decreasing the number of individuals who can harbor and spread an infectious agent, significantly decreasing the chances that susceptible individuals will become infected. This presents an important altruistic consideration: although many of us could survive infectious diseases for which we receive a vaccine (such as the flu), this is not true for everyone. Some individuals cannot receive the vaccine (e.g., the very young or immune compromised), and vaccination is never 100% effective. In other words, the susceptible, nonimmune individuals among us can benefit from the pervasive immunity of their neighbors.

However, there is a darker side to eradication and the end of universal vaccination. Over time, the number of people with no immunity to the disease will begin to rise, ending herd immunity. Vaccination for smallpox largely ended by the early to mid-1970s, leaving well over half of the current world population susceptible to the disease. This means that smallpox, or a weaponized version, is now considered a potential bioterrorism threat. In response, new and safer vaccines against smallpox are still being developed today, most of which go toward vaccinating U.S. military personnel thought to be at greatest risk of possible exposure.

In the United States and other industrialized nations, vaccines have eliminated a host of childhood diseases that were the cause of death for many young children just 50 years ago. Measles, mumps, chickenpox, whooping cough (pertussis), tetanus, diphtheria, and polio, once thought of as an inevitable part of childhood are now extremely rare or nonexistent in the United States because of current vaccination practices (Table 1-1). One can hardly estimate the savings to society resulting from the prevention of these diseases. Aside from suffering and mortality, the cost to treat these illnesses and their aftereffects or sequelae (such as paralysis, deafness, blindness, and mental retardation) is immense and dwarfs the costs of immunization. In fact, recent estimates suggest

TABLE 1-1	Cases of selected infectious disease in the United States before and after the introduction of effective vaccines		
	ANNUAL CASES/YR	CASES IN 2010	
Disease	**Prevaccine**	**Postvaccine**	**Reduction (%)**
Smallpox	48,164	0	100
Diphtheria	175,885	0	100
Rubeola (measles)	503,282	26	99.99
Mumps	152,209	2,612	98.28
Pertussis ("whooping cough")	147,271	27,550	81.29
Paralytic polio	16,316	0	100
Rubella (German measles)	47,745	5	99.99
Tetanus ("lockjaw")	1,314 (deaths)	26 (cases)	98.02
Invasive *Haemophilus influenzae*	20,000	3,151	84.25

SOURCE: Adapted from W. A. Orenstein et al., 2005. *Health Affairs* **24**:599 and CDC statistics of Notifiable Diseases.

that significant economic and human life benefits could be realized by simply scaling up the use of a few childhood vaccines in the poorest nations, which currently bear the brunt of the impact of these childhood infectious diseases. For example, it is estimated that childhood pneumonia alone, caused primarily by vaccine-preventable *Streptococcus pneumoniae* (aka, pneumococcus) and *Haemophilus influenzae* type b (aka, Hib), will account for 2.7 million childhood deaths in developing nations over the next decade if vaccine strategies in these regions remain unchanged.

Despite the many successes of vaccine programs, such as the eradication of smallpox, many vaccine challenges still remain. Perhaps the greatest current challenge is the design of effective vaccines for major killers such as malaria and AIDS. Using our increased understanding of the immune system, plus the tools of molecular and cellular biology, genomics, and proteomics, scientists will be better positioned to make progress toward preventing these and other emerging infectious diseases. A further issue is the fact that millions of children in developing countries die from diseases that are fully preventable by available, safe vaccines. High manufacturing costs, instability of the products, and cumbersome delivery problems keep these vaccines from reaching those who might benefit the most. This problem could be alleviated in many cases by development of future-generation vaccines that are inexpensive, heat stable, and administered without a needle. Finally, misinformation and myth surrounding vaccine efficacy and side effects continues to hamper many potentially life-saving vaccination programs (see Clinical Focus Box on p. 5).

Immunology Is About More Than Just Vaccines and Infectious Disease

For some diseases, immunization programs may be the best or even the only effective defense. At the top of this list are infec-

tious diseases that can cause serious illness or even death in unvaccinated individuals, especially those transmitted by microbes that also spread rapidly between hosts. However, vaccination is not the only way to prevent or treat infectious disease. First and foremost is preventing infection, where access to clean water, good hygiene practices, and nutrient-rich diets can all inhibit transmission of infectious agents. Second, some infectious diseases are self-limiting, easily treatable, and nonlethal for most individuals, making them unlikely targets for costly vaccination programs. These include the common cold, caused by the Rhinovirus, and cold sores that result from Herpes Simplex Virus infection. Finally, some infectious agents are just not amenable to vaccination. This could be due to a range of factors, such as the number of different molecular variants of the organism, the complexity of the regimen required to generate protective immunity, or an inability to establish the needed immunologic memory responses (more on this later).

One major breakthrough in the treatment of infectious disease came when the first antibiotics were introduced in the 1920s. Currently there are more than a hundred different antibiotics on the market, although most fall into just six or seven categories based on their mode of action. Antibiotics are chemical agents designed to destroy certain types of bacteria. They are ineffective against other types of infectious agents, as well as some bacterial species. One particularly worrying trend is the steady rise in antibiotic resistance among strains traditionally amenable to these drugs, making the design of next-generation antibiotics and new classes of drugs increasingly important. Although antiviral drugs are also available, most are not effective against many of the most common viruses, including influenza. This makes preventive vaccination the only real recourse against many debilitating infectious agents, even those that rarely cause mortality in healthy adults. For instance, because of the high mutation rate of the influenza virus, each year a new flu vaccine must

BOX 1-1

Vaccine Controversy: What's Truth and What's Myth?

Despite the record of worldwide success of vaccines in improving public health, some opponents claim that vaccines do more harm than good, pressing for elimination or curtailment of childhood vaccination programs. There is no dispute that vaccines represent unique safety issues, since they are administered to people who are healthy. Furthermore, there is general agreement that vaccines must be rigorously tested and regulated, and that the public must have access to clear and complete information about them. Although the claims of vaccine critics must be evaluated, many can be answered by careful and objective examination of records.

A recent example is the claim that vaccines given to infants and very young children may contribute to the rising incidence of autism. This began with the suggestion that thimerosal, a mercury-based additive used to inhibit bacterial growth in some vaccine preparations since the 1930s, was causing autism in children. In 1999 the U.S. Centers for Disease Control and Prevention (CDC) and the American Association of Pediatricians (AAP) released a joint recommendation that vaccine manufacturers begin to gradually phase out thimerosal use in vaccines. This recommendation was based on the increase in the number of vaccines given to infants and was aimed at keeping children at or below Environmental Protection Agency (EPA)–recommended maximums in mercury exposure. However, with the release of this recommendation, parent-led public advocacy groups began a media-fueled campaign to build a case demonstrating

what they believed was a link between vaccines and an epidemic of autism. These AAP recommendations and public fears led to a dramatic decline in the latter half of 1999 in U.S. newborns vaccinated for hepatitis B. To date, no credible study has shown a scientific link between thimerosal and autism. In fact, cases of autism in children have continued to rise since thimerosal was removed from all childhood vaccines in 2001. Despite evidence to the contrary, some still believe this claim.

A 1998 study appearing in *The Lancet*, a reputable British medical journal, further fueled these parent advocacy groups and anti-vaccine organizations. The article, published by Andrew Wakefield, claimed the measles-mumps-rubella (MMR) vaccine caused pervasive developmental disorders in children, including autism spectrum disorder. More than a decade of subsequent research has been unable to substantiate these claims, and 10 of the original 12 authors on the paper later withdrew their support for the conclusions of the study. In 2010, *The Lancet* retracted the original article when it was shown that the data in the study had been falsified to reach desired conclusions. Nonetheless, in the years between the original publication of the *Lancet* article and its retraction, this case is credited with decreasing rates of MMR vaccination from a high of 92% to a low of almost 60% in certain areas of the United Kingdom. The resulting expansion in the population of susceptible individuals led to endemic rates of measles and mumps infection, especially in several areas of Europe, and is credited with thou-

sands of extended hospitalizations and several deaths in infected children.

Why has there been such a strong urge to cling to the belief that childhood vaccines are linked with developmental disorders in children despite much scientific evidence to the contrary? One possibility lies in the timing of the two events. Based on current AAP recommendations, in the United States most children receive 14 different vaccines and a total of up to 26 shots by the age of 2. In 1983, children received less than half this number of vaccinations. Couple this with the onset of the first signs of autism and other developmental disorders in children, which can appear quite suddenly and peak around 2 years of age. This sharp rise in the number of vaccinations young children receive today and coincidence in timing of initial autism symptoms is credited with sparking these fears about childhood vaccines. Add to this the increasing drop in basic scientific literacy by the general public and the overabundance of ways to gather such information (accurate or not). As concerned parents search for answers, one can begin to see how even scientifically unsupported links could begin to take hold as families grapple with how to make intelligent public health risk assessments.

The notion that vaccines cause autism was rejected long ago by most scientists. Despite this, more work clearly needs to be done to bridge the gap between public perception and scientific understanding.

Gross, L. 2009. A broken trust: Lessons from the vaccine–autism wars. *PLoS Biology* **7**:e1000114.
Larson, H.J., et al. 2011. Addressing the vaccine confidence gap. *Lancet* **378**:526.

be prepared based on a prediction of the prominent genotypes likely to be encountered in the next season. Some years this vaccine is more effective than others. If and when a more lethal and unexpected pandemic strain arises, there will be a race between its spread and the manufacture and administration of a new vaccine. With the current ease of worldwide travel, present-day emergence of a pandemic strain of influenza could dwarf the devastation wrought by the 1918 flu pandemic, which left up to 50 million dead.

However, the eradication of infectious disease is not the only worthy goal of immunology research. As we will see later, exposure to infectious agents is part of our evolutionary history, and wiping out all of these creatures could potentially cause more harm than good, both for the host and the environment. Thanks to many technical advances allowing scientific discoveries to move efficiently from the bench to the bedside, clinicians can now manipulate the immune response in ways never before possible. For example, treatments to boost, inhibit, or redirect the specific efforts of immune cells are being applied to treat autoimmune disease, cancer, and allergy, as well as other chronic disorders. These efforts are already extending and saving lives. Likewise, a clearer understanding of immunity has highlighted the interconnected nature of body systems, providing unique insights into areas such as cell biology, human genetics, and metabolism. While a cure for AIDS and a vaccine to prevent HIV infection are still the primary targets for many scientists who study this disease, a great deal of basic science knowledge has been gleaned from the study of just this one virus and its interaction with the human immune system.

Immunity Involves Both Humoral and Cellular Components

Pasteur showed that vaccination worked, but he did not understand how. Some scientists believed that immune protection in vaccinated individuals was mediated by cells, while others postulated that a soluble agent delivered protection. The experimental work of Emil von Behring and Shibasaburo Kitasato in 1890 gave the first insights into the mechanism of immunity, earning von Behring the Nobel Prize in Physiology or Medicine in 1901 (Table 1-2). Von Behring and

TABLE 1-2	Nobel Prizes for immunologic research		
Year	**Recipient**	**Country**	**Research**
1901	Emil von Behring	Germany	Serum antitoxins
1905	Robert Koch	Germany	Cellular immunity to tuberculosis
1908	Elie Metchnikoff	Russia	Role of phagocytosis (Metchnikoff)
	Paul Ehrlich	Germany	and antitoxins (Ehrlich) in immunity
1913	Charles Richet	France	Anaphylaxis
1919	Jules Bordet	Belgium	Complement-mediated bacteriolysis
1930	Karl Landsteiner	United States	Discovery of human blood groups
1951	Max Theiler	South Africa	Development of yellow fever vaccine
1957	Daniel Bovet	Switzerland	Antihistamines
1960	F. Macfarlane Burnet	Australia	Discovery of acquired immunological
	Peter Medawar	Great Britain	tolerance
1972	Rodney R. Porter	Great Britain	Chemical structure of antibodies
	Gerald M. Edelman	United States	
1977	Rosalyn R. Yalow	United States	Development of radioimmunoassay
1980	George Snell	United States	Major histocompatibility complex
	Jean Dausset	France	
	Baruj Benacerraf	United States	
1984	Niels K. Jerne	Denmark	Immune regulatory theories (Jerne) and
	Cesar Milstein	Great Britain	technological advances in the development
	Georges E. Köhler	Germany	of monoclonal antibodies (Milstein and Köhler)
1987	Susumu Tonegawa	Japan	Gene rearrangement in antibody production
1991	E. Donnall Thomas	United States	Transplantation immunology
	Joseph Murray	United States	
1996	Peter C. Doherty	Australia	Role of major histocompatibility complex in
	Rolf M. Zinkernagel	Switzerland	antigen recognition by T cells
2002	Sydney Brenner	South Africa	Genetic regulation of organ development
	H. Robert Horvitz	United States	and cell death (apoptosis)
	J. E. Sulston	Great Britain	
2008	Harald zur Hausen	Germany	Role of HPV in causing cervical cancer
	Françoise Barré-Sinoussi	France	(Hausen) and the discovery of HIV
	Luc Montagnier	France	(Barré-Sinoussi and Montagnier)
2011	Jules Hoffman	France	Discovery of activating principles of innate
	Bruce Beutler	United States	immunity (Hoffman and Beutler) and role of
	Ralph Steinman	United States	dendritic cells in adaptive immunity (Steinman)

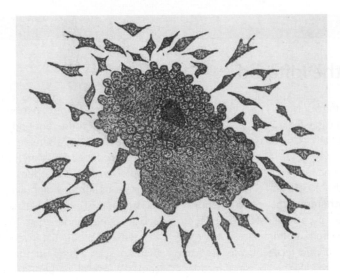 

FIGURE 1-3 Drawing by Elie Metchnikoff of phagocytic cells surrounding a foreign particle (left) and modern image of a phagocyte engulfing the bacteria that cause tuberculosis (right). Metchnikoff first described and named the process of phagocytosis, or ingestion of foreign matter by white blood cells. Today, phagocytic cells can be imaged in great detail using advanced microscopy techniques. *[Drawing reproduced by permission of The British Library:7616.h.19, Lectures on the Comparative Pathology of Inflammation delivered at the Pasteur Institute in 1891, translated by F. A. Starling and E. H. Starling, with plates by Il'ya Il'ich Mechnikov, 1893, p. 64, fig. 32. Photo courtesy Dr. Volker Brinkmann/Visuals Unlimited, Inc.]*

Kitasato demonstrated that serum—the liquid, noncellular component recovered from coagulated blood—from animals previously immunized with diphtheria could transfer the immune state to unimmunized animals.

In 1883, even before the discovery that a serum component could transfer immunity, Elie Metchnikoff, another Nobel Prize winner, demonstrated that cells also contribute to the immune state of an animal. He observed that certain white blood cells, which he termed *phagocytes*, ingested (phagocytosed) microorganisms and other foreign material (Figure 1-3, left). Noting that these phagocytic cells were more active in animals that had been immunized, Metchnikoff hypothesized that cells, rather than serum components, were the major effectors of immunity. The active phagocytic cells identified by Metchnikoff were likely blood monocytes and neutrophils (see Chapter 2), which can now be imaged using very sophisticated microscopic techniques (Figure 1-3, right).

Humoral Immunity

The debate over cells versus soluble mediators of immunity raged for decades. In search of the protective agent of immunity, various researchers in the early 1900s helped characterize the active immune component in blood serum. This soluble component could neutralize or precipitate toxins and could agglutinate (clump) bacteria. In each case, the component was named for the activity it exhibited: antitoxin, precipitin, and agglutinin, respectively. Initially, different serum components were thought to be responsible for each activity, but during the 1930s, mainly through the efforts of Elvin Kabat, a fraction of

serum first called gamma globulin (now **immunoglobulin**) was shown to be responsible for all these activities. The soluble active molecules in the immunoglobulin fraction of serum are now commonly referred to as **antibodies**. Because these antibodies were contained in body fluids (known at that time as the body *humors*), the immunologic events they participated in was called **humoral immunity**.

The observation of von Behring and Kitasato was quickly applied to clinical practice. **Antiserum**, the antibody-containing serum fraction from a pathogen-exposed individual, derived in this case from horses, was given to patients suffering from diphtheria and tetanus. A dramatic vignette of this application is described in the Clinical Focus box on page 8. Today there are still therapies that rely on transfer of immunoglobulins to protect susceptible individuals. For example, emergency use of immune serum, containing antibodies against snake or scorpion venoms, is a common practice for treating bite victims. This form of immune protection that is transferred between individuals is called **passive immunity** because the individual receiving it did not make his or her own immune response against the pathogen. Newborn infants benefit from passive immunity by the presence of maternal antibodies in their circulation. Passive immunity may also be used as a preventive (prophylaxis) to boost the immune potential of those with compromised immunity or who anticipate future exposure to a particular microbe.

While passive immunity can supply a quick solution, it is short-lived and limited, as the cells that produce these antibodies are not being transferred. On the other hand, administration

CLINICAL FOCUS

Passive Antibodies and the Iditarod

In 1890, immunologists Emil Behring and Shibasaburo Kitasato, working together in Berlin, reported an extraordinary experiment. After immunizing rabbits with tetanus and then collecting blood serum from these animals, they injected a small amount of immune serum (cell-free fluid) into the abdominal cavity of six mice. Twenty-four hours later, they infected the treated mice and untreated controls with live, virulent tetanus bacteria. All of the control mice died within 48 hours of infection, whereas the treated mice not only survived but showed no effects of infection. This landmark experiment demonstrated two important points. One, it showed that substances that could protect an animal against pathogens appeared in serum following immunization. Two, this work demonstrated that immunity could be passively acquired, or transferred from one animal to another by taking serum from an immune animal and injecting it into a nonimmune one. These and subsequent experiments did not go unnoticed. Both men eventually received titles (Behring became von Behring and Kitasato became Baron Kitasato). A few years later, in 1901, von Behring was

awarded the first Nobel Prize in Physiology or Medicine (see Table 1-2).

These early observations, and others, paved the way for the introduction of passive immunization into clinical practice. During the 1930s and 1940s, **passive immunotherapy**, the endowment of resistance to pathogens by transfer of antibodies from an immunized donor to an unimmunized recipient, was used to prevent or modify the course of measles and hepatitis A. Subsequently, clinical experience and advances in the technology of immunoglobulin preparation have made this approach a standard medical practice. Passive immunization based on the transfer of antibodies is widely used in the treatment of immunodeficiency and some autoimmune diseases. It is also used to protect individuals against anticipated exposure to infectious and toxic agents against which they have no immunity. Finally, passive immunization can be lifesaving during episodes of certain types of acute infection, such as following exposure to rabies virus.

Immunoglobulin for passive immunization is prepared from the pooled

plasma of thousands of donors. In effect, recipients of these antibody preparations are receiving a sample of the antibodies produced by many people to a broad diversity of pathogens—a gram of intravenous immune globulin (IVIG) contains about 10^{18} molecules of antibody and recognize more than 10^7 different antigens. A product derived from the blood of such a large number of donors carries a risk of harboring pathogenic agents, particularly viruses. This risk is minimized by modern-day production techniques. The manufacture of IVIG involves treatment with solvents, such as ethanol, and the use of detergents that are highly effective in inactivating viruses such as HIV and hepatitis. In addition to treatment against infectious disease, or acute situations, IVIG is also used today for treating some chronic diseases, including several forms of immune deficiency. In all cases, the transfer of passive immunity supplies only temporary protection.

One of the most famous instances of passive antibody therapy occurred in 1925, when an outbreak of diphtheria

---

of a vaccine or natural infection is said to engender active immunity in the host: the production of one's own immunity. The induction of active immunity can supply the individual with a renewable, long-lived protection from the specific infectious organism. As we discuss further below, this long-lived protection comes from memory cells, which provide protection for years or even decades after the initial exposure.

Cell-Mediated Immunity

A controversy developed between those who held to the concept of humoral immunity and those who agreed with Metchnikoff's concept of immunity imparted by specific cells, or **cell-mediated immunity**. The relative contributions of the two were widely debated at the time. It is now obvious that both are correct—the full immune response requires both cellular and humoral (soluble) components. Early studies of immune cells were hindered by the lack of genetically defined animal models and modern tissue culture tech-

niques, whereas early studies with serum took advantage of the ready availability of blood and established biochemical techniques to purify proteins. Information about cellular immunity therefore lagged behind a characterization of humoral immunity.

In a key experiment in the 1940s, Merrill Chase, working at The Rockefeller Institute, succeeded in conferring immunity against tuberculosis by transferring white blood cells between guinea pigs. Until that point, attempts to develop an effective vaccine or antibody therapy against tuberculosis had met with failure. Thus, Chase's demonstration helped to rekindle interest in cellular immunity. With the emergence of improved cell culture and transfer techniques in the 1950s, the lymphocyte was identified as the cell type responsible for both cellular and humoral immunity. Soon thereafter, experiments with chickens pioneered by Bruce Glick at Mississippi State University indicated the existence of two types of lymphocytes: **T lymphocytes (T cells)**, derived from

BOX 1-2

 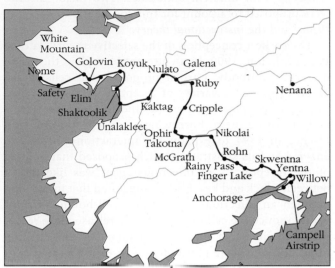

FIGURE 1
(left) Leonhard Seppala, the Norwegian who led a team of sled dogs in the 1925 diphtheria antibody run from Nenana to Nome, Alaska. (right) Map of the current route of the Iditarod Race, which commemorates this historic delivery of lifesaving antibody. [Source: Underwood & Underwood/Corbis.]

was diagnosed in what was then the remote outpost of Nome, Alaska. Lifesaving diphtheria-specific antibodies were available in Anchorage, but no roads were open and the weather was too dangerous for flight. History tells us that 20 mushers set up a dogsled relay to cover the almost 700 miles between Nenana, the end of the railroad run, and remote Nome. In this relay, two Norwegians and their dogs covered particularly critical territory and withstood blizzard conditions: Leonhard Seppala (Figure 1, left), who covered the most treacherous territory, and Gunnar Kaasen, who drove the final two legs in whiteout conditions, behind his lead dog Balto. Kaasen and Balto arrived in time to save many of the children in the town. To commemorate this heroic event, later that same year a statue of Balto was placed in Central Park, New York City, where it still stands today. This journey is memorialized every year in the running of the Iditarod sled dog race. A map showing the current route of this more than 1000-mile trek is shown in Figure 1, right.

the *thymus*, and **B lymphocytes (B cells)**, derived from the *bursa of* Fabricius in birds (an outgrowth of the cloaca). In a convenient twist of nomenclature that makes B and T cell origins easier to remember, the mammalian equivalent of the *bursa of Fabricius* is bone marrow, the home of developing B cells in mammals. *We now know that cellular immunity is imparted by T cells and that the antibodies produced by B cells confer humoral immunity.* The real controversy about the roles of humoral versus cellular immunity was resolved when the two systems were shown to be intertwined and it became clear that both are necessary for a complete immune response against most pathogens.

How Are Foreign Substances Recognized by the Immune System?

One of the great enigmas confronting early immunologists was what determines the specificity of the immune response for a particular foreign material, or **antigen**, the general term for any substance that elicits a specific response by B or T lymphocytes. Around 1900, Jules Bordet at the Pasteur Institute expanded the concept of immunity beyond infectious diseases, demonstrating that nonpathogenic substances, such as red blood cells from other species, could also serve as antigens. Serum from an animal that had been inoculated with noninfectious but otherwise foreign (nonself) material would nevertheless react with the injected material in a specific manner. The work of Karl Landsteiner and those who followed him showed that injecting an animal with almost any nonself organic chemical could induce production of antibodies that would bind specifically to the chemical. These studies demonstrated that antibodies have a capacity for an almost unlimited range of reactivity, including responses to compounds that had only recently been synthesized in the laboratory and are otherwise not found in nature! In addition, it was shown that molecules differing in the smallest detail, such as a single amino acid, could be distinguished by their reactivity with different antibodies. Two major theories were proposed to account for this specificity: the *selective theory* and the *instructional theory*.

The earliest conception of the selective theory dates to Paul Ehrlich in 1900. In an attempt to explain the origin of serum antibody, Ehrlich proposed that cells in the blood expressed a variety of receptors, which he called *side-chain receptors*, that could bind to infectious agents and inactivate them. Borrowing a concept used by Emil Fischer in 1894 to explain the interaction between an enzyme and its substrate, Ehrlich proposed that binding of the receptor to an infectious agent was like the fit between a lock and key. Ehrlich suggested that interaction between an infectious agent and a cell-bound receptor would induce the cell to produce and release more receptors with the same specificity (Figure 1-4). In Ehrlich's mind, the cells were pluripotent, expressing a number of

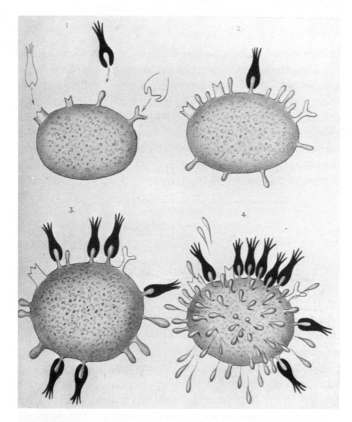

FIGURE 1-4 Representation of Paul Ehrlich's side chain theory to explain antibody formation. In Ehrlich's initial theory, the cell is pluripotent in that it expresses a number of different receptors or side chains, all with different specificities. If an antigen encounters this cell and has a good fit with one of its side chains, synthesis of that receptor is triggered and the receptor will be released. [*From Ehrlich's Croonian lecture of 1900 to the Royal Society.*]

different receptors, each of which could be individually "selected." According to Ehrlich's theory, the specificity of the receptor was determined in the host before its exposure to the foreign antigen, and therefore the antigen *selected* the appropriate receptor. Ultimately, most aspects of Ehrlich's theory would be proven correct, with the following minor refinement: instead of one cell making many receptors, *each cell makes many copies of just one membrane-bound receptor (one specificity)*. An army of cells, each with a different antigen specificity, is therefore required. *The selected B cell can be triggered to proliferate and to secrete many copies of these receptors in soluble form (now called antibodies) once it has been selected by antigen binding.*

In the 1930s and 1940s, the selective theory was challenged by various instructional theories. These theories held that antigen played a central role in determining the specificity of the antibody molecule. According to the instructional theorists, a particular antigen would serve as a template around which antibody would fold—sort of like an impression mold. The antibody molecule would thereby assume a configuration complementary to that of the antigen template. This concept was first postulated by Friedrich Breinl and Felix Haurowitz in about 1930 and redefined in the 1940s in terms of protein folding by Linus Pauling.

In the 1950s, selective theories resurfaced as a result of new experimental data. Through the insights of F. Macfarlane Burnet, Niels Jerne, and David Talmadge, this model was refined into a hypothesis that came to be known as the clonal selection theory. This hypothesis has been further refined and is now accepted as an underlying paradigm of modern immunology. According to this theory, an individual B or T lymphocyte expresses many copies of a membrane receptor that is specific for a single, distinct antigen. This unique receptor specificity is determined in the lymphocyte before it is exposed to the antigen. Binding of antigen to its specific receptor activates the cell, causing it to proliferate into a clone of daughter cells that have the same receptor specificity as the parent cell. The instructional theories were formally disproved in the 1960s, by which time information was emerging about the structure of protein, RNA, and DNA that would offer new insights into the vexing problem of how an individual could make antibodies against almost anything, sight unseen.

Overview Figure 1-5 presents a very basic scheme of clonal selection in the humoral (B cell) and cellular (T cell) branches of immunity. We now know that B cells produce antibodies, a soluble version of their receptor protein, which bind to foreign proteins, flagging them for destruction. T cells, which come in several different forms, also use their surface-bound T-cell receptors to sense antigen. These cells can perform a range of different functions once selected by antigen encounter, including the secretion of soluble compounds to aid other white blood cells (such as B lymphocytes) and the destruction of infected host cells.

An Outline for the Humoral and Cell-Mediated (Cellular) Branches of the Immune System

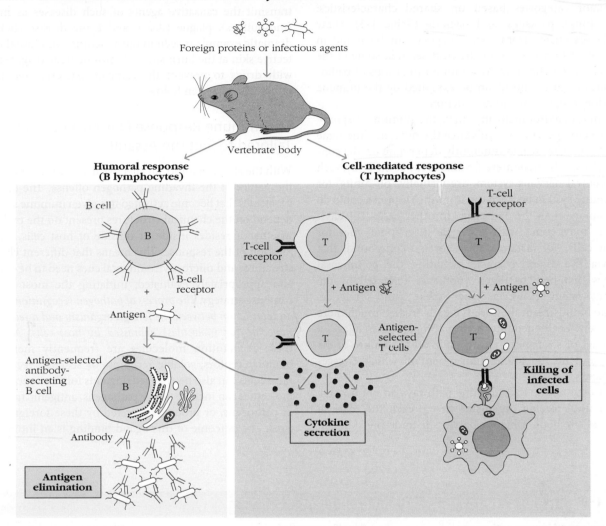

The humoral response involves interaction of B cells with foreign proteins, called antigens, and their differentiation into antibody-secreting cells. The secreted antibody binds to foreign proteins or infectious agents, helping to clear them from the body. The cell-mediated response involves various subpopulations of T lymphocytes, which can perform many functions, including the secretion of soluble messengers that help direct other cells of the immune system and direct killing of infected cells.

Important Concepts for Understanding the Mammalian Immune Response

Today, more than ever, we are beginning to understand on a molecular and cellular level how a vaccine or infection leads to the development of immunity. As highlighted by the historical studies described above, this involves a complex system of cells and soluble compounds that have evolved to protect us against an enormous range of invaders of all shapes, sizes, and chemical structures. In this section, we cover the range of organisms that challenge the immune system and several of the important new concepts that are unique hallmarks of how the immune system carries out this task.

Pathogens Come in Many Forms and Must First Breach Natural Barriers

Organisms causing disease are termed **pathogens**, and the process by which they induce illness in the host is called **pathogenesis**. The human pathogens can be grouped into four major categories based on shared characteristics: viruses, fungi, parasites, and bacteria (Table 1-3). Some example organisms from each category can be found in Figure 1-6. As we will see in the next section, some of the shared characteristics that are common to groups of pathogens, but not to the host, can be exploited by the immune system for recognition and destruction.

The microenvironment in which the immune response begins to emerge can also influence the outcome; the same pathogen may be treated differently depending on the context in which it is encountered. Some areas of the body, such as the central nervous system, are virtually "off limits" for the immune system because the immune response could do more damage than the pathogen. In other cases, the environment may come with inherent directional cues for immune cells. For instance, some foreign compounds that enter via the digestive tract, including the commensal microbes that help us digest food, are tolerated by the immune system. However, when these same foreigners enter the bloodstream they are typically treated much more aggressively. Each encounter with pathogen thus engages a distinct set of strategies that depends on the nature of the invader and on the microenvironment in which engagement occurs.

It is worth noting that immune pathways do not become engaged until foreign organisms first breach the physical barriers of the body. Obvious barriers include the skin and the mucous membranes. The acidity of the stomach contents, of the vagina, and of perspiration poses a further barrier to many organisms, which are unable to grow in low pH conditions. The importance of these barriers becomes obvious when they are surmounted. Animal bites can communicate rabies or tetanus, whereas insect puncture wounds can transmit the causative agents of such diseases as malaria (mosquitoes), plague (fleas), and Lyme disease (ticks). A dramatic example is seen in burn victims, who lose the protective skin at the burn site and must be treated aggressively with drugs to prevent the rampant bacterial and fungal infections that often follow.

The Immune Response Quickly Becomes Tailored to Suit the Assault

With the above in mind, an effective defense relies heavily on the nature of the invading pathogen offense. The cells and molecules that become activated in a given immune response depend on the chemical structures present on the pathogen, whether it resides inside or outside of host cells, and the location of the response. This means that different chemical structures and microenvironmental cues need to be detected and appropriately evaluated, initiating the most effective response strategy. *The process of pathogen recognition involves an interaction between the foreign organism and a recognition molecule (or molecules) expressed by host cells.* Although these recognition molecules are frequently membrane-bound receptors, soluble receptors or secreted recognition molecules can also be engaged. Ligands for these recognition molecules can include whole pathogens, antigenic fragments of pathogens, or products secreted by these foreign organisms. The outcome of this ligand binding is an intracellular

TABLE 1-3	Major categories of human pathogens	
Major groups of human pathogens	**Specific examples**	**Disease**
Viruses	*Poliovirus*	Poliomyelitis (Polio)
	Variola Virus	Smallpox
	Human Immunodeficiency Virus	AIDS
	Rubeola Virus	Measles
Fungi	*Candida albicans*	Candidiasis (Thrush)
	Tinea corporis	Ringworm
	Cryptococcus neoformans	Cryptococcal meningitis
Parasites	*Plasmodium species*	Malaria
	Leishmania major	Leishmaniasis
	Entamoeba histolytica	Amoebic colitis
Bacteria	*Mycobacterium tuberculosis*	Tuberculosis
	Bordetella pertussis	Whooping cough (pertussis)
	Vibrio cholerae	Cholera
	Borrelia burgdorferi	Lyme disease

(a) Virus: Rotavirus

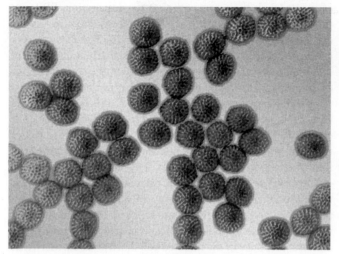

(b) Fungus: *Candida albicans*

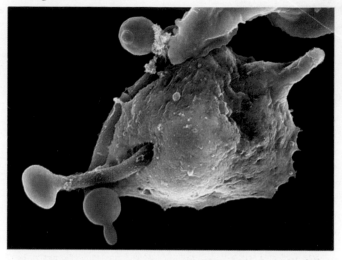

(c) Parasite: Filaria

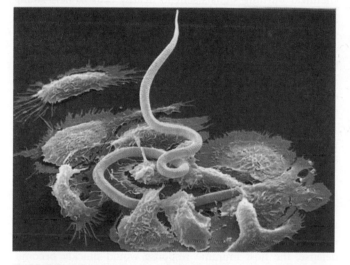

(d) Bacterium: *Mycobacterium tuberculosis*

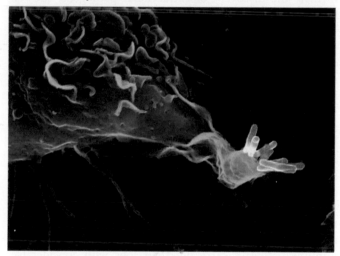

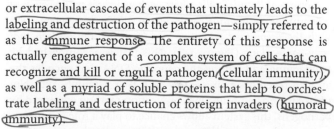

FIGURE 1-6 Pathogens representing the major categories of microorganisms causing human disease. (a) Viruses: Transmission electron micrograph of rotavirus, a major cause of infant diarrhea. Rotavirus accounts for approximately 1 million infant deaths per year in developing countries and hospitalization of about 50,000 infants per year in the United States. (b) Fungi: *Candida albicans,* a yeast inhabiting human mouth, throat, intestines, and genitourinary tract; *C. albicans* commonly causes an oral rash (thrush) or vaginitis in immunosuppressed individuals or in those taking antibiotics that kill normal bacterial flora. (c) Parasites: The larval form of filaria, a parasitic worm, being attacked by macrophages. Approximately 120 million persons worldwide have some form of filariasis. (d) Bacteria: *Mycobacterium tuberculosis,* the bacterium that causes tuberculosis, being ingested by a human macrophage. [(a) Dr. Gary Gaugler/Getty Images; (b) SPL/Photo Researchers; (c) Oliver Meckes/Nicole Ottawa/Eye of Science/Photo Researchers; (d) Max Planck Institute for Infection Biology/Dr. Volker Brinkmann.]

or extracellular cascade of events that ultimately leads to the labeling and destruction of the pathogen—simply referred to as the immune response. The entirety of this response is actually engagement of a complex system of cells that can recognize and kill or engulf a pathogen (cellular immunity), as well as a myriad of soluble proteins that help to orchestrate labeling and destruction of foreign invaders (humoral immunity).

The nature of the immune response will vary depending on the number and type of recognition molecules engaged. For instance, all viruses are tiny, obligate, intracellular pathogens that spend the majority of their life cycle residing inside host cells. An effective defense strategy must therefore involve identification of infected host cells along with recognition of the surface of the pathogen. This means that some immune cells must be capable of detecting changes that occur in a host cell after it becomes infected. This is achieved by a range of cytotoxic cells but especially **cytotoxic T lymphocytes (aka CTLs, or T_c cells),** a part of the cellular arm of immunity. In this case, recognition molecules positioned *inside* cells are key to the initial response. These intracellular receptors bind to viral proteins present in the cytosol and initiate an early warning system, alerting the cell to the presence of an invader.

Sacrifice of virally infected cells often becomes the only way to truly eradicate this type of pathogen. In general, this sacrifice is for the good of the whole organism, although in some instances it can cause disruptions to normal function. For example, the Human Immunodeficiency Virus (HIV) infects a type of T cell called a **T helper cell (T$_H$ cell)**. These cells are called helpers because they guide the behavior of other immune cells, including B cells, and are therefore pivotal for selecting the pathway taken by the immune response. Once too many of these cells are destroyed or otherwise rendered nonfunctional, many of the directional cues needed for a healthy immune response are missing and fighting all types of infections becomes problematic. As we discuss later in this chapter, the resulting immunodeficiency allows opportunistic infections to take hold and potentially kill the patient.

Similar but distinct immune mechanisms are deployed to mediate the discovery of extracellular pathogens, such as fungi, most bacteria, and some parasites. These rely primarily on cell surface or soluble recognition molecules that probe the extracellular spaces of the body. In this case, B cells and the antibodies they produce as a part of humoral immunity play major roles. For instance, antibodies can squeeze into spaces in the body where B cells themselves may not be able to reach, helping to identify pathogens hiding in these out-of-reach places. Large parasites present yet another problem; they are too big for phagocytic cells to envelop. In this case, cells that can deposit toxic substances or that can secrete products that induce expulsion (e.g., sneezing, coughing, vomiting) become a better strategy.

As we study the complexities of the mammalian immune response, it is worth remembering that a single solution does not exist for all pathogens. At the same time, these various immune pathways carry out their jobs with considerable overlap in structure and in function.

Pathogen Recognition Molecules Can Be Encoded in the Germline or Randomly Generated

As one might imagine, most pathogens express at least a few chemical structures that are not typically found in mammals. **Pathogen-associated molecular patterns** (or **PAMPs**) are common foreign structures that characterize whole groups of pathogens. It is these unique antigenic structures that the immune system frequently recognizes first. Animals, both invertebrates and vertebrates, have evolved to express several types of cell surface and soluble proteins that quickly recognize many of these PAMPs; a form of pathogen profiling. For example, encapsulated bacteria possess a polysaccharide coat with a unique chemical structure that is not found on other bacterial or human cells. White blood cells naturally express a variety of receptors, collectively referred to as **pattern recognition receptors (PRRs),** that specifically recognize these sugar residues, as well as other common foreign struc-

tures. When PRRs detect these chemical structures, a cascade of events labels the target pathogen for destruction. PPRs are proteins encoded in the genomic DNA and are always expressed by many different immune cells. These conserved, *germline-encoded* recognition molecules are thus a first line of defense for the quick detection of many of the typical chemical identifiers carried by the most common invaders.

A significant and powerful corollary to this is that it allows early categorizing or profiling of the sort of pathogen of concern. This is key to the subsequent immune response routes that will be followed, and therefore the fine tailoring of the immune response as it develops. For example, viruses frequently expose unique chemical structures only during their replication inside host cells. Many of these can be detected via intracellular receptors that bind exposed chemical moieties while still inside the host cell. This can trigger an immediate antiviral response in the infected cell that blocks further virus replication. At the same time, this initiates the secretion of chemical warning signals sent to nearby cells to help them guard against infection (a neighborhood watch system!). This early categorizing happens via a subtle tracking system that allows the immune response to make note of which recognition molecules were involved in the initial detection event and relay that information to subsequent responding immune cells, allowing the follow-up response to begin to focus attention on the likely type of assault underway.

Host-pathogen interactions are an ongoing arms race; pathogens evolve to express unique structures that avoid host detection, and the host germline-encoded recognition system co-evolves to match these new challenges. However, because pathogens generally have much shorter life cycles than their vertebrate hosts, and some utilize error-prone DNA polymerases to replicate their genomes, pathogens can evolve rapidly to evade host encoded recognition systems. If this were our only defense, the host immune response would quickly become obsolete thanks to these real-time pathogen avoidance strategies. How can the immune system prepare for this? How can our DNA encode a recognition system for things that change in random ways over time? Better yet, how do we build a system to recognize new chemical structures that may arise in the future?

Thankfully, the vertebrate immune system has evolved a clever, albeit resource intensive, response to this dilemma: to favor randomness in the design of some recognition molecules. This strategy, called **generation of diversity**, is employed only by developing B and T lymphocytes. *The result is a group of B and T cells where each expresses many copies of one unique recognition molecule, resulting in a population with the theoretical potential to respond to any antigen that may come along* (Figure 1-7). This feat is accomplished by rearranging and editing the genomic DNA that encodes the antigen receptors expressed by each B or T lymphocyte. Not unlike the error-prone DNA replication method employed by pathogens, this system allows chance to play a role in generating a menu of responding recognition molecules.

As one might imagine, however, this cutting and splicing of chromosomes is not without risk. Many B and T cells do not

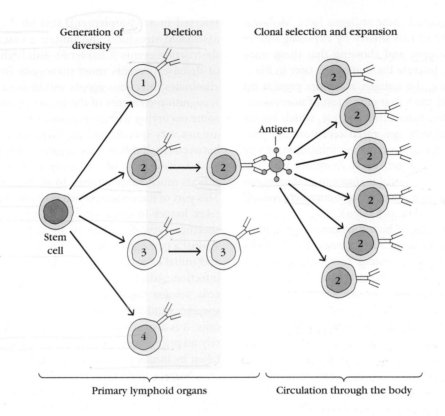

FIGURE 1-7 Generation of diversity and clonal selection in T and B lymphocytes. Maturation in T and B cells, which occurs in primary lymphoid organs (bone marrow for B cells and thymus for T cells) in the absence of antigen, produces cells with a committed antigenic specificity, each of which expresses many copies of surface receptor that binds to one particular antigen. Different clones of B cells (1, 2, 3, and 4) are illustrated in this figure. Cells that do not die or become deleted during this maturation and weeding-out process move into the circulation of the body and are available to interact with antigen. There, clonal selection occurs when one of these cells encounters its cognate or specific antigen. Clonal proliferation of an antigen-activated cell (number 2 or pink in this example) leads to many cells that can engage with and destroy the antigen, plus memory cells that can be called upon during a subsequent exposure. The B cells secrete antibody, a soluble form of the receptor, reactive with the activating antigen. Similar processes take place in the T-lymphocyte population, resulting in clones of memory T cells and effector T cells; the latter include activated T_H cells, which secrete cytokines that aid in the further development of adaptive immunity, and cytotoxic T lymphocytes (CTLs), which can kill infected host cells.

survive this DNA surgery or the quality control processes that follow, all of which take place in primary lymphoid organs: the thymus for T cells and bone marrow for B cells. Surviving cells move into the circulation of the body, where they are available if their specific, or *cognate*, antigen is encountered. When antigens bind to the surface receptors on these cells, they trigger clonal selection (see Figure 1-7). The ensuing proliferation of the selected clone of cells creates an army of cells, all with the same receptor and responsible for binding more of the same antigen, with the ultimate goal of destroying the pathogen in question. In B lymphocytes, these recognition molecules are **B-cell receptors** when they are surface structures and *antibodies* in their secreted form. In T lymphocytes, where no soluble form exists, they are **T-cell receptors**. In 1976 Susumu Tonegawa, then at The Basel Institute for Immunology in Switzerland, discovered the molecular mechanism behind the DNA recombination events that generate B-cell receptors and antibodies (Chapter 7 covers this in detail). This

was a true turning point in immunologic understanding; for this discovery he received widespread recognition, including the 1987 Nobel Prize in Physiology or Medicine (see Table 1-2).

Tolerance Ensures That the Immune System Avoids Destroying the Host

One consequence of generating random recognition receptors is that some could recognize and target the host. In order for this strategy to work effectively, the immune system must somehow avoid accidentally recognizing and destroying host tissues. This principle, which relies on self/nonself discrimination, is called tolerance, another hallmark of the immune response. The work credited with its illumination also resulted in a Nobel Prize in Physiology or Medicine, awarded to F. Macfarlane Burnet and Peter Medawar in 1960. Burnet was the first to propose that exposure to nonself antigens during certain stages of life could result in an immune system that ignored these antigens later. Medawar later proved the validity of this theory by exposing mouse embryos to foreign antigens and showing that these mice developed the ability to tolerate these antigens later in life.

To establish tolerance, the antigen receptors present on developing B and T cells must first pass a test of nonresponsiveness against host structures. This process, which begins shortly after these randomly generated receptors are produced, is achieved by the destruction or inhibition of any cells that have inadvertently generated receptors with the ability to harm the host. *Successful maintenance of tolerance ensures that the host always knows the difference between self and nonself* (usually referred to as *foreign*).

One recent re-envisioning of how tolerance is operationally maintained is called the danger hypothesis. This hypothesis suggests that the immune system constantly evaluates each new encounter more for its potential to be dangerous to the host than for whether it is self or not. For instance, cell death can have many causes, including natural homeostatic processes, mechanical damage, or infection. The former is a normal part of the everyday biological events in the body, and only requires a cleanup response to remove debris. The latter two, however, come with warning signs that include the release of intracellular contents, expression of cellular stress proteins, or pathogen-specific products. These pathogen or cell-associated stress compounds, sometimes referred to as danger signals, can engage specific host recognition molecules (e.g., PRRs) that deliver a signal to immune cells to get involved during these unnatural causes of cellular death.

One unintended consequence of robust self-tolerance is that the immune system frequently ignores cancerous cells that arise in the body, as long as these cells continue to express self structures that the immune system has been trained to ignore. Dysfunctional tolerance is at the root of most autoimmune diseases, discussed further at the end of this chapter and in greater detail in Chapter 16. As one might imagine, failures in the establishment or maintenance of tolerance can have devastating clinical outcomes.

The Immune Response Is Composed of Two Interconnected Arms: Innate Immunity and Adaptive Immunity

Although reference is made to "the immune system," it is important to appreciate that there are really two interconnected systems of immunity: innate and adaptive. These two systems collaborate to protect the body against foreign invaders. Innate immunity includes built-in molecular and cellular mechanisms that are encoded in the germline and are evolutionarily more primitive, aimed at preventing infection or quickly eliminating common invaders (Chapter 5). This includes physical and chemical barriers to infection, as well as the DNA-encoded receptors recognizing common chemical structures of many pathogens (see PRRs, above). In this case, rapid recognition and phagocytosis or destruction of the pathogen is the outcome. Innate immunity also includes a series of preexisting serum proteins, collectively

referred to as complement, that bind common pathogen-associated structures and initiate a cascade of labeling and destruction events (Chapter 6). This highly effective first line of defense prevents most pathogens from taking hold, or eliminates infectious agents within hours of encounter. The recognition elements of the innate immune system are fast, some occurring within seconds of a barrier breach, but they are not very specific and are therefore unable to distinguish between small differences in foreign antigens.

A second form of immunity, known as adaptive immunity, is much more attuned to subtle molecular differences. This part of the system, which relies on B and T lymphocytes, takes longer to come on board but is much more antigen specific. Typically, there is an adaptive immune response against a pathogen within 5 or 6 days after the barrier breach and initial exposure, followed by a gradual resolution of the infection. Adaptive immunity is slower partly because fewer cells possess the perfect receptor for the job: the antigen-specific, randomly generated receptors found on B and T cells. It is also slower because parts of the adaptive response rely on prior encounter and "categorizing" of antigens undertaken by innate processes. After antigen encounter, T and B lymphocytes undergo selection and proliferation, described earlier in the clonal selection theory of antigen specificity. Although slow to act, once these B and T cells have been selected and have honed their attack strategy, they become formidable opponents that can typically resolve the infection.

The adaptive arm of the immune response evolves in real time in response to infection and adapts (thus the name) to better recognize, eliminate, and remember the invading pathogen. Adaptive responses involve a complex and interconnected system of cells and chemical signals that come together to finish the job initiated during the innate immune response. The goal of all vaccines against infectious disease is to elicit the development of specific and long-lived adaptive responses, so that the vaccinated individual will be protected in the future when the real pathogen comes along. This arm of immunity is orchestrated mainly via B and T lymphocytes following engagement of their randomly generated antigen recognition receptors. How these receptors are generated is a fascinating story, covered in detail in Chapter 7 of this book. An explanation of how these cells develop to maturity (Chapters 9 and 10) and then work in the body to protect us from infection (Chapters 11-14) or sometimes fail us (Chapters 15-19) takes up the vast majority of this book.

The number of pages dedicated to discussing adaptive responses should not give the impression that this arm of the immune response is more important, or can work independently from, innate immunity. In fact, the full development of the adaptive response is dependent upon earlier innate pathways. The intricacies of their interconnections remain an area of intense study. The 2011 Nobel Prize in Physiology or Medicine was awarded to three scientists who helped clarify these two arms of the response: Bruce Beutler and Jules Hoffmann for discoveries related to the activation events important for innate immunity, and Ralph Steinman for his discovery of the role of dendritic cells in activating adaptive immune responses (see Table 1-2). Because innate pathways make first contact with pathogens, the cells and molecules involved in this arm of the response use information gathered from their early encounter with pathogen to help direct the process of adaptive immune development. *Adaptive immunity thus provides a second and more comprehensive line of defense, informed by the struggles undertaken by the innate system.* It is worth noting that some infections are, in fact, eliminated by innate immune mechanisms alone, especially those that remain localized and involve very low numbers of fairly benign foreign invaders. (Think of all those insect bites or splinters in life that introduce bacteria under the skin!) Table 1-4 compares the major characteristics that distinguish innate and adaptive immunity. Although for ease of discussion the immune system is typically divided into these two arms of the response, there is considerable overlap of the cells and mechanisms involved in each of these arms of immunity.

For innate and adaptive immunity to work together, these two systems must be able to communicate with one another. This communication is achieved by both cell-cell contact and by soluble messengers. Most of these soluble proteins are growth factor–like molecules known by the general name **cytokines**. Cytokines and cell surface ligands can bind with receptors found on responding cells and signal these cells to perform new functions, such as synthesis of other soluble factors or differentiation to a new cell type. A subset of these soluble signals are called **chemokines** because they have chemotactic activity, meaning they can recruit specific cells to the site. In this way, cytokines, chemokines, and other soluble factors produced by immune cells recruit or instruct cells and soluble proteins important for eradication of the pathogen from within the infection site. We've probably all felt this convergence in the form of swelling, heat, and tenderness at the site of exposure. These events are a part of a larger process collectively referred to as an **inflammatory response**, which is covered in detail in Chapter 15.

Adaptive Immune Responses Typically Generate Memory

One particularly significant and unique attribute of the adaptive arm of the immune response is **immunologic**

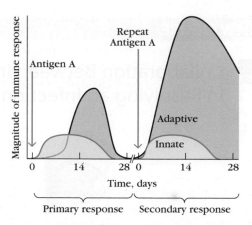

FIGURE 1-8 Differences in the primary and secondary response to injected antigen reflect the phenomenon of immunologic memory. When an animal is injected with an antigen, it produces a primary antibody response (dark blue) of low magnitude and short duration, peaking at about 10 to 20 days. At some later point, a second exposure to the same antigen results in a secondary response that is greater in magnitude, peaks in less time (1–4 days), and is more antigen specific than the primary response. Innate responses, which have no memory element and occur each time an antigen is encountered, are unchanged regardless of how frequently this antigen has been encountered in the past (light blue).

memory. This is the ability of the immune system to respond much more swiftly and with greater efficiency during a second exposure to the same pathogen. Unlike almost any other biological system, the vertebrate immune response has evolved not only the ability to learn from (adapt to) its encounters with foreign antigen in real time but also the ability to store this information for future use. During a first encounter with foreign antigen, adaptive immunity undergoes what is termed a **primary response** during which the key lymphocytes that will be used to eradicate the pathogen are clonally selected, honed, and enlisted to resolve the infection. As mentioned above, these cells incorporate messages received from the innate players into their tailored response to the specific pathogen.

All subsequent encounters with the same antigen or pathogen are referred to as the **secondary response** (Figure 1-8).

	Innate	Adaptive
Response time	Minutes to hours	Days
Specificity	Limited and fixed	Highly diverse; adapts to improve during the course of immune response
Response to repeat infection	Same each time	More rapid and effective with each subsequent exposure
Major components	Barriers (e.g., skin); phagocytes; pattern recognition molecules	T and B lymphocytes; antigen-specific receptors; antibodies

TABLE 1-4   Comparison of innate and adaptive immunity

Collaboration Between Innate and Adaptive Immunity in Resolving an Infection

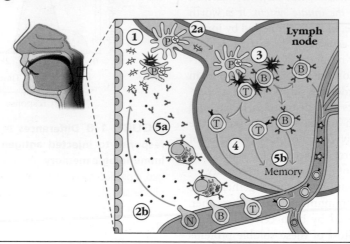

This very basic scheme shows the sequence of events that occurs during an immune response, highlighting interactions between innate and adaptive immunity. 1. Pathogens are introduced at a mucosal surface or breach in skin (bacteria entering the throat, in this case), where they are picked up by phagocytic cells (yellow). 2a. In this innate stage of the response, the phagocytic cell undergoes changes and carries pieces of bacteria to a local lymph node to help activate adaptive immunity. 2b. Meanwhile, at the site of infection resident phagocytes encountering antigen release chemokines and cytokines (black dots) that cause fluid influx and help recruit other immune cells to the site (inflammation). 3. In the lymph node, T (blue) and B (green) cells with appropriate receptor specificity are clonally selected when their surface receptors bind antigen that has entered the system, kicking off adaptive immunity. 4. Collaboration between T and B cells and continued antigen encounter occurs in the lymph node, driving lymphocyte proliferation and differentiation, generating cells that can very specifically identify and eradicate the pathogen. For example: 5a. B cells secrete antibodies specific for the antigen, which travels to the site of infection to help label and eradicate the pathogen. 5b. In addition to the cells that will destroy the pathogen here, memory T and B cells are generated in this primary response and will be available at the initiation of a secondary response, which will be much more rapid and antigen specific. (Abbreviations: T = T cell, B = B cell, P = phagocyte; N = neutrophil, a type of immune cell.)

During a secondary response, memory cells, kin of the final and most efficient B and T lymphocytes trained during the primary response, are re-enlisted to fight again. These cells begin almost immediately and pick up right where they left off, continuing to learn and improve their eradication strategy during each subsequent encounter with the same antigen. Depending on the antigen in question, memory cells can remain for decades after the conclusion of the primary response. Memory lymphocytes provide the means for subsequent responses that are so rapid, antigen-specific, and effective that when the same pathogen infects the body a second time, dispatch of the offending organism often occurs without symptoms. It is the remarkable property of memory that prevents us from catching many diseases a second time. Immunologic memory harbored by residual B and T lymphocytes is the foundation for vaccination, which uses crippled or killed pathogens as a safe way to "educate" the immune system to prepare it for later attacks by life-threatening pathogens.

Overview Figure 1-9 highlights the ways in which the innate and adaptive immune responses work together to resolve an infection. In this example, bacteria breach the mucosal lining of the throat, a skin or mucous barrier, where it is recognized and engulfed by a local phagocytic cell (step 1). As part of the innate immune system, the phagocytic cell releases cytokines and chemokines that attract other white blood cells to the site of infection, initiating inflammation (step 2b). That phagocytic cell may then travel to a local lymph node, the tissue where antigen and lymphocytes meet, carrying bacterial antigens to B and T lymphocytes (step 2a). Those lymphocytes with receptors that are specific for the antigen are selected, activated, and begin the adaptive immune response by proliferating (step 3). Activated T_H cells help to activate B cells, and clonal expansion of both types of lymphocyte occurs in the lymph node (step 4). This results in many T and B cells specific for the antigen, with the latter releasing antibodies that can attach to the intruder and direct its destruction (step 5a). The adaptive response leaves behind memory T and B cells available for a future, secondary encounter with this antigen (step 5b). It is worth noting that memory is a unique capacity that arises from adaptive responses; there is no memory component of innate immunity.

Sometimes, as is the case for some vaccines, one round of antigen encounter and adaptation is not enough to impart protective immunity from the pathogen in question. In many of these cases, immunity can develop after a second or even a third round of exposure to an antigen. It is these sorts of pathogens that necessitate the use of vaccine booster shots. Booster shots are nothing more than a second or third episode of exposure to the antigen, each driving a new round of adaptive events (secondary response) and refinements in the responding lymphocyte population. The aim is to hone these responses to a sufficient level to afford protection against the real pathogen at some future date.

The Good, Bad, and Ugly of the Immune System

The picture we've presented so far depicts the immune response as a multicomponent interactive system that always protects the host from invasion by all sorts of pathogens. However, failures of this system do occur. They can be dramatic and often garner a great deal of attention, despite the fact that they are generally rare. Certain clinical situations also pose unique challenges to the immune system, including tissue transplants between individuals (probably not part of any evolutionary plan!) and the development of cancer. In this section we briefly describe some examples of common failures and challenges to the development of healthy immune responses. Each of these clinical manifestations is covered in much greater detail in the concluding chapters of this book (Chapters 15–19).

Inappropriate or Dysfunctional Immune Responses Can Result in a Range of Disorders

Most instances of immune dysfunction or failure fall into one of the following three broad categories:

- **Hypersensitivity (including allergy):** overly zealous attacks on common benign but foreign antigens

- **Autoimmune Disease:** erroneous targeting of self-proteins or tissues by immune cells

- **Immune Deficiency:** insufficiency of the immune response to protect against infectious agents

A brief overview of these situations and some examples of each are presented below. At its most basic level, immune dysfunction occurs as a result of improper regulation that allows the immune system to either attack something it shouldn't or fail to attack something it should. Hypersensitivities, including allergy, and autoimmune disease are cases of the former, where the immune system attacks an improper target. As a result, the symptoms can manifest as *pathological inflammation*—an influx of immune cells and molecules that results in detrimental symptoms, including chronic inflammation and rampant tissue destruction. In contrast, immune deficiencies, caused by a failure to properly deploy the immune response, usually result in weakened or dysregulated immune responses that can allow pathogens to get the upper hand.

This is a good time to mention that the healthy immune response involves a balancing act between immune aggression and immune suppression pathways. While we rarely fail to consider erroneous attacks (autoimmunity) or failures to engage (immune deficiency) as dysfunctional, we sometimes forget to consider the significance of the suppressive side of the immune response. Imperfections in the inhibitory arm of the immune response, present as a check to balance all the immune attacks we constantly initiate, can be equally profound. Healthy immune responses must therefore be viewed as a delicate balance, spending much of the time with one foot on the brakes and one on the gas.

Hypersensitivity Reactions

Allergies and asthma are examples of hypersensitivity reactions. These result from inappropriate and overly active immune responses to common innocuous environmental antigens, such as pollen, food, or animal dander. The possibility that certain substances induce increased sensitivity (hypersensitivity) rather than protection was recognized in about 1902 by Charles Richet, who attempted to immunize dogs against the toxins of a type of jellyfish. He and his colleague Paul Portier observed that dogs exposed to sublethal doses of the toxin reacted almost instantly, and fatally, to a later challenge with even minute amounts of the same toxin. Richet concluded that a successful vaccination typically results in *phylaxis* (protection), whereas anaphylaxis (anti-protection)—an extreme, rapid, and often lethal overreaction of the immune response to something it has encountered before—can result in certain cases in which exposure to antigen is repeated. Richet received the Nobel Prize in 1913 for his discovery of the anaphylactic response (see Table 1-2). The term is used today to describe a severe, life-threatening, allergic response.

Fortunately, most hypersensitivity or allergic reactions in humans are not rapidly fatal. There are several different types of hypersensitivity reactions; some are caused by antibodies and others are the result of T-cell activity (see Chapter 15). However, most allergic or anaphylactic responses involve a type of antibody called *immunoglobulin E* (IgE). Binding of IgE to its specific antigen (allergen) induces the release of substances that cause irritation and inflammation, or the accumulation of cells and fluid at the site. When an allergic individual is exposed to an allergen, symptoms may include sneezing, wheezing (Figure 1-10), and difficulty in breathing (asthma); dermatitis or skin eruptions (hives); and, in more severe cases, strangulation due to constricted airways

The Hygiene Hypothesis

Worldwide, 300 million people suffer from asthma and approximately 250,000 people died from the disease in 2007 (see Chapter 15). As of 2009, in the United States alone, approximately 1 in 12 people (8.2%) are diagnosed with asthma. The most common reason for a trip to a hospital emergency room (ER) is an asthma attack, accounting for one-third of all visits. In addition to those treated in the ER, over 400,000 hospitalizations for asthma occurred in the United States in 2006, with an average stay of 3 to 4 days.

In the past 25 years, the prevalence of asthma in industrialized nations has doubled. This is coupled with an overall rise in other types of allergic disease during the same time frame. What accounts for this climb in asthma and allergy in the last few decades? One idea, called the hygiene hypothesis, suggests that a decrease in human exposure to environmental microbes has had adverse effects on the human immune system. The hypothesis suggests that several categories of allergic or inflammatory disease, all disorders caused by excessive immune activation, have become more prevalent in industrialized nations thanks to diminished exposure to particular classes of microbes following the widespread use of antibiotics, immunization programs, and overall hygienic practices in those countries. This idea was first proposed by D. P. Strachan and colleagues in an article published in 1989 suggesting a link between hay fever and household hygiene. More recently, this hypothesis has been expanded to include the view by some that it may be a contributing factor in many allergic diseases, several autoimmune disorders, and, more recently, inflammatory bowel disease.

What is the evidence supporting the hygiene hypothesis? The primary clinical support comes from studies that have shown a positive correlation between growing up under environmental conditions that favor microbe-rich (sometimes called "dirty") environments and a decreased incidence of allergy, especially asthma. To date, childhood exposure to cowsheds and farm animals, having several older siblings, attending day care early in life, or growing up in a developing nation have all been correlated with a decreased likelihood of developing allergies later in life. While viral exposures during childhood do not seem to favor protection, exposure to certain classes of bacteria and parasitic organisms may. Of late, the primary focus of attention has been on specific classes of parasitic worms (called helminthes), spawning New Age allergy therapies involving intentional exposure. This gives whole new meaning to the phrase "Go eat worms"!

What are the proposed immunologic mechanisms that might underlie this link between a lack of early-life microbial exposure and allergic disease? Current dogma supporting this hypothesis posits that millions of years of coevolution of microbes and humans has favored a system in which early exposure to a broad range of common environmental bugs helps set the immune system on a path of homeostatic balance between aggression and inhibition. Proponents of this immune regulation argument, sometimes referred to as the "old friends" hypothesis, suggest that antigens present on microbial organisms that have played a longstanding role in our evolutionary history (both pathogens and harmless microbes we ingest or that make up our historical flora) may engage with the pattern recognition receptors (PPRs) present on cells of our innate immune system, driving them to warn cells involved in adaptive responses to tone it down. This hypothesis posits that without early and regular exposure of our immune cells to these old friends and their antigens, the development of "normal" immune regulatory or homeostatic responses is thrown into disarray, setting us up for an immune system poised to overreact in the future.

Animal models of disease lend some support to this hypothesis and have helped immunologists probe this line of thinking. For instance, certain animals raised in partially or totally pathogen-free environments are more prone to type 1, or insulin-dependent, diabetes, an autoimmune disease caused by immune attack of pancreatic cells (see Chapter 16). The lower the infectious burden of exposure in these mice, the greater the incidence of diabetes. Animals specifically bred to carry enhanced genetic susceptibility favoring spontaneous development of diabetes (called NOD mice, for *non*-obese *d*iabetic) and treated with a variety of infectious agents can be protected from diabetes. Meanwhile, NOD mice maintained in pathogen-free housing almost uniformly develop diabetes. Much like this experimental model, susceptibility to asthma and most other allergies is known to run in families. Although all the genes linked to asthma have not yet been characterized, it is known that you have a 30% chance of developing the disease if one of your parents is a sufferer, and a 70% chance if both parents have asthma. While the jury may still be out concerning the verdict behind the hygiene hypothesis, animal and human studies clearly point to strong roles for both genes and environment in susceptibility to allergy. As data in support of this hypothesis continue to grow, the old saying concerning a dirty child—that "It's good for their immune system"—may actually hold true!

Centers for Disease Control and Prevention. 2012. CDC: Preventing Chronic Disease **9**: 110054.

Liu, A. H., and Murphy, J. R. 2003. Hygiene hypothesis: Fact or fiction? Journal of Allergy and Clinical Immunology **111**(3):471–478.

Okada, H., et al. 2010. The hygiene hypothesis for autoimmune and allergic diseases: An update. Clinical and Experimental Immunology **160**:1.

Sironi, M., and M. Clerici. 2010. The hygiene hypothesis: An evolutionary perspective. Microbes and Infection **12**:421.

FIGURE 1-10 Patient suffering from hay fever as a result of an allergic reaction. Such hypersensitivity reactions result from sensitization caused by previous exposure to an antigen in some individuals. In the allergic individual, histamines are released as a part of the hypersensitivity response and cause sneezing, runny nose, watery eyes, and such during each subsequent exposure to the antigen (now called an allergen) *[Source: Chris Rout/Alamy.]*

following extreme inflammation. A significant fraction of our health resources is expended to care for those suffering from allergies and asthma. One particularly interesting rationale to explain the unexpected rise in allergic disease is called the hygiene hypothesis and is discussed in the Clinical Focus on page 20.

Autoimmune Disease

Sometimes the immune system malfunctions and a breakdown in self-tolerance occurs. This could be caused by a sudden inability to distinguish between self and nonself or by a misinterpretation of a self-component as dangerous, causing an immune attack on host tissues. This condition, called autoimmunity, can result in a number of chronic debilitating diseases. The symptoms of autoimmunity differ, depending on which tissues or organs are under attack. For example, multiple sclerosis is due to an autoimmune attack on a protein in nerve sheaths in the brain and central nervous system that results in neuromuscular dysfunction. Crohn's disease is an attack on intestinal tissues that leads to destruction of gut epithelia and poor absorption of food. One of the most common autoimmune disorders, rheumatoid arthritis, results from an immune attack on joints of the hands, feet, arms, and legs.

Both genetic and environmental factors are likely involved in the development of most autoimmune diseases. However, the exact combination of genes and environmental exposures that favor the development of a particular autoimmune disease are difficult to pin down, and constitute very active areas of immunologic research. Recent discoveries in these areas and the search for improved treatments are all covered in greater detail in Chapter 16.

Immune Deficiency

In most cases, when a component of innate or adaptive immunity is absent or defective, the host suffers from some form of **immunodeficiency**. Some of these deficiencies produce major clinical effects, including death, while others are more minor or even difficult to detect. Immune deficiency can arise due to inherited genetic factors (called **primary immunodeficiencies**) or as a result of disruption/damage by chemical, physical, or biological agents (termed **secondary immunodeficiencies**). Both of these forms of immune deficiency are discussed in greater detail in Chapter 18.

The severity of the disease resulting from immune deficiency depends on the number and type of affected immune response components. A common type of primary immunodeficiency in North America is a selective immunodeficiency in which only one type of antibody, called *Immunoglobulin A* is lacking; the symptoms may be an increase in certain types of infections, or the deficiency may even go unnoticed. In contrast, a more rare but much more extreme deficiency, called **severe combined immunodeficiency (SCID)**, affects both B and T cells and basically wipes out adaptive immunity. When untreated, SCID frequently results in death from infection at an early age.

By far, the most common form of secondary immunodeficiency is Acquired Immune Deficiency Syndrome (AIDS), resulting from infection with Human Immunodeficiency Virus (HIV). As discussed further in Chapter 18, humans do not effectively recognize and eradicate this virus. Instead, a state of persistent infection occurs, with HIV hiding inside the genomes of T_H cells, its target cell type and the immune cell type that is critical to guiding the direction of the adaptive immune response. As the immune attack on the virus mounts, more and more of these T_H cells are lost. When the disease progresses to AIDS, so many T_H cells have been destroyed or otherwise rendered dysfunctional that a gradual collapse of the immune system occurs. It is estimated that at the end of 2010, more than 34 million people worldwide suffered from this disease (for more current numbers, see www.unaids.org), which if not treated can be fatal. For patients with access, certain anti-retroviral treatments can now prolong life with HIV almost indefinitely. However, there is neither a vaccine nor a cure for this disease.

It is important to note that many pervasive pathogens in our environment cause no problem for healthy individuals thanks to the immunity that develops following initial exposure. However, individuals with primary or secondary deficiencies in immune function become highly susceptible to disease caused by these ubiquitous microbes. For example, the fungus *Candida albicans*, present nearly everywhere and a nonissue for most individuals, can cause an irritating rash and a spreading infection in the mucosal surface of the mouth and vagina in patients suffering from immune deficiency (see Figure 1-6b). The resulting rash, called thrush, can sometimes be the first sign of immune dysfunction (Figure 1-11). If left unchecked, *C. albicans* can spread, causing systemic candidiasis, a life-threatening condition. Such

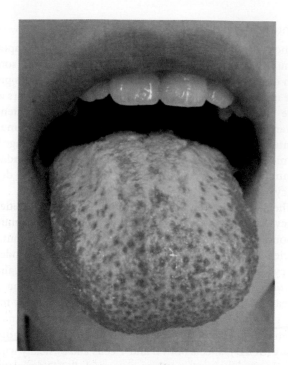

FIGURE 1-11 An immune deficient patient suffering from oral thrush due to opportunistic infection by _Candida albicans._ _[Creative Commons <http://en.wikipedia.org/wiki/File: Thrush.JPG]_

infections by ubiquitous microorganisms that cause no harm in an immune competent host but which are often observed only in cases of underlying immune deficiency, are termed **opportunistic infections**. Several rarely seen opportunistic infections identified in early AIDS patients were the first signs that these patients had seriously compromised immune systems, and helped scientists to identify the underlying cause. Unchecked opportunistic infections are still the main cause of death in AIDS patients.

The Immune Response Renders Tissue Transplantation Challenging

Normally, when the immune system encounters foreign cells, it responds strongly to rid the host of the presumed invader. However, in the case of transplantation, these cells or tissues from a donor may be the only possible treatment for life-threatening disease. For example, it is estimated that more than 70,000 persons in the United States alone would benefit from a kidney transplant. The fact that the immune system will attack and reject any transplanted organ that is nonself, or not a genetic match, raises a formidable barrier to this potentially lifesaving treatment, presenting a particularly unique challenge to clinicians who treat these patients. While the rejection of this transplant by the recipient's immune system may be seen as a "failure," in fact it is just a consequence of the immune system functioning properly. Normal tolerance processes governing self/nonself discrimination and immune engagement caused by danger signals (partially the result of the trauma caused by surgical transplantation) lead to the rapid influx of

immune cells and coordinated attacks on the new resident cells. Some of these transplant rejection responses can be suppressed using immune inhibitory drugs, but treatment with these drugs also suppresses general immune function, leaving the host susceptible to opportunistic infections.

Research related to transplantation studies has played a major role in the development of the field of Immunology. A Nobel Prize was awarded to Karl Landsteiner (mentioned earlier for his contributions to the concept of immune specificity) in 1930 for the discovery of the human ABO blood groups, a finding that allowed blood transfusions to be carried out safely. In 1980, G. Snell, J. Dausset, and B. Benacerraf were recognized for discovery of the major histocompatibility complex (MHC). These are the tissue antigens that differ most between non-genetically identical individuals, and are thus one of the primary targets of immune rejection of transplanted tissues. Finally, in 1991 E. D. Thomas and J. Murray were awarded Nobel prizes for treatment advances that paved the way for more clinically successful tissue transplants (see Table 1-2). Development of procedures that would allow a foreign organ or cells to be accepted without suppressing immunity to all antigens still remains a major goal, and a challenge, for immunologists today (see Chapter 16).

Cancer Presents a Unique Challenge to the Immune Response

Cancer, or **malignancy**, occurs in host cells when they begin to divide out of control. Since these cells are self in origin, self-tolerance mechanisms can inhibit the development of an immune response, making the detection and eradication of cancerous cells a continual challenge. That said, it is clear that many tumor cells do express unique or developmentally inappropriate proteins, making them potential targets for immune cell recognition and elimination, as well as targets for therapeutic intervention. However, as with many microbial pathogens, the increased genetic instability of these rapidly dividing cells gives them an advantage in terms of evading immune detection and elimination machinery.

We now know that the immune system actively participates in the detection and control of cancer in the body (see Chapter 19). The number of malignant disorders that arise in individuals with compromised immunity highlights the degree to which the immune system normally controls the development of cancer. Both innate and adaptive elements have been shown to be involved in this process, although adaptive immunity likely plays a more significant role. However, associations between inflammation and the development of cancer, as well as the degree to which cancerous cells evolve to become more aggressive and evasive under pressure from the immune system, have demonstrated that the immune response to cancer can have both healing and disease-inducing characteristics. As the mechanics of these elements are resolved in greater detail, there is hope that therapies can be designed to boost or maximize the anti-tumor effects of immune cells while dampening their tumor enhancing activities.

Our understanding of the immune system has clearly come a very long way in a fairly short time. Yet much still remains to be learned about the mammalian immune response and the ways in which this system interacts with other body systems. With this enhanced knowledge, the hope is that we will be better poised to design ways to modulate these immune pathways through intervention. This would allow us to develop more effective prevention and treatment strategies for cancer and other diseases that plague society today, not to mention preparing us to respond quickly to the new diseases or infectious agents that will undoubtedly arise in the future.

SUMMARY

- Immunity is the state of protection against foreign pathogens or substances (antigens).

- Vaccination is a means to prepare the immune system to effectively eradicate an infectious agent before it can cause disease, and its widespread use has saved many lives.

- Humoral immunity involves combating pathogens via antibodies, which are produced by B cells and can be found in bodily fluids. Antibodies can be transferred between individuals to provide passive immune protection.

- Cell-mediated immunity involves primarily antigen-specific T lymphocytes, which act to eradicate pathogens or otherwise aid other cells in inducing immunity.

- Pathogens fall into four major categories and come in many forms. The immune response quickly becomes tailored to the type of organism involved.

- The immune response relies on recognition molecules that can be germline encoded or randomly generated.

- The process of self-tolerance ensures that the immune system avoids destroying host tissue.

- The vertebrate immune response can be divided into two interconnected arms of immunity: innate and adaptive.

- Innate responses are the first line of defense, utilizing germline-encoded recognition molecules and phagocytic cells. Innate immunity is faster but less specific than adaptive responses, which take several days but are highly antigen specific.

- Innate and adaptive immunity operate cooperatively; activation of the innate immune response produces signals that stimulate and direct subsequent adaptive immune pathways.

- Adaptive immunity relies upon surface receptors, called B- and T-cell receptors, that are randomly generated by DNA rearrangements in developing B and T cells.

- Clonal selection is the process by which individual T and B lymphocytes are engaged by antigen and cloned to create a population of antigen-reactive cells.

- Memory cells are residual B and T cells that remain after antigen exposure and that pick up where they left off during a subsequent, or secondary, response.

- Dysfunctions of the immune system include common maladies such as allergies, asthma, and autoimmune disease (overly active or misdirected immune responses) as well as immune deficiency (insufficient immune responses).

- Transplanted tissues and cancer present unique challenges to clinicians, because the healthy immune system typically rejects or destroys nonself proteins, such as those encountered in most transplant situations, and tolerates self cells.

REFERENCES

Burnet, F. M. 1959. *The Clonal Selection Theory of Acquired Immunity.* Cambridge University Press, Cambridge, England.

Desour, L. 1922. *Pasteur and His Work* (translated by A. F. and B. H. Wedd). T. Fisher Unwin, London.

Kimbrell, D. A., and B. Beutler. 2001. The evolution and genetics of innate immunity. *Nature Reviews Genetics* 2:256.

Kindt, T. J., and J. D. Capra. 1984. *The Antibody Enigma.* Plenum Press, New York.

Landsteiner, K. 1947. *The Specificity of Serologic Reactions.* Harvard University Press, Cambridge, MA.

Medawar, P. B. 1958. *The Immunology of Transplantation: The Harvey Lectures, 1956–1957.* Academic Press, New York.

Metchnikoff, E. 1905. *Immunity in the Infectious Diseases.* Macmillan, New York.

Paul, W., ed. 2003. *Fundamental Immunology,* 5th ed. Lippincott Williams & Wilkins, Philadelphia.

Silverstein, A. M. 1979. History of immunology. Cellular versus humoral immunity: determinants and consequences of an epic 19th century battle. *Cellular immunology.* **48**:208.

Useful Web Sites

www.aai.org The Web site of the American Association of Immunologists contains a good deal of information of interest to immunologists.

www.ncbi.nlm.nih.gov/PubMed PubMed, the National Library of Medicine database of more than 9 million publications, is the world's most comprehensive bibliographic database for biological and biomedical literature. It is also a highly user-friendly site.

www.aaaai.org The American Academy of Allergy Asthma and Immunology site includes an extensive library of information about allergic diseases.

www.who.int/en The World Health Organization directs and coordinates health-related initiatives and collects worldwide health statistics data on behalf of the United Nations system.

www.cdc.gov Part of the United States Department of Health and Human Services, the Centers for Disease Control and Prevention coordinates health efforts in the United States and provides statistics on U.S. health and disease.

www.nobelprize.org/nobel_prizes/medicine/ laureates The official Web site of the Nobel Prize in Physiology or Medicine.

www.historyofvaccines.org A Web site run by The College of Physicians of Philadelphia with facts, articles, and timelines related to vaccine developments.

www.niaid.nih.gov The National Institute of Allergy and Infectious Disease is a branch of the U.S. National Institute of Health that specifically deals with research, funding, and statistics related to basic immunology, allergy, and infectious disease threats.

 STUDY QUESTIONS

1. Why was Jenner's vaccine superior to previous methods for conferring resistance to smallpox?

2. Did the treatment for rabies used by Pasteur confer active or passive immunity to the rabies virus? Is there any way to test this?

3. Infants immediately after birth are often at risk for infection with group B *Streptococcus*. A vaccine is proposed for administration to women of childbearing years. How can immunizing the mothers help the babies?

4. Indicate to which branch(es) of the immune system the following statements apply, using H for the humoral branch and CM for the cell-mediated branch. Some statements may apply to both branches (B).
   a. Involves B cells
   b. Involves T cells
   c. Responds to extracellular bacterial infection
   d. Involves secreted antibody
   e. Kills virus-infected self cells

5. Adaptive immunity exhibits several characteristic attributes, which are mediated by lymphocytes. List four attributes of adaptive immunity and briefly explain how they arise.

6. Name three features of a secondary immune response that distinguish it from a primary immune response.

7. Give examples of mild and severe consequences of immune dysfunction. What is the most common cause of immunodeficiency throughout the world today?

8. For each of the following statements, indicate whether the statement is true or false. If you think the statement is false, explain why.
   a. Booster shots are required because repeated exposure to an antigen builds a stronger immune response.
   b. The gene for the T cell receptor must be cut and spliced together before it can be expressed.
   c. Our bodies face the greatest onslaught from foreign invaders through our skin.
   d. Increased production of antibody in the immune system is driven by the presence of antigen.

   e. Innate immunity is deployed only during the primary response, and adaptive immunity begins during a secondary response.
   f. Autoimmunity and immunodeficiency are two different terms for the same set of general disorders.
   g. If you receive intravenous immunoglobulin to treat a snakebite, you will be protected from the venom of this snake in the future, but not venom from other types of snakes.
   h. Innate and adaptive immunity work collaboratively to mount an immune response against pathogens.
   i. The genomic sequences in our circulating T cells for encoding a T-cell receptor are the same as those our parents carry in their T cells.
   j. Both the innate and adaptive arms of the immune response will be capable of responding more efficiently during a secondary response.

9. What was the significance of the accidental re-inoculation of some chickens that Pasteur had previously exposed to the bacteria that causes cholera? Why do you think these chickens did not die after the first exposure to this bacterium?

10. Briefly describe the four major categories of pathogen. Which are likely to be the most homogenous in form and which the most diverse? Why?

11. Describe how the principle of herd immunity works to protect unvaccinated individuals. What characteristics of the pathogen or of the host do you think would most impact the degree to which this principle begins to take hold?

12. What is the difference between the discarded instructional theory for lymphocyte specificity and the selection theory, which is now the accepted explanation?

13. Compare and contrast innate and adaptive immunity by matching the following characteristics with the correct arm of immunity, using I for innate and A for adaptive:
   a. Is the first to engage upon initial encounter with antigen
   b. Is the most pathogen specific

c. Employs T and B lymphocytes

d. Adapts during the response

e. Responds identically during a first and second exposure to the same antigen

f. Responds more effectively during a subsequent exposure

g. Includes a memory component

h. Is the target of vaccination

i. Can involve the use of PAMP receptors

j. Involves antigen-specific receptors binding to pathogens

k. Can be mediated by antibodies

14. What is meant by the term *tolerance*? How do we become tolerant to the structures in our own bodies?

15. What is an antigen? An antibody? What is their relationship to one another?

16. How are PRRs different from B- or T-cell receptors? Which is most likely to be involved in innate immunity and which in adaptive immunity?

17. In general terms, what role do cytokines play in the development of immunity? How does this compare to chemokines?

2

Cells, Organs, and Microenvironments of the Immune System

A successful immune response to a pathogen depends on finely choreographed interactions among diverse cell types (see Figure 1-9): innate immune cells that mount a first line of defense against pathogen, antigen-presenting cells that communicate the infection to lymphoid cells, which coordinate the adaptive response and generate the memory cells, which prevent future infections. The coordination required for the development of a full immune response is made possible by the specialized anatomy and microanatomy of the immune system, which is dispersed throughout the body and organizes cells in time and space. **Primary lymphoid organs**—including the bone marrow and the thymus—regulate the development of immune cells from immature precursors. **Secondary lymphoid organs**—including the spleen, lymph nodes, and specialized sites in the gut and other mucosal tissues—coordinate the encounter of antigen with antigen-specific lymphocytes and their development into effector and memory cells. Blood vessels and lymphatic systems connect these organs, uniting them into a functional whole.

Remarkably, all functionally specialized, mature blood cells (red blood cells, granulocytes, macrophages, dendritic cells, and lymphocytes) arise from a single cell type, the **hematopoietic stem cell (HSC)** (Figure 2-1). The process by which HSCs differentiate into mature blood cells is called **hematopoiesis**. Two primary lymphoid organs are responsible for the development of stem cells into mature immune cells: the bone marrow, where HSCs reside and give rise to all cell types; and the thymus, where T cells complete their maturation. We will begin this chapter by describing the structure and function of each cell type that arises from HSCs, and the structure and function of both the bone marrow and thymus in the context of hematopoiesis and thymopoiesis. We will then describe the secondary lymphoid organs, where the immune response is initiated. The lymph nodes and spleen will be featured, but lymphoid tissue associated with mucosal layers will also be discussed. Four focused discussions are also

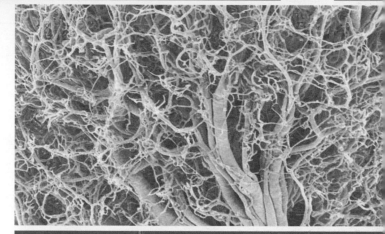

Scanning electron micrograph of blood vessels in a lymph node. *Susumu Nishinaga/ Photo Researchers*

- Cells of the Immune System
- Primary Lymphoid Organs—Where Immune Cells Develop
- Secondary Lymphoid Organs—Where the Immune Response Is Initiated

included in this chapter. Specifically, in two Classic Experiment Boxes, we describe the discovery of a second thymus and the history behind the identification of hematopoietic stem cells. In a Clinical Focus Box, we discuss the clinical use and promise of hematopoietic stem cells, and finally, in an Evolution Box, we describe some intriguing variations in the anatomy of the immune system among our vertebrate relatives.

Cells of the Immune System

Stem cells are defined by two capacities: (1) the ability to regenerate or "self-renew" and (2) the ability to differentiate into all diverse cell types. Embryonic stem cells have the capacity to generate every specialized cell type in an organism (in other words, they are *pluripotent*). Adult stem cells, in contrast, have the capacity to give rise to the diverse cell types that specify a particular tissue. Multiple adult organs harbor stem cells ("adult stem cells") that can give rise to mature tissue-specific cells. The HSC is considered the paradigmatic adult stem cell because it can differentiate into all the types of blood cells.

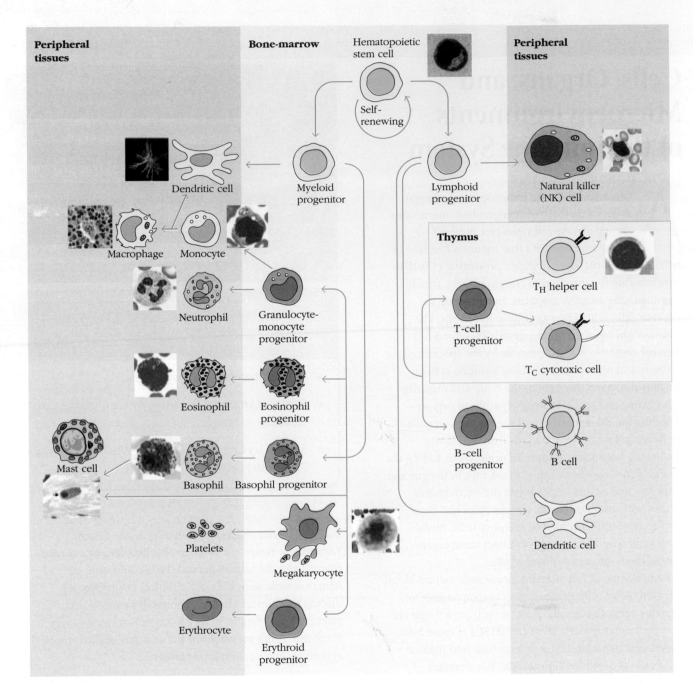

FIGURE 2-1 Hematopoiesis. Self-renewing hematopoietic stem cells give rise to lymphoid and myeloid progenitors. Most immune cells mature in the bone marrow and then travel to peripheral organs via the blood. Some, including mast cells and macrophages, undergo further maturation outside the bone marrow. T cells develop to maturity in the thymus.

Hematopoietic Stem Cells Have the Ability to Differentiate into Many Types of Blood Cells

HSCs are rare—fewer than one HSC is present per 5×10^4 cells in the bone marrow—and their numbers are strictly controlled by a balance of cell division, death, and differentiation.

Under conditions where the immune system is not being challenged by a pathogen (steady state or **homeostatic** conditions), most HSCs are quiescent. A small number divide, generating daughter cells. Some daughter cells retain the stem-cell characteristics of the mother cell—that is, they remain self-renewing and able to give rise to all blood cell types. Other daughter cells differentiate into *progenitor cells* that lose their self-renewal capacity and become progressively more committed to a particular blood cell lineage. As an organism ages, the number of HSCs decreases, demonstrating that there are limits to an HSC's self-renewal potential.

When there is an increased demand for hematopoiesis (e.g., during an infection or after chemotherapy), HSCs display an enormous proliferative capacity. This can be demonstrated in mice whose hematopoietic systems have been completely destroyed by a lethal dose of x-rays (950 rads). Such irradiated mice die within 10 days unless they are infused with normal bone marrow cells from a genetically identical mouse. Although a normal mouse has 3×10^8 bone marrow cells, infusion of only 10^4 to 10^5 bone marrow cells from a donor is sufficient to completely restore the hematopoietic system, which demonstrates the enormous capacity of HSCs for self-renewal. Our ability to identify and purify this tiny subpopulation has improved considerably, and investigators can now theoretically rescue irradiated animals with just a few purified stem cells, which give rise to progenitors that proliferate rapidly and populate the blood system relatively quickly.

Because of the rarity of HSCs and the challenges of culturing them in vitro, investigators initially found it very difficult to identify and isolate HSCs. The Classic Experiment Box on pages 29–31 describes experimental approaches that led to successful enrichment of HSCs. Briefly, the first successful efforts featured clever process-of-elimination strategies. Investigators reasoned that undifferentiated hematopoietic stem cells would not express surface markers specific for mature cells from the multiple blood lineages ("Lin" markers). They used several approaches to eliminate cells in the bone marrow that did express these markers (Lin$^+$ cells) and then examined the remaining (Lin$^-$) population for its potential to continually give rise to all blood cells over the long term. Lin$^-$ cells were, indeed, enriched for this potential. Other investigators took advantage of two technological developments that revolutionized immunological research—monoclonal antibodies and flow cytometry (see Chapter 20)—and identified surface proteins, including CD34, Sca-1, and c-Kit, that distinguished the rare hematopoietic stem cell population.

CLASSIC EXPERIMENT BOX 2-1

 ## Isolating Hematopoietic Stem Cells

By the 1960s researchers knew that hematopoietic stem cells (HSCs) existed and were a rare population in the bone marrow. However, they did not know what distinguished them from the other millions of cells that crowded the bone marrow. It was clear that in order to fully understand how these remarkable cells give rise to all other blood cells, and in order to harness this potential for clinical use, investigators would have to find a way to isolate HSCs.

But how do you find something that is very rare, whose only distinctive feature is its function—its ability to give rise to all blood cells? Investigators devised a variety of strategies that evolved rapidly with every technological advance, particularly with the advent of monoclonal antibodies and flow cytometry.

Regardless of what method one uses to try to isolate a cell, it is critically important to have a reliable experimental assay that can tell you that the cells that you have teased out are, indeed, the ones you are looking for. Fundamentally, in order to prove that you have enriched or purified an HSC, you have to show that it can proliferate and give rise to all blood cell types in an animal over the long term. Many of the assays that were originally established to show this are still in use. They include colony formation assays,

where the ability of individual cells to proliferate (and differentiate) is determined by looking for evidence of cell division either in vitro (on plates) or in vivo (in the spleens of irradiated mice). However, the best evidence for successful isolation of HSCs is the demonstration that they can restore the blood cells and immune system of a lethally irradiated animal, preventing its death. This can be done for mouse stem cells by injecting stem cell candidates into irradiated mice and determining if they confer survival and repopulate all blood cell types. The development of a mouse model that accepts human hematopoietic stem cells (the *SCID-hu* (man) mouse model) has greatly enhanced investigators' ability to verify the pluripotentiality of candidate human stem cell populations.

In the 1970s investigators did not have the ability to easily compare differences in protein and gene expression among single cells, so they had to try to distinguish cell types based on other physical and structural features. It wasn't until monoclonal antibodies were introduced into research repertoires that investigators could seriously consider purifying a stem cell. Monoclonal antibodies (described in Chapter 20) can be raised to virtually any protein, lipid, or carbohydrate. Monoclo-

nal antibodies can, themselves, be covalently modified with gold particles, enzymes, or fluorochromes in order to visualize their binding by microscopy.

In the early 1980s, investigators reasoned that HSCs were unlikely to express proteins specific for mature blood cells. Using monoclonal antibodies raised against multiple mature cells, they trapped and removed them from bone marrow cell suspensions, first via a process called panning, where the heterogeneous pool of cells was incubated with antibodies bound to a plastic and then those cells that did not stick were dislodged and poured off. The cells that did not stick to the antibodies were, indeed, enriched (in some cases by several thousandfold) in cells with HSC potential. This *negative selection* strategy against mature cell lineages continues to be useful and is now referred to as "lineage or Lin" selection; cells enriched by this method are referred to as "Lin$^-$." The panning process that yielded one of the first images of cells that included human hematopoietic stem cells is shown in Figure 1.

Investigators also worked to identify surface molecules that were specific to hematopoietic stem cells, so that they could *positively select* them from the diverse bone marrow cell types. The first protein that identified human HSCs, now known as

(continued)

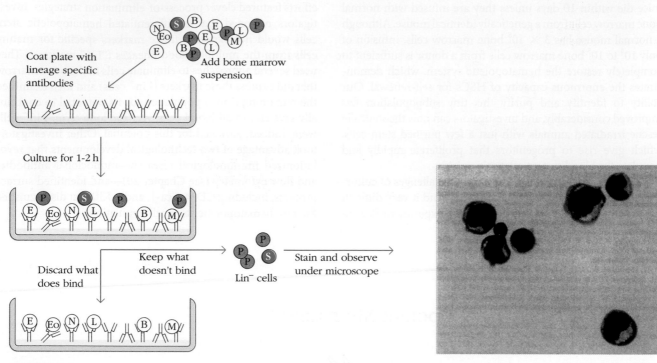

FIGURE 1

Panning for stem cells. Early approaches to isolate HSCs took advantage of antibodies that were raised against mature blood cells and a process called panning, where cells are incubated in plastic plates that are coated with antibodies. Specifically, investigators layered a suspension of bone marrow onto plastic plates coated with antibodies that could bind multiple different mature ("lineage positive") blood cells. They waited 1 to 2 hours and then washed the cells that did not stick (the immature, "lineage negative" blood cells) off the plate. Most cells—mature blood cells—did stick firmly to the plate. However, the cells that did not stick were enriched for hematopoietic stem cells. This process led to the first image of human bone marrow cells enriched for hematopoietic stem cells by panning. S = stem cell; P = progenitor cell; M = monocyte; B = basophil; N = neutrophil; Eo = eosinophil; L = lymphocyte; E = erythrocyte. [Emerson, S.G., Colin, A.S., Wang, E.A., Wong, G.G., Clark, S.C., and Nathan, D.G. Purification and Demonstration of Fetal Hematopoietic Progenitors and Demonstration of Recombinant Multipotential Colony-stimulating Activity. J. Clin. Invest., Vol. 76, Sept. 1985, 1286-1290. ©The American Society for Clinical Investigation.]

CD34, was identified with a monoclonal antibody raised against a tumor of primitive blood cell types (acute myeloid leukemia).

Although one can positively select cells from a diverse population using the panning procedures described above (or by using its more current variant where cells are applied to columns of resin-bound monoclonal antibodies or equivalents), the flow cytometer provides the most efficient way to pull out a rare population from a diverse group of cells. This machine, invented by the Herzenberg laboratory and its interdisciplinary team of scholars and inventors, has revolutionized immunology and clinical medicine. In a nutshell, it is a machine that allows one to identify, separate, and recover individual cells from a diverse pool of cells on the basis of the profiles of proteins and/or genes they express. Chapter 20 provides you with an under-the-

hood introduction to the remarkable technology. Briefly, cells from a heterogeneous suspension are tagged with ("stained with") monoclonal antibodies (or other molecules) that bind to distinct features and are coupled to distinct fluorochromes. These cells flow single file in front of lasers that excite the several fluorochromes, and the intensities of the multiple wavelengths given off by each individual cell are recorded. Cells that express specific antigens at desired levels (e.g., showing evidence for expression of CD34) can be physically separated from other cells and recovered for further studies.

In the late 1980s, Irv Weissman and his laboratories discovered that differences in expression of the Thy protein (a T-cell marker), and later the expression of Sca protein, differentiated mouse hematopoietic stem cells from more mature cells. His laboratory combined both negative and positive

selection techniques to develop one of the most efficient approaches for hematopoietic stem cell enrichment (Figure 2). As other surface molecules were identified, the approach was honed. Currently, HSCs are most frequently identified by their Lin$^-$ Sca-1$^+$ c-Kit$^+$ ("LSK") phenotype. It has become clear that even this subgroup, which represents less than 1% of bone marrow cells, is phenotypically and functionally heterogeneous. Other surface markers, including SLAM proteins, that can distinguish among these subpopulations continue to be identified. The synergy between technical developments and experimental strategies will undoubtedly continue so that, ultimately, we will be able to unambiguously identify, isolate, and manipulate what remains the holy grail of HSC investigations: the long-term stem cell that can both self-renew and give rise to all blood cell types.

BOX 2-1

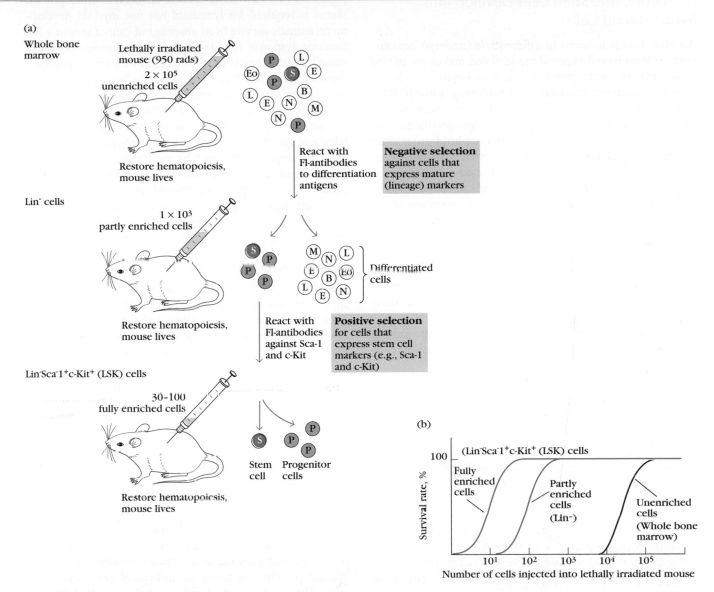

FIGURE 2

Current approaches for enrichment of the pluripotent stem cells from bone marrow. A schematic of the type of stem cell enrichment now routinely employed, but originated by Irv Weissman and colleagues. (a) Enrichment is accomplished by (1) removal (negative selection) of differentiated hematopoietic cells (white) from whole bone marrow after treatment with fluorescently labeled antibodies (Fl-antibodies) specific for membrane molecules expressed on differentiated (mature) lineages but absent from the undifferentiated stem cells (S) and progenitor cells (P), followed by (2) retention (positive selection) of cells within the resulting partly enriched preparation that bound to antibodies specific for Sca-1 and c-Kit, two early differentiation antigens. Cells that are enriched by removal of differentiated cells are referred to as "lineage-minus" or Lin$^-$ populations. Cells that are further enriched by positive selection according to Sca-1 and c-Kit expression are referred to as LSK (for Lin$^-$Sca-1$^+$c-Kit$^+$) cells. M = monocyte; B = basophil; N = neutrophil; Eo = eosinophil; L = lymphocyte; E = erythrocyte. (b) Enrichment of stem cell preparations is measured by their ability to restore hematopoiesis in lethally irradiated mice. Only animals in which hematopoiesis occurs survive. Progressive enrichment of stem cells (from whole bone marrow, to Lin populations, to LSK populations) is indicated by the decrease in the number of injected cells needed to restore hematopoiesis. A total enrichment of about 1000-fold is possible by this procedure.

Emerson, S. G., C. A. Sieff, E. A. Wang, G. G. Wong, S. C. Clark, and D. G. Nathan. 1985. Purification of fetal hematopoietic progenitors and demonstration of recombinant multipotential colony-stimulating activity. *Journal of Clinical Investigation* **76**:1286.

Shizuru, J. A., R. S. Negrin, and I. L. Weissman. 2005. Hematopoietic stem and progenitor cells: clinical and preclinical regeneration of the hematolymphoid system. *Annual Review of Medicine* **56**:509.

Weissman, I. L. 2000. Translating stem and progenitor cell biology to the clinic: Barriers and opportunities. *Science* **287**:1442.

Hematopoiesis Is the Process by Which Hematopoietic Stem Cells Develop into Mature Blood Cells

An HSC that is induced to differentiate (undergo hematopoiesis) loses its self-renewal capacity and makes one of two broad lineage commitment choices (see Figure 2-1). It can become a **common myeloid-erythroid progenitor (CMP)**, which gives rise to all red blood cells (the erythroid lineage), granulocytes, monocytes, and macrophages (the myeloid lineage), or it can become a **common lymphoid progenitor (CLP)**, which gives rise to B lymphocytes, T lymphocytes, and NK cells. Myeloid cells and NK cells are members of the innate immune system, and are the first cells to respond to infection or other insults. Lymphocytes are members of the adaptive immune response and generate a refined antigen-specific immune response that also gives rise to immune memory.

As HSCs progress along their chosen lineages, they lose the capacity to contribute to other cellular lineages. Interestingly, both myeloid and lymphoid lineages give rise to dendritic cells, antigen-presenting cells with diverse features and functions that play an important role in initiating adaptive immune responses. The concentration and frequency of immune cells in blood are listed in Table 2-1.

Regulation of Lineage Commitment during Hematopoiesis

Each step a hematopoietic stem cell takes toward commitment to a particular cellular lineage is accompanied by genetic changes. Multiple genes that specify lineage commitment have been identified. Many of these are transcriptional regulators. For instance, the transcription factor *GATA-2* is required for the development of all hematopoietic lineages; in its absence animals die during embryogenesis. Another transcriptional regulator, *Bmi-1*, is required for the self-renewal capacity of HSCs, and in its absence animals die within 2 months of birth because of the failure to repopulate their red and white blood

cells. *Ikaros* and *Notch* are both families of transcriptional regulators that have more specific effects on hematopoiesis. Ikaros is required for lymphoid but not myeloid development; animals survive in its absence but cannot mount a full immune response (i.e., they are severely *immunocompromised*). Notch1, one of four Notch family members, regulates the choice between T and B lymphocyte lineages (see Chapter 9). More master regulators of lineage commitment during hematopoiesis continue to be identified.

The rate of hematopoiesis, as well as the production and release of specific cell lineages, is also responsive to environmental changes experienced by an organism. For instance, infection can result in the release of cytokines that markedly enhance the development of myeloid cells, including neutrophils. Investigators have also recently shown that the release of mature cells from the bone marrow is responsive to circadian cycles and regulated by the sympathetic nervous system.

Distinguishing Blood Cells

Early investigators originally classified cells based on their appearance under a microscope, often with the help of dyes. Their observations were especially helpful in distinguishing myeloid from lymphoid lineages, granulocytes from macrophages, neutrophils from basophils and eosinophils. Hematoxylin and eosin (H&E) stains are still commonly used in combination to distinguish cell types in blood smears and tissues. They highlight intracellular differences because of their pH sensitivity and different affinities for charged macromolecules in a cell. Thus, the basic dye hematoxylin binds basophilic nucleic acids, staining them blue, and the acidic dye eosin binds eosinophilic proteins in granules and cytoplasm, staining them pink.

Microscopists drew astute inferences about cell function by detailed examination of the structure stained cells, as well as the behavior of live cells in solution. The advent of the flow cytometer in the 1980s revolutionized our understanding of cell subtypes by allowing us to evaluate multiple surface and internal proteins expressed by individual cells simultaneously. The development of ever more sophisticated fluorescent microscopy approaches to observe live cells in vitro and in vivo have allowed investigators to penetrate the complexities of the immune response in time and space. These advances coupled with abilities to manipulate cell function genetically have also revealed a remarkable diversity of cell types among myeloid and lymphoid cells, and continue to expose new functions and unexpected relationships among hematopoietic cells. Therefore, while our understanding of the cell subtypes is impressive, it is by no means complete.

Cells of the Myeloid Lineage Are the First Responders to Infection

Cells that arise from a common myeloid progenitor (CMP) include red blood cells (erythroid cells) as well as various types of white blood cells (myeloid cells such as granulocytes, monocytes, macrophages, and some dendritic cells). Myeloid

TABLE 2-1	Concentration and frequency of cells in human blood		
Cell type	**Cells/mm³**	**Total leukocytes (%)**
Red blood cells	5.0×10^6	
Platelets	2.5×10^5	
Leukocytes	7.3×10^3	
Neutrophil	$3.7–5.1 \times 10^3$	50–70
Lymphocyte	$1.5–3.0 \times 10^3$	20–40
Monocyte	$1–4.4 \times 10^2$	1–6
Eosinophil	$1–2.2 \times 10^2$	1–3
Basophil	$<1.3 \times 10^2$	<1

cells are the first to respond to the invasion of a pathogen and communicate the presence of an insult to cells of the lymphoid lineage (below). As we will see in Chapter 15, they also contribute to inflammatory diseases (asthma and allergy).

Granulocytes

Granulocytes are at the front lines of attack during an immune response and are considered part of the innate immune system. Granulocytes are white blood cells (leukocytes) that are classified as neutrophils, basophils, mast cells, or eosinophils on the basis of differences in cellular morphology and the staining of their characteristic cytoplasmic granules (Figure 2-2). All granulocytes have multilobed nuclei that make them visually distinctive and easily distinguishable from lymphocytes, whose nuclei are round. The cytoplasm of all granulocytes is replete with granules that are released in response to contact with pathogens. These granules contain a variety of proteins with distinct functions: Some damage pathogens directly; some regulate trafficking and activity of other white blood cells, including lymphocytes; and some contribute to the remodeling of tissues at the site of infection. See Table 2-2 for a partial list of granule proteins and their functions.

Neutrophils constitute the majority (50% to 70%) of circulating leukocytes (see Figure 2-2a) and are much more numerous than eosinophils (1%–3%), basophils (<1%), or mast cells (<1%). After differentiation in the bone marrow, neutrophils are released into the peripheral blood and circulate for 7 to 10 hours before migrating into the tissues, where they have a life span of only a few days. In response to many types of infections, the number of circulating neutrophils increases significantly and more are recruited to tissues, partially in response to cues the bone marrow receives to produce and release more myeloid cells. The resulting transient increase in the number of circulating neutrophils, called **leukocytosis**, is used medically as an indication of infection.

Neutrophils are recruited to the site of infection in response to inflammatory molecules (e.g., chemokines) generated by innate cells (including other neutrophils) that have engaged a pathogen. Once in tissues, neutrophils phagocytose (engulf) bacteria very effectively, and also secrete a range of proteins that have antimicrobial effects and tissue remodeling potential. Neutrophils are the dominant first responders to infection and the main cellular components of pus, where they accumulate at the end of their short lives. Although once considered a simple and "disposable" effector cell, the neutrophil has recently inspired renewed interest from investigations indicating that it may also regulate the adaptive immune response.

Basophils are nonphagocytic granulocytes (see Figure 2-2b) that contain large granules filled with basophilic proteins (i.e., they stain blue in standard H&E staining protocols). Basophils are relatively rare in the circulation, but can be very potent. In response to binding of circulating antibodies, basophils release the contents of their granules. Histamine, one of the best known proteins in basophilic granules, increases blood vessel permeability and smooth muscle activity. Basophils (and eosinophils, below) are critical to our response to parasites, particularly helminths (worms), but in areas where worm infection is less prevalent, histamines are best appreciated as the cause of allergy symptoms. Like neutrophils, basophils may also secrete cytokines that modulate the adaptive immune response.

Mast cells (see Figure 2-2c) are released from the bone marrow into the blood as undifferentiated cells; they mature only after they leave the blood. Mast cells can be found in a wide variety of tissues, including the skin, connective tissues of various organs, and mucosal epithelial tissue of the respiratory, genitourinary, and digestive tracts. Like circulating basophils, these cells have large numbers of cytoplasmic granules that contain histamine and other pharmacologically active substances. Mast cells also play an important role in the development of allergies.

Basophils and mast cells share many features and their relationship is not unequivocally understood. Some speculate that basophils are the blood-borne version of mast cells; others speculate that they have distinct origins and functions.

Eosinophils, like neutrophils, are motile phagocytic cells (see Figure 2-2d) that can migrate from the blood into the tissue spaces. Their phagocytic role is significantly less important than that of neutrophils, and it is thought that they play their most important role in the defense against multicellular parasitic organisms, including worms. They can be found clustering around invading worms, whose membranes are damaged by the activity of proteins released from eosinophilic granules. Like neutrophils and basophils, eosinophils may also secrete cytokines that regulate B and T lymphocytes, thereby influencing the adaptive immune response. In areas where parasites are less of a health problem, eosinophils are better appreciated as contributors to asthma and allergy symptoms.

Myeloid Antigen-Presenting Cells

Myeloid progenitors also give rise to a group of phagocytic cells (monocytes, macrophages, and dendritic cells) that have **professional antigen-presenting cell (APC)** function (Figure 2-3). Myeloid APCs are considered cellular bridges between the innate and adaptive immune systems because they make contact with a pathogen at the site of infection and communicate this encounter to T lymphocytes in the lymph node ("antigen presentation"). Each APC can respond to pathogens and secrete proteins that attract and activate other immune cells. Each can ingest pathogens via phagocytosis, digest pathogenic proteins into peptides, then present these peptide antigens on their membrane surfaces. Each can be induced to express a set of costimulatory molecules required for optimal activation of T lymphocytes. However, it is likely that each plays a distinct role during the immune response, depending on its locale and its ability to respond to pathogens. Dendritic cells, in particular, play a primary role in presenting antigen to—and activating—naïve T cells. Macrophages and neutrophils are especially efficient in removing both pathogen and damaged host cells, and can provide a first line of defense against pathogens.

(a) Neutrophil

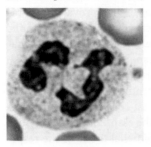
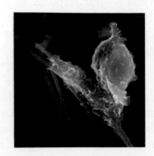
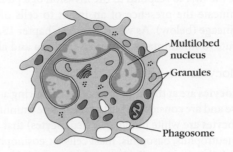

Multilobed nucleus

Granules

Phagosome

(b) Basophil

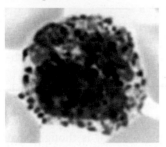
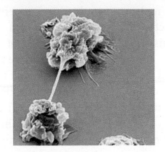
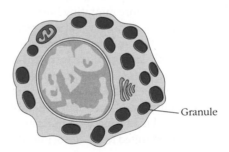

Glycogen

Granule

(c) Mast cell

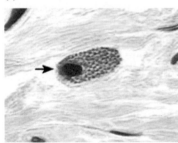

Granule

(d) Eosinophil

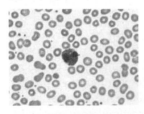
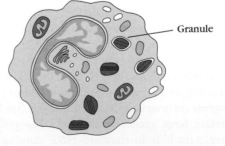

Granule

FIGURE 2-2 Examples of granulocytes. (a, b, c, d) Hematoxylin and eosin (H&E) stains of indicated cells in blood smears. (a, middle) Neutrophil engulfing bacteria visualized by scanning electron microscopy (SEM) and colorized digitally. (b, middle) SEM of activated granulocytes (colorized). Each image is accompanied by a cartoon depicting the typical morphology of the indicated granulocyte. Note differences in the shape of the nucleus and in the number, color, and shape of the cytoplasmic granules. [2-2a, left: Science Source/Getty Images; 2-2a, right: Creative Commons, http://es.wikipedia.org/wiki/Archivo:Neutrophil_with_anthrax_copy.jpg; 2-2b, left: Dr. Gladden Willis/Visuals Unlimited, Inc.; 2-2b, right: Steve Gschmeissner/Photo Researchers; 2-2c, left: Courtesy Gwen V. Childs, Ph.D., University of Arkansas for Medical Sciences; 2-2d, left; Pathpedia.com.]

TABLE 2-2 Examples of proteins contained in neutrophil, eosinophil, and basophil granules

Cell type	Molecule in granule	Examples	Function
Neutrophil	Proteases	*Elastase, Collagenase*	Tissue remodeling
	Antimicrobial proteins	*Defensins, lysozyme*	Direct harm to pathogens
	Protease inhibitors	*α1-anti-trypsin*	Regulation of proteases
	Histamine		Vasodilation, inflammation
Eosinophil	Cationic proteins	*EPO*	Induces formation of ROS
	Ribonucleases	*MBP*	Vasodilation, basophil degranulation
	Cytokines	*ECP, EDN*	Antiviral activity
	Chemokines	*IL-4, IL-10, IL-13, TNFα*	Modulation of adaptive immune responses
		RANTES, MIP-1α	Attract leukocytes
Basophil/Mast Cell	Cytokines	*IL-4, IL-13*	Modulation of adaptive immune response
	Lipid mediators	*Leukotrienes*	Regulation of inflammation
	Histamine		Vasodilation, smooth muscle activation

Monocytes make up about 5% to 10% of white blood cells and are a heterogeneous group of cells that migrate into tissues and differentiate into a diverse array of tissue-resident phagocytic cells, including macrophages and dendritic cells (see Figure 2-3a). During hematopoiesis in the bone marrow, granulocyte-monocyte progenitor cells differentiate into promonocytes, which leave the bone marrow and enter the blood, where they further differentiate into mature monocytes. Two broad categories of monocytes have recently been identified. *Inflammatory monocytes* enter tissues quickly in response to infection. *Patrolling monocytes*, a smaller group of cells that crawl slowly along blood vessels, provide a reservoir for tissue-resident monocytes in the absence of infection, and may quell rather than initiate immune responses.

Monocytes that migrate into tissues in response to infection can differentiate into specific tissue **macrophages**. Like monocytes, macrophages can play several different roles. Some macrophages are long-term residents in tissues and play an important role in regulating their repair and regeneration. Other macrophages participate in the innate immune response and undergo a number of key changes when they are stimulated by encounters with pathogens or tissue damage. These are referred to as *inflammatory macrophages* and play a dual role in the immune system as effective phagocytes that can contribute to the clearance of pathogens from a tissue, as well as antigen-presenting cells that can activate T lymphocytes. **Osteoclasts** in the bone, **microglial cells** in the central nervous system, and **alveolar macrophages** in the lung are tissue-specific examples of macrophages with these properties.

Activated, inflammatory macrophages are more effective than resting ones in eliminating potential pathogens for several reasons: They exhibit greater phagocytic activity, an increased ability to kill ingested microbes, increased secretion of inflammatory and cytotoxic mediators, and the ability to activate T cells. More will be said about the antimicrobial activities of macrophages in Chapter 5. Activated macrophages also function more effectively as antigen-presenting cells for helper T cells (T_H cells), which, in turn, regulate and enhance macrophage activity. Thus, macrophages and T_H cells facilitate each other's activation during the immune response.

Many macrophages also express receptors for certain classes of antibody. If an antigen (e.g., a bacterium) is coated with the appropriate antibody, the complex of antigen and antibody binds to antibody receptors on the macrophage membrane more readily than antigen alone and phagocytosis is enhanced. In one study, for example, the rate of phagocytosis of an antigen was 4000-fold higher in the presence of specific antibody to the antigen than in its absence. Thus, an antibody is an example of an **opsonin**, a molecule that binds an antigen marking it for recognition by immune cells. The modification of particulate antigens with opsonins (which come in a variety of forms) is called **opsonization**, a term from the Greek that literally means "to supply food" or "make tasty." Opsonization is traditionally described as a process that enhances phagocytosis of an antigen, but it serves multiple purposes that will be discussed in subsequent chapters.

Although most of the antigen ingested by macrophages is degraded and eliminated, early experiments with radiolabeled antigens demonstrated the presence of antigen peptides on the macrophage membrane. Although the macrophage is a very capable antigen-presenting cell, the dendritic cell is considered the most efficient activator of naïve T cells.

The discovery of the **dendritic cell (DC)** by Ralph Steinman in the mid 1970s resulted in awarding of the Nobel Prize in 2011. Dendritic cells are critical for the initiation of the immune response and acquired their name because they are covered with long membranous extensions that resemble the dendrites of nerve cells and extend and retract dynamically,

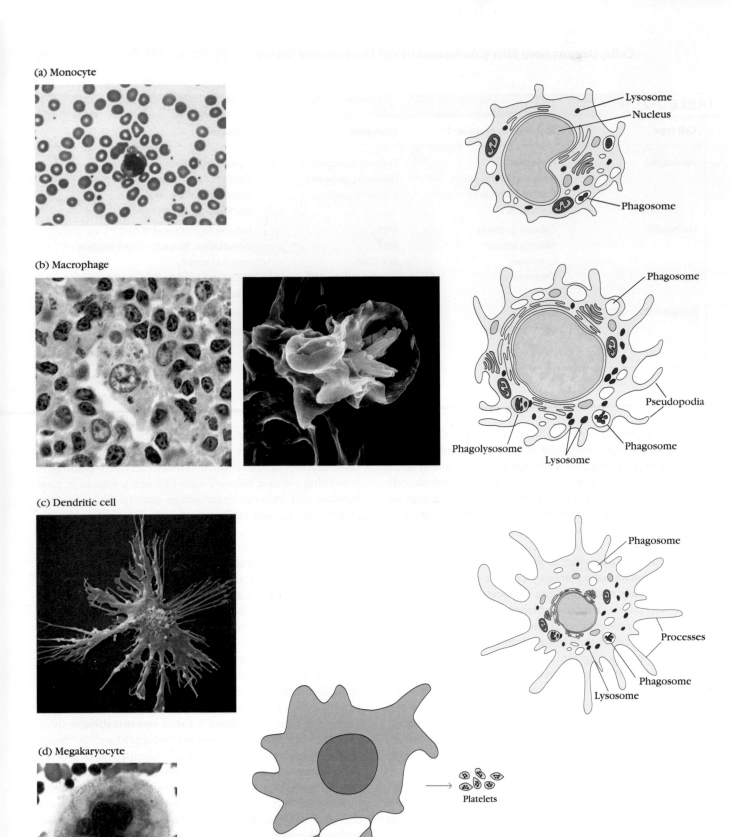

(a) Monocyte

Lysosome
Nucleus
Phagosome

(b) Macrophage

Phagosome
Pseudopodia
Phagolysosome
Lysosome
Phagosome

(c) Dendritic cell

Phagosome
Processes
Phagosome
Lysosome

(d) Megakaryocyte

Platelets
Megakaryocyte

FIGURE 2-3 Examples of monocytes, macrophages, dendritic cells, and megakaryocytes. (a, d) H&E stain of blood smear. (b) H&E stain of tissue section. (b, middle) SEM of macrophage engulfing mycobacteria (colorized). (c) SEM micrograph. Each image is accompanied by a cartoon depicting the typical morphology of the indicated cell. Note that macrophages are five- to tenfold larger than monocytes and contain more organelles, especially lysosomes. [2-3a, left: Pathpedia.com; 2-3b, left: Courtesy Dr. Thomas Caceci, Virginia-Maryland Regional College of Veterinary Medicine; 2-3b, middle: SPL/Photo Researchers; 2-3c, left: David Scharf/ Photo Researchers; 2-3d, left: Carolina Biological Supply Co./Visuals Unlimited.]

increasing the surface area available for browsing lymphocytes. They are more diverse a population of cells than once was thought, and seem to arise from both the myeloid and lymphoid lineages of hematopoietic cells. The functional distinctions among these diverse cells are still being clarified and are likely critically important in tailoring immune responses to distinct pathogens and targeting responding cells to distinct tissues.

Dendritic cells perform the distinct functions of antigen capture in one location and antigen presentation in another. Outside lymph nodes, immature forms of these cells monitor the body for signs of invasion by pathogens and capture intruding or foreign antigens. They process these antigens, then migrate to lymph nodes, where they present the antigen to naïve T cells, initiating the adaptive immune response.

When acting as sentinels in the periphery, immature dendritic cells take on their cargo of antigen in three ways. They engulf it by phagocytosis, internalize it by receptor-mediated endocytosis, or imbibe it by pinocytosis. Indeed, immature dendritic cells pinocytose fluid volumes of 1000 to 1500 μm^3 per hour, a volume that rivals that of the cell itself. Through a process of maturation, they shift from an antigen-capturing phenotype to one that is specialized for presentation of antigen to T cells. In making the transition, some attributes are lost and others are gained. Lost is the capacity for phagocytosis and large-scale pinocytosis. However, the ability to present antigen increases significantly, as does the expression of costimulatory molecules that are essential for the activation of naïve T cells. After activation, dendritic cells abandon residency in peripheral tissues, enter the blood or lymphatic circulation, and migrate to regions of the lymphoid organs, where T cells reside, and present antigen.

It is important to note that, although they share a name, **follicular dendritic cells** do not arise in bone marrow and have completely different functions from those described for the dendritic cells discussed above. Follicular dendritic cells do not function as antigen-presenting cells for T_H-cell activation. These dendritic cells were named for their exclusive location in organized structures of the lymph node called lymph follicles, which are rich in B cells. As discussed in Chapters 12 and 14, the interaction of B cells with follicular dendritic cells is an important step in the maturation and diversification of B cells.

It is clear that myeloid cells are not the only cells that can present antigen efficiently. As mentioned above, lymphoid-derived dendritic cells are fully capable APCs. In addition, activated B lymphocytes can act as professional antigen-presenting cells. B cells can internalize antigen very efficiently via their antigen-specific receptor, and can process and present antigenic peptides at the cell surface. Activated B cells also express the full complement of costimulatory molecules that are required to activate T cells. By presenting antigen directly to T cells, B cells efficiently solicit help, in the form of cytokines, that induces their differentiation into memory cells, as well as into antibody-producing cells (plasma cells).

Erythroid Cells

Cells of the erythroid lineage—**erythrocytes**, or red blood cells—also arise from a common myeloid precursor (sometimes referred to as a common myeloid-erythroid precursor). They contain high concentrations of hemoglobin, and circulate through blood vessels and capillaries delivering oxygen to surrounding cells and tissues. Damaged red blood cells can also release signals (free radicals) that induce innate immune activity. In mammals, erythrocytes are anuclear; their nucleated precursors (erythroblasts) extrude their nuclei in the bone marrow. However, the erythrocytes of almost all nonmammalian vertebrates (birds, fish, amphibians, and reptiles) retain their nuclei. Erythrocyte size and shape vary considerably across the animal kingdom—the largest red blood cells can be found among some amphibians, and the smallest among some deer species.

Megakaryocytes

Megakaryocytes are large myeloid cells that reside in the bone marrow and give rise to thousands of **platelets**, very small cells (or cell fragments) that circulate in the blood and participate in the formation of blood clots. Although platelets have some of the properties of independent cells, they do not have their own nuclei.

Cells of the Lymphoid Lineage Regulate the Adaptive Immune Response

Lymphocytes (Figure 2-4) are the principal cell players in the adaptive immune response. They represent 20% to 40% of circulating white blood cells and 99% of cells in the lymph. Lymphocytes can be broadly subdivided into three major populations on the basis of functional and phenotypic differences: B lymphocytes (B cells), T lymphocytes (T cells), and natural killer (NK) cells. In humans, approximately a trillion (10^{12}) lymphocytes circulate continuously through the blood and lymph and migrate into the tissue spaces and lymphoid organs. We briefly review the general characteristics and functions of each lymphocyte group and its subsets below.

Lymphocytes are relatively nondescript cells that are very difficult to distinguish morphologically. T and B cells, in particular, appear identical under a microscope. We therefore rely heavily on the signature of surface proteins they express to differentiate among lymphocyte subpopulations.

Surface proteins expressed by immune cells are often referred to by the **cluster of differentiation (CD** or *cluster of designation)* nomenclature. This nomenclature was established in 1982 by an international group of investigators who recognized that many of the new antibodies produced by laboratories all over the world (largely in response to the advent of monoclonal antibody technology) were seeing the same proteins. They therefore defined clusters of antibodies that appeared to be seeing the same protein and assigned a name—a cluster of differentiation or CD—to each group. Although originally designed to categorize the multiple antibodies, the

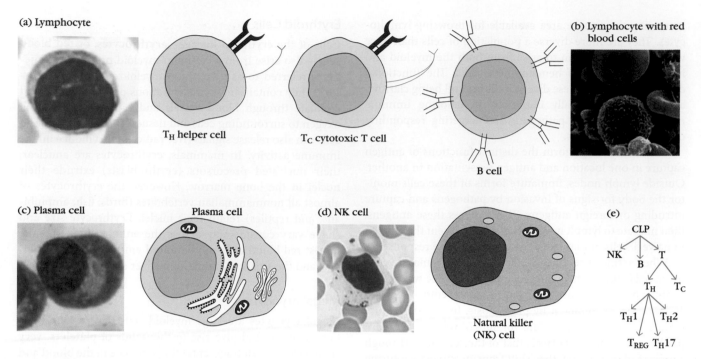

(a) Lymphocyte

T_H helper cell T_C cytotoxic T cell

B cell

(b) Lymphocyte with red blood cells

(c) Plasma cell   Plasma cell   (d) NK cell   Natural killer (NK) cell

(e)

CLP

NK   T

B

T_H T_C

T_H1 T_H2

T_{REG} T_H17

FIGURE 2-4 Examples of lymphocytes. (a, c, d) H&E stain of blood smear showing typical lymphocyte. Note that naïve B cells and T cells look identical by microscopy. (b) Scanning electron micrograph of lymphocytes and red blood cells. Cartoons depicting the typical morphology of the cells indicated accompany each image (including three different lymphocytes that would all have the same appearance). Note the enlarged area of cytoplasm of the plasma cell, which is occupied by an extensive network of endoplasmic reticulum and Golgi—an indication of the cell's dedication to antibody secretion. The NK cell also has more cytoplasm than a naïve lymphocyte; this is full of granules that are used to kill target cells. (e) A branch diagram that depicts the basic relationship among the lymphocyte subsets described in the text. [2-4a, left: Fred Hossler/ Visuals Unlimited; 2-4b: Creative Commons, http://commons.wikimedia.org/ wiki/File:SEM_blood_cells.jpg; 2-4c, left: Benjamin Koziner/Phototake; 2-4d: Courtesy Ira Ames, Ph.D., Dept. Cell & Developmental Biology, SUNY-Upstate Medical University.]

CD nomenclature is now firmly associated with specific surface proteins found on cells of many types. Table 2-3 lists some common CD molecules found on human and mouse lymphocytes. Note that the shift from use of a "common" name to the more standard "CD" name can take place slowly. (For example, investigators often still refer to the pan-T cell marker as "Thy-1" rather than CD90, and the costimulatory molecules as "B7-1" and "B7-2," rather than CD80 and CD86.) Appendix 1 lists over three hundred CD markers expressed by immune cells.

In addition to their CD surface signatures, each B or T cell also expresses an antigen-specific receptor (the **B cell receptor (BCR)** or the **T cell receptor (TCR)**, respectively) on its surface. Although the populations of B cells and T cells express a remarkable diversity of antigen receptors (more than a billion), all receptors on an individual cell's surface have identical structures and therefore have identical specificities for antigen. If a given lymphocyte divides to form two daughter cells, both daughters bear antigen receptors with antigen specificities identical to each other and to the parental cell from which they arose, and so will any descendants they produce. The resulting population of lymphocytes, all arising from the same founding lymphocyte, is a **clone**.

At any given moment, a human or a mouse will contain tens of thousands, perhaps a hundred thousand, distinct mature T- and B-cell clones, each distinguished by its own unique and identical cohort of antigen receptors. Mature B cells and T cells are ready to encounter antigen, but they are considered **naïve** until they do so. Contact with antigen induces naïve lymphocytes to proliferate and differentiate into both effector cells and memory cells. **Effector cells** carry out specific functions to combat the pathogen, while the **memory cells** persist in the host, and upon rechallenge with the same antigen mediate a response that is both quicker and greater in magnitude. The first encounter with antigen is termed a **primary response**, and the re-encounter a **secondary response**.

B Lymphocytes

The **B lymphocyte (B cell)** derived its letter designation from its site of maturation, in the *b*ursa of Fabricius in birds; the name turned out to be apt, as *b*one marrow is its major site of maturation in humans, mice, and many other mammals. Mature B cells are definitively distinguished from other lymphocytes and all other cells by their synthesis and

TABLE 2-3	Common CD markers used to distinguish functional lymphocyte subpopulations				
CD designation	**Function**	**B cell**	**T$_H$**	**T$_C$**	**NK cell**
CD2	Adhesion molecule; signal transduction	−	+	+	+
CD3	Signal transduction element of T-cell receptor	−	+	+	−
CD4	Adhesion molecule that binds to class II MHC molecules; signal transduction	−	+ (usually)	− (usually)	−
CD5	Unknown	+ (subset)	+	+	+
CD8	Adhesion molecule that binds to class I MHC molecules; signal transduction	−	− (usually)	+ (usually)	(variable)
CD16 (FcγRIII)	Low-affinity receptor for Fc region of IgG	−	−	−	+
CD19	Signal transduction; CD21 co-receptor	+	−	−	−
CD21 (CR2)	Receptor for complement (C3d and Epstein-Barr virus)	+	−	−	−
CD28	Receptor for costimulatory B7 molecule on antigen-presenting cells	−	+	+	−
CD32 (FcγRII)	Receptor for Fc region of IgG	+	−	−	−
CD35 (CR1)	Receptor for complement (C3b)	+	−	−	−
CD40	Signal transduction	+	−	−	−
CD45	Signal transduction	+	+	+	+
CD56	Adhesion molecule	−	−	−	+

Synonyms are shown in parentheses.

display of the B-cell receptor (BCR), a membrane-bound immunoglobulin (antibody) molecule that binds to antigen. Each B cell expresses a surface antibody with a unique specificity, and each of the approximately 1.5–3×10^5 molecules of surface antibody has identical binding sites for antigen. B lymphocytes also can improve their ability to bind antigen through a process known as somatic hypermutation and can generate antibodies of several different functional classes through a process known as class switching. Somatic hypermutation and class switching are covered in detail in Chapter 12.

Ultimately, activated B cells differentiate into effector cells known as **plasma cells** (see Figure 2-4c). Plasma cells lose expression of surface immunoglobulin and become highly specialized for secretion of antibody. A single cell is capable of secreting from a few hundred to more than a thousand molecules of antibody per second. Plasma cells do not divide and, although some long-lived populations of plasma cells are found in bone marrow, many die within 1 or 2 weeks.

T Lymphocytes

T lymphocytes (T cells) derive their letter designation from their site of maturation in the *t*hymus. Like the B cell, the T cell expresses a unique antigen-binding receptor called the T-cell receptor. However, unlike membrane-bound antibodies on B cells, which can recognize soluble or particulate antigen, T-cell receptors only recognize processed pieces of antigen (typically peptides) bound to cell membrane proteins called **major histocompatibility complex (MHC) molecules**. MHC molecules are genetically diverse glycoproteins found on cell membranes (their structure and function are covered in detail in Chapter 8). The ability of MHC molecules to form complexes with antigen allows cells to decorate their surfaces with internal (foreign and self) proteins, exposing them to browsing T cells. MHC comes in two versions: **class I MHC molecules**, which are expressed by nearly all nucleated cells of vertebrate species, and **class II MHC molecules**, which are expressed by professional antigen-presenting cells and a few other cell types during inflammation.

T lymphocytes are divided into two major cell types—**T helper (T_H) cells** and **T cytotoxic (T_C) cells**—that can be distinguished from one another by the presence of either **CD4** or **CD8** membrane glycoproteins on their surfaces. *T cells displaying CD4 generally function as T_H cells and recognize antigen in complex with MHC class II, whereas those displaying CD8 generally function as T_C cells and recognize antigen in complex with MHC class I.* The ratio of $CD4^+$ to $CD8^+$ T cells is approximately 2:1 in normal mouse and human peripheral blood. A change in this ratio is often an indicator of immunodeficiency disease (e.g., HIV), autoimmune diseases, and other disorders.

Naïve $CD8^+$ T cells browse the surfaces of antigen-presenting cells with their T-cell receptors. If and when they bind to an MHC-peptide complex, they become activated, proliferate, and differentiate into an effector cell called a **cytotoxic T lymphocyte (CTL)**. The CTL has a vital function in monitoring the cells of the body and eliminating any cells that display foreign antigen complexed with class I MHC, such as virus-infected cells, tumor cells, and cells of a foreign tissue graft. To proliferate and differentiate optimally, naïve $CD8^+$ T cells also need help from mature $CD4^+$ T cells.

Naïve $CD4^+$ T cells also browse the surfaces of antigen-presenting cells with their T-cell receptors. If and when they recognize an MHC-peptide complex, they can become activated and proliferate and differentiate into one of a variety of effector T cell subsets (see Figure 2-4e). **T helper type 1 ($T_H 1$) cells** regulate the immune response to intracellular pathogens, and **T helper type 2 ($T_H 2$) cells** regulate the response to many extracellular pathogens. Two additional T_H cell subsets have been recently identified. **T helper type 17 cells ($T_H 17$)**, so named because they secrete IL-17, play an important role in cell-mediated immunity and may help the defense against fungi. **T follicular helper cells (T_{FH})** play an important role in humoral immunity and regulate B-cell development in germinal centers. Which helper subtype dominates a response depends largely on what type of pathogen (intracellular versus extracellular, viral, bacterial, fungal, helminth) has infected an animal. Each of these $CD4^+$ T-cell subtypes produces a different set of cytokines that enable or "help" the activation of B cells, T_C cells, macrophages, and various other cells that participate in the immune response. The network of cytokines that regulate and are produced by these effector cells is described in detail in Chapter 11.

Another type of $CD4^+$ T cell, the **regulatory T cell (T_{REG})**, has the unique capacity to inhibit an immune response. These cells can arise during maturation in the thymus from autoreactive cells (natural T_{REG}), but also can be induced at the site of an immune response in an antigen-dependent manner (induced T_{REG}). They are identified by the presence of CD4 and CD25 on their surfaces, as well as the expression of the internal transcription factor FoxP3. T_{REG} cells are critical in helping us to quell autoreactive responses that have not been avoided via other mechanisms. In fact, mice depleted of T_{REG} cells are afflicted with a constellation of destructive self-reactive inflammatory reactions. However, T_{REG} cells may also play a role in limiting our normal T-cell response to a pathogen. $CD4^+$ and $CD8^+$ T-cell subpopulations may be even more diverse than currently described, and the field should expect identification of additional functional subtypes in the future.

Natural Killer Cells

Natural killer (NK) cells are lymphoid cells that are closely related to B and T cells. However, they do not express antigen-specific receptors and are considered part of the innate immune system. They are distinguished by the expression of a surface marker known as NK1.1, as well as the presence of cytotoxic granules. Once referred to as "large granular lymphocytes" because of their appearance under a microscope, NK cells constitute 5% to 10% of lymphocytes in human peripheral blood. They are efficient cell killers and attack a variety of abnormal cells, including some tumor cells and some cells infected with virus. They distinguish cells that should be killed from normal cells in a very clever way: by "recognizing" the absence of MHC class I, which is expressed by almost all normal cells, but is specifically down-regulated by some tumors and in response to some viral infections. How can cells recognize an absence? NK cells express a variety of receptors for self MHC class I that, when engaged, inhibit their ability to kill other cells. When NK cells encounter cells that have lost their MHC class I, these receptors are no longer engaged and can no longer inhibit the potent cytotoxic tendencies of the NK cell, which then releases its cytolytic granules and kills the abnormal target cell.

NK cells also express receptors for immunoglobulins and can therefore decorate themselves with antibodies that bind pathogens or proteins from pathogens on the surface of infected cells. This allows an NK cell to make a connection with a variety of target cells (independently of their MHC class I expression). Once the antibodies bring the NK cell in contact with target cells, the NK cell releases its granules and induces cell death. The mechanism of NK-cell cytotoxicity, the focus of much current experimental study, is described further in Chapter 13.

NKT Cells

Another type of cell in the lymphoid lineage, **NKT cells**, have received a great deal of recent attention and share features with both conventional T lymphocytes and NK cells. Like T cells, NKT cells have T-cell receptors (TCRs), and some express CD4. Unlike most T cells, however, the TCRs of NKT cells are not very diverse and recognize specific lipids and glycolipids presented by a molecule related to MHC proteins known as CD1. Like their innate immune relatives, NK cells, NKT cells have antibody receptors, as well as other receptors classically associated with NK cells. Activated NKT cells can release cytotoxic granules that kill target cells, but they can also release large quantities of cytokines that can both enhance and suppress the immune response. They appear to be involved in human asthma, but also may inhibit the development of autoimmunity and cancer. Understanding the exact role of NKT cells in immunity is one research priority.

Primary Lymphoid Organs—Where Immune Cells Develop

The ability of any stem cell to self-renew and differentiate depends on the structural organization and cellular function of specialized anatomic microenvironments known as **stem cell niches**. These sequestered regions are typically populated by a supportive network of stromal cells. Stem cell niche stromal cells express soluble and membrane-bound proteins that regulate cell survival, proliferation, differentiation, and trafficking. The organs that have microenvironments that support the differentiation of hematopoietic stem cells actually change over the course of embryonic development. However, by mid to late gestation, HSCs take up residence in the bone marrow, which remains the primary site of hematopoiesis throughout adult life. The bone marrow supports the maturation of all erythroid and myeloid cells and, in humans and mice, the maturation of B lymphocytes (as described in Chapter 10).

HSCs are also found in blood and may naturally recirculate between the bone marrow and other tissues. This observation has simplified the process used to transplant blood cell progenitors from donors into patients who are deficient (e.g., patients who have undergone chemotherapy). Whereas once it was always necessary to aspirate bone marrow from the donor—a painful process that requires anesthesia—it is now sometimes possible to use enriched hematopoietic precursors from donor blood, which is much more easily obtained (see the Clinical Focus Box 2-2 on pages 42–43).

Unlike B lymphocytes, T lymphocytes do not complete their maturation in the bone marrow. T lymphocyte precursors need to leave the bone marrow and travel to the unique microenvironments provided by the other primary lymphoid organ, the thymus, in order to develop into functional cells. The structure and function of the thymus will be discussed below and in more detail in Chapter 9.

The Bone Marrow Provides Niches for Hematopoietic Stem Cells to Self-Renew and Differentiate into Myeloid Cells and B Lymphocytes

The **bone marrow** is a primary lymphoid organ that supports self-renewal and differentiation of hematopoietic stem cells (HSCs) into mature blood cells. Although all bones contain marrow, the long bones (femur, humerus), hip bones (ileum), and sternum tend to be the most active sites of hematopoiesis. The bone marrow is not only responsible for the development and replenishment of blood cells, but it is also responsible for maintaining the pool of HSCs throughout the life of an adult vertebrate.

The adult bone marrow (Figure 2-5), the paradigmatic adult stem cell niche, contains several cell types that coordinate HSC development, including (1) *osteoblasts*, versatile cells that both generate bone and control the differentiation of HSCs, (2) *endothelial cells* that line the blood vessels and also regulate HSC differentiation, (3) *reticular cells* that send processes connecting cells to bone and blood vessels, and, unexpectedly, (4) *sympathetic neurons*, which can control the release of hematopoietic cells from the bone marrow. A microscopic cross-section reveals that the bone marrow is tightly packed with stromal cells and hematopoietic cells at every stage of differentiation. With age, however, fat cells gradually replace 50% or more of the bone marrow compartment, and the efficiency of hematopoiesis decreases.

The choices that an HSC makes depend largely on the environmental cues it receives. The bone marrow is packed with hematopoietic cells at all stages of development, but it is likely that the precursors of each myeloid and lymphoid subtype mature in distinct environmental micro-niches within the bone marrow. Our understanding of the microenvironments within the bone marrow that support specific stages of hematopoiesis is still developing. Evidence suggests, however, that the **endosteal niche** (the area directly surrounding the bone and in contact with bone-producing osteoblasts) and the **vascular niche** (the area directly surrounding the blood vessels and in contact with endothelial cells) play different roles (see Figure 2-5c). The endosteal niche appears to be occupied by quiescent HSCs in close association with osteoblasts that regulate stem cell proliferation. The vascular niche appears to be occupied by HSCs that have been mobilized to leave the endosteal niche to either differentiate or circulate. In addition, the more differentiated a cell is, the farther it appears to migrate from its supportive osteoblasts and the closer it moves to the more central regions of the bone. For example, the most immature B lymphocytes are found closest to the endosteum and osteoblasts, while the more mature B cells have moved into the more central sinuses of the bone marrow that are richly served by blood vessels.

Finally, it is important to recognize that the bone marrow is not only a site for lymphoid and myeloid development but is also a site to which fully mature myeloid and lymphoid cells can return. Mature antibody-secreting B cells (plasma cells) may even take up long-term residence in the bone marrow. Whole bone marrow transplants, therefore, do not simply include stem cells but also include mature, functional cells that can both help and hurt the transplant effort.

The Thymus Is a Primary Lymphoid Organ Where T Cells Mature

T cell development is not complete until the cells undergo selection in the **thymus** (Figure 2-6). The importance of the thymus in T-cell development was not recognized until the early 1960s, when J.F.A.P. Miller, an Australian biologist, worked against the power of popular assumptions to advance his idea that the thymus was something other than a graveyard for cells. It was an underappreciated organ, very large in prepubescent animals, that was thought by some to be detrimental to an organism, and by others to be an evolutionary dead-end. The cells that populated it—small, thin-rimmed,

Stem Cells—Clinical Uses and Potential

Stem cell transplantation holds great promise for the regeneration of diseased, damaged, or defective tissue. Hematopoietic stem cells are already used to restore hematopoietic function, and their use in the clinic is described below. However, rapid advances in stem cell research have raised the possibility that other stem cell types may soon be routinely employed for replacement of a variety of cells and tissues. Two properties of stem cells underlie their utility and promise. They have the capacity to give rise to lineages of differentiated cells, and they are self-renewing—each division of a stem cell creates at least one stem cell. If stem cells are classified according to their descent and developmental potential, three levels of stem cells can be recognized: pluripotent, multipotent, and unipotent.

Pluripotent stem cells can give rise to an entire organism. A fertilized egg, the zygote, is an example of such a cell. In humans, the initial divisions of the zygote and its descendants produce cells that are also pluripotent. In fact, identical twins develop when pluripotent cells separate and develop into genetically identical fetuses. *Multipotent stem cells* arise from embryonic stem cells and can give rise to a more limited range of cell types. Further differentiation of multipotent stem cells leads to the formation of *unipotent stem cells*, which can generate only the same cell type as themselves. (Note that "pluripotent" is often used to describe the hematopoietic stem cell. Within the context of blood cell lineages this is arguably true; however, it is probably strictly accurate to call the HSC a multipotent stem cell.)

Pluripotent cells, called *embryonic stem cells*, or simply *ES cells*, can be iso-lated from early embryos, and for many years it has been possible to grow mouse ES cells as cell lines in the laboratory. Strikingly, these cells can be induced to generate many different types of cells. Mouse ES cells have been shown to give rise to muscle cells, nerve cells, liver cells, pancreatic cells, and hematopoietic cells.

Advances have made it possible to grow lines of human pluripotent stem cells and, most recently, to induce differentiated human cells to become pluripotent stem cells. These are developments of considerable importance to the understanding of human development, and they also have great therapeutic potential. In vitro studies of the factors that determine or influence the development of human pluripotent stem cells along specific developmental paths are providing considerable insight into how cells differentiate into specialized cell types. This research is driven in part by the great potential for using pluripotent stem cells to generate cells and tissues that could replace diseased or damaged tissue. Success in this endeavor would be a major advance because transplantation medicine now depends entirely on donated organs and tissues, yet the need far exceeds the number of donations, and the need is increasing. Success in deriving cells, tissues, and organs from pluripotent stem cells could provide skin replacement for burn patients, heart muscle cells for those with chronic heart disease, pancreatic islet cells for patients with diabetes, and neurons for the treatment of Parkinson's disease or Alzheimer's disease.

The transplantation of HSCs is an important therapy for patients whose hematopoietic systems must be replaced. It has three major applications:

- Providing a functional immune system to individuals with a genetically determined immunodeficiency, such as severe combined immunodeficiency (SCID).

- Replacing a defective hematopoietic system with a functional one to cure patients with life-threatening nonmalignant genetic disorders in hematopoiesis, such as sickle-cell anemia or thalassemia.

- Restoring the hematopoietic system of cancer patients after treatment with doses of chemotherapeutic agents and radiation. This approach is particularly applicable to leukemias, including acute myeloid leukemia, which can be cured only by destroying the patient's own hematopoietic system—the source of the leukemia cells. Clinicians also hope that this approach can be used to facilitate treatment of solid tumors. High-dose radiation and cytotoxic regimens can be much more effective at killing solid tumors than therapies using more conventional doses of cytotoxic agents; however, they destroy the immune system, and stem cell transplantation makes it possible to recover from such drastic treatment.

Hematopoietic stem cells have extraordinary powers of regeneration. Experiments in mice indicate that as few as one HSC can completely restore the erythroid population and the immune system. In humans, for instance, as little as 10% of a donor's total volume of bone marrow can provide enough HSCs to completely restore the recipient's hematopoietic system. Once injected into a vein, HSCs enter the circulation and find their own way to

---

featureless cells called thymocytes—looked dull and inactive. However, Miller proved that the thymus was the all-important site for the maturation of T lymphocytes (see the Classic Experiment Box 2-3 on pages 46–47).

T-cell precursors, which still retain the ability to give rise to multiple hematopoietic cell types, travel via the blood from the bone marrow to the thymus. Immature T cells, known as **thymocytes** (thymus cells) because of their site of maturation, pass through defined developmental stages in specific thymic microenvironments as they mature into functional T cells. The thymus is a specialized environment where immature T cells generate unique antigen receptors (T cell receptors, or TCRs) and are then selected on the basis of their reactivity to self MHC-peptide complexes expressed

BOX 2-2

the bone marrow, where they begin the process of engraftment. In addition, HSCs can be preserved by freezing. This means that hematopoietic cells can be "banked." After collection, the cells are treated with a cryopreservative, frozen, and then stored for later use. When needed, the frozen preparation is thawed and infused into the patient, where it reconstitutes the hematopoietic system. This cell-freezing technology even makes it possible for individuals to store their own hematopoietic cells for transplantation to themselves at a later time. Currently, this procedure is used to allow cancer patients to donate cells before undergoing chemotherapy and radiation treatments, then later reconstitute their hematopoietic system using their own cells.

Transplantation of stem cell populations may be **autologous** (the recipient is also the donor), **syngeneic** (the donor is genetically identical; i.e., an identical twin of the recipient), or **allogeneic** (the donor and recipient are not genetically identical). In any transplantation procedure, genetic differences between donor and recipient can lead to immune-based rejection reactions. Aside from host rejection of transplanted tissue (host versus graft), lymphocytes conveyed to the recipient via the graft can attack the recipient's tissues, thereby causing **graft-versus-host disease (GVHD)**, a life-threatening affliction. In order to suppress rejection reactions, powerful immunosuppressive drugs must be used. Unfortunately, these drugs have serious side effects, and immunosuppression increases the patient's risk of infection and susceptibility to tumors. Consequently, HSC transplantation has the fewest complications when there is genetic identity between donor and recipient.

At one time, bone marrow transplantation was the only way to restore the hematopoietic system. However, both peripheral blood and umbilical cord blood are now also common sources of hematopoietic stem cells. These alternative sources of HSCs are attractive because the donor does not have to undergo anesthesia or the highly invasive procedure used to extract bone marrow. Although peripheral blood may replace marrow as a major source of hematopoietic stem cells for many applications, bone marrow transplantation still has some advantages (e.g., marrow may include stem cell subsets that are not as prevalent in blood). To obtain HSC-enriched preparations from peripheral blood, agents are used to induce increased numbers of circulating HSCs, and then the HSC-containing fraction is separated from the plasma and red blood cells in a process called leukapheresis. If necessary, further purification can be done to remove T cells and to enrich the CD34$^+$ population.

Umbilical cord blood contains an unusually high frequency of hematopoietic stem cells. Furthermore, it is obtained from placental tissue (the "afterbirth"), which is normally discarded. Consequently, umbilical cord blood has become an attractive source of cells for HSC transplantation. For reasons that remain incompletely understood, cord blood stem cell transplants do not engraft as reliably as peripheral blood stem cell transplants; however, grafts of cord blood cells produce GVHD less frequently than marrow grafts, probably because cord blood has fewer mature T cells.

Beyond its current applications in cancer treatment, autologous stem cell transplantation can also be useful for **gene therapy**, the introduction of a normal gene to correct a disorder caused by a defective gene. One of the most highly publicized gene therapy efforts—the introduction of the adenosine deaminase (ADA) gene to correct a form of severe combined immunodeficiency (SCID)—was performed successfully on hematopoietic stem cells. The therapy entails removing a sample of hematopoietic stem cells from a patient, inserting a functional gene to compensate for the defective one, and then reinjecting the engineered stem cells into the donor.

The advantage of using stem cells in gene therapy is that they are self-renewing. Consequently, at least in theory, patients would have to receive only a single injection of engineered stem cells. In contrast, gene therapy with engineered mature lymphocytes or other blood cells would require periodic injections because these cells are not capable of self-renewal. In the case of the SCID patients, hematopoietic stem cells were successfully infected with a retrovirus engineered to express the ADA gene. The cells were returned to the patients and did, indeed, correct the deficiency. Patients who previously could not generate lymphocytes to protect themselves from infection were able to generate normal cells and live relatively normal lives. Unfortunately, in a number of patients, the retrovirus used to introduce the ADA gene integrated into parts of the genome that resulted in leukemia. Investigators continue to work to improve the safety and efficiency of gene delivery; more successful gene therapy efforts are clearly in the future.

---

on the surface of thymic stromal cells. Those thymocytes whose T-cell receptors bind self MHC-peptide complexes with too high affinity are induced to die (**negative selection**), and those thymocytes that bind self MHC-peptides with an intermediate affinity undergo **positive selection**, resulting in their survival, maturation, and migration to the thymic medulla. Most thymocytes do not navigate the jour-

ney through the thymus successfully; in fact, it is estimated that 95% of thymocytes die in transit. The majority of cells die because they have too low an affinity for the self-antigen-MHC combinations that they encounter on the surface of thymic epithelial cells and fail to undergo positive selection.

These developmental events take place in several distinct thymic microenvironments (see Figure 2-6). T-cell precursors

(a)

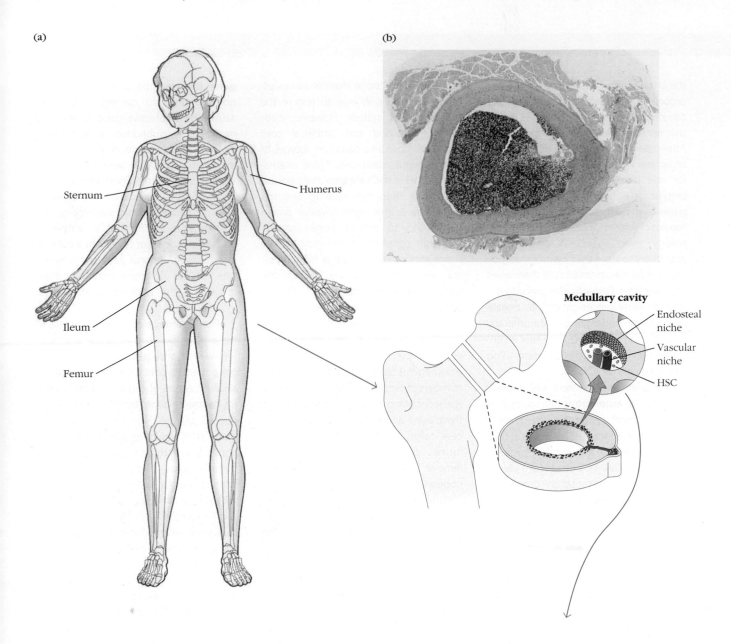

Sternum

Humerus

Ileum

Femur

(b)

Medullary cavity

Endosteal niche

Vascular niche

HSC

(c)

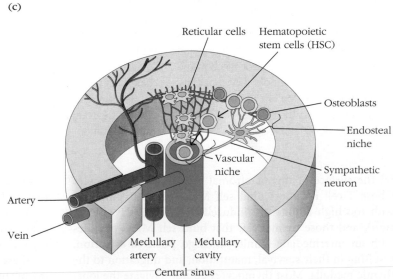

Reticular cells

Hematopoietic stem cells (HSC)

Osteoblasts

Endosteal niche

Vascular niche

Sympathetic neuron

Artery

Vein

Medullary artery

Medullary cavity

Central sinus

FIGURE 2-5 The bone marrow microenvironment. (a) Multiple bones support hematopoiesis, including the hip (ileum), femur, sternum, and humerus. (b) This figure shows a typical cross-section of a bone with a medullary (marrow) cavity. (c) Blood vessels (central sinus and medullary artery) run through the center of the bone and form a network of capillaries in close association with bone and bone surface (endosteum). Both the cells that line the blood vessels (endothelium) and the cells that line the bone (osteoblasts) generate niches that support hematopoietic stem cell (HSC) self-renewal and differentiation. The most immature cells appear to be associated with the endosteal (bone) niche; as they mature, they migrate toward the vascular (blood vessel) niche. Fully differentiated cells exit the marrow via blood vessels. [2-5b; Courtesy of Indiana University School of Medicine.]

(a)

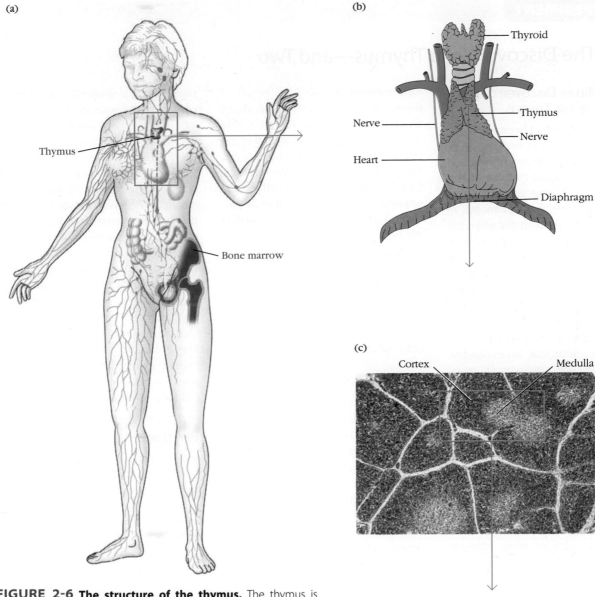

(b)

(c)

FIGURE 2-6 The structure of the thymus. The thymus is found just above the heart (a, b) and is largest prior to puberty, when it begins to shrink. Panel (c) depicts a stained thymus tissue section and (d) a cartoon of the microenvironments: the cortex, which is densely populated with DP immature thymocytes (blue) and the medulla, which is sparsely populated with SP mature thymocytes. These major regions are separated by the corticomedullary junction (CMJ), where cells enter from and exit to the bloodstream. The area between the cortex and the thymic capsule, the subcapsular cortex, is a site of much proliferation of the youngest (DN) thymocytes. The route taken by a typical thymocyte during its development from the DN to DP to SP stages is shown. Thymocytes are positively selected in the cortex. Autoreactive thymocytes are negatively selected in the medulla; some may also be negatively selected in the cortex. [2-6c: Dr. Gladden Willis/Getty Images.]

(d)

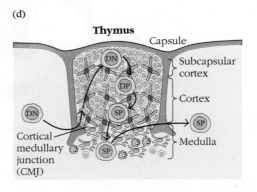

The Discovery of a Thymus—and Two

J.F.A.P. MILLER DISCOVERED THE FUNCTION OF THE THYMUS

In 1961, Miller, who had been investigating the thymus's role in leukemia, published a set of observations in *The Lancet* that challenged notions that, at best, this organ served as a cemetery for lymphocytes and, at worst, was detrimental to health (Figure 1). He noted that when this organ was removed in very young mice (in a process known as thymectomy), the subjects became susceptible to a variety of infections, failed to reject skin grafts, and died prematurely. On close examination of their circulating blood cells, they also appeared to be missing a type of cell that another investigator, James Gowans, had associated with cellular and humoral immune responses. Miller concluded that the thymus produced functional immune cells.

Several influential investigators could not repeat the data and questioned Miller's conclusions. Some speculated that the mouse strain he used was peculiar, others that his mice were exposed to too many pathogens and their troubles were secondary to infection. Dr. Miller responded to each of these criticisms experimentally, assessing the impact of thymectomy in different mouse strains and in germ-free facilities. His results were unequivocal, and his contention that this organ generated functional lymphocytes was vindicated. Elegant experiments by Dr. Miller, James Gowans, and others subsequently showed that the thymus produced a different type of lymphocyte than the bone marrow. This cell did not produce antibodies directly, but, instead, was required for optimal antibody production. It was called a T cell after the thymus, its organ of origin. Immature T cells are known as *thymocytes*. Miller is one of the few scientists credited with the discovery of the function of an entire organ.

A SECOND THYMUS

No one expected a new anatomical discovery in immunology in the 21st century. However, in 2006 Hans-Reimer Rodewald and his colleagues reported the existence of a second thymus in mice. The conventional thymus is a bi-lobed organ that sits in the thorax right above the heart. Rodewald and his colleagues discovered thymic tissue that sits in the neck, near the cervical vertebrae, of mice. This cervical thymic tissue is smaller in mass than the conventional thymus, consists of a single lobe or clusters of single lobes, and is populated by relatively more mature thymocytes. However, it contributes to T-cell development very effectively and clearly contributes to the mature T-cell repertoire. Rodewald's findings raise the possibility that some of our older observations and assumptions about thymic function need to be reexamined. In particular, studies based on thymectomy that indicated T cells could develop outside the thymus may need to be reassessed. The cells found may have come from this more obscure but functional thymic tissue. The evolutionary implications of this thymus are also interesting—thymi are found in the neck in several species, including the koala and kangaroo.

REFERENCES

Miller, J. F. 1961. Immunological function of the thymus. *Lancet* **2**:748–749.

Miller, J.F.A.P. 2002. The discovery of thymus function and of thymus-derived lymphocytes. *Immunological Reviews* **185**:7–14.

Rodewald, H-R. 2006. A second chance for the thymus. *Nature* **441**:942–943.

FIGURE 1
J.F.A.P. Miller (above) in 1961 and the first page (opposite) of his *Lancet* article (1961) describing his discovery of the function of the thymus. [The Walter and Eliza Hall Institute of Medical Research.]

BOX 2-3

748   SEPTEMBER 30, 1961   PRELIMINARY COMMUNICATIONS   THE LANCET

Preliminary Communications

IMMUNOLOGICAL FUNCTION OF THE THYMUS

It has been suggested that the thymus does not participate in immune reactions. This is because antibody formation has not been demonstrated in the normal thymus,[1] and because, even after intense antigenic stimulation plasma cells (the morphological expression of active antibody formation) and germinal centres have not been described in that organ.[2] Furthermore, thymectomy in the adult animal has had little or no significant effect on antibody production.[3]

On the other hand, there are certain clinical and experimental observations in man and other animals which suggest that the thymus may somehow be concerned in the control of immune responses. Thus, in acute infections, when presumably the need for antibody production is great, the thymus undergoes rapid involution; in patients with acquired agammaglobulinæmia the simultaneous occurrence of benign thymomas has been described,[4] and in fœtal or newborn animals, at a time when responsiveness to antigenic stimulation is deficient, the thymus is a very prominent organ.

The apparent contradiction between these two sets of observations may be partly explained by recent work,[5][6] which suggests that the thymus does not respond to circulating antigens because these cannot reach it owing to the existence of a barrier between the normal gland and the blood-stream. If the barrier is broken, for instance by local trauma, the histological reactions of antibody formation take place in the thymus.

In this laboratory, we have been interested in the role of the thymus in leukæmogenesis.[7] During this work it has become increasingly evident that the thymus at an early stage in life plays a very important part in the development of immunological response.

METHODS AND RESULTS

In the preliminary experiments mice of the C3H and Ak strains and of a cross between T_6 and Ak were used. The thymus was removed 1–16 hours after birth. Alternate littermates were used as sham-thymectomised controls—i.e., they underwent the full operative procedure, including excision of part of the sternum, but their thymuses were left intact. Mice in another group had thymectomy at 5 days of age. Wounds were closed with a continuous black silk suture and the baby mice were returned immediately to their mothers. No anti-

1. Fagreus, A. *J. Immunol.* 1948, **58**, 1.
2. Fagreus, A. *Acta med. scand.* 1948, suppl. 204, p. 3.
3. MacLean, L. D., Zak, S. J., Varco, R. L., Good, R. A. *Transplant. Bull.* 1957, **4**, 21.
4. Good, R. A., Varco, R. L. *Lancet*, 1955, i, 245.
5. Stoner, R. D., Hale, W. M. *J. Immunol.* 1955, **75**, 203.
6. Marshall, A. H. E., White, R. G. *Lancet*, 1961, i, 1030.
7. Miller, J. F. A. P. *Nature, Lond.* 1961, **191**, 248.

biotics were administered at any time either to the operated mice or to their mothers.

Mortality during and immediately after the operation ranged between 5 and 15% (excluding deaths due either to neglectful mothers or to cannibalism). Mortality in the thymectomised group was, however, higher between the 1st and 3rd month of life and was attributable mostly to common laboratory infections. This suggested that neonatally thymectomised mice were more susceptible to such infections than even sham-thymectomised littermate controls. When thymectomised and control groups were isolated from other experimental mice and kept under nearly pathogen-free conditions, the mortality in the thymectomised group was significantly reduced.

Absolute and differential white-cell counts were performed on tail blood at various intervals after thymectomy. The significant results of these estimations are summarised in fig. 1. In sham-thymectomised animals the lymphocyte/

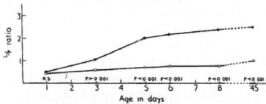

Fig. 1—Average lymphocyte:polymorph ratio of mice thymectomised in the neonatal period compared with sham-thymectomised controls. Statistical differences indicated.

   ○————○ thymectomised mice.
   ●————● sham-thymectomised mice.

polymorph ratio rose progressively in the first 8 days of life to reach the normal adult ratio of 2.5 ± 0.08. In the animals whose thymus was removed on the 1st day of life the ratio did not increase significantly and was only 1.0 ± 0.10 at 6 weeks of age.

Histological examination of lymph-nodes and spleens of thymectomised animals at 6 weeks of age revealed a conspicuous deficiency of germinal centres and only few plasma cells (figs. 2 and 3).

At 6 weeks of age, groups of thymectomised, sham-thymectomised, and entirely normal mice were subjected to skin grafting. Ak mice receiving C3H grafts and vice versa, and $(AkXT_6)F_1$ mice receiving C3H grafts. The median survival time of skin grafts in intact mice, sham-thymectomised mice, and mice thymectomised at 5 days of age ranged from 10 to 12 days. In more than 70% of mice whose thymus was removed on the 1st day of life the grafts were established and grew luxuriant crops of hair. Most of these grafts were tolerated for periods ranging from 6 weeks to 2 months and

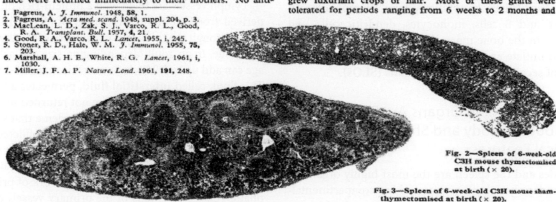

Fig. 2—Spleen of 6-week-old C3H mouse thymectomised at birth (× 20).

Fig. 3—Spleen of 6-week-old C3H mouse sham-thymectomised at birth (× 20).

enter the thymus in blood vessels at the *corticomedullary junction* between the *thymic cortex*, the outer portion of the organ, and the *thymic medulla*, the inner portion of the organ. At this stage thymocytes express neither CD4 nor CD8, markers associated with mature T cells. They are therefore called **double negative (DN) cells**. DN cells first travel to the region under the thymic capsule, a region referred to as the *subcapsular cortex,* where they proliferate and begin to generate their T-cell receptors. Thymocytes that successfully express TCRs begin to express both CD4 and CD8, becoming **double positive (DP) cells**, and populate the cortex, the site where most (85% or more) immature T cells are found. The cortex features a distinct set of stromal cells, *cortical thymic epithelial cells* (*cTECs*), whose long processes are perused by thymocytes testing the ability of their T-cell receptors to bind MHC-peptide complexes (Video 2-1). Thymocytes that survive selection move to the thymic medulla, where positively selected thymocytes encounter specialized stromal cells, *medullary thymic epithelial cells* (*mTECs*). Not only do mTECs support the final steps of thymocyte maturation, but they also have a unique ability to express proteins that are otherwise found exclusively in other organs. This allows them to negatively select a group of potentially very damaging, autoreactive T cells that could not be deleted in the cortex.[1]

Mature thymocytes, which express only CD4 or CD8 and are referred to as **single positive (SP)**, leave the thymus as they entered: via the blood vessels of the corticomedullary junction. Maturation is finalized in the periphery, where these new T cells (*recent thymic emigrants*) explore antigens presented in secondary lymphoid tissue, including spleen and lymph nodes.

Secondary Lymphoid Organs—Where the Immune Response Is Initiated

As just described, lymphocytes and myeloid cells develop to maturity in the primary lymphoid system: T lymphocytes in the thymus, and B cells, monocytes, dendritic cells, and granulocytes in the bone marrow. However, they encounter antigen and initiate an immune response in the microenvironments of **secondary lymphoid organs (SLOs)**.

Secondary Lymphoid Organs Are Distributed Throughout the Body and Share Some Anatomical Features

Lymph nodes and the spleen are the most highly organized of the secondary lymphoid organs and are compartmentalized from the rest of the body by a fibrous capsule. A somewhat less organized system of secondary lymphoid tissue, collectively referred to as mucosa-associated lymphoid tissue (MALT), is found associated with the linings of multiple organ systems, including the gastrointestinal (GI) and respiratory tracts. MALT includes tonsils, Peyer's patches (in the small intestine), and the appendix, as well as numerous lymphoid follicles within the lamina propria of the intestines and in the mucous membranes lining the upper airways, bronchi, and genitourinary tract (Figure 2-7).

Although secondary lymphoid organs vary in their location and degree of organization, they share key features. All SLOs include anatomically distinct regions of T-cell and B-cell activity, and all develop lymphoid follicles, which are highly organized microenvironments that are responsible for the development and selection of B cells that produce high-affinity antibodies.

Lymphoid Organs Are Connected to Each Other and to Infected Tissue by Two Different Circulatory Systems: Blood and Lymphatics

The immune cells are the most mobile cells in a body and use two different systems to traffic through tissues: the blood system and the lymphatic system. The blood has access to virtually every organ and tissue and is lined by endothelial cells that are very responsive to inflammatory signals. Hematopoietic cells can transit through the blood system—away from the heart via active pumping networks (arteries) and back to the heart via passive valve-based systems (veins) within minutes. Most lymphocytes enter secondary lymphoid organs via specialized blood vessels, and leave via the lymphatic system.

The lymphatic system is a network of thin walled vessels that play a major role in immune cell trafficking, including the travel of antigen and antigen-presenting cells to secondary lymphoid organs and the exit of lymphocytes from lymph nodes.

Lymph vessels are filled with a protein-rich fluid (**lymph**) derived from the fluid component of blood (**plasma**) that seeps through the thin walls of capillaries into the surrounding tissue. In an adult, depending on size and activity, seepage can add up to 2.9 liters or more during a 24-hour period. This fluid, called **interstitial fluid**, permeates all tissues and bathes all cells. If this fluid were not returned to the circulation, the tissue would swell, causing edema that would eventually become life threatening. We are not afflicted with such catastrophic edema because much of the fluid is returned to the blood through the walls of venules. The remainder of the interstitial fluid enters the delicate network of primary lymphatic vessels. The walls of the primary vessels consist of a single layer of loosely apposed endothelial cells. The porous architecture of the primary vessels allows fluids and even cells to enter the lymphatic network. Within these vessels, the fluid, now called lymph, flows into a series of progressively larger collecting vessels called **lymphatic vessels** (see Figures 2-7b and 2-7c).

---

[1]Note that some investigators describe positive selection as taking place in the cortex and negative selection solely in the medulla. However, several lines of evidence suggest that negative selection can also occur in the cortex, and we have adopted this perspective for this text.

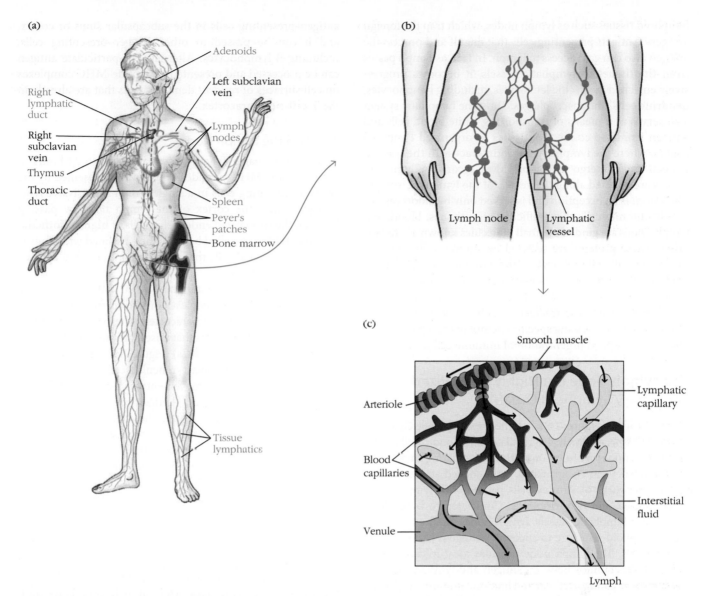

FIGURE 2-7 The human lymphoid system. The primary organs (bone marrow and thymus) are shown in red; secondary organs and tissues, in blue. These structurally and functionally diverse lymphoid organs and tissues are interconnected by the blood vessels (not shown) and lymphatic vessels (purple). Most of the body's lymphatics eventually drain into the thoracic duct, which empties into the left subclavian vein. However, the vessels draining the right arm and right side of the head (shaded blue) converge to form the right lymphatic duct, which empties into the right subclavian vein. The inset (b) shows the lymphatic vessels in more detail, and (c) shows the relationship between blood and lymphatic capillaries in tissue. The lymphatic capillaries pick up interstitial fluid, particulate and soluble proteins, as well as immune cells from the tissue surrounding the blood capillaries (see arrows). *[Part (a): Adapted from H. Lodish et al., 1995, Molecular Cell Biology, 3rd ed., Scientific American Books, New York.]*

All cells and fluid circulating in the lymph are ultimately returned to the blood system. The largest lymphatic vessel, the **thoracic duct**, empties into the left subclavian vein. It collects lymph from all of the body except the right arm and right side of the head. Lymph from these areas is collected into the *right lymphatic duct*, which drains into the right subclavian vein (see Figure 2-7a). By returning fluid lost from the blood, the lymphatic system ensures steady-state levels of fluid within the circulatory system.

The heart does not pump the lymph through the lymphatic system; instead, the slow, low-pressure flow of lymph is achieved by the movements of the surrounding muscles. Therefore, activity enhances lymph circulation. Importantly, a series of one-way valves along the lymphatic vessels ensures that lymph flows in only one direction.

When a foreign antigen gains entrance to the tissues, it is picked up by the lymphatic system (which drains all the tissues of the body) and is carried to various organized lymphoid tissues such as lymph nodes, which trap the foreign antigen. Antigen-presenting cells that engulf and process the antigen also can gain access to lymph. In fact, as lymph passes from the tissues to lymphatic vessels, it becomes progressively enriched in specific leukocytes, including lymphocytes, dendritic cells, and macrophages. Thus, the lymphatic system also serves as a means of transporting white blood cells and antigen from the connective tissues to organized lymphoid tissues, where the lymphocytes can interact with the trapped antigen and undergo activation. Most secondary lymphoid tissues are situated along the vessels of the lymphatic system. The spleen is an exception and is served only by blood vessels.

All immune cells that traffic through tissues, blood, and lymph nodes are guided by small molecules known as **chemokines**. These proteins are secreted by stromal cells, antigen-presenting cells, lymphocytes, and granulocytes, and form gradients that act as attractants and guides for other immune cells, which express an equally diverse set of receptors for these chemokines. The interaction between specific chemokines and cells expressing specific chemokine receptors allows for a highly refined organization of immune cell movements.

The Lymph Node Is a Highly Specialized Secondary Lymphoid Organ

Lymph nodes (Figure 2-8) are the most specialized SLOs. Unlike the spleen, which also regulates red blood cell flow and fate, lymph nodes are fully committed to regulating an immune response. They are encapsulated, bean-shaped structures that include networks of stromal cells packed with lymphocytes, macrophages, and dendritic cells. Connected to both blood vessels and lymphatic vessels, lymph nodes are the first organized lymphoid structure to encounter antigens that enter the tissue spaces. The lymph node provides ideal microenvironments for encounters between antigen and lymphocytes and productive, organized cellular and humoral immune responses.

Structurally, a lymph node can be divided into three roughly concentric regions: the cortex, the paracortex, and the medulla, each of which supports a distinct microenvironment (see Figure 2-8). The outermost layer, the **cortex**, contains lymphocytes (mostly B cells), macrophages, and follicular dendritic cells arranged in **follicles**. Beneath the cortex is the **paracortex**, which is populated largely by T lymphocytes and also contains dendritic cells that migrated from tissues to the node. The **medulla** is the innermost layer, and the site where lymphocytes exit (*egress*) the lymph node through the outgoing (*efferent*) lymphatics. It is more sparsely populated with lymphoid lineage cells, which include plasma cells that are actively secreting antibody molecules.

Antigen travels from infected tissue to the cortex of the lymph node via the incoming (*afferent*) lymphatic vessels, which pierce the capsule of a lymph node at numerous sites and empty lymph into the *subcapsular sinus* (see Figure 2-8b). It enters either in particulate form or is processed and presented as peptides on the surface of *migrating* antigen-presenting cells. Particulate antigen can be trapped by resident antigen-presenting cells in the subcapsular sinus or cortex, and it can be passed to other antigen-presenting cells, including B lymphocytes. Alternatively, particulate antigen can be processed and presented as peptide-MHC complexes on cell surfaces of *resident* dendritic cells that are already in the T-cell-rich paracortex.

T Cells in the Lymph Node

It takes every naïve T lymphocyte about 16 to 24 hours to browse all the MHC-peptide combinations presented by the antigen-presenting cells in a single lymph node. Naïve lymphocytes enter the cortex of the lymph node by passing between the specialized endothelial cells of **high endothelial venules (HEV)**, so-called because they are lined with unusually tall endothelial cells that give them a thickened appearance (Figure 2-9a).

Once naïve T cells enter the lymph node, they browse MHC-peptide antigen complexes on the surfaces of the dendritic cells present in the paracortex. The paracortex is traversed by a web of processes that arise from stromal cells called **fibroblast reticular cells (FRCs)** (Figure 2-9b). This network is referred to as the fibroblast reticular cell conduit system (FRCC) and guides T-cell movements via associated adhesion molecules and chemokines. Antigen-presenting cells also appear to wrap themselves around the conduits, giving circulating T cells ample opportunity to browse their surfaces as they are guided down the network. The presence of this specialized network elegantly enhances the probability that T cells will meet their specific MHC-peptide combination.

T cells that browse the lymph node but do not bind MHC-peptide combinations exit not via the blood, but via the *efferent lymphatics* in the medulla of the lymph node (see Figure 2-8). T cells whose TCRs do bind to an MHC-peptide complex on an antigen-presenting cell that they encounter in the lymph node will stop migrating and take up residence in the node for several days. Here it will proliferate and, depending on cues from the antigen-presenting cell itself, its progeny will differentiate into effector cells with a variety of functions. CD8$^+$ T cells gain the ability to kill target cells. CD4$^+$ T cells can differentiate into several different kinds of effector cells, including those that can further activate macrophages, CD8$^+$ T cells, and B cells.

B Cells in the Lymph Node

The lymph node is also the site where B cells are activated and differentiate into high-affinity antibody-secreting plasma cells. B cell activation requires both antigen engagement by the B-cell receptor (BCR) and direct contact with an activated CD4$^+$ T$_H$ cell. Both events are facilitated by the anatomy of the lymph node. Like T cells, B cells circulate through the blood and lymph and visit the lymph nodes on a daily basis, entering via the HEV. They respond to specific signals and chemokines that draw them not to the paracortex but to the lymph node follicle. Although they may initially take advantage of the FRCC for guidance, they

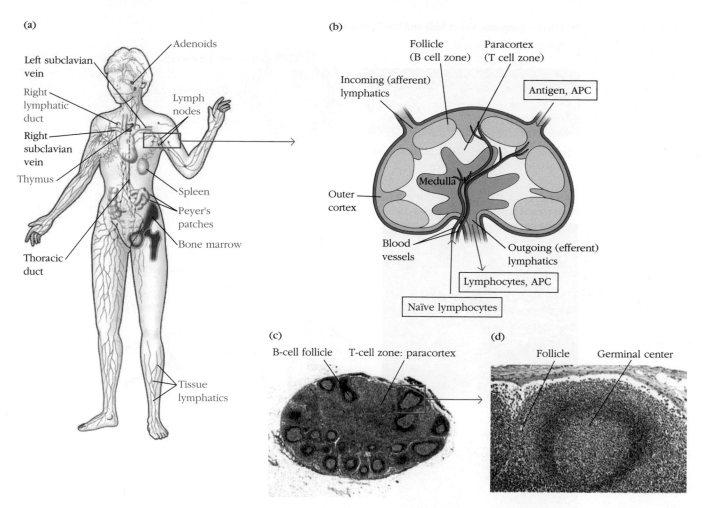

FIGURE 2-8 Structure of a lymph node. The microenvironments of the lymph node support distinct cell activities. (a) The lymph nodes are dispersed throughout the body and are connected by lymphatic vessels as well as blood vessels (not shown). (b) A drawing of the major features of a lymph node shows the major vessels that serve the organ: incoming (afferent) and outgoing (efferent) lymphatic vessels, and the arteries and veins. It also depicts the three major tissue layers: the outer cortex, the paracortex, and the innermost region, the medulla. Macrophages and dendritic cells, which trap antigen, are present in the cortex and paracortex. T cells are concentrated in the paracortex; B cells are primarily in the cortex, within follicles and germinal centers. The medulla is populated largely by antibody-producing plasma cells and is the site where cells exit via the efferent lymphatics. Naïve lymphocytes circulating in the blood enter the node via high endothelial venules (HEV) via a process called extravasation (see Advances Box 14-2). Antigen and some leukocytes, including antigen-presenting cells, enter via afferent lymphatic vessels. All cells exit via efferent lymphatic vessels. (c) This stained tissue section shows the cortex with a number of ovoid follicles, which is surrounded by the T-cell-rich paracortex. (d) A stained lymph node section showing a follicle that includes a germinal center (otherwise referred to as a secondary follicle). [2-8c: Dr. Gladden Willis/Getty Images; 2-8d: Image Source/Alamy.]

ultimately depend upon follicular dendritic cells (FDCs) for guidance (Figure 2-9c). FDCs are centrally important in maintaining follicular and germinal center structure and "presenting" antigen to differentiating B cells.

B cells differ from T cells in that their receptors can recognize free antigen. A B cell will typically meet its antigen in the follicle. If its BCR binds to antigen, the B cell becomes partially activated and engulfs and processes that antigen. As mentioned above, B cells, in fact, are specialized antigen-presenting cells that present processed peptide-MHC complexes on their surface to CD4$^+$ T$_H$ cells. Recent data show that B cells that have successfully engaged and processed antigen change their migration patterns and

move to the T-cell-rich paracortex, where they increase their chances of encountering an activated CD4$^+$ T$_H$ cell that will recognize the MHC-antigen complex they present. When they successfully engage this T$_H$ cell, they maintain contact for a number of hours, becoming fully activated and receiving signals that induce B cell proliferation (see Chapter 14).

Some activated B cells differentiate directly into an antibody-producing cell (plasma cell) but others re-enter the follicle to establish a germinal center. A follicle that develops a germinal center is sometimes referred to as a **secondary follicle**; a follicle without a germinal center is sometimes referred to as a **primary follicle**.

(a) Afferent lymphatic vessel   high endothelial venule

Lymphatic endothelial cell

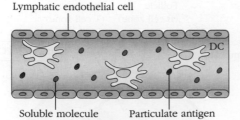

Soluble molecule   Particulate antigen

Blood endothelial cell

Perivascular sheath

Basal lamina

Naïve B cell   Naïve T cell

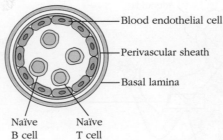

(b) Follicular reticular cell conduit system

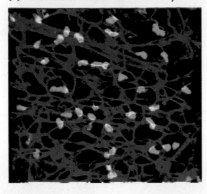

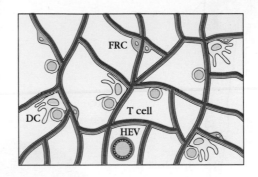

(c) Follicular dendritic cell

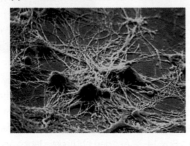

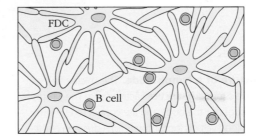

FIGURE 2-9 Features of lymph node microenvironments. The lymph node microenvironments are maintained and regulated by distinct cell types and structures. (a) The afferent lymphatics are the vessels through which dendritic cells, and particulate and soluble antigen, enter the lymph node. The high endothelial venules (HEVs) are the vessels through which naïve T and B cells enter the lymph node (via extravasation). (b) The paracortex is crisscrossed by processes and conduits formed by fibroblastic reticular cells (FRCs), which guide the migration of antigen-presenting cells and T cells, facilitating their interactions. The left panel shows an immunofluorescence microscopy image with the FRC shown in red and T cells in green. The right panel shows a cartoon of the network and cell participants. (c) The B-cell follicle contains a network of follicular dendritic cells (FDCs), which are shown as an SEM image as well as a cartoon. FDCs guide the movements and interactions of B cells. [2-9b, left: Courtesy of Stephanie Favre and Sanjiv A. Luther, University of Lausanne, Switzerland. 2-9c, left: Photograph courtesy of Mohey Eldin M. El Shikh.]

Germinal centers are remarkable substructures that facilitate the generation of B cells with increased receptor affinities. In the germinal center, an antigen-specific B cell clone will proliferate and undergo somatic hypermutation of the genes coding for their antigen receptors. Those receptors that retain the ability to bind antigen with the highest affinity survive and differentiate into plasma cells that travel to the medulla of the lymph node. Some will stay and release antibodies into the bloodstream; others will exit through the efferent lymphatics and take up residence in the bone marrow, where they will continue to release antibodies into circulation.

The initial activation of B cells and establishment of the germinal center take place within 4 to 7 days of the initial infection, but germinal centers remain active for 3 weeks or more (Chapter 12). Lymph nodes swell visibly and sometimes painfully, particularly during those first few days after infection. This swelling is due both to an increase in the number of lymphocytes induced to migrate into the node as well as the proliferation of antigen-specific T and B lymphocytes within the lobe.

The Generation of Memory T and B Cells in the Lymph Node

The interactions between T_H cells and APCs, and between activated T_H cells and activated B cells, results not only in the proliferation of antigen-specific lymphocytes and their functional differentiation, but also in the generation of memory T and B cells. Memory T and B cells can take up residence in secondary lymphoid tissues or can exit the lymph node and travel to and among tissues that first encountered the pathogen. Memory T cells that reside in secondary lymphoid organs are referred to as *central memory cells* and are distinct in phenotype and functional potential from *effector memory* T cells that circulate among tissues. Memory cell phenotype, locale, and activation requirements are an active area of investigation and will be discussed in in more detail in Chapters 11 and 12.

The Spleen Organizes the Immune Response Against Blood-Borne Pathogens

The **spleen**, situated high in the left side of the abdominal cavity, is a large, ovoid secondary lymphoid organ that plays a major role in mounting immune responses to antigens in the bloodstream (Figure 2-10). Whereas lymph nodes are specialized for encounters between lymphocytes and antigen drained from local tissues, the spleen specializes in filtering blood and trapping blood-borne antigens; thus, it is particularly important in the response to systemic infections. Unlike the lymph nodes, the spleen is not supplied by lymphatic vessels. Instead, blood-borne antigens and lymphocytes are carried into the spleen through the **splenic artery** and out via the **splenic vein**. Experiments with radioactively labeled lymphocytes show that more recirculating lymphocytes pass daily through the spleen than through all the lymph nodes combined.

The spleen is surrounded by a capsule from which a number of projections (**trabeculae**) extend, providing structural support. Two main microenvironmental compartments can be distinguished in splenic tissue: the **red pulp** and **white pulp**, which are separated by a specialized region called the **marginal zone** (see Figure 2-10d). The splenic red pulp consists of a network of sinusoids populated by red blood cells, macrophages, and some lymphocytes. It is the site where old and defective red blood cells are destroyed and removed; many of the macrophages within the red pulp contain engulfed red blood cells or iron-containing pigments from degraded hemoglobin. It is also the site where pathogens first gain access to the lymphoid-rich regions of the spleen, known as the white pulp. The splenic white pulp surrounds the branches of the splenic artery, and consists of the **periarteriolar lymphoid sheath (PALS)** populated by T lymphocytes as well as B-cell follicles. As in lymph nodes, germinal centers are generated within these follicles during an immune response. The marginal zone, which borders the white pulp, is populated by unique and specialized macrophages and B cells, which are the first line of defense against certain blood-borne pathogens.

Blood-borne antigens and lymphocytes enter the spleen through the splenic artery, and interact first with cells at the marginal zone. In the marginal zone, antigen is trapped and processed by dendritic cells, which travel to the PALS. Specialized, resident *marginal zone B cells* also bind antigen via complement receptors and convey it to the follicles. Migrating B and T lymphocytes in the blood enter sinuses in the marginal zone and migrate to the follicles and the PALS, respectively.

The events that initiate the adaptive immune response in the spleen are analogous to those that occur in the lymph node. Briefly, circulating naïve B cells encounter antigen in the follicles, and circulating naïve $CD8^+$ and $CD4^+$ T cells meet antigen as MHC-peptide complexes on the surface of dendritic cells in the T-cell zone (PALS). Once activated, $CD4^+$ T_H cells then provide help to B cells and $CD8^+$ T cells that have also encountered antigen. Some activated B cells, together with some T_H cells, migrate back into follicles and generate germinal centers.

It is unclear whether a reticular network operates as prominently within the spleen as it does in the lymph node. However, given that T cells, dendritic cells, and B cells find a way to interact efficiently within the spleen to initiate an immune response, it would not be surprising if a similar conduit system were present.

Although animals can lead a relatively healthy life without a spleen, its loss does have consequences. In children, in particular, **splenectomy** (the surgical removal of a spleen) can lead to overwhelming post-splenectomy infection (OPSI) syndrome characterized by systemic bacterial infections (sepsis) caused by primarily *Streptococcus pneumoniae*, *Neisseria meningitidis*, and *Haemophilus influenzae*. Although fewer adverse effects are experienced by adults, splenectomy can still lead to an increased vulnerability to blood-borne bacterial infections, underscoring the role the spleen plays in our immune response to pathogens that enter the circulation. It is also important to recognize that the spleen has other functions (e.g., in iron metabolism, thrombocyte storage, hematopoiesis) that will also be compromised if it is removed.

MALT Organizes the Response to Antigen That Enters Mucosal Tissues

Lymph nodes and the spleen are not the only organs that develop secondary lymphoid microenvironments. T- and B-cell zones and lymphoid follicles are also found in mucosal membranes that line the digestive, respiratory, and urogenital systems, as well as in the skin.

Mucosal membranes have a combined surface area of about 400 m^2 (nearly the size of a basketball court) and are the major sites of entry for most pathogens. These vulnerable membrane surfaces are defended by a group of organized lymphoid tissues known collectively as **mucosa-associated lymphoid tissue (MALT)**. Lymphoid tissue associated with different mucosal areas is sometimes given more specific names; for instance, the respiratory epithelium is referred to as **bronchus-associated lymphoid tissue (BALT)** or **nasal-associated lymphoid tissue (NALT)**, and that associated with the intestinal epithelium is referred to as **gut-associated lymphoid tissue (GALT)**.

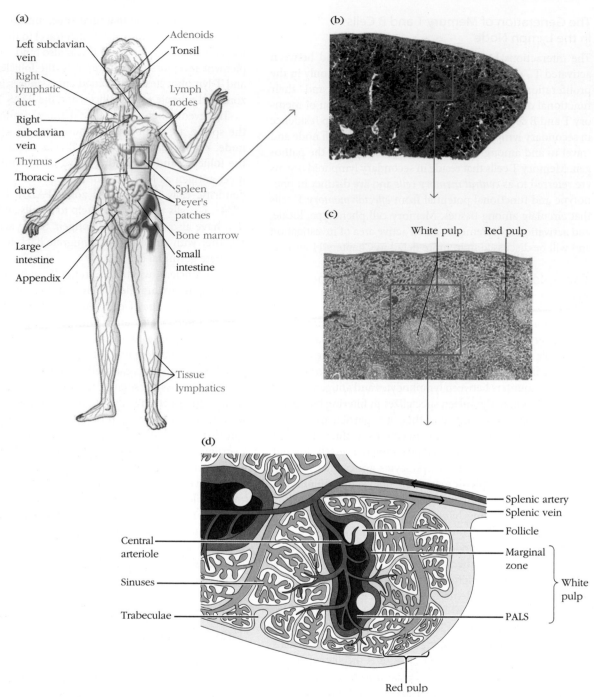

FIGURE 2-10 Structure of the spleen. (a) The spleen, which is about 5 inches long in human adults, is the largest secondary lymphoid organ. It is specialized for trapping blood-borne antigens. Panels (b) and (c) are stained tissue sections of the human spleen, showing the red pulp, white pulp, and follicles. These microenvironments are diagrammed schematically in (d). The splenic artery pierces the capsule and divides into progressively smaller arterioles, ending in vascular sinusoids that drain back into the splenic vein. The erythrocyte-filled red pulp surrounds the sinusoids. The white pulp forms a sleeve—the periarteriolar lymphoid sheath (PALS)—around the arterioles; this sheath contains numerous T cells. Closely associated with the PALS are the B-cell-rich lymphoid follicles that can develop into secondary follicles containing germinal centers. The marginal zone, a site of specialized macrophages and B cells, surrounds the PALS and separates it from the red pulp. [2-10b: Dr. Keith Wheeler/Photo Researchers; 2-10c: Biophoto Associates/Photo Researchers.]

The structure of GALT is well described and ranges from loose, barely organized clusters of lymphoid cells in the lamina propria of intestinal villi to well-organized structures such as the tonsils and adenoids (Waldeyer's tonsil ring), the appendix, and Peyer's patches, which are found within the intestinal lining and contain well-defined follicles and T-cell zones. As shown in Figure 2-11, lymphoid cells are found in various regions within the lining of the intestine. The outer mucosal epithelial layer contains **intraepithelial lymphocytes (IELs)**, many of which are T cells. The **lamina propria**, which lies under the epithelial

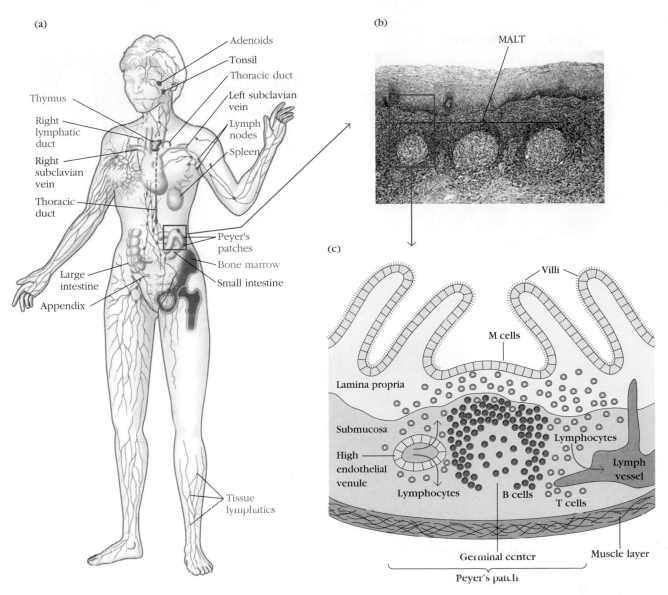

FIGURE 2-11 Mucosa-associated lymphoid tissue (MALT). (a) The Peyer's patch is a representative of the extensive MALT system that is found in the intestine. (b) A stained tissue cross-section of Peyer's patch lymphoid nodules in the intestinal submucosa is schematically diagrammed in (c). The intestinal lamina propria contains loose clusters of lymphoid cells and diffuse follicles. [2-11b: Dr. Gladden Willis/Visuals Unlimited.]

layer, contains large numbers of B cells, plasma cells, activated T cells, and macrophages in loose clusters. Microscopy has revealed more than 15,000 lymphoid follicles within the intestinal lamina propria of a healthy child. **Peyer's patches**, nodules of 30 to 40 lymphoid follicles, extend into the muscle layers that are just below the lamina propria. Like lymphoid follicles in other sites, those that compose Peyer's patches can develop into secondary follicles with germinal centers. The overall functional importance of MALT in the body's defense is underscored by its large population of antibody-producing plasma cells, whose number exceeds that of plasma cells in the spleen, lymph nodes, and bone marrow combined.

Some cellular structures and activities are unique to MALT. For instance, the epithelial cells of mucous membranes play an important role in delivering small samples of foreign antigen from the respiratory, digestive, and urogenital tracts to the underlying mucosa-associated lymphoid tissue. In the digestive tract, specialized **M cells** transport antigen across the epithelium (Figure 2-12). The structure of M cells is striking: they are flattened epithelial cells lacking the microvilli that characterize the rest of the mucosal epithelium. They have a deep invagination, or pocket, in the basolateral plasma membrane, which is filled with a cluster of B cells, T cells, and macrophages. Antigens in the intestinal lumen are endocytosed into vesicles that are transported from the luminal membrane to the underlying pocket membrane. The vesicles then fuse with the pocket membrane, delivering antigens to clusters of lymphocytes and antigen-presenting cells, the most important of which are dendritic cells, contained within the pocket. Antigen transported across the mucous membrane by M cells ultimately leads to the activation of B cells that differentiate and then secrete IgA. This class of antibody is concentrated in

(a)

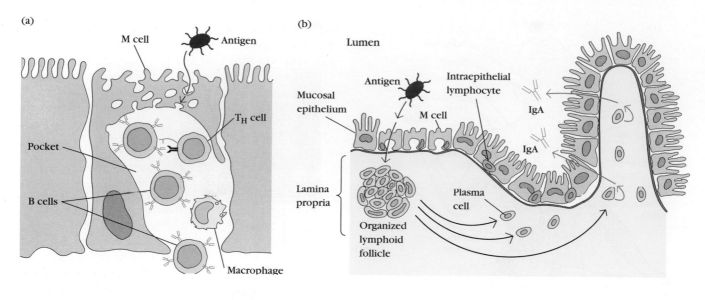

(b)

(c)

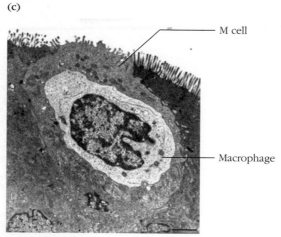

FIGURE 2-12 Structure of M cells and production of IgA at inductive sites. (a) M cells, situated in mucous membranes, endocytose antigen from the lumen of the digestive, respiratory, and urogenital tracts. The antigen is transported across the cell and released into the large basolateral pocket. (b) Antigen transported across the epithelial layer by M cells at an inductive site activates B cells in the underlying lymphoid follicles. The activated B cells differentiate into IgA-producing plasma cells, which migrate along the lamina propria, the layer under the mucosa.

The outer mucosal epithelial layer contains intraepithelial lymphocytes, of which many are T cells. (c) A stained section of mucosal lymphoid tissue (the Peyer's patch of the intestine) shows small, darkly stained intraepithelial lymphocytes encased by M cells, whose nuclei are labeled. Lymphocytes are also present in the lamina propria. [2-12c: Kucharzik, T., Lugering, N., Schmid, K.W., Schmidt, M.A., Stoll, R. Domschke, W. Human intestinal M cells exhibit enterocyte-like intermediate filaments. Gut 1998, 42: 54–52. doi: 10.1136/gut.42.1.54.]

secretions (e.g., milk) and is an important tool used by the body to combat many types of infection at mucosal sites.

The Skin Is an Innate Immune Barrier and Also Includes Lymphoid Tissue

The skin is the largest organ in the body and a critical anatomic barrier against pathogens. It also plays an important role in nonspecific (innate) defenses (Figure 2-13). The epidermal (outer) layer of the skin is composed largely of specialized epithelial cells called keratinocytes. These cells secrete a number of cytokines that may function to induce a local inflammatory

reaction. Scattered among the epithelial-cell matrix of the epidermis are Langerhans cells, skin-resident dendritic cells that internalize antigen by phagocytosis or endocytosis. These Langerhans cells undergo maturation and migrate from the epidermis to regional lymph nodes, where they function as potent activators of naïve T cells. In addition to Langerhans cells, the epidermis also contains **intraepidermal lymphocytes**, which are predominantly T cells; some immunologists believe that they play a role in combating infections that enter through the skin, a function for which they are well positioned. The underlying dermal layer of the skin also contains scattered lymphocytes, dendritic cells, monocytes, macrophages, and

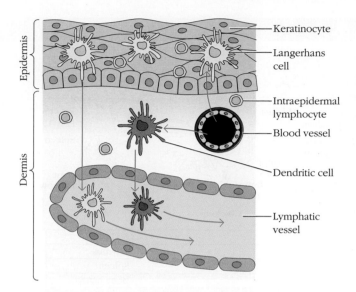

FIGURE 2-13 **The distribution of immune cells in the skin.** Langerhans cells reside in the outer layer, the epidermis. They travel to lymph nodes via the lymphatic vessels in the dermis, a layer of connective tissue below the epidermis. Dendritic cells also reside in the dermis and also can travel via the lymphatic vessels to lymph nodes when activated. White blood cells, including monocytes and lymphocytes travel to both layers of the skin via blood vessels, extravasating in the dermis, as shown.

may even include hematopoietic stem cells. Most skin lymphocytes appear to be either previously activated cells or memory cells, many of which traffic to and from local, draining lymph nodes that coordinate the responses to pathogens that have breached the skin barrier.

Tertiary Lymphoid Tissues Also Organize and Maintain an Immune Response

Tissues that are the sites of infection are referred to as **tertiary lymphoid tissue**. Lymphocytes activated by antigen in secondary lymphoid tissue can return to these organs (e.g., lung, liver, brain) as effector cells and can also reside there as memory cells. It also appears as if tertiary lymphoid tissues can generate defined microenvironments that organize the returning lymphoid cells. Investigators have recently found that the brain, for instance, establishes reticular systems that guide lymphocytes responding to chronic infection with the protozoan that causes toxoplasmosis. These observations together highlight the remarkable adaptability of the immune system, as well as the intimate relationship between anatomical structure and immune function. The conservation of structure/function relationships is also illustrated by the evolutionary relationships among immune systems and organs (see the Evolution Box 2-4 below on pages 57–59).

EVOLUTION   BOX 2-4

 ## Variations on Anatomical Themes

In Chapter 5, we will see that innate systems of immunity are found in vertebrates, invertebrates, and even in plants. Adaptive immunity, which depends on lymphocytes and is mediated by antibodies and T cells, only evolved in the subphylum Vertebrata. However, as shown in Figure 1, the kinds of lymphoid tissues seen in different orders of vertebrates differ dramatically.

All multicellular living creatures, plant and animal, have cellular and molecular systems of defense against pathogens. Vertebrates share an elaborate anatomical immune system that is compartmentalized in several predictable ways. For example, sites where immune cells develop (primary lymphoid organs) are separated from where they generate an immune response (secondary lymphoid tissue). Likewise, sites that support the development of B cells and myeloid cells, are separated from sites that generate mature T

cells. The spectrum of vertebrates ranges from the jawless fishes (Agnatha, the earliest lineages, which are represented by the lamprey eel and hagfish), to cartilaginous fish (sharks, rays), which represent the earliest lineages of jawed vertebrates (Gnathosomata), to bony fish, reptiles, amphibians, birds, and mammals, the most recently evolved vertebrates. If you view these groups as part of an evolutionary progression, you see that, in general, immune tissues and organs evolved by earlier orders have been retained as newer organs of immunity, such as lymph nodes, have appeared. For example, all vertebrates have gut-associated lymphoid tissue (GALT), but only jawed vertebrates have a well-developed thymus and spleen.

The adaptive immune system of jawed vertebrates emerged about 500 million years ago. T cells were the first cell population to express a diverse repertoire of anti-

gen receptors, and their appearance is directly and inextricably linked to the appearance of the thymus. This dependence is reflected in organisms today: all jawed vertebrates have a thymus, and the thymus is absolutely required for the development of T cells. Until recently, it was thought that jawless vertebrates (e.g., the lamprey eel) did not have any thymic tissue. However, recent studies indicate that even the lamprey supports the development of two distinct lymphocyte populations reminiscent of T and B cells, and may also harbor distinct thymic-like tissue in their gill regions (Figure 2). In contrast, their B-lymphocyte-like cells may develop in distinct regions associated with the kidneys and intestine (typhlosole). These findings suggest that jawed and jawless vertebrates share a common ancestor in which lymphocyte lineages and primary lymphoid organs were already compartmentalized.

(continued)

EVOLUTION   (continued)

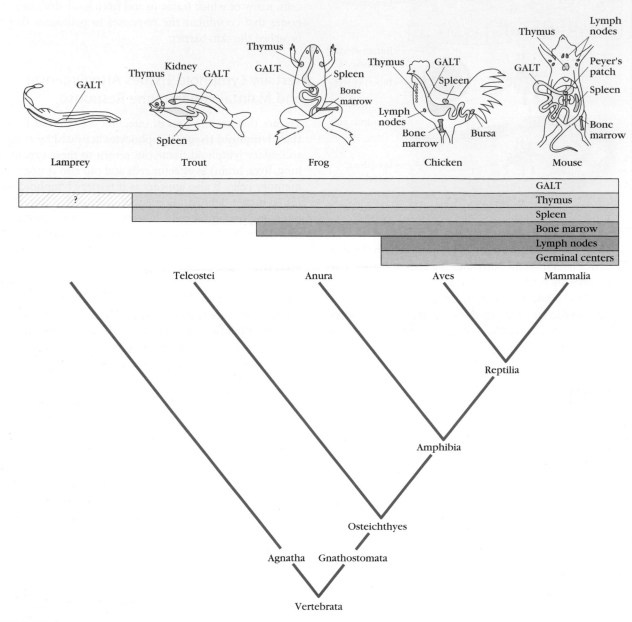

FIGURE 1

Evolutionary distribution of lymphoid tissues. The presence and location of lymphoid tissues in several major orders of vertebrates are shown. Although they are not shown in the diagram, cartilaginous fish such as sharks and rays have GALT, a thymus, and a spleen. Reptiles also have GALT, a thymus, and a spleen and may also have lymph nodes that participate in immunological reactions. The sites and nature of primary lymphoid tissues in reptiles are under investigation. Although jawless animals were thought not to have a thymus, recent data suggest that they might have thymic tissue and two types of lymphocytes analogous to B cells and T cells. [Adapted from Dupasquier and M. Flajnik, 2004, in Fundamental Immunology, 5th ed., W. E. Paul, ed., Lippincott-Raven, Philadelphia.]

Whereas T-cell development is inextricably linked to the presence of a thymus, B-cell development does not appear to be bound to one particular organ. Although the bone marrow is the site of B-cell development in many mammals, including mouse and human, it is not the site of B-cell development in all species. For instance, the anatomical sites of B-cell development shift over the course of development in sharks (from liver to kidney to spleen), in amphibians and reptiles (from liver and spleen to bone marrow). In contrast, B-cell development in bony fish takes place primarily within the kidney.

In birds, B cells complete their development in a lymphoid organ associated with the gut, the *bursa of Fabricius* (Figure 3). The gut is also a site of B-cell development in some mammals. In cattle and sheep, early in gestation, B cells mature in the fetal spleen. Later in gestation, however, this function is assumed by the

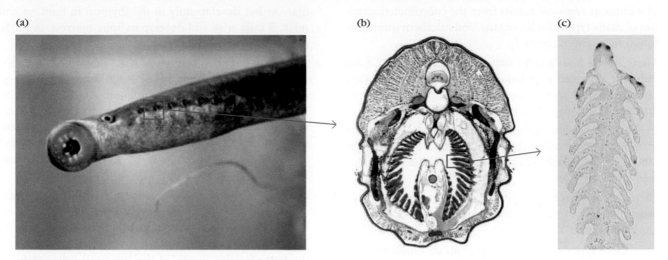

FIGURE 2
Thymic tissue in the lamprey eel. (a) Jawless vertebrates, including the lamprey eel (shown here in larval form), were thought not to have a thymus. Recent work suggests, however, that they generate two types of lymphocytes analogous to B and T cells and have thymic tissue at the tip of their gills (b). The thymic tissue shown in (c) is stained blue for a gene (CDA1) specific for lymphocytes found in lampreys. The thymus of jawed vertebrates arises from an area analogous to the gill region. [Left: Courtesy of Brian Morland, The Bellflask Ecological Survey Team; middle: Dr. Keith Wheeler/Science Photo Library/ Photo Researchers; right: Courtesy Thomas Boehm.]

patch of tissue embedded in the wall of the intestine called the ileal Peyer's patch, which contains a large number of B cells as well as T cells. The rabbit, too, uses gut-associated tissues, especially the appendix, as primary lymphoid tissue for important steps in the proliferation and diversification of B cells. Recent work suggests that even in animals that depend largely on the bone marrow to complete B cell maturation, lymphoid tissue in the gut can act as a primary lymphoid organ for generating mature B lymphocytes.

Secondary lymphoid organs are more variable in their numbers and location than primary lymphoid organs (e.g., rodents do not have tonsils, but do have well-developed lymphoid tissue at the base of the nose). The structural and functional features of these organs are, however, shared by most vertebrates with at least one interesting exception: pig lymph nodes exhibit a striking peculiarity—they are "inverted" anatomically, so that the medulla of the organ, where lymphocytes exit the organ, is on the

outside, and the cortex, where lymphocytes meet their antigen and proliferate, is on the inside. Lymphocyte egress is also inverted: they exit via blood vessels, rather than via efferent lymphatics. The adaptive advantages (if any) of these odd structural variations are unknown, but the example reminds us of the remarkable plasticity of structure/function relationships within biological structures and the creative opportunism of evolutionary processes.

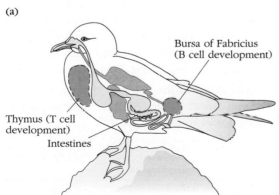

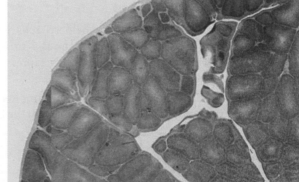

FIGURE 3
The avian bursa. (a) Like all vertebrates, birds have a thymus, spleen, and lymph nodes, and hematopoiesis occurs in their bone marrow. However, their B cells do not develop in the bone marrow. Rather they develop in an outpouching of the intestine, the bursa, which is located close to the cloaca (the common end of the intestinal and genital tracts in birds). A stained tissue section of the bursa and cloaca is shown in (b). [(b) Courtesy Dr. Thomas Caceci.]

- The immune response results from the coordinated activities of many types of cells, organs, and microenvironments found throughout the body.

- Many of the body's cells, tissues, and organs arise from different stem cell populations. All red and white blood cells develop from a pluripotent hematopoietic stem cell during a highly regulated process called hematopoiesis.

- In the adult vertebrate, hematopoiesis occurs in the bone marrow, a stem cell niche that supports both the self-renewal of stem cells and their differentiation into multiple blood cell types.

- Hematopoietic stem cells give rise to two main blood cell progenitors: common myeloid progenitors and common lymphoid progenitors.

- Four main types of cells develop from common myeloid progenitors: red blood cells (erythrocytes), monocytes (which give rise to macrophages and myeloid dendritic cells), granulocytes (which include the abundant neutrophils and less abundant basophils, eosinophils, and mast cells), and megakaryocytes.

- Macrophages and neutrophils are specialized for the phagocytosis and degradation of antigens. Macrophages also have the capacity to present antigen with MHC to T cells.

- Immature forms of dendritic cells have the capacity to capture antigen in one location, undergo maturation, and migrate to another location, where they present antigen to T cells. Dendritic cells are the most potent antigen-presenting cell for activating naïve T cells.

- There are three types of lymphoid cells: B cells, T cells, and natural killer (NK) cells. B and T cells are members of clonal populations distinguished by antigen receptors of unique specificity. B cells synthesize and display membrane antibody, and T cells synthesize and display T-cell receptors (TCRs). NK cells do not synthesize antigen-specific receptors; however a small population of TCR-expressing T cells have features of NK cells and are called NKT cells.

- T cells can be further subdivided into helper T cells, which typically express CD4 and see peptide bound to MHC class II, and cytotoxic T cells, which typically express CD8 and see peptide bound to MHC class I.

- Primary lymphoid organs are the sites where lymphocytes develop and mature. T cell precursors come from the bone marrow but develop fully in the thymus; in humans and mice, B cells arise and develop in bone marrow. In birds, B cells develop in the bursa.

- Secondary lymphoid organs provide sites where lymphocytes encounter antigen, become activated, and undergo clonal expansion and differentiation into effector cells. They include the lymph node, spleen, and more loosely organized sites distributed throughout the mucosal system (the mucosa-associated lymphoid tissue, or MALT) and are typically compartmentalized into T-cell areas and B-cell areas (follicles).

- Lymph nodes and the spleen are the most highly organized secondary lymphoid organs. T-cell and B-cell activity are separated into distinct microenvironments in both. T cells are found in the paracortex of the lymph nodes and the periarteriolar sheath of the spleen. B cells are organized into follicles in both organs.

- The spleen is the first line of defense against blood-borne pathogens. It also contains red blood cells and compartmentalizes them in the red pulp from lymphocytes in the white pulp. A specialized region of macrophages and B cells, the marginal zone, borders the white pulp.

- In secondary lymphoid tissue, T lymphocytes can develop into killer and helper effector cells. CD4$^+$ T cells can help B cells to differentiate into plasma cells, which secrete antibody or can also develop into effector cells that further activate macrophages and CD8$^+$ cytotoxic T cells. Both B and T cells also develop into long-lived memory cells.

- B cells undergo further maturation in germinal cells, a specialized substructure found in follicles after infection; here they can increase their affinity for antigen and undergo class switching.

- Lymphoid follicles and other organized lymphoid microenvironments are found associated with the mucosa, the skin, and even tertiary tissues at the site of infection.

- Mucosa-associated lymphoid tissue (MALT) is an important defense against infection at epithelial layers and includes a well-developed network of follicles and lymphoid microenvironments associated with the intestine (gut-associated lymphoid tissue, GALT).

- Unique cells found in the lining of the gut, M cells, are specialized to deliver antigen from the intestinal spaces to the lymphoid cells within the gut wall.

R E F E R E N C E S

Banchereau J., et al. 2000. Immunobiology of dendritic cells. *Annual Review of Immunology* **18**:767.

Bajenoff, M., J. G. Egen, H. Qi, A. Y. C. Huang, F. Castellino, and R. N. Germain. 2007. Highways, byways and breadcrumbs: Directing lymphocyte traffic in the lymph node. *Trends in Immunology* **28**:346–352.

Bajoghil, B., P. Guo, N. Aghaallaei, M. Hirano, C. Strohmeier, N. McCurley, D. E. Bockman, M. Schorpp, M. D. Cooper, and

T. Boehm. 2011. A thymus candidate in lampreys. *Nature* **47**:90–95.

Boehm, T., and C. E. Bleul. 2007. The evolutionary history of lymphoid organs. *Nature Immunology* **8**:131–135.

Borregaard, N. 2011. Neutrophils: From marrow to microbes. *Immunity* **33**:657–670.

Catron, D.M., A. A. Itano, K. A. Pape, D. L. Mueller, and M. K Jenkins. 2004. Visualizing the first 50 hr of the primary immune response to a soluble antigen. *Immunity* **21**:341–347.

Ema, H., et al. 2005. Quantification of self-renewal capacity in single hematopoietic stem cells from normal and Lnk-deficient mice. *Developmental Cell* **6**:907.

Falcone, F. H., D. Zillikens, and B. F.Gibbs. 2006. The 21st century renaissance of the basophil? Current insights into its role in allergic responses and innate immunity. *Experimental Dermatology* **15**:855–864.

Godfrey, D. I., II. R. MacDonald, M. Kronenberg, M. J. Smyth, and L. Van Kaer. 2004. NKT cells: What's in a name? *Nature Reviews Immunology* **4**:231.

Halin, C., J. R. Mora, C. Sumen, and U. H. von Andrian. In vivo imaging of lymphocyte trafficking. 2005 *Annual Review of Cell and Developmental Biology* **21**:581–603.

Kiel, M., and S. J. Morrison. 2006. Maintaining hematopoietic stem cells in the vascular niche. *Immunity* **25**:852–854.

Liu, Y. J. 2001. Dendritic cell subsets and lineages, and their functions in innate and adaptive immunity. *Cell* **106**:259.

Mebius, R. E., and G. Kraal. 2005. Structure and function of the spleen. *Nature Reviews Immunology*. **5**:606–616.

Mendez-Ferrer, S., D. Lucas, M. Battista, and P. S. Frenette. 2008. Haematopoietic stem cell release is regulated by circadian oscillations. *Nature* **452**:442 448.

Mikkola, H. K. A., and S. A. Orkin. 2006. The journey of developing hematopoietic stem cells. *Development* **133**:3733–3744.

Miller, J. F. 1961. Immunological function of the thymus. *Lancet* **2**:748–749.

Miller, J. F. A. P. 2002. The discovery of thymus function and of thymus-derived lymphocytes. *Immunological Reviews* **185**:7–14.

Moon, J. J., H. H. Chu, M. Pepper, S. J. McSorley, S. C. Jameson, R. M. Kedl, and M. K. Jenkins. 2007. Naïve CD4$^+$ T cell frequency varies for different epitopes and predicts repertoire diversity and response magnitude. *Immunity* **27**:203–213.

Pabst, R. 2007. Plasticity and heterogeneity of lymphoid organs: What are the criteria to call a lymphoid organ primary, secondary or tertiary? *Immunology Letters* **112**:1–8.

Piccirillo, C. A., and E. M. Shevach. 2004. Naturally occurring CD4+CD25+ immunoregulatory T cells: Central players in the arena of peripheral tolerance. *Seminars in Immunology* **16**:81.

Picker, L. J., and M. H. Siegelman. 1999. Lymphoid tissues and organs. In *Fundamental Immunology*, 4th ed. W. E. Paul, ed. Philadelphia: Lippincott-Raven, p. 145.

Rodewald, H-R. 2006. A second chance for the thymus. *Nature* **441**:942–943.

Rothenberg, M., and S. P. Hogan. 2006. The eosinophil. *Annual Review of Immunology* **24**:147–174.

Scadden, D. T. 2006. The stem-cell niche as an entity of action. *Nature* **441**:1075–1079.

Shizuru, J. A., R. S. Negrin, and I. L. Weissman. 2005. Hematopoietic stem and progenitor cells: Clinical and preclinical regeneration of the hematolymphoid system. *Annual Review of Medicine* **56**:509.

Weissman, I. L. 2000. Translating stem and progenitor cell biology to the clinic: Barriers and opportunities. *Science* **287**:1442.

Useful Web Sites

http://bio-alive.com/animations/anatomy.htm A collection of publicly available animations relevant to biology. Scroll through the list to find videos on blood and immune cells, immune responses, and links to interactive sites that reinforce your understanding of immune anatomy.

http://stemcells.nih.gov Links to information on stem cells, including basic and clinical information, and registries of available embryonic stem cells.

www.niaid.nih.gov/topics/immuneSystem/Pages/ structureImages.aspx A very accessible site about immune system function and structure.

www.immunity.com/cgi/content/full/21/3/341/ DC1 A pair of simulations that trace the activities of T cells, B cells, and dendritic cells in a lymph node. These movies are discussed in more detail in Chapter 14.

www.hhmi.org/research/investigators/cyster. html Investigator Jason Cyster's public website that includes many videos of B-cell activity in immune tissue.

www.hematologyatlas.com/principalpage.htm An interactive atlas of both normal and pathological human blood cells.

 STUDY QUESTIONS

RESEARCH FOCUS QUESTION Notch is a surface protein that regulates cell fate. When bound by its ligand, it releases and activates its intracellular region, which regulates new gene transcription. Investigators found that the phenotype of developing cells in the bone marrow differed dramatically when they overexpressed the active, intracellular portion of Notch. In particular, the frequency of BCR$^+$ cells plummeted, and the frequency of TCR$^+$ cells increased markedly. Interestingly, other investigators found that when you knocked out Notch, the phenotype of cells in the thymus changed: the frequency of BCR$^+$ cells increased and the frequency of TCR$^+$ cells decreased dramatically.

Propose a molecular model to explain these observations, and an experimental approach to begin testing your model.

CLINICAL FOCUS QUESTION The T and B cells that differentiate from hematopoietic stem cells recognize as self the bodies in which they differentiate. Suppose a woman donates HSCs to a genetically unrelated man whose hematopoietic system was totally destroyed by a combination of radiation and chemotherapy. Further, suppose that, although most of the donor HSCs differentiate into hematopoietic cells, some differentiate into cells of the pancreas, liver, and heart. Decide which of the following outcomes is likely and justify your choice.

a. The T cells that arise from the donor HSCs do not attack the pancreatic, heart, and liver cells that arose from donor cells but mount a GVH response against all of the other host cells.

b. The T cells that arise from the donor HSCs mount a GVH response against all of the host cells.

c. The T cells that arise from the donor HSCs attack the pancreatic, heart, and liver cells that arose from donor cells but fail to mount a GVH response against all of the other host cells.

d. The T cells that arise from the donor HSCs do not attack the pancreatic, heart, and liver cells that arose from donor cells and fail to mount a GVH response against all of the other host cells.

1. Explain why each of the following statements is false.

a. There are no mature T cells in the bone marrow.

b. The pluripotent stem cell is one of the most abundant cell types in the bone marrow.

c. There are no stem cells in blood.

d. Activation of macrophages increases their expression of class I MHC molecules, allowing them to present antigen to T$_H$ cells more effectively.

e. Mature B cells are closely associated with osteoblasts in the bone marrow.

f. Lymphoid follicles are present only in the spleen and lymph nodes.

g. The FRC guides B cells to follicles.

h. Infection has no influence on the rate of hematopoiesis.

i. Follicular dendritic cells can process and present antigen to T lymphocytes.

j. Dendritic cells arise only from the myeloid lineage.

k. All lymphoid cells have antigen-specific receptors on their membrane.

l. All vertebrates generate B lymphocytes in bone marrow.

m. All vertebrates have a thymus.

n. Jawless vertebrates do not have lymphocytes.

2. For each of the following sets of cells, state the closest common progenitor cell that gives rise to both cell types.

a. Dendritic cells and macrophages

b. Monocytes and neutrophils

c. T$_C$ cells and basophils

d. T$_H$ and B cells

3. List two primary and two secondary lymphoid organs and summarize their functions in the immune response.

4. What two primary characteristics distinguish hematopoietic stem cells from mature blood cells?

5. What are the two primary roles of the thymus?

6. At what age does the thymus reach its maximal size?
   a. During the first year of life
   b. Teenage years (puberty)
   c. Between 40 and 50 years of age
   d. After 70 years of age

7. Preparations enriched in hematopoietic stem cells are useful for research and clinical practice. What is the role of the SCID mouse in demonstrating the success of HSC enrichment?

8. Explain the difference between a monocyte and a macrophage.

9. What effect would removal of the bursa of Fabricius (bursectomy) have on chickens?

10. Indicate whether each of the following statements about the lymph node and spleen is true or false. If you think a statement is false, explain why.
    a. The lymph node filters antigens out of the blood.
    b. The paracortex is rich in T cells, and the periarteriolar lymphoid sheath (PALS) is rich in B cells.
    c. Only the lymph node contains germinal centers.
    d. The FRCC enhances T cell/APC interactions
    e. Afferent lymphatic vessels draining the tissue spaces enter the spleen.
    f. Lymph node but not spleen function is affected by a knockout of the Ikaros gene.

11. For each description below (1–14), select the appropriate cell type (a–p). Each cell type may be used once, more than once, or not at all.

Descriptions

1. Major cell type presenting antigen to naïve T cells

2. Phagocytic cell of the central nervous system

3. Granulocytic cells important in the body's defense against parasitic organisms

4. Give rise to red blood cells

5. Generally first cells to arrive at site of inflammation

6. Support maintenance of hematopoietic stem cells

7. Give rise to thymocytes

8. Circulating blood cells that differentiate into macrophages in the tissues

9. An antigen-presenting cell that arises from the same precursor as a T cell but not the same as a macrophage

10. Cells that are important in sampling antigens of the intestinal lumen

11. Granulocytic cells that release various pharmacologically active substances

12. White blood cells that play an important role in the development of allergies

13. Cells that can use antibodies to recognize their targets

14. Cells that express antigen-specific receptors

Cell Types
a. Common myeloid progenitor cells
b. Monocytes
c. Eosinophils
d. Dendritic cells
e. Natural killer (NK) cells
f. Mast cells
g. Neutrophils
h. M cells
i. Osteoblasts
j. Lymphocytes
k. NK1-T cell
l. Microglial cell
m. Myeloid dendritic cell
n. Hematopoietic stem cell
o. Lymphoid dendritic cell

Receptors and Signaling: B and T-Cell Receptors

The coordination of physiological functions throughout the body depends on the ability of individual cells to sense changes in their environment and to respond appropriately. One of the major routes by which a cell interprets its surroundings is through the binding of signaling molecules to cell-associated **receptor** proteins. A molecule that binds to a receptor is a **ligand**. Noncovalent binding of a ligand to its receptor may induce alterations in the receptor itself, in its polymerization state, and/or in the environment of that receptor. These changes act to transmit or **transduce** the ligand-binding signal into the interior of the cell, leading to alterations in cellular functions. In the nervous system, we would call the signaling molecules neurotransmitters; in the endocrine system, hormones. In the immune system, the foreign molecules that signal the presence of non-self entities are **antigens,** and the small molecules that communicate among the various populations of immune cells are **cytokines**. Specialized cytokines that induce chemo-attraction or -repulsion are termed **chemokines**.

In this chapter, we provide a general introduction to receptor-ligand binding and to the broad concepts and strategies that underlie signal transduction. We then focus specifically on the antigens and receptors of the adaptive immune system, introducing the B- and T-cell receptors and the intracellular signaling events that occur upon antigen binding.

Because the cells of the immune system are distributed throughout the body—with some resident in fixed tissues and others circulating through the various lymphoid tissues, the blood, and the lymphatics—the ability of these cells to communicate with and signal to one another via soluble cytokine and chemokine molecular messengers is essential to their function. In Chapter 4, we describe the signaling events that result when cytokines and chemokines bind to their cognate (matching) receptors. Chapter 5 includes a description of the pattern recognition receptors (PRRs) of the innate immune system and of the signaling events initiated by their antigen binding.

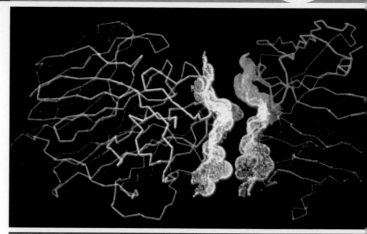

All signaling events begin with a complementary interaction between a ligand and a receptor. This figure depicts the molecular interaction between the variable regions of an antibody molecule (light and heavy chains are shown in blue and red, respectively) and the tip of the influenza hemagglutinin molecule, shown in yellow. [*Illustration based on x-ray crystallography data collected by P. M. Colman and W. R. Tulip from GJVH Nossal, 1993,* Scientific American *269(3):22.*]

- ■ Receptor-Ligand Interactions
- ■ Common Strategies Used in Many Signaling Pathways
- ■ Frequently Encountered Signaling Pathways
- ■ The Structure of Antibodies
- ■ Signal Transduction in B Cells
- ■ T-Cell Receptors and Signaling

The essential concepts that underlie cell signaling in the immune system can be summarized as follows:

- A cellular **signal** is any event that instructs a cell to change its metabolic or proliferative state.

- Signals are usually generated by the binding of a ligand to a complementary cell-bound receptor.

- A cell can become more or less susceptible to the actions of a ligand by increasing (up-regulating) or decreasing (down-regulating) the expression of the receptor for that ligand.

- The ligand may be a soluble molecule, or it may be a peptide, carbohydrate, or lipid presented on the surface of a cell.

- The ligand may travel long distances from its entry point through the body in either the bloodstream or the lymphatics before it reaches a cell bearing the relevant receptor.

- Ligand-receptor binding is noncovalent, although it may be of quite high affinity.

- Ligand binding to the receptor induces a molecular change in the receptor. This change may be in the form of a conformational alteration in the receptor, receptor dimerization or clustering, a change in the receptor's location in the membrane, or a covalent modification.

- Such receptor alterations set off cascades of intracellular events that include the activation of enzymes, and changes in the intracellular locations of important molecules.

- The end result of cellular signaling is often, but not always, a change in the transcriptional program of the target cell.

- Sometimes a cell must receive more than one signal through more than one receptor in order to effect a particular outcome.

- Integration of all the signals received by a cell occurs at the molecular level inside the recipient cell.

Receptor-Ligand Interactions

The antigen receptors of the adaptive immune system are transmembrane proteins localized at the plasma membrane. Ligand binding to its cognate receptor normally occurs via specific, noncovalent interactions between the ligand and the extracellular portion of the membrane receptor. Although a single lymphocyte expresses only one type of antigen receptor, it also may express many different receptor molecules for signals such as cytokines and chemokines, and therefore a healthy cell must integrate the signals from all the receptors that are occupied at any one time.

Receptor-Ligand Binding Occurs via Multiple Noncovalent Bonds

The surface of a receptor molecule binds to its complementary ligand surface by the same types of noncovalent chemical linkages that enzymes use to bind to their substrates. These include hydrogen and ionic bonds, and hydrophobic and van der Waals interactions. The key to a meaningful receptor-ligand interaction is that the sum total of these bonding interactions hold the two interacting surfaces together with sufficient binding energy, and for sufficient time, to allow a signal to pass from the ligand to the cell bearing the receptor. Because these noncovalent interactions

are individually weak, many such interactions are required to form a biologically significant receptor-ligand connection. Furthermore, since each of these noncovalent interactions operates only over a very short distance—generally about 1 Angstrom, (1 Å $= 10^{-10}$ meters)—a high-affinity receptor-ligand interaction depends on a very close "fit," or degree of complementarity, between the receptor and the ligand (Figure 3-1).

How Do We Quantitate the Strength of Receptor-ligand Interactions?

Consider a receptor, R, binding to a ligand, L. We can describe their binding reaction according to the following equation,

$$R + L \underset{k_{-1}}{\overset{k_1}{\rightleftharpoons}} RL \qquad \textbf{(Eq. 3-1)}$$

in which RL represents the bound receptor-ligand complex, k_1 is the forward or association rate constant, and k_{-1} is the reverse or dissociation rate constant.

The ratio of k_1/k_{-1} is equal to **Ka**, the **association constant** of the reaction, and is a measure of the **affinity** of the

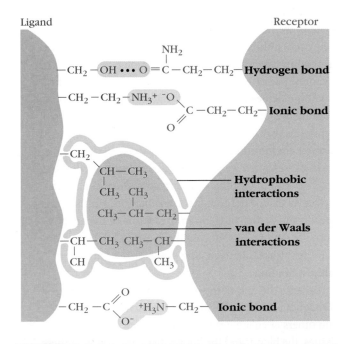

FIGURE 3-1 Receptor-ligand binding obeys the rules of chemistry. Receptors bind to ligands using the full-range of noncovalent bonding interactions, including ionic and hydrogen bonds and van der Waals and hydrophobic interactions. For signaling to occur, the bonds must be of sufficient strength to hold the ligand and receptor in close proximity long enough for downstream events to be initiated. In B- and T-cell signaling, activating interactions also require receptor clustering. In an aqueous environment, noncovalent interactions are weak and depend on close complementarity of the shapes of receptor and ligand.

receptor-ligand pair. The association constant is defined as the relationship between the concentration of reaction product, [RL], and the product of the concentrations of reactants, [R] multiplied by [L]. The units of affinity are therefore M^{-1}. *The higher the Ka, the higher the affinity of the interaction.*

$$Ka = \frac{[RL]}{[R][L]} \qquad \textbf{(Eq. 3-2)}$$

The reciprocal of the association constant, the **dissociation constant, Kd**, is often used to describe the interactions between receptors and ligands. It is defined as:

$$Kd = \frac{[R][L]}{[RL]} \qquad \textbf{(Eq. 3-3)}$$

The units of the dissociation constant are in molarity (M). Inspection of this equation reveals that when half of the receptor sites are filled with ligand—that is, [R] = [RL]—the Kd is equal to the concentration of free ligand [L]. *The lower the Kd, the higher the affinity of the interaction.*

For purposes of comparison, it is useful to consider that the Kd values of many enzyme-substrate interactions lie in the range of 10^{-3} M to 10^{-5} M. The Kd values of antigen-antibody interactions at the beginning of an immune response are normally on the order of 10^{-4} to 10^{-5} M. However, since the antibodies generated upon antigen stimulation are mutated and selected over the course of an immune response, antigen-antibody interactions late in an immune response may achieve a Kd as low as 10^{-12} M. Under these conditions, if an antigen is present in solution at a concentration as low as 10^{-12} M, half of the available antibody-binding sites will be filled. This is an extraordinarily strong interaction between receptor and ligand.

Interactions Between Receptors and Ligands Can Be Multivalent

Many biological receptors, including B-cell receptors, are **multivalent**—that is, they have more than one ligand binding site per molecule. When both receptors and ligands are multivalent—as occurs, for example, when a bivalent immunoglobulin receptor on the surface of a B cell binds to two, identical, repeated antigens on a bacterial surface—the overall binding interaction between the bacterial cell and the B-cell receptor is markedly enhanced compared with a similar, but univalent, interaction. In this way, multiple concurrent receptor-ligand interactions increase the strength of binding between two cell surfaces. Note, however, that binding via two identical receptor sites to two identical ligands on the same cell may be a little less than twice as firm as binding through a single receptor site. This is because the binding of both receptor sites to two ligands on a single antigen may somewhat strain the geometry of the binding at one or both of the sites and therefore slightly interfere with the "fit" of the individual interactions.

Much of the benefit of multivalency results from the fact that noncovalent-binding interactions are inherently

reversible; the ligand spends some of its time binding to the receptor, and some of its time in an unbound, or "off" state (Figure 3-2a). When more than one binding site is involved, it is less likely that all of the receptor sites will simultaneously be in the "off" state, and therefore that the receptor will release the ligand (Figures 3-2b and 3-2c).

The term **avidity** is used to describe the overall strength of the collective binding interactions that occur during multivalent binding. In the early phases of the adaptive immune response to multivalent antigens, B cells secrete IgM antibody, which has 10 available antigen binding sites per molecule. IgM is therefore capable of binding to antigens at a biologically significant level, even if the affinity of each individual binding site for its antigen is low, because of the avidity of the entire antigen-antibody interaction. Likewise, when a cell-bound receptor binds to a cell-bound ligand, their interaction may be functionally multivalent, even if the individual receptor molecules are monovalent, because multiple receptor molecules can cluster in the cell membrane and participate in the receptor-ligand interaction. In addition, *other co-receptor interactions* can also contribute to the overall avidity of the engagement between a cell and its antigen (see below).

The affinity of receptor-ligand interactions can be measured by techniques such as equilibrium dialysis or surface plasmon resonance (SPR). Both of these methods are described in Chapter 20.

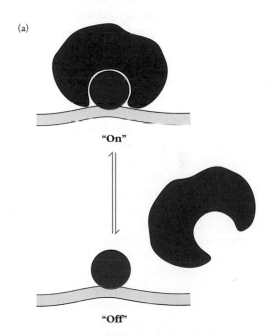

(a)

"On"

"Off"

FIGURE 3-2 Monovalent and bivalent binding. (a) Monovalent binding. The receptor exists in equilibrium with its ligand, represented here as a circle. Part of the time it is bound (binding is in the "on" state), and part of the time it is unbound (binding is in the "off" state). The ratio of the time spent in the "on" versus the "off" state determines the affinity of the receptor-ligand interaction and is related to the strength of the sum of the noncovalent binding interactions between the receptor and the ligand.

(b)

(c)

FIGURE 3-2 (continued) (b) Bivalent binding. Here, we visualize a bivalent receptor, binding to two sites on a single multivalent ligand. This could represent, for example, a bivalent antibody molecule binding to a bacterium with repeated antigens on its surface. In this rendering, both sites are occupied. (c) In this rendering, the bivalent receptor

has momentarily let go from one of its sites, but still holds onto the ligand with the other. Large proteins such as antibodies continuously vibrate and "breathe," resulting in such on-off kinetics. Bivalent or multivalent binding helps to ensure that, when one site momentarily releases the ligand, the interaction is not entirely lost, as it is in part (a).

Receptor and Ligand Expression Can Vary During the Course of an Immune Response

One of the most striking features of the molecular logic of immune responses is that receptors for some growth factors and cytokines are expressed only on an as-needed basis. Most lymphocytes, for example, express a hetero-dimeric (two-chain), low-affinity form of the receptor for the cytokine interleukin 2 (IL-2), which initiates a signal that promotes lymphocyte proliferation. (This cytokine and its receptor are covered in more detail in Chapter 4.) This low-affinity form of the receptor is unable to bind to IL-2 at physiological cytokine concentrations. However, when a lymphocyte is activated by antigen binding, the signal from the antigen-binding receptor causes an increase in the cell surface expression of a third chain of the IL-2 receptor (see Figure 3-3). Addition of this third chain converts the low-affinity form of the IL-2 receptor to a high-affinity form, capable of responding to the cytokine levels found in the lymphoid organs. The functional corollary of this is that only those lymphocytes that have already bound to antigen and have been stimulated by that binding have cytokine receptors of sufficiently high affinity to respond to the physiological cytokine concentrations. By waiting until it has interacted with its specific antigen before expressing the high-affinity cytokine receptor, the lymphocyte both conserves energy and prevents the accidental initiation of an immune response to an irrelevant antigen.

Local Concentrations of Cytokines and Other Ligands May Be Extremely High

When considering the interactions between receptors and their ligands, it is important to consider the anatomical environment in which these interactions are occurring. Lympho-

cytes and antigen-presenting cells engaged in an immune response spend significant amounts of time held together by multiple receptor-ligand interactions. Cytokine signals released by a T cell, for example, are received by a bound antigen-presenting cell at the interface between the cells, before the cytokine has time to diffuse into the tissue fluids. The secretion of cytokines into exactly the location where they are needed is facilitated by the ability of antigen receptor signaling to induce redistribution of the microtubule organizing center (MTOC) within the activated T cell. The MTOC reorientation, in turn, causes the redistribution of the secretory organelles (the Golgi body and secretory vesicles) within the T-cell cytoplasm, so that cytokines made by the T cell are secreted in the direction of the T-cell receptor, which, in turn, is binding to the antigen-presenting cell. The technical term for this phenomenon is the *vectorial (directional) redistribution of the secretory apparatus.* Thus, cytokines are secreted directly into the space between the activated T cell and antigen-presenting cell.

Vectorial redistribution of the secretory apparatus in antigen-presenting dendritic cells also occurs upon interaction with antigen-specific T cells. This ensures that cytokines, such as IL-12, which are secreted by dendritic cells, are efficiently delivered to antigen-recognizing T cells. Figure 3-4 shows the directional secretion of the cytokine IL-12 by an antigen-presenting dendritic cell engaged in the activation of a T cell.

The effector cells of the immune system, such as B cells, are also capable of presenting antigens on their cell surface for recognition by helper T cells. A B cell that has specifically recognized an antigen through its own receptor will process the antigen and express a particularly high concentration of those antigenic peptides bound to Major Histocompatibility Complex (MHC) molecules on its surface. A helper T cell, binding to an antigen presented on the surface of a B cell,

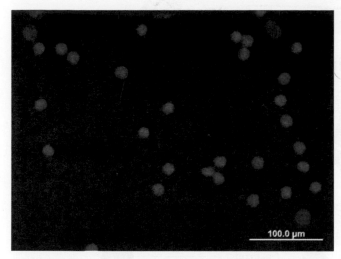

FIGURE 3-3 The expression of some cytokine receptors is upregulated following cell stimulation. White blood cells stained with DAPI, a stain that detects DNA (blue), were either untreated (left) or stimulated using the human T-cell mitogen phytohemagglutinin (right). Upon stimulation, the IL-2 receptor alpha (IL-2Rα) chain (yellow) is upregulated, increasing affinity of the IL-2 receptor for IL-2. A similar IL-2Rα upregulation is seen upon antigen stimulation. [Courtesy R&D Systems, Inc., Minneapolis, MN, USA.]

will therefore deliver a high concentration of cytokines directly to an antigen-specific, activated B cell. The local concentration of cytokines at the cellular interfaces can therefore be extremely high, much higher than in the tissue fluids generally.

Common Strategies Used in Many Signaling Pathways

A **signal transduction pathway** is the molecular route by which a ligand-receptor interaction is translated into a biochemical change within the affected cell. Communication of the ligand's message to the cell is initiated by the complementarity in structure between the ligand and its receptor, which results in a binding interaction of suffi-cient strength and duration to bring about a biochemical change in the receptor and/or its associated molecules. This biochemical change can, in turn, cause a diverse array of biochemical consequences in the cell (Overview Figure 3-5). The terms *upstream* and *downstream* are frequently used to describe elements of signaling pathways. The **upstream** components of a signaling pathway are those closest to the receptor; the **downstream** components are those closest to the effector molecules that determine the outcome of the pathway—for example, the transcription factors or enzymes whose activities are modified upon receipt of the signal.

As disparate and complex as some of these signaling pathways may be, many share common features. In this section, in order to provide a framework for analyzing the individual pathways employed by cells of the immune system, we

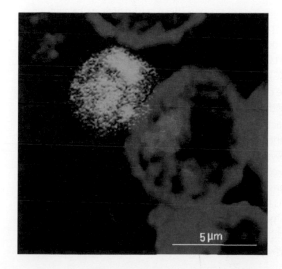

FIGURE 3-4 Polarized secretion of IL-12 (pink) by dendritic cells (blue) in the direction of a bound T cell (green). The higher magnification micrograph shows the secretion of packets of IL-12 through the dendritic cell membrane. [*J. Pulecio et al., 2010,* Journal of Experimental Medicine *207:2,719.*]

Concepts in lymphocyte signaling

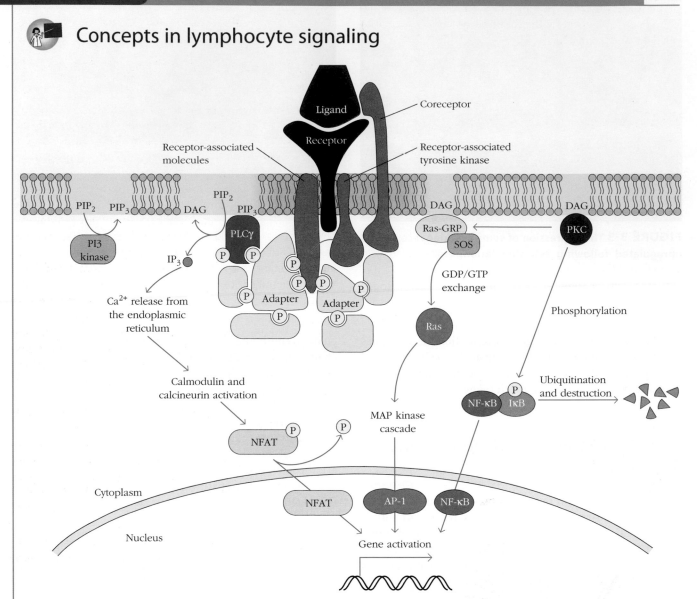

Ligand binding to receptors on a cell induces a variety of downstream effects, many of which culminate in transcription factor activation. Here we illustrate a few of the pathways that are addressed in this chapter. Binding of receptor to ligand induces clustering of receptors and signaling molecules into regions of the membrane referred to as lipid rafts (red). Receptor binding of ligand may be accompanied by binding of associated coreceptors to their own ligands, and causes the activation of receptor-associated tyrosine kinases, which phosphorylate receptor-associated proteins. Binding of downstream adapter molecules to the phosphate groups on adapter proteins creates a scaffold at the membrane that then enables activation of a variety of enzymes including phospholipase Cγ (PLCγ), PI3 kinase, and additional tyrosine kinases. PLCγ cleaves phosphatidyl inositol bisphosphate (PIP$_2$) to yield inositol trisphosphate (IP$_3$), which interacts with receptors on endoplasmic reticulum vesicles to

cause the release of calcium ions. These in turn activate calcineurin, which dephosphorylates the transcription factor NFAT, allowing it to enter the nucleus. Diacylglycerol (DAG), remaining in the membrane after PLCγ cleavage of PIP$_2$, binds and activates protein kinase C (PKC), which phosphorylates and activates enzymes leading to the destruction of the inhibitor of the transcription factor NF-κB. With the release of the inhibitor, NF-κB enters the nucleus and activates a series of genes important to the immune system. Binding of the adapter protein Ras-GRP to the signaling complex allows for the binding and activation of the Guanine nucleotide Exchange Factor (GEF) Son of Sevenless (SOS), which in turn initiates the phosphorylation cascade of the MAP kinase pathways. This leads to the entry of a third set of transcription factors into the nucleus, and activation of the transcription factor AP-1. (Many details that are explained in the text have been omitted from this figure for clarity.)

describe some of the strategies shared by many signaling pathways. The next section provides three examples of signaling pathways used by the immune system, as well as by other organ systems.

Ligand Binding Can Induce Conformational Changes in, and/or Clustering of, the Receptor

The first step necessary to the activation of a signaling pathway is that the binding of the ligand to its receptors in some way induces a physical or chemical change in the receptor itself, or in molecules associated with it. In the case of many growth factor receptors, ligand binding induces a conformational change in the receptor that results in receptor dimerization (Figure 3-6). Since the cytoplasmic regions of many growth factor receptors have tyrosine kinase activity, this results in the reciprocal phosphorylation of the cytoplasmic regions of each of the receptor molecules by its dimerization partner.

Other receptors undergo conformational changes upon ligand binding that result in higher orders of receptor polymerization. The receptors on B cells are membrane-bound forms of the antibody molecules that the B cell will eventually secrete. Two different types of antigen receptors exist on the surface of naïve B cells, which we term immunoglobulins M and D (IgM and IgD). Structural studies of the IgM form of the receptor have revealed that ligand binding induces a conformational change in a nonantigen binding part of the receptor, close to the membrane, that facilitates aggregation of the receptors into multimeric complexes and their subsequent movement into specialized membrane regions. T-cell receptors similarly cluster upon antigen binding. Whether or not they undergo a conformational change upon antigen binding remains controversial.

Some Receptors Require Receptor-Associated Molecules to Signal Cell Activation

In contrast to growth factor receptors, which have inducible enzyme activities built into the receptor molecule itself, B- and T-cell receptors have very short cytoplasmic components and therefore need help from intracellular receptor-associated molecules to bring about signal transduction. The Igα/Igβ (CD79α/β) heterodimer in B cells, and the hexameric CD3 complex in T cells are closely associated with their respective antigen receptors and are responsible for transmitting the signals initiated by ligand binding into the interior of the cell (Figure 3-7). Both of these complexes have a pair of long cytoplasmic tails that contain multiple copies of the **Immuno-receptor Tyrosine Activation Motif** or **ITAM**. ITAMs are recurrent sequence motifs found on many signaling proteins within the immune system, which contain tyrosines that become phosphorylated following signal transduction through the associated receptor. Phosphorylation of ITAM-tyrosine residues then allows docking of adapter molecules, thus facilitating initiation of the signaling cascade.

Other molecules associated with the B- or T-cell antigen receptors may also interact with the antigen or with other molecules on the pathogen's surface. For example, in the case of B cells, the CD19/CD21 complex binds to complement molecules covalently attached to the antigen (see Figure 3-7a). Similarly, CD4 and CD8 molecules on T cells bind to nonpolymorphic regions of the antigen-presenting MHC molecule and aid in signal transduction (see Figure 3-7b). Finally, the co-receptor CD28 on naïve T cells must interact with its ligands CD80 (B7-1) and CD86 (B7-2) for full T-cell activation to occur.

Ligand-Induced Receptor Clustering Can Alter Receptor Location

Ligand-induced clustering of B- and T-cell receptors slows down the rates of their diffusion within the planes of their respective cell membranes, and facilitates their movement into specialized regions of the lymphocyte membrane known as **lipid rafts**. These rafts are highly ordered, detergent-insoluble, cholesterol- and sphingolipid-rich membrane regions, populated by many molecules critical to receptor signaling. Moving receptors and co-receptors into the lipid rafts renders them susceptible to the action of enzymes associated with those rafts. Figure 3-8 shows how the raft-associated tyrosine kinase **Lyn** initiates the B-cell signaling cascade by phosphorylating the receptor-associated molecules Igα and Igβ. The tyrosine kinase **Lck** serves a similar role in the TCR signaling cascade.

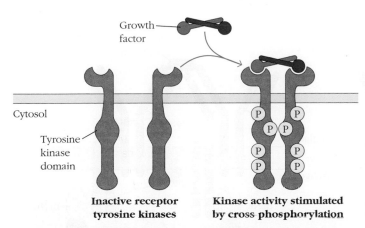

FIGURE 3-6 Some growth factors induce dimerization of their receptors, followed by reciprocal tyrosine phosphorylation of the receptor molecules. Many growth factor receptors possess tyrosine kinase activity in their cytoplasmic regions. Dimerization of the receptor occurs on binding of the relevant ligand and allows reciprocal phosphorylation of the dimerized receptor chains at multiple sites, thus initiating the signal cascade. As one example, stem cell factor binds to its receptor, c-kit (CD117) on the surface of bone marrow stromal cells.

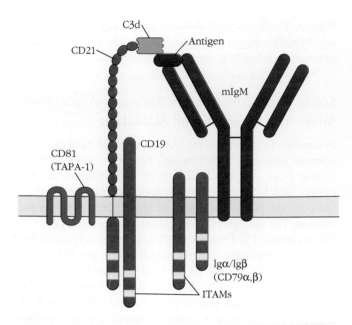
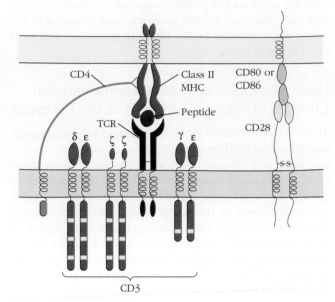

FIGURE 3-7 Both B- and T-cell receptors require receptor-associated molecules and co-receptors for signal transduction. (a) B-cell receptors require Igα/Igβ to transmit their signal. The CD21 co-receptor, which is associated with CD19, binds to the complement molecule C3d, which binds covalently to the antigen. The interaction between CD21 on the B cell, and C3d associated with the antigen, keeps the antigen in contact with the B-cell receptor, even when the antigen-BCR binding is relatively weak. The yellow bands on the cytoplasmic regions of the receptor-associated molecules indicate *I*mmuno-receptor *T*yrosine *A*ctivation *M*otifs (ITAMs). Phosphorylation of tyrosine residues in these motifs allows the binding of downstream adapter molecules and facilitates signal transduction from the receptors. Both CD19 and Igα/Igβ bear intracytoplasmic ITAMs and, along with CD81 (TAPA-1), participate in downstream signaling events. (b) T-cell receptors use CD3, a complex of δε, γε chains and a pair of ζ chain molecules or a ζν pair. CD4 and CD8 co-receptors bind to the non-polymorphic region of the MHC class II or class I molecules, respectively. This figure illustrates CD4 binding to MHC Class II. This binding secures the connection between the T cell and the antigen presenting cell, and also initiates a signal through CD4/8. The CD28 co-receptor provides another signal upon binding to CD80 or CD86. CD28 binding to costimulatory molecules on antigen-presenting cells is required for stimulation of naïve, but not memory, T cells.

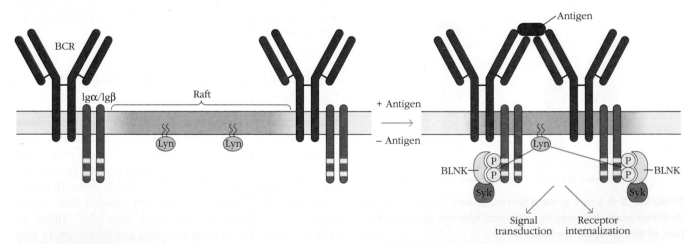
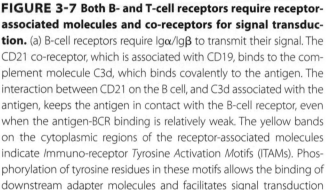

FIGURE 3-8 Lipid raft regions within membranes. In resting B cells, the B-cell receptor (BCR) is excluded from the lipid rafts, which are regions of the membrane high in cholesterol and rich in glycosphingolipids. The raft is populated by tyrosine kinase signaling molecules, such as Lyn. On binding to antigen, the BCR multimerizes (clusters), and the changes in the conformation of the BCR brought about by this multimerization increase the affinity of the BCR for the raft lipids. Movement of the BCR into the raft brings it into contact with the tyrosine kinase Lyn, which phosphorylates the receptor-associated proteins Igα/Igβ, thus initiating the activation cascade. A similar movement of TCRs into lipid rafts occurs upon T-cell activation. [*Adapted from S. K. Pierce, 2002,* Nature Reviews Immunology *2:96.*]

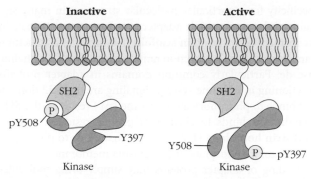

FIGURE 3-9 Activation of Src-family kinaes. Src-family kinases are maintained in an inactive closed configuration by the binding of a phosphorylated inhibitory tyrosine residue (pY508 in this example) with an SH2 domain in the same protein. Dephosphorylation of this tyrosine opens up the molecule, allowing substrate access to the enzymatic site. Opening up the kinase also allows phosphorylation of a different internal tyrosine (pY397), which further stimulates the Src kinase activity.

Tyrosine Phosphorylation is an Early Step in Many Signaling Pathways

Many signaling pathways, particularly those that signal cell growth or proliferation, are initiated with a tyrosine phosphorylation event. Many of the tyrosine kinases that initiate BCR and TCR activation belong to the family of enzymes known as the **Src-family kinases**. Src (pronounced "sark") family kinases have homology to a gene, *c-src*, that was first identified in birds. A constitutively active Rous sarcoma virus

homolog of these genes, *v-src*, was shown to induce a form of cancer, fibrosarcoma, in birds. The fact that a simple mutation in this viral form of a tyrosine kinase gene could result in the development of a tumor was the first hint that tyrosine kinase genes may be important in the regulation of cell proliferation and, indeed, in many aspects of cell signaling.

Src-family kinases are important in the earliest stages of activation of both T and B cells: Lck and Fyn are critical to T-cell activation, and Lyn, Fyn, and Blk play the corresponding initiating roles in B cells. Since inadvertent activation of these enzymes can lead to uncontrolled proliferation—a precursor to tumor formation—it is not surprising that their activity is tightly regulated in two different but interconnected ways.

Inactive Src-family tyrosine kinase enzymes exist in a closed conformation, in which a phosphorylated tyrosine is tightly bound to an internal **SH2 domain** (Src *homology 2* domain) (Figure 3-9, Table 3-1). (SH2 domains in proteins bind to phosphorylated tyrosine residues.) For as long as the inhibitory tyrosine is phosphorylated, the Src-family kinase remains folded in on itself and inactive. In lymphocytes, the tyrosine kinase enzyme Csk is responsible for maintaining phosphorylation of the inhibitory tyrosine. However, upon cell activation, a tyrosine phosphatase removes the inhibitory phosphate and the Src-family kinase opens up into a partially active conformation. Thus, the initiating event in signal transduction is often the movement of the tyrosine kinase into a region of the cell or the membrane populated by an appropriate, activating phosphatase and distant from the inhibitory kinase Csk. Full activity is then achieved when the Src-kinase phosphorylates itself on a second, activating tyrosine residue.

TABLE 3-1 Selected domains of adapter proteins and their binding target motifs

Domain of adapter protein	Binding specificity of adapter domain
Src-homology 2 (SH2)	Specific phosphotyrosine (pY)–containing motifs in the context of 3–6 amino acids located carboxy-terminal to the pY (An invariant arginine in the SH2 domain is required for pY binding.)
Src-homology 3 (SH3)	Proline-rich sequences in a left-handed polyproline helix (Proline residues are usually preceded by an aliphatic residue.)
Phosphotyrosine-binding (PTB)	pY-containing peptide motifs (DPXpY, where X is any amino acid) in the context of amino-terminal sequences
Pleckstrin homology (PH)	Specific phosphoinositides, which allow PH-containing proteins to respond to the generation of lipid second messengers generated by enzymes such as PI3 kinase
WW	Bind proline-rich sequences (Derive their name from two conserved tryptophan [W] residues 20–22 amino acids apart)
Cysteine-rich sequences (C1)	Diacylglycerol (DAG) (On association with DAG, the C1 domain exhibits increased affinity for the lipid membrane, promoting membrane recruitment of C1-containing proteins.)
Tyrosine kinase-binding (TKB)	Phosphotyrosine-binding domain divergent from typical SH2 and PTB domains (Consists of three structural motifs [a four-helix bundle, an EF hand, and a divergent SH2 domain], that together form an integrated phosphoprotein-recognition domain)
Proline-rich	Amino-acid sequence stretches rich in proline residues, able to bind modular domains including SH3 and WW domains
14-3-3- binding motifs	Phosphorylated serine residues in the context of one of the two motifs RSXpSXP and RXXXpSXp

[After G. A. Koretsky and P. S. Myung, 2001, Nature Reviews Immunology 1:95]

Tyrosine phosphorylation can bring about changes in a signaling pathway in more than one way. Sometimes, phosphorylation of a tyrosine residue can induce a conformational change in the phosphorylated protein itself, which can turn its enzymatic activity on or off. Alternatively, tyrosine phosphorylation of components of a receptor complex can permit other proteins to bind to it via their SH2 or phosphotyrosine binding domains (see Table 3-1), thereby altering their locations within the cell. Note that a phosphorylated tyrosine is often referred to as a "pY" residue.

Adapter Proteins Gather Members of Signaling Pathways

It is easy to imagine the cytoplasm as an ocean of macromolecules, with little structure or organization, in which proteins bump into one another at a frequency dependent only on their overall cytoplasmic concentrations. However, nothing could be further from the truth. The cytoplasm is, in fact, an intricately organized environment, in which three-dimensional arrays of proteins form and disperse in a manner directed by cellular signaling events. Many of these reversible interactions between proteins are mediated by **adapter proteins** as well as by interactions between members of signaling pathways with cytoskeletal components.

Adapter proteins have no intrinsic enzymatic or receptor function, nor do they act as transcription factors. Their function is to bind to specific motifs or domains on proteins or lipids, linking one to the other, bringing substrates within the range of enzymes, and generally mediating the redistribution of molecules within the cell. Table 3-1 lists a number of representative adapter domains along with their binding specificities. Adapter proteins are characterized by having multiple surface domains, each of which possesses a precise binding specificity for a particular molecular structure. In many signaling pathways, multiple adapter proteins may participate in the formation of a **protein scaffold** that provides a structural framework for the interaction among members of a signaling cascade. Particularly common domains in adapter proteins functioning in immune system signaling are the SH2 domain that binds to phosphorylated tyrosine (pY) residues, the **SH3 domain** that binds to clusters of proline residues, and the **pleckstrin homology (PH) domain** that binds to phosphatidyl inositol trisphosphate in the plasma membrane.

Binding of adapter proteins may simply bring molecules into contact with one another, so that, for example, an enzyme may act on its substrate. Alternatively, the binding of an adapter protein may induce a conformational change, which in turn can stabilize, destabilize, or activate the binding partner.

Signal transduction may induce conformational changes in proteins that in turn result in the uncovering of one or more protein domains with specific affinity for other proteins. Such interactions may be **homotypic** (interactions between identical domains) or **heterotypic** (interactions between different domains).

Phosphorylation on Serine and Threonine Residues is also a Common Step in Signaling Pathways

Whereas tyrosine phosphorylation is frequently seen at the initiation of a signaling cascade, the phosphorylation of proteins at serine and threonine residues (Figure 3-10) tends to occur at later steps in cell activation. Serine or threonine phosphorylation may serve to activate a phosphorylated enzyme, to induce the phosphorylated protein to interact with a different set of proteins, to alter the protein's location within the cell, to protect the protein from destruction, or, in

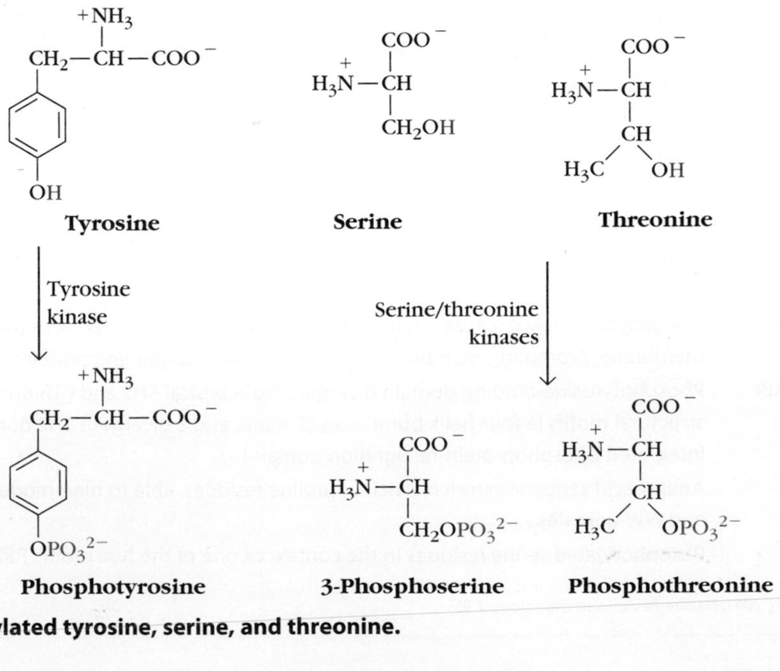

FIGURE 3-10 Phosphorylated tyrosine, serine, and threonine.

some important instances, to convert the phosphorylated protein into a target for proteasomal destruction.

Phosphorylation of Membrane Phospholipids Recruits PH Domain-Containing Proteins to the Cell Membrane

The phospholipid **Phosphatidyl Inositol bis-Phosphate (PIP$_2$)** is a component of the inner face of eukaryotic plasma membranes. However, PIP$_2$ is much more than a structural phospholipid; it also actively participates in cell signaling. The enzyme **Phosphatidyl Inositol-3-kinase (PI3 kinase)**, activated during the course of signaling through many immune receptors, phosphorylates PIP$_2$ to form **Phosphatidyl Inositol tris-Phosphate (PIP$_3$)** (Figure 3-11). PIP$_3$ remains in the membrane and serves as a binding site for proteins bearing pleckstrin homology (PH) domains (see Table 3-1). Thus, this lipid phosphorylation event serves to *move proteins from the cytosol to the membrane* by providing them with a binding site

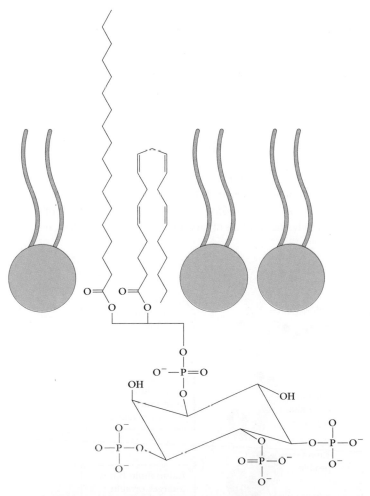

FIGURE 3-11 PIP$_2$ can be phosphorylated (red) by PI3 kinase to create PIP$_3$ during cellular activation. PIP$_3$ then creates binding sites for proteins with PH domains at the internal face of the plasma membrane.

at the inner membrane surface. Localization of proteins at the membrane brings them into contact with other members of a signaling cascade, allowing enzymes access to new substrates and enabling the modification of adapter proteins with the subsequent assembly of signaling complexes.

Signal-Induced PIP$_2$ Breakdown by PLC Causes an Increase in Cytoplasmic Calcium Ion Concentration

In addition to being phosphorylated by PI3 kinase, PIP$_2$ is also susceptible to a second type of signal-induced biochemical modification. A second enzyme (more correctly, a family of enzymes), **Phospholipase C (PLC)**, is also activated upon antigen signaling of lymphocytes (Figure 3-12). PLC hydrolyzes PIP$_2$, cleaving the sugar **inositol trisphosphate (IP$_3$)** from the **diacylglyerol (DAG)** backbone. The DAG remains in the membrane, where it binds and activates a number of important signaling enzymes. IP$_3$ is released into the cytoplasm, where it interacts with specific IP$_3$ receptors on the surface of endoplasmic reticulum vesicles, inducing the release of stored calcium ions (Ca^{2+}) into the cytoplasm. These calcium ions then bind several cellular proteins, changing their conformation and altering their activity.

The calcium ion's double positive charge and the size of its ionic radius render it particularly suitable for binding to many proteins; however, in the field of cell signaling, calmodulin (CaM) is unquestionably the most important calcium-binding protein. As shown in Figure 3-13a, CaM is a dumbbell-shaped protein, with two globular subunits separated by an α-helical rod. Each of the two subunits has two calcium binding sites, such that calmodulin has the capacity to bind four Ca^{2+} ions. On binding calcium, CaM undergoes a dramatic conformational change (Figure 3-13b), which allows it to bind to and activate a number of different cellular proteins.

As cytoplasmic Ca^{2+} ions are used up in binding to cytoplasmic proteins, and the free cytoplasmic Ca^{2+} ion concentration drops, Ca^{2+} channel proteins begin to assemble at the membrane. Eventually the channels open to allow more calcium to flood in from the extracellular fluid and complete the activation of the calcium-regulated proteins. The assembly of these so-called *store-operated calcium channel* proteins is prohibited in the presence of high intracellular calcium.

Many different types of ions are present in a cell, and it is reasonable to ask why eukaryotes should have evolved proteins with activities that are so responsive to intracellular Ca^{2+} ion concentrations. The answer to this question in part lies in the fact that it is relatively easy to alter the intracytoplasmic concentration of Ca^{2+} ions. The concentration of Ca^{2+} in the blood and tissue fluids is on the order of 1 mM, whereas the cytosolic concentration in a resting cell is closer to 100 nM—10,000 times lower than the extracellular concentration. This difference is maintained by an efficient (although energetically expensive) system of membrane pumps. Furthermore, higher-concentration Ca^{2+} stores exist

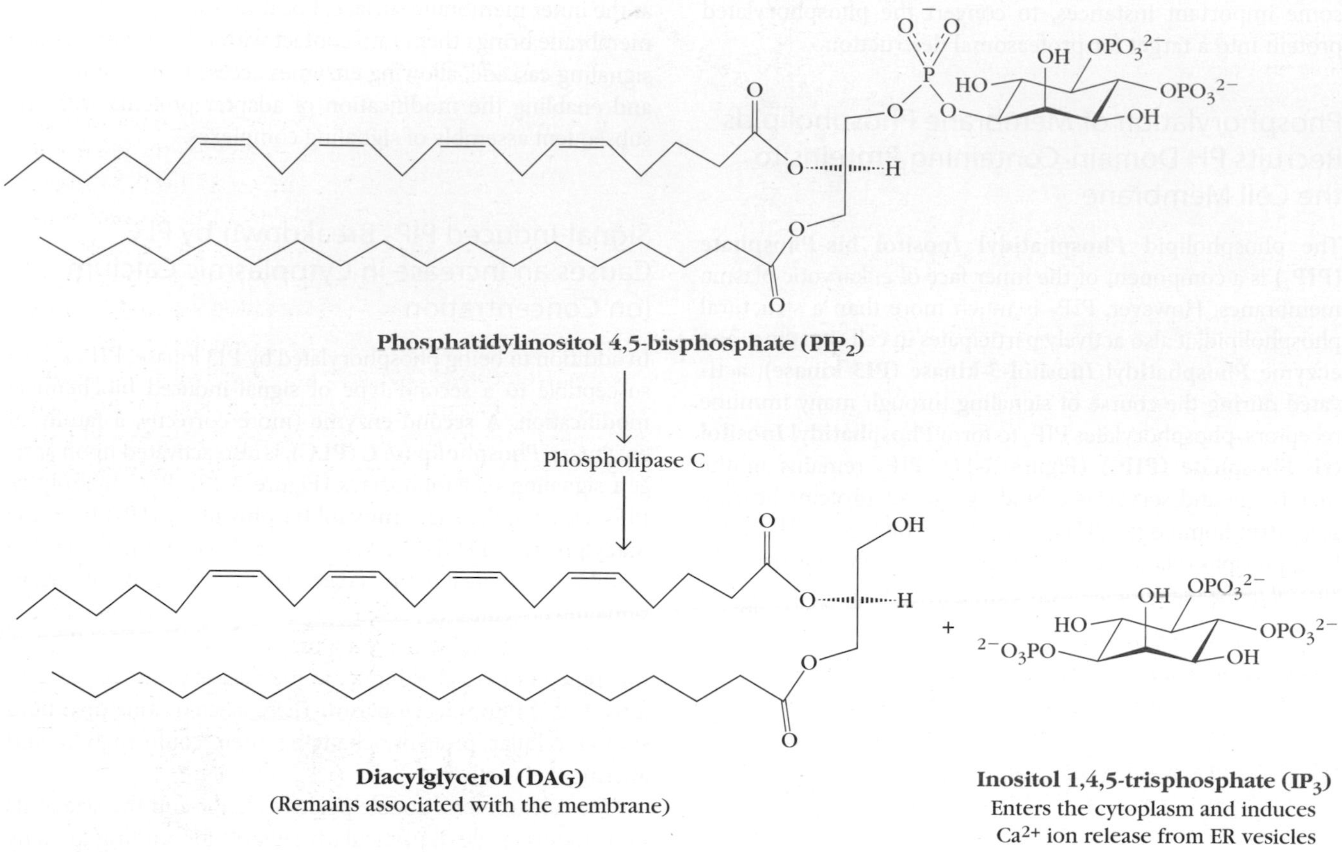

Phosphatidylinositol 4,5-bisphosphate (PIP$_2$)

Phospholipase C

Diacylglycerol (DAG)
(Remains associated with the membrane)

+

Inositol 1,4,5-trisphosphate (IP$_3$)
Enters the cytoplasm and induces
Ca^{2+} ion release from ER vesicles

FIGURE 3-12 PIP$_2$ can be hydrolyzed by PLC to create DAG and IP$_3$.

within intracellular vesicles associated with the endoplasmic reticulum and the mitochondria. Thus, any signaling event that opens up channels in the membrane of the endoplasmic reticulum, or in the plasma membrane, and allows for free flow of Ca^{2+} ions into the cytoplasm, facilitates a rapid rise in intracytoplasmic Ca^{2+} ion concentration.

Ubiquitination May Inhibit or Enhance Signal Transduction

The protein ubiquitin is a small, highly conserved, 76-residue, monomeric protein. Binding of the carboxy-terminal of ubiquitin to lysine residues of target proteins is often followed

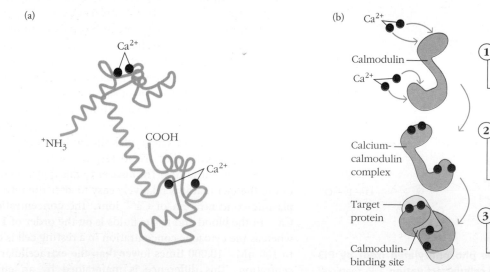

(a)

Ca^{2+}

$^+$NH$_3$

COOH

Ca^{2+}

(b)

Ca^{2+}

Calmodulin

Ca^{2+}

Calcium-calmodulin complex

Target protein

Calmodulin-binding site

① Calmodulin binds Ca^{2+} ions.

② Calmodulin changes conformation, becomes active.

③ Calmodulin binds a target protein.

FIGURE 3-13 The calcium-binding regulatory protein calmodulin undergoes a conformational shift on binding to calcium, which enables it to bind and activate other proteins. [Part (a) PDB ID 3CLN.]

by polymerization of a chain of ubiquitins onto selected lysine residues of the conjugated ubiquitin. This results in poly-ubiquitination of the target protein. In most instances, mono- or poly-ubiquitination serves as a mechanism by which the tagged proteins are targeted for destruction by the proteasome of the cell. However, we now know that ubiquitination on certain residues of proteins can also serve as an activating signal, and we will come across this alternative role of ubiquitin below as we learn about the activation of the NF-κB transcription factor, as well as in Chapters 4 and 5.

Frequently Encountered Signaling Pathways

Analysis of the biochemical routes by which molecular signals pass from cell surface receptors to the interior of the cell has revealed that a few signal transduction pathways are used repeatedly in the cellular responses to different ligands. These pathways are utilized, in slightly different forms, in multiple cellular and organ systems, and in many species. The end result of most of these pathways is an alteration in the transcriptional program of the cell. Here we briefly describe three such pathways of particular relevance to the adaptive immune system (see Overview Figure 3-5). Each of these pathways is triggered by antigen binding to the receptor and leads to the activation of a different family of transcription factors. The generation of these active transcription factors in turn initiates the up-regulation of a cascade of genes important to the immune response, including those encoding cytokines, antibodies, survival factors, and proliferative signals.

Although the conceptual framework of these signaling pathways is shared among many cell types, small modifications in the molecular intermediates and transcription factors involved allows for variations in the identities of the genes controlled by particular signals and for cell-type specific transcriptional control mechanisms. For example, the enzyme PLC exists in different forms; PLCγ1 is used in T cells and PLCγ2 in B cells. Similarly, the **Nuclear Factor of Activated T cells (NFAT)** family of transcription factors includes five members, some of which are expressed as differentially spliced variants, and each of which is affected by different arrays of molecular signals.

Furthermore, activation of transcription from some promoters requires the binding of multiple transcription factors. For example, full expression of the gene encoding interleukin 2 (IL-2) requires the binding of AP-1, NF-κB, *and* NFAT, in addition to several other transcription factors, to the interleukin promoter. In this way, the promoter can be partially activated in the presence of some of these transcription factors, but not fully active until all of them are in place. In the next section, we focus on the general principles that govern the control of these three pathways, from the receipt of an antigen signal, to the activation of a particular type of transcription factor.

The PLC Pathway Induces Calcium Release and PKC Activation

We have already met the essential elements of this pathway in our discussion of the enzymes that modulate PIP_2 upon cell activation (see Figure 3-12), and we now know that PLC breaks down PIP_2 to IP_3 and DAG. But how does PLC become activated in the first place? We use T cells here as our example.

As mentioned, T cells use the PLCγ1 form of the enzyme. On antigen stimulation, tyrosine phosphorylation of the ITAM residues of the CD3 receptor-associated complex results in the localization of the adapter protein LAT to the membrane (Figure 3-14). LAT in turn is phosphorylated, and PLCγ1 then binds to phosphorylated LAT. This binding ensures that PLCγ1 localizes to the cell membrane, the site of its substrate, PIP_2.

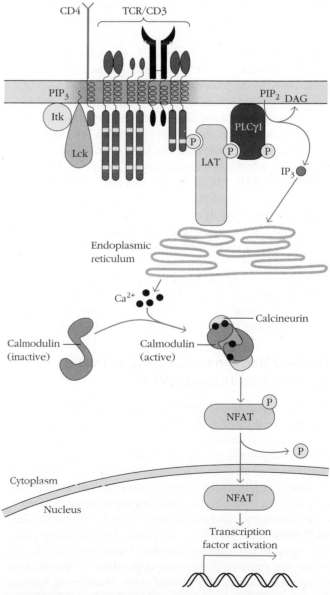

FIGURE 3-14 Activation of calcineurin by binding of the calcium-calmodulin complex induces NFAT dephosphorylation and nuclear entry.

Once located at the plasma membrane, PLCγ1 is further activated by tyrosine phosphorylation mediated by the receptor-activated kinase, Lck, as well as by a second tyrosine kinase, Itk. The phosphorylated and activated PLCγ1 then mediates cleavage of PIP_2 to IP_3 and DAG as described above.

How do these secondary mediators then function to bring about lymphocyte activation? Calcium ions, released from intracellular stores by IP_3, bind to the signaling intermediate, calmodulin. Each molecule of calmodulin binds to four calcium ions, and this binding results in a profound alternation in its conformation (see Figure 3-13). The calcium-calmodulin complex then binds and activates the phosphatase calcineurin, which dephosphorylates the transcription factor NFAT (see Figure 3-14). This dephosphorylation induces a conformational change in NFAT, revealing a nuclear localization sequence, which directs NFAT to enter the nucleus and activate the transcription of a number of important T-cell target genes. Genes activated by NFAT binding include that encoding interleukin 2, a cytokine pivotally important in the control of T cell proliferation.

The importance of the PLCγ1 pathway to T cell activation is illustrated by the profound immunosuppressant effects of the drug *cyclosporine*, which is used to treat both T-cell–mediated autoimmune disease and organ transplant rejection. In T cells, cyclosporine binds to the protein cyclophilin (immunophilin), and the cyclosporine/cyclophilin complex binds and inhibits calcineurin, effectively shutting down T-cell proliferation.

PIP_2 cleavage by PLC is a particularly efficient signaling strategy as both products of PIP_2 breakdown are active in subsequent cellular events. DAG, the second product of PIP_2 breakdown (see Figure 3-12) remains in the membrane, where it binds and activates enzymes of the **protein kinase C (PKC)** family. These kinases are serine/threonine kinases active in a variety of signaling pathways. Protein kinase Cθ, important in T-cell signaling, only requires DAG binding, whereas its relative, protein kinase C, must also bind Ca^{2+} ions for full activation. DAG is also implicated in the Ras pathway (described in the next section).

The Ras/Map Kinase Cascade Activates Transcription Through AP-1

The Ras signaling pathway was discovered initially after a mutated, viral form of the Ras protein was found to induce cancer in a rat model. **Ras** is a monomeric, GTP-binding protein (G protein, for short). When Ras binds to GTP, its conformation changes into an active state, in which it is capable of binding, and activating a number of serine/threonine kinases. However, the Ras protein possesses an intrinsic GTPase activity that hydrolyzes GTP to GDP, and the Ras–GDP form of the protein is incapable of transmitting a positive signal to downstream kinases (Figure 3-15). Activation of the Ras pathway is therefore dependent on the ability to maintain Ras in its GTP-bound state. Modulation between the active, GTP-bound form and the inactive, GDP-bound form is brought about by two families of enzymes. **Guanine-nucleotide Exchange Factors (GEFs)** *activate* Ras by induc-

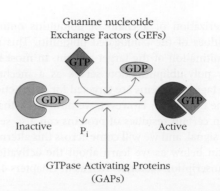

FIGURE 3-15 Small monomeric G proteins, such as Ras, alter conformation depending on whether they are bound to GTP or GDP. The GDP-bound form of small G proteins such as Ras (pale green) is inactive, and the GTP-bound form (dark green) is active. Small G proteins have intrinsic GTPase activity, which is further enhanced by GTPase Activating Proteins (GAPs). Guanine nucleotide Exchange Factors (GEFs) act in an opposite manner to release GDP and promote GTP binding.

ing it to release GDP and accept GTP; **GTPase Activating Proteins (GAPs)** *inhibit* the protein's activity by stimulating Ras's intrinsic ability to hydrolyze bound GTP.

Like the PLCγ pathway, the Ras pathway is initiated during both B and T lymphocyte activation and, again, we will use T cells as our example. DAG, released after PLCγ1-mediated PIP_2 cleavage, binds and activates Ras-GRP, an adapter protein that then recruits the GEF, *Son of Sevenless* (SOS). SOS binds to Ras, inducing it to bind GTP, at which point Ras gains the ability to bind and activate the first member of a cascade of serine/threonine kinase enzymes that phosphorylate and activate one another. Because the members of this cascade were first identified in experiments that studied the activation of cells by **mitogens** (agents that induce proliferation), it is referred to as the ***M*itogen *A*ctivated *P*rotein *K*inase** or **MAP kinase** cascade. The members of the cascade are held in close proximity to one another by the adapter protein KSR.

In the T-cell activation form of this cascade (illustrated in Figure 3-16), RasGTP binds to the MAP kinase kinase kinase (MAPKKK) Raf. Binding of RasGTP alters the conformation of Raf and stimulates its serine/threonine kinase activity. Raf then phosphorylates and activates the next enzyme in the relay, a MAP kinase kinase (MAPKK), in this case MEK. Activated MEK then phosphorylates its substrate, *E*xtracellular signal-*R*elated *K*inase, or Erk, a MAP kinase, which consequently gains the ability to pass through the nuclear membrane.

Once inside the nucleus, Erk phosphorylates and activates a transcription factor, Elk-1, which cooperates with a second protein, serum response factor (SRF), to activate the transcription of the *fos* gene. The Fos protein is also phosphorylated by Erk, and along with its partner, Jun, forms the master transcription factor, **AP-1**. Jun is phosphorylated and activated via a slightly different form of the MAP kinase pathway. AP-1 is another of the transcription factors that facilitates the transcription of the IL-2 gene.

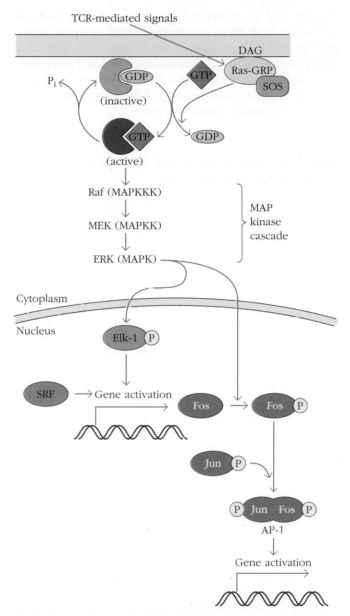

FIGURE 3-16 The Ras pathway involves a cascade of serine/threonine phosphorylations and culminates in the entry of the MAP kinase, Erk, into the nucleus where it phosphorylates the transcription factors Elk-1 and Fos.

The Ras pathway is an important component in many developmental and cell activation programs. Each pathway uses slightly different combinations of downstream protein kinases, but all adhere to the same general mode of passing the signal from the cell surface through to the nucleus via a cascade of phosphorylation reactions, with the resultant activation of a new transcriptional program.

PKC Activates the NF-κB Transcription Factor

NF-κB belongs to a family of heterodimeric transcription factors, and each dimer activates its own repertoire of promoters. In resting cells, NF-κB heterodimers are held in the

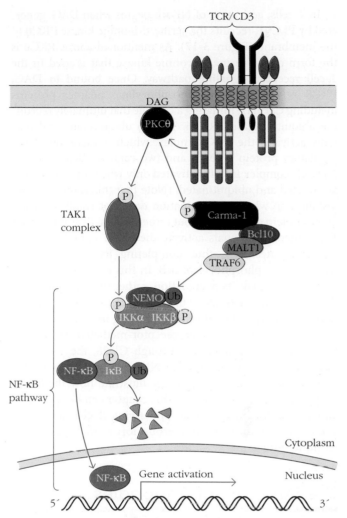

FIGURE 3-17 Phosphorylation and ubiquitination activates IKK, which phosphorylates and inactivates IκB, allowing translocation of NF-κB into the nucleus.

cytoplasm by binding to the *Inhibitor of NF-κB* (IκB) protein. Cell activation induces the phosphorylation of these inhibitory proteins by an **IκB kinase (IKK)** complex. The phosphorylated IκB protein is then targeted for proteasomal degradation, releasing the freed NF-κB to enter the nucleus and bind to the promoters of a whole array of immunologically important genes.

NF-κB is important in the transcription control of proteins needed for the proper functioning of many innate and adaptive immune system cell types; in general, NF-κB-mediated transcription is associated with proinflammatory and activation events, rather than with regulatory processes. Different types of cells and receptors use various forms of protein kinase activation as well as diverse combinations of adapter molecules in the pathway leading to NF-κB activation. However, all of these pathways culminate in the phosphorylation and subsequent destruction of the inhibitor of NF-κB, IκB. Below, we describe the pathway to NF-κB activation that is triggered by TCR antigen recognition.

In T cells, activation of NF-κB begins when DAG, generated by PLCγ1, recruits the serine/threonine kinase **PKCθ** to the membrane (Figure 3-17). As mentioned above, PKCθ is the form of the serine/threonine kinase that is used in the T-cell receptor signaling pathway. Once bound to DAG, PKCθ is activated and phosphorylates adapter proteins including Carma1, initiating a cascade that ultimately recruits the ubiquitin ligase TRAF6. TRAF6 ubiquitinates and partially activates the IKK complex, which is made up of the regulatory protein NEMO, and two catalytic IKK subunits. The IKK complex is fully activated only when it is both phosphorylated and ubiquitinated. (Note that this is one of those instances in which ubiquitination does not result in subsequent protein degradation, but rather in its activation.) Other T cell mediated signals activate the TAK1 complex, which phosphorylates IKK, thus completing its activation and allowing it to phosphorylate IκB. In this step, phosphorylation of IκB signals its degradation, rather than its activation. NF-κB is then free to move into the nucleus and activate the transcription of its target genes, including the IL2 gene.

In addition to this T-cell receptor-mediated pathway of NF-κB activation, signaling through CD28, the T-cell co-receptor, also positively controls NF-κB activation in T cells.

These three pathways together illustrate how the broad concepts introduced earlier in this chapter can be applied to understand the mechanisms underlying T-cell activation. These themes recur in numerous forms throughout the immune system.

The Structure of Antibodies

Having outlined some of the general concepts that underlie many different receptor-ligand interactions and signaling pathways, we now turn our attention more specifically to the antigen receptors of the adaptive immune system.

On stimulation with antigen, B cells secrete antibodies with antigen-binding sites identical to those on the B-cell membrane antigen receptor. The identity between the binding sites of the secreted antibody and the membrane-bound **B-cell receptor** (BCR) was first demonstrated by making reagents that bound to antibodies secreted by a particular clone of B cells and showing that those reagents also bound to the receptors on the cells that had secreted the antibodies. Since working with soluble proteins is significantly easier than manipulating membrane receptor proteins, the presence of a soluble form of the receptor greatly facilitated the characterization of B-cell receptor structure. Consequently, the basic biochemistry of the B-cell receptor was established long before that of the **T-cell receptor** (TCR), which, unlike the BCR, is not released in a secreted form. Box 3-1: Classic Experiment details the Nobel Prize–winning experiments that established the four-chain structure of the antibody molecule.

We begin this section with a discussion of antibody structure, and then go on to describe the B-cell membrane receptor and its associated signaling pathways. Secreted antibodies and their membrane-bound receptor forms belong to the immunoglobulin family of proteins. This large family of proteins, which includes both B- and T-cell receptors, adhesion molecules, some tyrosine kinases, and other immune receptors, is characterized by the presence of one or more **immunoglobulin domains**.

Antibodies Are Made Up of Multiple Immunoglobulin Domains

The immunoglobulin domain (Figure 3-18) is generated when a polypeptide chain folds into an organized series of antiparallel β-pleated strands. Within each domain, the β strands are arranged into a pair of β sheets that form a tertiary, compact domain. The number of strands per sheet varies among individual proteins. In antibody molecules, most immunoglobulin domains contain approximately 110 amino acids, and each β sheet contains three to five strands. The pair of β sheets within each domain are stabilized with respect to one another by an intrachain disulfide bond. Neighboring domains are connected to one another by a stretch of relatively unstructured polypeptide chain. Within the β strands, hydrophobic and hydrophilic amino acids alternate and their side chains are oriented perpendicular to the plane of the sheet. The hydrophobic amino acids on one sheet are oriented toward the opposite sheet, and the two sheets within each domain are therefore stabilized by hydrophobic interactions between the two sheets as well as by the covalent disulfide bond.

The immunoglobulin fold provides a perfect example of how structure determines and/or facilitates function. At the ends of each of the β sheets, more loosely folded polypeptide regions link one β strand to the next, and these loosely folded regions can accommodate a variety of amino acid side-chain lengths and structures without causing any disruption to the overall structure of the molecule. Hence, in the antibody molecule, the immunoglobulin fold is superbly adapted *to provide a single scaffold onto which multiple different binding sites can be built*, as the antigen-binding sites can simply be built into these loosely folded regions of the antigen-binding domains. These properties explain why the immunoglobulin domain has been used in so many proteins with recognition or adhesive functions. The essential domain structure provides a molecular backbone, while the loosely folded regions can be adapted to bind specifically to many adhesive or antigenic structures.

Indeed, the immunoglobulin domain structure is used by many proteins besides the BCR chains. As we will see later in this chapter, the T-cell receptor is also made up of repeating units of the immunoglobulin domain. Other proteins that use immunoglobulin domains include Fc receptors; the T-cell receptor accessory proteins CD2, CD4, CD8, and CD28; the receptor-associated proteins of both the TCR and the BCR; adhesion molecules; and others. Some of these immunoglobulin domain-containing proteins are illustrated in Figure 3-19. Each of these proteins is classified as a member

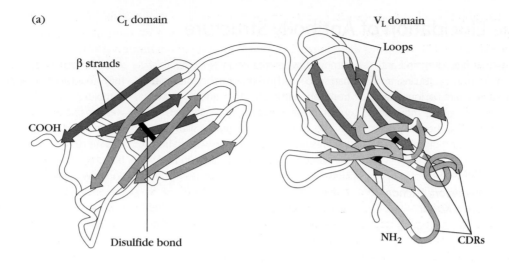

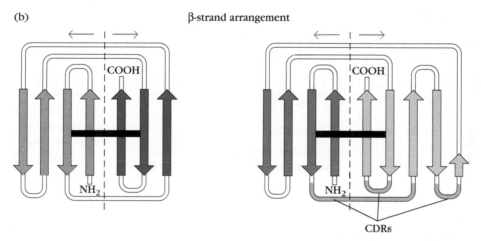

FIGURE 3-18 Diagram of the immunoglobulin fold structure of the antibody light chain variable (V_L) and constant (C_L) region domains. (a) The two β pleated sheets in each domain are held together by hydrophobic interactions and the conserved disulfide bond. The β strands that compose each sheet are shown in different colors. The three loops of each variable domain show considerable variation in length and amino acid sequence; these hypervariable regions (blue) make up the antigen-binding site. Hypervariable regions are usually called complementarity-determining regions (CDRs). Heavy-chain domains have the same characteristic structure. (b) The β-pleated sheets are opened out to reveal the relationship of the individual β strands and joining loops. Note that the variable domain contains two more β strands than the constant domain. *[Part (a) adapted from M. Schiffer et al., 1973, Biochemistry 12: 4620; reprinted with permission; part (b) adapted from A. F. Williams and A. N. Barclay, 1988, Annual Review of Immunology 6:381.]*

of the **immunoglobulin superfamily**, a term that is used to denote proteins derived from a common primordial gene encoding the basic domain structure.

Antibodies Share a Common Structure of Two Light Chains and Two Heavy Chains

All antibodies share a common structure of four polypeptide chains (Figure 3-20), consisting of two identical **light (L) chains** and two identical **heavy (H) chains**. Each light chain is bound to its partner heavy chain by a disulfide bond between corresponding cysteine residues, as well as by noncovalent interactions between the V_H and V_L domains and the C_H1 and C_L domains. These bonds enable the formation of a closely associated heterodimer (H-L). Multiple disulfide bridges link the two heavy chains together about halfway down their length, and the C-terminal parts of the two heavy chains also participate in noncovalent bonding interactions between corresponding domains.

As shown in Figure 3-21, the antibody molecule forms a Y shape with two identical antigen-binding regions at the tips of the Y. Each antigen-binding region is made up of amino acids derived from both the heavy- and the light-chain amino-terminal domains. The heavy and light chains both contribute two domains to each arm of the Y, with the non–antigen-binding domain of each chain serving to extend the antigen-binding arm. The base of the Y consists of the C-terminal domains of the antibody heavy chain.

The Elucidation of Antibody Structure

The experiments that first identified antibodies as serum immunoglobulins and then characterized their familiar four-chain structure represent some of the most elegant adaptations of protein chemistry ever used to solve a biological problem.

First Set of Experiments

Arne Tiselius, Pers Pederson, Michael Heidelberger, and Elvin Kabat

Since the late nineteenth century, it has been known that antibodies reside in the blood serum—that is, in that component of the blood which remains once cells and clotting proteins have been removed. However, the chemical nature of those antibodies remained a mystery until the experiments of Tiselius and Pederson of Sweden and Heidelberger and Kabat, in the United States, published in 1939. They made use of the fact that, when antibodies react with a multivalent protein antigen, they form a multimolecular cross-linked complex that falls out of solution. This process is known as immunoprecipitation (see Chapter 20 for modern uses of this technique). They immunized rabbits with the protein ovalbumin (the major component of egg whites), bled the rabbits to obtain an anti-ovalbumin antiserum, and then divided their antiserum into two aliquots. They subjected the first aliquot to electrophoresis, measuring the amount of protein that moved different distances from the origin in an electric field. The blue plot in Figure 1 depicts the four major protein subpopulations resolved by their technique. The first, and largest, is the albumin peak, the most abundant protein in serum, with responsibility for transporting lipids through the blood. They named the other peaks globulins. Two smaller peaks are denoted the α and β globulin peaks, and then a third globulin peak, γ globulin, clearly represents a set of proteins in high concentration in the serum.

However, the most notable part of the experiment occurred when the investigators mixed their second serum aliquot with ovalbumin, the antigen. The antibodies in the serum bound to ovalbumin in a multivalent complex, which fell out of solution into a precipitate. The precipitate was then removed by centrifugation. Now that they had succeeded in removing the antibodies from an antiserum, the question was, which protein peak would be affected?

The black plot in Figure 1 illustrates their results. Very little protein was lost from the albumin, or the α or β globulin peaks. However, immunoprecipitation resulted in a dramatic decrease in the size of the γ globulin peak, demonstrating that the majority of their anti-ovalbumin antibodies could be classified as γ globulins.

We now know that most antibodies of the IgG class are indeed found in the γ globulin class. However, antibodies of other classes are found in the α and β globulin peaks, which may account for the slight decrease in protein concentration found after immunoprecipitation in these other protein peaks.

Second Set of Experiments

Sir Rodney Porter, Gerald Edelman, and Alfred Nisonoff

Knowing the class of serum protein into which antibodies fall was a start, but immunochemists next needed to figure out what antibodies looked like. The fact that they could form precipitable multivalent complexes suggested that each antibody was capable of binding to more than one site on a multivalent antigen. But the scientists still did not know how many polypeptide chains made up an antibody molecule and how many antigen-binding sites were present in each molecule.

Two lines of experimentation conducted in a similar time frame on both sides of the Atlantic combined to provide the answers to these two questions. Ultracentrifugation experiments had placed the molecular weight of IgG antibody molecules at approximately 150,000 Daltons. Digestion of IgG with the enzyme papain produced three fragments, two of which were identical and a third which was clearly different (Figure 2). The third fragment, of approximately 50,000 Daltons, spontaneously formed crystals and was therefore named Fragment crystallizable, or Fc. By demonstrating that they could competitively inhibit the binding of antibodies to their antigen, the other two fragments were shown to retain the antigen-binding

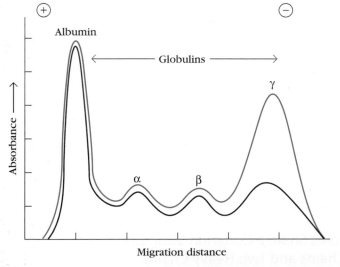

FIGURE 1

Experimental demonstration that most antibodies are in the γ-globulin fraction of serum proteins. After rabbits were immunized with ovalbumin (OVA), their antisera were pooled and electrophoresed, which separated the serum proteins according to their electric charge and mass. The blue line shows the electrophoretic pattern of untreated antiserum. The black line shows the pattern of antiserum that was first incubated with OVA to remove anti-OVA antibody and then subjected to electrophoresis. [Adapted from A. Tiselius and E. A. Kabat, 1939, Journal of Experimental Medicine 69:119, with copyright permission of the Rockefeller University Press.]

BOX 3-1

capacity of the original antibody. These fragments were therefore named Fab, or *Fragment antigen binding*. This experiment indicated that a single antibody molecule contained two antigen binding sites and a third part of the molecule that did not participate in the binding reaction.

Use of another proteolytic enzyme, pepsin, resulted in the formation of a single fragment of 100,000 Daltons, which contained two antigen-binding sites that were still held together in a bivalent molecule. Because the molecule acted as though it contained two Fab fragments, but clearly had an additional component that facilitated the combination of the two fragments into one molecule, it was named F(ab')$_2$. Pepsin digestion does not result in a recoverable Fc fragment, as it apparently digests it into small fragments. However, F(ab')$_2$ derivatives of antibodies are often used in experiments in which scientists wish to avoid artifacts resulting from binding of antibodies to Fc receptors on cell surfaces.

In the second set of experiments, investigators reduced the whole IgG molecule using β-mercaptoethanol, in order to break disulfide bonds, and alkylated the reduced product so that the disulfide bonds could not spontaneously reform. They then used a technique called gel filtration to separate and measure the size of the protein fragments generated by this reduction and alkylation. (Nowadays, we would use SDS PAGE gels to do this experiment.) In this way, it was shown that each IgG molecule contained two heavy chains, MW 50,000 and two light chains, MW 22,000.

Now the challenge was to combine the results of these experiments to create a consistent model of the antibody molecule. To do this, the scientists had to determine which of the chains was implicated in antigen binding, and which chains contributed to the crystallizable fragments. Immunologists often use immunological means to answer their questions, and this was no exception. They elected to use Fab and Fc fragments purified from rabbit IgG antibodies to immunize two separate goats. From these goats, they generated anti-Fab and anti-Fc antibodies, which they reacted, in separate experiments, with the heavy and light chains from the

reduction and alkylation experiments. The answer was immediately clear.

Anti-Fab antibodies bound to both heavy and light chains, and therefore the antigen binding site of the original rabbit IgG was made up of both heavy and light chain components. However, anti-Fc antibodies bound only to the heavy chains, but not to the light chains of the IgG molecule, demonstrating that the Fc part of the molecule was made up of heavy chains only. Finally, careful protein chemistry demonstrated that the amino-termini of the two chains resided in the Fab portion of the molecule. In this way, the familiar four-chain structure, with the binding sites at the amino-termini of the heavy and light chain pairs, was deduced from some classically elegant experiments.

In 1972, Sir Rodney Porter and Gerald Edelman were awarded the Nobel Prize in Physiology and Medicine for their work in discovering the structure of immunoglobulins.

Edelman, G. M., B. A. Cunningham, W. E. Gall, P. D. Gottlieb, U. Rutishauser, and M. J. Waxdal. 1969. The covalent structure of an entire gammaG immunoglobulin molecule. *Proceedings of the National Academy of Sciences of the United States of America* **63**:78–85.

Fleishman, J, B., R. H. Pain, and R. R. Porter. 1962 Sep. Reduction of gamma-globulins. *Archives of Biochemistry and Biophysics* **Suppl 1**:174–80.

Heidelberger, M., andK. O. Pedersen. 1937. The molecular weight of antibodies. *The Journal of Experimental Medicine* **65**:393–414.

Nisonoff, A., F. C. Wissler, and L. N. Lipman. 1960. Properties of the major component of a peptic digest of rabbit antibody. *Science* **132**:1770–1771.

Porter, R. R. (1972). Lecture for the Nobel Prize for physiology or medicine 1972: Structural studies of immunoglobulins. *Scandinavian Journal of Immunology* **34**(4):381–389.

Tiselius, A., and E. A. Kabat. 1939. An electrophoretic study of immune sera and purified antibody preparations. *The Journal of Experimental Medicine* **69**:119–131.

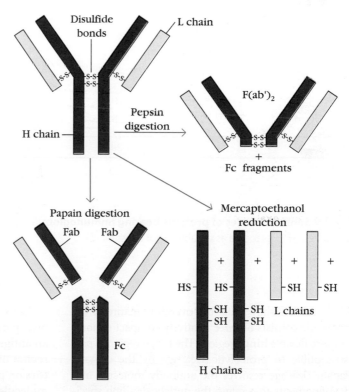

FIGURE 2

Prototype structure of IgG, showing chain structure and interchain disulfide bonds. The fragments produced by enzymatic digestion with pepsin or papain or by cleavage of the disulfide bonds with mercaptoethanol are indicated. Light (L) chains are in light blue, and heavy (H) chains are in dark blue.

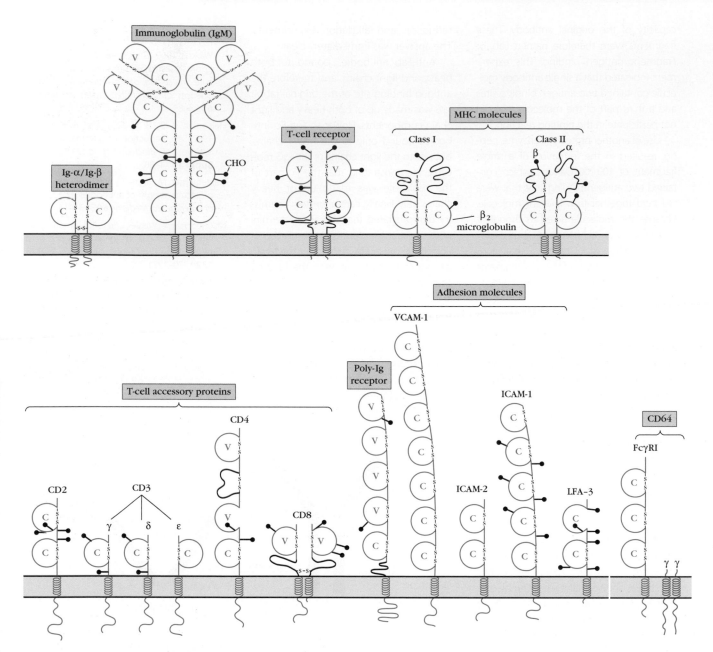

FIGURE 3-19 Some examples of proteins bearing immunoglobulin domains. Each immunoglobulin domain is depicted by a blue loop and, where relevant, is labeled as variable (V) or constant (C).

Figure 3-21 further shows that the overall structure of the antibody molecule consists of three relatively compact regions, joined by a more flexible **hinge** region. The hinge region is particularly susceptible to proteolytic cleavage by the enzyme **papain**. Papain cleavage resolves the antibody molecule into two identical *fragments* that retain the *antigen-binding* specificity of the original antibody (shown as **Fab regions** in Figure 3-21), and the remaining region of the molecule, which consists of the non antigen-binding portion. This latter region, which is identical for all antibodies of a given class, crystallizes easily and was thus called the **Fc region** (*fragment crystallizable*).

Each Fab region and Fc region of antibodies mediates its own particular functions during an antibody response to an antigen. The Fab regions bind to the antigen, and the Fc region of the antigen-coupled antibody binds to **Fc receptors** on phagocytic or cytolytic cells, or to immune effector molecules. In this way, antibodies serve as physiological bridges between an antigen present on a pathogen, and the cells or molecules that will ultimately destroy it. A family of Fc receptors exists; each Fc receptor is expressed on a different array of cells and binds to a different class of antibodies.

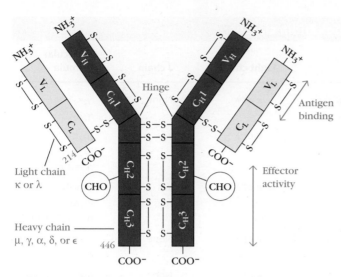

FIGURE 3-20 Schematic diagram of the structure of immunoglobulins derived from amino acid sequence analysis. Each heavy (dark blue) and light (light blue) chain in an immunoglobulin molecule contains an amino-terminal variable (V) region that consists of 100 to 110 amino acids and differs from one antibody to the next. The remainder of each chain in the molecule—the constant (C) regions—exhibits limited variation that defines the two light-chain subtypes and the five heavy-chain subclasses. Some heavy chains (γ, δ, and α) also contain a proline-rich hinge region. The amino-terminal portions, corresponding to the V regions, bind to antigen; effector functions are mediated by the carboxy-terminal domains. The μ and ε heavy chains, which lack a hinge region, contain an additional domain in the middle of the molecule. CHO denotes a carbohydrate group linked to the heavy chain.

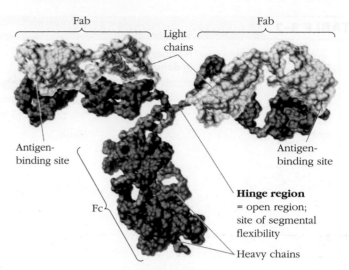

FIGURE 3-21 The three-dimensional structure of IgG. Clearly visible in this representation are the 110 amino acid immunoglobulin domains of which the molecule is constructed, along with the open hinge structure in the center of the molecule, which affords flexibility in binding to multivalent antigens. In IgG, the light chain contains two immunoglobulin domains, the heavy chain four. Fab = Antigen-binding portion of the antibody; contains paired V_L/V_H and C_L/C_H1 domains. Fc = Non–antigen-binding region of the antibody, with paired C_H2/C_H2 and C_H3/C_H3 domains. [Source: D. Sadava, H. C. Heller, G. H. Orians, W. K. Purves, and D. M. Hillis. *Life: The Science of Biology*, 7e (© 2004 Sinauer Associates, Inc., and W. H. Freeman & Co.), Figure 18-10 with modifications.]

There are Two Major Classes of Antibody Light Chains

Amino acid sequencing of antibody light chains revealed that the amino-terminal half (approximately 110 amino acids) of the light chain was extremely variable, whereas the sequence of the carboxyl-terminal half could be classified into one of two major sequence types. The N-terminal half of light chains is thus referred to as the **variable,** or V_L, region of the light chain, and the less variable part of the sequence is termed the **constant,** or C_L, region. The two major light chain constant region sequences are referred to as κ **(kappa)** or λ **(lambda) chains.** As more light-chain sequences were generated, it became apparent that the λ chain constant region sequences could be further subdivided into four subtypes—λ1, λ2, λ3, and λ4—based on amino acid substitutions at a few positions. In humans, the light chains are fairly evenly divided between the two light-chain classes; 60% of human light chains are κ whereas only 40% are λ. In mice, the situation is quite different: Only 5% of mouse light chains are of the λ light-chain type. All light chains have a molecular weight of approximately 22 kDa.

Further analysis of light-chain sequences demonstrated that, even within the variable regions of the light chain, there were regions of hypervariability. Since these **hypervariable** regions could be shown to interact with the bound antigen, they were renamed the **complementarity-determining regions,** or **CDRs.**

Those readers with expertise in genetics will immediately identify a problem: How is it possible to encode a single protein chain with some regions that are extremely diverse and other regions that are relatively constant, while maintaining that distinction across millions of years of evolutionary drift? The solution to this puzzle involves the encoding of a single antibody variable region in multiple genetic segments that are then joined together in different combinations in each individual antibody-producing cell. This unique mechanism will be fully described in Chapter 7.

There are Five Major Classes of Antibody Heavy Chains

Using antibodies directed toward the constant region of immunoglobulins and amino acid sequencing of immunoglobulins derived from plasmacytoma tumor cells, investigators discovered that the sequences of the heavy-chain constant regions fall into five basic patterns. These five basic sequences have been named with Greek letters: μ (mu), δ (delta), γ (gamma), ε (epsilon), and α (alpha). Each different heavy-chain constant region is referred to as an **isotype,** and the isotype of the heavy

TABLE 3-2   Chain composition of the five immunoglobulin classes

Class[*]	Heavy chain	Number of C_H Ig domains	Subclasses	Light chain	J chain	Molecular formula
IgG	γ	3	$\gamma1, \gamma2, \gamma3, \gamma4$ (human) $\gamma1, \gamma2a, \gamma2b, \gamma3$ (mouse)	κ or λ	None	$\gamma_2\kappa_2$ $\gamma_2\lambda_2$
IgM	μ	4	None	κ or λ	Yes	$(\mu_2\kappa_2)_n$ $(\mu_2\lambda_2)_n$ $n = 1$ or 5
IgA	α	3	$\alpha1, \alpha2$	κ or λ	Yes	$(\alpha_2\kappa_2)_n$ $(\alpha_2\lambda_2)_n$ $n = 1, 2, 3,$ or 4
IgE	ϵ	4	None	κ or λ	None	$\epsilon_2\kappa_2$ $\epsilon_2\kappa_2$
IgD	δ	3	None	κ or λ	None	$\delta_2\kappa_2$ $\delta_2\lambda_2$

[*]See Figure 3-22 for general structures of the five antibody classes.

chains of a given antibody molecule determines its **class**. Thus, antibodies with a heavy chain of the μ isotype are of the IgM class; those with a δ heavy chain are IgD; those with γ, IgG; those with ϵ, IgE; and those with α, IgA. The length of the constant region of the heavy chains is either 330 amino acid residues (for γ, δ, and α chains) or 440 amino acids (for μ and ϵ chains). Correspondingly, the molecular weights of the heavy chains vary according to their class. IgA, IgD, and IgG heavy chains weigh approximately 55 kDa, whereas IgM and IgE antibodies are approximately 20% heavier.

Minor differences in the amino acid sequences of groups of α and γ heavy chains led to further subclassification of these heavy chains into **sub-isotypes** and their corresponding antibodies therefore into **subclasses** (Table 3-2). There are two sub-isotypes of the α heavy chain, $\alpha1$ and $\alpha2$, and thus two IgA subclasses, IgA1 and IgA2. Similarly, there are four sub-isotypes of γ heavy chains, $\gamma1$, $\gamma2$, $\gamma3$, and $\gamma4$, with the corresponding formation of the four subclasses of IgG: IgG1, IgG2, IgG3, and IgG4. In mice, the four γ chain sub-isotypes are $\gamma1$, $\gamma2\alpha$, $\gamma2\beta$, and $\gamma3$, and the corresponding subclasses of mouse IgG antibodies are IgG1, IgG2a, IgG2b, and IgG3, respectively.

Further analysis revealed that the exact number, and precise positions of the disulfide bonds between the heavy chains of antibodies, vary among antibodies of different classes and subclasses (Figure 3-22).

Antibodies and Antibody Fragments Can Serve as Antigens

The essential principles of antibody structure were established prior to the development of the technology required to artificially generate monoclonal antibodies, and indeed much of the basic work of structure determination was completed before techniques were available for rapid DNA sequencing. As a source of homogenous antibodies, immunologists therefore turned to the protein products of antibody-secreting tumors. **Plasmacytomas** are tumors of plasma cells, the end-stage cell of B-cell differentiation, and the cells from that tumor are normally located in the bone marrow. When a single clone of plasma cells becomes cancerous, it is called a plasmacytoma for as long as it remains in a single bone. However, once it metastasizes into multiple bone marrow sites, the tumor is referred to as **multiple myeloma**. Plasmacytoma or myeloma tumors secrete large amounts of monoclonal antibodies into the serum and tissue fluids of the patients, and these antibodies can be purified in large quantities. Rather than secreting the whole antibody, some of these tumors will secrete only the light chains, or sometimes both the light chains and the whole antibodies, into the serum. The homogenous light chains secreted by these myeloma tumors are referred to as **Bence-Jones proteins**.

In the middle to late twentieth century, tumor-derived antibodies were used to generate a great deal of information regarding antibody structure and sequence. When a means was developed by which to generate these tumors artificially in mice, data from human tumors were supplemented with information derived from laboratory-generated murine cell lines, many of which are still in use to this day.

Both tumor-derived and purified serum antibodies were also used as antigens, and the anti-immunoglobulin antibodies so derived proved to be extraordinarily useful in the elucidation of antibody structure. Antibodies, or antibody fragments derived from one animal species (for example, a rabbit), can be injected into a second species

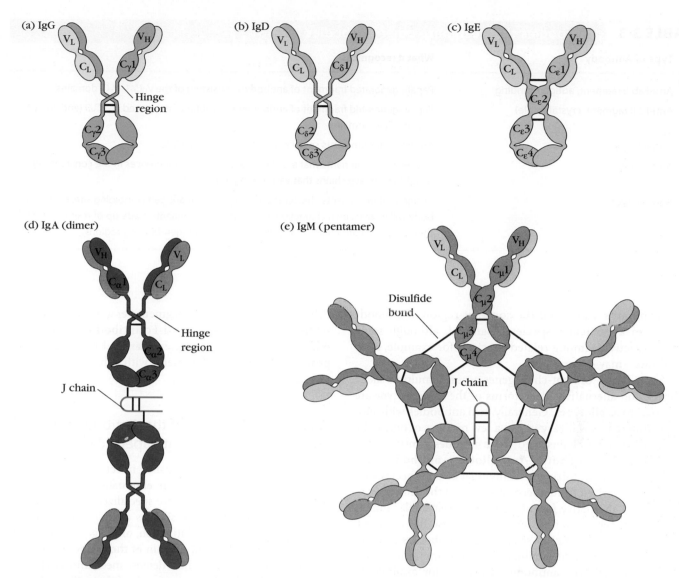

FIGURE 3-22 General structures of the five major classes of antibodies. Light chains are shown in lighter shades, disulfide bonds are indicated by thick black lines. Note that the IgG, IgA, and IgD heavy chains contain four domains and a hinge region, whereas the IgM and IgE heavy chains contain five domains but no hinge region. The polymeric forms of IgM and IgA contain a polypeptide, called the J chain, that is linked by two disulfide bonds to the Fc region in two different monomers. Serum IgM is always a pentamer; most serum IgA exists as a monomer, although dimers, trimers, and even tetramers are sometimes present. Not shown in these figures are intrachain disulfide bonds and disulfide bonds linking light and heavy chains (see Figure 3-23).

(for example, a goat), or even into another animal of the same species, in which the antibody or antibody fragment will serve as an antigen. Box 3-1 describes the series of experiments in which antibodies made against isolated proteolytic fragments of immunoglobulin molecules were used to help determine immunoglobulin structure. To understand these experiments, and many others that also use antibodies directed against whole immunoglobulins or against immunoglobulin fragments, it is important to have a good grasp of the nature of the antigenic determinants on immunoglobulin molecules. An **antigenic determinant** is defined as a region of an antigen that makes con-

tact with the antigen-combining region on an antibody. The different types of antibodies that recognize immunoglobulin determinants are described in Table 3-3.

Anti-isotype antibodies are directed against antigenic determinants present on the constant region of one particular heavy- or light-chain class of antibody, but not on any of the others. For example, an anti-isotype antibody may bind only to human μ heavy chains, but not to human δ, γ, α, or ε constant regions. Alternatively, it may bind to κ but not to λ light chains. Thus, an anti-isotype antibody binds only to a single antibody constant region class or subclass. Anti-isotype antibodies are usually specific for

TABLE 3-3   Antibodies that recognize other antibodies

Type of Antibody	What it recognizes
Anti-Fab (Fragment, antigen-binding)	Papain-generated fragment of antibodies consisting of the V_HC_H1–V_LC_L domains
Anti-Fc (Fragment, crystallizable)	Papain-generated fragment of antibodies consisting of the paired C_H2C_H3 (and, for IgM and IgE, C_H4) domains
Anti-isotype	Antigenic determinants specific to each heavy chain class
Anti-allotype	Antigenic determinants that are allele specific—small differences in the constant region of light and heavy chains that vary among individuals
Anti-idiotype	Antigenic determinants characteristic of a particular antigen combining site. Each antibody will have its own characteristic idiotypic determinants made up of residues from the heavy and light chains that contribute to the antigen-binding regions.

determinants present on the constant region of antibodies from a particular species of animal, although some cross-reactivity among related species—for example, mice and rats—may occur.

Some heavy- or light-chain genes occur in multiple allelic forms, and alternative allelic forms of the same isotype are referred to as **allotypes**. Generally, two antibodies with allotypic differences will vary in just a few residues of one of the immunoglobulin chains, and these residues constitute the **allotypic determinants**. **Anti-allotype antibodies** are generated by immunizing an individual of one species with antibodies derived from a second animal of the same species bearing an alternative form (allele) of the particular immunoglobulin gene. After an immune response has occurred, the investigator purifies or selects those antibodies that recognize the allotypic determinants from the immunized individual. Anti-allotype antibodies are used, for example, to trace particular cell populations in experiments in which B cells from a strain of animals bearing a particular allotype are transferred into a second strain that differs in immunoglobulin allotype.

Finally, antibodies directed against the antigen-binding site of a particular antibody are referred to as **anti-idiotypic antibodies**. They can be generated by immunizing an animal with large quantities of a purified monoclonal antibody, which, by definition, bears a single antigen-binding site. Those antibodies from the immunized animal that recognize the antigen-binding site of the original antibody can then be purified. Anti-idiotypic antibodies were used in the initial experiments that proved that the B-cell receptor had the same antigen-binding site as the antibody secreted by that B cell, and they are now used to follow the fate of B cells bearing a single receptor specificity in immuno-localization experiments.

As described above, treatment of whole antibody molecules with the enzyme papain cleaves the antibody molecule at the hinge region and releases two antigen-binding Fab fragments and one Fc fragment per antibody molecule. **Anti-Fab antibodies** and **anti-Fc antibodies** are made by immunizing a different species of animal from that which provided the antibody fragments with Fab or Fc fragments, respectively.

Each of the Domains of the Antibody Heavy and Light Chains Mediate Specific Functions

Antibodies protect the host against infection, by binding to pathogens and facilitating their elimination. Antibodies of different heavy-chain classes are specialized to mediate particular protective functions, such as complement activation (Chapter 6), pathogen agglutination, or phagocytosis (Chapter 13), and each different domain of the antibody molecule plays its own part in these host defense mechanisms. Here we briefly examine the structure of each of the antibody heavy-chain domains in turn, beginning with the C_H1 and C_L domains, which are those proximal to the V_H and V_L domains respectively (see Figure 3-20).

C_H1 and C_L Domains

As discussed above, the strength (avidity) of receptor binding to antigen is greatly enhanced by receptor multivalency. Antibodies have evolved to take advantage of this property by employing two antigen-binding sites, each of which can bind to individual determinants on multivalent antigens, such as are found on bacterial surfaces. The C_H1 and C_L domains serve to extend the antigen-binding arms of the antibody molecule, facilitating interactions with multivalent antigens and maximizing the ability of the antibody to bind to more than one site on a multivalent antigen. An interchain disulfide bond between these two domains holds them together and may facilitate the ability of some V_H-V_L pairs to hold on to one another and form a stable combining site.

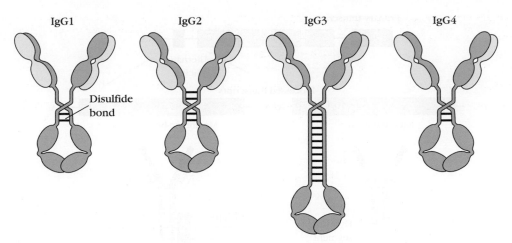

IgG1   IgG2   IgG3   IgG4

Disulfide bond

FIGURE 3-23 General structure of the four subclasses of human IgG, which differ in the number and arrangement of the inter-chain disulfide bonds (thick black lines) linking the heavy chains. A notable feature of human IgG3 is its 11 interchain disulfide bonds.

The Hinge Regions

The γ, δ, and α heavy chains contain an extended peptide sequence between the C_H1 and C_H2 domains that has no homology with the other domains (see Figures 3-20 and 3-22). This so-called **hinge region** is rich in proline residues, rendering it particularly flexible, and as a consequence, the two antigen-binding arms of IgG, IgD, and IgA antibodies can assume a wide variety of angles with respect to one another, which facilitates efficient antigen binding. The extended nature of the amino acid chain in the hinge region contributes to the vulnerability of this part of the molecule to protease cleavage, a vulnerability that was ingeniously exploited in the early experiments that characterized anti-body structure (see Box 3-1).

In addition to these proline residues, the hinge region also contains a number of cysteines, which participate in heavy-chain dimerization. The actual number of interchain disulfide bonds in the hinge region varies considerably among different heavy-chain classes and subclasses of anti-bodies (Figure 3-23) as well as between species. Lacking a hinge region, the heavy chains of IgE make their inter-heavy chain disulfide bonds between the C_H1 and C_H3 domains. In IgM, disulfide bonds bridge the pairs of heavy chains at the level of C_H3 and C_H4. Although μ and ε chains have no hinge regions, they do have an additional immunoglobulin domain that retains some hingelike qualities.

Carbohydrate Chains

The two C_H2 domains of α, δ, and γ chains and the two C_H3 domains of μ and ε chains are separated from their partner heavy-chain domains by oligosaccharide side chains that prevent the two heavy chains from nestling close to one another (Figure 3-24). As a result, the paired domains are significantly more accessible to the aqueous environment than other constant region domains. This accessibility is thought to contribute to the ability of IgM and IgG antibodies to bind to complement components. Immunoglobulins are in general quite extensively glycosylated, and some antibodies even have carbohydrates attached to their light chains.

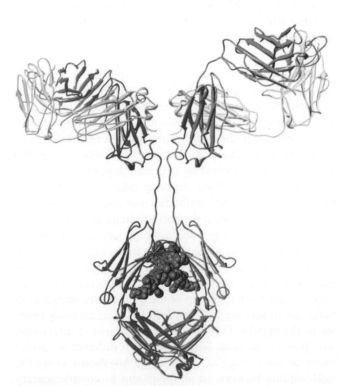

FIGURE 3-24 Carbohydrate residues, shown in pink, prevent close contact between the CH₂ domains of human IgG1. [PDB ID 1IGT.]

The Carboxy-Terminal Domains

The five classes of antibodies can be expressed as either membrane or secreted immunoglobulin. Secreted antibodies

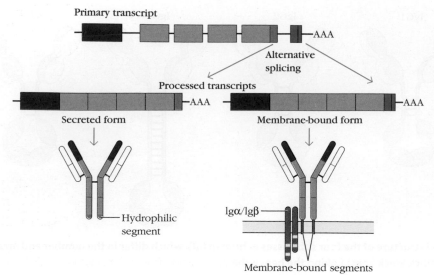

FIGURE 3-25 Membrane vs. secreted forms of immuno-globulin are created by alternative mRNA splicing. The dark blue boxes correspond to the mRNA transcript sections of the rear-ranged variable region sequences, and the light-blue boxes correspond to the parts of the mRNA transcript corresponding to the individual constant region domains of the heavy-chain genes. The IgM gene, shown here, contains one variable region and four constant region exons. The pink segment represents the transcript encoding the C-terminal portion of the secreted immunoglobulin. The green segments represent the transcript encoding the C-termi-nal portions of the membrane-bound immunoglobulin receptor, including the transmembrane and cytoplasmic regions. Alternative splicing creates the two different types of immunoglobulin mole-cules. Membrane-bound and secreted immunoglobulins on any one B cell share the identical antigen-binding regions and most of the heavy-chain sequences.

have a hydrophilic amino acid sequence of various lengths at the carboxyl terminus of the final C_H domain. In membrane-bound immunoglobulin receptors, this hydrophilic region is replaced by three regions (Figure 3-25):

- An extracellular, hydrophilic "spacer" sequence of approximately 26 amino acids
- A hydrophobic transmembrane segment of about 25 amino acids
- A very short, approximately three amino acid, cytoplasmic tail

B cells express different classes of membrane immuno-globulin at particular developmental stages and under differ-ent stimulatory conditions. Immature, pre-B cells express only membrane IgM. Membrane IgD co-expression along with IgM is one of the markers of differentiation to a fully mature B cell that has yet to encounter antigen. Following antigen stimulation, IgD is lost from the cell surface, and the constant region of the membrane and secreted immuno-globulin can switch to any one of the other isotypes. The antibody class secreted by antigen-stimulated B cells is determined by cytokines released by T cells and antigen presenting cells in the vicinity of the activated B cell. Anti-bodies of different heavy-chain classes have selective affini-ties for particular cell surface Fc receptors, as well as for components of the complement system. The effector func-tions of particular antibody classes are further discussed in Chapters 6 and 13.

X-ray Crystallography Has Been Used to Define the Structural Basis of Antigen-Antibody Binding

Crystallography has been used to explore the nature of antigen-antibody binding for a large number of antibodies and has demonstrated that either or both chains might pro-vide the majority of contact residues with any one antigen. Thus, some antibodies bind to the antigen mainly via con-tacts with heavy-chain variable region residues, with the light chain merely providing structural support to the bind-ing site; for other antigen-antibody interactions, the oppo-site is true. Still other antibodies use residues from both chains to directly contact the antigen. In most cases, though, contacts between antigen and antibody occur over a broad face and protrusions or depressions in the antibody surface are likely to be matched by complementary depressions or protrusions in the antigen. Figure 3-26 illustrates the bind-ing of an antibody to the tip of the influenza hemagglutinin molecule. In another well-studied case of an antibody bind-ing to the protein lysozyme, the surface area of interaction was shown to be quite large, ranging in different antibody-lysozyme pairs from 650 to 900 square Angstroms. Given the tight binding between an antibody and its complementary antigen, it should not be surprising that, at least in some cases, binding of antigen to antibody induces a conforma-tional change in the antibody, which can be visualized by x-ray crystallography (Figure 3-27).

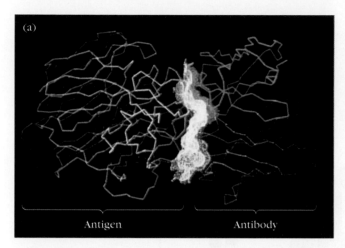 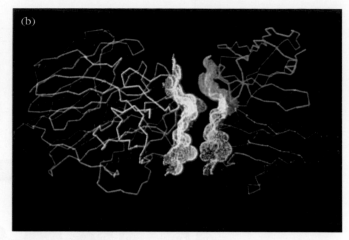

Antigen   Antibody

FIGURE 3-26 Computer simulation of an interaction between antibody and influenza virus antigen. (a) The antigen (yellow) is shown interacting with the antibody molecule; the variable region of the heavy chain is red, and the variable region of the light chain is blue. (b) The complementarity of the two molecules is revealed by separating the antigen from the antibody by 8 Å. *[Based on x-ray crystallography data collected by P. M. Colman and W. R. Tulip. From G.J.V.H. Nossal, 1993. Scientific American 269(3):22.]*

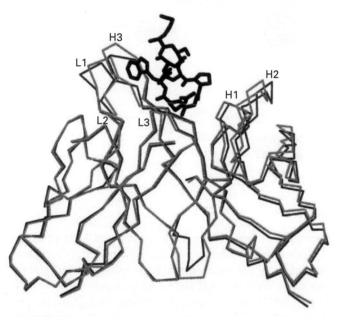

FIGURE 3-27 Conformational change can occur on binding of antigen to antibody. This figure shows a complex between a peptide derived from HIV protease and an Fab fragment from an anti-protease antibody. The peptide is shown in black. The red line shows the structure of the Fab fragment before it binds the peptide, and the blue line shows its structure after binding. There are significant conformational changes in the CDRs of the Fab on binding to the antigen. These are especially pronounced in the light chain CDR1 (L1) and the heavy chain CDR3 (H3). *[From J. Lescar et al., 1997, Journal of Molecular Biology 267:1207; courtesy of G. Bentley, Institute Pasteur.]*

Signal Transduction in B Cells

Having described the structure of antibody molecules and their membrane-bound form, the B-cell receptor (BCR), we now turn our attention to BCR function. Recall that the structure of the BCR is identical to that of the antibodies that the cell will secrete on antigen stimulation, with the exception of the C-terminal portion of the heavy chain; this portion is modified so as to anchor the receptor into the B-cell plasma membrane. Prior to antigen recognition, mature B cells residing in the secondary lymphoid tissues, such as the spleen or lymph nodes, express membrane-bound forms of both IgM and IgD.

As described above, the cytoplasmic tail of the BCR heavy chain is extremely short—only three amino acids—and so one of the puzzles that had to be solved by those exploring BCR-mediated antigen activation was how such a short cytoplasmic tail could efficiently pass a signal into the cytoplasm. This problem was solved when co-immunoprecipitation experiments revealed that each BCR molecule was noncovalently associated with a heterodimer, Igα/Igβ (see Figure 3-7a), that is responsible for transducing the antigen signal into the interior of the cell. Recall that Igα/Igβ chains contain ITAMs, which include tyrosine residues that become phosphorylated on activation through the receptor, and serve as docking residues for downstream signaling components. The BCR is therefore structurally and functionally divided into two components: a recognition component (the immunoglobulin receptor) and a signal transduction component (Igα/Igβ).

In previous sections of this chapter, we described some general principles of signal transduction and outlined three commonly employed signal transduction pathways (see Overview Figure 3-5). In this section, we show how those principles apply to the intracellular events that follow upon antigen-receptor interactions in B cells.

Antigen Binding Results in Docking of Adapter Molecules and Enzymes into the BCR-Igα/Igβ Membrane Complex

Antigen binding induces conformational alterations in the BCR, which expose regions in the Cμ4 domains of the

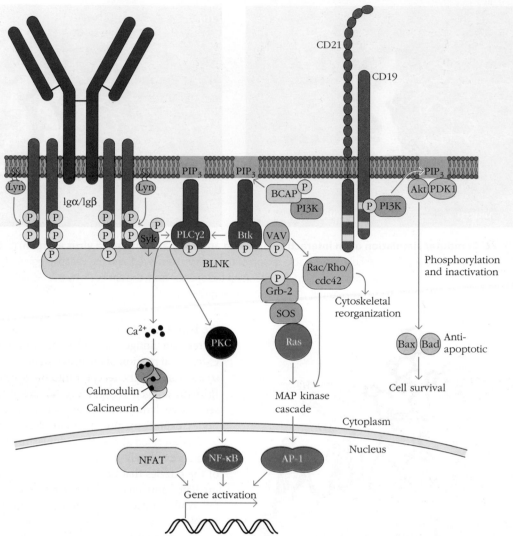

FIGURE 3-28 Signal transduction pathways emanating from the BCR. Antigen-mediated receptor clustering into the lipid raft regions of the membrane leads to src-family kinase phosphorylation of the co-receptors Igα/Igβ and CD19, the adapter proteins BLNK and BCAP, and the tyrosine kinase, Syk. BCAP and CD19 recruit PI3 kinase to the membrane with generation of PIP_3 and subsequent localization of PDK1 and Akt to the membrane. Phosphorylation by Akt enhances cell survival and leads to activation of the transcription factors NF-κB and CREB as described. Activation of the B cell isoform of PLC, PLCγ2 occurs on binding to the membrane-localized adapter, BLNK and phosphorylation by Syk, resulting in the generation of DAG and IP_3, with activation of the NFAT and NF-κB pathways as described in the text. MAP kinase pathways are activated through the binding of Grb2 to BLNK, with subsequent activation of Ras and through the activation of the GEF protein VAV which activates Rac. *[Adapted from M. E. Conley et al. 2009.* Annual Reviews of Immunology *27:199–227.]*

receptor heavy chains. Interactions between neighboring receptor molecules through these domains seed receptor oligomerization (formation of small clusters of antigen-receptor complexes) and subsequent movement of these receptor clusters into specialized lipid raft regions of the B-cell membrane. There, the ITAM residues of Igα/Igβ are brought into close proximity with the Src-family kinases Lyn (Figure 3-28), Fyn, and Blk. Tyrosine phosphorylation of the Igα/Igβ ITAM residues by these Src-family kinases, particularly Lyn, then provides attachment sites for the adapter protein BLNK and an additional tyrosine kinase, Syk, which is phosphorylated and activated by the Src-family kinases. Syk then phosphorylates BLNK, providing docking sites for multiple downstream components of the signaling pathway. The adapter protein BCAP and CD19, the B-cell co-receptor, are also phosphorylated by these tyrosine kinases and serve to recruit the enzyme PI3 kinase to the plasma membrane.

B Cells Use Many of the Downstream Signaling Pathways Described Above

The downstream signaling pathways used by the BCR will now be familiar. The tyrosine kinases Syk and Btk together phosphorylate and activate PLCγ2, which hydrolyzes PIP_2, as described above. The resultant increase in intracytoplasmic Ca^{2+} concentrations induces the activation of calcineurin and the movement of NFAT into the nucleus (see Figure 3-14). The other product of PIP_2 hydrolysis, DAG (see Figure 3-12), remains in the membrane and binds the B-cell isoform of protein kinase C, leading to the phosphorylation and release of the NF-κB inhibitor as described above for T cells (see Figure 3-17). This results in the nuclear localization and activation of NFκB.

Additional downstream effector pathways elicit the many other changes that take place upon B-cell activation. For example, PI3 kinase, now localized at the membrane, phosphorylates PIP_2 to PIP_3 (see Figure 3-11), allowing the recruitment of the PH domain–containing proteins PDK1 and Akt. On phosphorylation by the serine-threonine kinase PDK1, Akt promotes cell survival by phosphorylating and inactivating pro-apoptotic molecules such as Bax and Bad. It also phosphorylates and further activates the transcription factors NF-κB and CREB, which both support proliferation, differentiation, and survival functions of the activated B cells.

The MAP kinase pathway (see Figure 3-16) is also activated during B-cell activation. Grb2 attachment to BLNK brings about the binding of the GEF SOS, followed by the binding and activation of Ras, as described above. Similarly, Rac, another small monomeric G protein of the Ras family, is activated by binding to the GEF protein Vav. Ras and Rac are both small G proteins that act to initiate MAP kinase signaling pathways. As described above, activation of the MAP kinase pathway culminates in the expression of the phosphorylated transcription factor Elk. In B cells, Elk promotes the synthesis of the transcription factor Egr-1, which acts to induce alterations in the cell-surface expression of important adhesion molecules, and ultimately serves to aid B lymphocyte migration into and within the secondary lymphoid tissues. Downstream effectors from Rac promote actin polymerization, further facilitating B-cell motility.

In conclusion, antigen binding at the BCR leads to multiple changes in transcriptional activity, as well as in the localization and motility of B cells, which together result in their enhanced survival, proliferation, differentiation, and eventual antibody secretion. As exemplified in Box 3-2, defects in proteins involved in B-cell signaling can lead to immunodeficiencies.

CLINICAL FOCUS Box 3-2

 ## Defects in the B-Cell Signaling Protein Btk Lead to X-Linked Agammaglobulinemia

The characterization of the proteins necessary for B- and T-cell signaling opened up new avenues of exploration for clinicians working with patients suffering from immunodeficiency disorders. Clinicians and immunogeneticists now work closely together to diagnose and treat patients with immunodeficiencies, to the benefit of both the clinical and the basic sciences.

Characterization of the genes responsible for disorders of the immune system is complicated by the fact that antibody deficiencies may result from defective genes encoding either T- or B-cell proteins (since T cells provide helper factors necessary for B-cell antibody production), or even from mutations in genes encoding proteins in stromal cells important for healthy B-cell development in the bone marrow. However, no matter what the cause, all antibody deficiencies manifest clinically in increased susceptibility to bacterial infections, particularly those of the lung, intestines, and (in younger children) the ear.

In 1952, a pediatrician, Ogden Bruton, reported in the journal *Pediatrics* the case of an eight-year-old boy who suffered from multiple episodes of pneumonia. When the serum of the boy was subjected to electrophoresis, it was shown to be completely lacking in serum globulins, and his disease was therefore named agammaglobulinemia. This was the first immunodeficiency disease for which a laboratory finding explained the clinical symptoms, and the treatment that Bruton applied—administering subcutaneous injections of gamma globulin—is still used today. As similar cases were subsequently reported, it was noted that most of the pediatric cases of agammaglobulinemia occurred in boys, whereas when the disease was reported in adults, both men and women appeared to be similarly affected. Careful mapping of the disease susceptibility to the X chromosome resulted in the pediatric form of the disease being named XLA, for *X-linked* agammaglobulinemia.

With the characterization of the BCR signal transduction pathway components

in the 1980s and 1990s came the opportunity to define which proteins are damaged or lacking in particular immunodeficiency syndromes. In 1993, 41 years after the initial description of the disease, two groups independently reported that many cases of XLA resulted from mutations in a cytoplasmic tyrosine kinase called Bruton's tyrosine kinase, or Btk; at this point, we now know that fully 85% of patients affected with XLA have mutations in the *Btk* gene.

Btk is a member of the Tec family of cytoplasmic tyrosine kinases, which are predominantly expressed in hematopoietic cells. Tec family kinases share a C-terminal kinase domain, preceded by SH2 and SH3 domains, a proline-rich domain, and an amino-terminal PH domain, capable of binding to PIP_3 phospholipids generated by PI3 kinase activity. Btk is expressed in both B cells and platelets and is activated following signaling through the BCR, the pre-BCR (which is expressed in developing B cells), the IL-5 and IL-6 receptors, and also the CXCR4 chemokine receptor. Its involvement in pre-BCR signaling explains why

(continued)

children with XLA suffer from defective B-cell development.

Following activation, Btk moves to the inner side of the plasma membrane, where it is phosphorylated and partially activated (see Figure 3-28). Activation is completed when it autophosphorylates itself at a second phosphorylation site. Btk binds to the adapter protein BLNK, along with PLCγ2. Btk then phosphorylates and activates PLCγ2, leading, as described, to calcium flux and activation of the NF-κB and NFAT pathways. Btk therefore occupies a central position in B-cell activation, and it is no surprise that mutations in its gene result in such devastating consequences.

Over 600 different mutations have been identified in the *btk* gene, with the vast majority of these resulting from single base pair substitutions, or the insertion or deletion of less than five base pairs. As for other X-linked mutations that are lethal without medical intervention, XLA disease is maintained in the population by the generation of new mutations.

Patients with XLA are usually healthy in the neonatal (immediately after birth) period, when they still benefit from maternal antibodies. However, recurrent bacterial infections begin between the ages of 3 months and 18 months, and currently the mean age at diagnosis in North America is 3 years. XLA is a so-called "leaky" defect; almost all children with mutations in *btk* have some serum immunoglobulin, and a few B cells in the peripheral circulation. The prognosis for patients who are treated with regular doses of gamma globulin has improved dramatically over the last 25 years, with the use of prophylactic injections of gamma globulin.

The B cells in patients with XLA have a distinctive phenotype that can be used for diagnostic purposes. CD19 expression is low and variable in patients with XLA, whereas membrane IgM expression, normally variable in mature B cells, is relatively high and consistent in patients with XLA. This phenotype can be observed in Figure 1, which shows the flow cytometric profiles of a normal control individual (left two graphs) and a patient with a defective *btk* gene (right two graphs). In the top two plots, we note that the XLA patient has very few CD19[+] B cells compared with the control and that the levels of CD19 on the surface of those B cells that do exist are lower than those of the control cells. In the lower two plots, we note that, although there are fewer CD19[+] B cells overall in the *btk*-compromised patient, all of those CD19[+] B cells have relatively high levels of surface IgM, whereas the levels of membrane IgM are much more variable in the healthy controls.

Bruton, O. C. 1952. Agammaglobulinemia. *Pediatrics* **9**:722–728.

Conley, M. E., et al. 2009. Primary B cell immunodeficiencies: Comparisons and contrasts. *Annual Review of Immunology* **27**:199–227.

Tsukuda, S., et al. 1993. Deficient expression of a B cell cytoplasmic tyrosine kinase in human X-linked agammaglobulinemia. *Cell* **72**:279–290

Vetrie, D., et al. 1993. The gene involved in X-linked agammaglobulinemia is a member of the *src* family of protein tyrosine kinases. *Nature* **361**:226–233.

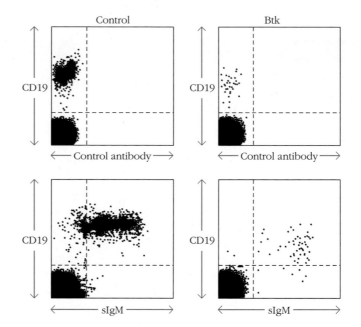

FIGURE 1
FACS profiles of a normal individual and an XLA patient. *[Adapted from Conley et al. 2009. Annual Review of Immunology 27:199.]*

B Cells Also Receive Signals Through Co-Receptors

The immunoglobulin receptor on the B cell membrane is noncovalently associated with three transmembrane molecules: CD19, CD21, and CD81 (TAPA-1) (see Figure 3-7a). Antigens are sometimes presented to the BCR already covalently bound to complement proteins, in particular to the complement component C3d. (The complement cascade is discussed in Chapter 6.) The B-cell co-receptor CD21 specifically binds to C3d, on C3d-coated antigens. This co-engagement of the BCR and CD21 brings the co-receptor and the BCR into close apposition with one another. When this happens, tyrosine residues on the cytoplasmic face of the co-receptor become phosphorylated by the same enzymes that phosphorylate the ITAMs on Igα/Igβ, providing sites of attachment for PI3 kinase. As illustrated in Figure 3-28, localization of PI3 kinase to the co-receptor enhances both cell survival and the alterations in the transcriptional program that accompany cell activation. CD19 also serves as an additional site of recruitment of PLCγ.

T-Cell Receptors and Signaling

T cells bind complex antigens made up of peptides located in the groove of membrane-bound MHC proteins. When the T-cell receptor makes contact with its MHC-peptide antigen on the surface of an antigen-presenting cell, the two cell membranes are brought into close apposition with one another. This adds an additional layer of complexity to the process of T-cell activation that finds no parallel in B-cell signaling. Notwithstanding this additional complexity, the events of T-cell activation still unfold according to a mix of the strategies described above, and bear many similarities to B-cell receptor signaling. Here, we briefly describe T-cell receptor structure and then turn to a characterization of the signaling routes through this receptor. In Box 3-3, we provide a description of the experiments that resulted in the isolation and characterization of the αβTCR.

The T-Cell Receptor is a Heterodimer with Variable and Constant Regions

There are two types of T-cell receptors, both of which are heterodimers (dimers made up of two different polypeptides). The majority of recirculating T cells bear αβ heterodimers, which bind to ligands made up of an antigenic peptide presented in a molecular groove on the surface of a type I or type II MHC molecule. A second subset of T cells instead expresses a heterodimeric T-cell receptor composed of a different pair of protein chains, termed γ and δ. T cells bearing γδ receptors have particular localization patterns (often in mucosal tissues) and some γδ T cells recognize different types of antigens from those bound by αβ T cells. Although some γδ T cells recognize conventional MHC-presented peptide antigens, other γδ T cells bind lipid or glycolipid moieties presented by noncanonical MHC molecules. Yet other γδ T-cell clones appear to recognize self-generated heat shock proteins or phosphoantigens derived from microbes. A unified theory of the precise nature of antigens recognized by γδ T cells remains elusive. However, this ability of γδ T cells to break the rules of MHC restriction may account for the evolution of a slight difference in the angle between the antigen-binding and constant regions of the T-cell receptor, which is apparent in an x-ray crystallographic analysis of the two types of receptor (Figure 3-29). Notwithstanding these functional differences in the αβ versus the γδ receptors, their essential biochemistry is quite similar.

For the remainder of this chapter, we focus on the structure and signaling of the dominant αβ TCR type, recognizing

that minor differences may exist between the two types of receptors and their downstream components.

Although the TCR is not an immunoglobulin per se, the TCR proteins are members of the immunoglobulin superfamily of proteins and therefore the domain structures of αβ and γδ TCR heterodimers are strikingly similar to those of the immunoglobulins (see Figure 3-19). The α chain has a molecular weight of 40–50 kDa, and the β chain's is 40–45 kDa. Like the antibody light chains, the TCR chains have two immunoglobulin-like domains, each of which contains an intrachain disulfide bond spanning 60 to 75 amino acids. The Cα domain of the TCR differs from most immunoglobulin domains in that it possesses only a single β sheet, rather than a pair, and the remainder of the sequence is more variably folded. The amino-terminal (variable) domain in both chains exhibits marked sequence variation, but the sequences of the remainder of each chain are conserved (constant). Each of the TCR variable domains has three hypervariable regions, which appear to be equivalent to the complementarity-determining regions (CDRs) in immunoglobulin light and heavy chains. A fourth hypervariable region on the TCRβ chain does not appear to contact antigen, and its functional significance is therefore uncertain.

At the C-terminal end of the constant domain, each TCR chain contains a short connecting sequence, in which a cysteine residue forms a disulfide link with the other chain of the heterodimer. C-terminal to this disulfide is a transmembrane region of 21 or 22 amino acids, which anchors each chain in the plasma membrane. The transmembrane domains of the TCR α and β chains are unusual in that they each contain positively charged amino acid residues that promote interaction with corresponding negatively charged residues on the chains of the signal-transducing CD3 complex. Finally, like BCRs, each TCR chain contains only a very short cytoplasmic tail at the carboxyl-terminal end.

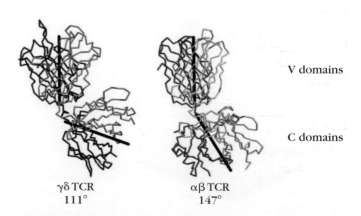

V domains

C domains

γδ TCR
111°

αβ TCR
147°

FIGURE 3-29 Comparison of the crystal structures of γδ and αβ TCRs. The difference (highlighted with black lines) in the elbow angle between the γδ and αβ forms of the TCR.

The Discovery of the αβ T-Cell Receptor

Once scientists had established that the BCR was simply a membrane-bound form of the secreted antibody, the elucidation of BCR structure became a significantly more tractable problem.

However, investigators engaged in characterizing the T-cell receptor (TCR) did not enjoy the same advantage, as the TCR is not secreted in soluble form. Understanding of TCR biochemistry

therefore lagged behind that of the BCR until the 1980s, when an important scientific breakthrough—the ability to make monoclonal antibodies from artificially constructed B-cell tumors, or

① Generation of a T cell hybridoma with known antigen specificity

Immunize mouse with ovalbumin (OVA)

Wait several days
Remove lymph nodes
Culture lymph node cell T cells with OVA

② Add polyethylene glycol to induce fusion of antigen-specific T cells with long-lived T cell line

③ Dilute out fused cells so that each well contains a single T cell hybridoma. Allow cells to divide to form clones and test each clone for its ability to secrete IL-2 when stimulated with ovalbumin peptides. Grow up individual clones of T cell hybridomas that recognize OVA.

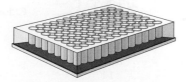

④ Production of antibodies that bind to TCR on the T cell hybridoma

Immunize a new mouse with cells from an OVA-specific T cell hybridoma

Wait several days
Isolate spleen

⑤ Fuse B cells from spleen with long-term B cell line

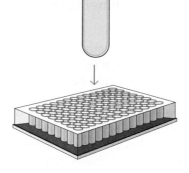

⑥ Selection and expansion of those long-term B cell clones that secrete monoclonal antibodies that bind to the T cell hybridoma

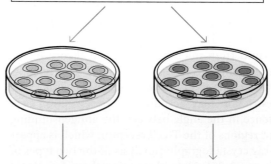

Collect monoclonal antibodies that bind to T cell lines

FIGURE 1
The generation of antibodies specific for the TCR

BOX 3-3

hybridomas—made the analysis of the TCR more technically feasible.

A hybridoma is a fusion product of two cells. B-cell hybridomas are generated by artificially fusing antibody-producing, short-lived lymphocytes with long-lived tumor cells in order to generate long-lived daughter cells secreting large amounts of monoclonal antibodies. (The term **monoclonal** refers to the fact that all of the cells in a given hybridoma culture are derived from the single clone of cells, and therefore carry the same DNA; details of the technology are described in Chapter 20.) Although this technique was first developed for the generation of long-lived B cells, scientists working in the laboratory of John Kappler and Philippa Marrack also applied it to T lymphocytes.

The researchers began by immunizing a mouse with the protein ovalbumin (OVA), allowing OVA-specific T cells to divide and differentiate for a few days, and then harvesting the lymph nodes from the immunized animal. To enrich their starting population with as many OVA-specific T cells as possible, they cultured the harvested lymph node cells in vitro with OVA for several hours (Figure 1, step 1).

After some time in culture, they fused these activated, OVA-specific T cells with cells derived from a T-cell tumor (Figure 1, step 2), thus generating a number of long-lived **T-cell hybridoma** cultures that recognized OVA peptides in the context of the MHC of the original mouse, an MHC allele called H-2^d. They then diluted out the fused cells in each culture, generating several T-cell hybridoma lines in which all the cells in an individual hybridoma line derived from the product of a single fusion event (Figure 1, step 3). This is referred to as cloning by limiting dilution. In this way, they isolated a T-cell hybridoma that expressed a TCR capable of recognizing a peptide from OVA, in the context of MHC Class 2 proteins from mice of the H-2^d strain.

These T cells could now be used as antigens and injected into a mouse (Figure 1, step 4). The spleen of this mouse was removed a few days later, and the mouse B cells were fused with B lineage tumor cells (Figure 1, step 5). After culturing to stabilize the hybrids, the investigators cloned the B-cell hybridomas and selected a B-cell hybridoma line that produced monoclonal antibodies that bound specifically to the

T-cell hybridoma (Figure 1, step 6). Most important, these antibodies interfered with the T cell's ability to recognize its cognate antigen (Figure 1, step 7). The fact that this monoclonal antibody inhibited TCR antigen binding suggested that the antibody was binding directly to the receptor, and competing with the antigen for TCR binding. They then used these antibodies to immunoprecipitate the TCR from detergent-solubilized membrane preparations and purify the TCR protein (Figure 1, step 8).

Concurrent with these experiments, the laboratories of Stephen Hedrick and Mark Davis at NIH and Tak Mak in Toronto had been making headway searching for the genes encoding the T-cell receptor. These experiments, as well as subsequent work by Susumu Tonegawa, which completed the identification of the TCR genes, are described in detail in Chapter 7.

Haskins et al. 1983. The major histocompatibility complex-restricted antigen receptor on T cells. I. Isolation with a monoclonal antibody. *The Journal of Experimental Medicine* **157**:1149–1169.

Haskins et al. 1984. The major histocompatibility complex-restricted antigen receptor on T cells. *Annual Review of Immunology* **2**:51–66.

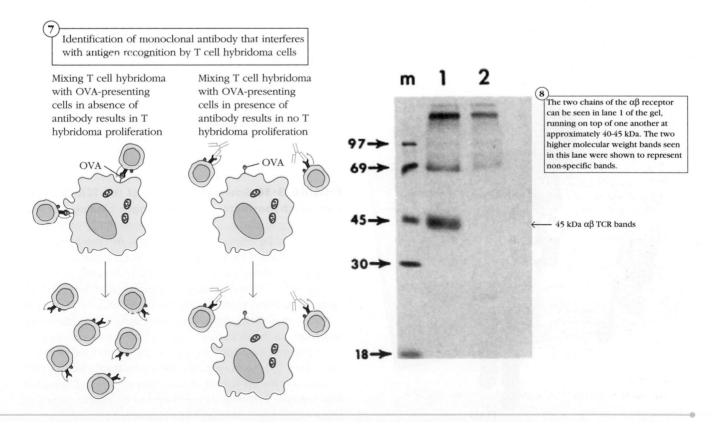

7 Identification of monoclonal antibody that interferes with antigen recognition by T cell hybridoma cells

Mixing T cell hybridoma with OVA-presenting cells in absence of antibody results in T hybridoma proliferation

Mixing T cell hybridoma with OVA-presenting cells in presence of antibody results in no T hybridoma proliferation

8 The two chains of the αβ receptor can be seen in lane 1 of the gel, running on top of one another at approximately 40-45 kDa. The two higher molecular weight bands seen in this lane were shown to represent non-specific bands.

45 kDa αβ TCR bands

The T-Cell Signal Transduction Complex Includes CD3

Just as B-cell signaling requires the participation of the Igα/Igβ signal transduction complex, signaling through the TCR depends on a complex of proteins referred to collectively as CD3 (Figure 3-30). The CD3 complex is made up of three dimers: a δε (delta epsilon) pair, a γε (gamma epsilon) pair, and a third pair that is made up either of two CD3ζ (zeta)

molecules or a ζη (zeta, eta) heterodimer. (Note that the CD3 γ and δ chains are different from the chains that make up the γδ TCR.) Like Igα and Igβ, the cytoplasmic tails of the CD3 molecules are studded with ITAM sequences that serve as docking sites for adapter proteins following activation-induced tyrosine phosphorylation. Each of the CD3 dimers contains negatively charged amino acids in its transmembrane domain that form ionic bonds with the positively charged residues on the intramembrane regions of the T-cell receptor.

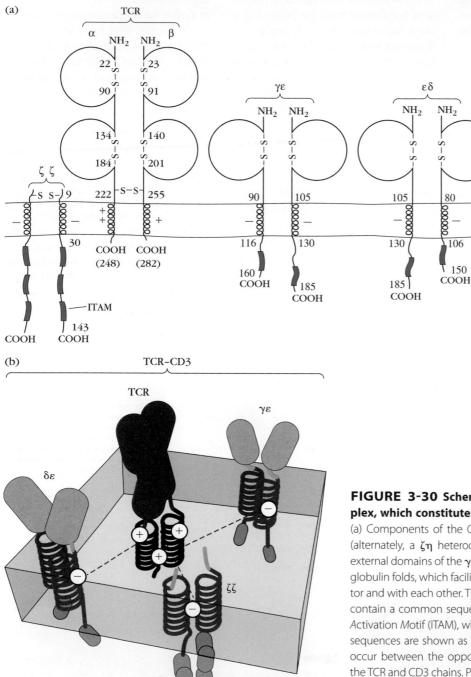

FIGURE 3-30 Schematic diagram of the TCR-CD3 complex, which constitutes the T-cell antigen-binding receptor.
(a) Components of the CD3 complex include the ζζ homodimers (alternately, a ζη heterodimer) plus γε and δε heterodimers. The external domains of the γ, δ, and ε chains of CD3 consist of immunoglobulin folds, which facilitates their interaction with the T-cell receptor and with each other. The long cytoplasmic tails of the CD3 chains contain a common sequence, the *Immunoreceptor Tyrosine-based Activation Motif* (ITAM), which functions in signal transduction; these sequences are shown as blue boxes. (b) Ionic interactions also may occur between the oppositely charged transmembrane regions in the TCR and CD3 chains. Proposed interactions among the CD3 components and the αβ TCR are shown. *[Adapted from M. E. Call and K. W. Wucherpfennig, 2004. Molecular mechanisms for the assembly of the T cell receptor-CD3 complex. Molecular Immunology 40:1295.]*

TABLE 3-4 Selected T cell accessory molecules participating in T-cell signal transduction

| Name | Ligand | FUNCTION | | |
		Adhesion	Signal transduction	Member of Ig superfamily
CD4	Class II MHC	+	+	+
CD8	Class I MHC	+	+	+
CD2 (LFA-2)	CD58 (LFA-3)	+	+	+
CD28	CD80, CD86	?	+	+
CTLA-4	CD80, CD86	?	+	−
CD45R	CD22	+	+	+
CD5	CD72	?	+	−

The T Cell Co-receptors CD4 and CD8 Also Bind the MHC

The T-cell receptor is noncovalently associated with a number of accessory molecules on the cell surface (Table 3-4). However, the only two such molecules that also recognize the MHC-peptide antigen are **CD4** and **CD8**. Recall that mature T cells can be subdivided into two populations according to their expression of CD4 or CD8 on the plasma membrane. $CD4^+$ T cells recognize peptides that are combined with class II MHC molecules, and function primarily as helper or regulatory T cells, whereas $CD8^+$ T cells recognize antigen that is expressed on the surface of class I MHC molecules, and function mainly as cytotoxic T cells.

CD4 is a 55 kDa monomeric membrane glycoprotein that contains four extracellular immunoglobulin-like domains (D_1–D_4), a hydrophobic transmembrane region, and a long cytoplasmic tail containing three serine residues that can be phosphorylated (Figure 3-31). CD8 takes the form of a disulfide-linked $\alpha\beta$ heterodimer or $\alpha\alpha$ homodimer. (These are not the same as the α and β chains that constitute the TCR heterodimer.) Both the α and β chains of CD8 are small glycoproteins of approximately 30 to 38 kDa. Each chain consists of a single, extracellular, immunoglobulin-like domain, a stalk region, a hydrophobic transmembrane region, and a cytoplasmic tail containing 25 to 27 residues, several of which can be phosphorylated.

The extracellular domains of CD4 and CD8 bind to conserved regions of MHC class II and MHC class I molecules respectively (see Figure 3-7b). The co-engagement of a single MHC molecule by both the TCR and its CD4 or CD8 co-receptor enhances the avidity of T-cell binding to its target. This co-engagement also brings the cytoplasmic domains of the TCR/CD3 and the respective co-receptor into close proximity, and it helps to initiate the cascade of intracellular events that activate a T cell.

Signaling through the antigen receptor, even when combined with that through CD4 or CD8, is insufficient to activate a T cell that has had no prior contact with antigen (a naïve T cell). A naïve T cell needs to be simultaneously signaled through the TCR and its co-receptor, CD28, in order to be activated. The TCR and CD28 molecules on a naïve T cell must co-engage the MHC-presented peptide and the CD28 ligand, CD80 (or CD86), respectively, on the antigen-presenting cell for full activation to occur. The signaling events mediated through CD28, which include the stimulation of interleukin 2 synthesis by the T cell, were alluded to above and are discussed fully in Chapter 11.

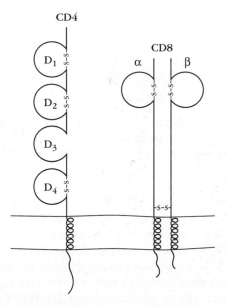

FIGURE 3-31 General structure of the CD4 and CD8 co-receptors; the Ig-like domains are shown as circles. CD8 takes the form of an $\alpha\beta$ heterodimer or an $\alpha\alpha$ homodimer. The monomeric CD4 molecule contains four Ig-fold domains; each chain in the CD8 molecule contains one.

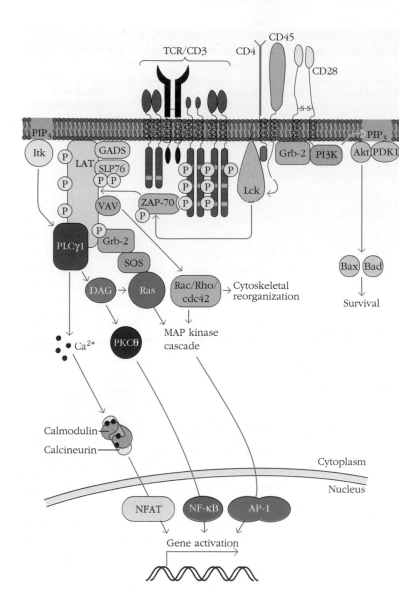

FIGURE 3-32 Signal transduction pathways emanating from the TCR. T cell antigen binding activates the src-family kinase Lck, which phosphorylates the kinase ZAP-70. ZAP-70 in turn phosphorylates the adapter molecules LAT, SLP76, and GADS which form a scaffold enabling the phosphorylation and activation of PLCγ1 and PKCθ with the consequent effects on transcription factor activation described in the text. The GEF proteins Vav and SOS are also activated on binding to LAT, leading to activation of the Ras/MAP kinase transcription factor pathway and the Rac/Rho/cdc42 pathway, leading to changes in cell shape and motility. PI3 kinase, translocated to the cytoplasmic side of CD28, forms PIP₃, inducing localization of the enzymes PDK1 and Akt to the membrane. This leads to further NF-κB activation and increased cell survival as described.

Lck is the First Tyrosine Kinase Activated in T Cell Signaling

When the T-cell receptor interacts with its cell-bound antigen, receptors, co-receptors, and signaling molecules cluster into the cholesterol-rich lipid rafts of the plasma membrane (Figure 3-32). The Src-family tyrosine kinase Lck is normally found associated with CD4 and CD8, and the association between Lck and CD4 is particularly close. Antigen-induced clustering of the receptor–co-receptor complex brings Lck into the vicinity of the membrane-associated tyrosine phosphatase, CD45, which removes the inhibitory phosphate group on Lck. Reciprocal phosphorylation by nearby Lck molecules at their activating tyrosine sites (see Figure 3-9) then induces Lck to phosphorylate CD3 ITAM residues. (Note that these early events parallel those induced in B cells by the Src-family kinase Lyn, which is also regulated by the phosphatase activity of CD45.)

Once the CD3 ITAMs are phosphorylated, a second tyrosine kinase, ZAP-70, docks at the phosphorylated tyrosine residues of the CD3ζ chains. ZAP-70 is activated by Lck-mediated phosphorylation and goes on to phosphorylate many adapter molecules including SLP-76 and LAT, as well as enzymes important in T-cell activation, such as PLCγ1.

T Cells Use Downstream Signaling Strategies Similar to Those of B Cells

Just as in B cells, signals initiating at the antigen receptor of T cells with tyrosine phosphorylation events are then fanned out to intracellular enzymes and transcription factors using a network of adapter molecules and enzymes.

In T cells, one of the earliest adapter molecules to be incorporated into the signaling complex is LAT (*Linker protein of Activated T cells*), a transmembrane protein associated with lipid rafts in the plasma membrane. Following TCR ligation, LAT is phosphorylated on multiple residues by ZAP-70, and these phosphorylated residues now provide docking sites for several important enzymes bearing SH2 domains, including PLCγ1 (see Figure 3-32). Phosphorylated LAT also binds to the adapter protein GADS, which is constitutively associated with the adapter SLP-76. This combination of adapter proteins is critical to T-cell receptor signaling, providing the structural framework for most downstream signaling events.

Many of those downstream events will now be familiar. PLCγ1, localized to the plasma membrane by binding to LAT, is further activated by tyrosine phosphorylation, mediated by the kinase Itk (which belongs to a family of kinases referred to as Tec kinases). As described earlier, PLCγ1 breaks down PIP$_2$, releasing IP$_3$, which induces the release of calcium and the activation of NFAT via calcineurin activation. The DAG created by PIP$_2$ hydrolysis binds, in T cells, to a specialized form of PKC called PKCθ (theta). As described above, this part of the signaling cascade similarly culminates in the degradation of the inhibitors of NF-κB and the translocation of the active transcription factor into the nucleus (see Figure 3-17).

Phosphorylated LAT also associates with the SH2 domain of Grb2, the now-familiar adapter molecule that brings in components of the Ras pathway to the signaling complex. Recall that Grb2 binds constitutively to SOS, the GEF that facilitates activation of the Ras pathway. In T cells, the Ras pathway is important both to the activation of the transcription factor AP-1, which functions to signal cytokine secretion, and to the passage of the signals that reorganize the actin cytoskeleton for directed cytokine release.

Thus, as for B cells, TCR-antigen binding leads to a multitude of consequences including transcription factor upregulation, reorganization of the cytoskeleton, and cytokine secretion. Again, as for B cells, T-cell signaling also affects the expression of adhesion molecules such as integrins on the cell surface, and chemokines, which has subsequent effects on cell localization.

Clearly the description of adaptive immune signaling offered in this chapter represents just the tip of the iceberg, and these signaling cascades contain more components and outcomes that can be alluded to in this brief outline. However, just as the adapter proteins provide a scaffold for the immune system to organize its signaling proteins, so we hope that this chapter has provided a similar scaffold for the organization of the reader's thoughts regarding these fascinating and complex processes.

SUMMARY

- Antigens bind to receptors via noncovalent bonding interactions.

- The interactions between antigens and receptors of the immune system are enhanced by simultaneous interactions between lymphocyte-expressed co-receptors and molecules on antigen-presenting cells or on complex antigens.

- Most receptor-antigen interactions are multivalent, and this multivalency significantly increases the avidity of the receptor-antigen binding interaction.

- Binding of antigen to receptor induces a signaling cascade in the receptor-bearing cell, which leads to alterations in the motility, adhesive properties, and transcriptional program of the activated cell.

- Antigen signaling is initiated in both T and B cells by antigen-mediated receptor clustering. The clustered receptors are located in specialized regions of the membranes called lipid rafts.

- The CD3 and Igα/Igβ proteins, which are T- and B-cell receptor-associated signal transduction elements, are phosphorylated on *Immunoreceptor Tyrosine Activation Motifs* (ITAMs), and these serve as docking points for adapter molecules.

- Downstream signaling enzymes and GEFs dock onto the adapter molecules and make contact with their substrates. These enzymes include phospholipase Cγ, which breaks down PIP$_2$ into DAG and IP$_3$. IP$_3$ interaction with ER-located receptors leads to release of intracellular calcium and activation of calcium-regulated proteins such as calcineurin phosphatase. Calcineurin dephosphorylates the transcription factor NFAT, allowing it access to the nucleus. DAG binds and activates protein kinase C, leading eventually to NF-κB activation. Docking of the adapter molecule Grb2 onto adapter proteins facilitates its binding to the GEF protein SOS and activation of the MAP kinase pathway, resulting in activation of the transcription factor AP-1.

- The antigen receptor on B cells is a membrane-bound form of the four-chain immunoglobulin molecule that the B cell secretes upon stimulation. An immunoglobulin molecule is commonly known as an antibody. Antibodies have two heavy and two light chains.

- The antibodies secreted by B cells upon stimulation are classified according to the amino acid sequence of the heavy chain, and antibodies of different classes perform different functions during an immune response.

- Antigen signaling in B cells proceeds according to signaling strategies shared among many cell types.

- The antigen receptor on T cells is not an immunoglobulin molecule, although its protein domains are classified as belonging to the immunoglobulin superfamily of proteins.

- Most T cells bear receptors made up of an αβ heterodimer that recognizes a complex antigen, made up of a short peptide inserted into a groove on the surface of a protein encoded by the major histocompatibility complex (MHC) of genes.

- Some T cells bear receptors made up of γδ receptor chains that recognize a different array of antigens.

- Antigen signaling in T cells shares many characteristic strategies with B cell signaling.

REFERENCES

Andreotti, A. H., P. L. Schwartzberg, R. E. Joseph, and L. J. Berg. 2010. T-cell signaling regulated by the Tec family kinase, Itk. *Cold Spring Harbor Perspectives in Biology* **2**:a002287.

Chaturvedi, A., Z. Siddiqui, F. Bayiroglu, and K. V. Rao. 2002. A GPI-linked isoform of the IgD receptor regulates resting B cell activation. *Nature Immunology* **3**:951.

Choudhuri, K., and M. L. Dustin. 2010. Signaling microdomains in T cells. *FEBS Letters* **584**:4823.

Edelman G. M., B. A. Cunningham, W. E. Gall, P. D. Gottlieb, U. Rutishauser, and M. J. Waxdal. 1969. The covalent structure of an entire gammaG immunoglobulin molecule. *Proceedings of the National Academy of Sciences U.S.A.* **63**:78.

Finlay, D., and D. Cantrell. The coordination of T-cell function by serine/threonine kinases. 2011. *Cold Spring Harbor Perspectives in Biology* **3**:a002261.

Fitzgerald, K. A., and Z. J. Chen. 2006. Sorting out Toll signals. *Cell* **125**:834.

Heidelberger, M., and F. E. Kendall. 1936. Quantitative studies on antibody purification : I. The dissociation of precipitates formed by Pneumococcus specific polysaccharides and homologous antibodies. *Journal of Experimental Medicine* **64**:161.

Heidelberger, M., and K. O. Pedersen. 1937. The Molecular Weight of Antibodies. *Journal of Experimental Medicine* **65**:393.

Kanneganti, T. D., M. Lamkanfi, and G. Nunez. 2007. Intracellular NOD-like receptors in host defense and disease. *Immunity* **27**:549.

Koretzky, G. A., and P. S. Myung. 2001. Positive and negative regulation of T-cell activation by adaptor proteins. *Nature Reviews Immunology* **1**:95.

Kurosaki, T. 2011. Regulation of BCR signaling. *Molecular Immunology* **48**:1287.

Love, P. E., and S. M. Hayes. 2010. ITAM-mediated signaling by the T-cell antigen receptor. *Cold Spring Harbor Perspectives in Biology* **2**:a002485.

Macian, F. 2005. NFAT proteins: Key regulators of T-cell development and function. *Nature Reviews Immunology* **5**:472.

Nisonoff, A., F. C. Wissler, and L. N. Lipman. 1960. Properties of the major component of a peptic digest of rabbit antibody. *Science* **132**:1770.

Palacios, E. H., and A. Weiss. 2004. Function of the Src-family kinases, Lck and Fyn, in T-cell development and activation. *Oncogene* **23**:7990.

Park, S. G., J. Schulze-Luehrman, M. S. Hayden, N. Hashimoto, W. Ogawa, M. Kasuga, and S. Ghos. 2009. The kinase PDK1 integrates T cell antigen receptor and CD28 coreceptor signaling to induce NF-kappaB and activate T cells. *Nature Immunology* **10**:158.

Porter, R. R. (1972). Lecture for the Nobel Prize for physiology or medicine 1972: Structural studies of immunoglobulins. *Scandinavian Journal of Immunology* **34**(4):381–389.

Tiselius, A. 1937. Electrophoresis of serum globulin. I. *The Biochemical Journal* **31**:313.

Tiselius, A., and E. A. Kabat. 1938. Electrophoresis of immune serum. *Science* **87**:416.

Tiselius, A., and E. A. Kabat. 1939. An electrophoretic study of immune sera and purified antibody preparations. *Journal of Experimental Medicine* **69**:119.

Wang, H., T. A. Kadlecek, B. B. Au-Yeung, H. E. Goodfellow, L. Y. Hsu, T. S. Freedman, and A. Weiss. 2010. ZAP-70: An essential kinase in T-cell signaling. *Cold Spring Harbor Perspectives in Biology* **2**:a002279.

Yamamoto, M., and K. Takeda. 2010. Current views of toll-like receptor signaling pathways. *Gastroenterology Research and Practice* **2010**:240365.

Useful Web Sites

www.genego.com This site offers a searchable database of metabolic and regulatory pathways.

www.nature.com/subject/cellsignaling This is a collection of original research articles, reviews and commentaries pertaining to cell signaling.

http://stke.sciencemag.org This is the Signal Transduction Knowledge Environment Web site, maintained by *Science* magazine. It is an excellent site, but parts of it are closed to those lacking a subscription.

www.signaling-gateway.org This gateway is powered by University of California at San Diego and is supported by Genentech and Nature. An excellent resource, complete with featured articles.

www.qiagen.com/geneglobe/pathwayview. aspx Qiagen. A useful commercial Web site.

www.biosignaling.com Part of Springer Science+Business Media.

www.youtube.com Many excellent videos and animations of signaling Web sites are available on YouTube, too numerous to mention here. Just type your pathways into a Web browser and go. But do be cognizant of the derivation of your video. Not all videos are accurate, so check your facts with the published literature.

www.hhmi.org/biointeractive/immunology/tcell. html Howard Hughes Medical Institute (HHMI) movie: *Cloning an Army of T Cells for Immune Defense.*

1. The NFAT family is a ubiquitous family of transcription factors.

   a. Under resting conditions, where is NFAT localized in a cell?
   b. Under activated conditions, where is NFAT localized in a cell?
   c. How is it released from its resting condition and permitted to relocalize?
   d. Immunosuppressant drugs such as cyclosporin act via inhibition of the calcineurin phosphatase. If NFAT is ubiquitous, how do you think these drugs might act with so few side effects on other signaling processes within the body?

2. In the early days of experiments designed to detect the T-cell receptor, several different research groups found that antibodies directed against immunoglobulin proteins appeared to bind to the T-cell receptor. Given what you know about the structure of immunoglobulins and the T-cell receptor, why is this not completely surprising?

3. True or false? Explain your answers.
   Interactions between receptors and ligands at the cell surface:

   a. are mediated by covalent interactions.
   b. can result in the creation of new covalent interactions within the cell.

4. Describe how the following experimental manipulations were used to determine antibody structure.

   a. Reduction and alkylation of the antibody molecule
   b. Enzymatic digestion of the antibody molecule
   c. Antibody detection of immunoglobulin fragments

5. What is an ITAM, and what proteins modify the ITAMs in Igα and Igβ?

6. Define an adapter protein. Describe how an interaction between proteins bearing SH2 and phosphorylated tyrosine (pY) groups helps to transduce a signal from the T-cell receptor to the ZAP-70 protein kinase.

7. IgM has ten antigen-binding sites per molecule, whereas IgG only has two. Would you expect IgM to be able to bind five times as many antigenic sites on a multivalent antigen as IgG? Why/why not?

8. You and another student are studying a cytokine receptor on a B cell that has a Kd of 10^{-6} M. You know that the cytokine receptor sites on the cell surface must be at least 50% occupied for the B cell to receive a cytokine signal from a helper T cell. Your lab partner measures the cytokine concentration in the blood of the experimental animal and detects a concentration of 10^{-7} M. She tells you that the effect you have been measuring could not possibly result from the cytokine you're studying. You disagree. Why?

9. Activation of Src-family kinases is the first step in several different types of signaling pathways. It therefore makes biological sense that the activity of this family of tyrosine kinases is regulated extremely tightly. Describe how phosphorylation of Src-family kinases can deliver both activating and inhibitory signals to Src kinases.

10. You have generated a T-cell clone in which the Src-family tyrosine kinase Lck is inactive. You stimulate that clone with its cognate antigenic peptide, presented on the appropriate MHC platform and test for interleukin 2 secretion, as a measure of T cell activation. Do you expect to see IL-2 secretion or not? Explain.

11. Name one protein shown to be defective in many cases of X-linked agammaglobulinemia, and describe how a reduction in the activity of this protein could lead to immunodeficiency.

12. The B- and T-cell receptor proteins have remarkably short intracytoplasmic regions of just a few amino acids. How can you reconcile this structural feature with the need to signal the presence of bound antigen to the interior of the cell?

13. Describe one way in which the structure of antibodies is superbly adapted to their function.

14. As a graduate student, your adviser has handed you a T-cell clone that appears to be constitutively (always) activated, although at a low level, even in the absence of antigenic stimulation, and he has asked you to figure out why. Your benchmate suggests you start by checking out the sequence of its *lck* gene, or the status of the Csk activity in the cell. You agree that those are good ideas. What is your reasoning?

Receptors and Signaling: Cytokines and Chemokines

The hundreds of millions of cells that comprise the vertebrate immune system are distributed throughout the body of the host (see Chapter 2). Some cells circulate through the blood and lymph systems, whereas others are sessile (remain in place) in the primary and secondary lymphoid tissues, the skin, and the mucosa of the respiratory, alimentary, and genito-urinary tracts. The key to success for such a widely dispersed organ system is the ability of its various components to communicate quickly and efficiently with one another, so that the right cells can home to the appropriate locations and take the necessary measures to destroy invading pathogens.

Molecules that communicate among cells of the immune system are referred to as **cytokines**. In general, cytokines are soluble molecules, although some also exist in membrane-bound forms. The interaction of a cytokine with its receptor on a target cell can cause changes in the expression of adhesion molecules and chemokine receptors on the target membrane, thus allowing it to move from one location to another. Cytokines can also signal an immune cell to increase or decrease the activity of particular enzymes or to change its transcriptional program, thereby altering and enhancing its effector functions. Finally, they can instruct a cell when to survive and when to die.

In an early attempt to classify cytokines, immunologists began numbering them in the order of their discovery, and naming them **interleukins**. This name reflects the fact that interleukins communicate between (Latin, *inter*) white blood cells (*leukocytes*). Examples include interleukin 1 (IL-1), secreted by macrophages, and interleukin 2 (IL-2), secreted by activated T cells. However, many cytokines that were named prior to this attempt at rationalizing nomenclature have resisted reclassification, and so students will come across cytokines such as Tumor Necrosis Factor or Interferons, that are also "interleukins" in all but name.

Although the term *cytokine* refers to all molecules that communicate among immune cells, the name **chemokine** is used specifically to describe that subpopulation of

Immunoglobulin family receptors | Hematopoietin-type receptors (class I) | Interferon-type receptors (class II)

TNF receptors | IL-17 receptors | Chemokine receptors

Cytokine and Chemokine Receptor Families

- General Properties of Cytokines and Chemokines
- Six Families of Cytokines and Associated Receptor Molecules
- Cytokine Antagonists
- Cytokine-Related Diseases
- Cytokine-Based Therapies

cytokines that share the specific purpose of mobilizing immune cells from one organ, or indeed, from one part of an organ, to another. Chemokines belong to the class of molecules called **chemoattractants**, molecules that attract cells by influencing the assembly, disassembly, and contractility of cytoskeleton proteins and the expression of cell-surface adhesion molecules. Chemokines attract cells with the appropriate chemokine receptors to regions where the chemokine concentration is highest. For example, chemokines are important in attracting cells of the innate immune system to the site of infection and inducing T cells to move toward antigen-presenting cells in the secondary lymphoid tissues. Leukocytes change their pattern of expression of chemokine receptors over the course of an immune response, first migrating to the secondary immune organs, in which they undergo differentiation to mature effector cells, and then moving out into the affected tissues to fight the infection, responding to different chemokine gradients with each movement. As we will learn in a later section, chemokines are also capable of instructing cells to alter their transcriptional programs.

The classification and nomenclature of chemokines is more logical than that of interleukins, and is based on their biochemical structures. Although chemokines technically fall under the umbrella classification of "cytokines," normal usage is evolving such that the term *chemokine* is used when referring to molecules that move immune cells from place to place, and the term *cytokine* is employed when referring to any other messenger molecule of the immune system.

Like all signaling molecules, cytokines can be further classified on the basis of the distance between the cell secreting the signaling ligand and the cell receiving that chemical signal. Cytokines that act on cells some distance away from the secreting cell, such that they must pass through the bloodstream before reaching their target, are referred to as **endocrine** (Figure 4-1). Those that act on cells near the secreting cell, such that the cytokine merely has to diffuse a few Ångstroms through tissue fluids or

across an immunological synapse, are referred to as **paracrine**. Sometimes, a cell needs to receive a signal through its own membrane receptors from a cytokine that it, itself, has secreted. This type of signaling is referred to as **autocrine**. Of note, the T-cell interleukin IL-2 acts effectively in all three modes. Unlike the classical hormones, such as insulin and glucagon, that generally act at long range in an endocrine fashion, many cytokines act over a short distance in an autocrine or paracrine fashion.

We begin this chapter with an introduction to the general properties of cytokines and chemokines followed by a discussion of the specific receptors and signaling pathways used by the six families of immune system cytokines and chemokines. Next, we describe the ways in which cytokine signaling can be regulated by antagonists. Finally, we turn to the role of cytokines and chemokines in disease and medicine.

General Properties of Cytokines and Chemokines

The activity of cytokines was first recognized in the mid-1960s, when supernatants derived from in vitro cultures of lymphocytes were found to contain soluble factors, usually proteins or glycoproteins, that could regulate proliferation, differentiation, and maturation of immune system cells. Production of these factors by cultured lymphocytes was induced by activation with antigens or with nonspecific **mitogens** (molecules inducing cell division, or mitosis). However, biochemical isolation and purification of cytokines was initially hampered because of their low concentrations in the culture supernatants and the absence of well-defined assay systems for individual cytokines.

The advent of hybridoma technology (see Chapter 20) allowed the production of artificially generated T-cell tumors that constitutively produced IL-2, allowing for its purification and characterization. Gene cloning techniques developed during the 1970s and 1980s then made it possible to generate pure cytokines by expressing the proteins from cloned genes derived from hybridomas or from normal leukocytes, after transfection into bacterial or yeast cells. Using these pure cytokine preparations, researchers were able to identify cell lines whose growth depended on the presence of a particular cytokine, thus providing them with biological cytokine assay systems. Since then, monoclonal antibodies specific for many cytokines have made it possible to develop rapid, quantitative, cytokine-specific immunoassays. ELISA assays measure the concentrations of cytokines in solution, Elispot assays quantitate the cytokines secreted by individual cells, and cytokine-specific antibodies can be used to identify cytokine-secreting cells using intracellular cytokine staining followed by flow cytometry or immuno-fluorescence microscopy (see Chapter 20).

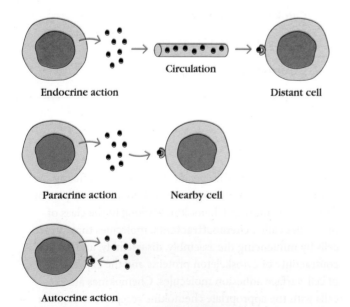

Endocrine action

Circulation

Distant cell

Paracrine action

Nearby cell

Autocrine action

FIGURE 4-1 Most immune system cytokines exhibit autocrine and/or paracrine action; fewer exhibit endocrine action.

Cytokines Mediate the Activation, Proliferation, and Differentiation of Target Cells

Cytokines bind to specific receptors on the membranes of target cells, triggering signal transduction pathways that ultimately alter enzyme activity and gene expression (Figure 4-2). The susceptibility of a target cell to a particular cytokine is determined by the presence of specific membrane receptors. In general, cytokines and their fully assembled receptors exhibit very high affinity for one another, with dissociation constants for cytokines and their receptors ranging from 10^{-8} to 10^{-12} M^{-1}. Because their receptor affinities are so high and because cytokines are often secreted in close proximity to their receptors, such that the cytokine concentration is not diluted by diffusion (as mentioned in Chapter 3), the secretion of very few cytokine molecules can mediate powerful biological effects.

Cytokines regulate the *intensity* and *duration* of the immune response by stimulating or inhibiting the activation, proliferation, and/or differentiation of various cells, by regulating the secretion of other cytokines or of antibodies, or in some cases by actually inducing programmed cell death in the target cell. In addition, cytokines can modulate the expression of various cell-surface receptors for chemokines, other cytokines, or even for themselves. Thus, the cytokines secreted by even a small number of antigen-activated lymphocytes can influence the activity of many different types of cells involved in the immune response.

Cytokines exhibit the attributes of pleiotropy, redundancy, synergism, antagonism, and cascade induction (Figure 4-3), which permit them to regulate cellular activity in a coordinated, interactive way. A cytokine that induces different biological effects depending on the nature of the target cells is said to have a **pleiotropic** action, whereas two or more cytokines that mediate similar functions are said to be **redundant**. Cytokine **synergy** occurs when the combined effect of two cytokines on cellular activity is greater than the additive effects of the individual cytokines. In some cases, the effects of one cytokine inhibit or **antagonize** the effects of another. **Cascade induction** occurs when the action of one cytokine on a target cell induces that cell to produce one or more additional cytokines.

Cytokines Have Numerous Biological Functions

Although a variety of cells can secrete cytokines that instruct the immune system, the principal producers are T_H cells, dendritic cells, and macrophages. Cytokines released from these cell types are capable of activating entire networks of interacting cells (Figure 4-4). Among the numerous physiological responses that require cytokine involvement are the generation of cellular and humoral immune responses, the induction of the inflammatory response, the regulation of hematopoiesis, and wound healing.

The total number of proteins with cytokine activity grows daily as research continues to uncover new ones. Table 4-1 summarizes the activities of some commonly encountered cytokines. An expanded list of cytokines can be found in Appendix II. Note, however, that many of the listed functions have been identified from analyses of the effects of recombinant cytokines, sometimes added alone to in vitro systems at nonphysiologic concentrations. In vivo, cytokines rarely, if ever, act alone. Instead, a target cell is exposed to a milieu containing a mixture of cytokines whose combined synergistic or antagonistic effects can have a wide variety of consequences. In addition, as we have learned, cytokines often induce the synthesis of other cytokines, resulting in cascades of activity.

Cytokines Can Elicit and Support the Activation of Specific T-Cell Subpopulations

As described in Chapters 2 and 11, helper T cells can be classified into subpopulations, each of which is responsible for the support of a different set of immune functions. For example, T_H1 cells secrete cytokines that promote the differentiation and activity of macrophages and cytotoxic T cells, thus leading to a primarily cytotoxic immune response, in which cells that have been infected with viruses and

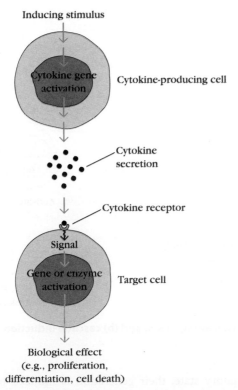

FIGURE 4-2 Overview of the induction and function of cytokines. An inducing stimulus, which may be an antigen or another cytokine, interacts with a receptor on one cell, inducing it to secrete cytokines that in turn act on receptors of a second cell, bringing about a biological consequence. In the case of IL-2, both cells may be antigen-activated T cells that secrete IL-2, which acts both on the secreting cell and on neighboring, activated T cells.

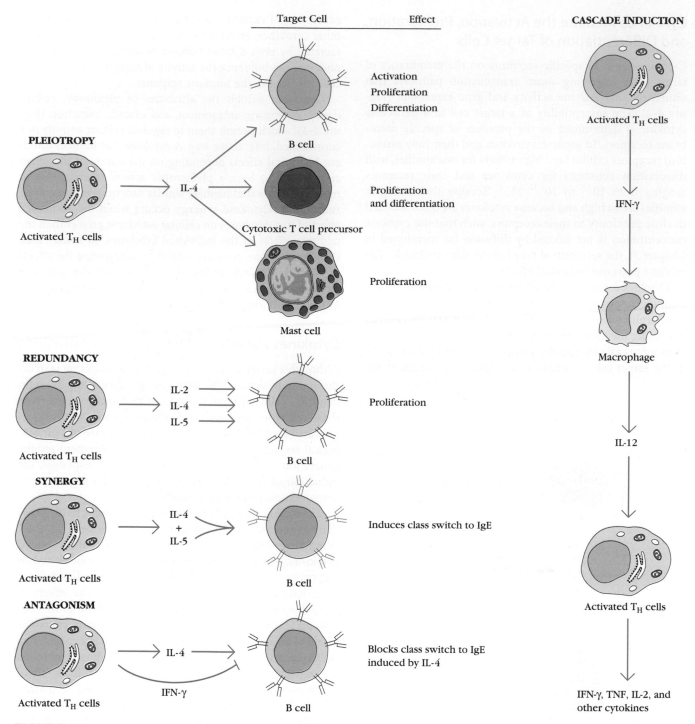

Target Cell   Effect

CASCADE INDUCTION

PLEIOTROPY

Activated T$_H$ cells

IL-4

B cell
Activation
Proliferation
Differentiation

Cytotoxic T cell precursor
Proliferation
and differentiation

Mast cell
Proliferation

Activated T$_H$ cells

IFN-γ

Macrophage

IL-12

REDUNDANCY

Activated T$_H$ cells

IL-2
IL-4
IL-5

B cell
Proliferation

SYNERGY

Activated T$_H$ cells

IL-4
+
IL-5

B cell
Induces class switch to IgE

Activated T$_H$ cells

ANTAGONISM

Activated T$_H$ cells

IL-4

IFN-γ

B cell
Blocks class switch to IgE
induced by IL-4

IFN-γ, TNF, IL-2, and
other cytokines

FIGURE 4-3 Cytokine attributes of (a) pleiotropy, redundancy, synergism, antagonism, and (b) cascade induction.

intracellular bacteria are recognized and destroyed. The cytokines IL-12 and interferon (IFN) γ induce T$_H$1 differentiation. In contrast, T$_H$2 cells activate B cells to make antibodies, which neutralize and bind extracellular pathogens, rendering them susceptible to phagocytosis and complement-mediated lysis. IL-4 and IL-5 support the generation of T$_H$2 cells. T$_H$17 cells promote the differentiation of activated macrophages and neutrophils, and support the

inflammatory state; their generation is induced by IL-17 and IL-23. The differentiation and activity of each distinctive T-cell subpopulation is therefore supported by the binding of different combinations of cytokines to T-cell surface receptors, with each cytokine combination inducing its own characteristic array of intracellular signals, and sending the helper T cell down a particular differentiation pathway.

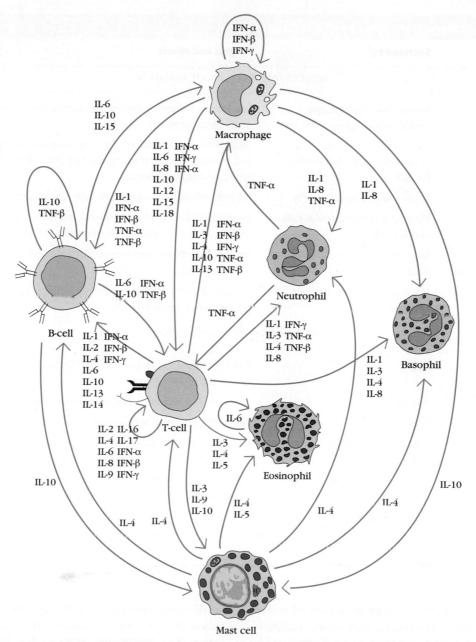

FIGURE 4-4 The cells of the immune system are subject to control by a network of cytokine actions.

Cell Activation May Alter the Expression of Receptors and Adhesion Molecules

The ability of cytokines to activate most, if not all, members of particular immune cell subpopulations appears to conflict with the established specificity of the immune system. What keeps cytokines from activating all T cells, for example, in a nonspecific fashion during the immune response?

In order for a cell to respond to a signaling molecule, it must express receptors for that molecule, and responsiveness to a molecular signal can thus be controlled by signal receptor expression. For example, antigen stimulation of a T cell induces alterations in the T-cell surface expression of chemokine receptors. Reception of chemokine signals

through these receptors therefore instructs *only those cells that have previously been activated by antigen* to migrate to nearby lymph nodes or to the spleen. Furthermore, activation-induced changes in the adhesion molecules that are expressed on the cell membrane ensures that stimulated cells migrate to, and then remain in, the location best suited to their function. T-cell activation by antigen also up-regulates the expression of the receptors for cytokines that provide proliferative signals, such as IL-2 (as described in Chapter 3), and also for differentiative cytokines such as IL-4. In this way, following antigen encounter, only those T cells that have been activated by antigen are primed to relocate and to receive the proliferative and differentiative signals they need to function as a mature immune effector cell. This

TABLE 4-1   Functional groups of selected cytokines[*]

Cytokine	Secreted by[†]	Targets and effects
SOME CYTOKINES OF INNATE IMMUNITY		
Interleukin 1 (IL-1)	Monocytes, macrophages, endothelial cells, epithelial cells	Vasculature (inflammation); hypothalamus (fever); liver (induction of acute phase proteins)
Tumor necrosis factor-α (TNF-α)	Macrophages, monocytes, neutrophils, activated T cells and NK cells	Vasculature (inflammation); liver (induction of acute phase proteins); loss of muscle, body fat (cachexia); induction of death in many cell types; neutrophil activation
Interleukin 12 (IL-12)	Macrophages, dendritic cells	NK cells; influences adaptive immunity (promotes T_H1 subset)
Interleukin 6 (IL-6)	Macrophages, endothelial cells, and T_H2 cells	Liver (induces acute phase proteins); influences adaptive immunity (proliferation and antibody secretion of B-cell lineage)
Interferon-α (IFN-α) (this is a family of molecules)	Macrophages dendritic cells, virus-infected cells	Induces an antiviral state in most nucleated cells; increases MHC Class I expression; activates NK cells
Interferon β (IFN-β)	Macrophages, dendritic cells, virus-infected cells	Induces an antiviral state in most nucleated cells; increases MHC Class I expression; activates NK cells
SOME CYTOKINES OF ADAPTIVE IMMUNITY		
Interleukin 2 (IL-2)	T cells	T-cell proliferation; can promote AICD. NK cell activation and proliferation; B-cell proliferation
Interleukin 4 (IL-4)	T_H2 cells, mast cells	Promotes T_H2 differentiation; isotype switch to IgE
Interleukin 5 (IL-5)	T_H2 cells	Eosinophil activation and generation
Transforming growth factor β (TGF-β)	T cells, macrophages, other cell types	Inhibits T-cell proliferation and effector functions; inhibits B-cell proliferation; promotes isotype switch to IgA; inhibits macrophages
Interferon γ (IFN-γ)	T_H1 cells, CD8$^+$ cells, NK cells	Activates macrophages; increases expression MHC Class I and Class II molecules; increases antigen presentation

[*]Many cytokines play roles in more than one functional category.

[†]Only the major cell types providing cytokines for the indicated activity are listed; other cell types may also have the capacity to synthesize the given cytokine. Activated cells generally secrete greater amounts of cytokine than unactivated cells.

pattern of activation-induced alteration in the cell surface expression of adhesion molecules, chemokine receptors, and cytokine receptors is a common strategy employed by the immune system.

Cytokines Are Concentrated Between Secreting and Target Cells

During the process of T-cell activation by an antigen-presenting dendritic cell, or of B-cell activation by a cognate T cell, the respective pairs of cells are held in close juxtaposition for many hours (see Chapter 14). Over that time period, the cells release cytokines that bind to relevant receptors on the partner cell surface, without ever entering the general circulation. Furthermore, during this period of close cell-cell contact, the secretory apparatus of the stimulating cell is oriented so that the cytokines are released right at the region of the cell membrane that is in closest contact with the recipient cell (see Figure 3-4). The close nature of the cell-cell interaction and

the directional release of cytokines by the secretory apparatus means that the effective concentration of cytokines in the region of the membrane receptors may be orders of magnitude higher than that experienced outside the contact region of the two cells. Thus, discussions of membrane receptor affinity and cytokine concentrations within tissue fluids must always take into account the biology of the responding system and the geography of the cell interactions involved. In addition, the half-life of cytokines in the bloodstream or other extracellular fluids into which they are secreted is usually very short, ensuring that cytokines usually act for only a limited time and over a short distance.

Signaling Through Multiple Receptors Can Fine Tune a Cellular Response

Cytokine and chemokine signaling in the immune response can be a strikingly complex and occasionally redundant affair. Effector molecules such as cytokines can bind to more than one receptor, and receptors can bind to more than one signaling molecule. Nowhere is the latter concept more clearly illustrated than in the chemokine system, in which approximately 20 receptors bind to close to 50 distinct chemokines (see Appendix III). Effector molecule signaling can also cooperate with signaling through antigen-specific receptors. Signals received through more than one receptor must then be integrated at the level of the biological response, with multiple pathways acting to tune up or tune down the expression of particular transcription factors or the activity of particular enzymes. Thus, the actual biological response mounted by a cell to a particular chemical signal depends not only on the nature of the individual receptor for that signal, but also on all of the downstream adapters and enzymes present in the recipient cell.

Six Families of Cytokines and Associated Receptor Molecules

In recent years, immunologists have enjoyed an explosion of information about new cytokines and cytokine receptors as a result of advances in genomic and proteomic analysis. Advances Box 4-1 describes a recently developed proteomic approach to the search for new, secreted cytokines and illustrates the manner in which a sophisticated appreciation of the molecular and cell biology of secretory pathways aids in the identification of new cytokines. The purpose of this chapter is not to provide an exhaustive list of cytokines and their receptors (see Appendices II and III for a comprehensive and current list of cytokines and chemokines), but rather to outline

---

ADVANCES   BOX 4-1

 ## Methods Used to Map the Secretome

The related approaches of genomics and proteomics provide scientists with tools they can use to assess the complex changes that occur in gene and protein expression induced by stimuli, such as antigen or cytokine stimulation. Vast arrays of information regarding the derivation and readout of genes in different cells and organisms, and the expression of particular proteins, can be analyzed and presented in ways not available to scientists just a few years ago.

Recently, the science of proteomics has been extended to address the mapping of proteins that are secreted by various cell types. The array of proteins secreted by a cell is referred to as its *secretome*, and the secretome can be more formally defined as the "proteins released by a cell, tissue, or organisms through classical and nonclassical secretion mechanisms."

Scientists first became interested in the concept of the secretome as a way to diagnose and identify various types of cancer. They reasoned that they could use the set of proteins secreted into the serum or other tissue fluids as a biological marker for spe-

cific tumor types. If particular proteins can be shown to be secreted at high concentration only under conditions of malignancy, then rapid and inexpensive tests can be developed that have the potential to screen for tumors at an early stage, when they are still amenable to treatment. Although such tumor-specific profiles of secreted proteins are surprisingly difficult to develop, given the range of mutations associated with the generation of a cancer, the ability to diagnose a tumor at an early stage using only a serum sample provides intense motivation, and many such attempts are ongoing.

The approaches used to define a set of cancer secretomes have since been applied to studies of many other, nonmalignant cell populations for which the description of a secretome would be a useful analytical tool. These populations include stem cells, cells of the immune system, and adipose cells. Given the diversity of cytokines that can be secreted by a single cell, and the manner in which the activities of cytokines can interact at the level of the target cell, cytokine biology is a superb target for such a global approach.

A recent secretome analysis (Botto et al., 2011) addressed the question of how the human cytomegalovirus induces the formation of new blood vessels (angiogenesis). Virus-free supernatant from virus-infected endothelial cells was found to induce angiogenesis. Secretome analysis of the infected endothelial cell supernatant revealed the presence of multiple cytokines, including IL-8, GM-CSF, and IL-6. The addition of a blocking anti-IL-6 antibody at the same time as the virus-free supernatant was then shown to inhibit its angiogenic activity, thus demonstrating that it was the IL-6 activity in the supernatant that was primarily responsible for inducing the new blood vessel growth.

One difficulty that is frequently encountered in trying to analyze the secretome of a particular type of cell is the need of many cells to grow in a tissue culture fluid supplemented with serum, which is itself a complex mixture of proteins. In this case it is important to distinguish between proteins released by the cells under study and those which were originally present in the serum. Several

(continued)

techniques are available to discriminate between secreted proteins and those from the tissue culture media, including adding inhibitors of secretion to some cultures and then comparing those proteins present in the culture supernatant in the presence and absence of inhibitors. Alternatively, culturing the cells in the presence of radioisotopes that only label newly synthesized proteins, such as ^{35}S methionine, can be used to distinguish these proteins from preexisting proteins in the culture medium.

In the case, such as that described above, that a cell line is being tested to determine whether it secretes a set of cytokines for which antibody assays already exist, two different types of multiplex measurements may be used (see Figure 1). Both of these approaches utilize antibodies to the array of cytokines to be analyzed, attached to some sort of solid phase support. This support may be glass, a membrane, or a set of beads, with each antibody attached to a bead of a different color. The sample of tissue culture fluid is added to the solid phase antibody, excess fluid is washed away, and then biotinylated antibodies are added. (Biotin, a small molecule, is used because it has an extremely high affinity for a protein, streptavidin, and is therefore used to couple two molecules together in assays such as these. For more details, see Chapter 20.) After antibody binding, the excess biotinylated antibodies are removed by washing and the cytokine concentrations are assessed by the addition of fluorescent streptavidin, which will bind to the biotin. A fluorescent signal indicates the presence of the cytokine in the sample, and the level of the signal reveals its concentration. Since each bead fluoresces at a different wavelength, the fluorescence associated with each cytokine can be distinguished.

Various bioinformatics tools have been developed that have particular application to secretome analysis. These include *SignalP*, which identifies the presence of signal peptides and also shows the location of signal peptide cleavage sites in bacterial and eukaryotic proteins. In addition, *SecretomeP* can be used for the pre-

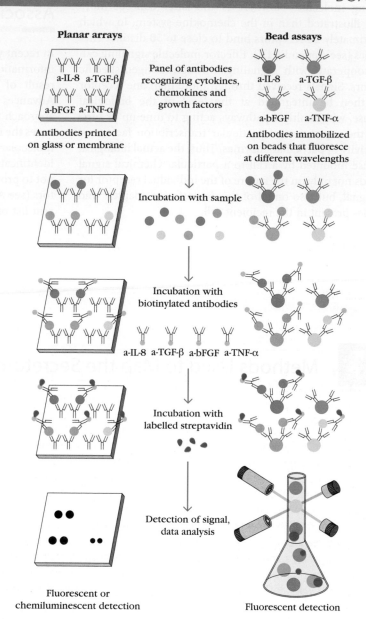

FIGURE 1

Principle of planar and bead-based multiplex detection and quantitation of cytokines, chemokines, growth factors, and other proteins. Assays use antibodies against (a-) various cytokines, and biotin (yellow)-streptavidin (green) conjugation. See text for details. *[Adapted from H. Skalnikova et al., Mapping of the secretome of primary isolates of mammalian cells, stem cells and derived cell lines, 2011, Proteomics 11:691.]*

diction of nonclassically secreted proteins. Several bioinformatics tools, including *TargetP* and *Protein Prowler*, use the protein sequence to predict its subcellular localization. Finally, *Ingenuity Pathway Analysis* allows the investigator to search for protein interaction partners and to

predict the involvement of the protein of interest in functional networks.

Botto, S., D. N. Streblow, V. DeFilippis, L. White, C. N. Kreklywich, P. P. Smith, and P. Caposio. (2011). IL-6 in human cytomegalovirus secretome promotes angiogenesis and survival of endothelial cells through the stimulation of survivin. *Blood* 117:352–361.

TABLE 4-2	Six Cytokine Families	
Family name	**Representative members of family**	**Comments**
Interleukin 1 family	IL-1α, IL-1β, IL-1Ra, IL-18, IL-33	IL-1 was the first noninterferon cytokine to be identified. Members of this family include important inflammatory mediators.
Hematopoietin (Class I cytokine) family	IL-2, IL-3, IL-4, IL-5, IL-6, IL-7, IL-12, IL-13, IL15, IL-21, IL-23, GM–CSF, G-CSF, Growth hormone, Prolactin, Erythropoietin/hematopoietin	This large family of small cytokine molecules exhibits striking sequence and functional diversity.
Interferon (Class II cytokine) family	IFN-α, IFN-β, IFN-γ, IL-10, IL-19, IL-20, IL-22, IL-24	While the IFNs have important roles in anti-viral responses, all are important modulators of immune responses.
Tumor Necrosis Factor family	TNF-α, TNF-β, CD40L, Fas (CD95), BAFF, APRIL, LTβ	Members of this family may be either soluble or membrane bound; they are involved in immune system development, effector functions, and homeostasis.
Interleukin 17 family	IL-17 (IL17-A), IL17B, C, D, and F	This is the most recently discovered family; members function to promote neutrophil accumulation and activation, and are proinflammatory.
Chemokines (see Appendix III)	IL-8, CCL19, CCL21, RANTES, CCL2 (MCP-1), CCL3 (MIP-1α)	All serve chemoattractant function.

some general principles of cytokine and receptor architecture and function that should then enable the reader to place any cytokine into its unique biological context.

Detailed studies of cytokine structure and function have revealed common features among families of cytokines. Cytokines are relatively small proteins and generally have a molecular mass of less than 30 kDa. Many are glycosylated, and glycosylation appears to contribute to cytokine stability, although not necessarily to cytokine activity. Cytokines characterized so far belong to one of six groups: the Interleukin 1 (IL-1) family, the Hematopoietin (Class I cytokine) family, the Interferon (Class II cytokine) family, the Tumor Necrosis Factor (TNF) family, the Interleukin 17 (IL-17) family, and the Chemokine family (Table 4-2). Each of these six families of cytokines, the receptors that engage them, and the signaling pathways that transduce the message received upon cytokine binding into the appropriate biological outcome are described in the following pages.

Cytokines of the IL-1 Family Promote Proinflammatory Signals

Cytokines of the **interleukin 1 (IL-1) family** are typically secreted very early in the immune response by dendritic cells and monocytes or macrophages. IL-1 secretion is stimulated by recognition of viral, parasitic, or bacterial antigens by innate immune receptors. IL-1 family members are generally *proinflammatory*, meaning that they induce an increase in the capillary permeability at the site of cytokine secretion, along with an amplification of the level of leukocyte migration into the infected tissues. In addition, IL-1 has systemic (whole body) effects and signals the liver to produce *acute*

phase proteins such as the Type I interferons (IFNs α and β), IL-6, and the chemokine CXCL8. These proteins further induce multiple protective effects, including the destruction of viral RNA and the generation of a systemic fever response (which helps to eliminate many temperature-sensitive bacterial strains). IL-1 also activates both T and B cells at the induction of the adaptive immune response.

Cytokines of the IL-1 Family

Members of the IL-1 cytokine and receptor family are shown in Figure 4-5. The canonical (most representative) members of the IL-1 family, IL-1α and IL-1β, are both synthesized as 31 kDa precursors, pro-IL-1α and pro-IL-1β. Pro IL-1α is biologically active, and often occurs in a membrane-bound form, whereas pro-IL-1β requires processing to the fully mature soluble molecule before it can function. Pro-IL-1α and β are both trimmed to their 17 kDa active forms by the proteolytic enzyme caspase-1 inside the secreting cell. Active caspase-1 is located in a complex set of proteins referred to as the *inflammasome* (see Chapter 5).

Other IL-1 family members, IL-18 and IL-33, have also been shown to be processed by caspase-1 in vitro (although there is ambiguity as to whether IL-33 requires this processing for full activity in vivo). IL-18 is related to IL-1, uses the same receptor family, and has a similar function; like IL-1, IL-18 is expressed by monocytes, macrophages, and dendritic cells and is secreted early in the immune response. In contrast, IL-33 is constitutively expressed in smooth muscle and in bronchial epithelia, and its expression can be induced by IL-1β and TNF-α in lung and skin fibroblasts. IL-33 has been shown to induce T_H2 cytokines that promote T-lymphocyte interactions with B cells, mast cells, and eosinophils. IL-33 has also been implicated in the pathology of diseases such as asthma and inflammatory airway and bowel diseases.

Two additional members of this cytokine family act as *natural inhibitors of IL-1 family function*. The soluble protein IL-1Ra (IL-1 *Receptor antagonist*) binds to the IL-1RI receptor, but prevents its interaction with its partner receptor chain, IL-1RAcP, thus rendering it incapable of transducing a signal to the interior of the cell. IL-1Ra therefore functions as an **antagonist** ligand of IL-1. IL-18BP adopts a different strategy of inhibition, binding to IL-18 in solution and preventing IL-18 from interacting productively with its receptor. The inhibitory effect of IL-18BP is enhanced by the further binding of IL-1F7 (see Figure 4-5b).

The IL-1 Family of Cytokine Receptors

The Interleukin 1 family of receptors includes the receptors for IL-1, IL-18, and IL-33. Both forms of IL-1—IL-1α and IL-1β—bind to the same receptors and mediate the same responses. Two different receptors for IL-1 are known, and both are members of the immunoglobulin superfamily of proteins (see Chapter 3). Only the type I IL-1R (IL-1RI), which is expressed on many cell types, is able to transduce a cellular signal; the type II IL-1R (IL-1RII) is limited to B cells and is inactive. For full functioning, the Type 1 IL-1R also requires the presence of an interacting accessory protein, IL-1RAcP (IL-1 *Receptor Accessory Protein* (see Figure 4-5a).

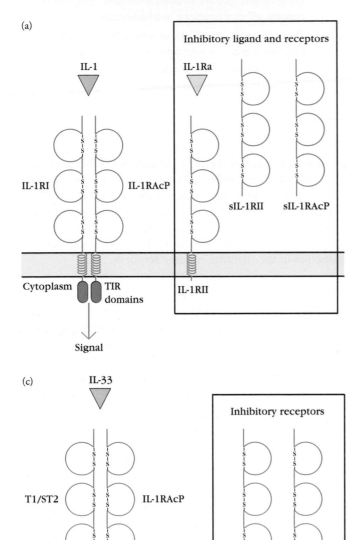

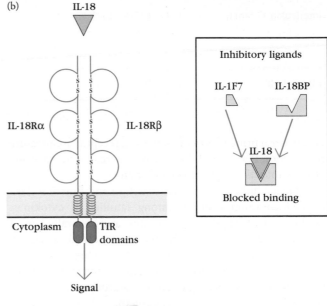

FIGURE 4-5 Ligands and receptors of the IL-1 family. (a) The two agonist ligands, IL-1α and IL-1β, are represented by IL-1 and the antagonist ligand by IL-1Ra. The IL-1 receptor, IL-1RI, has a long cytoplasmic domain and, along with IL-1RAcP, activates signal transduction pathways. IL-1Ra functions as an IL-1 inhibitor by binding to IL-1RI while not allowing interaction with IL-1RAcP. IL-1RII does not activate cells but functions as an IL-1 inhibitor both on the plasma membrane and in the cell microenvironment as a soluble receptor (sIL-1RII). IL-1RAcP can also inhibit IL-1 signals by cooperating with IL-1RII in binding IL-1 either on the plasma membrane or as a soluble molecule (sIL-1RAcP). (b) IL-18 binds to the IL-18Rα chain, and this complex then engages the IL-18Rβ chain to initiate intracellular signals. The soluble protein IL-18BP functions as an inhibitor of IL-18 by binding this ligand in the fluid phase, preventing interaction with the IL-18Rα chain. IL-1F7 appears to enhance the inhibitory effect of IL-18BP. (c) IL-33 binds to the T1/ST2 receptor, and this complex engages the IL-1RAcP as a co-receptor. A soluble form of ST2 (sST2) may function as an inhibitor of IL-33 by binding IL-33 in the cell microenvironment and sIL-1RAcP may enhance the inhibitory effects of sST2.

Note that both the IL-1RI and the IL-1RII receptor chains as well as the receptor accessory protein exist in both soluble and membrane-bound forms. However, *a full signal is transmitted only from the dimer of the membrane-bound forms of IL-1RI and IL-1RAcP.* The alternative, membrane-bound and soluble forms of IL-1 binding proteins, can "soak up" excess cytokine, but they are unable to transduce the interleukin signal. Thus, by secreting more or fewer of these inactive receptors, at different times during an immune response, the organism has the opportunity to fine-tune the cytokine signal by allowing the inactive and soluble receptors to compete with the signal-transducing receptor for available cytokine. This theme finds echoes in other immune system receptor families, and appears to be a frequently evolved strategy for controlling the strength of signals that give rise to important outcomes. In the case of IL-1, the ultimate result of successful IL-1 signaling is a global, proinflammatory state, and so the penalty paid by the host for an inappropriately strong IL-1 response would be physiologically significant and even potentially fatal.

The receptor for IL-18 is also a heterodimer, made up of IL-18Rα and IL-18Rβ. IL-33 is recognized by the IL-1RAcP in combination with a novel receptor protein, variously termed T1/ST-2 or IL-1RL1. As for IL-1, inhibitory receptors exist for IL-33 (see Figure 4-5c).

Signaling from IL-1 Receptors

Productive ligand binding to the extracellular portion of the IL-1 receptor leads to a conformational alteration in its cytoplasmic domain. This structural alteration in the receptor leads to a series of downstream signaling events (Figure 4-6). Most of the themes of these events will be familiar to the reader from Chapter 3, and we will encounter them again in the discussion of innate immune receptors in Chapter 5.

First, binding of the *adapter protein* **MyD88** to the occupied receptor allows recruitment to the receptor complex of one or more members of the **IL-1 Receptor Activated Kinase (IRAK)** protein family. One of these, IRAK-4, is activated by autophosphorylation and phosphorylates its fellow IRAKs, resulting in the generation of binding sites for *TNF Receptor Associated Factor 6* (TRAF6), which is associated with a ubiquitin-ligase complex capable of generating polyubiquitin chains. The IRAK-TRAF6 complex now dissociates from the receptor and interacts with a preformed cytosolic complex made up of the kinase *TGFβ—Associated Kinase 1* (TAK1) and two *TAK1-Binding proteins*, TAB1 and TAB2. Binding of polyubiquitin chains to the TAB proteins in the TAK1 complex activates it.

The TAK1 complex now performs two functions with which the reader should be familiar. It phosphorylates and activates the IKK complex, leading to the destruction of IκB and the resultant activation of the transcription factor NF-κB (see Figure 3-17). In addition, TRAF6 also plays a role in IKK activation by providing ubiquitination sites to which the NEMO component of IKK can bind, resulting in its further activation. TAK1 also activates downstream members

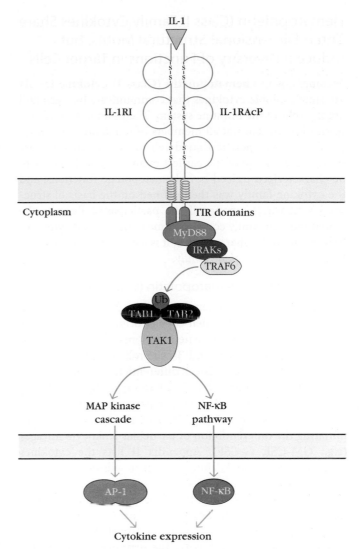

FIGURE 4-6 Signaling from members of the IL-1 receptor family. IL-1 binding to its receptor induces a conformational alteration in the receptor's Toll-IL-1R (TIR) domain that allows binding of the adapter protein MyD88 via its TIR domain. MyD88 recruits one or more IL-1 receptor activated kinases (IRAKs) to the receptor complex, which phosphorylate one another providing binding sites for TRAF6. The IRAK-TRAF6 complex dissociates from the receptor complex and interacts with the cytoplasmic protein TAK1 and its two binding proteins, TABs 1 and 2. TRAF6, together with a ubiquitin-ligase complex, catalyzes the generation of polyubiquitin chains that activate the TAK1 complex. TAK1 activates downstream events leading to the activation and nuclear localization of the transcription factor NF-κB. TAK1 also activates downstream members of the MAP kinase cascade, leading to activation of the AP-1 transcription factor.

of the MAP kinase cascade, which in turn activate the AP-1 transcription factor (see Figure 3-16). Binding of IL-1 family cytokines to their receptors thereby leads to a global alteration in the transcription patterns of the affected cells, which in turn results in the up-regulation of proinflammatory cytokines and adhesion molecules.

Hematopoietin (Class I) Family Cytokines Share Three-Dimensional Structural Motifs, but Induce a Diversity of Functions in Target Cells

Members of the **hematopoietin (Class I) cytokine family** are small, soluble cytokines that communicate between and among cells of the immune system. Their name is somewhat misleading in that not all members of this family are implicated in hematopoietic (blood-cell forming) functions per se. However, some of the earliest members of this family to be characterized indeed have hematopoietic functions, and the cytokine family was then defined on the basis of structural similarities among all the participants. Because the hematopoietin family contains some of the earliest cytokines to be structurally characterized, it is sometimes also referred to as the *Class I cytokine family*.

Cytokines of the Hematopoietin (Class I) Family

As more hematopoietin family members have been defined, it has become clear that their cellular origins and target cells are as diverse as their ultimate functions, which range from signaling the onset of T- and B-cell proliferation (e.g., IL-2), to signaling the onset of B-cell differentiation to plasma cells and antibody secretion (e.g., IL-6), to signaling the differentiation of a T helper cell along one particular differentiation pathway versus another (e.g., IL-4 vs. IL-12) and, finally, to initiating the differentiation of particular leukocyte lineages (e.g., GM-CSF, G-CSF). Appendix II lists the cytokines described in this book, along with their cells of origin, their target cells, and the functions they induce.

Significant homology in the three-dimensional structure of hematopoietin family cytokines defines them as members of a single protein family, despite a relatively high degree of amino acid sequence diversity. The defining structural feature of this class of cytokines is a four-helix bundle motif, organized into four anti-parallel helices (Figure 4-7). Members of this family can then be further subclassified on the basis of helical length. Cytokines such as IL-2, IL-4, and IL-3 typically have short helices of 8 to 10 residues in length. In contrast, the so-called long-chain cytokines, which include IL-6 and IL-12, typically have helical lengths of 10 to 20 residues.

The Hematopoietin or Class I Receptor Family

Most hematopoietin cytokine receptors include two types of protein domains: an immunoglobulin-like domain, made up of β sheets, as described in Chapter 3, and domains that bear structural homology to the FNIII domain of the extracellular matrix protein fibronectin. Binding sites for most cytokines are to be found in a structure made up of two, tandem (side-by-side) FNIII domains referred to as **Cytokine-binding Homology Regions (CHRs)**. As we will see, the CHR motif is common to cytokine receptors from several families.

A feature common to most of the Hematopoietin and Interferon cytokine receptor families is the presence of multiple subunits. Table 4-3 lists the three subfamilies of hematopoietin

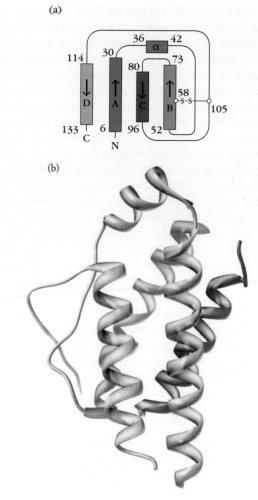

FIGURE 4-7 The four-helix bundle is the defining structural feature of the Hematopoietin family of cytokines. Structure of interleukin 2—the defining member of the Hematopoietin family—showing the four α-helices of the hematopoietin cytokines point in alternating directions. (a) Topographical representation of the primary structure of IL-2 showing α-helical regions (α and A-D) and connecting chains of the molecule. (b) Ribbon representation of the crystallographic structure of human Il-2. *[Part (b) PDB ID 1M47.]*

receptors, each subfamily being defined by a receptor subunit that is shared among all members of that family.

The γ-Chain Bearing, or IL-2 Receptor, Subfamily

Expression of a common γ chain defines the IL-2 receptor subfamily, which includes receptors for IL-2, IL-4, IL-7, IL-9, IL-15, and IL-21. The IL-2 and the IL-15 receptors are heterotrimers, consisting of a cytokine-specific α chain and two chains—β and γ—responsible for both signal transduction and cytokine recognition. The IL-2 receptor γ chain also functions as the signal-transducing subunit for the other receptors in this subfamily, which are all dimers. Congenital **X-linked severe combined immunodeficiency (XSCID)** results from a defect in the γ-chain gene, which maps to the X chromosome. The immunodeficiency observed in this disorder, which includes deficiencies in both T-cell and

TABLE 4-3	Subfamilies of hematopoietin family cytokine receptors share common subunits	
Common cytokine receptor subunit	**Cytokines recognized by receptors bearing that common subunit**	
γ	IL-2, IL-4, IL-7, IL-9, IL-15, IL-12	
β	IL-3, IL-5, GM-CSF	
gp130	IL-6, IL-11, LIF, OSM, CNTF, IL-27	

NK-cell activity, results from the loss of all the cytokine functions mediated by the IL-2 subfamily receptors.

The IL-2 receptor occurs in three forms, each exhibiting a different affinity for IL-2: the low-affinity monomeric IL-2Rα (CD25) (which can bind to IL-2, but is incapable of transducing a signal from it), the intermediate-affinity dimeric IL-2Rβγ (which is capable of signal transduction), and the high-affinity trimeric IL-2Rαβγ (which is responsible for most physiologically relevant IL-2 signaling) (Figure 4-8a). A recent x-ray crystallographic structure of the high-affinity trimeric form of the IL-2 receptor with an IL-2 molecule in its binding site reveals that IL-2 binds in a pocket formed by the β and γ chains

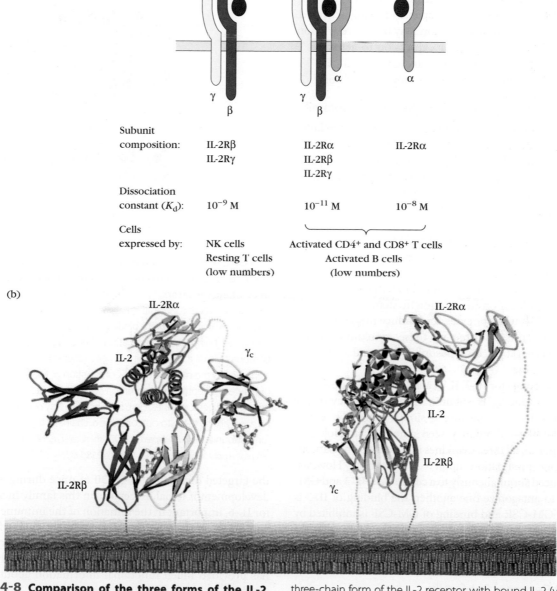

FIGURE 4-8 **Comparison of the three forms of the IL-2 receptor.** (a) Schematic of the three forms of the receptor and listing of dissociation constants and properties for each. Signal transduction is mediated by the β and γ chains, but all three chains are required for high-affinity binding of IL-2. (b) Three-dimensional structure of the three-chain form of the IL-2 receptor with bound IL-2 (views rotated by 90°). Note that the α chain completes the pocket to which IL-2 binds, accounting for the higher affinity of the trimeric form. [From X. Wang, M. Rickert, and K. C. Garcia, 2005, Structure of the quaternary complex of interleukin–2 with its α, β, and γc receptors. Science **310:1159.**]

(Figure 4-8b). Important additional contacts with the IL-2 ligand are contributed when the α chain is present, accounting for the higher affinity of binding by the trimer.

The expression of the three chains of the IL-2 receptor varies among cell types and in different activation states. The intermediate affinity (βγ) IL-2 receptors are expressed on resting T cells and on NK cells, whereas activated T and B cells express both the low-affinity (α) and the high-affinity (αβγ) receptor forms (see Figure 4-8a). Since there are approximately ten times as many low-affinity as high-affinity receptors on activated T cells (50,000 vs. 5000), one must ask what the function of the low-affinity receptor might be, and two possible ideas have been advanced. It may serve to concentrate IL-2 onto the recipient cell surface for passage to the high-affinity receptor. Conversely, it may reduce the local concentration of available IL-2, ensuring that only cells bearing the high-affinity receptor are capable of being activated. Whatever the answer to this question may be, the restriction of the high-affinity IL-2 receptor expression to activated T cells ensures that only antigen-activated CD4⁺ and CD8⁺ T cells will proliferate in response to physiologic levels of IL-2.

The β−Chain Bearing, or GM-CSF, Receptor Subfamily

Members of the GM-CSF receptor subfamily, which includes the receptors for IL-3, IL-5, and GM-CSF, share the β signaling subunit. Each of these cytokines binds with relatively low affinity to a unique, cytokine-specific receptor protein, the α subunit of a dimeric receptor. All three low-affinity subunits associate noncovalently with the common signal-transducing β subunit. The resulting αβ dimeric receptor has a higher affinity for the cytokine than the specific α chain alone, and is also capable of transducing a signal across the membrane upon cytokine binding (Figure 4-9a).

IL-3, IL-5, and GM-CSF exhibit redundant activities. IL-3 and GM-CSF both act on hematopoietic stem cells and progenitor cells, activate monocytes, and induce megakaryocyte differentiation, and all three of these cytokines induce eosinophil proliferation and basophil degranulation with release of histamine.

Since the receptors for IL-3, IL-5, and GM-CSF share a common signal-transducing β subunit, each of these cytokines would be expected to transduce a similar activation signal, accounting for the redundancy seen among their biological effects, and indeed, all three cytokines induce the same patterns of protein phosphorylation upon cell activation. However, when introduced simultaneously to a cell culture, IL-3 and GM-CSF appear to antagonize one another; the binding of IL-3 is inhibited by GM-CSF, and binding of GM-CSF is inhibited by IL-3. This antagonism is caused by competition for a limited number of β subunits available to associate with the cytokine-specific α subunits of the dimeric receptors (Figure 4-9b).

The gp130 Receptor Subfamily

The importance of the gp130 cytokine receptor family to the development and health of the individual is underscored by the results of deletion studies which have demonstrated that

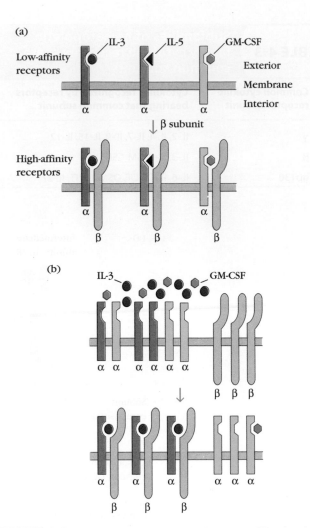

FIGURE 4-9 Interactions between cytokine-specific subunits and a common signal-transducing subunit of the β-chain family of cytokine receptors. (a) Schematic diagram of the low-affinity and high-affinity receptors for IL-3, IL-5, and GM-CSF. The cytokine-specific subunits exhibit low-affinity binding and cannot transduce an activation signal. Noncovalent association of each subunit with a common β subunit yields a high-affinity dimeric receptor that can transduce a signal across the membrane. (b) Competition of ligand-binding chains of different receptors for a common subunit can produce antagonistic effects between cytokines. Here binding of IL-3 by α subunits of the IL-3 receptor allows them to outcompete α chains of the GM-CSF receptor for β subunits. *[Part a adapted from T. Kishimoto et al., 1992, Interleukin-6 and its receptor: A paradigm for cytokines,* Science **258**:593.]

the targeted disruption of gp130 in mice during embryonic development is lethal. Receptors in this family include those for IL-6, important in the initiation of the immune response, and IL-12, critical for signaling differentiation of helper T cells along the T_H1 pathway. Targeted disruption of individual cytokine receptors, as well as of the cytokines themselves, has provided much functional information about signaling via these cytokine family members.

Cytokine specificity of the gp130 family of receptors is determined by the regulated expression of ligand-specific chains in dimers, or trimers, with the gp130 component. The gp130 subunit family of cytokine receptors is further subdivided into receptors specific for *monomeric cytokines, such as IL-6,* and those which bind the *dimeric cytokines, such as IL-12.*

Because the Hematopoietin (Class I) and Interferon (Class II) cytokine receptor families utilize similar signaling pathways, we will first describe the Interferons and their receptors and then consider the signaling pathways used by the two families together.

The Interferon (Class II) Cytokine Family Was the First to Be Discovered

In the late 1950s, investigators studying two different viral systems in two laboratories half a world apart almost simultaneously discovered interferons. Yasu-Ichi Nagano and Yashuhiko Kojima, Japanese virologists, were using a rabbit skin and testes tissue culture model to develop a vaccine against smallpox. They noted that immunization with a UV-inactivated form of the cowpox virus resulted in the localized inhibition of viral growth, following a subsequent injection of the same virus. Viral growth inhibition was restricted to a small area of skin close to the site of the original immunization, and the scientists postulated that the initial injection had resulted in the production of a "viral inhibitory factor." After showing that their "inhibitory factor" was not simply antibody, they published a series of papers about it. With hindsight, scientists now believe that their protective effect was mediated by interferons. However, the technical complexity of their system, and the fact that their papers were published in French, rather than in English, delayed the dissemination of their findings to the broader scientific community.

Meanwhile, in London, Alick Isaacs and Jean Lindenman were growing live influenza virus on chick egg chorioallantoic membranes (a method that is still used today), and noticed that exposure of their membranes to a heat-inactivated form of influenza interfered with subsequent growth of a live virus preparation on that surface preparation. They proved that the growth inhibition resulted from the production of an inhibiting molecule by the chick membrane. They named it "interferon" because of its ability to "interfere" with the growth of the live virus. Their more straightforward in vitro assay system enabled them to rapidly characterize the biological effects of the molecule involved, and they wrote a series of papers describing the biological effects of interferon(s) in the late 1950s. However, since interferons are active at very low concentrations, it was not until 1978 that they were produced in quantities sufficient for biochemical and crystallographic analysis. Since that time, investigators have shown that there are two major types of interferons, Types 1 and 2, and that Type 1 interferons can be subdivided into two subgroups.

Interferons

Type I interferons are composed of *Interferons* α, a family of about 20 related proteins, and *interferon-*β, which are secreted by activated macrophages and dendritic cells, as well as by virus-infected cells. Interferons α and β are also secreted by

virally infected cells after recognition of viral components by **pattern recognition receptors (PRRs)** located either at the cell surface, or inside the cell (see Chapter 5). Intracellular PRRs may interact with virally derived nucleic acids or with endocytosed viral particles. The secreted Type I interferons then interact in turn with membrane-bound interferon receptors on the surfaces of many different cell types. The results of their interaction with these receptors are discussed in detail in Chapter 5, but they include the induction of ribonucleases that destroy viral (and cellular) RNA, and the cessation of cellular protein synthesis. Thus, interferons prevent virally infected cells from replicating and from making new viral particles. However, they simultaneously inhibit normal cellular functions and destroy virally infected cells so that the infection cannot spread.

Type I interferons are dimers of 18 to 20 kDa polypeptides, predominantly helical in structure, and some members of this family are naturally glycosylated. Type I interferons are used in the treatment of a variety of human diseases, most notably hepatitis infections.

Type II interferon, otherwise known as *interferon-*γ, is produced by activated T and NK cells and is released as a dimer (Figure 4-10). Interferon-γ is a powerful modulator of

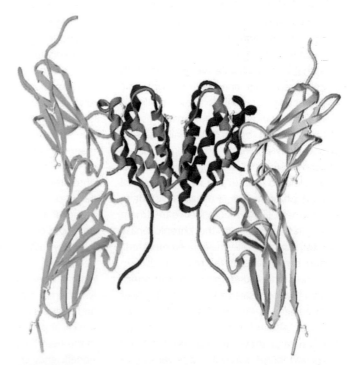

FIGURE 4-10 The complex between IFN-γ and the ligand-binding chains of its receptor. This model is based on the x-ray crystallographic analysis of a crystalline complex of interferon-γ (dark and light purple) bound to ligand-binding α chains of the receptors (green and yellow). Note that IFN-γ is shown in its native dimeric form; each member of the dimer engages the α chain of an IFN-γ receptor, thereby bringing about receptor dimerization and signal transduction. *[From M. R. Walter et al., 1995, Crystal structure of a complex between interferon- and its soluble high-affinity receptor. Nature* **376**:230, *courtesy M. Walter, University of Alabama.]*

CLINICAL FOCUS

Therapy with Interferons

Interferons are an extraordinary group of proteins with important effects on the immune system. Their actions affect both the adaptive and the innate arms of the immune system and include the induction of increases in the expression of both Class I and Class II MHC molecules and the augmentation of NK-cell activity. Cloning of the genes that encode IFN-α, IFN-β, and IFN-γ has made it possible for the biotechnology industry to produce large amounts of each of these interferons at costs that make their clinical use practical (Table 1).

IFN-α (also known by its trade names Roferon and Intron A) has been used for the treatment of hepatitis C and hepatitis B. It has also been found useful in a number of different applications in cancer therapy. A type of B-cell leukemia known as hairy-cell leukemia (because the cells are covered with fine, hairlike cytoplasmic projections) responds well to IFN-α. Chronic myelogenous leukemia, a disease characterized by increased numbers of granulocytes, begins with a slowly developing chronic phase that changes to an accelerated phase and terminates in a blast phase, which is usually resistant to treatment. IFN-α is an effective treatment for this leukemia in the chronic phase (70% response rates have been reported), and some patients (as many as 20% in some studies) undergo complete remission. Kaposi's sarcoma, the cancer most often seen in AIDS patients in the United States, also responds to treatment with IFN-α, and there are reports of a trend toward longer survival and fewer oppor-

tunistic infections in patients treated with this agent. Most of the effects mentioned above have been obtained in clinical studies that used IFN-α alone, but certain applications such as hepatitis C therapy commonly use it with an antiviral drug such as ribavirin. The clearance time of IFN-α is lengthened by using it in a form complexed with polyethylene glycol (PEG) called pegylated interferon.

IFN-β has emerged as the first drug capable of producing clinical improvement in multiple sclerosis (MS). Young adults are the primary target of this autoimmune neurologic disease, in which nerves in the central nervous system (CNS) undergo demyelination. This results in progressive neurologic dysfunction, leading to significant and, in many cases, severe disability. This disease is often characterized by periods of nonprogression and remission alternating with periods of relapse. Treatment with IFN-β provides longer periods of remission and reduces the severity of relapses. Furthermore, magnetic resonance imaging (MRI) studies of CNS damage in treated and untreated patients revealed that MS-induced damage was less severe in a group of IFN-β–treated patients than in untreated ones.

IFN-γ has been used, with varying degrees of success, to treat a variety of malignancies that include non-Hodgkin's lymphoma, cutaneous T-cell lymphoma, and multiple myeloma. A more successful clinical application of IFN-γ in the clinic is in the treatment of the hereditary immunodeficiency chronic granulomatous dis-

ease (CGD; see Chapter 18). CGD features a serious impairment of the ability of phagocytic cells to kill ingested microbes, and patients with CGD suffer recurring infections by a number of bacteria (*Staphylococcus aureus*, *Klebsiella*, *Pseudomonas*, and others) and fungi such as *Aspergillus* and *Candida*. Before interferon therapy, standard treatment for the disease included attempts to avoid infection, aggressive administration of antibiotics, and surgical drainage of abscesses. A failure to generate microbicidal oxidants (H_2O_2, superoxide, and others) is the basis of CGD, and the administration of IFN-γ significantly reverses this defect. Therapy of CGD patients with IFN-γ significantly reduces the incidence of infections. Also, the infections that are contracted are less severe, and the average number of days spent by patients in the hospital is reduced.

IFN-γ has also been shown to be effective in the treatment of osteopetrosis (*not osteoporosis*), a life-threatening congenital disorder characterized by overgrowth of bone that results in blindness and deafness. Another problem presented by this disease is that the buildup of bone reduces the amount of space available for bone marrow, and the decrease in hematopoiesis results in fewer red blood cells and anemia. The decreased generation of white blood cells causes an increased susceptibility to infection.

The use of interferons in clinical practice is likely to expand as more is learned about their effects in combination with other therapeutic agents.

---

the adaptive immune response, biasing T cell help toward the T_H1 type and inducing the activation of macrophages, with subsequent destruction of any intracellular pathogens and the differentiation of cytotoxic T cells. All three interferons increase the expression of MHC complex proteins on the surface of cells, thus enhancing their antigen-presentation capabilities.

Interferon-γ is used medically to bias the adaptive immune system toward a cytotoxic response in diseases such as leprosy and toxoplasmosis, in which antibody responses are less effective than those that destroy infected cells. Clinical Focus Box 4-2 describes additional roles of interferons in the clinic.

Minority members of the Interferon family of cytokines include IL-10, secreted by monocytes and by T, B, and

BOX 4-2

TABLE 1	Cytokine-based therapies in clinical use	
Agent	**Nature of agent**	**Clinical application**
Enbrel	Chimeric TNF-receptor/IgG constant region	Rheumatoid arthritis
Remicade or Humira	Monoclonal antibody against TNF-α receptor	Rheumatoid arthritis, Crohn's disease
Roferon	Interferon-α-2a[*]	Hepatitis B, Hairy-cell leukemia, Kaposi's sarcoma, Hepatitis C[†]
Intron A	Interferon-α–2b	Melanoma
Betaseron	Interferon-β–1b	Multiple sclerosis
Avonex	Interferon-β–1a	Multiple sclerosis
Actimmune	Interferon-γ–1b	Chronic granulomatous disease (CGD), Osteopetrosis
Neupogen	G-CSF (hematopoietic cytokine)	Stimulates production of neutrophils; reduction of infection in cancer patients treated with chemotherapy, AIDS patients
Leukine	GM-CSF (hematopoietic cytokine)	Stimulates production of myeloid cells after bone marrow transplantation
Neumega or Neulasta	Interleukin 11 (IL-11), a hematopoietic cytokine	Stimulates production of platelets
Epogen	Erythropoietin (hematopoietic cytokine)	Stimulates red-blood-cell production
Ankinra (kIneret)	Recombinant IL-1Ra	Rheumatoid arthritis
Daclizumab (Zenapax)	Humanized monoclonal antibody against IL-2R	Prevents rejection after transplantation
Basiliximab (Simulect)	Human/mouse chimeric monoclonal antibody against IL-2R	Prevents transplant rejection

[*]Interferon-α–2a is also licensed for veterinary use to combat feline leukemia.

[†] Normally used in combination with an antiviral drug (ribavirin) for hepatitis C treatment.

Although interferons, in common with other cytokines, are powerful modifiers of biological responses, the side effects accompanying their use are fortunately relatively mild. Typical side effects include flu-like symptoms, such as headache, fever, chills, and fatigue. These symptoms can largely be managed with acetaminophen (Tylenol) and diminish in intensity during continued treatment. Although interferon toxicity is usually not severe, treatment is sometimes associated with serious manifestations such as anemia and depressed platelet and white-blood-cell counts.

dendritic cells that regulates immune responses. IL-10 shares structural similarities with interferon-γ, and these similarities enable it to bind to the same class of receptors. In addition, a third class of interferons, the so-called interferon-λ, or *type III Interferon family*, was discovered in 2003. There are currently three members of this family: interferon-λ1 (IL-29), interferon-λ2 (IL-28A), and interferon-λ3 (IL-28B).

Like Type I interferons, the Type III interferons up-regulate the expression of genes controlling viral replication and host cell proliferation.

Interferon Receptors

Members of the Interferon receptor family are heterodimers that share similarly located, conserved cysteine residues with members of the Hematopoietin receptor family. Initially, only interferon-α, -β, and -γ were thought to be ligands for these receptors. However, recent work has shown that the receptor family consists of 12 receptor chains that, in their various assortments, bind no fewer than 27 different cytokines, including six members of the IL-10 family, 17 Type I interferons, one Type II interferon, and three members of the recently described interferon–λ family, including IL-28A, IL-28B, and IL-29.

The JAK-STAT Signaling Pathway

Early experiments in cytokine signaling demonstrated that a series of protein tyrosine phosphorylations rapidly followed the interaction of a cytokine with a receptor from the Class I or Class II cytokine receptor families. These results were initially puzzling, since the cytokine receptors lack the immunotyrosine activation motifs (ITAMs) characteristic of B- and T-cell receptors. However, studies of the molecular events triggered by binding of interferon gamma (IFN-γ) to its receptor shed light on the mode of signal transduction used by members of both the Hematopoietin and Interferon cytokine families.

In the absence of cytokine, the receptor subunits are associated only loosely with one another in the plane of the membrane, and the cytoplasmic region of each of the receptor subunits is associated noncovalently with inactive tyrosine kinases named **Janus Activated Kinases (JAKs)**. (Some members of this family of kinases retain their earlier name of **Tyk**, but share structural and functional properties with the JAK family of kinases.) The process of signal transduction from Class I and Class II cytokine receptors has been shown to proceed according to the following steps (Figure 4-11):

- Cytokine binding induces the association of the two separate cytokine receptor subunits and activation of the receptor-associated JAKs.

- The receptor-associated JAKs phosphorylate specific tyrosines in the receptor subunits.

- These phosphorylated tyrosine residues serve as docking sites for inactive transcription factors known as **Signal Transducers and Activators of Transcription (STATs)**.

- The inactive STATs are phosphorylated by JAK and Tyk kinases.

- Phosphorylated STAT transcription factors dimerize, binding to one another via SH2/phosphotyrosine interactions.

- Phosphorylation also results in a conformational change in the STAT dimer that reveals a nuclear localization signal.

- The STAT dimer translocates into the nucleus, where it initiates the transcription of specific genes.

Currently, we know of seven STAT proteins (STAT 1-4, 5A, 5B, and 6) and four JAK proteins (JAK 1–3 and Tyk2) in

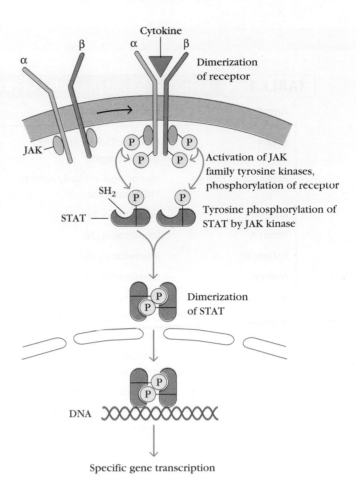

FIGURE 4-11 General model of signal transduction mediated by most Class I and Class II cytokine receptors. Binding of a cytokine induces dimerization of the receptor subunits, which leads to the activation of receptor subunit-associated JAK tyrosine kinases by reciprocal phosphorylation. Subsequently, the activated JAKs phosphorylate various tyrosine residues, resulting in the creation of docking sites for STATs on the receptor and the activation of one or more STAT transcription factors. The phosphorylated STATs dimerize and translocate to the nucleus, where they activate transcription of specific genes.

mammals. Specific STATs play essential roles in the signaling pathways of a wide variety of cytokines (Table 4-4).

Given the generality of this pathway among Class I and Class II cytokines, how does the immune system induce a specific response to each cytokine? First, there is exquisite specificity in the binding of cytokines to their receptors. Secondly, particular cytokine receptors are bound to specific partner JAK enzymes that in turn activate unique STAT transcription factors. Third, the transcriptional activity of activated STATs is specific because a particular STAT homodimer or heterodimer will only recognize certain sequence motifs and thus can interact only with the promoters of certain genes. Finally, only those target genes whose expression is permitted by a particular cell type can be activated within that variety of cell. For example, promoter regions in some cell types may be caught up in heterochromatin and

TABLE 4-4 STAT and JAK interaction with selected cytokine receptors during signal transduction

Each cytokine receptor must signal through a pair of Janus kinases. The JAKs may operate as either homo- or heterodimers.

Cytokine receptor	Janus kinase	STAT
IFN-α/-β	JAK 1, Tyk 2*	STATs 1 and 2
IFN-γ	JAK 1, JAK 2	STAT 1
IL-2	JAK 1, JAK 3	Mainly STATs 3 and 5. Also STAT 1.
IL-4	JAK 1, JAK 3	Mainly STAT 6. Also STAT 5.
IL-6	JAK 1, JAK 2	STAT 3
IL-7	JAK 1, JAK 3	STATs 5 and 3
IL-12	JAK 2, Tyk2	STATS 2, 3, 4, and 5
IL-15	JAK 1, JAK 3	STAT 5
IL-21	JAK 1, JAK 3	Mainly STATs 1 and 3; also STAT 5

* Despite its name, Tyk2 is also a Janus kinase.

inaccessible to transcription factors. In this way, the Class I cytokine IL-4 can induce one set of genes in T cells, another in B cells, and yet a third in eosinophils.

JAK-STAT pathways are not unique to the immune system. Among the many genes known to be regulated by mammalian STAT proteins are those encoding cell survival factors such as the Bcl-2 family members, those involved in cell proliferation such as *cyclin D1* and *myc*, and those implicated in angiogenesis or metastasis such as vascular endothelial growth factor, or *VEGF*.

At the close of cytokine signaling, negative regulators of the STAT pathway, such as *protein inhibitor of activated STAT (PIAS)*, *suppressor of cytokine signaling (SOCS)*, and protein tyrosine phosphatases are believed to be responsible for turning off JAK-STAT signaling and returning the cell to a quiescent state.

Members of the TNF Cytokine Family Can Signal Development, Activation, or Death

The **Tumor Necrosis Family (TNF) family** of cytokines regulates the development, effector function, and homeostasis of cells participating in the skeletal, neuronal, and immune systems, among others.

Cytokines of the TNF Family Can Be Soluble or Membrane Bound

TNF-related cytokines are unusual in that they are often firmly anchored into the cell membrane. Generally they are Type 2 transmembrane proteins with a short, intracytoplasmic N-terminal region, and a longer, extracellular C-terminal region. The extracellular region typically contains a canonical TNF-homology domain responsible for interaction with the cytokine receptors. Members of the TNF family can also act as soluble mediators, following cleavage of their extracellular

regions, and in some cases, the same cytokine exists in both soluble and membrane-bound forms.

There are two eponymous (having the same name as) members of the TNF family: TNF-α and TNF-β, though TNF-β is more commonly known as Lymphotoxin-α, or *LT-α*. Both of these are secreted as soluble proteins. TNF-α (frequently referred to simply as TNF) is a proinflammatory cytokine, produced primarily by activated macrophages, but also by other cell types including lymphocytes, fibroblasts, and keratinocytes (skin cells), in response to infection, inflammation, and environmental stressors. TNF elicits its biological effects by binding to its receptors, TNF-R1 or TNF-R2, which are described below. Lymphotoxin-α is produced by activated lymphocytes and can deliver a variety of signals. On binding to neutrophils, endothelial cells, and osteoclasts (bone cells), Lymphotoxin-α delivers activation signals; in other cells, binding of Lymphotoxin-α can lead to increased expression of MHC glycoproteins and of adhesion molecules.

We will also encounter five physiologically significant, membrane-bound members of the TNF cytokine family throughout this book. *Lymphotoxin-β*, a membrane-bound cytokine, is important in lymphocyte differentiation. We will learn about BAFF and APRIL in the context of B-cell development and homeostasis (Chapter 10). CD40L is a cytokine expressed on the surface of T cells that is required to signal for B-cell differentiation (Chapter 12). Fas ligand (FasL), or CD95L, induces apoptosis on binding to its cognate receptor, **Fas**, or CD95.

Whether membrane-bound or in soluble form, active cytokines of the TNF family assemble into trimers. Although in most cases they are homotrimeric, heterotrimeric cytokines do form between the TNF family members Lymphotoxin-α and Lymphotoxin-β and between APRIL and BAFF. Crystallographic analysis of TNF family members has revealed that

FIGURE 4-12 The TNF-family members act as trimers in vivo. *[PDB ID 1TNF]*

they have a conserved tertiary structure and fold into a β-sheet sandwich. The conserved residues direct the folding in the internal β strands that, in turn, promote the trimer formation (Figure 4-12).

TNF Receptors

Members of the TNF receptor superfamily are defined by the presence of *Cysteine-Rich Domains* (CRDs) in the extracellular, ligand-binding domain. Each CRD typically contains six cysteine residues, which form three disulfide bonded loops, and individual members of the superfamily can contain from one to six CRDs.

Although most TNF receptors are Type 1 membrane proteins (their N-terminals are outside the cell), a few family members are cleaved from the membrane to form soluble receptor variants. Alternatively, some lack a membrane anchoring domain at all, or are linked to the membrane only by covalently bound, glycolipid anchors. These soluble forms of TNF family receptors are known as "decoy receptors," as they are capable of intercepting the signal from the ligand before it can reach a cell, effectively blocking the signal. This is a theme that we have encountered before in our consideration of the IL-1 receptor family.

Signaling Through TNF Superfamily Receptors

The work of delineating the precise pathways of signaling through TNF family receptors is ongoing, and some impor-

tant questions still await resolution. One reason that these pathways have been so difficult to define is that the same receptor, TNF-R1, can transduce both activating and death-promoting signals, depending on the local cellular and molecular environment in which the signal is received, and investigators have yet to determine the trigger that shifts the signaling program from life to death. However, much is known about how each of these signaling pathways work, once that all-important decision has been made.

We will start by describing the proapoptotic (death-inducing) pathway that is initiated when the membrane-bound TNF family member FasL on one cell binds to a Fas receptor on a second cell, leading to death in the cell bearing the Fas receptor. With this as our foundation, we will then illustrate how the TNF-R1 receptor mediates both life- and death- promoting signals. Signaling through other TNF-R family members, such as CD40, BAFF, and April, will be described in later chapters in the context of the various immune responses in which they are involved.

Signaling Through the Fas Receptor

At the close of an immune response, when the pathogen is safely demolished and the immune system needs to eliminate the extra lymphocytes it has generated to deal with the invader, responding lymphocytes begin to express the TNF family receptor Fas on their cell surfaces. Fas, and its ligand FasL, are specialized members of the TNF receptor and the TNF cytokine families, respectively, and they work together to promote lymphocyte homeostasis. Mice with mutations in either the *fas* (mrl/lpr mice) or the *fasL* (gld mice) genes consequently suffer from severe lympho-proliferative disorders, indicative of their inability to eliminate lymphocytes that are no longer serving a useful purpose.

On interaction with other immune cells bearing FasL, the Fas receptor trimerizes and transduces a signal to the interior of the Fas-bearing cell that results in its elimination by **apoptosis**. Apoptosis, or **programmed cell death**, is a mechanism of cell death in which the cell dies from within and is fragmented into membrane-bound vesicles that can be rapidly phagocytosed by neighboring macrophages (Figure 4-13a). By using such well-controlled apoptotic pathways, the organism ensures that minimal inflammation is associated with the natural end of an immune response. Activation of the apoptotic pathway invokes the activation of caspases; these are proteases, bearing *Cysteine* residues at their active sites, which cleave after *ASPartic* acid residues.

Binding of Fas to FasL results in the clustering of the Fas receptors (Figure 4-13b). This, in turn, promotes interaction between their cytoplasmic regions, which include domains common to a number of proapoptotic signaling molecules called **death domains**. This type of interaction, between homologous protein domains expressing affinity for one another, is referred to as a *homotypic interaction*. As they bind to one another, the clustered Fas protein death domains incorporate death domains from the adapter

(a)

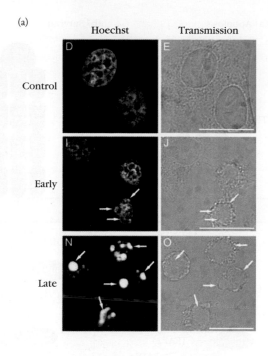

(b)

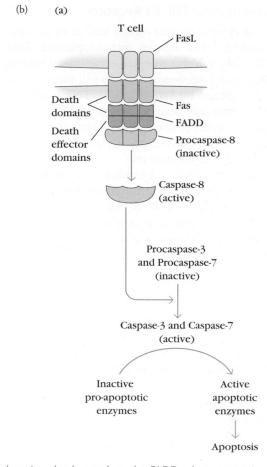

FIGURE 4-13 Apoptotic signaling through Fas receptors.
(a) Human HeLa cells were activated to undergo apoptosis through the Fas receptors. Hoechst-stained cells show the gradual condensation of nuclear DNA into membrane-bounded blebs, as the cell breaks up into vesicular packages that are recognized and phagocytosed by macrophages in the absence of inflammation. The same cells are also shown under transmission microscopy. Arrows show staining of the Nuclear Mitotic Apparatus protein, an early nuclear caspase target in apoptosis. (b) Signaling from Fas leads to apoptosis. Binding of FasL to Fas induces clustering of the Fas receptors and corresponding clustering of the Fas Death Domains (DDs). The DDs of the adapter protein FADD bind to the clustered Fas DDs via a homotypic interaction. Death effector domains also located on the FADD adapter proteins incorporate the DED domains of procaspase-8 into the membrane complex. Clustering of procaspase-8 induces cleavage of the pro domains of procaspase-8, leading to the release of the active caspase-8 protease. Caspase-8 cleaves the pro domains from the executioner caspases, caspase-3 and caspase-7, which in turn cleave and activate nucleases leading to the degradation of nuclear DNA. Caspase-8 also cleaves and activates the proapoptotic Bcl-2 family member protein, BID. [Taimen, Pekka and Kallajoki, Markku. NuMA and nuclear lamins behave differently in Fasmediated apoptosis J Cell Sci 2003 116:(3):571-583; Advance Online Publication December 11, 2002, doi:10.1242/jcs.00227. Reproduced with permission of Journal of Cell S]

protein FADD (*Fas-Associated Death Domain*-containing protein). FADD contains not only death domains, but also a related type of domain called a *Death Effector Domain* (DED). This, in turn, binds homotypically to the DED domains of procaspase-8, resulting in the clustering of procaspase-8 molecules. Procaspase-8 molecules contain the active caspase-8 enzyme, held in an inactive state by binding to prodomains.

The multimerization of procaspase-8 molecules results in mutual cleavage of their prodomains and induces capsase-8 activation. Caspase-8 then cleaves many target proteins

critical to the generation of apoptosis. The target proteins of caspase-8 include the executioner caspases, -3 and -7 (which cleave and activate nucleases leading to the degradation of nuclear DNA), and the proapoptotic Bcl-2 family member BID. The complex of Fas, FADD, and procaspase-8 is referred to as the **Death-Inducing Signaling Complex (DISC)**. The ultimate result of activation of this cascade is the condensation of nuclear material (see Figure 4-13a), the degradation of nuclear DNA into 240 base pair, nucleic acid fragments, and the subsequent breakdown of the cell into "easily digestible" membrane-bound fragments.

Signaling Through the TNF-R1 Receptor

The TNF-R1 receptor is present on the surface of all vertebrate cells and, like Fas, has an intracytoplasmic death domain (DD). Although this receptor is capable of binding to both TNF-α and Lymphotoxin-α, we will focus on the signaling that is elicited by TNF-α (TNF). TNF binding to the TNF-R1 receptor can lead to two very different outcomes: apoptosis (death) or survival (life). How it does so is still the focus of intensive investigation, but the story as it is unfolding is already a fascinating one.

The mechanism by which TNF binding leads to apoptosis is slightly different from that which follows Fas-FasL binding. Like FasL, binding of TNF to the TNF-R1 receptor induces trimerization of the receptor as well as an alteration in its conformation, and these together result in the binding of a DD-containing adapter molecule, in this case TRADD, to the internal face of the receptor molecule (Figure 4-14). The TRADD adapter molecule provides additional binding sites for the components RIP1 (a serine/threonine kinase, rather evocatively named *Rest In Peace 1*), which binds via its own DDs and TRAF2, the *TNF Receptor Associated Factor 2*. This is known as *complex I*. Intracellular localization experiments have shown that this complex can dissociate from the TNF-R at the membrane, and migrate to the cytoplasm where it binds to the now familiar DD-containing protein FADD. FADD recruits procaspase-8, as described above, resulting in the generation of an apoptotic signal. The proapoptotic cytoplasmic complex generated upon TNF-R1 receptor binding is shown in Figure 4-14a as *complex II*.

Counterintuitively, binding of this same molecule, TNF, to the same receptor, TNF-R1, can result in the delivery of survival as well as of proapoptotic signals. How can the same cytokine, acting through the same receptor, bring about two apparently opposing actions?

In the TNF-mediated survival pathway (Figure 4-14b), the generation of the original membrane complex appears to initiate in the same general manner as for the proapoptotic pathway. However, in the case of the prosurvival pathway, the TRADD-containing complex does not dissociate from the membrane, but rather remains at the cell surface and recruits a number of other components, including the ubiquitin ligases cIAP1 and cIAP2.

Once cIAP1 and cIAP2 join the TNF-R1 complex at the cell membrane, they recruit the LUBAC proteins, which attach linear ubiquitin chains to RIP1. Polyubiquitinated RIP1 then binds to the NEMO component of IKK as well as to TAK1, which is already complexed with its associated TAB proteins as described above. RIP1 and TAK1 activate the IKK complex, leading to IκB phosphorylation and destruction, and subsequent activation of NF-κB. Once NF-κB is fully activated, it turns on the expression of the cFLIP protein that then inhibits the activity of caspase-8. This effectively shuts down the antagonistic, proapoptotic pathway (Figure 4-14a). As previously described, the TAK1 complex also acts to activate the MAP kinase pathway, which further enhances survival signaling.

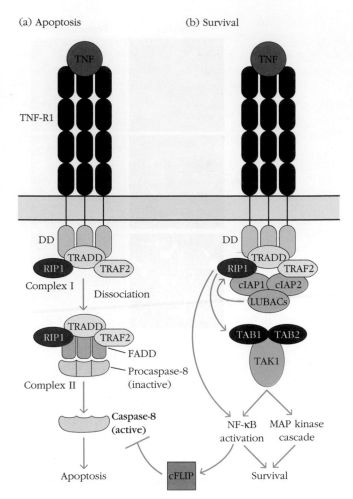

FIGURE 4-14 Signaling through TNF-R family receptors. Signaling through TNF-R family receptors can lead to pro- or antiapoptotic outcomes depending on the nature of the signal, the receptor, and the cellular context. (a) Apoptosis. Binding of TNF to TNF-R1 induces trimerization of the receptor and conformational alteration in its cytoplasmic domain, resulting in the recruitment of the DD-containing adapter molecule TRADD to the cytoplasmic face of the receptor. TRADD binds to the serine-threonine kinase RIP1 and the TNF receptor associated factor TRAF2. This complex of TRADD, RIP1, and TRAF2, known as complex I, dissociates from the receptor and migrates to the cytoplasm where it binds to the adapter protein FADD. FADD recruits procaspase-8, leading to apoptosis as described in Figure 4-13. The proapoptotic complex generated upon TNF-R1 receptor binding is shown as Complex II. (b). Survival. As for the apoptotic pathway, TNF ligation results in receptor trimerization, TRADD binding, and RIP1 recruitment. In this case, however, TRADD also recruits the ubiquitin ligases cIAP1 and cIAP2, which in turn bind to the proteins of the *l*inear *u*biquitin *a*ssembly *c*omplex (LUBAC) proteins. Polyubiquitination of RIP1 allows it to bind to the NEMO component of the IKK complex as well as to TAK1. TAK1 and RIP1 together activate the IKK complex, leading to IκB phosphorylation and destruction, and release of NF-κB to enter the nucleus. Among other prosurvival effects, NF-κB activates the transcription of the cFLIP protein, which inhibits caspase-8 action, thus tipping the scales in favor of survival. The TAK1 complex also activates MAP kinase signaling, which enhances cell survival.

The survival versus death decisions that are made at the level of the TNF-R1 receptor depend upon the outcome of the race between the generation of active caspase-8 on the one hand and the generation of the caspase-8 inhibitor cFLIP on the other. Although we now understand the molecular mechanisms that bring about the consequences of these decisions, we still have much to learn regarding how the cell integrates the signals received through TNF-R1 with other signals delivered to the cell in order to determine which of the two competing pathways will prevail. Since the generation of the membrane-bound complex that is capable of activating NF-κB is entirely dependent on interactions between various ubiquitinated proteins, it now appears that the life-death decision for a cell may be executed by a small protein previously thought to have only destructive intent.

The IL-17 Family Is a Recently Discovered, Proinflammatory Cytokine Cluster

The most recently described family of cytokines, the **IL-17 family**, includes interleukins 17A, 17B, 17C, 17D, and 17F. Signaling through most members of this family culminates in the generation of inflammation. IL-17 receptors are found on neutrophils, keratinocytes, and other nonlymphoid cells. Members of the IL-17 family therefore appear to occupy a location at the interface of innate and adaptive immunity. IL-17 cytokines do not share sequence similarity with other cytokines, but intriguingly the amino acid sequence of IL-17A is 58% identical to an open reading frame (ORF13) found in a T-cell–tropic herpesvirus. The significance of this sequence relationship is so far unknown; did the virus hijack the cytokine sequence for its own needs, or did the pilfering occur in the opposite direction?

IL-17 Cytokines

IL-17A, the first member of this family to be identified, is released by activated T cells and stimulates the production of factors that signal a proinflammatory state, including IL-6, CXCL8, and granulocyte colony-stimulating factor (G-CSF). As characterization of IL-17A and the T cells that secreted it progressed, it became clear that the T cells secreting this cytokine represent a new lineage, the T_H17 cell subset, which is currently the focus of intense investigation (see Chapter 11). Genomic sequencing has since led to the identification of a number of homologs of IL-17A (see Table 4-5). Most of the interleukins in the IL-17 family share the property of operating at the interface of innate and adaptive immunity, serving to coordinate the release of proinflammatory and neutrophil-mobilizing cytokines. However, IL-17E provides an exception to this general rule, instead promoting

TABLE 4-5	Expression and known functions of members of the extended IL-17 receptor family			
Family member	Other common names	Receptors	Expression by which cells	Main functions
IL-17A	IL-17 and CTL-8	IL-17RA and IL-17RC	T_H17 cells, CD8$^+$ T cells, γδ T cells, NK cells, and NKT cells	Autoimmune pathology, neutrophil recruitment, and immunity to extracellular pathogens
IL-17B	NA	IL-17RB	Cells of the GI tract, pancreas, and neurons	Proinflammatory activities?
IL-17C	NA	IL-17RE	Cells of the prostate and fetal kidney	Proinflammatory activities?
IL-17D	NA	Unknown	Cells of the muscles, brain, heart, lung, pancreas, and adipose tissue	Proinflammatory activities?
IL-17E	IL-25	IL-17RA and IL-17RB	Intraepithelial lymphocytes, lung epithelial cells, alveolar macrophages, eosinophils, basophils, NKT cells, T_H2 cells, mast cells, and cells of the gastrointestinal tract and uterus	Induces T_H2 responses and suppresses T_H17 responses
IL-17F	NA	IL-17RA and IL-17RC	T_H17 cells, CD8$^+$ T cells, γδ T cells, NK cells, and NKT cells	Neutrophil recruitment and immunity to extracellular pathogens
IL-17A/IL-17F heterodimer	NA	IL-17RA and IL-17RC	T_H17 cells, CD8$^+$ T cells, γδ T cells, NK cells, and NKT cells	Neutrophil recruitment and immunity to extracellular pathogens
vIL-17	ORF13	IL-17RA (and IL-17RC?)	*Herpesvirus saimiri*	Unknown

Adapted from Gaffen, S. L. 2009. Structure and signalling in the IL-17 receptor family. *Nature Reviews Immunology* **9**:556–567.

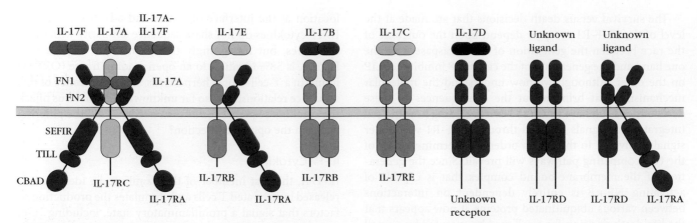

FIGURE 4-15 The IL-17 family of cytokines and their associated receptors. The cytokines that form the IL-17 family share a highly conserved structure, with four conserved cysteines. Only one of the proteins has so far been subject to x-ray analysis, which demonstrates that the structure is that of a "cysteine knot," a tightly folded protein that exists naturally as a dimer. The five proteins that make up the IL-17 receptor family are IL-17RA, IL-17RB, IL-17RC, IL-17RD, and IL-17RE. These are arranged into homo- and hetero- dimers and trimers to create the complete receptor molecules shown. Each receptor protein includes one or more fibronectin (FN) domains, as well as a cytoplasmic *SEF/IL-17R* (SEFIR) domain that is important in mediating downstream signaling events. The IL-17RA protein also includes a *TIR-like loop* domain (TILL), similar to that found in Toll-like receptors and IL-1 receptors, as well as a C/EBPβ activation domain, capable of interacting with the downstream transcription factor C/EBP. [*Adapted from S. Gaffen, 2009, Structure and signalling in the IL-17 receptor family,* Nature Reviews Immunology, *9:556.*]

the differentiation of the anti-inflammatory T$_H$2 subclass, while suppressing further T$_H$17 cell responses, in what amounts to a negative feedback loop.

In general, members of the IL-17 family exist as homodimers, but heterodimers of IL-17A and IL-17F have been described. Monomeric units of IL-17 family members range in molecular weight from 17.3 to 22.8 kDa, and crystallographic analysis has revealed that they share a structure that is primarily β sheet in nature, stabilized by intrachain disulfide bonds.

The IL-17 Family Receptors

The IL-17 receptor family is composed of five protein chains—IL-17RA, IL-17RB, IL-17RC, IL-17RD, and IL-17RE—which are variously arranged into homo- and hetero- dimeric and trimeric units to form the complete receptor molecules (Figure 4-15). Members of the IL-17 receptor family share fibronectin domains with the Hematopoietin and Interferon family cytokine receptors and are single transmembrane proteins. They all contain cytoplasmic *SEF/IL-17R* (*Similar Expression to Fibroblast growth factor interleukin 17 Receptor*, or SEFIR) domains, responsible for mediating the protein-protein interactions of the IL-17R signal transduction pathway. The IL-17RA chain also contains a *TIR-like loop* (TILL) domain, analogous to structures found in the Toll and IL-1 receptor molecules, as well as a *C/EBPβ activation domain* (CBAD), capable of activating the C/EBPβ transcription factor.

Signaling Through IL-17 Receptors

Analogous to signaling through IL-1 receptors, signaling through most IL-17 receptors results in an inflammatory response, and so it should not come as a surprise to learn that signaling through the IL-17 receptor results in activa-

tion of NF-κB, a hallmark transcription factor of inflammation. Details of the signaling pathways that emanate from the IL-17R are still being worked out, but Figure 4-16 illustrates the major features of our current knowledge.

1. *NF-κB activation via IL-17RA and IL-17RC.* Binding of IL-17A to the receptor molecules IL-17RA and IL-17RC results in the recruitment of the adapter protein ACT1 to the SEFIR domain. ACT1 binds other proteins, including TRAF3 and TRAF6, which then engage with the TAK1 complex. TAK1 activation results in the phosphorylation and inactivation of the inhibitor of NF-κB (IκB), allowing NF-κB activation and nuclear migration.

2. *Activation of MAP kinase pathway and cytokine mRNA stabilization.* Adapter proteins bound to the receptor also recruit components of the MAP kinase pathway, resulting in the activation of MAP kinases, including the extracellular signal-regulated kinase Erk1. Though unusual, it appears that the most important role of Erk1 in IL-17 signaling is not in the generation of phosphorylated transcription factors (as is the case for its involvement in TCR- and BCR-mediated cell signaling), but rather in *controlling the stability of cytokine mRNA transcripts.* Many of the target genes of IL-17 signaling are cytokines and chemokines whose transcription is up-regulated on receipt of an IL-17 signal. The levels of cytokine mRNA are controlled in part by binding of the cytoplasmic protein tristetraprolin to *AU-rich elements* (AREs) in the 3′-untranslated regions of mRNA transcripts. Tristetraprolin then delivers the cytokine mRNAs to the exosome complexes of the cells, where they are degraded. However, phosphorylation of tristetraprolin by MAP kinases inhibits

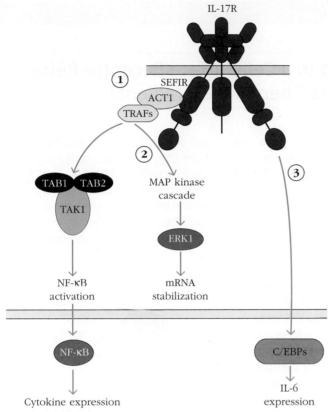

IL-17R

① SEFIR
ACT1
TRAFs

② MAP kinase cascade

③

TAB1 TAB2
TAK1

ERK1

NF-κB activation

mRNA stabilization

NF-κB

C/EBPs

Cytokine expression

IL-6 expression

FIGURE 4-16 Signaling from the IL-17 receptor. Binding of IL-17 to its receptor initiates three signaling pathways. (1) Binding of IL-17 to its receptor results in the recruitment of the adapter protein ACT1 to the cytoplasmic region of the receptor. ACT1 then serves as a docking point for TRAF proteins 3 and 6, which in turn recruit members of the TAK1 complex, consisting of the TAK1 kinase and TAK1 binding proteins. TAK1 activation results in the phosphorylation and activation of the IKK complex and resultant NF-κB activation, as described previously. (2) Adapter proteins bound to the SEFIR (*Similar Expression to Fibroblast growth factor interleukin 17 Receptor*) domain also recruit components of the MAP kinase pathway. The MAP kinase Erk1 phosphorylates the cytoplasmic protein tristetraprolin, and inhibits its ability to bind to AU-rich elements on mRNA encoding cytokines. Since tristetraprolin binding results in mRNA degradation, activation of this arm of the pathway results in enhancing the stability of cytokine mRNA. (3) IL-17 binding to its receptor also results in the activation of transcription factors of the C/EBP family, which promote the expression of the inflammatory cytokine IL-6.

its ability to recruit the degradative machinery and hence results in increased stability of the cytokine and chemokine mRNAs.

3. *Activation of the transcription factors C/EBPβ and C/EBPδ* In addition to up-regulation of NF-κB and members of the MAP kinase pathway, signaling through IL-17RA and IL-17RC proteins has also been shown to activate the transcription factors C/EBPβ and C/EBPδ, which promote expression of IL-6, one of the quintessential inflammatory cytokines.

Chemokines Direct the Migration of Leukocytes Through the Body

Chemokines are a structurally related family of small cytokines that bind to cell-surface receptors and induce the movement of leukocytes up a concentration gradient and toward the chemokine source. This soluble factor-directed cell movement is known as **chemotaxis**, and molecules that can elicit such movement are referred to as chemoattractants (Box 4-3). Some chemokines display innate affinity for the carbohydrates named glycosaminoglycans, located on the surfaces of endothelial cells, a property that enables them to bind to the inner surfaces of blood vessels and set up a cell-bound chemoattractant gradient along blood vessel walls, directing leukocyte movement.

Chemokine Structure

Chemokines are relatively low in molecular weight (7.5–12.5kDa) and structurally homologous. The tertiary structure of chemokines is constrained by a set of highly conserved disulfide bonds; the positions of the cysteine residues determine the classification of the chemokines into six different structural categories (Figure 4-17). Within any one category, chemokines may share 30% to 99% sequence identity.

The grouping of chemokines into the subclasses shown in Figure 4-17 has functional, as well as structural, significance. For example, the seven human CXC chemokines within the ELR subclass share the same receptor (CXCR2), attract neutrophils, are angiogenic, and have greater than 40% sequence identity. (A substance is *angiogenic* if it promotes the formation of new blood vessels; it is *angiostatic* if it prevents the formation of new blood vessels.) The non-ELR, CXCL chemokines CXCL9, CXCL10, and CXCL11, are also more than 40% identical to one another; however, this group is angiostatic, not angiogenic, and utilizes the CXCR4 receptor. Members of the two, structurally distinct CC groups are chemoattractants that attract monocytes and macrophages (although not neutrophils) to the site of infection. See Appendix III for a more comprehensive tabulation of chemokines and their immunologic roles.

Chemokine Receptors

In the 1950s, investigations of the mechanisms by which glucagon and adrenaline signaling led to an increase in the rate of glycogen metabolism revealed the existence of a class of receptors that threads through the membrane seven times and transduces the ligand signal via interactions with a polymeric GTP/GDP-binding "G protein." This class of **G-Protein–Coupled Receptors (GPCRs)** is used in the recognition of many types of signals, including those mediated by chemokines. Certain essential features of this pathway are conserved in all GPCR-type responses. (These larger, polymeric, seven membrane pass receptor-associated G proteins are different from the small, monomeric G proteins such as ras, which participate farther downstream in intracellular signaling pathways. Although both types of G proteins are

How Does Chemokine Binding to a Cell-Surface Receptor Result in Cellular Movement Along the Chemokine Gradient?

Chemotaxis is the mechanism by which the speed and direction of cell movement are controlled by a concentration gradient of signaling molecules. Receptors on the surface of responsive cells bind to chemotactic factors and direct the cell to move toward the source of factor secretion. Chemotaxis was recognized as a biological phenomenon as early as the 1880s, and is found in organisms as simple as bacteria, which use chemotaxis to locate sources of nutrition. However, only as scientists have developed the ability to analyze signaling pathways and to observe the movement of individual cells under in vivo conditions have we been able to approach a mechanistic understanding of the complex series of intracellular events that culminate in chemically directed cell movement in the immune system. Although we still do not understand all the details of this process, some general principles have begun to emerge.

We first address the means by which cells *sense* chemoattractant signals. We next develop an appreciation for how cells *move*, and finally we can describe a little of what we know regarding how they become *polarized* in order to direct their movement as specified by the chemotactic signal.

SIGNAL SENSING

Leukocytes recognize chemoattractant signals, or chemokines, using the G-protein–coupled receptors described in this chapter. Such receptors are located at the plasma membrane, and the cytoplasmic face of

these receptors is associated with a polymeric G protein, so called because of its affinity for guanosine phosphates. On binding to its chemokine ligand, the receptor alters its conformation, passing on that conformational change to its associated G protein. The bound G protein then loses affinity for guanosine diphosphate (GDP) and instead binds guanosine triphosphate (GTP), thus achieving its active conformation. Next, the G protein dissociates into two subunits, termed Gα-GTP and Gβγ. The signaling pathways emanating from each of these subunits are described in the text. It is difficult to overstate the effectiveness of these receptor signals in determining the direction and speed of cellular movement. For example, neutrophils can recognize and move in response to extremely shallow chemoattractant gradients, in which the concentration of the chemoattractant at the front of the gradient is as little as 2% higher than that at the back.

HOW CELLS MOVE

Figure 1 illustrates a model of amoeboid cellular movement, the mechanism used by leukocytes. Ameboid migration is a rapid type of cell movement; leukocytes and stem cells, which use this mode of migration, may move as quickly as 30 μm/minute. Cells routinely move over other cells and tissues (the substrata) by first forming cellular protrusions at their leading edge—the edge closest to the direction of movement. Depending on the cell type and the nature of the chemoattractant signal, these protrusions may be thin,

and almost like wires (filopodia), or protrude from the surface of the cell like a sheet (lamellipodia). The nucleus of the migrating cell lies behind these protrusions in the cell body, and at the posterior part of the cell lies a near-cylindrical tail, the uropod, which may be as long as 10 μm.

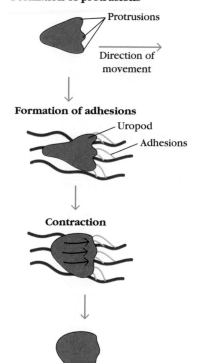

FIGURE 1

Schematic description of the classical migratory cycle. *[Adapted from P. Mrass et al., 2010, Cell-autonomous and environmental contributions to the interstitial migration of T cells, Seminars in Immunopathology* **32***:257.]*

The protrusions at the leading edge of a moving leukocyte form noncovalent attachments to the substratum using adhesive interactions between proteins on the cell protrusion and other proteins on the substratum. The substratum may, for example, be a capillary endothelial cell, when a white blood cell rolls along the inside of a blood capillary. Alternatively, during leukocyte movement within a lymph node, the adhesive interaction may occur between B or T lymphocyte surface integrins and proteins located on a cellular extension from a follicular reticular cell.

Once the migrating cell is temporarily attached to the substratum, contractions of the cell body mediated by actin and myosin, and resulting cytoskeletal rearrangements, bring the rest of the cell forward toward the leading edge and then the cycle of reaching forward with cellular protrusions and contraction of the rest of the cell in the direction of the leading edge can recur.

GENERATION OF POLARITY SIGNALS: HOW CELLS DECIDE ON THE DIRECTION OF MOVEMENT

Upon receipt of a chemotactic signal, leukocytes polarize very rapidly, and leukocyte locomotion is clearly visible within 30 seconds after the receipt of the chemokine by the cell.

Scientists initially hypothesized that movement of a leukocyte toward a particular chemoattractant may result from an asymmetric distribution of chemoattractant receptors on the cell surface, with a higher concentration of receptors localized to the leading edge of the cell. However, microscopic observations suggest that this initial hypothesis was incorrect. Instead, investigators found an asymmetrical distribution of signaling molecules and cytoskeletal components (Figure 2). Thus, although there is a uniform distribution of *receptors* on the cell surface, the distribution of *occupied receptors* is asymmetrical, which means that the *signal* received from chemokines on the side of the cell facing the chemokine source is stronger than that received elsewhere on the cell surface. The ultimate result of this asymmetrical signal from the occupied receptors is an internal redistribution of signaling pathway components and cytoskeletal elements; this in turn causes the cell to move toward the source of the signal.

Many of the signaling pathways that mediate this directional movement are already familiar to you from Chapter 3 and from this chapter. The $\beta\gamma$ dimer of the activated G protein recruits a particular subclass of PI3 kinase enzymes to the inner leaflet of the leading edge of the membrane, where it phosphorylates PIP2 and other phosphatidyl inositol phosphates. These phosphorylated lipids then serve as docking sites for PH-domain–containing proteins, including Akt, which in turn phosphorylates downstream effectors leading to actin polymerization. Depending on the cell type and the nature of the stimulation, the small GTPases Rac, Cdc42, and Rho are also implicated in the chemokine-mediated cytoskeletal modifications that are necessary for cell movement, and these are activated via tyrosine kinases in lymphocytes and by pathways that are as yet not fully characterized in other leukocyte cell types.

A cell can only move in response to a gradient if the trailing edge of the cell moves at the same rate as the leading edge, and the uropod of moving leukocytes is susceptible to its own set of G-protein–mediated signals. Following redistribution of the internal signaling components, the chemokine receptors at the trailing edge of the cell are coupled to a set of G-protein trimers different from those at the leading edge. Interaction of the uropod-localized *G-Protein–Coupled Receptors* (GPCRs) leads to the formation and contraction of actin-myosin complexes and subsequent retraction of the uropod.

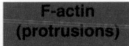

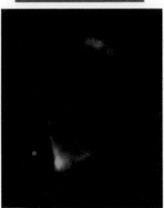

FIGURE 2
Polarization of neutrophils after receipt of a chemoattractant signal. A neutrophil exposed to a chemoattractant organizes filamentous actin at the protrusions on the leading edge of the cell (shown here in red) and an actino-myosin contractile complex at the uropod and along the sides of the cell (shown here stained green). [Fei Wang, The signaling mechanisms underlying cell polarity and chemotaxis. Cold-Spring Harbor Perspectives in Biology, October 2009, 1(4):a002980. doi: 10.1101/cshperspect.a002980. ? 2009 Cold Spring Harbor Laboratory Press, all rights reserved.]

Class	Structural signature	Names	Number (n) in class
CX3C	............CXXXC............C.........C.........	CX3CL1	1
Non-ELR CXC	............CX__C............C.........C.........	CXCL#	9
ELR CXC	...ELR....CX__C............C.........C.........	CXCL#	7
4C CC	............C___C............C.........C.........	CCL#	19
6C CC	............C___C......C....C.........C....C...	CCL#	5
C	....................C.......................C	XCL#	2

FIGURE 4-17 Disulfide bridges in chemokine structures. A schematic of the locations of cysteine residues in chemokines that shows how the locations of cysteines determine chemokine class. Chemokines are proteins of small molecular weight which share two, four, or six conserved cysteine residues at particular points in their sequence that form intrachain disulfide bonds. The number of cyste- ines as well as the positions of the disulfide bonds determine the subclass of these cytokines as shown. The overscores indicate the cysteines between which disulfide bonds are made. The naming of chemokines in part reflects the cysteine-determined class (see Appendix III). *[Adapted from W. E. Paul, 2008,* Fundamental Immunology, *6th ed. Lippincott Williams & Wilkins, Philadelphia, Figure 26.1]*

activated by GTP binding, their structures and functions are quite different).

The GPCRs are classified according to the type of chemokine they bind. For example, the CC receptors (CCRs) recognize CC chemokines, the CXCRs recognize CXCL chemokines, and so on. Chemokine receptors bind to their respective ligands quite tightly (Kd ≅ 10^{-9} M). Interestingly, the intrinsic specificity of the receptors is balanced by the capacity of many receptors to bind more than one chemokine from a particular family and of several chemokines to bind to more than one receptor. For example, the receptor CXCR2 recognizes seven different chemokines, and CCL5 can bind to both CCR3 and CCR5.

Signaling Through Chemokine Receptors

The cytoplasmic faces of seven membrane pass GPCRs associate with intracellular, trimeric **GTP-binding proteins** consisting of Gα, Gβ, and Gγ subunits (Figure 4-18a). When the receptor site is unoccupied, the Gα subunit of the trimeric G protein binds to GDP. Chemokine ligation to the receptor results in a conformational change that is transmitted to the G protein and in turn induces an exchange of GDP for GTP at the Gα binding site that is analogous to what occurs upon GTP binding to the small GTP-binding proteins such as Ras. This results in the dissociation of the G protein into a Gα-GTP monomer and a Gβγ dimer. In the case of the chemokine receptor associated G proteins, chemokine signaling is mediated by both the dissociated Gβγ dimer (which has no nucleotide binding site) and the Gα-GTP subunit.

Just as for the small G-protein–coupled pathways, the duration of signaling through the chemokine receptor is limited by the intrinsic GTPase activity of the Gα subunit, which in turn can be increased by **GTPase Activating**

Proteins (GAPs), also known as *Regulators of G-protein Signaling* (RGSs). Because the pathway is active only when the protein binds to GTP, GAPs *down-regulate* the activity mediated by the receptor (see Figure 3-15). Once the GTP in the Gα binding site is hydrolyzed to GDP, Gα re-associates with the Gβγ dimers, effectively terminating signaling. There are multiple subtypes of Gα and Gβ subunits that vary in representation between different cell types. Signaling through different subunits can give rise to different consequences, depending on the downstream pathways that are elicited. Just as for the small G protein, there are several different large polymeric G proteins, which vary in their cellular distribution and receptor partners.

Once released from its Gα partner, the Gβγ subunit of the trimeric G protein activates a variety of downstream effector molecules, including those of the Ras/MAP kinase pathway (path 1 in Figure 4-18). Full activation of MAP kinase is further facilitated by tyrosine phosphorylation mediated by Gα-GTP-activated tyrosine kinases (not shown). Activation of the Ras pathway culminates in the initiation of transcription as well as in up-regulation of integrin adhesion molecules on the cell membrane. Gβγ signaling also cooperates with signaling mediated by Gα-GTP to activate one isoform of phospholipase C, PLCβ, resulting in an increase in the activity of the transcription factor NF-κB (path 2).

The GαGTP complex also activates a signaling pathway that is initiated by the small G protein, Rho (path 3). This pathway leads to actin polymerization and the promotion of cell migration, so it is this third pathway that is responsible for the most commonly described aspect of chemokine signaling: cell movement (see also Box 4-3). Rho signaling is also instrumental in bringing about changes in the transcriptional program of the cell. Finally, a JAK associated with

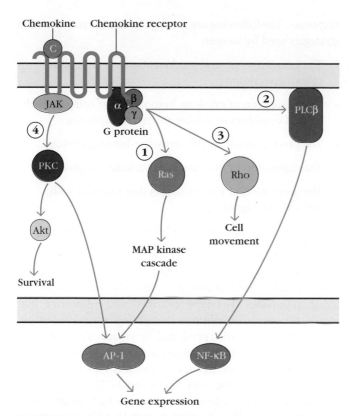

FIGURE 4-18 G-protein–coupled receptors interact with G proteins that transduce chemokine signals into the interior of the cell. Different chemokines can induce different signaling pathways. This figure therefore represents a composite of some of the most common pathways elicited by chemokine binding, which lead collectively to alterations in the transcriptional program, the enabling of cell movement, and changes in the adhesive properties of the signaled cell. (1) The Gβγ subunit binds to the adapter molecule Grb2, activating it and initiating the Ras signaling pathway that leads eventually to activation of MAP kinase and an alteration in the cell's transcriptional program, as shown in Figure 3-16. Ras pathway activation also leads eventually to activation of integrin adhesion molecules on the cell surface. Gα GTP simultaneously binds and activates a protein tyrosine kinase that phosphorylates and further activates MAP kinase (not shown). (2) Both GαGTP and Gβγ cooperate to activate PLCβ, which activates the NF-κB pathway. (3) GαGTP activates the small cytoplasmic G protein, Rho, initiating actin polymerization and cell movement. Other pathways emanate from Rho that lead to the activation of the transcription factor Serum Response Factor (SRF). (4) A JAK is stimulated by chemokine binding to the receptor, and it turns on the activity of PKC, leading eventually to the activation of the enzyme Akt. Akt affects cell survival by phosphorylating the proapoptotic genes Bax and Bad (not shown) and marking them for destruction and enhancing cell survival. It also phosphorylates and further activates the transcription factor NF-κB. JAK-mediated PKC activation can also lead to phosphorylation of the transcription factor Jun, and its dimerization with Fos to create the complete transcription factor AP-1.

the GPCR initiates signaling through PKC-mediated pathways that culminate in the activation of Akt and increased cell survival (path 4) as well as in further transcriptional alterations.

The quality of the response elicited by particular chemokines in different types of cells is dependent on the nature of the chemokine ligand, as well as on the signaling microenvironment, which in turn is determined by the range of G protein subtypes and regulatory molecules present in that cell. But from the complexity of signaling options available to the receptor, it is not difficult to see how binding of a chemokine molecule to its receptor can simultaneously bring about alterations in the location, the adhesion molecule binding capacity, and the transcriptional program of the chemokine-activated cell.

Cytokine Antagonists

A number of proteins that inhibit the biological activity of cytokines have been reported. These proteins act in one of two ways: either they bind directly to a cytokine receptor but fail to activate the cell, thus blocking the active cytokine from binding, or they bind directly to the cytokine itself, inhibiting its ability to bind to the cognate receptor. In this section, we describe some naturally occurring cytokine antagonists that modulate and refine the power of particular cytokine responses, as well as the ways in which various pathogens have hijacked cytokine responses to their own ends.

The IL-1 Receptor Antagonist Blocks the IL-1 Cytokine Receptor

The best-characterized cytokine inhibitor is the IL-1 receptor antagonist (IL-1Ra), which binds to the IL-1 receptor but does not elicit activation of the signaling pathway (see above). As previously described, ligation of IL-1Ra to the IL-1 receptor blocks the binding of both IL-1α and IL-1β, thus accounting for its inhibitory properties. IL-1Ra is synthesized by the same cells that secrete IL-1α and IL-1β, and its synthesis in the liver is up-regulated under inflammatory conditions, along with that of IL-1. Several animal and human models exist in which the levels of IL-1Ra are naturally reduced, and humans carrying an allele that decreases the expression of IL-1Ra suffer from arthritis and a variety of other autoimmune diseases. This observation suggests that the normal function of IL-1Ra is to provide for the host a means by which to modulate the numbers of receptors that are capable of mounting a physiological response to IL-1. Given the fiercely proinflammatory effects of IL-1, it makes sense that responses to this powerful cytokine should be carefully controlled. Indeed, we note quite often in biological systems that a process—if it has the potential to lead to deleterious consequences to the organism—is subject to several different means of regulation.

Recombinant IL-Ra has been used clinically, under the name of *anakrina*, for the treatment of rheumatoid arthritis. Investigations into how cells control the balance of IL-1 and IL-1Ra secretion are still ongoing, but preliminary studies suggest that the activation of different isoforms of PI3 kinase may play an important role in determining the relative amounts of IL-1 and IL-1Ra that are secreted by a stimulated monocyte.

Cytokine Antagonists Can Be Derived from Cleavage of the Cytokine Receptor

Some naturally occurring soluble antagonists arise from enzymatic cleavage of the extracellular domains of cytokine receptors. These soluble receptor components can compete with the membrane-bound receptor for cytokine binding and thus down-modulate the potential cytokine response. The best characterized of the soluble cytokine receptors consists of a segment containing the amino-terminal 192 amino acids of the IL-2Rα (CD25) subunit, which is released by proteolytic cleavage, forming a 45-kDa soluble IL-2 receptor (sIL-2R or sCD25). The shed receptor retains its ability to bind IL-2 and can therefore prevent the cytokine's productive interaction with the membrane-bound IL-2 receptor.

The origin of sIL-2R is still a matter for debate. Recently, regulatory T cells, which express high levels of CD25 on their membrane surfaces, have been shown to release sIL-2R upon activation. Since these T cells serve the function of down-regulating ongoing immune responses, it has been suggested that the soluble IL-2 receptors may serve the physiological function of soaking up excess IL-2 and thus reducing the amount of the cytokine that is available to irrelevant or even to competing effector T cells. The soluble IL-2 receptor is also found in the serum and bodily fluids of patients suffering from a number of hematologic malignancies (blood cell cancers), and high levels of sIL-2R in the blood correlate with a poor disease prognosis. However, the issue of whether the sIL-2R is released from the tumor cells per se, or whether it is released from regulatory T cells that may be acting to dampen the host anti-tumor response, has yet to be resolved.

Some Viruses Have Developed Strategies to Exploit Cytokine Activity

The cytokine antagonists described above derive from an organism's own immune system. However, as is so often the case in immunology, some pathogens have evolved ways in which to circumvent cytokine responses, by mimicking molecules and pathways used by the host. The evolution of anti-cytokine strategies by microbial pathogens provides biological evidence of the importance of cytokines in organizing and promoting effective antimicrobial immune responses. The following are among the various anti-cytokine strategies used by viruses:

- The generation of viral products that interfere with cytokine secretion
- The generation of cytokine homologs that compete with natural cytokines or inhibit anti-viral responses
- The production of soluble cytokine-binding proteins
- The expression of homologs of cytokine receptors
- The generation of viral products that interfere with intracellular signaling
- The induction of cytokine inhibitors in the host cell

Epstein-Barr virus (EBV), for example, produces an IL-10–like molecule (viral IL-10 or vIL-10) that binds to the IL-10 receptor. Just like host-derived IL-10, this viral homologue suppresses T_H1-type cell-mediated responses that would otherwise be effective in fighting a viral infection. Other cytokine mimics produced by viruses allow them to manipulate the immune response in alternative ways that aid the survival of the pathogen. For example, EBV produces an inducer of IL-1Ra, the host antagonist of IL-1. Poxviruses have also been shown to encode a soluble TNF-binding protein and a soluble IL-1-binding protein that block the ability of the bound cytokines to elicit a response. Since both TNF-α and IL-1 are critical to the early phases of an inflammatory, antiviral response, these soluble cytokine-binding proteins may allow the viruses an increased time window in which to replicate.

Yet other viruses produce molecules that inhibit the production of cytokines. One such example is the *cytokine response modifier* (Crm) protein of the cowpox virus, which inhibits the production of caspase-1 and hence prevents the processing of IL-1 precursor proteins. Finally, some viruses produce soluble chemokines and chemokine-binding proteins that interfere with normal immune cell trafficking, and allow the producing viruses and virally infected cells to evade an immune response. Table 4-6 lists a number of viral products that inhibit cytokines, chemokines, and their activities.

Cytokine-Related Diseases

Defects in the complex regulatory networks governing the expression of cytokines and cytokine receptors have been implicated in a number of diseases. Genetic defects in cytokines, their receptors, or the molecules involved in cytokine-directed signal transduction lead to immunodeficiencies such as those described in Chapter 18. Other defects in the cytokine network can cause an inability to defend against specific families of pathogens. For example, people with a defective receptor for IFN-γ are susceptible to mycobacterial

TABLE 4-6	Viruses use many different strategies to evade cytokine-mediated immune mechanisms
Virus	**Virally encoded proteins**
Epstein-Barr Virus (EBV), Cytomegalovirus	IL-10 homolog
Vaccinia virus, Variola virus	Soluble IL-1 receptors
Myxoma virus	Soluble IFN-γ receptor
Variola virus	Soluble TNF receptors
Adenovirus	RID complex proteins induce internalization of Fas receptor
Measles virus	Viral hemagglutinin binds to complement receptor, CD46, signaling disruption of IL-12 production and therefore inhibition of T_H1 pathway differentiation
Herpes simplex virus	Reverses translation block induced by Type 1 interferons
Adenovirus	Blocks interferon-induced JAK/STAT signaling

infections that rarely occur in the general population. In addition to the diseases rooted in genetic defects in cytokine activity, a number of other pathologic states result from overexpression or underexpression of cytokines or cytokine receptors. Several examples of these diseases are given below, followed by an account of therapies aimed at preventing the potential harm caused by cytokine activity.

Septic Shock Is Relatively Common and Potentially Lethal

Despite the widespread use of antibiotics, bacterial infections remain a major cause of septic shock, which may develop a few hours after infection by certain bacteria, including *Staphyloccocus aureus*, *E. coli*, *Klebsiella pneumoniae*, *Pseudomonas aeruginosa*, *Enterobacter aerogenes*, and *Neisseria meningitidis*.

Bacterial septic shock is one of the conditions that falls under the general heading of **sepsis**. Sepsis, in turn, may be caused not only by bacterial infection but also by trauma, injury, ischemia (decrease in blood supply to an organ or a tissue), and certain cancers. Sepsis is the most common cause of death in U.S. hospital intensive-care units and the 13th leading cause of death in the United States. A common feature of sepsis, whatever the underlying cause, is an overwhelming production of proinflammatory and fever-inducing cytokines such as TNF-α and IL-1β. The cytokine imbalance induces abnormal body temperature, alterations in the respiratory rate, and high white blood cell counts, followed by capillary leakage, tissue injury, widespread blood clotting, and lethal organ failure.

Bacterial septic shock often develops because bacterial cell wall **endotoxins** bind to innate immune system pathogen receptors, such as Toll-like receptors (see Chapter 5), on dendritic cells and macrophages, causing them to produce IL-1 and TNF-α at levels that lead to pathological capillary permeability and loss of blood pressure. A condition resembling bacterial septic shock can be produced in the absence of any bacterial infection simply by injecting mice with recombinant TNF-α. Several studies offer hope that neutralizing TNF-α or IL-1 activity with monoclonal antibodies or antagonists will prevent fatal shock from developing. In one such investigation, monoclonal antibodies to TNF-α protected animals from endotoxin-induced shock. In another, injection of a recombinant IL-1 receptor antagonist (IL-1Ra), which prevents binding of IL-1 to the IL-1 receptor as described above, resulted in a threefold reduction in mortality.

However, neutralization of TNF-α does not reverse the progression of septic shock in all cases, and antibodies against TNF-α give little benefit to patients with advanced disease. Recent studies in which the cytokine profiles of patients with septic shock were followed over time shed some light on this apparent paradox.

The increases in TNF-α and IL-1β occur rapidly in early sepsis, so neutralizing these cytokines is most beneficial early in the process. Indeed, in animal experiments, early intervention can prevent sepsis altogether. However, approximately 24 hours following the onset of sepsis, the levels of TNF-α and IL-1β fall dramatically, and other factors become more important. Cytokines critical in the later stages of sepsis include IL-6, MIF, and CCL-8. Sepsis remains an area of intense investigation, and clarification of the process involved in bacterial septic shock and other forms of sepsis can be expected to lead to advances in therapies for this major killer in the near future.

Bacterial Toxic Shock Is Caused by Superantigen Induction of T-Cell Cytokine Secretion

A variety of microorganisms produce toxins that act as **superantigens**. As described in Chapters 9 and 11, superantigens

Cytokines and Obesity

Even as those in the Third World suffer from malnutrition, among the leading current causes of disease and death in developed countries are obesity and its corollary, Type 2 diabetes.

Type 1, or juvenile onset, diabetes has long been known to have an autoimmune etiology (cause). Cells of the adaptive immune system kill the β cells of the Islets of Langerhans in the pancreas, leading to the complete absence of insulin in the diabetic patient, who must rely on exogenously delivered hormone to survive.

In Type 2 diabetes, the body fails to respond to insulin in an appropriate manner, and the cells fail to take in glucose from the blood and into the tissues. Just as in Type 1 diabetes, the presence of high levels of glucose in the blood and tissue fluids facilitates nonenzymatic glucose conversions to reactive carbohydrate derivatives such as glyoxals. These derivatives cross-link proteins and carbohydrates in the membranes and walls of blood vessels and neurons and within the extracellular matrix, leading to the familiar array of diabetic symptoms: poor peripheral circulation, plaque buildup in the arteries, and heart disease. However, we are now beginning to understand that Type 2, just like Type 1, diabetes is a disease closely related to the workings of the immune system.

In 1993, a seminal publication by Hotamisligil and colleagues made the link between inflammation and metabolic conditions such as Type 2 diabetes and obesity. These authors demonstrated that adipocytes (fat cells) constitutively express the proinflammatory cytokine TNF-α, and that TNF-α expression in adipocytes of obese animals is markedly increased. Later research demonstrated that this finding could be extended to humans. The adipose tissue of human subjects was found to constitutively express TNF-α, and blood levels of TNF-α fall after weight loss.

But is this increase in TNF-α expression affecting insulin sensitivity? On binding of insulin to its receptor, tyrosine kinase activity in the cytoplasmic region of the receptor is activated. This *Receptor Tyrosine Kinase* (RTK) then phosphorylates both itself (autophosphorylation) and some nearby proteins. The signaling cascade is initiated from the insulin receptor by the binding of adapter proteins such as Grb2 and IRS-1 via their SH2 domains to the phosphorylated tyrosine sites on the receptor molecule. We now know that TNF-α signaling in adipocytes inhibits the autophosphorylation of tyrosine residues of the insulin receptor and instead induces phosphorylation of serine residues of both the insulin receptor and the IRS-1 adapter. The serine phosphorylation inhibits any subsequent phosphorylation at tyrosine residues, and thus the passage of signals from insulin to the interior of the fat cell is prevented. Recently, the interleukin IL-6 has also been shown to inhibit insulin signal transduction in hepatocytes (liver cells) through a similar mechanism. The decrease in effectiveness of insulin signaling then becomes a self-reinforcing problem, as insulin signaling itself is anti-inflammatory, and so any decrease in the insulin signal can give rise to inflammatory side effects.

However, TNF-α and IL-6 are not the only cytokines implicated in the etiology of Type 2 diabetes. With the discovery and characterization of the proinflammatory cytokine family represented by IL-17, interest has arisen in the relationship between members of this family, obesity, and the control of fat cell metabolism. Since IL-6 is implicated in the differentiation of T lymphocytes to secrete IL-17, obesity and its associated inflammation tend to predispose an individual to secrete IL-17. However, again we find ourselves in a positive feedback loop, as IL-17 acts on monocytes to induce the further secretion of IL-6, thus ensuring the maintenance of an inflammatory state.

It is therefore clear that cytokine signaling plays a profound role in a disease that is emblematic of our time, and which is predicted to afflict close to 40% of the U.S. population by the middle of the next decade.

---

bind to MHC Class II molecules at a location in the MHC molecule that is outside the groove normally occupied by antigenic peptides (see Figure 11-6). They then bind to a part of the Vβ chain of the T-cell receptor that is outside the normal antigen-binding site, and this binding is sufficient to trigger T-cell activation. This means that a given superantigen can simultaneously activate all T cells bearing a particular Vβ domain. Because of their unique binding ability, superantigens can activate large numbers of T cells irrespective of the antigenic specificity of their canonical antigen-binding site.

Although less than 0.01% of T cells respond to a given conventional antigen (see Chapter 11), 5% or more of the T-cell population can respond to a given superantigen. Bacterial superantigens have been implicated as the causative agent of several diseases, such as bacterial toxic shock and food poisoning. Included among these bacterial superantigens are several enterotoxins, exfoliating toxins, and toxic shock syndrome toxin (TSST1) from *Staphylococcus aureus*; pyrogenic exotoxins from *Streptococcus pyrogenes*; and *Mycoplasma arthritidis* supernatant (MAS). The large number of T cells activated by these superantigens results in excessive production of cytokines. The TSST1, for example, has been shown to induce extremely high levels of TNF-α and IL-1β. As in bacterial septic shock, these elevated concentrations of cytokines can induce systemic reactions that include fever, widespread blood clotting, and shock.

In addition to those diseases described above, in which cytokines or their receptors are directly implicated, recent information has indicated the importance of cytokine involvement in the most important public health crisis currently afflicting the developed world: the increasing incidence of Type 2 diabetes. The roles of TNF-α and IL-6 in the induction and maintenance of this disease are described in Box 4-4.

Cytokine Activity Is Implicated in Lymphoid and Myeloid Cancers

Abnormalities in the production of cytokines or their receptors have been associated with some types of cancer. For example, abnormally high levels of IL-6 are secreted by cardiac myxoma (a benign heart tumor), myeloma and plasmacytoma cells, as well as cervical and bladder cancer cells. In myeloma and plasmacytoma cells, IL-6 appears to operate in an autocrine manner to stimulate cell proliferation. When monoclonal antibodies to IL-6 are added to in vitro cultures of myeloma cells, their growth is inhibited. In contrast, transgenic mice that express high levels of IL-6 have been found to exhibit a massive, fatal, plasma-cell proliferation, called plasmacytosis. In addition, as described above, high serum concentrations of the sIL-2R are found in patients suffering from various blood cell cancers, which may impede a vigorous anti-tumor response.

Cytokine Storms May Have Caused Many Deaths in the 1918 Spanish Influenza

Occasionally, a particularly virulent infection may induce the secretion of extremely high levels of cytokines, that then feed back on the immune cells to elicit yet more cytokines. Normally, these positive feedback loops represent effective modes of immune amplification; they are themselves usually kept in check by self-regulating immune mechanisms, such as the activation of regulatory T cells (see Chapter 11). However, some viruses cause a localized, exaggerated response, resulting in the secretion of extraordinarily high levels of cytokines. If this occurs in the lungs, for example, the localized swelling, inflammation, and increase in capillary permeability can lead to the accumulation of fluids and leukocytes that block the airways, thereby causing exacerbation of symptoms, or even death, before the cytokine levels can be controlled. It is unclear why some viruses induce these **cytokine storms** and others do not.

Historical documents detailing the symptoms of the 1918 Spanish influenza suggest that the massive fatalities associated with that pandemic most likely resulted from cytokine storms, and there is some evidence that the severe acute respiratory syndrome (SARS) epidemic of 1993 may have caused a similar, unregulated, immune cell cytokine secretion. Transplant surgeons also observe this phenomenon on occasions when leukocytes associated with a graft—often a bone marrow transplant—mount an immune response against the host. Effective treatments for patients undergoing cytokine storms are still being developed. Currently, patients are offered steroidal and nonsteroidal anti-inflammatory medications, but other drugs that are more specifically directed at the reduction of cytokine secretion and/or activity are being tested.

Cytokine-Based Therapies

The availability of purified cloned cytokines, monoclonal antibodies directed against cytokines, and soluble cytokine receptors offers the prospect of specific clinical therapies to modulate the immune response. Cytokines such as interferons (see Clinical Focus Box 4-2), colony-stimulating factors such as G-CSF, and IL-2 have all been used clinically. In addition, several reagents that specifically block the proinflammatory effects of TNF-α have proven to be therapeutically useful in certain diseases. Specifically, soluble TNF-α receptor (Enbrel) and monoclonal antibodies against TNF-α (Remicade and Humira) have been used to treat rheumatoid arthritis and ankylosing spondylitis in more than a million patients. These anti–TNF-α drugs reduce proinflammatory cytokine cascades; help to alleviate pain, stiffness, and joint swelling; and promote healing and tissue repair. In addition, as described above, the recombinant form of IL-1Ra— anakinra (Kineret)—has been shown to be relatively effective in the treatment of rheumatoid arthritis. Monoclonal antibodies directed against the α chain of the IL-2R— basiliximab (Simulect) and daclizumab (Zenapax)—are also in clinical use for the prevention of transplantation rejection reactions.

As powerful as these reagents may be, interfering with the normal course of the immune response is not without its own intrinsic hazards. Reduced cytokine activity brings with it an increased risk of infection and malignancy, and the frequency of lymphoma is higher in patients who are long-term users of the first generation of TNF-α blocking drugs.

In addition, the technical problems encountered in adapting cytokines for safe, routine medical use are far from trivial. As described above, during an immune response, interacting cells may produce extremely high local concentrations of cytokines in the vicinity of target cells, but achieving such high concentrations over a clinically significant time period, when cytokines must be administered systemically, is difficult. Furthermore, many cytokines have a very short half-life—recombinant human IL-2 has a half-life of only 7 to 10 minutes when administered intravenously—so frequent administration may be required. Finally, cytokines are extremely potent biological response modifiers, and they can cause unpredictable and undesirable side effects. The side effects from administration of recombinant IL-2, for example, range from mild (e.g., fever, chills, diarrhea, and weight gain) to serious (e.g., anemia, thrombocytopenia, shock, respiratory distress, and coma).

The use of cytokines and anti-cytokine therapies in clinical medicine holds great promise, and efforts to develop safe and effective cytokine-related strategies continue, particularly in those areas of medicine that have so far been resistant to more conventional approaches, such as inflammation, cancer, organ transplantation, and chronic allergic disease.

SUMMARY

- Cytokines are proteins that mediate the effector functions of the immune system.

- Most cytokines are soluble proteins, but some—for example, members of the TNF family, may be expressed in a membrane-bound form.

- Some cytokines are secreted following stimulation of the innate immune system (e.g., IL-1, TNF-α, CXCL8), whereas others are secreted by the T and B lymphocytes of the adaptive immune system (IL-2, IL-4, IL-17).

- Cytokines bind to receptors on the plasma membrane and elicit their effects through the activation of an intracellular signaling cascade.

- Cytokines can effect alterations in the differentiative, proliferative, and survival capacities of their target cells.

- Cytokines exhibit the properties of redundancy, pleiotropy, synergy, antagonism, and cascade induction.

- The levels of expression of cytokine receptors on the cell surface may change according to the activation status of a cell.

- There are six families of cytokines with associated receptors, distinguished on the basis of the structures of the cytokines and the receptor molecules, and on the nature of their signaling pathways.

- IL-1 family members interact with dimeric receptors to induce responses that are primarily proinflammatory. The physiological responses to some IL-1 family members are modulated by the presence of soluble forms of the receptors and soluble cytokine-binding proteins.

- The Hematopoietin (Class I cytokine) family is the largest family of cytokines, and members mediate diverse effects, including proliferation, differentiation, and antibody secretion. The Hematopoietin family members share a common, four-helix bundle structure.

- Receptors for cytokines from the Hematopoietin family are classified into three subgroups—the γ, β, or gp130 receptors—each of which shares a common signaling chain.

- The Interferon (Class II cytokine) family includes the Type I interferons (interferon α and interferon β), which were the first cytokines to be discovered and mediate early antiviral responses.

- Type II interferons (interferon γ) activate macrophages, interact with cells of the adaptive immune system and support the generation of T_H1 cells.

- The TNF family of cytokines act as trimers and may occur in either soluble or membrane-bound forms.

- FasL, a TNF family member, interacts with its receptor, Fas, to stimulate apoptosis in the recipient cell. This interaction is important at the close of the immune response.

- TNF interacts with the TNF-R1 receptor on the surface of the cell to induce either apoptosis or survival, depending on the physiological environment.

- The IL-17 family of cytokines has been defined quite recently, and its members are primarily proinflammatory in action.

- Chemokines act on GPCR-coupled receptors to promote chemoattraction, the movement of immune system cells into, within, and out of lymphoid organs.

- Naturally occurring and pathogen-derived inhibitors of cytokine function may modulate their activity in vivo.

- Levels of inflammatory cytokines such as IL-1, IL-6, and TNF may be increased in certain disease states such as rheumatoid arthritis, and such diseases are susceptible to treatment with drugs that inhibit cytokine activities.

REFERENCES

Abram, C. L., and C. A. Lowell. 2009. The ins and outs of leukocyte integrin signaling. *Annual Review of Immunology* **27**:339–362.

Ahmed, M., and S. L. Gaffen. 2010. IL-17 in obesity and adipogenesis. *Cytokine & Growth Factor Reviews* **21**:449–453.

Alcami, A. 2003. Structural basis of the herpesvirus M3-chemokine interaction. *Trends in Microbiology* **11**:191–192.

Arend, W.P., G. Palmer, and C. Gabay. 2008. IL-1, IL-18, and IL-33 families of cytokines. *Immunological Reviews* **223**:20–38.

Botto, S., et al. 2011. IL-6 in human cytomegalovirus secretome promotes angiogenesis and survival of endothelial cells through the stimulation of survivin. *Blood* **117**:352–361.

Boulanger, M. J., and K. C. Garcia. 2004. Shared cytokine signaling receptors: Structural insights from the gp130 system. *Advances in Protein Chemistry* **68**:107–146.

Crabtree, G. R., S. Gillis, K. A. Smith, and A. Munck. 1980. Mechanisms of glucocorticoid-induced immunosuppression: Inhibitory effects on expression of Fc receptors and production of T-cell growth factor. *Journal of Steroid Biochemistry* **12**:445–449.

Crispin, J. C., and G. C. Tsokos. 2009. Transcriptional regulation of IL-2 in health and autoimmunity. *Autoimmunity Reviews* **8**:190–195.

Dandona, P., A. Aljada, and A. Bandyopadhyay. 2004. Inflammation: The link between insulin resistance, obesity and diabetes. *Trends in Immunology* **25**:4–7.

Dandona, P., A. Aljada, A. Chaudhuri, P. Mohanty, and G. Rajesh. 2004. A novel view of metabolic syndrome. *Metabolic Syndrome and Related Disorders* **2**:2–8.

Eisenbarth, S. C., and R. A. Flavell. 2009. Innate instruction of adaptive immunity revisited: The inflammasome. *EMBO Molecular Medicine* **1**:92–98.

Fickenscher, H., et al. 2002. The interleukin-10 family of cytokines. *Trends in Immunology* **23**:89–96.

Gaffen, S. L. (2009). Structure and signalling in the IL-17 receptor family. *Nature Reviews. Immunology* **9**:556–567.

Gee, K., C. Guzzo, N. F. Che Mat, W. Ma, and A. Kumar. 2009. The IL-12 family of cytokines in infection, inflammation and autoimmune disorders. *Inflammation & Allergy Drug Targets* **8**:40–52.

Gillis, S., R. Mertelsmann, and M. A. Moore. 1981. T-cell growth factor (interleukin 2) control of T-lymphocyte proliferation: Possible involvement in leukemogenesis. *Transplantation Proceedings* **13**:1884–1890.

Guilherme, A., J. V. Virbasius, V. Puri, and M. P. Czech. 2008. Adipocyte dysfunctions linking obesity to insulin resistance and type 2 diabetes. *Nature Reviews. Molecular Cell Biology* **9**:367–377.

Horng, T., and G. S. Hotamisligil. 2011. Linking the inflammasome to obesity-related disease. *Nature Medicine* **17**:164–165.

Hotamisligil, G. S. 2003. Inflammatory pathways and insulin action. *International Journal of Obesity and Related Metabolic Disorders* **27**:Suppl 3, S53–55.

Hotamisligil, G. S., N. S. Shargill, and B. M. Spiegelman. 1993. Adipose expression of tumor necrosis factor-alpha: Direct role in obesity-linked insulin resistance. *Science* **259**:87–91.

Jones, L. L., and D. A. Vignali. 2011. Molecular interactions within the IL-6/IL-12 cytokine/receptor superfamily. *Immunologic Research* **51**:5–14.

Li, W. X. 2008. Canonical and non-canonical JAK-STAT signaling. *Trends in Cell Biology* **18**:545–551.

Micheau, M., and J. Tschopp. 2003. Induction of TNF Receptor 1-mediated apoptosis via two sequential signaling complexes. *Cell* **114**:181–190.

Raab, M., et al. 2010. T cell receptor "inside-out" pathway via signaling module SKAP1-RapL regulates T cell motility and interactions in lymph nodes. *Immunity* **32**:541–556.

Raman, D., T. Sobolik-Delmaire, and A. Richmond. 2011. Chemokines in health and disease. *Experimental Cell Research* **317**:5755–89.

Rochman, Y., R. Spolski, and W. J. Leonard. 2009. New insights into the regulation of T cells by gamma(c) family cytokines. *Nature Reviews. Immunology* **9**:480–490.

Rot, A., and U. H. von Andrian. 2004. Chemokines in innate and adaptive host defense: Basic chemokinese grammar for immune cells. *Annual Review of Immunollogy* **22**:891–928.

Sadler, A. J., and B. R. Williams. 2008. Interferon-inducible antiviral effectors. *Nature Reviews. Immunology* **8**:559–568.

Skalnikova, H., J. Motlik, S. J. Gadher, and H. Kovarova. Mapping of the secretome of primary isolates of mammalian cells, stem cells and derived cell lines. *Proteomics* **11**:691–708.

Skiniotis, G., P. J. Lupardus, M. Martick, T. Walz, T., and K. C. Garcia. 2008. Structural organization of a full-length gp130/LIF-R cytokine receptor transmembrane complex. *Molecular Cell* **31**:737–748.

Smith, K. A., P. E. Baker, S. Gillis, and F. W. Ruscetti. (1980). Functional and molecular characteristics of T-cell growth factor. *Molecular Immunology* **17**:579–589.

Stull, D., and S. Gillis. 1981. Constitutive production of interleukin 2 activity by a T cell hybridoma. *Journal of Immunology* **126**:1680–1683.

Walczak, H. 2011. TNF and ubiquitin at the crossroads of gene activation, cell death, inflammation and cancer. *Immunological Reviews* **244**:9–28.

Wellen, K. E., and G. S. Hotamisligil. 2003. Obesity-induced inflammatory changes in adipose tissue. *The Journal of Clinical Investigation* **112**:1785–1788.

Useful Web Sites

www.invitrogen.com

www.miltenyibiotec.com/cytokines

www.prospecbio.com/Cytokines

www.peprotech.com

www.rndsystems.com Many companies that sell recombinant cytokines or cytokine-related products provide useful information on their websites, or in print copy. The preceding are a few that are particularly helpful.

www.jakpathways.com/understandingjakpathways A useful animation of JAK-STAT signaling.

www.youtube.com/watch?v=ZUUfdP87Ssg&feature=related A rather wonderful movie of chemotaxis.

www.youtube.com/watch?v=EpC6G_DGqkI&feature=related A striking movie of a neutrophil chasing a bacterium.

www.youtube.com/watch?v=KiLJl3NwmpU An increasing number of medical animations are available on You Tube. This one shows a macrophage recognizing a pathogen and releasing cytokines in response.

www.netpath.org A curated set of pathways, with information on interacting proteins. Many interleukin pathways are included.

STUDY QUESTIONS

1. Distinguish between a hormone, a cytokine, a chemokine, and a growth factor. What functional attributes do they share, and what properties can be used to discriminate among them?

2. Measurement of the blood concentration of a particular cytokine reveals that it is rarely present above 10^{-10} M, even under the conditions of an ongoing immune response. However, when you measure the affinity of the cognate receptor, you discover that its dissociation constant is close to 10^{-8} M, implying that the receptor occupancy must rarely exceed 1%. How do you account for this discrepancy?

3. Describe how dimerization and phosphorylation of intracellular signaling molecules contribute to activation of cells by Type 1 cytokines.

4. Define the terms *pleiotropy*, *synergy*, *redundancy*, *antagonism*, and *cascade induction* as they apply to cytokine action.

5. How might receipt of a cytokine signal result in the alteration of the location of a lymphocyte?

6. Cytokines signaling through the Class I cytokine receptors can compete with one another, even though the recognition units of the receptors are different. Explain.

7. Describe one mechanism by which Type I interferons "interfere" with the production of new viral particles.

8. Signaling by tumor necrosis factor can paradoxically lead to cell activation or cell death. Explain how, by drawing diagrams of the relevant signaling pathways.

9. Describe two examples of ways in which vertebrates tune down the intensity of their own cytokine signaling.

10a. The cytokine IL-2 is capable of activating all T cells to proliferation and differentiation. How does the immune system ensure that only T cells that have been stimulated by antigen are susceptible to IL-2 signaling?

10b. The following diagram represents the results of a flow cytometry experiment in which mouse spleen cells were stained with antibodies directed against different components of the IL-2R. The more antibody that binds to the cells, the further they move along the relevant axis. The number of cells stained with fluorescein-conjugated anti-βγ IL-2R antibodies are shown along the x-axis of the flow cytometry plot, and cells that stain with phycoerythrin-labeled antibodies to the α subunit of the IL-2 receptor move along the Y-axis. We have drawn for your reference a circle that represents cells that stain with neither antibody.

On this plot, draw, as circles, and label where you would expect to find the populations representing unstimulated T cells and T cells after antigen activation, after treatment with the two fluorescent labels described above.

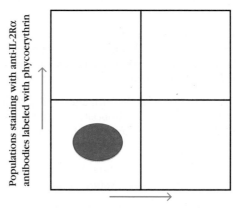

Populations staining with anti-IL-2Rβ and anti-IL-2Rγ antibodies labeled with fluorescein

Innate Immunity

Vertebrates are protected by both **innate immunity** and **adaptive immunity**. In contrast to adaptive immune responses, which take days to arise following exposure to antigens, innate immunity consists of the defenses against infection that are ready for immediate action when a host is attacked by a pathogen (viruses, bacteria, fungi, or parasites). The innate immune system includes anatomical barriers against infection—both physical and chemical—as well as cellular responses (Overview Figure 5-1). The main **physical barriers**—the body's first line of defense—are the epithelial layers of the skin and of the mucosal and glandular tissue surfaces connected to the body's openings; these epithelial barriers prevent infection by blocking pathogens from entering the body. **Chemical barriers** at these surfaces include specialized soluble substances that possess antimicrobial activity as well as acid pH. Pathogens that breach the physical and chemical barriers due to damage to or direct infection of the epithelial cell layer can survive in the extracellular spaces (some bacteria, fungi, and parasites) or they can infect cells (viruses and some bacteria and parasites), eventually replicating and possibly spreading to other parts of the body.

The **cellular innate immune responses** to invasion by an infectious agent that overcomes the initial epithelial barriers are rapid, typically beginning within minutes of invasion. These responses are triggered by cell surface or intracellular receptors that recognize conserved molecular components of pathogens. Some white blood cell types (macrophages and neutrophils) are activated to rapidly engulf and destroy extracellular microbes through the process of **phagocytosis**. Other receptors induce the production of proteins and other substances that have a variety of beneficial effects, including direct antimicrobial activity and the recruitment of fluid, cells, and molecules to sites of infection. This influx causes swelling and other physiological changes that collectively are called **inflammation.** Such local innate and inflammatory responses usually are beneficial for eliminating pathogens and damaged or dead cells and promoting healing. For example, increased levels of antimicrobial substances and

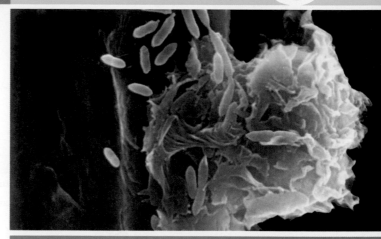

A macrophage (yellow) on the surface of a blood vessel (red) binds and phagocytoses bacteria (orange). [Dennis Kunkel Microscopy, Inc./Visuals Unlimited, Inc.]

- Anatomical Barriers to Infection
- Phagocytosis
- Induced Cellular Innate Responses
- Inflammatory Responses
- Natural Killer Cells
- Regulation and Evasion of Innate and Inflammatory Responses
- Interactions Between the Innate and Adaptive Immune Systems
- Ubiquity of Innate Immunity

phagocytic cells help to eliminate the pathogens, and dendritic cells take up pathogens for presentation to lymphocytes, activating adaptive immune responses. Natural killer cells recruited to the site can recognize and kill virus-infected, altered, or stressed cells. However, in some situations these innate and inflammatory responses can be harmful, leading to local or systemic consequences that can cause tissue damage and occasionally death. To prevent these potentially harmful responses, regulatory mechanisms have evolved that usually limit such adverse effects.

Despite the multiple layers of the innate immune system, some pathogens may evade the innate defenses. On call in vertebrates is the *adaptive immune system,* which counters infection with a specific tailor-made

OVERVIEW FIGURE 5-1

Innate Immunity

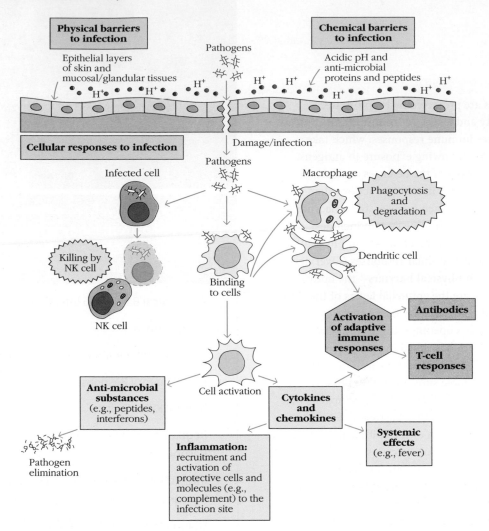

Key elements of innate immunity include the physical and chemical barriers that prevent infection, provided by the epithelial cell layers of the skin, mucosal tissues (e.g., gastrointestinal, respiratory, and urogenital tracts), and glandular tissues (e.g., salivary, lacrimal, and mammary glands). Once pathogens enter the body, such as through a breach in an epithelial layer, they are confronted by an array of cells with cell surface and intracellular receptors that recognize pathogen components and trigger a variety of cellular responses. Pathogen recognition by these receptors activates some cells to phagocytose and degrade the pathogen, and many cells are activated through their receptors to produce a variety of antimicrobial substances that kill pathogens, as well as cytokine and chemokine proteins that recruit cells, molecules, and fluid to the site of infection, leading to swelling and other symptoms collectively known as inflammation. The innate natural killer (NK) cells recognize and kill some virus-infected cells. Cytokines and chemokines can cause systemic effects that help to eliminate an infection, and also contribute—along with dendritic cells that carry and present pathogens to lymphocytes—to the activation of adaptive immune responses.

response to the attacking pathogen. This attack occurs in the form of B and T lymphocytes, which generate antibodies, and effector T cells that specifically recognize and neutralize or eliminate the invaders. Table 5-1 compares innate and adaptive immunity. While innate immunity is the most ancient form of defense, found in all multicellular plants and animals, adaptive immunity is a much more recent evolutionary invention, having arisen in vertebrates. In these animals, adaptive immunity complements a well-developed system of innate immune

TABLE 5-1	Innate and adaptive immunity	
Attribute	**Innate immunity**	**Adaptive immunity**
Response time	Minutes/hours	Days
Specificity	Specific for molecules and molecular patterns associated with pathogens and molecules produced by dead/damaged cells	Highly specific; discriminates between even minor differences in molecular structure of microbial or nonmicrobial molecules
Diversity	A limited number of conserved, germ line–encoded receptors	Highly diverse; a very large number of receptors arising from genetic recombination of receptor genes in each individual
Memory responses	Some (observed in invertebrate innate responses and mouse/human NK cells)	Persistent memory, with faster response of greater magnitude on subsequent exposure
Self/nonself discrimination	Perfect; no microbe-specific self/nonself patterns in host	Very good; occasional failures of discrimination result in autoimmune disease
Soluble components of blood	Many antimicrobial peptides, proteins, and other mediators	Antibodies and cytokines
Major cell types	Phagocytes (monocytes, macrophages, neutrophils), natural killer (NK) cells, other leukocytes, epithelial and endothelial cells	T cells, B cells, antigen-presenting cells

mechanisms that share important features with those of our invertebrate ancestors. A large and growing body of research has revealed that as innate and adaptive immunity have co-evolved in vertebrates, a high degree of interaction and interdependence has arisen between the two systems. Recognition by the innate immune system not only kicks off the adaptive immune response but also helps to ensure that the type of adaptive response generated will be effective for the invading pathogen.

This chapter describes the components of the innate immune system—physical barriers, chemical agents, and a battery of protective cellular responses carried out by numerous cell types—and illustrates how they act together to defend against infection. We conclude with an overview of innate immunity in plants and invertebrates.

Anatomical Barriers to Infection

The most obvious components of innate immunity are the external barriers to microbial invasion: the epithelial layers that insulate the body's interior from the pathogens of the exterior world. These epithelial barriers include the skin and the tissue surfaces connected to the body's openings: the mucous epithelial layers that line the respiratory, gastrointestinal, and urogenital tracts and the ducts of secretory glands such as the salivary, lacrimal, and mammary glands (which produce saliva, tears, and milk, respectively) (Figure 5-2). Skin and other epithelia provide a kind of living "plastic wrap" that encases and protects the inner domains of the

body from infection. But these anatomical barriers are more than just passive wrappers. They contribute to physical and mechanical processes that help the body shed pathogens and also generate active chemical and biochemical defenses by synthesizing and deploying molecules, including peptides and proteins, that have or induce antimicrobial activity.

Epithelial Barriers Prevent Pathogen Entry into the Body's Interior

The skin, the outermost physical barrier, consists of two distinct layers: a thin outer layer, the **epidermis**, and a thicker layer, the **dermis**. The epidermis contains several tiers of tightly packed epithelial cells; its outer layer consists of mostly dead cells filled with a waterproofing protein called keratin. The dermis is composed of connective tissue and contains blood vessels, hair follicles, sebaceous glands, sweat glands, and scattered myeloid leukocytes such as dendritic cells, macrophages, and mast cells. In place of skin, the respiratory, gastrointestinal, and urogenital tracts and the ducts of the salivary, lacrimal, and mammary glands are lined by strong barrier layers of epithelial cells stitched together by tight junctions that prevent pathogens from squeezing between them to enter the body.

A number of nonspecific physical and chemical defense mechanisms also contribute to preventing the entry of pathogens through the epithelia in these secretory tissues. For example, the secretions of these tissues (mucus, urine, saliva, tears, and milk) wash away potential invaders and also contain antibacterial and antiviral substances. Mucus, the viscous fluid secreted by specialized cells of the mucosal epithelial layers, entraps foreign microorganisms. In the

Organ or tissue	Innate mechanisms protecting skin/epithelium
Skin	Antimicrobial peptides, fatty acids in sebum
Mouth and upper alimentary canal	Enzymes, antimicrobial peptides, and sweeping of surface by directional flow of fluid toward stomach
Stomach	Low pH, digestive enzymes, antimicrobial peptides, fluid flow toward intestine
Small intestine	Digestive enzymes, antimicrobial peptides, fluid flow to large intestine
Large intestine	Normal intestinal flora compete with invading microbes, fluid/feces expelled from rectum
Airway and lungs	Cilia sweep mucus outward, coughing, sneezing expel mucus, macrophages in alveoli of lungs
Urogenital tract	Flushing by urine, aggregation by urinary mucins; low pH, anti-microbial peptides, proteins in vaginal secretions
Salivary, lacrimal, and mammary glands	Flushing by secretions; anti-microbial peptides and proteins in vaginal secretions

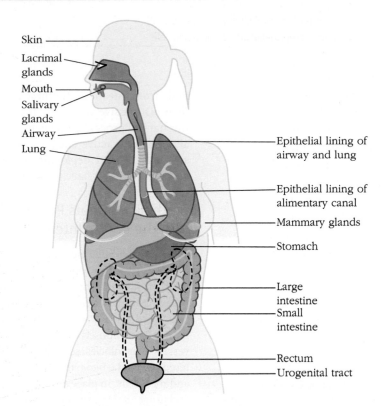

FIGURE 5-2 Skin and other epithelial barriers to infection. In addition to serving as physical barriers, the skin and mucosal and glandular epithelial layers are defended against microbial coloniza- tion by a variety of mechanisms: mechanical (cilia, fluid flow, smooth muscle contraction), chemical (pH, enzymes, antimicrobial peptides), and cellular (resident macrophages and dendritic cells).

lower respiratory tract, **cilia,** hairlike protrusions of the cell membrane, cover the epithelial cells. The synchronous move- ment of cilia propels mucus-entrapped microorganisms from these tracts. Coughing is a mechanical response that helps us get rid of excess mucus, with trapped microorganisms, that occurs in many respiratory infections. The flow of urine sweeps many bacteria from the urinary tract. With every meal, we ingest huge numbers of microorganisms, but they must run a gauntlet of defenses in the gastrointestinal tract that begins with the antimicrobial compounds in saliva and in the epithelia of the mouth and includes the hostile mix of digestive enzymes and acid found in the stomach. Similarly, the acidic pH of vaginal secretions is important in providing protection against bacterial and fungal pathogens.

Some organisms have evolved ways to evade these defenses of the epithelial barriers. For example, influenza

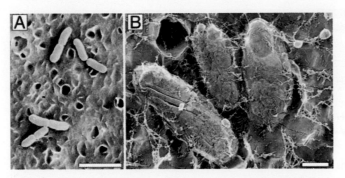

FIGURE 5-3 Electron micrograph of *E. coli* bacteria adhering to the surface of epithelial cells of the urinary tract. *E. coli* is an intestinal bacterial species that causes urinary tract infections affecting the bladder and kidneys. *[From N. Sharon and H. Lis, 1993, Scientific American 268(1):85; courtesy of K. Fujita; Matthew A. Mulvey, Joel D. Schilling, Juan J. Martinez, and Scott J. Hultgren, Bad bugs and beleaguered bladders: Interplay between uropathogenic Escherichia coli and innate host defenses. PNAS August 1, 2000 vol. 97 no. 16 8829-8835. Copyright 2000 National Academy of Sciences, U.S.A.]*

virus has a surface molecule that enables it to attach firmly to cells in mucous membranes of the respiratory tract, preventing the virus from being swept out by the ciliated epithelial cells. *Neisseria gonorrhoeae*, the bacteria that causes gonorrhea, binds to epithelial cells in the mucous membrane of the urogenital tract. Adherence of these and other bacteria to mucous membranes is generally mediated by hairlike protrusions on the bacteria called *fimbriae* or *pili* that have evolved the ability to bind to certain glycoproteins or glycolipids only expressed by epithelial cells of the mucous membrane of particular tissues (Figure 5-3).

Antimicrobial Proteins and Peptides Kill Would-be Invaders

To provide strong defense at these barrier layers, epithelial cells secrete a broad spectrum of proteins and peptides that provide protection against pathogens. The capacity of skin and other epithelia to produce a wide variety of antimicrobial agents on an ongoing basis is important for controlling the microbial populations on these surfaces, as breaks in these physical barriers from wounds provide routes of infection that would be readily exploited by pathogenic microbes if not defended by biochemical means.

Among the antimicrobial proteins produced by the skin and other epithelia in humans (Table 5-2), several are enzymes and binding proteins that kill or inhibit growth of bacterial and fungal cells. **Lysozyme** is an enzyme found in saliva, tears, and fluids of the respiratory tract that cleaves the peptidoglycan components of bacterial cell walls. *Lactoferrin* and *calprotectin* are two proteins that bind and sequester metal ions needed by bacteria and fungi, limiting their growth. Among the many antibacterial agents produced by human skin, recent research has identified *psoriasin*, a small protein of the S-100 family

(which also incudes calprotectin) with potent antibacterial activity against *Escherichia coli*, an enteric (intestinal) bacterial species. This finding answered a long-standing question: Why is human skin resistant to colonization by *E. coli* despite exposure to it from fecal matter resulting from lack of cleanliness or poor sanitation? As shown in Figure 5-4, incubation of *E. coli* on human skin for as little as 30 minutes kills the bacteria. Antimicrobial proteins can show some specificity toward particular pathogens. For example, psoriasin is a key antimicrobial for *E. coli* on the skin (and also on the tongue), but, as shown in Figure 5-4, it does not kill *Staphylococcus aureus* (a major cause of food poisoning and skin infections), whereas the related protein calprotectin kills *S. aureus* but not *E. coli*.

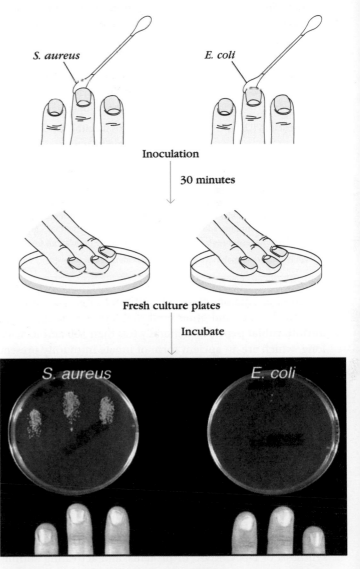

FIGURE 5-4 Psoriasin prevents colonization of the skin by *E. coli*. Skin secretes psoriasin, an antimicrobial protein that kills *E. coli*. Fingertips of a healthy human were inoculated with *Staphylococcus aureus* (*S. aureus*) and *E. coli*. After 30 minutes, the fingertips were pressed on a nutrient agar plate and the number of colonies of *S. aureus* and *E. coli* determined. Almost all of the inoculated *E. coli* were killed; most of the *S. aureus* survived. *[Photograph courtesy of Nature Immunology; from R. Gläser et al., 2005, Nature Immunology 6:57–64.]*

TABLE 5-2	Some human antimicrobial proteins and peptides at epithelial surfaces	
Proteins and peptides*	**Location**	**Antimicrobial activities**
Lysozyme	Mucosal/glandular secretions (e.g., tears, saliva, respiratory tract)	Cleaves glycosidic bonds of peptidoglycans in cell walls of bacteria, leading to lysis
Lactoferrin	Mucosal/glandular secretions (e.g., milk, intestine mucus, nasal/respiratory and urogenital tracts)	Binds and sequesters iron, limiting growth of bacteria and fungi; disrupts microbial membranes; limits infectivity of some viruses
Secretory leukocyte protease inhibitor	Skin, mucosal/glandular secretions (e.g., intestines, respiratory, and urogenital tracts, milk)	Blocks epithelial infection by bacteria, fungi, viruses; antimicrobial
S100 proteins, e.g.: - psoriasin - calprotectin	Skin, mucosal/glandular secretions (e.g., tears, saliva/tongue, intestine, nasal/ respiratory and urogenital tracts)	- Disrupts membranes, killing cells - Binds and sequesters divalent cations (e.g., manganese and zinc), limiting growth of bacteria and fungi
Defensins (α and β)	Skin, mucosal epithelia (e.g., mouth, intestine, nasal/respiratory tract, urogenital tract)	Disrupt membranes of bacteria, fungi, protozoan parasites, and viruses; additional toxic effects intracellularly; kill cells and disable viruses
Cathelicidin (LL37)**	Mucosal epithelia (e.g., respiratory tract, urogenital tract)	Disrupts membranes of bacteria; additional toxic effects intracellularly; kills cells.
Surfactant proteins SP-A, SP-D	Secretions of respiratory tract, other mucosal epithelia	Block bacterial surface components; promotes phagocytosis

*Examples listed in this table are all produced by cells in the epithelia of mucosal and glandular tissues; examples of prominent epithelial sites are listed. Most proteins and peptides are produced constitutively at these sites, but their production can also be increased by microbial or inflammatory stimuli. Many are also produced constitutively in neutrophils and stored in granules. In addition, synthesis and secretion of many of these molecules may be induced by microbial components during innate immune responses by various myeloid leukocyte populations (monocytes, macrophages, dendritic cells, and mast cells).

**While some mammals have multiple cathelicidins, humans have only one.

Another major class of antimicrobial components secreted by skin and other epithelial layers is comprised of **antimicrobial peptides**, generally less than 100 amino acids long, which are an ancient form of innate immunity present in vertebrates, invertebrates, plants, and even some fungi. The discovery that vertebrate skin produces antimicrobial proteins came from studies in frogs, where glands in the skin were shown to secrete peptides called *magainins* that have potent antimicrobial activity against bacteria, yeast, and protozoans. Antimicrobial peptides generally are cysteine-rich, cationic, and amphipathic (containing both hydrophilic and hydrophobic regions). Because of their positive charge and amphipathic nature, they interact with acidic phospholipids in lipid bilayers, disrupting membranes of bacteria, fungi, parasites, and viruses. They then can enter the microbes, where they have other toxic effects, such as inhibiting the synthesis of DNA, RNA, or proteins, and activating antimicrobial enzymes, resulting in cell death.

The main types of antimicrobial peptides found in humans are α- and β-defensins and cathelicidin. Human defensins kill a wide variety of bacteria, including *E. coli, S. aureus, Streptococcus pneumoniae, Pseudomonas aeruginosa,* and *Hemophilus influenzae.* Antimicrobial peptides also attack the lipoprotein envelope of enveloped viruses such as influenza virus and some herpesviruses. Defensins and cathelicidin LL-37 (the only cathelicidin expressed in humans) are secreted constitutively by epithelial cells in many tissues, as well as stored in neutrophil granules where they contribute to killing phagocytosed microbes. Recent studies have shown that human α-defensin antimicrobial peptides secreted into the gut by intestinal epithelial Paneth cells, located in the deep valleys (crypts) between the villi, are important for maintaining beneficial bacterial flora that are necessary for normal intestinal immune system functions. As we will see later, production of these antimicrobial peptides also can be induced in many epithelial and other cell types by the binding of microbial components to cellular receptors.

The epithelium of the respiratory tract secretes a variety of lubricating lipids and proteins called **surfactants**. Two surfactant proteins, SP-A and SP-D, which are present in the lungs as well as in the secretions of some other mucosal epithelia, are members of a class of microbe-binding proteins called **collectins**. SP-A and SP-D bind differentially to sets of carbohydrate, lipid, and protein components of microbial surfaces and help to prevent infection by blocking and modifying surface components and promoting pathogen clearance. For example, they differentially bind two alternating states of the lung pathogen *Klebsiella pneumonia* that differ in whether or not they are coated with a thick polysaccharide capsule: SP-A binds the complex polysaccharides coating many of the capsulated forms, while SP-D only binds the exposed cell wall lipopolysaccharide of the nonencapsulated form.

Phagocytosis

Despite the strong defenses of our protective epithelial layers, some pathogens have evolved strategies to penetrate these defenses, and epithelia may be disrupted by wounds, abrasions, and insect bites that may transmit pathogens. Once pathogens penetrate through the epithelial barrier layers into the tissue spaces of the body, an array of cellular membrane receptors and soluble proteins that recognize microbial components play the essential roles of detecting the pathogen and triggering effective defenses against it. Phagocytic cells make up the next line of defense against pathogens that have penetrated the epithelial cell barriers. Macrophages, neutrophils, and dendritic cells in tissues and monocytes in the blood are the main cell types that carry out phagocytosis—the cellular uptake (eating) of particulate materials such as bacteria—a key mechanism for eliminating pathogens. This major role of the cells attracted to the site of invading organisms is evolutionarily ancient, present in invertebrates as well as vertebrates. Elie Metchnikoff initially described the process of phagocytosis in the 1880s using cells from starfish (echinoderm invertebrates) similar to vertebrate white blood cells and ascribed to phagocytosis a major role in immunity. He was correct in this conclusion; we now know that defects in phagocytosis lead to severe immunodeficiency.

As described in Chapter 2, most tissues contain resident populations of macrophages that function as sentinels for the innate immune system. Through various cell surface receptors they recognize microbes such as bacteria, extend their plasma membrane to engulf them, and internalize them in phagosomes (endosomes resulting from phagocytosis, Figure 5-5). Lysosomes then fuse with the phagosomes, delivering agents that kill and degrade the microbes. Neutrophils are a second major type of phagocyte, usually recruited to sites of infection. Finally, dendritic cells also can bind and phagocytose microbes. As will be described later in this chapter and more extensively in Chapters 8 and 11, uptake and degradation of microbes by dendritic cells play key roles in the initiation of adaptive immune responses. In addition to triggering phagocytosis, various receptors on phagocytes recognize microbes and activate the production of a variety of molecules that contribute in other ways to eliminating infection, as will be described later in this chapter. A phagocyte's recognition of microbes and the responses that result are shown in Overview Figure 5-6.

Microbes are Recognized by Receptors on Phagocytic Cells

How does a phagocytic cell recognize microbes, triggering their phagocytosis? Phagocytes express on their surfaces a variety of receptors, some of which directly recognize specific conserved molecular components on the surfaces of microbes, such as cell wall components of bacteria and fungi. These conserved motifs, usually present in many copies on the surface of a bacterium, fungal cell, parasite, or virus particle, are called **pathogen-associated molecular patterns (PAMPs)**. Note that they can be expressed by microbes whether or not the microbes are pathogenic (cause

(a)

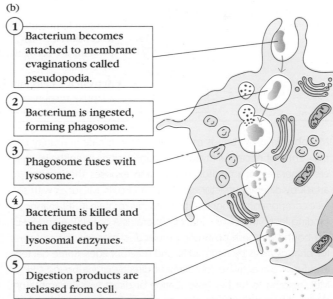

(b)

1. Bacterium becomes attached to membrane evaginations called pseudopodia.

2. Bacterium is ingested, forming phagosome.

3. Phagosome fuses with lysosome.

4. Bacterium is killed and then digested by lysosomal enzymes.

5. Digestion products are released from cell.

FIGURE 5-5 Phagocytosis. (a) Scanning electron micrograph of alveolar macrophage phagocytosis of *E. coli* bacteria on the outer surface of a blood vessel in the lung pleural cavity. (b) Steps in the phagocytosis of a bacterium. *[Part (a) from Dennis Kunkel Microscopy, Inc./Visuals Unlimited, Inc.]*

Effectors of Innate Immune Responses to Infection

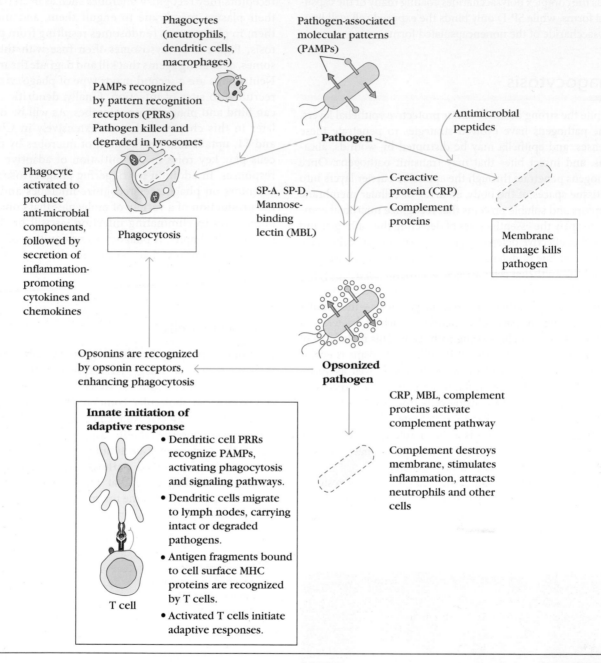

Phagocytes (neutrophils, dendritic cells, macrophages)

Pathogen-associated molecular patterns (PAMPs)

PAMPs recognized by pattern recognition receptors (PRRs) Pathogen killed and degraded in lysosomes

Antimicrobial peptides

Pathogen

Phagocyte activated to produce anti-microbial components, followed by secretion of inflammation-promoting cytokines and chemokines

Phagocytosis

SP-A, SP-D, Mannose-binding lectin (MBL)

C-reactive protein (CRP)

Complement proteins

Membrane damage kills pathogen

Opsonins are recognized by opsonin receptors, enhancing phagocytosis

Opsonized pathogen

CRP, MBL, complement proteins activate complement pathway

Complement destroys membrane, stimulates inflammation, attracts neutrophils and other cells

Innate initiation of adaptive response

- Dendritic cell PRRs recognize PAMPs, activating phagocytosis and signaling pathways.
- Dendritic cells migrate to lymph nodes, carrying intact or degraded pathogens.
- Antigen fragments bound to cell surface MHC proteins are recognized by T cells.
- Activated T cells initiate adaptive responses.

T cell

Microbial invasion brings many effectors of innate immunity into play. Entry of microbial invaders through lesions in epithelial barriers generates inflammatory signals and exposes the invaders to attack by different effector molecules and cells. Microbes with surface components recognized by C-reactive protein (CRP), mannose-binding lectin (MBL), or surfactant proteins A or D (SP-A and SP-D) are bound by these opsonizing molecules, marking the microbes for phagocytosis by neutrophils and macrophages. Some bacteria and fungi can activate complement directly, or via bound CRP or MBL, leading to further opsonization or direct lysis. Inflammatory signals cause phagocytes such as monocytes and neutrophils to bind to the walls of blood vessels, extravasate, and move to the site of infection, where they phagocytose and kill infecting microorganisms. Binding of microbes to receptors on phagocytes activates phagocytosis and production of additional antimicrobial and proinflammatory molecules that intensify the response, in part by recruiting more phagocytes and soluble mediators (CRP, MBL, and complement) from the bloodstream to the site of infection. *Inset:* Dendritic cells bind microbes via receptors and are activated to mature; they also internalize and degrade microbes. These dendritic cells migrate through lymphatic vessels to nearby lymph nodes, where they present antigen-derived peptides on their MHC proteins to T cells. Antigen-activated T cells then initiate adaptive immune responses against the pathogen. Cytokines produced during innate immune responses also support and direct the adaptive immune responses to infection.

TABLE 5-3	Human receptors that trigger phagocytosis	

Receptor type on phagocytes	Examples	Ligands
Pattern recognition receptors		**Microbial ligands (found on microbes)**
C-type lectin receptors (CLRs)	Mannose receptor	Mannans (bacteria, fungi, parasites)
	Dectin 1	β-glucans (fungi, some bacteria)
	DC-SIGN	Mannans (bacteria, fungi, parasites)
Scavenger receptors	SR-A	Lipopolysaccharide (LPS), lipoteichoic acid (LTA) (bacteria)
	SR-B	LTA, lipopeptides, diacylglycerides (bacteria), β-glucans (fungi)
Opsonin receptors		**Microbe-binding opsonins (soluble; bind to microbes)**
Collagen-domain receptor	CD91/calreticulin	Collectins SP-A, SP-D, MBL; L-ficolin; C1q
Complement receptors	CR1, CR3, CR4, CRIg, C1qRp	Complement components and fragments*
Immunoglobulin Fc receptors	FcαR	Specific IgA antibodies bound to antigen#
	FcγRs	Specific IgG antibodies bound to antigen;#
		C-reactive protein

* See Table 6-3 for specific complement components or fragments that are bound by individual receptors

Opsonization of antibody-bound antigens is an adaptive immune response clearance mechanism

disease); hence some researchers have started to use the more general term *microbe-associated molecular patterns* (*MAMPs*). The receptors that recognize PAMPs are called **pattern recognition receptors (PRRs)**. Some PRRs that bind microbes and trigger phagocytosis of the bound microbes are listed at the top of Table 5-3, along with the PAMPs they recognize. As we will see later, there are other PRRs that, after PAMP binding, do not activate phagocytosis but trigger other types of responses. Most PAMPs that induce phagocytosis are cell wall components, including complex carbohydrates such as mannans and β-glucans, lipopolysaccharides (LPS), other lipid-containing molecules, peptidoglycans, and surface proteins.

As shown in Figure 5-6, activation of phagocytosis can also occur indirectly, by phagocyte recognition of soluble proteins that have bound to microbial surfaces, thus enhancing phagocytosis, a process called **opsonization** (from the Greek word for "to make tasty"). Many of these soluble phagocytosis-enhancing proteins (called **opsonins**) also bind to conserved, repeating components on the surfaces of microbes such as carbohydrate structures, lipopolysaccharides, and viral proteins; hence they are sometimes referred to as soluble pattern-recognition proteins. Once bound to microbe surfaces, opsonins are recognized by membrane opsonin receptors on phagocytes, activating phagocytosis (see Table 5-3, bottom).

A variety of soluble proteins function as opsonins; many play other roles as well in innate immunity. For example, the two surfactant collectin proteins mentioned earlier, SP-A and SP-D, are found in the blood as well as in mucosal secretions

in the lungs and elsewhere, where they function as opsonins. After binding to microbes they are recognized by the CD91 opsonin receptor (see Table 5-3) and promote phagocytosis by alveolar and other macrophage populations. This function of SP-A and SP-D contributes to clearance of the fungal respiratory pathogen *Pneumocystis carinii*, a major cause of pneumonia in individuals with AIDS. **Mannose-binding lectin (MBL)**, a third collectin with opsonizing activity, is found in the blood and respiratory fluids. L-ficolin, a member of the **ficolin** family that is related to MBL and other collectins, is found in the blood, where it binds to acetylated sugars on microbes, including some streptococcal bacteria. The complement component C1q also functions as an opsonin, binding bacterial cell wall components such as lipopolysaccharides and some viral proteins.

MBL (and other collectins), ficolins, and C1q share structural features, including similar polymeric structures with collagen-like shafts, but have recognition regions with different binding specificities (Figure 5-7). As a result of their structural similarities, all are bound by the CD91 opsonin receptor (see Table 5-3) and activate pathogen phagocytosis. Another opsonin, **C-reactive protein (CRP)**, recognizes phosphorylcholine and carbohydrates on bacteria, fungi, and parasites, and is then bound by Fc receptors (FcRs) for IgG found on most phagocytes (see Chapter 3). Fc receptors also are important for the opsonizing activity of IgA antibodies and some IgG antibody subclasses. After binding specifically to antigens on microbe surfaces, the Fc regions of these antibodies can be recognized by specific FcRs, triggering phagocytosis. As an important mechanism by which the

(a) Mannose-binding lectin

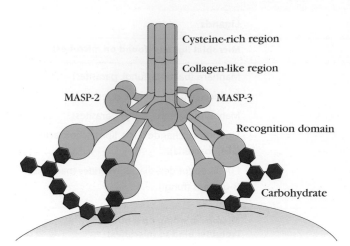

(c) C1 bound to LPS

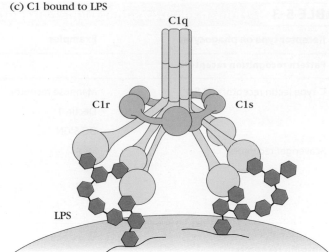

(b) H-ficolin

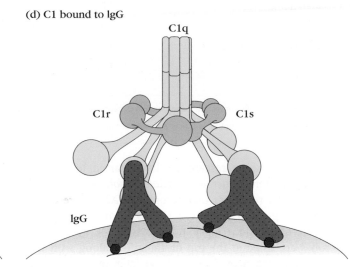

(d) C1 bound to IgG

FIGURE 5-7 Structures of opsonins. (a) Mannose-binding lectin (MBL), a collectin, and (b) H-ficolin are polymers of multiple polypeptide chains, each containing an N-terminal cysteine-rich region followed by a collagen-like region, an α-helical neck region, and a recognition domain. Both MBL and H-ficolin have bound MBL-associated serine proteases (MASPs), which become active after the recognition domains bind to specific carbohydrate residues on pathogen surfaces. The MASPs then activate the complement pathway. C1, the first component of complement, has a multimeric structure similar to that of MBL and H-ficolin. C1q can bind LPS on bacterial cell walls (c) or antibodies bound to antigens (d). After C1 binds to the pathogen-associated LPS or antibodies, the associated C1r and C1s subunit proteases, which are similar to MASPs, become activated and can initiate complement activation. *[Adapted from A. L. DeFranco, R. M. Locksley, and M. Robertson, 2007, Immunity (Primers in Biology), Sunderland, MA: Sinauer Associates.]*

adaptive immune response clears antigens, opsonization by antibodies (not shown in Figure 5-6) will be discussed in Chapter 13.

As mentioned above, among the most effective opsonins are several components of the complement system, which is described detail in Chapter 6. Present in both invertebrates and vertebrates, complement straddles both the innate and adaptive immune systems, indicating that it is ancient and important. In vertebrates, complement consists of more than 30 binding proteins and enzymes that function in a cascade of sequential activation steps. It can be triggered by several innate soluble pattern-recognition proteins (including the first complement component, C1q, and the structurally related lectins MBL and ficolins, C-reactive protein, and properdin), as well as by microbe-bound antibodies generated by the adaptive immune response. As we will see in Chapter 6, phagocytosis is one of many important antimicrobial effects resulting from complement activation. The importance of MBL's roles as both an opsonizer and an inducer of complement activation has been indicated by the effects of MBL deficiencies, which affect about 25% of the population. Individuals with MBL deficiencies are predisposed to severe respiratory tract infections, especially pneumococcal pneumonia. Interestingly, MBL deficiencies may be protective against tuberculosis, probably reflecting MBL's opsonizing role in enhancing the phagocytosis of *Mycobacterium tuberculosis*, the route by which it infects macrophages, potentially leading to tuberculosis.

Phagocytosed Microbes are Killed by Multiple Mechanisms

The binding of microbes—bacteria, fungi, protozoan parasites, and viruses—to phagocytes via pattern recognition receptors or opsonins and opsonin receptors (see Figure 5-6) activates signaling pathways. These signaling pathways trigger actin polymerization, resulting in membrane extensions around the microbe particles and their internalization, forming phagosomes (see Figure 5-5). The phagosomes then fuse with lysosomes and, in neutrophils, with preformed primary and secondary granules (see Figure 2-2a). The resulting phagolysosomes contain an arsenal of antimicrobial agents that then kill and degrade the internalized microbes. These agents include antimicrobial proteins and peptides (including defensins and cathelicidins), low pH, acid-activated hydrolytic enzymes (including lysozyme and proteases), and specialized molecules that mediate oxidative attack.

Oxidative attack on the phagocytosed microbes, which occurs in neutrophils, macrophages, and dendritic cells, employs highly toxic **reactive oxygen species (ROS)** and **reactive nitrogen species (RNS)**, which damage intracellular components (Figure 5-8). The reactive oxygen species are generated by the phagocytes' unique **NADPH oxidase enzyme complex** (also called **phagosome oxidase**), which is activated when microbes bind to the phagocytic receptors. The oxygen consumed by phagocytes to support ROS production by NADPH oxidase is provided by a metabolic process known as the **respiratory burst**, during which oxygen uptake by the cell increases severalfold. NADPH oxidase converts oxygen to superoxide ion ($\cdot O_2^-$); other ROS generated by the action of additional enzymes are hydrogen peroxide (H_2O_2), and hypochlorous acid (HClO), the active component of household bleach.

The generation of RNS requires the transcriptional activation of the gene for the enzyme **inducible nitric oxide synthase (iNOS, or NOS2)**—called that to distinguish it from related NO synthases in other tissues. Expression of iNOS is activated by microbial components binding to various PRRs. iNOS oxidizes L-arginine to yield L-citrulline and nitric oxide (NO), a potent antimicrobial agent. In combination with superoxide ion ($\cdot O_2^-$) generated by NADPH oxidase, NO produces an additional reactive nitrogen species, peroxynitrite (ONOO$^-$) and toxic S-nitrosothiols. Collectively the ROS and RNS are highly toxic to phagocytosed microbes due to the alteration of microbial molecules through oxidation, hydroxylation, chlorination, nitration, and S-nitrosylation, along with formation of sulfonic acids and destruction of iron-sulfur clusters in proteins. One example of how these oxidative species may be toxic to pathogens is the oxidation by ROS of cysteine sulfhydryls that are present in the active sites of many enzymes, inactivating the enzymes. ROS and RNS also can be released from activated neutrophils and macrophages and kill extracellular pathogens.

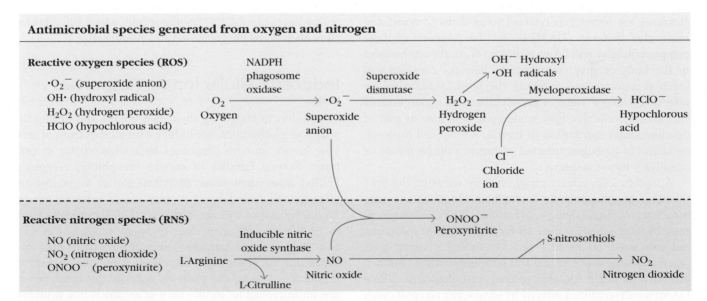

FIGURE 5-8 Generation of antimicrobial reactive oxygen and nitrogen species. In the cytoplasm of neutrophils, macrophages, and dendritic cells, several enzymes, including phagosome NADPH oxidase, transform molecular oxygen into highly reactive oxygen species (ROS) that have antimicrobial activity. One of the products of this pathway, superoxide anion, can interact with a reactive nitrogen species (RNS), generated by inducible nitric oxide synthase (iNOS) to produce peroxynitrite, another RNS. NO can also undergo oxidation to generate the RNS nitrogen dioxide.

Evidence from genetic defects in humans and mice highlights the critical roles of these reactive chemical species in microbial elimination by phagocytic cells. The importance to antimicrobial defense of phagosomal NADPH oxidase and its products, ROS and RNS, is illustrated by **chronic granulomatous disease (CGD)**. Patients afflicted with this disease have dramatically increased susceptibility to some fungal and bacterial infections, caused by defects in subunits of NADPH oxidase that destroy its ability to generate oxidizing species. In addition, studies with mice in which the genes encoding iNOS were knocked out have shown that nitric oxide and substances derived from it account for much of the antimicrobial activity of macrophages against bacteria, fungi, and parasites. These mice lost much of their usual ability to control infections by such intracellular pathogens as *M. tuberculosis* and *Leishmania major*, the intracellular protozoan parasite that causes leishmaniasis.

Phagocytosis Contributes to Cell Turnover and the Clearance of Dead Cells

Our discussion of phagocytosis thus far has focused on its essential roles in killing pathogens. As the body's main scavenger cells, macrophages also utilize their phagocytic receptors to take up and clear cellular debris, cells that have died from damage or toxic stimuli (necrotic cell death) or from apoptosis (programmed cell death), and aging red blood cells.

Considerable progress has been made in recent years in understanding the specific markers and receptors that trigger macrophage phagocytosis of dead, dying, and aging cells. Collectively the components of dead/dying cells and damaged tissues that are recognized by PRRs leading to their clearance are sometimes referred to as **damage-associated molecular patterns (DAMPs)**. As the presence of these components may also be an indicator of conditions harmful to the body or may contribute to harmful consequences (such as autoimmune diseases), the D in DAMP also can refer to a "Danger" signal. Phagocytosis is the major mode of clearance of cells that have undergone apoptosis as part of developmental remodeling of tissues, normal cell turnover, or killing of pathogen-infected or tumor cells by innate or adaptive immune responses.

Apoptotic cells attract phagocytes by releasing the lipid mediator lysophosphatidic acid, which functions as a chemoattractant. These dying cells facilitate their own phagocytosis by expressing on their surfaces an array of molecules not expressed on healthy cells, including phospholipids (such as phosphatidyl serine and lysophosphatidyl choline), proteins (annexin I), and altered carbohydrates. These DAMPs are recognized directly by phagocytic receptors such as the phosphatidyl serine receptor and scavenger receptor SR-A1. Other DAMPs are recognized by soluble pattern recognition molecules that function as opsonins, including the collectins MBL, SP-A, and SP-D mentioned earlier; various complement components; and the pentraxins C-reactive protein and serum amyloid protein. These opsonins are then recognized by opsonin receptors, activating phagocytosis and degradation of the apoptotic cells.

An important additional activity of macrophages in the spleen and those in the liver (known as Kupffer cells) is to recognize, phagocytose, and degrade aging and damaged red blood cells. As these cells age, novel molecules that are recognized by phagocytes accumulate in their plasma membrane. Phosphatidyl serine flips from the inner to the outer leaflet of the lipid bilayer and is recognized by phosphatidyl serine receptors on phagocytes. Modifications of erythrocyte membrane proteins have also been detected that may promote phagocytosis.

Obviously it is important for normal cells not to be phagocytosed, and accumulating evidence indicates that whether or not a cell is phagocytosed is controlled by sets of "eat me" signals—the altered membrane components (DAMPs) described above—and "don't eat me" signals expressed by normal cells. Young, healthy erythrocytes avoid being phagocytosed by not expressing "eat me" signals, such as phosphatidyl serine, and also by expressing a "don't eat me" signal, the protein CD47. CD47, expressed on many cell types throughout the body, is recognized by the SIRP-α receptor on macrophages, which transmits signals that inhibit phagocytosis. Recent studies have shown that tumors use elevated CD47 expression to evade tumor surveillance and phagocytic elimination by the immune system. Increased expression of CD47 on all or most human cancers is correlated with tumor progression, probably because the CD47 activates SIRP-1α-mediated inhibition of the phagocytosis of tumor cells by macrophages. This understanding of the role of CD47 in preventing phagocytosis is being used to develop novel therapies for certain cancers, such as using antibodies to block CD47 on tumor cells, which should then allow them to be phagocytosed and eliminated.

Induced Cellular Innate Responses

In addition to triggering their own uptake and killing by phagocytic cells, microbes induce a broad spectrum of cellular innate immune responses by a wide variety of cell types. Several families of pattern recognition receptors (PRRs) other than those described earlier as mediating phagocytosis (see Table 5-3) play major roles in innate immunity. As we will see in this section, these PRRs bind to PAMPs as well as to some endogenous (self) DAMPs and trigger signal-transduction pathways that turn on expression of genes with important functions in innate immunity. Among the proteins encoded by these genes are antimicrobial molecules such as antimicrobial peptides and interferons, chemokines and cytokines that recruit and activate other cells, enzymes such as iNOS (mentioned earlier) that generate antimicrobial molecules, and *proinflammatory mediators* (i.e., components that promote inflammation).

Cellular Pattern Recognition Receptors Activate Responses to Microbes and Cell Damage

Several families of cellular PRRs contribute to the activation of innate immune responses that combat infections. Some of these PRRs are expressed on the plasma membrane, while others are actually found *inside* our cells, either in endosomes/lysosomes or in the cytosol. This array of PRRs ensures that the cell can recognize PAMPs on both extracellular and intracellular pathogens. DAMPs released by cell and tissue damage also can be recognized by both cell surface and intracellular PRRs. Many cell types in the body express these PRRs, including all types of myeloid white blood cells (monocytes, macrophages, neutrophils, eosinophils, mast cells, basophils, dendritic cells) and subsets of the three types of lymphocytes (B cells, T cells, and NK cells). PRRs are also expressed by some other cell types, especially those commonly exposed to infectious agents; examples include the skin, mucosal and glandular epithelial cells, vascular endothelial cells that line the blood vessels, and fibroblasts and stromal support cells in various tissues. While it is unlikely that any single cell expresses all of these PRRs, subsets of the receptors are expressed by various cell subpopulations. The rest of this section describes the four main families of mammalian PRRs and the signaling pathways that they activate, leading to protective responses.

Toll-Like Receptors Recognize Many Types of Pathogen Molecules

Toll-like receptors (TLRs) were the first family of PRRs to be discovered and are still the best-characterized in terms of their structure, how they bind PAMPs and activate cells, and the extensive and varied set of innate immune responses that they induce.

Discovery of Invertebrate Toll and Vertebrate Toll-like Receptors

In the 1980s, researchers in Germany discovered that *Drosophila* fruit fly embryos could not establish a proper dorsal-ventral (back to front) axis if the gene encoding the Toll membrane protein is mutated. (The name "Toll" comes from German slang meaning "weird," referring to the mutant flies' bizarrely scrambled anatomy.) For their subsequent characterization of the *toll* and related homeobox genes and their role in the regulation of embryonic development, Christiane Nusslein-Volhard, Eric Wieschaus, and Edward B. Lewis were awarded the Nobel Prize in Physiology and Medicine in 1995. But what does this have to do with immune system receptors? Many mutations of the *toll* gene were generated, and in 1996 Jules Hoffman and Bruno Lemaitre discovered that mutations in *toll* made flies highly susceptible to lethal infection with *Aspergillus fumigatus*, a fungus to which wild-type flies were immune (Figure 5-9). This striking observation led to other studies showing that Toll and related proteins are involved in the activation of innate immune responses in invertebrates.

FIGURE 5-9 Impaired innate immunity in fruit flies with a mutation in the Toll pathway. Severe infection with the fungus *Aspergillus fumigatus* (yellow) results from a mutation in the signaling pathway downstream of the Toll pathway in *Drosophila* that normally activates production of the antimicrobial peptide drosomycin. *[Electron micrograph adapted from B. Lemaitre et al., 1996, Cell 86:973; courtesy of J. A. Hoffman, University of Strasbourg.]*

For his pivotal contributions to the study of innate immunity in *Drosophila*, Jules Hoffman was a co-winner of the 2011 Nobel Prize in Physiology and Medicine.

Characterization of the Toll protein surprisingly revealed that its cytoplasmic signaling domain was homologous to that of the vertebrate receptor for the cytokine IL-1 (IL-1R). Through a search for human proteins with cytoplasmic domains homologous to those of Toll and IL-1R, in 1997 Charles Janeway and Ruslan Medzhitov discovered a human gene for a protein similar to Toll that activated the expression of innate immunity genes in human cells. Appropriately, this and other vertebrate Toll relatives discovered soon thereafter were named *Toll-like receptors* (*TLRs*).

Through studies with mutant mice, in 1998 Bruce Beutler obtained the important proof that TLRs contribute to normal immune functions in mammals. Mice homozygous for a mutant form of a gene called *lps* were resistant to the harmful responses induced by lipolysaccharide (LPS; also known as endotoxin), a major component of the cell walls of *Gram-negative bacteria* (Figure 5-10). In humans, a buildup of endotoxin from severe bacterial infection can induce too strong of an innate immune response, causing septic shock, a life-threatening condition in which vital organs such as the

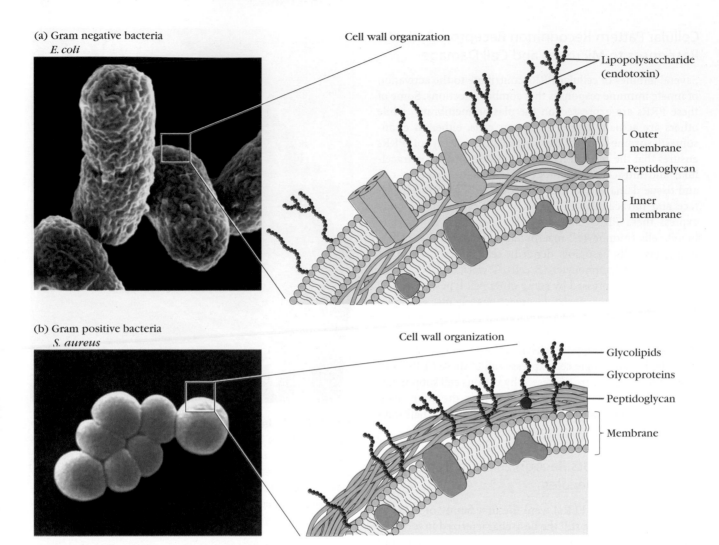

(a) Gram negative bacteria
E. coli

Cell wall organization

Lipopolysaccharide
(endotoxin)

Outer
membrane

Peptidoglycan

Inner
membrane

(b) Gram positive bacteria
S. aureus

Cell wall organization

Glycolipids

Glycoproteins

Peptidoglycan

Membrane

FIGURE 5-10 Cell wall components of Gram-negative and Gram-positive bacteria. Cell wall structures differ for Gram-negative (a) and Gram-positive (b) bacteria. Because of their thick peptidoglycan layer, Gram-positive bacteria retain the precipitate formed by the crystal violet and iodine reagents of the Gram stain, whereas the stain is easily washed out of the less-dense cell walls of Gram-negative bacteria. Thus the Gram stain identifies two distinct sets of bacterial genera that also differ significantly in other properties. *[Photographs: E. coli photograph from Gary Gaugler/ Visuals Unlimited. S. aureus photograph from Dr. Fred Hossler/Getty Images]*

brain, heart, kidney, and liver may fail. Each year, about 20,000 people die in the United States of septic shock caused by Gram-negative bacterial infections, so it was striking that some mutant strains of mice were resistant to fatal doses of LPS. Beutler found that the defective mouse *lps* gene encoded a mutant form of one TLR, TLR4, which differed from the normal form by a single amino acid so that it no longer was activated by LPS. This work provided an unequivocal demonstration that TLR4 is the cellular innate pattern recognition receptor that recognizes LPS and earned Beutler a share of the 2011 Nobel Prize.

Thus, this landmark series of experiments showed in rapid succession that invertebrates respond to pathogens, that they use receptors also found in vertebrates, and that one of these receptors is responsible for LPS-induced innate immune responses.

TLRs and Their Ligands

Intensive work over the last decade and a half has identified multiple TLR family members in mice and humans—as of 2011, 13 TLRs that function as PRRs have been identified in these species. TLRs 1-10 are conserved between mice and humans, although TLR10 is not functional in mice, while TLRs 11-13 are expressed in mice but not in humans. While TLRs have not been shown to be involved in vertebrate development, unlike in fruit flies, the set of TLRs present in a human or mouse can detect a wide variety of PAMPs from bacteria, viruses, fungi, and parasites, as well as DAMPs from damaged cells and tissues. Each TLR has a distinct repertoire of specificities for conserved PAMPs; the TLRs and some of their known PAMP ligands are listed in Table 5-4. Biochemical studies have revealed the structure of several TLRs and

TABLE 5-4 TLRs and their microbial ligands

TLRs*	Ligands	Microbes
TLR1	Triacyl lipopeptides	Mycobacteria and Gram-negative bacteria
TLR2	Peptidoglycans	Gram-positive bacteria
	GPI-linked proteins	Trypanosomes
	Lipoproteins	Mycobacteria and other bacteria
	Zymosan	Yeasts and other fungi
	Phosphatidlyserine	Schistosomes
TLR3	Double-stranded RNA (dsRNA)	Viruses
TLR4	LPS	Gram-negative bacteria
	F-protein	Respiratory syncytial virus (RSV)
	Mannans	Fungi
TLR5	Flagellin	Bacteria
TLR6	Diacyl lipopolypeptides	Mycobacteria and Gram-positive bacteria
	Zymosan	Yeasts and other fungi
TLR7	Single-stranded RNA (ssRNA)	Viruses
TLR8	Single-stranded RNA (ssRNA)	Viruses
TLR9	CpG unmethylated dinucleotides	Bacterial DNA
	Dinucleotides	
	Herpes virus components	Some herpesviruses
	Hemozoin	Malaria parasite heme byproduct
TLR10	Unknown	Unknown
TLR11	Unknown	Uropathogenic bacteria
	Profilin	Toxoplasma
TLR12	Unknown	Unknown
TLR13	Unknown	Vesicular stomatitis virus

*All function as homodimers except TLR1, 2, and 6, which form TLR2/1 and TLR2/6 heterodimers. Ligands indicated for TLR2 bind to both; ligands indicated for TLR1 bind to TLR2/1 dimers, and ligands indicated for TLR6 bind to TLR2/6 dimers.

how they bind their specific ligands. TLRs are membrane-spanning proteins that share a common structural element in their extracellular region called **leucine-rich repeats (LRRs)**; multiple LRRs make up the horseshoe-shaped extracellular ligand-binding domain of the TLR polypeptide chain (Figure 5-11a).

When TLRs bind their PAMP or DAMP ligands via their extracellular LRR domains, they are induced to dimerize. In most cases each TLR dimerizes with itself, forming a homodimer, but TLR2 forms heterodimers by pairing with either TLR1 or TLR6. How TLRs bind their ligands was not known until complexes of the extracellular LRR domain dimers with bound ligands were analyzed by x-ray crystallography. Structures of TLR2/1 with a bound lipopeptide and TLR3 with dsRNA are shown in Figure 5-11b; the characteristic "m"-shaped conformation of TLR dimers is apparent.

As shown in Figure 5-12, TLRs exist both on the plasma membrane and in the membranes of endosomes and lysosomes; their cellular location is tailored to enable them to respond optimally to the particular microbial ligands they recognize. TLRs that recognize PAMPs on the outer sur-

face of extracellular microbes (see Table 5-4) are found on the plasma membrane, where they can bind these PAMPs and induce responses. In contrast, TLRs that recognize internal microbial components that have to be exposed by the dismantling or degradation of endocytosed pathogens—nucleic acids in particular—are found in endosomes and lysosomes. Unique among the TLRs, TLR4 has been shown to move from the plasma membrane to endosomes/lysosomes after binding LPS or other PAMPs. As we will see below, it activates different signaling pathways from the two locations.

In addition to microbial ligands, TLRs also recognize DAMPs, endogenous (self) components released by dead/dying cells or damaged tissues. Among the DAMPS recognized by plasma membrane TLRs are heat shock and chromatin proteins, fragments of extracellular matrix components (such as fibronectin and hyaluronin), and oxidized low-density lipoprotein and amyloid-β. While self nucleic acids usually do not activate the intracellular PRR, under certain circumstances (such as when bound by anti-DNA or anti-chromatin antibodies in individuals with the autoimmune

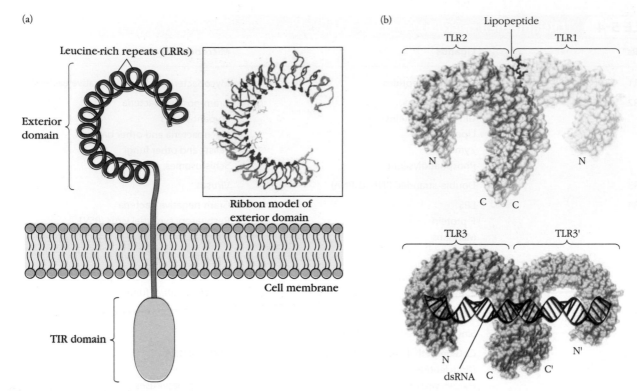

FIGURE 5-11 Toll-like receptor (TLR) structure and binding of PAMP ligands. (a) Structure of a TLR polypeptide chain. Each TLR polypeptide chain is made up of a ligand-binding exterior domain that contains many leucine-rich repeats (LRRs, repeating segments of 24-29 amino acids containing the sequence LxxLxLxx, where L is leucine and x is any amino acid), a membrane-spanning domain (blue), and an interior Toll/IL-1R (TIR) domain, which interacts with the TIR domains of other members of the TLR signal-transduction pathway. Two such polypeptide chains pair to form Toll/IL-1R (TLR) dimers, the form that binds ligands. (b) X-ray crystallographic structures of paired extracellular LRR domains of TLR dimers with bound ligands. Top: TLR2/1 dimer with a bound lipopeptide molecule. Bottom: TLR3/3 dimer with bound double-stranded (ds) RNA molecule. *[Part (b) from Jin, M.S., and Lee, J-O. 2008.* Immunity **29***:182. PDB IDs 2Z7X (top) and 3CIY (bottom).]*

disease lupus erythematosus) they can be endocytosed and activate endosomal/lysosomal TLRs, contributing to the disease (see Chapter 16).

Signaling Through TLRs: Overview

Given the wide variety of potential pathogens the innate immune system needs to recognize and combat, how does binding of a specific pathogen evoke an appropriate response for that pathogen? Signaling through TLRs utilizes many of the principles and some of the signaling molecules described in Chapters 3 and 4, along with some unique to pathways activated by TLRs (and by other PRRs, described below).

Detailed studies of the signaling pathways downstream of all of the TLRs, summarized schematically in Figure 5-13, have revealed that they include some shared components and activate expression of many of the same genes. An important example of a shared component is the transcription factor **NF-κB**. NF-κB is key for inducing many innate and inflammatory genes, including those encoding defensins; enzymes such as iNOS; chemokines; and cytokines such as the proinflammatory cytokines TNF-α, IL-1, and IL-6, produced by macrophages and dendritic cells. There are also TLR-specific signaling pathways and components

that induce the expression of subsets of proteins, some of which are particularly effective in combating the type of pathogen recognized by the particular TLR(s). An important example is expression of the potent antiviral Type 1 interferons, IFNs-α and -β, induced by pathways downstream of the TLRs that bind viral components. As described below, activation of the **interferon regulatory factors (IRFs)** is essential for inducing transcription of the genes encoding IFN-α and -β. Combinations of transcription factors contribute to inducing the expression of many of these genes; examples include combinations of NF-κB, IRFs, and/or transcription factors downstream of MAP kinase (MAPK) pathways (see Chapters 3 and 4), such as AP-1, that can be activated by signaling intermediates downstream of certain TLRs.

The particular signal transduction pathway(s) activated by a TLR dimer following PAMP binding are largely determined by the TLR and by the initial protein adaptor (see Chapter 3) that binds to the TLR's cytoplasmic domain. As shown in Figure 5-11a, this region is called the **TIR domain** (from Toll/IL-1 receptor), referring to the similarity noted earlier between the cytoplasmic domains of TLRs and IL-1 receptors (see Figure 4-5a). TIR domains of all TLR dimers serve as binding sites for the TIR domains of adaptors that activate the downstream

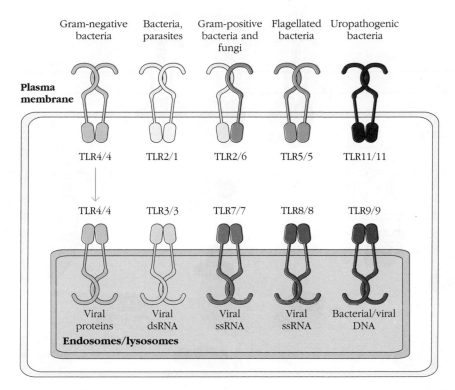

Gram-negative bacteria Bacteria, parasites Gram-positive bacteria and fungi Flagellated bacteria Uropathogenic bacteria

Plasma membrane

TLR4/4 TLR2/1 TLR2/6 TLR5/5 TLR11/11

TLR4/4 TLR3/3 TLR7/7 TLR8/8 TLR9/9

Viral proteins Viral dsRNA Viral ssRNA Viral ssRNA Bacterial/viral DNA

Endosomes/lysosomes

FIGURE 5-12 Cellular location of TLRs. TLRs that interact with extracellular ligands reside in the plasma membrane; TLRs that bind ligands generated from endocytosed microbes are localized to endosomes and/or lysosomes. Upon ligand binding, the TLR4/4 dimer moves from the plasma membrane to the endosomal/lysosomal compartment, where it can activate different signaling components.

signaling pathways. The two key adaptors are **MyD88** (*My*eloid *d*ifferentiation factor *88*) and **TRIF** (*TIR*-domain-containing adaptor-inducing *IF*N-β *f*actor). As shown in Figure 5-13, most TLRs, whether found on the cell surface or in endosomes/lysosomes, bind the adaptor protein MyD88 (activating MyD88-dependent signaling pathways). In contrast, TLR3 binds the alternative adaptor protein TRIF (activating TRIF-dependent signaling pathways). TLR4 is unique in binding both MyD88 (when it is in the plasma membrane, signaling its endocytosis) and TRIF (when it is in endosomes, after internalization). Figure 5-13 also shows the presence of two additional adaptors associated with most TLRs: TIRAP (*TIR*-domain containing *a*daptor *p*rotein) and TRAM (*TR*IF-related *a*daptor *m*olecule). They are TIR domain-containing adaptors that serve as sorting receptors: TIRAP helps to recruit MyD88 to TLRs 2/1, 2/6, and 4, and TRAM helps to recruit TRIF to both TLR3 and endosomal/lysosomal TLR4.

MyD88-Dependent Signaling Pathways

After associating with a TLR dimer following ligand binding, MyD88 initiates a signaling pathway that activates the NF-κB and MAPK pathways by essentially the same pathway as that activated by IL-1 (see Figure 4-6). As shown for plasma membrane TLRs 2/1, 4, and 5 in Figure 5-13, MyD88 recruits and activates several IRAK protein kinases, which then bind and activate **TRAF6**. TRAF6 ubiquitinates NEMO and TAB proteins, leading to the activation of **TAK1**, which then phosphorylates the IκB kinase (**IKK**) complex. Acti-

vated IKK then phosphorylates the inhibitory IκB subunit of NF-κB, releasing NF-κB to enter the nucleus and activate gene expression. TAK1 does double duty in this TLR signaling cascade. After separating from the IKK complex, it activates MAPK signaling pathways (see Chapter 3) that result in the activation of transcription factors including Fos and Jun, which make up the AP-1 dimer (see Figure 3-16).

In addition to activating NF-κB and MAPK pathways via the MyD88-dependent pathway, the endosomal TLRs 7, 8, and 9, which bind microbial nucleic acids (especially viral RNA and bacterial DNA), also trigger pathways that activate IRFs. As shown in Figure 5-13, when triggered by these TLRs, the MyD88/IRAK4/TRAF6 complex activates a complex containing TRAF3, IRAK1, and IKKα, leading to the phosphorylation, dimerization, activation, and nuclear localization of IRF7. IRF7 induces the transcription of genes for both Type 1 interferons, IFN-α and -β, which have potent antiviral activity. *Thus different TLRs may differentially activate distinct transcription factors (NF-κB, certain IRFs, and/or those activated by MAPK pathways), leading to variation in which genes are turned on to help protect us against the invading pathogens.*

TRIF-Dependent Signaling Pathways

For the two endosomal TLRs that recruit the TRIF adaptor instead of MyD88—TLR3 and endosomal TLR4—the downstream signaling pathways differ somewhat from those activated by MyD88 (see Figure 5-13). TRIF recruits the RIP1 protein kinase that in turn recruits and activates TRAF6,

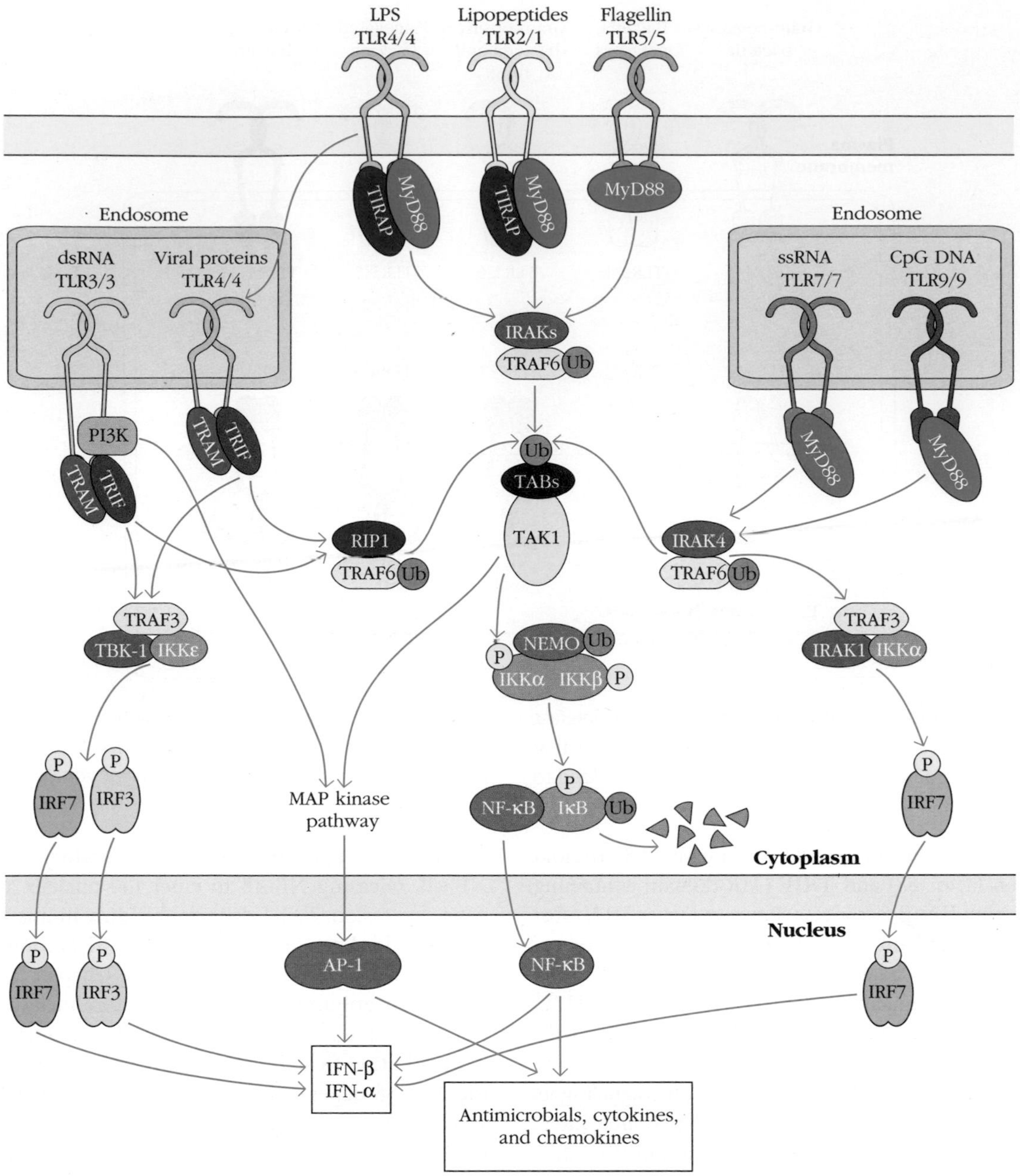

FIGURE 5-13 TLR signaling pathways. Signaling pathways downstream of TLRs are shown for three cell surface TLRs (TLR2/1, TLR4, and TLR5; also activated by TLR2/6 and TLR 11 [not shown]) and for endosomal/lysosomal TLRs TLR7 and TLR 9 (in endosome on right; also activated by TLR8 [not shown]) and TLR3 and TLR4 (in endosome on left). MyD88-dependent signaling pathways are activated by the TIRAP and MyD88 receptor-binding adaptors. TRIF-dependent signaling pathways are activated by the TRAM and TRIF receptor-binding adaptors. *Note:* Not all signaling components are shown. See text for details. The MAPK pathway is shown in Figure 3-16.

which then initiates the same steps as in the MyD88-dependent pathway. TLR3 also activates PI3K, which enhances MAPK pathway activation. In addition, TRIF activates TRAF3 and a complex of TBK-1 and IKKε, which phosphorylates and activates IRF7 and a different IRF—IRF3. IRF3 and IRF7 dimerize and move into the nucleus and (together with NF-κB and AP-1) induce the transcription of the IFN-α and -β genes. *Thus, all of the intracellular TLRs that bind viral PAMPs in an infected cell induce the synthesis*

and secretion of Type I interferons, which feed back to potently inhibit the replication of the virus in that infected cell.

C-Type Lectin Receptors Bind Carbohydrates on the Surfaces of Extracellular Pathogens

The second family of cell surface PRRs that activate innate and inflammatory responses is the **C-type lectin receptor (CLR)** family. CLRs are plasma membrane receptors expressed variably on monocytes, macrophages, dendritic cells, neutrophils, B cells, and T-cell subsets. CLRs generally recognize carbohydrate components of fungi, mycobacteria, viruses, parasites, and some allergens (peanut and dust mite proteins). Humans have at least 15 CLRs that function as PRRs, most of which recognize one or more specific sugar moieties such as mannose (e.g., the mannose receptor and DC-SIGN), fucose (e.g., Dectin-2 and DC-SIGN), and glucans (e.g., Dectin-1).

CLRs have a variety of functions. Unlike TLRs, which do not promote phagocytosis, some CLRs function as phagocytic receptors (see Table 5-3), and all CLRs trigger signaling pathways that activate transcription factors that induce effector gene expression. Some CLRs trigger signaling events that initially differ from TLR signaling but generally lead to downstream steps similar to the MyD88-dependent TLR pathways that activate the transcription factors NF-κB and AP-1. An example of this is Dectin-1, which binds cell wall β-1,3 glucans on mycobacteria, yeast, and other fungi (Fig-

ure 5-14). Dectin-1 contains a cytoplasmic domain with an ITAM similar to those in the signaling chains of B-cell and T-cell antigen receptors (see Chapter 3). After Dectin-1 binds a ligand, its ITAM is phosphorylated and then recognized by the tyrosine kinase Syk, also involved in the initial stages of B-cell activation (see Figure 3-28). Syk triggers MAPK pathways, leading to the activation of the transcription factor AP-1, and also activates CARD9, one of many signaling components with **Caspase recruitment domains (CARD**; see Table 3-1 and additional examples below). CARD9 forms a complex with additional signaling components, leading to the activation of IKK and the nuclear translocation of NF-κB as described above for TLR. NF-κB and AP-1 cooperate in inducing expression of inflammatory cytokines and IFN-β.

Signals through different PRRs can modulate each other's effects, enhancing or inhibiting expression of particular genes. As the signaling pathways downstream of some CLRs and TLRs are similar, signals coming into a cell from TLRs and

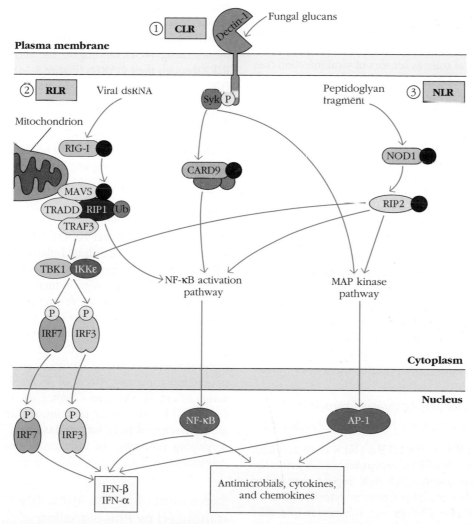

FIGURE 5-14 CLR, RLR, and NLR signaling pathways.
Signaling pathways downstream of selected non-TLR PRRs are shown for (1) the plasma membrane CLR Dectin-1 (2) the cytosolic RLR RIG-I and (3) the cytosolic NLR NOD1. *Note:* Not all signaling components are shown. CARD domains are shown in brown. See text for details. The MAPK pathway is shown in Figure 3-16 and the NF-κB pathway is described in Figure 5-13.

Dectin-1 (or related CLRs) that simultaneously recognize PAMPs on the same or different microbes can combine to enhance proinflammatory cytokine production. In addition, other signaling pathways downstream of Dectin-1 can activate and regulate different members of the NF-κB family, which control expression of genes for additional cytokines that regulate helper-T-cell differentiation, as we will see later. Other CLRs have different cytoplasmic domains, downstream signaling pathways, and outcomes. For example, the binding of PAMPs to one CLR, DC-SIGN, on dendritic cells curtails inflammation by inducing expression of the anti-inflammatory cytokine IL-10 or by reducing mRNA levels for proinflammatory cytokines. In some cases such modulatory effects are beneficial, as responses can be switched from potentially harmful inflammatory responses to the activation of protective T-cell populations. However, as is discussed later, pathogens can also take advantage of these control circuits to reduce responses that would contribute to their elimination.

Retinoic Acid-Inducible Gene-I-Like Receptors Bind Viral RNA in the Cytosol of Infected Cells

The **Retinoic acid-inducible gene-I-*like* receptors (RLRs)** are soluble PRRs that reside in the cytosol of many cell types, where they play critical roles as sensors of viral infection (see Figure 5-14). The three known RLRs (RIG-I, MDA5, and LGP2) are CARD-containing RNA helicases that recognize viral RNAs, usually from RNA viruses such as influenza, measles, and West Nile. These receptors appear to distinguish viral from cellular RNA on the basis of particular structural features not shared by normal cellular RNA, such as double-stranded regions of the RNA, virus-specific sequence motifs, and, in the case of RIG-I, a 5′ triphosphate modification that arises during viral RNA synthesis and processing. Upon viral RNA binding, RIG-I undergoes a conformational change that leads to binding via CARD-CARD interactions to its downstream adaptor molecule located in mitochondrial membranes, the mitochondria-associated MAVS (*m*itochondrial *a*nti-*v*iral *s*ignaling) protein. MAVS aggregates and recruits additional proteins, including the adaptor TRADD, TRAF3, and RIP1, which activate NEMO/IKKs that lead to the activation of IKK complexes, which activate IRFs 3 and 7 and the NF-κB pathway, leading to expression of IFNs α and β, other antimicrobials, chemokines, and proinflammatory cytokines.

Nod-Like Receptors are Activated by a Variety of PAMPs, DAMPs, and Other Harmful Substances

The final family of PRRs is the **NLRs.** (NLR is an acronym that stands for both **N**od-*like* **r**eceptors and **n**ucleotide **o**ligomerization **d**omain/*l*eucine-rich **r**epeat-containing **r**eceptors). The NLRs are a large family of cytosolic proteins activated by intracellular PAMPs and substances that alert cells to damage or danger (DAMPs and other harmful substances). They play major roles in activating beneficial innate immune and inflammatory responses, but, as we will see,

some NLRs also trigger inflammation that causes extensive tissue damage and disease. The human genome contains approximately 23 NLR genes, and the mouse genome up to 34. NLR proteins are divided into three major groups, based largely on their domain structure, as shown in Figure 5-15: NLRCs (which have *c*aspase *r*ecruitment *d*omains, CARD), NLRBs (some of which have *b*aculovirus *i*nhibitory *r*epeat, BIR, domains), and NLRPs (which have *py*rin *d*omains, PYD). The functions of many NLRs have not yet been well characterized; those of several NLRs are described below.

NOD1 and NOD2

The best-characterized members of the NLRC family are NOD1 and NOD2. Important PAMPs recognized by NOD1 and NOD2 are breakdown products (such as muramyl dipeptides) produced during the synthesis or degradation of cell wall peptidoglycans of either intracellular or extracellular bacteria—peptides from the latter must enter the cell to activate NODs. Studies in mice have shown that NOD1 also provides protection against the intracellular protozoan parasite *Trypanosoma cruzi*, which causes Chagas disease in humans, and that NOD2 activates responses to some viruses, including influenza. PAMP binding to the LRR regions of NODs initiates signaling by activating NOD binding to the serine/threonine kinase RIP2 through their CARDs (Figure 5-14). RIP2 then activates IKK, leading to nuclear translocation of NF-κB. RIP2 also activates MAPK pathways, leading to AP-1 activation. The activated NF-κB and AP-1 initiate transcription of inflammatory cytokines—including TNF-α, IL-1, and IL-6—and antimicrobial and other mediators. In addition, RIP2 activates the TRAF3/TBK1/IKKε complex, leading to phosphorylation of IRFs 3 and 7 and to production of IFNs-α and -β.

NLR Inflammasomes

Whereas the NOD1 and NOD2 NLRCs activate the transcription of genes encoding inflammatory cytokines and antimicrobial and other proteins, some other NLRs do not trigger signaling pathways inducing expression of genes involved in innate immune responses. Instead, these NLRs assemble with other proteins into a complex that activates proteases necessary for converting the inactive large precursor forms (procytokines) of IL-1 and IL-18 into the mature forms that are secreted by activated cells. Because of the very potent inflammatory effects of secreted IL-1 (and also to some extent IL-18) (see Chapter 4), these complexes of certain NLRs with other proteins, including key proteases, are now referred to as **inflammasomes**. The discovery and surprising properties of inflammasomes are described in Advances Box 5-1.

Expression of Innate Immunity Proteins is Induced by PRR Signaling

The PRR-activated signaling pathways described above induce the transcription of genes encoding an arsenal of

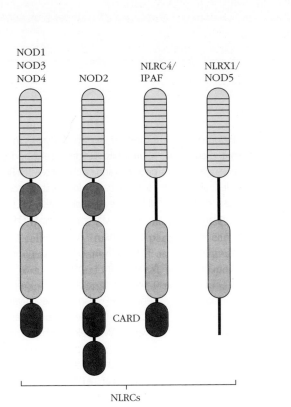

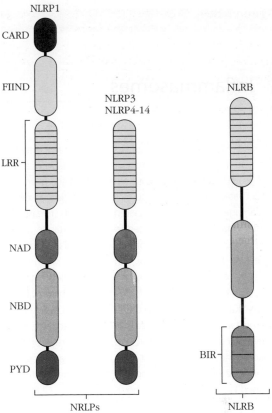

FIGURE 5-15 NLR families. NLRs are characterized by distinct protein domains. All NLRs have a leucine-rich repeat (LRR) domain, similar to those in TLRs, that functions in ligand binding in at least some NLRs, and a nuclear-binding domain (NBD). The three main classes are distinguished by their N-terminal domain (the lowest domains in the figure): NLRC receptors have CARD domains, NLRP receptors have pyrin domains (PYD), and NLRB receptors have BIR (*baculovirus inhibitor repeat*) domains (although the related protein IPAF does not). These domains function in protein-protein interactions, largely through homotypic domain interactions. *[Adapted from F. Martinon et al. 2007. Annual Review of Immunology 27:229; and E. Elinav et al. 2011. Immunity 34:665.]*

proteins that help us to mount protective responses. Some of the induced proteins are antimicrobial and directly combat pathogens, while others serve key roles in activating and enhancing innate and adaptive immune responses. There is tremendous variation in what proteins are made in response to pathogens, reflecting their PAMPs as well as the responding cell types and their arrays of PRRs. Some of the most common proteins and peptides that are secreted by cells following PAMP activation of PRRs and that contribute to innate and inflammatory responses are listed in Table 5-5.

Antimicrobial Peptides

Defensins and cathelicidins were mentioned earlier as being important in barrier protection, such as on the skin and the epithelial layers connected to the body's openings (see Table 5-2). Some cells and tissues constitutively (i.e., continually, without activation) express these peptides. For example, human intestinal Paneth epithelial cells constitutively express α-defensins and some β-defensins, and some defensins and the cathelicin LL-37 are constitutively synthesized and packaged in the granules of neutrophils, ready to

kill phagocytosed bacteria, fungi, viruses, and protozoan parasites. However, in some other cell types, such as mucosal and glandular epithelial cells, skin keratinocytes, and NK cells, the expression of these antimicrobial peptides is induced or enhanced by signaling through PRRs, in particular TLRs and the NOD NLRs. Macrophages do not produce these antimicrobial peptides following PRR activation; interestingly, however, there is an indirect pathway by which microbes induce cathelicidin in macrophages. Binding of microbial ligands to macrophage TLRs induces increased expression of receptors for vitamin D; binding of vitamin D to these receptors activates the macrophages to produce cathelicidin, which then can help the macrophages kill the pathogens.

Type I Interferons

Another major class of antimicrobial proteins transcriptionally induced directly by PRRs is the Type I interferons (IFN-α,β). Type I interferon production is generally activated by those cell surface TLRs and intracellular TLRs, RLRs, and NLRs that recognize viral nucleic acids and other

Inflammasomes

The cytokine IL-1 has been recognized as one of the potent inducers of inflammation. While it was known that the IL-1 gene is transcriptionally activated following exposure to pathogen, damage, or danger-associated molecules, additional steps were needed to generate the mature IL-1 protein from its large intracellular pro-IL-1 precursor. An enzyme, initially called IL-1 Converting Enzyme (ICE), now known as caspase-1, was shown to carry out the cleavage of pro-IL-1, but the enzyme itself existed in most cells as a large inactive precursor. The breakthrough in understanding how mature IL-1 is produced came in 2002 when Jurg Tschopp and others published biochemical studies show-

ing that activation of cells by LPS induced the formation of a large multiprotein aggregate containing an NLR and mature caspase-1 that cleaved pro-IL-1 into mature IL-1, allowing its release from cells. Because of the importance of IL-1 in promoting inflammation, Tschopp and his colleagues coined the term *inflammasome* for the large protein complex that activates caspase-1 to generate IL-1. Three NLRs (NLRP1, NLRP3, and NLRC4) have been shown to form inflammasomes that activate caspase-1 to cleave the large precursors of both IL-1 and IL-18, generating the mature proinflammatory cytokines.

The best-characterized inflammasome is the NLRP3 inflammasome, which is

expressed by monocytes, macrophages, neutrophils, dendritic cells, and some lymphocytes and epithelial cells. NLRP3 is a large complex containing multiple copies each of NLRP3, the adaptor protein ASC (which binds to NLRP3 via homotypic PYD:PYD domain interactions), and caspase-1 (Figure 1). NLRP3 can be activated in cells by a variety of components from bacteria, fungi, and some viruses. In addition to microbial components, NLRP3 can also be activated by nonmicrobial ("sterile") substances, including several DAMPs released by damaged tissues and cells, such as hyaluronan, β-amyloid (associated with Alzheimer's plaques), and extracellular ATP and glucose. Recent

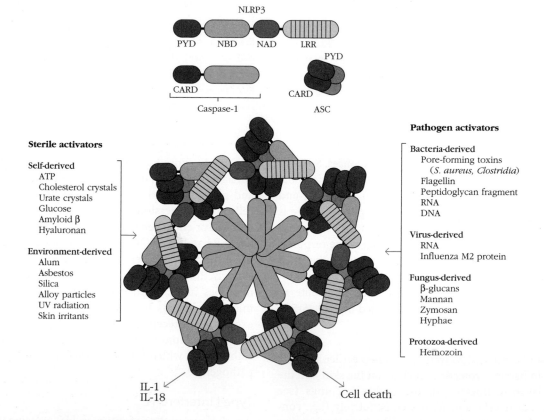

FIGURE 1

The NLRP3 inflammasome and its activators. Assembly of the NLRs, ASC, and caspase-1 due to homotypic domain interactions leads to the formation of a pentamer or heptamer structure: the inflammasome. Activators of the inflammasome are divided into two categories: sterile activators include non-microbial host- and environment-derived molecules, and pathogen-associated activators include PAMPs derived

from bacteria, virus, fungus, and protozoa. Activation of the inflammasome leads to maturation and secretion of IL-1 and IL-18 and, in some instances, to novel processes leading to inflammatory cell death. (Abbreviations: ASC, apoptosis-associated speck-like protein containing a caspase recruitment domain. For others, see legend to Figure 5-15 and text). [*Adapted from B. K. Davis et al. 2011. Annual Review of Immunology* **29***:707.*]

research has also implicated NLRP3 in mediating serious inflammatory conditions caused by an unusual class of harmful substances: crystals. Crystals of monosodium urate in individuals with hyperuricemia cause gout, an inflammatory joint condition, and inhalation of environmental silica or asbestos crystals cause the serious, often fatal, inflammatory lung conditions silicosis and asbestosis. When these crystals are phagocytosed, they damage lysosomal membranes, releasing lysosomal components into the cytosol. Similar effects may be responsible for the loosening (aseptic osteolysis) of artificial joints, caused by tiny metal alloy particles from the prosthesis that also activate NLRP3-mediated inflammation.

How these disparate PAMPs, DAMPs, crystals, and metal particles all activate the NLRP3 inflammasome is under intense investigation. However, to date there is no conclusive evidence for the binding of any ligands directly to NLRP3; thus it may not act as a PRR for PAMPs and DAMPs per se but instead may sense changes in the intracellular milieu resulting from exposure to these materials. Current models, summarized in Figure 2, involve the activation by these varied stimuli of a common set of intracellular signals, such as potassium ion efflux, reactive oxygen species (ROS), and/or leakage of lysosomal contents, all of which appear to induce NLRP3 inflammasome assembly and caspase-1 activation, leading to the processing and secretion of IL-1 and IL-18. It is clear, however, that inflammasomes usually work in tandem with other PRRs in a cell to generate these cytokines. Activation of many TLRs, CLRs, RLRs, and NOD NLRs by PAMPs or DAMPs primes the inflammatory response by inducing transcription of the IL-1 and IL-18 genes by the signaling pathways described above. Once activated by as yet unknown intracellular mediators, the NLRP3 inflammasome then processes the procytokine precursors to the mature, active IL-1 and IL-18, which then are released by the cell by an unknown atypical secretion process.

Somewhat less is known about the other inflammasomes, although they are also large multiprotein aggregates containing NLRs or related proteins and caspases and generate mature IL-1 and IL-18. The

inflammasome discovered by Tschopp contains the NLR protein now called NLRP1 (see Figure 5-15), which is expressed in a wide array of hematopoietic and other cell types. Its activation by the lethal toxin of *Bacillus anthracis* contributes to the lethality of anthrax. The NLRC4 (also called IPAF) inflammasome, expressed predominantly in hematopoietic cells, is activated by certain Gram-negative bacteria and by the cytosolic presence of bacterial flagellin. A fourth inflammasome utilizes the AIM2 protein, which does not contain an NBD and hence is not an NLR, but does contain a PYD domain and hence can form inflammasomes with ASC and caspase-1. Interestingly, AIM2, which is induced by interferons, has been shown recently to bind cytosolic double-stranded DNA, such as from intracellular viruses and bacteria, and also to generate mature IL-1 and IL-18.

The recent discovery of inflammasomes has provided the answer to one long-standing question: How is mature IL-1 generated in cells in response to innate and inflammatory stimuli? But other challenges remain, the most important being to identify the mechanism(s) by which inflammasomes are activated, so that their key caspase-1 enzyme becomes functional. In addition, as no functions are known yet for several other NLRs, more inflammasomes may yet be discovered.

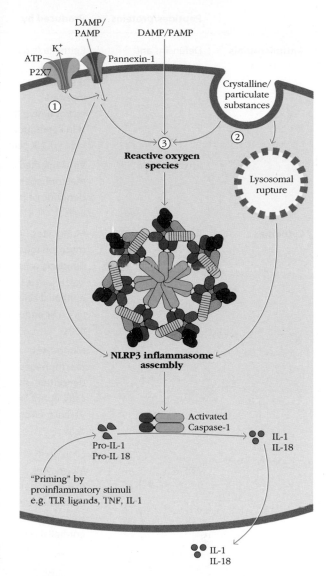

FIGURE 2

Models for NLRP3 inflammasome activation. Three major models for NLRP3 inflammasome activation have been proposed, which may not be mutually exclusive. Model 1: The NLRP3 activator, ATP, triggers P2X7-dependent pore formation by the pannexin-1 hemichannel, allowing potassium ion efflux and entry into the cytosol of extracellular DAMPs and/or PAMPs which then directly engage NLRP3. Model 2: Crystalline or particulate NLRP3 agonists are engulfed, and their physical characteristics lead to lysosomal rupture. Released lysosomal contents, including enzymes, then activate NLRP3 inflammasome in the cytoplasm by unknown mechanisms; one possibility is cathepsin-B-dependent processing of a direct NLRP3 ligand. Model 3: All DAMPs and PAMPs, including ATP and particulate/crystalline activators, induce the generation of reactive oxygen species (ROS). A ROS-dependent pathway triggers NLRP3 inflammasome complex formation. Following these initial activation steps, caspase-1 clustering in the inflammasome activates caspase-1, which then cleaves pro-IL-1 and pro-IL-18 precursors cytokines, generating IL-1 and IL-18, which are then secreted. In some instances the death of the cell may also be induced. [*Adapted from K. Schroder and J. Tschopp. 2010.* Cell *140:821.*]

TABLE 5-5   Secreted peptides and proteins induced by signaling through PRR

	Peptides/proteins	Produced by	Act on	Immune/inflammatory effects
Antimicrobials	Defensins and cathelicidin*	Epithelia (e.g., oro/nasal, respiratory, intestinal, reproductive tracts; skin keratinocytes, kidney); NK cells	Pathogens	Inhibit, kill
			Monocytes, immature dendritic cells, T cells	Chemoattractant; activate cytokine production
			Mast cells	Activate degranulation
	Interferons -α, -β	Virus-infected cells, macrophages, dendritic cells, NK cells	Virus-infected cells	Inhibit virus replication
			NK cells	Activate
			Macrophages, T cells	Regulate activity
Cytokines	IL-1	Monocytes, macrophages, dendritic cells, keratinocytes, epithelial cells, vascular endothelial cells	Lymphocytes	Enhances activity
			Bone marrow	Promotes neutrophil production
			Vascular endothelium	Activates; increases vascular permeability
			Liver	Induces acute-phase response
			Hypothalamus	Fever
	IL-6	Monocytes, macrophages, dendritic cells, NK cells, epithelial cells, vascular endothelial cells	Lymphocytes	Regulates activity
			Bone marrow	Promotes hematopoiesis → neutrophils
			Vascular endothelium	Activates; increases vascular permeability
			Liver	Induces acute-phase response
			Hypothalamus	Fever
	TNF-α	Monocytes, macrophages, dendritic cells, mast cells, NK cells, epithelial cells	Macrophages	Activates
			Vascular endothelium	Activates, increases vascular permeability, fluid loss, local blood clotting
			Liver	Induces acute-phase response
			Hypothalamus	Fever
			Tumors	Cytotoxic for many tumor cells
	GM-CSF	Macrophages, vascular endothelial cells	Bone marrow	Stimulates hematopoiesis → myeloid cells
	IL-12, IL-18	Monocytes, macrophages, dendritic cells	Naïve CD4 T cells	Induce T_H1 phenotype, IFN-γ production
			Naïve CD8 T cells, NK cells	Activate
	IL-10	Macrophages, dendritic and mast cells; NK, T, B cells	Macrophages, dendritic cells	Antagonizes inflammatory response, including production of IL-12 and T_H1 cells
Chemokines	Example: IL-8 (CXCL8) chemokine**	Macrophages, dendritic cells, vascular endothelial cells	Neutrophils, basophils, immature dendritic cells, T cells	Chemoattracts cells to infection site

* Defensins and cathelicidin LL-37 vary among tissues in expression and whether constitutive or inducible.

** Other chemokines that are induced by PRR activation of cells in certain tissues, including various epithelial layers, may also specifically recruit certain lymphoid and myeloid cells to that site. See text, Chapter 14, and Appendix III.

components and activate the IRF transcription factors. As summarized in Table 5-5, Type I interferons are produced in two situations. When virus infected, many cell types are induced to make IFN-α and/or IFN-β following binding of cytosolic viral PAMPs, such as nucleic acids, to their RLRs or NLRs (see Figure 5-14). In addition, many cells express cell surface TLRs recognizing viral PAMPs and/or internalize virus without necessarily being infected, allowing endo-somal TLRs to recognize viral components. Signaling from these TLRs activates the IRFs and IFN-α,β production. One particular type of dendritic cell, called the plasmacytoid dendritic cell (pDC) because of its shape, is a particularly effective producer of Type I IFNs. The pDCs endocytose virus that has bound to various cell surface proteins (including CLRs such as DC-SIGN, which, for example, binds HIV). TLRs 7 and 9 in the endosomes then are activated by viral PAMPs (ssRNA and CpG DNA, respectively), leading to IRF activation.

Type I interferons exert their antiviral and other effects by binding to a specific receptor, frequently called IFNAR (*IFN-alpha receptor*), expressed by most cell types (see Chapter 4). Similar to signaling pathways downstream of many cytokine receptors (see Table 4-3 and Figure 4-11), binding of IFN-α or -β activates IFNAR to recruit and activate specific JAKs (JAK1 and Tyk2), which then activate specific STATs (STATs 1 and 2; see Figure 5-16). STAT1/1 and STAT1/2 dimers then induce expression of various genes, including those for the three proteins that block viral replication: protein kinase R (PKR), which inhibits viral (and cellular) protein synthesis by inhibiting the elongation factor eIF2α; 2′,5′-oligoadenlyate A synthetase, which activates the ribonuclease RNase L that degrades viral mRNA; and Mx proteins, which inhibit both the transcription of viral genes into mRNAs and the assembly of virus particles. Reflecting the potent antiviral activities of Type I interferons, they are used to treat some viral infections, such as hepatitis B and C. In addition to their key

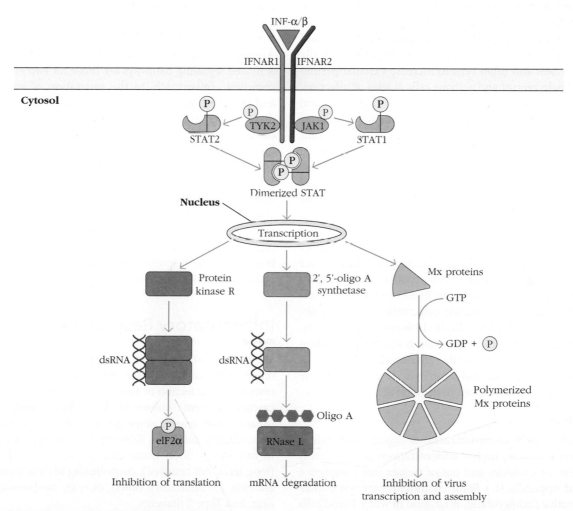

FIGURE 5-16 Major antiviral activities induced by Type I interferons. Interferons-α and -β bind to the IFNAR receptor, which then recruits and activates the JAK1 and TYK2 protein kinases. They bind and phosphorylate STATs 1 and 2, which dimerize, enter the nucleus, and stimulate expression of three proteins that activate antiviral effects. Protein kinase R (PKR) binds viral dsRNA and inhibits the activity of the eIF2α translation initiation factor. 2′,5′-oligoadenylate synthase synthesizes 2′,5′-oligoadenylate, which activates a ribonuclease, RNaseL, that degrades mRNAs. Mx proteins self-assemble into ringlike structures that inhibit viral replication and formation of new virus particles. *[Adapted from Fig. 3-46, A. DeFranco et al. 2007. Immunity (Primers in Biology). Sunderland, MA: Sinauer Associates.]*

roles in controlling viral infections, Type I interferons have other immune-related activities in that they activate NK cells and regulate activities of macrophages and T cells. Treatment with IFN-β has been shown to have beneficial effects in some forms of multiple sclerosis, a T-cell-mediated autoimmune disease with inflammatory involvement, probably by inhibiting production of certain proinflammatory T-cell-derived cytokines (see Chapter 16).

Cytokines

Among the proteins transcriptionally induced by PRR activation are several key cytokines, which—while not directly antimicrobial—activate and regulate a wide variety of cells and tissues involved in innate, inflammatory, and adaptive responses. As introduced in Chapter 4, cytokines function as the protein hormones of the immune system, produced in response to stimuli and acting on a variety of cellular targets. Several key examples of cytokines induced by PRR activation during innate immune responses are listed in Table 5-5, along with their effects on target cells and tissues.

Three of the most important cytokines are IL-1, IL-6, and TNF-α, the major proinflammatory cytokines that act locally on blood vessels and other cells to increase vascular permeability and help recruit and activate cells at sites of infection; they also have systemic effects (see below). IL-1, IL-6, and GM-CSF also feed back on bone marrow hematopoiesis to enhance production of neutrophils and other myeloid cells that will contribute to pathogen clearance. Activation of monocytes, macrophages, and dendritic cells through some TLRs also induces production of IL-12 and IL-18, cytokines that play key roles in inducing naïve helper T cells to become T_H1 cells, in particular by inducing production of IFN-γ. As will be discussed in Chapter 11, the hallmark cytokine of T_H1 cells is IFN-γ, which stimulates cell-mediated immunity and is an important macrophage-activating cytokine. Hence IL-12 and IL-18 are also considered proinflammatory. IL-10 is another important cytokine specifically induced by some TLRs in macrophages, dendritic cells, other myeloid cells, and subsets of T, B, and NK cells. IL-10 is anti-inflammatory, in that it inhibits macrophage activation and the production of proinflammatory cytokines by other myeloid cells. IL-10 levels increase over time and contribute to controlling the extent of inflammation-caused tissue damage.

Chemokines

These small protein **chemoattractants** (agents that induce cells to move toward higher concentrations of the agent) recruit cells into, within, and out of tissues (see Chapters 4 and 14 and Appendix III). Some chemokines are responsible for constitutive (homeostatic) migration of white blood cells throughout the body. Other chemokines, produced in response to PRR activation, have key roles in the early stages of immune and inflammatory responses in that they attract cells that contribute both to clearing the infection or damage and to amplifying the response. The first chemokine to be cloned, IL-8 (also called CXCL8), is produced in response to

activation—by PAMPs, DAMPs, or some cytokines—of a variety of cells at sites of infection or tissue damage, including macrophages, dendritic cells, epithelial cells, and vascular endothelial cells. One of IL-8's key roles occurs in the initial stages of infection or tissue damage; it serves as a chemoattractant for neutrophils, recruiting them to sites of infection. Other chemokines are specifically induced by PRR activation of epithelial cells in certain mucosal tissues and serve to recruit cells specifically to those sites, where they generate immune responses appropriate for clearing the invading pathogen. For example, B cells are recruited to the lamina propria, the lymphocyte-rich tissue under the intestinal epithelium, by two chemokines, CCL28 and CCL20, produced by intestinal epithelial cells activated by PAMP binding to their TLRs. These activated epithelial cells (as well as local dendritic cells activated by PAMPs) also produce cytokines that stimulate the B cells to produce IgA, the class of antibodies most effective in protecting against mucosal infections (see Chapter 13).

Enzymes: iNOS and COX2

Among other genes activated in many cell types by PRR-activated signaling pathways are those for two enzymes that contribute importantly to the generation of antimicrobial and proinflammatory mediators: inducible nitric oxide synthase (iNOS) and **cyclooxygenase 2 (COX2)**. As described earlier, the iNOS enzyme catalyzes an important step in the formation of nitric oxide, which kills phagocytosed microbes (see Figure 5-8). COX2, whose synthesis is induced by PRR activation in monocytes, macrophages, neutrophils, and mast cells, is key to converting the lipid intermediate arachidonic acid to prostaglandins, potent proinflammatory mediators. The next section provides an overview of the main processes of inflammatory responses initiated by innate immune responses. Inflammation will be covered in more detail in Chapter 15.

Inflammatory Responses

When the outer barriers of innate immunity—skin and other epithelial layers—are damaged, the resulting innate responses to infection or tissue injury can induce a complex cascade of events known as the **inflammatory response**. Inflammation may be acute (short-term effects contributing to combating infection, followed by healing)—for example, in response to local tissue damage—or it may be chronic (long term, not resolved), contributing to conditions such as arthritis, inflammatory bowel disease, cardiovascular disease, and Type 2 diabetes.

The hallmarks of a localized inflammatory response were first described by the Roman physician Celsus in the first century AD as *rubor et tumor cum calore et dolore* (redness and swelling with heat and pain). An additional mark of inflammation added in the second century by the physician Galen is loss of function (*functio laesa*). Today we know that these symptoms reflect an increase in vascular diameter (vasodilation), resulting in a rise of blood volume in the area. Higher blood volume heats the tissue and causes it to redden. Vascular permeability also increases, leading to leakage of fluid from the blood vessels, resulting in an accumulation of fluid (**edema**) that swells the tissue. Within a few hours, leukocytes also enter the tissue from the local blood vessels. These hallmark features of inflammatory responses reflect the activation of resident tissue cells—macrophages, mast cells, and dendritic cells—by PAMPs and DAMPs to release chemokines, cytokines, and other soluble mediators into the vicinity of the infection or wound. Recruited leukocytes are activated to phagocytose bacteria and debris and to amplify the response by producing additional mediators. Resolution of this acute inflammatory response includes the clearance of invading pathogens, dead cells, and damaged tissue; the activation of the systemic acute phase response and additional physiological responses, including the initiation of wound healing; and the induction of adaptive immune responses. However, if the infection or tissue damage is not resolved, it can lead to a chronic inflammatory state that can cause more local tissue damage and potentially have systemic consequences for the affected individual.

Inflammation Results from Innate Responses Triggered by Infection, Tissue Damage, or Harmful Substances

When there is local infection, tissue damage, or exposure to some harmful substances (such as asbestos or silica crystals in the lungs), sentinel cells residing in the epithelial layer—macrophages, mast cells, and dendritic cells—are activated by PAMPs, DAMPs, crystals, and so on to start phagocytosing the offending invaders (Figure 5-17). The cells are also activated to release innate immunity mediators that trigger a series of processes that collectively constitute the inflammatory response.

The recruitment of various leukocyte populations to the site of infection or damage is a critical early component of inflammatory responses. PRR signaling activates resident macrophages, dendritic cells, and mast cells to release the initial components of cellular innate immune responses, including the proinflammatory cytokines TNF-α, IL-1, and IL-6; chemokines; prostaglandins (following the induced expression of the COX2 enzyme); and histamine and other mediators released by mast cells. These factors act on the vascular endothelial cells of local blood vessels, increasing

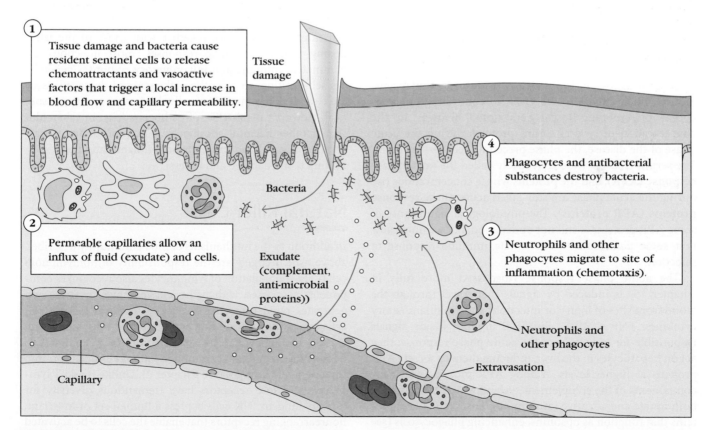

FIGURE 5-17 Initiation of a local inflammatory response. Bacterial entry through wounds activates initial innate immune mechanisms, including phagocytosis by and activation of resident cells, such as macrophages and dendritic cells. Recognition of bacteria by soluble and cellular pattern recognition molecules initiates an inflammatory response that recruits antimicrobial substances and phagocytes (first neutrophils and then monocytes) to the site of infection.

vascular permeability and the expression of **cell adhesion molecules (CAMs)** and chemokines such as IL-8. The affected epithelium is said to be inflamed or activated. Cells flowing through the local capillaries are induced by chemoattractants and adhesion molecule interactions to adhere to vascular endothelial cells in the inflamed region and pass through the walls of capillaries and into the tissue spaces, a process called **extravasation** that will be described in detail in Chapter 14. Neutrophils are the first to be recruited to a site of infection where they enhance local innate responses, followed by monocytes that differentiate into macrophages that participate in pathogen clearance and help initiate wound healing.

In addition to these key events at the site of infection or damage, the key cytokines made early in response to innate and inflammatory stimuli—TNF-α, IL-1, and IL-6—also have systemic effects, which will be described in more detail in Chapter 15. They induce fever (a protective response, as elevated body temperature inhibits replication of some pathogens) by inducing COX-2 expression, which activates prostaglandin synthesis, as mentioned above. Prostaglandin E2 (PGE2) acts on the hypothalamus (the brain center controlling body temperature), causing fever. These three proinflammatory cytokines also act on the liver, inducing the acute phase response, which contributes to the resolution of the inflammatory response.

Proteins of the Acute Phase Response Contribute to Innate Immunity and Inflammation

During the 1920s and 1930s, before the introduction of antibiotics, much attention was given to controlling pneumococcal pneumonia. Researchers noted changes in the concentration of several serum proteins during the acute phase of the disease, the phase preceding recovery or death. The serum changes were collectively called the **acute phase response (APR)**, and the proteins whose concentrations rise during the acute phase are still called **acute phase response proteins (APR proteins)**. The physiological significance of some APR proteins is still not understood, but we now know that some contribute to the innate immune response to infection.

The acute phase response (discussed more fully in Chapter 15) is induced by signals that travel through the blood from sites of injury or infection. The proinflammatory cytokines TNF-α, IL-1, and IL-6 are the major signals responsible for induction of the acute phase response; they act on hepatocytes in the liver, inducing them to secrete APR proteins at higher levels. Among APR proteins are many components of the complement system, which contribute to both innate and adaptive immune responses, and other proteins that function as opsonins, enhancing phagocytosis (see Table 5-3). As was mentioned briefly earlier and will be described in more detail in Chapter 6, complement components and fragments contribute to pathogen inactivation and clearance in a variety of ways, including serving as opsonins.

Several other proteins are produced at higher levels during the APR. Mannose-binding lectin (MBL), described earlier, is a collectin that recognizes mannose-containing molecular patterns on microbes and promotes phagocytosis by blood monocytes. When MBL binds to the surface of microorganisms it also initiates complement activation (see Chapter 6). Another important APR made in the liver is C-reactive protein (CRP), which belongs to a family of pentameric proteins called **pentraxins** that bind ligands in a calcium-dependent reaction. Among the ligands recognized by CRP are a polysaccharide found on the surface of pneumococcal bacteria and phosphorylcholine, which is present on the surface of many microbes. CRP is an opsonin and also activates a complement-mediated attack on the microbe. Circulating CRP levels are considered an indicator of the level of ongoing inflammation. Two other pentraxins, serum amyloid protein and PTX, similarly function as both opsonins and activators of the complement pathway. Also released into the blood at higher levels are the surfactant protein opsonins SP-A and SP-D, mentioned earlier as providing protection against lung infections, and a number of proteins that participate in or regulate the coagulation (clotting) pathway, such as fibrinogen. While most of these APR proteins are always present in the blood at low levels, their increased concentrations during the acute phase response provide enhanced protective functions during infections.

With the combined defenses mounted by innate and inflammatory responses, together with those of the later-arising adaptive immune responses, most infections are eliminated. Immune and inflammatory responses generally are self-limiting, so once the pathogen and damaged tissue are cleared, inflammation usually diminishes and the tissues heal. However, persistence of pathogens (e.g., in tuberculosis) or other harmful substances (e.g., monosodium urate crystals in gout) can cause chronic inflammation and continuing tissue damage and illness.

Natural Killer Cells

In addition to the mechanisms of innate immunity described above, which largely are the responsibility of nonlymphoid cell types, a population of lymphocytes also recognizes components associated with pathogens, damage, or stress and generates rapid protective responses. **Natural killer (NK) cells** constitute a third branch of lymphoid cells, along with B and T lymphocytes of the adaptive immune system, all differentiating from the common lymphoid progenitor into three separate lineages (see Chapter 2). Unlike B and T lymphocytes, whose receptors have tremendous diversity for foreign antigens, NK cells express a limited set of invariant, nonrearranging receptors that enable the cells to be activated by indicators of infection, cancer, or damage that are expressed by other cells. Again unlike B and T cells, which require days of activation, proliferation, and differentiation to generate their protective antibody and cell-mediated immune responses, NK cells are preprogrammed to respond immediately to appropriate stimuli, releasing from preformed secretory granules effector proteins that kill altered cells by inducing apoptosis. This mechanism of cell-mediated cytotoxicity is also carried out by cytotoxic T cells, which appear days later (see Chapter 13). NK-cell-mediated cytotoxic activity is enhanced by IFN-α produced early during virus infections, an example of positive feedback regulation in innate immunity. In addition to releasing cytotoxic mediators, activated NK cells also secrete cytokines—the proinflammatory cytokines IL-6 and TNF-α, as well as Type I IFNs and the Type II IFN, IFN-γ, a potent macrophage activator that also helps to activate and shape the adaptive response. Thus, along with the Type I IFNs produced by virus-infected cells or induced during innate responses by viral PAMPs, NK cells are an important part of the early innate response to viral infections (as well as to malignancy and other indicators of danger). *These early innate responses control the infection for the days to week it takes for the adaptive response (antibodies and cytotoxic T cells) to be generated.*

How do NK cells sense that our cells have become infected, malignant, or potentially harmful in other ways? As will be described more fully in Chapter 13, NK cells express a variety of novel receptors (collectively called NK receptors). Members of one group serve as *activating receptors* (of which more than 20 have been described in humans and mice) that have specificity for various cell surface ligands that serve as indicators of infection, cancer, or stress. While one of these activating ligands in mice has been shown to be a protein component of a virus (mouse cytomegalovirus), most NK activating receptors apparently have specificity not for a pathogen-associated component but for proteins specifically up-regulated on infected, malignant, or stressed cells, which serve as danger signals perceived by NK cells. NK cells also express TLRs, and binding of PAMPs or DAMPs can add to the activation signals.

As will be described later, to limit potential killing of normal cells in our body, NK cells also express *inhibitory receptors* that recognize membrane proteins (usually conventional MHC proteins) on normal healthy cells and inhibit NK-cell-mediated cytotoxic killing of those cells. Many virus-infected or tumor cells lose expression of their MHC proteins and thus do not send these inhibitory signals. Receiving an excess of activating signals compared to inhibitory signals tells an NK cell that a target cell is abnormal, and the NK cell is activated to kill the target cell. Thus NK cells are part of our innate sensing mechanisms that provide immediate protection, in this case recognizing and eliminating our own body's cells that have become harmful. Recent studies suggest that initial activation increases the population and/or activity of responding NK cells, which therefore would generate a greater response when exposed later to the same activating ligand(s). This resembles the immunological memory in the adaptive immune system and represents the best example of immunological memory in the vertebrate *innate* immune system.

Regulation and Evasion of Innate and Inflammatory Responses

Innate immune responses and the inflammatory responses they induce play key roles in clearing infections and healing infected or damaged tissues. The importance of some of the individual molecules involved in the generation of innate and inflammatory responses is dramatically demonstrated by the impact on human health of genetic defects and polymorphisms (genetic variants) that alter the expression or function of these molecules. Some of these defects and their clinical consequences are described in Clinical Focus box 5-2. As illustrated by these conditions, and by the many known roles (cited throughout this chapter) of innate and inflammatory mechanisms in protecting us against pathogens, these responses are essential to keeping us healthy. However, some other disorders show that innate and inflammatory responses can also be harmful, in that overproduction of various normally beneficial mediators and uncontrolled local or systemic responses can cause illness and even death. Therefore it is important that the occurrence and extent of innate and inflammatory responses be carefully regulated to optimize the beneficial responses and minimize the harmful responses.

Innate and Inflammatory Responses Can Be Harmful

To be optimally effective in keeping us healthy, innate and inflammatory responses should use their destructive mechanisms to eliminate pathogens and other harmful substances quickly and efficiently, without causing tissue damage or inhibiting the normal functioning of the body's systems. However, this does not always occur—a variety of conditions result from excessive or chronic innate and inflammatory responses.

The most dangerous of these conditions is **sepsis**, a systemic response to infection that includes fever, elevated heartbeat and breathing rate, low blood pressure, and compromised organ function due to circulatory defects. Several hundred thousand cases of sepsis occur annually in the United States, and mortality rates range from 20% to 50%, but sepsis can lead to **septic shock**, circulatory and respiratory collapse that has a 90% mortality rate. Sepsis results from **septicemia**, infections of the blood, in particular those involving Gram-negative bacteria such as *Salmonella*, although other pathogens can also cause sepsis.

The major cause of sepsis from Gram-negative bacteria is the cell wall component LPS (also known as endotoxin), which as we learned earlier is a ligand of TLR4. As we have seen, LPS is a highly potent activator of innate immune mediators, including the proinflammatory cytokines TNF-α, IL-1, and IL-6; chemokines; and antimicrobial components. Systemic infections activate blood cells including monocytes and neutrophils, vascular endothelial cells, and resident macrophages and other cells in the spleen, liver, and other

Genetic Defects in Components of Innate and Inflammatory Responses Associated with Disease

In combination with our rapidly expanding understanding of the mechanisms by which innate and inflammatory responses contribute to disease susceptibility and resistance, in recent years advances in human genetics have helped to identify a number of genetic defects that confer greater susceptibility to infectious and inflammatory diseases. The adverse effects of mutations in genes encoding essential components of innate and inflammatory processes highlight the critical roles of these proteins in keeping us healthy.

Since 2003, when the first mutations in innate immune components that predispose individuals to recurrent bacterial infections were discovered, a number of mutations interfering with the generation of protective innate immune responses have been identified. Two examples were mentioned earlier in this chapter: defects in NADPH oxidase, which cause chronic granulomatous disease, and MBL deficiencies, which predispose to respiratory infections. Also leading to defects in innate immunity are mutations in two proteins—MyD88 and IRAK4—required for the MyD88-dependent signaling pathway downstream of all TLRs except TLR3 (see Figure 5-13). Children with these defects suffer from severe invasive *S. pneumococcus* infections, some fatal, and

are also susceptible to *Staphylococcus aureus* and *Pseudomonas aeruginosa* (Figure 1a). The MyD88 mutations completely prevent cytokine and chemokine induction by ligands for TLRs 2/1, 2/6, 5, 7, and 8. Not surprisingly, the effects of these mutations are less significant for TLR3 (which activates TRIF signaling pathways instead of MyD88) and TLR4 (which activates both MyD88 and TRIF signaling pathways). That the MyD88 mutations do not leave these children more susceptible to a wider variety of pathogens probably reflects the induction of protective immunity by other PRRs as well as by the adaptive immune system. In fact, the children become less susceptible to infections as they get older (see Figure 1b), consistent with the buildup of adaptive immunological memory to these pathogens.

Other genetic defects with clinical consequences have been identified in the pathways by which Type 1 interferons (IFN-α,β) are induced by viral nucleic acid PAMPs and then block virus replication in infected cells. As highlighted in Figure 2, mutations that completely or partially block these pathways (red symbols) have been found in TLR3 and other components of the pathway that induce IFN-α,β. Mutations have also been found in TYK2 and STAT1, key components that activate the

antiviral effects of the interferons in infected cells. Interestingly, these mutations were all discovered in children presenting with herpes simplex virus (HSV) encephalopathy, a severe HSV infection of the central nervous system. Cells in the CNS express TLR3, and it may be that the mutations in these pathways severely disable innate responses that are critical to protection against CNS infection by this virus. Children with TYK2 and STAT1 mutations are also very susceptible to other infections, especially with mycobacteria, probably because macrophages must be activated by IFN-γ (which also utilizes TYK2 and STAT1 in its signaling pathway) to be able to kill these intracellular bacteria.

The final set of genetic defects associated with disease states to be discussed here involves the effects of genetic variants in NLRs (including inflammasomes) in promoting inflammatory diseases. Genome-wide genetic association studies have indicated that a number of allelic variants of TLRs and NLRs are associated with inflammatory disorders. Several variants are associated with inflammatory bowel disease, which includes ulcerative colitis and Crohn's disease. The most impressive genetic association was of Crohn's disease with mutations in NOD2 in or near its ligand-binding LRR region. As

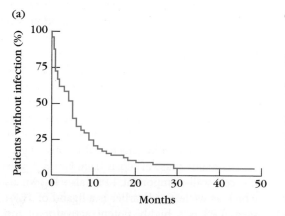

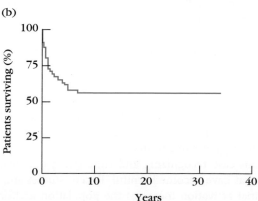

FIGURE 1
Severe bacterial infection and mortality among 60 children with MyD88 or IRAK4 deficiencies. (a) Decline in the percentage of children with these deficiencies who are asymptomatic reveals the incidence of the first severe bacterial infection during the first 50 months of life. (b) Survival curve of children with deficiencies shows reduced mortality after 5 years of age. *[Adapted from J-L. Casenova et al. 2011. Annual Review of Immunology* **29:**447*; based on data from C. Picard et al. 2010. Medicine* **89:**403*]*

BOX 5-2

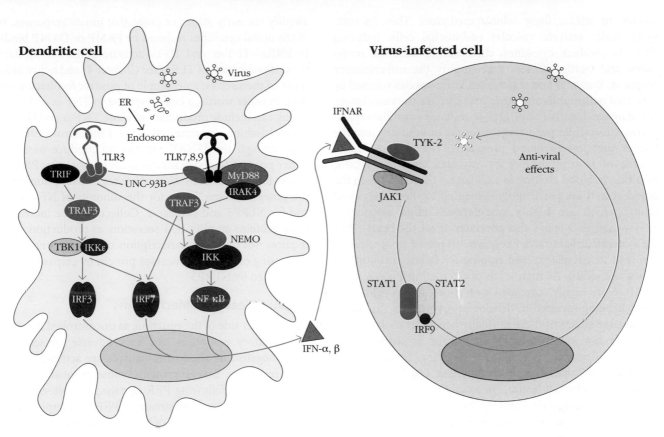

FIGURE 2

Genetic defects affecting the production and antiviral effects of IFN-α,β. This schematic shows some protein components involved in Type I IFN production by dendritic cells and the Type I IFN response in virus-infected cells. Proteins in which genetic mutations have been identified that result in defective functions and are associated with greater susceptibility to viral diseases are shown in red. Viruses are taken up by dendritic cells via specific receptors, and viral nucleic acids are detected by the various TLRs expressed in endosomes. Transport to endosomes of TLRs 3, 7, 8, and 9 is dependent on the ER protein UNC93B. Cytoplasmic signaling components activate transcription factors, including IRFs and NF-κB, leading to synthesis and secretion of IFN-α,β. TLR3, UNC93B, TRAF3, and NEMO deficiencies are associated with impaired IFN production, particularly during herpesvirus infection. The binding of IFN-α,β to their receptor induces the phosphorylation of JAK1 and TYK-2, activating the signal transduction proteins STAT-1, STAT-2, and IRF9. This complex translocates as a heterotrimer to the nucleus, where it acts as a transcriptional activator, binding to specific DNA response elements in the promoter region of IFN-inducible genes. TYK-2 and STAT-1 deficiencies are associated with impaired IFN responses.
*[Adapted from J. Bustamante et al., 2008, Current Opinion in Immunology **20**:39.]*

NOD2 is activated by bacterial cell wall fragments, investigators have hypothesized that intestinal epithelial cells with the mutant NOD2 PRRs are unable to activate adequate protective responses to gut bacteria and/or to maintain appropriate balance between normal commensals and pathogenic bacteria, and that these defective innate immune functions contribute to Crohn's disease pathogenesis. Consistent with this hypothesis, recent studies have found that intestinal Paneth cells from NOD2-defective individuals secrete reduced amounts of α-defensins which, as mentioned earlier, are essential for maintaining normal commensal gut flora.

Genetic variants of the NLRP3 inflammasome have also been shown to be associated with Crohn's disease and other inflammatory disorders. In fact, mutations in NLRP3 (originally called cryopyrin) have been show to be responsible for a set of autoinflammatory diseases (i.e., noninfectious inflammatory diseases affecting the body's tissues) collectively known as CAPS (for *cryopyrin-associated periodic fever syndromes*); one example is NOMID (*neonatal onset multisystem inflammatory disorder*). These devastating syndromes include many signs of systemic inflammation, including fever, rashes, arthritis, pain, and inflammation affecting the nervous system, with adverse effects on vision and hearing.

More than 70 inherited and novel mutations in NLRP3 associated with CAPS have been identified. Most are in the NRLP3 NBD domain, though some are in the LRR domain. What many have in common is their deregulating effect on NLRP3 activation of caspase-1, which may become constitutively active. Cells from NOMID patients have recently been shown to secrete higher levels of IL-1 and IL-18, both spontaneously and induced by PAMPs and DAMPs, promoting chronic inflammation. Another consequence of constitutive NLRP3 activation is the death of the activated cells, releasing DAMPs that lead to more inflammation. Fortunately, new therapeutic approaches that inhibit IL-1 activity seem to alleviate these symptoms in some patients.

tissues, to release these soluble mediators. They, in turn, systemically activate vascular endothelial cells, inducing them to produce cytokines, chemokines, adhesion molecules, and clotting factors that amplify the inflammatory response. Enzymes and reactive oxidative species released by activated neutrophils and other cells damage the vasculature; this damage, together with TNF-α–induced vasodilation and increased vascular permeability, results in fluid loss into the tissues that lowers blood pressure. TNF also stimulates release of clotting factors by vascular endothelial cells, resulting in blood clotting in capillaries. These effects on the blood vessels are particularly damaging to the kidneys and lungs, which are highly vascularized. High circulating TNF-α and IL-1 levels also adversely affect the heart. Thus the systemic inflammatory response triggered by septicemia can lead to circulatory and respiratory failure, resulting in shock and death. As high levels of circulating TNF-α, IL-1, and IL-6 are highly correlated with morbidity, considerable effort is being invested in developing treatments that block the adverse effects of these normally beneficial molecules.

While not as immediately dangerous as septic shock, chronic inflammatory responses resulting from ongoing activation of innate immune responses can have adverse consequences for our health. For example, a toxin from *Helicobacter pylori* bacteria damages the stomach by disrupting the junctions between gastric epithelial cells and also induces chronic inflammation that has been implicated in peptic ulcers and stomach cancer. Also, increasing evidence suggests that the noninfectious DAMPs cholesterol (as insoluble aggregates or crystals) and amyloid β contribute, respectively, to atherosclerosis (hardening of the arteries) and Alzheimer's disease. Other examples of harmful sterile (noninfectious) inflammatory responses discussed earlier—including gout, asbestosis, silicosis, and aseptic osteolysis—are induced, respectively, by crystals of monosodium urate, asbestos, and silica, and by metal alloy particles from artificial joint prostheses. These varied substances are all potent inflammatory stimuli due to their shared ability to activate the NLRP3 inflammasome, resulting in the release of the proinflammatory cytokines IL-1 and IL-18.

Innate and Inflammatory Responses are Regulated Both Positively and Negatively

As innate immune responses play essential roles in eliminating infections but also can be harmful when not adequately controlled, it is not surprising that many regulatory processes have evolved that either enhance or inhibit innate and inflammatory responses. These mechanisms control the induction, type, and duration of these responses, in most cases leading to the elimination of an infection without damaging tissues or causing illness.

Positive Feedback Mechanisms

Innate and inflammatory responses are increased by a variety of mechanisms to enhance their protective functions. To amplify the early stages of protective innate responses, two of the initial cytokines induced by PAMP or DAMP binding to PRRs—TNF-α and IL-1—activate pathways similar to those downstream of TLRs (see Chapter 4) and hence induce more of themselves, an example of positive feedback regulation. In other words, a cell's response to TNF-α or IL-1 signaling can include production of more TNF-α and IL-1. In a parallel fashion, in some cells Type I interferons can signal cells through the IFNAR receptor to produce more IFN. Other positive feedback mechanisms triggered by PRR activation include the activation of higher transcription rates of the genes for some TLRs, for the subunits of NF-κB itself, and for NLRP3 and caspase-1. Collectively the increases in these proteins amplify IL-1 secretion, as production of IL-1 requires not only the transcription of its gene but also the processing of its large precursor protein by caspase-1, which is activated by NLRP3.

Negative Feedback Mechanisms

On the other side of the equation, as uncontrolled innate and inflammatory responses can have adverse consequences, many negative feedback mechanisms are activated to limit these responses. Several proteins whose expression or activity is increased following PRR signaling inhibit steps in the signaling pathways downstream of PRRs. Examples include production of a short form of the adaptor MyD88 that inhibits normal MyD88 function, the activation of protein phosphatases that remove key activating phosphate groups on signaling intermediates, and the increased synthesis of IκB, the inhibitory subunit that keeps NF-κB in the cytoplasm. The activation of these and other intracellular negative feedback mechanisms can lead cells to become less responsive, limiting the extent of the innate immune response. In a well-studied example, when macrophages are exposed continually to the TLR4 ligand LPS, their initial production of antimicrobial and proinflammatory mediators is followed by the induction of inhibitors (including IκB and the short form of MyD88) that block the macrophages from continuing to respond to LPS. This state of unresponsiveness, called **LPS tolerance** (or endotoxin tolerance), reduces the possibility that continued exposure to LPS from a bacterial infection will cause septic shock.

Other feedback pathways inhibit the inflammatory effects of TNF-α and IL-1. Each induces production of a soluble version of its receptor or a receptor-like protein that binds the circulating cytokine molecules (see Chapter 4), preventing them from acting on other cells. In addition, the anti-inflammatory cytokine IL-10 is produced late in the macrophage response to PAMPs; it inhibits the production and effects of inflammatory cytokines and promotes wound healing. Another anti-inflammatory cytokine, TGF-β, is also produced by macrophages, dendritic cells, and other cells, especially after PRR activation by apoptotic cells. The inhibitory effects of TGF-β reduce the likelihood that DAMPs released by cells undergoing apoptosis will induce inflammatory responses. As a final example of a negative feedback

TABLE 5-6	Pathogen evasion of innate and inflammatory responses
Type of evasion	**Examples**
Avoid detection by PRRs	*Proteobacteria* flagellin has a mutation that prevents it from being recognized by TLR5.
	Helicobacter, Coxiella, and *Legionella* bacteria have altered LPS that is not recognized by TLR4.
	HTLV-1 virus p30 protein inhibits transcription and expression of TLR4.
	Several viruses (Ebola, influenza, Vaccinia) encode proteins that bind cytosolic viral dsRNA and prevent it from binding and activating RLR.
Block PRR signaling pathways, preventing activation of responses	Vaccinia virus protein A46R and several bacterial proteins have TIR domains that block MyD88 and TRIF from binding to TLRs.
	Several viruses block TBK1/IKK-activation of IRF3 and IRF7, required for IFN production.
	West Nile Virus NS1 protein inhibits NF-κB and IRF transport into the nucleus.
	Yersinia bacteria produce Yop proteins that inhibit inflammasome activity; the YopP protein inhibits transcription of the IL-1 gene.
Prevent killing or replication inhibition	*Salmonella* and *Listeria* bacteria rupture the phagosome membrane and escape to the cytosol.
	Mycobacteria tuberculosis blocks phagosome fusion with lysosomes.
	Vaccinia virus encodes a protein that binds to Type I IFNs and prevents them from binding to the IFN receptor.
	Hepatitis C virus protein NS3-4A and Vaccinia virus protein E3L bind protein kinase R and block IFN-mediated inhibition of protein synthesis.

loop, levels of circulating TNF-α and IL-1 that could potentially induce harmful inflammatory responses act via the hypothalamus to induce the adrenal medulla to secrete glucocorticoid hormones, which have a variety of potent anti-inflammatory effects, including inhibiting the production of TNF-α and IL-1 themselves.

Pathogens Have Evolved Mechanisms to Evade Innate and Inflammatory Responses

As it is advantageous to pathogens to evolve mechanisms that allow them to evade elimination by the immune system, many have acquired the ability to inhibit various innate and inflammatory signaling pathways and effector mechanisms that would clear them from the body. Most bacteria, viruses, and fungi replicate at high rates and, through mutation, may alter their components to avoid recognition or elimination by innate immune effector mechanisms. Other pathogens have evolved complex mechanisms that block normally effective innate clearance mechanisms. A strategy employed especially by viruses is to acquire genes from their hosts that have evolved and function as inhibitors of innate and inflammatory responses. Several strategies by which viruses evade immune responses triggered by interferons and other cytokines are

listed in Table 4-5. Examples of mechanisms by which pathogens more generally avoid detection by PRRs, activation of innate and inflammatory responses, or elimination by those responses are described in Table 5-6.

Interactions Between the Innate and Adaptive Immune Systems

The many layers of innate immunity are important to our health, as illustrated by the illnesses seen in individuals lacking one or another component of the innate immune system (see the Clinical Focus box and other examples mentioned earlier in this chapter). However, innate immunity is not sufficient to protect us fully from infectious diseases, in part because many pathogens have features that allow them to evade innate immune responses, as discussed above. Hence the antigen-specific responses generated by our powerful adaptive immune system are usually needed to resolve infections successfully. While our B and T lymphocytes are the key producers of adaptive response effector mechanisms—antibodies and cell-mediated immunity—it is becoming increasingly clear that our innate immune system plays important roles in helping to initiate and regulate adaptive immune responses so that they will be optimally effective. In addition, the adaptive immune system has co-opted several mechanisms by which the innate immune system eliminates pathogens, modifying them to enable antibodies to clear pathogens.

The Innate Immune System Activates and Regulates Adaptive Immune Responses

When pathogens invade our body, usually by penetrating our epithelial barriers, the innate immune system not only reacts quickly to begin to clear the invaders but also plays key roles in activating adaptive immune responses. While innate cells at the infection site (resident macrophages, dendritic cells, mast cells, and newly recruited neutrophils) are sensing the invading pathogens through their pattern recognition receptors and generating antimicrobial and pro-inflammatory responses that will slow down the infection, they are also initiating steps to bring the pathogens to the attention of lymphocytes and help activate responses that, days later, will generate the strong antigen-specific antibody and cell-mediated responses that will resolve the infection.

The first step in the generation of adaptive immune responses to pathogens is the delivery of the pathogen to lymphoid tissues where T and B cells can recognize it and respond. As is discussed in more detail in Chapter 14, dendritic cells (usually immature) that serve as sentinels in epithelial tissues bind microbes through various pattern-recognition receptors. The dendritic cells carry the bound microbes— either still attached to the cell surface or in phagosomes—via the lymphatic vessels to nearby secondary lymphoid tissues, such as the draining lymph nodes. There the dendritic cells can transfer or present the microbes or microbial components to other cells. In many cases the dendritic cell has internalized and degraded the microbe, and microbe-derived peptides come to the cell surface bound to its MHC class II proteins. Meanwhile, the binding of microbial PAMPs to the dendritic cell's PRRs also has activated the dendritic cell to mature. It now expresses higher levels of MHC class II proteins and has turned on expression of costimulatory membrane proteins, such as CD80 or CD86 (see Chapter 3), that are recognized by receptors on T_H cells. As a result of these processes of microbe binding, processing, and maturation, mature dendritic cells are the most effective antigen-presenting cells, particularly for the activation of naïve (not previously activated) T_H cells.

Pathogen-Specific Activation of T_H Cell Subsets

Activation of dendritic cells by binding of various PAMPs to PRRs has additional important consequences for adaptive immune responses. Depending on the nature of the pathogen and what PRRs and downstream signaling pathways are activated, dendritic cells are induced to secrete specific cytokines that influence what cytokines a naïve T_H (CD4$^+$) cell will produce after activation, thus determining its functional phenotype (Figure 5-18). These distinct phenotypes, or T_H cell subsets, will be discussed in detail in Chapter 11.

As one example, a Gram-negative bacteria may express PAMPs that bind to TLR4, TLR5, and/or the endosomal/lysosomal TLRs that bind bacterial nucleic acids. This binding stimulates dendritic cells to produce IL-12, which induces naïve T_H cells to become T_H1 cells (see Chapter 4). The hallmark cytokine of T_H1 cells is IFN-γ, which activates macrophages to kill intracellular bacteria, reinforces the T_H1 phenotype, contributes to the activation of virus-specific cytotoxic T cells, and, along with IL-2 (another T_H1 cytokine), helps to activate B cells.

In contrast, binding of PAMPs from fungi, Gram-positive bacteria, or helminths to TLR 2/1 or 2/6 activates production of IL-10 and inhibits secretion of IL-12. This combination of effects of stimulation through TLR2-containing receptors supports naïve T_H cell differentiation into T_H2 cells, which promote strong antibody responses that are more important for providing protection against these pathogens than are cell-mediated responses. Fungal PAMPs bind to the CLR Dectin-1, which activates the dendritic cell to produce the proinflammatory cytokines TNF-α, IL-6, and IL-23; the latter two induce the T_H to become T_H17 cells, which are characterized by the secretion of proinflammatory cytokines such as IL-17 and IL-6. These cytokines help recruit neutrophils to the site, which then phagocytose and kill the fungi.

It is important to recognize that the schematic shown in Figure 5-18 is oversimplified, as there are many complexities to the regulation of T_H phenotype not shown. For example, IL-4, a key cytokine for inducing formation of T_H2 cells, may be made by mast cells and/or basophils following PRR activation by certain pathogens, including helminth worms. The resulting T_H2 cells induce production of IgG and IgE antibodies that are particularly effective against some of these pathogens.

In summary, through their differential activation of specific PRRs, various pathogens induce dendritic and other innate cells to produce cytokines that activate T_H cells to acquire distinct cytokine-producing phenotypes. These distinct T_H subsets generally elicit the types of adaptive immune responses that will be most effective in eliminating the particular pathogen invaders. The molecular mechanisms by which naïve CD4$^+$ T_H cells are influenced to differentiate into various mature T_H subsets will be described in Chapter 11.

Pathogen-Specific Activation of T_C Cells

Recent research has also revealed key roles of dendritic cells in the activation of cytotoxic T-cell-mediated responses that are needed to kill cells in our body that have become harmful through infection with viruses or other intracellular pathogens or through malignant transformation. As was the case with naïve T_H (CD4$^+$) cells, naïve cytotoxic T_C (CD8$^+$) cells also are most effectively activated by mature dendritic cells (see Chapter 13). Again, dendritic cell maturation (also referred to as *licensing*) is usually induced by the binding of microbial components to a dendritic cell's PRRs. These mature/licensed DCs also are capable of novel pathways for processing internalized

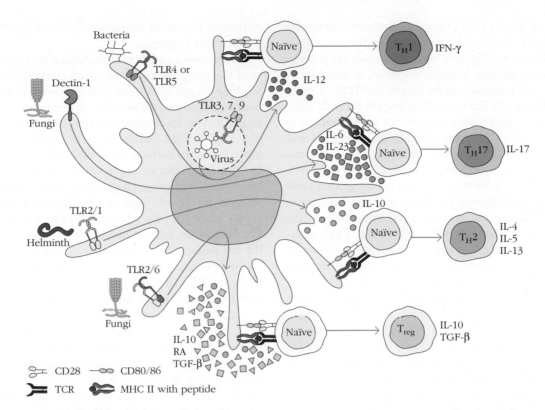

FIGURE 5-18 Differential signaling through dendritic cell PRRs influences helper T cell functions. Microbial PAMPs bind and activate distinct PRRs and signaling pathways that differentially induce production of various cytokines and other mediators, such as retinoic acid (RA). These cytokines interact with receptors on naïve CD4+ T cells that are in the process of being activated both by antigen-derived peptides bound to dendritic cell MHC class II proteins and by interactions with costimulatory molecules such as CD80 or CD86. Note that activation of dendritic cells by PAMP binding to TLRs stimulates increased expression of MHC class II and costimulatory proteins. The particular cytokine(s) to which a naïve CD4+ T cell is exposed induces that cell turn on genes for certain cytokines, determining that cell's functional phenotype (see text). [Adapted from B. Pulendran et al. 2010. Nature Immunology 11:647.]

pathogens so that peptides generated by endosomal/lysosomal degradation are carried to the cell surface bound to MHC class I proteins rather than to class II proteins (this process is called *cross-presentation* and explained in Chapter 8). These peptide/MHC class I complexes, together with costimulatory proteins such as CD80 or CD86 on the surface of the mature/licensed dendritic cell, activate the naïve T_C to give rise to cytotoxic T effector cells. As was the case with the activation of T_H cells, cytokines such as IL-12 and IFN-γ produced by innate cells (dendritic cells, macrophages, NK cells, or other cells) contribute to the differentiation and activation of T_C cells.

T-Independent Antigens

In addition to their indirect roles in promoting and regulating antibody and cell-mediated responses via their activation of DC and other innate cells, described above, TLRs also can be more directly involved in activating B and T cells. B cells express TLRs, and the binding of PAMPs to these TLRs activates signaling pathways that can add to or substitute for the signals from T_H cells normally required for B-cell activation.

One example has been well studied in mouse B cells. In combination with signals from the B cells' BCR after antigen binding, TLR4 binding of LPS (at low concentrations) can activate sufficient signals to induce the B cell to proliferate

and differentiate into antibody-secreting plasma cells without T_H cells. At high concentrations of LPS, TLR4-activated signals are sufficient to activate all B cells (polyclonal activation), regardless of their antigen-binding specificity; hence LPS has for many years been called a T-independent antigen (see Chapter 12). Human B cells do not express TLR4 and hence do not respond to LPS; however, they do express TLR9 and can be activated by microbial CpG DNA. The ability of TLR signals to replace T_H signals is often beneficial; co-binding of bacteria to both BCR and TLR on a cell may activate that cell more quickly than if it had to wait for signals from a T_H cell. Some T cells also express TLRs, which similarly function as costimulatory receptors to enhance protective responses. For example, mouse CD8+ T cells express TLR2; costimulation of CD8+ T cells by TLR2 ligands along with TCR recognition of peptide/MHC class I complexes reduces the T cells' need for costimulatory signals provided by dendritic cells and enhances their proliferation, survival, and functions.

Adjuvants Activate Innate Immune Responses to Increase the Effectiveness of Immunizations

Given these activating and potentiating effects of PRR ligands on adaptive immune responses, can they be used to enhance the efficacy of vaccines in promoting protective immunity against various pathogens? In fact, many of the materials—known as **adjuvants**—that have been shown over the years by trial and error to enhance immune responses in both laboratory animals and humans contain ligands for TLRs or other PRRs (see Chapter 17). For example, Complete Freund's Adjuvant, perhaps the most potent adjuvant for immunizations in experimental animals, is a combination of mineral oil and killed *Mycobacteria*. The mineral oil produces a slowly dispersing depot of antigen, a property of many effective adjuvants, while fragments of the bacteria's cell wall peptidoglycans serve as activating PAMPs. Alum (a precipitate of aluminum hydroxide and aluminum phosphate) is used in some human vaccines; it has recently been shown to activate the NRLP3 inflammasome, thereby enhancing IL-1 and IL-18 secretion and promoting inflammatory processes that enhance adaptive immune responses. As the responding innate cells do not make IL-12, immunizations with alum usually lead the activated T cells to become T_H2 cells, which promote strong antibody responses.

While many vaccines consist of killed or inactivated viruses or bacteria and hence contain their own PAMPS, which function as built-in adjuvants, some new vaccines consist of protein antigens that themselves are not very stimulatory to the immune system. Tumor antigens also tend to induce weak responses. Hence, considerable effort is being invested in developing new adjuvants based on current knowledge about PRRs. LPS is a highly potent adjuvant but generates too much inflammation to use; less harmful versions of LPS are being developed as potential adjuvants. The ability of the TLR3 ligand poly I:C (synthetic double-stranded RNA) to activate innate immunity and initiate inflammatory responses has inspired clinical trials to test its ability to enhance the effectiveness of weak vaccines, including antitumor vaccines. Vaccines using the TLR9 ligand CpG DNA (which mimics bacterial DNA) as an adjuvant are in clinical trials; there is great interest in CpG DNA as it preferentially elicits a T_H1 response, important for inducing cell-mediated immunity (see Figure 5-18). Other approaches for generating an effective vaccine for a pathogen protein include fusing the protein to a TLR ligand using genetic engineering. For example, fusions of pathogen proteins to the TLR5 ligand flagellin are currently being tested. These are all examples of how our expanding knowledge of the interactions between the innate and adaptive immune systems may contribute to the development of more effective vaccines.

Some Pathogen Clearance Mechanisms are Common to Both Innate and Adaptive Immune Responses

Adaptive immune responses—antibody responses in particular—have adopted and modified several effector functions by which the innate immune system eliminates antigen, so that they are also triggered by antibody binding to antigens. While some will be discussed in more detail in Chapters 6 and 13, several examples are mentioned briefly here as illustrations of important interactions between innate and adaptive immune responses.

As discussed earlier in this chapter, several soluble proteins that recognize microbial surface components—including SP-A, SP-D, and MBL—function as opsonins; when they are bound to microbial surfaces they are recognized by receptors on phagocytes, leading to enhanced phagocytosis. Antibodies of the IgG and IgA classes also serve as opsonins; after binding to microbial surfaces these antibodies can be recognized by Fc receptors that are expressed on macrophages and other leukocytes, triggering phagocytosis (see Chapter 3). One receptor for IgG Fc regions (FcγR), called CD16, is also expressed on NK cells, where it serves as an activating receptor. Thus, when IgG antibodies bind to foreign antigens on the cell surfaces (such as viral or tumor antigens), the IgG antibodies can be recognized by the FcγR on the NK cell, triggering NK-cell-mediated cytotoxic killing of the infected or malignant cell (see Chapter 13).

Finally, as mentioned earlier and described fully in Chapter 6, the complement pathway can be activated by both innate and antibody-mediated mechanisms. Components on microbe surfaces can be directly recognized by soluble pattern-recognition proteins, including MBL and C-reactive protein, leading to activation of the complement cascade. Similarly, when antibodies of the IgM class and certain IgG subclasses bind microbe surfaces, their Fc regions are recognized by the C1 component of complement, also triggering the complement cascade. As was shown in Figure 5-7, MBL and C1 are related structurally. Once the complement pathway is activated by any of these microbe-binding proteins, it generates a common set of protective activities; various complement components and fragments promote opsonization, lysis of membrane-bounded microbes, and generation of fragments that have proinflammatory and chemoattractant activities. Thus, the adaptive immune system makes good use of mechanisms that initially evolved to contribute to innate immunity, co-opting them for the elimination of pathogens.

Ubiquity of Innate Immunity

Determined searches among plant and invertebrate animal phyla for the signature proteins of the highly efficient adaptive immune system—antibodies, T cell receptors, and MHC proteins—have failed to find any homologs of these important vertebrate proteins. Yet without them multicellular organisms have managed to survive for hundreds of millions of years. The interior spaces of organisms as diverse as the tomato, fruit fly, and sea squirt (an early chordate, without a backbone) do not contain unchecked microbial populations. Careful studies of these and many other representatives of nonvertebrate phyla have found arrays of well-developed processes that carry out innate immune responses. The accumulating evidence leads to the conclusion that multiple immune mechanisms protect all multicellular organisms

TABLE 5-7 Immunity in multicellular organisms

Taxonomic group	Innate immunity (nonspecific)	Adaptive immunity (specific)	Invasion-induced protective enzymes and enzyme cascades	Phagocytosis	Antimicrobial peptides	Pattern recognition receptors	Lympho-cytes	Variable lympho-cyte receptors	Anti-bodies
Higher plants	+	−	+	−	+	+	−	−	−
Invertebrate animals									
Porifera (sponges)	+	−	?	+	+	+	−	−	−
Annelids (earthworms)	+	−	?	+	+	+	−	−	−
Arthropods (insects, crustaceans)	+	−	+	+	+	+	−	−	−
Vertebrate animals									
Jawless fish (hagfish, lamprey)	+	+	+	+	+	+	+	+	−
Elasmobranchs (cartilaginous fish; e.g., sharks, rays)	+	+	+	+	+	+	+	−	+
Bony fish (e.g., salmon, tuna)	+	+	+	+	+	+	+	−	+
Amphibians	+	+	+	+	+	+	+	−	+
Reptiles	+	+	+	+	+	+	+	−	+
Birds	+	+	+	+	+	+	+	−	+
Mammals	+	+	+	+	+	+	+	−	+

Sources: M. J. Flajnik and L. Du Pasquier, 2008, "Evolution of the Immune System," in *Fundamental Immunology* (6th ed.), W. E. Paul, ed., Philadelphia: Lippincott; J. H. Wong et al., 2007, "A review of defensins of diverse origins," *Current Protein and Peptide Science* **8**:446.

from microbial infection and exploitation (Table 5-7). Some of the innate immune system components occur across the plant and animal kingdoms. For example, as mentioned early in this chapter, virtually all plant and animal species, and even some fungi, have antimicrobial peptides similar to defensins. Most multicellular organisms have pattern-recognition receptors containing leucine-rich repeats (LRRs), although many organisms also have other families of PRRs. While the innate immune responses activated by these receptors in plants and invertebrates show both similarities and differences compared to those of vertebrates, innate immune response mechanisms are essential for the health and survival of these varied organisms.

Plants Rely on Innate Immune Responses to Combat Infections

Despite plants' tough outer protective barrier layers, such as bark and cuticle, and the cell walls surrounding each cell, plants can be infected by a wide variety of bacteria, fungi, and viruses, all of which must be combated by the plant innate immune system. Plants do not have phagocytes or other circulating cells that can be recruited to sites of infection to mount protective responses. Instead, they rely on local innate immune responses for protection against infection. As described in Box 5-3, some resemble innate responses of animals, while others are quite distinct.

Invertebrate and Vertebrate Innate Immune Responses Show Both Similarities and Differences

Additional innate immune mechanisms evolved in animals, and we vertebrates share a number of innate immunity features with invertebrates. PRRs (including relatives of *Drosophila* Toll and vertebrate TLRs) that have specificities for microbial carbohydrate and peptidoglycan PAMPs are found in organisms as primitive as sponges. Together with soluble opsonin proteins (including some related to complement components), some of these early invertebrate PRRs function in promoting phagocytosis, an early mechanism for clearing pathogens. Innate signaling has been well studied in *Drosophila*, and signaling proteins have been identified in the flies that are homologous to several of those downstream of vertebrate TLRs (including MyD88 and IRAK homologs). A difference is that fly Toll does not bind to PAMPs directly but is indirectly activated by the soluble protein product of an enzyme cascade triggered by pathogen binding to soluble

EVOLUTION

Plant Innate Immune Responses

In the plasma membrane under the cell wall, plant cells express pattern recognition receptors with LRR domains reminiscent of animal TLRs. These PRRs recognize what plant biologists refer to as microbe-associated molecular patterns (MAMPs), including bacterial flagellin, a highly conserved bacterial translation elongation factor, and various bacterial and fungal cell wall components (Figure 1). As is true for animal TLRs, some plant PRRs respond to danger-associated molecular patterns (DAMPs), which usually are created by pathogen enzymes that attack and fragment cell wall components. Some bacterial and fungal pathogens directly inject into plant cells toxin effector proteins that inhibit signaling through the plasma membrane PRRs. These toxins are recognized by a distinct class of LRR receptors in the cytoplasm called R proteins, which, like animal NLR proteins, have both LRR and nucleotide-binding domain domains. After ligand binding, plant PRRs activate signaling pathways and transcription factors distinct from those of

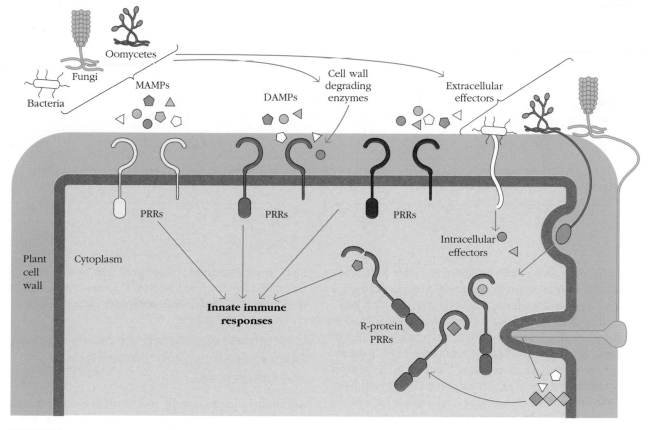

FIGURE 1
Activation of plant innate immune responses. Microbe-associated molecular patterns (MAMPs), danger-associated molecular patterns (DAMPs) generated by microbial enzymes (such as by degrading the cell wall), and microbial effectors (such as toxins) are recognized by plasma membrane LRR-containing pattern recognition receptors (PRRs). Microbial effectors that enter the cytoplasm are recognized by a class of PRRs called resistance (R) proteins. Recognition of MAMPs, DAMPs, and effectors by PRRs induces innate immune responses. [Adapted from T. Boller and G. Felix, 2009, Annual Review of Plant Biology **60**:379.]

BOX 5-3

vertebrate cells (plants do not have NF-κB or IRF homologs), triggering innate responses.

The primary protective innate immune response mechanisms of plants to infection are the generation of reactive oxygen and nitrogen species, elevation of internal pH, and induction of a variety of antimicrobial peptides (including defensins) and antimicrobial enzymes that can digest the walls of invading fungi (chitin-ases) or bacteria (α-1,3 glucanase). Plants may also be activated to produce organic molecules, such as phytoalexins, that have antibiotic activity. In some cases, the responses of plants to pathogens even goes beyond these protective substances to include structural responses. For example, to limit infection of leaves, PAMP binding to PRRs induces the epidermal guard cells that form the openings (stomata) involved in leaf gas exchange to close, preventing further invasion (Figure 2). Other protective mechanisms include the isolation of cells in the infected area by strengthening the walls of surrounding noninfected cells and the induced death (necrosis) of cells in the vicinity of the infection to prevent the infection from spreading to the rest of the plant. Mutations that disrupt any of these processes usually result in loss of the plant's resistance to a variety of pathogens.

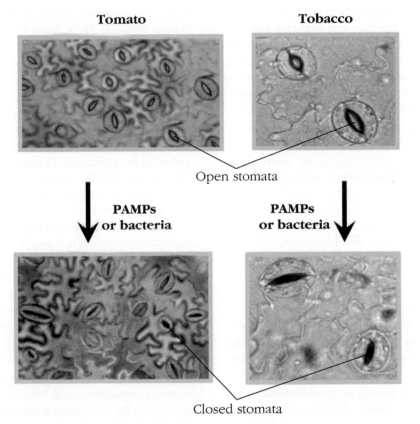

FIGURE 2
Induced closure of leaf stomata following exposure to bacterial PAMPs. In normal light conditions, the openings (stomata) formed by pairs of guard cells in the leaf epidermis are open (visible as the football-shaped openings in the upper pair of leaf photographs). However, as shown in the lower panels, exposure of the tobacco leaf to the bacterial plant pathogen *Pseudomonas syringae* and exposure of the tomato leaf to bacterial LPS induce the closure of the stomata (no openings visible). *[From M. Melotto et al. 2008. Annual Review of Phytopathology **46**:101–122.]*

pattern-recognition proteins. Through this pathway bacterial and fungal infections lead to the degradation of an IκB homolog and activation of NF-κB family members, Dif and Dorsal, which induce the production of drosomycin, an insect defensin, and other antimicrobial peptides. In addition to these and other pathways activated by PRRs, *Drosophila* and other arthropods employ other innate immune strategies, including the activation of phenoloxidase cascades that result in melanization—the deposition of a melanin clot around invading organisms that prevents their spread. Thus invertebrates and vertebrates have common as well as distinct innate immune response mechanisms.

SUMMARY

- The receptors of the innate immune system recognize conserved pathogen-associated molecular patterns (PAMPs), which are molecular motifs found in microbes, and damage-associated molecular patterns (DAMPs) from aging, damaged, or dead cells. Therefore these receptors are called pattern-recognition receptors (PRRs).

- The epithelial layers that insulate the interior of the body from outside pathogens—the skin and epithelial layers of the mucosal tracts and secretory glands—constitute an anatomical barrier that is highly effective in protecting against infection. These epithelial layers produce a variety of protective substances, including acidic pH, enzymes, binding proteins, and antimicrobial proteins and peptides.

- Phagocytosis—engulfment and internalization of particulate materials such as microbes—is mediated by receptors on phagocytes (monocytes, macrophages, neutrophils, and dendritic cells) that either directly recognize PAMPs on the surface of microbes or recognize soluble proteins (opsonins) that bind to the microbes. PAMP binding triggers microbe uptake into phagosomes, which fuse with lysosomes or prepackaged granules, leading to their killing through enzymatic degradation, antimicrobial proteins and peptides, and toxic effects of reactive oxygen species (ROS) and reactive nitrogen species (RNS).

- Families of PRRs (TLRs, CLRs, RLRs, and NLRs) recognize a wide variety of PAMP (and DAMP) ligands and trigger signaling pathways that activate genes encoding proteins that contribute to innate and inflammatory responses.

- Vertebrate Toll-like receptors (TLRs, homologous to the fruit fly Toll receptor) are dimers of chains with extracellular leucine-rich (LRR) domains that bind PAMPs and DAMPs. Of the 13 TLRs found in mice and humans, some are on the cell surface, while others are in endosomes and lysosomes, where each binds PAMPs revealed by the disassembly or degradation of pathogens, such as microbial nucleic acids.

- TLR binding of PAMPs activates signaling pathways; the particular pathway varies depending on the TLR and the adaptor protein that binds to its cytoplasmic TIR domain (either MyD88 or TRIF). The signaling pathways activate transcription factors NF-κB, various interferon regulating factors (IRFs), and transcription factors (such as AP-1) downstream of MAP kinase (MAPK) pathways.

- C-type lectin receptors (CLRs) comprise a heterogeneous family of cell surface PRRs that recognize cell wall components, largely sugars and polysaccharides, of bacteria and fungi. They trigger a variety of distinct signaling pathways that activate transcription factors, of which some are similar to those activated by TLRs.

- RIG-I-like receptors (RLRs) are RNA helicases that function as cytosolic PRRs. Most PAMPs that they recognize are viral double-stranded RNAs; after PAMP binding, RLRs trigger signaling pathways that activate IRFs and NF-κB.

- NOD-like receptors (NLRs) are a large family of cytosolic PRRs activated by intracellular PAMPs, DAMPs, and other harmful substances. The NOD NLRs bind intracellular microbial components such as cell wall fragments. Other NLRs, such as NLRP3, function as inflammasomes; they are activated by a wide variety of PAMPs, DAMPs, and crystals by sensing changes in the intracellular milieu. Inflammasomes then activate the caspase-1 protease, which cleaves the inactive large precursors of the proinflammatory cytokines IL-1 and IL-18, so that active cytokines can be released from cells.

- The signaling pathways downstream of PRRs activate expression of a variety of genes, including those for antimicrobial peptides, Type 1 interferons (potent antiviral agents), cytokines (including proinflammatory IL-1, TNF-α, and IL-6), chemokines, and enzymes that help to generate antimicrobial and inflammatory responses.

- Inflammatory responses are activated by the innate immune response to local infection or tissue damage—in particular, by the proinflammatory cytokines and certain chemokines that are produced. Key early components of inflammatory responses are increased vascular permeability, allowing soluble innate mediators to reach the infected or damaged site, and the recruitment of neutrophils and other leukocytes from the blood into the site.

- A later component of inflammatory responses is the acute phase response (APR), induced by certain proinflammatory cytokines (IL-1, TNF-α, and IL-6). The APR involves increased synthesis and secretion by the liver of several antimicrobial proteins, including MBL, CRP, and complement components, which activate a variety of processes that contribute to eliminating pathogens.

- Natural killer (NK) cells are lymphocytes with innate immune functions. They express a set of activating receptors that recognize surface components of the body's cells induced by infection, malignant transformation, or other stresses. Activated NK cells can kill the altered self cell and/or produce cytokines that help to induce adaptive immune responses to the altered cell.

- The importance of innate and inflammatory responses is demonstrated by the impact of a variety of genetic defects in humans. Defects in PRRs and signaling pathways activating innate responses lead to increased susceptibility to certain infections, while other defects that constitutively activate inflammasomes contribute to a variety of inflammatory disorders.

- As innate and inflammatory responses can be harmful as well as helpful, they are highly regulated by positive and negative feedback pathways.

- To escape elimination by innate immune responses, pathogens have evolved a wide array of strategies to block antimicrobial responses.

- There are many interactions between the innate and adaptive immune systems. The adaptive immune system has co-opted several pathogen-clearance mechanisms, such as opsonization and complement activation, so that they contribute to antibody-mediated pathogen elimination.

- B cells and some T cells express TLRs, which can serve as costimulatory receptors. PAMP binding helps activate these cells to generate adaptive immune responses. TLRs expressed by NK cells can serve as activating receptors.

- Dendritic cells are a key cellular bridge between innate and adaptive immunity. Microbial components acquired by dendritic cells via PRR binding during the innate response are brought from the site of infection to lymph nodes. Activation of a dendritic cell by PAMPs stimulates the cell to mature, so that it acquires the abilities to activate naïve T cells to become mature cytotoxic and helper T cells and to influence the sets of cytokines that the mature T_H will produce.

- Innate immunity appeared early during the evolution of multicellular organisms. While TLRs are unique to animals, PRRs with leucine-rich repeat (LRR) ligand-binding domains that induce innate immune responses are found in virtually all plants and animals.

REFERENCES

Areschoug, T., and S. Gordon. 2009. Scavenger receptors: role in innate immunity and microbial pathogenesis. *Cellular Microbiology* **11**:1160.

Beutler, B., and E. T. Rietschel. 2003. Innate immune sensing and its roots: The story of endotoxin. *Nature Reviews Immunology* **3**:169.

Bowie, A. G., and L. Unterholzner. 2008. Viral evasion and subversion of pattern-recognition receptor signalling. *Nature Reviews Immunology* **8**:911.

Bulet, P., et al. 2004. Anti-microbial peptides: from invertebrates to vertebrates. *Immunology Review* **198**:169–184.

Chroneos, Z. C., et al. 2010. Pulmonary surfactant: An immunological perspective. *Cellular Physiology and Biochemistry* **25**:13.

Coffman, R., et al. 2010. Vaccine adjuvants: Putting innate immunity to work. *Immunity* **33**:492–503.

Davis, B. K., et al. 2011. The inflammasome NLRs in immunity, inflammation, and associated diseases. *Annual Review of Immunology* **11**:707.

deFranco, A., et al. 2007. *Immunity (Primers in Biology)*. Sunderland, MA: Sinauer Associates.

Elinav, E., et al. 2011. Regulation of the antimicrobial response by NLR proteins. *Immunity* **34**:665.

Froy, O. 2005. Regulation of mammalian defensin expression by Toll-like receptor-dependent and independent signalling pathways. *Cellular Microbiology* **7**:1387.

Fukata, M., et al. 2009. Toll-like receptors (TLRs) and Nod-like receptors (NLRs) in inflammatory disorders. *Seminars in Immunology* **21**:242.

Geijtenbeek, T. B. H., and S. I. Gringhuis. 2009. Signaling through C-type lectin receptors: Shaping immune repertoires. *Nature Reviews Immunology* **9**:465.

Grimsley, C., and K. S. Ravichandran. 2003. Cues for apoptotic cell engulfment: Eat-me, don't-eat-me, and come-get-me signals. *Trends in Cellular Biology* **13**:648.

Herrin, B. R., and M. D. Cooper. 2010. Alternative adaptive immunity in jawless vertebrates. *Journal of Immunology* **185**:1367.

Kawai, T., and S. Akira. 2010. The role of pattern recognition receptors in innate immunity: Update on Toll-like receptors. *Nature Immunology* **11**:373.

Kerrigan, A. M., and G. D. Brown. 2009. C-type lectins and phagocytosis. *Immunobiology* **214**:562.

Lemaitre, B. 2004. The road to Toll. *Nature Reviews Immunology* **4**:521.

Litvack, M. L., and N. Palaniyar. 2010. Soluble innate immune pattern-recognition proteins for clearing dying cells and cellular components: Implications on exacerbating or resolving inflammation. *Innate Immunity* **16**:191.

Loo, Y.-M., and M. Gale. 2011. Immune signaling by RIG-I-like receptors. *Immunity* **34**:680.

Medzhitov, R., et al. 1997. A human homologue of the Toll protein signals activation of adaptive immunity. *Nature* **388**:394.

Nathan, C., and A. Ding. 2010. SnapShot: Reactive oxygen intermediates (ROI). *Cell* **140**:952.

Palm, N. W., and R. Medzhitov. 2009. Pattern recognition receptors and control of adaptive immunity. *Immunology Review* **227**:221.

Poltorak, A., et al. 1998. Defective LPS signaling in C3Hej and C57BL/10ScCr mice: Mutations in TLR4 gene. *Science* **282**:2085.

Salzman, N. H., et al. 2010. Enteric defensins are essential regulators of intestinal microbial ecology. *Nature Immunology* **11**:76.

Schroder, K., and J. Tschopp. 2010. The inflammasome. *Cell* **140**:821.

Sun, J. C., et al. 2011. NK cells and immune "memory." *Journal of Immunology* **186**:1891.

Takeuchi, O., and S. Akira. 2010. Pattern recognition receptors and inflammation. *Cell* **140**:805.

Tecle, T., et al. 2010. Defensins and cathelicidins in lung immunity. *Innate Immunity* **16**:151.

Tieu, D. T., et al. 2009. Alterations in epithelial barrier function and host defense responses in chronic rhinosinusitis. *Journal of Clinical Immunology* **124**:37.

van de Vosse, E., et al. 2009. Genetic deficiencies of innate immune signaling in human infectious disease. *The Lancet Infectious Diseases* **9**:688.

Willingham, S. B., et al. 2012. The CD47-signal regulatory protein alpha (SIRP1a) interaction is a therapeutic target for human solid tumors. *Proceedings of the National Academy of Sciences USA* **109**:6662.

Yu, M., and S. J. Levine. 2011. Toll-like receptor 3, RIG-I-like receptors and the NLRP3 inflammasome: Key modulators of innate immune responses to double-stranded viruses. *Cytokine & Growth Factor Reviews* **22**:63.

Useful Web Sites

www.biolegend.com/media_assets/.../InnateImmunityResourceGuide_3.pdf Presents a concise summary of innate immunity, including a detailed table on the cells of the innate immune system, including their locations in the body, functions, products, receptors, and modes of activation.

cpmcnet.columbia.edu/dept/curric-pathology/ pathology/pathology/pathoatlas/GP_I_menu. html Images are shown of the major inflammatory cells involved in acute and chronic inflammation, as well as examples of specific inflammatory diseases.

portal.systemsimmunology.org/portal/web/guest/ homepage Systems Approach to Immunology (systemsimmunology.org) is a large collaborative research program formed to study the mechanisms by which the immune system responds to infectious disease by inciting innate inflammatory reactions and instructing adaptive immune responses.

www.biomedcentral.com/1471-2172/9/7 Web site of The Innate Immunity Database, an NIH-funded multi-institutional project that assembled microarray data on expression levels of more than 200 genes in macrophages stimulated with a panel of TLR ligands. The database is intended to support systems biology studies of innate responses to pathogens.

www.immgen.org/index_content.html The Immunological Genome Project is a new cooperative effort for deep transcriptional profiling of all immune cell types.

www.ncbi.nlm.nih.gov/PubMed PubMed, the National Library of Medicine database of more than 15 million publications, is the world's most comprehensive bibliographic database for biological and biomedical literature. It is also highly user friendly, searchable by general or specific topics, authors, reviews, and so on. It is the best resource to use for finding the latest research articles on innate immunity or other topics in the biomedical sciences.

animaldiversity.ummz.umich.edu/site/index. html The Animal Diversity Web (ADW), based at the University of Michigan, is an excellent and comprehensive database of animal classification as well as a source of information on animal natural history and distribution. Here is a place to find information about animals that are not humans or mice.

en.wikipedia.org/wiki/Innate_immune_system The Wikipedia Web site presents a detailed summary of the innate immune system in animals and plants; it contains numerous figures and photographs illustrating various aspects of innate immunity, plus links to many references.

 STUDY QUESTIONS

CLINICAL FOCUS QUESTION What infections are unusually prevalent in individuals with genetic defects in TLRs or the MyD88-dependent TLR signaling pathway? in individuals with defects in pathways activating the production or antiviral activities of IFN-α,β? Why is it thought that these individuals aren't

susceptible to a wider range of diseases, and what evidence supports this hypothesis?

1. Use the following list to complete the statements that follow. Some terms may be used more than once or not at all.

Acute phase response

Antibodies

Arginine

Caspase-1

C-reactive protein (CRP)

Complement

Costimulatory molecules

Defensins

Dendritic cells

Ficolins

IL-1

Inflammasomes

iNOS

Interferons-α,β

IRFs

Lysozyme

Mannose-binding lectin (MBL)

MyD88

NADPH

NADPH phagosome oxidase

NF-κB

NK cells

NLRs

NO

O₂

PAMPs

Phagocytosis

Proinflammatory cytokines

PRRs

Psoriasin

RLRs

ROS

RNS

Surfactant proteins (SP) A, D

TRIF

T-cell receptors

TLR2

TLR3

TLR4

TLR7

TLR9

TNF-α

a. Examples of proteins and peptides with direct antimicrobial activity that are present on epithelial surfaces are _____, _____, and _____.

b. Soluble pattern-recognition proteins that function as opsonins, which enhance _____, include _____, _____, _____, and _____; of these, the ones that also activate complement are _____ and _____.

c. The enzyme _____ uses _____ to generate microbe-killing _____; one of these, plus the antimicrobial gas _____ generated by the _____ enzyme from the amino acid _____, are used to generate _____, which are also antimicrobial.

d. As components of innate immune responses, both _____ (secreted proteins) and _____ (a type of lymphocyte) defend against viral infection.

e. The _____ occurs when _____, such as _____ and _____, are generated by innate immune responses and act on the liver.

f. _____, receptors of innate immunity that detect _____, are encoded by germline genes, whereas the signature receptors of adaptive immunity, _____ and _____, are encoded by genes that require gene rearrangements during lymphocyte development to be expressed.

g. Among cell surface TLRs, _____ detects Gram-positive bacterial infections while _____ detects Gram-negative infections.

h. Some cells use the intracellular TLRs _____ and _____ to detect RNA virus infections and _____ to detect infections by bacteria and some DNA viruses.

i. _____ is unique among PRRs in that it functions both on the plasma membrane and in endosomes and binds both the _____ and _____ adaptor proteins.

j. _____ are receptors that detect cytosolic viral nucleic acids, while _____ include cytosolic receptors that detect intracellular bacterial cell wall components.

k. Key transcription factors for inducing expression of proteins involved in innate immune responses are _____ and _____.

l. The production of the key proinflammatory cytokine _____ is complex, as it requires transcriptional activation by signaling pathways downstream of _____ followed by cleavage of its large precursor protein by _____, which is activated by _____, members of the _____ family of innate receptors.

m. After maturation induced by binding of _____ to their _____, cells known as _____ become efficient activators of naïve helper and cytotoxic T cells.

n. _____ and _____ are innate immune system components common to both plants and animals.

2. What were the two experimental observations that first linked TLRs to innate immunity in vertebrates?

3. What are the hallmark characteristics of a localized inflammatory response? How are they induced by the early innate immune response at the site of infection, and how do these characteristics contribute to an effective innate immune response?

4. In vertebrates, innate immunity collaborates with adaptive immunity to protect the host. Discuss this collaboration, naming key points of interaction between the two systems. Include at least one example in which the adaptive immune response contributes to enhanced innate immunity.

5. As adaptive immunity evolved in vertebrates, the more ancient system of innate immunity was retained. Can you think of any disadvantages to having a dual system of immunity? Would you argue that either system is more essential?

ANALYZE THE DATA A variety of studies have shown that helminth worm parasites have developed numerous mechanisms for inhibiting innate and inflammatory responses that otherwise would contribute to their elimination. One particular parasite, the filarial nematode *Acanthocheilonema viteae*, has been shown to secrete a lipoglycoprotein, ES-62, that inhibits proinflammatory responses. Based on these initial observations, a group of investigators has been studying how ES-62 achieves this anti-inflammatory effect and whether ES-62 might inhibit the development of septic shock in individuals with sepsis. Interestingly, they found that ES-62 binds to TLR4 (but not to any other TLR). Based on what you have learned in this chapter and the information and data provided below from these investigators' recent paper (P. Puneet et al., 2011, *Nature Immunology* **12**:344), answer the following questions. (*Note:* The various methods used to obtain the data below are described in Chapter 20.)

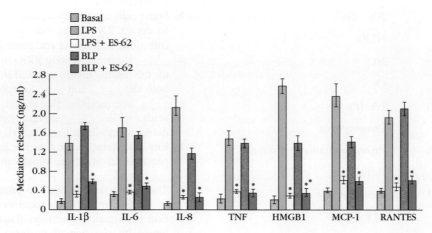

(a) Effects of ES-62 on proinflammatory mediator production

Macrophages from patients with sepsis were incubated with nothing ("Basal"), or with the TLR4 ligand LPS or the TLR2 lipoprotein ligand BLP with or without ES-62. After 24 hours, tissue culture fluids were harvested and assayed by ELISA for levels of the proin-flammatory cytokines IL-1, IL-6, and TNF-α; the chemokines IL-8 (CXCL8), MCP-1 (CCL2), and RANTES (CCL5); and the nuclear protein HMGB1 (which is released by activated or damaged cells and has proinflammatory activities).

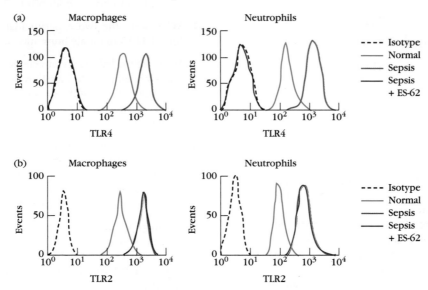

(b) Effects of ES-62 on cell surface expression of TLR4 and TLR2

Levels of TLR4 and TLR2 were measured on macrophages and neutrophils from normal people and from patients with sepsis incubated with or without ES-62 for 3 hours. Cells were stained with fluorescent antibodies to TLR4 or TLR2, or with control (nonreactive) antibodies with the same immunoglobulin heavy-chain class ("Isotype"). The stained cells were analyzed using a flow cytometer for the levels of their fluorescence, indicated on the X-axis. The fluorescence obtained with the Isotype control antibody represents no specific staining. The number of cells ("events") with different levels of fluorescence is indicated on the Y-axis.

(c) Effect of ES-62 on MyD88 levels

Cell extracts were prepared from macrophages and neutrophils from patients with sepsis that were incubated without (0) or for 3 or 6 hours with ES-62. The extract proteins were separated by size by SDS-PAGE and subjected to Western blots to assess levels of MyD88 and, as a control, the ubiquitous cellular protein α-tubulin. The gels were blotted onto filter paper and the filters stained with antibodies to MyD88 and to α-tubulin. Bound antibodies were visualized to reveal the levels of MyD88 and tubulin.

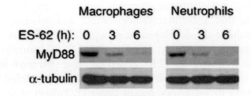

a. What is septic shock, and how is it induced by bacterial infections?

b. The investigators tested whether ES-62 inhibits the production of proinflammatory mediators by macrophages from normal people and from patients hospitalized with sepsis. Shown in panel (a) above are the results with macrophages from patients with sepsis; cells from normal individuals gave similar results. What effect does ES-62 have on the responses induced by the two PAMPs? Given that ES-62 binds to TLR4 but not TLR2, what do these results say about the mechanism of ES-62 inhibition?

The investigators compared the levels of TLR4 and TLR2 expressed by macrophages and neutrophils from normal people with those from patients with sepsis. As had been seen by others, panel (b) shows that cells from patients with sepsis express higher levels of these TLRs than do cells from normal individuals. Of what kind of feedback regulation is this an example? Why is this feedback response normally beneficial in individuals with limited infections but disadvantageous in individuals with sepsis?

In the same experiment shown in panel (b), some cells from patients with sepsis were incubated with ES-62. What effect does ES-62 have on TLR4 expression? on TLR2 expression?

The investigators also looked at the levels of MyD88 in macrophages and neutrophils from sepsis patients after no exposure or 3 or 6 hours of exposure to ES-62 (panel [c]). What does this Western blot show?

Returning to the initial finding that ES-62 inhibits the induction of proinflammatory mediators by PAMPs that activate either TLR4 or TLR2 [panel (a)], how might that finding be explained by the data shown in panels (b) and (c) (i.e., the effects of ES-62 on TLR and MyD88 expression)?

Given these results, do you think injection of ES-62 might be used to cure septic shock once it has occurred? Do you think ES-62 might prevent septic shock from occurring in patients with sepsis? Explain your answers.

The Complement System

The term **complement** (spelled with an *e*) refers to a set of serum proteins that cooperates with both the innate and the adaptive immune systems to eliminate blood and tissue pathogens. Like the components of the blood clotting system, complement proteins interact with one another in catalytic cascades. As mentioned in Chapters 3 and 5, various complement components bind and opsonize bacteria, rendering them susceptible to receptor-mediated phagocytosis by macrophages, which express membrane receptors for complement proteins. Other complement proteins elicit inflammatory responses, interface with components of the adaptive immune system, clear immune complexes from the serum, and/or eliminate apoptotic cells. Finally, a *Membrane Attack Complex (MAC)* assembled from complement proteins directly kills some pathogens by creating pores in microbial membranes. The biological importance of complement is emphasized both by the pathological consequences of mutations in the genes encoding complement proteins as well as by the broad range of strategies that have evolved in microorganisms to evade it.

Research on complement began in the 1890s when Jules Bordet at the Institut Pasteur in Paris showed that sheep antiserum to the bacterium *Vibrio cholerae* caused lysis (membrane destruction) of the bacteria, and that heating the antiserum destroyed its bacteriolytic activity. Surprisingly, the ability to lyse the bacteria was restored to the heated serum by adding fresh serum that contained no antibacterial antibodies. This finding led Bordet to reason that bacteriolysis required two different substances: the heat-stable specific antibodies that bound to the bacterial surface, and a second, heat-labile (sensitive) component responsible for the lytic activity.

In an effort to purify this second, nonspecific component, Bordet developed antibodies specific for red blood cells and used these, along with purified serum fractions, to identify those fractions that cooperated with the antibodies to induce lysis of the red blood cells (hemolysis). The famous immunologist Paul Ehrlich, working independently in Berlin, carried out similar

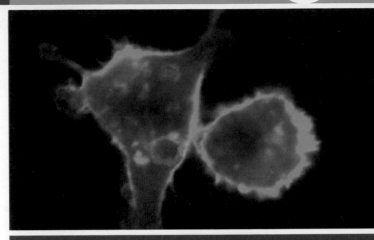

Sheep red blood cells (red) are phagocytosed by macrophages (green) after opsonization by complement.

- The Major Pathways of Complement Activation
- The Diverse Functions of Complement
- The Regulation of Complement Activity
- Complement Deficiencies
- Microbial Complement Evasion Strategies
- The Evolutionary Origins of the Complement System

experiments and coined the term *complement*, defining it as "the activity of blood serum that completes the action of antibody."

In the ensuing years, researchers have discovered that the action of complement is the result of interactions among a complex group of more than 30 glycoproteins. Most complement components are synthesized in the liver by hepatocytes, although some are also produced by other cell types, including blood monocytes, tissue macrophages, fibroblasts, and epithelial cells of the gastrointestinal and genitourinary tracts. Complement components constitute approximately 15% of the globulin protein fraction in plasma, and their combined concentration can be as high as 3 mg/ml. In addition, several of the regulatory components of the system exist on cell membranes, so the term *complement* therefore now embraces glycoproteins distributed among the blood plasma and cell membranes. Complement components can be classified into seven functional categories (Overview Figure 6-1):

Proteins Involved in the Complement System

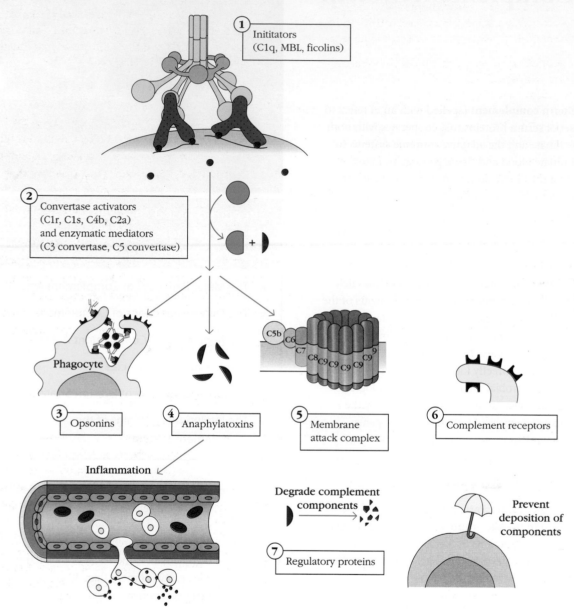

① Inititators
(C1q, MBL, ficolins)

② Convertase activators
(C1r, C1s, C4b, C2a)
and enzymatic mediators
(C3 convertase, C5 convertase)

Phagocyte

C5b
C6
C7
C8 C9 C9 C9 C9 C9 C9 C9

③ Opsonins

④ Anaphylatoxins

⑤ Membrane
attack complex

⑥ Complement receptors

Inflammation

Degrade complement
components

Prevent
deposition of
components

⑦ Regulatory proteins

(1) The complement pathways are initiated by proteins that bind to pathogens, either directly or via an antibody or other pathogen-specific protein. After a conformational change, (2) enzymatic mediators activate other enzymes that generate the central proteins of the complement cascade, the C3 and C5 convertases, which cleave C3 and C5, releasing active components that mediate all functions of complement, including (3) opsonization, (4) inflammation, and (5) the generation of the membrane attack complex (MAC). Effector complement proteins can label an antibody-antigen complex for phagocytosis (opsonins), initiate inflammation (anaphylatoxins), or bind to a pathogen and nucleate the formation of the MAC. Often, these effectors act through (6) complement receptors on phagocytic cells, granulocytes, or erythrocytes. (7) Regulatory proteins limit the effects of complement by promoting their degradation or preventing their binding to host cells.

1. *Initiator complement components.* These proteins initiate their respective complement cascades by binding to particular soluble or membrane-bound molecules. Once bound to their activating ligand, they undergo conformational alterations resulting in changes in their biological activity. The C1q complex, *Mannose Binding Lectin* (MBL), and the ficolins are examples of initiator complement components.

2. *Enzymatic mediators.* Several complement components, (e.g., C1r, C1s, MASP2, and factor B) are proteolytic enzymes that cleave and activate other members of the complement cascade. Some of these proteases are activated by binding to other macromolecules and undergoing a conformational change. Others are inactive until cleaved by another protease enzyme and are thus termed zymogens: proteins that are activated by proteolytic cleavage. The two enzyme complexes that cleave complement components C3 and C5, respectively, are called the C3 and C5 convertases and occupy places of central importance in the complement cascades.

3. *Membrane-binding components or opsonins.* Upon activation of the complement cascade, several proteins are cleaved into two fragments, each of which then takes on a particular role. For C3 and C4, the larger fragments, C3b and C4b, serve as **opsonins**, enhancing phagocytosis by binding to microbial cells and serving as binding tags for phagocytic cells bearing receptors for C3b or C4b. As a general rule, the larger fragment of a cleaved complement component is designated with the suffix "b," and the smaller with the suffix "a." However, there is one exception to this rule: the larger, enzymatically active form of the C2 component is named C2a.

4. *Inflammatory mediators.* Some small complement fragments act as inflammatory mediators. These fragments enhance the blood supply to the area in which they are released, by binding to receptors on endothelial cells lining the small blood vessels and inducing an increase in capillary diameter. They also attract other cells to the site of tissue damage. Because such effects can be harmful in excess, these fragments are called **anaphylatoxins**, meaning substances that cause **anaphylaxis** ("against protection"). Examples include C3a, C5a, and C4a.

5. *Membrane attack proteins.* The proteins of the **membrane attack complex (MAC)** insert into the cell membranes of invading microorganisms and punch holes that result in lysis of the pathogen. The complement components of the MAC are C5b, C6, C7, C8, and multiple copies of C9.

6. *Complement receptor proteins.* Receptor molecules on cell surfaces bind complement proteins and signal specific cell functions. For example, some complement receptors such as CR1 bind to complement components such as C3b on the surface of pathogens, triggering phagocytosis of the C3-bound pathogen. Binding of the complement component C5a to C5aR receptors on neutrophils stimulates neutrophil degranulation and inflammation. Complement receptors are named with "R," such as CR1, CR2, and C5aR.

7. *Regulatory complement components.* Host cells are protected from unintended complement-mediated lysis by the presence of membrane-bound as well as soluble regulatory proteins. These regulatory proteins include factor I, which degrades C3b, and Protectin, which inhibits the formation of the MAC on host cells.

This chapter describes the components of the complement system, their activation via three major pathways, the effector functions of the molecules of the complement cascade, as well as their interactions with other cellular and molecular components of innate and adaptive immunity. In addition, it addresses the mechanisms that regulate the activity of these complement components, the evasive strategies evolved by pathogens to avoid destruction by complement, and the evolution of the various complement proteins. This chapter's Classic Experiment tells the story of the scientist who discovered the alternative pathway of complement. In the Clinical Focus segment, we address various therapies that target elements of the complement cascades. Finally, an Advances Box describes some of the many tactics used by *Staphylococcus aureus* bacteria to escape complement-mediated destruction.

The Major Pathways of Complement Activation

Complement components represent some of the most evolutionarily ancient participants in the immune response. As viruses, parasites, and bacteria have attacked vertebrate hosts and learned to evade one aspect of the complement system, alternative pathways have evolved in an endless dance of microbial attack and host response. In the complex pathways that follow, we can gain a glimpse of the current state of what is actually an ongoing evolutionary struggle on the part of host organisms to combat microbial infection, while minimizing damage to their own cells.

The major initiation pathways of the complement cascade are shown in Figure 6-2. Although the initiating event of each of the three pathways of complement activation is different, they all converge in the generation of an enzyme complex capable of cleaving the C3 molecule into two fragments, C3a and C3b. The enzymes that accomplish this biochemical transformation are referred to as **C3 convertases**. As illustrated in Figure 6-2, the classical and lectin pathways use the dimer C4b2a for their C3 convertase activity, while the alternative pathway uses C3bBb to achieve the same end; however, the final result is the same: a dramatic increase in the concentration of C3b, a centrally located and multifunctional complement protein.

The second set of convertase enzymes of the cascade, **C5 convertases**, are formed by the addition of a C3b component

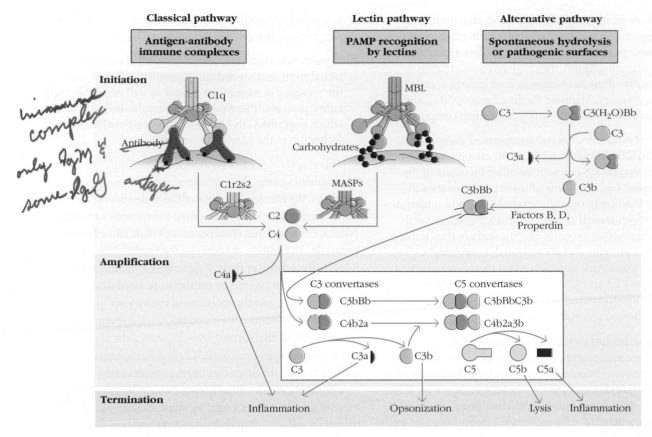

(handwritten margin notes: "immune complex only IgM & some IgG", "antigen")

FIGURE 6-2 The generation of C3 and C5 convertases by the three major pathways of complement activation. The classical pathway is initiated when C1q binds to antigen-antibody complexes. The antigen is shown here in dark red and the initiating antibody in green. The C1r enzymatic component of C1 (shown in blue) is then activated and cleaves C1s, which in turn cleaves C4 to C4a (an anaphylatoxin, bright red) and C4b. C4b attaches to the membrane, and binds C2, which is then cleaved by C1s to form C2a and C2b. (C2b is then acted upon further to become an inflammatory mediator.) C2a remains attached to C4b, forming the classical pathway C3 convertase (C4b2a). In the lectin pathway, mannose-binding lectin (MBL, green) binds specifically to conserved carbohydrate arrays on pathogens, activating the MBL-associated serine proteases (MASPs, blue). The MASPs cleave C2 and C4 generating the C3 convertase as in the classical pathway. In the alternative pathway, C3 undergoes spontaneous hydrolysis to C3(H_2O), which binds serum factor B. On binding to C3(H_2O), B is cleaved by serum factor D, and the resultant C3(H_2O)Bb complex forms a fluid phase C3 convertase. Some C3b, released after C3 cleavage by this complex, binds to microbial surfaces. There, it binds factor B, which is cleaved by factor D, forming the cell-bound alternative pathway C3 convertase, C3bBb. This complex is stabilized by properdin. The C5 convertases are formed by the addition of a C3b fragment to each of the C3 convertases.

to the C3 convertases. C5 convertases cleave C5 into the inflammatory mediator, C5a, and C5b, which is the initiating factor of the MAC.

We will now describe each of these pathways in more detail. The proteins involved in each of these pathways are listed in Table 6-1.

The Classical Pathway Is Initiated by Antibody Binding

The classical pathway of complement activation is considered part of the adaptive immune response since it begins with the formation of antigen-antibody complexes. These complexes may be soluble, or they may be formed when an antibody binds to antigenic determinants, or *epitopes*, situated on viral, fungal, parasitic, or bacterial cell membranes. Soluble antibody-antigen complexes are often referred to as *immune complexes*, and only complexes formed by IgM or certain subclasses of IgG antibodies are capable of activating the classical complement pathway (see Chapter 13). The initial stage of activation involves the complement components C1, C2, C3, and C4, which are present in plasma as zymogens.

The formation of an antigen-antibody complex induces conformational changes in the nonantigen-binding (Fc) portion of the antibody molecule. This conformational change exposes a binding site for the C1 component of complement. In serum, C1 exists as a macromolecular complex consisting of one molecule of C1q and two molecules each of the serine proteases, C1r and C1s, held together in a Ca^{2+}-stabilized complex (C1qr_2s_2) (Figure 6-3a). The C1q molecule itself is composed of 18 polypeptide chains that associate to form six collagen-like triple helical arms, the tips

TABLE 6-1 Proteins of the three major pathways of complement activation

Molecule	Biologically active fragments	Biological function	Active in which pathway
IgM, IgG		Binding to pathogen surface and initiating complement cascade	Classical pathway
Mannose-binding lectin, or ficolins		Binding to carbohydrates on microbial surface and initiating complement cascade	Lectin pathway
C1	C1q	Initiation of the classical pathway by binding Ig Binding to apoptotic blebs and initiating phagocytosis of apoptotic cells	Classical pathway
	(C1r)$_2$	Serine protease, cleaving C1r and C1s	
	(C1s)$_2$	Serine protease, cleaving C4 and C2	
MASP-1		MBL-Associated Serine Protease 1. MASP-2 appears to be functionally the more relevant MASP-protease.	Lectin pathway
MASP-2		Serine protease. In complex with MBL/ficolin and MASP-2, cleaves C4 and C2.	Lectin pathway
C2	C2a*	Serine protease. With C1 and C4b, is a C3 convertase.	Classical and lectin pathways
	C2b*	Inactive in complement pathway. Cleavage of C2b by plasmin releases C2 kinin, a peptide that stimulates vasodilation.	
C4	C4b	Binds microbial cell membrane via thioester bond. With C1 and C2a, is a C3 convertase.	Classical and lectin pathways
	C4a	Has weak anaphylatoxin (inflammatory) activity	
	C4c, C4d	Proteolytic cleavage products generated by factor I	
C3	C3a	Anaphylatoxin. Mediates inflammatory signals via C3aR.	
	C3b	Potent in opsonization, tagging immune complexes, pathogens, and apoptotic cells for phagocytosis	
		With C4b and C2a, forms the C5 convertase	Classical pathway
		With Bb, forms the C3 convertase	Tickover and properdin alternative pathways
		With Bb and one more molecule of C3b (C3bBb3b) acts as a C5 convertase	Tickover and properdin alternative pathways
	C3(H$_2$O)	C3 molecule in which the internal thioester bond has undergone hydrolysis With Bb, acts as a fluid phase C3 convertase.	Alternative pathway
	iC3b and C3f	Proteolytic fragments of C3b, generated by factor I. iC3b binds receptors CR3, CR4, and CRIg; CR2 binds weakly.	
	C3c and C3d or C3dg	Proteolytic fragments of iC3b generated by factor I. C3d/dg bind antigen and to CR2, facilitating antigen-binding to B cells. C3c binds CRIg on fixed tissue macrophages	
Factor B		Binds C3(H$_2$O) and is then cleaved by factor D into two fragments: Ba and Bb	Alternative pathway
	Ba	Smaller fragment of factor D-mediated cleavage of factor B. May inhibit proliferation of activated B cells.	
	Bb	Larger fragment of factor D-mediated cleavage of factor B	
		With C3(H$_2$O), acts as fluid phase C3 convertase	Alternative pathway
		With C3b, acts as cell-bound C3 convertase	Alternative pathway
		With two molecules of C3b, acts as C5 convertase	Alternative pathway
Factor D		Proteolytic enzyme that cleaves factor B into Ba and Bb only when it is bound to either C3(H$_2$O) or to C3b	Alternative pathway

(continued)

TABLE 6-1   (continued)

Factor P (properdin)		Stabilizes the C3bBb complex on microbial cell surface	Alternative pathway
C5	C5a	Anaphylatoxin binding to C5aR induces inflammation	
	C5b	Component of Membrane Attack Complex (MAC). Binds cell membrane and facilitates binding of other components of the MAC.	All
C6		Component of MAC. Stabilizes C5b. In the absence of C6, C5b is rapidly degraded.	All
C7		Component of MAC. Binds C5bC6 and induces conformational change allowing C7 to insert into interior of membrane.	All
C8		Component of MAC. Binds C5bC6C7 and creates a small pore in membrane.	All
C9		Component of MAC. 10–19 molecules of C9 bind C5bC6C7C8 and create large pore in membrane.	All

* Because C2a is the larger, active fragment of C2, some writers have tried to alter the nomenclature in order to make C2 conform to the convention that the "b" fragment is the larger, active fragment and the "a" fragment is smaller and may be an anaphylatoxin. However, this effort does not appear to be making headway. Note that the smaller fragment of C2, which we name C2b, is inactive in the complement pathway.

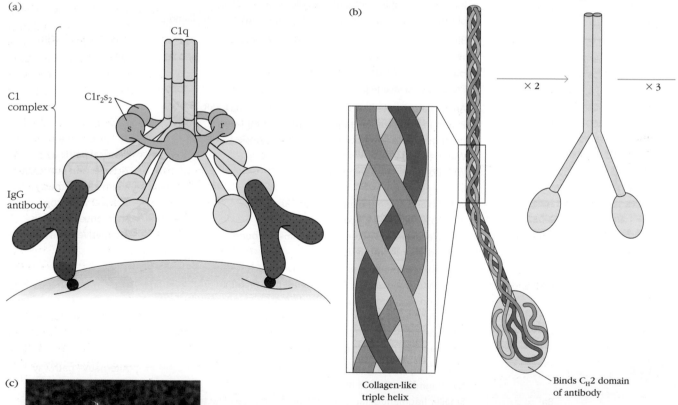

FIGURE 6-3 Structure of the C1 macromolecular complex. (a) C1q interacts with two molecules each of C1r and C1s to create the C1 complex. (b) The C1q molecule consists of 18 polypeptide chains in six collagen-like triple helices, each of which contains one A, B, and C chain, shown in three different shades of green. The head group of each triplet contains elements of all three chains. (c) Electron micrograph of C1q molecule showing stalk and six globular heads. The dark circle is a gold-labeled molecule of fibromodulin, a component of cartilage that binds the head groups of C1q. [Sjoberg, A., et al. September 16, 2005 The J. Biological Chem, Vol. 280, Issue 37, 32301-32308. ?2005 by the Merican Society for Biochemistry and Molecular Biology.]

(a)

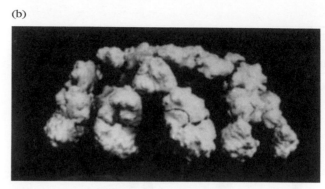

(b)

FIGURE 6-4 Models of pentameric IgM in planar form (a) and "staple" form (b). Several C1q-binding sites in the Fc region are accessible in the staple form, whereas none are exposed in the planar form. [*From A. Feinstein et al., 1981,* Monographs in Allergy *17:28, and 1981,* Annals of the New York Academy of Sciences *190:1104.*]

of which bind the C_H2 domain of the antigen-bound antibody molecule (Figure 6-3b, c).

Each C1 macromolecular complex must bind by its C1q globular heads to at least two Fc sites for a stable C1-antibody interaction to occur. Recall from Chapter 3 that serum IgM exists as a pentamer of the basic four-chain immunoglobulin structure. Circulating, nonantigen-bound IgM adopts a planar configuration (Figure 6-4a), in which the C1q-binding sites are not exposed. However, when pentameric IgM is bound to a multivalent antigen, it undergoes a conformational change, assuming the so-called "staple" configuration (Figure 6-4b), in which at least three binding sites for C1q are exposed. Thus, an IgM molecule engaged in an antibody-antigen complex can bind C1q, whereas circulating, nonantigen-bound IgM cannot.

In contrast to pentameric IgM, monomeric IgG contains only one C1q binding site per molecule, and the conformational changes IgG undergoes on antigen binding are much more subtle than those experienced by IgM. There is therefore a striking difference in the efficiency with which IgM and IgG are able to activate complement. Less than 10 molecules of IgM bound to a red blood cell can activate the classical complement pathway and induce lysis, whereas some 1000 molecules of cell-bound IgG are required to ensure the same result.

The intermediates in the classical activation pathway are depicted schematically in Figure 6-5. Proteins of the classical pathway are numbered in the order in which the proteins were discovered, which does not quite correspond with the order in which the proteins act in the pathway (a disconnect that has troubled generations of immunology students). Note that binding of one component to the next always induces either a conformational change or an enzymatic cleavage, which allows for the next reaction in the sequence to occur.

Binding of C1q to the C_H2 domains of the Fc regions of the antigen-complexed antibody molecule induces a conformational change in one of the C1r molecules that converts it to an active serine protease enzyme. This C1r molecule then cleaves and activates its partner C1r molecule. The two C1r proteases then cleave and activate the two C1s molecules (see Figure 6-5, part 1).

C1s has two substrates, C4 and C2. C4 is activated when C1s hydrolyzes a small fragment (C4a) from the amino terminus of one of its chains (see Figure 6-5, part 2). The C4b fragment attaches covalently to the target membrane surface in the vicinity of C1, and then binds C2. C4b binding to the membrane occurs when an unstable, internal thioester on C4b, exposed upon C4 cleavage, reacts with hydroxyl or amino groups of proteins or carbohydrates on the cell membrane. This reaction must occur quickly, otherwise the thioester C4b is further hydrolyzed and can no longer make a covalent bond with the cell surface (Figure 6-6). Approximately 90% of C4b is hydrolyzed before it can bind the cell surface.

On binding C4b, C2 becomes susceptible to cleavage by the neighboring C1s enzyme, and the smaller C2b fragment diffuses away, leaving behind an enzymatically active C4b2a complex. In this complex, C2a is the enzymatically active fragment, but it is only active when bound by C4b. This C4b2a complex is called C3 convertase, referring to its role in converting C3 into an active form. The smaller fragment generated by C4 cleavage, C4a, is an anaphylatoxin, and its function is described below.

The membrane-bound C3 convertase enzyme, C4b2a, now hydrolyzes C3, generating two unequal fragments; the small anaphylatoxin C3a and the pivotal fragment C3b. A single C3 convertase molecule can generate over 200 molecules of C3b, resulting in tremendous *amplification* at this step of the classical pathway.

OVERVIEW FIGURE   6-5

Intermediates in the Classical Pathway of Complement Activation up to the Formation of the C5 Convertase

1 C1q binds antigen-bound antibody, and induces a conformational change in one C1r molecule, activating it. This C1r then activates the second C1r and the two C1s molecules.

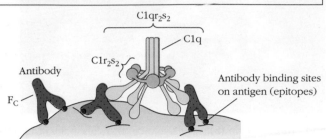

3 C3 convertase hydrolyzes many C3 molecules. Some combine with C3 convertase to form C5 convertase.

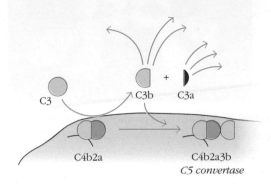

2 C1s cleaves C4 and C2. C4 is cleaved first and C4b binds to the membrane close to C1. C4b binds C2 and exposes it to the action of C1s. C1s cleaves C2, creating the C3 convertase, C4b2a.

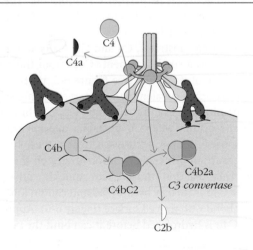

4 The C3b component of C5 convertase binds C5, permitting C4b2a to cleave C5.

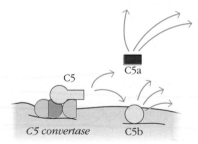

Antigenic determinants are shown in dark red, initiating components (antibodies and C1q) are shown in green, active enzymes are shown in blue, and anaphylatoxins in bright red.

The generation of C3b is centrally important to many of the actions of complement. Deficiencies of complement components that act prior to C3 cleavage leave the host extremely vulnerable to both infectious and autoimmune diseases, whereas deficiencies of components later in the pathway are generally of lesser consequence. This is because C3b acts in three important and different ways to protect the host.

First, in a manner very similar to that of C4b, C3b binds covalently to microbial surfaces, providing a molecular "tag" and thereby allowing phagocytic cells with C3b receptors to engulf the tagged microbes. This process is called **opsonization.** Second, C3b molecules can attach to the Fc portions of antibodies participating in soluble antigen-antibody complexes. These C3b-tagged immune complexes are bound by C3b receptors on phagocytes or red blood cells, and are

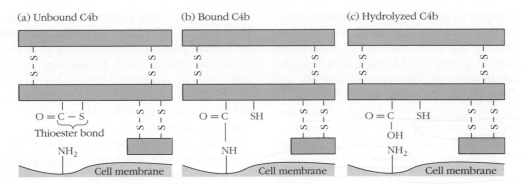

FIGURE 6-6 **Binding of C4b to the microbial membrane surface occurs through a thioester bond via an exposed amino or hydroxyl group.** (a) Both C3b and C4b contain highly reactive thioesters, which are subject to nucleophilic attack by hydroxyl or amino groups on cell membrane proteins and carbohy-drates. (b) Breakage of the thioester leads to the formation of covalent bonds between the membrane macromolecules and the complement components. (c) If this covalent bond formation does not occur quickly after generation of the C3b and C4b fragments, the thioester will be hydrolyzed to a nonreactive form.

either phagocytosed, or conveyed to the liver where they are destroyed.

Finally, some molecules of C3b bind the membrane-localized C4b2a enzyme to form the trimolecular, membrane-bound, C5 convertase complex C4b2a3b. The C3b component of this complex binds C5, and the complex then cleaves C5 into the two fragments: C5b and C5a (see Figure 6-5 part 4). *C4b2a3b is therefore the C5 convertase of the classical pathway.* This trio of tasks accomplished by the C3b molecule places it right at the center of complement attack pathways. As we will see, C5b goes on to form the MAC with C6, C7, C8, and C9.

The Lectin Pathway Is Initiated When Soluble Proteins Recognize Microbial Antigens

The lectin pathway, like the classical pathway, proceeds through the activation of a C3 convertase composed of C4b and C2a. However, instead of relying on antibodies to recognize the microbial threat and to initiate the complement activation process, this pathway uses lectins—proteins that recognize specific carbohydrate components primarily found on microbial surfaces—as its specific receptor molecules (Figure 6-7; see also Figure 5-7a). **Mannose-binding lectin (MBL)**, the first lectin demonstrated to be capable of initiating complement activation, binds close-knit arrays of mannose residues that are found on microbial surfaces such as those of *Salmonella*, *Listeria*, and *Neisseria* strains of bacteria; *Cryptococcus neoformans* and *Candida albicans* strains of fungi; and even the membranes of some viruses such as HIV-1 and respiratory syncytial virus. The complement pathway that it initiates is referred to as the **lectin pathway of complement activation**.

Further characterization of the sugars recognized by MBL demonstrated that MBL also recognizes structures in addition to mannose, including N-acetyl glucosamine, D-glucose, and L-fucose. All those sugars, including mannose, present their associated hydroxyl groups in a particular three-dimensional array. Thus, one can think of MBL as a classic pattern recognition receptor (see Chapter 5). Consistent with MBL's place at the beginning of an important immune cascade, many individuals with low levels of MBL suffer from repeated, serious bacterial infections.

MBL is constitutively expressed by the liver and, like C1q, which it structurally resembles, MBL belongs to the subclass of lectins known as collectins (see Chapter 5). More recently,

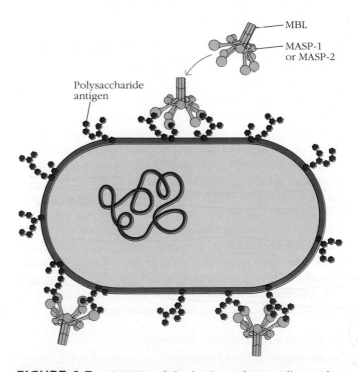

FIGURE 6-7 **Initiation of the lectin pathway relies on lectin receptor recognition of microbial cell surface carbohydrates.** Lectin receptors, such as MBL, bind microbial cell surface carbohydrates. Once attached to the carbohydrates, they bind the MASP family serine proteases, which cleave C2 and C4 with the formation of a lectin-pathway C3 convertase.

the ficolins (see Chapter 5), another family of proteins structurally related to the collectins, have been recognized as additional initiators of the lectin pathway of complement activation. L-ficolin, H-ficolin, and M-ficolin each bind specific types of carbohydrates on microbial surfaces. We will use MBL as our example in subsequent paragraphs.

MBL is associated in the serum with MBL-Associated Serine Proteases, or MASP proteins. Three MASP proteins—MASP1, MASP2, and MASP3—have been identified, but most studies of MASP function point to the MASP2 protein as being the most important actor in the next step of the MBL pathway.

MASP-2 is structurally related to the serine protease C1s, and can cleave both C2 and C4 (see Figure 6-2), giving rise to the C3 convertase, C4b2a, that we first encountered in our discussion of the classical pathway. Thus, the lectin pathway utilizes all the same components as the classical pathway with the single exception of the C1 complex. The soluble lectin receptor replaces the antibody as the antigen-recognizing component, and MASP proteins take the place of C1r and C1s in cleaving and activating the C3 convertase. Once the C3 convertase is formed, the reactions of the lectin pathway are the same as for the classical pathway; the C5 convertase of the lectin pathway, like that of the classical pathway, is also C4b2a3b.

The Alternative Pathway Is Initiated in Three Distinct Ways

Initiation of the **alternative pathway of complement activation** is independent of antibody-antigen interactions, and so this pathway, like the lectin pathway, is also considered to be part of the innate immune system. However, unlike the lectin pathway, the alternative pathway uses its own set of C3 and C5 convertases (see Figure 6-2). As we will see, the alternative pathway C3 convertase is made up of one molecule of C3b and one molecule unique to the alternative pathway, Bb. A second C3b is then added to make the alternative pathway C5 convertase.

Recent investigations have revealed that the alternative pathway can itself be initiated in three distinct ways. The first mode of initiation to be discovered, the "tickover" pathway, utilizes the four serum components C3, factor B, factor D, and properdin (see Table 6-1). The term *tickover* refers to the fact that C3 is constantly being made and spontaneously inactivated and is thus undergoing "tickover." Two additional modes of activation for the alternative pathway have also been identified: one is initiated by the protein **properdin**, and the other by proteases such as thrombin and kallikrein. The story of the discovery of properdin is addressed in Classic Experiment Box 6-1.

The Alternative Tickover Pathway

The **alternative tickover pathway** is initiated when C3, which is at high concentrations in serum, undergoes spontaneous hydrolysis at its internal thioester bond, yielding the molecule C3(H₂O). The conformation of C3(H₂O) has been demonstrated by spectrophotometric means to be different from that of the parent protein, C3. C3(H₂O) accounts for approximately 0.5% of plasma C3 and, in the presence of serum Mg^{2+},

binds another serum protein, factor B (Figure 6-8a). When bound to C3(H₂O), factor B becomes susceptible to cleavage by a serum protease, factor D. Factor D cleaves B, releasing a smaller Ba subunit, which diffuses away and leaves a catalytically active Bb subunit that remains bound to C3(H₂O).

This C3(H₂O)Bb complex, which at this point is still in the fluid phase (i.e., in the plasma, not bound to any cells), has C3 convertase activity. It rapidly cleaves many molecules of C3 into C3a and C3b (Figure 6-8b). This initiating C3 convertase is constantly being formed in plasma, and breaking down a few C3 molecules, but it is then just as rapidly degraded. However, if there is an infection present, some of

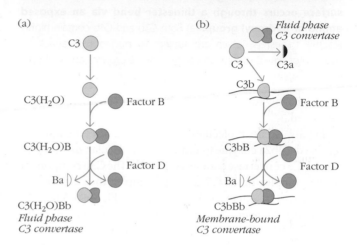

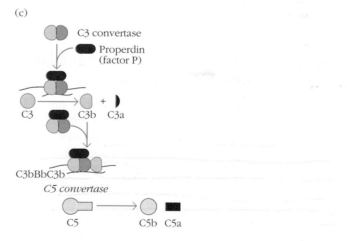

FIGURE 6-8 The initiation of the alternative tickover pathway of complement. (a) Spontaneous hydrolysis of soluble C3 to C3(H₂O) allows the altered conformation of C3(H₂O) to bind factor B, rendering it susceptible to cleavage by factor D. The resulting complex C3(H₂O)Bb forms a fluid phase convertase capable of cleaving C3 to C3a and C3b. (b) Some of the C3b molecules formed by the fluid phase convertase bind to cell membranes. C3b, like C3(H₂O) binds factor B in such a way as to make B susceptible to factor D-mediated cleavage. (c) However, membrane-bound C3bBb is unstable in the absence of properdin (factor P), which binds the C3bBb complex on the membrane and stabilizes it. Addition of a second C3b molecule to the C3bBb complex forms the C5 convertase, which is also stabilized by factor P.

the newly formed C3b molecules bind nearby microbial surfaces via their thioester linkages.

Since factor B is capable of binding to C3b as well as to $C3(H_2O)$, factor B now binds the newly attached C3b molecules on the microbial cell surface (see Figure 6-8b), and again becomes susceptible to cleavage by factor D, with the generation of C3bBb complexes. These C3bBb complexes are now located on the microbial membrane surface. Like the C4b2a complexes of the classical pathway, the cell-bound C3bBb complexes have C3 convertase activity, and this complex now takes over from the fluid phase $C3(H_2O)Bb$ as the predominant C3 convertase.

To be clear, there are two C3 convertases in the alternative tickover pathway: a fluid phase C3(H2O)Bb, which kicks off the pathway, and a membrane-bound C3bBb C3 convertase.

The cell-bound C3bBb C3 convertase is unstable until it is bound by properdin, a protein from the serum (Figure 6-8c). Once stabilized by properdin, these cell-associated, C3bBb C3 convertase complexes rapidly generate large amounts of C3b at the microbial surface; these, in turn, bind more factor B, facilitating its cleavage and activation and resulting in a dramatic amplification of the rate of C3b generation. This amplification pathway is rapid and, once the alternative pathway has been initiated, more than 2×10^6 molecules of C3b can be deposited on a microbial surface in less than 5 minutes.

Just as the C5 convertase of the classical and lectin pathways was formed by the addition of C3b to the C4b2a C3 convertase complex, so the C5 convertase of the alternative pathway is formed by the addition of C3b to the alternative pathway C3 convertase complex. The *C5 convertase* complex therefore has the composition *C3bBbC3b*, and, like the C3 convertase, is also stabilized by binding to factor P. Like the classical and lectin pathway C5 convertase, C3bBbC3b cleaves C5, which goes on to form the MAC (see Figure 6-8c).

The Alternative Properdin-Activated Pathway

In the previous section, we introduced properdin as a regulatory factor that stabilizes the C3bBb, membrane-bound C3 convertase. However, recent data suggest that, in addition to stabilizing the ongoing activity of the alternative pathway, *properdin may also serve to initiate it.*

In vitro experiments demonstrated that if properdin molecules were attached to an artificial surface and allowed to interact with purified complement components in the presence of Mg^{2+}, the immobilized properdin bound C3b and factor B (Figure 6-9). This bound factor B proved to be susceptible to cleavage by factor D, and the resultant *C3bPBb complex acted as an effective C3 convertase*, leading to the amplification process discussed above. Thus, it seemed that properdin could initiate activation of the alternative pathway on an artificial substrate.

However, proving that a set of reactions *can* occur in vitro does not necessarily mean that it actually *does* occur in vivo. Investigators next proved properdin's ability to bind specifically to certain microbes, including *Neisseria gonorrhoeae*, as

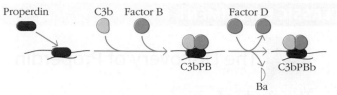

FIGURE 6-9 Initiation of the alternative pathway by specific, noncovalent binding of properdin to the target membrane. Properdin binds to components of microbial membranes, and stabilizes the binding of C3bBb complexes of the alternative complement pathway. The difference between this and the tickover pathways is that properdin binds first and initiates complement deposition on the membrane.

well as to apoptotic and necrotic cell surfaces. Once bound, properdin was indeed able to initiate the alternative pathway, as indicated above.

Support for the physiological relevance of this properdin-initiated pathway is provided by the observation that patients with a deficiency in properdin production are uniquely susceptible to meningococcal disease induced by the *Neisseria gonorrhoeae* bacterium. These findings suggest that properdin has the capacity to act as a pattern recognition receptor (PRR), specifically directing the activation of the alternative pathway onto the surface of *Neisseria* and other microbial cells.

Note that this pathway relies on the preexistence of low levels of C3b, which must be generated by mechanisms such as the tickover pathway. However, the specific binding of properdin to *Neisseria* membranes demonstrates how the properdin pathway can provide greater selectivity than that available from the nonspecific C3b binding of the tickover pathway.

The Alternative Protease-Activated Pathway

The complement and blood-coagulation pathways both use protease cleavage and conformational alterations to modify enzyme activities, as well as amplification of various steps of the pathways by feed-forward loops. Recently, some elegant work has revealed the existence of functional interactions as well as theoretical parallels between these two proteolytic cascades.

Several decades ago, it was shown that protein factors involved in blood clotting, such as thrombin, could cleave the complement components C3 and C5, in vitro, with the release of the active anaphylatoxins C3a and C5a. Since these cleavage reactions required relatively high thrombin concentrations, they were at first thought not to be physiologically meaningful. More recently, however, it has been demonstrated in a mouse disease model that initiation of the coagulation cascade may result in the cleavage of physiologically relevant amounts of C3 and C5 to produce C3a and C5a.

Specifically, in an immune-complex model of acute lung inflammation, thrombin was shown to be capable of cleaving C5, with the release of active C5a. This thrombin-mediated C5 cleavage was also demonstrated in C3 knockout mice, in which the canonical C5 convertases could not possibly have been generated. Thus, strong inflammatory reactions can result in the activation of at least a part of the alternative

The Discovery of Properdin

The study of the history of science shows us that scientists, like any other professionals, are often tempted to think about problems only in ways that are already well trodden and familiar. Science, like art and fashion, has its fads and its "in crowds." Indeed, sometimes those whose work moves in directions too different from that of the mainstream in their field have difficulty gaining credibility until the thinking of the rest of their colleagues catches up with their own. Such was the case for Louis Pillemer, the discoverer of properdin; by the time others in the burgeoning field of immunochemistry appreciated the power of his discovery, it was quite simply too late.

Louis Pillemer (Figure 1) was born in 1908 in Johannesburg, South Africa. In 1909 the family emigrated to the United States and settled in Kentucky. Pillemer completed his bachelor's degree at Duke University and started medical school at the same institution. However, in the middle of his third year, the emotional problems that would plague him for the rest of

his life surfaced for the first time, and he left. At that time, deep in the Depression, individuals in Kentucky who could pass an exam in the rudiments of medical care were encouraged to care for patients not otherwise served by a physician. Pillemer dutifully passed this examination and began to travel through Kentucky on horseback, visiting the sick and tendering whatever treatments were then available. In 1935, he quit this wandering life and entered graduate school at what was then Western Reserve University (now Case Western Reserve University).

There, Pillemer earned his Ph.D. and, except for some time at Harvard and a tour of duty at the Army Medical School in Washington D.C., he remained at Case Western Reserve for the rest of his life, developing a reputation as an excellent biochemist. Among his more noteworthy accomplishments were the first purifications of both tetanus and diphtheria toxins, which were used, along with killed pertussis organisms, in the development of the standard DPT vaccine.

After these successes, Pillemer turned his attention to the biochemistry of the complement system, which he had initially encountered during his graduate work. The antibody-mediated classical pathway had already been established. However, Pillemer was intrigued by some more recent experiments that had shown that mixing human serum with zymosan, an insoluble carbohydrate extract from yeast cell walls, at 37°C resulted in the selective loss of the vital third component of complement, C3. He was curious about the mechanism of this loss of C3, and initially interpreted the result to suggest that the C3 was being selectively adsorbed onto the zymosan surface. He reasoned that, if this were true, adsorption to zymosan might be used as a method to purify C3 from plasma. However, that first idea was proven incorrect, and instead he began to investigate whether the loss of C3 resulted from C3 cleavage that was occurring at the zymosan surface.

Pillemer succeeded in demonstrating that such was indeed the case and went on to show that C3 cleavage only happened when his experiments were run at pH 7.0 and at 37°C. This suggested to him that perhaps an enzyme in the serum was binding to the zymosan and causing the inactivation of the C3 component. Consistent with this hypothesis, when he mixed the serum and the zymosan at 17°C, no cleavage occurred. However, if he allowed the serum and zymosan to mix at 17°C and then warmed up the mixture to 37°C, the C3 was cleaved as effectively as if it had been incubated at 37°C all along.

Next, he incubated the serum and zymosan together at 17°C, spun down and removed the zymosan from the mixture, and then added fresh zymosan to the remaining serum containing the C3. He then raised the temperature to 37°C. Nothing happened. The C3 was untouched. Whatever enzymic activity in the serum was responsible for the breakdown of C3

FIGURE 1

Louis Pillemer, the discoverer of properdin. *[Presidential Address to the AAI, 1980, by Lepow; reprinted in J. Immunol. 125, 471, 1980. Fig.1. 1980. The American Association of Immunologists, Inc.]*

BOX 6-1

had been adsorbed by the zymosan and removed from the serum.

Pillemer concluded that a factor present in serum and adsorbed onto the zymosan was necessary for the cleavage of C3. With his students and collaborators, he purified this component and named it properdin, from the Latin *perdere* meaning "to destroy." His flow sheet for these initial experiments, and published as part of his landmark *Science* paper of 1954, is shown in Figure 2. He also identified a heat-labile factor in the serum that was required for C3 cleavage to occur.

Pillemer and colleagues went on to characterize properdin as a protein that represented less than 0.03% of serum proteins and whose activity was absolutely dependent on the presence of magnesium ions. In a brilliant series of experiments, Pillemer demonstrated the importance of his newly discovered factor in complement-related antibacterial and antiviral reactions, as well as its role in the disease known as paroxysmal nocturnal hemoglobinemia.

In light of today's knowledge, we can interpret exactly what was happening in his experiments. Properdin bound to the zymosan and stabilized the C3 convertase

of the alternative pathway, resulting in C3 cleavage. Indeed, experiments performed as recently as 2007 show that properdin binds to zymosan in a manner similar to its binding to *Neisseria* membranes.

Pillemer's discovery of properdin coupled with his purification of the tetanus and diphtheria toxins should have sealed his reputation as a world-class biochemist. Indeed, his findings were deemed of sufficient general interest at the time that they were publicized in the lay press as well as in the scientific media; articles and editorials about them appeared in the *New York Times, Time* magazine, and *Collier's*. Nor did Pillemer oversell his case. Scientists writing about this period are at pains to note that Pillemer did not make or broadcast any claims for his molecule beyond what appeared in the reviewed scientific literature.

However, in 1957 and 1958, scientist Robert Nelson offered an alternative explanation for Pillemer's findings. He pointed out that what Pillemer had described as a new protein could simply be a mixture of natural antibodies specific for zymosan. If that were the case, then all Pillemer had succeeded in doing was describing the classical pathway a

second time. Sensitive biochemical experiments indeed demonstrated the presence of low levels of antizymosan antibodies in properdin preparations, and the immunological community began to doubt the relevance of properdin to the complement activation that Pillemer described.

Pillemer was devastated. Never completely stable emotionally, Pillemer's "behavior became erratic, he occasionally abused alcohol and he appeared to be experimenting with drugs," according to his former graduate student, Irwin H. Lepow, who went on to become a distinguished immunologist in his own right. On August 31, 1957, right in the midst of the controversy over properdin, Pillemer died of acute barbiturate intoxication. His death was ruled a suicide, although nobody can know whether he was merely seeking short-term relief from stress and his death was therefore accidental, or whether he truly meant to bring an end to his life.

Subsequent experiments demonstrated that antibodies to zymosan could be removed from partially purified properdin without loss of the ability of the preparation to catalyze the cleavage of C3, thus confirming Pillemer's finding. Furthermore, the heat labile factor identified by Pillemer's experiments was identified as a previously unknown molecule, which was subsequently named factor B. By the late 1960s other laboratories entered the arena, and slowly Pillemer's discovery was confirmed and extended into what we now know as the alternative pathway of complement activation. It is one of the great tragedies of immunology that Pillemer did not live to enjoy the validation of his elegant work.

Lepow, I. H. 1980. Louis Pillemer, properdin and scientific controversy. Presidential address to American Association of Immunologists in Anaheim, CA, April 16, 1980. *Journal of Immunology* **125**:471–478.

Pillemer, L., et al. 1954. The properdin system and immunity. I. Demonstration and isolation of a new serum protein, properdin, and its role in immune phenomena. *Science* **120**:279–285.

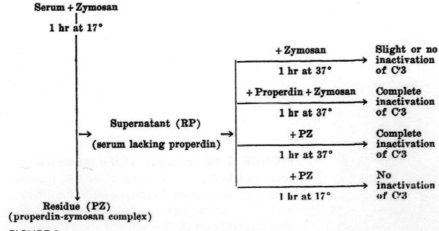

FIGURE 2
The flow sheet of Pillemer's experiments, which showed that a substance in serum that is adsorbed onto zymosan, the yeast cell wall extract, is capable of catalyzing the cleavage of C3. [*Pillemer, L., et al. The properdin system and immunity. I. Demonstration and isolation of a new serum protein, properdin, and its role in immune phenomena. Science. 1954 Aug 20;120(3112):279-85. Reprinted with permission from AAAS.*]

complement pathway, via the enzymes of the coagulation cascade. Given the proinflammatory roles of the anaphyla-toxins, this would result in further amplification of the inflammatory state. Additional experiments have since revealed that other coagulation pathway enzymes, such as plasmin, are capable of generating both C3a and C5a.

Furthermore, when blood platelets are activated during a clotting reaction, they release high concentrations of ATP and Ca²⁺ along with serine/threonine kinases. These enzymes act to phosphorylate extracellular proteins, includ-ing C3b. Phosphorylated C3b is less susceptible to proteo-lytic degradation than its unphosphorylated form, and thus, by this route, activation of the clotting cascade enhances all of the complement pathways.

The Three Complement Pathways Converge at the Formation of the C5 Convertase

The end result of the three initiation pathways is the formation of a C5 convertase. For the classical and lectin pathways, the C5 convertase has the composition C4b2a3b; for the alternative pathway, the C5 convertase has the formulation C3bBbC3b. (In

the thrombin-initiated pathway, the anaphylatoxin C5a is formed by cleavage of C5 by the blood clotting enzymes, but functionally meaningful C5b concentrations are not generated by this route.) However, *the end result of all types of C5 conver-tase activity is the same: the cleavage of the C5 molecule into two fragments, C5a and C5b.* The large C5b fragment is generated on the surface of the target cell or immune complex and pro-vides a binding site for the subsequent components of the MAC. However, the C5b component is extremely labile, is not covalently bound to the membrane like C3b and C4b, and is rapidly inactivated unless it is stabilized by the binding of C6.

C5 Initiates the Generation of the MAC

Up to this point in the complement cascades, all of the comple-ment reactions take place on the hydrophilic surfaces of microbes or on immune complexes in the fluid phase of blood, lymph, or tissues. In contrast, when C5b binds C6 and C7, the resulting complex undergoes a conformational change that exposes hydrophobic regions on the surface of the C7 compo-nent capable of inserting into the interior of the microbial membrane (Figure 6-10a). If, however, the reaction occurs on

(a)

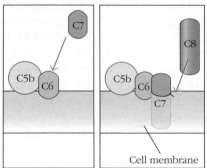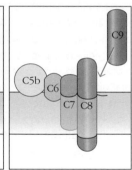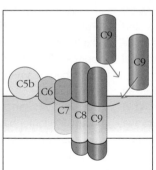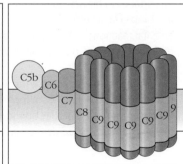

Cell membrane

(b)

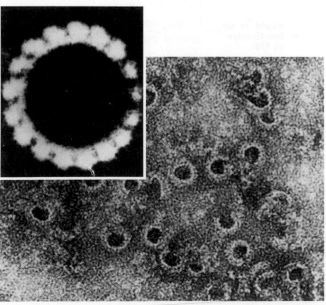

FIGURE 6-10 Formation of the membrane attack complex (MAC). (a) The formation of the MAC showing the addition of C6, C7, C8, and C9 components to the C5b compo-nent. (b) Photomicrograph of poly-C9 complex formed by in vitro polymerization of C9 (inset) and complement-induced lesions on the membrane of a red blood cell. These lesions result from formation of membrane-attack complexes. [Part b from E. R. Podack, 1986, Immunobiology of the Complement System, G. Ross, ed., Academic Press, Orlando, FL; part b inset from J. Humphrey and R. Dour-mashkin, 1969, The lesions in cell-membranes caused by complement, Advances in Immunology, 11:75.]

an immune complex or other noncellular activating surface, then the hydrophobic binding sites cannot anchor the complex and it is released. Released C5b67 complexes can insert into the membrane of nearby cells and mediate "innocent bystander" lysis. Under physiologic conditions, such lysis is usually minimized by regulatory proteins (see below).

C8 is made up of two peptide chains: C8β and C8αγ. Binding of C8β to the C5b67 complex induces a conformational change in the C8 dimer such that the hydrophobic domain of C8αγ can insert into the interior of the phospholipid membrane. The C5b678 complex can create a small pore, 10 Å in diameter, and formation of this pore can lead to lysis of red blood cells, but not of nucleated cells. The final step in the formation of the MAC is the binding and polymerization of C9 to the C5b678 complex. As many as 10 to 19 molecules of C9 can be bound and polymerized by a single C5b678 complex. During polymerization, the C9 molecules undergo a transition, so that they, too, can insert into the membrane. The completed MAC, which has a tubular form and functional pore diameter of 70 Å to 100 Å, consists of a C5b678 complex surrounded by a poly-C9 complex (Figure 6-10b). Loss of plasma membrane integrity leads irreversibly to cell death.

The Diverse Functions of Complement

Table 6-2 lists the main categories of complement function, each of which is discussed below. In addition to its long-understood role in antibody-induced lysis of microbes,

complement also has important functions in innate immunity, many of which are mediated by the soluble innate immune receptors such as MBL and the ficolins. The pivotal importance of C3b-mediated responses such as opsonization has been clearly demonstrated in C3$^{-/-}$ knockout animals, which display increased susceptibility to both viral and bacterial infections. In addition, recent experiments have explored the roles of various complement components at the interface of innate and adaptive immunity, and identified multiple mechanisms by which the release of active complement fragments acts to regulate the adaptive immune system. Complement also plays an important role in the contraction phase of the adaptive immune response, and recent work has even suggested that it is important in the elimination of excess synapses during the development of the nervous system. These various functions are described below.

Complement Receptors Connect Complement-Tagged Pathogens to Effector Cells

Many of the biological activities of the complement system depend on the binding of complement fragments to cell surface complement receptors. In addition, some complement receptors play an important role in regulating complement activity by mediating proteolysis of biologically active complement components. The levels of a number of the complement receptors are subject to regulation by aspects of the innate and adaptive immune systems. For example, activation of phagocytic cells by various agents, including the

TABLE 6-2	The three main classes of complement activity in the service of host defense
Activity	**Responsible complement component**
Innate defense against infection	
Lysis of bacterial and cell membranes	Membrane attack complex (C5b–C9)
Opsonization	Covalently bound C3b, C4b
Induction of inflammation and chemotaxis by anaphylatoxins	C3a, C4a, C5a (anaphylatoxins) and their receptors on leukocytes
Interface between innate and adaptive immunity	
Augmentation of antibody responses	C3b and C4b and their proteolyzed fragments bound to immune complexes and antigen; C3 receptors on immune cells
Enhancement of immunologic memory	C3b and C4b and their fragments bound to antigen and immune complexes; receptors for complement components on follicular dendritic cells
Enhancement of antigen presentation	MBL, C1q, C3b, C4b, and C5a
Potential effects on T cells	C3, C3a, C3b, C5a
Complement in the contraction phase of the immune response	
Clearance of immune complexes from tissues	C1, C2, C4; covalently bound fragments of C3 and C4
Clearance of apoptotic cells	C1q; covalently bound fragments of C3 and C4. Loss of CD46 triggers immune clearance.
Induction of regulatory T cells	CD46

This table has been considerably modified from the superb formulation of Mark Walport, Complement, *New England Journal of Medicine* **344**:1058–1066, Table 1.

TABLE 6-3   Receptors that bind complement components and their breakdown products

Receptor	Alternative name(s)	Ligand	Cell surface binding or expression	Function
CR1	CD35	C3b, C4b, C1q, iC3b	Erythrocytes, neutrophils, monocytes, macrophages, eosinophils, FDCs, B cells, and some T cells	Clearance of immune complexes, enhancement of phagocytosis, regulation of C3 breakdown
CR2	CD21, Epstein-Barr virus receptor	C3d, C3dg (human), C3d (mouse) iC3b	B cells and FDCs	Enhancement of B-cell activation, B-cell coreceptor, and retention of C3d-tagged immune complexes
CR3	CD11b/CD18, Mac-1	iC3b and factor H	Monocytes, macrophages, neutrophils, NK cells, eosinophils, FDCs, T cells	Binding to adhesion molecules on leukocytes, facilitates extravasation; iC3b binding enhances opsonization of immune complexes
CR4	CD11c/CD18	iC3b	Monocytes, macrophages, neutrophils, dendritic cells, NK cells, T cells	iC3b-mediated phagocytosis
CRIg	VSIG4	C3b, iC3b, and C3c	Fixed-tissue macrophages	iC3b-mediated phagocytosis and inhibition of alternative pathway
C1qR$_p$	CD93	C1q, MBL	Monocytes, neutrophils, endothelial cell, platelets, T cells	Induces T-cell activation; enhances phagocytosis
SIGN-R1	CD209	C1q	Marginal zone and lymph node macrophages	Enhances opsonization of bacteria by MZ macrophages
C3aR	None	C3a	Mast cells, basophils, granulocytes	Induces degranulation
C5aR	CD88	C5a	Mast cells, basophils, granulocytes, monocytes, macrophages, platelets, endothelial cells, T cells	Induces degranulation; chemoattraction; acts with IL-1β and/or TNF-α to induce acute phase response; induces respiratory burst in neutrophils
C5L2	None	C5a	Mast cells, basophils, immature dendritic cells	Uncertain, but most probably down-regulates proinflammatory effects of C5a

After Zipfel, P. F., and C. Skerka, 2009, Complement regulators and inhibitory proteins. *Nature Reviews Immunology* **10**:729–740; Kemper, C., and J. P. Atkinson, 2007, T-cell regulation: With complements from innate immunity, *Nature Reviews Immunology* **7**:9–18; Eggleton, P., A. J. Tenner, and K. B. M. Reid, 2000, C1q receptors, *Clinical and Experimental Immunology* **120**:406–412; Ohno, M., et al., 2000, A putative chemoattractant receptor, C5L2, is expressed in granulocyte and immature dendritic cells, but not in mature dendritic cells, *Molecular Immunology* **37**(8):407–412; and Kindt, T. J., B. A. Osborne, and R. A. Goldsby, 2006, *Kuby Immunology*, 6th ed., New York: W. H. Freeman, Table 7-4.

anaphylatoxins of the complement system, has been shown to increase the number of complement receptors by as much as tenfold.

Therefore, before we venture into a discussion of the biological functions of complement, we should first become familiar with the receptors for complement components and their activities. The complement receptors and their primary ligands are listed in Table 6-3. Where receptors have more than one name, we have offered both on first introduction, and subsequently refer to that receptor by the more common name.

CR1 (CD35) is expressed on both leukocytes and erythrocytes and binds with high affinity to C4b, C3b, and smaller C3b breakdown products. CR1 receptors on erythrocytes bind immune complexes and take them to the liver where they are picked up by phagocytes and cleared. Binding of complement-opsonized microbial cells via CR1 on phagocytes results in receptor-mediated phagocytosis and the secretion of proinflammatory molecules such as IL-1 and prostaglandins. CR1 on B cells mediates antigen uptake of C3b-bound antigen, leading to its degradation in the B-cell lysosomal system and subsequent presentation to T cells. This process makes CR1 an actor in the adaptive, as well as in the innate immune response. As we will see later, CR1 also mediates the protection of host cells against the ravages of complement attack, by serving as a cofactor for the destructive cleavage of C3b and C4b on host cell membranes by factor I, as well as acting as an accelerator of the decay of the C3 and C5 convertases.

C3b, either in solution or bound to the surface of cells, is subject to breakdown by endogenous proteases. **CD21** (CR2) is expressed on B cells and binds specifically to the

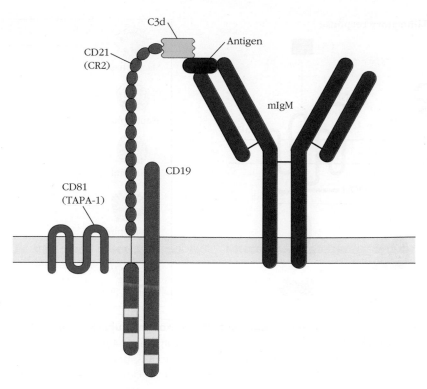

FIGURE 6-11 Coligation of antigen to B cells via IgM and CD21. The B-cell coreceptor is a complex of the three cell membrane molecules, CD21 (CR2), CD81 (TAPA-1), and CD19. The CD19 component is important in BCR antigen receptor signaling. Antigen that has been covalently bound to fragments of the C3 complement component is bound by both the immunoglobulin BCR and by the CD21 complement receptor, thus significantly increasing the avidity of the cell receptors for the antigen and allowing lower concentrations of antigen to trigger B-cell activation.

breakdown products of C3b: iC3b, C3d, and C3dg. Since C3b can form covalent bonds with antigens, the presence of CD21 on B cells enables the B cell to bind antigen via both the B-cell receptor and CD21 (Figure 6-11). This ability to simultaneously coengage antigen through two receptors has the effect of reducing the antigen concentration necessary for B-cell activation by up to a hundredfold (see Chapter 12). Deficiencies in CD21 have been identified in patients suffering from autoimmune diseases such as systemic lupus erythematosus (see below).

CR3 (a complex of CD11b and CD18) and CR4 (a complex of CD11c and CD18) are important in the phagocytosis of complement-coated antigens. CR3 and CR4 bind C3b and several of its breakdown products, including iC3b, C3c, and C3dg.

CRIg also binds C3b. It is expressed on macrophages resident in the fixed tissues, including on the Kupffer cells of the liver. Its importance in clearing C3b-opsonized antigens by facilitating their removal from circulation in the liver is emphasized by the finding that CRIg-deficient mice are unable to efficiently clear C3-opsonized particles. Animals with this deficiency are therefore subject to higher mortalities during infections.

C3aR, C5aR, and C5L2 are all members of the *G* protein coupled receptor (GPCR) family first encountered in Chapter 4. C3aR and C5aR mediate inflammatory functions after binding the small anaphylatoxins, C3a and C5a, respectively (Figure 6-12). The C5L2 receptor also binds C5a, is structurally similar to C5aR, and is expressed on some of the same cells. However, C5L2 is not functionally coupled to the G protein signaling pathway used by C5aR; instead, signaling through C5L2 appears to modulate C5a signaling through C5aR and C5L2 knockout animals express enhanced inflammatory responses upon C5a binding to its receptor.

More recently, a transmembrane lectin, SIGN-R1, able to bind C1q has been shown to be expressed on the surface of macrophages located in the marginal zone of the spleen. SIGN-R1 exists on the macrophage cell surface in aggregated form and is also able to bind carbohydrates present on the coat of the Gram-positive bacterium *Staphylococcus pneumoniae*. When SIGN-R1 binds to bacterial polysaccharides, the C1q-binding capacity is activated in the same, or in nearby SIGN-R1 molecules, eventually resulting in the opsonization of the bacterium with C3b. The opsonized bacteria are then released from these macrophages and bound by nearby phagocytes, B cells, or dendritic cells. This unusual mechanism explains a problem long noted with patients who have undergone splenectomy: an increased susceptibility to infection by *S. pneumoniae*.

Anaphylatoxins and inflammatory response

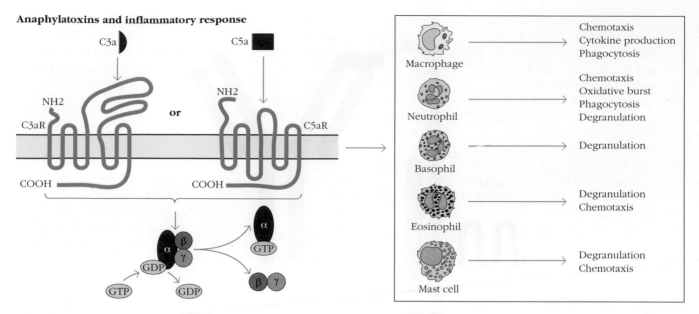

FIGURE 6-12 Binding of the anaphylatoxins C3a and C5a to the G-protein-coupled receptors C3aR and C5aR. The C3aR and C5aR receptors are members of the G-protein-coupled receptor family described in Chapter 4. Binding of the anaphylatoxins to these receptors stimulates the release of proinflammatory mediators from macrophages, neutrophils, basophils, eosinophils, and mast cells, as indicated. *[Adapted from J. R. Dunkelberger and W.-C. Song, 2010, Complement and its role in innate and adaptive immune responses,* Cell Research *20:34–50, Figure 3B.]*

Complement Enhances Host Defense Against Infection

Complement proteins actively engage in host defense against infection by forming the MAC, by opsonizing potentially pathogenic microbes, and by inducing an inflammatory response that helps to guide leukocytes to the site of infection.

MAC-Induced Cell Death

The first function of complement to be described was its role in inducing cell death following insertion of the MAC into target cell membranes. Early experiments on MAC formation used erythrocytes as the target membranes, and in this cellular system large pores involving 17 to 19 molecules of C9 were reported. Formation of these holes in the cell membrane (see Figure 6-10b) facilitated the free movement of small molecules and ions. The penetrated red blood cell membranes were unable to maintain osmotic integrity, and the cells lysed after massive influx of water from the extracellular fluid.

However, subsequent studies using nucleated eukaryotic cells indicated that smaller pores can be generated using only a few molecules of C9, and that death under these circumstances occurs via a type of apoptosis (programmed cell death), following calcium influx into the cytoplasm. Nuclear fragmentation, a hallmark of apoptotic death, was observed during MAC-induced lysis of nucleated cells, which further supported the notion that at least some MAC-targeted cells succumb to apoptosis. More careful observations indicated that the apoptosis induced by the MAC does not share all the biochemical characteristics normally associated with programmed cell death, and so this MAC-induced apoptosis has been termed *apoptotic necrosis.*

Killing eukaryotic cells with complement is actually quite difficult, as the membranes of such cells have a number of factors that act together to inactivate the complement proteins, and thus protect the host cells from collateral damage during a complement-mediated attack on infectious microorganisms (see below). However, when high concentrations of complement components are present, MACs can overwhelm the host defenses against MAC attack, and the resulting cell fragments, if present in sufficiently high concentrations, can induce autoimmunity. Indeed, complement-mediated damage is a problem in several autoimmune diseases, and the complement system is considered a target for therapeutic intervention in autoimmune syndromes.

Can a eukaryotic cell recover from a MAC attack? Well-documented studies have demonstrated that MACs can be removed from the cell surface, either by shedding MAC-containing membrane vesicles into the extracellular fluid, or by internalizing and degrading the MAC-containing vesicles in intracellular lysosomes. If the MAC is shed or internalized soon enough after its initial expression on the membrane, the cell can repair any membrane damage and restore its osmotic stability. An unfortunate corollary of this capacity to recover from MAC attack is that complement-mediated lysis directed by tumor-specific antibodies may be rendered ineffective by endocytosis or shedding of the MAC (see Chapter 19).

Different types of microorganisms are susceptible to complement-induced lysis to varying degrees. Antibody and complement play an important role in host defense against viruses and can be crucial both in containing viral spread during acute infection and in protecting against reinfection. Most enveloped viruses, including herpesviruses, orthomyxoviruses such as those causing measles and mumps, paramyxoviruses such as influenza, and retroviruses are susceptible to complement-mediated lysis.

However, some bacteria are not susceptible to MAC attack. Gram-positive bacteria efficiently repel complement assault, as the complement proteins cannot penetrate the bacterial cell wall to gain access to the membrane beyond. In contrast, when attacking Gram-negative bacteria, complement first permeabilizes the outer layer and, following destruction of the thin cell wall, lyses the inner bacterial membrane. In some cases, the MAC has been found to localize at regions of apposition of the inner and outer cell walls, and breaches them both simultaneously. *Neisseria*

meningitidis is a Gram-negative bacterium that is susceptible to MAC-induced lysis, and patients who are deficient in any of the complement components of the MAC are particularly vulnerable to potentially fatal meningitis caused by this bacterial species.

Promotion of Opsonization

As described in Chapter 5, the term *opsonization* refers to the capacity of antibodies and complement components (as well as other proteins) to coat dangerous antigens that can then be recognized by Fc receptors (for antibodies) or complement receptors (for complement components) on phagocytic cells. Binding of complement-coated antigen by phagocytic cells results in complement receptor-mediated phagocytosis and antigen destruction (Figure 6-13). In

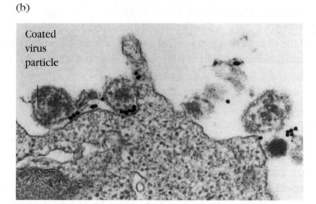

FIGURE 6-13 Opsonization of microbial cells by complement components and antibodies. (a) Phagocytosis is mediated by many different complement receptors on the surface of macrophages and neutrophils, including CR1, CR3, CR4, and CRIg. Phagocytes, using their Fc receptors, also bind to antigens opsonized by antibody binding. (b) Electron micrograph of Epstein-Barr virus coated with antibody and C3b and bound to the Fc and C3b receptor (CR1) on a B lymphocyte. *[Part b from N. R. Cooper and G. R. Nemerow, 1986,* Immunobiology of the Complement System, *G. Ross, ed., Orlando, FL: Academic Press.]*

addition, complement receptors on erythrocytes also serve to bind immune complexes, which are then transported to the liver for phagocytosis by macrophages (Figure 6-14). Although less visually compelling than MAC formation, opsonization may be the most physiologically important function assumed by complement components.

Opsonization with antibody and complement provides critical protection against viral infection. Antibody and complement can create a thick protein coat around a virus that neutralizes viral infectivity by preventing the virus from binding to receptors on the host cell. It also promotes phagocytosis via the complement receptors, followed by intracellular destruction of the digested particle.

Promotion of Inflammation

So far, we have focused on the roles of the larger products of complement factor fragmentation: C3b and C4b in opsonization and C5b in the formation of the MAC. However, the smaller fragments of C3, C4, and C5 cleavage—C3a, C4a, and C5a—are no less potent and mediate critically important events in immune responses, acting as anaphylatoxins or inducers of inflammation. We will focus here specifically on the activities of C3a and C5a.

C3a and C5a are structurally similar, small proteins (about 9 kDa, or 74–77 amino acid residues in size) that promote inflammation and also serve as chemoattractants for certain classes of leukocytes. C3a and C5a bind G-protein-coupled, activating receptors (C3aR for C3a, C5aR for C5a) on granulocytes, monocytes, macrophages, mast cells, endothelial cells, and some dendritic cells (see Figure 6-12). Binding of these anaphylatoxins to their receptors on some cells triggers a signaling cascade that leads to the secretion of soluble mediators such as IL-6 and TNF-α. These cytokines in turn induce localized increases in vascular permeability that enable leukocyte migration into the site of infection, and a concomitant increase in smooth muscle motility that helps to propel the released fluid to the site of damage. In addition, these proinflammatory mediators promote phagocytosis of offending pathogens and localized degranulation of granulocytes (neutrophils, basophils, and eosinophils) with the resultant release of a second round of inflammatory mediators, including histamines and prostaglandins. Inflammatory mediators expedite the movement of lymphocytes into neighboring lymph nodes where they are activated by the pathogen. This localized inflammatory response is further supported by systemic effects, such as fever, that decrease microbial viability.

The central role of the anaphylatoxins in the promotion of physiologically important inflammatory responses has brought them to the attention of clinicians seeking ways to tamp down the pathological levels of inflammation experienced by patients suffering from diseases such as rheumatoid arthritis and systemic lupus erythematosus (SLE). Clinical Focus Box 6-2 describes some approaches that investigators are taking to explore the complement system as a potential therapeutic target.

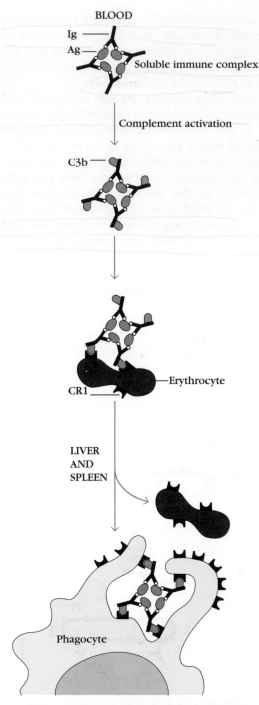

FIGURE 6-14 Clearance of circulating immune complexes by reaction with receptors for complement products on erythrocytes and removal of these complexes by receptors on macrophages in the liver and spleen. Because erythrocytes have fewer receptors than macrophages, the latter can strip the complexes from the erythrocytes as they pass through the liver or spleen. Deficiency in this process can lead to renal damage due to accumulation of immune complexes.

Complement Mediates the Interface Between Innate and Adaptive Immunities

Experiments in the past two or three decades have revealed multiple mechanisms by which components of the complement system modulate adaptive immunity. Many of these findings are extremely recent, and the study of how the binding of complement components and regulatory proteins may affect antigen-presenting cells, T cells, and B cells is still in its early phases.

Complement and Antigen-Presenting Cells

Dendritic cells (DCs), follicular dendritic cells (FDCs), and macrophages express all of the known complement receptors. When bound to antigens, MBL, C1q, C3b, and C4b are each capable of engaging their respective receptors on antigen-presenting cells during the process of antigen recognition, and signaling through their respective receptors acts to enhance antigen uptake.

In addition, signaling of antigen-presenting cells through the C5aR anaphylatoxin receptor has been shown to modulate their migration and affect interleukin production, particularly that of the cytokine IL-12. Production of IL-12 by an antigen-presenting cell normally skews the T-cell response toward the T_H1 phenotype (see Chapter 11). Since both induction and suppression of IL-12 production have been detected after activation of the anaphylatoxin receptors, depending upon the route of antigen delivery, the nature of the antigen, and the maturation status of the antigen-presenting cell, we must infer that many signaling pathways are being integrated to arrive at the eventual biological response to antigen and complement components.

Complement and B-Cell-Mediated Humoral Immunity

In 1972, Pepys showed that depleting mice of C3 impaired their T-dependent antigen-specific B-cell responses, thus implying that complement may participate in the initiation of the B-cell response. It now appears that this seminal observation was the first description of the complement receptor CD21 acting as a coreceptor in antigen recognition.

Recall that the C3b molecule is capable of covalently "tagging" antigen by interaction between its reactive thioester group and hydroxyl or amino groups on the surface of the antigen. As described in Chapters 3 and 12, the immunoglobulin receptor is expressed on the B-lymphocyte membrane along with the complement receptor CD21, the signaling protein CD19, and the transmembrane CD81 (tetraspanin, or TAPA-1) molecule (see Figure 6-11). CD21 is a receptor with specificity for both C3b and the products of C3b proteolytic breakdown, C3d (human) or C3dg (mouse). If C3b was originally bound to a microbial surface, the C3d or C3dg fragments remain bound to that surface after C3b proteolysis. A B cell bearing an immunoglobulin receptor specific for that microbe will therefore be able to bind the microbe with both its immunoglobulin receptor and with the associated CD21 complement receptor, thus enhancing the interaction between the B cell and antigen and reducing the amount of antigen that is required to signal a B-cell response by a factor of tenfold to a thousandfold.

Complement and T-Cell-Mediated Immunity

The mechanisms by which complement affects T-cell responses are not as well characterized as those for B cells and antigen-presenting cells. However, some recent, interesting results have been generated by the study of mice either genetically deficient in one or more complement components or following treatment with complement inhibitors. For example, mice lacking the C3 gene have reduced $CD4^+$ and $CD8^+$ T cell responses, implying that signaling through C3 enhances T-cell function. Furthermore, mice treated with inhibitors of C5aR signaling produced fewer antigen-specific $CD8^+$ T cells, following influenza infection, than wild-type mice, indicating that C5a may act as a costimulator during $CD8^+$ T-cell activation, possibly by increasing IL-12 production by antigen-presenting cells, as described above. Both these experiments demonstrate that signaling through complement components may have positive effects on T-cell-mediated adaptive immune responses.

The study of T-cell regulation by complement components is in its infancy, but these provocative findings suggest that there is much to learn and that complement may have more powerful effects on adaptive immunity than previously thought. As these effects are further explored, the question of whether complement is acting directly on the T cells, or indirectly, via effects on antigen-presenting cells, is an issue that will need to be clearly defined.

Complement Aids in the Contraction Phase of the Immune Response

At the close of an adaptive immune response, most of the lymphocytes that were generated in the initial proliferative phase undergo apoptosis (programmed cell death), leaving only a few antigen-specific cells behind to provide immunological memory (see Chapters 11 and 12). In addition, soluble antigen-antibody complexes may still be present in the bloodstream and immune organs. If autoimmune disease is to be avoided, these excess lymphocytes and immune complexes must be disposed of safely, without the induction of further inflammation, and complement components play a major role in these processes.

Disposal of Apoptotic Cells and Bodies

Apoptotic cells express the phospholipid **phosphatidyl serine** on the exterior surface of their plasma membranes. In healthy cells, this phospholipid is normally restricted to the cytoplasmic side of the membrane, and this change in its location serves to signal the immune system that the cell is undergoing programmed cell death. Surprisingly, experiments have also demonstrated the presence of nucleic acids on the surface membrane of apoptotic cells. Nuclear fragmentation and

CLINICAL FOCUS

The Complement System as a Therapeutic Target

The involvement of complement-derived anaphylatoxins in inflammation makes complement an interesting therapeutic target in the treatment of inflammatory diseases, such as arthritis. In addition, autoimmune diseases that potentially result in complement-mediated damage to host cells, such as multiple sclerosis and age-related macular degeneration, are also potential candidates for complement-focused therapeutic interventions.

The last three decades have seen the crystallization and molecular characterization of several complement components, a necessary precursor to the development of designer drugs targeted to specific proteins. At the time of writing, however, just two therapies directed at interfering with complement components are approved for clinical use, although at least one more is currently in clinical trials.

Purified C1 esterase inhibitor is currently in use in the clinic as a therapeutic option for the treatment of hereditary angioedema. However, its beneficial effects are most likely the result of the inhibition of excess esterase activation external to the complement system per se.

The other successful complement-related treatment is more specifically directed toward a dysregulation of the complement cascade. Paroxysmal noctur-

nal hemoglobinuria (PNH) manifests as increased fragility of erythrocytes, leading to chronic hemolytic anemia, pancytopenia (loss of blood cells of all types), and venous thrombosis (formation of blood clots). The name PNH derives from the presence of hemoglobin in the urine, most commonly observed in the first urine passed after a night's sleep. The cause of PNH is a general defect in the synthesis of cell-surface proteins, which affects the expression of two regulators of complement: DAF (CD55) and Protectin (CD59).

DAF and Protectin are cell-surface proteins that function as inhibitors of complement-mediated cell lysis, acting at different stages of the process. DAF induces dissociation and inactivation of the C3 convertases of the classical, lectin, and alternative pathways (see Figure 6-16). Protectin acts later in the pathway by binding to the C5b678 complex and inhibiting C9 binding, thereby preventing formation of the membrane pores that destroy the cell under attack. Deficiency in these proteins leads to increased sensitivity of host cells to complement-mediated lysis and is associated with a high risk of thrombosis.

Protectin and DAF are attached to the cell membrane via glycosylphosphatidyl inositol (GPI) anchors, rather than by

stretches of hydrophobic amino acids, as is the case for many membrane proteins. Patients with PNH lack an enzyme, *Phosphatidyl Inositol Glycan Class A* (PIGA), that attaches the GPI anchors to the appropriate proteins. Patients thus lack the expression of DAF and Protectin needed to protect erythrocytes against lysis. The term *paroxysmal* refers to the fact that episodes of erythrocyte lysis are often triggered by stress or infection, which result in increased deposition of C3b on host cells. The gene encoding the PIGA enzyme is X-linked.

The defect identified in PNH occurs early in the enzymatic pathway leading to formation of a GPI anchor and resides in the *pig-a* gene. Transfection of cells from PNH patients with an intact *pig-a* gene restores the resistance of the cells to host complement lysis. Further genetic analysis revealed that the defect is not encoded in the germline genome (and therefore is not transmissible to offspring), but rather is the result of mutations that occurred within the hematopoietic stem cells themselves, such that any one individual may have both normal and affected cells. Those patients who are most adversely affected display a preferential expansion of the affected cell populations.

PNH is a chronic disease with a mean survival time between 10 and 15 years.

DNA cleavage follow quickly after outer membrane phosphatidyl serine expression. Once apoptosis begins, the dying cell is broken down into membrane-bound vesicles termed *apoptotic bodies*, which express phosphatidyl serine and/or surface-bound DNA on their exterior surfaces. Recent work has demonstrated that C1q binds specifically to DNA.

The complement component C1q specifically binds apoptotic bodies and assists in their clearance. When light-sensitive keratinocytes (skin cells) are treated with UVB to initiate apoptosis and then stained with anti-C1q antibody, C1q staining is restricted to the apoptotic blebs, where DNA, exposed on apoptotic membranes, is specifically recognized by C1q (Figure 6-15). With C1q deposition, the classical pathway is initiated and the apoptotic cells are then opsonized by C3b. This, in turn, leads to clearance of the apoptotic cells by phagocytes.

In the absence of C1q, the apoptotic membrane blebs are released from the dying cells as apoptotic bodies, which can then act as antigens and initiate autoimmune responses. As a consequence, mice deficient in C1q show increased mortality and higher titers of auto-antibodies than control mice and also display an increased frequency of glomerulonephritis, an autoimmune kidney disease. Analysis of the kidneys of C1q knockout mice reveal deposition of immune complexes as well as significant numbers of apoptotic bodies.

Disposal of Immune Complexes

As mentioned earlier, the coating of soluble immune complexes with C3b facilitates their binding by CR1 on erythrocytes. Although red blood cells express lower levels of CR1

The most common causes of mortality in PNH are venous thromboses affecting hepatic veins and progressive bone marrow failure.

A breakthrough in treatment of PNH was reported in 2004 using a humanized monoclonal antibody that targets complement component C5 and thus inhibits the terminal steps of the complement cascade and formation of the membrane-attack complex. This antibody—eculizumab (Soliris)—was infused into patients, who were then monitored for the loss of red blood cells. A dramatic improvement was seen in patients during a 12-week period of treatment with eculizumab (Figure 1). Treatment of PNH patients with eculizumab relieves hemoglobinuria, reverses the kidney damage resulting from high levels of protein in the urine, and significantly reduces the frequency of thromboses. In 2007, eculizumab was approved for the treatment of PNH in the general population.

Since the control of infections with Meningococcal bacteria (*Neisseria meningitidis*) relies upon an intact and functioning membrane attack complex (MAC), patients being treated with eculizumab are routinely vaccinated against meningococcus.

The pathology of PNH underscores the potential danger to the host that is inherent in the activation of the complement system. Complex systems of regulation are necessary to protect host cells from the activated complement complexes generated to lyse intruders, and alterations in the expression or effectiveness of these regulators has the potential to result in a pathological outcome.

The centrality of C3 within the complement cascades may suggest that it would be an excellent target for therapeutic intervention in complement-mediated disease. However, its very position at the crossroads of the three pathways means that interference with C3 activity places patients at increased risk for infectious diseases normally curtailed by the activities of the innate as well as the adaptive immune systems, and so systemic drugs specifically targeting C3 may prove to have too high a risk-benefit ratio.

However, in one success story, targeted to a particular organ system, the drug Compstatin was developed as a result of phage display experiments designed to discover compounds that bound to C3. Chemical refinements based on structural and computational studies followed, and a Compstatin derivative, POT-4, is now in clinical trials for the treatment of age-related macular degeneration, a progressive and debilitating eye disease that leads inexorably and quickly to total blindness. Because POT-4 can be delivered directly into the vitreous humor of the eye, systemic side effects on the patient's immune system are minimized.

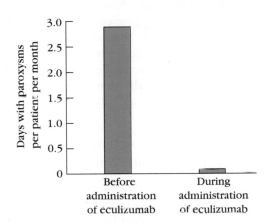

FIGURE 1
Treatment of PNH patients with eculizumab relieves hemoglobinuria. The number of days with paroxysm (onset of attack) per patient per month in the month prior to treatment (left bar) and for a 12-week period of treatment with eculizumab (right bar) is shown. [*From P. Hillmen, et al., 2004. Eculizumab in patients with paroxysmal nocturnal hemoglobinuria. New England Journal of Medicine **350**:6, 552–559.*]

(100–1000 molecules per cell, depending on the age of the cell and the genetic constitution of the donor) than do granulocytes (5×10^4 per cell), there are about 1000 erythrocytes for every white blood cell, and therefore erythrocytes account for about 90% of the CR1 in blood. Erythrocytes also therefore play an important role in clearing C3b-coated immune complexes by conveying them to the liver and spleen where the immune complexes are stripped from the red blood cells and phagocytosed (see Figure 6-14).

The importance of the complement system in immune complex clearance is emphasized by the finding that patients with SLE, the autoimmune disease, have high concentrations of immune complexes in their serum that are deposited in the tissues. Complement is activated by these tissue-deposited immune complexes, and pathological inflammation is induced in the affected tissues (see Chapter 15).

Since complement activation is implicated in the pathogenesis of SLE, it may therefore seem paradoxical that the incidence of SLE is highly correlated to *C4 deficiency.* Indeed, 90% of individuals who completely lack C4 develop SLE. The resolution to this paradox lies in the fact that deficiencies in the early components of complement lead to a reduction in the levels of C3b that are deposited on the immune complexes. This reduction, in turn, inhibits their clearance by C3b-mediated opsonization and allows for the activation of the later inflammatory and cytolytic phases of complement activation.

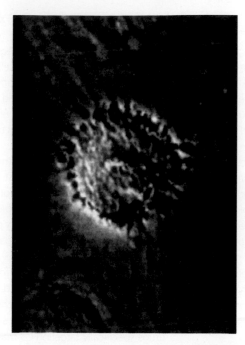

FIGURE 6-15 Binding of C1q to apoptotic keratinocytes (skin cells) is limited to surface apoptotic blebs. Apoptosis was induced in human skin cells by exposure to ultraviolet light. The cells were then incubated in human serum and stained with an antibody to C1q. This is an image derived from confocal microscopy that shows the C1q staining in green, clearly localizing to the apoptotic blebs. The cell itself is visualized by phase contrast microscopy, and the image was generated by merging the phase contrast and immunofluorescent frames. [Korb and Ahearn. 1997. J. Immunol. 158: 4525–4528.]

Complement Mediates CNS Synapse Elimination

Complement has also been shown to play an important role outside the boundaries of the immune system. In the developing nervous system, growing neurons first form a relatively high number of connections (synapses) with one another; the number of those synapses is then pruned as the nervous system matures. Scientists studying the development of the mouse eye have demonstrated that animals deficient in C1q or C3 fail to eliminate these early, excess synapses, and display anatomical abnormalities in the visual nervous system, indicating that complement may play an important role in the process of *synaptic remodeling*.

In healthy animals, the expression of complement components in mouse neurons is up-regulated early in development, in response to signals provided by immature astrocytes (glial cells that assist in the function and maintenance of the nervous system). Complement component expression is then down-regulated as the animals mature and astrocyte function is dampened. However, patients suffering from glaucoma have been shown to display erroneously high

levels of complement components in the adult retina during the early phases of the disease, suggesting that inappropriate, complement-mediated synaptic pruning may be a causal factor in that disease. C1q up-regulation has also been observed in animal models of both ALS and Alzheimer's disease, potentially implicating inappropriate complement activation in the etiology of some clinically important neurodegenerative illnesses.

The Regulation of Complement Activity

All biological systems with the potential to damage the host, be they metabolic pathways, cytotoxic cells, or enzyme cascades, are subject to rigorous regulatory mechanisms, and the complement system is no exception to this general rule. Especially in light of the potent positive feedback mechanisms and the absence of antigen specificity of the alternative pathway, the host must ensure that the destructive potential of complement proteins is confined to the appropriate pathogen surfaces and that collateral damage to healthy host tissue is minimized.

Below, we discuss the different mechanisms by which the host protects itself against the potential ravages of inadvertent complement activation.

Complement Activity Is Passively Regulated by Protein Stability and Cell-Surface Composition

Protection of vertebrate host cells against complement-mediated damage is achieved by both general, passive and specific, active regulatory mechanisms. The *relative instability* of many complement components is the first means by which the host protects itself against inadvertent complement activation. For example, the C3 convertase of the alternative pathway, C3bBbC3b, has a half-life of only 5 minutes before it breaks down, unless it is stabilized by reaction with properdin. A second passive regulatory mechanism depends upon the difference in the *cell surface carbohydrate composition of host versus microbial cells.* For example, fluid phase proteases that destroy C3b bind much more effectively to host cells, which bear high levels of sialic acid, than to microbes that have significantly lower levels of this sugar. Hence, any C3b that happens to alight on a host cell is likely to be destroyed before it can effect significant damage.

In addition to these more passive environmental brakes on inappropriate complement activation, a series of active regulatory proteins act to inhibit, destroy, or tune down the activity of complement proteins and their active fragments. The stages at which complement activity is subject to regulation are illustrated in Figure 6-16, and the regulatory proteins are listed in Table 6-4.

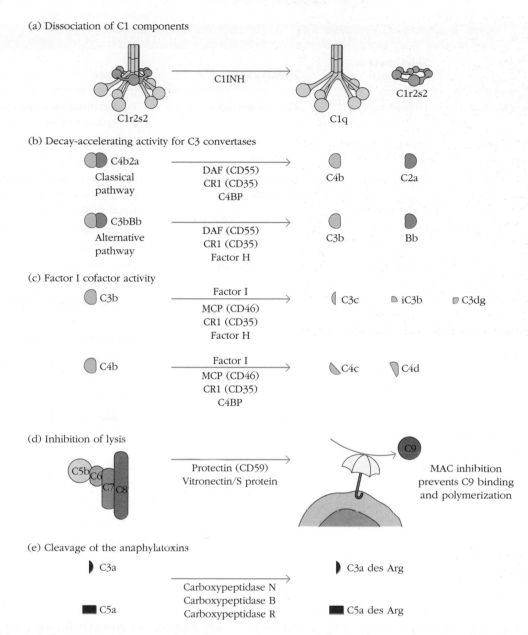

(a) Dissociation of C1 components

C1r2s2 — C1INH → C1q + C1r2s2

(b) Decay-accelerating activity for C3 convertases

C4b2a Classical pathway — DAF (CD55) CR1 (CD35) C4BP → C4b + C2a

C3bBb Alternative pathway — DAF (CD55) CR1 (CD35) Factor H → C3b + Bb

(c) Factor I cofactor activity

C3b — Factor I / MCP (CD46) / CR1 (CD35) / Factor H → C3c + iC3b + C3dg

C4b — Factor I / MCP (CD46) / CR1 (CD35) / C4BP → C4c + C4d

(d) Inhibition of lysis

C5b C6 C7 C8 — Protectin (CD59) Vitronectin/S protein → C9 MAC inhibition prevents C9 binding and polymerization

(e) Cleavage of the anaphylatoxins

C3a — Carboxypeptidase N / Carboxypeptidase B / Carboxypeptidase R → C3a des Arg

C5a → C5a des Arg

FIGURE 6-16 The regulation of complement activity. The various stages at which complement activity is subject to regulation (see text for details). [Adapted from J. R. Dunkelberger and W.-C. Song, 2010. Cell Research **20**:34–50.]

The C1 Inhibitor, C1INH, Promotes Dissociation of C1 Components

C1INH, the C1 inhibitor, is a plasma protein that binds in the active site of serine proteases, effectively poisoning them. C1INH belongs to the class of proteins called *ser*ine *protease in*hibitors (serpins), and it acts by forming a complex with the C1 proteases, C1r2s2, causing them to dissociate from C1q and preventing further activation of C4 or C2 (see Figure 6-16a). C1INH inhibits both C3b and the serine protease MASP2. It is the only plasma protease capable of inhibiting the initiation of both the classical and lectin complement pathways. Its presence in plasma serves to control the time period during which they can remain active.

Decay Accelerating Factors Promote Decay of C3 Convertases

Since the reaction catalyzed by the C3 convertase enzymes is the major amplification step in complement activation, the generation and lifetimes of the C3 convertases C4b2a and C3bBb are subject to rigorous control. Several membrane-bound factors accelerate the decay of the C4b2a on the surface of host cells, including *d*ecay *a*ccelerating *fac*tor, or DAF (CD55), CR1, and C4BP (*C4 b*inding *p*rotein)

| TABLE 6-4 | Proteins involved in the regulation of complement activity |

Protein	Fluid phase or membrane	Pathway affected	Function
C1 inhibitor (C1INH)	Fluid phase	Classical and lectin	Induces dissociation and inhibition of $C1r_2s_2$ from C1q; serine protease inhibitor
Decay Accelerating Factor (DAF) CD55	Membrane bound	Classical, alternative, and lectin	Accelerates dissociation of C4b2a and C3bBb C3 convertases
CR1 (CD35)	Membrane bound	Classical, alternative, and lectin	Blocks formation of, or accelerates dissociation of, the C3 convertases C4b2a and C3bBb by binding C4b or C3b
			Cofactor for factor I in C3b and C4b degradation on host cell surface
C4BP	Soluble	Classical and lectin	Blocks formation of, or accelerates dissociation of, C4b2a C3 convertase
			Cofactor for factor I in C4b degradation
Factor H	Soluble	Alternative	Blocks formation of, or accelerates dissociation of, C3bBb C3 convertase
			Cofactor for factor I in C3b degradation
Factor I	Soluble	Classical, alternative, and lectin	Serine protease: cleaves C4b and C3b using cofactors shown in Figure 6-16
Membrane cofactor of proteolysis, MCP (CD46)	Membrane bound	Classical, alternative, and lectin	Cofactor for factor I in degradation of C3b and C4b
S protein or Vitronectin	Soluble	All pathways	Binds soluble C5b67 and prevents insertion into host cell membrane
Protectin (CD59)	Membrane bound	All pathways	Binds C5b678 on host cells, blocking binding of C9 and the formation of the MAC complex
Carboxypeptidases N, B, and R	Soluble	Anaphylatoxins produced by all pathways	Cleave and inactivate the anaphylatoxins C3a and C5a

(see Figure 6-16b). These decay accelerating factors cooperate to *accelerate the breakdown of the C4b2a* complex into its separate components. The enzymatically active C2a diffuses away, and the residual membrane-bound C4b is degraded by another regulatory protein, factor I (see Figure 6-16c).

In the alternative pathway, DAF and CR1 function in a similar fashion and are joined by factor H in separating the C3b component of the alternative pathway C3 convertase from its partner, Bb. Again, Bb diffuses away, and residual C3b is degraded (see Figure 6-16b).

Whereas DAF and CR1 are membrane-bound components and their expression is therefore restricted to host cells, factor H and C4BP are soluble regulatory complement components. Host-specific function of factor H is ensured by its binding to polyanions such as sialic acid and heparin, essential components of eukaryotic, but not prokaryotic, cell surfaces. Similarly, C4BP is preferentially bound by host cell membrane proteoglycans such as heparan sulfate. In this

way, host cells are protected from deposition of complement components, whereas it is permitted on the membranes of microbial invaders that lack DAF and CR1 expression and fail to bind factor H or C4BP.

Factor I Degrades C3b and C4b

Factor H, C4BP, and CR1 also figure in a second and potentially more critical pathway of complement regulation: that catalyzed by factor I. Factor I is a soluble, constitutively (always) active serine protease that can cleave membrane-associated C3b and C4b into inactive fragments (Figure 6-16c).

However, if factor I is indeed soluble, constitutively active, and designed to destroy C3b and C4b, one might wonder how the complement cascades described above can ever succeed in destroying invading microbes? The answer is that factor I requires the presence of cofactors in order to function and that two of these cofactors, *Membrane Cofactor of Proteolysis* (MCP, or CD46) and CR1, are found on the surface of host cells, but not on the surfaces of the microbial cells (see Figure 6-16c). The other two cofactors, factor H and C4BP, described above, bind host cell-surface peptidoglycans. Hence, cleavage of membrane-bound C3b on host cells is conducted by factor I in collaboration with the membrane-bound host cell proteins, CD46 and CR1, and the soluble cofactor H. Similarly, cleavage of membrane-bound C4b is effected again by factor I, this time in collaboration with membrane-bound CD46 and CR1 and soluble cofactor C4BP. Since these membrane-bound, or membrane-binding cofactors are not found on microbial cells, C3b and C4b are thus allowed to remain on microbial cells and effect their specific functions.

CD46 has recently been implicated as a factor in the control of apoptosis of dying T cells. When a T cell commits to apoptosis, it expresses DNA on its cell membrane that binds circulating C1q, as described above. It then begins to shed CD46 from the cell surface. Only after CD46 is lost can progression of the classical pathway occur, with increased C3b binding and phagocytosis of the apoptotic T cells. As the cell continues to deteriorate, it may also begin to release nucleotides and other molecules into the tissue fluids that serve as chemotactic signals for phagocytes.

Protectin Inhibits the MAC Attack

In the case of a particularly robust antibody response, or of an inflammatory response accompanied by extensive complement activation, inappropriate assembly of MAC complexes on healthy host cells can potentially occur, and mechanisms have evolved to prevent the resulting inadvertent host cell destruction. A host cell surface protein, **Protectin** (CD59), binds any C5b678 complexes that may be deposited on host cells and prevents their insertion into the host cell membrane (see Figure 6-16d). Protectin also blocks further C9 addition to developing MACs. In addition, the soluble *complement S protein*, otherwise known as *Vitronectin*, binds any fluid phase C5b67 complexes released from microbial cells, preventing their insertion into host cell membranes.

Carboxypeptidases Can Inactivate the Anaphylatoxins C3a and C5a

Anaphylatoxin activity is regulated by cleavage of the C-terminal arginine residues from both C3a and C5a by serum carboxypeptidases, resulting in rapid inactivation of the anaphylatoxin activity (Figure 6-16e). Carboxypeptidases are a general class of enzymes that remove amino acids from the carboxyl termini of proteins; the specific enzymes that mediate the control of anaphylatoxin concentrations are carboxypeptidases N, B, and R. These enzymes remove arginine residues from the carboxyl termini of C3a and C5a to form the so-called *des-Arg* ("without Arginine"), inactive forms of the molecules. In addition, as mentioned above, binding of C5a by C5L2 also serves to modulate the inflammatory activity of C5a.

Complement Deficiencies

Genetic deficiencies have been described for each of the complement components. Homozygous deficiencies in any of the early components of the classical pathway (C1q, C1r, C1s, C4, and C2) result in similar symptoms, notably a marked increase in immune-complex diseases such as SLE, glomerulonephritis, and vasculitis. The effects of these deficiencies highlight the importance of the early complement reactions in generating C3b and C3b's role in the solubilization and clearance of immune complexes. In addition, as described above, C1q has been shown to bind apoptotic blebs. In the absence of C1q binding, cells bearing these apoptotic blebs, or the blebs themselves, can act as autoantigens and lead to the development of autoimmune diseases such as SLE. Individuals with deficiencies in the early complement components may also suffer from recurrent infections by pyogenic (pus-forming) bacteria such as streptococci and staphylococci. These organisms are Gram-positive and therefore are normally resistant to the lytic effects of the MAC. Nevertheless, the early complement components can help to mitigate such infections by mediating a localized inflammatory response and opsonizing the bacteria.

A deficiency in MBL, the first component of the lectin pathway, has been shown to be relatively common, and results in serious pyrogenic (fever-inducing) infections in babies and children. Children with MBL deficiency suffer from respiratory tract infections. MBL deficiency is also found with a frequency two to three times higher in SLE patients than in normal subjects, and certain mutant forms of MBP are found to be prevalent in chronic carriers of hepatitis B. Deficiencies in factor D and properdin—early components of the alternative pathway—appear to be associated with *Neisseria* infections but not with immune-complex disease.

People with C3 deficiencies display the most severe clinical manifestations of any of the complement deficiency patients, reflecting the central role of C3 in opsonization and in the formation of the MAC. The first person identified with a C3 deficiency was a child suffering from frequent, severe bacterial infections leading to meningitis, bronchitis, and pneumonia, who was initially diagnosed with agammaglobulinemia. After tests revealed normal immunoglobulin levels, a deficiency in C3 was discovered. This case highlighted the critical function of the complement system in converting a humoral antibody response into an effective defense mechanism. The majority of people with C3 deficiency have recurrent bacterial infections and may also present with immune-complex diseases.

Levels of C4 (one of the proteins operative very early in the classical pathway) vary considerably in the population. Specifically, the genes encoding C4 are located in the major histocompatibility locus (see Chapter 8), and the number of C4 genes may vary among individuals from two to six. Low gene copy numbers are associated with lower levels of C4 in plasma and with a correspondingly higher incidence of SLE, for the reasons described above. Patients with complete deficiencies of the components of the classical pathway such as C4 suffer more frequent infections with bacteria such as *S. pneumoniae*, *Haemophilus influenza*, and *N. meningitidis*. However, even patients with low copy numbers of C4 appear to be relatively well protected against such infections. Interestingly, C4 exists in two isoforms: C4A and C4B. C4B is more effective in binding to the surfaces of the three bacterial species mentioned above, and homozygous C4B deficiency has been shown to result in a slightly higher incidence of infection than is experienced by individuals with C4A deficiency, or those who are heterozygous for the two isoforms.

Individuals with deficiencies in members of the terminal complement cascade are more likely than members of the general population to suffer from meningococcal infections, indicating that cytolysis by complement components C5-C9 is of particular relevance to the control of *N. meningitidis*. This has resulted in the release of public health guidelines that highlight the need for vaccinations against *N. meningitidis* for individuals deficient in the terminal complement components.

Deficiencies of complement regulatory proteins have also been reported. As described above, C1INH, the C1 inhibitor, regulates activation of the classical pathway by preventing excessive C4 and C2 activation by C1. However, as a serine protease inhibitor, it also controls two serine proteases in the blood clotting system. Therefore, patients with C1INH deficiency suffer from a complex disorder that includes excessive production of vasoactive mediators (molecules that control blood vessel diameter and integrity), which in turn leads to tissue swelling and extracellular fluid accumulation. The resultant clinical condition is referred to as hereditary angioedema. It presents clinically as localized tissue edema that often follows trauma, but sometimes occurs with no known cause. The edema can be in subcutaneous tissues; within the bowel, where it causes abdominal pain; or in the upper respiratory tract, where it can result in fatal obstruction of the airway. C1INH deficiency is an autosomal dominant condition with a frequency of 1 in 1000 in the human population.

Studies in humans and experimental animals with homozygous deficiencies in complement components have provided important information regarding the roles of individual complement components in immunity. These initial observations have been significantly enhanced by studies using knockout mice, genetically engineered to lack expression of specific complement components. Investigations of in vivo complement activity in these animals has allowed dissection of the complex system of complement proteins and the assignment of precise biologic roles to each.

Microbial Complement Evasion Strategies

The importance of complement in host defense is clearly illustrated by the number and variety of strategies that have evolved in microbes, enabling them to evade complement attack (see Table 6-5). Gram-positive bacteria have developed thick cell walls and capsules that enable them to shrug off the insertion of MAC complexes, while others escape into intracellular vacuoles to escape immune detection. However, these two general strategies are energy intensive for the microbe, and many microbes have adopted other tactics to escape complement-mediated destruction.

In Advances Box 6-3, we describe the multiple ways in which *Staphylococcus aureus*, a major human pathogen, protects itself against complement-mediated destruction. In this

TABLE 6-5 Some microbial complement evasion strategies	
Complement evasion strategy	**Example**
Interference with antibody-complement interaction	Antibody depletion by Staphylococcal protein A
	Removal of IgG by Staphylokinase
Binding and inactivation of complement proteins	*S. aureus* protein SCIN binds and inactivates the C3bBb C3 convertase
	Parasite protein C2 receptor trispanning protein disrupts the binding between C2 and C4
Protease-mediated destruction of complement component	Elastase and alkaline phosphatase from *Pseudomonas* degrade C1q and C3/C3b
	ScpA and ScpB from *Streptococcus* degrade C5a
Microbial mimicry of complement regulatory components	*Streptococcus pyogenes* M proteins bind C4BP and factor H to the cell surface, accelerating the decay of C3 convertases bound to the bacterial surface
	Variola and *Vaccinia* viruses express proteins that act as cofactors for factor I in degrading C3b and C4b

section, we address complement microbial evasion at a more conceptual level, categorizing the approaches that microbes have evolved to elude this effector arm of the immune response.

Some Pathogens Interfere with the First Step of Immunoglobulin-Mediated Complement Activation

Many classes of viruses act at the beginning of the classical complement pathway by synthesizing proteins and glycoproteins that specifically bind the Fc regions of antibodies, thus preventing complement binding and the generation of the classical pathway reactions. Some viruses also effect enhanced clearance of antibody-antigen complexes from the surfaces of virus-infected cells and/or manufacture proteins that induce rapid internalization of viral protein-antibody complexes.

Microbial Proteins Bind and Inactivate Complement Proteins

Since highly specific protein-protein interactions between complement components are central to the functioning of the cascade, it is logical that microbes have evolved mechanisms that interfere with some of these binding reactions. Advances Box 6-3 describes those used by *S. aureus*, but the strategy of inhibiting the interactions between complement proteins is not restricted to bacteria. The production of molecules that inhibit the interactions between complement components has also been detected in certain human parasites. Specifically, a protein generated by some species of both *Schistosoma* and *Trypanosoma*, the complement C2 receptor trispanning protein, disrupts the interaction between C2a and C4b and thus prevents the generation of the classical pathway C3 convertase.

Microbial Proteases Destroy Complement Proteins

Some microbes produce proteases that destroy complement components. This strategy is utilized primarily by bacteria, and numerous bacterial proteases exist that are capable of digesting a variety of complement components. For example, elastase and alkaline protease proteins from *Pseudomonas aeruginosa* target the degradation of C1q and C3/C3b, and two proteases derived from Streptococcal bacteria, scpA and scpB, specifically attack the anaphylatoxin C5a. Streptococcal pyrogenic exotoxin B was found to degrade the complement regulator properdin, with resultant destabilization of the alternative pathway C3 convertase on the bacterial surface.

Fungi can also inactivate complement proteins. The opportunistic human pathogen *Aspergillus fumigatus* secretes an alkaline protease, Alp1, that is capable of cleaving C3, C4, C5, and C1q, as well as IgG. Since this pathogen tends to attack patients who are already immunocompromised, its capacity to further damage the immune system is particularly troubling to the clinician.

Some Microbes Mimic or Bind Complement Regulatory Proteins

Several microbial species have exploited the presence of cell-surface or soluble regulators of complement activity for their own purposes, either by mimicking the effects of these regulators, or by acquiring them directly. Indeed, recent work suggests that sequestration of host-derived regulators of complement activity may be one of the most widely utilized microbial mechanisms for complement evasion.

Many microbes have developed the ability to bind one or another of the fluid phase inhibitors of complement, C4BP or factor H. For example, *Streptococcus pyogenes*, an important human pathogen, expresses a family of proteins called "M proteins" capable of binding to C4BP and factor H. Expression of these regulatory proteins on the bacterial surface results in the inhibition of the later steps of complement fixation.

More surprisingly, some microbes appear to acquire membrane-bound regulators from host cells. *Helicobacter pylori*, obtained from patients suffering from gastric ulcers, was found to stain positive for Protectin, a potent host inhibitor of MAC complex formation. Protectin is normally attached to the host membrane by a glycosylphosphatidyl inositol anchor, and this anchor must be transferable in some way from the host to the bacterial cell membrane.

Many viruses have evolved the capacity to produce proteins that closely mimic the structure and function of complement regulatory proteins. For example, the *Variola* (smallpox) and *Vaccinia* (cowpox) viruses express complement inhibitory proteins that bind C3b and C4b and serve as cofactors for factor I, thus preventing complement activation at the viral membrane. In addition to generating complement regulatory proteins, it should be noted that some viruses actively induce regulatory complement components within the cells that harbor them. Other viruses camouflage themselves during budding from the host cell by hiding within regions of the host membrane on which regulatory complement components are expressed. And finally, some viruses mimic eukaryotic membranes by incorporating high levels of sialic acid into the viral membrane, thus facilitating the binding of the cofactors for factor I that normally bind only to host cell membranes. This therefore results in the inhibition of complement activation on the viral surface.

The Evolutionary Origins of the Complement System

Although the complement system was initially characterized by its ability to convert antibody binding into pathogen lysis, studies of complement evolution have identified the

Staphylococcus aureus Employs Diverse Methods to Evade Destruction by the Complement System

Staphylococcus aureus has developed an impressive variety of mechanisms that inhibit both the classical and the alternative pathways of complement activation.

S. aureus is an encapsulated, Gram-positive bacterial strain. The capsular polysaccharide itself provides the bacterium with some mechanical inhibition of opsonization (1 in Figure 1). Although complement factors can assemble on the cell-wall surface underneath the capsule, most are then inaccessible to the complement receptors on phagocyte surfaces. However, if the C3 convertases C4b2a and C3bBb do succeed in assembling at the bacterial surface, they are then bound by a small, 9.8 kDa protein called *Staphylococcus* Complement *IN*hibitor (SCIN). Blockage of further convertase activity prevents amplification of

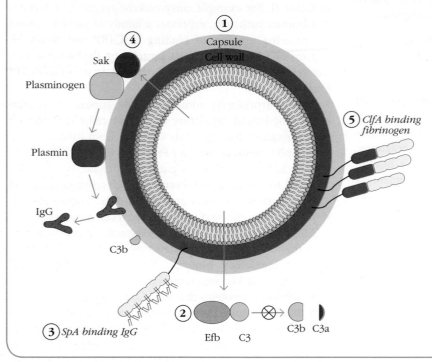

FIGURE 1

Mechanisms by which *S. aureus* avoids opsonophagocytosis. (1) The capsular polysaccharides denies access of neutrophils to opsonized bacteria. (2) The extracellular fibrinogen binding protein (Efb) binds C3, preventing it from reaching the cell surface and inhibiting further activation of the complement cascade. (3) Protein A (SpA) binds IgG in a conformation that does not permit Fc receptor binding. (4) Staphylokinase (Sak), secreted by the bacterium, activates plasminogen, a protease capable of cleaving and inactivating IgG and C3b. (5) Clumping factor A binds factor I and localizes it to the microbial surface, where it cleaves and inactivates any C3b that binds there. [*Adapted from Foster, T. J., 2005. Immune evasion by Staphylococci. Nature Reviews Microbiology **3**: 948–958, Figure 3.*]

components of the lytic MAC as relatively late additions to the animal genome. Long before the emergence of the adaptive immune system, complement components provided invertebrate organisms with an important advance on the peptide-based humoral immunity previously available to them (see Chapter 5). Activation of complement cascades in invertebrates and early vertebrates most likely culminated in opsonization and phagocytosis by primitive hemocytes. It appears probable that, among the non-MAC complement components, the proteins of the alternative pathway were the first to appear, followed by those of the lectin recognition systems. A fully fledged MAC emerged only around the same time as the appearance of the adaptive immune system (see Figure 6-17 and Table 6-6).

Genomic analysis has classified complement components into five gene families, each of which possesses unique domain structures that have enabled investigators to trace their phylogenetic origins. The first of these gene families encodes C1q, mannose-binding lectin (MBL), and ficolins. Genes for prototypical C1q molecules have been identified in species as primitive as lampreys (the jawless fishlike vertebrate; Phylum Chordata, Subphylum Vertebrata), sea urchins (Phylum Echinodermata, Subphylum Echinozoa), and ascidians (Phylum Chordata, Subphylum Urochordata). Analysis of C1q gene clusters in different species has demonstrated that C1q genes appeared prior to the generation of immunoglobulin genes, and that at least some of these C1q proteins bind to specific carbohydrates, implying that they

BOX 6-3

the complement cascade. A second protein secreted by *S. aureus*, the *E*xtracellular *F*ibrinogen-*B*inding protein (Efb), binds C3, thus preventing its deposition on the bacterial cell surface (2). But if the immune system succeeds in depositing C3, in spite of all these evasive actions by the bacterium, *S. aureus* has evolved the capacity to manufacture other proteins to destroy it. Recently, a metalloproteinase enzyme, aureolysin, has been shown to be secreted by *S. aureus*. Aureolysin cleaves C3 in such a manner that it is then further degraded and inactivated by host regulatory proteins.

Other mechanisms interfere with the ability of antibodies to initiate the complement cascade. *S*taphylococcal protein *A* (SpA) is anchored into the carbohydrate cell wall, and possesses four or five extracellular domains, each of which is capable of binding to the Fc portion of a molecule of IgG, thus effectively blocking effector functions of the antibody molecules, including complement activation (3). In addition, *S. aureus* secretes an enzyme called *S*taphylokinase (SAK) that attaches to the bacterial surface and binds to plasminogen, the host cell plasma zymogen, activating it upon binding (4). The activated plasminogen has a serine protease activity that cleaves and releases any surface-bound IgG and C3b.

The *S. aureus* protein Sbi, which is found both in association with the bacterial cell and in solution, contains two immunoglobulin-binding domains and two other domains that bind to the C3 complement component. The cell-associated Sbi has been demonstrated to be capable of preventing Fc-mediated protective functions such as opsonization and complement activation, by blocking complement, or Fc receptor binding, to the IgG constant regions. In addition, the portion of the Sbi protein that is released into the extracellular fluid has also been shown to be protective for the bacterium. Although the precise nature of the protective mechanism is still being investigated, it is thought to result from the depletion of complement components in the vicinity of the bacterium.

S. aureus has also evolved mechanisms that take advantage of host complement control strategies. A bacterial cell-surface protein, *C*lumping factor *A* (ClfA) is able to bind factor I, thus localizing it to the bacterial surface, where it cleaves and inactivates any C3b that has bound there (5). Another complement inhibitory protein secreted by *S. aureus* acts in a slightly different way. The *C*hemotaxis *I*nhibitory *P*rotein of *S. aureus* (CHIPs) binds two chemotactic receptors on neutrophils, the C5a receptor, and

the formylated peptide receptor, thus effectively eliminating the capacity of neutrophils to respond to C5a-mediated chemotactic signals. Other Staphylococcal proteins are active in the inhibition of neutrophil binding to the membrane surfaces of the endothelial cells lining blood capillaries, and thus retard the extravasation of neutrophils in the vicinity of a local infection.

The variety and effectiveness of these inhibitory mechanisms that have evolved in just one strain of bacteria emphasize for us the importance of the complement system in the normal control of bacterial infections, and the complexity that researchers and clinicians face when attempting to develop the next generation of antibacterial therapies.

Foster, T. J. 2005. Immune evasion by staphylococci. *Nature Reviews Microbiology* **3**:948–958.

Laarman, A., F. Milder, J. van Strijp, and S. Rooijakkers. 2010. Complement inhibition by gram-positive pathogens: Molecular mechanisms and therapeutic implications. *Journal of Molecular Medicine (Berlin)* **88**:115–120.

Serruto, D., R. Rappuoli, M. Scarselli, P. Gros, and J. A. van Strijp. 2010. Molecular mechanisms of complement evasion: Learning from staphylococci and meningococci. *Nature Reviews Microbiology* **8**:393–399.

are potentially able to discriminate between host and pathogen. Thus, C1q may have expressed the capacity to recognize foreign molecules at a very early point in time, independent of antibody binding and in a manner similar to that of MBL. Mannose-binding lectins have been identified in lampreys, thus placing the origin of the MBLs at least as far back as the early vertebrates. Indeed, functional lectin pathways have been characterized in ascidians. However, although genes encoding MBL-like proteins have been characterized in ascidian genomes, the nature of their relationship to vertebrate MBL proteins is still unclear.

The next three gene families to be considered all encode proteins that are implicated in the cleavage of the C3, C4, and C5 components of complement. The first of these families is composed of *factor B (Bf) and the serine protease C2*; the second family comprises the serine proteases *MASP-1, MASP-2, MASP-3, C1r, and C1s*; and the third family contains the C3 family members *C3, C4, and C5*, which are not themselves proteases, but which share a common internal thioester and are subject to protease cleavage and activation.

Studies of a number of invertebrate species have indicated that the genome of each species contains only a single copy representative of each of these three gene families. In contrast, one or more *gene duplication events* have occurred within each gene family in most vertebrates, suggesting that the gene duplications found in vertebrates probably occurred in the early stages of jawed vertebrate evolution. Genes for each of the Bf, C3, and MASP families have been identified from all

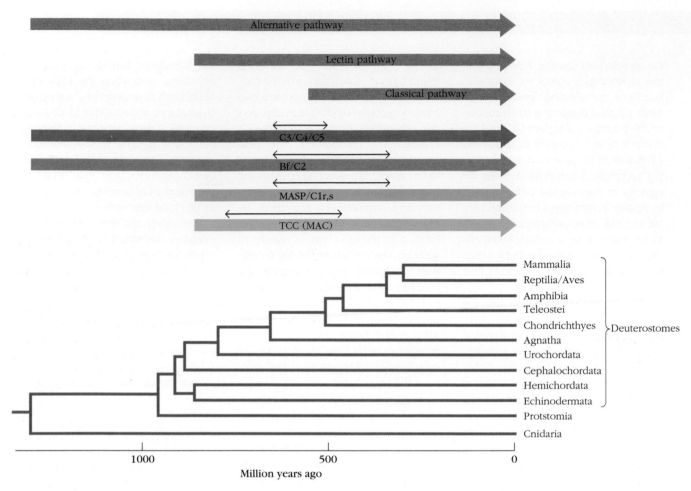

FIGURE 6-17 Evolution of complement components. The appearance of each of the three major complement pathways is illustrated by gray arrows, and the timing of the gene duplications that gave rise to the classical complement pathway is indicated by double-headed arrows. [Adapted from Nonaka, M., and A. Kimura, 2006, Genomic view of the evolution of the complement system, Immunogenetics 58:701.]

TABLE 6-6	Complement system pathways in the major groups of deuterostome animals				
Animal group	**Alternate pathways**	**Classical pathway**	**Lectin pathway**	**TCC complex**	**Antibodies present?**
Mammals	+	+	+	+	+
Birds	+	+	+	+	+
Reptiles	+	+	+	+	+
Amphibians	+	+	+	+	+
Teleost fish	+	+	+	+	+
Cartilaginous fish	+	+	+	+	+
Agnathan fish	+	−	+	?	−
Tunicates	+	−	+	?	−
Echinoderms	+	−	?	?	−

After J. O. Sunyer et al., 2003. Evolution of complement as an effector system in innate and adaptive immunity. *Immunologic Research* **27**:549–564, Figure 2.

invertebrate deuterostomes that have so far been analyzed, with the single exception of the MASP family gene, which is missing from the sea urchin (Phylum Echinodermata) genome.

The situation in protostomes is somewhat more complicated. Whereas members of the C3 and Bf families have been identified in some very early protostomes—for example, the horseshoe crab and the carpet-shell clam (Phyla: Arthropoda and Mollusca respectively)—analysis of whole genome sequences of insects such as *Drosophila melanogaster* (Arthropoda) or the nematode *Caenorhabditis elegans* (Nematoda) have failed to locate any proto-complement protein sequences. This contrasts starkly with the presence of all three family genes in sea anemones and corals (Phylum, Cnidaria), and suggests that the ability to encode proteins of the complement system has been lost from the genome of the common, model-system protostomes *D. melanogaster* and *C. elegans*. The absence of genes encoding any of the later, cytolytic components of the complement cascade in protostomes and cnidarians supports the hypothesis that the proto-complement systems of cnidarians and protostomes function by opsonization.

The proteins encoded by the fifth gene family, C6, C7, C8α, C8βγ, and C9, together make up the terminal complement complex (TCC). They share a unique domain structure that enables molecular phylogenetic analysis. In particular, the mammalian TCCs, C6, C7, C8α, C8βγ, and C9 all share the *MAC/perforin* (MACPF) domain, in addition to other domains common among two or more members of this protein group. Mammals, birds, and amphibians share all TCC genes (with the exception of the bird C9 gene, which is missing from the chicken genome sequence). Although the full complement of MAC proteins has not yet been documented in cartilaginous fishes such as the sharks (the earliest known organisms to possess an adaptive immune system), a gene encoding a C8α subunit orthologous to mammalian C8α has been cloned and characterized. Furthermore, the serum of sharks has been known for decades to express hemolytic activity, and microscopic analysis of the pores formed on target cell surfaces reveals a transmembrane pore structure indistinguishable from that induced by a MAC attack. It thus seems likely that the functionality of the MAC emerged not long after the adaptive immune system.

Genomic analysis has traced the origin of the TCC genes back to before the divergence of the urochordates, cephalochordates, and vertebrates (see Figure 6-17) as ascidians (urochordates) and amphioxus (cephalochordates) possess primitive copies of TCC genes. However, the ascidian and amphioxus TCC proteins could not have functioned in a manner similar to that of later vertebrate species because they lack the domain responsible for interacting with the C5 protein. Intriguingly, toxins of the venomous sea anemone, which express a very high hemolytic potential, are closest in structure to the TCCs of the complement pathway.

In summary, complement evolution is based on the diversification and successive duplications of members of five gene families that evolved first in response to the need for microbial recognition and opsonization and then responded to the appearance of the adaptive immune system by additional gene duplication and functional adaptation to form the complement system as we know it today.

SUMMARY

- The complement system comprises a group of serum proteins, many of which circulate in inactive forms that must first be cleaved or undergo conformational changes prior to activation.

- Complement proteins include initiator molecules, enzymatic mediators, membrane-binding components or opsonins, inflammatory mediators, membrane attack proteins, complement receptor proteins, and regulatory components.

- Complement activation occurs by three pathways—classical, lectin, and alternative—which converge in a common sequence of events leading to membrane lysis.

- The classical pathway is initiated by antibodies of the IgM or IgG classes binding to a multivalent antigen. This allows binding of the first component of complement, C1q, and begins the process of complement deposition.

- The lectin pathway is initiated by binding of lectins such as mannose-binding lectin or members of the ficolin family to microbial surface carbohydrates.

- The alternative pathway is initiated when the third component of complement, C3, undergoes nonspecific breakdown. C3b is formed and adheres to a cell surface. Inadvertent initiation is prevented by the presence of control proteins on host cell membranes.

- The end result of the initiating sequence of all three pathways is the generation of enzymes that cleave C3 into C3a and C3b, and C5 into C5a and C5b.

- C3b opsonizes microbial cells and immune complexes, rendering them suitable for phagocytosis.

- Activation of the terminal components of the complement cascade C5b, C6, C7, C8 and C9 results in the deposition of a membrane attack complex (MAC) onto the microbial cell membrane. This complex introduces large pores in the membrane, preventing it from maintaining osmotic integrity and resulting in the death of the cell.

- Binding of the complement component C1q to apoptotic cells, apoptotic bodies, and immune complexes results in their opsonization with C3b and subsequent phagocytosis.

- Patients deficient in the early complement components do not clear apoptotic cells efficiently and suffer disproportionally from autoimmune disease such as SLE.

- In addition to acting during host defense against infection, complement proteins also bind to receptors on the surfaces of antigen-presenting cells, T cells, and B cells, inducing interleukin production and aiding in their activation.

- A system of regulatory complement proteins ensures that inadvertent complement activation on the surface of host cells is prevented and controlled, by deactivating complement components on the surface of host cells, and ensuring that regulatory proteins are bound specifically to host, but not to microbial cell membranes.

- Patients suffering from complement deficiencies often present with immune complex disorders and suffer disproportionally from infections by encapsulated bacteria such as *Neisseria meningitidis*. Animal models exist for most complement deficiencies.

- Underscoring the importance of complement in the immune system, a broad variety of complement evasion strategies has evolved in viruses, bacteria, fungi, and parasites, including mimicking regulatory proteins, interfering with the interactions between antibodies and complement components, or between proteins of the complement pathways, or by destruction of the complement components.

- The genes encoding complement components belong to five families. Genes of the alternative pathway components appear first in evolution, and those encoding the terminal complement components appear last. Thus, prior to the emergence of adaptive immunity, complement served its protective functions by mediating phagocytosis.

REFERENCES

Alexander, J. J., A. J. Anderson, S. R. Barnum, B. Stevens, B., and A. J. Tenner. 2008. The complement cascade: Yin-Yang in neuroinflammation—neuro-protection and -degeneration. *Journal of Neurochemistry* 107:1169–1187.

Al-Sharif, W. Z., J. O. Sunyer, J. D. Lambris, and L. C. Smith. 1998. Sea urchin coelomocytes specifically express a homologue of the complement component C3. *Journal of Immunology* 160:2983–2997.

Azumi, K., et al. 2003. Genomic analysis of immunity in a Urochordate and the emergence of the vertebrate immune system: "Waiting for Godot." *Immunogenetics* 55:570–581.

Behnsen, J., et al. 2010. Secreted *Aspergillus fumigatus* protease Alp1 degrades human complement proteins C3, C4, and C5. *Infection and Immunity* 78:3585–3594.

Blom, A. M., T. Hallstrom, and K. Riesbeck, K. 2009. Complement evasion strategies of pathogens—acquisition of inhibitors and beyond. *Molecular Immunology* 46:2808–2817.

Botto, M., et al. 1998. Homozygous C1q deficiency causes glomerulonephritis associated with multiple apoptotic bodies. *Nature Genetics* 19:56–59.

Dunkelberger, J. R., and W. C. Song. 2010. Complement and its role in innate and adaptive immune responses. *Cell Research* 20:34–50.

Elward, K., et al. 2005. CD46 plays a key role in tailoring innate immune recognition of apoptotic and necrotic cells. *Journal of Biological Chemistry* 280:36342–36354.

Fearon, D. T., and M. C. Carroll. 2000. Regulation of B lymphocyte responses to foreign and self-antigens by the CD19/CD21 complex. *Annual Review of Immunology* 18:393–422.

Gros, P., F. J. Milder, and B. J. Janssen. 2008. Complement driven by conformational changes. *Nature Reviews Immunology* 8:48–58.

Jensen, J. A., E. Festa, D. S. Smith, and M. Cayer. 1981. The complement system of the nurse shark: hemolytic and comparative characteristics. *Science* 214:566–569.

Kang, Y. S., et al. 2006. A dominant complement fixation pathway for pneumococcal polysaccharides initiated by SIGN-R1 interacting with C1q. Cell 125:47–58.

Kemper, C., and J. P. Atkinson. 2007. T-cell regulation: With complements from innate immunity. *Nature Reviews Immunology* 7:9–18.

Kemper, C., and D. E. Hourcade. 2008. Properdin: New roles in pattern recognition and target clearance. *Molecular Immunology* 45:4048–4056.

Lambris, J. D., D. Ricklin, and B. V. Geisbrecht. 2008. Complement evasion by human pathogens. *Nature Reviews Microbiology* 6:132–142.

Litvack, M. L. and N. Palaniyar. 2010. Soluble innate immune pattern-recognition proteins for clearing dying cells and cellular components: Implications on exacerbating or resolving inflammation. *Innate Immunity* 16:191–200.

Longhi, M. P., C. L. Harris, B. P. Morgan, and A. Gallimore, A. 2006. Holding T cells in check: A new role for complement regulators? *Trends in Immunology* 27:102–108.

Markiewski, M. M., B. Nilsson, K. N. Ekdahl, T. E. Mollnes, and J. D. Lambris. 2007. Complement and coagulation: Strangers or partners in crime? *Trends in Immunology* 28:184–192.

Miller, D. J., et al. 2007. The innate immune repertoire in Cnidaria: Ancestral complexity and stochastic gene loss. *Genome Biology* 8:R59.

Nonaka, M., and A. Kimura. 2006. Genomic view of the evolution of the complement system. *Immunogenetics* 58:701–713.

Price, J. D., et al. 2005. Induction of a regulatory phenotype in human CD4+ T cells by streptococcal M protein. *Journal of Immunology* 175:677–684.

Ricklin, D., and J. D. Lambris. 2008. Compstatin: A complement inhibitor on its way to clinical application. *Advances in Experimental and Medical Biology* 632:273–292.

Roozendaal, R., and M. C. Carroll. 2007. Complement receptors CD21 and CD35 in humoral immunity. *Immunological Reviews* **219**:157–166.

Rosen, A. M., and B. Stevens. 2010. The role of the classical complement cascade in synapse loss during development and glaucoma. *Advances in Experimental and Medical Biology* **703**:75–93.

Stevens, B., et al. 2007. The classical complement cascade mediates CNS synapse elimination. *Cell* **131**:1164–1178.

Sunyer, J. O., et al. 2003. Evolution of complement as an effector system in innate and adaptive immunity. *Immunologic Research* **27**:549–564.

Suzuki, M. M., N. Satoh, and M. Nonaka. 2002. C6-like and C3-like molecules from the cephalochordate, amphioxus, suggest a cytolytic complement system in invertebrates. *Journal of Molecular Evolution* **54**:671–679.

Ward, P. A. 2009. Functions of C5a receptors. *Journal of Molecular Medicine* **87**:375–378.

Zhu, Y., S. Thangamani, B. Ho, and J. L. Ding. 2005. The ancient origin of the complement system. *European Molecular Biology Organization Journal* **24**:382–394.

Zipfel, P. F., and C. Skerka. 2009. Complement regulators and inhibitory proteins. *Nature Reviews Immunology* **9**:729–740.

Useful Web Sites

www.complement-genetics.uni-mainz.de The Complement Genetics Homepage from the University of Mainz gives chromosomal locations and information on genetic deficiencies of complement proteins.

www.cehs.siu.edu/fix/medmicro/cfix.htm A clever graphic representation of the basic assay for complement activity using red blood cell lysis, from D. Fix at the University of Southern Illinois, Carbondale.

www.youtube.com/watch?v=y2ep6j5kHUc

www.youtube.com/watch?v=AljaiJV4m2g There are a number of animations of the complement cascade available on the Internet. These links are to clear animations.

STUDY QUESTIONS

1. Indicate whether each of the following statements is true or false. If you think a statement is false, explain why.

   a. A single molecule of bound IgM can activate the C1q component of the classical complement pathway.

   b. The enzymes that cleave C3 and C4 are referred to as convertases.

   c. C3a and C3b are fragments of C3 that are generated by proteolytic cleavage mediated by two different enzyme complexes.

   d. Nucleated cells tend to be more resistant to complement-mediated lysis than red blood cells.

   e. Enveloped viruses cannot be lysed by complement because their outer envelopes are resistant to pore formation by the membrane attack complex (MAC).

   f. MBL has a function in the lectin pathway analogous to that of IgM in the classical pathway, and MASP-1 and MASP-2 take on functions analogous to C1 components.

2. Explain why serum IgM cannot activate complement prior to antigen binding.

3. Genetic deficiencies have been described in patients for all of the complement components except factor B. Particularly severe consequences result from a deficiency in C3. Describe the consequences of an absence of C3 for each of the following:

   a. Initiation of the classical and alternate pathways

   b. Clearance of immune complexes

   c. Phagocytosis of infectious bacteria

4. Describe three ways in which complement acts to protect the host during an infection.

5. Complement activation can occur via the classical, alternative, or lectin pathway.

   a. How do the three pathways differ in the substances that can initiate activation?

   b. Which parts of the overall activation sequence differ among the three pathways, and which parts are similar?

   c. How does the host ensure that inadvertent activation of the alternative pathway on its own healthy cells does not lead to autoimmune destruction?

6. Briefly explain the mechanism of action of the following complement regulatory proteins. Indicate which pathway(s) each protein regulates.

   a. C1 inhibitor (C1INH)

   b. C4b-binding protein (C4bBP)

   c. Decay-accelerating factor (DAF)

   d. Membrane cofactor of proteolysis protein, MCP (CD46)

   e. Protectin (CD59)

   f. Carboxypeptidase N

7. Explain why complement deficiencies in the early components of complement give rise to immune-complex-mediated disorders such as systemic lupus erythematosus.

8. You have prepared knockout mice with mutations in the genes that encode various complement components. Each knockout strain cannot express one of the complement components listed across the top of the table below. Predict the effect of each mutation on the steps in complement activation and on the complement effector functions indicated in the table using the following symbols: NE = no effect; D = process/function decreased but not abolished; A = process/function abolished.

	Complement component knocked out						
	C1q	**C4**	**C3**	**C5**	**C9**	**Factor B**	**MASP-2**
Formation of classical pathway C3 convertase							
Formation of alternative pathway C3 convertase							
Formation of classical pathway C5 convertase							
Formation of lectin pathway C3 convertase							
C3b-mediated opsonization							
Neutrophil chemotaxis and inflammation							
Cell lysis							

CLINICAL FOCUS QUESTION

As shown in the figure below, two flow cytometric histograms of red blood cells were obtained from a patient, stained with antibodies toward CD59 (Protectin) (part A) and CD55 (Decay Accelerating Factor or DAF) (part B). On the right of each histogram is a large population of cells expressing relatively high levels of each antigen (population I). Populations II express midrange levels of the two antigens, and populations III express levels of antigen below statistical detectability. (Detecting laser voltages on flow cytometers are normally set such that negative staining populations show levels of fluorescence below 10^1 on the lower logarithmic scale. Note also the creep of cells up the 10^0 axis on each plot.)

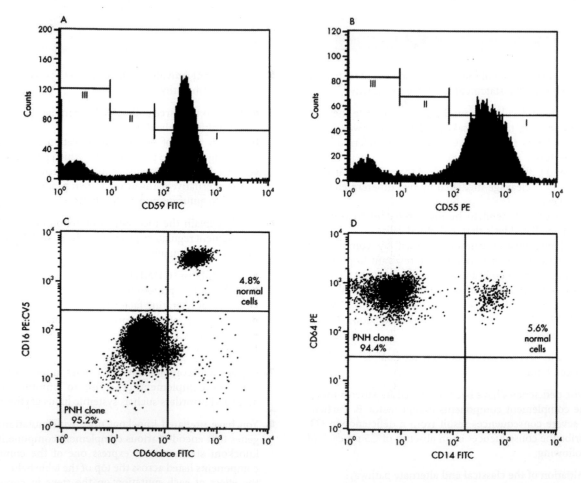

a. Can you offer an explanation as to why this patient may have red blood cells expressing low levels of these two particular antigens? From what disease do you think this patient is suffering?

b. Why do you think a single patient can generate red blood cells expressing three different levels of CD59 and CD55?

c. Now look at parts C and D of the same figure, which display the expression of two other markers on red blood cells from the same patient. In these flow cytometric profiles, the investigators have shown you quadrant markers to indicate whether the cells they are investigating are considered to be positive or negative for the proteins labeled on the axes. For example, the cell population high and right in part C is considered to express both CD66abce and CD16, whereas the cell population low and left is negative for both of these markers.

Using your answer for part (a) as a starting point, what can you deduce about the biochemistry of the membrane proteins CD16, CD66abce, CD46, and CD14?

ANALYZE THE DATA

a. In the figure below, a flow cytometric histogram describes the numbers of human Jurkat T cells expressing low and high levels of CD46 after treatment with an apoptosis-inducing reagent. On the figure, label the cell population that you think is undergoing apoptosis and explain your reasoning.

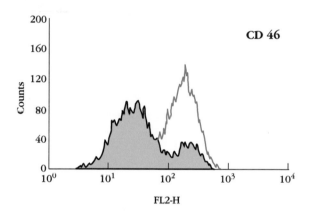

On the flow cytometric dot plot shown below, indicate the following and explain your labeling:
a. Where you would expect to find a healthy T-cell population?
b. Where you would expect to find a T-cell population undergoing apoptosis?

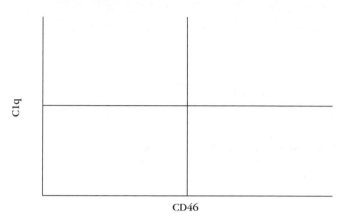

The Organization and Expression of Lymphocyte Receptor Genes

To protect its host, the immune system must recognize a vast array of rapidly evolving microorganisms. To accomplish this, it must generate a diverse and flexible repertoire of receptor molecules, while minimizing the expression of receptors that recognize self antigens. As described in Chapter 3, each B or T lymphocyte expresses a unique antigen-specific receptor. When these receptors bind to their corresponding antigens under the appropriate conditions (described in Chapters 11 and 12), T and B lymphocytes proliferate and differentiate into effector cells that eliminate the microbial threat (Chapter 13).

In Chapter 3, we described the biochemistry of the T and B lymphocyte receptors and the secreted antibodies formed by B lymphocytes following antigen stimulation. We also outlined the experiments which demonstrated that secreted antibodies are identical in antigen-binding specificity to the B-cell receptors of the secreting cell. In this chapter, we address the question of how an organism can encode and express receptors capable of recognizing a constantly evolving universe of microbial threats using a finite amount of genetic information.

The production of specific lymphocyte receptors employs a number of genetic mechanisms that are unique to the immune system. In 1987, Susumu Tonegawa won the Nobel Prize for Physiology or Medicine "for his discovery of the genetic principle for generation of antibody diversity," a discovery that challenged the fundamental concept that one gene encoded one polypeptide chain. Tonegawa and his colleagues showed that *the antibody light chain was encoded in the germ line by not one but three families of gene segments separated by kilobases of DNA* (Figure 7-1). (The germ-line DNA is the genetic information encoded in the sperm and egg, which can be passed on to future generations.) Their work demonstrated that two DNA segments, one from each family, are conjoined, only in B lymphocytes, to create the mature form of the light-chain *variable* region of the immunoglobulin (Ig) gene. A third segment encodes the *constant* region of the gene. Different combinations of

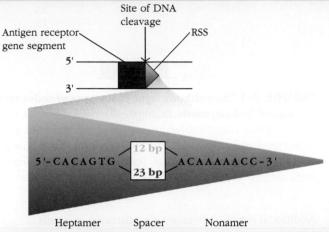

The Recombination Signal Sequence (RSS) serves as the site of DNA cleavage. The RSS is composed of conserved heptamer and nonamer sequences, separated by a spacer region of 12 or 23 bp, which is conserved in length, but not in sequence. Cleavage occurs at the junction of the heptamer and the variable region coding segment. *[Adapted from D. G. Schatz and Y. Ji, 2011, Recombination centres and the orchestration of V(D)J recombination, Nature Reviews. Immunology **11**:251–263.]*

- The Puzzle of Immunoglobulin Gene Structure
- Multigene Organization of Ig Genes
- The Mechanism of V(D)J Recombination
- B-Cell Receptor Expression
- T-Cell Receptor Genes and Expression

segments are used in each B cell, to create the diverse repertoire of light-chain receptor genes. Subsequent experiments have shown that all of the B- and T-cell receptor genes are assembled from multiple gene segments by similar rearrangements.

We describe below the unique genetic arrangements of T- and B-cell receptor gene segments, and the mechanisms by which they are rearranged and expressed. We will address here only those mechanisms that shape the receptor repertoires of mouse and human **naïve** B and T cells, which have not yet been exposed to antigen.

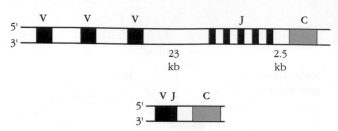

FIGURE 7-1 The antibody light-chain gene encodes three families of DNA segments. During B-cell development, one V segment and one J segment (which encode contiguous parts of the light-chain variable region) join together with the C (constant) region to form the gene for the antibody light chain. This gene rearrangement occurs in the DNA, prior to gene transcription into mRNA.

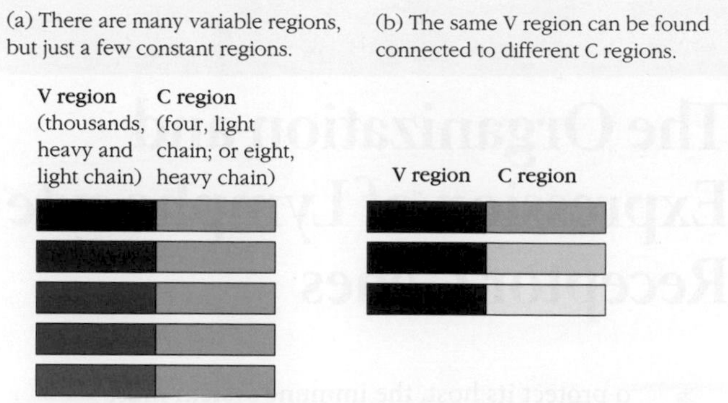

Additional layers of diversity are generated in B cells following antigen exposure and those will be addressed in Chapter 12. These powerful mechanisms include the generation of antibodies of different heavy-chain classes, each capable of mediating a discrete set of effector functions, as well as the creation of modified antigen-binding regions by somatic hypermutation, followed by antigen-driven selection. Both of these processes are triggered in human and mouse B cells only after antigen contact.

The Puzzle of Immunoglobulin Gene Structure

The immune system relies on a vast array of **B-cell receptors (BCRs)** that possess the ability to bind specifically to a correspondingly large number of potential pathogens. The first indication of the immense size of the antibody repertoire was provided by immunologists using synthetic molecules to stimulate antibody production. They discovered that antibodies can discriminate between small synthetic molecules differing in as little as the position of an amino or hydroxyl group on a phenyl ring. If the immune system can discriminate between small molecules that it had presumably never encountered during evolutionary selection and that differ in such subtle ways, then, it was reasoned, the number of potential antibodies must be very large indeed. A series of experiments conducted in the late 1970s and early 1980s estimated the number of different BCRs generated in a normal mouse immune system to be at least 10^7, but we now know that estimate was many orders of magnitude too small.

Investigators trying to make sense of Ig genetics were also faced with an additional puzzle: protein sequencing of mouse and human antibody heavy and light chains revealed that the first (amino terminal) 110 amino acids of antibody heavy and light chains are extremely variable among different antibody molecules. This region was therefore defined as the **variable (V) region** of the antibody molecule. In contrast, the remainder of the light and heavy chains could be classified into one

FIGURE 7-2 Early sequencing studies indicated that the light chain may be encoded in more than one segment. (a) Many Ig variable regions in both heavy and light chains could be found in association with few constant regions. (b) The same variable region can be found in contiguity with several different heavy-chain constant regions.

of only four sequences (light chain) or eight sequences (heavy chain) (Figure 7-2a) and was, therefore, named the **constant (C) region**. This raised an intriguing question: If each of 10^5 to 10^7 antibodies is encoded by a separate gene, how could the constant region part of the gene remain constant in sequence in the face of evolutionary drift?

Furthermore, antibodies could be found in which the same antibody variable (V) region was associated with more than one heavy-chain constant (C) region (Figure 7-2b), lending further support to the possibility that the expression of the variable and constant regions of each antibody chain were independently controlled. (These additional constant regions are generated by the antigen-induced process of class switch recombination, discussed in Chapter 12. In the current context, the only point of relevance is the independence with which the variable and constant regions appear to be expressed.) It was rapidly becoming clear that the solution to the antibody gene puzzle was more complex than had previously been imagined.

Investigators Proposed Two Early Theoretical Models of Antibody Genetics

Classical **germ-line theories** of antibody diversity suggested that the genetic information for each antibody is encoded, in its entirety, within the germ-line genome. This means that the genes encoding the entire sequence of every antibody heavy or light chain that the animal could ever make must be present in every cell. However, a quick calculation is sufficient to demonstrate that if there are 10^7 or more antibodies, each of which requires approximately 2,000 nucleotides, this would require massive expenditure of genetic information—indeed, more DNA would be required to encode the receptors of the immune system than is available to the organism. Although arguments were initially made that the dedication of a considerable fraction of the genome to the immune system may represent a reasonable evolutionary strategy, it became clear, as estimates of the size of the antibody repertoire were revised upward, that there simply was not enough DNA to go around and new ideas must be found.

In 1965, William Dreyer and J. Claude Bennett proposed that antibody heavy and light chains are each encoded in two separate segments in the germ-line genome and that one of each of the V region- and C region-encoding segments are brought together in B-cell DNA to form complete antibody heavy and light-chain genes. The idea that DNA in somatic cells might engage in recombinatorial activity was revolutionary. However, germ-line theorists began to modify their ideas to embrace this possibility.

Others suggested the equally innovative idea that the number of variable region genes in the genome might be extremely limited, and proposed the **somatic hypermutation theory**. According to this hypothesis, a limited number of antibody genes is acted upon by unknown mutational mechanisms in somatic cells to generate a diverse receptor repertoire in mature B lymphocytes. This latter idea had the advantage of explaining how a large repertoire of antibodies could be generated from a relatively small number of genes, but the disadvantage that such a process, like the somatic cell gene recombination suggested by others, had never been observed. There was additional argument over whether mutation would occur prior to, or only after, antigen contact and, therefore, whether mutation was responsible for generation of the so-called "primary repertoire" that exists prior to antigen binding by the B cell.

Heated debate continued between the proponents of modified germ-line versus somatic mutation theories throughout the early 1970s, until a seminal set of experiments revealed that both sides were correct. We now know that each antibody molecule is encoded by multiple, germ-line, variable-region gene segments, which are rearranged differently in each naïve immune cell to produce a diverse primary receptor repertoire. These rearranged genes are then further acted upon after antigen encounter by somatic hypermutation and antigenic selection, resulting in an expanded and exquisitely honed repertoire of antigen-specific B cells (see Chapter 12).

Breakthrough Experiments Revealed That Multiple Gene Segments Encode the Light Chain

From the mid 1970s until the mid 1980s, a small group of brilliant immunologists completed a series of experiments that fundamentally altered the way in which scientists think about the genetics of immune receptor molecules. The first breakthrough occurred when Susumu Tonegawa showed, as Dreyer and Bennett had predicted, that multiple gene segments encode the antibody light chain. His achievement is all the more impressive because the modern tools of molecular biology were not yet available. The experiment he performed with Nobumichi Hozumi is described in Classic Experiment Box 7-1.

CLASSIC EXPERIMENT   BOX 7-1

 ## Hozumi and Tonegawa's Experiment: DNA Recombination Occurs in immunoglobulin Genes in Somatic Cells

The paradigm-shifting experiment of Hozumi and Tonegawa was designed to determine whether the DNA encoding Ig light-chain constant and variable regions existed in separate segments in non antibody-producing cells, such that a single constant region gene could associate with different variable region genes in different B cells. They hypothesized that this might be the case, because of Ig amino acid sequencing data showing that the sequences of the constant regions of many Ig light chains were identical. They reasoned that if multiple copies of a constant region gene existed, then each copy would be expected to accumulate silent or neutral mutations over time. The most likely explanation for the finding of a single constant region amino acid sequence arranged in tandem with many different variable regions was, therefore, that a single constant region gene segment was cooperating with multiple variable region gene segments in different cells or situations to generate the light-chain gene. However, this notion was heretical to a generation of scientists brought up on the concept that one gene encodes one polypeptide chain.

Scientists are accustomed to thinking about the concept of alternative RNA splicing, wherein different proteins may be encoded by the same piece of DNA by differentially using particular RNA segments, cut and spliced following transcription of a long precursor RNA transcript. However, this experiment asked an entirely new question: *Can a piece of DNA change its place on a chromosome in a somatic cell?*

To serve as their source of germ-line DNA, Hozumi and Tonegawa used DNA from an organ in which Ig genes would not be expressed, *embryonic liver* in this case. (Sperm and egg DNA would have been much more difficult to obtain.) For B-cell DNA, they used DNA from an antibody-producing plasma cell tumor line, MOPC 321, which secretes fully functional κ light chains. They separately cut both sets of DNA with the same restriction enzyme and used radioactive probes to determine *the sizes of the DNA fragments* on which the variable and constant regions of the light-chain gene were found in the two sets of cells.

(continued)

Below, we describe how Hozumi and Tonegawa conducted their experiment, and indicate how the experimental protocol would be modified using the reagents and techniques available to a modern molecular biologist. We then display their data as it would appear both in the modern gel format as well as in its original form. Hozumi and Tonegawa:

a. Purified genomic DNA from embryonic liver cells and from the MOPC 321 tumor cell line.

b. Cut the two sets of genomic DNA with a restriction endonuclease (BamHI) and separated the DNA fragments using electrophoresis. (Today we would use a polyacrylamide gel and electrophorese submicrogram quantities of samples over the course of a few hours; Hozumi and Tonegawa used a foot-long agarose gel that needed 2 liters of agarose. They loaded 5 mg of DNA and ran the gel for 3 days.)

c. Made ^{125}I-labeled nucleic acid probes specific for the two regions of mouse κ light chain. One of Tonegawa's probes was a full-length radiolabeled piece of mRNA encoding the entire κ-chain sequence. The other probe was the 3' half of the sequence, which would hybridize to the constant, but not to the variable region of the κ-chain gene. Today, making enzymatically labeled, stable DNA probes for any particular sequence is a safe and relatively straightforward task. In Hozumi and Tonegawa's day, this task was significantly more challenging.

d. Probed the nuclease-generated DNA fragments to determine the size of the fragments carrying the variable and constant region sequences. We would normally use a Southern blot procedure, electrophoresing the fragments, blotting the gel with nitrocellulose paper, and then probing the paper with enzyme-labeled fragments complementary to the sequences of interest, developing the blots with

luminescent or fluorescent substrates. Hozumi and Tonegawa's approach was much more time-consuming. They cut the gel into about 30 slices, melted the agarose, and separately eluted the DNA from each slice. To each DNA sample, they added radiolabeled RNA,

allowed it to anneal, and removed the un-annealed RNA using RNase. The radioactivity remaining in each fraction was then plotted against the size of the DNA in the slice.

Figure 1 shows the results that would be obtained from a modern-day Southern

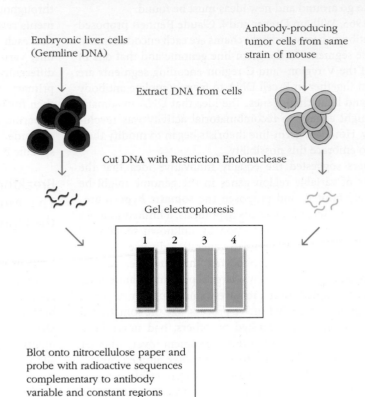

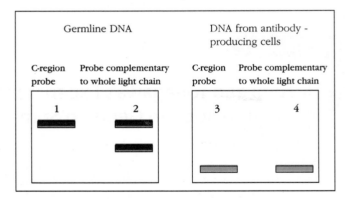

FIGURE 1

The κ light-chain gene is formed by DNA recombination between variable and constant region gene segments. DNA from embryonic liver cells was used as a source of germ-line DNA, and DNA from a B-cell tumor cell line was used as an example of DNA from antibody-producing cells. In embryonic liver, the DNA sequences encoding the variable and constant regions, respectively, were located on different restriction endonuclease fragments. However, these two sequences were co-located on a single restriction fragment in the myeloma DNA. (See text for details of the experiment.)

blot of Tonegawa's fragments. The germ-line DNA blot probed with the constant region mRNA sequence probe (lane 1) shows only a single band. This indicates that the restriction endonuclease is not cutting in the middle of the light-chain constant region, but rather that the whole constant region sequence is encoded within a single fragment.

In contrast, probing the blot with the whole light-chain sequence (V and C, lane 2) yields two bands. Since the difference between lanes 1 and 2 is the presence of the variable region sequence in the probe used for lane 2, the presence of two bands in lane 2 indicates that the restriction endonuclease used to generate the DNA fragments is cutting somewhere in between the variable and constant regions in the germ-line DNA, such that the variable region lies on a DNA fragment distinct from that bearing the constant region.

Analyzing the blots from the plasma cell tumor, we first note that the large fragment containing the constant region DNA and the midsized fragment bearing the variable region DNA have disappeared. Both the variable and constant region gene segments are now located on smaller fragments. This implies that a new pair of restriction endonuclease sites now forms the boundaries of each of the constant and variable region fragments. *Right away, we can tell that the DNA environment around the light-chain genes changes as the B cell differentiates.*

Next, we note that the sizes of the DNA fragments on which the constant and variable region gene segments are located are apparently the same. This implies, although it does not yet prove, that the movement of the constant and/or variable region gene segments has brought them into close proximity with one another, such that the constant and variable region gene segments co-locate on the same fragment. An alternative explanation is that they have both altered their locations and the similarity in the size of the fragments is coincidental. DNA sequencing supported the former

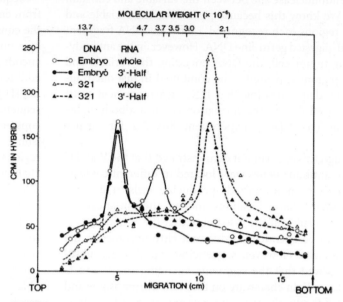

FIGURE 2
The original data from Hozumi and Tonegawa's classic experiment proving immunoglobulin gene recombination occurs in B cells. DNA from the sources shown was run out on an agarose gel for 3 days at 4°C. The gel was sliced, the DNA eluted, and then each sample was hybridized with radiolabeled probes for either the constant region of the κ light chain, or the whole chain. The plot shows the amount of radioactivity in each fraction as a function of the migration distance, which reflects the molecular weight of the DNA fragment. (See text for details.) *[From N. Hozumi and S. Tonegawa, 1976,* Evidence for somatic rearrangement of immunoglobulin genes coding for variable and constant regions, *Proceedings of the National Academy of Sciences USA* **73***:3628–3632.]*

interpretation: *as the B cell differentiates, the variable and constant region gene segments are moved from distant regions of the chromosome into close apposition with one another.*

It is all too easy, with modern molecular biology technologies, to forget what a tour de force this experiment actually represented. Few of the reagents or pieces of apparatus we commonly encounter in the molecular biology laboratory today were available to Hozumi and Tonegawa— they had to make their own.

The original paper speaks of purifying their own restriction endonuclease (BamHI) from a bacterial sample obtained from a colleague, and of performing the electrophoresis at 4°C for 3 days. Furthermore, technical difficulties precluded their being able to make a good 5′ (V-region) probe. They therefore probed each sample

with a 3′ (C-region) probe, as well as one that bound to the whole light chain, and then inferred 5′ (V-region) binding by subtraction. It was an extraordinary piece of work. Their original data are shown in Figure 2.

What happened to the DNA on the alternative allele that did not encode the tumor cell secreted light chain? Subsequent analysis of the DNA from this tumor showed that the DNA from both chromosomes had undergone rearrangement. Hozumi and Tonegawa were fortunate that the variable regions used by both rearrangements were close to one another and so the fragment patterns overlapped.

Hozumi, N., and S. Tonegawa. 1976. Evidence for somatic rearrangement of Ig genes coding for variable and constant regions. *Proceedings of the National Academy of Sciences USA* **73**:3628–3632.

Tonegawa's experiment showed that, in the DNA from non-antibody-producing, embryonic liver cells, there is a BamHI endonuclease site between the variable and constant regions. We know this because probes for the variable and constant regions each recognized a different DNA fragment in BamHI-digested germ-line DNA. However, in the antibody-producing tumor cell, the DNA encoding the variable and constant regions appeared to be combined in just one fragment, thus demonstrating that DNA rearrangement must have occurred during the formation of an antibody light-chain gene (see Classic Experiment Box 7-1 for further details).

Although this experiment demonstrated that the V and C regions of antibody genes were located in different contexts in the DNA of non-antibody-producing cells, it did not speak to their relative locations; indeed, the initial experiment did not rule out the possibility that the V and C fragments could be encoded on different chromosomes in the embryonic cells. However, sequencing experiments subsequently showed that the segments encoding the V and C genes of the κ light chains are on the same chromosome and that, in non B cells, the two segments are separated by a long non-coding DNA sequence.

The impact of this result on the biological community was profound. For the first time, DNA was shown to be cut and rejoined during the process of cell differentiation. This experiment not only provided the experimental proof of Dreyer and Bennett's prediction; it also paved the way for the next surprising finding.

Tonegawa's experiment had identified the V and C segments encoding the kappa light chain. However, when scientists in his group sequenced the antibody light-chain DNA, they encountered another unexpected result. As expected, the embryonic (unrearranged) V region segment had, at its 5′ terminus, a short hydrophobic leader sequence, a common feature of membrane proteins necessary to guide the nascent protein chain into the membrane (Figure 7-3a). A 93 bp sequence of non-coding DNA separated the leader sequence from a long stretch of DNA that encoded the first 97 amino acids of the V region. But the light-chain V region

domain is approximately 110 amino acids long. Where was the coding information for the remaining 13 amino acids?

Sequencing of the light-chain constant region fragment from embryonic DNA provided the answer. Upstream from the constant region coding sequence, and separated from it by a non-coding DNA segment of 1250 bp, were the 39 bp encoding the remaining 13 amino acids of the V region. This additional light-chain coding segment was named the **joining (J) gene segment** (shown in red in Figure 7-3a). Further sequencing of mouse and human light-chain variable and constant region genes confirmed Tonegawa's second, astonishing finding. Not only are the variable and constant regions encoded in two separate segments, but the light-chain variable regions themselves are encoded in two separate gene segments, the V and J segments, that are made contiguous only in antibody-producing cells.

The heavy-chain variable region gene was then shown to require yet a third genetic segment. Adopting a similar strategy of cloning and sequencing Ig heavy-chain genes, Lee Hood's group identified a germ-line **Heavy-chain Variable Region (V_H) gene** fragment that encoded amino acid residues 1–101 of the antibody heavy chain and a second fragment that included a **Heavy-chain Joining gene segment (J_H)** that determined the sequence of amino acid residues 107–123. Neither of these segments contained the DNA sequence necessary to encode residues 102–106 of the heavy chain. Significantly, these missing residues were included in the third complementarity-determining region of the antibody heavy chain, **CDR3**, which provides contact residues in the binding of most antigens (see Chapter 3). Gene segments encoding this part of the antibody heavy chain were eventually located 5′ of the J region in mouse embryonic DNA by Hood and colleagues (Figure 7-3b). The importance of the contribution made by this gene segment to the diversity of antibody specificities is denoted by its name; the **Diversity (D) region**. (Because there is no D region in the light chain, immunologists usually drop the subscript denoting the heavy chain.)

Thus, the variable region of the heavy chain of the antibody molecule is encoded by three discrete gene segments,

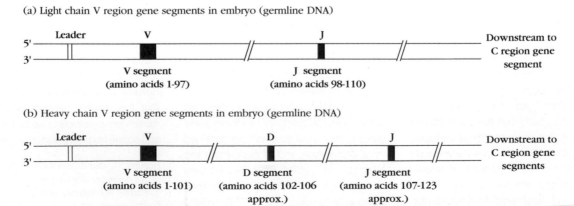

(a) Light chain V region gene segments in embryo (germline DNA)

V segment
(amino acids 1-97)

J segment
(amino acids 98-110)

(b) Heavy chain V region gene segments in embryo (germline DNA)

V segment
(amino acids 1-101)

D segment
(amino acids 102-106
approx.)

J segment
(amino acids 107-123
approx.)

FIGURE 7-3 Antibody light chains are encoded in two segments—V and J, whereas antibody heavy chains are encoded in three segments—V, D, and J.

TABLE 7-1	Chromosomal locations of immunoglobulin genes in humans and mice	
Gene	**Human chromosome**	**Mouse chromosome**
λ light chain	22	16
κ light chain	2	6
Heavy chain	14	12

and the variable region of the light chain by two segments, in the germ-line genome. These segments are brought together by a process of DNA recombination that occurs only in the B lymphocyte lineage to create the complete variable region gene. Furthermore, the DNA at the junction between the V and J segments for light chains, and at the VD and DJ junctions in heavy chains, accounts for the extraordinary diversity that was first observed by Kabat and Wu in the CDR3 regions of both heavy and light chains (see Chapter 3). Below, we will describe how further genetic diversity is generated at these junctions by additional processes unique to immune system genetics.

Multigene Organization of Ig Genes

Recall that Ig proteins consist of two identical heavy chains and two identical light chains (see Chapter 3). The light chains can be either **kappa (κ) light chains** or **lambda (λ) light chains**. The heavy-chain, kappa, and lambda gene families are each encoded on separate chromosomes (Table 7-1).

Kappa Light-Chain Genes Include V, J, and C Segments

The mouse Ig$_\kappa$ locus spans 3.5 Mb and includes 91 potentially functional V$_\kappa$ genes, which have been grouped into 18 V gene families based on sequence homology (Figure 7-4a). (If two sequences are greater than 80% identical, they are classified as belonging to the same V gene family.) These V gene families can be further grouped into V gene clans, based again on sequence homology. Each V$_\kappa$ segment includes the leader exon encoding the signal peptide. Individual V$_\kappa$ segments are separated by non-coding gaps of 5 to 100 kb. The transcriptional orientation of particular V$_\kappa$ segments may be in the same or in the opposite direction as the constant region segment. The relative orientations of the variable and constant region segments do not affect the use of the segments, but do alter some details of the recombinational mechanism that creates the complete light-chain gene, as will be discussed later.

Downstream of the V$_\kappa$ region cluster are four functional J$_\kappa$ segments and one pseudogene, or other nonfunctional open reading frame (Table 7-2). A similar arrangement is found in the human V$_\kappa$ locus, although the numbers of V and J gene segments vary. A single C$_\kappa$ segment is found downstream of the J region, and all kappa light-chain constant regions are encoded by this segment.

Lambda Light-Chain Genes Pair Each J Segment with a Particular C Segment

The mouse Ig$_\lambda$ locus spans a region of approximately 190 kb. Lambda light chains are only found in 5% of mouse antibodies because of a deletional event in the mouse genome that has eliminated most of the lambda light-chain variable region segments. It was therefore not surprising to discover that

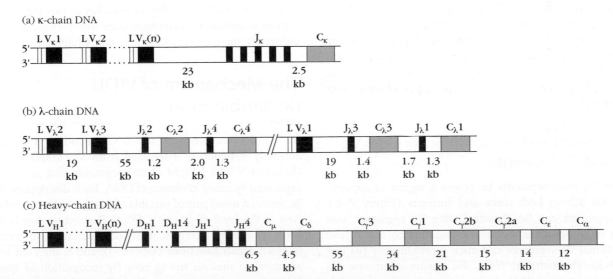

FIGURE 7-4 Organization of immunoglobulin germ-line gene segments in the mouse. The (a) κ light chain and (b) λ light chains are encoded by V, J, and C gene segments. The (c) heavy chain is encoded by V, D, J, and C gene segments. The distances in kilobases (kb) separating the various gene segments in mouse germ-line DNA are shown below each diagram.

TABLE 7-2   Immunoglobulin variable region gene numbers in humans and mice

	Human	Mouse
V_κ	34–48 functional; 8 ORFs; 30 pseudogenes.	91 functional; 9 ORFs; 60 pseudogenes
J_κ	5 functional; multiple alleles	4 functional; 1 ORF
V_λ	33 functional; 6 ORFs; 36 pseudogenes	3–8 functional
J_λ	5 functional; 2 ORFs	4 functional; 1 pseudogene
V_H	38–44 functional; 4 ORFs; 79 pseudogenes	101 functional; 69 pseudogenes*
D	23 functional; 4 ORFs	19 functional; 7 ORFs; 6 pseudogenes
J_H	6 functional; 3 pseudogenes; several different alleles found	4 functional; again, multiple haplotypes

* Gene numbers of the mouse V_H locus refer to the sequenced chromosome 14 of the C57 BL/6 mouse.

A germ-line gene is considered to be functional if the coding region has an open reading frame (ORF) without a stop codon, and if there is no described defect in the splicing sites, RSS, and/or regulatory elements.

A germ-line entity is considered to be an ORF if the coding region has an open reading frame but alterations have been described in the splicing sites, RSSs, and/or regulatory elements, and/or changes of conserved amino acids have been suggested by the investigators to lead to incorrect folding and/or the entity is an orphon, a nonfunctional gene located outside the main chromosomal locus.

A germ-line entity is considered to be a pseudogene if the coding region has a stop codon(s) and/or a frameshift mutation. In the case of V gene segments, these mutations may be either in the V gene coding sequence, or in the leader sequence.

SOURCE: Gene numbers and definitions summarized from the International Immunogenetics Information System Web site: http://imgt.org. Accessed November 16, 2011.

there are only three fully functional V_λ gene segments, although this number varies somewhat by strain (see Table 7-2). Interestingly, there are also three fully functional λ chain constant regions, each one associated with its own J region segment (Figure 7-4b). The J-C_λ4 segments are not expressed because of a splice site defect.

Since recombination of *Ig* gene segments always occurs in the downstream direction (V to J), the location of the V_λ1 variable region sequence upstream of the JC_λ3 and JC_λ1 segments but downstream from JC_λ2 means that V_λ1 is always expressed with either JC_λ3 or JC_λ1 but never with JC_λ2. For the same reason, V_λ2 is usually associated with the JC_λ2 pairing, although occasional pairings of V_λ2 with the 190 kb distant JC_λ1 segments have been observed.

In humans, 40% of light chains are of the λ type and about 30 V_λ-chain gene segments are used in mature antibody light chains (see Table 7-2). Downstream from the human V_λ locus is a series of seven JC_λ pairs, of which four pairs are fully functional.

Heavy-Chain Gene Organization Includes V_H, D, J_H, and C_H Segments

Multiple V_H gene segments lie across a region of approximately 3.0 Mb in both mice and humans (Figure 7-4c). These segments can be classified, like V_κ segments, into families of homologous sequences. Humans express at least 38 functional V_H segments and mice approximately 101 (see Table 7-2). Downstream from the cluster of mouse V_H region segments is an 80 kb region containing approximately 14 D regions (the actual number varies among different

mouse strains). Just 0.7 kb downstream of the most 3′ D segment is the J_H region cluster, which contains four functional J_H regions. A further gap separates the last J_H segment from the first constant region exon, C_μ1.

The human V_H locus has a similar arrangement with approximately 30 functional D segments, and 6 functional J_H segments. Human D regions can be read in all three reading frames, whereas mouse D regions are mainly read in reading frame 1, because of the presence of stop codons in reading frames 2 and 3 in most mouse D regions.

The eight constant regions of antibody heavy chains are encoded in a span of 200 kb of DNA downstream from the J_H locus, as illustrated in Figure 7-4c. Recall that the constant regions of antibodies determine their heavy-chain class and, ultimately, their effector functions (see Chapter 13).

The Mechanism of V(D)J Recombination

In V(D)J recombination, the DNA encoding a complete antibody V-region is assembled from V, D, and J (heavy chain) or V and J (light chain) segments that are initially separated by many kilobases of DNA. Each developing B cell generates a novel pair of variable region genes by recombination at the level of genomic DNA. Recombination is catalyzed by a set of enzymes, many of which are also involved in DNA repair functions (Table 7-3), and is directed to the appropriate sites on the Ig gene by recognition of specific DNA sequence motifs called **Recombination Signal Sequences (RSSs)**. These sequences ensure that one of each

TABLE 7-3	Proteins involved in V(D)J recombination
Protein	**Function**
RAG-1/2	Lymphoid-specific complex of two proteins that catalyze DNA strand breakage and rejoin to form signal and coding joints
TdT	Lymphoid-specific protein that adds N region nucleotides to the joints between gene segments in the Ig heavy chain and at all joints between TCR gene segments
HMG1/2 proteins	Stabilize binding of RAG1/2 to Recombination Signal Sequences (RSSs), particularly to the 23-bp RSS; stabilize bend introduced into the 23-bp spacer DNA by the RAG1/2 proteins
Ku70 and Ku80 heterodimers	Binds DNA coding and signal ends and holds them in protein-DNA complex
DNA PKcs	In complex with Ku proteins, recruits and phosphorylates Artemis
Artemis	Opens the coding end hairpins
XRCC4	Stabilizes and activates DNA ligase IV
DNA ligase IV	In complex with XRCC4, and Cernunnos ligates DNA ends
Cernunnos	With XRCC4, activates DNA ligase IV

type of segment (V and J for the light chain, or V, D, and J for the heavy chain) is included in the recombined variable region gene. During cleavage and ligation of the segments, the DNA is edited in various ways, adding further variability to the recombined gene.

Recombination Is Directed by Signal Sequences

If recombination is to occur in the DNA of every lymphocyte, then a mechanism must exist to ensure that it only occurs in antibody and T-cell receptor genes, and that the moving DNA segments end up exactly where they should be in the genome. Otherwise, dire consequences, including malignancy, can ensue.

In the late 1970s, investigators working with light-chain genes, described two blocks of conserved sequences—a **nonamer** (a set of nine base pairs) and a **heptamer** (a set of seven base pairs)—that are highly conserved and occur in the noncoding regions upstream of each J segment. The heptamer appeared to end exactly at the J region coding sequence. Further sequencing showed that the same motif was repeated in an inverted manner on the downstream side of the V region coding sequences, again with the heptamer sequence ending flush with the V-region gene segment (Figure 7-5a).

A further noteworthy feature of these sequences was the presence of a spacer sequence of either 12 or 23 bp between the heptamer and the nonamer. The significance of the spacer lengths was clear; they represented one, or two, turns

(a) Nucleotide sequence of RSSs

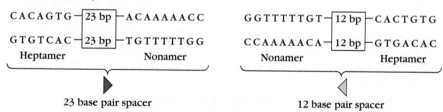

(b) Location of RSSs in germ-line immunoglobulin DNA

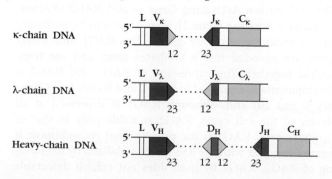

FIGURE 7-5 Two conserved sequences in light-chain and heavy-chain DNA function as recombination signal sequences (RSSs). (a) Both signal sequences consist of a conserved heptamer and conserved AT-rich nonamer; these are separated by nonconserved spacers of 12 or 23 bp. (b) The two types of RSS have characteristic locations within λ-chain, κ-chain, and heavy-chain germ-line DNA. During DNA rearrangement, gene segments adjacent to the 12-bp RSS can join only with segments adjacent to the 23-bp RSS.

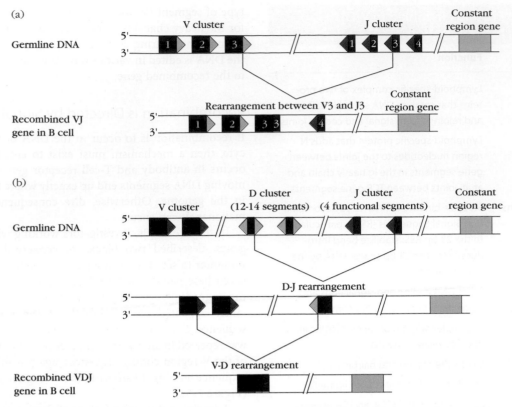

FIGURE 7-6 Recombination between gene segments is required to generate a complete light chain gene. (a) Recombination between a V region (in this case, V3) and a J region (in this case J3) generates a single VC light-chain gene in each B cell. The recombinase enzymes recognize the RSS downstream of the V region (orange triangle) and upstream of the J region (brown triangle). In every case, an RSS with a 12-bp (one-turn) spacer is paired with an RSS with a 23-bp (two-turn) spacer. This ensures that there is no inadvertent V-V or J-J joining. (b) Recombination of V (blue), D (purple), and J (red) segments creates a complete heavy-chain variable region gene. Again, the recombinase enzyme recognizes the RSS sequences downstream of the V region, up- and downstream of the D region, and upstream of the J region, pairing 23-bp spacers with 12-bp spacers.

of the double helix. Thus, the spacer sequence ensures that the ends of the nonamer and heptamer closest to the spacers would be on the same side of the double helix and, thus, accessible to binding by the same enzyme. The investigators correctly concluded that they had discovered the signal that directs recombination between the V and J gene segments and termed this heptamer 12/23 nonamer motif, the recombination signal sequence (RSS).

To summarize, the RSS consists of three elements:

- An absolutely conserved, 7-bp (heptamer) consensus sequence 5'-CACAGTG –3'
- A less conserved spacer of either 12 or 23 bp
- A second conserved, 9-bp (nonamer) consensus sequence 5'-ACAAAAACC-3'

In the heavy-chain gene segments, a similar pattern was noted. The spacer regions separating the heptamer and nonamer pairs were 23 bp in length following the V and preceding the J segments and 12 bp in length before and after the D segments. The relative locations of the 12 and 23 base pair spacers (Figure 7-5b) suggested that the VDJ recombinase enzyme is designed to bring together one sequence with a 12-bp spacer with one sequence with a 23-bp spacer, something we now know to be the case.

Figures 7-6a and 7-6b illustrate the generation of complete light-chain and heavy-chain genes from individual V and J and V, D, and J segments, respectively.

Gene Segments Are Joined by the RAG1/2 Recombinase

Two proteins, encoded by closely linked genes, **RAG1 (Recombination Activating Gene 1)** and **RAG2 (Recombination Activating Gene 2)**, were shown to be necessary for recombining antibody genes. The *RAG1* and *RAG2* genes are encoded just 8 kilobases apart and are transcribed together. The expression of RAG1 and RAG2 is developmentally regulated in both T and B cells (see Chapters 9 and 10) and, although RAG1 is expressed at all phases of the cell cycle, RAG2 is stable only in G_0- or G_1-phase cells. RAG1 is the predominant recombinase; it forms a complex with the RSS that is stabilized by the binding of RAG2. RAG2 by itself does not exhibit detectable RSS binding activity.

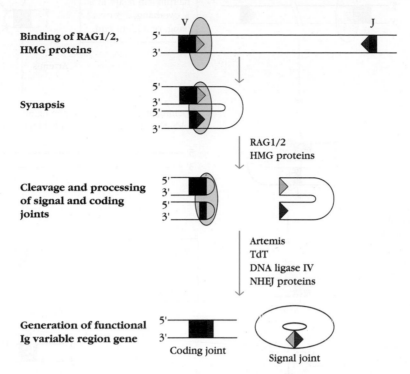

FIGURE 7-7 Overview of recombination of immunoglobulin variable region genes. The RAG1/2 complex (represented together by the green oval) binds the RSSs and catalyzes recombination. Other enzymes fill in or cleave nucleotides at the coding end, and ligase completes the process. See text for details. [Adapted from Krangel, M. Nature Immunology 4, p. 625. 2003]

Only three of the proteins implicated in V(D)J recombination are unique to lymphocytes: RAG1, RAG2, and **Terminal deoxynucleotidyl Transferase (TdT)**, which is responsible for the generation of additional diversity in the CDR3 region of the antibody heavy chain (as we will see below). TdT is expressed only in developing lymphocytes and adds untemplated "N" nucleotides to the free 3′ termini of coding ends following their cleavage by RAG1/2 recombinases.

Other enzymes participating in the recombination process are not lymphoid specific. Whereas binding of the RSS by RAG1/2 can occur in the absence of any other proteins, other cellular factors, most of which are part of the *Non-Homologous End Joining (NHEJ)* pathway of DNA repair are necessary for completion of V(D)J recombination. These other non-lymphocyte-specific proteins known to participate in V(D)J joining are described in Table 7-3.

V(D)J Recombination Results in a Functional Ig Variable Region Gene

The process of V(D)J recombination occurs in several phases (Figure 7-7). The end product of each successful rearrangement is an intact Ig gene, in which V and J (light chains) or V, D, and J (heavy chains) segments are brought together to create a complete heavy or light chain gene. The new joints in the antibody V region gene, created by this recombination process are referred to as **coding joints**. The joints between the heptamers from the RSSs are referred to as **signal joints**.

The first phase of this process, DNA recognition and cleavage, is catalyzed by the RAG1/2 proteins. The second phase, end processing and joining, requires a more complex set of enzymatic activities in addition to RAG1/2, including Artemis, TdT, DNA ligase IV, and other NHEJ proteins. The individual steps involved in the process of recombination between V_κ and J_κ segments are shown sequentially in Figure 7-8.

Step 1 *Recognition of the heptamer-nonamer Recombination Signal Sequence (RSS) by the RAG1/RAG2 enzyme complex.* The RAG1/2 recombinase forms a complex with the heptamer-nonamer RSSs contiguous with the two gene segments to be joined. Complex formation is initiated by recognition of the nonamer RSS sequences by RAG1 and the 12-23 rule is followed during this binding.

Step 2 *One-strand cleavage at the junction of the coding and signal sequences.* The RAG1/2 proteins then perform one of their unique functions: the creation of a single-strand nick, 5′ of the heptameric signal sequence on the coding strand of each V segment and a similar nick on the non-coding strand exactly at the heptamer-J region junction. (Figure 7-8 shows this process for the V segment only.)

Step 3 *Formation of V and J region hairpins and blunt signal ends.* The free 3′ hydroxyl group at the end of the coding strand of the V_κ segment now attacks the phosphate group on the opposite, non-coding V_κ strand, forming a new covalent bond *across* the double helix and yielding a DNA hairpin structure on the V-segment side of the break (coding end). Simultaneously, a blunt DNA end is formed at

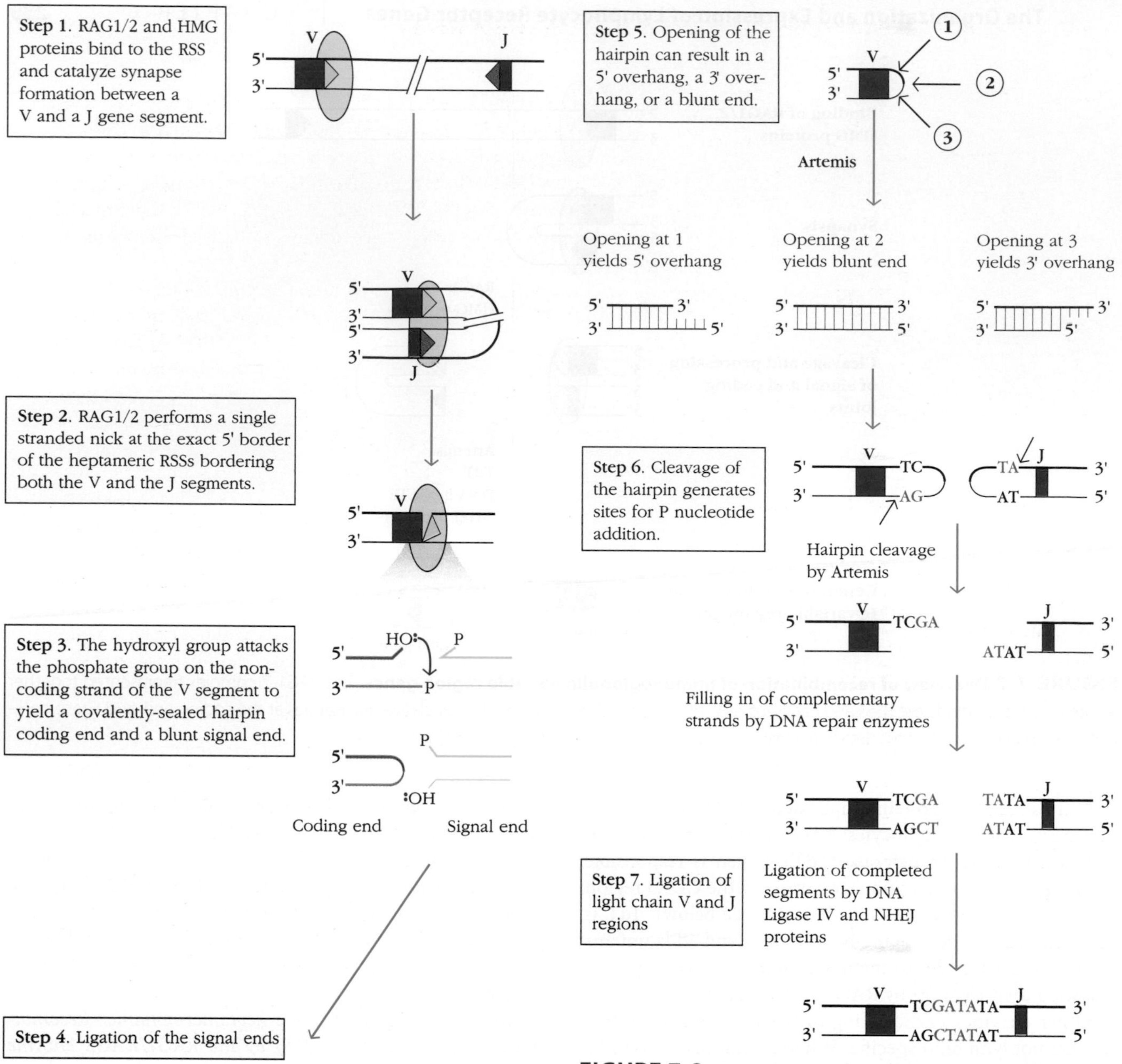

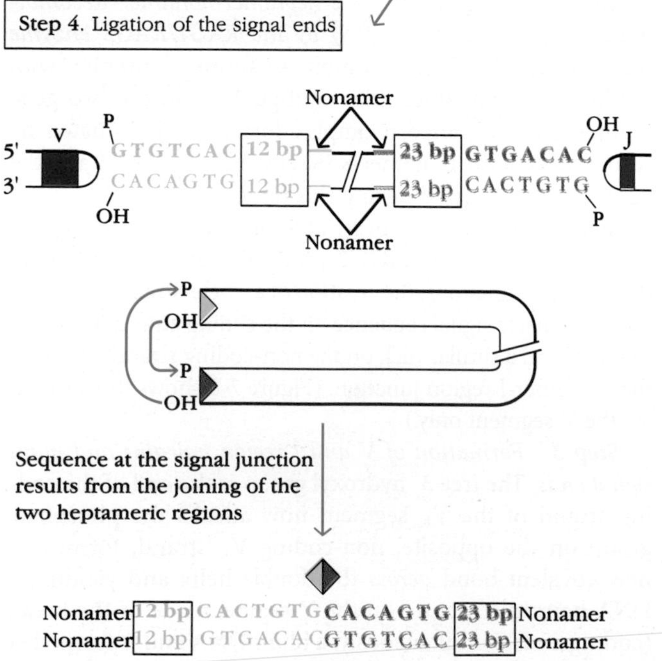

Step 1. RAG1/2 and HMG proteins bind to the RSS and catalyze synapse formation between a V and a J gene segment.

Step 2. RAG1/2 performs a single stranded nick at the exact 5' border of the heptameric RSSs bordering both the V and the J segments.

Step 3. The hydroxyl group attacks the phosphate group on the non-coding strand of the V segment to yield a covalently-sealed hairpin coding end and a blunt signal end.

Coding end   Signal end

Step 4. Ligation of the signal ends

Nonamer

Nonamer

Sequence at the signal junction results from the joining of the two heptameric regions

Nonamer | 12 bp | CACTGTG **CACAGTG** | 23 bp | Nonamer
Nonamer | 12 bp | GTGACAC **GTGTCAC** | 23 bp | Nonamer

Step 5. Opening of the hairpin can result in a 5' overhang, a 3' overhang, or a blunt end.

Artemis

Opening at 1 yields 5' overhang

Opening at 2 yields blunt end

Opening at 3 yields 3' overhang

Step 6. Cleavage of the hairpin generates sites for P nucleotide addition.

Hairpin cleavage by Artemis

Filling in of complementary strands by DNA repair enzymes

Step 7. Ligation of light chain V and J regions

Ligation of completed segments by DNA Ligase IV and NHEJ proteins

FIGURE 7-8 The mechanism of V(D)J recombination. 1. The RAG1/2 complex (pale green oval) binds to the RSSs and catalyzes synapse formation. The coding (5' → 3') strand of DNA is drawn in a thick line, the non-coding (3' → 5') strand in a thin line. For steps 1 to 5 we show only the events associated with the V$_\kappa$ region gene segment, although the single-strand cleavage, hairpin formation, and templated nucleotide addition simultaneously occur at the borders of the V$_\kappa$ and the J$_\kappa$ segments. In this example, the V and J regions are encoded in the same direction on the chromosome, and so the DNA encoding the RSSs and the intervening DNA is released into the nucleus as a circular episome and will be lost on cell division. The DNA that was on the coding strand of the V region prior to rearrangement is emboldened. The actual signal joint is between the residues that were in contiguity with the V and J regions, respectively. Only the heptamer sequence is written out to preserve clarity. Nucleotides encoded in the germ-line genome are shown in black; P nucleotides are in blue, and non-templated nucleotides added by TdT at heavy chain VD and DJ joints are shown in red. Steps 8, 9, and 10, shown on the facing page, only occur in heavy chain loci. (See text for details.)

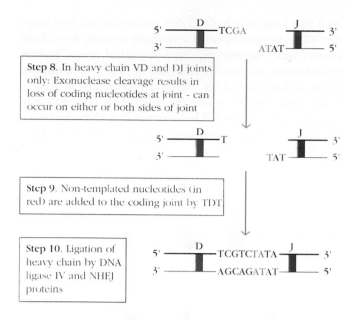

Step 8. In heavy chain VD and DJ joints only: Exonuclease cleavage results in loss of coding nucleotides at joint - can occur on either or both sides of joint

Step 9. Non-templated nucleotides (in red) are added to the coding joint by TDT

Step 10. Ligation of heavy chain by DNA ligase IV and NHEJ proteins

the edge of the heptameric signal sequence. The same process occurs simultaneously on the J_κ side of the incipient joint. At this stage, the RAG1/2 proteins and HMG proteins are still associated with the coding and signal ends of both the V and J segments in a postcleavage complex.

Step 4 Ligation of the signal ends. DNA ligase IV then ligates the free blunt ends to form the signal joint. The involvement of particular enzymes in this process was deduced from observations of V(D)J recombination in natural and artificially generated systems lacking one or more enzymes (see Table 7-3).

Step 5 Hairpin cleavage. The hairpins at the ends of the V and J regions are now opened in one of three ways. The identical bond that was formed by the reaction described in step 3 above, may be reopened to create a blunt end at the coding joint. Alternatively, the hairpin may be opened asymmetrically on the "top" or on the "bottom" strand, to yield a 5′ or a 3′ overhang, respectively. A 3′ overhang is more common in in vivo experiments. Hairpin opening is catalyzed by **Artemis**, a member of the NHEJ pathway.

Step 6 Overhang extension, leading to palindromic nucleotides. In Ig light-chain rearrangements, the resulting overhangs can act as substrates for extension DNA repair enzymes, leading to double stranded **palindromic (P) nucleotides** at the coding joint. For example, the top row of bases in the V region in the 5′ to 3′ direction reads TCGA. Reading backward on the bottom strand from the point of ligation also yields TCGA. The palindromic nature of the bases at this joint is a direct function of an asymmetric hairpin opening reaction. P nucleotide addition can also occur at both the VD and DJ joints of the heavy-chain gene segments, but, as described below, other processes can intervene to add further diversity at the V_H-D and D-J_H junctions.

Step 7 Ligation of light-chain V and J Segments. Members of the NHEJ pathway repair both the signal and the coding joints, but the precise roles of each, and potentially other enzymes in this process, have yet to be fully characterized.

During B-cell development, Ig heavy-chain genes are rearranged first, followed by the light-chain genes. This temporal dissociation of the two processes enables two additional diversifying mechanisms to act on heavy-chain V region segments. The enzymes responsible for these mechanisms are usually turned off before light-chain rearrangements begin.

Comparative sequence analysis of germ-line and mature B-cell Ig genes demonstrated that both loss of templated nucleotides (nucleotides found in the germ-line DNA) and addition of untemplated nucleotides (nucleotides not found in germ-line DNA) could be identified in heavy-chain sequences. Two distinct enzyme-catalyzed activities are responsible for these findings.

Step 8 Exonuclease trimming. An exonuclease activity, which has yet to be identified, trims back the edges of the V region DNA joints. Since the RAG proteins themselves can trim DNA near a 3′ flap, it is possible that the RAG proteins may cut off some of the lost nucleotides. Alternatively, Artemis has also been shown to have exonuclease, as well as endonuclease activity, and could be the enzyme responsible for the V(D)J-associated exonuclease function. Exonuclease trimming does not necessarily occur in sets of three nucleotides, and so can lead to out-of-phase joining. V segment sequences in which trimming has caused the loss of the correct reading frame for the chain cannot encode antibody molecules, and such rearrangements are said to be **unproductive**. As mentioned above, such exonuclease trimming is more common at the two heavy-chain V gene joints (V-D and D-J) than at the light-chain V-J joint. In cases where trimming is extensive, it can lead to the loss of the entire D region as well as the elimination of any P nucleotides formed as a result of asymmetric hairpin cleavage.

Step 9 N nucleotide addition. **Non-templated (N) nucleotides** are added by TdT to the coding joints of heavy-chain genes after hairpin cleavage. This enzyme can add up to 20 nucleotides to each side of the joint. The two ends are held together throughout this process by the enzyme complex, and again, loss of the correct phase may occur if nucleotides are not added in the correct multiples of three required to preserve the reading frame.

Step 10 Ligation and repair of the heavy-chain gene. This occurs as for the light-chain genes.

In describing V(D)J recombination, investigators must explain not only the mechanism of RSS recognition, cleavage, and ligation but also address the question of how two RSSs, located many kilobases distant from one another in the linear DNA sequence, are brought into close apposition. Furthermore, once successful recombination has occurred on one heavy-chain and one light-chain allele, this information must be communicated to the homologous chromosome, so that the other alleles can be silenced. How does this occur?

Recent research indicates that both the structure and the location of recombinationally active, Ig V region DNA change significantly as B-cell development proceeds, and that some of these changes are signaled by epigenetic alterations in the chromatin structure, mediated by specific methylation reactions on associated histone residues. DNA close to the nuclear membrane (pericentric DNA) does not include recombinationally active V regions; rather, DNA undergoing recombination moves away from the nuclear membrane, toward the center of the nucleus. The structure of the chromatin undergoing recombination also alters,

such that long stretches of DNA are condensed into loops that allow the recombining sequences to come into closer contact with one another. Once successful recombination has occurred, the inactive allele has been shown to migrate to the pericentric regions. Thus, the recombination process is occurring within an actively modulating nucleoplasmic context, and is controlled by enzymes that alter chromatin structure, in addition to those that cleave and recombine the DNA.

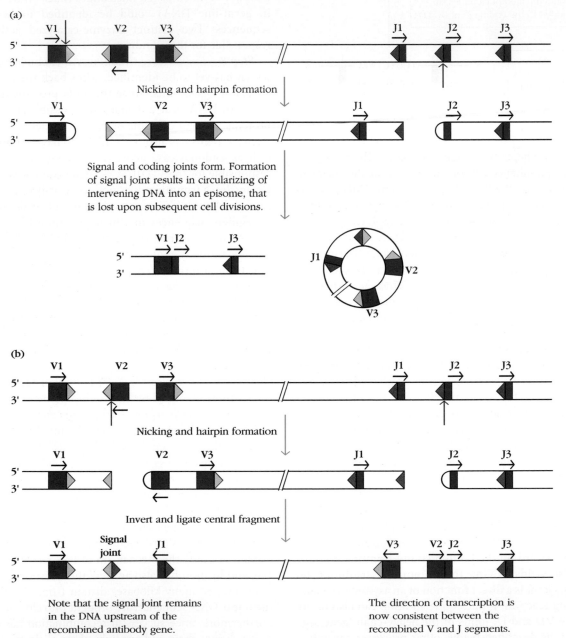

FIGURE 7-9 Recombination can occur between DNA segments aligned in the same, or opposite, transcriptional direction on the chromosome. (a) Recombination is shown occurring between V1 and J2, which are both encoded in the same transcriptional orientation, from left to right. The intervening DNA is

excised as a circular episome. (b) Recombination is shown occurring between V2 and J2, which are encoded in opposite transcriptional orientations. In this case, the DNA containing the signal joint remains inverted in the DNA upstream of the recombined pair.

TABLE 7-4	Combinatorial antibody diversity in humans		
Nature of segment	**Number of heavy-chain segments (estimated)**	**Number of κ-chain segments (estimated)**	**Number of λ-chain segments (estimated)**
V	41	41	33
D	23		
J	6	5	5
Possible number of combinations	$41 \times 23 \times 6 = 5658$	$41 \times 5 = 205$	$30 \times 5 = 165$
Possible number of heavy-light chain combinations in the human $= 5658 \times (205 + 165) = 2.09 \times 10^6$			

V(D)J Recombination Can Occur between Segments Transcribed in Either the Same or Opposite Directions

Sequencing of chromosomal fragments containing multiple antibody V region genes showed that transcription of some V region genes occurs in a direction opposite from that of downstream D and J segments. However, work with artificial recombinase substrates has demonstrated that the same general mechanism of recombination is used, regardless of the transcriptional direction of the gene segments involved. When recombination occurs between two gene segments that are transcribed in the same direction, the intervening DNA is deleted and lost (Figure 7-9a). When they are transcribed in opposite directions, the DNA that was located between the V and J segments is retained, in an inverted orientation, in the DNA upstream of the rearranged VJ region (Figure 7-9b).

Five Mechanisms Generate Antibody Diversity in Naïve B Cells

This description of the complex and sophisticated apparatus by which Ig genes are created allows us to understand how such an immensely diversified antibody repertoire can be generated from a finite amount of genetic material. To summarize, the diversity of the naïve BCR repertoire is shaped by the following mechanisms:

1. *Multiple gene segments* exist at heavy (V, D, and J) and light-chain (V and J) loci. These can be combined with one another to provide extensive combinatorial diversity (Table 7-4).
2. *P nucleotide addition* results when the DNA hairpin at the coding joint is cleaved asymmetrically. Filling in the single-stranded DNA piece resulting from this asymmetric cleavage generates a short palindromic sequence.
3. *Exonuclease trimming* sometimes occurs at the VDJ and VJ junctions, causing loss of nucleotides.
4. *Non-templated N nucleotide addition* in heavy chains results from TdT activity. Mechanisms 2, 3, and 4 give rise to extra diversity at the junctions between gene segments, which contribute to CDR3.

In addition to these four mechanisms for generating antibody diversity that operate on individual heavy- or light-chain variable segments, the combination of different heavy and light-chain pairs to form a complete antibody molecule provides further opportunities for increasing the number of available antibody combining sites (Table 7-4).

5. *Combinatorial diversity:* The same heavy chain can combine with different light chains, and vice versa.

These five mechanisms are responsible for the creation of the diverse repertoire of BCRs and antibodies that is available to the organism before any contact with pathogens or antigen has occurred. Following antigenic stimulation, B cells are able to use yet another mechanism, unique to the immune system, to further diversify and refine the antigen-specific receptors and antibodies: *somatic hypermutation.* As described in Chapter 12, a specialized enzyme complex targets the genes encoding the variable regions of Ig genes *only in those B cells that have undergone antigen-specific activation in the presence of T-cell help.* B cells are exposed to successive cycles of mutation at the BCR loci, followed by antigen-mediated selection in specialized regions of the lymph node and spleen. The end result of this process is that the average affinity of antigen-specific BCRs and antibodies formed at the end of an immune response is considerably higher than that at its instigation. This process is referred to as **affinity maturation**.

This section has described the process of the generation of the primary Ig variable region repertoire as it occurs in humans and rodents. Although the same principles apply to most vertebrate species, different species have evolved their own variations. For example, the process of gene conversion is used in chickens, and some species, such as sheep and cows, use somatic hypermutation in the generation of the primary as well as the antigen-experienced repertoire.

In Evolution Box 7-2, we describe the evolution of this system of recombined lymphocyte receptors, addressing the current hypothesis that the key event was the introduction of the *RAG1/2* gene segment into the early vertebrate genome as a transposon.

EVOLUTION

Evolution of Recombined Lymphocyte Receptors

Scientists are slowly beginning to understand how the process of V(D)J recombination may have evolved. With that understanding has come an appreciation for the fact that the BCRs and TCRs of the vertebrate adaptive immune system may not be the only immune receptor molecules generated by recombinatorial genetics.

Ig-like receptors have been identified in species as ancient as the earliest jawed vertebrates. However, extensive analyses of the only surviving jawless vertebrates (Agnathans), the hagfish and lamprey, show no evidence of TCR or BCR V(D)J segments, RAG1/2 genes, or genes encoding elements of a primitive *Major Histo-Compatibility Complex (MHC)* system. This suggests that the RSS-based, recombinatorially generated set of adaptive immune receptors first emerged in a common ancestor of jawed vertebrates, most probably more than 500 million years ago (Figure 1).

The driving force in the development of the adaptive immune system was most probably the incorporation of an ancestor of the RAG1 genes into the ancestral genome in the form of a transposable element. This hypothesis is supported by several observations and experiments:

- The DNA-binding region of RAG1 is strikingly homologous to that of known transposable elements and is evolutionarily related to members of the *Transib* transposase family, currently expressed in species such as fruit flies, mosquitoes, silkworm, and the red flour beetle.
- The inverted repeats in the RSSs of the Ig and TCR gene segments are structurally similar to those found in other transposons.
- The mechanism of action of the RAG proteins involves formation of a DNA hairpin intermediate reminiscent of the action of certain transposases. For example, another phylogenetically con-

served transposon in flies, the HERMES transposon, has been shown to induce a double-strand break via the hairpin-formation mechanism described above for V(D)J recombination.

- RAG proteins have the ability to transpose an RSS-containing segment of DNA to an unrelated target DNA in vitro.

Current thinking supports the hypothesis that the appearance of the transposon in a primordial antigen receptor gene might have separated the receptor gene into two or more pieces. This hypothesis is supported by the presence, in lower chordates, of genes that are related to BCR and TCR V regions, and that therefore could

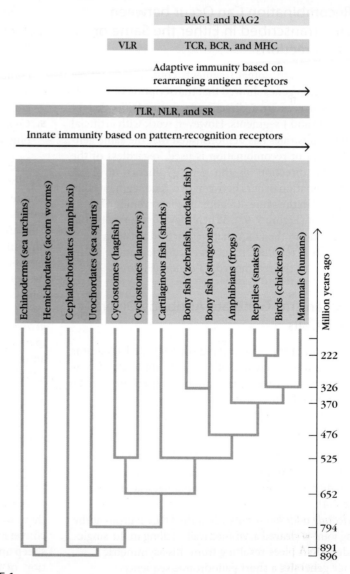

FIGURE 1

The emergence of the RSS-dependent recombination-based adaptive immune system coincides with the first appearance of the jawed vertebrates. SR = scavenger receptor. [*Adapted from Figure 1, M. F. Flajnik and M. Kasahara, 2010, Origin and evolution of the adaptive immune system: Genetic events and selective pressures,* Nature Reviews Genetics **11**:47–59.]

BOX 7-2

have provided the substrate for the transposon invasion. When the transposon subsequently jumped out of the gene, the RSSs would have been left behind. Multiple such rounds of transposition followed by gene duplication would have given rise to the TCR and BCR loci we know today. Interestingly, our knowledge of phylogeny suggests that two waves of gene duplications occurred at the time of vertebrate origin, right around the time of the proposed transposon entry.

Much still remains unknown about the evolution of the adaptive immune receptor molecules. What was the nature of the primordial antigen receptor that was the target for the first transposition? At what stage did the RAG2 gene become associated with the RAG1 component? And how do we account for the evolution of the Ig receptors of species such as the shark, in which fully rearranged V(D)J genes exist in the germ-line genome?

Recent evidence has suggested that the recombinatorial strategy for generating antigen receptors is not limited to the RSS-based Ig and TCR systems of vertebrates. Jawless fish, such as lampreys and hagfish, possess lymphocyte-like cells that can be stimulated to divide and differentiate. Moreover, these cells release specific agglutinins after immunization with antigens, and higher serum levels of these agglutinins are secreted after a secondary, than after a primary, immunization, suggesting that immunological memory exists in these fish. These lymphocyte-like cells therefore appear to express all the hallmarks of adaptive immunity, and yet the agglutinins they secrete do not appear to have an Ig-like structure.

Analysis of the agglutinins has revealed that activated lamprey "lymphocytes" express abundant quantities of *leucine-rich-repeat* (LRR)-containing proteins. LRRs are protein motifs that are frequently associated with protein-protein recognition, and such motifs have already been encountered in the context of the Toll-like receptors described in Chapter 5. The lamprey agglutinin receptors are gener-

ated by recombination of gene segments during lymphocyte development, and the LRR receptor repertoires of these fish are remarkably diverse. These receptors have therefore been named *variable lymphocyte receptor* (VLR) molecules.

Figure 2a shows an example of the arrangement of the protein modules in one of these VLR molecules. At the amino terminal of the protein is an invariant signal peptide, followed by a 27–34 residue *N-Terminal Leucine-Rich Repeat* (LRRNT). This is followed by the first of several

24-residue *Leucine-Rich Repeat* (LRR) modules, LRR1, connected to a series of up to eight 24-residue LRR modules with *variable* sequences (LRRVs). At the C-terminal of the molecule, a 24-residue end *LRRV* segment (LRRVe) is attached to a short 16-residue *connecting peptide* (CP), which culminates in a 48–63 residue *C-terminal LRR* (LRRCT). The VLR molecules are attached to the lymphocyte membrane by an invariant stalk, rich in threonine and proline residues, connecting to a *glycosyl-phosphatidyl-inositol*

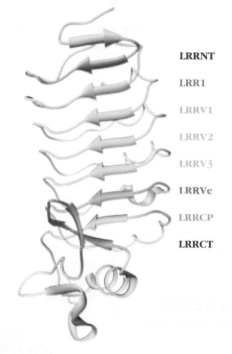

(a)

Amino acids/cassette

SP | LRRNT | LRR1 | LRRV (0-8) | LRRVe | CP | LRRCT — Stalk — GPI — HP

27-34   24   24   24   16   48-63

(b)

5' LRR cassettes   3' LRR cassettes

Functional VLR

LRRNT
LRR1
LRRV1
LRRV2
LRRV3
LRRVe
LRRCP
LRRCT

FIGURE 2
Agnathans such as lampreys and hagfish have evolved a recombination-based system of recognition molecules that does not depend on RSSs. Agnathans, such as lamprey and hagfish, have lymphocyte-like cells which carry antigen receptors that are remarkably diverse and are created by a system of recombination of small gene cassettes. (a) The modules that make up a complete VLR receptor. (See text for details.) (b) The assembly of a complete VLR gene from the basic skeleton germ-line and flanking cassettes. (c) The three-dimensional structure of a VLR, showing the repeating arrangement of the randomly recombined LRR protein modules. [*Adapted from Figures 1 and 2, Herrin, B. R., and M. D. Cooper. 2010. Journal of Immunology **185:**1367.*]

(continued)

(GPI) anchor and a hydrophobic peptide. Lymphocyte activation leads to phospholipase cleavage at the GPI anchor, enabling soluble forms of the VLRs to be released from the lymphocyte following antigen stimulation.

The assembly process that generates the completed protein occurs only in lymphocytes. In the germ-line DNA, LRR cassettes flank the skeleton *LRR* genes, which consist initially only of sequences coding for parts of the LRRNT and LRRCT, separated by non-coding DNA (Figure 2b). During lymphocyte development, the non-coding sequence is replaced with variable LRRs. Gene segments are copied from one part of the genome into another in a one-way process, similar to gene conversion. Each lamprey lymphocyte

expresses a unique VLR gene from a single allele, and the diversity of the repertoire is limited only by the number of lymphocytes.

Figure 2c shows a cartoon of the structure of one of these receptors, generated from an x-ray crystallographic model. Comparison of the sequences of several of these primarily β-stranded structures has shown that their sequence diversity is concentrated on the concave (left) side of the molecule and can be attributed to the inherent diversity of the LRR cassettes.

Did the two LRR- and V(D)J-based, recombinatorially derived sets of immune receptors exist side by side in an ancestral species? This we do not know. Agnathans other than lampreys and hagfish were extinguished 400 million years ago, and

we have access only to the fossil remains of ostracoderms (agnathans with dermal skeletons), which are thought to be the ancestors of the gnathostomes. One theory would hold that the VLR recombinatorial system evolved in an ancestor common only to hagfish and lampreys, and could have been causal in their ability to survive when other agnathan species succumbed to environmental insult. Alternatively, the VLR and Ig-based systems may have co-existed for a time, with lymphocytes bearing both types of receptors, until one of the systems was lost.

Combinatorial genetics in the generation of the vertebrate antigen receptors may not be as unique a mechanism as immunologists had previously thought.

B-Cell Receptor Expression

The expression of a receptor on the surface of a B cell is the end result of a complex and tightly regulated series of events. First, the cell must ensure that the various gene recombination events culminate in productive rearrangements of both the heavy- and light-chain loci. Second, only one heavy-chain and one light-chain allele must be expressed in each B cell. Finally, the receptor must be tested to ensure it does not bind self antigens, in order to protect the host against the generation of an autoimmune response.

Allelic Exclusion Ensures That Each B Cell Synthesizes Only One Heavy Chain and One Light Chain

The random nature of Ig heavy- and light-chain gene rearrangement means that more than one heavy-light-chain pair could potentially be expressed on the surface of individual B cells. Furthermore, since each heavy chain could potentially combine with both light chains and vice versa, this could result in the creation of B cells bearing a variety of different antigen-binding sites. Whereas this opportunity to increase the number of available receptors may initially sound advantageous to the organism, in practice the presence of more than one receptor per B cell creates prohibitive difficulties for those mechanisms that protect against the generation of autoimmunity. The mechanism by which B cells ensure that only one heavy- and one light-chain allele are transcribed and translated is referred to as **allelic exclusion**.

The rearrangement of *Ig* genes occurs in an ordered way, and begins with recombination at one of the two homologous chromosomes carrying the heavy-chain loci. The production of a complete heavy chain and its expression on the B-cell surface in concert with a surrogate light chain, made up of two proteins, VpreB and λ5 (Figure 7-10), signals the end of heavy-chain gene rearrangement and, thus, only one antibody heavy chain is allowed to complete the rearrangement process. The mechanism by which this allelic exclusion occurs is still under investigation. However, we do know that following successful arrangement at one allele, the V_H gene locus on the other chromosomes is methylated and recruited to heterochromatin.

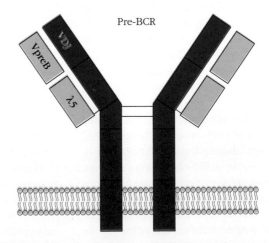

FIGURE 7-10 Ig heavy-chain expression on the cell surface in concert with VpreB and λ5 signals, via Igα/Igβ, the end of heavy-chain rearrangement.

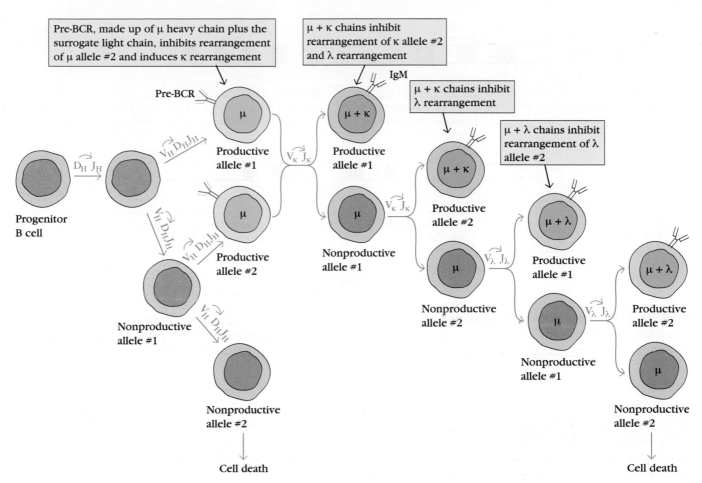

FIGURE 7-11 The generation of a functional immunoglobulin receptor requires productive rearrangement at heavy- and light-chain alleles. Heavy-chain genes rearrange first, and once a productive heavy-chain gene rearrangement occurs, the pre-BCR containing the μ protein product prevents rearrangement of the other heavy-chain allele and initiates light-chain rearrangement. In the mouse, rearrangement of κ light-chain genes precedes rearrangement of the λ genes, as shown here. In humans, either κ or λ rearrangement can proceed once a productive heavy-chain rearrangement has occurred. Formation of IgM inhibits further light-chain gene rearrangement. If a nonproductive rearrangement occurs for one allele, then the cell attempts rearrangement of the other allele. *[Adapted from G. D. Yancopoulos and F. W. Alt, 1986, Regulation of the assembly and expression of variable-region genes, Annual Review of Immunology 4:339.]*

If the first attempt at a heavy-chain rearrangement is unproductive (i.e., results in the formation of out-of-frame joints caused by P and N nucleotide addition or by exonuclease trimming at the joint), rearrangement will initiate at the second heavy-chain allele. If this is also unsuccessful, the B cell will die. Once a complete heavy chain is expressed, light-chain rearrangement begins. In mice, light-chain rearrangement always begins on one of the κ alleles, and continues until a productive light-chain rearrangement is completed. In humans, it may begin at either a κ or a λ light-chain locus. Because successive arrangements (using upstream V regions and downstream J regions) can occur on the same light-chain chromosome (see below) and there are four alleles (two κ and two λ) to choose from, once a B cell has made a complete heavy chain, it will normally progress to maturity. Successful completion of light-chain rearrangement results in the expression of the BCR on the cell surface, and this signals the end of further *Ig* receptor gene rearrangement events (Figure 7-11).

The process of generating a fully functional set of B cells is energetically expensive for the organism because of the amount of waste involved. On average, two out of three attempts at the first heavy-chain chromosomal locus will result in an unproductively rearranged heavy chain, and the same is true for the rearrangement process at the second heavy-chain locus. Therefore the probability of successfully generating a functional heavy chain is only 1/3 (at the first allele) + {2/3 (probability the first rearrangement was not successful) × 1/3 (probability that the second rearrangement is successful)} = 0.55 or 55%.

Receptor Editing of Potentially Autoreactive Receptors Occurs in Light Chains

The immune system can still make some changes even after the completed receptor is on the cell surface. If the receptor is found to be auto-reactive, the cell can swap it out for

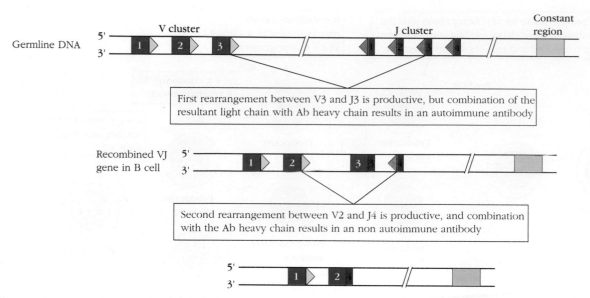

FIGURE 7-12 κ light-chain receptor editing. In the event of a productive light-chain rearrangement that leads to the formation of an auto-antibody, additional rounds of rearrangement may occur between upstream V_κ and downstream J_κ gene segments. Here, the primary rearrangement between V3 and J3 is edited out with a secondary rearrangement between V2 and J4.

a different one, in a process termed **receptor editing**. Since the receptor's binding site contains elements of both the heavy and the light chain, changing just one of these is often sufficient to alter the specificity, and light-chain receptor editing has been shown to occur quite frequently.

In this process, the DNA rearrangement machinery is switched back on after the first completed receptor in the immature B cell has been found to have autoimmune specificity. For example, as shown in Figure 7-12, the initial rearrangement of V3 to J3 may result in an Ig receptor molecule that bound to a self antigen. The B cell can then rearrange a light-chain gene at the second κ-chain allele, it can rearrange a λ-chain allele, or it can edit the original κ-chain rearrangement. All of these choices have been demonstrated in different B cells. However, the term *receptor editing* refers to the process in which a B cell uses the same allele more than once, and engages in a second rearrangement process in which a variable region gene segment upstream of the original segment (in this case V2) is recombined with a J segment downstream of the initially utilized J region (in this case, J4).

Switching out the rearranged V_H segments is less common. Since rearrangement of the heavy-chain variable region deletes the RSSs on either side of the D region sequences, it was at first thought to be impossible. However, recent advances have demonstrated that cryptic RSSs exist that can be used in heavy-chain editing, and so some measure of heavy-chain receptor editing does occur.

Ig Gene Transcription Is Tightly Regulated

Immunoglobulin genes are expressed only in cells of the B lymphocyte lineage, and within this lineage, expression of *Ig*

genes is developmentally regulated (see Chapter 10). The control of *Ig* gene transcription is therefore necessarily complex, although it involves the familiar paradigm of transcription factors that bind to sequence elements in the DNA (Table 7-5 and Figure 7-13).

Expression of the Ig genes requires the coordination of two types of cis regulatory elements: promoters and enhancers. Like other promoters, the Ig heavy and light-chain gene promoters contain a highly conserved, AT-rich sequence called the **TATA box**, which serves as the binding site for RNA polymerase. RNA polymerase II starts transcribing the DNA from the initiation site, about 25 base pairs (bp) downstream of the TATA box. Each V_H and V_L gene segment has a promoter located just upstream from its leader sequence. The promoters upstream of Ig V_H and V_L gene segments bind to RNA polymerase II quite weakly, and the rate of transcription of V_H and V_L coding regions is very low in unrearranged germ-line DNA. Rearrangement of Ig genes brings the promoters close enough to enhancers located in the intron sequences and in regions downstream of the C region genes for them to influence the V region promoters. Consequently, the rate of transcription is increased by a factor of 10^4 upon V gene rearrangement.

However, although the transcription of unrearranged genes is low, it is not absent in developing B cells; evidence now suggests that, as the Ig genes move out of the pericentric regions of the nucleus and assume more euchromatic configurations prior to recombination, a low level of transcription occurs which indicates to the recombination machinery that they are in a state of readiness for rearrangement to occur.

Elements	Function	Location(s) in Ig genes	Symbol in Figure 7-13
Promoters	Promote initiation of transcription of neighboring gene in a specific direction. Bind RNA polymerase and direct the formation of the pre-initiation complex.	Upstream of V_H, D_μ, V_κ, $V_\lambda 1$, and $V_\lambda 2$. Additional promoters are found upstream of all constant regions.	Arrow denotes direction of transcription.
Enhancers	Stimulate transcription of associated genes/gene segments when bound by transcription factors (TFs). Can function at variable distance from promoter, and in either orientation.	Intronic enhancers lie between J_H and C_μ and between J_κ and $C\kappa$. 3' α enhancers lie to the 3' end of the C_α gene and may be part of a locus control region. Light-chain enhancers lie 3' of C_κ and C_λ genes.	
Locus control regions	Collection of smaller elements that may each have enhancer function.	Complex regulatory region, RR, 3' of murine $C\alpha$ gene. Contains four enhancers in two separate units.	

TABLE 7-5 Some of the cis regulatory elements that control immunoglobulin gene transcription

The J_κ cluster and each of the D genes of the heavy-chain locus are also preceded by promoters. These D and J_κ promoters allow for transient transcription of DJ and J segments, which appears to be important for efficient VDJ rearrangement. So far, there is no evidence of conventional promoter sequences upstream of any of the J_λ loci, although some germ-line transcription of J_λ sequences has been detected.

The control of Ig gene transcription depends on the interplay between a large number of transcription factors that bind to the promoter and enhancer regions of the Ig genes, and binding of different combinations of these proteins to

promoters and enhancers can lead to very different outcomes. In Chapter 10, we discuss how transcription of Ig genes is controlled by varying combinations of transcription factors during B cell development.

Occasionally, the powerful B-cell Ig enhancers can engage in activities that are less than benign. Translocation of the *c-myc* oncogene close to the Ig enhancer regions results in constitutive expression of the c-Myc protein and the generation of an aggressive, highly proliferative B-cell lymphoma. Similarly, the translocation of the *bcl-2* gene close to the Ig enhancer leads to suspension of programmed cell death in B cells, resulting in follicular B-cell lymphoma.

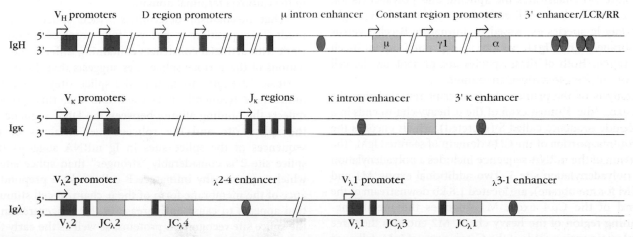

FIGURE 7-13 Locations of the cis regulatory control elements on immunoglobulin genes. The expression of Ig genes is controlled by transcription factor binding to promoter and enhancer regions. Promoters are designated by rectangular arrows and enhancer regions by green ovals. Accessibility to recombination is often preceded by low levels of transcription from downstream D and J segments.

Mature B Cells Express Both IgM and IgD Antibodies by a Process That Involves mRNA Splicing

As described above, the same Ig heavy-chain variable region can be expressed in association with more than one heavy-chain constant region. Furthermore, each of the heavy-chain constant regions can be synthesized as both membrane-bound and secreted forms.

Immature B cells in the bone marrow express only membrane IgM receptors. However, as the B cells mature, they also place IgD receptors on their cell surface. These IgD receptors bear the identical antigen-binding V region, but carry a δ, rather than a μ constant region. In this section, we describe how cells create the membrane-bound and secreted forms of IgM and IgD by RNA splicing. The generation of antibodies belonging to the heavy-chain classes other than IgM and IgD occurs by an additional DNA recombination process, referred to as **class switch recombination (CSR)**, and results in the loss of the DNA encoding the μ and δ gene segments. CSR is described in Chapter 12, because it is a process that occurs only after antigen-mediated activation of B cells.

The primary heavy-chain transcript in a B cell that has not yet been exposed to antigen encodes both the Cμ and the Cδ constant regions, and is approximately 15 kb in length. The determination of whether a cell will synthesize the membrane-bound or secreted forms of μ and/or δ heavy chains is made by nuclear proteins that select particular choices of polyadenylation and RNA splicing sites. During mRNA processing, polyadenylation sites signal for the cleavage and addition of a poly-A sequence at a particular site on the primary transcript, that then becomes the 3′ terminus of the mature, processed mRNA. (Poly-A tails are characteristically found at the 3′ termini of many eukaryotic mRNA species.) A nuclear complex called the **spliceosome** determines the sites at which the primary transcript will be cleaved and spliced to form the mature mRNA.

Figure 7-14 illustrates the splicing and polyadenylation choices made when a cell generates membrane bound **IgM** (the only Ig species expressed by immature B cells) versus membrane-bound **IgD**, which is expressed upon B-cell maturation. Both of these species are present on the cell membrane prior to antigen encounter.

Analysis of the primary Ig transcript revealed that the Cμ4 exon (the 3′-most exon of the μ heavy chain) ends in a nucleotide sequence called S (secreted), which encodes the hydrophilic portion of the CH4 domain of secreted IgM. The S portion of the mRNA sequence includes a polyadenylation site, polyadenylation site 1. Two additional exons, M1 and M2 (M for membrane), are located 1.8 kb downstream of the 3′ end of the Cμ4 exon. M1 encodes the membrane-spanning region of the heavy chain. M2 encodes the three cytoplasmic amino acids at the C-terminus of IgM, followed by polyadenylation site 2 (see Figure 7-14a). Production of

mRNA encoding the membrane-bound form of the μ chain occurs when cleavage of the primary transcript and addition of the poly-A tail occurs at polyadenylation site 2 (see Figure 7-14b). RNA splicing then removes the S sequence at the 3′ end of the Cμ4 exon, and joins the remainder of the Cμ4 exon to the M1 and M2 exons. The RNA sequence between the M1 and M2 exons is also spliced out, resulting in the production of mRNA encoding the membrane-bound form of the IgM heavy chain. This is the only splicing pattern that occurs in *immature* B cells.

As the B cell matures, it begins to express membrane-bound IgD in addition to membrane IgM (see Figure 7-14c). The exons encoding the membrane-bound and secreted forms of IgD are arranged similarly to those of IgM, with polyadenylation sites 3 and 4 at the 3′ termini of the sequences encoding the secreted and membrane-bound forms of IgD respectively. If polyadenylation occurs at site 4, and the Cμ RNA is spliced out, the cell will make the membrane-bound δ chain, and express IgD as well as IgM on the cell surface. Since *mature* B cells bear both IgM and IgD on their surfaces, it is clear that both splicing patterns can occur simultaneously.

Following antigen stimulation, the B cell begins to generate the secreted form of IgM (Figure 7-15a). Polyadenylation shifts to site 1, and the mRNA sequences downstream of site 1 are degraded. (Similarly, the use of splice site 3 generates secreted IgD, although this is not shown for purposes of clarity.) The secreted form of IgD is used less frequently than the secreted form of the other antibody heavy-chain classes, and so splice site 3 is used relatively rarely.

At the protein level, the carboxyl-terminus of the membrane-bound form of the Ig μ chain consists of a sequence of about 40 amino acids, which is composed of a hydrophilic segment that extends outside the cell, a hydrophobic transmembrane segment, and a very short hydrophilic segment at the carboxyl end that extends into the cytoplasm (Figure 7-15b). In the secreted form, these 40 amino acids are replaced by a hydrophilic sequence of about 20 amino acids in the carboxyl terminal domain.

What mechanism determines which splice sites are used? This question is not yet fully resolved. However, work with artificial constructs containing different combinations of the various splice sites suggests that there is an intrinsic "strength" to each of the splice sites that determines how frequently it is used, and that this strength reflects the binding affinity between the DNA sequence at the splice site and the spliceosome. Analysis of the sequences of the splice sites in Ig mRNA suggests that splice site 2 is considerably "stronger" than splice site 1, which explains why immature B cells make a preponderance of the membrane form of the μ chain. B-cell stimulation appears to cause an increase in the concentration of the splice site recognition proteins, as well as the early termination of transcription prior to the M1 and M2 sequences, thereby facilitating the use of splice site 1 and

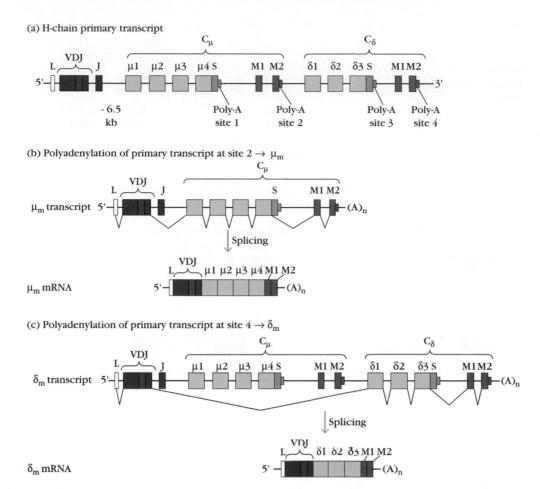

FIGURE 7-14 Differential expression of membrane forms of μ and δ heavy chains by alternative RNA processing. (a) Structure of rearranged heavy-chain gene showing C_μ and C_δ exons and poly-adenylation sites. (b) Structure of μ_m primary transcript and μ_m mRNA resulting from polyadenylation at site 2, followed by RNA splicing. (c) Structure of δ_m primary transcript and δ_m mRNA resulting from polyadenylation at site 4 and RNA splicing. Both processing pathways can proceed in any given B cell.

generation of the secreted form of the protein. A similar mechanism may control the differential splicing of the membrane-bound and secreted forms of the δ chain.

T-Cell Receptor Genes and Expression

The initial characterization of the BCR molecule was facilitated by the fact that the secreted Ig product of the B cell shares the antigen-binding region with the membrane receptor. The secreted Ig proteins could be used as antigens to generate antibodies in other species that recognized the B-cell surface receptor and these anti-receptor antibodies could then be employed as reagents to isolate and characterize the BCR. In addition, a set of monoclonal B-cell tumors secreting high concentrations of soluble antibodies was available.

Investigators attempting to purify and analyze the *T-cell receptor* (TCR) enjoyed no such advantages, and therefore it should not be surprising to learn that several decades

elapsed between the characterization of the BCR and that of its T-cell counterpart. Indeed, characterization of the TCR was so difficult and vexing that one of the distinguished writers for the scientific journal *Nature*, Miranda Robertson, referred to it as "The Hunting of the Snark," in reference to the Lewis Carroll poem of the same name, which described the search for a mythical creature.

In Chapter 3, we described the TCR protein. Below, we describe the parallel story of the discovery of the TCR genes.

Understanding the Protein Structure of the TCR Was Critical to the Process of Discovering the Genes

In Chapter 3 we described how investigators were able to use a monoclonal antibody (mAb) specific for the TCR to purify and characterize it as an αβ heterodimeric protein. Each chain was shown to consist of one variable and one constant region domain, and both chains contained regions arranged

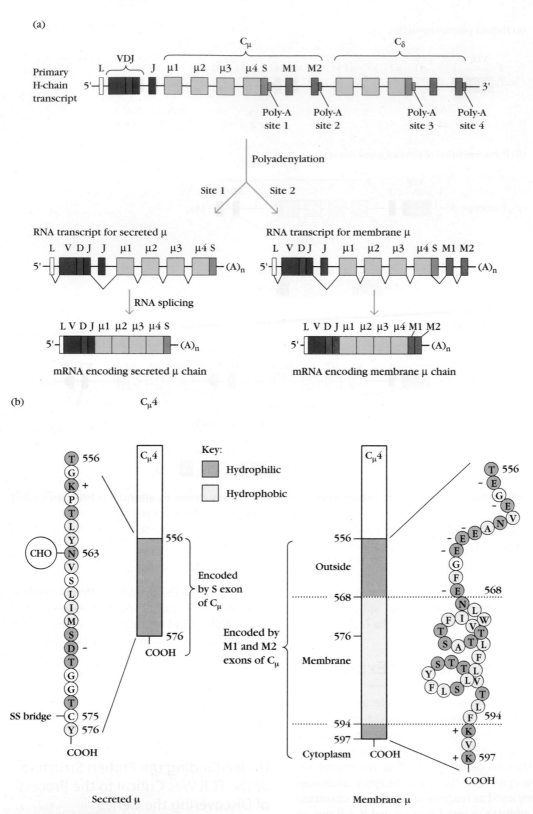

FIGURE 7-15 Differential expression of the secreted and membrane-bound forms of immunoglobulin μ chains is regulated by alternative RNA processing. (a) Structure of the primary transcript of a rearranged heavy chain, showing the C_μ exons and poly-A sites. Polyadenylation of the primary transcript at either site 1 or site 2, and subsequent splicing (indicated by V-shaped lines) generates mRNAs encoding either secreted or membrane μ chains, as shown. (b) Amino acid sequences of the carboxyl terminal ends of secreted and membrane μ heavy chains. Residues are described by the single-letter amino acid code. Hydrophilic (pink) and hydrophobic (yellow) residues and regions are shown, charged amino acids are indicated with a + or −, and the N-linked glycosylation site on the secreted form is labeled "CHO."

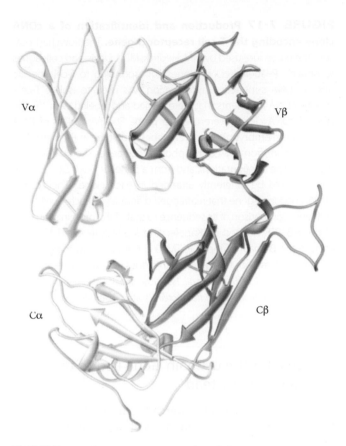

FIGURE 7-16 The heterodimeric αβ T-cell receptor. The alpha chain is shown here in green and the beta chain in blue. *[PDB ID 1TCR.]*

in the characteristic Ig fold (Figure 7-16). The presence of these variable and constant regions provided clues as to the potential arrangement of the genes that encode the TCR proteins, and enabled researchers to design the experiments that ultimately led to the characterization of the TCR genes.

The β-Chain Gene Was Discovered Simultaneously in Two Different Laboratories

Two different research groups published papers in the same issue of *Nature* in 1984 that described the discovery of the TCR genes in mice and in humans respectively. Mark Davis, Stephen Hedrick, and their collaborators utilized a uniquely creative approach to isolate the receptor genes from several mouse T-cell hybridoma cell lines, which we describe below. The other group, headed by Tak Mak, used more classical methods of genetic analysis to isolate the TCR genes from human cell lines.

The Davis-Hedrick group strategy was based on four assumptions about TCR genes:

- TCR genes will be expressed in T cells but not B cells.
- Since the genes encode a membrane-bound receptor, the transcribed mRNA will be found associated with membrane-bound polyribosomes.

- Like Ig genes, TCR genes will code for a variable and a constant region.
- Like Ig genes, the genes that encode the TCR will undergo rearrangement in T cells.

Prior experiments had revealed that only about 2% of the genes expressed by lymphocytes differed between T and B cells. Furthermore, only a small proportion (about 3%) of T-cell mRNA had been shown to be associated with membrane-bound polyribosomes. Davis and Hedrick reasoned that if they could generate ^{32}P-labeled *DNA* copies (cDNA) of the membrane-bound polyribosomal fraction of antigen-specific T-cell mRNA, and remove from that population those sequences that were also expressed in B cells, they would be left with labeled, T-cell-derived cDNA that would be greatly enriched in sequences encoding the TCR (Figure 7-17).

They therefore performed the following steps:

- Extract membrane-bound polysomal mRNA (RNA sequences that encode membrane-bound proteins) from several T hybridoma lines.
- Reverse transcribe the polysomal mRNA in the presence of ^{32}P-labeled nucleotides, to generate radioactive cDNA copies of each of the mRNA species.
- Mix the ^{32}P-labeled T-cell cDNA with B-cell mRNA and remove all the cDNA that hybridized with B-cell mRNA. (This leaves behind only the T-cell mRNA that encodes proteins not expressed in B cells.)
- Recover the "T-cell-specific cDNAs" and use them to identify hybridizing DNA clones from a T-cell specific library.
- Radioactively label those clones that hybridized to the "T-cell-specific" cDNAs, and use cloned probes individually to probe Southern blots of DNA from liver cells and B-cell lymphomas (which would not be expected to rearrange TCR genes) and DNA from different T-cell hybridomas (which would be expected to show different patterns of rearrangements in different T-cell clones).

Probing with the TM90 cDNA probe (see bottom of Figure 7-17) resulted in a pattern of bands that did not vary depending on the origin of the cellular DNA. This indicated that TM90 recognizes DNA found in T cells but not B cells, that does *not* rearrange in different ways in different cells, and therefore is probably not complementary to a TCR gene. In contrast, the TM86 cDNA probe hybridized to bands in the T-cell DNA of different sizes than in either the liver or the B-cell DNA. Furthermore, TM86 *showed different patterns of hybridization in different T-cell hybridomas*, indicating that the sequence recognized by the TM86 probe is found in varying contexts within the DNA of different hybridomas. Note that this probe also shows the same pattern of bands in the liver cells and B cells. This is consistent with the notion that the genes encoding the TCR recombine in T cells, but not in any other cell type. These results indicated

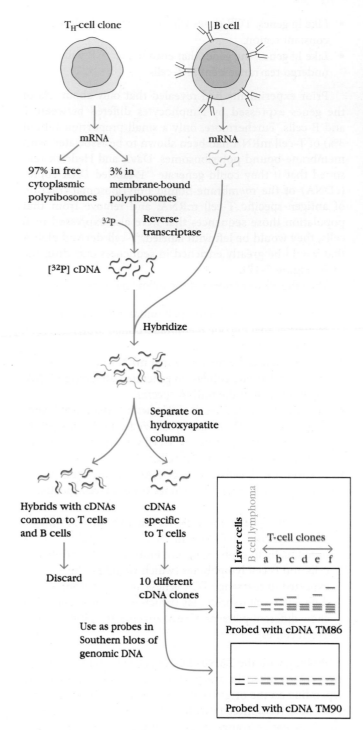

T$_H$-cell clone

B cell

mRNA

mRNA

97% in free cytoplasmic polyribosomes

3% in membrane-bound polyribosomes

32p

Reverse transcriptase

[^{32}P] cDNA

Hybridize

Separate on hydroxyapatite column

Hybrids with cDNAs common to T cells and B cells

cDNAs specific to T cells

Discard

10 different cDNA clones

Use as probes in Southern blots of genomic DNA

Liver cells

B cell lymphoma

T-cell clones

a b c d e f

Probed with cDNA TM86

Probed with cDNA TM90

FIGURE 7-17 Production and identification of a cDNA clone encoding the T-cell receptor β gene. The flowchart outlines the procedure used by S. M. Hedrick, M. M. Davis, and colleagues to obtain ^{32}P-labeled cDNA clones corresponding to T-cell-specific mRNAs. DNA subtractive hybridization was used to isolate T-cell-specific cDNA fragments that were cloned and labeled with ^{32}P. The labeled cDNA clones were used to probe Southern blots of DNA isolated from embryonic liver (TCR genes should be in the germ-line configuration), from a B-cell lymphoma (TCR genes should also be in the germ-line configuration), and from a panel of T-cell clones. (TCR genes should be differently arranged in each clone.) Their clone, TM86, encodes a gene that rearranged differently in each T-cell clone analyzed. Comparison of its sequence to that of the protein sequence of the αβ TCR isolated by Kappler and Marrack revealed TM86 to encode the β chain of the TCR. TM90 cDNA identified the gene for another T cell membrane molecule that does not undergo rearrangement. [Based on S. M. Hedrick et al., 1984, Isolation of cDNA clones encoding T cell-specific membrane-associated proteins, Nature **308**:149.]

A Search for the α-Chain Gene Led to the γ-Chain Gene Instead

Focus now switched to the search for the TCR α chain. But here, immunology took one of its strange and wonderful twists.

Tonegawa's lab, this time using the subtractive hybridization approach pioneered by Hedrick and Davis, succeeded in cloning a gene that appeared at first to have all the hallmarks of the TCR α chain gene. It was expressed in T cells, but not in B cells; it was rearranged in T cell clones; and on sequencing, it revealed regions corresponding to a signal peptide, two Ig family domains, a transmembrane region, and a short cytoplasmic peptide. Furthermore, the predicted molecular weight of the encoded chain appeared to be very close to that of the α-chain protein. Tonegawa's lab initially suggested that "while direct evidence is yet to be produced, it is very likely the pHDS4/203 codes for the α subunit of the T cell receptor."

However, biochemical analysis of the TCR α and β chain proteins by another laboratory had clearly demonstrated that both the α and the β chains of the TCR heterodimer were glycosylated. A search for potential sequences corresponding to sites of carbohydrate attachment on Tonegawa's putative α-chain sequence came up short. It therefore seemed that the gene Tonegawa's lab had isolated was not the α chain, but something very similar. Further work demonstrated that they had discovered a gene encoding an hitherto unknown receptor that contained both variable and constant regions and recombined only in T cells. They named this gene (and the receptor chain it encoded) γ.

Davis and colleagues next isolated a genomic sequence expressed in T and not B cells that encoded a protein with four potential N-linked glycosylation sites, and a molecular weight consistent with that of the α chain of the TCR. With

that the TM86 cDNA probe was binding to DNA that is rearranged specifically only in T cells and thus most probably encodes a gene for a TCR protein. In this way the first mouse TCR gene was isolated.

Comparison of the DNA sequence of Mak's human clone with the amino acid sequence from a human hybridoma confirmed that Davis and Hedrick et al., and Mak et al., had isolated the gene for the mouse and human forms of the TCR β gene, respectively.

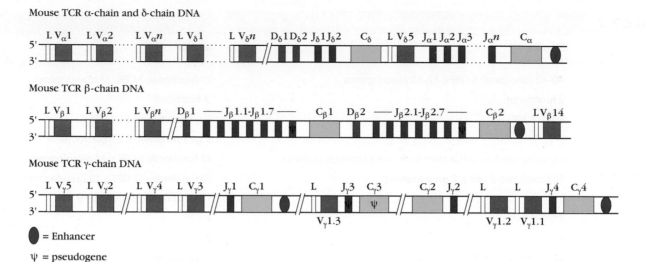

Mouse TCR α-chain and δ-chain DNA

Mouse TCR β-chain DNA

Mouse TCR γ-chain DNA

⬤ = Enhancer

ψ = pseudogene

FIGURE 7-18 Germ-line organization of the mouse TCR α-, β-, γ-, and δ- chain gene segments. Each C gene segment is composed of a series of exons and introns, which are not shown. The organization of TCR gene segments in humans is similar. Approximate numbers of the gene segments are shown in Table 7-7. *[Adapted* *from D. Raulet, 1989, The structure, function, and molecular genetics of the gamma/delta T cell receptor,* Annual Review of Immunology *7:175; and M. M. Davis, 1990, T cell receptor gene diversity and selection,* Annual Review of Biochemistry *59:475.]*

the genes for the α and β chains of the TCR fully characterized, those attempting to understand TCR genetics were left with a series of intriguing questions. Was the γ chain expressed by the same T cells that expressed the αβ receptor? Did the γ chain that Tonegawa had discovered also exist as a heterodimer? If so, what was its partner, and who would find it first?

After 3 years of searching, Chien and colleagues, working in the Davis lab, described a fourth TCR gene that they named δ. Intriguingly, the V_δ chain gene was found to lie downstream of the previously described V_α genes and just 5' to the $J_\alpha C_\alpha$ coding regions. Rearrangement of this locus was found to occur very early in thymic differentiation, and expression of δ-chain RNA paralleled that of the expression of the γ chain in thymic subpopulations, occurring in the same T cells. Most striking about the location of the δ genes was the observation that, if a T cell rearranged the α chain, the intervening δ genes would be lost (Figure 7-18). Study of the T cells expressing these γ,δ receptors demonstrated that γ,δ-bearing T cells represented an entirely new T-cell subset

with an antigen repertoire distinct from that of α,β T cells and a different anatomical distribution within the lymphoid system in the host.

TCR Genes Undergo a Process of Rearrangement Very Similar to That of Ig Genes

Table 7-6 shows the chromosomal locations of the TCR receptor genes in mice and humans and Figure 7-18 shows the arrangement of the TCR genes in the mouse genome. The four TCR loci are organized in the germ line in a manner remarkably similar to the multigene organization of the Ig loci, shown in Figure 7-4, and as in the case of Ig genes, functional TCR genes are produced by rearrangements of V and J segments in the α- and γ-chain families and between V, D, and J segments in the β- and δ-chain genes.

The approximate numbers of different V, D, and J segments in each of the TCR gene families of mice and humans are shown in Table 7-7. Interestingly, some of the V_α and V_δ gene segments have been shown to be used in both α and δ TCR chains, which further adds to the diversity of the TCR repertoire.

Rearrangement of TCR variable region genes follows the same general outline as that of Ig genes. One point of contrast however is that, whereas Ig light chains incorporate few N nucleotides because of the developmental down-regulation of expression of the TdT enzyme in B cells by the time of light-chain rearrangement, N nucleotides are seen at similar frequencies in all of the TCR chains.

The α- and γ-chain variable genes, like the Ig light-chain gene, are generated from one V and one J segment. The TCR β and δ chain variable region genes are assembled from V,

TABLE 7-6	Chromosomal locations of TCR genes in humans and mice	
Gene	**Human**	**Mouse**
α chain	14	14
β chain	7	6
γ chain	7	13
δ chain	14	14

TABLE 7-7	TCR gene segments in humans and mice	
	Human	**Mouse**
V_β	40–42 functional; 6 ORFs; 12–13 pseudogenes	21 functional; 1 ORF; 12 pseudogenes
D_β	2 functional	2 functional
J_β	6 $J_\beta 1$; 7 $J_\beta 2$	5 $J_\beta 1$ functional; 2 $J_\beta 1$ ORFs
		6 $J_\beta 2$ functional; 1 $J_\beta 2$ pseudogene
V_α	43; some used in conjunction with α or δ constant regions	71 functional; 2 ORFs; 14 pseudogenes
J_α	50 functional; 8 ORFs; 3 pseudogenes	39 functional; 12 ORFs; 10 pseudogenes
V_δ	3*	15 functional; 1 pseudogene
D_δ	3	2
J_δ	4	2
V_γ	4–6 functional; 3 ORFs; 5 pseudogenes	7 functional
J_γ	5	4

* Additional gene segments (5 for human and 9 for mouse) have been found able to associate with downstream δ or α gene segments.

A germ-line gene is considered to be functional if the coding region has an ORF without a stop codon and, if there is no described defect in the splicing sites, RSS and/or regulatory elements.

A germ-line entity is considered to be an ORF if the coding region has an open reading frame but alterations have been described in the splicing sites, RSSs, and/or regulatory elements, and/or changes of conserved amino acids have been suggested by the authors to lead to incorrect folding and/or the entity is an orphan.

A germ-line entity is considered to be a pseudogene if the coding region has a stop codon(s) and/or a frameshift mutation. In the case of V gene segments, these mutations may be either in the V gene coding sequence, or in the leader sequence.

SOURCE: Gene numbers and definitions summarized from the International Immunogenetics Information System Web site: http://imgt.org. Accessed November 16, 2011.

D, and J segments, like the Ig heavy chains. In contrast to the Ig genes, however, the TCR families do not have a functionally diverse set of C regions. There is one constant region gene for each of the α and δ gene families, and although there is more than one gene encoding each of the C_β and C_γ regions, they are highly homologous and do not appear to differ in function.

Animals deficient in RAG1/2 or the other enzymes affecting Ig gene recombination, such as Artemis or TdT, are similarly deficient in the generation of both their BCRs and their TCRs, and so it is thought that the essential features of the rearrangement processes are similar in B and T cells (Clinical Focus Box 7-3). The heptamer-nonamer recombination signal sequences are found at the 3′ termini of the V sequences, the 5′ and 3′ termini of the D sequences, and the 5′ termini of the J sequences, just as for the Ig variable region gene segments.

Recall that the 12-23 rule specifies that recombination can only occur between gene segments contiguous in one case with an RSS containing a 12 (one-turn) bp spacer and in the other with an RSS containing 23 (two-turn) bp spacer (see Figure 7-5). By this rule, recombination between V_H and J_H immunoglobulin gene segments is disallowed, and hence all properly rearranged heavy-chain V regions must have all three segments represented.

In contrast, the D regions of the TCR β and δ chain genes are bordered on the one side by a 12-bp spacer and on the other by a 23-bp spacer (Figure 7-19). By the 12-23 rule,

V and J segments from the β and δ gene families can theoretically recombine directly, without an intervening D segment. Furthermore, the rule would also allow for D-D segment recombination. Do these unusual recombinations occur in vivo?

In the case of the β-chain gene, the answer appears to be that they do not. Evidence obtained from sequencing cDNA fragments containing precisely known V, D, and J sequences has demonstrated that functioning TCR β-chain genes always contain one each of the V, D, and J segments. However, the situation is not quite as simple for the TCR δ-chain genes.

Scientists noted that many TCR V_δ genes were considerably longer than TCR V_β genes; furthermore, the increased sequence length seemed primarily to occur in those parts of the gene encoding the CDR3 residues that make antigen contact. Subsequent analysis showed that many TCR V_δ genes have incorporated not one but two D-region segments, and this additional segment is primarily responsible for the observed increase in gene length. In some V_δ genes, further length was added to the TCR δ-chain gene by N nucleotide addition at the D-D joint.

In contrast to the dramatic levels of diversity facilitated by the incorporation of N nucleotides in all chains of the TCR, TCRs do not appear to undergo somatic hypermutation following antigen contact at any appreciable rate.

Table 7-8 compares the mechanisms of the generation and expression of diversity in BCR versus TCR molecules.

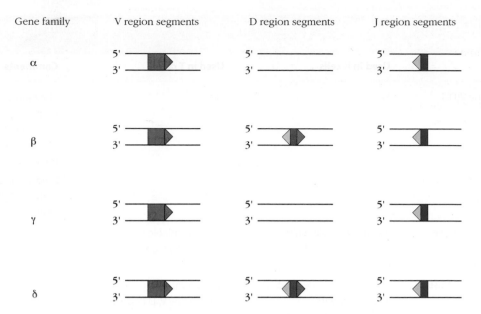

FIGURE 7-19 The locations of the 12-bp and 23-bp RSS spacers in TCR genes. Note that the spacers on either side of the D_β and D_δ gene segments are different, allowing for the occurrence of recombined genes bearing two copies of the D_δ. Duplication of D_β segments has not been observed in vivo.

TCR Expression Is Controlled by Allelic Exclusion

Allelic exclusion occurs among TCR genes almost as efficiently as among BCR genes, with one exception: some fraction of T cells bear more than one TCR α chain. However, it would be unlikely, although not impossible, that a T cell bearing two α chains would be able to recognize more than one antigen, given the complexity of T-cell antigen recognition. In Chapter 8, we explain how T cells recognize an antigen made up of a foreign peptide presented to the immune system in a specialized groove in a *Major Histocompatibility Complex* (MHC) antigen. As we will learn in Chapter 9, TCRs are tested twice within the thymus to determine their functionality before the T cells are released into the periphery. First, T cells are tested to ensure that the α and β chains can pair into a functioning TCR that is capable of recognizing self MHC antigens bearing self peptides with low to moderate affinity (*positive selection*). This is necessary because the T cell must have some minimal capacity to recognize self MHC if it is to be able to bind to a peptide presented on the surface of the MHC antigen. Then, T cells bearing receptors that recognize self peptides in association with MHC antigens at high affinity are eliminated in order to eliminate the generation of autoreactive receptors (*negative selection*). The likelihood that a single TCR β chain could pair with more than one TCR α chain and meet all these stringent criteria is extremely small. The presence of even one αβ TCR capable of binding self antigens at high affinity would result in the elimination of the cell, irrespective of the specificity of the receptor formed using an alternative α chain.

Although B and T cells use very similar mechanisms for variable region gene rearrangements, complete rearrangement of Ig genes does not occur in T cells and complete rearrangement of TCR genes does not occur in B cells. Not only is the recombinase enzyme system differently regulated in each lymphocyte lineage, but chromatin is uniquely reconfigured in B cells and T cells to allow the recombinase access only to the appropriate specific antigen receptor genes.

TCR Gene Expression Is Tightly Regulated

As at Ig loci, expression of TCR genes requires the coordinated use of transcriptional promoters and enhancers that serve at specific stages in T-cell development to make the receptor chromatin open and accessible for recombination (see Chapter 9). T cells begin to rearrange the TCR β gene at the pro-T-cell stage. The mouse TCR_β locus (see Figure 7-18) consists of approximately 40 V_β genes located upstream of 2 distinct $D_\beta J_\beta$ clusters. Immediately upstream of each of the two D_β genes is a promoter region ($PD_\beta 1$ and $PD_\beta 2$). These promoters require the activity of a single enhancer, E_β, located at the 3′ end of the locus. Activation of the enhancer serves to facilitate the opening of the chromatin throughout the $D_\beta J_\beta$ locus, and activation of each of the two PD_β promoters localizes the rearrangement to their own set of DJ regions.

Several TCR enhancer-binding proteins have been identified and, as for B-cell Ig genes, some of these are shared among cells of many different types. Among the T-cell-specific proteins is GATA-3, which binds in a sequence-specific manner to the enhancer elements of all four TCR genes. Tissue- and stage-specific expression of enhancer-binding proteins at the TCR loci helps to facilitate locus accessibility that in turn allows the complex processes of recombination and transcription of the TCR genes and, eventually, cell-surface expression of the TCR proteins.

TABLE 7-8   Comparison of the mechanisms of the generation and expression of diversity among B-cell and T-cell receptor molecules

Mechanism	Used in B cells	Used in T cells	Comments
Multiple germ-line V(D)J genes	Yes	Yes	The mouse V_λ locus has undergone a severe contraction, and, therefore, only 5% of mouse light chains are of the λ type.
			J region diversity is notably higher in TCR α-chain genes than in other TCR or Ig genes
"Light-chain" segment use	κ and λ variable regions encoded by V and J segments	α and γ variable regions encoded by V and J segments	
"Heavy-chain" segment use	V_H regions encoded by V, D and J segments	β and δ variable regions encoded by V, D and J segments	
Absolute dependence on RAG1/2 expression	Yes	Yes	
Junctional diversity: P nucleotides and N-nucleotide addition	Yes	Yes	Many fewer N nucleotides found in Ig light chains because of developmental regulation of TdT
Multiple D regions per recombined chain	Not in Ig heavy chains	Present in TCR δ, but not TCR β chains	The presence of two D segments allows an additional site for N nucleotide addition.
Allelic exclusion of receptor gene expression	Absolute	Allelic exclusion of TCR α genes is not absolute.	
On activation, secretes product with the same binding site as the receptor	Yes	TCR is not found in secreted form.	
Nature of constant region determines function	Yes; constant region of secreted antibody product determines its function. Constant region of membrane receptor anchors receptor in membrane and connects with signal transduction complex	No secreted product. Constant region of membrane receptor anchors receptor in membrane and connects with signal transduction complex.	
Receptor genes undergo somatic hypermutation following antigen stimulation	Yes	No	

BOX 7-3

CLINICAL FOCUS

Some Immunodeficiencies Result from Impaired Receptor Gene Recombination

Since the RAG1/2 enzymes are responsible for both BCR and TCR gene rearrangements, mutations in the genes encoding RAG1 or RAG2 have catastrophic consequences for the immune system, resulting in patients with severe combined immunodeficiency, or SCID. The RAG1/2 genes are located on human chromosome 11 (i.e., an autosome), so such mutations are inherited in an autosomally recessive manner.

Babies born with nonfunctional RAG1 or RAG2 genes have essentially no circulating B or T cells, although they do express normal levels of natural killer (NK) cells and their myeloid and erythroid cells are normal in number and function. Because infants receive antibodies passively from the maternal circulation, the first manifestation of this disease is the complete absence of T-cell function, and hence such infants suffer from severe, recurrent infections with fungi and viruses that would normally be combated by T cells in a healthy neonate. SCID used to be inevitably fatal within the first few months of life, unless babies were delivered directly into a sterile environment. The life span of a SCID patient can be prolonged by preventing contact with all potentially harmful microorganisms. Air must be filtered, all food must be sterilized, and there must be no direct contact with other people. Such isolation is feasible only as a temporary measure, pending treatment.

Nowadays, babies diagnosed sufficiently early with RAG1/2 deficiency can be successfully treated with bone marrow transplantation. If a suitable donor can be found, and transplantation is performed early in the patient's first year of life, the chances of the patient surviving to live a normal life are 97% or greater.

Patients who carry mutations resulting in partially active or impaired RAG1 or RAG2 genes are diagnosed with Omenn syndrome. Omenn syndrome patients have no circulating B cells and abnormal lymph node architecture with deficiencies in the B-cell zones of the lymph nodes. Although some T cells are present, they are oligoclonal (derived from a very few precursors and hence have very few different receptors), and these T cells tend to be inappropriately activated. They do not respond normally to mitogens or to antigens in vitro. This condition, like RAG1/2 deficiency, is invariably fatal unless corrected by bone marrow transplantation.

If deficiency in RAG1/2 genes causes such a catastrophic loss of immune function, it makes sense that loss of any of the other genes implicated in V(D)J recombination would also result in a SCID phenotype. In 1998, it was determined that deficiency of the *Artemis* DNA repair enzyme also results in the loss of T- and B-cell function, in the face of normal NK cell activity. Loss of *Artemis* results in a SCID syndrome referred to as Athabascan SCID. Since *Artemis* is necessary for normal DNA repair, as well as V(D)J recombination, patients with Athabascan SCID also suffer from increased radiation sensitivity of skin fibroblasts and bone marrow cells. Similarly, human patients suffering from a deficiency in DNA ligase IV present with an immunodeficiency that affects T, B, and NK cells. Since DNA ligase IV is implicated in DNA repair functions outside the immune system, such patients also suffer from generalized chromosomal instability, radiosensitivity, and developmental and growth retardation.

A SCID defect in mice, first noted and characterized by Prof. Melvin Bosma, results from a nonsense mutation in the gene encoding the DNA-PKcs. Although a similar recessive mutation has been found in Arabian horses, few, if any, human cases of SCID have been reported carrying this mutation.

SUMMARY

- The antigen receptor on the surface of a B cell is an Ig with the same binding site as the antibody the B cell's progeny will eventually secrete.

- The number of different antibody molecules that can be made by a single individual mouse or human is vast: close to 10^{13-14}. This number is achieved by combinatorial association of multiple gene segments termed V (variable) and J (joining), which together encode the variable region of the light chain of the Ig molecule and V, D (diversity), and J, which encode the variable region of the heavy chain of the Ig molecule.

- V(D)J recombination occurs at the level of the DNA and is catalyzed by the lymphoid-specific enzymes RAG1 and RAG2.

- RAG1 and RAG2 recognize Recombination Signal Sequences (RSSs) that are contiguous with the coding regions. RSSs consist of a conserved heptamer and a conserved nonamer sequence, separated by a spacer region conserved in length, but not completely in sequence. Spacer regions have a length corresponding to one turn of the double helix (12-bp spacer) or two turns of the double helix (23-bp spacer). RAG1/2 catalyzes recombination between segments bordered by different spacers.

- At the sites of recombination, hairpin cleavage can be asymmetric, resulting in P nucleotides, and in the heavy-chain genes, non-templated nucleotide addition can result in N nucleotides. Exonuclease nibbling can reduce the number of N and P nucleotides in the final receptor gene products.

- The variable regions of TCRs are generated by the same general mechanism. Most T cells have receptors of the $\alpha\beta$ type. The variable regions of α chains are made up of V and J segments, whereas those of β chains have V, D, and J

regions. Other T cells have $\gamma\delta$ receptors. The variable regions of γ chains, like those of α chains, have V and J segments. δ chains are encoded within the TCRα locus, and their variable regions are made up of V, D, and J segments. δ but not β chains can express more than one D region per chain.

REFERENCES

Alder, M. N., et al. 2008. Antibody responses of variable lymphocyte receptors in the lamprey. *Nature Immunology* **9**:319–327.

Calame, K., et al. 1980. Mouse Cmu heavy chain immunoglobulin gene segment contains three intervening sequences separating domains. *Nature* **284**:452–455.

Capone, M., et al. 1993. TCR beta and TCR alpha gene enhancers confer tissue- and stage-specificity on V(D)J recombination events. *EMBO Journal* **12**:4335–4346.

Cesari, I. M., and M. Weigert. 1973. Mouse lambda-chain sequences. *Proceedings of the National Academy of Sciences of the United States of America* **70**:2112–2116.

Chien, Y., et al. 1984. A third type of murine T-cell receptor gene. *Nature* **312**:31–35.

Chien, Y. H., N. R. Gascoigne, J. Kavaler, N. E. Lee, and M. M. Davis. 1984. Somatic recombination in a murine T-cell receptor gene. *Nature* **309**:322–326.

Chun, J. J., D. G. Schatz, M. A. Oettinger, R. Jaenisch, R., and D. Baltimore. 1991. The recombination activating gene-1 (RAG-1) transcript is present in the murine central nervous system. *Cell* **64**:189–200.

Clatworthy, A. E., M. a. Valencia-Burton, J. E. Haber, and M. A. Oettinger. 2005. The MRE11-RAD50-XRS2 complex, in addition to other non-homologous end-joining factors, is required for V(D)J joining in yeast. *Journal of Biological Chemistry* **280**:20,247–20,252.

Cooper, M. D., and M. N. Alder. 2006. The evolution of adaptive immune systems. *Cell* **124**:815–822.

Davis, M. M., et al. 1980. An immunoglobulin heavy-chain gene is formed by at least two recombinational events. *Nature* **283**:733–739.

Davis, M. M., Y. H. Chien, N. R. Gascoigne, and S. M. Hedrick. 1984. A murine T cell receptor gene complex: Isolation, structure and rearrangement. *Immunological Reviews* **81**:235–258.

Early, P., H. Huang, M. Davis, K. Calame, and L. Hood. 1980. An immunoglobulin heavy chain variable region gene is generated from three segments of DNA: VH, D and JH. *Cell* **19**:981–992.

Early, P., et al. 1980. Two mRNAs can be produced from a single immunoglobulin mu gene by alternative RNA processing pathways. *Cell* **20**:313–319.

Early, P.W., M. M. Davis, D. B. Kaback, N. Davidson, and L. Hood. 1979. Immunoglobulin heavy chain gene organization in

mice: Analysis of a myeloma genomic clone containing variable and alpha constant regions. *Proceedings of the National Academy of Sciences of the United States of America* **76**:857–861.

Ernst, P., and S. T. Smale. 1995. Combinatorial regulation of transcription II: The immunoglobulin mu heavy chain gene. *Immunity* **2**:427–438.

Flajnik, M. F., and M. Kasahara. 2004. Origin and evolution of the adaptive immune system: Genetic events and selective pressures. *Nature Reviews Genetics* **11**:47–59.

Gascoigne, N. R., Y. Chien, D. M. Becker, J. Kavaler, and M. M. Davis. 1984. Genomic organization and sequence of T-cell receptor beta-chain constant- and joining-region genes. *Nature* **310**:387–391.

Gearhart, P. J. 2004. The birth of molecular immunology. *Journal of Immunology* **173**:4259.

Gilfillan, S., A. Dierich, M. Lemeur, C. Benoist, and D. Mathis. 1993. Mice lacking TdT: Mature animals with an immature lymphocyte repertoire. *Science* **261**:1175–1178.

Hedrick, S. M., D. I. Cohen, E. A. Nielsen, and M. M. Davis. 1984. Isolation of cDNA clones encoding T cell-specific membrane-associated proteins. *Nature* **308**:149–153.

Hedrick, S. M., E. A. Nielsen, J. Kavaler, D. I. Cohen, and M. M. Davis. 1984. Sequence relationships between putative T-cell receptor polypeptides and immunoglobulins. *Nature* **308**:153–158.

Herrin, B. R., and M. D. Cooper. 2010. Alternative adaptive immunity in jawless vertebrates. *Journal of Immunology* **185**:1367.

Hood, L., et al. 1981. Two types of DNA rearrangements in immunoglobulin genes. *Cold Spring Harbor Symposia on Quantitative Biology* **45**(Pt 2):887–898.

Hozumi, N., and S. Tonegawa. 1976. Evidence for somatic rearrangement of immunoglobulin genes coding for variable and constant regions. *Proceedings of the National Academy of Sciences of the United States of America* **73**:3628–3632.

Ju, Z., et al. 2007. Evidence for physical interaction between the immunoglobulin heavy chain variable region and the 3′ regulatory region. *Journal of Biological Chemistry* **282**:35,169–35,178.

Kasahara, M., T. Suzuki, and L. D. Pasquier. 2004. On the origins of the adaptive immune system: Novel insights from invertebrates and cold-blooded vertebrates. *Trends in Immunology* **25**:105–111.

Kavaler, J., M. M. Davis, and Y. Chien. 1984. Localization of a T-cell receptor diversity-region element. *Nature* **310**:421–423.

Koralov, S. B., T. I. Novobrantseva, J. Konigsmann, A. Ehlich, and K. Rajewsky. 2006. Antibody repertoires generated by VH replacement and direct VH to JH joining. *Immunity* **25**:43–53.

Landau, N. R., D. G. Schatz, M. Rosa, and D. Baltimore. 1987. Increased frequency of N-region insertion in a murine pre-B-cell line infected with a terminal deoxynucleotidyl transferase retroviral expression vector. *Molecular and Cellular Biology* **7**:3237–3243.

Mansilla-Soto, J., and P. Cortes. 2003. VDJ recombination: Artemis and its in vivo role in hairpin opening. *Journal of Experimental Medicine* **197**:543–547.

Oestreich, K. J., et al. 2006. Regulation of TCR beta gene assembly by a promoter/enhancer holocomplex. *Immunity* **24**:381–391.

Oettinger, M. A., D. G. Schatz, C. Gorka, and D. Baltimore. 1990. RAG-1 and RAG-2, adjacent genes that synergistically activate V(D)J recombination. *Science* **248**:1517–1523.

Pancer, Z., and M. D. Cooper. 2006. The evolution of adaptive immunity. *Annual Review of Immunology* **24**:497–518.

Rivera-Munoz, P., et al. 2007. DNA repair and the immune system: From V(D)J recombination to aging lymphocytes. *European Journal of Immunology* **37**(Suppl 1):S71–82.

Rooney, S., et al. 2003. Defective DNA repair and increased genomic instability in Artemis-deficient murine cells. *Journal of Experimental Medicine* **197**:553–565.

Saito, H., et al. 1984. Complete primary structure of a heterodimeric T-cell receptor deduced from cDNA sequences. *Nature* **309**:757–762.

Saito, H., et al. 1984. A third rearranged and expressed gene in a clone of cytotoxic T lymphocytes. *Nature* **312**:36–40.

Schatz, D. G., and D. Baltimore. 1988. Stable expression of immunoglobulin gene V(D)J recombinase activity by gene transfer into 3T3 fibroblasts. *Cell* **53**:107–115.

Schatz, D. G., and Y. Ji. 2011 Recombination centres and the orchestration of V(D)J recombination. *Nature Reviews Immunology* **11**:251–263.

Schatz, D. G., M. A. Oettinger, and D. Baltimore. 1989. The V(D)J recombination activating gene, RAG-1. *Cell* **59**:1035–1048.

Tonegawa, S., C. Brack, N. Hozumi, G. Matthyssens, and R. Schuller. 1977. Dynamics of immunoglobulin genes. *Immunological Reviews* **36**:73–94.

Tonegawa, S., C. Brack, N. Hozumi, and V. Pirrotta. 1978. Organization of immunoglobulin genes. *Cold Spring Harbor Symposia in Quantitative Biology* **42**(Pt 2):921–931.

Tonegawa, S., C. Brack, N. Hozumi, and R. Schuller. 1977. Cloning of an immunoglobulin variable region gene from mouse embryo. *Proceedings of the National Academy of Sciences of the United States of America* **74**:3518–3522.

Tonegawa, S., N. Hozumi, G. Matthyssens, and R. Schuller. 1977. Somatic changes in the content and context of immunoglobulin genes. *Cold Spring Harbor Symposia in Quantitative Biology* **41**(Pt 2):877–889.

Wei, X. C., et al. 2005. Characterization of the proximal enhancer element and transcriptional regulatory factors for murine recombination activating gene-2. *European Journal of Immunology* **35**:612–621.

Weigert, M. G., I. M. Cesari, S. J. Yonkovich, and M. Cohn. 1970. Variability in the lambda light chain sequences of mouse antibody. *Nature* **228**:1045–1047.

Zha, S., F. W. Alt, H. L. Cheng, J. W. Brush, and G. Li. 2007. Defective DNA repair and increased genomic instability in Cernunnos-XLF-deficient murine ES cells. *Proceedings of the National Academy of Sciences of the United States of America* **104**:4518–4523.

Zhang, Z. 2007. VH replacement in mice and humans. *Trends in Immunology* **28**:132–137.

Useful Web Sites

http://imgt.org The creators of this website monitor the literature for information on the numbers of genes encoding Ig and TCRs in a variety of species, and regularly update information regarding the numbers of gene segments that have been sequenced. Also includes information on other gene families related to the immune system, such as the MHC.

http://cellular-immunity.blogspot.com/2007/12/vdj-recombination.html A clear and brief explanation of V(D)J recombination, with clickable links.

http://highered.mcgraw-hill.com/sites/0072556781/student_view0/chapter32/animation_quiz_2.html A clear, even if rather simplistic, animation of V(D)J recombination, which offers a "big picture" of the process. May be useful to view before launching into more detailed reading. Note that this animation speaks of there being five constant regions, rather than the eight referred to in this chapter; this is because two of the heavy-chain classes—γ and α—can be divided into subclasses, which are ignored in this animation.

http://users.rcn.com/jkimball.ma.ultranet/BiologyPages/A/AgReceptorDiversity.html A useful review of antigen receptor diversity with clickable links. Do not be put off by minor differences between this website and this chapter in the numbers of gene segments identified as belonging to particular families—these numbers are constantly being updated, and they vary among different individuals and inbred strains of animals.

http://biophilessurf.info/immuno.html A useful collection of databases pertaining to immunological topics, including RAG1 mutations that give rise to immunodeficiencies.

www.ncbi.nlm.nih.gov The National Center for Biotechnology Information (NCBI) site offers library tools as well as numerous sequence analysis and protein structure resources (under the "Resources" tab) pertinent to immunology.

STUDY QUESTIONS

1. Indicate whether each of the following statements is true or false. If you think a statement is false, explain why.

   a. In generating a B-cell receptor gene, V_κ segments sometimes join to C_λ segments.

   b. In generating a T-cell receptor gene, V_α segments sometimes join to C_δ.

   c. The switch in constant region use from to IgM to IgD is mediated by DNA rearrangements.

   d. Although each B cell carries two alleles encoding the immunoglobulin heavy and light chains, only one allele is expressed.

   e. Like the variable regions of the heavy chain of the B-cell Ig receptor, TCR variable genes are all encoded in three segments.

2. Explain why a V_H segment cannot join directly with a J_H segment in heavy-chain gene rearrangement.

3. For each incomplete statement below, select the phrase(s) that correctly completes the statement. More than one choice may be correct.

   a. Recombination of Ig gene segments serves to:
      (1) Promote Ig diversification
      (2) Assemble a complete Ig coding sequence
      (3) Allow changes in coding information during B-cell maturation
      (4) Increase the affinity of Ig for antibody
      (5) All of the above

   b. Kappa and lambda light-chain genes:
      (1) Are located on the same chromosome
      (2) Associate with only one type of heavy chain
      (3) Can be expressed by the same B cell
      (4) All of the above
      (5) None of the above

   c. Generation of combinatorial diversity within the variable regions of Ig involves
      (1) mRNA splicing
      (2) DNA rearrangement
      (3) Recombination signal sequences
      (4) One-turn/two-turn joining rule
      (5) Switch sites

   d. The mechanism that permits Ig to be synthesized in either a membrane-bound or secreted form is:
      (1) Allelic exclusion
      (2) Codominant expression
      (3) Class switch recombination
      (4) The one-turn/two-turn joining rule
      (5) Differential RNA processing

   e. During Ig V_H recombination, the processes that contribute to additional diversity at the third complementarity-determining region of Ig variable regions include:
      (1) Introduction of the D_H gene segments into the heavy-chain V gene
      (2) mRNA splicing of the membrane form of the C region of the constant chain
      (3) Exonuclease cleavage of the ends of the gene segments
      (4) P nucleotide addition
      (5) N nucleotide addition

4. You have been given a cloned mouse myeloma cell line that secretes IgG with the molecular formula $\gamma_2\lambda_2$. Both the heavy and light chains in this cell line are encoded by genes derived from allele 1 (i.e., the first of the two homologous alleles encoding each type of chain). Indicate the form(s) in which each of the genes listed below would occur in this cell line using the following symbols: G = germ-line form; P = productively rearranged form; NP = nonproductively rearranged form. State the reason for your choice in each case.

   (a) Heavy-chain allele 1
   (b) Heavy-chain allele 2
   (c) κ-chain allele 1
   (d) κ-chain allele 2
   (e) λ-chain allele 1
   (f) λ-chain allele 2

5. You have identified a B-cell lymphoma that has made nonproductive rearrangements for both heavy-chain alleles. What is the arrangement of its light-chain DNA? Why?

6. The random addition of nucleotides by TdT is a wasteful and potentially risky evolutionary strategy. State why it may be disadvantageous to the organism and why, therefore, you think it is sufficiently useful to have been retained during vertebrate evolution.

7. Are there any differences in the genetic strategies used to generate complete V genes in T and BCRs?

8. What known features of the TCR did Hedrick, Davis, and colleagues use in their quest to isolate the TCR genes?

9. The following figure describes the end of a V region sequence and the beginning of the D region sequence to which is it about to be joined. Arrows mark the cleavage points where the RAG1/RAG2 complex will make the cut and recombination will be targeted.

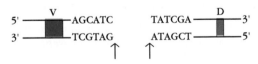

   a. Is this a heavy-chain or a light-chain sequence? How do you know?

   b. What DNA sequence structure would you find just downstream of the AGCATC sequence immediately adjacent to the 3′ end of the V segment?

   c. Here is one possible VD joint structure formed after recombination between these two gene segments:

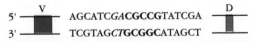

      (1) Which residue(s) MAY be P-region nucleotide(s)?
      (2) Can we know for certain that this residue is a P-region nucleotide?
      (3) Which residues *must* have been added by TdT, and therefore *must* be N-region nucleotides?
      (4) Can we know for certain whether a residue is an N-region nucleotide?

10. Challenge data question. In Classic Experiment Box 7.1, Figure 1, we see the pattern of bands that Hozumi and Tonegawa's experiment would have revealed, using the particular cell line that they used. Below, see a pair of gels that represent the results of a hypothetical experiment performed using the same general protocol. In this hypothetical experiment, our probes correspond to either the V region or the C region. Furthermore, the investigators used a different tumor cell line and a different restriction endonuclease.

a. Why are there two C-region bands in the germ-line blot?
b. Why are these two bands still present in the myeloma blot, and why are there two bands recognized by the V-region probe on the myeloma blot?
c. From this plot, would you hypothesize that the cell had achieved a successful arrangement at the first κ allele? Why or why not?
d. How would you prove your answer to part c?

The Major Histocompatibility Complex and Antigen Presentation

Although both T and B cells use surface molecules to recognize antigen, they accomplish this in very different ways. In contrast to antibodies or B-cell receptors, which can recognize an antigen alone, T-cell receptors only recognize pieces of antigen that are positioned on the surface of other cells. These antigen pieces are held within the binding groove of a cell surface protein called the **major histocompatibility complex (MHC) molecule**, encoded by a cluster of genes collectively called the MHC locus. These fragments are generated inside the cell following antigen digestion, and the complex of the antigenic peptide plus MHC molecule then appears on the cell surface. MHC molecules thus act as a cell surface vessel for holding and displaying fragments of antigen so that approaching T cells can engage with this molecular complex via their T-cell receptors.

The MHC got its name from the fact that the genes in this region encode proteins that determine whether a tissue transplanted between two individuals will be accepted or rejected. The pioneering work of Benacerraf, Dausset, and Snell helped to characterize the functions controlled by the MHC, specficially, organ transplant fate and the immune responses to antigen, resulting in the 1980 Nobel Prize in Medicine and Physiology for the trio (see Table 1-2). Follow-up studies by Rolf Zinkernagel and Peter Doherty illustrated that the proteins encoded by these genes play a seminal role in adaptive immunity by showing that T cells recognize MHC proteins as well as antigen. Structural studies done by Don Wiley and others showed that different MHC proteins bind and present different antigen fragments. There are many alleles of most MHC genes, and the specific alleles one inherits play a significant role in susceptibility to disease, including the development of autoimmunity. The mechanisms by which this family of molecules exerts such a strong influence on the development of immunity to nearly all types of antigens has become a major theme in immunology, and has taken

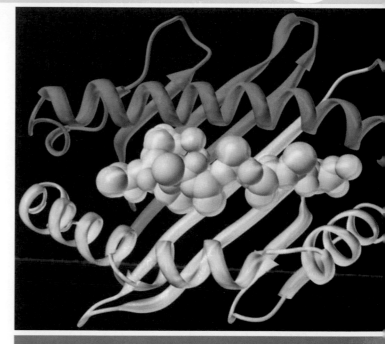

Ribbon diagram of a human MHC class I molecule (blue) with a space-filled peptide (orange) held in the binding groove.

- The Structure and Function of MHC Molecules
- General Organization and Inheritance of the MHC
- The Role of the MHC and Expression Patterns
- The Endogenous Pathway of Antigen Processing and Presentation
- The Exogenous Pathway of Antigen Processing and Presentation
- Cross-Presentation of Exogenous Antigens
- Presentation of Nonpeptide Antigens

the study of the MHC far beyond its origins in the field of transplantation biology.

There are two main classes of MHC molecules: class I and class II. These two molecules are very similar in their final quaternary structure, although they differ in how they create these shapes via primary through quaternary protein arrangements. Class I and class II MHC molecules also differ in terms of which cells express them and in the source of the antigens they present to T cells. Class I molecules are present on all nucleated cells in the body and specialize in presenting antigens that originate from the cytosol, such as viral proteins. These are presented to $CD8^+$ T cells, which recognize and kill cells expressing such intracellular antigens. In contrast, class II MHC molecules are expressed almost exclusively on a subset of leukocytes called **antigen-presenting cells (APCs)** and specialize in presenting antigens from extracellular spaces that have been engulfed by these cells, such as fungi and extracellular bacteria. Once expressed on the cell surface, the MHC class II molecule presents the antigenic peptide to $CD4^+$ T cells, which then become activated and go on to stimulate immunity directed primarily toward destroying extracellular invaders.

This chapter will begin by discussing the structure of MHC molecules, followed by the genetic organization of the DNA region that encodes these proteins and inheritance patterns. We then turn to the function of the MHC in regulating immunity, including reference to several seminal studies in these areas. In this section we also discuss regulation of MHC expression. This is followed by a detailed discussion of the cellular pathways that lead to antigen degradation and association with each type of MHC molecule (**antigen processing**) and the appearance of these MHC-peptide complexes on the cell surface for recognition by T cells (**antigen presentation**). The role of particular APCs in these processes will also be presented (no pun intended!). The chapter concludes with a discussion of unique processing and presentation pathways, such as cross-presentation and the handling of nonpeptide antigens by the immune system.

The Structure and Function of MHC Molecules

Class I and class II MHC molecules are membrane-bound glycoproteins that are closely related in both structure and function. Both classes of MHC molecule have been isolated and purified, and the three-dimensional structures of their extracellular domains have been resolved by x-ray crystallography. These membrane glycoproteins function as highly specialized antigen-presenting molecules with grooves that form unusually stable complexes with peptide ligands, displaying them on the cell surface for recognition by T cells via T-cell receptor (TCR) engagement. In contrast, class III MHC molecules are a group of unrelated proteins that do not share structural similarity or function with class I and II molecules, although many of them do participate in other aspects of the immune response.

Class I Molecules Have a Glycoprotein Heavy Chain and a Small Protein Light Chain

Two polypeptides assemble to form a single class I MHC molecule: a 45-kilodalton (kDa) α *chain* and a 12-kDa

$β_2$-**microglobulin** molecule (Figure 8-1, left). The α chain is organized into three external domains (α1, α2, and α3), each approximately 90 amino acids long; a transmembrane domain of about 25 hydrophobic amino acids followed by a short stretch of charged (hydrophilic) amino acids; and a cytoplasmic anchor segment of 30 amino acids. Its companion, $β_2$-microglobulin, is similar in size and organization to the α3 domain. $β_2$-microglobulin does not contain a transmembrane region and is noncovalently bound to the MHC class I α chain. Sequence data reveal strong homology between the α3 domain of MHC class I, $β_2$-microglobulin, and the constant-region domains found in immunoglobulins.

The α1 and α2 domains interact to form a platform of eight antiparallel β strands spanned by two long α-helical regions (Figure 8-2a). The structure forms a deep groove, or cleft, with the long α helices as sides and the β strands of the β sheet as the bottom (Figure 8-2b). This *peptide-binding groove* is located on the top surface of the class I MHC molecule, and it is large enough to bind a peptide of 8 to 10 amino acids. During the x-ray crystallographic analysis of class I molecules, small noncovalently associated peptides that had co-crystallized with the protein were found in the groove. The big surprise came when these peptides were later discovered to be processed self-proteins bound to this deep groove and not the foreign antigens that were expected.

The α3 domain and $β_2$-microglobulin are organized into two β pleated sheets each formed by antiparallel β strands of amino acids. As described in Chapter 3, this structure, known as the immunoglobulin fold, is characteristic of immunoglobulin domains. Because of this structural similarity, which is not surprising given the considerable sequence similarity with the immunoglobulin constant regions, class I MHC molecules and $β_2$-microglobulin are classified as members of the immunoglobulin superfamily (see Figure 3-19). The α3 domain appears to be highly conserved among class I MHC molecules and contains a sequence that interacts strongly with the CD8 cell surface molecule found on T_C cells.

All three molecules (class I α chain, $β_2$-microglobulin, and a peptide) are essential to the proper folding and expression of the MHC-peptide complex on the cell surface. This is demonstrated in vitro using Daudi tumor cells, which are unable to synthesize $β_2$-microglobulin. These tumor cells produce class I MHC α chains but do not express them on the cell membrane. However, if Daudi cells are transfected with a functional gene encoding $β_2$-microglobulin, class I molecules will appear on the cell surface.

Class II Molecules Have Two Non-Identical Glycoprotein Chains

Class II MHC molecules contain two different polypeptide chains, a 33-kDa α chain and a 28-kDa β chain, which associate by noncovalent interactions (see Figure 8-1, right). Like class I α chains, class II MHC molecules are membrane-bound glycoproteins that contain external domains, a transmembrane segment, and a cytoplasmic anchor segment. Each

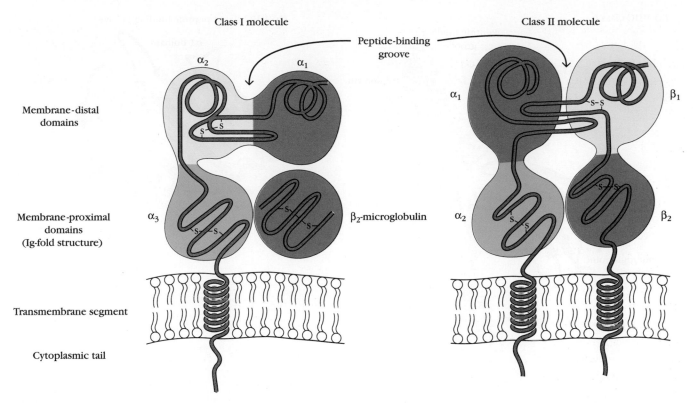

FIGURE 8-1 Schematic diagrams of class I and class II MHC molecules showing the external domains, transmembrane segments, and cytoplasmic tails. The peptide-binding groove is formed by the membrane-distal domains in both class I and class II molecules (blue). The membrane-proximal domains (green and red) possess the basic immunoglobulin-fold structure; thus, both class I and class II MHC molecules are classified as members of the immunoglobulin superfamily.

chain in a class II molecule contains two external domains: α1 and α2 domains in one chain and β1 and β2 domains in the other (Figure 8-2c). The membrane-proximal α2 and β2 domains, like the membrane-proximal α3/β2-microglobulin domains of class I MHC molecules, bear sequence similarity to the immunoglobulin-fold structure. For this reason, class II MHC molecules are also classified in the immunoglobulin superfamily. The α1 and β1 domains form the peptide-binding groove for processed antigen (Figure 8-2d). Although similar to the peptide-binding groove of MHC class I, this groove is thus formed by the association of two separate chains.

X-ray crystallographic analysis reveals the similarity between these two classes of molecule, strikingly apparent when class I and class II molecules are superimposed (Figure 8-3). Interestingly, despite the fact that these two structures are encoded quite differentially (one chain *versus* two), the final quaternary structure is similar and retains the same overall function: the ability to bind antigen and present it to T cells. The peptide-binding groove of class II molecules, like that found in class I molecules, is composed of a floor of eight antiparallel β strands and sides of antiparallel α helices, where peptides typically ranging from 13 to 18 amino acids can bind. The class II molecule lacks the conserved residues in the class I molecule that bind to the terminal amino acids of short peptides, and therefore forms more of an open pocket. In this way, class I presents more of a socket-like opening, whereas class II pos-

sesses an open-ended groove. The functional consequences of these differences in fine structure will be explored below.

Class I and II Molecules Exhibit Polymorphism in the Region That Binds to Peptides

Several hundred different allelic variants of class I and II MHC molecules have been identified in humans. Any one individual, however, expresses only a small number of these molecules—up to six different class I molecules and 12 or more different class II molecules. Yet this limited number of MHC molecules must be able to present an enormous array of different antigenic peptides to T cells, permitting the immune system to respond specifically to a wide variety of antigenic challenges. Thus, peptide binding by class I and II molecules does not exhibit the fine specificity characteristic of antigen binding by antibodies and T-cell receptors. Instead, *a given MHC molecule can bind numerous different peptides, and some peptides can bind to several different MHC molecules.* Because of this broad specificity, the binding between a peptide and an MHC molecule is often referred to as "promiscuous." This promiscuity of peptide binding allows many different peptides to match up with the MHC binding groove and for exchange of peptides to happen on occasion, unlike the relatively stable, high-affinity cognate interactions of antibodies and T-cell receptors with their specific ligands.

(a) MHC class I

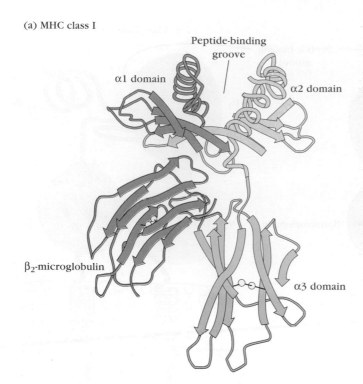

(b) MHC class I peptide-binding groove

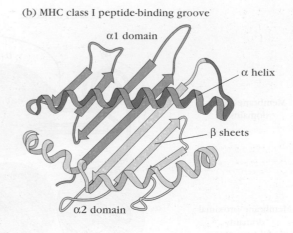

(c) MHC class II

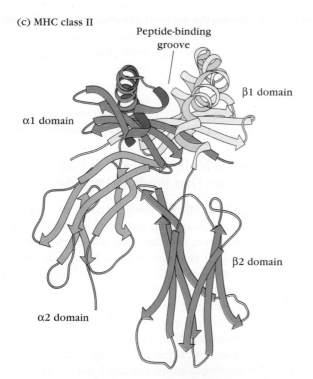

(d) MHC class II peptide-binding groove

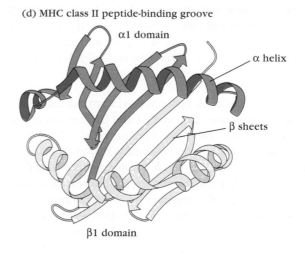

FIGURE 8-2 Representations of the three-dimensional structure of the external domains of human MHC class I and class II molecules based on x-ray crystallographic analysis. (a, c) Side views of class I and II, respectively, in which the β strands are depicted as thick arrows and the α helices as spiral ribbons. Disulfide bonds are shown as two interconnected spheres. The α1 and α2 domains of class I and the α1 and β1 domains of class II interact to form the peptide-binding groove. Note the immunoglobulin-fold structure of each membrane proximal domain, including the β₂-microglobulin molecule. (b, d) The α1 (dark blue) and α2 domains (light blue) of class I and the α1 (dark blue) and β1 (light blue) domains of class II as viewed from the top, showing the peptide-binding grooves, consisting of a base of antiparallel β strands and sides of α helices for each molecule. This groove can accommodate peptides of 8 to 10 residues for class I and 13 to 18 residues for class II.

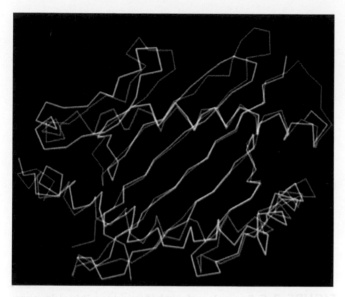

FIGURE 8-3 **The peptide-binding groove of a human class II MHC molecule (blue), superimposed over the corresponding regions of a human class I MHC molecule (red).** Overlapping regions are shown in white. These two molecules create very similar binding grooves such that most of the structural differences (long stretches, kinks, bends, etc.) lie outside the peptide-binding regions. *[From J. H. Brown et al., 1993, Nature 364:33.]*

Given the similarities in the structures of the peptide-binding grooves in class I and II MHC molecules, it is not surprising that they exhibit some common peptide-binding features (Table 8-1). In both types of MHC molecules, peptide ligands are held in a largely extended conformation that runs the length of the groove. The peptide-binding groove in class I molecules is blocked at both ends, whereas the ends of the groove are open in class II molecules (Figure 8-4). As a result of this difference, class I molecules bind peptides that typically contain 8 to 10 amino acid residues, whereas the open groove of class II molecules accommodates slightly longer peptides of 13 to 18 amino acids. Another difference is that class I binding requires that the peptide contain specific amino acid residues near the N and C termini; there is

no such requirement for class II peptide binding. The association of peptide and MHC molecule is very stable under physiologic conditions (K_d values range from ~ 10^{-6} to 10^{-10}). Thus, most of the MHC molecules expressed on the membrane of a cell will be associated with a peptide.

Class I MHC-Peptide Interaction

Class I MHC molecules bind peptides and present them to $CD8^+$ T cells. These peptides are often derived from endogenous intracellular proteins that are digested in the cytosol. The peptides are then transported from the cytosol into the endoplasmic reticulum (ER), where they interact with class I MHC molecules. This process, known as the cytosolic or endogenous processing pathway, is discussed in detail later in this chapter.

Each human or mouse cell can express several unique class I MHC molecules, each with slightly different rules for peptide binding. Because a single nucleated cell expresses about 10^5 copies of each of these unique class I molecules, each with its own peptide promiscuity rules, many different peptides will be expressed simultaneously on the surface of a nucleated cell. This means that although many of the peptide fragments of a given foreign antigen will be presented to $CD8^+$ T cells, the set of MHC class I alleles inherited by each individual will determine which specific peptide fragments from a larger protein get presented.

The bound peptides isolated from different class I molecules have two distinguishing features: they are 8 to 10 amino acids in length, most commonly 9, and they contain specific amino acid residues at key locations in the sequence. The ability of an individual class I MHC molecule to bind to a diverse spectrum of peptides is due to the presence of the same or similar amino acid residues at these key positions along the peptides (Figure 8-5). Because these amino acid residues *anchor* the peptide into the groove of the MHC molecule, they are called **anchor residues**. The side chains of the anchor residues in the peptide are complementary with surface features of the binding groove of the class I MHC molecule. The amino acid residues lining the binding sites vary among different class I allelic variants, and this is what determines the chemical

TABLE 8-1	Peptide binding by class I and class II MHC molecules	
	Class I molecules	**Class II molecules**
Peptide-binding domain	$\alpha1/\alpha2$	$\alpha1/\beta1$
Nature of peptide-binding groove	Closed at both ends	Open at both ends
General size of bound peptides	8–10 amino acids	13–18 amino acids
Peptide motifs involved in binding to MHC molecule	Anchor residues at both ends of peptide; generally hydrophobic carboxyl-terminal anchor	Conserved residues distributed along the length of the peptide
Nature of bound peptide	Extended structure in which both ends interact with MHC groove but middle arches up away from MHC molecule	Extended structure that is held at a constant elevation above the floor of the MHC groove

(a)

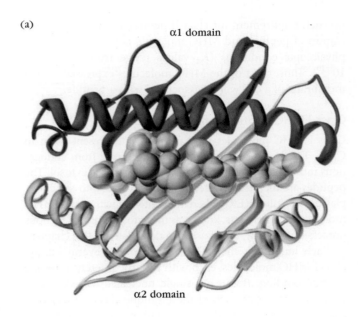

α1 domain

α2 domain

(b)

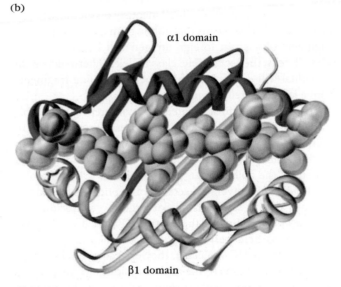

α1 domain

β1 domain

FIGURE 8-4 MHC class I and class II molecules with bound peptides. (a) Ribbon model of human class I molecule HLA-A2 (α1 in dark blue, α2 in light blue) with a peptide from the HIV-1 envelope protein GP120 (space-filling, orange) in the binding groove. (b) Ribbon model of human class II molecules HLA-DR1 with the DRα chain shown in dark blue and the DRβ chain in light blue. The peptide (space-filling, orange) in the binding groove is from influenza hemagglutinin (amino acid residues 306–318). *[Part (a) PDB ID 1HHG, part (b) PDB ID 1DLH.]*

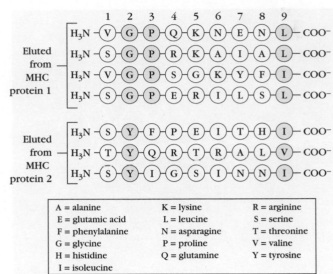

FIGURE 8-5 Examples of anchor residues (blue) in non-americ peptides eluted from two different class I MHC molecules. Anchor residues, at positions 2/3 and 9, that interact with the class I MHC molecule tend to be hydrophobic amino acids. The two MHC proteins bind peptides with different anchor residues. *[Data from V. H. Engelhard, 1994, Current Opinion in Immunology 6:13.]*

are approximately 10^5 copies of each class I protein per cell, it is estimated that each of the 2000 distinct peptides is presented with a frequency of between 100 and 4000 copies per cell. Evidence suggests that even a single MHC-peptide complex may be sufficient to target a cell for recognition and lysis by a cytotoxic T lymphocyte with a receptor specific for that target structure.

All peptides examined to date that bind to class I molecules contain a carboxyl-terminal anchor (e.g., see position 9 in Figure 8-5). These anchors are generally hydrophobic residues (e.g., leucine, isoleucine), although a few charged amino acids have been reported. Besides the anchor residue found at the carboxyl terminus, another anchor is often found at the second, or second and third, positions at the amino-terminal end of the peptide. In general, any peptide of the correct length that contains the same or chemically similar anchor residues will bind to the same class I MHC molecule. Knowledge of these key positions and the chemical restrictions for amino acids at these positions may allow for more nuanced clinical design studies, such as future vaccines targeted at eliciting protective immunity to particular pathogens.

X-ray crystallographic analyses of class I MHC-peptide complexes have revealed how the peptide-binding groove in a given MHC molecule can interact stably with a broad spectrum of different peptides. The anchor residues at both ends of the peptide are buried within the binding groove, thereby holding the peptide firmly in place (see Figure 8-4a). Nonameric peptides are bound preferentially. The main contacts between class I MHC molecules and the peptides they bind

identity of the anchor residues that can interact with a given class I molecule.

In a critical study of peptide binding by MHC molecules, peptides bound by two different class I MHC proteins were released chemically and analyzed by high-performance liquid chromatography (HPLC) mass spectrometry. More than 2000 distinct peptides were found among the peptide ligands released from these two class I MHC molecules. Since there

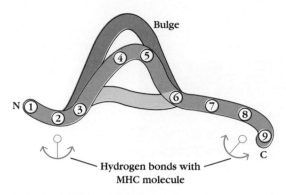

FIGURE 8-6 Conformation of peptides bound to class I MHC molecules. Schematic diagram depicting, in a side view, the conformational variance in MHC class I-bound peptides of different lengths. Longer peptides bulge in the middle, presumably interacting more with TCR, whereas shorter peptides lay flat in the groove. Contact with the MHC molecule is by hydrogen bonds to anchor residues 2 and/or 3 and 9. [Adapted from P. Parham, 1992, Nature 360:300, © 1992 Macmillan Magazines Limited]

involve residue 2 at the amino-terminal end and residue 9 at the carboxyl terminus of the peptide. Between the anchors, the peptide arches away from the floor of the groove in the middle (Figure 8-6), allowing peptides that are slightly longer or shorter to be accommodated. Amino acids in the center of the peptide that arch away from the MHC molecule are more exposed and presumably can interact more directly with the T-cell receptor.

Class II MHC-Peptide Interaction

Class II MHC molecules bind and present peptides to CD4[+] T cells. Like class I molecules, MHC class II molecules can bind a variety of peptides. These peptides are typically derived from exogenous proteins (either self or nonself) that are degraded within the exogenous processing pathway, described later in this chapter. Most of the peptides associated with class II MHC molecules are derived from self membrane-bound proteins or foreign proteins internalized by phagocytosis or by receptor-mediated endocytosis and then processed through the exogenous pathway.

Peptides recovered from class II MHC-peptide complexes generally contain 13 to 18 amino acid residues, somewhat longer than the nonameric peptides that most commonly bind to class I molecules. The peptide-binding groove in class II molecules is open at both ends (see Figure 8-4b), allowing longer peptides to extend beyond the ends, like an extra-long hot dog in a bun. Peptides bound to class II MHC molecules maintain a roughly constant elevation on the floor of the binding groove, another feature that distinguishes peptide binding to class I and class II molecules.

Peptide-binding studies and structural data for class II molecules indicate that a central core of 13 amino acids determines the ability of a peptide to bind class II. Longer peptides may be accommodated within the class II groove,

but the binding characteristics are determined by the central 13 residues. The peptides that bind to a particular class II molecule often have internal conserved sequence motifs, but unlike class I–binding peptides, they appear to lack conserved anchor residues (see Table 8-1). Instead, hydrogen bonds between the backbone of the peptide and the class II molecule are distributed throughout the binding site rather than being clustered predominantly at the ends of the site as is seen in class I–bound peptides. Peptides that bind to class II MHC molecules contain an internal sequence of 7 to 10 amino acids that provide the major contact points. Generally, this sequence has an aromatic or hydrophobic residue at the amino terminus and three additional hydrophobic residues in the middle portion and carboxyl-terminal end of the peptide. In addition, over 30% of the peptides eluted from class II molecules contain a proline residue at position 2 and another cluster of prolines at the carboxyl-terminal end. This relative flexibility contributes to peptide binding promiscuity.

General Organization and Inheritance of the MHC

Every vertebrate species studied to date possesses the tightly linked cluster of genes that constitute the MHC. As we have just discussed, MHC molecules have the important job of deciding which fragments of a foreign antigen will be "seen" by the host T cells. In general terms, MHC molecules face a similar ligand binding challenge to that faced, collectively, by B-cell and T-cell receptors: they must be able to bind a wide variety of antigens, and they must do so with relatively strong affinity. However, these immunologically relevant molecules meet this challenge using very different strategies. Although B- and T-cell receptor diversity is generated through genomic rearrangement and gene editing (described in Chapter 7), MHC molecules have opted for a combination of peptide binding promiscuity (discussed above) and the expression of several different MHC molecules on every cell (described below). Using this clever combined strategy, the immune system has evolved a way of maximizing the chances that many different regions, or *epitopes*, of an antigen will be recognized.

Studies of the MHC gene cluster originated when it was found that the rejection of foreign tissue transplanted between individuals in a species was the result of an immune response mounted against cell surface molecules, now called **histocompatibility antigens**. In the mid-1930s, Peter Gorer, who was using inbred strains of mice to identify blood-group antigens, identified four groups of genes that encode blood-cell antigens. He designated these I through IV. Work carried out in the 1940s and 1950s by Gorer and George Snell established that antigens encoded by the genes in the group designated as II took part in the rejection of transplanted tumors and other tissue. Snell called these *histocompatibility genes*; their current designation as histocompatibility-2 (H-2, or MHC) genes in the mouse was in reference to Gorer's original group II blood-cell antigens.

Mouse H-2 complex

Complex	H-2						
MHC class	I	II		III			I
Region	K	IA	IE	S			D
Gene products	H-2K	IA αβ	IE αβ	C' proteins	TNF-α Lymphotoxin-α	H-2D	H-2L*

*Not present in all haplotypes

Human HLA complex

Complex	HLA							
MHC class	II			III		I		
Region	DP	DQ	DR	C4, C2, BF		B	C	A
Gene products	DP αβ	DQ αβ	DR αβ	C' proteins	TNF-α Lymphotoxin-α	HLA-B	HLA-C	HLA-A

FIGURE 8-7 Comparison of the organization of the major histocompatibility complex (MHC) in mouse and human. The MHC is referred to as the H-2 complex in mice and as the HLA complex in humans. In both species, the MHC is organized into a number of regions encoding class I (pink), class II (blue), and class III (green) gene products. The class I and II gene products shown in this figure are considered to be the classical MHC molecules. The class III gene products include other immune function–related compounds such as the complement proteins (C') and tumor necrosis factors (TNF-α and Lymphotoxin α).

The MHC Locus Encodes Three Major Classes of Molecules

The major histocompatibility complex is a collection of genes arrayed within a long continuous stretch of DNA on chromosome 6 in humans and on chromosome 17 in mice. The MHC is referred to as the **human leukocyte antigen (HLA) complex** in humans and as the **H-2 complex** in mice, the two species in which these regions have been most studied. Although the arrangement of genes is somewhat different in the two species, in both cases the MHC genes are organized into regions encoding three classes of molecules (Figure 8-7):

- **Class I MHC genes** encode glycoproteins expressed on the surface of nearly all nucleated cells; the major function of the class I gene products is presentation of endogenous peptide antigens to CD8$^+$ T cells.

- **Class II MHC genes** encode glycoproteins expressed predominantly on APCs (macrophages, dendritic cells, and B cells), where they primarily present exogenous antigenic peptides to CD4$^+$ T cells.

- **Class III MHC genes** encode several different proteins, some with immune functions, including components of the complement system and molecules involved in inflammation.

Class I MHC molecules were the first discovered and are expressed in the widest range of cell types. Unlike in the human, this region in mouse is noncontinuous, interrupted by the class II and III regions (Figure 8-8). Recall that there are two chains to the MHC class I molecule: the more variable and antigen-binding α chain and the common β-2-microglobulin chain. The α chain molecules are encoded by the K and D regions in mice, with an additional L region found in some strains, and by the A, B, and C loci in humans. β2-microglobulin is encoded by a gene outside the MHC. Collectively, these are referred to as *classical class I molecules*; all posses the functional capability of presenting protein fragments of antigen to T cells.

Additional genes or groups of genes within the class I region of both mouse and human encode *nonclassical class I* molecules that are expressed only in specific cell types and have more specialized functions. Some appear to play a role in self/nonself discrimination. One example is the class I HLA-G molecule. These are present on fetal cells at the maternal-fetal interface and are credited with inhibiting rejection by maternal CD8$^+$ T cells by protecting the fetus from identification as foreign, which may occur when paternally derived antigens begin to appear on the developing fetus.

Class II MHC molecules are encoded by the IA and IE regions in mice and by the DP, DQ, and DR regions in humans (see Figures 8-7 and 8-8). The terminology is somewhat confusing, since the D region in mice encodes *class I* MHC molecules, whereas DP, DQ, and DR in humans refers to *class II* genes and molecules. Recall that class II molecules are composed of two chains, both of which interact with

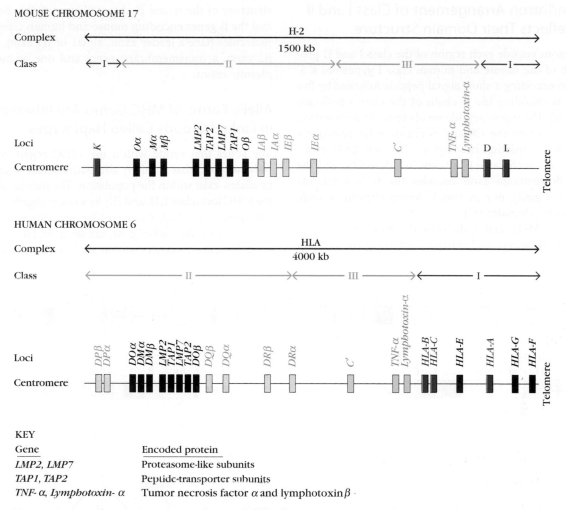

KEY

Gene	Encoded protein
LMP2, LMP7	Proteasome-like subunits
TAP1, TAP2	Peptide-transporter subunits
TNF-α, Lymphotoxin-α	Tumor necrosis factor α and lymphotoxin β

FIGURE 8-8 Simplified map of the mouse and human MHC loci. The MHC class I genes are colored red, MHC class II genes are colored blue, and genes in MHC III are colored green. Classical class I genes are labeled in red, classical class II genes are labeled in blue, and the nonclassical MHC genes are labeled in black. The concept of classical and nonclassical does not apply to class III. Only some of the non-classical class I genes are shown; the functions of only some of their proteins are known.

antigen. The class II region of the MHC locus encodes both the α chain and the β chain of a particular class II MHC molecule, and in some cases multiple genes are present for either or both chains. For example, individuals can inherit up to four functional DR β-chain genes, and all of these are expressed simultaneously in the cell. This allows any DR α-chain gene product to pair with any DR β chain product. Since the antigen-binding groove of class II is formed by a combination of the α and β chains, this creates several unique antigen-presenting DR molecules on the cell.

As with the class I loci, additional nonclassical class II molecules with specialized immune functions are encoded within this region. For instance, human nonclassical class II genes designated *DM* and *DO* have been identified. The *DM* genes encode a class II–like molecule (HLA-DM) that facilitates the loading of antigenic peptides into class II MHC molecules. Class II DO molecules, which are expressed only in the thymus and on mature B cells, have been shown to serve as regulators of class II antigen processing.

Class I and II MHC molecules have common structural features, and both have roles in antigen processing and presentation. By contrast, the class III MHC region encodes several molecules that are critical to immune function but have little in common with class I or II molecules. Class III products include the complement components C4, C2, and factor B (see Chapter 6), as well as several inflammatory cytokines, including the two tumor necrosis factor proteins (TNF-α and Lymphotoxin-α [TNF-β]). Allelic variants of some of these class III MHC gene products have been linked to certain diseases. For example, polymorphisms within the TNF-α gene, which encodes a cytokine involved in many immune processes (see Chapter 4), have been linked to susceptibility to certain infectious diseases and some forms of autoimmunity, including Crohn's disease and rheumatic arthritis. Despite its differences from class I and class II genes, the linked class III gene cluster is conserved in all species with an MHC region, suggesting similar evolutionary pressures have come to bear on this cluster of genes.

The Exon/Intron Arrangement of Class I and II Genes Reflects Their Domain Structure

Separate exons encode each region of the class I and II proteins. Each of the mouse and human class I genes has a 5′ leader exon encoding a short signal peptide followed by five or six exons encoding the α chain of the class I molecule (Figure 8-9a). The signal peptide serves to facilitate insertion of the α chain into the ER and is removed by proteolytic enzymes after translation is complete. The next three exons encode the extracellular α1, α2, and α3 domains, and the following downstream exon encodes the transmembrane (T_m) region. Finally, one or two 3′-terminal exons encode the cytoplasmic domains (C).

Like class I MHC genes, the class II genes are organized into a series of exons and introns mirroring the domain structure of the α and β chains (Figure 8-9b). Both the α and the β genes encoding mouse and human class II MHC molecules have a leader exon, an α1 or β1 exon, an α2 or β2 exon, a transmembrane exon, and one or more cytoplasmic exons.

Allelic Forms of MHC Genes Are Inherited in Linked Groups Called Haplotypes

The genes that reside within the MHC region are highly **polymorphic**; that is, many alternative forms of each gene, or **alleles**, exist within the population. The individual genes of the MHC loci (class I, II, and III) lie so close together that their inheritance is linked. Crossover, or recombination between genes, is more likely when genes are far apart. For instance, the recombination frequency within the H-2 complex (i.e., the

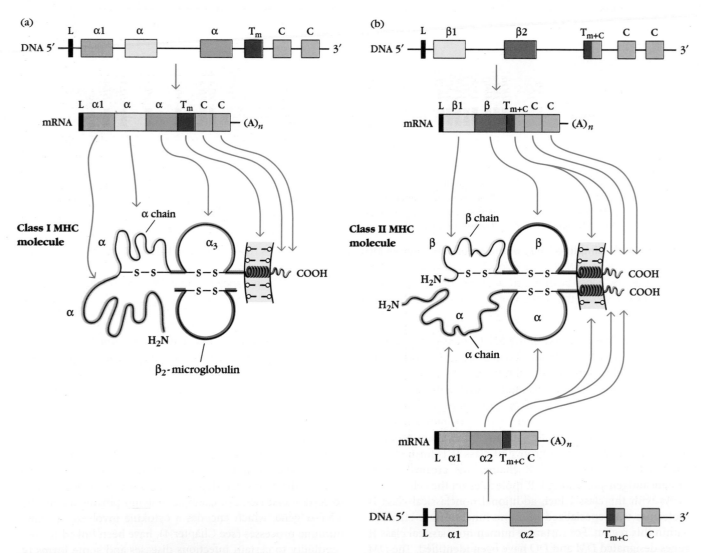

FIGURE 8-9 Schematic diagram of (a) class I and (b) class II MHC genes, mRNA transcripts, and protein molecules. There is strong correspondence between exons and the domains in the gene products of MHC molecules. Note that the mRNA transcripts are spliced to remove the intron sequences. Each exon, with the exception of the leader (L) exon, encodes a separate domain of an MHC molecule. The leader peptide is removed in a post-translational reaction before the molecule is expressed on the cell surface. The gene encoding β_2-microglobulin is located on a different chromosome in both human and mouse. T_m = transmembrane; C = cytoplasmic.

frequency of chromosome crossover events during meiosis, indicative of the distance between given genes) is only 0.5%. Thus, crossover occurs only once in every 200 meiotic cycles. For this reason, most individuals inherit all the alleles encoded by these genes as a set (known as linkage disequilibrium). This set of linked alleles is referred to as a **haplotype**. An individual inherits one haplotype from the mother and one haplotype from the father, or two sets of alleles.

In outbred populations, such as humans, the offspring are generally heterozygous at the MHC locus, with different alleles contributed by each of the parents. If, however, mice are inbred, each H-2 locus becomes homozygous because the maternal and paternal haplotypes are identical, and all offspring begin to express identical MHC molecules. Certain mouse strains have been intentionally inbred in this manner and are employed as prototype strains. The MHC haplotype expressed by each of these strains is designated by an arbitrary italic superscript (e.g., H-2^a, H-2^b). These designations refer to the entire set of inherited H-2 alleles within a strain without having to list the specific allele at each locus individually (Table 8-2). Different inbred strains may share the same set of alleles, or MHC haplotype, with another strain (i.e., CBA, AKR, and C3H) but will differ in genes outside the H-2 complex.

Detailed analysis of the H-2 complex in mice has been made possible by the development of congenic H-2 strains that differ only at the MHC locus. Inbred mouse strains are said to be **syngeneic**, or identical at all genetic loci. Two strains are **congenic** if they are genetically identical *except* at a single genetic region. Any phenotypic differences that can be detected between congenic strains is therefore related to the genetic region that distinguishes the two strains. Congenic strains that are identical with each other except at the

MHC can be produced by a series of crosses, backcrosses, and selections between two inbred strains that differ at the MHC. A frequently used congenic strain, designated B10.A, is derived from B10 mice (which is H-2^b) genetically manipulated to possess the H-2^a haplotype at the MHC locus. Recombination within the H-2 region of congenic mouse strains then allows the study of individual MHC genes and their products. Examples of these are included in the list in Table 8-2. For example, the B10.A (2R) strain has all the MHC genes from the *a* haplotype except for the D region, which is derived from the H-2^b parent.

MHC Molecules Are Codominantly Expressed

The genes within the MHC locus exhibit a **codominant** form of expression, meaning that both maternal and paternal gene products (from both haplotypes) are expressed at the same time and in the same cells. Therefore, if two mice from inbred strains possessing different MHC haplotypes are mated, the F$_1$ generation inherits both parental haplotypes and will express all these MHC alleles. For example, if an H-2^b strain is crossed with an H-2^k strain, then the F$_1$ generation inherits both parental sets of alleles and is said to be H-2$^{b/k}$ (Figure 8-10a). Because such an F$_1$ generation expresses the MHC proteins of both parental strains on its cells, it is said to be histocompatible with both parental strains. This means offspring are able to accept grafts from either parental source, each of which expresses MHC alleles viewed as "self" (Figure 8-10b). However, neither of the inbred parental strains can accept a graft from its F$_1$ offspring because half of the MHC molecules (those coming from the *other* parent) will be viewed as "nonself," or foreign, and thus subject to recognition and rejection by the immune system.

TABLE 8-2 H-2 haplotypes of some mouse strains

Prototype strain	Other strains with the same haplotype	Haplotype	K	IA	IE	S	D
CBA	AKR, C3H, B10.BR, C57BR	k	k	k	k	k	k
DBA/2	BALB/c, NZB, SEA, YBR	d	d	d	d	d	d
C57BL/10 (B10)	C57BL/6, C57L, C3H.SW, LP, 129	b	b	b	b	b	b
A	A/He, A/Sn, A/Wy, B10.A	a	k	k	k	d	d
B10.A(2R)*		h2	k	k	k	d	b
B10.A(3R)		i3	b	b	k	d	d
B10A.(4R)		h4	k	k	b	b	b
A.SW	B10.S, SJL	s	s	s	s	s	s
A.TL		t1	s	k	k	k	d
DBA/1	STOLI, B10.Q, BDP	q	q	q	q	q	q

*The R designates a recombinant haplotype, in this case between the H-2^a and H-2^b types. Gene contribution from the *a* strain is shown in yellow and from the *b* strain in red.

(a) Mating of inbred mouse strains with different MHC haplotypes

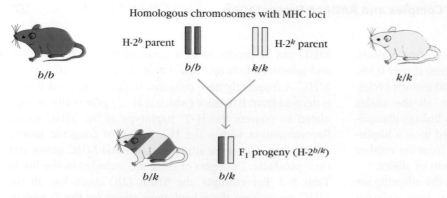

Homologous chromosomes with MHC loci

H-2^b parent   H-2^k parent

b/b   k/k

b/b   k/k

F$_1$ progeny (H-2$^{b/k}$)

b/k   b/k

(b) Skin transplantation between inbred mouse strains with same or different MHC haplotypes

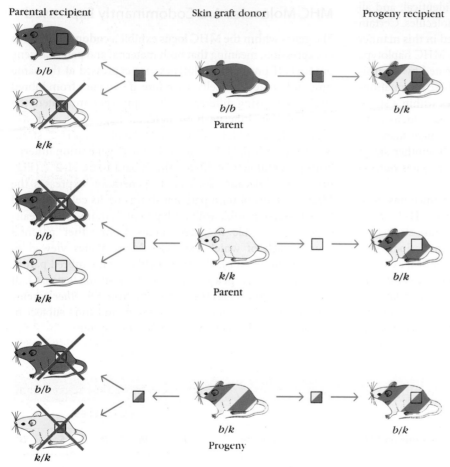

Parental recipient   Skin graft donor   Progeny recipient

b/b

k/k   b/b
Parent   b/k

b/b

k/k   k/k
Parent   b/k

b/b

k/k   b/k
Progeny   b/k

(c) Inheritance of HLA haplotypes in a typical human family

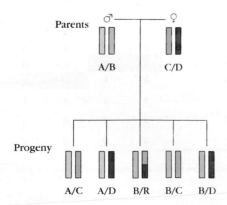

Parents   ♂ ⎯⎯⎯ ♀

A/B   C/D

Progeny

A/C   A/D   B/R   B/C   B/D

FIGURE 8-10 Illustration of inheritance of MHC haplotypes in inbred mouse strains. (a) The letters b/b designate a mouse homozygous for the H-2^b MHC haplotype, k/k homozygous for the H-2^k haplotype, and b/k a heterozygote. Because the MHC genes are closely linked and inherited as a set, the MHC haplotype of F$_1$ progeny from the mating of two different inbred strains can be predicted easily. (b) Acceptance or rejection (X) of skin grafts is controlled by the MHC type of the inbred mice. The progeny of the cross between two inbred strains with different MHC haplotypes (H-2^b and H-2^k) will express both haplotypes (H-2$^{b/k}$) and will accept grafts from either parent and from one another. However, neither parent strain will accept grafts from the offspring. (c) Inheritance of HLA haplotypes in a hypothetical human family. For ease, the human paternal HLA haplotypes are arbitrarily designated A and B, maternal C and D. Note that a new haplotype, R (recombination), can arise from rare recombination of a parental haplotype (maternal shown here).

In an outbred population such as humans, each individual is generally heterozygous at each locus. The human HLA complex is highly polymorphic, and multiple alleles of each class I and class II gene exist. However, as with mice, the human MHC genes are closely linked and usually inherited as a haplotype. When the father and mother have different haplotypes, as in the example shown in Figure 8-10c, there is a one-in-four chance that siblings will inherit the same paternal *and* maternal haplotypes and therefore will be histocompatible (i.e., genetically identical at their MHC loci) with each other; none of the offspring will be fully histocompatible with the parents.

Although the rate of recombination by crossover is low within the HLA complex, it still contributes significantly to the diversity of the loci in human populations. Genetic recombination can generate new allelic combinations, or haplotypes (see haplotype R in Figure 8-10c), and the high number of intervening generations since the appearance of humans as a species has allowed extensive recombination. As a result of recombination and other mechanisms for generating mutations, it is rare for any two unrelated individuals to have identical sets of HLA genes. This makes transplantation between individuals who are not identical twins quite challenging! To address this, clinicians begin by looking for family members who will be at least partially histocompatible with the patient, or they rely on donor databases to look for an MHC match. Even with partial matches, physicians still need to administer heavy doses of immunosuppressive drugs to inhibit the strong rejection responses that typically follow tissue transplantation due to differences in the MHC proteins (see Chapter 16).

Class I and Class II Molecules Exhibit Diversity at Both the Individual and Species Levels

As noted earlier, any particular MHC molecule can bind many different peptides (called promiscuity), which gives the host an advantage in responding to pathogens. Rather than relying on just one gene for this task, the MHC region has evolved to include multiple genetic loci encoding proteins with the same function. In humans, HLA-A, B, or C molecules can all present peptides to $CD8^+$ T cells and HLA-DP, DQ, or DR molecules can present to $CD4^+$ T cells. The MHC region is thus said to be **polygenic** because it contains multiple genes with the same function but with slightly different structures. Since the MHC alleles are also codominantly expressed, heterozygous individuals will express the gene products encoded by both alleles at each MHC gene locus. In a fully heterozygous individual this amounts to 6 unique classical class I molecules on each nucleated cell. An F_1 mouse, for example, expresses the K, D, and L class I molecules from each parent (six different class I MHC molecules) on the surface of each of its nucleated cells (Figure 8-11). The expression of so many individual class I MHC molecules allows each cell to display a large number of different peptides in the peptide-binding grooves of its MHC molecules.

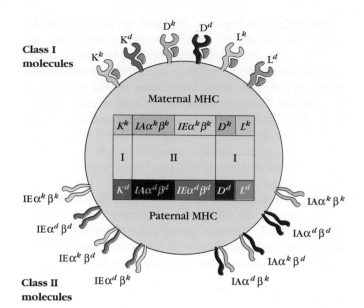

FIGURE 8-11   Diagram illustrating the various MHC molecules expressed on antigen-presenting cells of a heterozygous H-2$^{k/d}$ mouse. Both the maternal and paternal MHC genes are expressed (codominant expression). Because the class II molecules are heterodimers, new molecules containing one maternal-derived and one paternal-derived chain are also produced, increasing the diversity of MHC class II molecules on the cell surface. The β_2-microglobulin component of class I molecules (pink) is encoded by a nonpolymorphic gene on a separate chromosome and may be derived from either parent.

MHC class II molecules have even greater potential for diversity. Each of the classical class II MHC molecules is composed of two different polypeptide chains encoded by different loci. Therefore, a heterozygous individual can express α-β combinations that originate from the same chromosome (maternal only or paternal only) as well as class II molecules arising from unique chain pairing derived from separate chromosomes (new maternal-paternal α-β combinations). For example, an H-2^k mouse expresses IA^k and IE^k class II molecules, whereas an H-2^d mouse expresses IA^d and IE^d molecules. The F_1 progeny resulting from crosses of these two strains express four parental class II molecules (identical to their parents) and also four new molecules that are mixtures from their parents, containing one parent's α chain and the other parent's β chain (as shown in Figure 8-11). Since the human MHC contains three classical class II genes (*DP, DQ,* and *DR*), a heterozygous individual expresses six class II molecules identical to the parents and six molecules containing new α and β chain combinations. The number of different class II molecules expressed by an individual can be further increased by the occasional presence of multiple β-chain genes in mice and humans, and in humans by the presence of multiple α-chain genes. The diversity generated by these new MHC molecules likely increases the number of different antigenic peptides that can be presented and is therefore advantageous to the organism in fighting infection.

The variety of peptides displayed by MHC molecules echoes the diversity of antigens bound by antibodies and T-cell receptors. This evolutionary pressure to diversify comes from the fact that both need to be able to interact with antigen fragments they have never before seen, or that may not yet have evolved. However, the strategy for generating diversity within MHC molecules and the antigen receptors on T and B cells is not the same. Antibodies and T-cell receptors are generated by several somatic processes, including gene rearrangement and the somatic mutation of rearranged genes (see Chapters 7 and 12). Thus, the generation of T- and B-cell receptors is dynamic, changing over time within an individual. By contrast, the MHC molecules expressed by an individual are fixed. However, promiscuity of antigen binding ensures that even "new" proteins are likely to contain some fragments that can associate with any given MHC molecule.

TABLE 8-3	Genetic diversity of MHC loci in the human population

MHC CLASS I	
HLA locus	**Number of allotypes (proteins)**
A	1448
B	1988
C	1119
E	3
F	4
G	16

MHC CLASS II	
HLA locus	**Number of allotypes (proteins)**
DMA	4
DMB	7
DOA	3
DOB	5
DPA1	17
DPB1	134
DQA1	47
DQB1	126
DRA	2
DRB1	860
DRB3	46
DRB4	8
DRB5	17

Source: Data obtained from http://hla.alleles.org, a Web site maintained by the HLA Informatics Group based at the Anthony Nolan Trust in the United Kingdom, with up-to-date information on the numbers of HLA alleles and proteins.

Collectively, this builds in enormous flexibility within the host for responding to unexpected environmental changes that might arise in the future—an elegant evolutionary strategy.

Countering this limitation on the range of peptides that can be presented by any one individual is the vast array of peptides that can be presented at the species level, thanks to the diversity of the MHC in any outbred population. The MHC possesses an extraordinarily large number of different alleles at each locus and is one of the most polymorphic genetic complexes known in higher vertebrates. These alleles differ in their DNA sequences from one individual to another by 5% to 10%. The number of amino acid differences between MHC alleles can be quite significant, with up to 20 amino acid residues contributing to the unique structural nature of each allele. Analysis of human HLA class I genes as of July 2012 reveals approximately 2013 A alleles, 2605 B alleles, and 1551 C alleles (Table 8-3 shows the number of protein products; not all alleles encode expressed proteins). In mice, the polymorphism is similarly enormous. The human class II genes are also highly polymorphic, and in some cases, different individuals can even inherit different numbers of genes. The number of HLA-DR β-chain genes (DRB) may vary from 2 to 9 in different haplotypes, and over 1200 alleles of DRB genes alone have been reported. Interestingly, the DRA gene is highly conserved, with only seven different alleles and two proteins identified. Current estimates of actual polymorphism in the human MHC are likely on the low side because the earliest and most detailed data have primarily concentrated on populations of European descent.

This enormous polymorphism results in a tremendous diversity of MHC molecules within a species. Using the numbers given above for the allelic forms of human HLA-A, -B, and -C, we can calculate a theoretical number of combinations that can exist, which is upward of *1.7 billion* different class I haplotypes possible in the population. If class II loci are considered, the numbers are even more staggering, with over 10^{15} different class II combinations. Because each haplotype contains *both* class I and class II genes, multiplication of the numbers gives a total of more than 1.7×10^{24} possible combinations of these class I and II alleles within the entire human population!

Some evidence suggests that a reduction in MHC polymorphism within a species may predispose that species to infectious disease (see Evolution Box 8-1). In one example, cheetahs and certain other wild cats, such as Florida panthers, that have been shown to be highly susceptible to viral disease also have very limited MHC polymorphism. It is postulated that the present cheetah population arose from a limited breeding population, or genetic bottleneck, causing a loss of MHC diversity. This increased susceptibility of cheetahs to various viruses may result from a reduction in the number of different MHC molecules available to the species as a whole and a corresponding limitation on the range of processed antigens with which these MHC molecules can interact. As a corollary, this suggests that the high level of MHC polymorphism that has been observed in many outbred species, including humans, may provide a survival advantage by supplying a broad range of MHC molecules and thus a broad range of presentable antigens.

BOX 8-1

 The Sweet Smell of Diversity

As early as the mid-1970s, mate choice in mice was shown to be influenced by genes at the MHC (H-2) locus. This was followed by investigations in other rodents, fish, and birds, all with similar conclusions. Thanks to results from several studies conducted in the last 15 years, it looks like humans could be added to this list. However, questions remain as to the precise evolutionary pressures, the mechanism, and the magnitude of this effect, among others, in influencing mate choice in humans.

In terms of evolutionary pressure, local pathogens play a significant role in maintaining MHC diversity in the population and in selecting for specific alleles. As we now appreciate, this is because the MHC influences immune responsiveness. Due to its role in selecting the peptide fragments that will be presented, the inheritance of specific alleles at particular loci can predispose individuals to either enhanced susceptibility or resistance to specific infectious agents and immune disorders (see Clinical Focus Box 8-2). Over long time periods, endemic local pathogens exert evolutionary pressures that drive higher- or lower-than-expected rates of certain MHC alleles in a population, as well as overrepresentation of other resistance-associated genes.

The degree of diversity at the MHC locus clearly influences susceptibility to disease in populations; witness the enhanced viral susceptibility seen in cheetahs (see page 274) and some devastating human stories. For instance, the European introduction of the smallpox virus to New World populations is credited with wiping out large Native American groups. This may be due to a lack of past evolutionary pressure for conservation of resistance-associated MHC alleles, which would therefore be rare or nonexistent in this population, as well as to a lack of any individuals with immunity from prior infection.

But how does an individual evaluate how well a potential mate will contribute "new" MHC alleles to one's offspring, leading to greater diversity and the potential for enhanced fitness? The primary candidate is odor: the MHC is known to influence odor in many vertebrate species. For example, the urine of mice from distinct MHC congenic lines can be distinguished by both humans and rodents. Mice show a distinct preference for mating with animals that carry MHC alleles that are *dissimilar* to their own. In terms of maximizing the range of peptides that can be presented, this makes clear evolutionary sense, as increased diversity at the MHC should increase the number of different pathogenic peptides that can be "seen" by the immune system, increasing the likelihood of effective anti-pathogen responses. To back this up, the advantage of overall MHC diversity has been shown experimentally in mice, where most simulated epidemic experiments have found a survival advantage for H-2 heterozygous animals over their homozygous counterparts. In humans, research in HIV-infected individuals has shown that extended survival and a slower progression to AIDS are correlated with full heterozygosity at the HLA class I locus, as well as absence of certain AIDS-associated HLA-B and -C alleles. This specific link to class I is not surprising in light of the key role of CD8$^+$ T cells in combating viral infections.

Human studies of attraction and mating also point to preferences for individuals with MHC dissimilar alleles. In one key study involving what is commonly known as the "sweaty T-shirt test," college-age volunteers were asked to rate their preference or sexual attraction to the odor of T-shirts worn by individuals of the opposite sex. In general, both males and females preferred the odor of T-shirts worn by individuals with dissimilar HLA types. The one key exception was seen in women who were concurrently using an estrogen-based birth control pill; they instead showed a preference for the odors of MHC-*similar* individuals, suggesting that this hormone not only interferes with this response but that it potentially shifts the outcome.

But how? you ask. Soluble forms of MHC molecules have been found in many bodily fluids, including urine, saliva, sweat, and plasma. However, these molecules are unlikely to be small or volatile enough to account for direct olfactory detection. Other hypotheses for how MHC might influence odor include via olfactory recognition of natural ligands or specific volatile peptides carried by MHC molecules, or by MHC-driven differences in natural flora, which are also known to impact body odor. The recent finding that some polymorphic olfactory receptor genes are closely linked to the MHC may also help explain this apparent association with mating preference. To date, the jury is still out on which mechanism(s) likely account(s) for MHC-specific odor differences.

As one might imagine, large methodological challenges are inherent in asking questions related to mate choice in humans, where differing levels of outbreeding as well as social and cultural factors influence the outcome. Nonetheless, it appears that like most primate species tested so far, we humans may also have the capacity to use smell as an evolutionary strategy for promoting a robust immune response in our offspring.

J. Havlicek, and S. C. Roberts. 2009. MHC-correlated mate choice in humans: A review. *Psychoneuroendocrinology* **34**:497–512.

C. Wedekind et al. 1995. MHC-dependent mate preferences in humans. *Proceedings. Biological Science* **260**:245–249.

C. Wedekind et al. 1997. Body odor preferences in men and women: Do they aim for specific MHC combinations or simply heterozygosity? *Proceedings. Biological Science* **264**:1471–1479.

Although some individuals within a species may not be able to develop an immune response to a given pathogen and therefore will be susceptible to infection, extreme polymorphism ensures that at least some members of a species will be resistant to that disease. In this way, MHC diversity at the population level may protect the species as a whole from extinction via a wide range of infectious diseases.

MHC Polymorphism Has Functional Relevance

Although the sequence divergence among alleles of the MHC within a species is very high, this variation is not randomly distributed along the entire polypeptide chain. Instead, polymorphism in the MHC is clustered in short stretches, largely within the membrane-distal α1 and α2 domains of class I molecules (Figure 8-12a). Similar patterns of diversity are observed in the α1 and β1 domains of class II molecules.

Structural comparisons have located the polymorphic residues within the three-dimensional structure of the membrane-distal domains in class I and class II MHC molecules and have related allelic differences to functional differences (Figure 8-12b). For example, of 17 amino acids previously shown to display significant polymorphism among HLA-A molecules, 15 were shown by x-ray crystallographic analysis to be in the peptide-binding groove of this molecule. The location of so many polymorphic amino acids within the binding site for processed antigen strongly suggests that allelic differences contribute to the observed differences in the ability of MHC molecules to interact with a given peptide ligand. Polymorphisms that lie outside these regions and might affect basic domain folding are rare. This clustering of polymorphisms around regions that make contact with antigen also suggest possible reasons why certain MHC genes or haplotypes can become associated with certain diseases (see Clinical Focus Box 8-2).

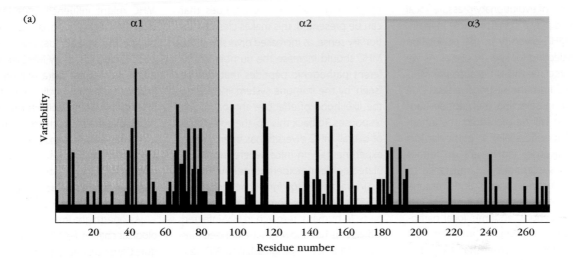

(a)

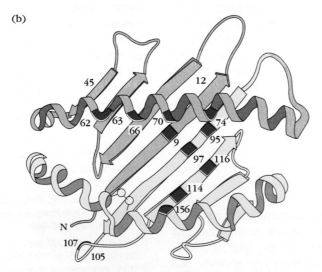

(b)

FIGURE 8-12 Variability in the amino acid sequences of allelic class I HLA molecules. (a) In the external domains, most of the variable residues are in the membrane-distal α1 and α2 domains. (b) Location of polymorphic amino acid residues (red) in the α1/α2 domain of a human class I MHC molecule. *[Part (a) adapted from R. Sodoyer et al., 1984, EMBO Journal 3:879, reprinted by permission of Oxford University Press; part (b) adapted, with permission, from P. Parham, 1989, Nature 342:617, © 1989 Macmillan Magazines Limited.]*

MHC Alleles and Susceptibility to Certain Diseases

Some HLA alleles occur at a much higher frequency in people suffering from certain diseases than in the general population. The diseases associated with particular MHC alleles include auto-immune disorders, certain viral diseases, disorders of the complement system, some neurologic disorders, and several different allergies. In humans, the association between an HLA allele and a given disease may be quantified by determining the frequency of that allele expressed by individuals afflicted with the disease, then comparing these data with the frequency of the same allele in the general population. Such a comparison allows calculation of an individual's **relative risk (RR)**.

$$RR = \frac{\text{frequency of disease in the allele}^{+}\text{ group}}{\text{frequency of disease in the allele}^{-}\text{ group}} \quad (Eq.\ 8.1)$$

A RR value of 1 means that the HLA allele is expressed with the same frequency in disease-afflicted and general populations, indicating that this allele confers no increased risk for the disease. A RR value substantially above 1 indicates an association between the HLA allele and the disease. For example, individuals with the HLA-B27 allele are 90 times more likely (RR = 90) to develop the autoimmune disease ankylosing spondylitis, an inflammatory disease of vertebral joints characterized by destruction of cartilage, than are individuals who lack this HLA-B allele. Other disease associations with significantly high RR include HLA-DQB1 and narcolepsy (RR = 130) and HLA-DQ2 with celiac disease (RR=50), an allergy to gluten. A few HLA alleles have also been linked to relative protection from disease or clinical progres-

sion. This is seen in the case of individuals inheriting HLA-B57, which is associated with greater viral control and a slower progression to AIDS in HIV-infected individuals.

When the associations between MHC alleles and disease are weak, reflected by low RR values, it is possible that multiple genes influence susceptibility, of which only one lies within the MHC locus. The genetic origins of several autoimmune diseases, such as multiple sclerosis (associated with DR2; RR = 5) and rheumatoid arthritis (associated with DR4; RR = 10) have been studied in depth. The observation that these diseases are not inherited by simple Mendelian segregation of MHC alleles can be seen clearly in identical twins, where both inherit the same MHC risk factor, but frequently only one develops the disease. This finding suggests that multiple genetic factors plus one or more environmental factors are at play in development of this disease. As we highlight in Chapter 16, this combined role for genes and the environment in the development of autoimmunity is not uncommon.

The existence of an association between an MHC allele and a disease should not be interpreted to imply that the expression of the allele has caused the disease. The relationship between MHC alleles and development of disease is more complex, partly thanks to the genetic phenomenon of linkage disequilibrium. The fact that some of the class I MHC alleles are in linkage disequilibrium with the class II and class III alleles can make their contribution to disease susceptibility appear more pronounced than it actually is. If, for example, DR4 contrib-

utes to risk of a disease and it also occurs frequently in combination with HLA-A3 because of linkage disequilibrium, then A3 would incorrectly appear to be associated with the disease. Improved genomic mapping techniques make it possible to analyze the fine linkage between genes within the MHC and various diseases more fully and to assess the contributions from other loci. In the case of ankylosing spondylitis, for example, it has been suggested that alleles of *TNF-α* and *Lymphotoxin α* may produce protein variants that are involved in destruction of cartilage, and these alleles happen to be linked to certain HLA-B alleles. In the case of HLA-B57 and AIDS progression, the allele has been linked more directly to disease. It is believed that this class I allele is particularly efficient at presenting important components of the virus to circulating T_C cells, leading to increased destruction of virally infected cells and delayed disease progression.

Other hypotheses have also been offered to account for a direct role of particular MHC alleles in disease susceptibility. In some rare cases, certain allelic forms of MHC genes can encode molecules that are recognized as receptors by viruses or bacterial toxins, leading to increased susceptibility in the individuals who inherit these alleles. As will be explored further in Chapters 15 through 19, in many cases complex interactions between multiple genes, frequently including the MHC, and particular environmental factors are required to create a bias toward the development of certain diseases.

The preceding discussion points to additional parallels between MHC molecules and lymphocyte antigen receptors. The somatic hypermutations seen in B-cell receptor genes is also not randomly arrayed within the molecule, but instead is clustered in the regions most likely to interact directly with peptide (see Chapter 12), providing yet another example of how the immune system has solved a similar functional dilemma using a very different strategy.

The Role of the MHC and Expression Patterns

As we have just discussed, several genetic features help ensure a diversity of MHC molecules in outbred populations, including polygeny, polymorphism, and codominant expression. All this attention paid to maximizing the number of different binding grooves suggests that variety within the MHC plays an important role in survival (see Clinical Focus Box 8-2). In fact, in addition to fighting infection, MHC expression throughout the body plays a key role in maintaining homeostasis and health even when no foreign antigen is present.

Although the presentation of MHC molecules complexed with foreign antigen to T cells garners much attention (and space in this book!), most MHC molecules spend their lives presenting other things, and often to other cells. There are several reasons why an MHC molecule on the surface of a cell is important. In general, these include the following:

- To display self class I to demonstrate that the cell is healthy
- To display foreign peptide in class I to show that the cell is infected and to engage with T_C cells
- To display a self-peptide in class I and II to test developing T cells for autoreactivity (primary lymphoid organs)
- To display a self-peptide in class I and II to maintain tolerance to self-proteins (secondary lymphoid organs)
- To display a foreign peptide in class II to show the body is infected and activate T_H cells

It is worth noting that although some of these instances lead to immune activation, some do not. It is also important to note that the cell type, tissue location, and timing of expression vary for each of these situations. For instance, developing T cells first encounter MHC class I and II molecules presenting self peptides in the thymus, where these signals are designed to *inhibit* the ability of T cells to attack self-structures later (see Chapter 9). Others occur only after certain types of exposure, such as during an immune response to extracellular pathogens or when cells in the body are infected with viral invaders, and is designed to *activate* T cells to act against the pathogen (see Chapter 11).

To help set the stage for this discussion, we turn now to the where, when, and why of MHC expression. This is followed by a description of the how, detailing the pathways that lead to peptide placement in the binding groove of each class of molecule. As we will see in greater detail in the following sections, the type of MHC molecule(s) expressed by a given cell is linked to the role of the cell and the function of each class of molecule.

MHC Molecules Present Both Intracellular and Extracellular Antigens

In very general terms, the job of MHC class I molecules is to collect and present antigens that come from *intracellular* locations. This is a form of ongoing surveillance of the internal happenings of the cell—in essence, a window for displaying on the surface of the cell snippets of what is occurring inside. Basically, all cells in the body need this form of check and balance, and this shows up in the fairly ubiquitous nature of MHC class I expression in the body (there are a few notable exceptions, which we will touch on later). Often, nothing

other than normal cellular processes are occurring in the cytosol, and in these instances our cells present self-peptides in the groove of MHC class I molecules. The expresssion of self-MHC class I (with self peptides) signals that a cell is healthy; absence of self-MHC class I (as can occur in virus-infected and tumor cells) targets that cell for killing by NK cells (see Chapter 13). When foreign proteins are present in the cytosol and begin to appear in the groove of MHC class I on the cell surface, this alerts $CD8^+$ T cells to the presence of this unwelcome visitor, targeting the cell for destruction. In this case, the cell is called a **target cell** because it becomes a target for lysis by cytotoxic T cells.

Conversely, MHC class II molecules primarily display peptides that have come from the *extracellular* spaces of the body. Since this sampling of extracellular contents is not a form of policing that all cells need to perform, only specialized leukocytes posses this ability. These cells are collectively referred to as antigen-presenting cells (APCs), because their job is to present extracellular antigen to T cells, charging them with the ultimate job of coordinating the elimination of this extracellular invader. While all cells in the body express MHC class I proteins and can present peptides from foreign intracellular antigens, the term APC is usually reserved for MHC class II-expressing cells.

MHC Class I Expression Is Found Throughout the Body

In general, classical class I MHC molecules are expressed constitutively on almost all nucleated cells of the body. However, the level of expression differs among different cell types, with the highest levels of class I molecules found on the surface of lymphocytes. These molecules may constitute approximately 1% of the total plasma membrane proteins, or some 5×10^5 MHC class I molecules per lymphocyte. In contrast, some other cells, such as fibroblasts, muscle cells, liver hepatocytes, and some neural cells, express very low to undetectable levels of class I MHC molecules. This low-level expression on liver cells may contribute to the relative success of liver transplants by reducing the likelihood of graft rejection by T_c cells of the recipient. A few cell types (e.g., subsets of neurons and sperm cells at certain stages of differentiation) appear to lack class I MHC molecules altogether. Nucleated cells without class I expression are, however, very rare. Non-nucleated cells, such as red blood cells in mammals, do not generally express any MHC molecules.

In normal, healthy cells, class I molecules on the surface of the cell will display self-peptides resulting from normal turnover of self-proteins inside the cell. In cells infected by a virus, viral peptides as well as self-peptides will be displayed. Therefore, a single virus-infected cell can be envisioned as having various class I molecules on its membrane, some displaying a subset of viral peptides derived from the viral proteins being manufactured within. Because of individual allelic differences in the peptide-binding grooves of the class I MHC molecules, different individuals within a species will have the ability to bind and present different sets of viral peptides. In addition to virally infected cells, altered self cells such as cancer cells, aging body cells, or cells from an allogeneic graft (i.e., from a genetically-different individual), also can serve as target cells due to their expression of foreign MHC proteins and can be lysed by T_C cells. The importance of constitutive expression of class I is highlighted by the response of natural killer (NK) cells to somatic cells that lack MHC class I, as can occur during some viral infections. NK cells can kill a cell that has stopped expressing MHC class I on the surface, presumably because this suggests that the cell is no longer healthy or has been altered by the presence of an intracellular invader.

Expression of MHC Class II Molecules Is Primarily Restricted to Antigen-Presenting Cells

MHC class II molecules are found on a much more restricted set of cells than class I, and sometimes only after an inducing event. As mentioned above, cells that display peptides associated with class II MHC molecules to $CD4^+$ T_H cells are called antigen-presenting cells (APCs), and these cells are primarily certain types of leukocytes. APCs are specialized for their ability to alert the immune system to the presence of an invader and drive the activation of T cell responses.

Among the various APCs, marked differences in the level of MHC class II expression have been observed. In some cases, class II expression depends on the cell's differentiation stage or level of activation (such as in macrophages; see below). APC activation usually occurs following interaction with a pathogen and/or via cytokine signaling, which then induces significant increases in MHC class II expression.

A variety of cells can function as bona fide APCs. Their distinguishing feature is their ability to express class II MHC molecules and to deliver a costimulatory, or second activating signal, to T cells. Three cell types are known to have these characteristics and are thus often referred to as *professional antigen-presenting cells* (*pAPCs*): dendritic cells, macrophages, and B lymphocytes. These cells differ from one another in their mechanisms of antigen uptake, in whether they constitutively express class II MHC molecules, and in their costimulatory activity, as follows:

- Dendritic cells are generally viewed as the most effective of the APCs. Because these cells constitutively express a

high levels of class II MHC molecules and have inherent costimulatory activity, they can activate naïve T_H cells.

- Macrophages must be activated (e.g., via TLR signaling) before they express class II MHC molecules or costimulatory membrane molecules such as CD80/86.

- B cells constitutively express class II MHC molecules and posses antigen-specific surface receptors, making them particularly efficient at capturing and presenting their cognate antigen. However, they must be activated by, for example, antigen, cytokines, or pathogen-associated molecular patterns (PAMPs), before they express the costimulatory molecules required for activating naïve T_H cells.

Several other cell types, classified as nonprofessional APCs, can be induced to express class II MHC molecules and/or a costimulatory signal under certain conditions (Table 8-4). These cells can be deputized for professional antigen presentation for short periods and in particular situations, such as during a sustained inflammatory response.

MHC Expression Can Change with Changing Conditions

As noted above, MHC class I is constitutively expressed by most cells in the body, whereas class II is only expressed under certain conditions and in a very limited number of cell types. The different roles of the two molecules help explain this, and suggest that in certain instances specific changes in MHC expression may prove advantageous. The MHC locus can respond to both positive and negative regulatory pressures. For instance, MHC class I production can be disrupted or depressed by some pathogens. MHC class II expression on APCs is already quite variable, and the microenvironment surrounding an APC can further modulate expression, usually enhancing the expression of these molecules. APCs in particular are conditioned to respond to local cues leading to their activation and heightened MHC expression, enabling them to arm other cells in the body for battle. As one might imagine, this activation and arming of APCs must be carefully regulated, lest these cells orchestrate unwanted aggressive maneuvers against self components or benign companions, as occurs in clinical conditions such as autoimmunity or allergy, respectively. The mechanisms driving these changes in expression are described below.

Genetic Regulatory Components

The presence of internal or external triggers, such as intracellular invaders or cytokines (see below), can induce a signal transduction cascade that leads to changes in MHC gene expression. Research aimed at understanding the mechanism of control of MHC expression has been advanced by the now complete sequence of the mouse genome. Both class I and class II MHC genes are flanked by 5′ promoter sequences that bind sequence-specific transcription factors. The promoter

TABLE 8-4	Antigen-presenting cells	
Professional antigen-presenting cells	**Nonprofessional antigen-presenting cells**	
Dendritic cells (several types)	Fibroblasts (skin)	Thymic epithelial cells
Macrophages	Glial cells (brain)	Thyroid epithelial cells
B cells	Pancreatic beta cells	Vascular endothelial cells

motifs and the transcription factors that bind to these motifs have been identified for a number of MHC genes, with examples of regulation mediated by both positive and negative elements. For example, a class II MHC transcriptional activator called CIITA (also known as *class II, major histocompatibility complex, transactivator*) and another transcription factor called RFX, have both been shown to activate the promoter of class II MHC genes. Defects in these transcription factors cause one form of *bare lymphocyte syndrome*. Patients with this disorder lack class II MHC molecules on their cells and suffer from severe immunodeficiency, highlighting the central role of class II molecules in T-cell maturation and activation.

Viral Interference

One clear example of negative regulation of MHC comes from viruses that interfere with MHC class I expression and thus avoid easy detection by CD8[+] T cells. These viruses include human cytomegalovirus (CMV), hepatitis B virus (HBV), and adenovirus 12 (Ad12). In some cases, reduced expression of class I MHC molecules is due to decreased levels of a component needed for peptide transport or MHC class I assembly rather than decreased transcription. For example, in the case of cytomegalovirus infection, a viral protein binds to β_2-microglobulin, preventing assembly of class I MHC molecules and their transport to the plasma membrane. Adenovirus 12 infection causes a pronounced decrease in transcription of the transporter genes (*TAP1* and *TAP2*). As described in the following section on antigen processing, the *TAP* gene products play an important role in peptide transport from the cytoplasm into the rough endoplasmic reticulum (RER). Blocking of *TAP* gene expression inhibits peptide transport; as a result, class I MHC molecules cannot assemble with β_2-microglobulin or be transported to the cell membrane. These observations are especially important because decreased expression of class I MHC molecules, by whatever mechanism, is likely to help viruses evade the immune response. Lower levels of class I decrease the likelihood that virus-infected cells can display viral peptide complexes on their surface and become targets for CTL-mediated destruction.

Cytokine-Mediated Signaling

The expression of MHC molecules is externally regulated by various cytokines. Leading among these are the interferons (α, β, and γ) and the tumor necrosis factors (TNF-α and Lymphotoxin-α), each of which have been shown to increase expression of class I MHC molecules on cells. Typically, phagocytic cells that are involved in innate responses, or locally infected cells, are the first to produce these MHC-regulating cytokines. In particular, IFN-α (produced by a cell following viral or bacterial infection) and TNF-α (secreted by APCs after activation) are frequently the first cytokines to kick off an MHC class I up-regulation event. In the later stages, interferon gamma (IFN-γ), secreted by activated T_H cells as well as other cell types, also contributes to increased MHC expression.

Binding of these cytokines to their respective receptors induces intracellular signaling cascades that activate transcription factors and alter expression patterns. These factors bind to their target promoter sequences and coordinate increased transcription of the genes encoding the class I α chain, β_2-microglobulin, and other proteins involved in antigen processing and presentation. IFN-γ has been shown to induce expression of the class II transcriptional activator (CIITA), thereby indirectly increasing expression of class II MHC molecules on a variety of cells, including non-APCs (e.g., skin keratinocytes, intestinal epithelial cells, vascular endothelium, placental cells, and pancreatic beta cells).

Other cytokines influence MHC expression only in certain cell types. For example, IL-4 increases expression of class II molecules in resting B cells, turning them into more efficient APCs. Conversely, expression of class II molecules by B cells is down-regulated by IFN-γ. Corticosteroids and prostaglandins can also decrease expression of class II MHC molecules. These naturally occurring, membrane-permeable compounds bind to intracellular receptors and are some of the most potent suppressors of adaptive immunity, primarily based on this ability to inhibit MHC expression. This property is exploited in many clinical settings, where these compounds are used as treatments to suppress overly zealous immune events, such as in allergic responses or during transplant rejection (see Chapters 15 and 16).

Class II MHC Alleles Play a Critical Role in Immune Responsiveness

MHC haplotype plays a strong role in the outcome of an immune response, as these alleles determine which fragments of protein will be presented. Recall that class II MHC molecules present foreign antigen to CD4[+] T_H cells, which go on to activate B cells to produce serum antibodies. Early studies by Benacerraf in which guinea pigs were immunized with simple synthetic antigens were the first to show that the ability of an animal to mount an immune response, as measured by the production of serum antibodies, is determined by its MHC haplotype. Later experiments by H. McDevitt, M. Sela, and their colleagues used congenic (matched at the MHC locus but not elsewhere) and recombinant congenic mouse strains to specifically map the control of *immune responsiveness* to class II MHC genes. In early reports, the genes responsible for this phenotype were designated *Ir* or immune response genes; retaining the initial *I*, mouse class II products are now called IA and IE. We now know that the dependence of immune responsiveness on the genes within the class II MHC reflects the central role of these molecules in determining which fragments of foreign proteins are presented as antigen to T_H cells.

Two explanations have been proposed to account for this variability in immune responsiveness observed among different haplotypes. According to the **determinant-selection model**, different class II MHC molecules differ in their ability to bind particular processed antigens. In the end, some peptides may be more crucial to eliminating the pathogen than others. A separate hypothesis, termed **holes-in-the-repertoire model**, postulates that T cells bearing receptors that recognize certain foreign antigens which happen to closely resemble self-antigens may be eliminated during T-cell development, leaving the organism without these cells/receptors for future responses to foreign molecules. Since the T-cell response to an antigen involves a trimolecular complex of the T-cell receptor, an antigenic peptide, and an MHC molecule (discussed in detail in Chapter 11), both models appear correct. That is, the absence of an MHC molecule that can bind and present a particular peptide, or the absence of T-cell receptors that can recognize a given MHC-peptide molecule complex, both result in decreased immune responsiveness to a given foreign substance, and can account for the observed relationship between MHC haplotype and the ability to respond to particular exogenous antigens.

T Cells Are Restricted to Recognizing Peptides Presented in the Context of Self-MHC Alleles

In the 1970s a series of experiments were carried out to further explore the relationship between the MHC and immune response. These investigations contributed two very crucial discoveries: (1) that both $CD4^+$ and $CD8^+$ T cells can recognize antigen only when it is presented in the groove of an MHC molecule, and (2) that the MHC haplotype of the APC and the T cell must *match*. This happens naturally in the host, where T cells develop alongside host APCs, both expressing only that individual's MHC molecules (see Chapter 9). This constraint is referred to as **self-MHC restriction**, which refers to the dual specificity of T cells for self MHC as well as for foreign antigen.

A. Rosenthal and E. Shevach showed that antigen-specific proliferation of T_H cells occurs only in response to antigen presented by macrophages of the same MHC haplotype as the T cells recognizing the antigen. In their experimental system, guinea pig macrophages from strain 2 were initially incubated with an antigen (Figure 8-13). After the "antigen-pulsed" macrophages had processed the antigen and presented it on their surface, they were mixed with T cells from the same strain (strain 2), a different strain (strain 13), or their F_1 progeny (2 X 13), and the magnitude of T-cell proliferation in response to the antigen-pulsed macrophages was measured. The results of these experiments showed that strain-2 antigen-pulsed macrophages activated T cells from strain-2 and F_1 mice but not T cells from strain-13 animals. Similarly, strain-13 antigen-pulsed macrophages activated strain-13 and F_1 T cells but not strain-2 T cells. Subsequently, congenic and recombinant congenic strains of mice, which differed from each other only in selected regions of the MHC were used as the source of macrophages and T cells. These experiments confirmed that the $CD4^+$ T_H cell is activated and proliferates only in the presence of antigen-pulsed macrophages that share class II MHC alleles. Thus, antigen recognition by the $CD4^+$ T_H cell is said to be *class II MHC restricted*. In 1974, Zinkernagel and Doherty similarly demonstrated the self-MHC class I restriction of $CD8^+$ T cells (see Classic Experiment Box 8-3). Over two decades later, the pair was awarded the Nobel Prize in Medicine for their

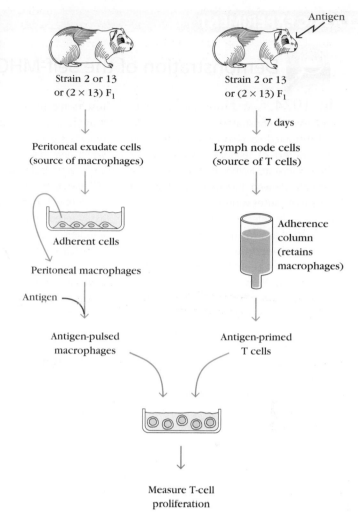

Antigen-primed T cell	Antigen-pulsed macrophages		
	Strain 2	Strain 13	(2×13) F_1
Strain 2	+	−	+
Strain 13	−	+	+
(2×13) F_1	+	+	+

FIGURE 8-13 Experimental demonstration of self-MHC restriction of T_H cells. Peritoneal exudate cells from strain 2, strain 13, or (2 × 13) F_1 guinea pigs were incubated in plastic petri dishes, allowing enrichment of macrophages, which are adherent cells. The peritoneal macrophages were then incubated with antigen. These "antigen-pulsed" macrophages were incubated in vitro with T cells from strain 2, strain 13, or (2 × 13) F_1 guinea pigs, and the degree of T-cell proliferation was assessed (+ vs −). The results indicated that T_H cells could proliferate only in response to antigen presented by macrophages that shared MHC alleles. *[Adapted from A. Rosenthal and E. Shevach, 1974, Journal of Experimental Medicine 138:1194, by copyright permission of the Rockefeller University Press.]*

seminal studies (see Table 1-2 for a list of Nobel Prizes related to Immunology). As we will see in Chapter 9, self-MHC restriction is the result of T-cell development in the thymus, where only T cells that recognize self-MHC are selected for survival.

BOX 8-3

 ## Demonstration of the Self-MHC Restriction of CD8$^+$ T Cells

In 1974, R. M. Zinkernagel and P. C. Doherty demonstrated that CD8$^+$ T cells are restricted to recognizing antigen in the context of self-MHC molecules. For this seminal study, they took advantage of recently derived inbred mice strains, a virus that causes widespread neurological damage in infected animals, and new assays that allowed quantification of cytotoxic T cell responses.

In their experiments, mice were first immunized with lymphocytic choriomeningitis virus (LCMV), a pathogen that

results in central nervous system inflammation and neurological damage. Several days later, the animals' spleen cells, which included T$_C$ cells specific for the virus, were isolated and incubated with LCMV-infected target cells of the same or different haplotype (Figure 1). The assay relied on measuring the release of a radioisotope (^{51}Cr) from labeled target cells (called a chromium release assay; see Figure 13-18). They found that the T$_C$ cells killed syngeneic virus-infected target cells (or cells with matched MHC

alleles) but not uninfected cells or infected cells from a donor that shared no MHC alleles with these cytotoxic cells.

Later studies with congenic and recombinant congenic strains showed that the T$_C$ cell and the virus-infected target cell must specifically share class I molecules encoded by the K or D regions of the MHC. Thus, antigen recognition by CD8$^+$ cytotoxic T cells is *class I MHC restricted*. In 1996, Doherty and Zinkernagel were awarded the Nobel Prize in Medicine for their major contribution to understanding the role of the MHC in cell-mediated immunity.

Before this discovery, histocompatibility antigens were mainly blamed for transplantation rejection, but their importance in the everyday process of cellular immunity was not appreciated. Zinkernagel and Doherty's discovery is all the more revolutionary when one takes into consideration that little was known about cellular immunity and no T-cell receptor had yet been discovered!

This milestone in immunologic understanding set the stage for the development of two key models of how T cells respond to foreign antigen: altered self and dual recognition. The altered self hypothesis posited that histocompatibility molecules that associate with foreign particles, such as viruses, may appear to be altered forms of self-proteins. The dual recognition model proposed that T cells must be capable of simultaneous recognition of both the foreign substance and these self-histocompatibility molecules. We now appreciate the validity of both of these models, and the role of MHC restriction in the immune response to both microorganisms and allogeneic transplants.

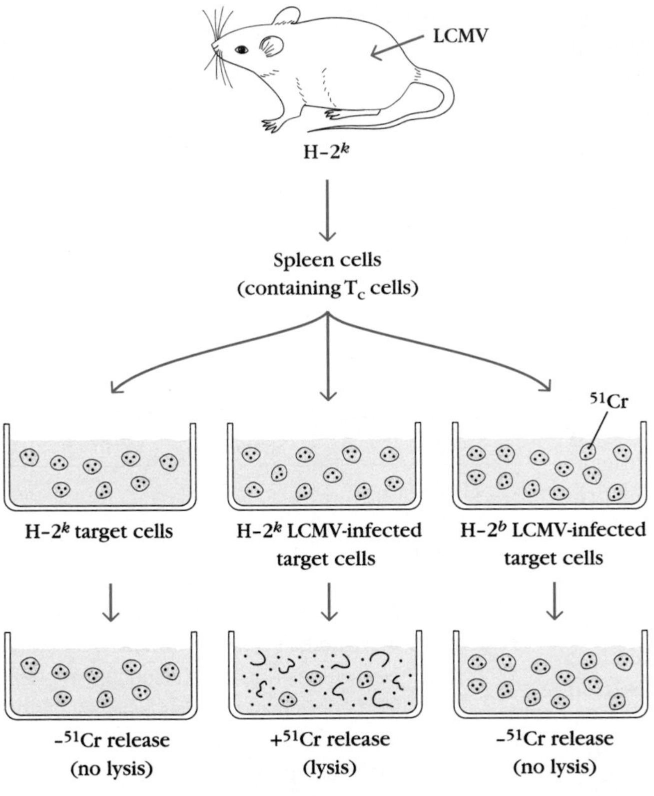

Doherty, P. C., and R. M. Zinkernagel. 1975. H-2 compatibility is required for T-cell mediated lysis of target cells infected with lymphocytic choriomeningitis virus. *Journal of Experimental Medicine* **141**:502.

LCMV

H-2^k

Spleen cells
(containing T$_C$ cells)

^{51}Cr

H-2^k target cells

H-2^k LCMV-infected
target cells

H-2^b LCMV-infected
target cells

$-^{51}$Cr release
(no lysis)

$+^{51}$Cr release
(lysis)

$-^{51}$Cr release
(no lysis)

FIGURE 1
Classic experiment of Zinkernagel and Doherty demonstrating that antigen recognition by T$_C$ cells exhibits MHC restriction. H-2^k mice were primed with the lymphocytic choriomeningitis virus (LCMV) to induce cytotoxic T lymphocytes (CTLs) specific for the virus. Spleen cells from this LCMV-primed mouse were then added to target cells of different H-2 haplotypes that were intracellularly labeled with ^{51}Cr (black dots) and either infected or not with the LCMV. CTL-mediated killing of the target cells, as measured by the release of ^{51}Cr into the culture supernatant, occurred only if the target cells were infected with LCMV and had the same MHC haplotype as the CTLs. *[Adapted from P. C. Doherty and R. M. Zinkernagel, 1975,* Journal of Experimental Medicine *141:502.]*

Processing of Antigen Is Required for Recognition by T Cells

In the 1980s, K. Ziegler and E. R. Unanue showed that an intracellular processing step by APCs was required to activate T cells. These researchers observed that T_H-cell activation by bacterial protein antigens was prevented by treating the APCs with paraformaldehyde prior to antigen exposure (in essence, killing the cells and immobilizing the antigens in the membrane; see Figure 8-14a). However, if the APCs were first allowed to ingest the antigen and were fixed with paraformaldehyde 1 to 3 hours later, T_H-cell activation still occurred (Figure 8-14b). During that interval of 1 to 3 hours, the APCs had taken up the antigen, processed it into peptide fragments, and displayed these fragments on the cell membrane in a form capable of activating T cells.

Subsequent experiments by R. P. Shimonkevitz showed that internalization and processing could be bypassed if APCs were exposed to already digested peptide fragments instead of the native antigen (Figure 8-14c). In these experi-

ments, APCs were treated with glutaraldehyde (this chemical, like paraformaldehyde, fixes the cell, rendering it metabolically inactive) and then incubated with native ovalbumin or with ovalbumin that had been subjected to partial enzymatic digestion. The digested ovalbumin was able to interact with the glutaraldehyde-fixed APCs, thereby activating ovalbumin-specific T_H cells. However, the native ovalbumin failed to do so. These results suggest that antigen processing requires the digestion of the protein into peptides that can be recognized by the ovalbumin-specific T_H cells.

At about the same time, several investigators, including W. Gerhard, A. Townsend, and their colleagues, began to identify the proteins of influenza virus that were recognized by T_C cells. Contrary to their expectations, they found that internal proteins of the virus, such as polymerase and nucleocapsid proteins, were often recognized by T_C cells better than the more exposed envelope proteins found on the surface of the virus. Moreover, Townsend's work revealed that T_C cells recognized short linear peptide sequences of influenza proteins.

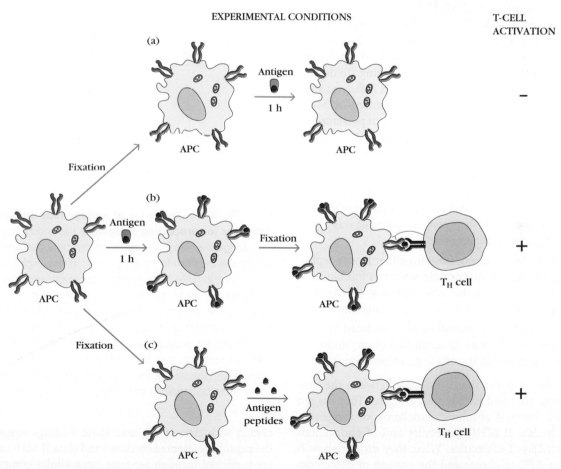

EXPERIMENTAL CONDITIONS

T-CELL ACTIVATION

FIGURE 8-14 Experimental demonstration that antigen processing is necessary for T_H-cell activation. (a) When antigen-presenting cells (APCs) are fixed before exposure to antigen, they are unable to activate T_H cells. (b) In contrast, APCs fixed at least 1 hour after antigen exposure can activate T_H cells. (This simplified figure does not show costimulatory molecules needed for T-cell activation.) (c) When APCs are fixed before antigen exposure and incubated with peptide digests of the antigen (rather than native antigen), they also can activate T_H cells. T_H-cell activation is determined by measuring a specific T_H-cell response (e.g., cytokine secretion).

In fact, when noninfected target cells were incubated in vitro with synthetic peptides corresponding to only short sequences of internal influenza proteins, these cells could be recognized by T_C cells and subsequently lysed just as well as target cells that had been infected with whole, live influenza virus. These findings, along with those presented below, suggest that *antigen processing is a metabolic process that digests proteins into peptides, which can then be displayed on the cell surface together with a class I or class II MHC molecule.*

Evidence Suggests Different Antigen Processing and Presentation Pathways

The immune system typically uses different pathways to eliminate intracellular and extracellular antigens. As a general rule, endogenous antigens (those generated within the cell) are processed in the **cytosolic** or **endogenous pathway** and presented on the membrane with class I MHC molecules. Exogenous antigens (those taken up from the extracellular environment by endocytosis) are typically processed in the **exogenous pathway** and presented on the membrane with class II MHC molecules (Figure 8-15).

Experiments carried out by L. A. Morrison and T. J. Braciale provided an excellent demonstration that the antigenic peptides presented by class I and class II MHC molecules follow different routes during antigen processing. To do this, they used a set of T-cell clones specific for an influenza virus antigen; some recognized the antigen presented by MHC class I, and others recognized the *same antigen* presented by MHC class II. Examining the T-cell responses, they derived the following general principles about the two pathways:

- Class I presentation requires internal synthesis of virus protein, as shown by the requirement that the target cell be infected by live virus, and by the inhibition of class I presentation observed when protein synthesis was blocked by the inhibitor emetine.

- Class II presentation can occur with either live or replication-incompetent virus; protein synthesis inhibitors had no effect, indicating that new protein synthesis is not a necessary condition of class II presentation.

- Class II, but not class I, presentation is inhibited by treatment of the cells with an agent that blocks endocytic processing within the cell (e.g., chloroquine).

These studies support the distinction between the processing of exogenous and endogenous antigens. They suggest a preferential, but not absolute, association of exogenous antigens with class II MHC molecules and of endogenous antigens with class I molecules. What they do not show is that both the APC of choice and the antigen of choice can influence the outcome, as will be seen later when these rules are subverted during cross-presentation. In the experiments described above, association of viral antigen with class I MHC molecules required replication of the influenza virus and viral protein synthesis within the target cells, but asso-

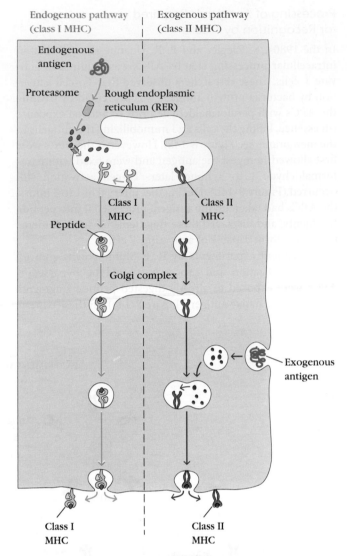

FIGURE 8-15 Overview of endogenous and exogenous pathways for processing antigen. In the endogenous pathway (left), antigens are degraded by the proteasome, converting proteins into smaller peptides. In the exogenous pathway (right), extracellular antigens are engulfed into endocytic compartments where they are degraded by acidic pH-dependent endosomal and lysosomal enzymes. The antigenic peptides from proteasome cleavage and those from endocytic compartments associate with class I or class II MHC molecules respectively, and the MHC-peptide complexes are then transported to the cell membrane. It should be noted that the ultimate fate of most peptides in the cell is neither of these pathways but rather to be degraded completely into amino acids.

ciation with class II did not. These findings suggested that the peptides presented by class I and class II MHC molecules are trafficked through separate intracellular compartments; class I MHC molecules interact with peptides derived from cytosolic degradation of endogenously synthesized proteins, class II molecules with peptides derived from endocytic degradation of exogenous antigens. The next two sections examine these two pathways in detail.

The Endogenous Pathway of Antigen Processing and Presentation

In eukaryotic cells, protein levels are carefully regulated. Every protein is subject to continuous turnover and is degraded at a rate that is generally expressed in terms of its half-life. Some proteins (e.g., transcription factors, cyclins, and key metabolic enzymes) have very short half-lives. Denatured, misfolded, or otherwise abnormal proteins also are degraded rapidly. Defective ribosomal products are polypeptides that are synthesized with imperfections and constitute a large part of the products that are rapidly degraded. The average half-life for cellular proteins is about 2 days, but many are degraded within 10 minutes. The consequence of steady turnover of both normal and defective proteins is a constant deluge of degradation products within a cell. Most will be degraded to their constituent amino acids and recycled, but some persist in the cytosol as peptides. The cell samples these peptides and presents some on the plasma membrane in association with class I MHC molecules, where cells of the immune system can sample these peptides to survey for foreign proteins. The pathway by which these endogenous peptides are generated for presentation with class I MHC molecules utilizes mechanisms similar to those involved in the normal turnover of intracellular proteins, but exactly how particular proteins are selected for degradation and peptide presentation still remains unclear.

Peptides Are Generated by Protease Complexes Called Proteasomes

Intracellular proteins are degraded into short peptides by a cytosolic proteolytic system present in all cells, called the proteasome (Figure 8-16a). The large (20S) proteasome is composed of 14 subunits arrayed in a barrel-like structure of symmetrical rings. Many proteins are targeted for proteolysis when a small protein called **ubiquitin** is attached to them. These ubiquitin-protein conjugates enter the proteasome complex, consisting of the 20S base and an attached 19S regulatory component, through a narrow channel at the 19S end. The proteasome complex cleaves peptide bonds in an ATP-dependent process. Degradation of ubiquitin-protein complexes is thought to occur within the central hollow of the proteasome.

The immune system also utilizes this general pathway of protein degradation to produce small peptides for presentation by class I MHC molecules. In addition to the standard 20S proteasomes resident in all cells, a distinct proteasome of the same size can be found in pAPCs and the cells of infected tissues. This distinct proteasome, called the **immunoproteasome**, has some unique components that can be induced by exposure to interferon-γ or TNF-α (Figure 8-16b). *LMP2* and *LMP7*, genes that are located within the class I region (see Figure 8-8) and are responsive to these cytokines, encode replacement catalytic protein subunits that convert

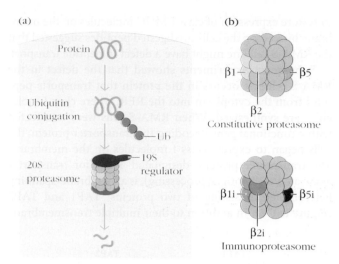

FIGURE 8-16 Cytosolic proteolytic system for degradation of intracellular proteins. (a) Endogenous proteins may be targeted for degradation by ubiquitin conjugation. These proteins are degraded by the 26S proteasome complex, which includes the 20S constitutive proteasome and a 19S regulator. (b) In activated APCs, several proteins in the constitutive proteasome (β1, β2, and β5) are replaced by proteins encoded by the LMP genes and specific to the immunoproteasome (β1i, β2i, and β5i). This immunoproteasome has increased proteolytic efficiency for creating peptides that can assemble with MHC class I molecules. [Adapted from M. Groettrup et al. 2010. *Proteasomes in immune cells: More than peptide producers.* Nature Reviews. Immunology 10:73–78. doi:10.1038/nri2687]

standard proteasomes into immunoproteasomes, increasing the production of peptides that bind efficiently to MHC class I proteins. The immunoproteasome turns over more rapidly than a standard proteasome, possibly because the increased level of protein degradation in its presence may have consequences beyond the targeting of infected cells. It is possible that in some cases autoimmunity results from increased processing of self-proteins in cells with high levels of immunoproteasomes.

Peptides Are Transported from the Cytosol to the RER

Insight into the cytosolic processing pathway came from studies of cell lines with defects in peptide presentation by class I MHC molecules. One such mutant cell line, called RMA-S, expresses about 5% of the normal levels of class I MHC molecules on its membrane. Although RMA-S cells synthesize normal levels of class I α chain and β₂-microglobulin, few class I MHC complexes appear on the membrane. A clue to the mutation in the RMA-S cell line was the discovery by Townsend and his colleagues that "feeding" these cells peptides restored their level of membrane-associated class I MHC molecules to normal. These investigators suggested that peptides might be required to stabilize the interaction between the class I α chain and β₂-microglobulin. The ability to restore expression of class I MHC molecules on the membrane by feeding the cells predigested peptides suggested that the RMA-S cell line might have a defect in peptide transport.

Subsequent experiments showed that the defect in the RMA-S cell line occurs in the protein that transports peptides from the cytoplasm into the RER, where class I molecules are synthesized. When RMA-S cells were transfected with a functional gene encoding the transporter protein, the cells began to express class I molecules on the membrane. The transporter protein, designated **TAP** (for *transporter associated with antigen processing*), is a membrane-spanning heterodimer consisting of two proteins: TAP1 and TAP2 (Figure 8-17a). In addition to their multiple transmembrane

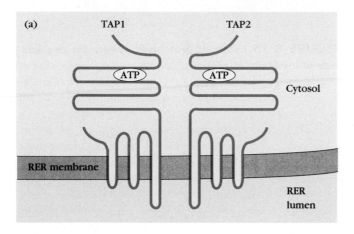

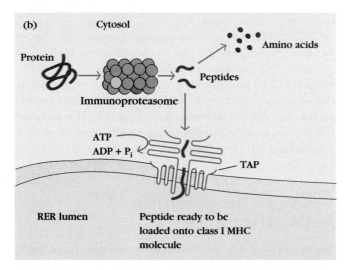

FIGURE 8-17 TAP (transporter associated with *antigen processing*). (a) Schematic diagram of TAP, a heterodimer anchored in the membrane of the rough endoplasmic reticulum (RER). The two chains are encoded by *TAP1* and *TAP2*. The cytosolic domain in each TAP subunit contains an ATP-binding site, and peptide transport depends on the hydrolysis of ATP. (b) In the cytosol, association of β1i, β2i, and β5i (colored spheres) with a proteasome changes its catalytic specificity to favor production of peptides that bind to class I MHC molecules. These peptides are translocated by TAP into the RER lumen, where, in a process mediated by several other proteins, they will associate with class I MHC molecules.

segments, the TAP1 and TAP2 proteins each have a domain projecting into the lumen of the RER and an ATP-binding domain that projects into the cytosol. Both TAP1 and TAP2 belong to the family of ATP-binding cassette proteins found in the membranes of many cells, including bacteria. These proteins mediate ATP-dependent transport of amino acids, sugars, ions, and peptides.

Peptides generated in the cytosol by the proteasome are translocated by TAP into the RER by a process that requires the hydrolysis of ATP (Figure 8-17b). TAP has affinity for peptides containing 8 to 16 amino acids. The optimal peptide length for class I MHC binding is around 9 amino acids, and longer peptides are trimmed by enzymes present in the ER, such as **ERAP** (*endoplasmic reticulum aminopeptidase*). In addition, TAP appears to favor peptides with hydrophobic or basic carboxyl-terminal amino acids, the preferred anchor residues for class I MHC molecules. Thus, TAP is optimized to transport peptides that are most likely to interact with class I MHC molecules. The *TAP1* and *TAP2* genes map within the class II MHC region, adjacent to the *LMP2* and *LMP7* genes, and different allelic forms of these genes exist within the population. TAP deficiencies can lead to a disease syndrome that has aspects of both immunodeficiency and autoimmunity (see Clinical Focus Box 8-4).

Chaperones Aid Peptide Assembly with MHC Class I Molecules

Like other proteins destined for the plasma membrane, the α chain and β2-microglobulin components of the class I MHC molecule are synthesized on ribosomes on the RER. Assembly of these components into a stable class I MHC molecular complex that can exit the RER requires the presence of a peptide in the binding groove of the class I molecule. The assembly process involves several steps and includes the participation of *molecular chaperones* that facilitate the folding of polypeptides.

The first molecular chaperone involved in class I MHC assembly is **calnexin**, a resident membrane protein of the ER. ERp57, a protein with enzymatic activity, and calnexin associate with the free class I α chain and promote its folding (Figure 8-18). When β2-microglobulin binds to the α chain, calnexin is released and the class I molecule associates with the chaperone **calreticulin** and with **tapasin**. Tapasin (*TAP-associated protein*) brings the TAP transporter into proximity with the class I molecule and allows it to acquire an antigenic peptide. The TAP protein promotes peptide capture by the class I molecule before the peptides are exposed to the luminal environment of the RER.

Exoproteases in the ER will act on peptides not associated with class I MHC molecules. One ER aminopeptidase, ERAP1, removes the amino-terminal residue from peptides to achieve optimum class I binding size (see Figure 8-18). ERAP1 has little affinity for peptides shorter than eight amino acids in length. As a consequence of productive

Deficiencies in TAP Can Lead to Bare Lymphocyte Syndrome

A relatively rare condition known as bare lymphocyte syndrome (BLS) has been recognized for more than 20 years. The lymphocytes in BLS patients express MHC molecules at below-normal levels and, in some cases, not at all. Type 1 BLS is caused by a deficiency in MHC class I molecules; in type 2 BLS, expression of class II molecules is impaired. The pathogenesis of type 1 BLS underscores the importance of the class I family of MHC molecules in their dual roles of preventing autoimmunity as well as defending against pathogens.

One study identified a group of patients with type 1 BLS caused by defects in *TAP1* or *TAP2* genes. As described in this chapter, TAP proteins are necessary for the loading of peptides onto class I molecules, a step that is essential for expression of class I MHC molecules on the cell surface. Lymphocytes in individuals with TAP deficiency express levels of class I molecules significantly lower than those of normal controls. Other cellular abnormalities include increased numbers of NK and $\gamma\delta$ T cells and decreased levels of CD8$^+$ $\alpha\beta$ T cells. As we will see, the disease manifestations are reasonably well explained by these deviations in the levels of certain cells involved in immune function.

In early life, the TAP-deficient individual suffers frequent bacterial infections of the upper respiratory tract and in the second decade begins to experience chronic infection of the lungs. It is thought that a postnasal-drip syndrome common in younger patients promotes the bacterial lung infections in later life. Noteworthy is the absence of susceptibility to severe viral infection, which is common in immunodeficiencies with T-cell involvement (see Chapter 18). Bronchiectasis (dilation of the bronchial tubes) often occurs, and

recurring infections can lead to lung damage that may be fatal. The most characteristic mark of the deficiency is the occurrence of necrotizing skin lesions on the extremities and the midface (Figure 1). These lesions ulcerate and may cause disfigurement. The skin lesions are probably due to activated NK cells and $\gamma\delta$ T cells; NK cells were isolated from biopsied skin from several patients, supporting this possibility. Normally, the activity of NK cells is limited through the action of killer-cell-inhibitory receptors (KIRs), which deliver a negative signal to the NK cell following interaction with class I molecules (see Chapter 13). The lack of class I molecules in BLS patients with TAP deficiency explains the excessive activity of the NK cells.

Activation of NK cells further explains the absence of severe viral infections, which are limited by NK and $\gamma\delta$ T cells.

The best treatment for the characteristic lung infections appears to be a combination of antibiotics and intravenous immunoglobulin. Attempts to limit the skin disease by immunosuppressive regimens, such as steroid treatment or cytotoxic agents, can lead to exacerbation of lesions and are therefore contraindicated. Mutations in the promoter region of *TAP* that prevent expression of the gene were found for several patients, suggesting the possibility of gene therapy, but the cellular distribution of class I is so widespread that it is not clear what cells would need to be corrected to alleviate all symptoms.

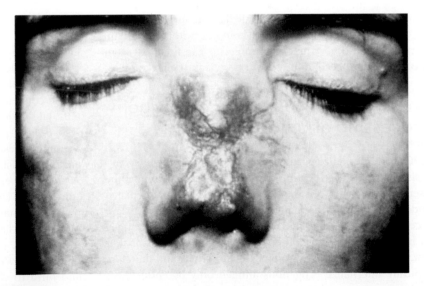

FIGURE 1
Necrotizing granulomatous lesions in the midface of a patient with TAP-deficiency syndrome. TAP deficiency leads to a condition with symptoms characteristic of autoimmunity, such as the skin lesions that appear on the extremities and the midface, as well as immunodeficiency that causes chronic sinusitis, leading to recurrent lung infection. *[From S. D. Gadola et al., 1999, Lancet 354:1598, and 2000, Clinical and Experimental Immunology 121:173.]*

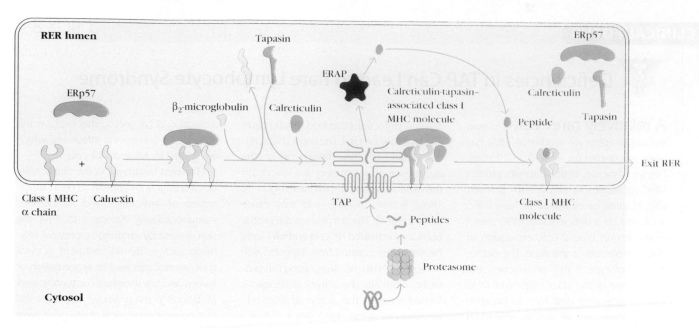

FIGURE 8-18 Assembly and stabilization of class I MHC molecules. Within the rough endoplasmic reticulum (RER) membrane, a newly synthesized class I α chain associates with calnexin, a molecular chaperone, and ERp57 until β₂-microglobulin binds to the α chain. The binding of β₂-microglobulin releases calnexin and allows binding to calreticulin and to tapasin, which is associated with the peptide transporter TAP. This association promotes binding of an antigenic peptide. Antigens in the ER can be further processed via exopeptidases such as ERAP1, producing fragments ideally suited for binding to class I. Peptide association stabilizes the class I molecule-peptide complex, allowing it to be transported from the RER to the plasma membrane.

peptide binding, the class I molecule displays increased stability and can dissociate from the complex with calreticulin, tapasin, and ERp57. The class I molecule can then exit from the RER and proceed to the cell surface via the Golgi complex.

The Exogenous Pathway of Antigen Processing and Presentation

APCs can internalize particulate material by simple phagocytosis (also called "cell eating"), where material is engulfed by pseudopods of the cell membrane, or by receptor-mediated endocytosis, where the material first binds to specific surface receptors. Macrophages and dendritic cells internalize antigen by both processes. Most other APCs, whether professional or not, demonstrate little or no phagocytic activity and therefore typically internalize exogenous antigen only by endocytosis (either receptor-mediated endocytosis or by pinocytosis, "cell drinking"). B cells, for example, internalize antigen very effectively by receptor-mediated endocytosis using their antigen-specific membrane immunoglobulin as the receptor.

Peptides Are Generated from Internalized Antigens in Endocytic Vesicles

Once an antigen is internalized, it is degraded into peptides within compartments of the endocytic processing pathway.

As the experiment shown in Figure 8-14 demonstrated, internalized antigen takes 1 to 3 hours to traverse the endocytic pathway and appear at the cell surface in the form of class II MHC-peptide complexes. The endocytic antigen processing pathway appears to involve several increasingly acidic compartments, including early endosomes (pH 6.0–6.5); late endosomes, or endolysosomes (pH 4.5–5.0); and lysosomes (pH 4.5). Internalized antigen progresses through these compartments, encountering hydrolytic enzymes and a lower pH in each compartment (Figure 8-19). Antigen-presenting cells have a unique form of late endosome, the MHC class II-containing compartment (MIIC), in which final protein degradation and peptide loading into MHC class II proteins occurs. Within the compartments of the endocytic pathway, antigen is degraded into oligopeptides of about 13 to 18 residues that meet up with and bind to class II MHC molecules in late endosomes. Because the hydrolytic enzymes are optimally active under acidic conditions (low pH), antigen processing can be inhibited by chemical agents that increase the pH of the compartments (e.g., chloroquine) as well as by protease inhibitors (e.g., leupeptin).

The mechanism by which internalized antigen moves from one endocytic compartment to the next has not been conclusively demonstrated. It has been suggested that early endosomes from the periphery move inward to become late endosomes and eventually lysosomes. Alternatively, small

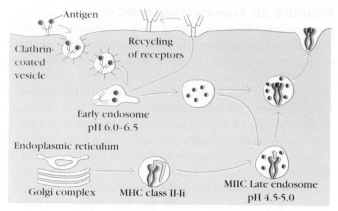

FIGURE 8-19 Generation of antigenic peptides in the exogenous processing pathway. Internalized exogenous antigen moves through several acidic compartments ending in specialized MIIC late endosomes, where it is degraded into peptide fragments which associate with class II MHC molecules transported in vesicles from the Golgi complex. The cell shown here is a B cell, which internalizes antigen by receptor-mediated endocytosis, with the membrane-bound antibody functioning as an antigen-specific receptor.

transport vesicles may carry antigens from one compartment to the next. Eventually the endocytic compartments, or portions of them, return to the cell periphery, where they fuse with the plasma membrane. In this way, the surface receptors are recycled.

The Invariant Chain Guides Transport of Class II MHC Molecules to Endocytic Vesicles

Since APCs express both class I and class II MHC molecules, some mechanism must exist to prevent class II MHC molecules from binding to the antigenic peptides destined for the class I molecules. When class II MHC molecules are synthesized within the RER, these class II αβ chains associate with a protein called the **invariant chain (Ii, CD74)**. This conserved, non-MHC encoded protein interacts with the class II peptide-binding groove preventing any endogenously derived peptides from binding while the class II molecule is within the RER (Figure 8-20a). The invariant chain also appears to be involved in the folding of the class II α and β chains, their exit from the RER, and the subsequent routing of class II molecules to the endocytic processing pathway from the trans-Golgi network.

The role of the invariant chain in the routing of class II molecules has been demonstrated in transfection experiments with cells that lack both the genes encoding class II MHC molecules and the invariant chain. Immunofluorescent labeling of these cells transfected only with class II MHC genes revealed that, in the absence of invariant chain, class II molecules remain primarily in

the ER and do not transit past the cis-Golgi. However, in cells transfected with both the class II MHC genes and the *Ii* gene, the class II molecules were localized in the cytoplasmic vesicular structures of the endocytic pathway. The invariant chain contains sorting signals in its cytoplasmic tail that direct the transport of the class II MHC complex from the trans-Golgi network to the endocytic compartments.

Peptides Assemble with Class II MHC Molecules by Displacing CLIP

Recent experiments indicate that most class II MHC-invariant chain complexes are transported from the RER, where they are formed, through the Golgi complex and trans-Golgi network, and then through the endocytic pathway, moving from early endosomes to the MIIC late endosomal compartment in some cases. As the proteolytic activity increases in each successive compartment, the invariant chain is gradually degraded. However, a short fragment of the invariant chain termed **CLIP** (for *cl*ass II–associated *i*nvariant chain *p*eptide) remains bound to the class II molecule after the majority of the invariant chain has been cleaved within the endosomal compartment. Like antigenic peptide, CLIP physically occupies the peptide-binding groove of the class II MHC molecule, preventing any premature binding of antigen-derived peptide (Figure 8-20b).

A nonclassical class II MHC molecule called HLA-DM is required to catalyze the exchange of CLIP with antigenic peptides (see Figure 8-20a). The *DM*α and *DM*β genes are located near the *TAP* and *LMP* genes in the MHC complex of humans, with similar genes in mice (see Figure 8-8). Like other class II MHC molecules, HLA-DM is a heterodimer of α and β chains. However, unlike other class II molecules it is relatively nonpolymorphic and is not normally expressed at the cell membrane but is found predominantly within the endosomal compartment. HLA-DM has been found to associate with the class II MHC β chain and to function in removing or "editing" peptides, including CLIP, that associate transiently with the binding groove of classical class II molecules. Peptides that make especially strong molecular interactions with class II MHC, creating long-lived complexes, are harder for HLA-DM to displace, and thus become the repertoire of peptides that ultimately make it to the cell surface as MHC-peptide complexes.

As with class I MHC molecules, peptide binding is required to maintain the structure and stability of class II MHC molecules. Once a peptide has bound, the class II MHC-peptide complex is transported to the plasma membrane, where the neutral pH appears to enable the complex to assume a compact, stable form. Peptide is bound so strongly in this compact form that it is difficult to replace a class II–bound peptide on the membrane with another peptide at physiologic conditions.

(a)

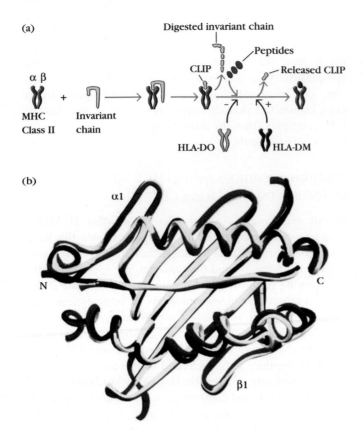

(b)

FIGURE 8-20 Assembly of class II MHC molecules. (a) Within the rough endoplasmic reticulum, a newly synthesized class II MHC molecule binds an invariant chain. The bound invariant chain prevents premature binding of peptides to the class II molecule and helps to direct the complex to endocytic compartments containing peptides derived from exogenous antigens. Digestion of the invariant chain leaves CLIP, a small fragment remaining in the binding groove of the class II MHC molecule. HLA-DM, a nonclassical MHC class II molecule present within the MIIC compartment, mediates exchange of antigenic peptides for CLIP. The nonclassical class II molecule HLA-DO may act as a negative regulator of class II antigen processing by binding to HLA-DM and inhibiting its role in the dissociation of CLIP from class II molecules. (b) Comparison of three-dimensional structures showing the binding groove of HLA class II molecules (α1, β1) containing different antigenic peptides or CLIP. The red lines show HLA-DR4 complexed with collagen II peptide, yellow lines are HLA-DR1 with influenza hemagglutinin peptide, and dark blue lines are HLA-DR3 associated with CLIP. (N indicates the amino-terminus and C the carboxyl-terminus of the peptides.) No major differences in the structures of the class II molecules or in the conformation of the bound peptides are seen. This comparison shows that CLIP binds the class II molecule in a manner identical to that of antigenic peptides. *[Part b from A. Dessen et al., 1997, Immunity 7:473–481; courtesy of Don Wiley, Harvard University.]*

One additional nonclassical member of the class II MHC family, HLA-DO, like DM, is another relatively nonpolymorphic class II molecule. Unlike other class II molecules, its expression is not induced by IFN-γ. HLA-DO has been found to act as a negative regulator of antigen binding, modulating the function of HLA-DM and changing the repertoire of peptides that preferentially bind to classical class II molecules. In cells that express both DO and DM, these two molecules strongly associate in the ER and maintain this interaction all the way to the endosomal compartments. Although this interaction has been recognized for many years, the function of this negative regulator and the impact of this changed peptide repertoire is only now being resolved.

Originally observed only in B cells and the thymus, the cellular expression profile of HLA-DO has recently been expanded to dendritic cells (DCs), where it is thought to play a role in the maintenance of self-tolerance (discussed further in Chapter 16). This phenomenon was studied by L. K. Denzin and colleagues using diabetes-prone mice engineered to express human HLA-DO. DCs are known to be essential in the establishment of self-tolerance, as well as in the presentation of autoantigens to self-reactive T cells. Self-reactive T cells develop in these non-obese diabetic animals, and these cells ultimately destroy pancreatic β cells, causing

diabetes. In this study, the development of diabetes was blocked by the presence of the HLA-DO transgene in DCs. In addition, using specific monoclonal antibodies, they showed that the repertoire of peptides being presented to the autoreactive T cells was significantly altered, resulting in a reduced efficiency in presenting key self-peptides. This and other work suggest that normal DO expression may play an important role in modulating DM behavior to ensure the presentation of a self-peptide repertoire that encourages tolerance to self-antigens. Interestingly, in wild-type mice and humans, DO expression is down-regulated following DC activation by antigen, releasing HLA-DM to carry out its normal function of encouraging the presentation of a diverse array of peptides, many of which will presumably be derived from the foreign proteins that stimulated these APCs.

Figure 8-21 illustrates the endogenous pathway (left side) and compares it with the separate exogenous pathway (right), described above. Whether an antigenic peptide associates with class I or with class II molecules is partially dictated by the mode of entry into the cell, either exogenous or endogenous, by the site of processing, and by the cell type. However, in the next section, we will see that these assignments are not absolute and that exogenous antigens may, in some APCs, be presented by class I antigens via a unique pathway.

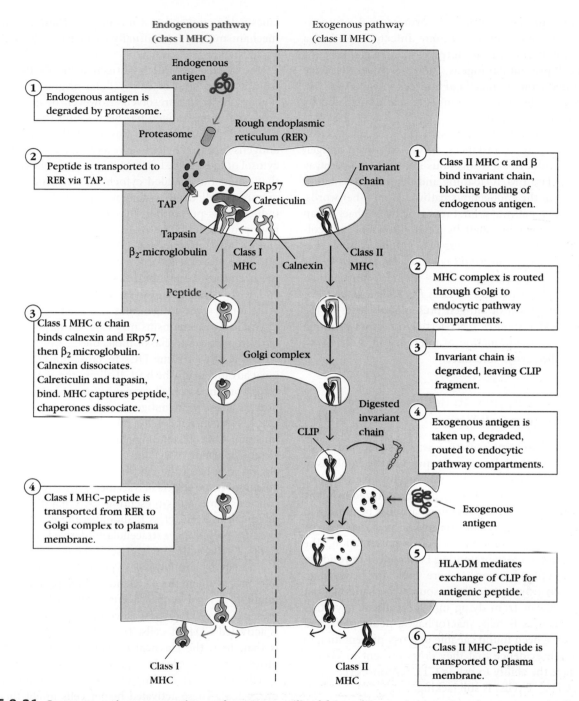

FIGURE 8-21 Separate antigen-presenting pathways are utilized for endogenous (green) and exogenous (red) antigens. The mode of antigen entry into cells and the site of antigen processing determine whether antigenic peptides associate with class I MHC molecules in the rough endoplasmic reticulum or with class II molecules in endocytic compartments.

Cross-Presentation of Exogenous Antigens

We have seen that after pAPCs internalize extracellular antigen they can process and present these antigens via the exogenous pathway. Because pAPCs also express costimulatory molecules, engagement of CD4$^+$ T cells with these class II MHC-peptide complexes can lead to

activation of T helper responses. On the other hand, infected cells will generally process and present cytosolic peptides via the endogenous pathway, leading to class I MHC-peptide complexes on their surface. However, unless these infected cells are pAPCs, they will not express the costimulatory molecules necessary to activate naïve CD8$^+$ T cells. This leaves us with a dilemma: how does the immune system activate naïve CD8$^+$ T cells to eliminate *intracellular* microbes unless a professional antigen-presenting cell happens to become infected? And if a pAPC is not harboring an intracellular infection, how does this cell present pathogens, such as viruses that have been engulfed from *extracellular* sources in ways that will activate the needed CTL responses (usually mediated by the endogenous pathway)?

The answer to this dilemma is a process called **cross-presentation**. In some instances, APCs will divert antigen obtained by endocytosis (exogenous antigen) to a pathway that leads to class I MHC loading and peptide presentation to CTLs (like in the endogenous pathway)—in other words, crossing the two pathways. First reported by Michael Bevan and later described in detail by Peter Cresswell and colleagues, *the phenomenon of cross-presentation requires that internalized antigens that would normally be handled by the exogenous pathway leading to class II MHC presentation somehow become redirected to a class I peptide loading pathway*. When this form of antigen presentation leads to the activation of CTL responses, it is referred to as **cross-priming**; when it leads to the induction of tolerance in these CD8$^+$ T cells it is called **cross-tolerance**.

Dendritic Cells Appear to Be the Primary Cross-Presenting Cell Type

The concept of cross-presentation was first recognized as early as 1976, but the relevant cell types and mechanisms of action remained a mystery until more recently. New studies, focused on subsets of DCs that are now recognized to be the most efficient of the cross-presenters, are just beginning to shed light on this process. For example, in vivo depletion or inactivation of DCs compromises cross-presentation. The most potent of the cross-presenting DCs appear to reside in secondary lymphoid organs, where they are believed to receive antigen from extracellular sources by handoff from migrating APCs or from dying infected cells. A few other cells types, such as B cells, macrophages, neutrophils, and mast cells, have been found capable of cross-presentation in vitro, although evidence that they cross-present in vivo or that they have the ability to prime CTL responses (i.e., activate naïve CD8$^+$ T cells) is still lacking.

Mechanisms and Functions of Cross-Presentation

Two possible models have been proposed for how DCs accomplish cross-presentation. The first hypothesizes that cross-presenting cells possess special antigen-processing machinery that allows loading of exogenously derived peptides onto class I MHC molecules. The second theory postulates specialized endocytosis machinery that can send internalized antigen directly to an organelle (such as a phagosome or early endosome), where peptides from those antigens are then loaded onto class I MHC mole-

cules using conventional machinery. These two proposed mechanisms are not mutually exclusive, and evidence for both exists.

The means by which antigen achieves the crossover from its exogenous or extracellular origins to the endogenous pathway is unresolved. However, in many instances cross-presented antigen has been found to enter the cytoplasm. In the early 1990s, it was shown that bead-conjugated antigens captured by cross-presenting DCs could reach the cytosol of these cells. Later, M. L. Lin and colleagues showed that exogenously added cytochrome c, which causes programmed cell death when present in the cytosol, could induce cross-presenting DCs to undergo apoptosis. It is now proposed that retro-translocation, or movement of endocytosed proteins *out* of endocytic compartments and into the cytosol, can occur via TAP molecules present in these endocytic membranes—a sort of backflow out of endocytic vesicles. However, TAP-independent cross-presentation of some antigens has also been observed, suggesting multiple mechanisms may exist for targeting externally derived peptides to class I molecules.

The ability of some DCs to cross-present antigens has great advantage for the host. It allows these APCs to capture virus from the extracellular environment or from dying cells, process these viral antigens, and activate CTLs that can attack virus-infected cells, inhibiting further spread of the infection. One outstanding question remains: if this pathway is so important, why don't all APCs cross-present? The answer may lie in the fact that cross-presenting cells could quickly become targets of lysis themselves. Another dilemma concerns how cross-presenting DCs handle self-peptides presented in MHC class I. Shouldn't these cells break tolerance by presenting extracellularly-derived self-peptides to CD8$^+$ T cells, activating anti-self CTL responses?

To help explain how cross-presentation is regulated and tolerance is maintained, scientists have proposed that DCs might first need to be "licensed" before they can cross-present. The cell type postulated to supply this licensing role is activated CD4$^+$ T cells. The way this is believed to work is that, first, the classical exogenous pathway of antigen processing in DCs leads to presentation of antigen to CD4$^+$ T cells via class II, leading to activation of these cells (Figure 8-22a). These activated helper cells might then return the favor by inducing costimulatory molecules in the DC and by cytokine secretion (e.g., IL-2), supplying a "second opinion" that, respectively, licenses the DC to present internalized antigens via MHC class I and helps activate naïve CD8$^+$ T cells (Figure 8-22b). This requirement for licensing by a T$_H$ cell could help avoid accidental induction of CTLs to nonpathogenic antigens or self-proteins.

If this is the case, any *unlicensed* DCs that cross-present antigens may serve the opposite and equally important purpose. Since they have not received the go-ahead from the T$_H$ cell to activate CTL responses, they may instead induce tolerance in the CD8$^+$ T cells they encounter, helping to dampen reactivity to these antigens.

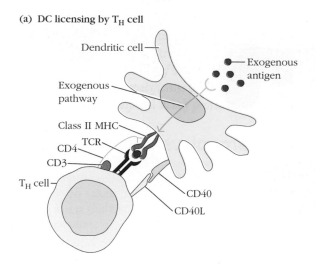

(a) DC licensing by T_H cell

Dendritic cell

Exogenous antigen

Exogenous pathway

Class II MHC

TCR

CD4

CD3

T_H cell

CD40

CD40L

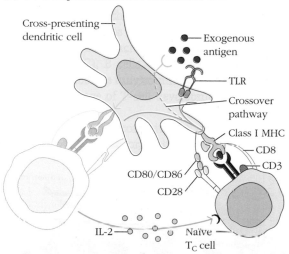

(b) DC cross-presentation and activation of CTL

Cross-presenting dendritic cell

Exogenous antigen

TLR

Crossover pathway

Class I MHC

CD8

CD3

CD80/CD86

CD28

IL-2

Naïve T_C cell

FIGURE 8-22 Activation of naïve Tc cells by exogenous antigen requires DC licensing and cross-presentation. (a) Dendritic cells (DCs) first internalize and process antigen through the exogenous pathway, presenting to CD4$^+$ T_H cells via MHC class II molecules and activating these cells through, among other things, CD40-CD40L engagement. (b) These activated T_H cells can then serve as a bridge to help activate CTL responses; they provide local IL-2 and they in turn license the DC to cross-present internalized antigen in MHC class I, up-regulate costimulatory molecules, and down-regulate their inhibitory counterparts. DC licensing creates an ideal situation for the stimulation of antigen-specific CD8$^+$ T cell responses. When the TLRs on these DCs are engaged, this further activates these cells, providing added encouragement for cross-presentation. Dashed arrows indicate antigen directed for cross-presentation. *[Adapted from Kurts et al., 2010,* Nature Reviews Immunology *10:403.]*

Presentation of Nonpeptide Antigens

To this point, the discussion of the presentation of antigens has been limited to protein antigens and their presentation by classical class I and II MHC molecules. It is well known that some nonprotein antigens are also recognized by T cells, and in the 1980s T-cell proliferation was detected in the presence of nonprotein antigens derived from infectious agents. Many reports now indicate that various types of T cells (expressing γδ as well as αβ T–cell receptors) can react against lipid antigens, such as mycolic acid, derived from well-known pathogens, such as *Mycobacterium tuberculosis*. These antigens are presented by members of the CD1 family of nonclassical class I molecules.

CD1 molecules share structural similarity to classical MHC class I but overlap functionally with MHC class II. Five human CD1 genes are known; all are encoded on a chromosome separate from the classical class I molecules and display very limited polymorphism. Much like classical MHC class I, CD1 proteins are formed by a transmembrane heavy chain, composed of three extracellular α domains, which associates noncovalently with β$_2$-microglobulin. In terms of trafficking and expression profile, however, most CD1 molecules resemble MHC class II proteins, moving intracellularly to endosomal compartments, where they associate with exogenous antigen. Like MHC class II molecules, CD1 proteins are expressed by many immune cell types, including thymocytes, B cells, and DCs, although some members of the family have also been found on hepatocytes and epithelial cells.

Many different lipid or lipid-linked structures have been found to associate non-covalently with CD1 molecules. In general, most of these are glycolipid or lipoprotein antigens, where the lipid moiety fits into deep pockets within the CD1 binding groove and the hydrophilic head group remains exposed, allowing recognition by T cells. Crystal structures have demonstrated that CD1 contains a binding groove that is both deeper and narrower than classical MHC molecules. These grooves are lined with nonpolar amino acids, which can easily accommodate lipid structures. Some of these pockets correspond roughly with the antigen-binding pockets of classical MHC class I and allow antigen access to a deep groove through a narrow opening—more like a foot sliding into a shoe (Figure 8-23a). Figure 8-23b illustrates a comparison of the binding groove of CD1b holding a lipid antigen with a classical class I molecule complexed with peptide antigen. The relatively nonspecific manner of antigen association with the CD1 binding groove, which relies primarily on many hydrophobic interactions, probably accounts for the diversity of self- and foreign antigens that can be presented by these nonclassical class I molecules. A variety of T cells are known to bind these nonclassical MHC molecules.

It is now hypothesized that short-chain self-lipids with relatively low affinity are loaded onto CD1 molecules in the ER, shortly after translation, and allow proper CD1 protein folding. These self-antigen loaded CD1 molecules then travel to the cell surface, where in some cases exogenous lipids may be exchanged with these low-affinity self-antigens. Like with MHC class II molecules, after endocytosis and movement to lower pH environments, CD1 proteins are believed to

(a)

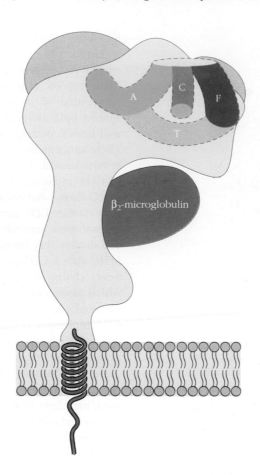

(b)

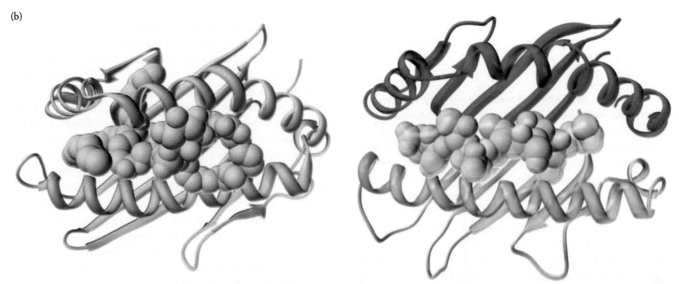

FIGURE 8-23 Lipid antigen binding to the CD1 molecule. (a) Cartoon of the deep binding pockets (A, C, and F) and the deep binding tunnel (T) that accommodate lipid antigens in a foot-in-shoe fashion in the members of the CD1 family of nonclassical class I molecules. (b) Ribbon diagram of the binding groove or pocket of human CD1b complexed with lipid (left) compared with a classical class I molecule binding peptide antigen (right). *[Part b PDB IDs 1GZQ (left) and 1HHG (right).]*

exchange their self peptide, low-affinity binding partners for exogenously derived long-chain, high-affinity lipid antigens resident in phagolysosomes or late endosomes. These newly loaded CD1 molecules then return to the cell surface for recognition by CD1-restricted T cells. Recently constructed transgenic mice that express human CD1 molecules will likely lead the way to new murine models aimed at better defining the role of CD1 restricted T cells in protection from infection.

SUMMARY

- The major histocompatibility complex (MHC) encodes class I and II molecules, which function in antigen presentation to T cells, and class III molecules, which have diverse functions.

- Class I MHC molecules consist of a large glycoprotein α chain encoded by an MHC gene, and β_2-microglobulin, a protein with a single domain that is encoded elsewhere.

- Class II MHC molecules are composed of two noncovalently associated glycoproteins, the α and β chains, encoded by separate MHC genes.

- MHC genes are tightly linked and generally inherited as a unit from parents; these linked units are called haplotypes.

- MHC genes are polymorphic (many alleles exist for each gene in the population), polygenic (several different MHC genes exist in an individual), and codominantly expressed (both maternal and paternal copies).

- MHC alleles influence the fragments of antigen that are presented to the immune system, thereby influencing susceptibility to a number of diseases.

- Class I molecules are expressed on most nucleated cells; class II molecules are restricted to B cells, macrophages, and dendritic cells (pAPCs).

- In most cases, class I molecules present processed endogenous antigen to $CD8^+$ T_C cells and class II molecules present processed exogenous antigen to $CD4^+$ T_H cells.

- Endogenous antigens are degraded into peptides within the cytosol by proteasomes, assemble with class I molecules in the RER, and are presented on the membrane to $CD8^+$ T_C cells. This is the endogenous processing and presentation pathway.

- Exogenous antigens are internalized and degraded within the acidic endocytic compartments and subsequently combine with class II molecules for presentation to $CD4^+$ T_H cells. This is the exogenous processing and presentation pathway.

- Peptide binding to class II molecules involves replacing a fragment of invariant chain in the binding groove by a process catalyzed by nonclassical MHC molecule HLA-DM.

- In some cases, exogenous antigens in certain cell types (mainly DCs) can gain access to class I presentation pathways in a process called cross-presentation.

- Presentation of nonpeptide (lipid and lipid-linked) antigens derived from pathogens involves the nonclassical class I–like CD1 molecules.

REFERENCES

Amigorena, S., and A. Savina. 2010. Intracellular mechanisms of antigen cross presentation in dendritic cells. *Current Opinion in Immunology* **22**:109.

Brown, J. H., et al. 1993. Three-dimensional structure of the human class II histocompatibility antigen HLA-DR1. *Nature* **364**:33.

Doherty, P. C., and R. M. Zinkernagel. 1975. H-2 compatibility is required for T-cell mediated lysis of target cells infected with lymphocytic choriomeningitis virus. *Journal of Experimental Medicine* **141**:502.

Fahrer, A. M., et al. 2001. A genomic view of immunology. *Nature* **409**:836.

Gadola, S. D., et al. 2000. TAP deficiency syndrome. *Clinical and Experimental Immunology* **121**:173.

Horton, R., et al. 2004. Gene map of the extended human MHC. *Nature Reviews Genetics* **5**:1038.

International Human Genome Sequencing Consortium. 2001. Initial sequencing and analysis of the human genome. *Nature* **409**:860.

Kelley, J., et al. 2005. Comparative genomics of major histocompatibility complexes. *Immunogenetics* **56**:683.

Kurts, C., B. W. S. Robinson, and P. A. Knolle. 2010. Cross-priming in health and disease. *Nature Reviews in Immunology* **10**:403.

Li, X. C., and M. Raghavan. 2010. Structure and function of major histocompatibility complex class I antigens. *Current Opinions in Organ Transplantation* **15**:499.

Madden, D. R. 1995. The three-dimensional structure of peptide-MHC complexes. *Annual Review of Immunology* **13**:587.

Margulies, D. 1999. The major histocompatibility complex. In *Fundamental Immunology*, 4th ed. W. E. Paul, ed. Lippincott-Raven, Philadelphia.

Moody, D. B., D. M. Zajonc, and I. A. Wilson. 2005. Anatomy of CD1-lipid antigen complexes. *Nature Reviews in Immunology* **5**:387.

Rock, K. L., et al. 2004. Post-proteosomal antigen processing for major histocompatibility complex class I presentation. *Nature Immunology* **5**:670.

Rothenberg, B. E., and J. R. Voland. 1996. Beta 2 knockout mice develop parenchymal iron overload: A putative role for class I genes of the major histocompatibility complex in iron metabolism. *Proceedings of the National Academy of Science U.S.A.* **93**:1529.

Rouas-Freiss, N., et al. 1997. Direct evidence to support the role of HLA-G in protecting the fetus from maternal uterine natural killer cytolysis. *Proceedings of the National Academy of Science USA.* **94**:11520.

Strominger, J. L. 2010. An alternative path for antigen presentation: Group 1 CD1 proteins. *Journal of Immunology* **184**(7):3303

Wearsch, P. A., and P. Cresswell. 2008. The quality control of MHC class I peptide loading. *Current Opinion in Cell Biology* **20**(6):624.

Yaneva, R., et al. 2010. Peptide binding to MHC class I and II proteins: New avenues from new methods. *Molecular Immunology* **47**:649.

Yi, W., et al. 2010. Targeted regulation of self-peptide presentation prevents type I diabetes in mice without disrupting general immunocompetence. *Journal of Clinical Investigation* **120**(4):1324.

Useful Web Sites

www.bioscience.org This *Frontiers in Bioscience* site includes in the Research Tools section a database that lists gene knockouts. Information is available on studies of the consequences of targeted disruption of MHC molecules and other component molecules, including β_2-microglobulin and the class II invariant chain.

www.bshi.org.uk The British Society for Histocompatibility and Immunogenetics home page contains information on tissue typing and transplantation and links to worldwide sites related to the MHC.

www.nature.com/nrg/journal/v5/n12/poster/ MHCmap/index.html A poster with an extended gene map of the human MHC from the Horton et al. reference cited above.

www.ebi.ac.uk/imgt/hla The International ImMunoGeneTics (IMGT) database section contains links to sites with information about HLA gene structure and genetics and up-to-date listings and sequences for all HLA alleles officially recognized by the World Health Organization HLA nomenclature committee.

www.hla.alleles.org A Web site maintained by the HLA Informatics Group based at the Anthony Nolan Trust in the United Kingdom, with up-to-date information on the numbers of HLA alleles and proteins.

 S T U D Y   Q U E S T I O N S

CLINICAL FOCUS QUESTION Patients with TAP deficiency have partial immunodeficiency as well as autoimmune manifestations. How do the profiles for patients' immune cells explain the partial immunodeficiency? Why is it difficult to design a gene therapy treatment for this disease, despite the fact that a single gene defect is implicated?

1. Indicate whether each of the following statements is true or false. If you think a statement is false, explain why.

   a. A monoclonal antibody specific for β_2-microglobulin can be used to detect both class I MHC K and D molecules on the surface of cells.

   b. Antigen-presenting cells express both class I and class II MHC molecules on their membranes.

   c. Class III MHC genes encode membrane-bound proteins.

   d. In outbred populations, an individual is more likely to be histocompatible with one of its parents than with its siblings.

   e. Class II MHC molecules typically bind to longer peptides than do class I molecules.

   f. All nucleated cells express class I MHC molecules.

   g. The majority of the peptides displayed by class I and class II MHC molecules on cells are derived from self-proteins.

2. You cross a BALB/c (H-2^d) mouse with a CBA (H-2^k) mouse. What MHC molecules will the F_1 progeny express on (a) its liver cells and (b) its macrophages?

3. To carry out studies on the structure and function of the class I MHC molecule K^b and the class II MHC molecule IA^b, you decide to transfect the genes encoding these proteins into a mouse fibroblast cell line (L cell) derived from the C3H strain (H-2^k). L cells do not normally function as antigen-presenting cells. In the following table, indicate which of the listed MHC molecules will ($+$) or will not ($-$) be expressed on the membrane of the transfected L cells.

Transfected gene	MHC molecules expressed on the membrane of the transfected L cells					
	D^k	D^b	K^k	K^b	IA^k	IA^b
None						
K^b						
$IA\alpha^b$						
$IA\beta^b$						
$IA\alpha^b$ and $IA\beta^b$						

4. The SJL mouse strain, which has the H-2^s haplotype, has a deletion of the $IE\alpha$ locus.

   a. List the classical MHC molecules that are expressed on the membrane of macrophages from SJL mice.

   b. If the class II $IE\alpha$ and $IE\beta$ genes from an H-2^k strain are transfected into SJL macrophages, what additional classical MHC molecules would be expressed on the transfected macrophages?

5. Draw diagrams illustrating the general structure, including the domains, of class I MHC molecules, class II MHC molecules, and membrane-bound antibody on B cells. Label each chain and the domains within it, the antigen-binding regions, and regions that have the immunoglobulin-fold structure.

6. One of the characteristic features of the MHC is the large number of different alleles at each locus.

   a. Where are most of the polymorphic amino acid residues located in MHC molecules? What is the significance of this location?

   b. How is MHC polymorphism thought to be generated?

7. As a student in an immunology laboratory class, you have been given spleen cells from a mouse immunized with the LCM virus (LCMV). You determine the antigen-specific

functional activity of these cells with two different assays. In assay 1, the spleen cells are incubated with macrophages that have been briefly exposed to LCMV; the production of interleukin 2 (IL-2) is a positive response. In assay 2, the spleen cells are incubated with LCMV-infected target cells; lysis of the target cells represents a positive response in this assay. The results of the assays using macrophages and target cells of different haplotypes are presented in the table below. Note that the experiment has been set up in a way to exclude alloreactive responses (reactions against nonself MHC molecules).

a. The activity of which cell population is detected in each of the two assays?

b. The functional activity of which MHC molecules is detected in each of the two assays?

c. From the results of this experiment, which MHC molecules are required, in addition to the LCM virus, for specific reactivity of the spleen cells in each of the two assays?

d. What additional experiments could you perform to unambiguously confirm the MHC molecules required for antigen-specific reactivity of the spleen cells?

e. Which of the mouse strains listed in the table could have been the source of the immunized spleen cells tested in the functional assays? Give your reasons.

Mouse strain used as source of macrophages and target cells	MHC haplotype of macrophages and virus-infected target cells				Response of spleen cells	
	K	IA	IE	D	IL-2 production in response to LCMV-pulsed macrophages (assay 1)	Lysis of LCMV-infected cells (assay 2)
C3H	k	k	k	k	+	−
BALB/C	d	d	d	d	−	+
(BALB/C X B10.A)F₁	d/k	d/k	d/k	d/k	+	+
A.TL	s	k	k	d	+	+
B10.A (3R)	b	b	b	d	−	+
B10.A (4R)	k	k	—	b	+	−

8. A T_C-cell clone recognizes a particular measles virus peptide when it is presented by H-2D^b. Another MHC molecule has a peptide-binding groove identical to the one in H-2D^b but differs from H-2D^b at several other amino acids in the α1 domain. Predict whether the second MHC molecule could present this measles virus peptide to the T_C-cell clone. Briefly explain your answer.

9. Human red blood cells are not nucleated and do not express any MHC molecules. Why is this property fortuitous for blood transfusions?

10. The hypothetical allelic combination *HLA-A99* and *HLA-B276* carries a relative risk of 200 for a rare, and yet unnamed, disease that is fatal to preadolescent children.

a. Will every individual with *A99/B276* contract the disease?

b. Will everyone with the disease have the *A99/B276* combination?

c. How frequently will the *A99/B276* allelic combination be observed in the general population? Do you think that this combination will be more or less common than predicted by the frequency of the two individual alleles? Why?

11. Explain the difference between the terms *antigen-presenting cell* and *target cell*, as they are commonly used in immunology.

12. Define the following terms:

a. Self-MHC restriction
b. Antigen processing
c. Endogenous antigen
d. Exogenous antigen
e. Anchor residues
f. Immunoproteasome

13. Ignoring cross-presentation, indicate whether each of the cell components or processes listed is involved in the processing and presentation of exogenous antigens (EX), endogenous antigens (EN), or both (B). Briefly explain the function of each item.

a. _____ Class I MHC molecules
b. _____ Class II MHC molecules
c. _____ Invariant (Ii) chains
d. _____ Lysosomal hydrolases
e. _____ TAP1 and TAP2 proteins
f. _____ Transport of vesicles from the RER to the Golgi complex
g. _____ Proteasomes
h. _____ Phagocytosis or endocytosis
i. _____ Calnexin
j. _____ CLIP
k. _____ Tapasin

14. Antigen-presenting cells have been shown to present lysozyme peptide 46–61 together with the class II IA^k molecule. When CD4$^+$ T_H cells are incubated with APCs and native lysozyme or the synthetic lysozyme peptide 46–61, T_H-cell activation occurs.

a. If chloroquine is added to the incubation mixture, presentation of the native protein is inhibited, but the peptide continues to induce T_H-cell activation. Explain why this occurs.

b. If chloroquine addition is delayed for 3 hours, presentation of the native protein is not inhibited. Explain why this occurs.

15. Cells that can present antigen to T_H cells have been classified into two groups: professional and nonprofessional APCs.

a. Name the three types of professional APCs. For each type, indicate whether it expresses class II MHC molecules and a costimulatory signal constitutively or must be activated before doing so.

b. Give three examples of nonprofessional APCs. When are these cells most likely to function in antigen presentation?

16. Predict whether T_H-cell proliferation or CTL-mediated cytolysis of target cells will occur with the following mixtures of cells. The $CD4^+$ T_H cells are from lysozyme-primed mice, and the $CD8^+$ CTLs are from influenza-infected mice. Use R to indicate a response and NR to indicate no response.

   a. _____ H-2^k T_H cells + lysozyme-pulsed H-2^k macrophages

   b. _____ H-2^k T_H cells + lysozyme-pulsed H-$2^{b/k}$ macrophages

   c. _____ H-2^k T_H cells + lysozyme-primed H-2^d macrophages

   d. _____ H-2^k CTLs + influenza-infected H-2^k macrophages

   e. _____ H-2^k CTLs + influenza-infected H-2^d macrophages

   f. _____ H-2^d CTLs + influenza-infected H-$2^{d/k}$ macrophages

17. Molecules of the CD1 family were recently shown to present nonpeptide antigens.

   a. What is a major source of nonpeptide antigens?

   b. Why are CD1 molecules not classified as members of the MHC family even though they associate with β_2-microglobulin?

   c. What evidence suggests that the CD1 pathway is different from that utilized by classical class I MHC molecules?

18. A slide of macrophages was stained by immunofluorescence using a monoclonal antibody for the TAP1/TAP2 complex. Which of the following intracellular compartments would exhibit positive staining with this antibody?

   a. Cell surface

   b. Endoplasmic reticulum

   c. Golgi apparatus

   d. The cells would not be positive because the TAP1/TAP2 complex is not expressed in macrophages.

19. HLA determinants are used not only for tissue typing of transplant organs, but also as one set of markers in paternity testing. Given the following phenotypes, which of the potential fathers is most likely the actual biological father? Indicate why each could or could not be the biological father.

	HLA-A	HLA-B	HLA-C
Offspring	A3, 43	B54, 59	C5, 8
Biological Mother	A3, 11	B59, 78	C8, 8
Potential Father 1	A3, 33	B54, 27	C5, 5
Potential Father 2	A11, 43	B54, 27	C5, 6
Potential Father 3	A11, 33	B59, 26	C6, 8

20. Assuming the greatest degree of heterozygosity, what is the maximum number of different classical HLA class I molecules that one individual can employ to present antigens to $CD8^+$ T cells? Assuming no extra beta chain genes, what is the maximum number of different class II molecules (isoforms) that one individual can use to present antigen to $CD4^+$ T cells?

21. Define the following terms and give examples using the human HLA locus: polygeny, polymorphism, and codominant expression. How exactly does each contribute to ensuring that a diversity of antigens can be presented by each individual?

22. What is the cellular phenotype that results from inactivation or mutation in both copies of the gene for the invariant chain?

23. In general terms, describe the process of cross-presentation. Which cell types are most likely to be involved in cross-presentation, and what unique role does this process play in the activation of naïve $CD8^+$ T cells?

ANALYZE THE DATA Smith and colleagues (2002, *Journal of Immunology* 169:3105) examined the ability of two peptides to bind two different allotypes—L^d and L^q—of the mouse class I MHC molecule. They looked at (a) the binding of a murine cytomegalovirus peptide (MCMV; amino acid sequence YPH-FMPTNL) and (b) a synthesized peptide, tum⁻ P91A 14–22 (TQNHRALDL). Before and after pulsing the cells with the target peptides, the investigators measured the amount of peptide-free MHC molecules on the surface of cells expressing either L^d, L^q, or a mutant L^d in which tryptophan had been mutated to arginine at amino acid 97 (L^d W97R). The ability of the peptide to decrease the relative number of open forms on the cell surface reflects increased peptide binding to the class I MHC molecules. Answer the following questions based on the data and what you have learned from reading this book.

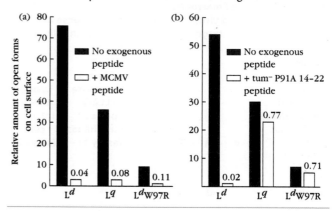

   a. Are there allotypic differences in the binding of peptide by native class I MHC molecules on the cell surface?

   b. Is there a difference in the binding of MCMV peptide to L^d after a tryptophan (W) to arginine (R) mutation in the L^d molecule at position 97? Explain your answer.

   c. Is there a difference in the binding of tum² P91A 14–22 peptide to L^d after a tryptophan (W) to arginine (R) mutation in the L^d molecule at position 97? Explain your answer.

   d. If you wanted to successfully induce a $CD8^+$ T-cell response against tum² P91A 14–22 peptide, would you inject a mouse that expresses L^q or L^d? Explain your answer.

   e. The T cells generated against the MCMV peptide have a different specificity than the T cells generated against the tum² P91A 14–22 peptide when L^d is the restricting MHC molecule. Explain how different peptides can bind the same L^d molecule yet restrict/present peptide to T cells with different antigen specificities.

T-Cell Development

Thymocytes in the cortex of the thymus.

- ■ Early Thymocyte Development
- ■ Positive and Negative Selection
- ■ Lineage Commitment
- ■ Exit from the Thymus and Final Maturation
- ■ Other Mechanisms That Maintain Self-Tolerance
- ■ Apoptosis

T cells initiate the adaptive immune response by interacting, via their T-cell receptors, with MHC/peptide complexes on antigen-presenting cells that have been exposed to pathogens. The range of pathogens to which we are exposed is considerable, and, as described in Chapter 7, vertebrates have evolved a remarkable mechanism to generate a comparable range of T-cell (and B-cell) receptor specificities. Each of the several million T cells circulating in the body expresses a distinct T-cell receptor. The generation of this diverse population, with its diverse receptor repertoire, takes place in the thymus, an organ both required for and dedicated to the development of T cells. Immature cells entering the thymus from the bone marrow express no mature lymphocyte features and no antigen receptors; those leaving the thymus are mature T cells that are diverse in their receptor specificities, and are both tolerant to self and restricted to self-MHC.

How do we generate such a diverse self-restricted and self-tolerant group of T cells? This question has intrigued immunologists for decades and has inspired and required much experimental ingenuity. The innovation has paid off and, although our understanding is not comprehensive, we now have a fundamental appreciation of the remarkable strategies taken by the thymus, the nursery of immature T cells or **thymocytes**, to produce a functional, safe, and useful T-cell repertoire.

In this chapter we describe these strategies and divide T-cell development into two clusters of events: *early thymocyte development*, during which a dizzyingly diverse TCR⁺ population of immature T cells is generated, and *selection events* that depend on TCR interactions to shape this population so that only those cells that are self-restricted and self-tolerant will leave to populate the periphery (Overview Figure 9-1). Early thymocyte development is T-cell receptor independent and brings cells through uncommitted CD4⁻CD8⁻ (double negative, DN) stages to the T-cell receptor-expressing,

CD4⁺CD8⁺ (double positive, DP) stage. The specific events in this early stage include:

1. commitment of hematopoietic precursors to the T cell lineage,
2. the initiation of antigen receptor gene rearrangements, and
3. the selection and expansion of cells that have successfully rearranged one of their T-cell receptor genes (β-selection).

The second phase of T-cell development is largely dependent on T-cell receptor interactions and brings cells to maturity from the CD4⁺CD8⁺ stage to the CD4⁺ or CD8⁺ single positive (SP) stage. The events in this last phase of development include:

1. **positive selection**, selection *for* those cells whose T-cell receptors respond to self-MHC,
2. **negative selection**, selection *against* those cells whose T-cell receptors react strongly to self-peptide/MHC combinations, and
3. **lineage commitment**, commitment of thymocytes to effector cell lineages, including CD4⁺ helper or CD8⁺ cytotoxic populations.

Development of T cells in the mouse

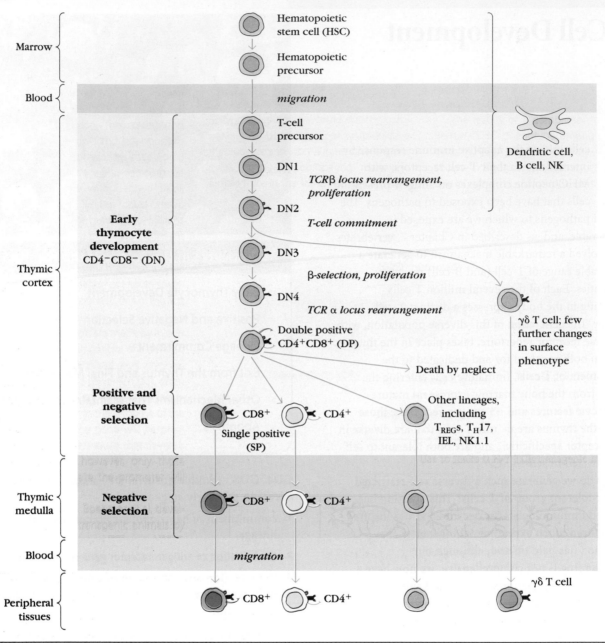

T-cell precursors from the bone marrow travel to the thymus via the bloodstream, undergo development to mature T cells, and are exported to the periphery, where they can undergo antigen-induced activation and differentiation into effector cells and memory cells. Each stage of development occurs in a specific microenvironment and is characterized by specific intracellular events and distinctive cell-surface markers. The most immature thymocytes are CD4−CD8− (double negative, DN) and pass through several stages (DN1-DN4) during which they commit to the T-cell lineage and begin to rearrange their T-cell receptor (TCR) gene loci. Those that successfully rearrange their TCRβ chain proliferate, initiate rearrangement of their TCRα chains, and become CD4+CD8+ (double positive, DP) thymocytes, which dominate the

thymus. DP thymocytes undergo negative and positive selection in the thymic cortex (see text and Overview Figure 9-5 for details). Positively selected thymocytes continue to mature and migrate to the medulla, where they are subject to another round of negative selection to self-antigens that include tissue-specific proteins. Mature T cells express either CD4 or CD8 (single positive, SP) and leave the thymus with the potential to initiate an immune response. Although most thymocytes develop into conventional TCRαβ CD4+ or CD8+ T cells, some DN and DP thymocyte cells develop into other cell lineages, including lymphoid dendritic cells, TCRγδ T cells, natural killer T cells (NKT), regulatory T cells (TREG), and intraepithelial lymphocytes (IELs), each of which has a distinct function (see text).

Understanding T-cell development has been a challenge for students and immunologists alike, but it helps to keep in mind the central purpose of the process: *to generate a large population of T cells that express a diverse array of receptor specificities that can interact with self-MHC, but do not respond to self proteins.*

Early Thymocyte Development

T-cell development occurs in the thymus and begins with the arrival of small numbers of lymphoid precursors migrating from the bone marrow and blood into the thymus, where they proliferate, differentiate, and undergo selection processes that result in the development of mature T cells.

The precise identity of the bone marrow cell precursor that gives rise to T cells is still debated. However, it is clear that this precursor, which is directed to the thymus via chemokine receptors, retains the potential to give rise to more than one type of cell, including natural killer (NK) cells, dendritic cells (DC), B cells, and even myeloid cells. This precursor only becomes fully committed to the T-cell lineage in the late DN2 stage of T-cell development (see Overview Figure 9-1).

Studies revealed that commitment to the T-cell lineage was dependent on a receptor, **Notch**, which had been classically associated with embryonic cell development. Notch, in fact, regulates the decision of a lymphoid precursor to become a T versus a B lymphocyte. When a constitutively active version of Notch1, one of four versions of Notch, is overexpressed in hematopoietic cells, T cells rather than B cells develop in the bone marrow. Reciprocally, when the Notch1 gene is knocked out among hematopoietic precursors, B cells rather than T cells develop in the thymus (!).

The importance of Notch in T-cell commitment was underscored by an in vitro system for studying T-cell development. Until recently, we thought that development of T cells from their hematopoietic precursors required the intact architecture of the thymus. Whereas B-cell development could be achieved in vitro using a preparation of bone marrow stem cells grown on stromal cells in the presence of appropriate cytokines, in vitro studies of T-cell development once could only be performed using intact fetal thymic organ culture systems. In 2002, J. C. Zuniga-Pfluker and colleagues demonstrated that T cells could be induced to develop when bone marrow stem cells were cultured on a stromal cell line that expressed a ligand for Notch. As shown in Figure 9-2, growth of hematopoietic stem cells (HSCs) on stromal cells that express Notch ligand drives the development of these multipotent stem cells to the T-cell lineage. This assay system has been invaluable in defining interactions that control early T-cell development and have helped investigators reveal, for instance, that the transcription factor GATA-3 is a critical participant in Notch-mediated T-cell commitment.

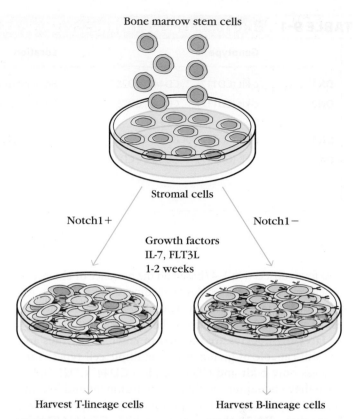

FIGURE 9-2 Development of T cells from hematopoietic stem cells on bone marrow stromal cells expressing the Notch ligand. Investigators can induce lymphoid development from hematopoietic stem cells in vitro using a combination of stromal cell lines and soluble cytokines and growth factors, as indicated. Investigators discovered that Notch signaling was the key to inducing development to the T- rather than B-lymphocyte lineage. After transfecting the stromal cell line with a gene encoding the Notch ligand, lymphoid precursors would adopt the T-cell lineage. Otherwise, they would develop into B cells. *[Adapted from J. C. Zuniga-Pflucker, 2002, Nature Reviews Immunology **4**:67–72.]*

Thymocytes Progress through Four Double-Negative Stages

T-cell development is elegantly organized, spatially and temporally. Different stages of development take place in distinct microenvironments that provide membrane-bound and soluble signals that regulate maturation. After arriving in the thymus from the bone marrow via blood vessels at the cortico-medullary boundary, T-cell precursors encounter Notch ligands, which are abundantly expressed by the thymic epithelium. Recall that T-cell precursors first travel to the outer cortex where they slowly proliferate, then they pass through the thymic medulla before exiting at the cortico-medullary junction (see Figure 2-6). During the time it takes cells to develop in the thymus (1 to 3 weeks), thymocytes pass through a series of stages defined by changes in their cell surface phenotype (see Overview Figure 9-1). The earliest T cells lack detectable CD4 and CD8 and are therefore referred to as **double-negative (DN)** cells. DN T cells can be subdivided

TABLE 9-1   Double-negative thymocyte development

	Genotype	Location	Description
DN1	c-kit (CD117)$^{++}$, CD44$^+$, CD25$^-$	Bone marrow to thymus	Migration to thymus
DN2	c-kit (CD117)$^{++}$, CD44$^+$, CD25$^+$	Subcapsular cortex	TCRγ, δ, and β chain rearrangement; T-cell lineage commitment
DN3	c-kit (CD117)$^+$, CD44$^-$, CD25$^+$	Subcapsular cortex	Expression of pre-TCR; β selection
DN4	c-kit (CD117)$^{low/-}$, CD44$^-$, CD25$^-$	Subcapsular cortex to cortex	Proliferation, allelic exclusion of β-chain locus; α-chain locus rearrangement begins; becomes DP thymocyte

into four subsets (DN1-4) based on the presence or absence of other cell surface molecules, including **c-kit** (CD117), the receptor for stem cell growth factor; **CD44**, an adhesion molecule; and **CD25**, the α chain of the IL-2 receptor (Table 9-1).

DN1 thymocytes are the first to enter the thymus and are still capable of giving rise to multiple cell types. They express only c-kit and CD44 (c-kit^{++}CD44$^+$CD25$^-$), but once they encounter the thymic environment and become resident in the cortex, they proliferate and express CD25, becoming **DN2** thymocytes (c-kit^{++}CD44$^+$CD25$^+$). During this critical stage of development, the genes for the TCR γ, δ, and β chains begin to rearrange; however, the TCR α locus does not rearrange, presumably because the region of DNA encoding TCRα genes is not yet accessible to the recombinase machinery. At the late DN2 stage, T-cell precursors fully commit to the T-cell lineage and reduce expression of both c-kit and CD44. Cells in transition from the DN2 to **DN3** (c-kit$^+$CD44$^-$CD25$^+$) stages continue rearrangement of the TCRγ, TCRδ, and TCRβ chains and make the first major decision in T-cell development: whether to join the TCRγδ or TCRαβ T-cell lineage. Those DN3 T cells that successfully rearrange their β chain and therefore commit to the TCRαβ T-cell lineage lose expression of CD25, halt proliferation, and enter the final phase of their DN stage of development, **DN4** (c-kit$^{low/-}$CD44$^-$CD25$^-$), which mature directly into CD4$^+$CD8$^+$ DP thymocytes.

Thymocytes Can Express Either TCRαβ or TCRγδ Receptors

Vertebrates generate two broad categories of T cells: those that express TCRα and β receptor chains and those that express TCRγ and δ receptor chains. TCRαβ cells are the dominant participants in the adaptive immune response in secondary lymphoid organs; however, TCRγδ cells also play an important role, particularly in protecting our mucosal tissues from outside infection. Both types of cells are generated in the thymus, but how does a cell make the decision to become one or the other? To a large extent, the choice to

become a γδ or αβ T cell is dictated by when and how fast the genes that code for each of the four receptor chains successfully rearrange.

Recall from Chapter 7 that TCR genes are generated by the shuffling (rearrangement) of V and J (and sometimes D) segments, an event responsible for the vast diversity of receptor specificities. Rearrangement of the β, γ, and δ loci begins during the DN2 stage. To become an αβ T cell, a cell must generate a TCRβ chain—an event that depends on a single in-frame VDJ rearrangement event. To become a γδ cell, however, a thymocyte must generate two functional proteins that depend on two separate in-frame rearrangement events. Probability favors the former fate and, in fact, T cells are at least three times as likely to become TCRαβ cells than TCRγδ cells.

TCRγδ T-cell generation is also regulated developmentally. They are the first T cells that arise during fetal development, and provide a very important protective function perhaps even prior to birth. Studies show, for instance, that γδ T cells are required to protect very young mice against the protozoal pathogen that causes coccidiosis. However, production of γδ T cells declines after birth, and the TCRγδ T-cell population represents only 0.5% of all mature thymocytes in the periphery of an adult animal (Figure 9-3).

Most TCRγδ T cells are quite distinct in phenotype and function from conventional TCRαβ T cells. Most do not go through the DP stage of thymocyte development and leave the thymus as mature DN T cells. Many emerge from the thymus with the ability to secrete cytokines, a capacity gained by most TCRαβ cells only after they encounter antigen in secondary lymphoid tissues. The TCRγδ T-cell population also expresses receptors that are not as diverse as TCRαβ T cells, and many appear to recognize unconventional antigens, including lipids associated with unconventional MHC molecules. Many take up long-term residence in mucosal tissues and skin and join innate immune cells in providing a first line of attack against invading microbes, as well as the response to cellular stress.

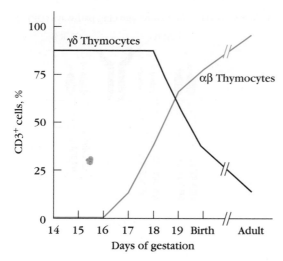

FIGURE 9-3 Time course of appearance of γδ thymocytes and αβ thymocytes during mouse fetal development. The graph shows the percentage of cells in the thymus that are double negative (CD4⁻CD8⁻) and bear the γδ T-cell receptor (black) or are double positive (CD4⁺CD8⁺) and bear the αβ T-cell receptor (blue). Fetal animals generate more γδ T cells than αβ T cells, but the proportion of γδ T cells generated drops off dramatically after birth. This early dominance of TCRγδ cells may have adaptive value: a large portion of these cells express nondiverse TCR specificities for common pathogen proteins and can mount a quick defense before the more traditional adaptive immune system has fully developed.

For the rest of this chapter, we will focus on the development of TCRαβ T cells, which make up the vast majority of T lymphocytes in the body and are continually being generated, even after puberty, when the thymus shrinks considerably in size (*involutes*).

DN Thymocytes Undergo β-Selection, Which Results in Proliferation and Differentiation

Double-negative (DN) thymocytes that have successfully rearranged their TCRβ chains are valuable, and are identified and expanded via a process known as β-**selection** (see Overview Figure 9-1). This process involves a protein that is uniquely expressed at this stage of development, a 33-kDa invariant glycoprotein known as the **pre–Tα chain**. Pre-Tα acts as a surrogate for the real TCRα chain, which has yet to rearrange, and assembles with a successfully rearranged and translated β chain, as well as CD3 complex proteins. This precursor TCR/CD3 complex is known as the **pre-TCR** (Figure 9-4a) and acts as a sensor by initiating a signal transduction pathway. The signaling that the pre-TCR complex initiates is dependent on many of the same T-cell specific kinases used by a mature TCR (see Chapter 3), but does not appear to be dependent on ligand binding. In fact, little if any of the complex is expressed on the cell surface; rather, successful assembly of the complex may be sufficient to acti-

vate the signaling events. As we will learn in Chapter 10, B cells also express an immature receptor complex (the pre-BCR) that is coded by distinct genes, but is fully analogous to the pre-TCR.

Pre-TCR signaling results in the following cascade of events:

1. Maturation to the DN4 stage (c-kit$^{low/-}$CD44⁻CD25⁻)
2. Rapid proliferation in the subcapsular cortex
3. Suppression of further rearrangement of TCR β-chain genes, resulting in allelic exclusion of the β-chain locus
4. Development to the CD4⁺CD8⁺ **double-positive (DP)** stage
5. Cessation of proliferation
6. Initiation of TCRα chain rearrangement

It is important to note that the proliferative phase prior to α chain rearrangement enhances receptor diversity considerably by generating clones of cells with the same TCR β-chain rearrangement. Each of the cells within a clone can then rearrange a different α-chain gene, thereby generating an even more diverse population than if the original cell had undergone rearrangement at both the β- and α-chain loci prior to proliferation. TCR α-chain gene rearrangement does not begin until double-positive thymocytes stop proliferating.

Most T cells fully rearrange and express a TCRβ chain from only one of their two TCR alleles, a phenomenon known as **allelic exclusion**. Allelic exclusion is the result of inhibition of further rearrangement at the other TCRβ allele (which must be fully rearranged to be expressed). This can be accomplished by reducing RAG expression so no more rearrangement can occur, as well as by making the locus inaccessible to further RAG interaction via more permanent changes in chromatin packaging. The details of the mechanisms responsible for this shutdown are still being investigated. However, negative feedback signals from a successfully assembled pre-TCRβ/pre-Tα complex during β-selection clearly have a significant influence. Other events, including the proliferative burst that follows β-selection, which dilutes RAG protein levels, can also play a role.

RAG levels continue to change after β-selection. They are restored after the proliferative burst and allow TCRα rearrangement to occur. They decrease once again after expression of a successfully assembled TCRαβ dimer.*

---

*The mechanisms that turn off rearrangement at this (DP) stage of development are not as efficient as those that turn off additional TCRβ chain rearrangement. In fact, both TCRα alleles successfully rearrange more frequently than we originally supposed. This means that many T cells actually express two TCR specificities: an interesting reality check that complicates a fundamentalist interpretation of the clonal selection hypothesis.

(a) Pre-TCR (DN3, DN4, DP)

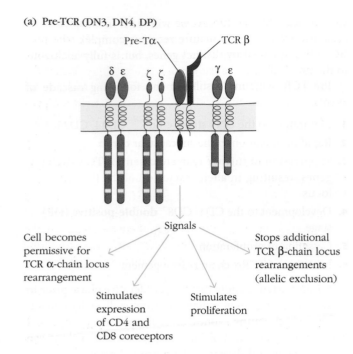

Cell becomes permissive for TCR α-chain locus rearrangement

Stimulates expression of CD4 and CD8 coreceptors

Stimulates proliferation

Stops additional TCR β-chain locus rearrangements (allelic exclusion)

(b) Mature αβTCR, immature signaling pathways (DP)

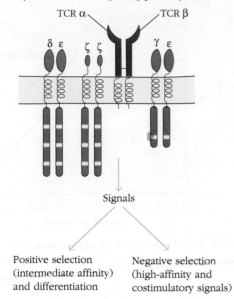

Positive selection (intermediate affinity) and differentiation

Negative selection (high-affinity and costimulatory signals)

FIGURE 9-4 Changes in the structure and activity of the T-cell receptor through T-cell development. (a) The pre-TCR. The pre-TCR is assembled during the DN stage of development when a successfully rearranged TCRβ chain dimerizes with the nonvariant pre-Tα chain. Like the mature αβTCR dimer, the pre-TCR is noncovalently associated with the CD3 complex. Assembly of this complex results in intracellular signals that induce a variety of processes, including the maturation to the DP stage. (b) Mature αβTCR expressed at the DP stage of development. Once a DP thymocyte has successfully rearranged a TCRα chain, it will dimerize with the TCRβ, replacing the pre-Tα chain. This mature αβTCR generates the signals that lead to either positive or negative selection (differentiation or death, respectively), depending on the affinity of the interaction. Note that the TCRα chain has a shorter intracellular region than the pre-Tα chain and cannot generate intracellular signals independently. (c) Mature αβTCR expressed at the SP stage. Although the αβTCR/CD3 complex expressed by mature SP T cells is structurally the same as that expressed by DP thymocytes, the signals it generates are distinct. It responds to high-affinity engagement not by dying, but by initiating cell proliferation, activation, and the expression of effector functions. Low-affinity signals generate survival signals. The basis for the differences in signals generated by DP and SP TCR complexes is still unknown.

(c) Mature αβTCR, mature signaling pathways (SP)

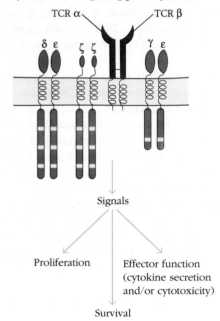

Proliferation

Effector function (cytokine secretion and/or cytotoxicity)

Survival

Positive and Negative Selection

CD4+CD8+ (DP) thymocytes, small, nonproliferating cells that reside in the thymic cortex, are the most abundant subpopulation in the thymus, comprising more than 80% of cells. Most important, they are the first subpopulation of thymocytes that express a fully mature surface TCRαβ/CD3 complex and are therefore the primary targets of thymic selection (see Figure 9-1). Thymic selection shapes the TCR repertoire of DP thymocytes based on the affinity of their T-cell receptors for the MHC/peptides they encounter as they browse the thymic cortex.

Once a young double-positive (DP) thymocyte successfully rearranges and expresses a TCRα chain, this chain will associate with the already produced TCRβ chain, taking the place of the surrogate pre-TCRα chain, which is no longer actively expressed (Figure 9-4b). At this point, several "goals" have been accomplished by the thymus: hematopoietic cell precursors have expanded in the subcapsular cortex, committed to the T-cell lineage, and rearranged a set of TCR genes. They have also "chosen" to become a TCRαβ or TCRγδ T cell. This TCRαβ population now expresses both CD4 and CD8, and is ready for the second stage of T-cell development: selection.

Why is thymic selection necessary? The most distinctive property of the mature T cells is that they recognize only foreign antigen combined with self-MHC molecules. However, randomly generated TCRs will certainly have no inherent affinity for "foreign antigen plus self-MHC molecules." They could just as well recognize foreign MHC/peptide combinations, which would not be useful, or self-MHC/self-peptide combinations, which could be dangerous.

Two distinct selection processes are required:

- Positive selection, which selects for those thymocytes bearing receptors capable of binding self-MHC molecules, resulting in **MHC restriction**

- Negative selection, which selects against thymocytes bearing high-affinity receptors for self-MHC/peptide complexes, resulting in **self-tolerance**.

Because only self-peptides are presented in the thymus, and only in association with self-MHC molecules, these two selection processes ensure that surviving thymocytes express TCRs that have low affinity for self-peptides in self-MHC. On the other hand, the processes do not guarantee that the T cells generated will bear receptors with high affinity for any specific self-MHC/foreign peptide combination. The ability of the immune system to respond to a foreign antigen depends on the probability that one of the millions of T cells that survives selection will bind one of the many MHC/peptide combinations expressed by an antigen-presenting cell that has processed pathogenic proteins outside the thymus.

The vast majority of DP thymocytes (~98%) never meet the selection criteria and die by apoptosis within the thymus (Overview Figure 9-5). The bulk of DP thymocyte death (~95%) occurs among thymocytes that fail positive selection because their receptors do not specifically recognize self-MHC molecules. These cells do not receive survival signals through their TCRs, and die by a process known as **death by neglect**. A small percentage of cells (2%–5%) are eliminated by negative selection. Only 2% to 5% of DP thymocytes actually exit the thymus as mature T cells.

Below, we briefly describe the experimental evidence for thymic involvement in MHC restriction. We also describe what is currently known about the positive and negative selection processes, as well as the models that have been advanced to explain what is known as "the thymic paradox."

Thymocytes "Learn" MHC Restriction in the Thymus

Early evidence for the role of the thymus in selection of the T-cell repertoire came from chimeric mouse experiments performed by R. M. Zinkernagel and his colleagues. Recall that Zinkernagel won the Nobel Prize for showing that mature T cells are MHC restricted (i.e., they need to recog-

nize antigen in the context of the MHC of their host to respond; see Classic Experiment in Chapter 8). However, he and his colleagues were curious to know how T cells became so "restricted." They considered two possibilities: Either the T cell and APC simply had to have matching MHC types (i.e., an "A" strain T cell had to see an "A" strain antigen-presenting cell) or T cells, regardless of their own MHC type, "learned" the MHC type of their host sometime during development. Zinkernagel and colleagues thought that such learning could take place in the thymus, the T-cell nursery. To determine if T cells could be "taught" to recognize the host MHC, they removed the thymus (thymectomized) and irradiated (A × B) F_1 mice so they had no functional immune system (Figure 9-6). They then reconstituted the hematopoietic cells with an intravenous infusion of F_1 bone marrow cells, but replaced the thymus with one from a B-type mouse. (To be certain that the thymus graft did not contain any mature T cells, they irradiated it before transplantation.)

In this experimental system, T-cell progenitors from the (A × B) F_1 bone marrow would mature within a thymus that expresses only B-haplotype MHC molecules on its stromal cells. Would these (A × B) F_1 T cells now be MHC restricted to the B haplotype of the thymus in which they developed? Or, because they expressed both A and B MHC, would they be able to recognize both A and B MHC haplotypes? To answer this question, the investigators infected the chimeric mice with lymphocytic choriomeningitis virus (LCMV, the antigen) and removed the immunized, mature splenic T cells to see which LCMV-infected target cells (APCs) they could kill. They tested them against infected APCs from strain A, strain B, and strain A × B mice. As shown in Figure 9-6, T cells from the chimeric mice could only lyse LCMV-infected target cells from strain B mice. Thus, the MHC haplotype of the thymus in which T cells develop determines their MHC restriction. T cells "learned" which MHC haplotype they are restricted to during their early days in the thymus. Although once we referred to this process as "T-cell education," we now know that it is the consequence of a brutal selection process.

T Cells Undergo Positive and Negative Selection

In the thymus, thymocytes come into contact with thymic epithelial cells that express high levels of class I and class II MHC molecules on their surface. These self-MHC molecules present self-peptides, which are typically derived from intracellular or extracellular proteins that are degraded in the normal course of cellular metabolism. DP thymocytes undergo positive and negative selection, depending on the signals they receive when they encounter self-MHC/self-peptide combinations with their TCRs.

While working to understand negative and positive selection, it helps to picture an early model that was advanced by several seminal immunologists (see Overview

Positive and Negative Selection of Thymocytes in the Thymus

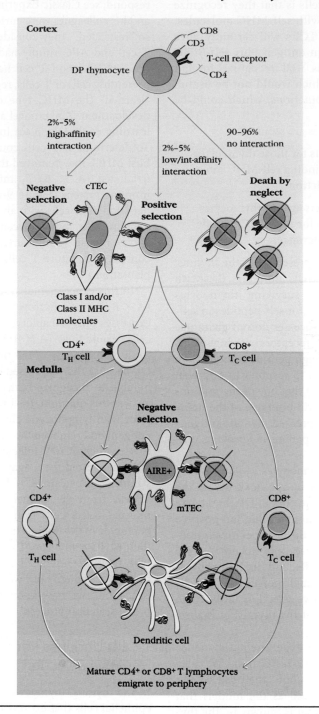

Thymic selection involves multiple interactions of DP and SP thymocytes with both the cortical and medullary thymic stromal cells, as well as dendritic cells and macrophages. Selection results in a mature T-cell population that is both self-MHC restricted and self-tolerant. DP thymocytes that express new TCRαβ dimers browse the MHC/peptide complexes expressed by the *cortical* thymic epithelial cells (cTECs). The large majority of DP thymocytes die in the cortex by neglect because of their failure to bind MHC/peptide combinations with sufficient affinity. The small percentage whose TCRs bind MHC/peptide with high affinity die by clonal deletion (negative selection). Those DP thymocytes whose receptors bind to MHC/peptide with intermediate affinity are positively selected and mature to single-positive (CD4+ or CD8+) T lymphocytes. These migrate to the medulla, where they are exposed to AIRE+ *medullary* thymic epithelial cells (mTECs), which express tissue-specific antigens and can mediate negative selection. Medullary dendritic cells can acquire mTEC antigens by engulfing mTECs, and mediate negative selection (particularly of MHC Class II restricted CD4+ thymocytes).

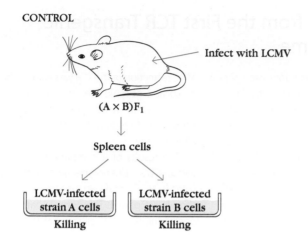

CONTROL

Infect with LCMV

$(A \times B)F_1$

Spleen cells

LCMV-infected strain A cells
Killing

LCMV-infected strain B cells
Killing

EXPERIMENT

$(A \times B)\ F_1\ (H\text{-}2^{a/b})$

① Thymectomy
② Lethal x-irradiation

Strain B thymus graft (H-2^b)
$(A \times B)\ F_1$ hematopoietic stem cells (H-$2^{a/b}$)
Infect with LCMV

Spleen cells

LCMV-infected strain A cells
No killing

LCMV-infected strain B cells
Killing

FIGURE 9-6 Experimental demonstration that the thymus selects for maturation only those T cells whose T-cell receptors recognize antigen presented on target cells with the haplotype of the thymus. Control (A × B) F_1 mice infected with LCMV produce mature T cells that are able to kill both infected-strain A cells and infected-strain B cells, demonstrating that these T cells are restricted to MHC molecules expressed by both strain A (H-2^a) and strain B (H-2^b). In order to determine the involvement of the thymus in the restriction specificity of T cells, investigators grafted thymectomized and lethally irradiated (A × B) F_1 (H-$2^{a/b}$) mice with a strain-B (H-2^b) thymus and reconstituted it with (A × B) F_1 bone marrow cells. After infection with LCMV, the CD8$^+$ cytotoxic (CTL) cells from this reconstituted mouse were assayed for their ability to kill ^{51}Cr-labeled strain A or strain B target cells infected with LCMV. Only strain B target cells were lysed, suggesting that the H-2^b grafted thymus had selected for maturation only those T cells that could recognize antigen combined with H-2^b MHC molecules.

Figure 9-5). Although we now know that some aspects of this model are too simplistic, its core principles remain relevant and provide a very useful framework for understanding thymic selection. Briefly, DP thymocytes can be considered to have one of three fates, depending on the affinity of their new T-cell receptors for the MHC/self-peptide combinations that they encounter: If their newly generated TCRs do not bind to any of the MHC/self-peptides they encounter on stromal cells, they will die by neglect. If they bind too strongly to MHC/self-peptide complexes they encounter, they will be negatively selected. If they bind with a low, "just right" affinity to MHC/self-peptide complexes, they will be positively selected and mature to the single-positive stage of development. (We will also see that cells that are positively selected in the cortex undergo a second, very important round of negative selection in the medulla, following lineage commitment to CD4$^+$ or CD8$^+$ cells.) We review the origins of the model and discuss some of the modifications that have been suggested by recent data below. See Classic Experiment Box 9-1 for experimental evidence in support of this model from the earliest TCR transgenic mouse.

Thymic stromal cells, including epithelial cells, macrophages, and dendritic cells, play essential roles in positive and negative selection and should be part of our visualization of thymic selection events. Young DP thymocytes are in intimate contact with these cells and "browse" the MHC/self-peptides displayed on their surfaces. Each of these cell types has the capacity to express high levels of class I and class II MHC proteins, and typically have extended processes that contact many developing thymocytes. Some also express costimulatory ligands, including CD80 (B7-1) and CD86 (B7-2). As the thymocytes migrate through the thymus, they encounter multiple different stromal cell surfaces, and have the opportunity to bind many different MHC/peptide combinations.

Positive Selection Ensures MHC Restriction

If a CD4$^+$CD8$^+$ thymocyte recognizes a self-MHC/peptide complex on the cortical epithelial cells that they browse, it will undergo positive selection, a process that induces both survival and differentiation of DP thymocytes. Remarkably, the majority of newly generated thymocytes do not successfully engage the MHC/peptides they encounter with their TCRs. Either they have not generated a functional TCRαβ combination or the combination does not bind MHC/peptide complexes with sufficient affinity. These cells have "failed" the positive selection test and die by apoptosis within 3 to 4 days. TCR/MHC/self-peptide interactions that initiate positive selection are thought to be at least three times lower in intensity than interactions that initiate negative selecting signals (Figure 9-7).

The signal cascade generated by positive selecting TCR interactions has not yet been fully characterized. We do know that many of the signaling molecules required for

Insights about Thymic Selection from the First TCR Transgenic Mouse Have Stood the Test of Time

In the late 1980s Harald von Boehmer and colleagues published a series of seminal experiments that provided direct and compelling evidence for the influence of T-cell receptor interactions on positive and negative selection. They recognized that in order to understand how positive and negative selection worked, they would need to be able to trace and compare the fate of thymocytes with defined T-cell receptor specificities. But how could one do that if every one of the hundreds of millions of normal thymocytes expresses a different receptor? The researchers decided to develop a system in which all thymocytes expressed a T-cell receptor of known specificity, and thereby generated one of the first TCR transgenic mice.

Transgenic animals are made by injecting a gene under the control of a defined promoter into a fertilized egg (zygote) (see Chapter 20). The gene—in fact, often many copies of the gene—will be incorporated randomly into the genome of the zygote. Therefore, the gene will be present in all cells of the mouse that develop from that zygote; however, only those cells that can activate the promoter will express the gene.

The team of von Boehmer et al. developed the first TCR transgenic animals by isolating fully rearranged TCRα and TCRβ genes of a mature CD8⁺ T-cell line that was known to be specific for an antigen expressed only in male mice (the "H-Y" antigen) in the context of H-2D^b Class I MHC molecules (Figure 1). They generated a genetic construct that included both rearranged genes under the control of a T-cell specific promoter and injected this into a mouse zygote. Since the receptor transgenes were already rearranged, other TCR gene rearrangements were suppressed in the transgenic mice (the phenomenon of allelic exclusion; see text and Chapter 7); therefore, a very high percentage of the thymocytes in the transgenic mice expressed the TCRαβ combination specific for the H-Y/H-2D^b. (It turns out that not all thymocytes expressed this TCR combination. Why not?

Recall that allelic exclusion does not operate well for the TCRα locus in particular.)

The team's choice of specificities was very clever. Because all thymocytes expressed an anti-male antigen TCR specificity, the researchers could directly compare the phenotype of autoreactive thymocytes and non-autoreactive thymocytes simply by comparing male and female mice in the same litter. Because the

MHC restriction of the TCR was known, they could also observe the influence of MHC haplotype on thymocyte development simply by breeding the mice to other inbred strains.

The results of a comparison of CD4⁺ versus CD8⁺ phenotype among thymocytes in H-2D^b male versus female mice is shown in Figure 1. The primary data from one of the first publications are shown in

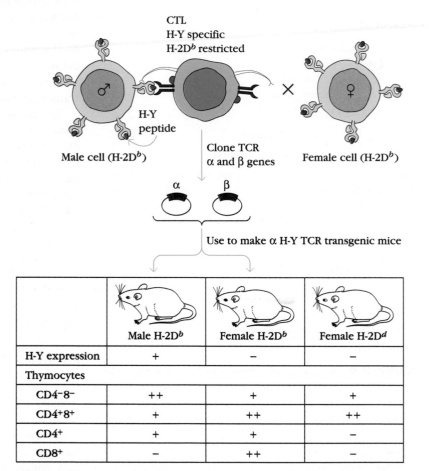

	Male H-2D^b	Female H-2D^b	Female H-2D^d
H-Y expression	+	−	−
Thymocytes			
CD4⁻8⁻	++	+	+
CD4⁺8⁺	+	++	++
CD4⁺	+	+	−
CD8⁺	−	++	−

FIGURE 1

Experimental demonstration that negative selection of thymocytes requires both self-antigen and self-MHC, and positive selection requires self-MHC. In this experiment, H-2D^b male and female transgenics were generated by injecting TCR transgenes specific for H-Y antigen plus the D^b MHC molecule into zygotes. The H-Y antigen is expressed only in males. H-2D^d mice were also generated by backcrossing these transgenics onto an H-2^d strain (e.g., BALB/c). FACS analysis of thymocytes from the transgenics showed that mature CD8⁺ T cells expressing the transgene were absent in the male H-2D^b mice but present in the female H-2D^b mice, suggesting that thymocytes that bind with high affinity to a self antigen (in this case, H-Y antigen in the male mice) are deleted during thymic development. Results also show that DP but not CD8⁺ thymocytes develop in female H-2D^d mice, which do not express the proper self-MHC. These findings indicate that positive selection and maturation require self-MHC interactions. [Adapted from H. von Boehmer and P. Kisielow, 1990, Science **248**:1369.]

BOX 9-1

Figure 2 and depict flow cytometric data of CD4 and CD8 expression of thymocytes. The flow cytometric profile (see Chapter 20) of a typical wild-type thymus is shown in Figure 2a and can be used for comparison. In this profile, the CD8-expression status of a cell is shown on the X-axis, the CD4-expression status on the Y-axis. The data are quantified in the upper right corner of the profile. As you can see, more than 80% of normal thymocytes are DP, 10% to 15% are CD4 single positive, 3% to 5% are CD8 single-positive cells, and only a small percentage of cells are DN. How do the CD4 versus CD8 phenotypes of male and female H-Y TCR transgenic mice (Figure 2b) differ, and what does this tell you?

Thymocytes in female mice are in all stages of development: DN, DP, and SP. In fact, they seem to have an abundance of mature CD8[+] SP thymocytes. We'll come back to this below. The male mice, however, have few if any anti-H-Y TCR transgenic DP thymocytes. These data show that in an environment where DP thymocytes are exposed to the antigen for which they are specific (in this case the male H-Y antigen), they are eliminated. This result was fully consistent with the concept of negative selection and showed that self-reactive cells were removed from the developing repertoire, perhaps at the DP stage. However, the mice offered insights into many more aspects of thymic selection.

Evidence for the role of TCR interactions in positive selection, for instance, came

from a different experiment in which the investigators asked what would happen if the "correct" restricting element, H-2D[b], was not present in the mouse. To do this they performed backcrosses of their TCR transgenic mice to mice that expressed a different H-2 haplotype, an H-2D[d] mouse. (For those who want to brush up on their genetics, determine what crosses you would make to do this, and include a plan for assessing the genotype of your mice.) Figure 2c shows the CD4 versus CD8 phenotype of H-Y TCR transgenic thymocytes from female H-2D[b/d] mice and H-2D[d] mice. What did they find?

H-2D[d] females had no mature T cells, indicating that in the absence of the MHC haplotype to which the TCR was restricted, thymocytes could not mature beyond the DP stage. These data provided direct evidence for the necessity of a TCR/self-MHC interaction for thymocyte maturation (positive selection) to occur.

These "simple" experiments also revealed a third feature of thymic development and initiated a controversy that continues to this day. Return to Figure 2b and take a look at the female mouse data. Unlike wild-type thymocytes, which typically include 3%- to -5% CD8[+] T cells, 46% of developing thymocytes in the anti-H-Y TCR transgenic were mature CD8[+] cells. What did this mean? Recall that the T-cell line that the TCR genes came from originally was a CD8[+] T-cell line, restricted to Class I MHC (H-2D[b]). These data showed that the MHC restriction specificity

of the TCR (in this case, a restriction for Class I) influenced its CD4 versus CD8 lineage choice. In other words, the results suggested that thymocytes with new TCRs that preferentially bound Class II would become CD4[+] SP mature T cells, and thymocytes with TCRs that preferentially bound Class I would develop into CD8[+] SP mature T cells.

The fundamental conclusions of this now classic set of experiments have stood the test of time and have been supported by many different, equally clever experiments: (1) *Negative selection* of autoreactive cells results in their elimination at the DP stage and beyond. (It is important to note that it later became clear that the promoter driving these first H-Y TCR transgenes permitted earlier-than-normal expression of the TCR, which resulted in elimination of autoreactive cells at an earlier-than-usual point in development. Experiments now indicate that negative selection by clonal deletion typically occurs after positive selection as cells transit between the DP and SP stages of development.) (2) *Positive selection* involves a low affinity interaction between thymocyte TCR and self-MHC/ peptide interactions. (3) *Commitment to the CD4 versus CD8 lineage is determined by the preference of the TCR for MHC Class I versus Class II.* How this last event happens remains a topic of intense controversy (see text).

von Boehmer, H., H. Teh, and P. Kisielow. 1989. The thymus selects the useful, neglects the useless and destroys the harmful. *Immunology Today* **10**:57–61.

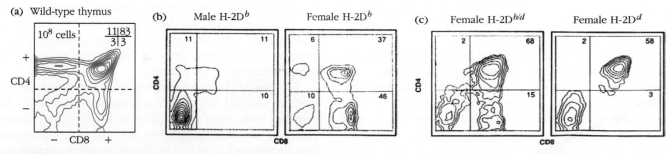

FIGURE 2

Primary data from experiments summarized in Figure 1. The relative staining by an anti-CD8 antibody is shown on the X-axis, and the relative staining by an anti-CD4 antibody appears on the Y-axis. DP, DN, CD4 SP, and CD8 SP phenotypes are divided into quadrants, and percentages are given by quadrant. (a) CD4 versus CD8 phenotype of a wild-type thymus. (b) CD4 versus CD8 phenotype of thymocytes from H-Y TCR transgenic H-2[b] male and female. (c) CD4 versus CD8 phenotype of thymocytes from H-Y TCR transgenic H-2[b/d] and H-2[d/d] females. *[Parts (b) and (c) from H. von Boehmer and P. Kisielow, 1990, Science* **248**:*1369.]*

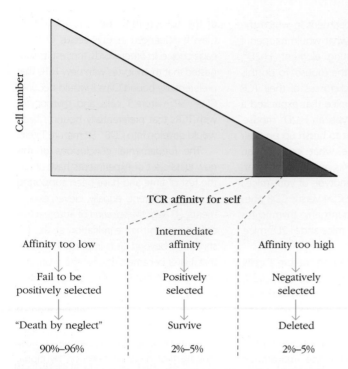

FIGURE 9-7 **Relationship between TCR affinity and selection.** This figure schematically illustrates the association between thymocytes' fate and the affinity of their TCR for self-MHC/peptide complexes that they encounter in the thymus. Fewer than 5% of thymocytes produce TCRs that bind to MHC/peptide complexes with high affinity. Most of these will be deleted by negative selection (some will become regulatory T cells and other specialized cell types). More than 90% generate TCRs that either do not bind to MHC/peptide complexes or bind them with very low affinity. These die by neglect. Fewer than 5% generate TCRs that bind with just the right intermediate affinity to self-MHC/peptide complexes. These will survive and mature.

positive selection are also involved in T-cell receptor mediated activation of mature T cells (see Chapter 3), an event, however, that requires higher-affinity TCR interactions. We still do not fully understand how low-affinity positive selecting signals initiate maturation to the single-positive stage, or how they differ from pro-apoptotic negative selecting signals, which are also TCR mediated.

What is the adaptive value of positive selection? Why didn't we evolve a system that negatively selects but leaves the T-cell receptor repertoire otherwise intact? Some have suggested that the paring down of the repertoire is important to increasing the efficiency of negative selection, as well as to increasing the efficiency of peripheral T-cell responses. Without positive selection, the system would be cluttered with a great many cells that are unlikely to recognize anything, and reduce the probability that a T cell will find "its" antigen in a reasonable period of time. This is a reasonable speculation; however, positive selection may also offer more subtle advantages that have often inspired discussion in the field.

Negative Selection (Central Tolerance) Ensures Self-Tolerance

Autoreactive CD4$^+$CD8$^+$ thymocytes with high-affinity receptors for self-MHC/self-peptide combinations are potentially dangerous to an organism, and many are killed by negative selection in the thymus. In fact, errors in the negative selection process are responsible for a host of autoimmune disorders, including Type 1 diabetes (see Clinical Focus). Negative selection is defined broadly as any process that rids a repertoire of autoreactive clones and is responsible for central tolerance. It is likely that most negative selection occurs via a process known as **clonal deletion**, where high-affinity TCR interactions directly induce apoptotic signals.

Clonal deletion of DP thymocytes appears to be optimally mediated by the same cells (APCs) and same interactions (high-affinity TCR engagement coupled with costimulatory signals) that activate mature T cells. Why strong TCR signals result in death of immature T cells, but proliferation and differentiation of mature T cells (see Figure 9-4c) remains an active area of investigation.

Thymic dendritic cells and macrophages, which are found in multiple areas of the thymus, clearly have the ideal features to mediate negative selection, but interestingly, so do medullary epithelial cells, which express high levels of the costimulatory ligands CD80 and CD86 as well as a unique transcription factor that allows them to present tissue-specific antigens (see below). Therefore, both the cortex and the medulla have the potential to induce negative selection.

Do We Delete Thymocytes Reactive to Tissue-Specific Antigens?

We have seen that MHC molecules can present peptides derived from both endogenous and exogenous proteins on the surface of antigen-presenting cells (see Chapter 8). However, only a fraction of cell types—thymocytes, stromal cells, macrophages, and other antigen-presenting cells—reside in the thymus, so we would expect these cells to produce only a subset of the proteins encoded in the genome. How, then, can the thymus possibly get rid of developing T cells that are autoreactive to tissue-specific antigens—for example, proteins specific to the brain, the liver, the kidney, or the pancreas?

This question bothered immunologists for a long time. For a bit, we assumed that other mechanisms of tolerance in the periphery took care of autoreactivity to tissue-specific proteins, but investigations in the late 1990s surprised us all and revealed that the thymus had an extraordinary capacity to express and present proteins from all over the body. Bruno Kyewski and colleagues showed that this capacity was a unique feature of thymic medullary epithelial cells. Diane Mathis, Cristoph Benoist, and their colleagues went on to show that some medullary epithelial cells express a unique protein, **AIRE**, that allows cells

How Do T Cells That Cause Type 1 Diabetes Escape Negative Selection?

We have an extensive appreciation of the pain caused by autoimmune disease and its clinical progression, but although we have also gained a deeper understanding of the mechanisms behind immune tolerance, we still know little about the precise origins of autoimmune disorders. Indisputably, the primary cause of autoimmune disease is the escape of self-reactive lymphocytes—B cells, T cells, or both—from negative selection. Surprisingly little is known about the reasons for this escape, or the specificity and phenotype of the escapees that cause even the best-characterized disorders, including Type I diabetes (T1D), multiple sclerosis (MS), rheumatoid arthritis (RA), and systemic lupus erythematosus (SLE).

Recent work on a mouse strain, the non-obese diabetic or NOD mouse, which is markedly susceptible to Type 1 diabetes, has shed light on the features of the self-reactive T cells that cause the disease, and revealed some interesting reasons for their escape from central tolerance. T1D is a T-cell mediated autoimmune disease caused by T cells that kill insulin-producing beta islet cells in the pancreas. Many of the autoreactive T cells actually recognize a specific peptide from the insulin protein itself.

When Emil Unanue and colleagues examined the fine specificity of T cells that had infiltrated the islet cells in diabetes-prone mice, they found two types of cells: T cells that could respond to the insulin peptide that was generated via intracellular MHC Class II processing pathways by dendritic cells ("Type A" T cells), and an unexpected population of T cells that responded to the insulin peptide only if it were added exogenously to the MHC Class II molecules on dendritic cells ("Type B" T cells). The investigators speculate that the endogenously processed MHC/insulin peptide combination and the exogenously formed MHC/insulin peptide complex differ in conformation and therefore activate different T-cell receptor specificities.

Perhaps more important, these observations suggest an elegant and precise reason for the escape of autoreactive T cells from the thymus. Dendritic cells in the pancreas, but not other tissues, appear uniquely capable of forming MHC Class II/insulin peptide from exogenous sources, presumably because they are in an environment replete with extracellular insulin secreted by the beta islet cells. Thymic medullary epithelial cells or dendritic cells would not have this capacity, and therefore "Type B" T cells would sneak through the negative selection process in the thymus.

Why "Type A" T cells also escape is less clear. They are autoreactive to more conventionally processed insulin/MHC Class II complexes, which are clearly present on thymic medullary cells. Perhaps Type A cells expressed too low an affinity for insulin/MHC Class II peptides in the thymus, but were inspired by the high levels of insulin and inflammatory signals in a diabetic islet? Perhaps they escaped negative selection because the level of insulin peptide expression on medullary epithelium was too low in the NOD mouse strain? In fact, recent data from both mice and humans suggest that the level of insulin expression significantly influences the progression of disease and the efficiency of negative selection in the thymus.

Understandably, most current therapies for autoimmune disease focus on inhibiting the secondary but most proximal cause of autoimmune disease: the peripheral activation of autoreactive T and B cells that escaped negative selection. However, by defining the molecular reasons for self-reactive lymphocyte escape, we may ultimately find a way to develop therapeutic approaches to correct negative selection defects, too.

---

to express, process, and present proteins that are ordinarily only found in specific organs. This group discovered AIRE while studying the molecular basis for autoimmunity by examining human disease. They traced a mutation that caused autoimmune polyendocrinopathy syndrome 1 (APS1, otherwise known as APECED, see Chapter 18) to this gene, and called it AIRE for "*autoimmune regulator*." The title of the paper that announced their discovery, "Projection of an Immunological Self Shadow within the Thymus by the AIRE Protein," aptly described AIRE's powerful function.

AIRE's mechanism of action is still under investigation. It has classic features of a transcription factor and may be part of a transcriptional complex that facilitates expression of tissue-specific genes by regulating not only translation, but also chromatin packing. Thus, it allows medullary epithelial cells to express proteins not ordinarily found in the thymus, process them, and present them in MHC molecules. This is particularly useful for the presentation of MHC Class I peptides and negative selection of CD8[+] thymocytes. However, neighboring dendritic cells and macrophages are also thought to phagocytose medullary epithelial cells and so can then present their protein contents in MHC Class II molecules, and mediate negative selection of CD4[+] thymocytes (see Overview Figure 9-5).

Other Mechanisms of Central Tolerance

Other mechanisms of thymic negative selection (central tolerance) that do not necessarily involve cell death have been proposed. They include clonal arrest, where thymocytes that express autoreactive T-cell receptors are prevented from maturation; clonal anergy, where autoreactive cells are inactivated, rather than deleted; and clonal editing, where autoreactive cells are given a second or third chance to rearrange a TCRα gene. Each of these mechanisms has some experimental support, but clonal deletion is probably the most common mechanism responsible for thymic negative selection. The generation of regulatory T cells from autoreactive thymocytes can certainly be considered of importance among central tolerance mechanisms and will be discussed below.

Superantigens

Superantigens, viral or bacterial proteins that bind simultaneously to the Vβ domain of a T-cell receptor and to the α chain of a Class II MHC molecule (outside the peptide binding groove), can induce activation of all T cells that express that particular family of Vβ chains (see Figure 11-5). Superantigens are also expressed in the thymus of mice and humans and influence thymocyte maturation. Because they mimic high-affinity TCR interactions, superantigens will induce the negative selection of all DP thymocytes whose receptors possess Vβ domains targeted by the superantigen. However, because we continually generate T cells with a wide range of T-cell receptor specificities, this loss does not appear to have major clinical consequences.

The Selection Paradox: Why Don't We Delete All Cells We Positively Select?

Full understanding of the process of positive and negative selection requires an appreciation of the following paradox: If positive selection allows only thymocytes reactive with self-MHC molecules to survive, and negative selection eliminates self-MHC-reactive thymocytes, then why are any T cells allowed to mature? Other factors must operate to prevent these two MHC-dependent processes from eliminating the entire repertoire of MHC-restricted T cells.

The most straightforward model advanced to explain the paradox of MHC-dependent positive and negative selection is the **affinity hypothesis,**[*] which asserts that differences in the *strength* of TCR-mediated signals received by thymo-

cytes undergoing positive and negative selection determine the outcome of the interaction. Double-positive thymocytes that receive low-affinity signals would be positively selected, those that received high-affinity (autoreactive) signals would be negatively selected, and those that received no signal at all would die by neglect (see Figure 9-7 and Overview Figure 9-5).

The OT-I TCR transgenic mouse, developed by Kristin Hogquist and colleagues, is a superb model for the study of thymic selection using TCRs and MHC/peptides with known affinities. Not only is its peptide/MHC specificity known precisely (it binds a chicken ovalbumin peptide in the context of the H-2K[b] MHC Class I molecule), but its affinity for a range of peptides that vary in sequence has also been defined. To determine the influence of affinity on T-cell development, investigators also took advantage of two other immunological tools: (1) the fetal thymic organ culture (FTOC) system, where thymic development can be followed in vitro (Figure 9-8a), and (2) mice defective in MHC Class I antigen processing. Several different mutations that lead to defective MHC Class I presentation exist. An experimental scheme using the TAP-1 knock out (TAP-1[−]), a mutant mouse where newly generated peptides cannot access newly generated MHC Class I molecules (see Chapter 8), is depicted in Figure 9-8b. These mice cannot load their MHC Class I with endogenous peptides; they are expressed on the cell surface but are "empty" and do not hold their shape. However, they can be loaded with exogenous peptides, which stabilize their conformation. Therefore, by incubating fetal thymic organs from these mutant mice with peptides of choice, investigators were able to control the type of peptide presented by the MHC Class I.

What did investigators find? As expected, OT-I[+]/TAP-1[+] fetal thymic organ cultures produced more CD8[+] cells than wild-type FTOC. This is because virtually all of their thymocytes expressed a receptor that bound to the variety of MHC Class I (H-2K[b])/self-peptide complexes expressed by the normal stroma with intermediate affinity, generating a signal that results in positive selection (see Classic Experiment Box 9-1). Also, as expected, CD8[+] mature T cells failed to develop in OT-I[+]/TAP-1[−] thymic organ cultures because there was no normal MHC Class I and, therefore, no TCR signal—a failure of positive selection, resulting in death by neglect (see Figure 9-8b, top row). However, when a low-affinity peptide was added to the OT-I[+]/TAP-1[−] culture, MHC Class I was able to load itself with peptide and assume a normal conformation (see Figure 9-8b, middle row). Positive selection occurred, and CD8[+] T cells developed successfully. Addition of high-affinity peptides induced a strong TCR signal and negative selection, resulting in deletion of DP cells in OT-I[+]/TAP-1[−] thymic organ cultures (see Figure 9-8b, bottom row). Interestingly, low concentrations of high-affinity peptides also induced positive selection, although the cells that developed were not fully functional. These results and many others provided clear support for the affinity model, showing that *the*

---

[*]Note that a similar version of this hypothesis is also sometimes referred to as the avidity hypothesis. Avidity and affinity do have distinct meanings, which can differ for different investigators. For the sake of simplicity, we consider *affinity* to mean the strength of interaction between the TCR and its MHC/peptide ligand. For a more nuanced discussion of the differences between affinity and avidity and their influence on thymic development, see the review by Kyewski listed in the reference section of this chapter.

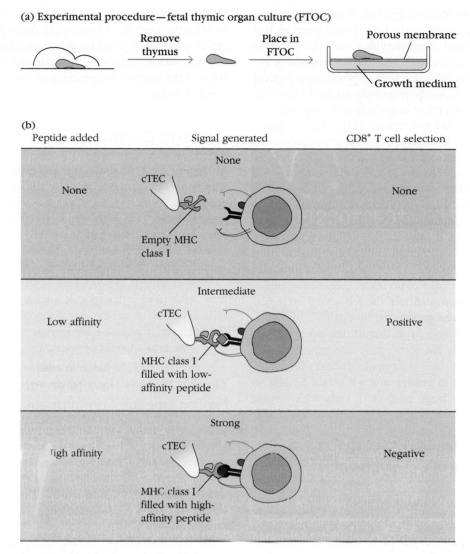

(a) Experimental procedure—fetal thymic organ culture (FTOC)

Remove thymus → Place in FTOC →

Porous membrane

Growth medium

(b)

| Peptide added | Signal generated | CD8+ T cell selection |

None — None (cTEC, Empty MHC class I) — None

Low affinity — Intermediate (cTEC, MHC class I filled with low-affinity peptide) — Positive

High affinity — Strong (cTEC, MHC class I filled with high-affinity peptide) — Negative

FIGURE 9-8 Role of TCR affinity for peptide in thymic selection. Fetal thymocyte populations have not yet undergone positive and negative selection and can be easily manipulated to study the development and selection of single-positive (CD4+CD8− and CD4−CD8+) T cells. (a) Outline of the experimental procedure for in vitro fetal thymic organ culture (FTOC). Fetal thymi are cultured on a filter disc at the interface between medium and air. Reagents added to the medium are absorbed by the thymi. In this experiment, peptide is added to the medium of FTOC from TAP-1 knockout (TAP-1−) mice, which are unable to form peptide-MHC Class I complexes unless peptide is added exogenously to the culture medium. (b) The development and selection of CD8+CD4− class I–restricted T cells depends on TCR peptide–MHC class I interactions. The mice used in this study were transgenic for OT-I TCR α and β chains, which recognize an ovalbumin peptide when associated with MHC Class I; the proportion of CD8+ T cells that develop in these thymi is higher than normal because all thymocytes in the mouse express the MHC Class I restricted specificity. Different peptide/MHC complexes interact with the TCR with different affinities. Varying the peptide added to FTOC from OT-I+/TAP-1− mice revealed that peptides that interact with low affinities (middle row) resulted in positive selection, and levels of CD4−CD8+ cells that approached that seen in the OT-I+/TAP-1+ mice. Peptides that interact with high affinities (bottom row) cause negative selection, and fewer CD4−CD8+ T cells appeared.

strength of TCR/MHC/peptide interaction does, indeed, influence cell fate.

The OT-I system has also been used to estimate the range of TCR affinities that define positive versus negative selection outcomes. It appears as if negative selection occurs at affinities that are threefold higher than those that induce positive selection.

An Alternative Model Can Explain the Thymic Selection Paradox

Philippa Marrack, John Kappler, and their colleagues thoughtfully offered an alternative possibility to explain the "thymic paradox"—that is, why we don't negatively select all cells that we positively select. These investigators advanced the **altered peptide model**, a suggestion that cortical epithelium, which induces positive selection, makes peptides that are unique and distinct from peptides made by all other cells, including the thymic cells that induce negative selection. Thus, those thymocytes selected on these unique peptide/MHC complexes would not be negatively selected when they browse the medulla and other negatively selecting cells.

Initial experiments did not support this model; however, advances in our ability to analyze peptides presented on cell surfaces in detail suggests the model has merit. Cortical epithelial cells may, indeed, process peptides differently, and are likely to present a different array of peptides to developing T cells. Recall that peptides are processed intracellularly by several mechanisms, including the activity of the proteosome (see Figure 8-16). It appears that components of the proteosome expressed by cortical epithelial cells (the "thymoproteasome") are unique. The thymic epithelium may also express unique versions of proteases (e.g., cathepsins).

The altered peptide and affinity models are by no means mutually exclusive. Investigations firmly establish that affinity plays an important role in distinguishing positive from negative selection. That the cortex may also generate novel peptides simply underscores the possibility that the thymus evolved multiple ways to ensure that we would be able to generate a sufficiently large pool of T cells with diverse specificities.

Do Positive and Negative Selection Occur at the Same Stage of Development, or in Sequence?

Strictly interpreted, the affinity model assumes that positive and negative selection operate on exactly the same target population (DP thymocytes) and in the same microenvironment of the thymus (the thymic cortex). Although it is clear that thymocytes undergo positive selection in the cortex, as we have seen above, the medulla is the site of negative selection to tissue-specific antigens. This compartmentalization of function suggests some thymocytes are positively selected and initiate a maturation program prior to negative selection. That this can happen is supported by observations showing that "semi-mature" thymocytes—those that have been positively selected and are in transition from the DP to the SP stages—are excellent targets of negative selection.

Most likely, thymocytes can be negatively selected at more than one point of development (illustrated in Overview Figure 9-5). Those thymocytes that bind with high affinity as they browse MHC/peptide complexes in the cortex are negatively selected, and those that receive positively selecting signals are given permission not only to mature, but to migrate to the cortical medullary boundary and ultimately to the medulla. In the medulla, semi-mature thymocytes can again be subject to negative selection if they interact with high affinity to tissue-specific antigens.

Migration of positively selected thymocytes to the medulla is now known to be dependent on expression of the CCR7 chemokine receptor: CCR7 deficient thymocytes fail to enter the medulla. Interestingly, they still mature and are exported to the periphery, but there they cause autoimmunity—an observation that, again, underscores the central importance of the medulla in removing autoreactive cells from the T-cell repertoire.

Lineage Commitment

As thymocytes are being screened on the basis of their TCR affinity for self-antigens, they are also being guided in their lineage decisions. Specifically, a positively selected double-positive thymocyte must decide whether to join the CD8$^+$ cytotoxic T-cell lineage or the CD4$^+$ helper T-cell lineage. Lineage commitment requires changes in genomic organization and gene expression that result in (1) silencing of one coreceptor gene (CD4 or CD8) as well as (2) expression of genes associated with a specific lineage function. Immunologists continue to debate about the cellular and molecular mechanisms responsible for lineage commitment. We will review the most current perspectives.

Several Models Have Been Proposed to Explain Lineage Commitment

When early studies with TCR transgenic mice revealed that affinity for MHC Class I versus Class II preference dictated the CD8$^+$ versus CD4$^+$ fate of developing thymocytes (see Classic Experiment Box 9-1, investigators advanced two simple testable models to explain how developing T cells "matched" their TCR preference with their coreceptor expression.

In the **instructive model**, TCR/CD4 and TCR/CD8 coengagement generates unique signals that directly initiate distinct developmental programs (Figure 9-9a). For example, if a thymocyte randomly generated a TCR with an affinity for MHC Class I, the TCR and CD8 would bind MHC Class I together, and generate a signal that specifically initiated a program that silenced CD4 expression and induced expression of genes specific for cytotoxic T-cell lineage function. Likewise, TCR/CD4 coengagement would generate a unique signal that initiated CD8 silencing and the helper T-cell developmental program.

In the **stochastic model**, a positively selected thymocyte randomly down-regulates CD4 or CD8 (Figure 9-9b). Only those cells that express the "correct" coreceptor—the ones that can coengage MHC with the TCR—generate a TCR signal strong enough to survive to mature. In this model, TCR/CD4 and TCR/CD8 coengagement does not necessarily generate distinct signals. Unfortunately, studies that followed the consequences of such mismatches confounded researchers by providing evidence in support of both models! Clearly they were too simplistic.

Further experiments challenged some of the assumptions on which these early efforts were based and indicated that

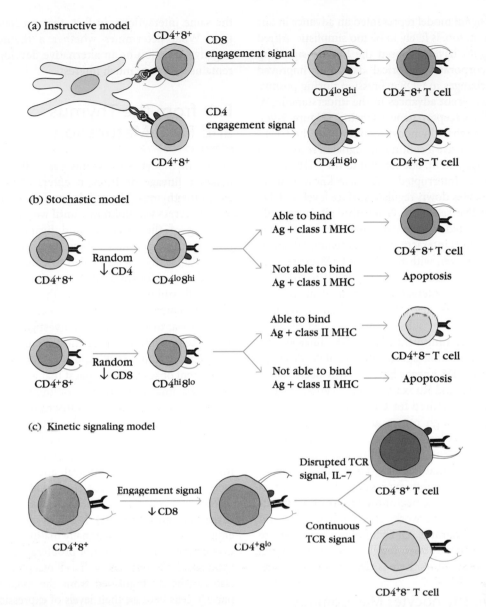

FIGURE 9-9 Proposed models of lineage commitment, the decision of double-positive thymocytes to become helper CD4$^+$ or cytotoxic CD8$^+$ T cells. (a) According to the instructive model, interaction of a coreceptor with the MHC molecule for which it is specific results in down-regulation of the other coreceptor. (b) According to the stochastic model, down-regulation of CD4 or CD8 is a random process. (c) According to the kinetic signaling model, the decision to commit to the CD4 or CD8 lineage is based on the continuity of the TCR signal that a thymocyte receives. Positive selection results in down-regulation of CD8 on all thymocytes. This will not alter the intensity of a TCR/CD4/MHC Class II signal, and cells receiving this signal will continue development to the CD4 SP lineage. However, down-regulation of CD8 diminishes (interrupts) a TCR/CD8/MHC Class I signal, an experience that sends a cell toward the CD8 lineage. IL-7 signals are required to "seal" CD8 lineage commitment.

the strength and duration of the T-cell receptor/coreceptor signal experienced by a thymocyte play a more important role in dictating cell fate than its specificity for MHC Class I or Class II. In fact, thymocytes whose TCR had a known preference of MHC Class II could be coaxed into the CD8$^+$ T-cell fate simply by weakening the CD4/Class II interaction (e.g., by mutating kinases downstream of TCR signaling so that they signaled more or less effectively). By inhibiting TCR signaling, investigators could coax cells that normally commit to the CD4 lineage to become CD8$^+$ cells. By enhancing TCR signaling, investigators could coax MHC Class I restricted cells to become CD4$^+$. These data suggested that stronger positive selecting TCR signals resulted in CD4 lineage commitment, and weaker positive selecting TCR signals resulted in CD8 lineage commitment. They were consistent with the observation that the intracellular tail of CD4 interacts more effectively with the tyrosine kinase lck than the intracellular tail of CD8. Therefore, a TCR/CD4 coengagement is likely to generate stronger signals than TCR/CD8 and result in CD4$^+$ T-cell commitment.

The *strength of signal* model represented an advance in our understanding, but it, too, is likely to be too simplistic. Alfred Singer and colleagues have proposed the **kinetic signaling model**, which incorporates historical data, our improved understanding of changes in CD8 expression during positive selection, as well as recent advances in the understanding of the complexity of coreceptor gene silencing (Figure 9-9c). Briefly, they propose that thymocytes commit to the CD4$^+$ T-cell lineage if they receive a continuous signal in response to TCR/coreceptor engagement, but commit to the CD8 lineage if the TCR signal is interrupted. We now know that all CD4$^+$CD8$^+$ thymocytes down-regulate surface levels of CD8 in response to positive selection. Given this response, only MHC Class II restricted T cells will maintain continuous TCR/CD4/MHC Class II interaction, and therefore develop to the CD4 lineage. However, with the loss of CD8 expression, MHC Class I restricted T cells will lose the ability to maintain TCR/CD8/MHC Class I interactions. Singer et al. provide evidence that these thymocytes, interrupted from their TCR/coreceptor engagement, are subsequently rescued by IL-7, which facilitates their commitment to the CD8$^+$ lineage.

Transcription factors that specifically regulate development to the CD4 and CD8 lineage have recently been identified. At present, ThPok and Runx3 are taking center stage as transcriptional factors required for CD4 and CD8 commitment, respectively, although others also play a role. Both ThPok and Runx3 act, at least in part, by suppressing genes involved in differentiation to the other lineage. ThPok inhibits expression of genes that regulate CD8 differentiation, including Runx3. Runx3 inhibits expression of CD4, itself, as well as ThPok. This is just the beginning of our understanding of the transcriptional networks that regulate CD4 and CD8 developmental programs. Likely to play a role, too, are miRNAs; some have already been implicated in CD8 single-positive T-cell differentiation. There is clearly more to come.

Double-Positive Thymocytes May Commit to Other Types of Lymphocytes

Small populations of DP thymocytes can also commit to other T-cell types, including the NK T cell, regulatory T cell, and intraepithelial lymphocyte (IEL) lineages. **NKT cells** (which include mature cells that express only CD4 and cells that have lost both CD4 and CD8) play a role in innate immunity (see Chapter 5) and express a TCR that includes an invariant TCRα chain (Vα14). For this reason, NKT cells are sometimes referred to as iNKT cells for *i*nvariant. Their invariant TCR interacts not with classical MHC, but with the related molecule CD1, which presents glycolipids, not peptides (see Chapter 8). **Intraepithelial lymphocytes** (IEL), most of which are CD8$^+$, also have features of innate immune cells and patrol mucosal surfaces. *Regulatory T cells* (T_{REG}), another CD4$^+$ subset discussed in more detail below, quench adaptive immune reactions. All three subpopulations are thought to develop from DP thymocytes in response to autoreactive, high-affinity TCR interactions—

the same interactions, in fact, that mediate negative selection. What determines whether a thymocyte undergoes negative selection or an alternative developmental pathway remains a topic of much interest.

Exit from the Thymus and Final Maturation

Once a thymocyte successfully passes through selection and makes a lineage decision, it enters a quiescent state and leaves the thymus. The cellular and molecular basis for thymocyte egress was unknown until we gained a better understanding of the receptors and ligands that regulate cell migration. As mentioned above, one set of investigators discovered that thymocytes failed to enter the medulla if they were deficient for the CCR7 chemokine receptor. However, these cells still were able to leave the thymus, indicating that other receptors control thymic exit. The identity of this receptor was discovered when investigators found that few if any T cells made it out of the thymus in a *sphingosine-1-phosphate receptor (SIPR)* knockout mouse.

Current observations suggest that a cascade of events controls these final stages of maturation: positive selecting TCR signals up-regulate Foxo1, a transcription factor that controls expression of several genes related to T-cell function. Foxo1 regulates expression of Klf2, which, in turn, up-regulates SIPR. Foxo1 also up-regulates both IL-7R, which helps to maintain mature T-cell survival, and CCR7, the chemokine receptor that directs mature T-cell traffic to the lymph nodes.

Mature T cells that exit the thymus are referred to as **recent thymic emigrants (RTEs)**. It is now clear that recent thymic emigrants are not as functionally mature as most naïve T cells in the periphery: they do not proliferate or secrete cytokines as vigorously in response to T-cell receptor stimulation. RTEs also can be distinguished from the majority of peripheral naïve T cells because their levels of expression of several surface proteins (e.g., the maturation markers HSA and Qa-2, as well as the IL-7 receptor) are more similar to their immature thymocyte ancestors than their fully mature T-cell descendants. Investigators are particularly interested in RTEs because they are an important source of mature T cells in individuals who are *lymphopenic* (i.e., have a reduced pool of functional lymphocytes), including people who have undergone chemotherapy, newborns, and the aged. Although research is ongoing, studies suggest that the final maturation of these cells is influenced by their interactions with both MHC and non-MHC ligands in secondary lymphoid organs.

Other Mechanisms That Maintain Self-Tolerance

As we have seen, negative selection of thymocytes can rid a developing T-cell repertoire of cells that express a high affinity for both ubiquitous self-antigens and, thanks to the activity of AIRE, tissue-specific antigens. However, negative selection in the thymus is not perfect. Autoreactive T cells do escape, either because they have too low an affinity for self to induce clonal deletion, or they happen not to have browsed the "right" tissue-specific antigen/MHC combination. The body has evolved several other mechanisms to avoid autoimmunity, including what has become a major focus of interest for immunologists: the development in both the thymus and the periphery of a fascinating group of cells now known as regulatory T cells.

T_{REG} Cells Negatively Regulate Immune Responses

Regulatory T cells (T_{REG} cells) can inhibit the proliferation of other T-cell populations in vitro, effectively suppressing autoreactive immune responses. They express on their surface CD4 as well as CD25, the α chain of the IL-2 receptor. However, T_{REG} cells are more definitively identified by their expression of a master transcriptional regulator, FoxP3, the expression of which is necessary and sufficient to induce differentiation to the T_{REG} lineage.

T_{REG} cells can develop in the thymus and appear to represent an alternative fate for autoreactive T cells. As we have seen, most thymocytes that express receptors with high affinity for self-antigen die via negative selection. However, a small fraction appear to commit to the regulatory T-cell lineage and leave the thymus to patrol the body and thwart autoimmune reactions. What determines whether a self-reactive thymocyte dies or differentiates into a T_{REG} cell is still an open question. Investigators are currently trying to understand if the choice is made based on subtle differences in affinity for self or on differences in their maturation state when they receive a high-affinity signal. Recent work suggests that T_{REG}s develop in a unique microenvironmental niche within the thymus, and that the available space for developing cells in this niche is limited. These findings suggest that thymocytes that commit to the regulatory T-cell lineage are likely to receive unique stimulatory signals. These *natural* T_{REG}s share the periphery with *induced* T_{REG}s that can develop from conventional mature T cells that are exposed to TGF-β and IL-10 cytokines (see Chapter 11).

Animal studies show that members of the FoxP3$^+$ T_{REG} population inhibit the development of autoimmune diseases such as experimentally induced inflammatory bowel disease, experimental allergic encephalitis, and autoimmune diabetes. Suppression by these regulatory cells is antigen specific because it depends on activation through the T-cell receptor. Exactly how T_{REG}s quench responses is still debated, although they probably do so via a variety of means: directly inhibiting an antigen-presenting cell's ability to activate T cells, directly killing T cells, indirectly inhibiting T-cell activity by secreting inhibitory cytokines IL-10 and TGF-β, and/or depleting the local environment of stimulatory cytokines such as IL-2 (Figure 9-10).

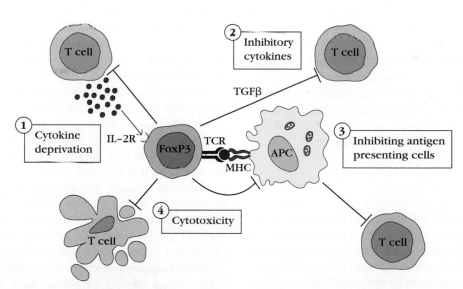

FIGURE 9-10 How regulatory T cells inactivate traditional T cells. Some possible mechanisms of T_{REG} activity are illustrated in this schematic. These may all contribute to quelling immune responses in vivo. (1) *Cytokine deprivation.* T_{REG}s express relatively high levels of high-affinity IL-2 receptors and can compete for the cytokines that activated T cells need to survive and proliferate. (2) *Cytokine inhibition.* T_{REG}s secrete several cytokines, including IL-10 and TGF-β, which bind receptors on activated T cells and reduce signaling activity. (3) *Inhibition of antigen-presenting cells.* T_{REG}s can interact directly with MHC Class II expressing antigen-presenting cells and inhibit their maturation, leaving them less able to activate T cells. (4) *Cytotoxicity.* T_{REG}s can also display cytotoxic function and kill cells by secreting perforin and granzyme.

The existence of regulatory T cells that specifically suppress immune responses has clinical implications. Depletion or inhibition of T_{REG} cells before immunization may enhance immune responses to conventional vaccines. Elimination of T cells that suppress responses to tumor antigens may also facilitate the development of antitumor immunity. Conversely, increasing the suppressive activity of regulatory T-cell populations could be useful in the treatment of allergic or autoimmune diseases. The ability to increase the activity of regulatory T-cell populations might also be useful in suppressing organ and tissue rejection. Investigators doing future work on this regulatory cell population will seek deeper insights into the mechanisms by which members of FoxP3^{+} T-cell populations regulate immune responses. They will also make determined efforts to discover ways the activities of these populations can be increased to diminish unwanted immune responses and decreased to promote desirable ones.

Peripheral Mechanisms of Tolerance also Protect Against Autoreactive Thymocytes

The body has evolved several other mechanisms to manage the autoreactive escapee in the periphery. Briefly, many antigens are "hidden" from autoreactive T cells because only a subset of cells (professional APCs) express the right costimulatory molecules needed to initiate the immune response. Autoreactive naïve T cells that can see an MHC/self-peptide combination on a nonprofessional antigen-presenting cell will not receive the correct costimulatory signals and will not divide or differentiate. For example, if a thymocyte specific for a peptide made by a kidney cell escaped from the thymus, it will not be activated unless that peptide were presented on a professional antigen-presenting cell. Kidney cells do not express the costimulatory ligands required for activating a CD4^{+} or a CD8^{+} T cell. In fact, a high-affinity interaction with MHC/peptide combinations on the surface of kidney cell, in the absence of costimulatory ligands, could result in T-cell anergy—another peripheral tolerance mechanism that is described in more detail in Chapter 11. The clinical consequences of failures of central and peripheral tolerance will be discussed in Chapter 16.

Apoptosis

As we have seen here, and will also see in Chapter 10, apoptosis features prominently during T- and B-lymphocyte development. T-cell development is a particularly expensive process for the host. An estimated 95% to 98% of all thymocytes do not mature—most die by apoptosis within the thymus either because they fail to make a productive TCR gene rearrangement or because they fail to interact with self-MHC. Some (2%–5%) die, also via apoptosis, because they are negatively selected.

Not only does apoptosis shape the developing lymphocyte repertoire; it also regulates immune cell homeostasis by returning myeloid, T-, and B-cell populations to their appro-priate levels after bursts of infection-inspired proliferation, and it is the means by which a cell dies after being targeted by cytotoxic T cells or NK cells. Understanding its fundamental biological features is an important part of an overall understanding of immune cell function. Because apoptosis is such a critical part of lymphocyte development, we will take the opportunity in this chapter to briefly describe the apoptotic process and how it is regulated.

Apoptosis Allows Cells to Die without Triggering an Inflammatory Response

Apoptosis, or **programmed cell death**, is an energy-dependent process by which a cell brings about its own demise. It is most often contrasted with **necrosis**, a form of cell death arising from injury or toxicity (Figure 9-11a). In necrosis, injured cells swell and burst, releasing their contents and possibly triggering a damaging inflammatory response. Apoptotic cells, in contrast, dismantle their contents without disrupting their membranes and induce neighboring cells to engulf them before they can release any inflammatory material. Morphologic changes associated with apoptosis include a pronounced decrease in cell volume; modification of the cytoskeleton, which results in membrane blebbing; a condensation of the chromatin; and degradation of the DNA into fragments (Figure 9-11b).

Different Stimuli Initiate Apoptosis, but All Activate Caspases

Although a lymphocyte can be signaled to die in several ways, all apoptotic signals ultimately activate a class of proteases called **caspases**. Caspase activation is common to almost all death pathways known in both vertebrates and invertebrates, demonstrating that apoptosis is an ancient process. Many of its molecular participants have been conserved through evolution (Table 9-2).

The role of caspases was first revealed by studies of developmentally programmed cell death in the nematode *Caenorhabditis elegans*, where the death of cells was shown to be totally dependent on the activity of a gene that encoded a cysteine protease with specificity for aspartic acid residues. We now know that mammals have at least 14 cysteine-*asp*artic prote*ases*, or *caspases*, and all apoptotic cell deaths require the activity of at least a subset of these molecules.

How do caspases trigger cell death? Apoptotic pathways typically involve two classes of caspases, both of which are maintained in an inactive state until apoptosis is initiated. **Initiator caspases** are activated by a death stimulus and, in turn, activate **effector caspases**, which execute the death program by (1) cleaving critical targets necessary for cell survival (e.g., cytoskeletal proteins) as well as (2) activating other enzymes that carry out the dismantling of the cell. Ultimately, the catalytic cascade initiated by caspases induces death via an orderly disassembly of intracellular molecules. It also results in the packaging of cell contents into vesicles that are ultimately engulfed.

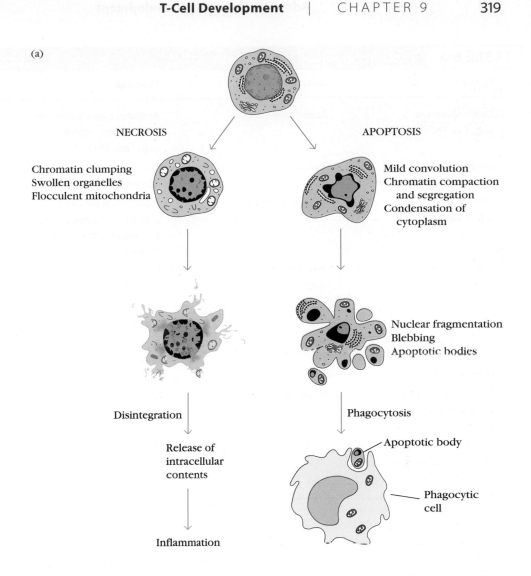

(a)

NECROSIS

APOPTOSIS

Chromatin clumping
Swollen organelles
Flocculent mitochondria

Mild convolution
Chromatin compaction
and segregation
Condensation of
cytoplasm

Nuclear fragmentation
Blebbing
Apoptotic bodies

Disintegration

Phagocytosis

Release of
intracellular
contents

Apoptotic body

Phagocytic
cell

Inflammation

(b)   Normal   Apoptotic

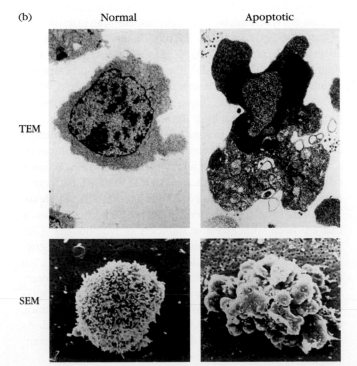

TEM

SEM

FIGURE 9-11 Morphological changes that occur during apoptosis. (a) Comparison of morphologic changes that occur in apoptosis versus necrosis. Apoptosis, which results in the programmed cell death of hematopoietic cells, does not induce a local inflammatory response. In contrast, necrosis, the process that leads to death of injured cells, results in release of the cells' contents, which may induce a local inflammatory response. (b) Microscopic images of apoptotic cells. Transmission and scanning electron micrographs (TEM and SEM) of normal and apoptotic thymocytes, as indicated. *[Part (b) from B. A. Osborne and S. Smith, 1997, Journal of NIH Research **9**:35; courtesy of B. A. Osborne, University of Massachusetts at Amherst.]*

TABLE 9-2	Proteins involved in apoptosis			
Protein	**Location in cell**	**Function**	**Role in apoptosis**
Death receptors (e.g., Fas, TNFR)	Membrane	Activates caspase cascade after binding ligand (e.g., FasL, TNF)	Promotes
Initiator caspase (e.g., caspase-8, caspase-9)	Cytosol	Protease; cleaves and activates effector caspases	Promotes
Effector caspase (e.g., caspase-3)	Cytosol, nucleus	Protease; cleaves and activates enzymes, cleaves and disassembles structural proteins	Promotes
Cytochrome c	Intermembrane space, mitochondria	Participates in formation of apoptosome	Promotes
Apaf-1	Cytosol	Participates in formation of apoptosome	Promotes
Anti-apoptotic Bcl-2 family members (e.g., Bcl-2, Bcl-x_L)	Mitochondria, ER	Regulates cytochrome c release	Inhibits
Pro-apoptotic Bcl-2 family members (e.g., Bax, Bak)	Mitochondria	Regulates cytochrome c release; opposes Bcl-2, Bcl-x_L	Promotes
BH3 proteins (e.g., Bim, Bid, PUMA)	Cytosol and mitochondria	Opposes activity of anti-apoptotic Bcl-2 family members at mitochondria	Promotes

Stimuli that lead to apoptosis can be divided into two categories based on how they activate the effector caspases. Signals such as radiation, stress, and growth factor withdrawal induce caspase activation and apoptosis via the *intrinsic pathway* that is mediated by mitochondrial molecules, while membrane-bound or soluble ligands that bind to membrane receptors stimulate caspase activation and apoptosis via the *extrinsic pathway*.

Apoptosis of Peripheral T Cells Is Mediated by the Extrinsic (Fas) Pathway

Outside of the thymus, most of the TCR-mediated apoptosis of mature T cells is induced via the extrinsic pathway through membrane-associated *death receptors*, including Fas (CD95). Repeated or persistent stimulation of peripheral T cells via their T-cell receptors results in the co-expression of both Fas and Fas ligand (FasL) on T cells, ultimately leading to Fas/FasL-mediated death. Apoptosis induced by antigen receptor activation is called *activation-induced cell death* (*AICD*) and is a major homeostatic mechanism, helping to reduce the pool of activated T cells after antigen is cleared and helping to remove stray autoreactive T cells that are stimulated by self-antigens.

The signaling cascade initiated by Fas (and by all death receptors) leads directly to the activation of the initiator caspase, caspase-8. When FasL binds Fas, procaspase-8, an inactive form of caspase-8, is recruited to the intracellular tail of Fas by an adaptor molecule called FADD (*Fas-a*ssociated protein with *death d*omain) (Figure 9-12a). The aggregation of procaspase-8 results in its cleavage and conversion to active caspase-8, which will then activate effector caspase-3, which in turn initiates the proteolytic cascade that leads to the death of the cell.

The importance of Fas and FasL in the removal of mature activated T cells is underscored by abnormalities in *lpr/lpr* mice, a naturally occurring loss of function mutation in Fas. When T cells become activated in these mice, the Fas/FasL pathway is not operative, and stimulated T cells continue to proliferate. These mice spontaneously develop strikingly large lymph nodes that are filled with excessive numbers of lymphocytes. Ultimately, they develop autoimmune disease, clearly demonstrating the consequences of a failure to delete activated T cells. An additional spontaneous mutant, *gld/gld*, has the complementary loss of function. These mice lack functional FasL and display abnormalities very similar to those found in the *lpr/lpr* mice. Recently, humans with defects in Fas have been reported. As expected, these individuals display characteristics of autoimmune disease (see Clinical Focus Box 9-3).

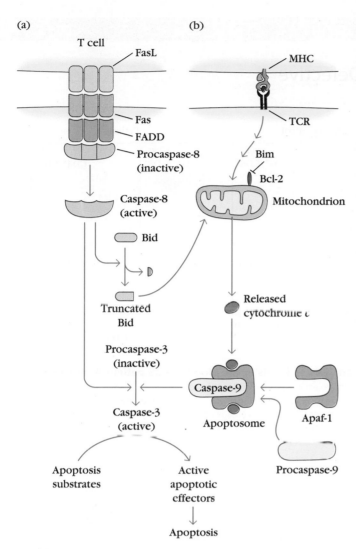

FIGURE 9-12 Two pathways to apoptosis in T cells.
(a) Activated peripheral T cells are induced to express high levels of Fas and FasL and stimulate the extrinsic apoptotic pathway. FasL induces the trimerization of Fas on a neighboring cell. FasL can also engage Fas on the same cell, resulting in a self-induced death signal. Trimerization of Fas leads to the recruitment of FADD, which leads in turn to the cleavage of associated molecules of procaspase-8 to form active caspase-8. Caspase-8 cleaves procaspase-3, producing active caspase-3, which results in the death of the cell. Caspase-8 can also cleave the Bcl-2 family member Bid to a truncated form that can activate the mitochondrial death pathway. (b) Other signals, such as the engagement of the TCR by peptide-MHC complexes on an APC during T-cell development, result in the activation of the intrinsic, mitochondrial death pathway. A key feature of this pathway is the release of cytochrome *c* from the inner mitochondrial membrane into the cytosol, an event that is regulated by Bcl-2 family members. Cytochrome *c* interacts with Apaf-1 and subsequently with procaspase-9 to form the active apoptosome. The apoptosome initiates the cleavage of procaspase-3, producing active caspase-3, which initiates the execution phase of apoptosis by proteolysis of substances whose cleavage commits the cell to apoptosis. [*Adapted in part from S. H. Kaufmann and M. O. Hengartner, 2001,* Trends in Cell Biology *11:526.*]

Fas and FasL are members of a family of related receptor/ligands including tumor necrosis factor (TNF) and its ligand, TNFR (*tumor necrosis factor receptor*), which can also induce apoptosis via the activation of caspase-8 followed by the activation of effector caspases such as caspase-3 (see Figure 4-14).

TCR-Mediated Negative Selection in the Thymus Induces the Intrinsic (Mitochondria-Mediated) Apoptotic Pathway

Fas/FasL interactions do not appear to play a central role in negative selection in the thymus. Instead, TCR-mediated negative-selecting signals in the thymus induce a route to apoptosis in which mitochondria play a central role (Figure 9-12b). In mitochondria-dependent (intrinsic) apoptotic pathways, cytochrome *c*, which normally resides between the inner and outer mitochondrial membranes, leaks into the cytosol. The release of cytochrome *c* is regulated by several different protein families. In thymocytes, Bim, a Bcl-2 family member, and Nur77, an orphan nuclear receptor, both play a role.

In the cytosol, cytochrome *c* binds to a protein known as Apaf-1 (*apoptotic protease-activating factor 1*), which then oligomerizes. Binding of oligomeric Apaf-1 to procaspase-9 results in its transformation to active caspase-9, an initiator caspase. The cytochrome *c*/Apaf-1/caspase-9 complex, called the **apoptosome**, proteolytically cleaves procaspase-3, generating active caspase-3, which initiates the cascade of reactions that kills the cell.

Interestingly, although over 95% of thymocytes die during development, apoptotic thymocytes are very difficult to find in a normal thymus. Investigators showed that if you inhibit the activity of thymic macrophages, apoptotic thymocytes are abundantly evident. This experiment dramatically underscored both the importance and the efficiency of phagocytes in clearing dying cells from the thymus.

Cell death induced by Fas/FasL is swift, with rapid activation of the caspase cascade leading to cell death in 2 to 4 hours. On the other hand, TCR-induced negative selection appears to involve a more circuitous pathway, requiring the activation of several processes, including mitochondrial membrane failure, the release of cytochrome *c*, and the formation of the apoptosome before caspases become involved. Consequently, TCR-mediated negative selection can take as long as 8 to 10 hours.

Bcl-2 Family Members Can Inhibit or Induce Apoptosis

The *Bcl-2* (*B-cell lymphoma 2*) family of genes, which regulate cytochrome *c* release, plays a prominent role in immune cell physiology. They include both anti-apoptotic and pro-apoptotic proteins that can insert into the mitochondrial membrane. Although their precise mechanism of action is still

CLINICAL FOCUS

Failure of Apoptosis Causes Defective Lymphocyte Homeostasis

The maintenance of appropriate numbers of various types of lymphocytes is extremely important to an effective immune system. One of the most important elements in this regulation is apoptosis mediated by the Fas/FasL system. The following excerpts from medical histories show what can happen when this key regulatory mechanism fails.

Patient A: A woman, now 43, has had a long history of immunologic imbalances and other medical problems. By age 2, she was diagnosed with Canale-Smith syndrome (CSS), a severe enlargement of such lymphoid tissues as lymph nodes (lymphadenopathy) and spleen (splenomegaly). Biopsy of lymph nodes showed that, in common with many other CSS patients, she had greatly increased numbers of lymphocytes. She had reduced numbers of platelets (thrombocytopenia) and, because her red blood cells were being lysed, she was anemic (hemolytic anemia). The reduction in numbers of platelets and the lysis of red blood cells could be traced to the action of circulating antibodies that reacted with these host components. At age 21, she was diagnosed with grossly enlarged pelvic lymph nodes that had to be removed. Ten years later, she was again found to have an enlarged abdominal mass, which on surgical removal turned out to be a half-pound lymph node aggregate. She has continued to have mild lymphadenopathy and, typical of these patients, the lymphocyte populations of enlarged nodes had elevated numbers of T cells (87% as opposed to a

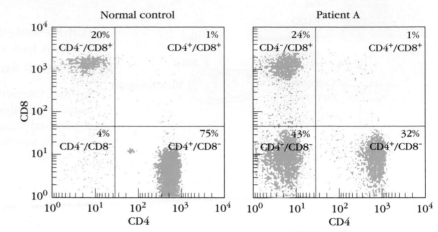

FIGURE 1

Flow-cytometric analysis of T cells in the blood of CSS patient A and a control subject. Mature T cells are either CD4+ or CD8+. Although almost all of the T cells in the control subject are CD4+ or CD8+, the CSS patient shows high numbers of double-negative T cells (43%), which express neither CD4 nor CD8. The percentage of each category of T cells is indicated in the quadrants. [*Adapted from M. D. Drappa et al., 1996,* New England Journal of Medicine **335**:1643.]

normal range of 48%–67% T cells). Examination of these cells by flow cytometry and fluorescent antibody staining revealed an excess of double-negative T cells (Figure 1). Also, like many patients with CSS, she has had cancer: breast cancer at age 22 and skin cancer at ages 22 and 41.

Patient B: A man who was eventually diagnosed with CSS had severe lymphadenopathy and splenomegaly as an infant and child. He was treated from age 4 to age 12 with corticosteroids and the immunosuppressive drug mercaptopurine. These appeared to help, and the swelling of lymphoid tissues became milder during adolescence and adulthood. At age 42, he died of liver cancer.

Patient C: An 8-year-old boy, the son of patient B, was also afflicted with CSS and showed elevated T-cell counts and severe lymphadenopathy at the age of 7 months. At age 2 his spleen became so enlarged that it had to be removed. He also developed hemolytic anemia and thrombocytopenia. However, although he continued to have elevated T-cell counts, the severity of his hemolytic anemia and thrombocytopenia have so far been controlled by treatment with methotrexate, a DNA-synthesis-inhibiting drug used for immunosuppression and cancer chemotherapy.

Recognition of the serious consequences of a failure to regulate the number of lymphocytes, as exemplified by

controversial, the balance of pro-apoptotic and anti-apoptotic family members influences a cell's response to stress. The Bcl-2 family also includes an important, third group of proteins, the *BH3 family*, all of which do not insert in the mitochondrial membrane, but instead inhibit the anti-apoptotic Bcl-2 family members, and so are ultimately pro-apoptotic.

Intrinsic apoptotic stimuli act, in part, by activating BH3 family members, which are often sequestered from the mito-

chondria in a healthy cell. For example, TCR-mediated negative selection activates Bim, a BH3 family protein that facilitates cytochrome *c* release (see Figure 9-12b). However, even the extrinsic pathway can activate BH3 family members. For instance, caspase-8 generated by the Fas pathway cleaves the BH3 family member Bid, releasing it from sequestration and allowing it to activate the mitochondrial death pathway.

BOX 9-3

these case histories, emerged from detailed study of several children whose enlarged lymphoid tissues attracted medical attention. In each of these cases of CSS, examination revealed grossly enlarged lymph nodes that were 1 cm to 2 cm in girth and sometimes large enough to distort the local anatomy. In four of a group of five children who were studied intensively, the spleens were so massive that they had to be removed.

Even though the clinical picture in CSS can vary from person to person, with some individuals suffering severe chronic affliction and others only sporadic episodes of illness, there is a common feature: a failure of activated lymphocytes to undergo Fas-mediated apoptosis. Isolation and sequencing of Fas genes from a number of patients and more than a hundred controls reveals that CSS patients are heterozygous ($fas^{+/-}$) at the *fas* locus and thus carry one copy of a defective *fas* gene. A comparison of Fas-mediated cell death in T cells from normal controls who do not carry mutant Fas genes with death induced in T cells from CSS patients shows a marked defect in Fas-induced death (Figure 2). Characterization of the Fas genes so far seen in CSS patients reveals that they have mutations in or around the region encoding the death-inducing domain (the "death domain") of this protein. Such mutations result in the production of Fas protein that lacks biological activity but still competes with normal Fas molecules for interactions with essential components of the Fas-mediated death pathway. Other mutations have been found in the extracellular domain of Fas, often associated with milder forms of CSS or no disease at all.

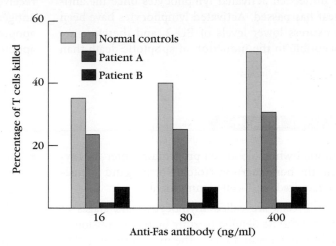

FIGURE 2

Fas-mediated killing takes place when Fas is cross-linked by FasL, its normal ligand, or by treatment with anti-Fas antibody, which artificially cross-links Fas molecules. This experiment shows the percentage of dead T cells after induction of apoptosis in T cells from patients and controls by cross-linking Fas with increasing amounts of an anti-Fas monoclonal antibody. T cells from the Canale-Smith patients (A and B) are resistant to Fas-mediated death. *[Adapted from M. D. Drappa et al., 1996, New England Journal of Medicine* **335:**1643.]

A number of research groups have conducted detailed clinical studies of CSS patients, and the following general observations have been made:

- The cell populations of the blood and lymphoid tissues of CSS patients show dramatic elevations (5-fold to as much as 20-fold) in the numbers of lymphocytes of all sorts, including T cells, B cells, and NK cells.

- Most of the patients have elevated levels of one or more classes of immunoglobulin (hyper-gammaglobulinemia).

- Immune hyperactivity is responsible for such autoimmune phenomena as the production of autoantibodies against red blood cells, resulting in hemolytic anemia, and a depression in platelet counts due to the activity of antiplatelet autoantibodies.

These observations establish the importance of the death-mediated regulation of lymphocyte populations in lymphocyte homeostasis. Such cell death is necessary because the immune response to antigen results in a sudden and dramatic increase in the populations of responding clones of lymphocytes and temporarily distorts the representation of these clones in the repertoire. In the absence of cell death, the periodic stimulation of lymphocytes that occurs in the normal course of life would result in progressively increasing, and ultimately unsustainable, lymphocyte levels. As CSS demonstrates, without the essential culling of lymphocytes by apoptosis, severe and life-threatening disease can result.

The signature member of this gene family, *bcl-2*, was discovered near the breakpoint of a chromosomal translocation in a cancer known as human B-cell lymphoma. The translocation moved the *bcl-2* gene into the immunoglobulin heavy-chain locus, resulting in transcription of *bcl-2* along with the immunoglobulin gene, with consequent overproduction of the encoded Bcl-2 protein by the lymphoma cells. The resulting high levels of Bcl-2 resulted in the accumula-

tion of long-lived B cells and contributed to their transformation into cancerous lymphoma cells.

Bcl-2 levels play an important role in regulating the normal life span of various hematopoietic cell lineages, including lymphocytes. A normal adult has about 5 liters of blood, with about 2000 lymphocytes/mm³, for a total of about 10^{11} circulating lymphocytes. During acute infection, the lymphocyte count increases fourfold or more. The immune system cannot sustain such a massive increase in cell numbers for an extended period, so the system must eliminate unneeded activated lymphocytes once the antigenic threat has passed. Activated lymphocytes have been found to express lower levels of Bcl-2, and therefore are more susceptible to the induction of apoptotic death than

are naïve lymphocytes or memory cells. However, if the lymphocytes continue to be activated by antigen, signals received during activation block the apoptotic signal. As antigen levels subside, the levels of the signals that block apoptosis diminish, and the lymphocytes begin to die by apoptosis.

S U M M A R Y

- Uncommitted white blood cell progenitors enter the thymus from the bone marrow. Notch/Notch ligand interactions are required for T-cell commitment.

- Immature T cells are called thymocytes, and stages of development can be defined broadly by the expression of the coreceptors CD4 and CD8. The most immature thymocytes express neither coreceptor and are referred to as CD4$^-$CD8$^-$ or double-negative (DN) cells. These progress to the CD4$^+$CD8$^+$ (double positive or DP) stage, which in turn mature to the CD4$^+$CD8$^-$ or CD4$^-$CD8$^+$ (single positive or SP) stages.

- DN thymocytes progress through four stages of development (DN1-DN4) defined by CD44, CD25, and c-Kit expression. During these stages they proliferate and rearrange the TCRβ, δ, and γ antigen receptor genes.

- Thymocytes that rearrange δ and γ receptor genes successfully mature to the TCRγδ lineage.

- Those thymocytes that successfully rearrange the β receptor chain are detected by a process called β-selection. β-selection is initiated by assembly of the TCRβ protein with a surrogate, invariant TCRα chain and CD3 complex proteins. Assembly of this pre-TCR results in commitment to the TCRαβ lineage, another burst of proliferation, maturation to the DP stage of development, and initiation of TCRα rearrangement.

- DP thymocytes are the most abundant subpopulation in the thymus and are the first cells to express a mature TCRαβ receptor. They are the targets of positive and negative selection, which are responsible for self-tolerance and self-MHC restriction, respectively.

- Positive and negative selection are regulated by the affinity of a DP thymocyte's TCR for MHC/self-peptide complexes expressed by the thymic epithelium.

- High-affinity TCR/MHC/peptide interactions result in negative selection, typically by initiating apoptosis (clonal deletion).

- Low/intermediate-affinity signals result in positive selection and initiate a maturation program to the helper CD4$^+$ or cytotoxic CD8$^+$ single-positive lineages.

- The large majority of thymocytes (95%) do not interact with any MHC/self-peptides expressed by the thymic epithelium and die by neglect.

- The distinction among TCR signaling cascades that activate positive versus negative selection is not fully understood but may involve differences in MAPK and Ca^{2+} signaling.

- Positively selected DP thymocytes have to decide whether to become helper CD4$^+$ or cytotoxic CD8$^+$ T cells. This process, called lineage commitment, appears to depend on continuity of the signals positively selected DP thymocytes receive through their TCR.

- Thymocytes whose TCR preferentially interacts with MHC Class II generate a continuous signal that initiates a CD4$^+$ helper T-cell developmental program.

- Thymocytes whose TCR preferentially interacts with MHC Class I cannot generate a continuous signal because CD8 surface levels are down-regulated in response to positive selection. The interruption in signaling, followed by further stimulation, initiates a CD8 developmental program.

- Transcription factors Th-Pok and Runx3 play important roles in CD4$^+$ helper cell and CD8$^+$ cytotoxic cell commitment. They work in part by reciprocally regulating each other.

- Positively selected thymocytes migrate from the thymic cortex to the thymic medulla via interactions with the CCR7 chemokine receptor. These cells, in transition from the DP to the SP stage of development, are negatively selected for tissue-specific antigens in the medulla. This is the second opportunity to remove autoreactive T cells from the developing repertoire.

- The thymic cortex and thymic medulla carry out distinct functions in the thymus. Medullary epithelial cells, but not cortical epithelial cells, express the transcription factor AIRE, which is responsible for their unique capacity to express tissue-specific antigens.

- Fully mature thymocytes exit the thymus via interactions via the sphingosine-1-phosphate receptor (S1PR) and undergo final functional maturation in peripheral lymphoid tissues.

- A small percentage of cells also develop within the thymus to other cell lineages, including NK T cells, IELs, and regulatory T cells.

- Regulatory T cells that develop in the thymus are called natural T$_{REG}$s. They share the periphery with regulatory T cells that develop from conventional T cells (induced T$_{REG}$s) and play an important role in inhibiting autoimmune responses.

■ Apoptosis, a process of cell death that is internally initiated and highly regulated, has a major influence on the shape of the T-cell repertoire. Thymocytes that do not pass β-selection (and do not successfully express a TCRγδ complex) die by apoptosis; 95% of developing DP thymocytes also die by apoptosis, the vast majority because they do not receive sufficient TCR signaling (death by neglect) and a smaller population because they receive too high a level of TCR signaling (negative selection).

■ The process by which a cell undergoes apoptosis signaling is evolutionarily conserved and inspired by many different receptor interactions in most if not all cell types in the body.

■ Apoptotic stimuli either activate death receptors (extrinsic pathway) or mitochondrial cytochrome *c* release (intrinsic pathway). Both pathways ultimately activate effector caspases, which finalize cell death events.

■ Apoptotic cells are rapidly engulfed by neighboring phagocytes and, for this reason, are difficult to detect in living tissues.

■ Bcl-2 family members include both anti-apoptotic and pro-apoptotic family members that reside in the mitochondrial membrane and regulate cytochrome *c* release. A third set of family members, the pro-apoptotic BH3 proteins, can be activated by apoptotic stimuli, favoring a shift in balance of Bcl-2 family member activity at the mitochondrial membrane toward cytochrome *c* release.

REFERENCES

Alam, S., et al. 1999. Qualitative and quantitative differences in T cell receptor binding of agonist and antagonist ligands. *Immunity* **10**:227–237.

Anderson, M., et al. 2002. Projection of an immunological self shadow within the thymus by the aire protein. *Science* **298**:1395–1401.

Baldwin, T., K. Hogquist, and S. Jameson. 2004. The fourth way? Harnessing aggressive tendencies in the thymus. Journal of Immunology **173**:6515–6520.

Baldwin, T., M. Sandau, S. Jameson, and K. Hogquist. 2005. The timing of TCR alpha expression critically influences T cell development and selection. *Journal of Experimental Medicine* **202**:111–121.

Bouillet, P., et al. 2002. BH3-only Bcl-2 family member Bim is required for apoptosis of autoreactive thymocytes. *Nature* **415**:922–926.

Carpenter, A., and R. Bosselut. 2010. Decision checkpoints in the thymus. *Nature Immunology* **11**:666–673.

Caton, A., et al. 2004. CD4(+) CD25(+) regulatory T cell selection. *Annals of the New York Academy of Sciences* **1029**:101–114.

Ceredig, R., and T. Rolink. 2002. A positive look at double-negative thymocytes. *Nature Reviews Immunology* **2**:888–897.

Drennan, M., D. Elewaut, and K. Hogquist. 2009. Thymic emigration: sphingosine-1-phosphate receptor-1-dependent models and beyond. *European Journal of Immunology* **39**:925–930.

Gascoigne, N. 2010. CD8+ thymocyte differentiation: T cell two-step. *Nature Immunology* **11**:189–190.

Germain, R. 2002. T-cell development and the CD4-CD8 lineage decision. *Nature Reviews. Immunology* **2**:309–322.

Germain, R. 2008. Special regulatory T-cell review: A rose by any other name: from suppressor T cells to T$_{REG}$s, approbation to unbridled enthusiasm. *Immunology* **123**:20–27.

He, X., et al. 2005. The zinc finger transcription factor Th-POK regulates CD4 versus CD8 T-cell lineage commitment. *Nature* **433**:826–833.

Hogquist, K., T. Baldwin, and S. Jameson. 2005. Central tolerance: Learning self-control in the thymus. *Nature Reviews. Immunology* **5**:772–782.

Hogquist, K., M. Gavin, and M. Bevan. 1993. Positive selection of CD8+ T cells induced by major histocompatibility complex binding peptides in fetal thymic organ culture. *Journal of Experimental Medicine* **177**:1469–1473.

Hogquist, K., S. Jameson, and M. Bevan. 1995. Strong agonist ligands for the T cell receptor do not mediate positive selection of functional CD8+ T cells. *Immunity* **3**:79–86.

Hogquist, K., et al. 1994. T cell receptor antagonist peptides induce positive selection. *Cell* **76**:17–27.

Kappes, D., and X. He. 2006. ole of the transcription factor Th-POK in CD4:CD8 lineage commitment. *Immunological Reviews* **209**:237–252.

Kisielow, P., H. Blüthmann, U. Staerz, M. Steinmetz, and H. von Boehmer, H. 1988. Tolerance in T-cell-receptor transgenic mice involves deletion of nonmature CD4+8+ thymocytes. *Nature* **333**:742–746.

Klein, L., M. Hinterberger, G. Wirnsberger, and B. Kyewski. 2009. Antigen presentation in the thymus for positive selection and central tolerance induction. Nature Reviews. Immunology **9**:833–844.

Kyewski, B., and P. Peterson. 2010. Aire: Master of many trades. *Cell* **140**:24–26.

Li, R., and D. Page. 2001. Requirement for a complex array of costimulators in the negative selection of autoreactive thymocytes in vivo. *Journal of Immunology* 166:6050–6056.

Marrack, P., L. Ignatowicz, J. Kappler, J. Boymel, and J. Freed. 1993. Comparison of peptides bound to spleen and thymus class II. *Journal of Experimental Medicine* 178:2173–2183.

Marrack, P., J. McCormack, and J. Kappler. 1989. Presentation of antigen, foreign major histocompatibility complex proteins and self by thymus cortical epithelium. *Nature* 338:503–505.

Mathis, D., and C. Benoist. 2009. Aire. *Annual Review of Immunology* 27:287–312.

McNeil, L., T. Starr, and K. Hogquist. 2005. A requirement for sustained ERK signaling during thymocyte positive selection in vivo. *Proceedings of the National Academy of Sciences of the United States of America* 102:13,574–13,579.

Mohan, J., et al. 2010. Unique autoreactive T cells recognize insulin peptides generated within the islets of Langerhans in autoimmune diabetes. *Nature Immunology* 11:350–354.

Page, D., L. Kane, J. Allison, and S. Hedrick. 1993. Two signals are required for negative selection of CD4+CD8+ thymocytes. *Journal of Immunology* 151:1868–1880.

Park, J., et al. 2010. Signaling by intrathymic cytokines, not T cell antigen receptors, specifies CD8 lineage choice and promotes the differentiation of cytotoxic-lineage T cells. *Nature Immunology* 11:257–264.

Pui, J., et al. 1999. Notch1 expression in early lymphopoiesis influences B versus T lineage determination. *Immunity* 11:299–308.

Punt, J., B. Osborne, Y. Takahama, S. Sharrow, and A. Singer. 1994. Negative selection of CD4+CD8+ thymocytes by T cell receptor-induced apoptosis requires a costimulatory signal that can be provided by CD28. *Journal of Experimental Medicine* 179:709–713.

Sambandam, A., et al. 2005. Notch signaling controls the generation and differentiation of early T lineage progenitors. *Nature Immunology* 6:663–670.

Schmitt, T., and J. Zúñiga-Pflücker. 2002. Induction of T cell development from hematopoietic progenitor cells by delta-like-1 in vitro. *Immunity* 17:749–756.

Singer, A. 2010. Molecular and cellular basis of T cell lineage commitment: An overview. *Seminars in Immunology* 22:253.

Stadinski, B., et al. 2010. Diabetogenic T cells recognize insulin bound to IAg7 in an unexpected, weakly binding register. *Proceedings of the National Academy of Sciences of the United States of America* 107:10,978–10,983.

Starr, T., S. Jameson, and K. Hogquist. 2003. Positive and negative selection of T cells. *Annual Review of Immunology* 21:139–176.

Sun, G., et al. 2005. The zinc finger protein cKrox directs CD4 lineage differentiation during intrathymic T cell positive selection. *Nature Immunology* 6:373–381.

Teh, H., et al. 1988. Thymic major histocompatibility complex antigens and the alpha beta T-cell receptor determine the CD4/CD8 phenotype of T cells. *Nature* 335:229–233.

Uematsu, Y., et al. 1988. In transgenic mice the introduced functional T cell receptor beta gene prevents expression of endogenous beta genes. *Cell* 52:831–841.

Venanzi, E., C. Benoist, and D. Mathis. 2004. Good riddance: Thymocyte clonal deletion prevents autoimmunity. *Current Opinion in Immunology* 16:197–202.

von Boehmer, H., H. Teh, and P. Kisielow. 1989. The thymus selects the useful, neglects the useless and destroys the harmful. *Immunology Today* 10:57–61.

Wang, L., and R. Bosselut. 2009. CD4-CD8 lineage differentiation: Thpok-ing into the nucleus. *Journal of Immunology* 183:2903–2910.

Wang, L., et al. 2008. Distinct functions for the transcription factors GATA-3 and ThPOK during intrathymic differentiation of CD4(+) T cells. *Nature Immunology* 9:1122–1130.

Wilson, A., H. MacDonald, and F. Radtke. 2001. Notch 1-deficient common lymphoid precursors adopt a B cell fate in the thymus. *Journal of Experimental Medicine* 194:1003–1012.

Zúñiga-Pflücker, J. 2004. T-cell development made simple. *Nature Reviews. Immunology* 4:67–72.

Useful Web Sites

http://www.bio.davidson.edu/courses/immunology/Flash/Main.htm A Dr. Victor Lemas–generated animation of positive and negative selection events in the thymus, used at Davidson College.

http://www.bio.davidson.edu/courses/movies.html A full list of animations assembled by and in many cases generated by individuals associated with Davidson College.

http://bmc.med.utoronto.ca/student_video_gallery.php?title=images/stories/videos/Janice_Yau&h=360&w=640&m4v=1 Another animation of T-cell development generated as part of a master's project by Janice Yau. This is from an impressive site that features multiple video projects generated by master's students in the University of Toronto at Mississauga's Biomedical Communications Graduate Program.

http://bio-alive.com/categories/apoptosis.htm Videos of cells undergoing apoptosis.

www.celldeath.de/encyclo/aporev/aporev.htm Detailed online review of apoptosis biology.

www.ncbi.nlm.nih.gov/pmc/articles/PMC2008650 Access to original Kerr, Wyllie, and Currie *British Journal of Cancer* 1972 article where the term *apoptosis* was coined.

CLINICAL FOCUS QUESTIONS

1. The susceptibility to autoimmune diseases often has a genetic basis and has been linked to many different gene loci. Identify three possible genes other than Fas and AIRE (whose connection to autoimmunity have been explicitly described in this chapter) that could be involved in an increased susceptibility to autoimmune disease. Explain your reasoning.

2. Susceptibility to many autoimmune diseases has been linked to MHC gene variants. One of the best examples of such a linkage is provided by multiple sclerosis (MS), a human autoimmune disease caused by autoreactive T cells whose activity ultimately damages the myelin sheaths around neurons. Susceptibility to MS has been consistently associated with variants in the HLA-DR2 gene. Although this link was first recognized in 1972, we still don't fully understand the basis for this susceptibility. One perspective on the reasons for the link between MHC variations and autoimmune disease was offered in a recent review article. The authors state, "The mechanisms underlying MHC association in autoimmune disease are not clearly understood. One long-held view suggests a breakdown in immunological tolerance to self-antigens through aberrant class II presentation of self or foreign peptides to autoreactive T lymphocytes. Thus, it seems likely that specific MHC class II alleles determine the targeting of particular autoantigens resulting in disease-specific associations." (Fernando, M. M. A., et al. 2008 Defining the role of the MHC in autoimmunity: A review and pooled analysis. *PLoS Genet* 4:4: e1000024. doi:10.1371/journal.pgen.1000024.)
   a. Paraphrase this perspective using your own words. What, specifically, might the authors mean by "aberrant class II presentation . . . to autoreactive T lymphocytes"?
   b. (**Advanced question.**) Although this speculation has some merit, it does not resolve all questions. Why? Pose one question that this explanation inspires or does not answer.
   c. (**Very advanced question.**) Offer one addition to the explanation (or an alternative) that helps resolve the question you posed above.

3. Over a period of several years, a group of children and adolescents are regularly dosed with compound X, a lifesaving drug. However, in addition to its beneficial effects, it was found that this drug interferes with Fas-mediated signaling.
   a. What clinical manifestations of this side effect of compound X might be seen in these patients?
   b. If white blood cells from an affected patient are stained with a fluorescein-labeled anti-CD4 and a phycoerythrin-labeled anti-CD8 antibody, what might be seen in the flow-cytometric analysis of these cells? What pattern would be expected if the same procedure were performed on white blood cells from a healthy control?

STUDY QUESTIONS

1. Each of the following statements is false. Correct them (and explain your correction[s]).
   Knockout mice lacking class I MHC molecules fail to produce CD4⁺ mature thymocytes.
   β-selection initiates negative selection.

   Negative selection to tissue-specific antigens occurs in the cortex of the thymus.
   Most thymocytes die in the thymus because they are autoreactive.
   The extrinsic pathway of apoptosis never activates the intrinsic pathway.
   Cytochrome *c* is an important downstream molecule in the extrinsic apoptotic pathway.
   Bcl-2 enhances the activity of Bax and therefore inhibits apoptosis.

2. Whereas the majority of T cells in our bodies express an αβ TCR, up to 5% of T cells express the γδ TCR instead. Explain the difference in antigen recognition between these two cell types.

3. You have fluorescein-labeled anti-CD4 and phycoerythrin-labeled anti-CD8. You use these antibodies to stain thymocytes and lymph-node cells from normal mice and from RAG-1 knockout mice. In the forms below, draw the FACS plots that you would expect.

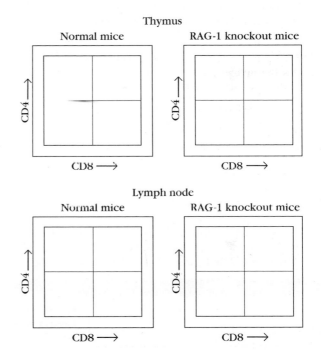

4. What stages of T-cell development (DN1, DN2, DN3, DN4, DP, CD4 SP, or CD8 SP) would be affected in mice with the following genetic modifications? Justify your answers.
   a. Mice that do not express MHC Class II.
   b. Mice that do not express AIRE.
   c. Mice that do not express the TCRα chain.

5. You stain thymocytes with PE conjugated anti-CD3 and FITC conjugated anti-TCRβ. Most cells stain with both. However, you find a proportion of cells that stain with neither antibody. You also find a small population that stain with anti-CD3, but not with anti-TCRβ. What thymocyte populations might each of these populations represent?

6. You immunize an H-2^k mouse with chicken ovalbumin (a protein against which the mouse will generate an immune response) and isolate a CD4$^+$ mature T cell specific for an ovalbumin peptide. You clone the αβ TCR genes from this cell line and use them to prepare transgenic mice with the H-2^k or H-2^d haplotype.

   a. What approach can you use to distinguish immature thymocytes from mature CD4$^+$ thymocytes in the transgenic mice?

   b. Would thymocytes from a TCR transgenic mouse of the H-2^k background have a proportion of CD4$^+$ thymocytes that is higher, lower, or the same as a wild-type mouse?

   c. Would thymocytes from a TCR transgenic mouse of the H-2^d background have a proportion of CD4$^+$ thymocytes that is higher, lower, or the same as a wild-type mouse?

   Speculate and explain your answer.

   d. You find a way to "make" the medullary epithelium of an H-2^k TCR transgenic mouse express and present the ovalbumin peptide for which your T cell is specific. What would the CD4 versus CD8 profile of a TCR transgenic thymus look like in these mice?

   You also find a way to "make" the cortical epithelium express this ovalbumin peptide in its MHC Class II dimer. What might the CD4 versus CD8 profile of this TCR transgenic thymus look like?

7. In his classic chimeric mouse experiments, Zinkernagel took bone marrow from a mouse of one MHC haplotype (mouse 1) and a thymus from a mouse of another MHC haplotype (mouse 2) and transplanted them into a third mouse, which was thymectomized and lethally irradiated. He then immunized this reconstituted mouse with the lymphocytic choriomeningitis virus (LCMV) and examined the activity of the mature T cells isolated from the spleen and lymph nodes of the mouse.

   He was specifically interested to see if the mature CD8$^+$ T cells in these mice could kill target cells infected with LCMV with the MHC haplotype of mouse 1, 2, or 3. The results of two such experiments using H-2^b strain C57BL/6 mice and H-2^d strain BALB/c mice as bone marrow and thymus donors, respectively, are shown in the following table.

Experiment	Bone marrow donor	Thymus donor	Thymectomized x-irradiated recipient	Lysis of LCMV-infected target cells		
				H-2^d	H-2^k	H-2^b
A	C57BL/6 (H-2^b)	BALB/c (H-2^d)	C57BL/6 × BALB/c	+	−	−
B	BALB/c (H-2^d)	C57BL/6 (H-2^b)	C57BL/6 × BALB/c	−	−	+

   a. Why were the H-2^b target cells not lysed in experiment A but lysed in experiment B?

   b. Why were the H-2^k target cells not lysed in either experiment?

8. You have a CD8$^+$ CTL clone (from an H-2^k mouse) that has a T-cell receptor specific for the H-Y antigen. You clone the αβ TCR genes from this cloned cell line and use them to prepare transgenic mice with the H-2^k or H-2^d haplotype.

   a. How can you distinguish the immature thymocytes from the mature CD8$^+$ thymocytes in the transgenic mice?

   b. For each of the following transgenic mice, indicate with (1) or (2) whether the mouse would have immature double-positive and mature CD8$^+$ thymocytes bearing the transgenic T-cell receptor: H-2^k female, H-2^k male, H-2^d female, H-2^d male.

   c. Explain your answers for the H-2^k transgenics.

   d. Explain your answers for the H-2^d transgenics.

9. To demonstrate positive thymic selection experimentally, researchers analyzed the thymocytes from normal H-2^b mice, which have a deletion of the class II *IE* gene, and from H-2^b mice in which the class II *IA* gene had been knocked out.

   a. What MHC molecules would you find on antigen-presenting cells from the normal H-2^b mice?

   b. What MHC molecules would you find on antigen-presenting cells from the IA knockout H-2^b mice?

   c. Would you expect to find CD4$^+$ T cells, CD8$^+$ T cells, or both in each type of mouse? Why?

10. You wish to determine the percentage of various types of thymocytes in a sample of cells from mouse thymus using the indirect immunofluorescence method.

   a. You first stain the sample with goat anti-CD3 (primary antibody) and then with rabbit FITC-labeled antigoat Ig (secondary antibody), which emits a green color. Analysis of the stained sample by flow cytometry indicates that 70% of the cells are stained. Based on this result, how many of the thymus cells in your sample are expressing antigen-binding receptors on their surface? Would all be expressing the same type of receptor? Explain your answer. What are the remaining unstained cells likely to be?

   b. You then separate the CD3$^+$ cells with the fluorescence-activated cell sorter (FACS) and restain them. In this case, the primary antibody is hamster anti-CD4, and the secondary antibody is rabbit PE-labeled antihamster Ig, which emits a red color. Analysis of the stained CD3$^+$ cells shows that 80% of them are stained. From this result, can you determine how many T$_C$ cells are present in this sample? If yes, then how many T$_C$ cells are there? If no, what additional experiment would you perform to determine the number of T$_C$ cells that are present?

B-Cell Development

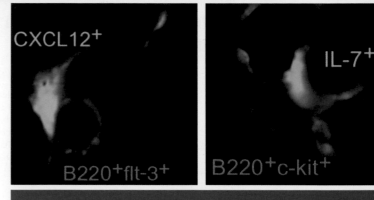

CXCL12+

B220+flt-3+

IL-7+

B220+c-kit+

- The Site of Hematopoiesis

- B-Cell Development in the Bone Marrow

- The Development of B-1 and Marginal-Zone B Cells

- Comparison of B- and T-Cell Development

Millions of B lymphocytes are generated in the bone marrow every day and exported to the periphery. The rapid and unceasing generation of new B cells occurs in a carefully regulated sequence of events. Cell transfer experiments, in which genetically marked donor **hematopoietic stem cells** (HSCs) are injected into an unmarked recipient, have indicated that B-cell development from HSC to mature B cell takes from 1 to 2 weeks; donor-derived mature B cells can be detected in the recipient by 2 weeks following transfer of HSCs into recipient mice.

B-cell development begins in the bone marrow with the asymmetric division of an HSC and continues through a series of progressively more differentiated progenitor stages to the production of **common lymphoid progenitors** (CLPs), which can give rise to either B cells or T cells (see Overview Figure 10-1). Progenitor cells destined to become T cells migrate to the thymus where they complete their maturation (see Chapter 9); the majority of those that remain in the bone marrow become B cells. As differentiation proceeds, the developing B cell expresses on its cell surface a precisely calibrated sequence of *cell-surface receptor* and *adhesion molecules*. Some of the signals received from these receptors induce the *differentiation* of the developing B cell; others trigger its *proliferation* at particular stages of development and yet others direct its movements within the bone marrow environment. These signals collectively allow differentiation of the CLP through the early B-cell stages to form the **immature B cell** that leaves the marrow to complete its differentiation in the spleen. For the investigator, the expression of different cell-surface molecules at each stage of B-cell maturation provides an invaluable experimental tool with which to recognize and isolate B cells poised at discrete points in their development.

The primary function of mature B cells is to secrete antibodies that protect the host against pathogens, and so one major focus of those studying B-cell differentiation is the analysis of the *timing and order of rearrangement and expression of immunoglobulin receptor heavy- and light-chain genes*. Recall from Chapter 7 that immunoglobulin gene rearrangements begin with heavy-chain D to J_H gene-segment rearrangement, followed by the stitching together of the μ heavy-chain V_H and DJ_H segments. These rearrangements culminate in the cell-surface expression of the *pre-B-cell receptor* during the *pre-B-cell* stage, in which the rearranged heavy chain is expressed in combination with the surrogate light chain. Rearrangement of the light chain is initiated after several rounds of division of cells bearing the pre-BCR.

Like T cells, developing B cells must solve the problem of creating a repertoire of receptors capable of recognizing an extensive array of antigens, while ensuring that self-reactive B cells are either eliminated by apoptosis or rendered functionally unreactive or *anergic*. However, unlike T-cell receptors, B-cell receptors are not constrained by the need to be MHC restricted. Again, unlike T-cell maturation, B-cell development is almost complete by the time the B cell leaves the bone marrow; in mammalian systems there is no thymic equivalent in which B-cell development is accomplished. Instead, immature B cells are released to

B-Cell Development Begins in the Bone Marrow and Is Completed in the Periphery

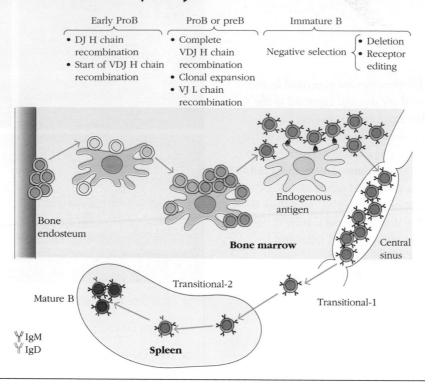

B-cell development begins with a hematopoietic stem cell (HSC) and passes through progressively more delimited progenitor-cell stages until it reaches the pro-B cell stage. At this stage, the precursor cell is irreversibly committed to the B-cell lineage and the recombination of the immunoglobulin genes begins. Once the completed immunoglobulin is expressed on the cell surface, the immature B cell, now a transitional B cell, leaves the bone marrow to complete its maturation in the spleen.

the periphery, where they complete their developmental program in the spleen.

In this chapter, we will follow B-cell development from its earliest stages in the primary lymphoid organs to the generation of fully mature B cells in the secondary lymphoid tissues. As for T cells, multiple B-cell subsets exist, and we will briefly address how the process of differentiation of the minority B-1 and *marginal-zone* (MZ) B-cell subsets differs from the developmental program followed by the predominant B-2 B-cell subset. We will conclude with a brief comparison of the maturational processes of T and B lymphocytes.

The Site of Hematopoiesis

In adult animals, **hematopoiesis**, the generation of blood cells, occurs in the bone marrow; the HSCs in the marrow are the source of all blood cells of the erythroid, myeloid, and lymphoid lineages (Chapter 2). Various non-hematopoietic cells in the bone marrow express cell-surface molecules and secrete hormones that guide hematopoietic cell development. Developing lymphocytes move within the bone marrow as they mature, thus interacting with different populations of cells and signals at various developmental stages. However, fetal animals face particular challenges to their developing immune systems; how can they generate blood cells when their bones are still not yet fully developed?

The Site of B-Cell Generation Changes during Gestation

Hematopoiesis is a complex process in the adult animal, and during fetal maturation additional challenges must be met. Red blood cells must be quickly generated de novo in order to provide the embryo with sufficient oxygen, and HSCs must proliferate at a rate sufficient to populate the adult as well as provide for the hematopoietic needs of the maturing fetus. Furthermore, since the bone marrow appears relatively late in

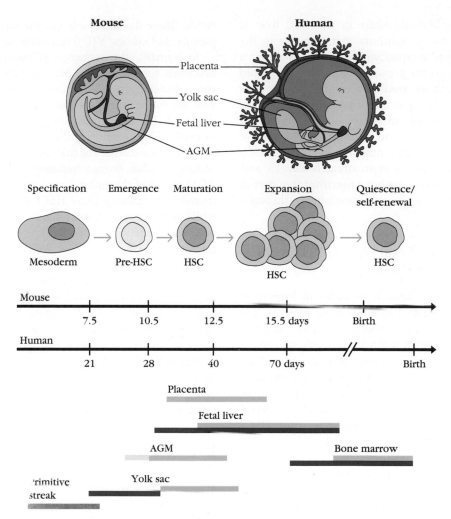

FIGURE 10-2 The anatomy and timing of the earliest stages in hematopoiesis. (a) Blood-cell precursors are initially found in the yolk sac (yellow), then spread to the placenta fetal liver (pink), and aorta-gonad-mesonephros (AGM) region (green), before finding their adult home in the bone marrow. The mouse embryo is shown at 11 days of gestation; the human embryo at the equivalent 5 weeks of gestation. (b) In the embryo, cells within the primitive streak mesodermal tissue adopt either hematopoietic (blood-cell forming) or vascular (blood-vessel forming) fates. Those destined to become blood cells emerge first as rapidly proliferating pre-HSCs and eventually mature into relatively quiescent hematopoietic stem cells that populate the bone marrow. The colored bars in the timeline illustrate the ages at which the various murine and human hematopoietic sites are active. Mesoderm (gray); generation of fetal HSC (yellow); active hematopoietic differentiation (red); emergence of functional, adult-type HSCs (blue). *[Adapted from H. K. Mikkola and S. H. Orkin, 2006, The journey of developing hematopoietic stem cells, Development 133:3733–3744, Figures 1 and 2.]*

development, the whole process of blood-cell generation must shift location several times before moving into its final home.

The gestation period for mice is 19 to 21 days. Hematopoiesis begins, in the mouse, around 7 days post fertilization (Figure 10-2) when precursor cells in the **yolk sac** begin differentiating to form primitive, nucleated, erythroid cells that carry the oxygen the embryo needs for early development. *Fetal HSCs* capable of generating all blood-cell types can be detected in the early a*orta-gonad-m*esonephros *(AGM) region* on day 8, when the fetal heart starts beating. On day 10, *mature HSCs* capable of completely repopulating the hematopoietic system of irradiated adult mice can be isolated from the AGM, and by day 11 they can be found in the *yolk sac, placenta,* and

fetal liver. Between days 11.5 and 12.5, there is rapid expansion of the placental HSC pool, at the end of which time the placenta holds more HSCs than either the AGM or the yolk sac. By day 13.5, the number of HSCs in the placenta begins to decrease while the HSC pool in the liver continues to expand.

As the mouse embryo completes its development, the predominant site of HSC generation remains within the fetal liver, but some hematopoiesis can be detected in the spleen in the perinatal (around the time of birth) period. The number of fetal liver HSCs in mice reaches a maximum of approximately 1000 by days 15.5 to 16.5 of embryonic development, after which it starts to decline. Within the fetal liver, HSCs differentiate to form progenitor cells. At the earliest time points, hematopoiesis in the fetal liver is dominated by erythroid progenitors that give rise to the true, enucleated mature erythrocytes, in order to ensure a steady oxygen supply to the growing embryo, and myeloid and lymphoid progenitors gradually emerge. **Pre-B cells (precursor-B cells)**, defined as cells that express immunoglobulin in their cytoplasm but not on their surfaces, are first observed at day 13 of gestation, and surface IgM-positive B cells are present in detectable numbers by day 17. HSCs first seed the bone marrow at approximately day 15, and over a period of a few weeks, the bone marrow takes over as the main site of B-cell development, remaining so throughout post-natal life.

Hematopoiesis in the Fetal Liver Differs from That in the Adult Bone Marrow

Developing B cells in the fetal liver differ in important ways from their counterparts in adult bone marrow. The liver is the primary site of B-cell generation in the fetus, and provides the neonatal animal with the cells it needs to populate its nascent immune system. In order to accomplish this, hematopoietic stem cells and their progeny must undergo a phase of rapid proliferation, and fetal liver HSCs, as well as their daughter cells, undergo several rounds of cell division over a short time. In contrast, HSCs derived from the bone marrow of a healthy adult animal are relatively quiescent.

B cells generated from fetal liver precursors are predominantly **B-1 B cells**, which will be described more fully in Chapter 12. Briefly, B-1 B cells are primarily located in the body (specifically the peritoneal and pleural) cavities. They are therefore well-positioned to protect the gut and the lungs, which are the major ports of entry of microbes in the fetus and neonate. Antibodies secreted by B-1 B cells are broadly cross-reactive; many bind to carbohydrate antigens expressed by a number of microbial species. Since terminal *d*eoxynucleotidyl *t*ransferase (TdT) is minimally expressed at this point in ontogeny, and the RAG1/2 recombinase proteins appear not to use the full range of V, D, and J region gene segments at this stage in embryonic development, the immunoglobulin receptors of B-1 B cells express minimal receptor diversity. In expressing an oligoclonal (few, as opposed to many, clones) repertoire of B-cell receptors that bind to a limited number of carbohydrate antigens shared among many microbes, B-1 B cells occupy a functional niche that bridges the innate and adaptive immune systems. We will describe B-1 B-cell development further at the end of this chapter.

Over a period of 2 to 4 weeks after birth, the process of hematopoiesis in mice shifts from the fetal liver and spleen to the bone marrow, where it continues throughout adulthood. The B-1 B-cell population represents an exception to this general rule, as it is self-renewing in the periphery. This means that new daughter B-1 B cells are generated continually from preexisting B-1 B cells in the peritoneal and pleural cavities, and in those other parts of the body in which B-1 B cells

reside. These daughter cells use the same receptors as their parents, and no new V(D)J recombinase activity is required.

In humans, the sequence of events is similar to that described for the mouse, but the time frame is obviously somewhat elongated. Blood-cell precursors first appear in the yolk sac in the third week of embryonic development, but these cells, like their analogues in the mouse provide primarily erythroid progenitors and are not capable of generating all subsets of blood cells. The first cells capable of entirely repopulating an adult human hematopoietic system arise in the AGM region of the embryo and/or the yolk sac. By the third month of pregnancy, these HSCs migrate to the fetal liver, which then becomes responsible for the majority of hematopoiesis in the fetus. By the fourth month of pregnancy, HSCs migrate to the bone marrow, which gradually assumes the hematopoietic role from the fetal liver until, by the time of birth, it is the primary generative organ for blood cells. Prior to puberty in humans, most of the bones of the skeleton are hematopoietically active, but by the age of 18 years only the vertebrae, ribs, sternum, skull, pelvis, and parts of the humerus and femur retain hematopoietic potential.

Just as B-cell development in the fetus and neonate differs from that in the adult, so does B-cell hematopoiesis in the aging animal. Clinical Focus Box 10-1 describes some aspects of B-cell development that alter as humans age.

B-Cell Development in the Bone Marrow

In Chapter 2 (Figure 2-5), we presented the structure of bone and bone marrow. The bone marrow microenvironment is a complex, three-dimensional structure with distinctive cellular niches which are specialized to influence the development of the cell populations that mature there. A dense network of fenestrated (leaky) thin-walled blood vessels—the bone marrow sinusoids—permeates the marrow, allowing the passage of newly formed blood cells to the periphery and facilitating blood circulation through the marrow.

In addition to serving as a source of hematopoietic stem cells, bone marrow also contains stem cells that can differentiate into adipocytes (fat cells), chondrocytes (cartilage cells), osteocytes (bone cells), myocytes (muscle cells), and potentially other types of cells as well. Each of these different classes of stem cells requires specific sets of factors, secreted by particular *bone marrow stromal cells* to enable their proper differentiation.

What are bone marrow stromal cells? The term *stroma* derives from the Greek for *mattress*, and a **stromal cell** is a general term that describes a large adherent cell that supports the growth of other cells. During B-cell development, bone marrow stromal cells fulfill two functions. First, by interacting with adhesion molecules on the surfaces of HSCs and progenitor cells, stromal cells retain the developing cell populations in the specific bone marrow niches where they can receive the appropriate molecular signals required for

Box 10-1

 # B-Cell Development in the Aging Individual

People of retirement age and older represent a greater segment of the population than they used to, and these older individuals expect to remain active and productive members of society. However, physicians and immunologists have long known that the elderly are more susceptible to infection than are young men and women, and that vaccinations are less effective in older individuals. In this feature, we explore the differences in B-cell development between younger and older vertebrates, which may account for some of these immunological disparities between adult and older individuals.

Aging individuals display deficiencies in many aspects of B-cell function, including a poor antibody response to vaccination, inefficient generation of memory B cells, and an increase in the expression of autoimmune disorders. Does this reflect defective functioning only in the mature antigen-responsive B-cell population, or does it result from problems manifested during earlier stages in B-cell development? Current research demonstrates that aging individuals display a range of shortcomings in developing B cells.

Experiments employing reciprocal bone marrow chimeras—in which aging HSCs were transplanted into young recipients or HSCs from young mice were injected into aging recipients—have shown that the suboptimal process of B-cell development in aging individuals results from deficiencies in both the aging stem cells and in the supporting stromal cells. For example, bone marrow stromal cells from aging mice secrete lower levels of IL-7 than do stromal cells from younger animals, suggesting an environmental defect in the aging bone marrow. However, study of isolated, aging B-cell progenitors reveals that they also respond less efficiently to IL-7 than do B cells from younger mice, and so the IL-7 response in aging individuals is affected at both the secretory and recipient-cell levels.

Indeed, the problems encountered by developing B cells from aging indi-

viduals start at the very beginning of their developmental program. The epigenetic regulation of HSC genes in aging mice is compromised, resulting in diminished levels of HSC self-renewal. Furthermore, the balance between the production of myeloid versus lymphoid progenitors is shifted in older individuals, with down-regulation of genes associated with lymphoid specification and a correspondingly enhanced expression of genes specifying myeloid development. The net effect of these changes in the HSC population is a reduction with age in the numbers of early B-cell progenitors, which is reflected in a decrease in the numbers of pro- and pre-B-cell precursors at all stages of development.

Detailed studies of the expression of particular genes important in B-cell development demonstrate that the expression of important transcription factors, such as those encoded by the E2A gene, is reduced in older animals. Furthermore, the *Rag* genes, as well as the gene encoding the surrogate light-chain component, λ5, are down-regulated in older animals compared with young adults, resulting in a reduction in the bone marrow output of immature B cells.

Multiple mechanisms therefore help to explain why the numbers of B cells released from the bone marrow are smaller in aging than in younger individuals. But is the antigen recognition capacity—the quality—as well as the quantity of B cells different between the two populations? In particular, do B cells from aging mice express a repertoire of receptors similar to those obtained from younger animals? The answer to this question has come from the development of techniques that enable a global assessment of repertoire diversity. Study of the sizes and sequences of CDR3 regions from large numbers of human B cells suggests that in aging individuals the size of the repertoire (the number of different B-cell receptors an individual expresses) is drastically diminished, and that this decrease

in repertoire diversity correlates with a reduction in the health of the aging patient.

The mechanisms for this age-related repertoire truncation appear to be complex. A decrease in output of immature B2 cells from the bone marrow could provide the opportunity for B-1 B cells to increase their share of the peripheral B-cell niche, and as is well appreciated, B-1 B cells have a less diverse receptor repertoire than do B-2 B cells. A lifetime of generating memory cells may also result in an individual having less room in B-cell follicles for newly formed B cells to enter, and a decreased concentration of the homeostatic regulatory cytokines may make it more difficult for primary B cells to compete with their more robust memory counterparts. Clearly, this is an area of increasing clinical interest as the average age of the population of the developed world continues to increase.

REFERENCES

Cancro, M. P., et al. 2009. B cells and aging: Molecules and mechanisms. *Trends in Immunology* **30:**313–318.

Dorshkin, K., E. Montecino-Rodriguez, and R. A. Signer. 2009. The ageing immune system: Is it ever too old to become young again? *Nature Reviews Immunology* **9:**57–62.

Goodnow, C. C. 1992. Transgenic mice and analysis of B-cell tolerance. *Annual Reviews Immunology* **10:**489–518.

Labrie, J. E., 3rd, A. P. Sah, D. M. Allman, M. P. Cancro, and R. M. Gerstein. 2004. Bone marrow microenvironmental changes underlie reduced RAG-mediated recombination and B cell generation in aged mice. *Journal of Experimental Medicine* **200:**411–423.

Nemazee, D. A., and K. Bürki. 1989, February. Clonal deletion of B lymphocytes in a transgenic mouse bearing anti-MHC class I antibody genes. *Nature* **337:**562–566. doi:10.1038/337562a0

Van der Put, E., E. M. Sherwood, B. B. Blomberg, and R. L. Riley. 2003. Aged mice exhibit distinct B cell precursor phenotypes differing in activation, proliferation and apoptosis. *Experimental Gerontology* **38:**1137–1147.

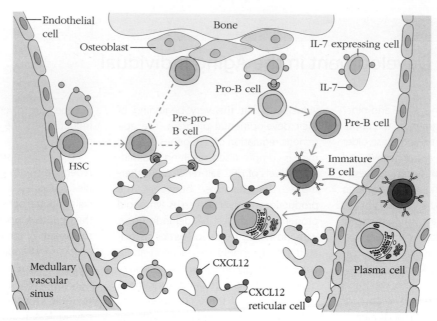

FIGURE 10-3 HSCs and B-cell progenitors make contact with different sets of bone marrow cells as they progress through their developmental program. HSCs begin their developmental program close to the osteoblasts (top). An HSC is also shown entering from the blood (left-hand side), illustrating the fact that HSCs are capable of recirculation in the adult animal. Progenitor cells then move to gain contact with CXCL12-expressing stromal cells, where they mature into pre-pro B cells. By the time differentiation has progressed to the pro-B-cell stage, the developing cell has moved to receive signals from IL-7-producing stromal cells. After leaving the IL-7-expressing stromal cell, the pre-B cell completes its differentiation and leaves the bone marrow as an immature B cell. CXCL-12 is shown in purple; IL-7 in blue. [*Adapted from T. Nagasawa, 2006, February, Microenvironmental niches in the bone marrow required for B-cell development,* Nature Reviews Immunology *6:107–116. doi:10.1038/nri1780*]

their further differentiation. Second, diverse populations of stromal cells express different cytokines. At various points in their development, progenitor and precursor B cells must interact with stromal cells secreting particular cytokines, and thus the developing B cells move in an orderly progression from location to location within the bone marrow. This progression is guided by chemokines secreted by particular stromal cell populations. For example, HSCs begin their life in close contact with osteoblasts located close to the lining of the endosteal (bone marrow) cavity. Once differentiated to the pre-pro-B-cell stage, the developing B cells require signals from the chemokine CXCL12, which is secreted by a specialized set of stromal cells, in order to progress to the pro-B-cell stage. Pro-B cells then require signaling from the cytokine IL-7, which is secreted by yet another stromal cell subset (Figure 10-3). Many of these stromal cell factors serve to induce the expression of specialized transcription factors important in B-cell development.

The Stages of Hematopoiesis Are Defined by Cell-Surface Markers, Transcription-Factor Expression, and Immunoglobulin Gene Rearrangements

Full characterization of a developmental pathway requires that scientists understand the phenotypic and functional characteristics of each cell type in that pathway, as well as the molecular signals and transcription factors that drive differentiation at each stage. Cells at particular stages of differentiation can be characterized by their surface molecules, which include cell-surface antigens, adhesion molecules, and receptors for chemokines and cytokines. They are also defined by the array of active transcription factors that determine which genes are expressed at each step in the developmental process. Finally, in the case of B cells, the developmental stages are defined by the status of the rearranging heavy- and light-chain immunoglobulin genes. B-cell development is not yet completely understood; however, most of the important cellular intermediates have been defined, and developmental immunologists are gradually filling in the gaps in our knowledge.

Investigators delineating the path of B-lymphocyte differentiation employed three general experimental strategies. First, they generated *antibodies against molecules (antigens or markers) present on the surface of bone marrow cells.* They then determined which of these molecules were present at the same time as other antigens, and which combinations of antigens appeared to define unique cell types (Figure 10-4a). In addition, *culturing cells* in vitro that bear known cell-surface antigens, followed by *flow cytometric analysis* of the daughter cells generated in culture, enabled them to describe the sequential expression of particular combinations of cell-surface molecules (Figure 10-4b).

(a)

Characterization of progenitors bearing
different sets of cell surface molecules.

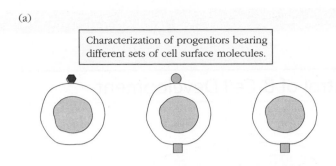

(b)

Determining sequence of marker expression by culturing cells
from each stage. Culturing cells with the red antigen gives rise to
daughter cells of both types. Culturing cells with only the blue
antigen gives rise to the cells bearing both blue and green
antigens, but never cells bearing the red antigens. Therefore, we
can sequence the three cell types in this way.

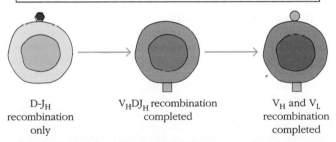

(c)

Sequencing of different antigen-bearing cells by analyzing each
population for the stage of V(D)J rearrangements in heavy and
light chains.

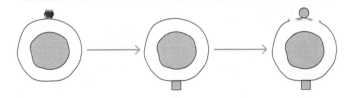

D-J_H
recombination
only

$V_H DJ_H$ recombination
completed

V_H and V_L
recombination
completed

(d)

Knocking out particular transcription factors
(TFs) stops development at particular points.

Knocking out TF1 leaves only this population. Therefore
TF1 is required to progress to V_HD recombination.

Knocking out TF2 leaves these two populations. Therefore
TF2 is required for progression to light chain rearrangement.

(e)

Placing GFP under the control of the TF2 promoter reveals that TF2
expression occurs in these two cell populations. Clearly, it is turned
on during the end of the blue stage, and is needed for progression
to the blue and green stage of development.

$V_H DJ_H$
recombination
completed

V_H and V_L
recombination
completed

FIGURE 10-4 Experimental approaches to the staging and characterization of B-cell progenitors. In this figure, the different icons do not represent specific antigens, but are used in order to illustrate the principle of the experiment. Investigators delineated the stage of B-cell development by using flow cytometry to characterize the cell-surface expression of developmental markers and molecular biology to correlate the expression of specific markers with the stage of immunoglobulin gene rearrangement. The requirement for transcription factor activity at each step was determined using both knockout and knockin genetic approaches. (See text for details.)

Second, by sorting cells bearing particular combinations of cell-surface markers, and *analyzing those cell populations for the occurrence of immunoglobulin gene rearrangements*, scientists were able to confirm the staging of the appearance of particular developmental antigens. For example, an antigen that appears on a cell in which no variable region gene rearrangement has taken place is clearly expressed very early in B-cell differentiation. Similarly, an antigen present on the surface of a cell that has rearranged heavy-chain, but not light-chain, genes defines a stage in B-cell differentiation later than that defined by the marker described above, whereas an antigen present on a cell that has undergone both heavy-chain and light-chain rear-

rangement characterizes a very late stage in B-cell development. In this way, cell-surface markers were defined that could serve as indicators of particular steps in B-cell differentiation (Figure 10-4c).

Third, investigators use the power of **knockout genetics** to determine the effects on B-cell development of eliminating the expression of particular genes, such as those encoding particular transcription factors (Figure 10-4d). For example, knocking out a gene encoding a particular transcription factor eliminates an animal's ability to complete V_H to DJ_H recombination while still allowing D to J_H rearrangement. This tells us that the particular transcription factor in question is not necessary for stages of B-cell differentiation

ADVANCES

The Role of miRNAs in the Control of B-Cell Development

Geneticists have long known that only a small fraction of chromosomal DNA specifies protein sequences, and early papers relegated the nonprotein-coding DNA segments to the somewhat igno- miniously described status of "junk DNA." In 1993, however, scientists studying the genome of the nematode *C. elegans* described groundbreaking investigations of some of the nonprotein-coding sequences that they had identified as hav- ing been transcribed but not translated. They showed that these primary tran- scripts were processed into small pieces of RNA, 18 to 30 nucleotides in length, that were capable of exerting control over the level of expression of mRNA.

The biosynthesis of these micro-RNAs follows a similar form in eukaryotes as diverse as *C. elegans* and humans (Figure 1). Fully capped and polyadenylated RNAs (pri-microRNAs) are synthesized by RNA polymerase II and are then cleaved into a hairpin-shaped 70- to 100-nucleotide pre- microRNA by the nuclear RNAase Drosha, which works in tandem with a second double-stranded RNA-binding protein, DGCR8. The cleaved pre-micro-RNA is then exported to the cytoplasm, where a second RNAase, Dicer, acting in associa- tion with two other proteins, processes it to an 18- to 30-nucleotide miRNA duplex, consisting of the mature miRNA and its anti-sense strand. In a final step, the mature miRNA, now single stranded, asso- ciates with a protein complex called the RNA-induced silencing complex, or RISC.

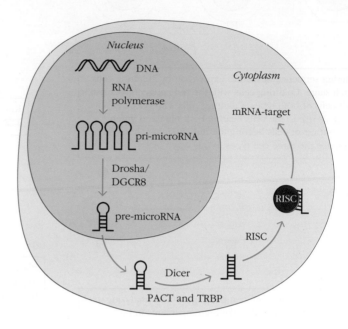

FIGURE 1

The generation of functioning microRNAs (miRNAs). Just like mRNA, miRNA species are transcribed as long, capped and polyadenylated RNA species (**pri-miRNA**) by RNA polymerase II. They are then cleaved by a nuclear RNase, Drosha, into a hairpin shaped nucleotide precursor molecule, termed a **pre-miRNA**. Drosha works in a protein complex with the protein DGCR8 (DiGeorge syndrome chromosomal region 8). Pre-miRNAs are then exported to the cytoplasm where a second ribonuclease, **Dicer**, in association with the proteins **PACT** and **TRBP** processes the pre-miRNA into a 19 to 24 nucleotide miRNA duplex, by removing the terminal loop. Next, a protein complex called RISC (*RNA-Induced Silencing Complex*) binds to one of the two strands of the duplex. The strand of miRNA that binds to RISC is the mature miRNA, and it drives the RISC enzyme to the target mRNA, resulting in mRNA silencing and/or destruction. *[Adapted from Vasilatou et al., 2009. The role of microRNAs in normal and malignant hematopoiesis. European Journal of Hematology, 84, 1 to 16. Figure 1.]*

The mature miRNA operates by com- plementary binding of a so-called "seed" region of 6 to 8 nucleotides at its 5' end to a region on its target mRNA. Once the miRNA has bound, three things can hap- pen: the target mRNA can be directly *tar-* *geted for cleavage*; the mRNA can be *destabilized*; or *translation from the mRNA can be repressed*. Furthermore, we now know that a single miRNA can target the synthesis of many proteins, and each mRNA can be the target of more than one

---

prior to D to J_H rearrangement, but it is required for one or more stages, starting with V_H to DJ_H recombination.

One drawback of the knockout approach, however, is that it only defines the first stage in differentiation at which the transcription factor is required. More recent variations have exploited **knockin genetics** (Figure 10-4e) to generate ani- mals that express fluorescent markers under the control of

the transcription factor promoters, so that every point at which the transcription factor is expressed can be delin- eated. The staging of transcription factor expression is then correlated with the expression of both cell-surface markers and immunoglobulin gene rearrangement. The sequence of B-cell development described in the next few sections has been elucidated using a combination of these strategies.

miRNA, thus adding to both the flexibility and the complexity of this mode of control over gene expression.

But does regulation by miRNAs operate during B-cell development? Several lines of evidence suggest that the answer to this question is an unequivocal "yes." From a theoretical standpoint, it is clear that the developmental changes that occur as B cells mature require rapid changes in the concentrations of such important proteins as transcription factors and pro- and anti-apoptotic molecules, among other regulatory proteins. The need for such rapid alterations in protein concentrations can be met efficiently by the type of post-transcriptional control mechanisms mediated by miRNAs.

Conditional loss of the gene encoding the Dicer nuclease destroys all capacity to synthesize mature miRNAs. Ablation of Dicer in early B-cell progenitors resulted in a developmental block at the pro- to pre-B-cell transition. In these experiments, the *pro-apoptotic molecule Bim* was expressed at higher concentrations in Dicer-ablated than in normal B-cell progenitors. Sequences in the 3′ untranslated region of the Bim gene were found to be complementary to miRNAs of the 17~92 family, suggesting that members of this family normally down-regulate Bim at this stage of development, enabling B cells to pass through this transition. The 17~92 family of miRNAs was also shown to affect expression of TdT and hence N-sequence addition. Other alterations in immunoglobulin gene expression were also observed in the absence of 17~92 miRNAs, including an increase in the expression of sterile transcripts. These data collectively demonstrate that *miRNAs are important in the control of the pro- to pre-B-cell transition* and affect both the expression of pro-apoptotic molecules and the nature of the Ig repertoire. As one would predict from these collective results, animals with increased expression of miR-17~92 family members express lower levels of Bim and suffer from a lympho-proliferative disorder and autoimmune disease.

Members of the 17~92 miRNA family were also found to control the levels of the Pten protein, which acts as an inhibitor of the pro-survival PI3 kinase and Akt signaling pathway. An increase in the levels of the miR-17~92 molecules allows for greater destruction of the Pten mRNA, resulting in increased cell survival and a corresponding increase in the number of lymphocytes available to proliferate.

Other investigators have addressed the question of which miRNAs are expressed at different stages of B-cell development. A combination of genomic (*in silico* analyses) and more classical molecular biological approaches have identified several miRNA species and/or families that are implicated in the control of B-cell development.

Perhaps the most well studied miRNA is miR-150, which is highly expressed in mature and resting B cells but not in their progenitors. The miR-150 molecule has been shown to depress the level of expression of the transcription factor c-Myb, known to be essential for the control of B-cell development. As might have been predicted, the pattern of c-Myb expression in lymphocyte development is complementary to that of miR-150, in that c-Myb is highly expressed in lymphoid progenitors and is down-regulated upon their maturation; in addition, transcriptional analysis confirmed that miR-150 is an important factor in the regulation of the levels of the c-Myb transcription factor during B-cell development.

MiR-150 is also implicated in B-1/B-2 lineage specification. B-cell-specific deletion of the *c-myb* gene stops B-cell development at the pro- to pre-B transition and also leads to the complete disappearance of the B-1 subset of B cells. If miR-150 is responsible for down-modulating the levels of c-Myb in vivo, then it might be predicted that a deficiency in miR-150 would result in an opposite phenotype to that expressed by a c-Myb deficient animal, and such was indeed found to be the case. Deficiency of miR-150 was found to result in an expansion of the B-1-B-cell pool, with a resulting increase in the levels of IgM antibody secretion.

The study of miRNA control of mammalian gene expression and lymphocyte development is still in its infancy, but the ability to manipulate and isolate cells at discrete stages in the developmental sequence provides a particularly tractable system in which to analyze the range of actions of different families of miRNAs. This field will undoubtedly be one to watch.

REFERENCES

Baltimore et al, 2008. MicroRNAs: new regulators of immune cell development and function. *Nature Immunology* **9:**839–845.

Koralov, S. B., et al. 2008. Dicer ablation affects antibody diversity and cell survival in the B lymphocyte lineage. *Cell* **132:**860–874.

Vasilatou, D., S. Papageorgiou, V. Pappa, E. Papageorgious, and J. Dervenoulas. 2009. The role of microRNAs in normal and malignant hematopoiesis. *European Journal of Haematology* **84:**1–16.

Xiao, C., and K. Rajewsky. 2009. MicroRNA control in the immune system: Basic principles. *Cell* **136:**26–36.

Recently, attention has begun to focus on small molecular weight miRNA species, which have profound effects on the stability of mRNA and hence on the expression of particular proteins. In Advances Box 10-2, we describe the effects of some of the miRNAs that have recently been shown to affect B-cell differentiation. This is currently an extremely active research area.

The Earliest Steps in Lymphocyte Differentiation Culminate in the Generation of a Common Lymphoid Progenitor

In this section, we will describe the process by which an HSC in the bone marrow develops into a CLP. Unless otherwise specified, the developmental pathway we describe refers to

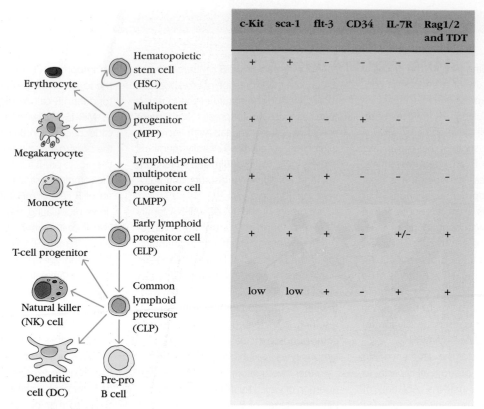

	c-Kit	sca-1	flt-3	CD34	IL-7R	Rag1/2 and TDT
Hematopoietic stem cell (HSC)	+	+	−	−	−	−
Multipotent progenitor (MPP)	+	+	−	+	−	−
Lymphoid-primed multipotent progenitor cell (LMPP)	+	+	+	−	−	−
Early lymphoid progenitor cell (ELP)	+	+	+	−	+/−	+
Common lymphoid precursor (CLP)	low	low	+	−	+	+

FIGURE 10-5 Expression of cell-surface markers on HSC and lymphoid progenitor cells. The maturation of HSCs into lymphoid progenitors, and the progressive loss of the ability to differentiate into other blood-cell lineages can be followed by the expression of the cell-surface markers as well as by the acquisition of RAG and TdT activity.

that followed by the predominant B-2-B-cell population. Specific aspects of development that differ among the various B-cell subsets will be addressed toward the end of this chapter.

HSCs

HSCs are both **self-renewing** (they can divide to create identical copies of the parent cell) and **multipotential** (they can divide to form daughter cells that are more differentiated than the parent cell and that can develop along distinct blood-cell lineages), and can give rise to all cells of the blood. HSCs maintain a relatively large number of genes in a so-called "primed" state, and an individual HSC may possess primed genes characteristic of multiple cell lineages. *Primed chromatin* is associated with lower-than-usual numbers of nucleosomes, is more accessible to enzyme activity than the majority of chromatin in the cell, and shows histone methylation and acetylation patterns characteristic of active chromatin. Depending on the environmental stimuli to which any given HSC is exposed, transcription factors may drive the cell down a number of possible developmental pathways. During the differentiation process that follows, primed chromatin regions containing genes that are not needed for the selected developmental pathway are shut down.

In HSCs bound for a B-cell fate, the transcription factors **Ikaros, Purine box factor 1 (PU.1)**, and **E2A** participate in the earliest stages of B-lineage development. Ikaros recruits chromatin remodeling complexes to particular regions in the DNA and ensures the accessibility of genes necessary for B-cell development. PU.1 presides over a leukocytic "balancing act"; low levels of PU.1 favor lymphoid differentiation, whereas cells expressing higher levels of PU.1 veer off to a myeloid fate. The level of PU.1 protein expressed is in turn regulated by the transcriptional repressor Gfi1, which downregulates the expression of PU.1 to the levels necessary for progression down the B-cell pathway. E2A expression contributes to the maintenance of the HSC pool by participating in the regulation of cell cycle control in this population.

As noted in Chapter 9, HSCs express the cell-surface molecule **c-Kit (CD117)**, which is the receptor for *stem cell factor* **(SCF)**. SCF is a cytokine that exists in both membrane-bound and soluble forms and the SCF-c-Kit interaction is critical for the development, in adult animals, of *multipotential progenitor cells* **(MPPs)**. Membrane-bound SCF plays a role in retaining the HSCs and its daughter progenitor cells in the appropriate environmental niches in the bone marrow. HSCs also express the *stem cell associated antigen-1* **(Sca-1)**. Both c-Kit and Sca-1 are expressed in parallel on early progenitor cells, and the levels of their expression drop as the cells commit to a particular cell lineage (Figure 10-5). HSCs are often described as being **Lin⁻**, a designation that refers to the fact that they have no "lineage markers" characteristic of a particular blood-cell subpopulation.

MPPs

MPPs generated on receipt of SCF/c-Kit signaling lose the capacity for extensive self-renewal, but retain the potential to differentiate into several different hematopoietic lineages. MPP cells retain the expression of c-Kit and Sca-1 and transiently express the molecule CD34. Indeed, antibodies to CD34 are used clinically to isolate cells at this stage in hematopoiesis. MPPs also express the chemokine receptor CXCR4, which enables them to bind the stromal cell-derived chemokine CXCL12. The interaction between CXCL12 and CXCR4 is important in ensuring that the progenitor cell occupies the correct niche within the bone marrow (see Figure 10-3).

LMPPs

The progenitor cell on its way to becoming a B cell then begins to express the *fms-related tyrosine kinase 3 receptor* (**flt-3**). Flt-3 binds to the membrane-bound flt-3 ligand on bone marrow stromal cells and signals the progenitor cell to begin synthesizing the **IL-7 receptor** (IL-7R, CD127). Flt-3 is expressed on B-cell progenitors from this point until the pro-B stage and acts synergistically with IL-7R to promote the growth of cells bearing flt-3 and IL-7R. The expression of flt-3 on the surface of the developing cell marks the loss of the potential of the MPP cell to develop into erythrocytes or megakaryocytes, and therefore characterizes a new level of cell commitment; however, this progenitor still retains the capacity to develop along either the *myeloid* or the *lymphoid* pathways.

These cells, now c-Ki$^-$, Sca-1$^+$, and flt-3$^+$, are termed **lymphoid-primed, multipotential progenitors (LMPPs)** (see Figure 10-5). As they become further committed to the lymphoid lineage, levels of the stem-cell antigens c-Kit and Sca-1 fall, and cells destined to become lymphocytes begin to express **RAG1/2** and **terminal deoxynucleotidyl transferase (TdT)** (Chapter 7). Expression of the genes encoding RAG1/2, TdT, IL-7R, and the B-cell-specific transcription factor EBF1 are all up-regulated at the end of this stage.

ELPs

Expression of RAG1/2 defines the cell as an **early lymphoid progenitor cell (ELP)**. A subset of ELPs migrates out of the bone marrow to seed the thymus and serve as the T-cell progenitors (discussed in Chapter 9). The rest of the ELPs remain in the bone marrow as B-cell progenitors. On these cells, the levels of the early c-Kit and Sca-1 antigens decrease as the levels of the IL-7R increase, and the ELP now develops into a CLP.

CLPs

At the CLP stage, the progenitor on its way to B-cell commitment still retains the potential to mature along the NK, conventional DC, or T-cell lineages. At this point in development, signals received through IL-7R promote cell survival and enhance the production of EBF-1 and other transcription factors that are required for later steps in the B-cell differentiation pathway. Signaling through the IL-7R occurs via pathways familiar from Chapter 4. Specifically, an IL-7R-mediated JAK-STAT pathway induces the up-regulation of the anti-apoptotic molecule Mcl1. Signaling through IL-7R also results in the up-regulation of the *C-myc* and *N-myc* genes, which signal the cell proliferation characteristic of the later, pro-B-cell stage.

CLPs are c-Kit^{low}, Sca-1^{low}, and IL-7R$^+$ and have lost myeloid potential. However, as a CLP destined to differentiate along the B cell pathway matures, the chromatin containing the immunoglobulin locus becomes increasingly accessible and the developing lymphocyte approaches the point at which it is irrevocably committed to the B-cell lineage.

The Later Steps of B-Cell Development Result in Commitment to the B-Cell Phenotype

Figure 10-6 illustrates the expression of cell-surface markers, and the patterns of rearrangement of immunoglobulin heavy-chain and light-chain genes starting at the pre-pro B-cell stage of development. The stages of B-cell differentiation have been defined by more than one group of scientists and, as a result, two systems of nomenclature are in common use. The first, and most widely used, is the Basel nomenclature (pre-pro, pro, pre-B, immature B) developed by Melchers and colleagues. The second (A, B, C, C′, D, E) is that defined by Hardy et al., and the process by which this system of classification of B-cell development was established is described in detail in Classic Experiment Box 10-3.

Pre-Pro B Cells

With the acquisition of the B-cell lineage-specific marker *B220 (CD45R)*, and the expression of increasing levels of the transcription factor EBF1, the developing cell enters the pre-pro-B-cell stage. EBF1 is an important transcription factor in lymphoid development, and therefore *transcription of the Ebf1 gene is itself under the control of multiple transcription factors* (Figure 10-7). These each bind at distinct promoter regions, and hence the level of transcription of the *Ebf1* gene can vary considerably depending on the combination of controlling factors present at any particular developmental stage. At the pre-pro-B-cell stage, *EBF-1*, along with E2A, binds to the immunoglobulin gene, promoting *accessibility of the D-J$_H$ locus* and preparing the cells for the first step of Ig gene recombination. *EBF-1* is also essential to the full expression of many B-cell proteins, including Igα,Igβ (CD79α,β), and the genes encoding the pre-B-cell receptor, which will be expressed when heavy-chain VDJ recombination is complete.

Pre-pro B cells remain in contact with CXCL-12-secreting stromal cells in the bone marrow. However, the onset of D to J$_H$ gene recombination classifies the cell as an early pro-B cell, and at this stage the developing cell moves within the bone marrow, seeking contact with IL-7 secreting stromal cells.

	Cell stage	Status of Ig genes	Surface Ig receptor expression	c-Kit	IL-7R	CD25	CD19	B220 (CD45R)
○	Pre-pro (Fr. A)[1]	GL[2]	None	–	lo	–	–	+
◔	Early pro (Fr. B)	DJ$_H$	None	lo	lo/+	–	+	+
◑	Late pro (Fr. C)	Some V$_H$DJ$_H$	None	lo	+	+/–	+	+
	1st checkpoint							
Ⓨ	Large Pre (Fr. C')	V$_H$DJ$_H$	Pre-BCR	–	+	+	+	+
Ⓨ	Small Pre (Fr. D)	V$_H$DJ$_H$ V$_L$J$_L$ re-arrange-ment begins	Decreasing levels of Pre-BCR	–	+	+	+	+
Ⓨ	Immature B (Fr. E)	V$_H$DJ$_H$ V$_L$J$_L$	IgM	–	+	+	+	+
	2nd checkpoint							

Ⓨ Pre-BCR
Ⓨ BCR
[1] Labeled fractions refer to the "Hardy nomenclature," described in the Classic Experiment Box 10-1.
[2] GL = germ line arrangement of heavy and/or light chain V region segments

FIGURE 10-6 Immunoglobulin gene rearrangements and expression of marker proteins during B-cell development. The expression of selected marker proteins is correlated with the extent of Ig gene rearrangement during B-cell development from the pre-pro B cell to the immature T1 B-cell stage. (See text for details.) [Adapted from K. Samitas, J. Lötvall, and A. Bossios, B cells: From early development to regulating allergic diseases, Archivum Immunologiae et Therapiae Experimentalis **58**:209–225, Figure 1.]

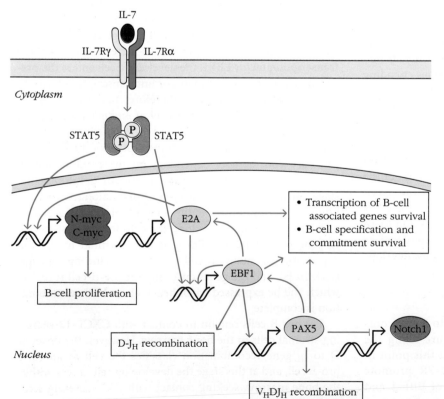

FIGURE 10-7 The interplay of transcription factors during early B-cell development. Dimerization and activation of the transcription factor STAT5 is stimulated by IL-7 binding to its receptor. STAT5 stimulates B-cell proliferation by activating the proliferative control proteins N-myc and C-myc. STAT5 collaborates with E2A proteins to promote the expression of early B-cell factor 1 (EBF1). EBF1 in turn promotes the expression of PAX5, and together the E2A proteins, EBF1, and PAX5 activate many genes leading to B-cell lineage specification and commitment. PAX5 and EBF1 both participate in positive feedback loops that enhance the levels of both EBF1 and PAX5 transcription. [Adapted from B. L. Kee, 2009, E and ID proteins branch out, Nature Reviews Immunology **9**:175–184.]

BOX 10-3

The Stages of B-Cell Development: Characterization of the Hardy Fractions

Richard Hardy's laboratory was one of the first to combine flow cytometry and molecular biology in experiments designed to analyze lymphocyte maturation. In this feature, we describe what those researchers did and how they generated a model of the sequencing of the stages of B-cell development from their data.

When Hardy and colleagues began their characterization of B-cell lineage development in the early 1990s, prior work using molecular analysis of long term bone marrow cell lines had already established the sequential rearrangement of heavy-chain and light-chain immunoglobulin genes. In addition, the expression of a number of cell-surface markers on bone marrow cells had been measured, and several of these antigens had been shown to be co-expressed with B220 (CD45R), which had already been established as a B-cell differentiation antigen. Hardy's approach was to characterize the sequence of expression of those antigenic markers that were found on the same cells as B220. The hypothesis was that some of these markers may be expressed on early B-cell progenitors and might therefore help to generate a scheme of B-cell development. In order to place cells expressing different combinations of markers into a developmental lineage, Hardy then sorted cells bearing each combination of his selected markers, and placed them into co-cultures with a bone marrow stromal cell line. After defined times in culture, he harvested the hematopoietic cells and re-characterized their surface marker expression.

The markers used in these experiments included B220 (CD45R) and CD43 (leukosialin), which had previously been shown to be expressed on granulocytes and all T cells, but was not present on mature B cells, with the exception of plasma cells. In addition, their experiments employed antibodies directed against *Heat Stable Antigen*, or HSA (CD24) and BP-1, an antigen on bone

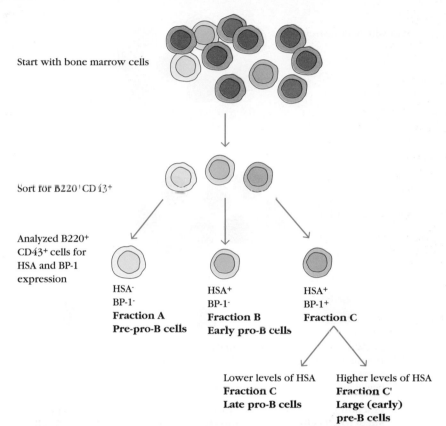

Start with bone marrow cells

Sort for B220⁺CD43⁺

Analyzed B220⁺ CD43⁺ cells for HSA and BP-1 expression

HSA⁻ BP-1⁻	HSA⁺ BP-1⁻	HSA⁺ BP-1⁺
Fraction A	**Fraction B**	**Fraction C**
Pre-pro-B cells	**Early pro-B cells**	

Lower levels of HSA **Fraction C** **Late pro-B cells**

Higher levels of HSA **Fraction C'** **Large (early) pre-B cells**

FIGURE 1
The isolation of Hardy's fractions. A, B, C, C'. Bone marrow cells were sorted for cells bearing B220 and CD43 and then analyzed for their expression of the cell-surface markers HSA and BP-1.

marrow cells. Both HSA and BP-1 had been previously shown to be differentially expressed at varying stages of lymphoid differentiation.

The first set of experiments analyzed those cells bearing both B220 and CD43 for the levels of their expression of HSA and BP-1 (Figures 1 and 2). Flow cytometry plots demonstrated that the B220⁺CD43⁺ cells neatly resolved into three discrete subpopulations. The first, labeled A in Figure 1, expressed neither HSA, nor BP-1. The second, labeled B, expressed HSA, but not BP-1, and the third expressed both of these antigens. Analysis of Ig gene rearrangements in these populations revealed that no gene rearrangements occurred in fraction A but that D to J_H gene segment rearrangements had begun in fraction B. Subsequent work has shown that V_H to DJ_H

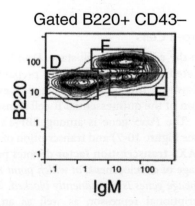

Gated B220+ CD43–

FIGURE 2
Flow cytometric characterization of the early developmental stages of B cells. (See text for details.) *[Hardy et al., J. Exp. Med. 173, 1213–1225. May 1991. By permission of Rockefeller University Press]*

(continued)

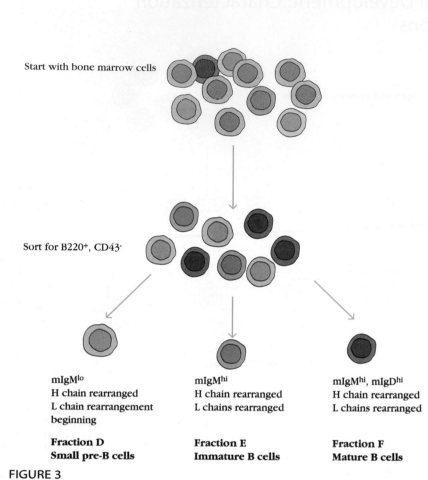

Start with bone marrow cells

Sort for B220⁺, CD43⁻

mIgM^{lo}
H chain rearranged
L chain rearrangement
beginning

Fraction D
Small pre-B cells

mIgM^{hi}
H chain rearranged
L chains rearranged

Fraction E
Immature B cells

mIgM^{hi}, mIgD^{hi}
H chain rearranged
L chains rearranged

Fraction F
Mature B cells

FIGURE 3
The isolation of Hardy's fractions D, E, and F. Bone marrow cells were sorted for cells bearing B220, but not CD43 (recognized by monoclonal antibody S7) and then analyzed for cell-surface expression of IgM and IgD.

rearrangements occur in fraction C (although at the time, the method of analysis that Hardy and colleagues used failed to reveal this second type of rearrangement).

So at this point, we know that three distinct types of B-cell precursors express both B220 and CD43, and can be discriminated on the basis of their levels of the further two antigens, HSA and BP-1. Fraction A corresponds to what we now know as pre-pro B cells, fraction B to early pro-B cells, and fraction C to late pro-B cells.

Culture of fraction C cells yielded cells that expressed membrane (m) IgM; similarly, culture of fraction B cells also yielded daughter cells expressing mIgM, but at a lower frequency than fraction C cells, suggesting that cells in fraction C were further along the differentiation pathway to mIgM⁺ B cells. Furthermore, the three different fractions displayed differential dependence on the need to adhere to the stromal cell layer. Cells from fraction A required stromal cell contact for survival. Fraction B cells survived best in contact with the stromal cells, but were able to survive in a culture in which they were separated from the stromal cells by a semipermeable membrane. Under these conditions, they could still receive soluble factors generated by the stromal cells, but were prevented from generating adhesive interactions with stromal cell-surface-bound growth factors. Fraction C cells

Pro-B Cells

In the early pro-B-cell stage, D to J$_H$ recombination is completed and the cell begins to prepare for V to DJ$_H$ joining. However, this final recombination event awaits the expression of the quintessential B-cell transcription factor, PAX5.

The *Pax5* gene is among EBF-1's transcriptional targets (see Figure 10-7) and transcription of genes controlled by the **PAX5 transcription factor** denotes passage to the pro-B-cell stage of development, *at which point the expression of non-B-lineage genes is permanently blocked.* PAX5 can act as a transcriptional repressor, as well as an activator, and blocks *Notch-1* gene expression, thereby terminating any residual potential of the pro-B cell to develop along the T-cell lineage. Many important B-cell genes are turned on at this stage, under the control of PAX5 and other transcription factors. Among these is the gene encoding **CD19**, which we first encountered in

Chapter 3, as one of the components of the B-cell co-receptor. CD19 is considered a quintessential B-cell marker and is often used as such in flow cytometry experiments.

Once the PAX5 protein is expressed, a mutual reinforcement occurs between PAX5 and EBF-1 expression, as illustrated in Figure 10-7, with each transcription factor serving to enhance the expression of the other. The PAX5 protein continues to be expressed in mature B cells until the B cell commits to a plasma-cell fate following antigenic stimulation (see Chapter 12). The higher levels of EBF1 expression induced by PAX5 also allow for an increase in the level of IL-7R expression.

PAX5 promotes V$_H$ to D recombination by contracting the Ig$_H$ locus, thus bringing the distant V$_H$ gene segments closer to the D-J$_H$ region. B cells deficient in PAX5 permit D to J$_H$ Ig gene rearrangement, but do not allow recombination of

BOX 10-3

survived and proliferated in the absence of stromal cell contact. Analysis of the factors secreted by the stromal cells that were necessary for survival and proliferation of the fraction B and C cells revealed one of them to be interleukin 7.

Hence, using the criteria of Ig gene rearrangements and phenotypic analysis of cultured cell populations, Hardy and colleagues were able to place the three fractions in sequence; cells of fraction A gave rise to cells of fraction B, which in turn mature into cells of fraction C.

Careful analysis of the contour graph of fraction C reveals that it, in turn, can be subdivided on the basis of the levels of expression of HSA. That population of cells bearing higher levels of HSA, as well as BP-1, is now defined as fraction C', which corresponds to early or large pre-B cells.

Hardy and colleagues next turned their attention to those cells that expressed B220, but had lost CD43, and measured their cell-surface expression of mIgM (Figure 3). Three populations of cells were again evident, which they labeled D, E, and F (Figure 4). Cells belonging to fraction D expressed zero to low levels of mIgM, showed complete heavy-chain rearrangement, and some light-chain rearrangement and correspond to small pre-B cells. Cells belonging to fraction E displayed high levels of mIgM as well as of B220, complete heavy-chain

rearrangement, and most of the cells in that fraction also displayed light-chain gene rearrangement. Fraction E cells are thus immature B cells ready to leave the bone marrow. Subsequent further characterization of fraction F cells showed that, in addition to surface IgM, these cells also bear surface IgD and therefore represent fully mature B cells, presumably recirculating through the bone marrow.

Thus, Hardy's experiments revealed that the pool of progenitor and precursor B cells in the bone marrow represents a complex mixture of cells at different stages of development, with varying requirements for stromal cell contact and interleukin support.

These elegant experiments still had one more story to tell that did not appear in the original paper, but which emerged in later publications. Single-cell PCR analysis of fraction C cells showed that many of them had nonproductive rearrangements on both heavy-chain chromosomes (see Figure 2). In contrast, all the cells from the C' fraction demonstrated productive rearrangements on one of the heavy-chain chromosomes. Fraction C cells therefore represent B cells that have been unable to productively rearrange one of their heavy-chain genes and that will therefore eventually die by apoptosis. The C' fraction also included the highest proportion of cells in cycle of any of the B220$^+$ B-cell stages in

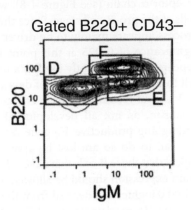

Gated B220+ CD43−

FIGURE 4
Flow cytometric characterization of the later developmental stages of B cells. (See text for details.) [Hardy et al., J. Exp. Med. 173, 1213–1225. May 1991. By permission of Rockefeller University Press]

the marrow. This is consistent with the notion that the new heavy chain is associating with the surrogate light chain at the C' stage and the pre-B-cell receptor complex is expressed on the cell surface, triggering the period of clonal expansion of B cells described in this chapter.

REFERENCES

Hardy, R. R., et al. 1991. Resolution and characterization of pro-B and pre-pro-B cell stages in normal mouse bone marrow. *Journal of Experimental Medicine* **173**:1213–1225.

Hardy, R. R., P. W. Kincade, and K. Dorshkind. 2007. The protean nature of cells in the B lymphocyte lineage. *Immunity* **26**:703–714.

the V_H to the D Ig gene segment, indicating that PAX5 is essential to the second step of Ig gene rearrangement.

Expression of the signaling components of the B-cell receptor, Igα and Igβ, also begins at the pro-B-cell stage and the Igα,Igβ signaling complex is briefly placed on the cell surface in complex with the chaperone protein calnexin. Although this Igα,Igβ complex has been referred to as a "pro-BCR," no ligand has yet been established for it, nor do we yet understand the importance of any signaling that may emanate from it.

Also during the pro-B-cell stage, c-Kit is once more turned on briefly, enabling the cell to receive signals from stem cell factor. By the beginning of the pre-B-cell stage of development, expression of c-Kit is irreversibly turned back off. By the *late pro-B-cell stage*, most cells have initiated V_H to DJ_H Ig gene segment recombination, which is completed by the onset of the early pre-B-cell stage.

Pre-B Cells

During the pre-B-cell stage, the cell expresses a **pre-B-cell receptor** composed of the rearranged heavy chain, complexed with the **VpreB** and λ5 components of the **surrogate light chain** (see Figure 7-10). The appearance of this pre-B-cell receptor signals the entry of the developing B cell into the large, or **early pre-B-cell phase**. As we learned in Chapter 7, the expression of the heavy chain at the cell surface is necessary for the termination of further heavy-chain rearrangement and ensures allelic exclusion of the Ig heavy-chain genes. Animals deficient in the expression of either the pre-B-cell receptor, or of the signaling components Igα,Igβ, fail to progress to the pre-B-cell stage.

Signaling through the pre-B-cell receptor induces a few rounds of proliferation in the pre-B cell. This proliferative phase correlates in time with the expression on the pre-B-cell surface of CD25, the α chain of the high-affinity IL-2 receptor α chain (see Figure 4-8), which first appears on B cells at the pro-B-cell stage. Since the pre-B-cell proliferative process appears mainly to be driven by IL-7, the functional significance of CD25 at this point is unclear. However, its appearance is frequently used as a marker of the late pro-B-cell to early pre-B-cell stage of development.

V_H gene recombination is energetically expensive to the organism, as not all developing B cells are successful in undergoing productive V_H gene rearrangements, and those that fail to do so are lost by apoptosis. It therefore seems logical that those B cells that have achieved productive heavy-chain expression should be allowed to proliferate. Each individual daughter cell derived from this proliferative process is then free to participate in a different light-chain rearrangement event. Most individuals will therefore have multiple B cells expressing precisely the same heavy-chain rearrangement but each with a different light chain, and many different receptor specificities can thereby be generated from each successful heavy-chain rearrangement. Recall from Chapter 9 that an analogous process of β chain rearrangement, followed by proliferation prior to α rearrangement, occurs in T cells.

If the pre-B-cell receptor cannot be displayed on the cell surface because of non-productive V_HDJ_H gene rearrangements, B-cell development is halted and the cell is lost to apoptosis. This stage in B-cell development is therefore referred to as the **pre-B-cell (1st) checkpoint** (see Figure 10-6). Progress through this checkpoint depends on some type of signaling event through the pre-B-cell receptor, and recent evidence suggests that this is mediated via interactions between arginine-rich regions in the non-immunoglobulin portion of the λ5 component of the surrogate light chain, or in the CDR3 regions of some heavy chains, with negatively charged molecules on the surface of the stromal cells.

Pre-B-cell receptor signaling induces the transient downregulation of RAG1/2 and the loss of TdT activity. Together, these events ensure that, as soon as one heavy-chain gene has successfully rearranged, no further heavy-chain recombination is possible. This results in the phenomenon of **allelic exclusion**, whereby the genes of only one of the two heavy-chain alleles can be expressed in a single B cell. As a result of this pre-B-cell receptor signaling, the chromatin at the unused heavy-chain locus undergoes a number of physical changes that render it incapable of participating in further rearrangement events. Recall that IL-7 provided one of the signals that brought the V_H, D, and J_H loci into close apposition with one another at the beginning of V_HDJ_H recombination. A reduction in IL-7 signaling at the pre-B-cell stage now reverses that initial locus contraction, resulting in the physical separation of the V_H, D, and J_H gene segments in the unrearranged heavy-chain locus. This *decontraction* is then followed by deacetylation events that deactivate the unused heavy-chain locus and return it to a *heterochromatic* (inactive, closed) configuration.

Surrogate light-chain expression is also terminated by a negative feedback round of signaling through the pre-B-cell receptor. At the end of pre-BCR-signaled cell proliferation, the pre-B-cell receptor is lost from the surface, and this signals entry into the small or **late pre-B-cell stage**. At this point, light-chain rearrangement is initiated with the re-expression of the *Rag1/2* genes. Very little TdT activity remains at this stage, and therefore N region addition occurs less frequently in light chains than in heavy chains. In the mouse, light-chain rearrangement begins on one of the κ chain chromosomes, followed by the other. If neither κ chain rearrangement is successful, rearrangement is then successively attempted on each of the λ chain chromosomes. In humans, rearrangement is initiated randomly at either the κ or the λ loci.

Once a light-chain gene rearrangement has been successfully completed, the IgM receptor is expressed on the cell surface, signaling entry into the immature B-cell stage. If the attempts at light-chain immunoglobulin gene rearrangement are not successful, the nascent cell is eventually lost at the immature B-cell (2nd) checkpoint (see Figure 10-6). However, given the availability of four separate chromosomes on which to attempt rearrangement, and the opportunity for light-chain editing in the case of unproductive rearrangement, most pre-B cells that have successfully rearranged their heavy chains will progress to the formation of an immature B cell.

Immature B Cells in the Bone Marrow Are Exquisitely Sensitive to Tolerance Induction

Immature B cells bear a functional receptor in the form of membrane IgM, but have not yet begun to express any other class of immunoglobulin. They continue to express B220, CD25, IL-7R, and CD19.

Once the functional BCR is assembled on the B-cell membrane, the receptor must be tested for its ability to bind to self antigens, in order to ensure that as few as possible autoreactive B cells emerge from the bone marrow. Those immature B cells that are found to bear autoreactive receptors undergo one of three fates; some are lost from the repertoire prior to leaving the bone marrow, by the BCR-mediated apoptotic process of **clonal deletion**. The loss of B cells bearing self-reactive receptors *within the bone marrow* is referred to as **central tolerance**. Other autoreactive B cells reactivate their RAG genes to initiate the process of light-chain **receptor editing** (see Chapter 7). Some autoreactive B cells that recognize soluble self antigens within the bone marrow may survive to escape the bone marrow environment, but become **anergic**, or unresponsive, to any further antigenic stimuli.

The concept of negative selection of lymphocytes bearing autoreactive receptors should be familiar from the discussion of T-cell tolerance in Chapter 9. However, functional differences between T cells and B cells mean that the selection processes against autoreactive B cells are different from those that protect against the emergence of autoreactive T cells, and indeed can be somewhat less stringent. Since stimulation of B-2 B cells requires T-cell help, an autoreactive B cell cannot respond to antigen with antibody production unless there is also an autoreactive T cell that can provide the necessary cytokines and costimulation (Chapter 12). Thus, it is quite possible that most individuals carry significant numbers of autoreactive B cells within their mature B-cell repertoires that are never activated.

There are also mechanistic differences between the modes of negative selection among B and T cells. At this point, no equivalent of the AIRE protein has been shown to exist for B-cell selection, and so B-cell negative selection within the bone marrow is more limited with respect to the available tolerogenic antigenic specificities than is T-cell negative selection in the thymus.

Many, but Not All, Self-Reactive B Cells Are Deleted within the Bone Marrow

Our understanding of how the immune system eliminates or neutralizes autoreactive threats has been facilitated by the development of transgenic animals which express both deliberately introduced auto-antigens and the receptors that recognize them. It has long been established that cross-linking the IgM receptors of immature B cells in vitro (performed experimentally by treating the cells with antibodies against the receptor μ chain) results in death by apoptosis. In contrast, performing the same experiments with mature B cells, bearing both IgM and IgD receptors, results in activation. David Nemazee and colleagues set out to test whether the apoptotic response of immature B cells in vitro reflected what happens in the bone marrow in vivo when an immature B cell meets a self antigen.

Nemazee et al.'s approach was conceptually simple, although experimentally complex, particularly for the time period in which the work was done (1989). They generated mice transgenic for both a heavy and a light chain specific for the MHC molecule H-2K^k. All the B cells in this mouse therefore made only anti-H-2K^k antibodies. If immature B cells undergo selection to prevent autoimmunity, these cells would be selected against in a mouse that expresses the H-2K^k gene for MHC. By appropriate breeding, they introduced the immunoglobulin H-2K^k-specific transgenes into mice bearing two different MHC genotypes.

In the first group of mice (Figure 10-8a), which bore H-2K^d but no H-2K^k antigens, they were able to detect the transgenic antibody at high frequency on the surface of B cells and at high concentration in the serum (Table 10-1). This makes sense, as the transgenic antibody would be unable to bind the H-2K^d molecules and so the B cells that produce it would not be negatively selected. However, when they bred these animals with mice of the H-2K^k type (Figure 10-8b), no membrane-bound or secreted anti-H-2K^k antibodies could be detected, suggesting that all immature B cells bearing the potentially autoimmune receptor antibodies had been deleted in the bone marrow. This deletion occurred via induction of apoptosis in the autoimmune cells.

Interestingly, in the H-2K$^{k/d}$ mice, not all B cells bearing the autoimmune transgenes were deleted, even though all B cells in this mouse should bear the anti-H-2K^k receptor

(Figure 10-8c). Closer examination revealed that some of the residual transgene-expressing B cells in the bone marrow had undergone *light-chain receptor editing* (see Chapter 7), changing their antigen specificity so they no longer bound the H-2K^k antigen. Recent experiments suggest that in vivo, a significant fraction of potentially autoimmune B cells undergo receptor editing (or even V_H gene replacement (Chapter 7), and successfully generate acceptable BCRs prior to release from the bone marrow.

In normal animals, not all potentially autoimmune B cells are lost to clonal deletion or altered via receptor editing or V_H gene replacement within the bone marrow, however; some are released to the periphery and subject to further rounds of selection.

B Cells Exported from the Bone Marrow Are Still Functionally Immature

Once the B cell expresses IgM on its membrane (mIgM), it is referred to as an *immature B cell*. This B cell is ready for export to the spleen, where it completes its developmental program. Immature B cells have a short half-life, in part as a result of expressing low levels of the anti-apoptotic molecules Bcl-2 and Bcl-xl. They also express high levels of the cell-surface molecule, Fas, which is capable of transmitting a death signal when bound by its ligand (see Chapters 4 and 9). Immature B cells are exquisitely susceptible to tolerance induction, and if they encounter a self antigen at this stage of development, the B cells will re-express the RAG1 and RAG2 genes and edit their light-chain genes. If receptor editing fails to yield a suitable receptor, the cell undergoes apoptosis.

The study of B-cell development in the periphery, like that in the bone marrow, has benefited significantly from the ingenious application of flow cytometry, which has enabled the classification of immature B cells into two subpopulations of **transitional B cells (T1, T2)**. These transitional B cells act sequentially as the precursors to the fully mature B cell.

T1 and T2 Transitional B Cells

T1 and T2 transitional B cells were characterized initially on the basis of their cell-surface expression of immunoglobulin receptors and membrane markers (Table 10-2). T1 cells are mIgM^{hi}, mIgD$^{-/lo}$, CD21$^-$, CD23$^-$, CD24$^+$, and CD93$^+$. T2 cells differ from T1 cells in having higher levels of mIgD and in expressing CD21 (the complement receptor and B-cell co-receptor; see Figure 3-7) and CD23. T2 cells also express BAFF-R, the receptor for the B-cell survival factor **BAFF**, whose expression is dependent on signals received through the BCR. As B cells differentiate from the transitional T2 state to full maturity, they raise their levels of mIgD still further, while reducing the expression of mIgM. They also cease to express CD24 and CD93.

T1 cells that have been labeled and transferred into recipient mice develop into T2 cells. Similarly, both transitional B-cell subpopulations have been demonstrated to have the

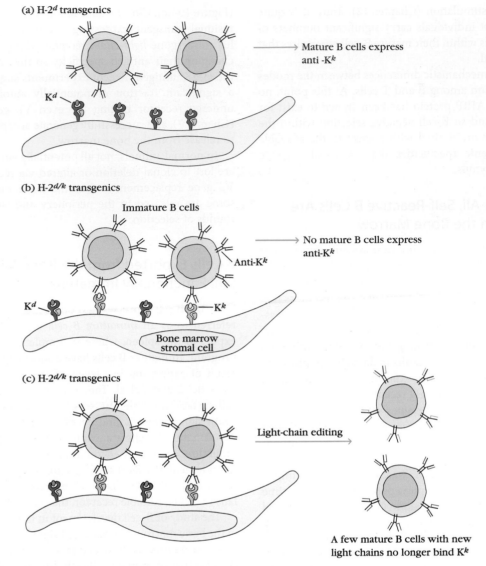

(a) H-2^d transgenics

→ Mature B cells express anti -K^k

K^d

(b) H-2$^{d/k}$ transgenics

Immature B cells

Anti-K^k

→ No mature B cells express anti-K^k

K^d K^k

Bone marrow stromal cell

(c) H-2$^{d/k}$ transgenics

Light-chain editing

A few mature B cells with new light chains no longer bind K^k

FIGURE 10-8 Experimental evidence for negative selection (clonal deletion) of self-reactive B cells during maturation in the bone marrow. The presence or absence of mature peripheral B cells expressing a transgene-encoded IgM against the H-2 class I molecule K^k was determined in H-2^d mice (a) and H-2$^{d/k}$ mice (b) and (c). (a) In the H-2^d transgenics, the immature B cells did not bind to a self antigen and consequently went on to mature, so that splenic B cells expressed the transgene-encoded anti-K^k as membrane Ig. (b) In the H-2$^{d/k}$ transgenics, many of the immature B cells that recognized the self antigen K^k were deleted by negative selection. (c) More detailed analysis of the H-2$^{d/k}$ transgenics revealed a few peripheral cells that expressed the transgene-encoded μ chain but a different light chain. Apparently, a few immature B cells underwent light-chain editing, so they no longer bound the K^k molecule and consequently escaped negative selection. *[Adapted from D. A. Nemazee and K. Burki, 1989, Nature 337:562; S. L. Tiegs et al., 1993, Journal of Experimental Medicine 177:1009.]*

TABLE 10-1   Expression of transgene encoding IgM antibody to H-2^k class I MHC molecules

		Expression of transgene	
Experimental animal	Number of animals tested	As membrane Ab	As secreted Ab (μg/ml)
Nontransgenics	13	(−)	<0.3
H-2^d transgenics	7	(+)	93.0
H-2$^{d/k}$ transgenics	6	(−)	<0.3

[Adapted from D. A. Nemazee and K. Bürki, 1989, February, Clonal deletion of B lymphocytes in a transgenic mouse bearing anti-MHC class I antibody genes, Nature 337:562–566. doi:10.1038/337562a0.]

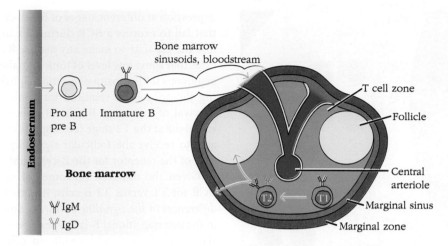

FIGURE 10-9 T2, but not T1, transitional immature cells can enter the B-cell follicles and recirculate. Immature B cells leave the bone marrow as T1 transitional immature B cells. They enter the spleen from the bloodstream through the marginal sinuses, percolating into the T-cell zones, and differentiating into T2 transitional B cells, which gain the ability to enter the B-cell follicles and recirculate. There, the T2 cells complete their differentiation into mature follicular B-2 cells. Marginal zone cells have also been shown to derive from T2 cells. *[Adapted from J. B. Chung, M. Silverman, and J. G. Monroe, 2003, Transitional B cells: Stem by step towards immune competence, Trends in Immunology 24:343–348, Figure 1; and R. E. Meblus, and G. Kraal, 2005, Structure and function of the spleen. Nature Reviews Immunology 5:606–616, Figure 1.]*

capacity to differentiate into mature B cells. These experiments have therefore proven that the order of the developmental sequence progresses from T1 to T2 to mature B cell. The time in transit of a T1 cell to a mature B cell has been measured to be approximately 3 to 4 days. Most T1 B cells differentiate to T2 cells within the spleen, but a minority of about 25% of transitional B cells emerge from the bone marrow already in the T2 state. The increased level of maturity of T2 cells correlates with changes in the expression of chemokine and cytokine receptors, such that T2 cells, but not T1 cells, are capable of

recirculating among the blood, lymph nodes, and spleen; and T2 cells, but not T1 cells, can enter B-cell follicles.

Figure 10-9 shows the path of the developing transitional B cell as it leaves the bone marrow and enters the spleen through the central arteriole, which deposits it in the marginal sinuses, just inside the outer marginal zone. (In humans, the anatomy of the spleen is slightly different, and the cells arrive in the spleen in a peri-follicular zone.) From there, the T1 B cell percolates through to the T-cell zone, where some fraction of T1 cells will mature into the T2 state. T2 B cells are then able to enter the follicles or the marginal zone where they complete their developmental program into fully mature, recirculating B lymphocytes.

In Chapter 7, we learned that mature B cells bear on their surfaces two classes of membrane-bound immunoglobulins—IgM and IgD—and that the expression of mIgD along with mIgM requires carefully regulated mRNA splicing events. It is at the point of transit between the T1 and T2 stages of development that we observe the onset of these splicing capabilities. Mature B cells bear almost 10 times more mIgD than mIgM, and so mIgD expression results in significant up-regulation in the number of B-cell immunoglobulin receptors.

The effect of strong BCR engagement with a multivalent, or membrane-bound, antigen depends on the maturational status of the transitional B cell (Figure 10-10 and Table 10-3). Self-reactive T1 B cells are eliminated by apoptosis in response to a strong antigenic signal, in a process reminiscent of thymocyte negative selection, leading to peripheral tolerance; recent experiments have suggested that in healthy adults, fully 55% to 75% of immature B cells are lost by this process. In contrast, once the B cell has matured into a T2 transitional B cell, it becomes resistant to antigen-induced apoptosis, reminiscent of thymocytes that have reached the single-positive stage of development. This

			Mature
Marker	**T1**	**T2**	**B-2 cells**
mIgM	High	High	Intermediate
mIgD	−/Low	Intermediate	High
CD24	+	+	−
CD93	+	+	−
CD21	−	+	+
CD23	−	+	+
BAFF receptor (BAFF-R)	+/−	+	+

TABLE 10-2 Surface marker expression on transitional T1 and T2 and mature B-2 B cells

Note: CD93 is defined by the AA4 monoclonal antibody. CD24 is otherwise known as the Heat Stable Antigen (HSA). CD23 is a low-affinity receptor for IgE. CD21 is a receptor for complement and part of the B-cell co-receptor.

After D. Allman and S. Pillai. 2008. Peripheral B cell subsets. Current Opinion in Immunology 20:149–157, and others.

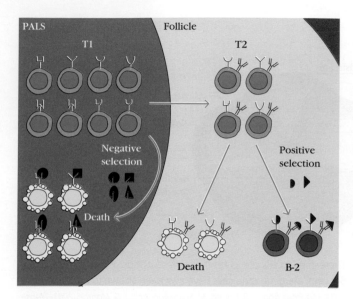

FIGURE 10-10 Transitional B cells bound for a follicular fate undergo positive and negative selection in the spleen. T1 transitional B cells, which recognize antigen with high affinity in the spleen, are eliminated by negative selection, and never reach the splenic follicles. Those T1 cells that escape negative selection enter the follicles and differentiate into T2 B cells. In the follicles, their BCRs interact with an unknown molecule(s) that deliver(s) a stimulatory survival signal. Transitional cells that have received this survival signal up-regulate their BAFF receptors (positive selection). Those T2 B cells that fail to receive the stimulatory signal (or that fail to receive a BAFF survival signal) die in the spleen. Selecting antigens are shown as violet shapes; T1 and T2 cells are green. White cells represent dead cells that were either negatively selected or failed positive selection. [Adapted from T. T. Su et al., 2004, Signaling in transitional type 2 B cells is critical for peripheral B-cell development, Immunological Reviews **197**:161–178, Figure 3.]

resistance to receptor-induced cell death results in part from the fact that T2 B cells have increased their expression of the anti-apoptotic molecule Bcl-xl.

Using conditional knockout genetic techniques (see Chapter 20), animals can be generated that lack Ig receptor

expression at different stages of B-cell development. Animals that fail to express a BCR during the immature B-cell stage lose the capacity to make any mature B cells at all. This indicates that some low level of tonic signaling through the BCR, analogous to the positive selection signal needed by developing thymocytes, is required for continued generation and survival of immature B cells. B cells unable to receive this signal die at the T2 stage (see Figure 10-10). Those T2 B cells able to receive the follicular signal up-regulate the expression of the receptor for the B-cell survival factor BAFF.

Given the different outcomes of signaling through the BCR for T1 versus T2 B cells, it is clear that there must be *differences in the signaling pathways* downstream of the BCR in the two transitional B-cell types. Specifically, BCR-mediated signaling of T1 B cells results in calcium release without significant production of diacylglycerol, and provides an apoptotic signal. In contrast, receipt of BCR signals by T2 B cells induces both an increase in the concentration of intracytoplasmic calcium and in diacylglycerol production. This combination of intracellular second messengers delivers both maturational and survival signals to the cell and suggests the involvement of a diacylglycerol-activated protein kinase in survival signaling (see Chapter 3).

But what causes this difference in the signal transduction pathways between T1 and T2 cells? A partial answer to this question appears to lie in differences in the composition of the lipid membranes of the two types of cells. Immature T1 B cells contain approximately half as much cholesterol as their more mature counterparts, and this reduction in cholesterol levels appears to prevent efficient clustering of the B-cell receptor into lipid rafts upon BCR stimulation. This may cause a reduction of the strength of BCR signaling in T1 versus T2 immature B cells.

The development of B cells through the transitional phase is absolutely dependent on signaling through the **BAFF receptor (BAFF-R)**. BAFF-R expression is first detected in T1 B cells and increases steadily thereafter. BAFF is then required constitutively throughout the life of mature B cells. Signaling through the BAFF/BAFF-R axis promotes survival of transitional B cells by inducing the synthesis of anti-apoptotic factors such as Bcl-2, Bcl-xl, and Mcl-1, as well as by interfering with the function of the pro-apoptotic molecule Bim (see Figure 9-12).

The discovery of a B-cell survival signal mediated by BAFF/BAFF-R interactions, and distinct from survival signals emanating from the BCR, extends our thinking about how B cells are selected for survival in the periphery. In the presence of high levels of BAFF, B cells that may not otherwise receive sufficient quantities of survival signals via the BCR may survive a selection process that would otherwise eliminate them. In this way, BAFF can provide plasticity and flexibility in the process of B-cell deletion. However, this may be accomplished at the cost of maintaining potentially autoreactive cells.

T3 B Cells Are Primarily Self-Reactive and Anergic

Transitional T3 B cells were first characterized in the blood and lymphoid organs by flow cytometry and were described

TABLE 10-3	Responses to strong BCR signaling in T1, T2, and mature B-2 B cells		
Nature of Response	**T1**	**T2**	**Mature B-2 B cells**
Formation of lipid rafts	+/−	+	+
Increase in cytoplasmic Ca²⁺ ion concentrations	+	+	+
Increase in diacylglycerol concentrations	−	+	+
Induction of Bcl-xl	−	+	+
Induction of apoptosis	+	−	−

as being CD93$^+$mIgM^{low}CD23$^+$. The function of CD93 is so far unknown; CD23 is a low-affinity receptor for some classes of immunoglobulin, and the two markers were used in these experiments only to identify the cell populations, and not because of any particular reasons relevant to their functionality. Recent experiments have suggested that the T3 population may represent B cells that have been rendered anergic by contact with soluble self antigen but have not yet been eliminated from the B-cell repertoire.

A transgenic system developed by Goodnow and colleagues first placed the concept of B-cell anergy, or unresponsiveness, onto a firm experimental footing. Anergic lymphocytes clearly recognize their antigens, as shown by the identification of low levels of molecular signals generated within the cells after binding to antigen. However, rather than being activated by antigen contact, anergic B cells fail to divide, differentiate, or secrete antibody after stimulation, and many die a short time after receipt of the antigenic signal.

Goodnow et al. developed the two groups of transgenic mice illustrated in Figure 10-11a. One group of mice carried a *hen egg-white lysozyme* (HEL) transgene linked to a metallothionine promoter, which placed transcription of the *HEL* gene under the control of zinc levels in the animals' diet. This allowed the investigators to alter the levels of soluble *HEL* expressed in the experimental animals by changing the concentration of zinc in their food. Under these experimental conditions, *HEL* was expressed in the periphery of the animal, but not in the bone marrow. The other group of transgenic mice carried rearranged immunoglobulin heavy-chain and light-chain transgenes encoding anti-HEL antibody; in these transgenic mice, the rearranged anti-HEL transgene is expressed by 60% to 90% of the mature peripheral B cells. Goodnow then mated the two groups of transgenics to produce "double-transgenic" offspring carrying both the HEL and anti-HEL transgenes (Figure 10-11b) and asked what effect *peripheral HEL* expression would have on antibody expression by B cells bearing the anti-HEL antibodies. They found that the double-transgenic mice continued to generate mature, peripheral B cells bearing anti-HEL membrane immunoglobulin of both the IgM and IgD classes, indicating that the B cells had fully matured. However, these B cells were functionally nonresponsive, or anergic.

Flow-cytometric analysis of B cells from the double-transgenic mice showed that, although large numbers of anergic anti-HEL cells were present, they expressed membrane IgM at levels about 20-fold lower than anti-HEL single transgenics (Figure 10-11b). When these mice were given an immunizing dose of HEL, few anti-HEL plasma cells were induced and the serum anti-HEL titer was very low (Table 10-4). Furthermore, when antigen was presented to these anergic B cells in the presence of T-cell help, many of the anergic B cells responded by undergoing apoptosis. Additional analysis of the anergic B cells demonstrated that they had a shorter half-life than normal B cells and appeared to be excluded from the B-cell follicles in the lymph nodes and spleen. These properties were dependent on the continuing presence of

antigen, as the B cell half-lives were restored to normal lengths upon adoptive transfer of the transgene-bearing B cells to an animal that was not expressing *HEL*. The anergic response appears to be generated in vivo when immature B cells meet a soluble self antigen.

More recent experiments have focused on defining the differences between the signal transduction events leading to anergy versus activation. Anergic B cells show much less antigen-induced tyrosine phosphorylation of signaling molecules, when compared with their nonanergic counterparts and antigen-stimulated calcium release from storage vesicles into the cytoplasm of the anergic B cells was also dramatically reduced. Anergic B cells also require higher levels of the cytokine BAFF for continued survival, and it is likely that their reduced half-lives result from unsuccessful competition with normal B cells for limiting amounts of this survival molecule. One of the outcomes of BAFF signaling is a reduction in the cytoplasmic levels of the pro-apoptotic molecule BIM; as might be expected, anergic B cells show higher-than-normal levels of BIM and a correspondingly increased susceptibility to apoptosis.

The conclusion from these experiments is that mechanisms exist, even after the B cells have exited the bone marrow and entered the periphery, that minimize the risk that B cells make antibodies to soluble self proteins expressed outside the bone marrow. B cells reactive to such proteins respond to receptor stimulation in the absence of appropriate T-cell help by anergy and eventual apoptosis.

What might be the function of these anergic cells? One possibility is that they serve to absorb excess self antigens that might otherwise be able to deliver activating signals to high-affinity B cells and thus lead to autoimmune reactions. Another is that they represent cells destined for apoptosis that do not yet display the characteristic microanatomy of apoptotic cells. Yet another is that these cells will eventually develop into B regulatory cells (see Chapter 12). As is so often the case in the immune system, it is more than possible that all of these functions are subsumed within this intriguing cell population, which remains the subject of intensive current investigation.

Mature, Primary B-2 B Cells Migrate to the Lymphoid Follicles

Fully mature B cells express high levels of IgD and intermediate levels of IgM on their cell surfaces (see Table 10-2). Mature B cells recirculate between the blood and the lymphoid organs, entering the B-cell follicles in the lymph nodes and spleen, and responding to antigen encounter in the presence of T-cell help with antibody production (Chapter 12). Approximately 10 million to 20 million B cells are produced in the bone marrow of the mouse each day, but only about 10% of this number ever take up residence in the periphery and only 1% to 3% will ever enter the recirculating follicular B-2 B-cell pool. Some of these cells are lost to the process of clonal deletion, but others are perfectly harmless B cells that nonetheless fail to thrive. Experimental depletion of the mature B-cell population, either chemically or by irradiation,

(a)

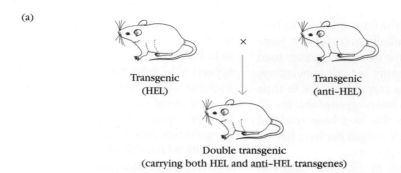

(b)

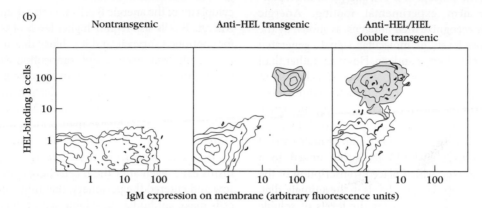

FIGURE 10-11 Goodnow's experimental system for demonstrating clonal anergy in mature peripheral B cells. (a) Production of double-transgenic mice carrying transgenes encoding HEL (hen egg-white lysozyme) and anti-HEL antibody. (b) Flow-cytometric analysis of peripheral B cells that bind HEL compared with membrane IgM levels. The number of B cells binding HEL was measured by determining how many cells bound fluorescently labeled HEL. Levels of membrane IgM were determined by incubating the cells with anti-mouse IgM antibody labeled with a fluorescent label different from that used to label HEL. Measurement of the fluorescence emitted from this label indicated the level of membrane IgM expressed by the B cells. The nontransgenics (*left*) had many B cells that expressed high levels of surface IgM but almost no B cells that bound HEL above the background level of 1. Both anti-HEL transgenics (*middle*) and anti-HEL/HEL double transgenics (*right*) had large numbers of B cells that bound *HEL* (blue), although the level of membrane IgM was about 20-fold lower in the double transgenics. The data in Table 10-4 indicate that the B cells expressing anti-HEL in the double transgenics cannot mount a humoral response to *HEL*.

followed by in vivo reconstitution, results in rapid replenishment of the B-cell follicular pool. This suggests that the follicular B-cell niches have a designated capacity and that once full they turn away additional B cells. Most probably, the mechanism for this homeostatic control of B-cell numbers relies on competition for survival factors, particularly BAFF and its related proteins.

Experiments using conditional RAG2 knockout animals, in which all new B-cell development was prevented in otherwise healthy adult animals, indicate that follicular B-2 B cells have a half-life of approximately 4.5 months. In contrast, since B-1 B cells can self-renew in the periphery, their numbers are unaffected in this experimental knockout animal.

TABLE 10-4	Expression of anti-HEL transgene by mature peripheral B cells in single- and double-transgenic mice			
Experimental Group	**HEL level**	**Membrane anti-HEL Ig**	**Anti-HEL PFC/spleen***	**Anti-HEL serum titer**
Anti-HEL single transgenics	None	+	High	High
Anti-HEL/HEL double transgenics	10^{-9} M	+	Low	Low

* Experimental animals were immunized with hen egg-white lysozyme (HEL). Several days later hemolytic plaque assays for the number of plasma cells secreting anti-HEL antibody were performed and the serum anti-HEL titers were determined. PFC = plaque-forming cells.

Adapted from Goodnow, C. C., 1992, Annual Review of Immunology *10:489.*

The Development of B-1 and Marginal-Zone B Cells

This chapter has so far focused on the development of those B cells that belong to the best characterized B-cell subpopulation, **B-2 B cells** (or *follicular B cells*). Mature B-2 B cells recirculate between the blood and the lymphoid organs, and can be found in large numbers in the B-cell follicles of the lymph nodes and spleen. However, other subsets of B cells have been recognized that perform distinct functions, occupy distinct anatomical locations, and pursue different developmental programs. This section of the chapter will therefore address the development and function of B-1 B cells and of marginal-zone B cells (see Figure 10-12 for a comparison of the properties of these three cell types).

As described more fully in Chapter 12, B-1 B cells generate antibodies against antigens shared by many bacterial species and may do so even in the absence of antigenic stimulation. They are the source of the so-called natural antibodies: serum IgM antibodies that provide a first line of protection against invasion by many types of microorganisms. Marginal-zone B cells take their name from their location in the outer zones of the white pulp of the spleen. They are the first B cells encountered by blood-borne antigens

entering the spleen and, like B-1 B cells, mainly (although not exclusively) produce broadly cross-reactive antibodies of the IgM class. Both B-1 and marginal-zone B cells can generate antibodies in the absence of T-cell help, although the addition of helper T cells enhances antibody secretion and allows for some degree of heavy-chain class switching.

B-1 B Cells Are Derived from a Separate Developmental Lineage

B-1 B cells are phenotypically and functionally distinct from B-2 B cells in a number of important ways. They occupy *different anatomical niches* from B-2 B cells, constituting 30% to 50% of the B cells in the pleural and peritoneal cavities of mice, and representing about 1 million cells in each space. A similar number of B-1 B cells can also be found in the spleen, but there they represent a much smaller fraction (around 2%) of the splenic B-cell population. B-1 B cells have only a relatively *limited receptor repertoire*, and their receptors tend to be directed toward the recognition of commonly expressed *microbial carbohydrate antigens*. These broadly cross-reactive, low-affinity antigen receptors expressing minimal repertoire diversity are reminiscent of the *pathogen-associated molecular pattern* (PAMP) receptors of the innate immune system

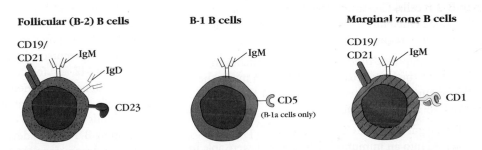

Attribute	Follicular (B-2) B cells	B-1 B cells	Marginal zone B cells
Major sites	Secondary lymphoid organs	Peritoneal and pleural cavities	Marginal zones of spleen
Source of new B cells	From precursors in bone marrow	Self-renewing (division of existing B-1 cells)	Long-lived May be self-renewing
V-region diversity	Highly diverse	Restricted diversity	Somewhat restricted
Somatic hypermutation	Yes	No	Unclear
Requirements for T-cell help	Yes	No	Variable
Isotypes produced	High levels of IgG	High levels of IgM	Primarily IgM; some IgG
Response to carbohydrate antigens	Possibly	Yes	Yes
Response to protein antigens	Yes	Possibly	Yes
Memory	Yes	Very little or none	Unknown
Surface IgD on mature B cells	Present on naïve B cells	Little or none	Little or none

FIGURE 10-12 The three major populations of mature B cells in the periphery. The cell-surface properties and functions of B-2, B-1, and marginal-zone (MZ) cells are shown. Conventional B-2 cells were so named because they develop after B-1 B cells.

(Chapter 5), and B-1 B cells are thus considered to play a role that bridges those of the innate and adaptive immune systems.

In contrast with the T-1 transitional B-2 B cells, which undergo apoptosis upon antigen challenge, transitional B-1 cells undergo apoptosis *unless* they interact with self antigens. Work in a number of different transgenic systems has suggested that relatively strong BCR engagement by self antigens provides a *positive selection* rather than a *negative selection* survival signal for B-1 B cells. In contrast with B-2 cells and marginal-zone cells (see below), B-1 B cells do not require interaction with BAFF during the transitional stage of development.

For many years, the preeminent issue debated by those interested in the development of B-1 B cells was whether they constituted a separate developmental lineage, or whether they derived from the same progenitors as B-2 B cells. This controversy has since been resolved in favor of the assertion that B-1 and B-2 B cells derive from distinct lineages of progenitor cells. Several lines of evidence support this conjecture:

- B-1 B cells appear before B-2 B cells during ontogenic development. B cells generated from the AGM region and the liver in the fetus have a cell-surface phenotype characteristic of B-1 B cells and secrete natural IgM antibodies without the need for deliberate immunization. B-1 B cells may be generated early in development in order to protect the fetus from commonly encountered bacterial pathogens.

- B-1 B cells display much more limited V region diversity than B-2 B cells. Generated at a point in ontogeny before TdT can be efficiently activated, many B-1 cells lack any evidence of N region addition.

- B-1 B cells populate different anatomical niches in the mouse than the later-arising B-2 B cells, in much the same manner that has been described for the fetally derived γδ T cells (Chapter 9).

- B-cell progenitors of the CD19$^+$CD45R$^{low/-}$ phenotype transferred into an immunodeficient mouse were able to repopulate the B-1, but not the B-2 B-cell compartments. Conversely, CD19$^+$CD45R^{hi} B-cell progenitors gave rise to B-2, but not to B-1, daughter cells in an immunodeficient recipient mouse, supporting the notion that the two subclasses derive from different lineages of progenitor cells.

- Whereas B-2 B cells must be constantly replenished by the emergence of newly generated cells from the bone marrow, B-1 B cells are constantly regenerated in the periphery of the animal. Bone marrow ablation therefore leaves the mouse with a depleted B-2 pool, but with a fully functional B-1 population.

There are very few absolutes in biology, and so it should be clearly stated that it is probable that not all B-1 B cells in an adult animal are derived from fetal liver precursors and that some replenishment of B-1 precursors occurs in the adult bone marrow. Not all B-1 B cells are restricted to IgM production, and the antibody production of some B-1 B cells does benefit from the provision of T-cell help. But notwithstanding these variations on the general themes we have elaborated, it remains the case that evidence clearly supports the notion that B-1 B cells are derived from a different progenitor cell lineage from B-2 B cells.

Marginal-Zone Cells Share Phenotypic and Functional Characteristics with B-1 B Cells and Arise at the T2 Stage

The term *marginal zone* refers to the fact that marginal zone (MZ) B cells are located in the *outer regions of the white pulp of the spleen* (Figure 10-13). The spreading out of the fast-moving arterial flow into the marginal sinuses results in a decrease in the rate of blood flow and allows blood-borne antigens to interact with cells resident within the marginal zone. Indeed, MZ B cells appear to be specialized for recognizing blood-borne antigens. They are capable of responding to both protein and carbohydrate antigens, and evidence suggests that some, but maybe not all, MZ B cells can do so without the need for help from T cells.

MZ cells are characterized by relatively high levels of membrane IgM, and the complement receptor/B-cell co-receptor CD21 (see Chapters 3 and 6), but low levels of membrane IgD and the Fc receptor CD23. They also display phospholipid receptors and adhesion molecules that enable them to make adhesive interactions with other cells within the marginal zone that hold them in place. MZ cells are long lived and may self-renew in the periphery.

What drives an immature B cell down the MZ pathway? Phenotypic characterization of bone marrow and peripheral immature B-cell populations, combined with labeling studies that define precursor/daughter-cell relationships have suggested that MZ and follicular B-2 B cells both derive from the T2 transitional population (see Figure 10-9). MZ cells, like B-2 B cells, are also reliant on the B-cell survival factor BAFF, which binds to the MZ B cells through the receptor BR3. Like developing B-1 B cells, developing MZ cells appear to require relatively strong signaling through the BCR in order to survive. Surprisingly, unlike any other B-cell subset so far described, the differentiation of MZ B cells also requires signaling through ligands of the Notch pathway (see Chapter 9). Loss of the Notch 2 receptor or deletion of the Notch 2 ligand, Delta-like 1 (Dl-1), results in the selective deletion of MZ B cells. Both MZ and B-1 B cell populations are enriched in cells that express self antigen-specific receptors, and so relatively strong signaling of T2 cells by binding of self antigens through the BCR is also necessary for MZ B-cell differentiation.

Comparison of B- and T-Cell Development

In closing this chapter, it is instructive to consider the many points of comparison between the development of the two arms of the adaptive immune system, B cells and T cells (Table 10-5). Both cell lineages have their beginnings in the fetus and the neonate. In the neonate, γδ T cells and B-1 B cells are dispatched to their own particular peripheral niches,

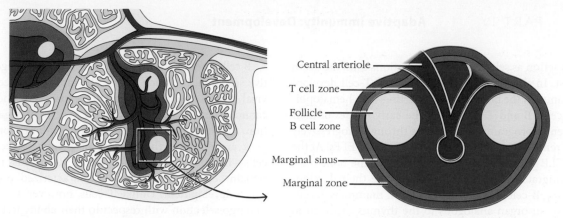

FIGURE 10-13 The relative locations of the marginal zone and the follicles in the spleen. This figure shows a cross-section of the spleen, displaying the anatomical relationships between the central arteriole, the T-cell and B-cell zones, and the marginal zone.

TABLE 10-5 Comparison between T-cell and B-cell development

Structure or process	B cells	T cells
Development begins in the bone marrow.	+	+
Development continues in the thymus.	−	+
Ig heavy chain or TCRβ chain many gene rearrangement begins with D-J and continues with V-DJ recombination.	+	+
The H chain (BCR) or β chain (TCR) is expressed with a surrogate form of the light chain on the cell surface. Signaling from this pre-BCR or pre-TCR is necessary for development to continue.	+	+
Signaling through the pre-B- or pre-TCRs results in proliferation.	+	+
The κ/λ (BCR) or α (TCR) chain bears only V and J segments.	+	+
Signaling from the completed receptor is necessary for survival (positive selection).	+	+
Positive selection requires recognition of self components.	+/− True for the minority B-1 B-cell subset. Ligand for B-2 and MZ subsets unclear.	+ Low-affinity binding of self MHC and self peptide in the thymus is necessary for positive selection.
Receptor editing of κ/λ (BCR) or α (TCR) chain modifies autoreactive specificities.	+	Has been shown to occur, but rarely used as a mechanism of escaping autoreactivity.
Immature cells bearing high affinity autoreactive receptors are eliminated by apoptosis (negative selection).	+	+
Negative selection involves ectopic expression of auto-antigens in the primary lymphoid organs.	−	+ Auto-antigens are expressed in the thymus under the influence of the AIRE transcription factor.
Negative selection involves recognition of MHC-presented peptides.	−	+
Heavy or TCRβ chain allelic exclusion.	+	+
Light (κ/λ) or TCRα chain allelic exclusion.	+	Many T cells display more than one TCRα chain.
90% of lymphocytes are lost prior to export to the periphery.	+	+
Development is completed in the periphery to allow for tolerance to antigens not expressed in the primary lymphoid organs	+	+

Note: The comparisons in this table most accurately refer to the predominant αβ T and follicular B-2 cell subsets, although many of the statements made are equally valid for the minority lymphocyte subsets.

there to function as self-renewing populations until the death of the host. In the adult animal, B-cell and T-cell development continue in the bone marrow, starting with hematopoietic stem cells. B and T cells share the early phases of their developmental programs, as they pass through progressively more differentiated stages as MPPs, LMPPs, and ELPs. At the ELP and CLP stages, T-cell progenitors leave the bone marrow, and migrate to the thymus to complete their development, leaving B-cell progenitors behind. In mammals, B cells do not have an organ analogous to the thymus in which to develop into mature, functioning cells, although birds do possess such an organ—the bursa of Fabricius. (Many students are unaware that it is from this organ, and not from the bone marrow, that the "B" in B cells originates.)

With the initiation of V(D)J recombination, the B cell irreversibly commits to its lineage, and begins the process of receptor rearrangement, followed by B-cell selection and differentiation that will culminate in the formation of a complete repertoire of functioning peripheral B cells.

B cells and T cells must both pass through stages of positive selection, in which those cells capable of receiving survival signals are retained at the expense of those which cannot. They must also survive the process of negative selection, in which lymphocytes with high affinity for self antigens are deleted. The process of positive selection in B cells—its mechanism, and the BCR ligands involved—remains one of the least well characterized processes in B-cell development. Unlike T cells, however, B cells do not undergo selection with respect to their ability to bind to self MHC antigens.

With the expression of high levels of IgD on the cell surface and the necessary adhesion molecules to direct their recirculation, development of the mature, follicular B-2 cell is complete and, for a few weeks to months, it will recirculate, ready for antigen contact in the context of T-cell help and subsequent differentiation to antibody production. For the final, antigen-stimulated stages of B-cell differentiation, the reader is directed to Chapter 12.

SUMMARY

- Hematopoiesis in the embryo generates rapidly dividing hematopoietic stem cells that populate the blood system of the animal, providing red blood cells that supply the oxygen needs of the fetus, and generating the early precursors of other blood-cell lineages. Early sites of blood-cell development include the yolk sac and the placenta, as well as the aorta-gonad-metanephros (AGM) region and the fetal liver.

- The earliest B-cell progenitors in the fetal liver supply B-1 B cells that migrate into the pleural and peritoneal cavities and remain self-renewing throughout the life of the animal.

- In adult animals, hematopoiesis occurs in the bone marrow.

- Progenitor stages of the conventional B-2 subset are defined by the presence of particular cell-surface markers, which include chemokine and lymphokine receptors and proteins involved in adhesive interactions.

- Progenitor stages are also defined by the status of immunoglobulin V region gene rearrangements. The heavy-chain V genes rearrange first, with D to J_H recombination occurring initially, followed by V_H to DJ_H recombination.

- The heavy chain is then expressed on the cell surface in combination with the surrogate light chain, which is made up of VpreB and λ5. Together they form the pre-B-cell receptor, which is expressed on the cell surface along with the Igα,Igβ signaling complex.

- Signaling through the B-cell receptor stops V_H gene rearrangement and calls for a few rounds of cell division. This allows multiple B cells to use the same, successfully rearranged heavy chain in combination with many different light chains.

- After light-chain rearrangement, and the expression of the completed immunoglobulin receptor on the cell surface, immature B cells specific for self antigens present in the bone marrow are deleted by apoptosis.

- Immature B cells emerge from the bone marrow as transitional 1 (T1) B cells and circulate to the spleen. Interaction with self antigens in the spleen can give rise to apoptosis.

- T1 B-2 B cells then enter the follicles, where the level of IgD expression increases, they become T2 cells, and then mature into either a follicular B-2 cell (a conventional B cell) or a marginal-zone (MZ) B cell.

- The three major subsets of B cells differ according to their site of generation, their sites of maturation, their anatomical niches in the adult, the antigens to the which they respond, their need for T-cell help in antibody production, the diversity of their immunoglobulin repertoires, and their abilities to undergo somatic hypermutation and memory generation following antigenic stimulation.

- Like T cells, developing B cells must undergo both positive and negative selection. Unlike T cells, B cells need not be selected for their ability to recognize antigens in the context of MHC antigens, nor is there a primary immune organ aside from the bone marrow specialized for their maturation. There is no equivalent of the AIRE protein that has yet been discovered to provide for ectopic expression of antigens in the bone marrow in order to facilitate clonal deletion of self antigen-specific B cells.

REFERENCES

Allman, D., and Pillai, S. 2008. Peripheral B cell subsets. *Current Opinion in Immunology* **20**:149–157.

Cambier, J. C., S. B. Gauld, K. T. Merrell, and B. J. Vilen. 2007. B-cell anergy: From transgenic models to naturally occurring anergic B cells? *Nature Reviews Immunology* **7**:633–643.

Carsetti, R., M. M. Rosado, and H. Wardmann, H. 2004. Peripheral development of B cells in mouse and man. *Immunological Reviews* **197**:179–191.

Casola, S. 2007. Control of peripheral B-cell development. *Current Opinions in Immunology* **19**:143–149.

Chung, J. B., R. A. Sater, M. L. Fields, J. Erikson, and J. G. Monroe. 2002. CD23 defines two distinct subsets of immature B cells which differ in their responses to T cell help signals. *International Immunology* **14**:157–166.

Dorshkind, K., and E. Montecino-Rodriguez. 2007. Fetal B-cell lymphopoiesis and the emergence of B-1-cell potential. *Nature Reviews Immunology* **7**:213–219.

Dzierzak, E., and N. A. Speck. 2008. Of lineage and legacy: The development of mammalian hematopoietic stem cells. *Nature Immunology* **9**:129–136.

Fuxa, M., and J. A. Skok. 2007. Transcriptional regulation in early B cell development. *Current Opinions in Immunology* **19**:129 136.

Hoek, K. L., et al. 2006. Transitional B cell fate is associated with developmental stage-specific regulation of diacylglycerol and calcium signaling upon B cell receptor engagement. *Journal of Immunology* **177**:5405–5413.

Kee, B. L. 2009. E and ID proteins branch out. *Nature Reviews Immunology* **9**:175–184.

Koralov, S. B., et al. 2008. Dicer ablation affects antibody diversity and cell survival in the B lymphocyte lineage. *Cell* **132**:860–874.

Kurosaki, T., H. Shinohara, and Y. Baba. 2010. B cell signaling and fate decision. *Annual Review of Immunology* **28**:21–55.

Liao, D. 2009. Emerging roles of the EBF family of transcription factors in tumor suppression. *Molecular Cancer Research* **7**:1893–1901.

Mackay, F., and P. Schneider. 2009. Cracking the BAFF code. *Nature Reviews Immunology* **9**:491–502.

Malin, S., et al. 2010. Role of STAT5 in controlling cell survival and immunoglobulin gene recombination during pro-B cell development. *Nature Immunology* **11**:171–179.

Mikkola, H. K., and S. H. Orkin. 2006. The journey of developing hematopoietic stem cells. *Development* **133**:3733–3744.

Monroe, J. G., and K. Dorshkind. 2007. Fate decisions regulating bone marrow and peripheral B lymphocyte development. *Advances in Immunology* **95**:1–50.

Nagasawa, T. 2006. Microenvironmental niches in the bone marrow required for B-cell development. *Nature Reviews Immunology* **6**:107–116.

Nemazee, D. 2006. Receptor editing in lymphocyte development and central tolerance. *Nature Reviews Immunology* **6**:728–740.

Nutt, S. L., and Kee, B. L. 2007. The transcriptional regulation of B cell lineage commitment. *Immunity* **26**:715–725.

Palis, J., et al. 2001. Spatial and temporal emergence of high proliferative potential hematopoietic precursors during murine embryogenesis. *Proceedings of the National Academy of Sciences of the United States of America* **98**:4528–4533.

Perez-Vera, P., A. Reyes-Leon, and E. M. Fuentes-Panana. 2011. Signaling proteins and transcription factors in normal and malignant early B cell development. *Bone Marrow Research* **2011**:502,751.

Pillai, S., and A. Cariappa. 2009. The follicular versus marginal zone B lymphocyte cell fate decision. *Nature Reviews Immunology* **9**:767–777.

Rajewsky, K., and H. von Boehmer. 2008. Lymphocyte development: Overview. *Current Opinions in Immunology* **20**:127–130.

Ramirez, J., K. Lukin, and J. Hagman. 2010. From hematopoietic progenitors to B cells: Mechanisms of lineage restriction and commitment. *Current Opinions in Immunology* **22**:177–184.

Srivastava, B., W. J. Quinn, 3rd, K. Hazard, J. Erikson, and D. Allman. 2005. Characterization of marginal zone B cell precursors. *Journal of Experimental Medicine* **202**:1225–1234.

Tokoyoda, K., T. Egawa, T. Sugiyama, B. I. Choi, and T. Nagasawa, T. 2004. Cellular niches controlling B lymphocyte behavior within bone marrow during development. *Immunity* **20**:707–718.

Tung, J. W., and L. A. Herzenberg. 2007. Unraveling B-1 progenitors. *Current Opinions in Immunology* **19**:150–155.

von Boehmer, H., and F. Melchers. 2010. Checkpoints in lymphocyte development and autoimmune disease. *Nature Immunology* **11**:14–20.

Whitlock, C. A., and O. Witte. 1982. Long-term culture of B lymphocytes and their precursors from murine bone marrow. *Proceedings of the National Academy of Sciences of the United States of America* **79**:3608–3612.

Xiao, C., and K. Rajewsky. 2009. MicroRNA control in the immune system: Basic principles. *Cell* **136**:26–36.

Yin, T., and L. Li. 2006. The stem cell niches in bone. *Journal of Clinical Investigation* **116**:1195–1201.

Useful Web sites

www.bio.davidson.edu/courses/immunology/Flash/Bcellmat.html An unusual animation of B-cell development.

1. You wish to study the development of B-1 B cells in the absence of the other two major B-cell subsets. You have a recipient Rag1$^{-/-}$ mouse that you have already repopulated with T cells. What would you choose to be your source of B-1 progenitors and why? Which anatomical sites would you expect to harvest the B-1 B cells from?

2. Describe the phenotypic and functional differences between T1 and T2 immature B cells.

3. Following expression of the pre-B-cell receptor on the progenitor B-cell surface, the B cell undergoes a few rounds of cell division. What purpose does this round of division serve in the development of the B-cell repertoire?

4. Immature B cells bearing potentially autoimmune receptors can be managed in three ways to minimize the probability of disease. Describe these three strategies, noting whether they are shared by T-cell progenitors.

5. You suspect that a new transcription factor is expressed at the pre-pro-B-cell stage of development. How would you test your hypothesis? What is the status of heavy-and light-chain rearrangement at this stage of development and how would you test it?

6. How would you determine whether a particular stage of B-cell development occurs in association with a stromal cell that expresses CXCL12?

7. Describe the order in which B-cell receptor genes undergo rearrangement, indicating at what steps you might expect to see the B cell express one or both chains on the cell surface. In what sense(s) does this gene rearrangement process mimic the analogous progression in αβ T cells, and in what ways do the two processes differ?

ANALYZE THE DATA The two columns of data in the following figure below are flow cytometric plots that describe the levels of the antigens denoted on the x and y axes. The left column represents the antigens present on spleen (part A) and bone marrow (parts B) from wild-type (genetically normal) animals. The plots represent all lymphocytes in the spleen (part A) or B-cell progenitor and precursor cells in the bone marrow (part B). The right column shows the same plots from animals in which the *Dicer* gene has been knocked out. As you recall, the *Dicer* gene is required for the maturation of controlling miRNAs.

a. For each pair of plots, describe the differences in the cell populations, indicating whether the differences reflect losses or gains in particular developing B-cell populations.
b. At what point(s) in B-cell development do you think miRNAs are functioning?

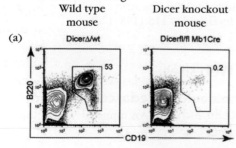

(a)

Wild type mouse   Dicer knockout mouse

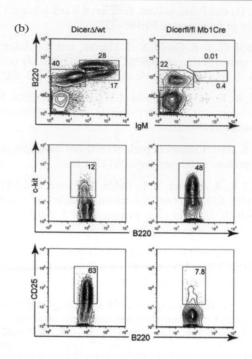

(b)

ANALYZE THE DATA The following figure is derived from the same paper as those above. In this case the data are expressed as histograms, in which the y axis represents the number of cells binding the molecule shown on the x axis, Annexin V. Annexin V binds to phosphatidyl serine on the outer leaflet of cell membranes. Phosphatidyl serine is found on the outer leaflet only in cells about to undergo apoptosis. The top two panels represent cells from a wild-type animal, and the bottom two panels represent cells from animals in which the *Dicer* gene has been knocked out.

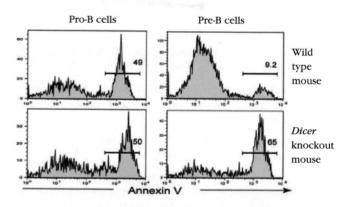

a. Does the presence of Dicer have an effect on the fraction of pro-B cells undergoing apoptosis? Explain your reasoning.
b. Does the presence of Dicer have an effect on the fraction of pre-B cells undergoing apoptosis? Again explain your reasoning.
c. Describe one function that you now think miRNAs fulfill in B-cell development.

T-Cell Activation, Differentiation, and Memory

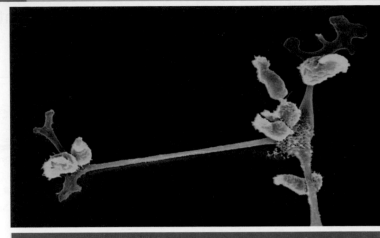

Dendritic cell (orange) interacting with T cells (green). [M. Rohde, HZI, Braunschweig, Germany.]

- T-Cell Activation and the Two Signal Hypothesis
- T-Cell Differentiation
- T-Cell Memory

The interaction between a naïve T cell and an antigen-presenting cell (APC) is *the* initiating event of the adaptive immune response. Prior to this, the innate immune system has been alerted at the site of infection or tissue damage, and APCs, typically dendritic cells, have been activated via their pattern recognition receptors. These cells may have engulfed extracellular (or opsonized intracellular) pathogens, or they may have been infected by an intracellular pathogen. In either case, they have processed and presented peptides from these pathogens in complex with surface MHC class I and class II molecules, and have made their way to a local (draining) lymph node and/or the spleen. The APCs have taken up residence in the T-cell zones of the lymph node or spleen to join networks of other cells that are continually scanned by roving naïve CD8$^+$ and CD4$^+$ T cells, which recognize MHC class I-peptide and MHC class II-peptide complexes, respectively.

We have seen that each mature T cell expresses a unique antigen receptor that has been assembled via random gene rearrangement during T-cell development in the thymus (Chapter 9). Because developing T cells undergo selection events within the thymus, each mature, naïve T cell is tolerant to self antigens, and restricted to self-MHC (Chapter 9). Some naïve T cells have committed to the CD8$^+$ cytotoxic T-cell lineage, some to the CD4$^+$ helper T-cell lineage. When a naïve CD8$^+$ or CD4$^+$ T cell binds tightly to an MHC-peptide complex expressed by an activated dendritic cell, it becomes activated by signals generated through the TCR (see Chapter 3). These signals, in concert with signals from other factors that we will describe below, stimulate the T cell to proliferate and differentiate into an effector cell.

As you know, naïve CD8$^+$ T cells become cytotoxic cells in response to engagement of MHC class I-peptide combinations. Although we refer to the activation of CD8$^+$ T cells in this chapter, we will discuss their effector functions in detail in Chapter 13. Naïve CD4$^+$ T cells become helper cells in response to engagement of MHC class II-peptide combinations (Overview Figure 11-1).

This chapter focuses on this event, which is critical for the development of both humoral and cell-mediated immunity, as well as the development of B-cell and CD8$^+$ T-cell memory. As discussed below, CD4$^+$ T cells can differentiate into a surprising number of distinct helper subsets, each of which has a different function in combating infection.

In this chapter, we briefly review the cellular and molecular events that activate T cells and then deepen your understanding of the costimulatory interactions that play an important role in determining the outcome of T cell-APC interactions. We then discuss the outcomes of naïve T-cell activation—the development of effector and memory T cells—focusing primarily on the different fates and functions of the CD4$^+$ helper T-cell subsets that drive the adaptive response.

Which helper subset a naïve CD4$^+$ T cell becomes depends on the types of signals (e.g., cytokines, costimulatory signals) they receive from the dendritic cells they engage via their TCRs. And as described in Chapter 5, the signals dendritic cells are able to deliver depend in large part on the pathogen to which they have been exposed (see Figure 5-18). Investigators are still working to understand all the variables involved in determining the lineage choices of T helper cells, but we introduce you to the current thinking.

OVERVIEW FIGURE

 T-Cell Activation and Differentiation

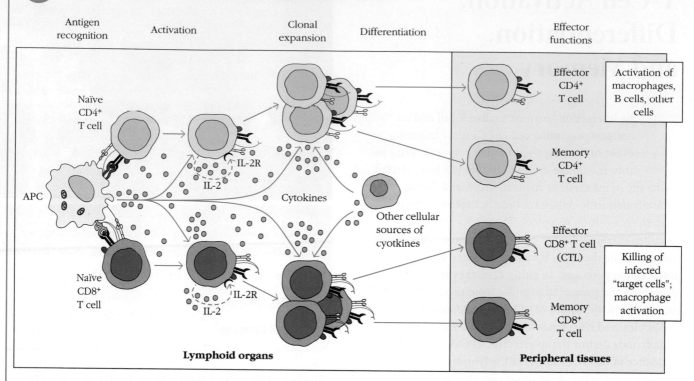

Activation of a naïve T cell in a secondary lymphoid organ results in the generation of effector and memory T cells. Activation requires several receptor-ligand interactions between the T cell and a dendritic cell, as well as signals through cytokines produced by the activating APC, as well as other supportive cells in the lymphoid organ. Effector CD4$^+$ T cells become helper T cells (T$_H$) and secrete cytokines that enhance the activity of many other immune cells. Effector CD8$^+$ T cells are cytotoxic cells (T$_C$) that kill infected cells.

We also describe the known functions of the specialized helper cells, focusing on T$_H$1, T$_H$2, T$_H$17, T$_{FH}$, and T$_{REG}$ cells. Finally, we close the chapter with a discussion of T-cell memory, which is dependent on CD4$^+$ T cell help, and describe both what is known and what is currently under investigation.

A Classic Experiment box and Clinical Focus box are offered as a pair and describe the basic research behind the discovery of the costimulatory molecule CD28, an essential participant in naïve T-cell activation, and then the development of a molecular therapy for autoimmune diseases that takes advantage of what we know about the biology of costimulation. These boxes, together, illustrate the powerful connections between basic research and clinical development, which underlie *translational research*, an effort to bring bench scientific discovery to the "bedside" that has captured the imagination of many biomedical investigators.

The Advances box describes a more recent effort to figure out precisely how many T-cell receptors must be engaged to initiate T-cell activation. The answer was initially surprising, yet in hindsight may not be surprising at all. The final Clinical Focus box discusses how a disease, an "experiment of nature," has helped us to better understand the basic biology and physiological function of the effector cells introduced in this chapter.

T-Cell Activation and the Two-Signal Hypothesis

CD4$^+$ and CD8$^+$ T cells leave the thymus and enter the circulation as resting cells in the G$_0$ stage of the cell cycle. These **naïve** T cells are mature, but they have not yet encountered antigen. Their chromatin is condensed, they have very little cytoplasm, and they exhibit little transcriptional activity.

However, they are mobile cells and recirculate continually among the blood, lymph, and secondary lymphoid tissues, including lymph nodes, browsing for antigen. It is estimated that each naïve T cell recirculates from blood through lymph nodes and back again every 12 to 24 hours. Because only about 1 in 10^5 naïve T cells is likely to be specific for any given antigen, this large-scale recirculation increases the chances that a T cell will encounter appropriate antigen.

If a naïve T cell does not bind any of the MHC-peptide complexes encountered as it browses the surfaces of stromal cells of a lymph node, it exits through the efferent lymphatics, ultimately draining into the thoracic duct and rejoining the blood (see Chapter 2). However, if a naïve T cell does encounter an APC expressing an MHC-peptide to which it can bind, it will initiate an activation program that produces a diverse array of cells that orchestrate efforts to clear infection.

Recall from Chapter 3 that a successful T cell-APC interaction results in the stable organization of signaling molecules into an immune synapse (Figure 11-2). The TCR/MHC-peptide complexes and coreceptors are aggregated in the central part of this synapse (central supramolecular activating complex, or cSMAC). The intrinsic affinity between the TCR and MHC-peptide surfaces is quite low (K_d ranges from 10^{-4} M to 10^{-7} M) and is stabilized by the activity of several molecules which together increase the **avidity** (the combined affinity of all cell-cell interactions) of the cellular interaction. The coreceptors CD4 and CD8, which are found in the cSMAC, stabilize the interaction between TCR and MHC by binding MHC class II and MHC class I molecules, respectively. Interactions between adhesion molecules and their ligands (e.g., LFA-1/ICAM-1 and CD2/LFA-3) help to sustain the signals generated by allowing long-term cell interactions. These molecules are organized around the central aggregate, forming the peripheral or "p" SMAC.

However, even the increased functional avidity offered by coreceptors and adhesion molecules is still not sufficient to fully activate a T cell. Interactions between costimulatory receptors on T cells (e.g., CD28) and costimulatory ligands on dendritic cells (e.g., CD80/86) provide a second, required signal. In addition, as you will see below, a third set of signals, provided by local cytokines (Signal 3), directs T-cell differentiation into distinct effector cell types.

Costimulatory Signals Are Required for Optimal T-Cell Activation and Proliferation

What evidence pointed to a requirement for a second, costimulatory signal? In 1987, Helen Quill and Ron Schwartz recognized that, in the absence of functional APCs, isolated high affinity TCR-MHC interactions actually led to T-cell non-responsiveness rather than activation—a phenomenon they called T-cell **anergy**. Their studies led to the simple but powerful notion that not one but two signals were required for full T-cell activation: *Signal 1* is provided by antigen-specific TCR engagement (which can be enhanced by coreceptors and adhesion molecules), and *Signal 2* is provided by contact with

a costimulatory ligand, which can only be expressed by a functional APC. When a T cell receives both Signal 1 and Signal 2, it will be activated to produce cytokines that enhance entry into cell cycle and proliferation (Figure 11-3).

It is now known that Signal 2 results from an interaction between specific **costimulatory receptors** on T cells and costimulatory ligands on dendritic cells (Table 11-1). Recall from Chapter 5 that dendritic cells and other APCs become activated by antigen binding to PRRs, to express costimulatory ligands (e.g., CD80 and CD86) and produce cytokines that enhance their ability to activate T cells. CD28 is the most commonly cited example of a costimulatory receptor, but other related molecules that provide costimulatory signals during T-cell activation have since been identified and are also described below. Because these molecules enhance TCR signaling, they are collectively referred to as "positive" costimulatory receptors and ligands.

Negative costimulatory receptors, which inhibit TCR signaling, have also been identified. Although our understanding of their specific functions is incomplete, as a group these play important roles in (1) maintaining peripheral T-cell tolerance and (2) reducing inflammation both after the natural course of an infection and during responses to chronic infection. As you can imagine, the expression and activity of negative and positive costimulatory molecules must be carefully regulated temporally and spatially. Naïve T cells, for example, do not express negative costimulatory receptors, allowing them to be activated in secondary lymphoid tissue during the initiation of an immune response. On the other hand, effector T cells up-regulate negative costimulatory receptors at the end of an immune response, when proliferation is no longer advantageous. However, these generalizations belie the complexity of regulation of this highly important costimulatory network, and investigators are still working to understand the details. Below we introduce aspects of the structure, function, expression, and, when known, regulation of several positive and negative costimulators.

Positive Costimulatory Receptors: CD28

CD28, a 44 kDa glycoprotein expressed as a homodimer, was the first costimulatory molecule to be discovered (see Classic Experiment Box 11-1). Expressed by all naïve and activated human and murine CD4$^+$ T cells, all murine CD8$^+$ T cells, and, interestingly, only 50% of human CD8$^+$ T cells, it markedly enhances TCR-induced proliferation and survival by cooperating with T-cell receptor signals to induce expression of the pro-proliferative cytokine IL-2 and the prosurvival bcl-2 family member, bcl-xL.

CD28 binds to two distinct ligands of the B7 family of proteins: CD80 (B7-1) and CD86 (B7-2). These are members of the immunoglobulin superfamily, which have similar extracellular domains. Interestingly, their intracellular regions differ, suggesting that they might not simply act as passive ligands; rather, they may have the ability to generate signals that influence the APC, a view that has some experimental support.

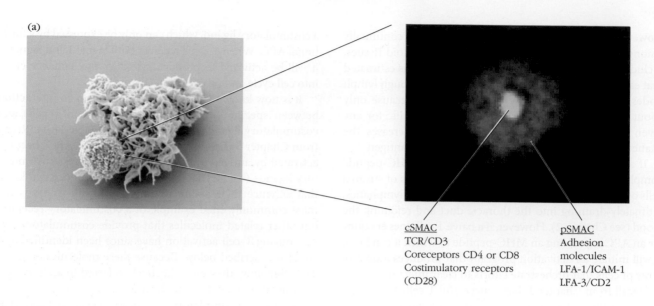

(a)

cSMAC
TCR/CD3
Coreceptors CD4 or CD8
Costimulatory receptors
(CD28)

pSMAC
Adhesion
molecules
LFA-1/ICAM-1
LFA-3/CD2

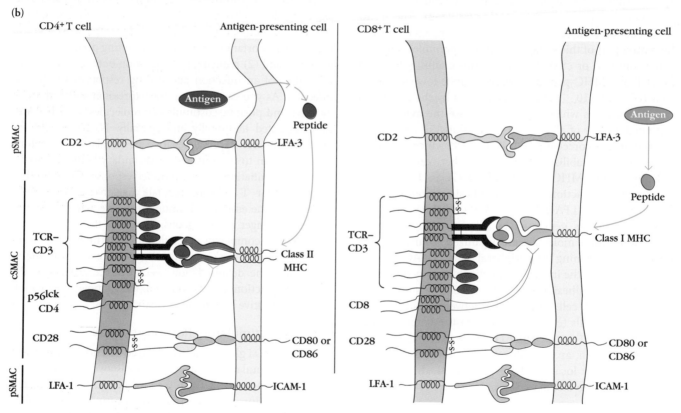

FIGURE 11-2 Surface interactions responsible for T-cell activation. (a) A successful T-cell/dendritic-cell interaction results in the organization of signaling molecules into an immune synapse. A scanning electron micrograph (left) shows the binding of a T cell (artificially colored yellow) and dendritic cell (artificially colored blue). A fluorescent micrograph (right) shows a cross-section of the immune synapse, where the TCR is stained with fluorescein (green) and adhesion molecules (specifically LFA-1) are stained with phycoerythrin (red). Other molecules that can be found in the central part of the synapse (central *supramolecular activation complex* [cSMAC]) and the *peripheral* part of the synapse (pSMAC) are listed. (b) The interactions between a CD4+ (left) or CD8+ (right) T-cell and its activating dendritic cell. A dendritic cell (to the right of each diagram) can engulf an antigen and present peptide associated with MHC class II to a CD4+ T cell or can load internal peptides into MHC class I and present the combination to a CD8+ T cell. Binding of the TCR to MHC-peptide complexes is enhanced by the binding of coreceptors CD4 and CD8 to MHC class II and class I, respectively. CD28 interactions with CD80/86 provide the required costimulatory signals. Adhesion molecule interactions, two of which (LFA-1/ICAM-1, LFA-3/CD2) are depicted, markedly strengthen the connection between the T cell and APC or target cell so that signals can be sustained. *[(a) right: Michael L. Dustin, J Clin Invest. 2002; 109(2): 155, Fig.1. doi:10.1172/JCI14842. Left: Dr. Olivier Schwartz, Institut Pasteur/Photo Researchers.]*

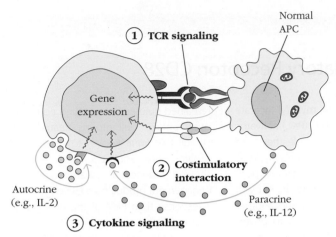

FIGURE 11-3 Three signals are required for activation of a naïve T cell. The TCR/MHC-peptide interaction, along with CD4 and CD8 coreceptors and adhesion molecules, provide Signal 1. Co-stimulation by a separate set of molecules, including CD28 (or ICOS, not shown) provide Signal 2. Together, Signal 1 and Signal 2 initiate a signal transduction cascade that results in activation of transcription factors and cytokines (Signal 3) that direct T-cell proliferation (IL-2) and differentiation (polarizing cytokines). Cytokines can act in an *autocrine* manner, by stimulating the same cells that produce them, or in a *paracrine* manner, by stimulating neighboring cells.

Although most T cells express CD28, most cells in the body do not express its ligands. In fact, only professional APCs have the capacity to express CD80/86. Mature dendritic cells, the best activator of naïve T cells, appear to con-stitutively express CD80/86, and macrophages and B cells have the capacity to up-regulate CD80/86 after they are activated by an encounter with pathogen (see Chapter 5).

Positive Costimulatory Receptors: ICOS

Since the discovery of CD28, several other structurally related receptors have been identified. Like CD28, the closely related *inducible costimulator* (ICOS) provides positive costimulation for T-cell activation. However, rather than binding CD80 and CD86, ICOS binds to another member of the growing B7 family, ICOS-ligand (ICOS-L), which is also expressed on a subset of activated APCs.

Differences in patterns of expression of CD28 and ICOS indicate that these positive costimulatory molecules play distinct roles in T-cell activation. Unlike CD28, ICOS is not expressed on naïve T cells; rather, it is expressed on memory and effector T cells. Investigations suggest that *CD28 plays a key costimulatory role during the initiation of activation and ICOS plays a key role in maintaining the activity of already differentiated effector and memory T cells.*

Negative Costimulatory Receptors: CTLA-4

The discovery of CTLA-4 (CD152), the second member of the CD28 family to be identified, caused a stir. Although closely related in structure to CD28 and also capable of binding both CD80 and CD86, CTLA-4 did not act as a positive costimulator. Instead, it antagonized T-cell activating signals and is now referred to as a negative costimulatory receptor.

CTLA-4 is not expressed constitutively on resting T cells. Rather, it is induced within 24 hours after activation of a

TABLE 11-1 T-cell costimulatory molecules and their ligands

Costimulatory receptor on T cell	Costimulatory ligand	Activity
Positive costimulation		
CD28	CD80 (B7-1) or CD86 (B7-2) *Expressed by professional APCs, (and medullary thymic epithelium)*	Activation of naïve T cells
ICOS	ICOS-L *Expressed by B cells, some APCs, and T cells*	Maintenance of activity of differentiated T cells; a feature of T-/B-cell interactions
Negative costimulation		
CTLA-4	CD80 (B7-1) or CD86 (B7-2) *Expressed by professional APCs (and medullary thymic epithelium)*	Negative regulation of the immune response (e.g., maintaining peripheral T-cell tolerance; reducing inflammation; contracting T-cell pool after infection is cleared)
PD-1	PD-L1 or PD-L2 *Expressed by professional APCs, some T and B cells, and tumor cells*	Negative regulation of the immune response, regulation of T_{REG} differentiation
BTLA	HVEM *Expressed by some APCs, T and B cells*	Negative regulation of the immune response, regulation of T_{REG} differentiation (?)

CLASSIC EXPERIMENT

 ## Discovery of the First Costimulatory Receptor: CD28

In 1989, Navy immunologist Carl June took the first step toward filling (and revealing) the considerable gaps in our understanding of T-cell proliferation and activation by introducing a new actor: CD28. CD28 had been recently identified as a dimeric glycoprotein expressed on all human CD4$^+$ T cells and half of human CD8$^+$ T cells, and preliminary data suggested that it enhanced T-cell activation. June and his colleagues specifically wondered if CD28 might be related to the Signal 2 that was known to be provided by APCs (sometimes referred to as "accessory cells" in older literature).

June and his colleagues isolated T cells from human blood by density gradient centrifugation and by depleting a T-cell-enriched population of cells that did not express CD28 (negative selection). They then measured the response of these CD28$^+$ T cells to TCR stimulation in the presence or absence of CD28 engagement (Figure 1a).

To mimic the TCR-MHC interaction, June and colleagues used either monoclonal antibodies to the CD3 complex or the mitogen phorbol myristyl acetate (PMA), a protein kinase C (PKC) activator. To engage the CD28 molecule, they used an anti-CD28 monoclonal antibody. They included two negative controls: one population that was cultured in growth medium with no additives and another population that was exposed to phytohaemagglutinin (PHA), which was known to activate T cells only in the presence of APCs. This last control was clever and would be used to demonstrate that the researchers' isolated populations were not contaminated with APCs, which would express their own sets of ligands and confound interpretation.

June and colleagues measured the proliferation of each of these populations by measuring incorporation of a radioactive (tritiated) nucleotide, [^{3}H]Uridine, which is incorporated by cells that are synthesizing new RNA (and, hence, are showing signs of activation). Their results were striking, particularly when responses to PMA were examined (Figure 1b). As expected, T cells grown without stimulation or with incomplete stimulation (PHA) remained quiescent, exhibiting no nucleotide uptake. Cell groups treated with stimuli that were known to cause T-cell proliferation—anti-CD3 and PMA—showed evidence of activation, with CD3 engagement producing relatively more of a response than PMA at the time points examined.

The cells treated with anti-CD28 only were just as quiescent as the negative control samples, indicating that engagement of CD28, alone, could not induce activation. However, when CD28 was engaged at the same time cells were exposed to PMA, incorporation of [^{3}H]Uridine increased markedly. T cells cotreated with anti-CD28 and anti-CD3 also took up more [^{3}H]Uridine than those treated with anti-CD3 alone.

What new RNA were these cells producing? Using Northern blot analysis and functional assays (bioassays) to characterize the cytokines in culture supernatant of the stimulated cells, June and colleagues went on to show that CD28 stimulation induced anti-CD3 stimulated T cells to produce higher levels of cytokines

naïve T cell and peaks in expression within 2 to 3 days poststimulation. Peak surface levels of CTLA-4 are lower than peak CD28 levels, but because it binds CD80 and CD86 with markedly higher affinity, CTLA-4 competes very favorably with CD28. Interestingly, CTLA-4 expression levels increase in proportion to the amount of CD28 costimulation, suggesting that CTLA-4 acts to "put the brakes on" the proproliferative influence of TCR-CD28 engagement. The importance of this inhibitory function is underscored by the phenotype of CTLA-4 knockout mice, whose T cells proliferate without control, leading to lymphadenopathy (greatly enlarged lymph nodes), splenomegaly (enlarged spleen), autoimmunity, and death within 3 to 4 weeks after birth.

Negative Costimulatory Receptors: PD-1 and BTLA

PD-1 (CD279) and *B and T lymphocyte attenuator* (BTLA (CD272)) are relatively new additions to our list of negative costimulatory receptors. Although more distantly related to CD28 family members than CTLA-4, they also inhibit TCR-mediated T-cell activation. *program death-1* (PD-1) is expressed by both B and T cells and binds to two ligands, PD-L1 (B7-H1) and PD-L2 (B7-DC), which are also members of the CD80/86 family. PD-L2 is expressed predominantly on APCs; however, PD-L1 is expressed more broadly and may help to mediate T-cell tolerance in nonlymphoid tissues. Recent data suggest that PD-L1/PD-L1/2 interactions regulate the differentiation of regulatory T cells.

BTLA is more broadly expressed: not only has it been found on conventional T$_H$ cells, as well as γδ T cells and regulatory T cells, but it is also expressed on NK cells, some macrophages and dendritic cells, and most highly on B cells. Interestingly, BTLA's primary ligand appears not to be a B7 family member, but a TNF receptor family member known as *herpesvirus-entry mediator* (HVEM), which is also expressed on many cell types. Studies on the role of this interesting costimulatory receptor-ligand pair are ongoing, but there are indications that BTLA-HVEM interactions also play a role in down-regulating inflammatory and autoimmune responses.

As the genome continues to be explored, additional costimulatory molecules—both negative and positive in influence—are likely to be identified. Understanding their

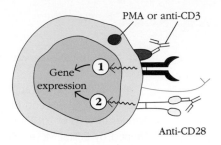

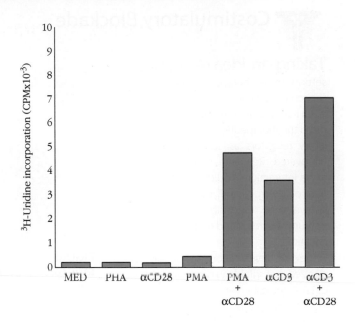

FIGURE 1
(a) June's experimental setup used anti-CD3 monoclonal antibodies or a mitogen, PMA, to provide Signal 1, and anti-CD28 antibodies to provide Signal 2. (b) June measured [³H]Uridine incorporation, an indicator of RNA synthesis, in response to various treatments. Addition of stimulating anti-CD28 antibody increased RNA synthesis in response to activation by PMA or anti-CD3. *[Part (b) adapted from C. Thompson et al., 1989, CD28 activation pathway regulates the production of multiple T-cell-derived lymphokines/cytokines. Proceedings of the National Academy of Sciences of the United States of America 86:1333.]*

involved in antiviral, anti-tumor, and proliferative activity, generating an increase in T-cell immune response.

When this paper was published, June did not know the identity of the natural ligand for CD28, or even if the homodimer could be activated in a natural immune context. We now know that CD28 binds to CD80/86 (B7), providing the critical Signal 2 during naïve T-cell activation, a signal required for optimal up-regulation of IL-2 and the IL-2 receptor. Finding this second switch capable of modulating T-cell activation was only the beginning of a landslide of discoveries of additional costimulatory signals—positive and negative—involved in T-cell activation, and the recognition that T cells are even more subtly perceptive than we once appreciated.

Based on a contribution by Harper Hubbeling, Haverford College, 2011.

Thompson, C., et al. 1989. CD28 activation pathway regulates the production of multiple T-cell-derived lymphokines/cytokines. *Proceedings of the National Academy of Sciences of the United States of America* **86**(4):1333–1337.

regulation and function will continue to occupy the attention of the immunological community and has already provided the clinical community with new tools for manipulating the immune response during transplantation and disease (see Clinical Focus Box 11-2).

Clonal Anergy Results if a Costimulatory Signal Is Absent

Experiments with cultured cells show that if a resting T cell's TCR is engaged (Signal 1) in the absence of a suitable costimulatory signal (Signal 2), that T cell will become unresponsive to subsequent stimulation, a state referred to as anergy. There is good evidence that both CD4⁺ and CD8⁺ T cells can be anergized, but most studies of anergy have been conducted with CD4⁺ T_H cells.

Anergy can be demonstrated in vitro with systems designed to engage the TCR in the absence of costimulatory molecule engagement. For instance, T cells specific for a MHC-peptide complex can be induced to proliferate in vitro by incubation with activated APCs that express both the appropriate MHC-peptide combination and CD80/86. However, glutaraldehyde-fixed APCs, which express MHC class II-peptide complexes, but cannot be induced to express CD80/86, render T cells unresponsive (Figure 11-4a). These anergic T cells are no longer able to secrete cytokines or proliferate in response to subsequent stimulation (Figure 11-4b). T-cell anergy can also be induced in vitro by incubating T cells with normal, CD80/86-expressing APCs in the presence of the Fab portion of anti-CD28, which blocks the interaction of CD28 with CD80/86 (Figure 11-4c).

In vivo, the requirement for costimulatory ligands to fully activate a T cell decreases the probability that circulating autoreactive T cells will be activated and become dangerous. For instance, a naïve T cell expressing a T-cell receptor specific for an MHC class I insulin-peptide complex would be rendered nonresponsive if it encountered a β-islet cell expressing this MHC class I-peptide complex. Why? β-islet cells cannot be induced to express costimulatory ligands, and the encounter would result in T-cell anergy, preventing an immune attack on these insulin-producing cells.

Costimulatory Blockade

Taking an idea from the research setting to a therapeutic reality—from "bench to bedside"—is a dream for many, but a reality for few. Immunologists recognized the therapeutic promise of costimulatory receptors as soon as they were discovered. Reagents that would block these interactions could block the activation of destructive T cells, which were known to be responsible for many autoimmune disorders and for transplantation rejection.

Several investigators specifically recognized that soluble versions of CD28 and CTLA-4 could be very valuable. By blocking the interaction between costimulatory receptors and their CD80/86 ligands, soluble CD28 and CTLA-4 could inhibit destructive T-cell responses (e.g., those involved in autoimmune disease or transplantation rejection). Couple this idea with technological advances in protein design and you had a novel reagent—and a potential drug.

In the early 1990s, at least two groups converted human CTLA-4 into a soluble protein by fusing the extracellular domain of human CTLA-4 with the F_c portion of an IgG1 antibody (Figure 1). The F_c portion enhances (1) the ability to manipulate a protein during production by taking advantage of F_c binding as well as (2) the distribution of the reagent in an organism (the F_c portion confers some of the antibodies' tissue distribution behaviors). The F_c portion is modified so that it does not

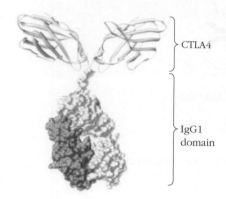

FIGURE 1

Structure of CTLA-4 Ig The extracellular domain of human CTLA-4 is fused to a modified F_c region of human IgG1. *[PDB IDs 1IGY and 3OSK.]*

bind to F_c receptors and cause unintended cytotoxicity. Using this new protein, Peter Linsley et al. found that their soluble CTLA-4-Ig bound CD80/86 with higher affinity than CD28-Ig, and could therefore more potently block costimulation.

The development of CTLA-4 Ig as a drug did not proceed without difficulty. Originally designed and tested as a treatment for T-cell-mediated transplantation rejection, it did not originally perform as well as expected. However, hope in its utility was revived when it showed significant promise as a treatment for rheumatoid arthritis. The marketing of CTLA-4 Ig was not without controversy, either. At least two groups claimed patent rights, and communication between companies

and basic researchers was not always smooth.

In late 2005, the FDA approved the use of CTLA-4 Ig (abatacept) for rheumatoid arthritis (RA). As of 2012 it is marketed as Orencia by Bristol Meyers Squibb, which shares patent royalties with Repligen Corporation. Abatacept shows promise in delaying the joint damage seen with RA, and clinical trials are also underway to evaluate its potential to ameliorate psoriasis, lupus, Type 1 diabetes, and more.

This true bench-to-bedside story is still not finished and continues to be informed by basic researchers' growing knowledge of the complexities of costimulation. Investigators have already developed a modified version of CTLA-4 Ig that differs in two amino acids in the CD80/86 binding region and exhibits a higher affinity for CD86, which may play a more important role in transplantation rejection. This drug is also showing promise in clinical trials and may result in fewer side effects than standard immunosuppressant therapies (e.g., cyclosporine), which are not specific for T cells. It is also important to recognize that CTLA-4 Ig not only blocks positive costimulatory reactions, but also has the potential to inhibit negative ones and in some circumstances could lead to enhancement of T-cell activity. Researchers will continue to draw from basic and clinical knowledge to determine how best to use the drug and to improve its design for enhanced safety and efficacy.

Interactions between negative costimulatory receptors and ligands can also induce anergy. This phenomenon, which would typically only apply to activated T cells that have up-regulated negative costimulatory receptors, could help curb T-cell proliferation when antigen is cleared. Unfortunately, negative costimulation may also contribute to the T cell "exhaustion" during chronic infection, such as that caused by mycobacteria, HIV, hepatitis virus, and more. T cells specific for these pathogens express high levels of PD-1 and BTLA, and are functionally anergic. Some recent therapies, in fact, are designed to block this interaction and allow T cells to reactivate.

Although anergy is a well-established phenomenon, the precise biochemical pathways that regulate this state of nonresponsiveness are still not fully understood. During the past few years, microarray analyses (see Chapter 20) have identified several key enzymes expressed by anergic T cells, including ubiquitin ligases that appear to target key components of the TCR signaling pathway for degradation by the proteasome.

Cytokines Provide Signal 3

We have now seen that naïve T cells will initiate activation when they are costimulated by engagement with both

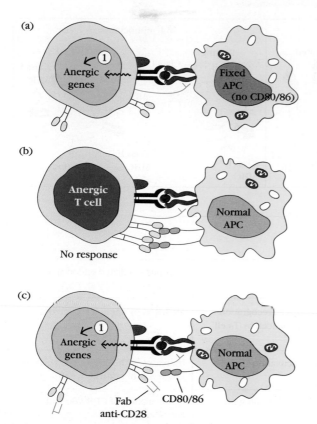

FIGURE 11-4 Experimental demonstration of clonal anergy versus clonal expansion. (a) Only Signal 1 is generated when resting T cells are incubated with glutaraldehyde-fixed APCs, which cannot be stimulated to up-regulate the costimulatory ligand CD80/86. (b) Anergic cells cannot respond to subsequent challenge, even when APCs can engage costimulatory receptors. (c) Anergy can also be induced when naïve T cells are incubated with normal APCs in the presence of the Fab portion of anti-CD28, which blocks interaction with CD80/86.

MHC-peptide complexes and costimulatory ligands on dendritic cells. However, the consequences and extent of T-cell activation are also critically shaped by the activity of soluble cytokines produced by both APCs and T cells. These assisting cytokines are referred to, by some, as *Signal 3* (see Figure 11-3).

Cytokines bind surface cytokine receptors, stimulating a cascade of intracellular signals that can enhance both proliferation and/or survival (see Chapter 4). IL-2 is one of the best-known cytokines involved in T-cell activation and plays a key role in inducing optimal T-cell proliferation, particularly when antigen and/or costimulatory ligands are limiting. Costimulatory signals induce transcription of genes for both IL-2 and the α chain (CD25) of the high-affinity IL-2 receptor. Signals also enhance the stability of the IL-2 mRNA. The combined increase in IL-2 transcription and improved IL-2 mRNA stability results in a 100-fold increase in IL-2 production by the activated T cell. Secretion of IL-2 and its subsequent binding to the high-affinity IL-2 receptor induces activated naïve T cells to proliferate vigorously.

As we will see below, Signal 3 is also supplied by other cytokines (produced by APCs, T cells, NK cells, and others), known as *polarizing cytokines*, that play central roles not just in enhancing proliferation, but also in determining what types of effector cells naïve T cells will become.

Antigen-Presenting Cells Have Characteristic Costimulatory Properties

Which cells are capable of providing both Signal 1 and Signal 2 to a naïve T cell? Although almost all cells in the body express MHC class I, only professional APCs—dendritic cells, activated macrophages, and activated B cells—express the high levels of class II MHC molecules that are required for T-cell activation. Importantly, these same professional APCs are among the only two cell types capable of expressing costimulatory ligands. (The only other cell type known to have this capacity is the thymic epithelial cell; see Chapter 9.) Professional APCs are more diverse in function and origin than originally imagined, and each subpopulation differs both in the ability to display antigen and in the expression of costimulatory ligands (Figure 11-5).

In the early stages of an immune response in secondary lymphoid organs, T cells encounter two main types of professional APCs: the dendritic cell and the activated B cell. Mature dendritic cells that have been activated by microbial components via their pattern recognition receptors (PRRs) are present throughout the T-cell zones. They express antigenic peptides in complex with high levels of MHC class I and II molecules. They also express costimulatory ligands and are the most potent activators of naïve CD4$^+$ and CD8$^+$ T cells.

Resting B cells residing in the follicles gain the capacity to activate T cells after they bind antigen through their B-cell receptor (BCR). This engagement stimulates the up-regulation of MHC class II and costimulatory CD80/86, allowing the B cell to present antigen to CD4$^+$ T cells they encounter at the border between the follicle and T-cell zone. Because of their unique ability to internalize antigen (e.g., pathogens) via specific BCRs and present them in MHC class II, B cells are most efficient at activating CD4$^+$ T cells that are specific for the same pathogen for which they are specific. This situation serves the immune response very well, focusing the attention of antigen-specific CD4$^+$ T cells activated in the T-cell zone on B cells activated by the same antigen in the neighboring follicle. The pairing of B cells with their helper T cells occurs at the junction between the B- and T-cell zones (see Chapter 14) and allows T cells to deliver the help required for B-cell proliferation, differentiation, and memory generation (see Chapter 12).

Macrophages are also found in secondary lymphoid organs, but can activate T cells in a wide range of other peripheral tissues. They also must be activated to reveal their full antigen-presenting capacity. They up-regulate MHC molecules and costimulatory ligands in response to interactions with pathogens, as well as in response to cytokines produced by other cells, including IFN-γ.

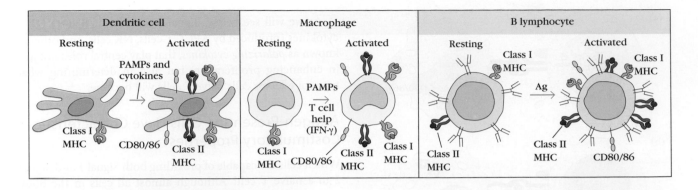

Antigen-presenting cell	Dendritic cell	Macrophage	B cell
Antigen uptake	Endocytosis Phagocytosis	Phagocytosis	Receptor-mediated endocytosis
Activation	Mediated by pattern recognition receptors	Mediated by pattern recognition receptors and enhanced by T-cell help	Mediated by antigen recognition
MHC Class II expression	Increases with activation (may express low levels constitutively)	Increases with activation	Increases with activation (expresses low levels of constitutively)
Costimulatory activity	Up-regulation of CD80/86 with activation	Up-regulation of CD80/86 with activation	Up-regulation of CD80/86 with activation
T-cell activation	Naïve T cells Effector T cells Memory T cells	Effector T cells Memory T cells	Effector T cells Memory T cells
Location	**Resting:** *Circulation peripheral tissues* **Activated:** *SLOs (T-cell zones) Tertiary tissues*	**Resting:** *Circulation peripheral tissues* **Activated:** *SLOs (subcapsular cortex of lymph node, marginal zones of spleen) Peripheral tissues*	**Resting:** *Circulation SLOs (follicles)* **Activated:** *SLOs (B cell/T-cell zone interface, germinal centers, and marginal zones)*

FIGURE 11-5 Differences in the properties of professional antigen-presenting cells that induce T-cell activation. This figure describes general features of three major classes of professional APCs. Dendritic cells are the best activators of naïve T cells. This may be due, in part, to relatively high levels of expression of MHC and co-stimulatory molecules when they are mature and activated. Activated B cells interact most efficiently with differentiated T_H cells that are specific for the same antigen that activated them. Macrophages play several different roles, processing and distributing antigen in second-ary lymphoid tissues (SLOs) as well as interacting with effector cells in the periphery. It is important to recognize that the distinctions shown are rules of thumb only. Functions among the APC classes overlap, and the field now recognizes different subsets within each major group of APC, each of which may act independently on different T-cell subsets. This diversity may be a consequence of activation by different innate immune receptors or may reflect the existence of independent cell lineages. Note that activation of effector and memory T cells is not as dependent on costimulatory interactions.

It turns out that there are several different dendritic cell and macrophage subtypes; however, their functions are still being clarified. Some are likely to activate or induce differentiation of specific effector T cells that travel to the site of infection, and some may be involved in reactivating memory cells that reside in tissues. Others may help to quell, rather than to activate, responses. Full understanding awaits more research.

Superantigens Are a Special Class of T-Cell Activators

Superantigens are viral or bacterial proteins that bind simultaneously to specific Vβ regions of T-cell receptors and to the α chain of class II MHC molecules. Vβ regions are encoded by over 20 different Vβ genes in mice and 65 different genes in humans. Each superantigen displays a

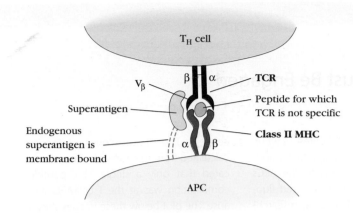

FIGURE 11-6 Superantigen-mediated cross-linkage of T-cell receptor and class II MHC molecules. A superantigen binds to all TCRs bearing a particular Vβ sequence regardless of their antigen specificity. Exogenous superantigens are soluble secreted bacterial proteins, including various exotoxins. Endogenous super-antigens are membrane-embedded proteins produced by certain viruses; they include Mls antigens encoded by mouse mammary tumor virus.

"specificity" for one of these Vβ versions, which can be expressed by up to 5% of T cells, regardless of their antigen specificity. This clamp-like connection mimics a strong TCR-MHC interaction and induces activation, bypassing the need for TCR antigen specificity (Figure 11-6). Superantigen binding, however, does not bypass the need for costimulation; professional APCs are still required for full T-cell activation by these microbial proteins.

Both endogenous superantigens and exogenous superantigens have been identified. *Exogenous* superantigens are soluble proteins secreted by bacteria. Among them are a variety of **exotoxins** secreted by Gram-positive bacteria, such as staphylococcal enterotoxins, toxic shock syndrome toxin, and exfoliative dermatitis toxin. Each of these exogenous superantigens binds particular Vβ sequences in T-cell receptors (Table 11-2) and cross-links the TCR to a class II MHC molecule.

Endogenous superantigens are cell-membrane proteins generated by specific viral genes that have integrated into mammalian genomes. One group, encoded by mouse mammary tumor virus (MTV), a retrovirus that is integrated into the DNA of certain inbred mouse strains, produces proteins called **minor lymphocyte-stimulating (Mls) determinants**, which bind particular Vβ sequences in T-cell receptors and cross-link the TCR to class II MHC molecules. Four Mls superantigens, originating from distinct MTV strains, have been identified.

Because superantigens bind outside the TCR antigen-binding cleft, any T cell expressing that particular Vβ sequence will be activated by a corresponding superantigen. Hence, the activation is polyclonal and can result in massive T-cell activation, resulting in overproduction of T_H-cell cytokines and systemic toxicity. Food poisoning induced by staphylococcal enterotoxins and toxic shock induced by toxic shock syndrome toxin are two examples of disorders caused by superantigen-induced cytokine overproduction.

Given that superantigens result in T-cell activation and proliferation, what adaptive value could they possibly have for the pathogens that make them? The answer is not clear, but there is evidence that such antigen-nonspecific T-cell

---

TABLE 11-2   Exogenous superantigens and their Vβ specificity

Superantigen	Disease*	Vβ SPECIFICITY Mouse	Vβ SPECIFICITY Human
Staphylococcal enterotoxins			
SEA	Food poisoning	1, 3, 10, 11, 12, 17	nd
SEB	Food poisoning	3, 8.1, 8.2, 8.3	3, 12, 14, 15, 17, 20
SEC1	Food poisoning	7, 8.2, 8.3, 11	12
SEC2	Food poisoning	8.2, 10	12, 13, 14, 15, 17, 20
SEC3	Food poisoning	7, 8.2	5, 12
SED	Food poisoning	3, 7, 8.3, 11, 17	5, 12
SEE	Food poisoning	11, 15, 17	5.1, 6.1–6.3, 8, 18
Toxic shock syndrome toxin (TSST1)	Toxic shock syndrome	15, 16	2
Exfoliative dermatitis toxin (ExFT)	Scalded skin syndrome	10, 11, 15	2
Mycoplasma arthritidis supernatant (MAS)	Arthritis, shock	6, 8.1–8.3	nd
Streptococcal pyrogenic exotoxins (SPE-A, B, C, D)	Rheumatic fever, shock	nd	nd

*Disease results from infection by bacteria that produce the indicated superantigens.

ADVANCES

How Many TCR Complexes Must Be Engaged to Trigger T-Cell Activation?

Until single-cell imaging techniques were developed, indirect methods were used to calculate how many ligands a T cell must recognize in order to be activated. Mark Davis and colleagues at Stanford School of Medicine approached this question using an acutely sensitive, two-part microscopic visualization technique (Figure 1). APCs were cultured (pulsed) briefly with peptide bound to a biotin molecule. When APCs are exposed to soluble peptides a small number will exchange it with a peptide bound to an MHC molecule on the cell surface. The number of peptides that actually bound to MHC could be determined in this system by adding a fluorescent (phycoerythrin) streptavidin conjugate, which binds the biotin of the biotinylated peptide and can be detected and quantified by fluorescent microscopy.

The response of T cells specific for the complex could also be quantified using another fluorescent tool: fura-2, a dye that can enter cells and fluoresces when intracellular calcium is released. By adding fura-2 "loaded" T cells to the APCs bound to varying numbers of red fluorescent peptide (see Figure 1a), researchers could determine (1) if a T cell was activated and (2) how many molecules of peptide were present at the point of contact between the T cell and the APC—in other words, how many TCR/MHC-biotinylated peptide complexes were required to stimulate the release of intracellular calcium.

One set of data from these experiments is illustrated in Figure 1b. An activated, fura-2-loaded T cell (blue) is shown interacting with an APC in the upper left. The fluorescent micrograph of the peptide at the junction of the T cell and APC is shown in the top right image. The intensity of the red fluorescence varies with the number of peptides bound. (The image is "stretched" artificially because of the computer program used to quantify fluores-

cence.) The fluorescence intensity calculated from this particular image indicated that only a single MHC-peptide combination was at the T cell-APC synapse. The plot below these images shows the intensity of fura-2 fluorescence (i.e., the increase in intracellular Ca^{2+}) over time after T cell-APC engagement. The initial spike is an indicator that this single MHC-peptide could inspire robust Ca^{2+} signals.

The investigators quantified many interactions in this way and definitively concluded that a single MHC-peptide combination could stimulate significant Ca^{2+} release. Maximal Ca^{2+} release was achieved in CD4[+] T cells when as few as 10 TCR complexes were engaged. Similar results were obtained with CD8[+] T cells.

Irvine, D. J., M. A. Purbhoo, M. Krogsgaard, and M. M. Davis. 2002. Direct observation of ligand recognition by T cells. *Nature* **419**:845–849.

---

activation and inflammation hampers the development of a coordinated antigen-specific response that would most effectively thwart infection. Some speculate that the large-scale proliferation and cytokine production that occur harm the cells and microenvironments that are required to start a normal response; others argue that these events induce T-cell tolerance to the pathogen.

T-Cell Differentiation

How does an interaction between a naïve T cell and a dendritic cell result in the generation of cells with effector functions? An activating, costimulatory interaction between a naïve T cell and an APC typically lasts 6 to 8 hours, a period that permits the development of a cascade of signaling events (see Chapter 3) that alter gene programs and induce differentiation into a variety of distinct effector and memory cell subtypes. Just a few TCR-MHC interactions (as described in Advances Box 11-3) stimulates a signaling cascade that,

in combination with costimulation and signals received by soluble cytokines, culminate in the activation of "effector" molecules that regulate (1) *cell survival*, (2) *cell cycle entry*, and, as we shall see below, (3) *cell differentiation*.

In 1 to 2 days after successful engagement with a dendritic cell in the T-cell zone of a secondary lymphoid organ, a naïve T cell will enlarge into a blast cell and begin undergoing repeated rounds of cell division. Signals 1 plus 2 induce up-regulation of expression and activity of pro-survival genes (e.g., bcl-2), as well as the transcription of genes for both IL-2 and the α chain (CD25) of the high-affinity IL-2 receptor (Figure 11-7). The combined effect on a naïve T cell is activation and robust proliferation. Activated T cells divide 2 to 3 times per day for 4 to 5 days, generating a clone of progeny cells that differentiate into memory and effector T-cell populations.

Activated T cells and their progeny gain unique functional abilities, becoming *effector* helper or cytotoxic T cells that indirectly and directly act to clear pathogen. CD8[+] cytotoxic T cells leave the secondary lymphoid tissues and circulate

BOX 11-3

(a)

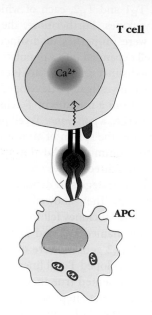

(b)

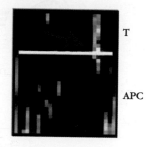

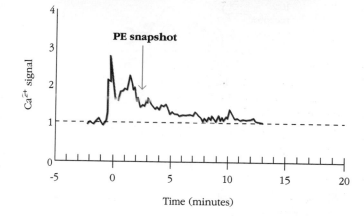

FIGURE 1

Engagement of a single T-cell receptor can induce Ca²⁺ signals in a T cell. (a) The experimental approach taken by the investigators to determine the ability of a single TCR-MHC interaction to generate Ca²⁺ signals. See text for details. (b) One example of original data on Ca²⁺ signaling generated from a single T cell that has engaged a single MHC-peptide ligand on the surface of an APC. The interaction between the T cell

and APC is captured by bright field microscopy (top left) and by high-resolution fluorescence microscopy (top right), where the arrow points to the single TCR/MHC-peptide interaction. The Ca²⁺ signal generated within the T cell over a 20-minute period is depicted in the graph and measured using software that quantifies the intensity of fura-2 fluorescence, which increases with a rise in cytosolic Ca²⁺ levels. [*Irvine, D.J. et al. 2002 Nature 419:845–849.*]

---

to sites of infection where they bind and kill infected cells. CD4⁺ helper T cells secrete cytokines that orchestrate the activity of several other cell types, including B cells, macrophages, and other T cells. Some CD4⁺ T cells, particularly those that "help" B cells and those that generate lymphocyte memory, stay within secondary lymphoid tissue to continue to regulate the generation of the response. Others return to the sites of infection and enhance the activity of macrophages and cytotoxic cells.

Effector cells tend to be short-lived and have life spans that range from a few days to a few weeks. However, the progeny of an activated T cell can also become long-lived memory T cells that reside in secondary and tertiary tissues for significant periods of time, providing protection against a secondary infection.

Effector T cells come in more varieties than was originally anticipated, and each subset plays a specific and important role in the immune response. The first effector cell distinction to be recognized was between CD8⁺ T cells and CD4⁺ T cells. Activated CD8⁺ T cells acquire the ability to induce

the death of target cells, becoming "killer" or "cytotoxic" T lymphocytes (CTL, or T_C). Because cytotoxic CD8⁺ T cells recognize peptide bound to MHC class I, which is expressed by almost all cells in the body, they are perfectly poised to clear the body of cells that have been internally infected by the pathogen that resulted in their activation. On the other hand, activated CD4⁺ T cells (helper T cells or T_H) acquire the ability to secrete factors that enhance the activation and proliferation of other cells. Specifically, they regulate activation and antibody production of B cells, enhance the phagocytic, anti-microbial, cytolytic, and antigen-presenting capacity of macrophages, and are required for the development of B-cell and CD8⁺ T-cell memory.

As immunologists developed and adopted more tools to distinguish proteins expressed and secreted by helper T cells, it became clear that CD4⁺ helper T cells were particularly diverse, differentiating into several different subtypes, each of which secretes a signature set of cytokines. The cytokines secreted by CD4⁺ T_H cells can either act directly on the same cell that produced them (acting in an *autocrine* fashion) or

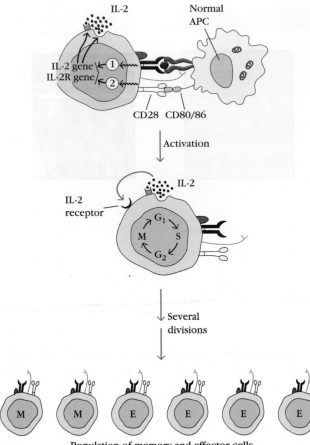

FIGURE 11-7 Activation of a naïve T cell up-regulates expression of IL-2 and the high-affinity IL-2 receptor. Signal 1 and Signal 2 cooperate to enhance transcription and stability of mRNA for IL-2 and IL-2R. Secreted IL-2 will bind the IL-2R, which generates signals that enhance the entry of the T cell into the cell cycle and initiates several rounds of proliferation. Most of the cells differentiate into effector helper or cytotoxic cells, but some will differentiate into effector or central memory cells.

can bind to receptors and act on cells in the vicinity (acting in a *paracrine* fashion). Below we describe the major features and functions of the best-characterized CD4$^+$ helper T-cell subsets. Although CD8$^+$ cytotoxic T cells also secrete cytokines, and there are indications that CD8$^+$ T cells may also differentiate into more than one killer subtype, the diversity of CD8$^+$ T-cell effector functions are clearly more restricted than those of CD4$^+$ T$_H$ cells. The generation and activity of CTL cells are described in more detail in Chapter 13.

Helper T Cells Can Be Divided into Distinct Subsets

Tim Mosmann, Robert Coffman, and colleagues can be credited with one of the earliest experiments definitively demonstrating that helper CD4$^+$ T cells were more heterogeneous in phenotype and function than originally supposed. Earlier investigations, showing that helper T cells produced a diverse

array of cytokines, hinted at this possibility. However, Mosmann and Coffman definitively identified two distinct functional subgroups, T$_H$1 and T$_H$2, each of which produced a different set of cytokines. Furthermore, they showed that these differences were properties of distinct T-cell clones: each activated T cell expanded into a population of effector T cells that secreted a distinct array of cytokines.

Specifically, these investigators developed over 50 individual T-cell clones from a mixture of T cells with different receptor specificities (i.e., a *polyclonal* T-cell population) that had been isolated from the spleen of an immunized mouse. At a time when the community did not have the tools to identify most cytokines directly, these researchers developed clever (and elaborate) strategies to define each clone's cytokine secretion pattern. They showed that the cytokines secreted by each of the 50 clones fell into one of two broad categories, which they named T$_H$1 and T$_H$2.

Because T$_H$1 and T$_H$2 subsets were originally identified from in vitro studies of cloned T-cell lines, some researchers doubted that they represented true in vivo subpopulations. They suggested instead that these subsets might represent different maturational stages of a single lineage. In addition, the initial failure to locate either subset in humans led some to believe that T$_H$1, T$_H$2, and other subsets of T helper cells were not present in this species. Further research corrected these views. In many in vivo systems, the full commitment of populations of T cells to either a T$_H$1 or T$_H$2 phenotype occurs late in an immune response. Hence, it was difficult to find clear T$_H$1 or T$_H$2 subsets in studies employing healthy human subjects, where these cells would not have developed. Indeed, T$_H$1 and T$_H$2 populations in T cells were ultimately isolated from humans during chronic infectious disease or chronic episodes of allergy, and studies in humans and mice definitively confirmed their lineage independence.

With the benefit of new tools and technology, we now have a more detailed understanding of the panels of cytokines that each group produces. The **T$_H$1 subset** secretes IL-2, IFN-γ, and Lymphotoxin-α (TNF-β), and is responsible for many classic cell-mediated functions, including activation of cytotoxic T lymphocytes and macrophages. The **T$_H$2 subset** secretes IL-4, IL-5, IL-6, IL-9, IL-10, and IL-13, and regulates B-cell activity and differentiation.

These experiments set the stage for the discovery over the last decade that CD4$^+$ T cells can adopt not just two but at least five distinct effector fates after activation. The T$_H$1 and T$_H$2 subpopulations have been joined definitively by the T$_H$17 and T$_{REG}$ subpopulations, each of which produces a distinct cytokine profile and regulates distinct activities within the body. Yet another subpopulation, T follicular helper cells (T$_{FH}$), has recently been characterized, and has achieved membership among the major helper subsets. More are bound to reveal themselves, although it will be important to determine whether each represents distinct subgroups or variants within the major subgroups.

In retrospect, we probably should have anticipated the heterogeneity of helper responses, which allows an organism to "tailor" a response to a particular type of pathogen. The type of effector T_H cell that a naïve T cell (also called a T_H0 cell) becomes depends largely on the kind of infection that occurs. For example, extracellular bacterial infections result in the differentiation of activated $CD4^+$ T cells into T_H2 cells, which help to activate B cells to secrete antibodies that can opsonize bacteria and neutralize the toxins they produce. On the other hand, infection by an intracellular virus or bacterium induces differentiation of $CD4^+$ T cells into T_H1 helpers that enhance the cytolytic activity of macrophages and $CD8^+$ T cells, which can then kill infected cells. Responses to fungi stimulate the differentiation of different helper responses than responses to worms, and so on. The reality is, of course, more complex. Infections evoke the differentiation of more than one helper subtype, some of which have overlapping roles. What regulates the differentiation of each effector subset and what function each subset plays are still being actively investigated. We describe the fundamentals of our current understanding below.

The Differentiation of T Helper Cell Subsets Is Regulated by Polarizing Cytokines

As you know, T-cell activation requires TCR and costimulatory receptor engagement, both of which are supplied by an activated APC. It is now clear that the functional fate of activated T cells is determined by signals they receive from additional cytokines generated during the response. These cytokines (Signal 3) are referred to as *polarizing* cytokines because they are responsible for guiding a helper T cell toward one of several different effector fates. For example, T cells that are activated in the presence of IL-12 and IFN-γ tend to differentiate, or polarize, to the T_H1 lineage, whereas T cells that are activated in the presence of IL-4 and IL-6 polarize to the T_H2 lineage.

Polarizing cytokines can be generated by the stimulating APC itself, or by neighboring immune cells that have also been activated by antigen. Which cytokines are produced during an immune response depends on (1) the cell of origin (DC, macrophage, B cell, NK cell, etc.), (2) its maturation and activation status, (3) which pathogens and other inflammatory mediators it encounters, and (4) in what tissue environment it encounters that pathogen. Innate interactions therefore have a critical role in shaping adaptive responses (see Figure 5-18). Specifically, by influencing APC secretions and the surface and the microenvironmental landscape that a T cell encounters, innate immune responses directly influence the functional fate of helper T cells.

Recall from Chapter 5 that APCs and other innate immune cells are activated by interaction with pathogens bearing pathogen-associated molecular patterns (PAMPs). These PAMPs bind PRRs, including, but not limited to Toll-like receptors (TLRs). PRR interactions activate dendritic cells by stimulating the up-regulation of MHC and costimulatory proteins. They also determine the type of cytokine(s) that dendritic cells and other immune cells will secrete.

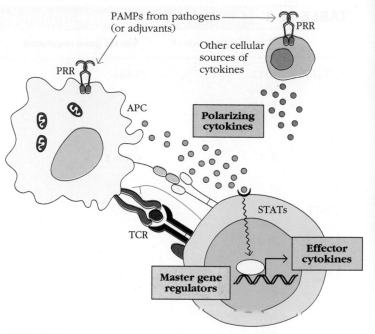

FIGURE 11-8 General events and factors that drive T_H subset polarization. Interaction of pathogen with pattern recognition receptors (PRRs) on dendritic cells and other neighboring immune cells determines which polarizing cytokines are produced and, hence, into which T helper subset a naïve cell will differentiate. In general, polarizing cytokines that arise from dendritic cells or other neighboring cells interact with cytokine receptors and generate signals that induce transcription of unique master gene regulators. These master regulators, in turn, regulate expression of various genes, including effector cytokines, which define the function of each subset.

Hence, PRR signals regulate the fate a T cell will adopt following activation (Figure 11-8).

For example, double-stranded RNA, a product of many viruses, binds TLR3 receptors on dendritic cells, initiating a signaling cascade that results in production of IL-12, which directly promotes T_H1 differentiation. On the other hand, worms stimulate PRRs on innate immune cells, including mast cells, which generate IL-4. IL-4 directly promotes polarization of activated T cells to the T_H2 subset, which coordinates the IgE response to helminths (see Figure 11-8). In this case, the main polarizing cytokine is not made by the activating dendritic cell, but is generated by a neighboring immune cell. The pathogen interactions that give rise to the polarizing cytokines that drive T helper cell differentiation are often complex and a very active area of research.

Adjuvants, which have been used for decades to enhance the efficacy of vaccines, are now understood to exert their influence on the innate immune system by regulating the expression of costimulatory ligands and cytokines by APCs, events that ultimately shape the consequences of T-cell activation. PAMPs and cytokines such as IL-12, produced by APCs themselves, are considered natural adjuvants. Dead mycobacteria, which clearly activate many PRRs, have long

TABLE 11-3	Regulation and function of T helper subtypes			
	Polarizing cytokines	**Master gene regulators**	**Effector cytokines**	**Functions**
T_H1	IL-12 IFN-γ IL-18	T-Bet	IFN-γ TNF	Enhances APC activity Enhances T_C activation Protects against intracellular pathogens Involved in delayed type hypersensitivity, autoimmunity
T_H2	IL-4	GATA-3	IL-4 IL-5 IL-13	Protects against extracellular pathogens (particularly IgE responses) Involved in allergy
T_H17	TGF-β IL-6 (IL-23)	RORγ	IL-17A IL-17F IL-22	Protects against some fungal and bacterial infections Contributes to inflammation, autoimmunity
T_{REG}	TGF-β IL-2	FoxP3	IL-10 TGF-β	Inhibits inflammation
T_{FH}	IL-6 IL-21	Bcl-6	IL-4 IL-21	B cell help in follicles and germinal centers

been used as a very potent adjuvant for immune responses in mice. Very few adjuvants are approved for human vaccination, but given our new and evolving understanding of the molecules that stimulate PRRs and the consequences of that stimulation, investigators expect that we will one day be able to shape the effector response to vaccine antigens by varying the adjuvants—natural and/or synthetic—included in vaccine preparations (see Chapter 17).

Effector T Helper Cell Subsets Are Distinguished by Three Properties

Each helper T-cell subset is defined by an array of features, the details of which can rapidly overwhelm a new (or old) student of immunology. Understanding these specifics is an important first step to clarifying the role each subset plays in resolving infection and causing disease. Having a reference to them is also helpful when deciphering primary literature describing advances. However, some generalizations provide a useful conceptual framework for organizing some of these details. Consider the following:

- Each of the major T helper cell subsets is characterized by (1) a distinct set of *polarizing cytokines* that induce the expression of (2) a *master gene regulator* that regulates expression of (3) a signature set of *effector cytokines* the T-cell population produces once it is fully differentiated (see Figure 11-8 and Table 11-3).

- Which effector subset an activated helper cell becomes depends on the quality and quantity of signals its naïve cell precursor receives from APCs in a secondary lymphoid organ; that activity, in turn, depends on the

nature of the pathogen the APC encountered at the site of infection.

- Broadly speaking, T_H1 and T_H17 cells regulate cell-mediated immunity (CD8$^+$ T cells and macrophages) and T_H2 and T_{FH} cells regulate humoral immunity (B cells). However, it is important to recognize that all CD4$^+$ effector T-cell subsets may have the potential to provide help to B cells. T_H1 and T_H17 subsets generally encourage B cells to produce antibodies that contribute to cell-mediated immunity (e.g., isotypes like IgG2a that can "arm" NK cells for cytotoxicity; see Chapter 13). T_H2 cells encourage B cells to produce antibodies that mediate the clearance of extracellular pathogens (e.g., isotypes like IgE that induce the release of molecules that harm extracellular parasites).

- Helper T-cell subsets often "cross-regulate" each other. The cytokines they secrete typically enhance their own differentiation and expansion and inhibit commitment to other helper T-cell lineages. This is particularly true of the T_H1 and T_H2 pair, as well as the T_H17 and T_{REG} pair.

- Helper cell lineages may not be fixed; some subsets can assume the cytokine secretion profile of other subsets if exposed to a different set of cytokines, particularly early in the differentiation process.

- The precise biological function and sites of differentiation and activity of each subset continue to be actively investigated. Much remains unknown.

We start our discussion of helper cell subset characteristics with the first two subsets to be identified: T_H1 and T_H2 cells. They provide an illustrative example of the features that

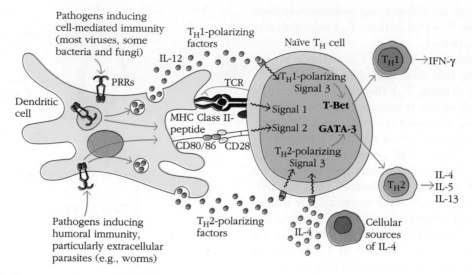

FIGURE 11-9 Regulation of T$_H$1 and T$_H$2 subset differentiation. This figure depicts some of the cellular events that drive T$_H$1 and T$_H$2 lineage commitment in more detail. Intracellular pathogens activate a cascade of signals that polarize cells to the T$_H$1 lineage. For example, viruses interact with PRRs (e.g., TLR-3) that induce dendritic cells to generate IL-12. This binds to receptors on naïve T cells, activating a signal transduction pathway mediated by STAT4 that induces expression of the transcription factor T-Bet. T-Bet, in turn, activates expression of effector cytokines, including IFN-γ, which define the T$_H$1 subset's functional capacities (and can also enhance T$_H$1 polarization) On the other hand, extracellular pathogens activate signal cascades that can polarize naïve T cells to the T$_H$2 lineage. Parasitic worms interact with PRRs on neighboring immune cells (such as mast cells, basophils, or germinal center B cells), triggering the release of the signature T$_H$2 polarizing cytokine IL-4. This interacts with receptors on T cells that activate STAT6, up-regulating expression of the transcriptional regulator GATA-3. GATA-3, in turn, induces expression of the T$_H$2 effector cytokines, including IL-4, IL-5, and IL-13. *[Adapted from M. L. Kapsenberg, 2003, Dendritic-cell control of pathogen-driven T-cell polarization, Nature Reviews Immunology 3:984.]*

distinguish T helper cells as well as the relationship between subsets. We follow this section with summaries of what is currently understood about the more recently characterized helper subsets: T$_H$17, T$_{FH}$, and T$_{REG}$.

The Differentiation and Function of T$_H$1 and T$_H$2 Cells

The key polarizing cytokines that induce differentiation of naïve T cells into T$_H$1 cells are IL-12, IL-18, and IFN-γ (Figure 11-9). IL-12 is produced by dendritic cells after an encounter with pathogens via PRRs (e.g., TLR4, TLR3). It is also up-regulated in response to IFN-γ, which is generated by activated T cells and activated NK cells. IL-18, which is also produced by dendritic cells, promotes proliferation of developing T$_H$1 cells and enhances their own production of IFN-γ. These polarizing cytokines trigger signaling pathways that up-regulate the expression of the master gene regulator T-Bet. This master transcription factor induces commitment to the T$_H$1 lineage, inducing expression of the signature T$_H$1 effector cytokines, including IFN-γ and TNF.

IFN-γ is a particularly potent effector cytokine. It activates macrophages, stimulating these cells to increase microbicidal activity, up-regulate the level of class II MHC, and, as mentioned above, secrete cytokines such as IL-12, which further enhance T$_H$1 differentiation. IFN-γ secretion also induces antibody class switching in B cells to IgG classes (such as IgG2a in the mouse) that support phagocytosis and fixation of complement. Finally, IFN-γ secretion promotes

the differentiation of fully cytotoxic T$_C$ cells from CD8$^+$ precursors by activating the dendritic cells that engage naïve T$_C$ cells. These combined effects make the T$_H$1 subset particularly suited to respond to viral infections and other intracellular pathogens. They also contribute to the pathological effects of T$_H$1 cells, which are also involved in the delayed type hypersensitivity response to poison ivy (see Chapter 15).

Just as differentiation to the T$_H$1 subset is promoted by IL-12 and IFN-γ, differentiation to the T$_H$2 subset is promoted by a defining polarizing cytokine, IL-4 (see Figure 11-9). Exposing naïve helper T cells to IL-4 at the beginning of an immune response causes them to differentiate into T$_H$2 cells. Interestingly, T$_H$2 development is greatly favored over T$_H$1 development. Even in the presence of IFN-γ and IL-12, naïve T cells will differentiate into T$_H$2 effectors if IL-4 is present. IL-4 triggers a signaling pathway within the T cell that up-regulates the master gene regulator GATA3, which, in turn, regulates expression of T$_H$2-specific cytokines, including IL-4, IL-5, and IL-13.

Dendritic cells do not make IL-4, so from where does it come? Mast cells, basophils, and NKT cells can be induced to make IL-4 after exposure to pathogens and could influence helper T cell fate in the periphery. Germinal-center B cells and T$_{FH}$ cells can also produce IL-4, which could influence helper T-cell polarization in the lymph nodes and spleen. And T$_H$2 cells themselves are an excellent source of additional IL-4 that can enhance polarization events.

Investigators, however, are still working to definitively identify the source of the IL-4 that initiates T_H2 polarization in secondary lymphoid tissues.

The effector cytokines produced by T_H2 cells help to clear extracellular parasitic infections, including those caused by helminths. IL-4, the defining T_H2 effector cytokine, acts on both B cells and eosinophils. It induces eosinophil differentiation, activation, and migration and promotes B-cell activation and class switching to IgE. These effects act synergistically because eosinophils express IgE receptors (FcεR) which, when cross-linked, release inflammatory mediators that are particularly good at attacking roundworms. Thus, IgE antibody can form a bridge between the worm and the eosinophil, binding to worm antigens via its variable regions and binding to FcεR via its constant region. IgE antibodies also mediate allergic reactions, and the role of T_H2 activity in these pathological responses is described in Chapter 15.

IL-5 can also induce B-cell class switching to IgG subclasses that do not activate the complement pathway (e.g., IgG1 in mice), and IL-13 functions largely overlap with IL-4. Finally, IL-4 itself can suppress the expansion of T_H1-cell populations.

T_H1 and T_H2 Cross-regulation

The major effector cytokines produced by T_H1 and T_H2 subsets (IFN-γ and IL-4, respectively) not only influence the overall immune response, but also influence the helper T cell subsets. First, they promote the growth (and in some cases even the polarization) of the subset that produces them; second, they inhibit the development and activity of the opposite subset, an effect known as *cross-regulation*. For instance, IFN-γ (secreted by the T_H1 subset) inhibits proliferation of the T_H2 subset, and IL-4 (secreted by the T_H2 subset) downregulates the secretion of IL-12 by APCs, thereby inhibiting T_H1 differentiation. Furthermore, IL-4 enhances T_H2 cell development by making T_H cells less susceptible to the T_H1 promoting cytokine signals (and vice versa).

Similarly, these cytokines have opposing effects on target cells other than T_H subsets. In mice, where the T_H1 and T_H2 subsets have been studied most extensively, the cytokines have distinct effects on the type of antibody made by B cells. Recall from Chapter 3 (and see Chapter 13) that the antibody isotype IgG2a enhances cell-mediated immunity by arming NKT cells, whereas the isotypes IgG1 and IgE enhance humoral immunity by their activities on extracellular pathogens. IFN-γ secreted by the T_H1 subset promotes IgG2a production by B cells but inhibits IgG1 and IgE production. On the other hand, IL-4 secreted by the T_H2 subset promotes production of IgG1 and IgE and suppresses production of IgG2a. Thus, these effects on antibody production are consistent with T_H1 and T_H2 subsets' overall tendencies to promote cell-mediated versus humoral immunity, respectively.

IL-10 secreted by T_H2 cells also inhibits (cross-regulates) T_H1 cell development, but not directly. Instead, IL-10 acts on monocytes and macrophages, interfering with their ability to activate the T_H1 subset by abrogating their activation,

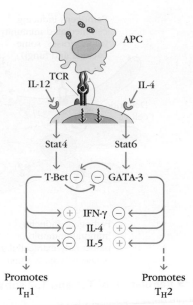

FIGURE 11-10 Cross-regulation of T helper cell subsets by transcriptional regulators. GATA-3 and T-Bet reciprocally regulate differentiation of T_H1 and T_H2 lineages. IL-12 promotes the expression of the T_H1-defining transcription factor, T-Bet, which induces expression of T_H1 effector cytokines, including IFN-γ. At the same time, T-Bet represses the expression of the T_H2 defining master transcriptional regulator, GATA-3, as well as expression of the effector cytokines IL-4 and IL-5. Reciprocally, IL-4 promotes expression of GATA-3, which up-regulates the synthesis of IL-4 and IL-5, and at the same time represses the expression of T-Bet and the T_H1 effector cytokine IFN-γ. *[Adapted from J. Rengarajan, S. Szabo, and L. Glimcher, 2000, Transcriptional regulation of Th1/Th2 polarization, Immunology Today 21:479.]*

specifically by (1) inhibiting expression of class II MHC molecules, (2) suppressing production of bactericidal metabolites (e.g., nitric oxide) and various inflammatory mediators (e.g., IL-1, IL-6, IL-8, GM-CSF, G-CSF, and TNF-β), and even by (3) inducing apoptosis.

The master regulators T-Bet and GATA-3 also play an important role in cross-regulation. (Figure 11-10) Specifically, the expression of T-Bet drives cells to differentiate into T_H1 cells and suppresses their differentiation along the T_H2 pathway. Expression of GATA-3 does the opposite, promoting the development of naïve T cells into T_H2 cells while suppressing their differentiation into T_H1 cells. Consequently, cytokine signals that induce one of these transcription factors sets in motion a chain of events that represses the other.

T_H17 Cells

The discovery of the **T_H17 subset** of T cells, which, like the T_H1 subset, is involved in cell-mediated immunity, arose in part from a recognition that IL-12, one of the polarizing cytokines that induces T_H1 development, was a member of a family of cytokines that shared a subunit (p40) with IL-23. The p40 knockout mouse was a favorite model for studying the importance of T_H1 cells because, in the absence of IL-12, it failed to generate T_H1 cells. However, once it was understood

T Helper Subset Differentiation

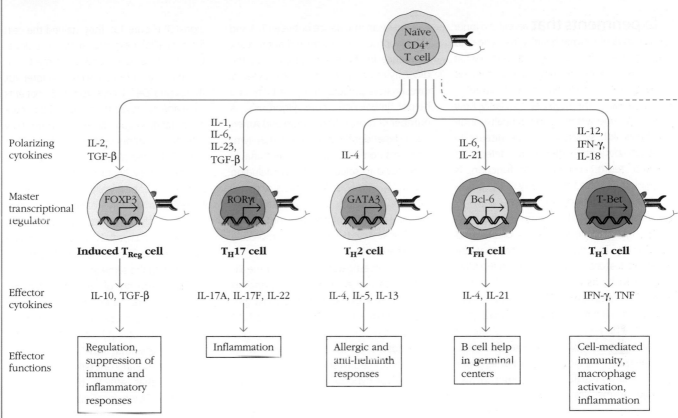

This figure synthesizes current information about the distinguishing features of T helper subset differentiation and activity. Polarizing cytokines, master transcriptional regulators, effector cytokines, and broad functions in health and disease are depicted for each of the major helper subsets. Neither cross-regulation nor the potential plasticity in differentiation among subsets is depicted, but both are described in the text. [Adapted from S. L. Swain, K. K. McKinstry, and T. M. Strutt, Expanding roles for CD41 T cells in immunity to viruses, Nature Reviews. Immunology 12:136–148.]

that p40 was also required for IL-23 production, investigators quickly realized that the results from these mice were no longer unambiguous. In fact, it became clear that these mice also failed to produce a T-cell subset that required IL-23. Some of the functions originally attributed to IL-12-induced T_H1 cells were actually carried out by an IL-23-induced T helper cell subpopulation now referred to as T_H17 cells.

T_H17 cells are generated when naïve T cells are activated in the presence of IL-6 and TGF-β, the key polarizing cytokine for iT_{REG} differentiation (Overview Figure 11-11). IL-23 also plays a role in finalizing the commitment to the T_H17 fate and is induced in APCs by interactions with PAMPs including fungal wall components, with TLR2 and the CLR Dectin-1. Like T_H1 and T_H2 differentiation, T_H17 cell differentiation is also controlled by a master transcriptional regulator whose expression is induced by polarizing cytokines. In this case the master regulator is the orphan steroid receptor RORγt,

which also plays a role in T-cell development. RORγt knock-out mice have reduced severity of experimental autoimmune encephalitis (EAE, a mouse model of multiple sclerosis) apparently because of the reduction in T_H17 cells.

T_H17 cells are so named because they produce IL-17A, a cytokine associated with chronic inflammatory and autoimmune responses, including those that result in inflammatory bowel disease, arthritis, and multiple sclerosis. T_H17 cells are the dominant inflammatory cell type associated with chronic autoimmune disorders. They also produce IL-17F and IL-22, cytokines associated with tissue inflammation. We have only begun to understand the true physiological function of T_H17 cells, which in healthy humans have been found at epithelial surfaces (e.g., lung and gut) and may play a role in warding off fungal and extracellular bacterial infections (see Clinical Focus Box 11-4). However, a full appreciation of their biological role awaits further investigations.

CLINICAL FOCUS

What a Disease Reveals about the Physiological Role of T$_H$17 Cells

Experiments that reveal the inner workings of normal healthy immune cells have offered invaluable insights into what can and does go wrong. However, at times we need "experiments of nature"—the diseases themselves—to clarify how the immune system works in healthy individuals. Job syndrome, a rare disease in which patients suffer from defects in bone, teeth, and immune function, is helping us solve the mystery of the physiological function of T$_H$17 cells.

People with Job syndrome suffer from recurrent infections, typically of the lung and skin. The painful abscesses that are often a feature of the disease and the trials endured by its patients explain its name, which comes from a biblical story of a man (Job) who is subject to horrific hardship as a test of his piety. Another cardinal feature of the disease, elevated IgE levels in serum, is the basis for its more formal name, *Hyper IgE Syndrome* (HIES). HIES comes in two major forms: Job syndrome, the most common, is also referred to as Type 1 HIES. Patients with Type 2 HIES, which we will not discuss here, do not have trouble with bone or dental development.

The abundance of IgE originally suggested to some investigators who were savvy about the roles of helper T-cell subsets that Type 1 HIES symptoms may be

caused by an imbalance between T$_H$1 and T$_H$2 responses. Initial work did not reveal a difference in these activities, but as the diversity of T helper subsets was revealed, investigators pursued the possibility of a helper imbalance more vigorously. A number of groups from Japan and America independently discovered that lymphocytes from HIES patients were unable to respond to select cytokines, including IL-6, IL-10, and IL-23. An analysis of genes that could explain this signaling failure revealed a dominant negative mutation in STAT3, a key cytokine signaling molecule (see Chapter 4).

These investigators knew from the literature that the absence of signaling through these cytokines suggested a specific problem with polarization to the T$_H$17 helper subset. Indeed, circulating T$_H$17 cells were absent in HIES patients with STAT3 mutations. The investigators then directly tested their hypothesis that this absence was due to a failure of CD4$^+$ T cells to polarize normally. Specifically, they removed T cells from blood, and exposed them to conditions that would ordinarily polarize them to the T$_H$17 lineage: T-cell receptor stimulation (Signal 1) in the presence of costimulatory signals (CD28, Signal 2) and cytokines known to drive human T$_H$17 differentiation (IL-1, IL-6, TGF-β, and IL-23,

Signal 3) (Figure 1a). They stained the cells for intracellular expression of T$_H$17's signature cytokine, IL-17, and performed flow cytometry (Figure 1b; also see Chapter 20). Although CD4$^+$ T cells from HIES patients were able to develop into other helper lineages (and, as you can see from the flow cytometry contour plots, were also able to make IFN-γ, indicating they could become T$_H$1 cells), they could not be induced to secrete IL-17. Specifically, whereas 18.3% of T cells from healthy patients that were subject to these conditions stained with antibodies against IL-17, 0.05% (essentially none) of the T cells from HIES patients stained with the same antibodies.

These striking observations suggest that the recurrent infections HIES patients experience are caused at least in part by the absence of T$_H$17 cells. Reciprocally, they indicate that T$_H$17 cells play an important role in controlling the type of infection that afflicts these patients, including *Staphylococcus aureus* skin infections and pneumonia. The critical role of T$_H$17 in controlling bacterial and fungal infections at epithelial surfaces has been supported by studies in mice, too.

J. D. Milner et al. 2008. Impaired T$_H$17 cell differentiation in subjects with autosomal dominant Hyper-IgE syndrome. *Nature* **452**:773–776.

(Induced) T$_{REG}$ Cells

Another major CD4$^+$ T-cell subset negatively regulates T-cell responses and plays a critically important role in peripheral tolerance by limiting autoimmune T-cell activity. This subset of T cells, designated *induced* **T$_{REG}$ (iT$_{REG}$) cells,** is similar in function to the *n*atural T$_{REG}$ cells (nT$_{REG}$s) that originate from the thymus (see Chapter 9). Induced T$_{REG}$ cells, however, do not arise in the thymus, but from naïve T cells that are activated in secondary lymphoid tissue in the presence of TGF-β (see Overview Figure 11-11). TGF-β induces expression of FoxP3, the master transcriptional regulator responsible for iT$_{REG}$ commitment. The iT$_{REG}$ cells secrete the effector cytokines IL-10 and TGF-β, which down-regulate inflammation via their inhibitory effects on APCs, and can also exert their suppressive func-

tion by interacting directly with T cells. The depletion of iT$_{REG}$ cells in otherwise healthy animals leads to multiple autoimmune outbreaks, revealing that even healthy organisms are continually warding off autoimmune responses. Recent data also indicate that iT$_{REG}$ cells are critically important for maintaining a mother's tolerance to her fetus.

T$_H$17 and T$_{REG}$ Cross-Regulation

Just as T$_H$1 and T$_H$2 cells reciprocally regulate each other, T$_{REG}$ and T$_H$17 cells also cross-regulate each other. TGF-β induces T$_{REG}$ differentiation; however, when accompanied by IL-6, TGF-β induces T$_H$17 differentiation. Specifically TGF-β appears to up-regulate both FoxP3 and RORγ (which control T$_{REG}$ and T$_H$17 differentiation, respectively). In combination with signals generated by IL-6,

BOX 11-4

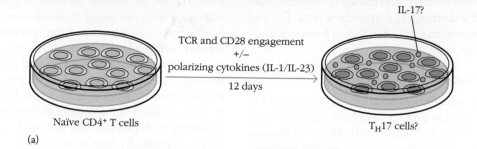

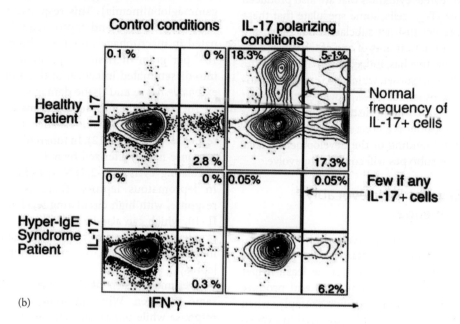

FIGURE 1

CD4⁺ T cells from Hyper-IgE patients do not differentiate into T$_H$17 cells. (a) Conditions used to polarize CD4⁺ T cells to the T$_H$17 lineage in vitro. The ability of the cells to make IL-17 (a feature of T$_H$17 cells) or IFN-γ (a property of T$_H$1, not T$_H$17 cells) after they were polarized was assessed by staining for intracellular cytokines. (b) Flow cytometric analysis of intracellular staining. The boxes outlined in red indicate the quadrant where IL-17⁺ (T$_H$17) cells would appear. [Irvine, D.J. et al. 2002. Nature 419: 845–849. Reprinted by permission from Macmillan Publishers]

signals generated by TGF-β inhibit FoxP3 expression, letting RORγ dominate and induce T$_H$17 development. The T$_H$17 versus iT$_{REG}$ relationship may be very adaptive. At rest, a healthy organism may favor the development of an anti-inflammatory iT$_{REG}$ population, which would be reinforced by the iT$_{REG}$ cell's own production of TGF-β. Inflammation, however, would induce the generation of acute response proteins, including IL-6. In the presence of IL-6, TGF-β activity would shift development of T cells away from iT$_{REG}$s toward the pro-inflammatory T$_H$17, so a proper defense could be mounted.

T$_{FH}$ Cells

Follicular helper T (T$_{FH}$) cells are a very recent addition to the helper T-cell subset family. Whether they represent an independent lineage or a developmental option for other helper lineages remains controversial. Like T$_H$2 cells, T$_{FH}$ cells play a central role in mediating B-cell help and are found in B-cell follicles and germinal centers. However, the effector cytokines secreted by follicular helper T cells differ partially from those secreted by T$_H$2 cells.

Cytokines that polarize activated T cells toward the T$_{FH}$ lineage include IL-6 and IL-21. These polarizing cytokines induce the expression of Bcl-6, a transcriptional repressor that is thought to be T$_{FH}$'s master transcriptional regulator (see Overview Figure 11-11). Cross-regulation is also a feature of T$_{FH}$ function: Bcl-6 expression inhibits T-bet, GATA3, and RORγt expression, thus inhibiting T$_H$1, T$_H$2, and T$_H$17 differentiation, respectively, while inducing T$_{FH}$ polarization.

Although both T_{FH} and T_H2 cells secrete IL-4, T_{FH} cells are best characterized by their secretion of IL-21, which induces B-cell differentiation. Interestingly, they can also produce IFN-γ (the defining T_H1 cytokine). How T_{FH} and T_H2 cells divide responsibilities for inducing B-cell antibody production is still an open question.

Other Potential Helper T-Cell Subsets

Other T-cell subsets with distinct polarizing requirements and unique cytokine secretion profiles have been identified (e.g., T_H9 cells, which secrete IL-9 and IL-10). However, because these subpopulations secrete cytokines that are also produced by T_H1, T_H2, T_H17, or iT_{REG} cells, some speculate that these cell types do not represent distinct subclasses but rather are developmental or functional variants of one of the major subpopulations. This perspective has, indeed, been applied to the follicular helper T-cell (T_{FH}) subset, which also expresses cytokines shared by several other subtypes. However, this subset has a distinct gene signature and a distinct master regulator (bcl-6), so most now consider it a bona fide independent lineage. Clearly, our understanding of the developmental relationship among effector subtypes will continue to evolve.

Helper T Cells May Not Be Irrevocably Committed to a Lineage

Investigations now suggest that the relationship among T_H cell subpopulations may be more plastic than previously suspected. At early stages in differentiation, at least, helper cells may be able to shift their commitment and produce another subset's signature cytokine(s). For example, when exposed to IL-12, young T_H2 cells can be induced to express the signature T_H1 cell cytokine, IFN-γ. Young T_H1 cells can also be induced to express the signature T_H2 cytokine, IL-4, under T_H2 polarizing conditions.

Interestingly, T_H1 and T_H2 cells do not seem able to adopt T_H17 or iT_{REG} characteristics. On the other hand, T_H17 and iT_{REG} cells are more flexible and can adopt the cytokine expression profiles of other subsets, including T_H1 and T_H2. The T_H17 subset may be the most unstable or "plastic" lineage and can be induced to express IFN-γ and IL-4, depending on environmental input. Some iT_{REG} cells can also be induced to express IFN-γ, and some can be redirected toward a T_H17 phenotype if exposed to IL-6 and TGF-β. This fluidity among subsets makes it difficult to definitively establish the independence of helper cell lineages. In fact, some of the emerging subgroups may be variants of T_H1, T_H2, T_H17, and iT_{REG} subsets that have been exposed to other polarizing environments.

Helper T-Cell Subsets Play Critical Roles in Immune Health and Disease

Studies in both mice and humans show that the balance of activity among T-cell subsets can significantly influence the outcome of the immune response. A now classic illustration of the influence of T-cell subset balance on disease outcome is provided by leprosy, which is caused by *Mycobacterium*

leprae, an intracellular pathogen that can survive within the phagosomes of macrophages. Leprosy is not a single clinical entity; rather, the disease presents as a spectrum of clinical responses, with two major forms of disease, tuberculoid and lepromatous, at each end of the spectrum. In **tuberculoid leprosy**, cell-mediated immune responses destroy most of the mycobacteria. Although skin and peripheral nerves are damaged in tuberculoid leprosy, it progresses slowly and patients usually survive. In contrast, **lepromatous leprosy** is characterized by a humoral response; cell-mediated immunity is depressed. The humoral response sometimes results in markedly high levels of immunoglobulin (hypergammaglobulinemia). This response is not as effective in inhibiting disease, and mycobacteria are widely disseminated in macrophages, often reaching numbers as high as 10^{10} per gram of tissue. Lepromatous leprosy progresses into disseminated infection of the bone and cartilage with extensive nerve and tissue damage.

The development of lepromatous or tuberculoid leprosy depends in part on the balance between T_H1 and T_H2 responses (Figure 11-12). In tuberculoid leprosy, the immune response is characterized by a T_H1-type response and high circulating levels of IL-2, IFN-γ, and Lymphotoxin-α (LT-α). In lepromatous leprosy, there is a T_H2-type immune response, with high circulating levels of IL-4 and IL-5 (and IL-10, which can also be made by T_H2 cells). This cytokine profile explains the diminished cell-mediated immunity and increased production of serum antibody in lepromatous leprosy.

Presumably each of these patients was exposed to the same pathogen. Why did some develop an effective T_H1 response while others did not? Studies suggest that genetic differences among human hosts play a role. For example, differences in susceptibility may correlate with individual differences in the expression of PRRs (TLR1 and TLR2) expressed

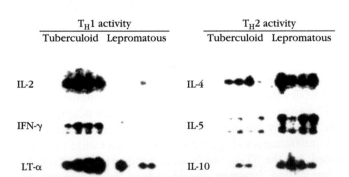

FIGURE 11-12 Correlation between type of leprosy and relative T_H1 or T_H2 activity. Messenger RNA isolated from lesions from tuberculoid and lepromatous leprosy patients was analyzed by Southern blotting using the cytokine probes indicated. Cytokines characteristic of T_H1 cells (IFN-γ and TNF-β, for instance) predominate in the tuberculoid patients, whereas cytokines characteristic of T_H2 cells (IL-4) predominate in the lepromatous patients. *[From P. A. Sieling and R. L. Modlin, 1994, Cytokine patterns at the site of mycobacterial infection,* Immunobiology 191:378.]

by innate cells. This makes sense given that interactions between pathogen and innate immune cells determine the cytokine environment that influences the outcome of T-cell polarization. Differences in TLR expression or activity could alter the quality or quantity of cytokines produced.

Progression of HIV infection to AIDS may also be influenced by T-cell subset balance. Early in the disease, T_H1 activity is high, but as AIDS progresses, some have suggested that a shift may occur from a T_H1-like to a T_H2-like response, which is less effective at controlling viral infection. In addition, some pathogens may "purposely" influence the activity of the T_H subsets. The Epstein-Barr virus, for example, produces a homolog (mimic) of human IL-10 called viral *IL-10* (vIL-10). Like cellular IL-10, vIL-10 tends to suppress T_H1 activity by inhibiting the polarizing activation of macrophages. Some researchers have speculated that vIL-10 may reduce the cell-mediated response to the Epstein-Barr virus, thus conferring a survival advantage on the virus.

T_H17 cells first received attention because of their association with chronic autoimmune disease. Mice that were unable to make IL-23, a cytokine that contributes to T_H17 polarization, were remarkably resistant to autoimmunity. T_H17 cells and their defining effector cytokine, IL-17, are often found in inflamed tissue associated with rheumatoid arthritis, inflammatory bowel disease, multiple sclerosis, psoriasis, and asthma. However, the role T_H17 cells played in protecting organisms from infection was not immediately obvious. Studies of individuals with an autosomal dominant form of a disease known as Hyper-IgE syndrome or Job syndrome confirmed indications from mice that T_H17 cells were important in controlling infections by extracellular bacteria and fungi (see Clinical Focus Box 11-4).

These disorders and those described in Chapters 15 and 16 are just some examples of the influence of helper T-cell subsets on disease development. It is important to recognize that our current perspectives of the roles of helper subsets in disease and health are still simplistic. Our developing appreciation of the complexity of the interplay among subsets will improve and add more subtlety to our explanations in the future.

T-Cell Memory

T-cell activation results in a proliferative burst, effector cell generation, and then a dramatic contraction of cell number. At least 90% of effector cells die by apoptosis after pathogen is cleared, leaving behind an all-important population of antigen-specific **memory T cells**. Memory T cells are generally long-lived and quiescent, but respond with heightened reactivity to a subsequent challenge with the same antigen. This **secondary immune response** is both faster and more robust, and hence more effective than a primary response.

Until recently, memory cells were difficult to distinguish from effector T cells and naïve T cells by phenotype, and for some time they were defined best by function. Like naïve T cells, most memory T cells are resting in the G_0 stage of the cell cycle, but they appear to have less stringent requirements

for activation than naïve T cells. For example, naïve T cells are activated almost exclusively by dendritic cells, whereas memory T cells can be activated by macrophages, dendritic cells, or B cells. Memory cells express different patterns of surface adhesion molecules and costimulatory receptors that allow them to interact effectively with a broader spectrum of APCs. They also appear to be more sensitive to stimulation and respond more quickly. This may, in part, be due to their ability to regulate gene expression more readily because of differences in epigenetic organization that occurred during their formation. Finally, memory cells display recirculation patterns that differ from those of naïve or effector T cells. Some stay for long periods of time in the lymph node and other secondary lymphoid organs, some travel to and remain in tertiary immune tissues where the original infection occurred, anticipating the possibility of another infection with the antigen to which they are specific.

As our ability to identify different cell surface proteins improved, so has our ability to identify and distinguish memory cell subpopulations. We now broadly distinguish two subsets of memory T cells, **central memory T cells (T_{CM})** and **effector memory T cells (T_{EM})**, on the basis of their location, their patterns of expression of surface markers, and, to some extent, their function. Recent work also reveals a great deal of diversity within these subsets, whose relationship is still being clarified. We describe some useful generalizations below, and close with the many questions that remain.

Naïve, Effector, and Memory T Cells Display Broad Differences in Surface Protein Expression

Three surfaces markers have been used to broadly distinguish naïve, effector, and memory T cells: CD44, which increases in response to activation signals; CD62L, an adhesion protein; and CCR7, a chemokine receptor. Both CD62L and CCR7 are involved in homing to secondary lymphoid organs (Table 11-4). Naïve T cells express low levels of CD44, reflecting their unactivated state, and high levels of the adhesion molecule CD62L, directing them to the lymph node or spleen. In contrast, effector helper and cytotoxic T cells have the reciprocal phenotype. They express high levels of CD44, indicating that they have received TCR signals, and low levels of CD62L, which prevents them from recirculating to secondary lymphoid tissue, allowing them to thoroughly probe sites of infection in the periphery.

Both types of memory T cells also tend to express CD44, indicating that they are *antigen experienced* (i.e., have received signals through their TCR). Like naïve T cells, central memory cells (T_{CM}) express CD62L and the chemokine receptor, CCR7, consistent with their residence in secondary lymphoid organs. Effector memory cells (T_{EM}), which are found in a variety of tissues, can express varying levels of CD62L depending on their locale; however, they do not express CCR7, reflecting their travels through and residence in nonlymphoid tissues. Other markers have been used to distinguish subtypes of memory cells, but these still provide a useful starting point for gauging the status of a T cell.

TABLE 11-4	Surface proteins that are used to distinguish naïve, effector, and memory T cells		
Cell type	CD44	CD62L	CCR7
Naïve T cell	low	+	+
Effector T cell	+	low	−
Effector memory T cell	+	variable	−
Central memory T cell	+	+	+

T_CM and T_EM Are Distinguished by Their Locale and Commitment to Effector Function

A small proportion (<10%) of the progeny of a naïve cell that has proliferated robustly in response to antigen differentiates into T_{CM} and T_{EM} cells. In general, these two subsets are distinguished by where they reside as well as their level of commitment to a specific effector cell fate.

In general, T_{CM} cells reside in and travel between secondary lymphoid tissues. They live longer and have the capacity to undergo more divisions than their T_{EM} counterparts. When they reencounter their cognate pathogen in secondary lymphoid tissue, they are rapidly activated and have the capacity to differentiate into a variety of effector T-cell subtypes, depending on the cytokine environment.

On the other hand, T_{EM} cells travel to and between tertiary tissues (including skin, lung, liver, and intestine). They are arguably better situated to contribute to the first line of defense against reinfection because they have already committed to an effector lineage during the primary response and exhibit their effector functions quite rapidly after reactivation by their cognate pathogen.

It is important to note that some of these generalizations may not hold up to scientific scrutiny. For instance, some

investigators contest the definitive distinction between T_{CM} and T_{EM} cells and emphasize the diversity and continuum of variations among these subtypes. Hence, memory cell biology is one of the most active fields of investigation, one that is critical to our ability to develop the best vaccines.

How and When Do Memory Cells Arise?

Current work suggests that memory cells arise very early in the course of an immune response (e.g., within 3 days), but their cell of origin remains controversial. Some investigations suggest that memory cells arise as soon as naïve T cells are activated. Others suggest that memory cells arise from more fully differentiated naïve T cells. Still others raise the intriguing possibility that naïve T-cell activation generates a "memory stem cell" that is self-renewing and gives rise to memory effector cell populations. These models are not mutually exclusive, and it is possible that memory cells can arise at several different stages of T-cell activation throughout a primary response.

The relationship between T_{CM} and T_{EM} cells is also debated. They may originate independently from naïve and effector cells, respectively, or may give rise to each other. Studies suggest, in fact, that T_{CM} cells arise from T_{EM} cells, and one possible model of relationships is shown in Figure 11-13. Here, investigators speculate that T_{CM} cells arise prior to T_{EM} cells, from cells at an earlier stage of differentiation into effector (helper or cytotoxic) T cells. T_{EM} cells arise late, and also may develop from fully differentiated effector cells. The model also suggests that effector cells can replenish central memory cells.

It should be stressed, however, that several other models have also been advanced. For instance, recent work suggests interactions experienced by effector cells determines their T_{CM} versus T_{EM} fate. Effector cells that interact with B cells may preferentially develop into central and not effector memory T cells. New models may also need to incorporate intriguing recent observations, including the possibilities that (1) memory

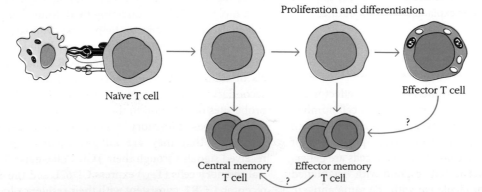

Proliferation and differentiation

Naïve T cell — Effector T cell

Central memory T cell   Effector memory T cell

FIGURE 11-13 One possible model for the development of central and effector memory T cells. This model, only one of several that have been advanced, suggests that central memory cells arise early after naïve T-cell activation, perhaps from the first divisions. Effector memory T cells may arise later, after the progeny have divided more and have assumed at least some

effector cell features. The model also includes the possibilities that (1) some effector memory T cells arise from fully differentiated effector T cells and (2) effector memory T cells can develop into central memory T cells. [*Adapted from D. Gray, 2002, A role for antigen in the maintenance of immunological memory, Nature Reviews Immunology 2:60.*]

cells arise from the asymmetric cell division of activated T cells, where one daughter cell becomes an effector cell, and another contributes to the memory pool, and (2) that T-cell activation generates a self-renewing memory stem cell population that provides a long-term source of memory T cells.

What Signals Induce Memory Cell Commitment?

Most investigators agree that T-cell help is critical to generating long-lasting memory. For instance, CD8$^+$ T cells can be activated in the absence of CD4$^+$ T-cell help, but this "helpless" activation event does not yield long-lived memory CD8$^+$ T cells. The relative importance of other variables in driving memory development is still under investigation. Although intensity of T-cell receptor engagement was thought to be a factor in memory cell commitment, recent data suggest that even low-affinity interactions can generate memory T cells. All studies, however, appear consistent with the recognition that the more proliferation a response inspires, the better the memory pool.

Do Memory Cells Reflect the Heterogeneity of Effector Cells Generated during a Primary Response?

We have seen that naïve T cells differentiate into a wide variety of effector T-cell subpopulations, largely determined by the cytokine signals they receive during activation. Studies indicate that the memory cell response is also very diverse, in terms of both the T-cell receptor specificities and the array

of cytokines produced. However, the cellular origin of this diversity is still under investigation. Specifically, does this diverse memory response strictly reflect the functional effector diversity generated during the primary response? Or does it develop anew from central memory T cells responding to different environmental cues during rechallenge? The answer is likely to be "Both," but investigations continue.

Are There Differences between CD4$^+$ and CD8$^+$ Memory T Cells?

The simple answer is "Maybe." Memory CD8$^+$ T cells are clearly more prevalent than memory CD4$^+$ T cells. This is partly because CD8$^+$ T cells proliferate more robustly and therefore generate proportionately more memory T cells. It may also be due to differences in the life span of memory T cells: CD4$^+$ memory T cells may not be as long-lived as CD8$^+$ memory T cells.

How Are Memory Cells Maintained over Many Years?

Whether memory cells can persist for years in the absence of antigen remains controversial, although evidence seems to favor the possibility that they do. Regardless, it does seem that memory persistence depends on the input of cytokines that induce occasional divisions, a process known as homeostatic proliferation, which maintains the pool size by balancing apoptotic events with cell division. Both IL-7 and IL-15 appear important in enhancing homeostatic proliferation, but CD4$^+$ and CD8$^+$ memory T-cell requirements may differ.

SUMMARY

- T-cell activation is the central event in the initiation of the adaptive immune responses. It results from the interaction in a secondary lymphoid tissue between a naïve T cell and an APC, specifically a dendritic cell.

- Activation of naïve T cells leads to the differentiation of effector cells, which regulate the response to pathogen, and of long-lived memory T cells, which coordinate the stronger and quicker response to future infections by the same pathogen.

- Three distinct signals are required to induce naïve T-cell activation, proliferation, and differentiation. Signal 1 is generated by the interaction of the TCR-CD3 complex with an MHC-peptide complex on a dendritic cell. Signal 2 is a costimulatory signal provided by the interaction between molecules of the B7 family expressed by APC with the positive costimulatory molecules CD28 or ICOS expressed by T cells. Signal 3 is provided by soluble cytokines and plays a key role in determining the type of effector cell that a T cell becomes.

- In the absence of a costimulatory signal (Signal 2), T-cell receptor engagement results in T-cell inactivity or clonal anergy.

- CD8$^+$ T cells, which recognize MHC class I-peptide complexes that are expressed by virtually all cells in the body, become cytotoxic (T$_C$) cells with the capacity to kill infected cells.

- CD4$^+$ T cells, which recognize MHC class II-peptide complexes that are expressed by professional APCs, become helper (T$_H$) cells, secreting cytokines that regulate (positively and negatively) cells that clear infection, including B cells, macrophages, and other T cells.

- CD4$^+$ T cells differentiate into at least five main subpopulations of effector cells: T$_H$1, T$_H$2, T$_H$17, iT$_{REG}$, and T$_{FH}$. Each subpopulation is characterized by (1) a unique set of polarizing cytokines that initiate differentiation, (2) a unique master transcriptional regulator that regulates the production of helper-cell-specific genes, and (3) a distinct set of effector cytokines that they secrete to regulate the immune response.

- T$_H$1 and T$_H$17 cells generally enhance cell-mediated immunity and inflammatory responses. T$_H$2 and T$_{FH}$ cells enhance humoral immunity and antibody production, and induced T$_{REG}$ cells inhibit T-cell responses.

- Helper T-cell subsets have also been associated with disease and play a role in the development of autoimmunity and allergy.

- Memory T cells, which are more easily activated than naïve cells, are responsible for secondary responses.

- The generation of memory B cells as well as CD4$^+$ and CD8$^+$ memory T cells requires T-cell help.

- Two types of memory T cells have been described. Central memory (T_{CM}) cells are longer lived, reside in secondary lymphoid tissues, and can differentiate into several different effector T cells. Effector memory (T_{EM}) cells populate the sites of infection (tertiary tissues) and immediately reexpress their original effector function after reexposure to antigen.

REFERENCES

Ahmed, R., M. Bevan, S. Reiner, and D. Fearon. 2009. The precursors of memory: Models and controversies. *Nature Reviews. Immunology* 9:662–668.

Gattinoni, L., et al. 2011. A human memory T cell subset with stem cell-like properties. *Nature Medicine* 17:1290–1297.

Hoyer, K. K., W. F. Kuswanto, E. Gallo, and A. K. Abbas. 2009. Distinct roles of helper T-cell subsets in systemic autoimmune disease. *Blood* 113:389–395.

Jameson, S., and D. Masopust. 2009. Diversity in T cell memory: An embarrassment of riches. *Immunity* 31:859–871.

Jelley-Gibbs, D., T. Strutt, K. McKinstry, and S. Swain. Influencing the fates of CD4$^+$ T cells on the path to memory: Lessons from influenza. *Immunology and Cell Biology* 86:343–352.

Kaiko, G. E., J. C. Horvat, K. W. Beagley, and P.M. Hansbro. 2007. Immunological decision making: How does the immune system decide to mount a helper T-cell response? *Immunology* 123: 326–338.

Kapsenberg, M. L. 2003. Dendritic cell control of pathogen-driven T-cell polarization. *Nature Reviews Immunology* 3: 984–993.

Kassiotis, G., and A. O'Garra. 2009. Establishing the follicular helper identity. *Immunity* 31:450–452.

Khoury, S., and M. Sayegh. 2004. The roles of the new negative T cell costimulatory pathways in regulating autoimmunity. *Immunity* 20:529–538.

King, C. 2009. New insights into the differentiation and function of T follicular helper cells. *Nature Reviews. Immunology* 9:757–766.

Korn, T., E. Bettelli, M. Oukka, and V. Kuchroo. 2009. IL-17 and T_H17 Cells. *Annual Review of Immunology* 27:485–517.

Lefrancois, L., and D. Masopust. 2009. The road not taken: Memory T cell fate "decisions." *Nature Immunology* 10:369–370.

Linsley, P., and S. Nadler. 2009. The clinical utility of inhibiting CD28-mediated costimulation. *Immunological Reviews* 229:307–321.

Malissen, B. 2009. Revisiting the follicular helper T cell paradigm. *Nature Immunology* 10:371–372.

Mazzoni, A., and D. Segal. 2004. Controlling the Toll road to dendritic cell polarization. *Journal of Leukocyte Biology* 75:721–730.

McGhee, J. 2005. The world of T_H1/T_H2 subsets: First proof. *Journal of Immunology* 175:3–4.

Miossec, P., T. Korn, and V.K. Kuchroo. 2009. Interleukin 17 and Type 17 helper T cells. *New England Journal of Medicine* 361:888–898.

Mosmann, T., H. Cherwinski, M. Bond, M. Giedlin, and R. Coffman. 1986. Two types of murine helper T cell clone. I. Definition according to profiles of lymphokine activities and secreted proteins. *Journal of Immunology* 136:2348–2357.

Murphy, K., C. Nelson, and J. Sed. 2006. Balancing co-stimulation and inhibition with BTLA and HVEM. *Nature Reviews Immunology* 6:671–681.

Palmer, M. T., and C. T. Weaver. 2010. Autoimmunity: Increasing suspects in the CD4$^+$ T cell lineup. *Nature Immunology* 11:36–40.

Pepper, M., and M. K. Jenkins. 2011. Origins of CD4$^+$ effector and memory T cells. *Nature Immunology* 12:467–471.

Readinger, J., K. Mueller, A. Venegas, R. Horai, and P. Schwartzberg. 2009. Tec kinases regulate T-lymphocyte development and function: New insights into the roles of Itk and Rlk/Txk. *Immunological Reviews* 228:93–114.

Reiner, S. 2008. Inducing the T cell fates required for immunity. *Immunological Research* 42:160–165.

Riley, J. 2009. PD-1 signaling in primary T cells. *Immunology Reviews* 229:114–125.

Rudd, C., A. Taylor, and H. Schneider. 2009. CD28 and CTLA-4 coreceptor expression and signal transduction. *Immunological Reviews* 229:12–26.

Sallusto, F., and A. Lanzavecchia. 2009. Heterogeneity of CD4$^+$ memory T cells: Functional modules for tailored immunity. *European Journal of Immunology* 39:2076–2082.

Sharpe, A. 2009. Mechanisms of costimulation. *Immunological Reviews* 229:5–11.

Smith-Garvin, J., G. Koretzky, and M. Jordan. 2009. T cell activation. *Annual Review of Immunology* 27:591–619.

Thomas, R. 2004. Signal 3 and its role in autoimmunity. *Arthritis Research & Therapy* 6:26–27.

Thompson, C., et al. 1989. CD28 activation pathway regulates the production of multiple T-cell-derived lymphokines/cytokines. *Proceedings of the National Academy of Sciences of the United States of America* 86:1333–1337.

Wang, S., and L. Chen. 2004. T lymphocyte co-signaling pathways of the B7-CD28 family. *Cellular & Molecular Immunology* **1**:37–42.

Weaver, C., and R. Hatton. 2009. Interplay between the T$_H$17 and T$_{REG}$ cell lineages: A (co)evolutionary perspective. *Nature Reviews. Immunology* **9**:883–889.

Yu, D., et al. 2009. The transcriptional repressor Bcl-6 directs T follicular helper cell lineage commitment. *Immunity* **31**:457–468.

Zhou, L., M. Chong, and D. Littman. 2009. Plasticity of CD4$^+$ T cell lineage differentiation. *Immunity* **30**:646–655.

Zhu, J., and W. Paul. 2010. Heterogeneity and plasticity of T helper cells. *Cell Research* **20**:4–12.

Useful Web Sites

The following are examples of well-organized Web sites developed by undergraduate students, graduate students, and teachers of immunology who have done an excellent job of simplifying a complex topic: helper T-cell subset differentiation. The Web sites may not be continually updated, so, as always with Internet sources, double-check the date and the accuracy of the information.

http://wenliang.myweb.uga.edu/mystudy/ immunology/ScienceOfImmunology/Biological process(immuneresponses).html These Web notes from a former graduate student at the University of Georgia provide a rudimentary, but accurate and accessible, description of the T-cell subsets.

http://microbewiki.kenyon.edu/index.php/Host_ Dependency_of_Mycobacterium_leprae You will find here a posting from MicrobeWiki, "a student-edited microbiology resource" originating from Kenyon College.

http://users.rcn.com/jkimball.ma.ultranet/ BiologyPages/T/T$_H$1_T$_H$2.html#Types_of_ Helper_T_Cells This selection is from immunologist and teacher John Kimball's online version of his textbook.

STUDY QUESTIONS

CLINICAL AND EXPERIMENTAL FOCUS QUESTION Multiple sclerosis is an autoimmune disease in which T$_H$ cells participate in the destruction of the protective myelin sheath around neurons in the central nervous system. Each person with this disease has different symptoms, depending on which neurons are affected, but the disease can be very disabling. Recent work in a mouse model of this disease suggests that transplantation of cell precursors of neurons may be a good therapy. Although these immature cells may work because they can develop into neuronal cells that replace the lost myelin sheath, some investigators realized that they play another, perhaps even more important role. These scientists showed that the neuronal cell precursors secrete a cytokine called *Leukemia Inhibiting Factor* (LIF). In fact, the administration of this factor, alone, ameliorated symptoms.

These investigators were curious to know if this cytokine had an effect on T-cell activity. They added LIF to cultures of (normal) T cells that were being stimulated under different polarizing conditions (i.e., TCR and CD28 engagement in the presence of cytokines that drive differentiation to distinct T helper subsets). They stained T cells for cytokine production and analyzed the results by flow cytometry. The data below show their results from normal mouse T cells polarized to the lineage indicated in the absence (top, control) or presence (bottom) of LIF.

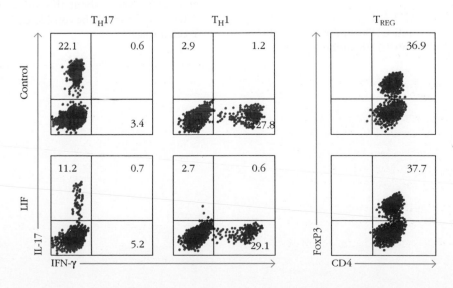

Which T helper lineage(s) is(are) most affected by the addition of LIF? Explain your answer. Why might these results explain the beneficial effect of LIF on the disease? Knowing what you know now about the molecular events that influence T-cell differentiation, speculate on the molecular basis for the activity of LIF.

1. Which of the following conditions would lead to T-cell anergy?

   a. A naïve T-cell interaction with a dendritic cell in the presence of CTLA-4 Ig.

   b. A naïve T cell stimulated with antibodies that bind both the TCR and CD28.

   c. A naïve T cell stimulated with antibodies that bind only the TCR.

   d. A naïve T cell stimulated with antibodies that bind only CD28.

2. A virus enters a cut in the skin of a mouse and infects dendritic cells, stimulating a variety of PRRs both on and within dendritic cells that induce it to produce IL-12. The mouse subsequently mounts an immune response that successfully clears the infection. Which of the following statements is(are) likely to be true about the immune response that occurred? Correct any that are false.

   a. The infected dendritic cells up-regulated CD80/CD86 and MHC class II.

   b. The dendritic cells encountered and activated naïve T cells in the skin of the mouse.

   c. Naïve T cells activated by these dendritic cells generated signals that released internal Ca^{2+} stores.

   d. Naïve T cells activated by these dendritic cells were polarized to the T_H2 lineage.

   e. Only effector memory T cells were made in this mouse.

3. Your lab acquires mice that do not have the GATA-3 gene (GATA-3 knockout mice). You discover that this mouse has a difficult time clearing helminth (worm) infections. Why might this be?

4. You isolate naïve T cells from your own blood and want to polarize them to the T_H1 lineage in vitro. You can use any of the following reagents to do this. Which would you choose?

Anti-TCR antibody	CTLA-4 Ig	IL-12
IL-4	anti-CD80 antibody	IL-17
IFNγ	anti-CD28 antibody	

5. The following sentences are all false. Identify the error(s) and correct.

   a. Macrophages activate naïve T cells better than dendritic cells.

   b. ICOS enhances T-cell activation and is called a negative coreceptor.

   c. Virtually all cells in the body express costimulatory ligands.

   d. CD28 is the only costimulatory receptor that binds to B7 family members.

   e. Signal 3 is provided by negative costimulatory receptors.

   f. Toxic shock syndrome is an example of an autoimmune disease.

   g. Superantigens mimic TCR-MHC class I interactions.

   h. $CD4^+$ T cells interact with MHC class I on $CD8^+$ T cells.

   i. Naïve T cells produce IFN-γ.

   j. T-bet and GATA-3 are effector cytokines.

   k. Polarizing cytokines are only produced by APCs.

   l. Bcl-6 is involved in the delivery of costimulatory signals.

   m. T_H17 and T_{FH} cell subsets are the major sources of B-cell help.

   n. iT_{REG} cells enhance inflammatory disease.

   o. Effector cytokines act exclusively on T cells.

   p. Central memory T cells tend to reside in the site of infection.

   q. Like naïve T cells, effector memory cells express CCR7.

6. Like T_H1 and T_H2 cells, T_H17 and T_{REG} cells cross-regulate each other. Which of these two statements about this cross-regulation is (are) true? Correct either, if false.

   a. TGF-β is a polarizing cytokine that stimulates up-regulation of each of the master transcriptional regulators that polarize T cells to the T_H17 and T_{REG} lineages.

   b. IL-6 inhibits polarization to the T_{REG} lineage by inhibiting expression of RORγ.

7. A new effector T-cell subset, T_H9, has been recently identified. It secretes IL-9 and IL-10 and appears to play a role in the protection against intestinal worm infection. What other information about this subset would help you to determine if it should be considered an independent helper T-cell lineage?

B-Cell Activation, Differentiation, and Memory Generation

The function of a B cell is to secrete antibodies capable of binding to any organism or molecule that poses a threat to the host. The secreted antibodies have antigen-binding sites identical to those of the receptor molecules on the B cell surface. Antibodies belong to the class of proteins known as immunoglobulins, and once secreted, they can protect the host against the pathogenic effects of invading viruses, bacteria, or parasites in a variety of ways, as described in Chapters 1 and 13.

Our current understanding of B-cell clonal selection, activation, proliferation, and deletion finds its beginnings in a theoretical paper in the *Australian Journal of Science*, written by Sir Frank MacFarlane Burnet (Figure 12-1) over the course of a single weekend in 1957. In this paper, built on prior work by Neils Jerne, David Talmage, Peter Medawar and others, Burnet laid out the **Clonal Selection Hypothesis**, which provided the conceptual underpinnings for the entire field of immunology (see Figure 1-7). The Clonal Selection Hypothesis suggested for the first time that the receptor molecule on the lymphocyte surface and the antibody products secreted by that cell had identical antigen-binding specificities. Furthermore, it posited that stimulation of a single B cell would result in the generation of a clone of cells having the identical receptor specificity as the original cell, and that these clones would migrate to, and function within, the secondary lymphoid organs. The daughter cells within each clone would not only be able to secrete large amounts of antibody to neutralize the pathogen, but some progeny cells would also remain viable within the organism and available to neutralize a secondary infection by the same pathogen.

In other words, in one brilliant paper Professor Burnet gave generations of immunologists the basis for thinking about B-cell receptor diversity, lymphocyte trafficking, and immune memory—all this at a time when practitioners in the field were still unsure of the differences between T and B cells.

Burnet's paper went on to predict the generation of the vast array of antibody specificities known to exist today. It is difficult for us now, in the second decade of the twenty-first century, to even begin to appreciate the prescience of

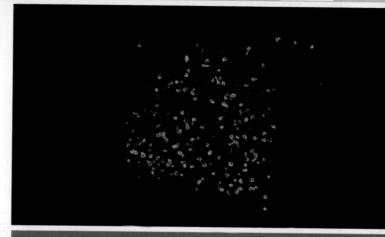

The germinal center of the lymph node contains B cells (green) in the dark zone and follicular dendritic cells (red) in the light zone. Naïve B cells (blue) define the follicular mantle zone. *[Reprinted from Victora, G.D. et al., 2010, Germinal center dynamics revealed by multiphoton microscopy with a photoactivatable fluorescent reporter, Cell 143:592–605. Copyright 2010 with permission from Elsevier.]*

- T-Dependent B-Cell Responses
- T-Independent B Cell Responses
- Negative Regulation of B Cells

Burnet's suggestion that "The theory requires at some stage in early embryonic development *a genetic process for which there is no available precedent* [italics added]. In some way, we have to picture a 'randomization' of the coding responsible for part of the specification of gamma globulin molecules, so that after several cell generations, in early mesenchymal cells, there are specifications in the genomes for virtually every variant that can exist as a gamma globulin molecule." Just four years after the eludication of the double helical structure of DNA, Burnet was postulating that there are movable genetic segments!

The paper ends with a final flourish, in which Burnet predicted the need for clonal deletion in the developing lymphocyte repertoire, in order to eliminate B cells bearing receptors with specificity for self antigens. His formulation of the Clonal Selection Hypothesis, along with the brilliant experimental work he performed with others on the generation of immunological tolerance, resulted in

FIGURE 12-1 Sir MacFarlane Burnet, 1899–1985. Professor McFarlane Burnet, the author of the Clonal Selection Hypothesis, shared the 1960 Nobel Prize for Physiology or Medicine with Sir Peter Medawar of Britain for "the discovery of acquired immunological tolerance." *[Sir Macfarlane Burnet 1960–61 by Sir William Dargie. Oil on composition board. Collection: National Portrait Gallery, Canberra. Purchased 1999]*

his being awarded the Nobel Prize for Physiology or Medicine in 1960, along with Sir Peter Medawar. The essential tenets of the Clonal Selection Hypothesis are summarized below, and a visual representation of the events he predicted are shown in Figure 12-2.

- Immature B lymphocytes bear *i*mmunoglobulin (Ig) receptors on their cell surfaces. All receptors on a single B cell have identical specificity for antigen.

- Upon antigen stimulation, the B cell will mature and migrate to the lymphoid organs. There it will replicate. Its clonal descendants will bear the same receptor as the parental B cell and secrete antibodies with an identical specificity for antigen.

- At the close of the immune response, more B cells bearing receptors for the stimulating antigen will remain in the host than were present before the antigenic challenge. These memory B cells will then be capable of mounting an enhanced secondary response.

- B cells with receptors for self antigens are deleted during embryonic development.

In summary, each B cell bears a single type of Ig receptor generated by the processes described in Chapter 7,

and on stimulation will create a clone of cells bearing the same antigen receptor as the original B cell. Many B cells bearing receptors with specificity for self antigens are eliminated from the primary B-cell repertoire as described in Chapter 10. In this chapter, we describe what happens when mature B cells, located in the peripheral lymphoid organs, encounter antigen.

The two major types of B-cell responses are elicited by structurally distinct types of antigens. The first type of response that we will describe is generated upon recognition of protein antigens and requires the participation of CD4$^+$ helper T cells. Because T cells are involved, this class of B-cell response is therefore known as a **T-*d*ependent (TD) response**, and it is mediated by **B-2 B cells** binding to **TD antigens**. Because B-2 B cells represent the majority of B cells, we will routinely refer to the B-2 B-cell subset simply as "B cells," and distinguish the other B-cell subclasses by their particular names as B-1, marginal zone, or B-10 B cells.

The T-dependent response requires two distinct signals. The first is generated when a multivalent antigen binds and cross-links *m*embrane *i*mmunoglobulin receptors (mIg) (Figure 12-3a). The second signal is provided by an activated T cell, which binds to the B cell both through its antigen receptor and via a separate interaction between **CD40** on the B cell and **CD40L** (CD154) on the activated T$_H$ cell. The bound T cell then delivers cytokines and other signals to its partner B cell to complete the activation process.

The second type of response, which we will describe later in this chapter, is directed toward multivalent or highly polymerized antigens, and does not require T-cell help. This type of response is referred to as a **T-independent response**, and the antigens that elicit such responses are *T-i*ndependent (TI) antigens. One class of TI antigens (**TI-1 antigens**), exemplified by the lipopolysaccharide moiety of Gram-negative bacteria, interacts with the B cell via both mIg and innate immune receptors. TI-1 antigens are mitogenic (induce proliferation) for most B cells at high concentrations, as a result of their ability to bind to *p*attern *r*ecognition *r*eceptors (PRRs) on the surface of the B cell. However, at lower concentrations they activate only those B cells that bind antigen with their Ig receptors (Figure 12-3b). The other class of TI antigens, **TI-2 antigens**, includes highly repetitive antigens, such as bacterial capsular polysaccharides. These antigens are not inherently mitogenic, but their ability to cross-link a large fraction of the Ig receptors on the surface of a B cell allows them to deliver an activation signal in the absence of T-cell help. Many TI-2 antigens are specifically and covalently bound by the complement component C3d, and mice depleted of C3d mount poor responses to TI-2 antigens (Figure 12-3c).

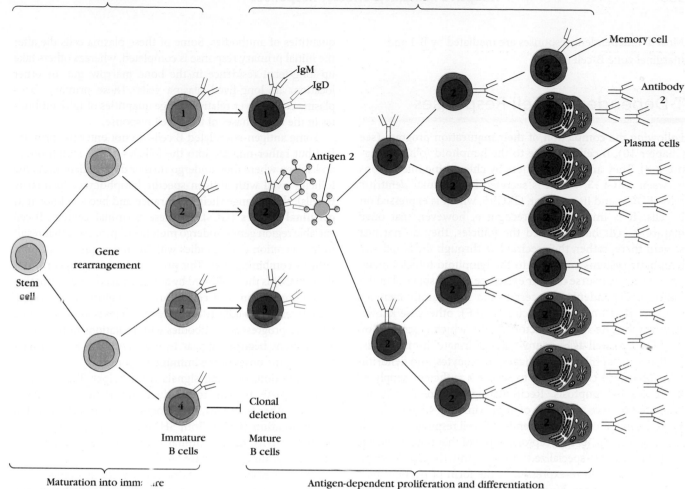

IgM
IgD

Antigen 2

Memory cell

Antibody 2

Plasma cells

Gene rearrangement

Stem cell

Clonal deletion

Immature B cells

Mature B cells

Maturation into immature committed B cell

Antigen-dependent proliferation and differentiation into plasma and memory cells

FIGURE 12-2 Maturation and clonal selection of B lymphocytes. B-cell maturation, which occurs in the absence of antigen, first produces immature B cells bearing IgM receptors. Each B cell bears receptors of one specificity only. Any B cell with receptors specific for antigens expressed in the bone marrow are deleted at the immature B cells stage (indicated by clone 4). Those B cells that do not express self-reactive receptors mature to express both IgM and IgD receptors and are released into the periphery, where they recirculate among the blood, lymph, and lymphoid organs. Clonal selection occurs when an antigen binds to a B cell with a receptor specific for that antigen. Clonal expansion of an antigen-activated B cell (number 2 in this example) leads to a clone of effector B cells and memory B cells; all cells in the expanded clone are specific for the original antigen. The effector, plasma cells secrete antibody reactive with the activating antigen.

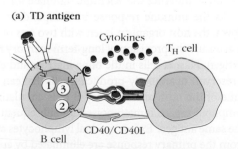
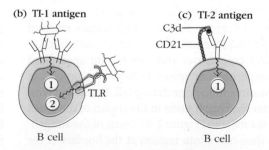

(a) TD antigen

Cytokines
T_H cell

CD40/CD40L
B cell

(b) TI-1 antigen

TLR
B cell

(c) TI-2 antigen

C3d
CD21

B cell

FIGURE 12-3 Different types of antigens signal through different receptor units. (a) T-dependent (TD) antigens bind to the Ig receptor of B cells. Some of the antigen is processed and presented to helper T cells. T cells bind to the MHC-peptide antigen, and deliver further activating signals to the B cell via interaction between CD40L (on the T cells) and CD40 on the B cells. In addition, T cells secrete activating cytokines, such as IL-2 and IL-4, which are recognized by receptors on the B-cell surface. Cytokines deliver differentiation, proliferation, and survival signals to the B cells. (b) Type 1 T-independent (TI-1) antigens bind to B cells through both Ig and innate immune receptors. For example, LPS from Gram negative organisms binds to B cells via both mIg and TLR4, resulting in signaling from both receptors. (c) Type 2 T-independent (TI-2) antigens are frequently bound by C3d complement components and cross-link both mIg and CD21 receptors on B cells. Cross-linking of between 12 and 16 Ig receptors has been shown to be sufficient to deliver an activating signal.

Most T-independent responses are mediated by B-1 and marginal zone B-cell types.

T-Dependent B-Cell Responses

Following the completion of their maturation program (see Chapter 10), B cells migrate to the lymphoid follicles (Figure 2-8). They are directed there by chemokine interactions between CXCL13, which is secreted by *follicular dendritic cells* (FDCs) and its receptor, CXCR5, which is expressed on B cells. It is important to recognize, however, that once mature B cells have reached the follicles, they do not just remain there; rather, they recirculate through the blood and lymphatic systems and back to the lymphoid follicles many times over the course of their existence. B-cell survival in the follicle is dependent on access to the TNF-family member cytokine *B-cell activation factor* (BAFF), otherwise known as *B lymphocyte survival factor* (BLyS), which is secreted by the FDCs, as well as by many types of innate immune cells, such as neutrophils, macrophages, monocytes, and dendritic cells. Mature B cells unable to secure a sufficient supply of BAFF die by apoptosis. Recirculating, mature B cells are estimated to have a half-life of approximately 4.5 months.

At the initiation of a T-dependent B-cell response, the B cell binds antigen via its Ig receptors. Some of that bound antigen is internalized into specialized vesicles within the B cells, where it is processed and re-expressed in the form of peptides presented in the antigen-binding groove of class II MHC molecules (see Figure 8-19). T cells that have been previously activated by an encounter with antigen-bearing dendritic cells now bind to the MHC-presented peptide on the surface of the B cell. This induces a vectorial redistribution of the T cell's secretory apparatus, such that it releases its activating cytokines directly into the T-cell/B-cell interface (as described in Chapter 4). Other interactions also occur between accessory molecules on the T- and B-cell surfaces that enhance their binding and provide further signals to the pair of cells. For example, CD40L on the T cell binds to CD40 on the B cell (see Figure 12-3a) and the CD28 coreceptor on the T cell engages with CD80 and CD86 on the surface of the B cell.

The B cell then integrates the activating signals received through its antigen receptor with those from its various cytokine receptors and coreceptors and moves into specialized regions of the lymph node or spleen to begin the process of differentiation into an antibody-secreting cell. Recall that lymphocytes first enter the lymph node in the region of the T-cell zone, just inside the follicles (Figure 2-8). Some of the antigen-activated B cells then move into regions at the borders of the T-cell and B-cell areas, where they differentiate into clusters of activated B cells known as **primary foci** (see Overview Figure 12-4). There, they complete their differentiation into plasma cells. At around 4 days post stimulation, when this differentiation is complete, they migrate into the medullary cord regions of the lymph node, or to parts of the red pulp of the spleen close to the T-cell zones (Figure 2-10), where they secrete large quantities of antibodies. Some of these plasma cells die after the initial primary response is completed, whereas others take up long-term residence in the bone marrow, gut, or other locations as long-lived plasma cells. These primary focus plasma cells deliver relatively large quantities of IgM antibodies in the early phases of the B-cell response.

Some antigen-stimulated B cells do not enter the primary foci but rather migrate into the follicles of the lymph nodes and spleen where they undergo further differentiation. As the follicles swell with antigen-specific lymphocytes (primarily B cells), they change their appearance and become known as **germinal centers (GCs)**. In these germinal centers, B-cell variable region genes undergo mutational processes that result in the secretion of antibodies with altered sequences in their antigen-combining sites. This process of **somatic hypermutation (SHM)** is then followed by antigen selection, culminating in the production of B cells bearing receptors and secreting antibodies whose affinity for antigen increases as the immune response progresses. Antibodies with mutations in their variable regions begin to appear in the circulation 6 to 10 days following the onset of the immune response.

In addition, ongoing signals from helper T cells direct their cognate B cells to produce antibodies of isotypes (classes) other than IgM, in a process known as **class switch recombination (CSR)**. Both SHM and CSR are dependent on the activity of a germinal center enzyme: *activation-induced cytidine deaminase* (AID).

The end result of this extraordinary set of events is the production of high-affinity antibody molecules that eliminate the pathogenic threat by one or more of the means described in Chapter 13. At the close of the primary immune response, **memory B cells** remain that are the daughters (and more distant progeny) of those cells that were stimulated during the primary response. Many of these progeny B cells now carry mutated and selected *B-cell receptors* (BCRs). On secondary exposure to the same antigen, these memory B cells will be stimulated more quickly and will deliver higher-affinity and heavy-chain class-switched antibodies, resulting in faster elimination of microbial pathogens. This improved memory response underlies the scientific rationale for vaccination.

As the immune response to T-dependent antigens winds down, the host organism is left with two sets of long-lived cells that ensure the provision of long-term memory responses to the antigen. Plasma cells in the bone marrow and elsewhere create a reservoir of antibody-producing cells that can last for the lifetime of the host, and memory B cells circulating through the lymphoid organs remain poised for subsequent stimulation by the same antigen. Excess residual lymphocytes and plasma cells from the primary response are eliminated by apoptosis.

T-Dependent Antigens Require T-Cell Help to Generate an Antibody Response

The initial experiments proving that B cells required "help" from T cells in order to complete their differentiation were performed by Miller, Mitchell, Mitchison, and

B-Cell Activation

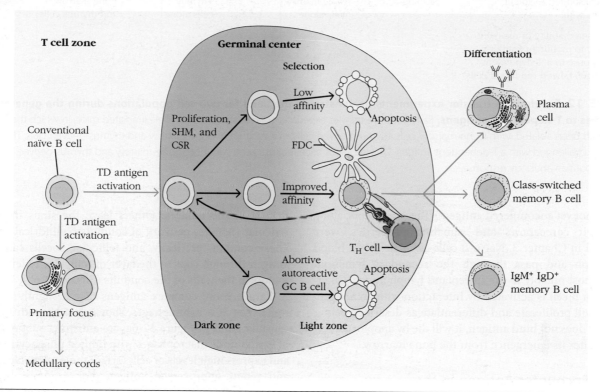

B cells engaged in T-dependent activation first encounter antigen-specific T cells outside of the B-cell follicles. Some activated B cells differentiate into antibody-producing plasma cells in primary foci that lie outside the follicles, and then migrate to the medullary cords of the lymph node, or the bone marrow, where they continue to secrete antigen-specific antibodies. Other antigen-activated B cells enter the follicle where they divide and differentiate. As the follicle fills with proliferating B cells, it develops into a germinal center, characterized by a light zone, in which follicular dendritic cells reside, and a dark zone in which the cell density is particularly high. Within the germinal center, the immunoglobulin genes undergo class switching, in which μ constant regions are replaced by constant regions of other isotypes. The variable region is subject to somatic hypermutation, and the mutated variable regions are subject to antigen-mediated selection within the germinal centers. Low-affinity and autoimmune receptor-bearing B cells die, and those B cells with enhanced receptors leave the germinal centers for the periphery. (See the text for details.)

others in the 1960s, using the technique of adoptive transfer. In an adoptive transfer experiment, a mouse is lethally irradiated in order to eliminate its immune system; the investigator then attempts to reconstitute the ability to generate an immune response by adding back purified cells of different types from genetically identical mouse donors.

Using this technique, investigators showed that, in order for a mouse to produce antibodies against a protein antigen, it must receive cells derived from both the bone marrow and from the thymus of a healthy donor animal. Neither thymus-derived nor bone marrow-derived cells were capable of reconstituting the responding animal's immune system on their own (Figure 12-5). We now know, of course, that the thymus-derived cells active in this response were mature helper T cells,

whereas the bone marrow-derived, antibody-producing cells were mature B cells, recirculating through the bone marrow. In this way it was demonstrated that the antibody response to protein antigens required both B cells and T cells.

Having introduced the major players and briefly described the geography of the landscape in which the cells are operating, we will now sequentially step through a T-dependent B-cell response.

Antigen Recognition by Mature B Cells Provides a Survival Signal

The first step in antibody production is antigen recognition by the Ig receptors on the surface of a naïve B cell (a B cell

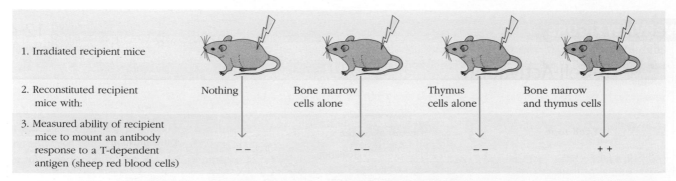

FIGURE 12-5 Adoptive transfer experiments demonstrated the need for two cell populations during the generation of antibodies to T-dependent antigens. Early adoptive transfer experiments reconstituted syngeneic irradiated mice, in which the immune system had been ablated, with bone-marrow cells, thymus-derived cells, or a mixture of bone-marrow and thymus-derived cells. These mice were then challenged with a T-dependent antigen. Only recipient mice reconstituted with both bone-marrow and thymus-derived cells were able to mount an antibody response.

that has not yet encountered antigen). These receptors, along with their coreceptors and signaling properties, were described in Chapter 3. Naïve B cells circulate in the blood and lymph and pass through the secondary lymphoid organs, most notably to the spleen and lymph nodes (Chapter 2). If a B cell is activated by interaction with an antigen, the cell will proliferate and differentiate as described below. If a B cell does not bind antigen, it will die by apoptosis a few months after its emergence from the bone marrow.

B Cells Encounter Antigen in the Lymph Nodes and Spleen

When antigen is introduced into the body, it becomes concentrated in various peripheral lymphoid organs. Blood-borne antigen is filtered by the spleen, whereas antigen from tissue spaces drained by the lymphatic system is filtered by regional lymph nodes (or lymphoid nodules in the gut). Here, we will focus on antigen presentation to B cells in the lymph nodes.

Antigen enters the lymph nodes either alone or associated with antigen-transporting cells. Recall that B cells are capable of recognizing antigenic determinants on native, unprocessed antigens, whereas T cells must recognize antigen as a peptide presented in the context of MHC antigens. In addition, as described in Chapter 6, antigen is often covalently modified with complement fragments, and complement receptors on B cells play an important role in binding complement-coupled antigen with sufficient affinity to trigger B-cell activation.

The mechanism of B-cell antigen acquisition varies according to the size of the antigen. Small, soluble antigens can be directly acquired from the lymphatic circulation by follicular B cells, without the intervention of any other cells. These antigens enter the lymph node via the afferent lymph and pass into the subcapsular sinus region (Figure 12-6a,b). Some small antigens may diffuse between the **subcapsular sinus (SCS) macrophages** that line the sinus to reach the

B cells in the follicles. Others leave the sinus through a reticular (netlike) network of conduits (cylindrical vessels). These conduits are leaky, and follicular B cells can access antigen through gaps in the layer of follicular reticular cells that form the walls of the conduits (Figure 12-6b).

Larger, more complex antigens take a slightly different route. The SCS macrophages, shown just beneath the subcapsular sinus in Figure 12-6a, are a distinct subpopulation of lymph node macrophages with limited phagocytic ability, and express high levels of cell-surface molecules able to bind and retain unprocessed antigen. For example, bacteria, viruses, particulates, and other complex antigens that have been covalently linked to complement components are held by complement receptors on the surfaces of these macrophages. Antigen-specific B cells within the follicles can acquire the antigens directly from the macrophages and become activated. In addition, since SCS macrophages, B cells, and FDCs all bear high levels of complement receptors, it is quite possible that antigen may be passed between the three subtypes of cells until it is finally recognized by a B cell with the matching Ig receptor. Other classes of cells within the lymph node can also present unprocessed antigens to B cells. For example, a population of dendritic cells that is located close to the high endothelial venules in lymph nodes has also been shown to be capable of presenting unprocessed antigen to B cells.

Any discussion of antigen presentation to B cells would be incomplete without consideration of the role of *f*ollicular *d*endritic cells (FDCs). Early electron microscope investigations of lymph-node-derived cells revealed the dendrites of FDCs to be studded with antigen-antibody complexes called **iccosomes** (Figure 12-6c). Further analysis showed that these complexes are retained on the surface of the FDC through interaction either with complement or with Fc receptors. Because of the high surface density of antigen on FDCs, many scientists at first posited that iccosomes' primary role may be in antigen presentation to naïve B cells. However, current evidence suggests that their main function

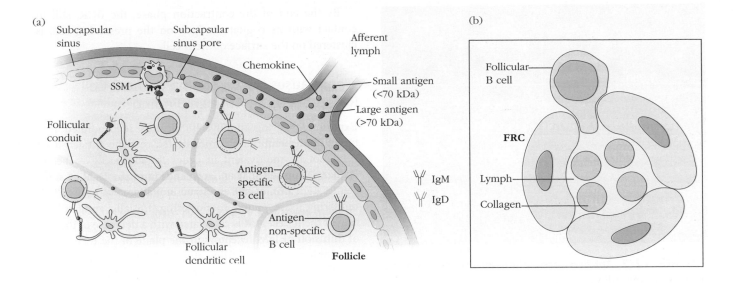

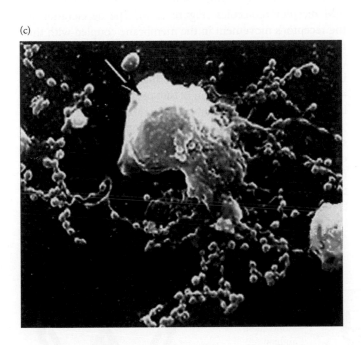

FIGURE 12-6 Antigen presentation to follicular B cells in the lymph node. (a) Lymphatic fluid containing antigens (red) and cytokines and chemokines (blue) reaches the lymph node through the afferent lymph vessel and enters the subcapsular sinus (SCS) region. The SCS region is lined with a porous border of SCS macrophages (SSMs) that prevent the free diffusion of the lymph fluid into the lymph node. Larger antigens and signaling molecules are bound by surface receptors on the SCS macrophages and then presented directly to B cells. Smaller antigens, and chemokines less than approximately 70 kDa in molecular weight, access B cells in the follicles either by direct diffusion through pores in the SCS, or by passage through conduits emanating from the sinus. (b) These conduits are formed by follicular reticular cells (FRCs) wrapped around collagen fibers, and B cells are able to access their contents through pores in the sides of the conduits. (See the text for further details.) (c) A follicular dendritic cell with its dendrites studded with iccosomes. Follicular dendritic cells bind antigen-antibody complexes via complement or Fc receptors on their cell membranes. These complexes are visible in the electron microscope, as iccosomes. *[Parts a and b adapted from Harwood, N.E., and Batista, F.D., 2009, The antigen expressway: follicular conduits carry antigen to B cells, Immunity 30:177–189. Part (c) Courtesy Andras K. Szakal PhD.]*

is to provide a reservoir of antigen for B cells to bind as they undergo mutation, selection, and differentiation along the path to a memory phenotype, rather than playing the primary role in the initial antigen presentation process. In addition, FDCs secrete survival factors that ensure the survival of B cells within the lymph node.

B-Cell Recognition of Cell-Bound Antigen Results in Membrane Spreading

Prior to contact with antigen, the majority of BCRs are expressed in monomeric, bivalent form on the cell surface. Interaction of these monomeric receptors on the surface of the B cell with multivalent, cell-bound antigens then induces

a rather spectacular cellular response. First, a cluster of BCRs and their cognate antigens forms at the initial site of contact. Next, the B cell rapidly spreads over the target membrane, before contracting back.

This spreading reaction can be clearly seen in Figure 12-7, in which the cell-bound antigen was modeled using a protein incorporated into a planar lipid bilayer. Membrane spreading occurs 2 to 4 minutes after antigen contact, during which time micro-clusters of antigen-receptor interactions can be seen by time-lapse fluorescence microscopy at the membrane-lipid interface. After reaching a maximum surface area of approximately 25 μm², the area of contact between the cell and the artificial lipid membrane begins to contract, and the antigen-receptor complex is gathered into a central, defined

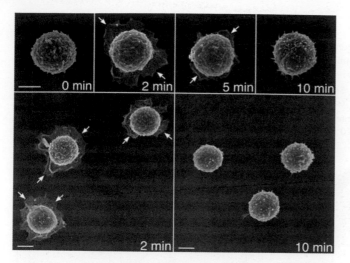

FIGURE 12-7 Antigen recognition by the BCR triggers membrane spreading. Transgenic B cells expressing a BCR specific for *hen egg lysozyme* (HEL) were settled onto planar lipid bilayers containing HEL and were incubated for the time period shown, followed by fixation and visualization with scanning electron microscopy. At 2 to 4 minutes, the B-cell membrane can clearly be seen spreading over the surface of the planar lipid bilayer (arrows). By 5 minutes, the membrane begins to contract again, after which the BCR molecules are found clustered on the cell surface. (See the text for details.) *[Fleire, S.J., Goldman, J.P., Carrasco Y.R., Weber, M., Bray, D. and Batista, F.D. B cell ligand discrimination through a spreading and contraction response. Science 5 May 2006, Vol. 312, no. 5774, pp. 738–741. © 2006 The American Association for the Advancement of Science]*

cluster with an area of approximately 16 μm². The contraction phase takes an additional 5 to 7 minutes. The effectiveness of the ultimate antibody response has been shown to correlate with the extent of this spreading reaction.

By the end of the contraction phase, the BCR, still in contact with its cognate antigen on the presenting cell, is clustered on the surface of the B cell.

What Causes the Clustering of the B-Cell Receptors Upon Antigen Binding?

Experiments using monovalent ligands incorporated into target lipid membranes demonstrated that the antigen itself need not be multivalent in order for clustering to begin. Rather, it appears that antigen binding to the BCR at sufficiently high affinity causes a structural alteration in the BCR, rendering it susceptible to clustering. Initially, antigen ligation of the Ig receptor is associated with a decrease in the rate of diffusion of IgM receptors in the plane of the membrane. In addition, antigen binding appears to induce a conformational change in the Cμ4 constant region domains of occupied IgM membrane receptors. These two changes result in an increase in the tendency of the bound IgM receptors to bind to the corresponding domains of other antigen-bound IgM receptor molecules (Figure 12-8). The deceleration in the receptor's movement in the membrane coupled with this receptor oligomerization occurs whether or not the Igα,Igβ (CD79α,β) signaling components of the receptor are present, indicating that the receptor clustering does not require signaling events. It is now thought that this BCR clustering may be the initiating event in antigen signaling through the BCR.

Upon oligomerization, the BCR complex moves transiently into parts of the membrane characterized by highly ordered, detergent-insoluble, sphingolipid- and cholesterol-rich regions, colloquially designated as **lipid rafts**. Recall that association of the BCR with the lipid rafts then brings the ITAMs of the Igα and Igβ components into close apposition with the raft-tethered src-family member tyrosine

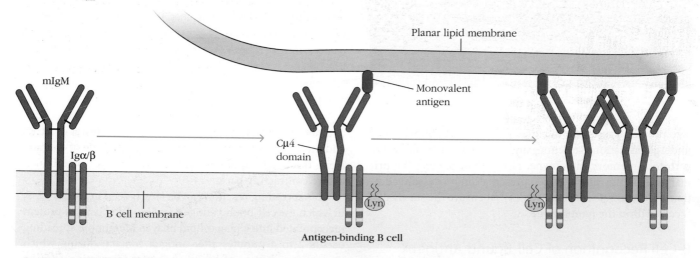

FIGURE 12-8 Antigen binding induces a conformational change in the Cμ4 domain and subsequent oligomerization of antigen-bound IgM molecules in the plane of the membrane. When monovalent antigens floating in an artificial bilayer encounter IgM receptors, they bind and induce a conformational change in the Cμ4 domain of the IgM heavy chain. This conformational change facilitates oligomerization of the IgM receptors in the plane of the membrane. (See text for details). *[Adapted from P. Tolar et al., 2009, The molecular assembly and organization of signaling active B-cell receptor oligomers, Immunological Reviews **232**:34–41, Figure 2.]*

kinase, Lyn. This sets off the B-cell signaling cascade shown in Figure 3-28.

Antigen Receptor Clustering Induces Internalization and Antigen Presentation by the B Cell

Antigen binding by the B cell activates the signaling cascade described in Figure 3-28, and this cascade induces several changes within the cell. One of these changes is the formation around the BCR of clathrin-coated pits that facilitate the internalization of the receptor-antigen complexes. Indeed, we now know that most of the antigen-occupied BCR molecules are internalized, leaving just a few copies of the receptor on the cell surface to serve as signaling scaffolds. Experiments have shown that those antigen-bound BCR-Igα,Igβ receptor complexes that remain at the cell surface and provide the B-cell signaling function are more highly phosphorylated than are those receptors that are internalized.

Once internalized, the antigen is processed as described in detail in Chapter 8. This results in an increased expression of peptide-loaded class II MHC molecules on the B-cell surface, as well as an up-regulation of the expression of the costimulatory molecules CD80 and CD86. Recall that CD80 and CD86 both bind to the T-cell costimulatory molecule CD28. These changes in the expression of CD80 and CD86, which occur within 1 to 2 hours after antigen recognition, therefore prepare the B cells for their subsequent interactions with T cells.

A B cell that has taken up its specific antigen via receptor-mediated endocytosis is up to 10,000-fold more efficient in presenting antigen to cognate T cells than is a nonantigen-specific B cell that can only acquire the same antigen by nonspecific pinocytotic mechanisms. Effectively, this means that *the only B cells that process and present antigen to T cells in a physiologically relevant context are those B cells with the capacity to make antibody to that antigen.*

Now that the B cell is ready to present antigen to T cells, we must ask how a B cell, with a receptor that is normally expressed at extremely low frequency within the receptor repertoire, can possibly find and bind to a T cell specific for the same antigen, which is also present at low frequency among all the available T cells?

Activated B Cells Migrate to Find Antigen-Specific T Cells

We owe a great deal of our understanding of the movement of cells and antigens through the lymph nodes to recent advances in imaging cells within their biological context. Specifically, in some recent experiments described in Chapter 14, lymph nodes have been brought outside the body of living, anesthetized animals, and lymphocyte circulation through these nodes has been studied using fluorescently tagged T, B, and antigen-presenting cells. In these experiments, the blood and lymphatic circulation to the lymph node is preserved, and the lymph node is gently lifted onto a warmed, humidified microscope stage for observation. Sometimes the fluorescent tags are built into the genome of the animal by placing fluorescent proteins, such as *green fluorescent protein* (GFP), under the control of specific promoters that are active only in specific lymphocyte subsets. In other experiments, the animal is injected with fluorescent antibodies just prior to being anesthetized. This technique is known as **intravital fluorescence microscopy** (see Chapters 14 and 20) and has contributed hugely to our understanding of the process of lymphocyte activation in the lymph nodes.

Investigators have used immunohistochemical, immunofluorescence, and the aforementioned intravital imaging techniques to visualize particular cell populations during the course of an ongoing immune response. This has, in some cases, required the generation of transgenic mice in which all, or most, B and T cells have defined antigen specificity. All of these techniques are described in Chapter 20. What follows is a distillation of the information gained from a large number of such experiments using various antigen and transgenic model systems.

As we have learned, B cells pick up their antigen in the follicular regions of the lymph node or spleen. By approximately 2 hours post antigen contact, the B cell has internalized and processed its antigen and expressed antigenic peptides on its surface in the context of class II MHC antigens. In response to antigen recognition by the BCR, the B cell begins to express the chemokine receptor CCR7. CCR7 binds to the chemokines CCL19 and CCL21, which are secreted by stromal cells in the T-cell zones of secondary lymphoid organs. The B cell still also expresses CXCR5, which binds CXCL13, expressed by FDCs in the B-cell follicles. Because they express receptors for both T- and B-cell-zone-derived chemokines, *by approximately 6 hours post stimulation, antigen-engaged B cells move to the boundary of the B-cell and T-cell zones.* Figure 12-9a shows these various lymph node zones in diagrammatic form. Activated B cells also up-regulate the expression of receptors for cytokines released by activated T cells, allowing B cells to receive the signals to proliferate, to differentiate, and also to initiate an anti-apoptotic program. The receptor for IL-4, an important B-cell-specific cytokine, can be detected as early as 6 hours post antigen contact and reaches maximal levels at 72 hours.

Antigen-stimulated B cells move about within the T-cell zone of the lymph node until they make contact with an antigen-specific T cell, an event that most probably occurs over the 24 to 48 hours post antigen contact. By this time, many of the important cell-surface changes characteristic of B lymphocyte activation can be observed, such as increased levels of class II MHC antigens and of the two costimulatory molecules, CD80 and CD86. Once contact with an activated T cell is made, these antigen-stimulated B cells engage with their conjugate T-cell partners over extended periods of time, ranging from a few minutes to several hours. During this period, the activated helper T cell also expresses CD40L (CD154), a cell-membrane-bound member of the TNF receptor family, which interacts with CD40 (a TNF family member)

(a)

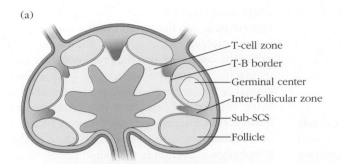

- T-cell zone
- T-B border
- Germinal center
- Inter-follicular zone
- Sub-SCS
- Follicle

(b)

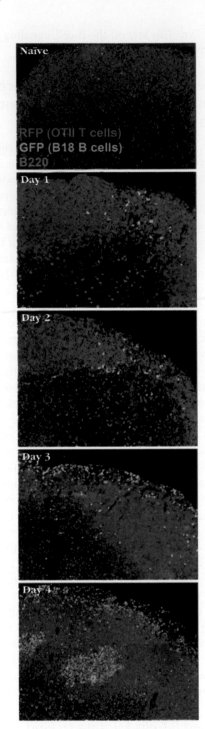

FIGURE 12-9 Movement of antigen-specific T and B cells within the lymph node after antigen encounter. (a) Section of the lymph node in diagrammatic form, showing the various regions of the lymph node. (b) The locations of antigen-specific B cells (green) and antigen-specific T cells (red) within the lymph node were visualized at specified times after antigen stimulation. The antigen used in the experiment was *nitro-phenacetyl* (NP), conjugated to *oval*bumin (OVA). Fluorescently labeled NP-specific B cells and OVA-specific T cells were transferred into a recipient animal, which was immunized with NP-OVA in the footpads 2 days after cell transfer. Draining popliteal lymph nodes were excised at the stated times. Anti-B220 (blue) was used to identify the B-cell follicles. The times refer to the days (D) post immunization. D1 and D2: Antigen-specific B-T cell pairs could be found both at the border of the T- and B-cell zones and in the interfollicular regions. D3: T cells have begun to enter the follicle, and many B cells can be seen just underneath the subcapsular sinus of the lymph node. D4: B cells have taken up residence in the follicle, and the formation of the germinal center can be seen. [*Fleire, S.J., Goldman, J.P., Carrasco Y.R., Weber, M., Bray, D. and Batista, F.D. B cell ligand discrimination through a spreading and contraction response.* Science 5 May 2006, Vol. 312, no. 5774, pp. 738–741. ©2006 The American Association for the Advancement of Science]

(see Chapter 4) on the activated B cell. The CD40-CD40L interactions, coupled with BCR-mediated antigen recognition and interleukin signaling, elicit all of the downstream activation pathways shown in Figure 3-28, and together drive the B cell into the proliferative phase of the cell cycle.

After this period of intense communication between the activated T and B cells, activated B cells then down-regulate CCR7 while maintaining high levels of expression of CXCR5. This allows them to leave the T-cell areas and to migrate into the inter- and outer-follicular regions. The activated B cells appear to spend 1 or 2 days in the regions to the outside and in between the follicles before finally entering the follicle around 4 days post immunization (Figure 12-9b).

In the experiment depicted in Figure 12-9b, a transgenic mouse has been immunized with the T-dependent antigen *nitro*phenacetyl-*oval*bumin (NP-OVA). Antigen-specific B cells express GFP, and are therefore labeled green; antigen-specific T cells are labeled red; and the B-cell follicles are stained blue. In the naïve animal, T cells can be seen clearly localized in the T-cell zones, and occasional green B cells can be seen scattered throughout the follicles. With time post immunization, B and T cells can be seen migrating first into the interfollicular regions between the follicles (days 1 and 2 post immunization). By day 3, B and T cells can be found clustered in the outer- as well as the interfollicular regions, and by 4 days post immunization most, but not all, of the B cells have entered the follicles and the beginning of germinal center development can be seen.

Interestingly, B cells derived from mice deficient in CXCR5 or CXCL13 behave in a fashion that is not as different from those derived from normal mice as might have been expected, and this observation led to the discovery of a second ligand-receptor pair that induced the same migrating behavior in B cells. EBI2-ligand is recognized by a G-protein-coupled receptor, *EBV-i*nduced *l*igand *2* receptor (EBI2), that is present on all mature B cells. The level of expression of EBI2 increases markedly and quickly following B-cell activation, and interaction of EBI2 with its ligand results in the movement of B cells into the outer- and interfollicular regions in a manner analogous to that instructed by CXCR5 interaction with CXCL13. Biochemical experiments suggest that the ligand for EBI2 is lipid in nature, and investigators are still working to determine the nature of the cell types that express the ligand, the dynamics of its expression, and the way in which the two sets of chemoattractant signals interact to ensure the correct movement of the antigen-stimulated B cell through the lymph node.

The outer- and interfollicular regions of the lymph node contain high numbers of SCS macrophages, dendritic cells, and a cell population referred to as *m*arginal *r*eticular *c*ells (MRCs), and the roles of each of these cell populations in the migratory behavior of antigen-activated B cells is another area of active investigation.

Activated B Cells Move Either into the Extrafollicular Space or into the Follicles to Form Germinal Centers

Over the first few days after their interaction with T cells, daughter cells of the proliferating B cell elect one of two fates. Some of the stimulated daughter cells in the extrafollicular spaces migrate into the borders of the T-cells zone, form a primary focus, and differentiate into **plasmablasts**. Plasmablasts are B cells that can still divide and present antigen to T cells, but they have already begun to secrete antibody. Plasmablasts and plasma cells in the primary focus secrete measurable levels of IgM by approximately 4 days after antigen contact and are responsible for the earliest manifestations of the antibody response.

Other daughter cells generated by stimulation of the original B cells enter the B-cell follicles, as shown in Figure 12-9b in the bottom panel. There, they divide rapidly and undergo further differentiation. This movement of B cells into the follicles is facilitated by the expression of the transcription factor Bcl-6, which represses the expression of EBI-2 and enables the B cells to leave the outer- and interfollicular regions. As the B cells differentiate under the influence of **follicular helper T cells** (T$_{FH}$) (see Chapter 11), the follicle becomes larger and more dense, developing into a germinal center. (The germinal center reaction will be discussed below.)

The precise nature of the mechanisms responsible for determining which cells enter the primary focus and which

elect to enter the follicle and establish a germinal center reaction are unclear at this time and may be stochastic. Below, we first describe the processes that lead to primary focus formation and then go on to discuss the extraordinary fate that awaits those B cells that enter the follicles, and eventually develop into germinal center B cells.

Plasma Cells Form within the Primary Focus

Plasma cells are essentially Ig-producing machines. Their surface Ig levels are close to zero, and they are no longer capable of being further stimulated by antigen, or of presenting antigen to responsive T cells. An activated B cell in the extrafollicular regions of the lymph node that is initiating a program of differentiation toward the plasma cell endpoint begins to divide rapidly and to secrete low levels of Ig. As the cell divides, it decreases levels of membrane Ig, MHC proteins, and CD80/86, and increases its rate of Ig secretion to become a plasmablast. Eventually, the cell reaches a stage of terminal differentiation in which it is no longer capable of cell division and has achieved a maximal rate of Ig secretion: the **plasma cell**. Plasma cells are found within the first 5 to 6 days of an immune response in the medullary cord region of the lymph node.

Most primary focus plasma cells have short half-lives, dying by apoptosis within 5 to 10 days of their generation, and for many years it was thought that all plasma cells from the primary foci endured this fate. However, recent experiments have suggested that some of these plasma cells may migrate to the bone marrow or to other locations within the body, where they provide long-lasting Ig memory.

There is currently a great deal of interest in the transcription factors that control whether antigen-stimulated B cells differentiate along the plasma cell or, alternatively, the germinal center route. Scientists now understand that these transcription factors are linked in a mutually regulatory network (Figure 12-10).

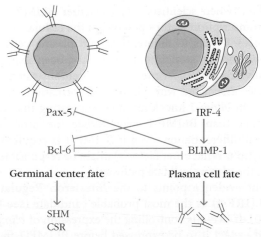

Germinal center fate   Plasma cell fate

SHM
CSR

FIGURE 12-10 A regulatory network of transcription factors controls the germinal center B cell/plasma cell decision point. The transcription factors that control germinal center B cell versus plasma cell states of differentiation are related to one another through a mutually regulatory network. (See the text for details).

Transcription factors that favor the generation of proliferating germinal center B cells, Pax-5 and Bcl-6, actively repress the expression of the plasma cell transcription factor BLIMP-1. Indeed, *bcl-6* gene knockout animals are incapable of forming germinal centers, a factor that demonstrates the centrality of Bcl-6 in determination of the germinal center phenotype. Immune responses from these knockout mice also fail to undergo affinity maturation and class switch recombination, processes that depends on the normal functioning of germinal centers.

In wild-type animals, as illustrated in Figure 12-10, induction of BLIMP-1 and IRF-4 results in the inhibition of Bcl-6 and Pax-5 expression. This antigen-induced expression of two alternative sets of transcription factors that control the decision between two potential lymphocyte cell fates is reminiscent of the induction of T-Bet and GATA-3 and their respective effects on T_H1 and T_H2 induction (Figure 11-10). In both cases, induction of the transcription factor(s) that initiate one differentiation pathway results in the direct inhibition of the synthesis and activity of the transcription factors supporting the alternative pathway.

The transcription factor **B-lymphocyte-induced maturation protein 1 (BLIMP-1)** had long been considered to be the master regulator of plasma cell differentiation. BLIMP-1 knockout animals lack all capacity to generate plasma cells, while enforced expression of BLIMP-1 is sufficient to promote plasma cell differentiation. In wild-type animals, BLIMP-1 expression peaks at the plasma cell stage of B-cell differentiation, and the highest levels of BLIMP-1 expression are seen in long-lived plasma cells in the bone marrow and primary focus plasma cells in the secondary lymphoid organs. Expression of BLIMP-1 also decreases the levels of MHC expression on the surface of the cell, consistent with the inability of plasma cells to engage in presentation of antigen to T cells. BLIMP-1 also promotes the alternative splicing of Ig mRNA, which enables the generation of mature transcripts encoding the secreted form of antibodies (see Figure 7-15).

However, when scientists asked whether BLIMP-1 was the *first* transcription factor operative in the cellular process that induces B cells to differentiate into plasma cells, the answer was not as straightforward as had been expected. Specifically, recent experiments have demonstrated that some genes associated with plasma-cell formation are still turned on in *blimp-1* knockout mice, suggesting that some factor other than BLIMP-1 may *initiate* the process of plasma cell differentiation, even if BLIMP-1 is required for its completion. What other factor(s) might be responsible for initiating the plasma-cell fate pathway?

Current evidence points to the **Interferon Regulatory Factor 4 (IRF-4)** as the most probable candidate (see Figure 12-10). If IRF-4 is controlling the expression of *blimp-1*, we would expect it to be expressed before BLIMP-1 in the developing B cell, and to control the expression of the *blimp-1* gene. Such is indeed the case. The *irf-4* gene is expressed before *blimp-1*, and the IRF-4 protein binds to elements upstream of the *blimp-1* gene, up-regulating its transcription. Furthermore IRF-4 down-regulates the expression of the genes encoding both Pax-5 and Bcl-6. Finally, the concentration of IRF-4 varies in B cells at different stages of differentiation in a manner consistent with it being a determining factor in plasma cell differentiation. Thus, it appears that IRF-4 may initiate differentiation to the plasma cell stage. But is it required, or merely a useful adjunct to the process of plasma-cell differentiation? For the answer to this question, we must turn to knockout experiments.

Experiments with IRF-4-deficient mice have shown that they completely lack Ig-secreting plasma cells; conversely, and as for BLIMP-1, overexpression of IRF-4 promotes plasma cell differentiation. IRF-4 has been shown to bind to Ig enhancer regions and is important in the generation of high levels of Ig secretion. Thus, it would appear that IRF-4 may be *a* determining factor, or possibly *the* determining factor, in driving a B cell toward the plasma cell, rather than the germinal center phenotype.

The generation of the plasma cells in the primary focus is a critically important step in the antibody response, and the rate of secretion of Ig by these cells is extraordinarily high. At the peak of an antibody response, a single antibody-secreting cell can release as much as 0.3 ng of Ig per hour, with fully 30% of its cellular protein synthesis devoted to the generation of secreted antibody molecules. Therefore, these cells provide high levels of antibody that can bind antigen, neutralizing it and/or opsonizing it for phagocytosis by macrophages.

However, the antigen-binding affinities of the antibodies secreted by the primary focus have not yet been optimized by the related processes of somatic hypermutation and antigen selection, and their affinity for the antigen is often relatively low. In the case of a particularly virulent infection, the antibodies secreted by the cells of the primary focus may therefore serve primarily to contain microbial numbers until the high-affinity antibodies generated by the germinal center reaction are released. It is within the germinal center that the next extraordinary events in B-cell differentiation will occur.

Other Activated B Cells Move into the Follicles and Initiate a Germinal Center Response

As described above, 4 or 5 days after T-dependent B-cell activation occurs at the border between the T-cell zone and the follicles, some antigen-specific B cells enter the follicles. There, they undergo rapid proliferation, resulting in the formation of large clusters of antigen-specific B cells. The follicle, which contains both T_{FH} and follicular dendritic cells (FDCs), becomes larger as the entering antigen-specific cells proliferate, and the teeming follicle in which an immune response is actively ongoing is referred to as a germinal center (GC) (Figure 12-11). Interaction between CD40 on the B cell and its ligand, CD40L, on the T cell is necessary for germinal center formation, and no germinal centers are generated in mice in which the genes for either CD40 or CD40L have been deleted.

In the germinal centers, B cells undergo a period of intense proliferation, and then their Ig genes are subjected to some of

(a)

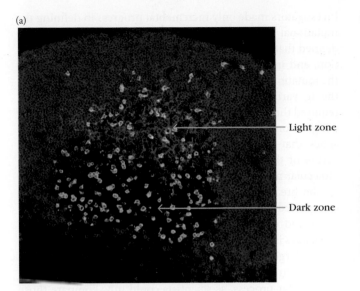

(b)

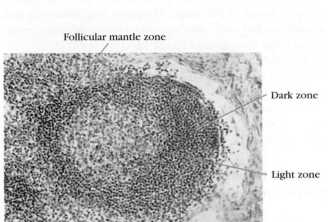

FIGURE 12-11 The germinal center. (a) A mouse lymph node germinal center 7 days after secondary immunization, in which 10% of the germinal center B cells express green fluorescent proteins (and are therefore labeled green). These proteins are seen most intensely in the dark zone of the germinal center, but are also present less frequently in the light zone. Follicular dendritic cells, labeled red with an antibody to the FDC marker CD35, mark the light zone of the germinal center. The blue staining marks IgD-bearing B cells not specific for the immunizing antigen. (b) Histochemical staining of a germinal center, illustrating the remarkably high centroblast cell density within the dark zone. *[(a) Reprinted from Victora, G.D. et al., 2010, Germinal center dynamics revealed by multiphoton microscopy with a photoactivatable fluorescent reporter, Cell 143:592–605. Copyright 2010 with permission from Elsevier. (b) Courtesy Dr. Roger C. Wagner, Professor Emeritus of Biological Sciences, University of Delaware]*

the most extraordinary processes in biology. First, the Ig variable (but not the constant) region genes undergo extremely high rates of mutagenesis, on the order of one mutation per thousand base pairs per generation. (Contrast this with the background cosmic ray-induced mutation rate of one mutation per hundred million base pairs per generation.) The mutated genes are then expressed, and the encoded Ig receptors are tested to see if their affinities for antigen have been altered for the better. Those B cells bearing higher-affinity receptors than those on the parental cells are selected for further rounds of mutation and selection, while those with no, or decreased, affinity for antigen are allowed to die by apoptosis.

Second, also within the germinal center, interleukin signals from follicular T cells drive the process of Ig CSR: the replacement of the μ constant region gene segments with segments encoding other classes of constant regions. We are just beginning to understand the molecular biology of these processes, which take place in a rapidly modulating anatomical environment. Given the active mutational and selection events that occur within the germinal center, it should come as no surprise that it has been referred to by some as a "Darwinian Microcosm."[1] Below, we will describe the biology of the germinal center and then look in turn at the two genetic processes that are unique to the immune system: somatic hypermutation and the Ig class switch.

[1]Garnett Kelsoe. (1998). V(D)J hypermutation and DNA mismatch repair: Vexed by fixation. *Proceedings of the National Academy of Sciences of the United States of America* **95**:6576.

Germinal Center Formation

When the activated B cells enter the B-cell follicles, they proliferate in an environment provided by a network of FDCs and T_{FH} cells. T_{FH} cells were described in detail in Chapter 11, and both FDCs and T_{FH}s provide survival signals to the rapidly dividing germinal center B cells.

B-cell proliferation results initially in a decrease in IgD expression on the cell surface, although IgM expression is maintained. As the GC expands, nonresponding IgD$^+$ B cells are displaced into the border region of the follicle, forming a corona of naïve B cells referred to as the **follicular mantle zone**. This is illustrated rather dramatically in Figure 12-11a as a blue border to the germinal center and is also illustrated in Figure 12-11b.

Germinal center formation requires a number of cytokines, including Lymphotoxin-α (TNF-β), that are produced by the FDCs, as well as by the T_{FH} and potentially by other cells within the lymph node. These soluble signaling molecules help to repress genes such as *blimp-1*, and promote the expression of genes such as *AID*, which have important roles in the developing germinal centers.

Dark- and Light-Zone Development

As the germinal center matures, two zones become visible: the light zone and the dark zone (see Figure 12-11). The "darkness" of the dark zone results from a dense packing of rapidly proliferating B cells (centroblasts), whereas the "lighter" nature of the light zone results from the less dense distribution of B cells (centrocytes) within a network of FDCs; indeed, FDCs are largely absent from the dark zone. (Note that in Figure 12-11a, not all the B cells have been labeled, in order to allow for clarity in the image. The green dark zone is, in fact, densely packed with dividing centroblasts, a characteristic that is better visualized in Figure 12-11b.)

The distribution of B cells between the two zones is dependent on their relative levels of expression of the chemokine receptors CXCR4 and CXCR5, which interact with chemokines preferentially expressed in each of the two zones. CXCR4 is up-regulated on centroblasts, whereas CXCR5 has higher expression on centrocytes.

Somatic Hypermutation and Affinity Selection Occur within the Germinal Center

The term *somatic hypermutation* describes one of the most extraordinary processes in biology. The word *somatic* tells us that the mutational processes are occurring outside of the germ line (egg and sperm) cells. *Hypermutation* alludes to the fact that the mutational processes occur extremely rapidly. Somatic hypermutation (SHM) in mice and humans occurs only following antigen contact, affects only the variable regions of the antibody heavy and light chains, and requires the engagement of T cells, which must be able to interact with B cells via CD40-CD40L binding.

The possibility that SHM of Ig genes might play a role in antibody diversification was first implied by amino acid sequencing studies conducted in the 1970s by Cesari, Weigert, Cohn, and others. Their studies focused on antibodies produced by mouse myeloma tumors expressing λ light chains. The mouse λ locus has been severely truncated, and as a result, mice have very few different λ chain variable regions. These investigators were therefore able to compare each of their myeloma light-chain sequences to a known germ-line sequence. They showed that the myeloma tumors expressed point mutations that were restricted to the variable regions of the λ chains and clustered in the complementarity determining regions. With the advent of nucleic acid sequencing technology, these data were confirmed and extended by others who showed that SHM affects the variable regions, but not the constant regions, of both the heavy and the light chains of Igs; that the frequency of somatic hypermutations increases with time post immunization and further; and that somatic hypermutation, followed by antigen selection, results in the secretion of antibodies whose affinity for the immunizing antigen increases with time post immunization.

These experiments defined the process of SHM, but left open the question of *where* it occurred. A series of papers published in the early 1990s, and representing a technical tour de force, proved that somatic hypermutation occurs within the germinal centers of an active lymph node (see Classic Experiment Box 12-1).

Once investigators knew *where* mutation was occurring, their next question, inevitably, was *how*? This question presented huge experimental challenges, and for almost 30 years

investigators made only incremental progress in defining the mutational mechanism. Certain nucleotide sequences were defined that appeared to be particularly susceptible to mutation, and immunogeneticists were able to demonstrate that the mutational process appeared to be remarkably focused on the Ig variable regions. Other studies showed that SHM required the involvement of T cells, that it occurred only after antigen contact, and that mutation appeared to affect only genes that were being rapidly transcribed. However, the nature of the enzymes that mediate the response, and the molecular mechanisms involved, remained obscure.

The breakthrough came, as so often happens, from an unexpected direction. In experiments designed to detect the genes and enzymes that mediate CSR, Takasu Honjo and colleagues isolated a cell line in which the antibody genes did not undergo CSR. This type of finding is often the crucial first step in identifying those proteins that are implicated in a particular process, as scientists could now compare those cells that could and could not undertake CSR and look for differences. However, importantly, they then noticed that *the antibody genes in these cell lines were also devoid of somatic mutations in the variable regions*. Could it be that the same enzyme was implicated in both SHM and CSR?

Members of Honjo's laboratory quickly isolated the gene that had been mutated in this cell line and proved that it encoded the enzyme **Activation-Induced Cytidine Deaminase (AID)**, which was subsequently shown to be necessary for both somatic hypermutation and CSR. But now the scientists were faced with the new questions: How does AID induce hypermutation, and how is it restricted only to antibody variable regions?

AID-Mediated Somatic Hypermutation

Although all the details of the SHM mechanism are not yet completely understood, several parts of the process are now well established. AID-induced loss of the amino group from cytidine residues located in mutation hot spots results in the formation of uridine, as shown in Figure 12-12. This creates a U-G mismatch. Several alternative mechanisms then come into play that participate in the resolution of the original mismatch, leading to the creation of shorter or longer stretches of mutated DNA (Figure 12-13).

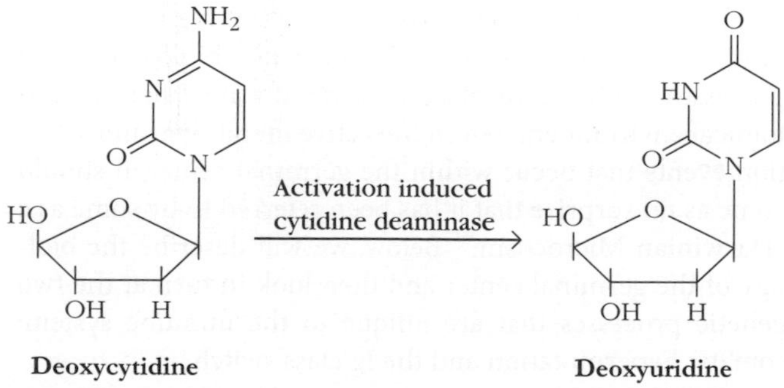

FIGURE 12-12 Activation-induced Cytidine deaminase (AID) mediates the deamination of cytidine and the formation of uridine.

BOX 12-1

Experimental Proof That Somatic Hypermutation and Antigen-Induced Selection Occurred within the Germinal Centers

In a series of papers written in the early 1990s, the laboratory of Garnett Kelsoe conclusively demonstrated that the phenomena of somatic hypermutation and antigen-induced selection occurred within the germinal center regions of the secondary lymphoid organs. The researchers made use of the fact that the immune response to the T-dependent hapten-protein antigen—4-hydroxy-3-nitrophenyl-acetyl (NP)-hemocyanin—generates an immune response that uses a predictable combination of heavy- and light-chain variable regions. This enabled them to study the changes in antibody genes as the immune response progressed, whereas a similar study of a response that used, for example, a hundred or more different B-cell clones would have been technically impossible.

The researchers generated serial tissue sections from lymph nodes at different time points during an immune response to the T-dependent antigen. The tissue samples were derived from the primary foci as well as from the germinal centers of antigen-responding lymph nodes. By scraping the primary foci or germinal center cells off the slides using microdissection tools, isolating the mRNA, and then subjecting the mRNA to RT *polymerase chain reaction* (PCR) and nucleic acid sequencing, mRNA from 100–300 B cells per sample were obtained. The results can be summarized as follows:

- *Each focus and germinal center is initiated by seeding with one to six antigen-specific B cells.* Statistical analysis of the immunoglobulin variable region sequences obtained from each sample showed that each focus or germinal center contained few B cells at the onset of the response. With time, the number of distinct clones of B cells per focus or germinal center decreased.

- *Somatic hypermutation is restricted to the germinal center B cells.* Nucleic acid sequencing of immunoglobulin variable regions obtained from B cells in both primary foci and germinal centers demonstrated that foci-derived sequences did not show any evidence of mutation. In contrast, mutations could be detected in germinal center-derived, V-region genes and accumulated as the response progressed.

- *With time post immunization, mutations in the complementarity-determining regions are selected for; mutations in framework regions are selected against.* Somatic hypermutation is an antigen-driven process, unlike the other mechanisms for the generation of antibody diversity. As discussed in Chapter 7, the variable region of an immunoglobulin gene can be divided into framework and complementarity-determining regions. Jacobs and colleagues analyzed what fraction of the mutations they observed were present in framework versus complementarity-determining regions. Those mutations affecting framework residues might be expected to be either neutral with respect to antigen binding or, if they affected the folding of the antibody variable region, to be deleterious. In contrast, those in the complementarity-determining regions of the gene, which encode the antigen-binding region of the antibody molecule, have the potential to be advantageous. Jacob and Kelsoe asked whether selective mechanisms were at play that increased the frequency of B cells with enhanced antigen-binding capacity. Their results showed that, whereas at 8 days post immunization, mutations were randomly distributed throughout the Ig variable region gene, by the end of the primary immune response, a greater fraction of the mutations was located in the complementarity determining regions, despite the fact that these regions comprised only 21% of the V region sequence.

- *Affinity measurements of antibodies generated in vitro from PCR-amplified samples derived from foci and germinal center cells showed that mutated and selected antibodies displayed increased affinity for antigen.*

Thus, not only did Kelsoe's lab demonstrate that the germinal center was the site of somatic hypermutation; their experiments also showed that mutations in the complementarity-determining regions that led to increased affinity for antigen were positively selected.

Jacob, J., C. Miller, and G. Kelsoe. 1992. In situ studies of the antigen-driven somatic hypermutation of immunoglobulin genes. *Immunology and Cell Biology* **70:**145–152.

Jacob, J., and G. Kelsoe. 1992. In situ studies of the primary immune response to (4-hydroxy-3-nitrophenyl)acetyl. II. A common clonal origin for periarteriolar lymphoid sheath-associated foci and germinal centers. *Journal of Experimental Medicine* **176:**679–687.

Briefly, the simplest mechanism is the interpretation of the deoxyuridine as a deoxythymidine by the DNA replication apparatus. In this case, one of the daughter cells would have an A-T pair instead of the original G-C pair found in the parent cell (Figure 12-13, left). Alternatively, the mismatched uridine could be excised by a DNA uridine glycosylase enzyme. Error-prone polymerases would then fill the gap as part of the cell's short-patch base excision repair mechanism (Figure 12-13, middle). Third, *mismatch repair* (MMR) mechanisms could be invoked that result in the excision of a longer stretch of DNA surrounding the mismatch. The excised strand could then be repaired by error-prone DNA polymerases, such as DNA polymerase η, leading to a lengthier series of mutations in the region of the original mismatch (Figure 12-13, right).

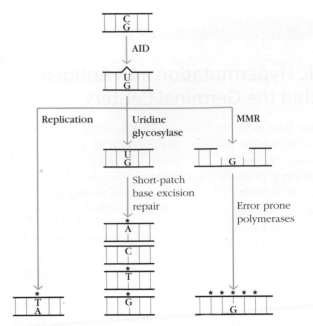

FIGURE 12-13 The generation of somatic cell mutations in Ig genes by AID. AID deaminates a cytidine residue, creating a uridine-guanosine (U-G) mismatch. Resolution of this mismatch may be facilitated by any one of several pathways that may compete with one another. On the left, the deoxyuridine is interpreted by the DNA replication machinery as if it were a deoxythymidine, resulting in the creation of an A-T pair in place of the original G-C pair in one of the daughter cells. In the center example, the mismatched uridine has been excised, most probably by one of the uridine DNA glycosylase enzymes, leaving an abasic site. The uridine can then be replaced by any of the four bases, in a reaction known as short-patch base excision repair, which can be catalyzed by one of a number of error-prone polymerases. Finally, as shown on the right, mismatch repair enzymes can detect the mismatch and excise a longer stretch of the DNA surrounding the U-G couple. Error-prone polymerases are then recruited to the hypermutable site by *proliferating cell nuclear antigen* (PCNA), and these polymerases can introduce a number of mutations around the original mismatch. Thus, depending on the repair mechanism, a mutation may occur only in the originally altered base or in one or more bases surrounding it. [Adapted from J. U. Peled et al., 2008, The Biochemistry of Somatic Hypermutation. Annual Reviews Immunology *26*:481–511.]

Mutational Apparatus Targeting

Since the rate of hypermutation is orders of magnitude higher in Ig variable region DNA than in other genes in germinal center B cells, some mechanism must exist to direct the mutational machinery to the correct chromosomal location. In addition, a variation on this mechanism should also direct AID to those parts of the constant region genes that are recognized during CSR.

When DNA is being actively transcribed, localized regions of DNA become transiently detached from their partner strands. The presence of single-stranded DNA appears to be necessary for SHM to occur, and the number of mutations that accumulate in Ig DNA is roughly proportional to the rate of Ig transcription. However, genes other than Igs are transcribed in germinal center B cells, as are the constant regions of the immunoglobulin genes, and they do not undergo mutation. Clearly this distinction alone does not account for the fact that mutations concentrate almost exclusively in the variable regions of Ig heavy and light chains, and something other than active transcription must be targeting AID to the variable regions.

Careful analysis of Ig variable region sequences revealed that some sequence motifs were more likely than others to be targeted by the mutational apparatus, and these are referred to as **mutational hot spots**. In particular, it was noted that the sequence motif DGYW/WRCH was frequently targeted for mutation at the underlined G-C pair. (In this description of the sequence, the four nucleotides after the backslash represent the inverted complement of the DGYW sequence, such that the underlined G and C are paired with one another.) The code used to describe the targeted hot spot sequence is as follows:

$$D = A/G/T$$

$$Y = C/T$$

$$R = A/G$$

$$W = A/T$$

$$H = T/C/A$$

The DGYW motif is also found frequently in the class switch regions, and so it appears to be an important sequence that directs AID binding to certain parts of the DNA.

Antigen-Induced Selection of B Cells with Higher Affinity

In this section, we will describe how those B cells with higher-affinity antigen receptors successfully compete with their lower-affinity counterparts for survival and proliferation signals delivered to them by T$_{FH}$ cells in the germinal center.

Seminal experiments using intravital fluorescence microscopy in transgenic mouse models have enabled investigators to address the question of how higher-affinity B cells are allowed to survive, while their lower-affinity precursors and cousins succumb to apoptosis. In these mice, antigen-specific T and B cells as well as FDCs were labeled with different fluorescent dyes, and the movement of B and T cells during an ongoing germinal center reaction was observed with time-lapse fluorescence microscopy. Investigators observed that B cells spend a relatively short amount of time in the light zones of the germinal centers, where they must take the opportunity to come into contact with T$_{FH}$ cells, in order to receive T-cell-derived growth and survival signals. They further noticed that there are many more B cells than T$_{FH}$ cells in the light zones, which indicated that B cells may have to compete with one another for the privilege of interacting with T cells.

A well-accepted current model suggests that B cells bearing higher-affinity receptors capture and process antigen more effectively than do competing B cells with lower-affinity receptors. The more antigen that a B cell processes, the more antigen it will present on its class II MHC molecules for recognition by cognate T_{FH} cells. Therefore, a B cell that has undergone an advantageous mutation in the antigen-binding region of its Ig genes will be better able to interact with T_{FH} cells than will its competitors, and therefore it will receive more proliferative and survival signals from those T_{FH} cells.

Additional experiments have indicated that B-cell competition for antigen within the germinal center may be more direct than previously thought, as some B cells were observed actually stripping antigen from other B cells. This suggests that *higher-affinity B cells actually steal antigen from their lower-affinity counterparts*. In this way B cells with higher-affinity receptors present more antigen to T cells, and enjoy better interactions with them that lead directly to enhanced survival. In the absence of positive survival signals, lower-affinity B cells in the germinal center undergo apoptosis.

Genetic dissection of this process has shown that signals delivered from the T cell through CD40 on the B-cell surface provide an indispensable component of the signal that T cells deliver to successfully competing B cells. In addition, BCR-induced PI3 kinase activation (see Chapter 3) in germinal center B cells results in the activation of the serine/threonine kinase Akt. Akt is a pleiotropic (has many effects) kinase, which not only promotes cell survival and inhibits apoptotic proteins, but also promotes the degradation of p53, thus allowing the GC B cells to cycle.

Thus, those B cells that bind, process, and present more antigen to T cells will win out over those B cells that cannot express antigen as effectively. Low-affinity B cells will lose out in the competition for antigen, and will die because of the absence of T-cell-mediated survival signals.

Since mutation is a random process, some B cells may acquire self-reactive receptors. One proposed mechanism for the destruction of such mutation-generated, self-specific B cells relies on the fact that self molecules, such as serum proteins, will be expressed at extremely high concentrations in the lymph nodes, whereas foreign antigens will be expressed at much lower levels. One might therefore expect that all the BCRs on a self-reactive B cell will be occupied. Full receptor occupancy leads to rapid internalization of the vast majority of the BCRs on the cell surface. In follicular B cells, such loss of cell surface BCR expression would result, in turn, in the loss of signaling through the BCR, exit from the cell cycle, and the induction of apoptosis. In contrast, on B cells specific for foreign antigens, only a relatively small proportion of their receptors will be occupied by antigen and therefore, on these cells, enough BCRs will remain on the cell surface to provide the scaffold for the signaling cascade.

Inevitably, some self-specific B cells will escape into the periphery. Recall, however, that these B cells will lack cognate T-cell help and, in the absence of stimulation, will simply be lost by neglect.

As described above, both SHM and CSR depend on the activity of the AID enzyme. However, analysis of mutated forms of AID has demonstrated that different parts of the same molecule catalyze the two different processes. Furthermore, mutations in different parts of the molecule lead to different immunodeficiency states.

Specifically, investigators analyzed the structure of AID genes isolated from a series of patients with Hyper IgM syndrome. In this immunodeficiency disease, patients generate only IgM antibodies that fail to undergo either SHM or CSR. Such patients suffer from recurrent, severe infections, thus emphasizing the physiological importance of both the mutational and the class switching processes to the fully functioning immune system. However, some unusual patients were found to have AID genes with premature stop codons near the 3' end. These patients generated antibodies that could undergo SHM, but were severely compromised in their ability to perform CSR. Such individuals displayed only mild symptoms, and their immunodeficiency was often not diagnosed until adulthood. This implies that different sections of the AID protein are necessary for SHM and for CSR and further suggests that the ability to generate high-affinity antibodies may be more functionally relevant than the capacity to synthesize antibodies of different heavy-chain classes.

Class Switch Recombination Occurs Within the Germinal Center after Antigen Contact

In Chapter 7, we noted that naïve B cells could simultaneously express both mIgM and mIgD: both proteins are encoded on the same long transcript, and the decision to translate μ (IgM) versus δ (IgD) heavy chains is made at the level of RNA splicing. In contrast, the decision to switch from the expression of IgM to expression of any of the other classes of antibodies is made at the level of DNA recombination, and the process by which it occurs is referred to as class switch recombination (CSR). The switch to the expression of any heavy-chain class other than δ results in the irreversible loss of the intervening DNA.

The Ig heavy-chain locus is approximately 200 kb in length. The formation of γ, ε, and α heavy-chain genes requires cutting and rejoining of the heavy-chain DNA (Figure 12-14) in such a way that the desired constant region lies directly downstream from the rearranged VDJ region. Class switching occurs by the induction of recombination between donor and acceptor **switch (S) regions** located 2 kb to 3 kb upstream from each C_H region (except for C_δ). The *donor S region* is the S region upstream from the antibody heavy-chain constant region gene expressed prior to the class switch (which is normally μ, except for those instances in which a B cell undergoes more than one class switch). The *acceptor S region* is the S region upstream of the antibody heavy-chain constant region that the B cell will express next. Switch regions consist of tandem repeats of short, G-rich sequences, 20 bp to 80 bp in length, that differ slightly for

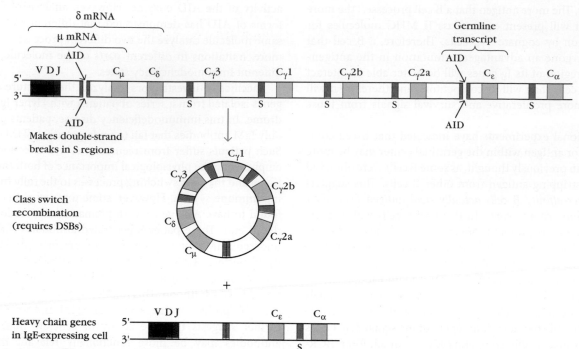

FIGURE 12-14 Class switch recombination from a Cμ to a Cε heavy-chain constant region gene. The activation induced cytidine deaminase (AID) enzyme initiates CSR by deaminating cytidine residues within the switch (S) regions upstream of Cμ and Cε on both strands. This leads to the formation of double-strand breaks within both S regions that are then resolved by DNA repair mechanisms, with the loss of the intervening DNA sequence, as described in the text. [Adapted from J. Stavnezer et al., 2008, Mechanism and Regulation of Class Switch Recombination. Annual Review of Immunology 26:261–292.]

each isotype and contain targeting sites for AID. In CSR, genetic analysis has indicated that the critical sequence motif required for AID binding is a pair of WGCW overlapping motifs on the top and bottom strand, where W represents either adenine or thymine. (Note that WGCW merely represents a subset of the group of sequences described above as DGYW.) The overall length of the switch regions vary from 1 kb to 10 kb, and CSR can occur anywhere within or near the S regions.

Signals for Class Switch Recombination

B cells must receive costimulatory signals from CD40 or, occasionally, B-cell Toll-like receptors in order to engage in CSR. The importance of CD40-CD40L interactions in the mediation of CSR is illustrated in patients suffering from **X-linked Hyper-IgM syndrome**, an immunodeficiency disorder in which T_H cells fail to express CD40L. Patients suffering from this disorder express IgM, but no other isotypes. Such patients also fail to form germinal centers, they fail to generate memory cell populations, and their antibodies do not show evidence of SHM.

The cytokine signal received by the B cell determines which class of Ig it will make (see Table 12-1). These cytokines signal the B cells to induce transcription from *germline promoters* located upstream (5′) of the respective donor

and acceptor switch regions. The resulting germ-line transcripts do not encode proteins and are therefore referred to as sterile RNAs. Importantly, no CSR can occur in the absence of this transcriptional activity, which is probably important in the creation of localized regions of single-stranded DNA recognized by AID. The germ-line promoters express the appropriate cytokine-responsive elements. For example, germ-line γ1 and ε promoters, which are induced by IL-4, have binding sites for the IL-4-induced transcriptional activator Stat6. But how does a B cell that receives a signal from IL-4 know whether to switch to IgG1 or IgE? The answer to this, and similar questions, is not yet clear, and there is still much to learn regarding the details of the signals that differentially regulate switching to particular classes of antibodies.

The Molecular Mechanism of Class Switch Recombination

CSR occurs by an end-joining mechanism and, like SHM, the process is initiated by AID. AID deaminates several cytosines within both the donor and acceptor S sites that have been previously activated as a result of cytokine signaling. DNA uridine glycosylase enzymes remove the U, created by the deamination of cytidine, and then apurinic/apyrimidinic endonucleases nick the DNA backbone at the abasic sites,

TABLE 12-1	Specific cytokines signal B cells to undergo CSR to different heavy-chain classes
Cytokine signal	**Isotype synthesized by target B cell**
IL-4	IgG1, IgE
TGF-β	IgA, IgG2b
IL-5	IgA
IFN-γ	IgG3, IgG2a

creating single-strand breaks at multiple points in the donor and acceptor S sites. Mismatch DNA repair enzymes then convert the single-strand DNA breaks into double-strand breaks. In the final step of the process, the cell's double-strand break repair machinery steps in and ligates the two switch regions, resulting in the excision of the intervening sequence. The process of CSR can occur more than once during the lifetime of the cell. For example, an initial CSR event can switch the cell from making IgM to synthesizing IgG1, and a second CSR can switch it to making IgE or IgA.

Most Newly Generated B Cells Are Lost at the End of the Primary Immune Response

Between 14 and 18 days after its initiation, the primary immune response winds down, and the immune system is faced with a problem of excess. Rapid proliferation of antigen-specific B cells over the course of the immune response leads to the generation of expanded clones of cells, and if all the cells from each clone were allowed to survive at the close of every immune response, there would soon be no room for new B cells emerging from the bone marrow and seeking to circulate through the lymphoid follicles and receive their survival signals.

Although we know that most B cells are lost by apoptosis at the end of the immune response, the exact mechanism by which this occurs has not yet been fully characterized. As antigen levels wane, the balance between survival signals and death signals experienced by the B cell in the lymph node may be tipped in favor of apoptosis. Certainly, the B cell no longer receives survival signals via antigen binding to the BCR. In addition, animals deficient in the genes encoding Fas have excess numbers of B cells, implicating the Fas-FasL interaction in the control of B-cell numbers. But we do not yet know the whole story. This is an area of active research, and a number of important questions remain to be answered:

- Does the cessation of B-cell signaling simply lead to a decrease in anti-apoptotic proteins, such that the cell is no longer protected from these death-inducing signals? Or is a more active switch engaged that leads to the cell's demise?

- What roles do residual antigen, follicular dendritic cells, and T-cell signaling play in the survival of B cells after the primary response is over?

- Since some cells will survive this primary response as either memory B cells or long-lived plasma cells, how are B cells within a single clone selected for survival versus cell death? Is the decision made earlier or later during the course of the response, and is it random or directed in some as yet unknown way?

Some Germinal Center Cells Complete Their Maturation as Plasma Cells

At some point in the ongoing immune response, approximately 5 to 15 days after antigen encounter, a fraction of germinal center B cells will begin to up-regulate IRF-4 expression, heralding the beginning of their differentiation into antibody-secreting plasma cells. As described earlier, IRF-4 expression then induces the generation of the transcriptional repressor, BLIMP-1, which down-regulates those genes important to B-cell proliferation, CSR, and SHM, and up-regulates the rate of the synthesis and secretion of Ig genes. As the germinal center B cell differentiates into a fully mature plasma cell, it reduces the level of expression of the chemokine receptor CXCR5, which has been responsible for retaining it within the germinal center. Instead the nascent plasma cell begins to express CXCR4, which enables it to leave the lymph node and circulate within the peripheral tissues. These germinal center-derived plasma cells differ from those generated from the primary focus in that their Ig genes have undergone both SHM and CSR, and hence the antibodies they secrete will be of high affinity and may be class switched.

For many years, it was thought that plasma cells localized primarily in the medullary cords of the lymph nodes (the inner parts of the kidney-shaped lymph nodes), or the red pulp in the spleen, and that they were relatively short-lived. However in the past decade or so, we have come to understand that plasma cells can home to several other locations and that the 10% to 20% of plasma cells that home to the bone marrow can be very long-lived. Indeed, smallpox-specific serum antibodies have been identified 75 or more years after immunization with smallpox vaccine, suggesting that the plasma cells secreting these antibodies may persist for the lifetime of the host. We now know that these long-lived plasma cells derive from both the plasma cells of the primary focus as well as from the B cells that have passed through the germinal center and have undergone CSR and SHM.

Within the bone marrow, the niches occupied by fully differentiated plasma cells differ from those inhabited by developing B cells (Figure 12-15). Experiments using in vitro culture techniques to determine the survival requirements for long-lived bone marrow plasma cells have highlighted the need for CXCL12 (recognized by CXCR4 on the plasma cell) and the TNF family cytokine member APRIL (recognized

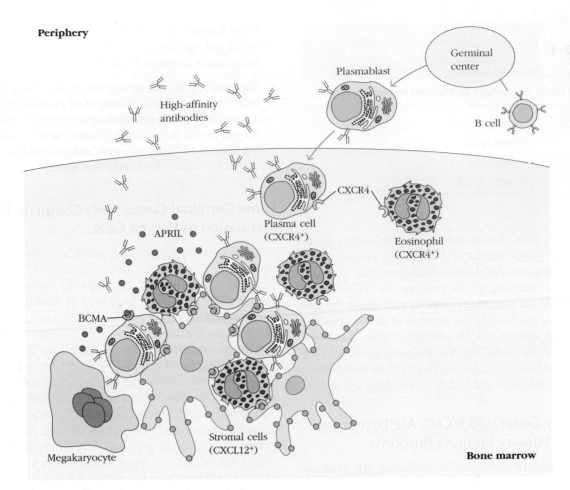

FIGURE 12-15 The bone marrow niche occupied by plasma cells is supported by eosinophils and megakaryocytes, as well as by mesenchymal stromal cells. Plasmablasts and plasma cells that have passed through the germinal center reaction and entered the circulation take up residence in the bone marrow, where they seek niches adjacent to eosinophils and megakaryocytes, as well as to the traditional mesenchymally derived stromal cells. The eosinophils and megakaryocytes provide the long-lived plasma cell with the survival factor APRIL, a TNF family member recognized by the receptor BCMA, whereas the stromal cells release CXCL12, recognized by the receptor CXCR4 on plasma cells.

by the plasma cell APRIL receptor, BCMA). CXCL12 is produced in the bone marrow by mesenchymal stromal cells, and APRIL by both eosinophils and megakaryocytes. Characterization of the bone marrow plasma cell niche continues to be a rapidly advancing area of research.

Other plasma cells, generated in the lymphoid tissues of the gut, remain associated with the mucosal tissues and secrete large amounts of IgA. IgA-secreting plasma cells are generated both in the Peyer's patches, areas of lymphoid concentration within the gut tissues, or in isolated lymphoid follicles in the lamina propria of the gut (see Figures 2-11 and 2-12). These gut-associated, IgA-secreting plasma cells have significant functional differences from the IgG-secreting plasma cells in other lymphoid tissues, and share some important characteristics with cells of the granulocyte-monocyte lineages. In particular, they produce the antimicrobial mediators TNF-α and inducible nitric oxide synthase (iNOS), molecules normally associated with monocyte and granulocyte activation. In order to continue producing these mediators, gut IgA-producing plasma cells must remain in contact with gut stromal cells and be subject to microbial costimulation. Deletion of TNF-α and iNOS in B-lineage cells resulted in poor clearance of gut pathogens and has been associated with a concomitant reduction in IgA synthesis.

B-Cell Memory Provides a Rapid and Strong Response to Secondary Infection

The first recorded concept of immunological memory appears in the writings of Thucydides, around 430 BCE. Describing the plague in Athens he wrote, "Only those who had recovered from the plague could nurse the sick, because they would not contract the disease." Implicit in this statement was the fact that those who had suffered and recovered had immunity to the plague, and were therefore able to mount a stronger and faster response to future infection.

Figure 12-16 illustrates the classic conception of a memory immune response. The index of immune responsiveness

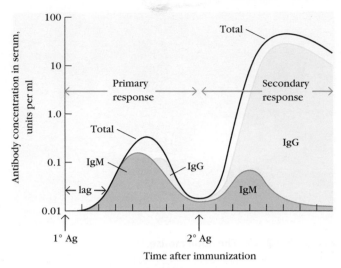

FIGURE 12-16 Concentration and isotype of serum antibody following primary and secondary immunization with antigen. The antibody concentrations are plotted on a logarithmic scale. The time units are not specified because the kinetics differ somewhat with type of antigen, administration route, presence or absence of adjuvant, and the species or strain of animal.

in this figure is the production of serum antibody. The primary response is characterized by a lag period, which reflects the time required for the division and differentiation of B cells within the primary foci and their movement into the germinal center. The B cells of the primary foci then release IgM, and after a short delay B cells that have migrated into the germinal center join the response with the concomitant release of both IgM and IgG. As the primary response draws to a close, somatically hypermutated receptors make their first appearance, and then a selected subpopulation of B cells leaves the germinal centers and enters the memory B-cell compartment.

The secondary response to the antigen is both faster and stronger than the first. Antigen-specific B-cell division has already occurred, and an expanded set of memory B cells bearing high-affinity receptors is available for immediate differentiation to high-affinity IgG secretion. Interestingly, somatically

hypermutated B cells can undergo further hypermutation with additional antigen exposure, such that the average affinity of antigen-specific antibodies increases through the third, and even a fourth, immunization with the same antigen.

The actual kinetics of both primary and secondary B-cell responses varies with the nature, dose, and route of administration of the antigen. For example, immunization of mice with an antigen such as sheep red blood cells typically results in a lag phase of 4 to 5 days before antibody is reliably detected in serum, and peak serum antibody levels are attained by 7 to 10 days. In contrast, the lag phase for soluble protein antigens is a little longer, often lasting about a week, and peak serum titers do not occur until around 14 days.

What properties distinguish memory from naïve B cells (Table 12-2)? Memory and naïve B cells share the characteristics of being nonsecreting cells that need further antigen stimulation prior to expansion and differentiation to plasma cells. However, memory B cells differ from their naïve counterparts in terms of their sensitivity to antigen activation; memory B cells are able to respond to lower concentrations of antigen than primary naïve B cells can, as a consequence of having generated higher-affinity Ig receptors during the process of SHM. They also respond more quickly to antigen activation, and therefore there is a shorter lag between antigen contact and antibody synthesis. Memory B cells do not express IgD and usually carry Ig receptors that are class switched (although some IgM-bearing memory cells may persist). Memory B cells have longer half-lives than primary B cells and recirculate among the lymphoid tissues.

Despite the fact that scientists have known about the existence of these cells for decades, surprisingly little is known about the factors that control their generation and maintenance. Questions regarding memory B cells that still await resolution include the following:

- At what stage in B-cell differentiation is the B-cell fate determined, and what are the signals that control that fate? It seems clear that at some point in the maturation process of a B cell, individual daughter cells of the same stimulated clone of primary B cells will be selected for memory generation (life) or death. What mechanism

TABLE 12-2	**Functional differences between primary and secondary B cells**	
	Naïve B cell	**Memory B cell**
Lag period after antigen administration	4–7 days (depends on antigen)	1–3 days (depends on antigen)
Time of peak response	7–10 days	3–5 days
Magnitude of peak antibody response	Varies, depending on antigen	Generally 10 to 1000 times higher than primary response
Antibody isotype produced	IgM predominates in early primary response	IgG predominates (IgA in the mucosal tissues)
Antigens	Thymus independent and thymus dependent	Primarily thymus dependent
Antibody affinity	Low	High
Life span of cells	Short-lived (days to weeks)	Long-lived, up to life span of animal host
Recirculation	Yes	Yes

allows for the differential survival of members of the same clone of B cells? Exciting, ongoing experiments that study asymmetric cell division in antigen-stimulated B cells are beginning to unravel the threads of this particular immunological knot.

- Where, in the body, do memory cells bearing differentially class-switched receptors reside? In mice, some memory B cells localize to the marginal zones of the spleen, and others have been located in the follicles; in humans, memory cells appear to localize to the marginal zones and to sub- and intraepithelial surfaces. Interestingly, during an ongoing infection or in an individual suffering from an autoimmune disease, ectopic reservoirs of organized lymphoid tissues can develop in multiple organs not usually associated with immune responsiveness, including the kidneys, pancreas, lungs, thyroid, and so on. The contributions of such ectopic tissues, as well as of organized lymphoid organs within the bone marrow, gut, and other mucosal tissues, to memory responses have yet to be investigated.

- How does the balance of pro- and anti-apoptotic molecules differ between primary and memory B cells, and are particular cytokine signals required for persistent memory B-cell survival?

- Are there differences in the efficiency of signal transduction in the two kinds of cells that may help to explain the variations in their lag times to antibody synthesis?

T-Independent B-Cell Responses

Not all antibody-producing responses require the participation of T cells, and certain subsets of B cells have evolved mechanisms so as to respond with antibody production to particular classes of antigens, without T-cell help (see Figure 12-3). Antigens capable of eliciting T-independent antibody responses tend to have polyvalent, repeating determinants that are shared among many microbial species. Many of these antigens are recognized by B cells of the B-1 as well as the marginal zone subsets (discussed below). B-1 B cells secrete mainly IgM antibodies that are not subject to SHM. Because of the shared nature of their antigens and their oligoclonal antibody response, B-1 B cells may be considered to belong to an "innate-like" category of lymphocytes. Our understanding of the physiology of MZ is still developing, but it is clear that B cells of this unusual subset of cells may receive help from cell types other than T cells, as is discussed below (see Advances Box 12-2).

T-Independent Antigens Stimulate Antibody Production without the Need for T-Cell Help

The nude (nu/nu) mouse is one of the most bizarre accidents of nature. Devoid of body hair, its ears seem oversized, and it appears absurdly vulnerable (Figure 12-17). These mice have

FIGURE 12-17   The nude mouse. Nude mice have a mutation in the Foxn1 gene that results in hair loss (alopecia) and interferes with thymic development, such that they possess very few T cells and only a thymic rudiment. Nude mice provided useful early models for the exploration of a T-cell-depleted immune response. [*David A. Nortcott/Corbis*]

a mutation in the gene for the transcription factor Foxn1 that, in addition to affecting hair growth, also results in athymia: mice and humans with this mutation have only thymic rudiments and possess few mature recirculating T cells.

Using nude mice as well as mice whose thymuses were surgically removed early in life (referred to as neonatally thymectomized), immunologists showed that most protein antigens fail to elicit an antibody response in such animals even while the response to many carbohydrate antigens was unaffected. Those antigens capable of generating antibody responses in athymic mice were referred to as T-independent antigens because they did not require T-cell help to generate an antibody response. Since T-cell interactions are required for the induction of AID, TI antigens elicit predominantly low-affinity antibody responses from B cells expressing only the IgM isotype. We now know that T-independent antigens fall into two subclasses (Table 12-3).

TI-1 Antigens

The first class of *T-i*ndependent antigens (TI-1 antigens) is exemplified by the bacterial *lipopolysaccharide* (LPS). TI-1 antigens bind to innate immune receptors on the surface of all B cells (including the majority B-2 B-cell population), and are capable, at high-antigen doses, of being mitogenic for all B cells bearing the responding innate receptors. Since B-cell stimulation in this instance is occurring through the innate receptor (TLR4), only a small minority of the antibodies produced will be able to bind directly to the TI-1 antigen. The huge in vivo polyclonal TI-1 responses generated in response to high levels of Gram-negative organisms can be catastrophic for an individual and are associated with the phenomenon we know as *septic shock* (see Chapter 15).

At lower doses, the innate immune receptors are unable to bind sufficient antigen to be stimulatory for the B cell. However,

TABLE 12-3 Properties of thymus-dependent and thymus-independent antigens

Property	TD antigens	TI antigens Type 1	Type 2
Chemical nature	Soluble protein	Bacterial cell-wall components (e.g., LPS)	Polymeric protein antigens; capsular polysaccharides
Humoral response			
Isotype switching	Yes	No	Limited
Affinity maturation	Yes	No	No
Immunologic memory	Yes	No	No
Polyclonal activation	No	Yes (high doses)	No

in those B cells that bind to the TI-1 antigen through their Ig receptors, the TI-1 antigen may still be able to cross-link the Ig and innate receptors, thereby eliciting an oligoclonal (few clones) B-cell response that remains independent of T-cell involvement. Under these circumstances, all the secreted antibodies will be specific for the TI-1 antigen.

Given the number of commensal bacteria in the gut, many of which bear these TI-1 antigens, why aren't our gut B cells in a constant state of activation? Part of the answer to this question lies in the anatomy of the gut. The gut is lined by a mucous layer, which prevents access of most commensal bacteria to gut-resident lymphocytes. Second, regulatory T cells in the gut help to tone down the immune response, thus preventing the inflammation that would ensue were organisms to mount a constant immune response to commensal organisms.

TI-2 Antigens

Unlike TI-1 antigens, type 2 T-independent antigens (TI-2 antigens), such as capsular bacterial polysaccharides or polymeric flagellin, are not mitogenic at high concentrations. Their capacity to activate B cells in the absence of T-cell help results from their ability to present antigenic determinants in a flexible and remarkably multivalent array, which causes extensive cross-linking of the BCR. In addition, most naturally occurring TI-2 antigens are characterized by the ability to bind to the complement fragments, C3d and C3dg (see Chapter 6). As a result, TI-2 antigens are capable of activating B cells by cross-linking the BCR and CD21 receptors on the B-cell surface.

TI-2 antigens can only partially stimulate B cells in the complete absence of help from other cells. Monocytes, macrophages, and dendritic cells have been shown to facilitate B-cell responses to TI-2 antigens by expressing a molecule known as **BAFF**—a membrane-bound homolog of tumor necrosis factor—to which mature B cells bind though the TACI (*trans-membrane activator* and *CAML interactor*) receptor. This interaction activates important transcription factors that promote B-cell survival, maturation, and antibody secretion.

T cells can also enhance B-cell activation by producing cytokines that push TI-2 antigen-activated B cells from activation to antibody production. Recognition by B cell-bound receptors of these T-cell cytokines may stimulate the B cell to secrete antibody classes other than IgM in response to TI-2 antigen stimulation. Unlike TI-1 antigens, TI-2 antigens cannot stimulate immature B cells and do not act as polyclonal activators.

Two Novel Subclasses of B Cells Mediate the Response to T-Independent Antigens

All B cells bear Ig receptors and secrete antibodies, but recent research has demonstrated that there are multiple subpopulations of B cells varying in locations, phenotypes, and functions (Table 12-4). Some of these subpopulations (the *transitional* B-cell populations, T1 and T2) represent temporal stages of B-cell development that occur after the B cell leaves the bone marrow. An additional transitional subset, T3, appears to represent an anergic subpopulation of B cells (see Chapter 10). Others (B-1a, B-1b, B-2, and marginal zone B cells) represent different subpopulations of mature B cells, each characterized by a preferential location and range of functional capacities. The T-dependent B-cell response described above is conducted by the B-2 cell population, also known as follicular B cells.

The presence of B cells with properties distinct from those of the majority B-2 subset was first suggested when scientists noticed that B cells responding to T-independent antigens differed in several important ways from B cells recognizing T-dependent antigens. Those B cells with specificity for T-independent antigens can be divided into two major subtypes that differ in aspects of their development, their anatomical location, and their cell surface markers. We will describe each of them in turn.

B-1 B Cells

Like γδ intraepithelial T cells, **B-1 B cells** occupy a functional niche midway between those of the innate and adaptive

New Ideas on B-Cell Help: Not All Cells That Help B Cells Make Antibodies Are T Cells

Since the classical experiments of Miller, Mitchell, Mitchison, and others in the 1960s, immunologists investigating the sources of help for B-cell activation and antibody production have been conditioned to think in terms of T lymphocytes. However, we are gradually coming to appreciate that cells other than T lymphocytes may promote B-cell activation and survival. We now know that mesenchymal stromal cells and eosinophils cooperate in ensuring plasma cell survival in the bone marrow. Recently, in what can be considered a major shift in our thinking about aiding B-cell function, Puga and colleagues have demonstrated that marginal zone (MZ) B cells may receive help in activation, somatic hypermutation (SHM), class-switch recombination (CSR), and antibody secretion from an unexpected source: neutrophils.

Since many cells are capable of exerting effects in vitro that are not replicated when tested in vivo, we first note that the numbers and/or activity of MZ B cells are diminished in patients with reduced

numbers of neutrophils (neutropenic individuals) or with other neutrophil disorders. Such patients show reduced levels of IgM, IgG, and IgA antibodies to microbial TI antigens normally recognized by MZ B cells, whereas the amounts of these Igs with specificity for T-dependent (TD) antigens were normal. Furthermore, in cases where somatic hypermutation could be assessed, patients with neutrophil disorders showed less evidence of V-region mutations in MZ B cells than did healthy control subjects. These findings all suggest that neutrophils may play an important role in the survival and activation of MZ B cells in vivo, a suggestion borne out by the careful microscopic and in vitro culture experiments described by Puga et al.

Neutrophils were observed colonizing the regions of the spleen around the MZ, even in the absence of infection or inflammation. Although microscopic observation and flow cytometry failed to detect major morphological distinctions between the MZ splenic neutrophils and

those in the circulation, subsequent functional assays showed that the splenic neutrophils were able to induce IgM secretion by splenic MZ cells, even in the absence of prior neutrophil activation by antigen. Surprisingly, the splenic neutrophils were more effective in the induction of IgM secretion by MZ B cells than $CD4^+$ T cells (Figure 1). Culture of the

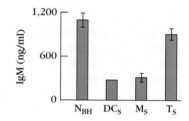

FIGURE 1

N_{BH}s enhance IgM secretion by MZ B cells even more effectively than do helper T cells. Splenic MZ B cells were cultured for 6 days with N_{BH} cells, splenic dendritic cells (DC_s), macrophages (M_s), or $CD4^+$ T cells (T_s). The concentration of IgM released into the culture supernatant was assessed by *enzyme-linked immunosorbent assay* (ELISA, see Chapter 20).

immune systems; many representatives of both cell types are located outside the classical secondary lymphoid tissues, and the repertoires of their antigen receptors are less diverse than those of other, more conventional adaptive immune lymphocyte subsets. Both γδ intraepithelial T cells and B-1 B cells are self-renewing in the periphery—their populations are maintained without the need to be continuously reseeded from bone-marrow-derived precursors. Finally, both of these subsets respond rapidly to antigen challenge with relatively low-affinity responses.

The existence of the B-1 B-cell subpopulation was first described in 1983, when the lab of Leonard and Leonore Herzenberg discovered a set of B cells bearing the **CD5 antigen**, whose expression had previously been thought to be restricted to T cells. These $CD5^+$ B cells were termed B-1 B cells to distinguish them from the conventional B-2 B-cell population. (The numbering reflects the order of appearance of the two B-cell subsets in ontogeny.)

In humans and mice, B-1 B cells make up only about 5% of B cells, although in some species such as rabbits and cattle,

B-1-like cells represent the major subset. However, even in humans and mice, B-1 cells predominate in the pleural and peritoneal cavities, and it is probable that their major function is to protect these body cavities from bacterial infection. Because of their priority in the B-cell developmental sequence, B-1 B cells are found in relatively high numbers in fetal and neonatal life. Cells having the functional characteristics of B-1 cells but lacking expression of the CD5 molecule were identified at a later stage and termed **B-1b B cells**.

Because the lack of T-cell involvement in their stimulation means that AID is never activated, B-1 B cells secrete antibodies of relatively low affinity that are primarily of the IgM class. Since B-1 B cells derive from a limited number of B-cell clones generated early in ontogeny, the antibodies they secrete are also significantly less diverse than the antibodies secreted by B-2 B cells. Antibodies secreted by B-1 B cells have evolved to recognize antigenic determinants expressed by gut and respiratory system bacteria and are primarily directed toward such common repeated antigens as phosphatidyl choline (a component of pneumococcal cell walls), lipopolysaccharide, splenic neutrophils, termed B helper neutrophils, or N_{BH} by the authors, with $CD4^+$ T cells resulted in *suppression* of T-cell proliferation following stimulation with anti-CD3 and IL-2. This shows that the neutrophils are not stimulating B cells indirectly by first activating T cells, and also allows for the possibility that the neutrophils may be selectively biasing splenic B cells toward T-independent responses.

Using qPCR, Puga and colleagues demonstrated the presence of *aicda* mRNA (encoding the AID protein) in MZ B cells. They further showed that culturing B cells in medium conditioned by the growth of N_{BH} cells in it (N_{BH}-conditioned medium) resulted in the up-regulation of *aicda* mRNA expression in those B cells. The amount of *aicda* mRNA in the MZ B cells was greater than in naïve B cells, but not as much as in B-2 B cells engaged in an active germinal center reaction. Culturing MZ B cells with N_{BH} cells, or with N_{BH}-conditioned medium, resulted in both SHM and CSR and confirmed that N_{BH}-derived help was sufficient to drive both processes in the MZ B cells (Figure 2).

What form does this neutrophil-derived help take? Is it purely in the form of soluble factors, or does cell-cell contact also play a role? The answer to this question appears to be *both*. Comparing the mRNA and protein expression, as well as the cell surface presence of a variety of helper B-cell factors between the circulating, inactive neutrophils (N_C) and the N_{BH}s, Puga et al. showed that the helper B-cell neutrophils produced higher quantities of soluble BAFF and APRIL, as well as of IL-21 than did the N_C

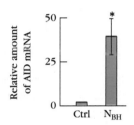

FIGURE 2

Quantitative RT PCR analysis of the expression of the *aicda* gene in MZ B cells cultured for 2 days in control medium (left) or in medium conditioned by N_{BH} cells (right). The relative amounts of mRNA encoding the AID protein were assessed by qPCR. Gene expression was normalized to the expression of pax5 mRNA. Culture of MZ B cells in N_{BH}-conditioned medium was shown to enhance the expression of the *aicda* gene by approximately 35-fold.

cells. They also expressed more BAFF on their membranes. Investigators reduced the amount of each of these cytokines that was available to B cells by adding soluble forms of the cytokine receptors to in vitro B cell cultures. The soluble receptors act by blocking B cell recognition. Under these conditions, IgM secretion, and CSR to IgG and IgA were both significantly reduced, thus demonstrating the relevance of the three cytokine signals to MZ B cell activation. The authors also note that N_{BH} cells express surface CD40L, although at lower levels than those found on $CD4^+$ T cells.

Thus, at least for MZ B cells, the role played by helper neutrophils in B-cell activation may be even more relevant than that played by helper T cells. Paradigm shifts such as this one enliven a field in delightful and stimulating ways.

Source: I. Puga, et al. 2011. B cell-helper neutrophils stimulate the diversification and production of immunoglobulin in the marginal zone of the spleen. *Nature Immunology* **13**:170–180.

and influenza viruses. B-1 B cells represent the majority component of the B-cell responses to TI antigens.

Very early in the history of immunology, investigators noted that the serum of unimmunized mice and humans contains so-called *natural IgM antibodies* that bind a broad spectrum of antigens with relatively low affinity. These antibodies derive mainly from B-1 B cells, and their presence in unimmunized animals suggests that B-1 B cells may exist in a partially activated state and constitutively secrete low levels of natural antibodies. In addition, a relatively high frequency of these IgM antibodies display autoimmune reactivities, although their affinities for self antigens are sufficiently low that they rarely induce disease. Low-level interactions with self antigens during development may therefore be important in the development of B-1 B-cell function and in the maintenance of their partially activated phenotype.

Although antibody secretion by B-1 B cells is not dependent on T-cell help, it can be enhanced in the presence of T-cell cytokines. Indeed, recent data suggests that, in the presence of T-cell help, B-1b cells may express certain attri-butes of B-2 cells, such as Ig class switching (with the resultant production of IgA antibodies), SHM, and the generation of a long-lasting antibody response.

Marginal Zone B Cells

The second class of B cells capable of responding to TI antigens is the **marginal zone (MZ)** subset, which resides in the marginal zone of the spleen (Figure 12-18). As for B-1 B cells, maintenance of physiological levels of MZ B cells appears to depend on their capacity to receive low-level signals through the BCR. Again, like B-1 B cells, MZ B cells have the capacity for self-renewal in the periphery and do not need to be constantly replenished from bone marrow precursors. MZ B cells derive originally from the transitional T2 B-cell population, and it has recently been shown that the Notch2 signaling system plays a role in sending B cells down the MZ pathway. This is notable because Notch signaling was previously thought to be restricted in its activity among lymphocytes to T cells. MZ B cells bear unusually high levels of CD21 (CR2), enabling

TABLE 12-4   Functional differences among mature B-cell subsets

Attribute	Conventional B2 B cells	B-1 B cells	Marginal zone B cells
Major sites	Secondary lymphoid organs	Pleural and peritoneal cavity; also spleen	Marginal zones of spleen in mice; primates also have MZ cells in other locations
V region diversity	Highly diverse	More restricted diversity	Moderate diversity
Rapidity of antibody response	Slow; plasma cells appear 7–10 days post stimulation	Rapid; plasma cells appear as early as 3 days after stimulation	Rapid; plasma cells appear as early as 3 days after stimulation
Surface IgD?	High levels of IgD	Low levels of IgD	Low levels of IgD
Somatic hypermutation	Yes	No	Yes in primates; possibly in rodents
Requirements for help from other cell types	Provided by T cells	No, although T and other cells can enhance response	Dendritic cells and neutrophils can enhance response (see Box 12-2)
Participate in germinal center reaction?	Yes	No	Possibly, although with slower kinetics than follicular B cells
Isotypes produced	All isotypes	Predominantly IgM	Predominantly IgM
Immunological memory	Yes	Very little	Yes; source of IgM-producing memory cells

them to bind very efficiently to antigens covalently conjugated to C3d or C3dg. Since one characteristic of TI-2 antigens is an enhanced tendency to bind C3d, MZ B cells are particularly important in the host protection against pathogens bearing TI-2 antigens.

MZ B cells are specialized to respond to blood-borne antigens that enter the immune system via the splenic MZ. Antigen stimulation of MZ B cells results in their movement from the MZ to the bridging channels and red pulp of the spleen, where they undergo a burst of proliferation, forming foci of plasmablasts not unlike the primary foci formed in the lymph nodes upon antigen challenge. These cells produce high levels of antigen-specific IgM within 3 to 4 days after antigenic stimulation. Very recent work has suggested that MZ B cells may be helped in antibody secretion, SHM, and CSR not by T cells (as we have come to expect), but instead by neutrophils, which we normally think of as participants in the innate arm, rather than the adaptive arm, of immunity. These experiments are described in Advances Box 12-2.

Some of the characteristics of MZ B cells, as well as their local environments, differ between rodents and primates. Of particular interest is the possibility that the variable regions

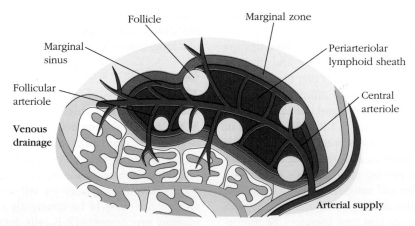

FIGURE 12-18 Longitudinal section through the mouse spleen, showing its blood supply and the location of lymphoid populations. The periarteriolar lymphoid sheath (PALS), containing the T-cell zone, can be seen surrounding the central arteriole, outside of which are the lymphoid follicles. Outside the follicles lies the marginal sinus, and the marginal zone separates the sinus from the red pulp of the spleen. The anatomy of the marginal regions varies somewhat between primate and rodent spleens. [Adapted from S. Pillai et al., 2005, Marginal Zone B Cells. Annual Review of Immunology 23:161–196, Figure 1.]

of antibodies derived from human MZ B cells may undergo SHM in the absence of an obvious germinal center reaction, and conceivably even in the absence of antigenic stimulation. Furthermore, in rodents, B cells with the characteristics of MZ cells are restricted to the MZs of the spleen, whereas in primates they can be found in other peripheral lymphoid tissues such as the tonsils.

Negative Regulation of B Cells

Up until this point, we have been discussing how B cells are activated, and the functional correlates of that activation. However, antigen stimulation of B cells results in a proliferative response that is as rapid as any observed in vertebrate organisms; an activated lymphocyte may divide once every 6 hours. Control mechanisms have therefore evolved to ensure that B-cell proliferation is slowed down once sufficient specific B cells have been generated and that most of the B cells enter into an apoptotic program once the pathogen has been eliminated. In this section, we will address negative regulation of B-cell activation that is mediated through two different molecules on the B-cell surface.

Negative Signaling through CD22 Shuts Down Unnecessary BCR Signaling

In addition to CD19/CD21 and CD81, the BCR of resting B cells is also associated with an additional transmembrane molecule, CD22. CD22 bears an **Immunoreceptor Tyrosine-based Inhibitory Motif (ITIM)**, similar in structure to the ITAM motifs introduced in Chapter 3, but mediating inhibitory, rather than activating functions. Activation of B cells results in phosphorylation of the ITIM, thus allowing association of the SHP-1 tyrosine phosphatase with the cytoplasmic tail of CD22. SHP-1 can then strip activating phosphates from the tyrosine of neighboring signaling complexes.

For as long as the BCR signaling pathway is being activated by antigen engagement, phosphate groups are reattached to the tyrosine residues of adapter molecules and other signaling intermediates as fast as the phosphatases can strip them off. However, once antigen levels begin to decrease, receptor-associated tyrosine kinase activity slows down, and signaling through CD22 can then induce the removal of any residual activating phosphates. CD22 thus functions as a negative regulator of B-cell activation, and its presence and activity ensure that signaling from the BCR is shut down when antigen is no longer bound to the BCR. Consistent with this negative feedback role, levels of B-cell activation are elevated in CD22 knockout mice, and aging CD22 knockout animals have increased levels of autoimmunity.

CD22 is a cell-surface receptor molecule that recognizes N-glycolyl neuraminic acid residues on serum glycoproteins and other cell surfaces and can thus double as an adhesion molecule. It is expressed in mature B cells that bear both mIgM and mIgD Ig receptors.

Negative Signaling through the FcγRIIb Receptor Inhibits B-Cell Activation

It has long been known that the presence of circulating, specific, antigen-IgG complexes is inhibitory for further B-cell activation, and this phenomenon has now been explained at the molecular level by the characterization of the **FcγRIIb** receptor (also known as CD32). FcγRIIb recognizes immune complexes containing IgG and, like CD22, bears a cytoplasmic ITIM domain. Co-ligation of the B cell's FcγRIIb receptor molecules with the BCR by a specific antigen-antibody immune complex results in activation of the FcγRIIb signaling cascade, and phosphorylation of FcγRIIb's ITIM. The phosphorylated ITIM serves as a docking site for the inositol phosphatase SHIP, which binds to the ITIM via its SH2 domain. SHIP then hydrolyzes PIP_3 to PIP_2, thus interfering with the membrane localization of the important signaling molecules Btk and PLCγ2 (see Chapter 3) and causing the effective abrogation of B-cell signaling. Signaling through FcγRIIb also results in decreased phosphorylation of CD19 and reduced recruitment of PI3 kinase to the membrane.

Negative signaling through the FcγRIIb receptor makes intuitive sense, as the presence of immune complexes containing the antigen for which a B cell is specific signals the presence of high levels of antigen-specific antibody, and hence a reduced need for further B-cell differentiation.

B-10 B Cells Act as Negative Regulators by Secreting IL-10

Recently, an unusual population of B cells has been discovered that appears to be capable of negatively regulating potentially inflammatory immune responses by secreting the cytokine IL-10 upon stimulation.

Working with two mouse autoimmune models, investigators demonstrated that B cells capable of secreting the immunoregulatory cytokine IL-10 could alleviate the symptoms of a mouse suffering from a form of the antibody-mediated autoimmune disease multiple sclerosis. Recall from Chapter 11 that IL-10 is a cytokine normally associated with regulatory T cells. It has pleiotropic effects on other immune system cells, which include the suppression of T-cell production of the cytokines IL-2, IL-5, and TNF-α. Furthermore, IL-10 interacts with antigen-presenting cells in such a manner as to reduce the cell surface expression of MHC antigens. The finding that B cells could be capable of secreting this immunoregulatory cytokine represents the first indication that they, as distinct from T cells, might have the capacity to down-regulate the function of other immune system cells. A small population of splenic B cells appears to account for almost all of the B-cell-derived IL-10. However, at this point, we do not know whether this IL-10 secreting B-cell population truly represents a single developmental B-cell lineage. For example, B cells producing IL-10 have been identified among both B-1 and B-2

B-cell populations. In addition, some, but not all B cells producing IL-10 bear markers typical of the transitional T2 B-cell subset.

Importantly, all B-10 B cells appear to demonstrate the capacity to secrete a diverse repertoire of antibodies, with specificities for both foreign and auto-antigens. It is thought that the function of these cells may be to limit and control inflammation during the course of an ongoing immune response. A great deal of work is still required to tease out the lineage relationships among these various IL-10-secreting and other B-cell subpopulations.

SUMMARY

- The Clonal Selection Hypothesis states that each B cell bears a single antigen receptor. When antigen interacts with a B-cell receptor (BCR) specific for that antigen, it induces proliferation of that B cell to form a clone of cells bearing the identical specificity. Cells from this clone secrete antibodies having the same specificity as the antigen receptor. At the close of the immune response, there will be more cells bearing that specificity than existed prior to the response, and these so-called "memory B cells" will generate a faster and higher response upon secondary antigen exposure. Those cells bearing receptors with specificity for self antigens are eliminated from the B-cell repertoire during development.

- The B-cell response to some antigens requires help from T cells. These T-dependent responses are mounted by B-2 or follicular B cells. The response to some other antigens can occur in the absence of T-cell help. These T-independent responses are primarily mounted by B-1 or marginal zone (MZ) B cells.

- Mature B-2 B cells migrate to the lymphoid follicles under the influence of the chemokine CXCL13. Survival of mature B-2 B cells depends upon access to the survival factor BAFF.

- B cells may acquire antigens directly. Some antigens enter the lymph node by squeezing between the subcapsular sinus (SCS) macrophages that line the lymph node, whereas others enter via a leaky network of conduits that are sampled by the follicular B cells. Yet others are taken up first by the SCS macrophages or follicular dendritic cells and passed on to the B cells.

- When the BCR recognizes its cognate antigen, the receptors on the B-cell membrane briefly spread over the antigen surface, and then contract, resulting in B-cell receptor clustering. This represents the earliest phase of B-cell activation.

- Receptor clustering results in internalization of the receptor-antigen complex followed by antigen presentation by the B cell to the T cell. The T-cell/B-cell interaction occurs primarily at the border between the T- and B-cell zones of the lymph node.

- T cells interact with their cognate B cells by binding to the processed antigen with their T-cell receptor (TCR), as well as by interactions between T cell CD28 with B cell CD80 and CD86, and between T cell CD40L and B cell CD40. These interactions facilitate the directional secretion of T-cell cytokines that are necessary for full B-cell activation.

- Following stimulation of primary B cells at the T-cell/B-cell border within the lymph node, some B cells differentiate quickly into plasma cells that form primary foci and secrete an initial wave of IgM antibodies. This requires the up-regulation of the plasma cell transcription factors IRF4 and BLIMP-1.

- Other B cells from the antigen-stimulate clones migrate to the primary follicles and form germinal centers.

- Within germinal centers, B-cell differentiation continues, with the generation of somatically hypermutated receptors that are then subject to antigen selection, with the eventual generation of high-affinity antibodies.

- Somatic hypermutation (SHM) affects certain sequences called mutational hot spots, in the variable region genes of antibody molecules.

- Also within the germinal center, the constant region of the heavy chain of the antibody genes undergoes class switch recombination (CSR). This results in the formation of antibodies bearing mutated variable regions and constant regions other than μ. CSR occurs between switch regions upstream of each heavy chain constant region gene (except for the δ constant region).

- Both somatic hypermutation and class switch recombination are mediated by the enzyme activation-induced cytidine deaminase (AID), followed by DNA repair.

- At the close of the B-cell response, long-term memory exists in two forms. Recirculating B memory cells must be reactivated by antigen in order to yield a higher, faster, and stronger response than the primary response. In addition, long-lived plasma cells residing in the bone marrow and other locations continually secrete antibodies and ensure that antibodies to commonly encountered antigens constantly circulate within the blood.

- T-independent responses generated by B-1 and marginal zone (MZ) B cells give rise to relatively low-affinity, primarily IgM antibodies.

- TI-1 antigens interact with B cells via both the BCR and innate immune receptors, whereas TI-2 receptors are highly polymerized antigens that do not have an intrinsic, mitogenic activity. Both types of T-independent responses are enhanced by interactions with other cell types, including T cells, macrophages, and monocytes, and possibly, neutrophils.

- B-10 B cells release the interleukin IL-10 upon antigenic stimulation, and may serve to reduce inflammation during an ongoing immune response.

- B cells can be negatively signaled through CD22 and FcγRIIb—a receptor that recognizes the presence of IgG-containing immune complexes in the blood.

REFERENCES

Ada, G. L., and P. Byrt. 1969. Specific inactivation of antigen-reactive cells with 125I-labelled antigen. *Nature* **222**:1291–1292.

Allen, C. D., T. Okada, and J. G. Cyster. 2007. Germinal-center organization and cellular dynamics. *Immunity* **27**:190–202.

Allen, C. D., T. Okada, H. L. Tang, and J. G. Cyster. 2007. Imaging of germinal center selection events during affinity maturation. *Science* **315**:528–531.

Berek, C., G. M. Griffiths, and C. Milstein. 1985. Molecular events during maturation of the immune response to oxazolone. *Nature* **316**:412–418.

Burnet, F. M. 1957. A modification of Jerne's theory of antibody production using the concept of clonal selection. *The Australian Journal of Science* **20**:67–69.

Catron, D. M., et al. 2004. Visualizing the first 50 hr of the primary immune response to a soluble antigen. *Immunity* **21**:341–347.

Dal Porto, J. M., et al. 2004. B cell antigen receptor signaling 101. *Molecular Immunology* **41**:599–613.

Dykstra, M., et al.2003. Location is everything: Lipid rafts and immune cell signaling. *Annual Review of Immunology* **21**:457–481.

Eisen, H. N., and G. W. Siskind. 1964. Variations in affinities of antibodies during the immune response. *Biochemistry* **3**:996–1008.

Flajnik, M. F., and M. Kasahara. 2010. Origin and evolution of the adaptive immune system: Genetic events and selective pressures. *Nature Reviews Genetics* **11**:47–59.

Fleire, S. J., et al. 2006. B cell ligand discrimination through a spreading and contraction response. *Science* **312**:738–741.

Germain, R. N., et al. 2008. Making friends in out-of-the-way places: How cells of the immune system get together and how they conduct their business as revealed by intravital imaging. *Immunological Reviews* **221**:163–181.

Haas, K. M., J. C. Poe, D. A. Steeber, and T. F. Tedder, T.F. 2005. B-1a and B-1b cells exhibit distinct developmental requirements and have unique functional roles in innate and adaptive immunity to S. pneumoniae. *Immunity* **23**:7–18.

Hannum, L. G., A. M. Haberman, S. M. Anderson, and M. J. Shlomchik. 2000. Germinal center initiation, variable gene region hypermutation, and mutant B cell selection without detectable immune complexes on follicular dendritic cells. *Journal of Experimental Medicine* **192**:931–942.

Harwood, N. E., and F. D. Batista. 2009. The antigen expressway: Follicular conduits carry antigen to B cells. *Immunity* **30**:177–179.

Hayakawa, K., R. R. Hardy, D. R. Parks, and L. A. Herzenberg. 1983. The "Ly-1 B" cell subpopulation in normal immunodefective, and autoimmune mice. *Journal of Experimental Medicine* **157**:202–218.

Jacob, J., and G. Kelsoe. 1992. In situ studies of the primary immune response to (4-hydroxy-3-nitrophenyl)acetyl. II. A common clonal origin for periarteriolar lymphoid sheath-associated foci and germinal centers. *Journal of Experimental Medicine* **176**:679–687.

Jacob, J., J. Przylepa, C. Miller, and G. Kelsoe. 1993. In situ studies of the primary immune response to (4-hydroxy-3-nitrophenyl)acetyl. III. The kinetics of V region mutation and selection in germinal center B cells. *Journal of Experimental Medicine* **178**:1293–1307.

Kohler, G., and C. Milstein. 1975. Continuous cultures of fused cells secreting antibody of predefined specificity. *Nature* **256**:495–497.

Koshland, M. E. 1975. Structure and function of the J chain. *Advances in Immunology* **20**:41–69.

LeBien, T. W., and T. F. Tedder. 2008. B lymphocytes: How they develop and function. *Blood* **112**:1570–1580.

Martin, F., and J. F. Kearney. 2002. Marginal-zone B cells. *Nature Reviews Immunology* **2**:323–335.

Mitchell, G. F., and J. F. Miller. 1968. Cell to cell interaction in the immune response. II. The source of hemolysin-forming cells in irradiated mice given bone marrow and thymus or thoracic duct lymphocytes. *Journal of Experimental Medicine* **128**:821–837.

Muramatsu, M., et al. 2007. Discovery of activation-induced cytidine deaminase, the engraver of antibody memory. *Advances in Immunology* **94**:1–36.

Pape, K. A., D. M. Catron, A. A. Itano, and M. K. Jenkins. 2007. The humoral immune response is initiated in lymph nodes by B cells that acquire soluble antigen directly in the follicles. *Immunity* **26**:491–502.

Raff, M. C., M. Feldmann, and S. De Petris. 1973. Monospecificity of bone marrow-derived lymphocytes. *Journal of Experimental Medicine* **137**:1024–1030.

Schneider, P. 2005. The role of APRIL and BAFF in lymphocyte activation. *Current Opinion in Immunology* **17**:282–289.

Shapiro-Shelef, M., and K. Calame. 2005. Regulation of plasma-cell development. *Nature Reviews Immunology* 5:230–242.

Sixt, M., et al. 2005. The conduit system transports soluble antigens from the afferent lymph to resident dendritic cells in the T cell area of the lymph node. *Immunity* 22:19–29.

Weigert, M. G., I. M. Cesari, S. J. Yonkovich, and M. Cohn. 1970. Variability in the lambda light chain sequences of mouse antibody. *Nature* 228:1045–1047.

Weill, J. C., S. Weller, and C. A. Reynaud. 2009. Human marginal zone B cells. *Annual Review of Immunology* 27:267–285.

Useful Websites

http://bio-alive.com/seminars/immunology. htm This Bio Alive Web site is an excellent source of lectures by accomplished immunologists that can be downloaded for personal use. The series contains lectures pertinent to this chapter, including seminars entitled *Immunological Memory*; *Regulating B Cell Immunity*; *Moviemaking and Modeling* (this one by Ronald Germain, one of the pioneers of the application of sophisticated imaging technology to the study of the immune system); *Molecular Mechanisms of Leukocyte Migration*; and *Somatic Hypermutation*.

www.sciencedirect.com/science/article/pii/ S1074761304002389 Two movies from a paper by D. M. Catron et al. (2004, Visualizing the first 50 hr of the primary immune response to a soluble antigen, *Immunity* 21:341–347) that is discussed in Chapter 14 can be found at this Web site. The first portrays the movements of T cells and dendritic cells prior to antigen entry into the observed lymph node. The second shows the movements of T cells, B cells, and dendritic cells through a nearby lymph node, following injection of antigen into the footpad of a mouse.

STUDY QUESTIONS

1. Name one distinguishing feature of TI-1 and TI-2 antigens.

2. In the following flow cytometric dot plots, draw a circle where you expect to see the designated cell populations.

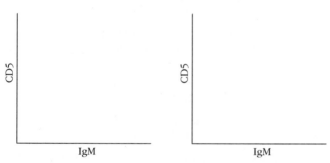

(a) B2 (follicular) B cells   (b) B-1 B cells

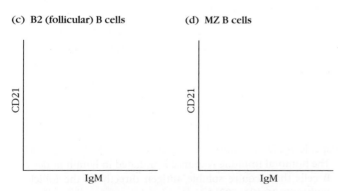

(c) B2 (follicular) B cells   (d) MZ B cells

3. You have generated a knockout mouse that does not express the CD40 antigen. On stimulation with a T-dependent antigen, do you expect this mouse to be able to:

   a. secrete IgG antibodies specific for the antigen?
   b. secrete antigen-specific antibodies that have undergone somatic hypermutation?

4. You are looking at lymph node sections of a mouse that has had its Lymphotoxin-α gene disrupted, and you are comparing it with sections from a mouse with an intact Lymphotoxin-α gene. Both mice were immunized approximately 10 days previously with a T-dependent antigen. What is the most striking feature that is missing in your knockout animal?

5. You have immunized two mice with a T-dependent antigen. One of them is a wild-type mouse, and the other belongs to the same strain as the first but is a knockout mouse that does not have the gene encoding activation-induced cytidine deaminase (AID). Upon primary immunization, both mice express similar titer (concentrations) of IgM antibodies. You allow the mice to rest for 6 weeks and then reimmunize. Do you expect the secondary antibodies to be similar or different between the two animals? Explain your answer.

6. Draw the biochemical reaction catalyzed by AID, and describe how subsequent DNA repair mechanisms can lead to somatic hypermutation (SHM) and to class switch recombination (CSR).

7. Describe one mechanism by which higher-affinity B cells may gain a selective survival advantage within the germinal center.

8. The presence of circulating antigen-antibody complexes has been shown to result in the down-regulation of B-cell activation. State why this would be advantageous to the organism, and offer one mechanism by which the down-regulation may occur.

9. Recent evidence suggests that not all regulatory cytokines secreted by lymphocytes derive from T cells. Explain.

10. Describe three mechanisms by which antigen can enter the lymph node and make contact with the B-cell receptor.

Effector Responses: Cell- and Antibody-Mediated Immunity

I n previous chapters, we have focused on how the immune response is initiated. Here, we will finally describe how the targets of the immune system—pathogen, infected cells, and even tumor cells—are actually cleared from the body.

We have already seen how the innate immune system initiates the response to pathogen and alerts the adaptive immune system to the presence and nature of that pathogen (Chapter 5). In Chapters 9 through 12, we described the development and activation of the antigen-specific cells of the adaptive immune system, B and T lymphocytes. You were also introduced to the differentiation and activity of helper T lymphocytes, a type of effector cell that regulates the activity and function of cytotoxic T cells, B cells, and other antigen-presenting cells. Here we focus on the effector cells and molecules of both the cell-mediated and the humoral (antibody-mediated) immune responses that directly rid the organism of pathogens and abnormal cells. These effector responses are arguably the most important manifestations of the immune response: they protect the host from infection and rid the host of pathogens that have breached defenses.

The effector functions of cell-mediated and humoral branches of the immune system assume different, although overlapping, roles in clearing infection from a host. The effector molecules of the humoral branch are antibodies, the secreted version of the highly specific receptor on the surface of B cells. Antibodies secreted into extracellular spaces are exquisitely antigen specific and have several methods at their disposal to rid a body of pathogen. How an antibody contributes to clearing infection depends on its isotype, which determines whether it can recruit complement (recall Chapter 6). The isotype also determines which receptors an antibody can bind. Antibody-binding receptors, which bind to the constant regions of antibodies and are therefore called Fc receptors or FcRs, determine which cells an antibody can recruit to aid in its destructive mission, as well as the tissues to which it can gain entry.

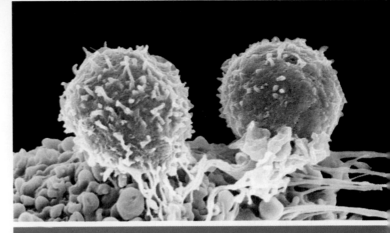

Two cytotoxic T lymphocytes bind to a tumor-specific antigen on the surface of a cancer cell, deliver the "kiss of death," and induce apoptosis. *[Steve Gschmeissner/Photo Researchers]*

- Antibody-Mediated Effector Functions
- Cell-Mediated Effector Responses
- Experimental Assessment of Cell-Mediated Cytotoxicity

If antibodies were the only agents of immunity, intracellular pathogens, which occupy spaces that antibody cannot access, would likely escape the immune system. Fortunately there is another effector branch of our immune system, cell-mediated immunity, which detects and kills cells that harbor intracellular pathogens. Cell-mediated immunity consists of both helper T cells (CD4$^+$ T$_H$) and several types of cytotoxic cells. As you have seen in Chapter 11, T$_H$ cells exert their effector functions indirectly, by contributing to the activation of antigen-presenting cells, B cells, and cytotoxic T cells via receptor-ligand interactions and soluble cytokines and chemokines. On the other hand, cytotoxic cells exert their effector functions directly, by attacking infected cells and, in some cases, the pathogens themselves.

Effector cytotoxic cells arise from both the adaptive and innate immune systems and, therefore, include both antigen-specific and -nonspecific cells. Antigen-nonspecific (innate immune) cells that contribute to the clearance of infected cells include NK cells and nonlymphoid cell types such as macrophages, neutrophils, and eosinophils. Antigen-specific cytotoxic cells include $CD8^+$ *T lymphocytes* (CTLs or T_C cells), as well as the $CD4^+$ NKT cell subpopulation, which, although derived from the T-cell lineage, displays some useful features of innate immune cell types, too. Populations of cytotoxic $CD4^+$ T_H cells have also been described and may contribute to delayed-type hypersensitivity. The discussion of DTH reactions and the role of $CD4^+$ T cells in their orchestration appears in Chapter 15.

The humoral and cell-mediated immune systems also cooperate effectively. Cells such as macrophages, NK cells, neutrophils, and eosinophils all express Fc receptors, which induce phagocytosis of antibody-antigen complexes as well as direct killing of target cells via a process known as antibody-dependent cell-mediated cytotoxicity.

It is also important to note that the cell-mediated immune system plays a role in recognizing and eliminating not only infected cells, but tumor cells, which often have undergone genetic modifications that lead to surface expression of antigens not typical of normal cells.

In this chapter we will first focus on antibody-mediated activities, describing not only their ability to fix complement but also the multiple functions they acquire by interacting with cells expressing Fc receptors. We will then describe cytotoxic effector mechanisms mediated by cells of the innate and adaptive immune system. We will close with a discussion of experimental assays of cytotoxicity.

A Clinical Focus feature and a Classic Experiment feature are also included in this chapter. The first describes how knowledge of antibody structure and function has led to new and successful disease therapies. The second introduces recent data showing that lymphocytes are not the only cell type that can develop a memory; natural killer cells, members of the innate immune system, also appear to have this capacity.

Antibody-Mediated Effector Functions

Antibodies generated by activated B cells (see Chapter 12) protect the body against the pathogen in several ways: they can **neutralize** pathogen by binding to and blocking receptors that the pathogen uses to gain entry into a cell; they can

opsonize pathogen by binding and recruiting phagocytic cells; and they can fix **complement** by binding to pathogen and initiating the complement cascade, which punctures cell membranes (see Chapter 6). In addition, they can cooperate with the cellular branch of the immune system by binding and recruiting the activities of cytotoxic cells, specifically natural killer (NK) cells, in a process called **antibody-dependent cell-mediated cytotoxicity** (ADCC) (Figure 13-1 and Table 13-1). The ability of antibodies to mediate these responses is dependent not only on their antigen specificity, but also on their isotype (see Chapter 3 and Table 13-2), which determines whether an antibody will "fix" complement, to which Fc receptors they will bind, and therefore, which cellular effects they will have. *Collectively, these mechanisms result in the destruction of pathogen, either directly or indirectly, by stimulating phagocytosis and digestion of pathogen-antibody complexes, directly destroying pathogenic cell membranes, or inducing suicide of the infected cell.*

Antibodies Mediate the Clearance and Destruction of Pathogen in a Variety of Ways

Antibodies are versatile effector molecules that play a direct role in resolving infection by (1) blocking pathogen entry into cells (neutralization) and (2) recruiting cytotoxic molecules and cytotoxic cells to kill pathogen (via complement fixation, opsonization, and ADCC).

Neutralization

Most viruses and some bacteria gain entry into a cell by binding specifically to one or more cell-surface proteins, stimulating endocytosis. For example, the *human immunodeficiency virus* (HIV) expresses a coat protein, gp120, that specifically binds the CD4 coreceptor. The influenza virus expresses a protein, hemagglutinin, that binds to sugar-modified proteins on epithelial cells. Antibodies that bind such proteins are particularly potent effector molecules because they can prevent a pathogen from ever initiating an infection. They are referred to as **neutralizing antibodies**. Pathogens disarmed by neutralizing antibodies are typically phagocytosed by macrophages.

Neutralizing antibodies can also block entry of toxins into cells. For example, tetanus toxoid, a product of the *Clostridium tetani* bacteria, is a neurotoxin that can result in uncontrolled muscle contraction and death. The tetanus vaccine contains an inactive version of this toxin that stimulates B-cell production of anti-tetanus toxoid antibodies. These antibodies bind and potently inhibit the entry of the toxin into nerve cells. Neutralizing antibodies have also been raised against snake venom toxins and are effective therapies for some snakebites.

Physiologically, neutralization is the most effective mode of protection against infection. However, because microbes proliferate rapidly, they can generate genetic

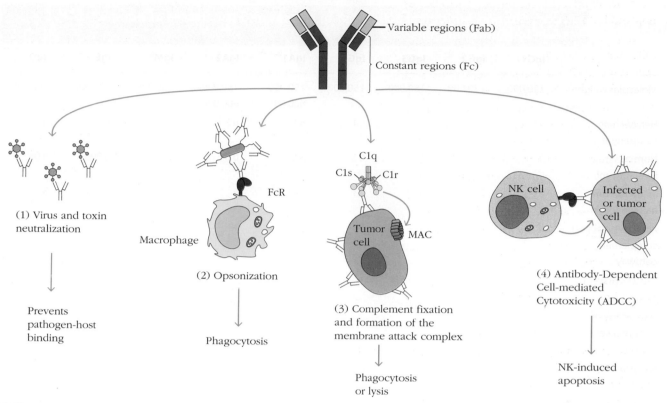

FIGURE 13-1 Four broad categories of antibody effector functions. Antibody binding can enhance the clearance of pathogens by (1) binding to proteins that pathogens use to infect cells (neutralization), (2) interacting with Fc receptors on the surface of phagocytic cells and inducing internalization and degradation (opso-nization), (3) recruiting complement proteins that can directly kill pathogen or enhance its phagocytosis, and (4) binding to FcRs on cytotoxic cells (e.g., NK cells) and directing their activity to infected cells, tumor cells, and/or pathogens (antibody-dependent cell-medi-ated cytotoxicity [ADCC]).

variants able to evade neutralizing antibodies. For example, influenza viral variants expressing modified hemaggluti-nins that do not bind circulating antibodies have a distinct reproductive advantage and can rapidly take over an infec-tion. HIV is one of the most adept viruses at evading anti-

body responses. In theory, gp120 is an ideal target for neutralizing antibodies; however, few individuals generate such antibodies. This is in part because the region of gp120 that binds CD4 is concealed from the body until just before viral entry.

TABLE 13-1 Ways antibodies can protect the host against invasion by pathogens

Mode of antibody-mediated protection	How it works
Neutralization	Antibody binds to sites on the pathogen or a toxin that interact with host proteins, masking them, and inhibiting entry of that pathogen or toxin into the host. Antibody-pathogen com-plexes are then eliminated, often after phagocytosis.
Opsonization	Antibody binds to pathogen and is then bound by Fc receptors on phagocytic cells. The antibody-antigen binding to FcRs induces internalization and destruction by the phagocytic cell.
Complement fixation	The antibody-antigen complex becomes bound by complement components in serum and is either phagocytosed via cells expressing C3 receptors or lysed as a result of pore formation by the complement components C7, C8, and C9.
Antibody-dependent cell-mediated cytotoxicity	Antibody-antigen complexes are bound by Fc receptors on NK cells and granulocytes, thus directing the cytotoxicity of these cells toward the antigen targeted by the antibody (e.g., viral proteins on the surface of an infected cell), and inducing apoptosis of the target cell.

TABLE 13-2	Properties and biological activities* of classes and subclasses of human serum immunoglobulins								
	IgG1	**IgG2**	**IgG3**	**IgG4[†]**	**IgA1**	**IgA2**	**IgM[§]**	**IgE**	**IgD**
Molecular weight[‡]	150,000	150,000	150,000	150,000	150,000–600,000	150,000–600,000	900,000	190,000	150,000
Heavy-chain component	$\gamma1$	$\gamma2$	$\gamma3$	$\gamma4$	$\alpha1$	$\alpha2$	μ	ε	δ
Normal serum level (mg/ml)	9	3	1	0.5	3.0	0.5	1.5	0.0003	0.03
In vivo serum half-life (days)	23	23	8	23	6	6	5	2.5	3
Activates classical complement pathway	+	+/−	++	−	−	−	++	−	−
Crosses placenta	+	+/−	+	+	−	−	−	−	−
Present on membrane of mature B cells	−	−	−	−	−	−	+	−	+
Binds to Fc receptors of phagocytes	++	+/−	++	+	−	−	?	−	−
Mucosal transport	−	−	−	−	++	++	+	−	−
Induces mast cell degranulation	−	−	−	−	−	−	−	+	−

*Activity levels are indicated as follows: ++ = high; + = moderate; +/− = minimal; − = none; ? = questionable.
[†]Note that mice do not make IgG4 but generate two different versions of IgG2 (IgG2a and IgG2b).
[‡]IgG, IgE, and IgD always exist as monomers; IgA can exist as a monomer, dimer, trimer, or tetramer. Membrane-bound IgM is a monomer, but secreted IgM in serum is a pentamer.
[§]IgM is the first isotype produced by the neonate and during a primary immune response.

Opsonization

From the Greek word for "to make tasty," **opsonization** refers to the ability of antibodies to promote and/or enhance the engulfment of antigens by phagocytes. Opsonizing antibodies coat antigen and interact with Fc receptors expressed by phagocytic cells.

Thus, phagocytic cells of the innate immune system (e.g., macrophages and neutrophils) have two sets of receptors capable of binding pathogens: they can bind them directly using innate immune receptors (pattern recognition receptors such as Toll-like receptors), or they can bind antigens bound in immune complexes with antibodies, via Fc receptors.

Innate immune receptors are important in activating antigen-presenting cells and shaping the signals that they will convey to lymphocytes (see Chapter 5 and Chapter 11). Fc receptors also generate signals and regulate the effector responses of innate immune cells (see below). In the case of opsonization, binding of pathogen (antigen)-antibody complexes to an Fc receptor on phagocytes will induce internal-ization of the complex and internal digestion of the pathogen in lysosomes.

Complement Fixation

As we saw in Chapter 6, antigen-antibody complexes can also induce a complement cascade. Specifically, when antibodies that associate with complement bind to the surface of bacteria and some (enveloped) viruses, they can initiate a cascade of reactions that results in the generation of the *membrane attack complex* (MAC), perforating pathogen membranes and killing the microbe. This capacity to stimulate complement-mediated membrane damage is referred to as *fixing complement* and is a property of specific antibody isotypes, including some IgGs and IgM.

Activation of the early part of the complement cascade can also protect the host by opsonizing pathogens in an antibody-independent manner. This complement function does not require the later components (C4–C9) of the complement cascade. Instead, the complement protein fragment C3b, which is produced early on in the complement reaction, binds to the surface of the pathogen and is recognized by the many different blood cells that have C3b receptors. Binding of C3b on the surface of a pathogen by a macrophage results in phagocytosis and destruction of that pathogen by mechanisms similar to those described above for antibody-mediated phagocytosis. Interestingly, binding of C3b-bound pathogen by red blood cells results in the delivery of the pathogens to the liver or spleen, where the pathogen is removed from the red blood cell, without affecting the viability of the red blood cell, followed by phagocytosis of the pathogen by a macrophage.

Antibody-Dependent Cell-Mediated Cytotoxicity

Antibodies can recruit the activities of multiple cytotoxic cells, including NK cells (lymphoid members of the innate immune system) and granulocytes (myeloid members of the innate immune system). Unlike cytotoxic T lymphocytes, these cells do not express antigen-specific T-cell receptors. However, they do express Fc receptors (specifically, FcγRIII) and can use these to "arm" themselves with antibodies that allow them to "adopt" an antigen specificity. Antibodies bound to NK cells then direct the cytotoxic activity to a specific cellular target—for example, an infected cell that expresses viral proteins on its surface or a tumor cell that expresses tumor-specific proteins on its surface (see Figure 13-1). This phenomenon, referred to as ADCC, is under intense investigation because of its therapeutic potential. Monoclonal antibodies (mAbs) developed as cancer treatments are thought to act in part by taking advantage of ADCC mechanisms and directing NK cytotoxicity toward tumor cell targets (see Clinical Focus Box 13-1). As with complement fixation, the ability to mediate ADCC is antibody-isotype dependent, and murine IgG2a and human IgG1 are most potent in this capacity (see below).

Antibody Isotypes Mediate Different Effector Functions

As you learned in Chapters 3 and 12, activated B cells develop into plasma cells, which are remarkably productive antibody-producing cellular factories. Antibody specificity is defined by their Fab regions, and their isotype or class is defined by their Fc region (see Figure 3-21). Depending on the type of T cell help they received during activation, as well as the cytokines to which they were exposed, B cells will secrete one of four major classes of antibody: IgM, IgG, IgA, or IgE. Although immature B cells express IgD, little if any is secreted and its function remains a bit of a mystery. The structure of each of these antibodies is reviewed in detail in Chapter 3. Each plays a distinct role in thwarting and clearing infection, as described below and depicted in Figure 13-2.

Effector Functions of IgM Antibodies

IgM antibodies are the first class of antibodies to be produced during a primary immune response. They tend to be low-affinity antibodies because the B cells that produce them have not gone through affinity maturation. However, because they are often pentavalent (bound together in groups of five) they can be very effective in binding pathogen. Interestingly, most circulating IgM antibodies come not from conventional B cells but from B-1 cells, which constitutively produce IgM antibodies that appear to protect us from common pathogens, particularly those that can come from the gut and other mucosal areas.

IgM antibodies are very good at fixing complement and efficiently induce lysis of pathogens to which they bind. They are also good at forming dense antibody-pathogen complexes that are efficiently engulfed by macrophages.

Effector Functions of IgG Antibodies

IgG antibodies are the most common antibody isotype in the serum. They are also the most diverse, and include several subclasses (IgG1, IgG2, IgG3, and IgG4 in humans; IgG1, IgG2a, IgG2b, and IgG3 in mice), each of which has distinct effector capabilities. All of the IgG variants bind to Fc receptors and can enhance phagocytosis by macrophages (opsonization). Human IgG1 and IgG3 are particularly good at fixing complement. Mouse IgG2a and human IgG1 are particularly good at mediating ADCC by NK cells.

Specific IgG isotypes are also associated with disease pathologies. For example, in a mouse lupus model, IgG2a and IgG2b isotypes dominate. In human lupus, IgG1 and IgG3 are involved in anti-DNA reactions, and IgG2 and IgG3 are associated with kidney dysfunction. Investigators are making excellent use of our growing knowledge of IgG effector functions and are tailoring the isotypes of mAbs used in therapies to enhance their effect. For example, IgG1 antibodies are most commonly used in tumor therapy because they can both fix complement and mediate ADCC by NK cells. However, recent data suggest that IgG2 is a potent mediator of ADCC by myeloid cells (including neutrophils) and may offer some therapeutic advantages.

Effector Functions of IgA Antibodies

Although IgA antibodies are also found in circulation, they are the major isotype found in secretions, including mucus in the gut, milk from mammary glands, tears, and saliva. In these secretions, IgA can neutralize both toxins and pathogens, continually interacting with the resident (commensal) bacteria that colonize our mucosal surfaces and preventing them from entering the bloodstream. Because IgA cannot fix complement, these interactions do not induce inflammation, which is advantageous given that IgA acts continually on antigens and pathogens that typically pose no threat.

IgA's half-life in secretions is relatively long because the amino acid sequence of the Fc region is resistant to many of the proteases that are present. Interestingly, several microbes, including those that cause gonorrhea and strep infections, produce proteases that do degrade IgA, allowing them to evade this form of immune protection.

IgA does exist in monomeric form (particularly in circulation) but also forms dimers and polymers in mucosal tissues. The dimeric and polymeric forms bind to receptors

Monoclonal Antibodies in the Treatment of Cancer

Cancer is a disease in which a single cell multiplies rapidly and uncontrollably to form a tumor. As the tumor cells replicate, the DNA repair functions that normally ensure faithful replication of DNA sequence during cell division may begin to fail, and new mutations begin to accumulate, leading to a worsening prognosis. Cancer remains one of the worst killers in the developed world, and despite decades of intensive and creative work by research scientists and clinicians, a diagnosis of cancer is still followed all too frequently by a shortened life span, preceded by a course of weakening and toxic treatment.

Attempts to develop effective anticancer drug strategies have had to address two main problems. First, because many cancer cells are derived from self and do not "look" too different from healthy cells, any drug that binds to the cancer cell is likely also to bind and enter the patient's other, healthy cells. The trick is to find a molecule on the surface of a cancer cell that is either not expressed in normal cells, or is expressed at much lower concentrations in normal cells compared to cancer cells, and then to design a drug that will preferentially bind to the malignant cell. Second, because cancer cells and normal dividing cells use the same metabolic strategies to generate ATP, any untargeted drug that affects the metabolism or growth of a cancer cell will also affect normal dividing cells.

Most of the first generation of anticancer drugs, however, are focused on inhibiting cancer cell division. Drugs such as daunorubicin, cis-platin, and others enter all cells of the body and interfere with the process of cell division. As mentioned above, such drugs will also target cells that divide normally as part of their day-to-day function, and the patient's well-being will be compromised not only by the tumor they carry, but also by the treatment they endure. Cells of the alimentary tract, skin, hair follicles, and immune system all divide more frequently than others such as the liver, kidney, and brain and

tend to be more sensitive to these drugs. This is why conventional chemotherapy leads to hair loss, digestive distress, and impaired immune function.

The development of monoclonal antibodies, with their extraordinary capacity to target a single 6- to 8-amino-acid length of a protein was seized upon by scientists as a potential opportunity to specifically target tumor cells, while leaving the neighboring healthy cells undamaged. Immunotherapeutics, the production and refinement of mAbs that will specifically target cancer cells for death and healthy neighbors, is a burgeoning field. Now part of the therapeutic arsenal, mAbs have met with some success. Some examples of current reagents and their mechanisms of action are described below.

How Do Antibodies Kill Cancer Cells?

- *Competitive binding to cell surface receptors and preventing their activation.* Antibody binding to molecules on the surface of a cell can block interactions that may be necessary to allow growth of that cell. For example, many tumor cells become dependent for their growth on binding of specific growth factors such as *E*pidermal *g*rowth *f*actor (EGF). Several of the mAbs currently in clinical use, such as erbitux, bind to the *EGF* receptor (EGFR) and interfere with its function. Herceptin is an example of an extremely successful drug that targets the Her-2 EGFR on advanced breast cancer cells. All EGFRs must dimerize on binding to EGF so that they can send their growth signal to the nucleus. Herceptin interferes with this dimerization reaction and, so, interrupts the EGFR signal.

- *Interference with the generation of new blood vessels.* Another group of drugs, including bevacizumab (Avastin), targets activity of *v*ascular *e*ndothelial *g*rowth *f*actor (VEGF). VEGF interacts with *VEGF* receptors (VEGFR1 and VEGFR2) located on blood vessels. Binding stimulates the formation of

new blood vessels that can be used to supply the nascent tumor cells with nutrients. Avastin binds to VEGF and prevents it from interacting with its receptors, thereby inhibiting the development of new blood vessels and compromising tumor growth.

- *Direct binding to receptors with induction of apoptosis.* Some mAbs, such as rituximab (Rituxan), may kill their target cells by binding to the surface of the cell and mimicking the binding of natural ligands. Binding of rituximab to follicular B lymphoma cells results in interference with cell cycle regulation and the induction of apoptosis.

- *Induction of ADCC.* Binding of certain isotypes of antibodies to the surface of a cell results in the induction of *anti*body-*d*ependent *c*ell-*m*ediated *c*ytotoxicity (ADCC). If a tumor cell becomes coated with antibodies of human IgG1 or IgG3 subclasses, FcγRIII molecules on the surface of natural killer (NK) cells and other immune cells will bind to the Fc regions of these antibodies. Once bound to the NK cell, the tumor cell becomes the target of secreted perforin and granzymes, which induce apoptosis in the affected cell. In addition, macrophages will recognize the Fc portion of antibody clusters of the IgG1 isotype that are bound to a cell surface, and be induced to phagocytose the antibody-cell complex and/or release toxic metabolites.

- *Induction of complement fixation.* IgG3 and IgG1 antibody binding to the surface of a cell can also result in the attachment (fixation) of a set of serum proteins known as the complement cascade to the cell surface. By a complex set of reactions detailed in Chapter 6, this can result in either the dissolution of the cell following disruption of the integrity of its membrane, or enhanced phagocytosis of that cell by macrophages.

- *Delivery of toxins.* Very early on in the development of mAbs as clinical tools,

BOX 13-1

scientists began modifying them with toxins such as radioactive isotopes and cytotoxic reagents. The isotopes ^{90}Y and ^{131}I are both capable of delivering short-range cytotoxic doses of radiation to anything bound by the conjugated antibody and the immuno-conjugates Zevalin and Bexxar were approved for use in humans for the treatment of relapsed or refractory non-Hodgkin's lymphoma in 2002 and 2003, respectively. Other toxins that have been studied in the lab and/or used in the clinic include modified bacterial toxins such as *Pseudomonas* exotoxin A, which has been conjugated to an mAb directed toward the CD22 molecule and has proven promising in the treatment of chemoresistant hairy cell leukemia. Ricin, another bacterial toxin, has also been successfully conjugated with antibodies directed toward CD22 and CD19, and these molecules are currently being developed for the treatment of pediatric and acute lymphoblastic leukemia.

Currently, scientists are working to improve the biological efficacy of the antibodies employed in cancer treatment by subjecting proven antibodies to mutation in vitro followed by selection using phage display technology to improve the affinity and selectivity of binding to tumor cells. Others are combining multiple antibodies with different specificity to the same cell in a single treatment, or combining mAbs with more conventional chemotherapeutic approaches. The Human Genome Project has provided a wealth of information regarding the number and structure of cell-surface signaling molecules, and it is also being exploited in the design of novel drugs and toxins, including antibodies and antibody fragments.

For further reading on this topic, see the references below. Also keep an eye on the news. Immunotherapeutics is one of the fastest-moving and most exciting areas of research at the confluence of the laboratory and the clinic. New ideas arise every day, and new reagents are likely to follow.

Rivera, F., et al. 2008. Current situation of panitumumab, matauzumab, nimotuzumab and zalutumumab. *Acta Oncologia* **47**:9–19

Strome, S. E., et al. 2007. A mechanistic perspective of monoclonal antibodies in cancer beyond target-related effects. *The Oncologist* **12**:1084–1095.

TABLE 1 — Some monoclonal antibodies in clinical use

mAb product (trade name)	Nature of antibody	Target (antibody specificity)	Modification of antibody	Treatment
Rituximab, (Rituxan)	Chimeric	CD20 (mouse B-cell antigen)	None	Relapsed or refractory non-Hodgkin's lymphoma
Trastuzumab (Herceptin)	Humanized	Human epidermal growth factor receptor 2 (HER-2)	None	HER-2 receptor positive advanced breast cancers
Alemtuzumab (Campath)	Humanized	CD52 (an antigen on many types of leukocytes)	None	Chronic lymphocytic leukemia
Bevacizumab (Avastin)	Humanized	Vascular endothelial growth factor (VEGF)	None	Colorectal cancer
Cetuximab (Erbitux)	Chimeric	EGFR	None	Colorectal cancer
Panitumumab (Vectibix)	Human	EGFR	None	Colorectal cancer
Ibritumomab	Mouse	CD20	None	Relapsed or refractory non-Hodgkin's lymphoma
Ibritumomab tiuxetan (Zevalin)	Mouse	CD20	^{90}Y	Relapsed or refractory non-Hodgkin's lymphoma
Tositumomab (Bexxar)	Mouse	CD20	^{131}I	Relapsed or refractory non-Hodgkin's lymphoma
Gemtuzumab ozogamicin (Mylotarg)	Humanized	CD33 (glycoprotein antigens on myeloid progenitor cells and monocytes)	Attached to an anti-tumor agent that cleaves double-stranded DNA at specific sequences	Acute myelogenous leukemia
Epratuzumab	Humanized	CD22 (glycoprotein antigens on mature and neoplastic B cells)	None	Relapsed or refractory non-Hodgkin's lymphoma

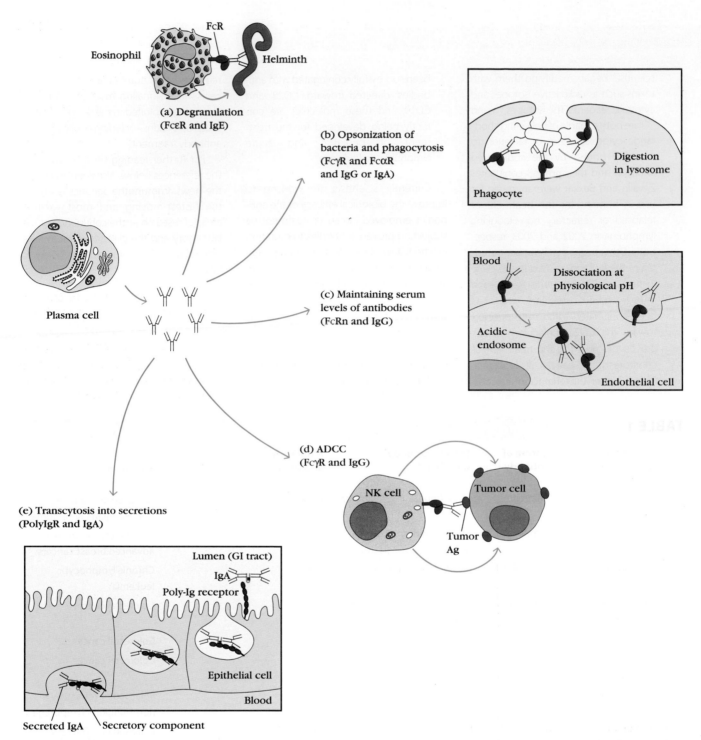

FIGURE 13-2 Functions of Fc receptors. Fc receptors (FcRs) come in a variety of types and are expressed by many different cell types. Some of their major functions are illustrated here. (a) When bound by IgE-pathogen (e.g., worm) complexes, FcεRs expressed by granulocytes can induce the release of histamine and proteases (degranulation). (b) When bound by antibody-pathogen complexes, FcαR and FcγR can induce macrophage activation and phagocytosis. (c) The neonatal FcR (FcRn) binds antibody that has been nonspecifically engulfed by endothelial cells and returns the antibody, intact, to the blood. (d) When bound by IgG antibodies coating infected or tumor cells, FcγRs activate the cytolytic activity of natural (NK) cells. (e) *PolyIg* receptors (PolyIgR) expressed by the inner (basolateral) surface of epithelial cells (facing the blood) will bind dimers and multimers of IgA and IgM antibodies and transfer them through the cell to their apical (outer) surface and into the lumen of an organ (e.g., the GI tract). This is a process referred to as transcytosis and is responsible for the accumulation of antibodies in bodily secretions.

on epithelial cells, an event that triggers endocytosis and transport of the molecules from the basolateral (inner) to the apical (outer) sides of the epithelial cell and into the tissue lumen (see Figure 13-2).

Two subclasses of IgA—IgA1 and IgA2—are found in humans. IgA1 is more prevalent in the serum (and typically monomeric), and IgA2 is more prevalent in secretions. Both isotypes can mediate ADCC by binding FcRs on NK T cells, and both can trigger degranulation of granulocytes.

Effector Functions of IgE Antibodies

IgE antibodies are best known for their role in allergy and asthma. However, research suggests that they play an important role in protection against parasitic helminths (worms) and protozoa. They are made in very small quantities but have a very potent effect, inducing degranulation of eosinophils and basophils, and release of molecules such as histamine that do permanent damage to a large pathogen.

Fc Receptors Mediate Many Effector Functions of Antibodies

Although several antibody isotypes have a direct effect on the viability of a pathogen by initiating complement-mediated lysis, many of the effector functions of the antibodies described above depend on their ability to interact with receptors on cells in the innate and adaptive immune systems as well as cells in various epithelial and endothelial tissues. These antibody receptors were originally discovered over 40 years ago by early antibody biochemists, who referred to them as Fc receptors because they bound specifically to the Fc portion of the antibody.

FcRs allow nonspecific immune cells to take advantage of the exquisite specificity of antibodies to focus their cellular functions on specific antigens and pathogens. They also provide a physical bridge between the humoral and cell-mediated immune systems. When bound by antibody alone, FcRs do not trigger signals; however, when cross-linked by multiple antibody-antigen complexes, FcRs initiate effector responses (see Figure 13-2).

The diversity of FcRs, most of which are members of the immunoglobulin superfamily (Figure 13-3), is now fully appreciated. Each differs by (1) the antibody class (isotype) to which they bind (e.g., FcγRs bind to several IgG isotypes, FcεRs bind to the IgE isotype, and FcαRs bind to IgA isotypes), (2) the cells that express them (macrophages, granulocytes, NK cells, T and B cells), and (3) their signaling properties. Together these properties define a distinct effector function that an antibody can inspire (Table 13-3).

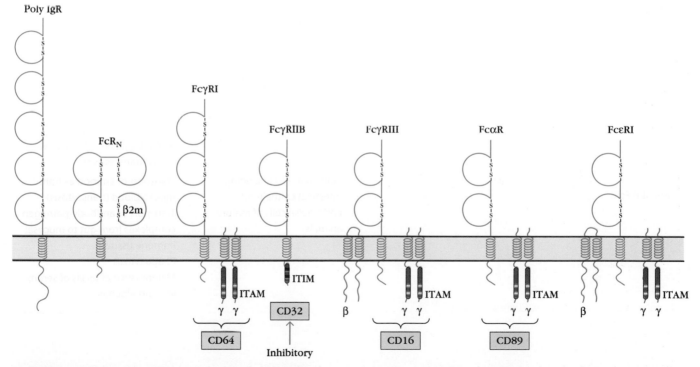

FIGURE 13-3 The structure of a number of human Fc receptors. The antibody-binding polypeptides are shown in blue and, where present, accessory signal-transducing polypeptides are shown in green. Most FcRs are activating receptors and are associated with signaling proteins that contain the *I*mmuno-receptor *T*yrosine *A*ctivation *M*otif (ITAM) or include ITAM(s) in their own intracellular regions. FcγRIIB is an inhibiting FcR and includes an *I*mmuno-receptor *T*yrosine *I*nhibition *M*otif (ITIM) in its intracellular region. All FcRs shown in this figure are members of the immunoglobulin supergene family; their extracellular regions include two or more immunoglobulin folds. These molecules are expressed on the surface of various cell types as cell-surface antigens and, as indicated in the figure, some have been assigned cluster of differentiation (CD) designations (for clusters of differentiation, see Appendix 1). *[Adapted from M. Daeron, 1999, Fc Receptors, in The Antibodies, M. Zanetti and J. D. Capra, eds., vol. 5, p. 53, Amsterdam: Harwood Academic Publishers.]*

TABLE 13-3	Expression and function of FcRs		
FcR	**Isotypes they bind**	**Cells that express them**	**Function**
FcγRI (CD64)	IgG2a in mice, IgG1 and IgG3 in humans High-affinity receptor	Dendritic cells, monocytes, macrophages, granulocytes, B lymphocytes	Phagocytosis Cell activation
FcγRII (CD32)	IgG	Dendritic cells, monocytes, macrophages, granulocytes, B lymphocytes, some immature lymphocytes	Inhibitory receptor Traps antigen-antibody complexes in germinal center Abrogates B-cell activation
FcγRIII (CD16) Humans generate two versions: FcγRIIIA (CD16a) and FcγRIIIB (CD16b)	IgG1, IgG2a, and IgG2b in mouse; IgG1 in human Low-affinity receptor Only FcR that binds mouse IgG1	Dendritic cells, monocytes, macrophages, granulocytes, B lymphocytes Only FcR expressed by NK cells	ADCC Cell activation
FcγRIV (in mouse, with some similarity to human FcγRIIIA and/or human FcεRI)	IgG2a and IgG2b in mice; IgG1 in humans Intermediate affinity receptor, although exhibits higher affinity for human IgG1 than FcgRIIA.	Monocytes, macrophages, granulocytes Not on lymphocytes	ADCC Cell activation
FcεRI	IgE	Eosinophils, basophils, mast cells	Degranulation of granulocytes, including eosinophils, basophils, mast cells
FcεRII (CD23)	IgE (low affinity)	B lymphocytes	Regulation of B-cell production of IgE Transport of IgE-antigen complexes to B-cell follicles
FcαRI (CD89)	IgA	Dendritic cells, monocytes, macrophages, granulocytes, some liver cells	Phagocytosis Cell activation ADCC
pIgR	IgA and IgM	Multiple epithelial cells	Transport of antibody from blood to the lumens of GI, respiratory, and reproductive tracts (transcytosis)
FcRn (neonatal FcR)	IgG	Epithelial cells (including intestinal epithelium) Endothelial cells of mature animals	Transport of antibodies from milk to blood (transcytosis) Transport of antibody-pathogen complexes from gut to mucosal immune tissue Phagocytosis Maintenance of levels of serum IgG and albumin

For example, antibody-antigen complexes binding to FcεRs on eosinophils and other granulocytes trigger the release of histamine and proteases from granules (degranulation). FcγRs and FcαRs on macrophages trigger a signaling cascade that induces the internalization of the pathogen into a phagocyte, which can destroy it via a variety of mechanisms, including oxidative damage, enzyme digestion, and membrane-disrupting effects of antibacterial peptides. "*Neonatal*" *FcR* (FcRn)

molecules on cells that line the bloodstream help maintain antibody levels in serum by protecting them from degradation. Other FcγRs expressed by cytotoxic cells, including NK cells, trigger a cascade that results in the release of granule contents that will cause the death of targeted cells (ADCC). And FcαRs expressed by epithelial cells trigger the passage of antibodies from one side of a cell to another—from the bloodstream to the gut, for instance.

Interestingly, some FcRs do not stimulate effector responses but, rather, inhibit cell activity. Indeed, a cell can express varying levels of both activating and inhibiting FcRs. The cell's response is determined by the balance between the positive and negative signals it receives when antibody-antigen complexes bind these two types of receptors. We will discuss these events in more detail below, where we describe the function of individual classes of FcRs.

FcR Signaling

A single antibody will not activate an Fc receptor; instead multiple FcRs need to be cross-linked or oligomerized by binding to the multiple antibodies coating a single pathogen. This cross-linking results in the generation of either a positive or a negative signal that enhances or suppresses the effector function of that cell.

Whether engagement of an FcR generates positive or negative signals depends on whether that FcR is associated with an ITAM or an ITIM, respectively (see Figure 13-3). Although some activating FcRs include their own ITAM motif in their cytoplasmic region, many do not and, instead, associate at the membrane with a common gamma (γ) chain, which provides the services of an ITAM motif to the stimulating receptor complex.

The ITAM amino acid sequence is a common activating motif found in many signaling complexes. Recall that ITAM motifs are found in the cytoplasmic domains of TCR and BCR antigen-receptor associated molecules (CD3 molecules, Igα, and Igβ; see Chapter 3). They are also features of NK cell receptors. What makes these motifs so useful? As you know from Chapter 3, they act as substrates for tyrosine kinases, which are recruited after receptor engagement. Figure 13-4 provides an example of signaling pathways downstream of one activating FcγR (FcγRIII) and illustrates their similarity to BCR signaling. Antibody-antigen engagement recruits tyrosine kinases lyn (or lck), phosphorylating ITAMs, which, in this case, are expressed by an associated pair of common γ chains. The tyrosine kinase Syk is recruited to the phosphorylated ITAMs and phosphorylates the Btk kinase, which enhances the activity of PLCγ. PLCγ

(a)

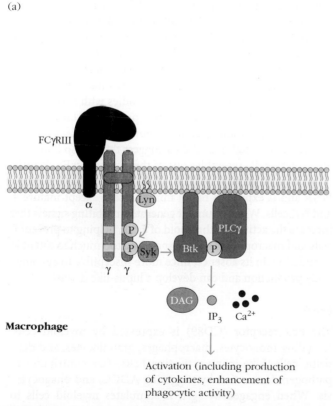

Macrophage

Activation (including production
of cytokines, enhancement of
phagocytic activity)

(b)

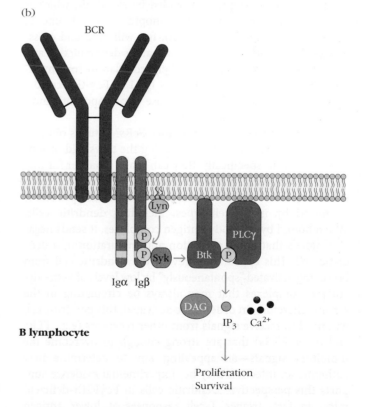

B lymphocyte

Proliferation
Survival

FIGURE 13-4 FcR signaling. When bound by antibody-antigen complexes, activating Fc receptors oligomerize and transmit signals (a) that are very similar to those generated by antigen binding to the B-cell receptor (b and Figure 3-28). This is a very simplified schematic illustrating some of these similarities. Briefly, oligomerization of receptors recruits a tyrosine kinase (e.g., lyn), which phosphorylates the ITAMs associated with γ-chains (in the case of FcRs) or Igα and Igβ (in the case of B cells). Phosphorylation allows the tyrosine kinase Syk to dock, bringing it into proximity with other activating molecules including adapter proteins (not shown) and Btk, another kinase. Btk can activate the phospholipase PLCγ, which generates two important second messengers, DAG and IP3. These and other signals initiated by receptor binding result in cell activation and effector functions (some of which are illustrated in Figure 13-2). Inhibitory FcR signaling (not shown) recruits phosphatase activity that can antagonize these signals by removing the activating phosphate groups.

generates the potent second messengers DAG and Ca^{2+}. These events and others that are not illustrated in this simple diagram result in activation of the FcR-expressing cell. In this case, the cell is a macrophage, and the signaling pathways stimulate and enhance its phagocytic capacity, and can induce expression and secretion of cytokines (e.g., the inflammatory cytokines TNFα, IL-1, and IL-6). Although the general signaling pathways are similar, the precise consequences of FcR signaling differ depending on the receptor that is activated, the versions of downstream signaling enzymes that are present, and the genes that are accessible in the specific cell type that is stimulated.

ITIMs are found in the intracellular regions of the FcγRII receptor, the main inhibitory receptor among the FcR family. As you may remember from Chapter 9, ITIMs are also found in the intracellular region of the inhibitory costimulatory molecule CTLA-4 on T cells. These motifs associate not with kinases, but with phosphatases, such as SHIP, which help to assemble complexes that inhibit signaling cascades, in part by dephosphorylating what was phosphorylated by activating signals (see Figure 13-4).

The role of activating Fc receptors is not difficult to appreciate. A classic example is provided by the FcεR, which is expressed on mast cells and eosinophils. Once bound to pathogen (e.g., a parasitic worm), IgE will bind and cross-link FcεR, activating a cascade that causes degranulation and the release of histamine and other proinflammatory molecules. Similarly, binding of human IgG1 to FcγRI receptors on NK cells will stimulate the release of pro-apoptotic molecules from NK cells (see below).

But what is the role of inhibitory FcRs? It turns out that they are critically important in tuning the threshold of activation of a cell. Specifically, they can make it harder for that cell to be activated by low-level signals. FcγRIIB, the main inhibitory FcR, binds to several different IgG isotypes and is expressed by many cell types, including dendritic cells. When bound by antibody-antigen complexes, it sends negative signals that block activation and maturation of a dendritic cell. This appears to prevent the dendritic cell from becoming activated "spontaneously" by low levels of antibody-antigen complexes that may always be circulating in the serum. Therefore, in order to be activated, this dendritic cell will need to receive signals from other receptors (e.g., TLRs and other PRRs) that are strong enough to overcome the inhibitory signals—an appealing way to determine how authentic an infection may be. Experimental evidence supports this perspective. Dendritic cells in FcγRIIB-deficient mice, in fact, trigger T-cell responses at lower antigen thresholds. FcγRIIB deficient mice also develop autoimmunity, suggesting that negative signals play an important role in maintaining immune tolerance.

Interestingly, cells can co-express both inhibitory and activating FcRs and tune their response according to their integration of positive and negative signals. Which FcRs a cell expresses depends on the signals that the cell receives during an immune response. For instance, TGFβ and IL-4

up-regulate the expression of inhibitory FcRs, whereas inflammatory cytokines (TNF and IFNγ) up-regulate the expression of some activating FcRs.

FcγRs

FcγRs are the most diverse group of FcRs and are the main mediators of antibody functions in the body. Expressed by a wide range of cells, they bind antibodies of the IgG class, often several different IgG subclasses. Most are activating receptors and can induce phagocytosis if expressed by a macrophage as well as degranulation if expressed by cytotoxic cells.

The four families of FcγR (FcγRI [CD64], FcγRII [CD32], FcγRIII [CD16], and FcγRIV) are conserved across mammals. Three of these, FcγRI, FcγRIII, and FcγRIV, are activating receptors and vary in their binding affinities and cellular expression. FcγRI binds to antibodies with high affinity, and prefers to bind to IgG2a in mouse, and IgG1 and IgG3 in humans. FcγRIII binds with lower affinity to several IgG isotypes. Most immune cells, except mature T cells, express both of these receptors. NK cells only express FcγRIII.

FcγRIV was discovered only recently. Investigators wishing to study the role of FcγRs in immune responses developed mice where the two major FcγRs, RI and RIII, were absent. They were puzzled when IgG antibodies still exhibited effector functions, but recognized that this probably meant mice expressed other, as yet unidentified, FcγRs. Indeed, they found a new protein, FcγRIV, expressed exclusively on innate immune cells (including dendritic cells, macrophages, and neutrophils), which bound IgG antibodies with high affinity. This receptor also shows an isotype preference, binding best to murine IgG2a and IgG2b. Its human equivalent may be FcγRIIIA, although recent data suggest that it also binds IgE and functionally resembles one of the human FcεRs (FcεRI).

As we saw above, FcγRIIB (CD32) is the major inhibiting FcγR and is expressed on all immune cells except mature T and NK cells. When bound, it generates inhibiting signals that increase the activation threshold of a cell, helping to prevent B cells and macrophages from responding too much to too little stimulation. In its absence, mice more susceptible to autoantibody production and can develop a lupus-like disease.

FcαR

The Fcα receptor (CD89) is expressed by myeloid cells, including monocytes, macrophages, granulocytes, and dendritic cells. Like many activating FcRs, Fcα contributes to pathogen destruction by triggering ADCC, and phagocytosis. When engaged by IgA, it stimulates myeloid cells to release inflammatory cytokines and generate superoxide free radicals, which help kill internalized pathogens. It has also been found in a soluble form, which binds circulating IgA. IgA can also be bound by the polyIgR described below.

FcεR

FcεRs are expressed by granulocytes, including mast cells, eosinophils, and basophils and, when bound by IgE, trigger a signaling cascade that releases histamine, proteases, and other inflammatory mediators. They are most often associated with allergy symptoms, which arise when an antigen induces an IgE response. Their physiologic function has been contested, but most agree that they evolved as part of our immune response to parasitic worms. Two types of FcεR are recognized: the high-affinity FcεRI that is similar in structure to most activating FcRs and is expressed by mast cells, eosinophils, and basophils; and the lower-affinity FcεRII (CD23) that is a member of the C-type lectin family, not the immunoglobulin supergene family. This low-affinity form is expressed by B cells and eosinophils, and its function is still under investigation (see Chapter 15).

pIgR

The *poly*meric *i*mmunoglobulin *r*eceptor (PolyIgR) is expressed by epithelial cells and has a unique function. It initiates the transport of IgA and IgM from blood to the lumen (the inside) of multiple tissues, including the GI, respiratory, and reproductive tracts (see Figure 13-2e). This is the FcR responsible for carrying antibodies into tears and milk, and for populating our gut mucosa with IgA antibodies that protect us from the invasion of microbes and toxins that we ingest.

FcRn

Although the *n*eonatal *Fc receptor* (FcRn) can be considered another FcγR, it is unique in structure and function. Related to MHC class I, it, too, associates with β₂-microglobulin. Expressed on the surface of many different cell types, including epithelial and endothelial cells, it binds not only to IgG, but also to albumin, inhibiting the degradation of both molecules. FcRn is responsible for carrying antibodies ingested from mother's milk across epithelial cells of an infant's intestine to the bloodstream—the opposite direction of the antibodies carried by polyIgR. Even though its expression levels drop after an animal is weaned, FcRn still plays an important role in managing intestinal infections and can transport antibody-pathogen complexes from the intestinal lumen into mucosal immune tissue where antigen can be processed and presented to T cells. FcRn can also augment the role of pIgR (which carries IgA into organ secretions) and carry IgG antibodies into milk and GI and respiratory secretions. When expressed on phagocytes, it can also play a more conventional role: stimulating internalization and destruction of pathogens.

Perhaps, however, one of the most important roles of FcRn in adult animals is to maintain levels of circulating IgG and albumin. Endothelial cells can take proteins up nonspecifically by pinocytosis. Under ordinary circumstances these proteins would be degraded in acidic lysosomes. However, endothelial cells also express FcRns, which bind tightly to antibodies and albumin within the acid environment of the lysosomes. They sort these important proteins into vesicles that are carried back to the blood-

stream, where the higher pH of blood permits their release (see Figure 13-2c).

Visualizing Antibody and FcR Effector Responses

We can now visualize effector antibody activity in the context of an immune response. In summary, after infection B cells encounter pathogen and their proteins in the secondary lymphoid organs. Those that bind antigen are activated, and if they receive T-cell help differentiate into antibody-producing plasma cells, either immediately or after undergoing somatic mutation and isotype (class) switching in the germinal center. The isotypes produced by plasma cells depend on the timing of the response (early versus late) and the nature of the pathogen. A single infection can induce the production of antibodies with the same specificity but multiple isotypes.

Plasma cells release antibody into the bloodstream from a number of vantage points, including the medulla of a lymph node, the sinuses of a spleen, and the bone marrow. IgM antibodies, produced early in the response, may find pathogen in the bloodstream and initiate the complement cascade, lysing that pathogen directly. It can also be carried by polyIgRs to the lumens of other tissues and bind pathogens there, generating antibody-antigen complexes that can be engulfed by macrophages. IgG antibodies, produced later in the response, will not only fix complement and opsonize pathogen, but can be transported into tissues to interact with innate cells that express FcRs (e.g., neutrophils), enhancing their inflammatory activity. Whereas extracellular pathogens bias B-cell class switching to isotypes that can fix complement and opsonize pathogen, intracellular pathogens bias B-cell class switching to antibodies that can recruit cytotoxic cells, including NK T cells, which empty their apoptosis-inspiring contents and kill an infected cell (ADCC). This arm of the immune system is the topic of the next section of the chapter.

Cell-Mediated Effector Responses

Cytotoxic effector cells include the CD8⁺ cytotoxic T cells of the adaptive immune system, NKT lymphocytes, which bridge the innate and adaptive immune systems, and NK cells, which were once associated strictly with the innate immune system but share intriguing functional features with their adaptive immune lymphocyte relatives. CTL, NKT, and NK effectors all induce cell death by triggering apoptosis in their target cells. Not only do these cytotoxic cells eliminate targets infected with intracellular pathogens (virus and bacteria), but they also play a critical role in eliminating tumor cells and cells that have been stressed by extreme temperatures or trauma (Table 13-4). NK cells also play a less desirable role in rejecting cells from allogeneic organ transplants. Essentially, the cell-mediated immune response is prepared to recognize and attack any cell that exhibits "non-self" or "altered-self" characteristics.

TABLE 13-4 How cytotoxic lymphocytes kill

Cell type	Effector molecules produced	Mechanism of killing
CTL (typically CD8+ T cell)	Cytotoxins (perforins and granzymes), IFN-γ, TNF, Fas ligand (FASL)	Cytotoxic granule release and FASL-FAS interactions
NK T cell	IFN-γ, IL-4, GMCSF, IL-2, TNF	FASL interactions predominantly; can activate NK cells indirectly
NK cell	Cytotoxins (perforins and granzymes), IFN-γ, TNF, Fas ligand (FASL)	Cytotoxic granule release and FASL-FAS interactions

Cytotoxic T Lymphocytes Recognize and Kill Infected or Tumor Cells Via T-Cell Receptor Activation

The importance of cytotoxic T lymphocytes in the cell-mediated immune response to pathogens is underscored by the pathologies that afflict those defective in T-cell generation. Children with DiGeorge syndrome are born without a thymus and therefore lack the T-cell component of the cell-mediated immune system. They are generally able to cope with extracellular bacterial infections, but cannot effectively eliminate intracellular pathogens such as viruses, intracellular bacteria, and fungi. The severity of the cell-mediated immunodeficiency in these children is such that even the attenuated virus present in a vaccine, which is capable of only limited growth in normal individuals, can produce life-threatening infections.

T-cell-mediated immune responses can be divided into two major categories according to the different effector populations that are mobilized: cytotoxic T cells and helper T cells. The differentiation and activity of helper T cell subsets was discussed in detail in Chapter 11. Here, we focus on the immune response mediated by **cytotoxic T lymphocytes (T$_c$ cells or CTLs)**, which can be divided into two phases. In the first phase, naïve T$_C$ cells undergo activation and differentiation into functional effector CTLs within secondary lymphoid tissue. In the second phase, effector CTLs recognize class I MHC-antigen complexes on specific target cells in the periphery, an event that ultimately induces the destruction of the target cells. CTLs are predominantly CD8+ and are therefore class I MHC restricted. Since virtually all nucleated cells in the body express class I MHC molecules, a CTL can recognize and eliminate almost any cell in the body that displays the specific antigen recognized by that CTL in the context of a class I MHC molecule.

As you have learned from Chapter 11, differentiation of a naïve CD8+ T cell into an effector cytotoxic T cell involves interaction with class I MHC-peptide complexes on activated dendritic cell as well as help from T$_H$1 or T$_H$17 effector CD4+ T cells. This help is provided both indirectly, via the ability of CD4+ T cells to enhance dendritic cell function, and directly, via the release of cytokines by helper CD4+ T cells.

Effector CTL Generation from CTL Precursors

Naïve T$_C$ cells are incapable of killing target cells and are therefore also referred to as **CTL precursors (CTL-Ps)** to denote their functionally immature state. Only after a CTL-P has been activated does the cell differentiate into a functional CTL with cytotoxic activity. The threshold to activate CTLs from CTL-Ps is high compared to that of other effector T cells and appears to require at least three signals (Figure 13-5):

- An antigen-specific signal transmitted by the TCR complex upon recognition of a class I MHC-peptide complex on a "licensed" APC (*licensing* is explained below)

- A costimulatory signal transmitted by the CD28-CD80/86 (B7) interaction of the CTL-P and the licensed APC

- A signal induced by the interaction of IL-2 with the high-affinity IL-2 receptor, resulting in proliferation and differentiation of the antigen-activated CTL-P into effector CTLs (IL-2 can be generated by helper T cells as well as by the CD8+ T cell itself.)

Differentiation of a CTL-P is initiated by its recognition of antigen on an APC presented in the context of a class I MHC molecule, but other factors are also required. For CTL-Ps to mature into cytotoxic cells they need to interact with a *licensed* APC. An APC can be licensed in several ways—by a CD4+ helper T cell (T$_H$) of the T$_H$1 or T$_H$17 subtype—or by engagement with microbial products, which activate Toll-like receptors. There is no absolute requirement for CD4+ T-cell help at this interval.

In contrast, CD4+ T cell help is absolutely required for the development of effective CD8+ T-cell memory and for an optimal proliferative burst of functional T cells. This was demonstrated by examining the behavior of CTLs in class II MHC deficient mice that could not generate CD4+ T cells. These mice generated functional CTLs in response to a primary infection, and even developed cells with memory phenotypes. However, the CTLs were not able to mount a secondary response, demonstrating that in the absence of CD4+ T cell help, functional memory CTLs did not develop.

What does a CD4+ T cell provide that is so important for optimal activation of a CD8+ T cell? Recall from Chapter 11

(a) Sequential

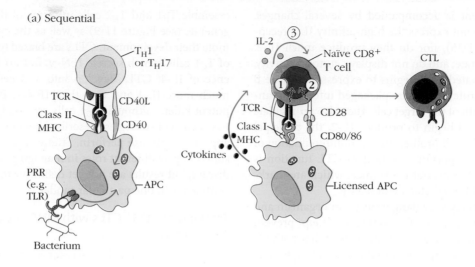

(b) Simultaneous

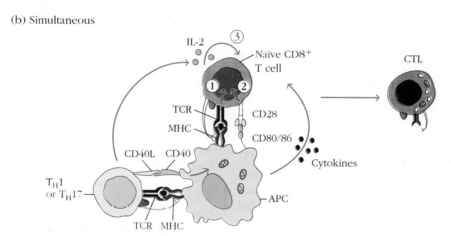

FIGURE 13-5 Generation of effector CTLs. The differentiation of a naïve CD8$^+$ T cell (a precursor CTL) into a functional CTL requires several events. The precursor CTL must engage class I MHC-peptide complexes and costimulatory ligands on a "licensed" antigen-presenting cell (dendritic cell). Licensing of dendritic cells occurs either through engagement with an activated, CD40L$^+$ helper T cell (e.g., a T$_H$1 or T$_H$17 helper cell) or through engagement with pathogens (e.g., via Toll-like receptors). For optimal proliferation and memory generation, CD4$^+$ T-cell help is required. CD4$^+$ T-cell help can be given prior to CTL engagement of the dendritic cell as depicted in (a). Alternatively, it can be given at the same time as the precursor CTL engages the dendritic cell (b). Simultaneous engagement of a dendritic cell by both helper CD4$^+$ and precursor CD8$^+$ T cells would allow optimal delivery of IL-2 generated by helper CD4$^+$ T cells. In response to TCR stimulation, precursor CTLs begin to up-regulate expression of the IL-2 receptor (and begin to make more of their own IL-2). IL-2 is critical for successful differentiation into a functional CTL.

that helper T cells generate cytokines (e.g., IFN-γ) that activate antigen-presenting cells. However, activated helper T cells also express *CD40 ligand* (CD40L, also known as CD154), a member of the TNF family of proteins, which provides an all-important costimulatory signal to the APC. CD40L interacts with CD40, a TNF receptor family member expressed by activated professional APCs. When engaged, CD40 initiates a signaling cascade within the APC that increases the expression of costimulatory ligands (CD80 and CD86), chemokines, and cytokines, significantly enhancing the APC's ability to activate the CD8$^+$ T cell (see Figure 13-5).

Investigators envision an interaction among three cells that results in CD8$^+$ T-cell activation: the T$_H$ cell, which interacts with and further activates an APC, which, in turn, interacts with and activates the CTL-P cell (see Figure 13-5a). In fact, fluorescent-imaging studies have provided direct evidence for the formation of dendritic cell/CD4$^+$ T-cell/CD8$^+$ T-cell complexes during the T-cell response to viral antigen (see Figure 13-5b and Chapter 14).

It is important to realize, however, that the interactions between the dendritic cell and the two different T cells may not have to be concurrent; rather, a dendritic cell "licensed" by a T$_H$ cell could retain its capacity to activate CTL-Ps for some period after the T$_H$ cell disengages. In this case, specific T$_H$ cells could move on to activate other dendritic cells, amplifying the activation response.

CTL development is accompanied by several changes. Precursor CTLs do not express the high-affinity IL-2 receptor alpha chain (CD25), nor do they produce much IL-2. They do not proliferate, and do not display cytotoxic activity. A successfully activated CTL begins to express granzyme B and perforin, the proteins that are packaged into lytic granules and induce death of the target cell. They also up-regulate the IL-2 receptor and begin to produce IL-2, the principal cytokine required for full proliferation and differentiation of effector CTLs. The importance of IL-2 in CTL function is underscored in IL-2 knockout (KO) mice, which are markedly deficient in CTL-mediated cytotoxicity.

CTLs are potentially very dangerous to an organism and the stringent requirements for activation help prevent unwanted destruction. Thus, the requirement that both T_H and T_C cells recognize antigen before the T_C cell is optimally activated provides a safeguard against inappropriate self-reactivity by cytotoxic cells. The fact that the IL-2 receptor is not expressed until after a naïve $CD8^+$ T cell has been activated by T-cell receptor engagement also ensures that only the antigen-specific CTL-Ps proliferate and become cytotoxic.

The Importance of Cross-Presentation in CTL Activation

Regardless of whether activation of a $CD8^+$ T cell occurs via sequential or simultaneous cellular interactions, the APC involved must be able to present antigens from the infecting pathogen to both MHC class II restricted $CD4^+$ T_H cells and MHC class I restricted T_C cells. This capacity initially puzzled immunologists. Strictly speaking, shouldn't antigens from intracellular pathogens (the principal target of CTLs) only be presented by MHC class I? Even more importantly, what if the viral pathogen doesn't even infect dendritic cells (surely, many of them won't)? How can it process and present antigens from microbes that don't naturally infect antigen-presenting cells? The phenomenon of antigen cross-presentation (see Chapter 8) as well as the ability of dendritic cells to endocytose antigen helped resolve these questions.

Recall that cross-presentation is the capacity of dendritic cells to process and present exogenous antigens in both class I and class II MHC molecules. Thus, even if dendritic cells weren't infected by a pathogen, they could engulf it, process it, and present antigens in both MHC class II (by traditional routes) and MHC class I (by cross-presentation). Those that were infected would, of course, have the capacity to process and present antigens in MHC class I via both traditional and cross-presenting routes.

T_C1 and T_C2: Two Types of Effector CTLs

As you know from Chapter 11, $CD4^+$ effector T helper cells can differentiate into several subtypes, each of which respond to and secrete a distinct panel of cytokines. Effector $CD8^+$ cytotoxic cells are not as diverse, but can develop into two distinct subtypes: T_C1 and T_C2. These subtypes loosely resemble T_H1 and T_H2 cells in terms of the cytokines they generate (see Figure 11-9) as well as the cytokines that promote their development. CTLs are biased toward production of T_C1 cells, which secrete IFN-γ, but no IL-4. In the presence of IL-4, CTLs develop into T_C2 cells, which secrete much more IL-4 and IL-5 than IFN-γ. Both subtypes are potent killers, although T_C1 cells can use both perforin and Fas-mediated strategies (described below), whereas Tc2 cells appear to only use perforin. Studies suggest that these two subtypes play different roles in managing (and exacerbating) disease, but results are not yet definitive to generate enough confidence in one particular model.

Tracking of $CD8^+$ CTLs with MHC Tetramer Technology

To fully understand the basis for a successful (and unsuccessful) immune response, immunologists realized that they would need to be able to track the behavior of antigen-specific lymphocytes in an organism. This was initially very difficult because the frequency of antigen-specific lymphocytes is quite low and ways to identify cells with known specificity were crude at best. Even though antigen-specific cells proliferated in response to antigen, they still represented a minority of all lymphocytes in circulation, and their activities were very difficult to tease apart from the activities of nonspecific lymphocytes. Identifying rare antigen-specific memory cells was essentially impossible. A novel and clever technology, based on the use of **MHC tetramers**, was developed in the 1990s and made it possible to identify, isolate, and follow the behavior of lymphocytes specific for an antigen of choice. Now used routinely, MHC tetramer staining has been an invaluable tool in our efforts to understand the complexities of lymphocyte behavior in space and time.

MHC tetramers are laboratory-generated complexes of four identical MHC molecules bound to four peptides and linked to a fluorescent molecule (Figure 13-6). MHC class I tetramers were generated first, but now MHC class II tetramers are also available. The MHC-peptide combination used to generate these complexes depends on the antigen of interest. For example, one laboratory has developed tetramers that include four HLA-A2 molecules (human MHC class I) complexed with a peptide from HIV that is known to stimulate a CTL response. A given MHC class I tetramer-peptide complex will bind only those $CD8^+$ T cells with TCRs specific for that particular peptide-MHC complex. Thus, when a particular MHC-peptide tetramer is added to a diverse T-cell population, cells that bear TCRs specific for the tetramer will bind and become fluorescently labeled. These cells can then be traced using fluorescence microscopy and flow cytometry. With these technologies, it is now possible to determine the number, locale, and phenotype of cells in a population that have TCRs specific for that particular antigen (that particular MHC-peptide combination) before, during, and after an infection. Flow cytometry is sensitive enough to detect antigen-specific T cells even when their frequency in the $CD8^+$ population is as low as 0.1%.

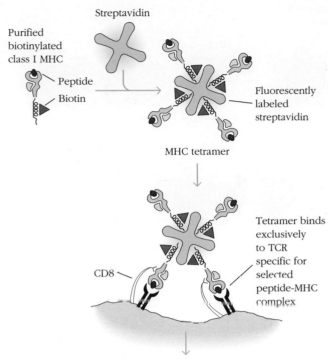

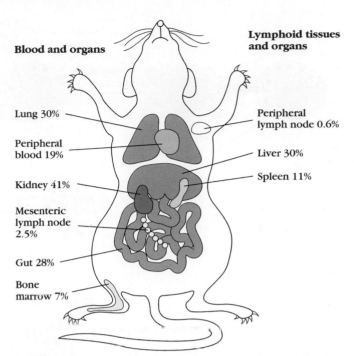

FIGURE 13-7 Localizing antigen-specific CD8+ T-cell populations in vivo. Mice were infected with vesicular stomatitis virus (VSV), and during the course of the acute stage of the infection cell populations were isolated from the tissues indicated. These cells were then incubated with fluorescent tetramers containing VSV-peptide-MHC complexes. Flow-cytometric analysis allowed determination of the percentages of CD8+ T cells that were VSV specific in each of the populations examined. [*Adapted from P. Klenerman, V. Cerundolo, and P. R. Dunbar, 2002, Tracking T cells with tetramers: New tales from new tools.* Nature Reviews Immunology *2:263.*]

Antigen specific cells now fluoresce and
signal measured by flow cytometer
and/or fluorescent microscopy

FIGURE 13-6 MHC tetramers. A homogeneous population of peptide-bound class I MHC molecules (e.g., HLA-A1 bound to an HIV-derived peptide) is conjugated to biotin and mixed with fluorescently labeled streptavidin. In the ideal situation, four biotinylated MHC-peptide complexes bind to the high-affinity binding sites of fluorescent streptavidin to form a tetramer. Addition of the now fluorescent tetramer to a population of T cells results in binding of the fluorescent tetramer only to those CD8+ T cells with TCRs that are specific for the peptide-MHC complexes of the tetramer. The subpopulation of T cells that are specific for the target antigen are now fluorescently tagged, making them readily detectable by flow cytometry. [*Adapted in part from P. Klenerman, V. Cerundolo, and P. R. Dunbar, 2002, Tracking T cells with tetramers: New tales from new tools,* Nature Reviews Immunology *2:263.*]

With this powerful tool one can directly measure the increase in antigen-specific CD8+ T cells in response to exposure to pathogens such as viruses or cancer-associated antigens, and trace their tissue distribution. For instance, researchers infected mice with vesicular stomatitis virus (VSV) and, using tetramer technology, examined all organs for the presence of CD8+ cells specific for a VSV-derived peptide-MHC complex. This study demonstrated that during acute infection with VSV, VSV-specific CD8+ cells migrate away from the lymphoid system and are distributed widely, with large numbers found in liver and kidney, but antigen-specific cells virtually everywhere (Figure 13-7). Follow-up studies show that, regardless of the original site of infection, effector CD8+ T cells distribute themselves throughout the body.

Tetramer-based studies have been exceptionally useful in tracing the phenotype and fate of not just pathogen-specific T cells, but also autoreactive T cells. For example, different MHC/insulin-peptide tetramers were used to identify insulin-reactive CD8+ T cells in diabetic patients. Although these studies are ongoing, some have shown that CD8+ T cells at the earliest stages of the disease respond to a different insulin peptide than CD8+ T cells at later stages of the disease. The investigators speculate that insulin therapy itself may influence which autoreactive CD8+ T cells become dominant in the disease.

How CTLs Kill Cells

A CTL can kill a target in two major ways: either via the directional release of granule contents or via a Fas-FasL membrane signaling interaction. Rather than inducing cell lysis, both of these processes induce the target cell to undergo apoptosis, typically within a few hours of contact with the cytotoxic cell.

Regardless of which method is employed, CTL killing involves a carefully orchestrated sequence of events that begins when the attacking cell binds to the target cell (Figure 13-8) and forms a cell-cell conjugate, an event so intimate

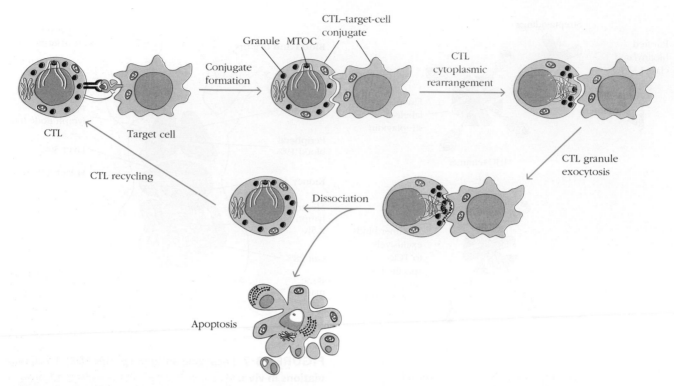

FIGURE 13-8 Stages in CTL-mediated killing of target cells. T-cell receptors on a CTL interact with processed antigen-class I MHC complexes on an appropriate target cell, leading to formation of a CTL-target cell conjugate. The *microtubule organizing center* (MTOC) polarizes to the site of interaction, repositioning the Golgi stacks and granules toward the point of contact with the target cell, where the granules' contents are released by exocytosis. After dissociation of the conjugate, the CTL is recycled and the target cell dies by apoptosis. *[Adapted from P. A. Henkart, 1985, Mechanism of lymphocyte-mediated cytotoxicity, Annual Review of Immunology 3:31.]*

that it is sometimes referred to as the "kiss of death." Formation of a CTL-target cell conjugate is followed within several minutes by a Ca^{2+}-dependent, energy-requiring step in which, ultimately, the CTL induces death of the target cell. The CTL then dissociates from the target cell and goes on to bind another target cell.

The specific signaling events involved in establishing the CTL-target interaction are very similar to those associated with an activating T cell-APC interaction. The TCR-CD3 membrane complex on a CTL recognizes antigen in association with class I MHC molecules on a target cell, an event that triggers the development of a highly organized immunological synapse (see Chapter 10) that is characterized by a central ring of TCR molecules, surrounded by a peripheral ring of adhesion molecules, formed primarily by interactions between the integrin receptor LFA-1 on the CTL membrane and the *intracellular adhesion molecules* (ICAMs) on the target-cell membrane.

TCR signals directly enhance the adhesion between killer and target by converting LFA-1 from a folded, low-affinity state to an extended, high-affinity state (Figure 13-9a). This change is a consequence of "inside-out signaling" where a signal cascade (generated by the TCR but also by some cytokine and chemokine receptors) act on the intracellular portion of LFA-1, inducing a conformational change that straightens out the extracellular region of LFA so that its high-affinity face is accessible to ICAM. LFA-1 persists in its high-affinity state for only 5 to 10 minutes after antigen-mediated activation, and then it returns to the low-affinity state. This downshift in LFA-1 affinity may facilitate dissociation of the CTL from the target cell.

The importance of TCR signals in promoting CTL adhesion was demonstrated by an experiment where purified ICAM protein was coated onto plastic and the ability of resting CTLs versus TCR-stimulated CTLs to adhere was measured. As you can see from Figure 13-9b, more than 10 times the number of TCR-stimulated CTLs (versus unstimulated CTLs) bound to the ICAM-coated plastic. Antibodies that could block the interaction between LFA-1 and ICAM abrogated the effect of TCR stimulation, showing that the adhesion was, in fact, LFA-1/ICAM specific.

Granzyme and Perforin Mediated Cytolysis   Many CTLs initiate killing of their targets via the delivery of pro-apoptotic molecules. These molecules are packaged within granules that can be visualized by microscopy (Figure 13-10). Investigators originally isolated CTL granules by subcellular fractionation and showed that they could induce target-cell damage directly. Analysis of their contents revealed 65-kDa monomers of a pore-forming protein called **perforin** and

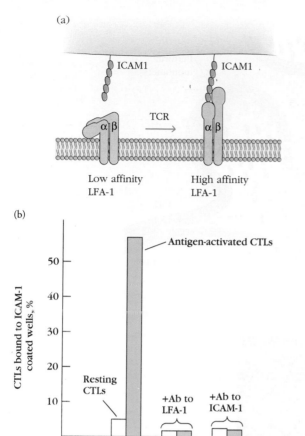

FIGURE 13-9 Effect of antigen activation on the ability of CTLs to bind to the intercellular cell adhesion molecule ICAM-1. (a) TCR signals induce a conformational change in LFA-1 molecules from a folded state to an extended state that allows them to bind to ICAM with high affinity. (b) The importance of TCR signaling in inducing LFA-1 mediated adhesion is illustrated by an experiment in which resting mouse CTLs were first incubated with anti-CD3 antibodies. Cross-linkage of CD3 molecules on the CTL membrane by anti-CD3 has the same activating effect as interaction with antigen–class I MHC complexes on a target cell. Adhesion was assayed by binding radiolabeled CTLs to microwells coated with ICAM-1. Antigen activation increased CTL binding to ICAM-1 more than 10-fold. The presence of excess monoclonal antibody to LFA-1 or ICAM-1 in the microwell abolished binding, demonstrating that both molecules are necessary for adhesion. *[Part (b) based on M. L. Dustin and T. A. Springer, 1989, T-cell receptor cross-linking transiently stimulates adhesiveness through LFA-1, Nature 341:619.]*

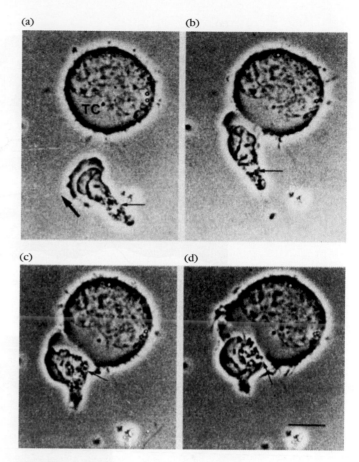

FIGURE 13-10 Formation of a conjugate between a CTL and a target cell and reorientation of CTL cytoplasmic granules as recorded by time-lapse photography. (a) A motile mouse CTL (thin arrow) approaches an appropriate target cell (TC). The thick arrow indicates direction of movement. (b) Initial contact of the CTL and target cell has occurred. (c) Within 2 minutes of initial contact, the membrane-contact region has broadened and the rearrangement of dark cytoplasmic granules within the CTL (thin arrow) is underway. (d) Further movement of dark granules toward the target cell is evident 10 minutes after initial contact. *[From J. R. Yanelli et al., 1986, Reorientation and fusion of cytotoxic T lymphocyte granules after interaction with target cells as determined by high resolution cinemicrography, Journal of Immunology 136:377.]*

several serine proteases called **granzymes (fragmentins)**. CTL-Ps lack cytoplasmic granules and perforin; however, once activated, they begin to form cytoplasmic granules that include perforin monomers and granzyme molecules.

Almost immediately after conjugate formation, CTL granules containing granzyme and perforin are brought to the site of interaction between a killer and target (see Figure 13-10) via the activity of the *microtubule organizing center* (MTOC), which polarizes to this synapse in response to TCR stimulation. The vesicles fuse with the outer membrane and release perforin monomers and granzyme proteases into the space at the junction between the two cells.

As the perforin monomers contact the target-cell membrane, they undergo a conformational change, exposing an amphipathic domain that inserts into the target-cell membrane; the monomers then polymerize (in the presence of Ca^{2+}) to form cylindrical pores with an internal diameter of 5 to 20 nm (Figure 13-11a). A large number of perforin pores are visible on the target-cell membrane in the region of conjugate formation (Figure 13-11b). Perhaps not surprisingly, perforin exhibits some sequence homology with the terminal C9 component of the complement system, and the membrane pores formed by perforin are similar to those observed in complement-mediated lysis. The importance of

(a)

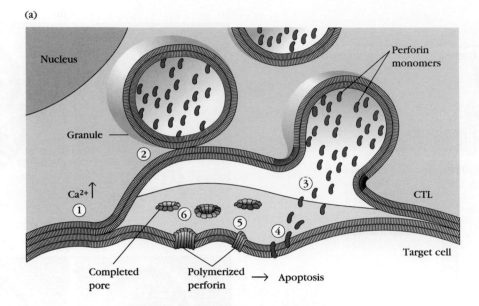

(b)

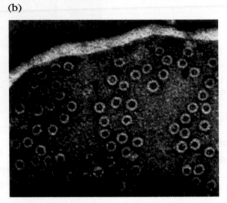

FIGURE 13-11 CTL-mediated pore formation in target-cell membrane. (a) In this model, a rise in intracellular Ca^{2+} triggered by CTL-target cell interaction (1) induces exocytosis, in which the granules fuse with the CTL cell membrane (2) and release monomeric perforin into the small space between the two cells (3). The released perforin monomers undergo a Ca^{2+}-induced conformational change that allows them to insert into the target-cell membrane (4). In the presence of Ca^{2+}, the monomers polymerize within the membrane (5), forming cylindrical pores (6). (b) Electron micrograph of perforin pores on the surface of a rabbit erythrocyte target cell. The arrow indicates a single pore. *[Part a adapted from J. D. E. Young and Z. A. Cohn, 1988, How killer cells kill, Scientific American **258**(1):38; part b from E. R. Podack and G. Dennert, 1983, Assembly of two types of tubules with putative cytolytic function by cloned natural killer cells, Nature **301**:442.]*

perforin to CTL-mediated killing is demonstrated in perforin-deficient KO mice, which are unable to eliminate *lymphocytic choriomeningitis virus* (LCMV) from the body even though they mount a significant CD8[+] immune response to virally infected cells.

Although granzyme B was once thought to gain entry into the target cell via surface perforin pores, it is now thought that it enters via endocytic processes. Many target cells have a molecule known as the mannose 6-phosphate receptor on their surface that also binds to granzyme B. Granzyme B-mannose 6-phosphate receptor complexes are internalized and appear inside vesicles. Perforin internalized at the same time then forms pores that release granzyme B from the vesicle into the cytoplasm of the target cell.

Regardless of the mechanism of entry, once in the cytoplasm, granzyme initiates a cascade of reactions that result in the fragmentation of the target-cell DNA into oligomers of 200 base pairs (bp); this type of DNA fragmentation is typical of apoptosis. Granzyme proteases do not directly mediate DNA fragmentation. Rather, they activate an apoptotic pathway within the target cell. This apoptotic process does not require mRNA or protein synthesis in either the CTL or the target cell. Within 5 minutes of CTL contact, target cells begin to exhibit DNA fragmentation. Interestingly, viral DNA within infected target cells has also been shown to be fragmented during this process. This observation shows that CTL-mediated killing not only kills virus-infected cells but also can destroy the viral DNA in those cells directly. The rapid onset of DNA fragmentation after CTL contact may prevent continued viral replication and assembly in the period before the target cell is destroyed.

How do CTLs protect themselves from their own perforin and granzyme activity? Studies show that CTLs are more resistant to the activities of granzyme and perforin than their targets. The strategies they use are not fully understood, but investigators have found that CTLs express high levels of serine protease inhibitors (serpins), which protect them from granzyme B activity. In support of the role of serpins in protecting CTLs from their own pro-apoptotic functions, CTLs do not survive in mice deficient for one of the most common serpins expressed by $CD8^+$ T cells, Spi-1.

Fas-FasL Mediated Cytolysis Some potent CTL lines have been shown to lack perforin and granzymes. In these cases, cytotoxicity is mediated by **Fas (CD95)**. As described in Chapter 3, this transmembrane protein, which is a member of the TNF-receptor family, can deliver a death signal when cross-linked by its natural ligand, a member of the tumor necrosis factor family called **Fas ligand (FasL)**. FasL is found on the membrane of CTLs, and the interaction of FasL with Fas on a target cell triggers apoptosis.

Fas mutations lead to multiple disorders in both mice and humans. Mice that are homozygous for the *lpr* mutation express little or no Fas on their cells, and have remarkably large lymph nodes. In more rigorous terms, they are afflicted with a lymphoproliferative disease characterized by the accumulation of mature, activated T and B lymphocytes in their lymph nodes. CTLs in *lpr* mutant mice can be induced to express FasL; however, these CTLs cannot kill targets, which express no Fas. These mice also develop autoimmune diseases, presumably because they kill autoreactive lymphocytes that are activated during an immune response. A very similar disorder, now known as *autoimmune lymphoproliferative syndrome* (ALPS or Canale-Smith syndrome), has been identified in human patients who have genetic defects in Fas, Fas ligand, or Fas signaling pathways.

The critical importance of both the perforin and the Fas-FasL systems in CTL-mediated cytolysis was revealed by experiments with two types of mutant mice: perforin KO mice and the Fas deficient *lpr* strain (Figure 13-12). Investigators wanted to determine the relative importance of each of these molecules and developed a system where they generated CTLs by co-culturing T cells from $H-2^b$ mice with killed target cells from $H-2^k$ mice. CTLs generated in such *mixed lymphocyte reactions* (see below) reacted strongly against allogeneic MHC and could kill $H-2^k$-expressing cells. With this system, investigators first asked if CTLs could kill target cells generated from *lpr* mice (which expressed no Fas). In fact, they could, demonstrating that Fas-FasL interactions were not absolutely required for CTL activity. Next they asked if perforin was required and generated CTLs from perforin KO mice. They found that these cells could also kill targets, demonstrating that perforin wasn't

absolutely required. Finally, they asked if killing occurred in the absence of both perforin *and* Fas-FasL interactions. They generated CTL cells from perforin KO mice and determined their ability to kill target cells from the Fas-deficient mice. No killing was observed. These and other observations indicate that CTLs employ these and only these two methods—perforin-mediated and Fas-mediated apoptosis—to kill their targets.

Both the perforin and Fas-FasL killing strategies activate a signaling pathway in the target cell that induces apoptosis (Figure 13-13). A central feature of cell death by apoptosis is the involvement of the **caspase** family of cysteine proteases, which cleave after an aspartic acid residue. The term *caspase* incorporates each of these elements (*c*ysteine, *asp*artate, prote*ase*). Normally, caspases are present in the cell as inactive proenzymes—procaspases—that require proteolytic cleavage for conversion to the active forms. Cleavage of a procaspase produces an active caspase, which cleaves other procaspases, thereby activating their proteolytic activity, which results in a cascade of events that systematically disassembles the cell—the hallmark of apoptosis. More than a dozen different caspases have been found, each with its own specificity. They are typically divided into two categories: those that initiate the caspase cascade (initiator caspases) and those that directly initiate apoptosis (effector caspases). See Chapter 9 for a more thorough discussion of the apoptotic process.

What strategies do CTLs use to initiate caspase activation? The engagement of Fas on a target cell by Fas ligand on a CTL first induces the activation of an initiator caspase in the target cell. Fas is associated with a protein known as *Fas-a*ssociated protein with *d*eath *d*omain (FADD), which in turn associates with procaspase-8 (see Figure 13-13a). Upon Fas cross-linking, procaspase-8 is converted to caspase-8 and initiates an apoptotic caspase cascade.

The granzymes introduced into the target cell by a CTL are proteolytic and have several targets (see Figure 13-13b). They can directly cleave procaspase-3, an event that appears to only partially activate this effector caspase. They can also cleave bid, which induces mitochondrial release of cytochrome c and, ultimately, the activation of caspase-9. Both activities appear to be required for optimal pro-apoptotic activity.

The end result of both perforin-granzyme and Fas-mediated pathways is, therefore, the activation of dormant death pathways that are present in the target cell. As one immunologist has so aptly put it, CTLs don't so much kill target cells as persuade them to commit suicide.

Natural Killer Cells Recognize and Kill Infected Cells and Tumor Cells by Their Absence of MHC Class I

Another immune cell subtype, NK cells, initiates apoptotic pathways in target cells using very similar mechanisms but very different receptors. NK cells were discovered essentially

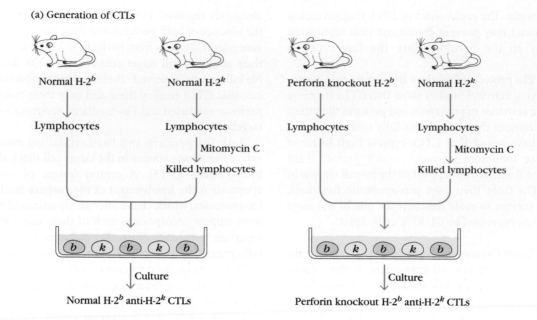

(a) Generation of CTLs

(b) Interaction of CTLs with Fas⁺ and Fas⁻ targets

	Target cells	
CTLs	Normal H-2^k	*lpr* mutant H-2^k (no Fas)
Normal H-2^b anti-H-2^k	Killed	Killed
Perforin knockout H-2^b anti-H-2^k	Killed	Survive

FIGURE 13-12 Experimental demonstration that CTLs use Fas and perforin pathways. (a) Lymphocytes were harvested from mice of H-2^b and H-2^k MHC haplotypes. H-2^k haplotype cells were killed by treatment with mitomycin C and co-cultured with H-2^b haplotype cells to stimulate the generation of anti-H-2^k CTLs. If the H-2^b lymphocytes were derived from normal mice, they gave rise to CTLs that had both perforin and Fas ligand. If the CTLs were raised by stimulation of lymphocytes from perforin knockout (KO) mice, they expressed Fas ligand but not perforin. (b) Interaction of CTLs with Fas⁺ and Fas⁻ targets. Normal H-2^b anti-H-2^k CTLs that express both Fas ligand and perforin kill normal H-2^k target cells and H-2^k *lpr* mutant cells, which do not express Fas⁺. In contrast, H-2^b anti-H-2^k CTLs from perforin KO mice kill Fas1 normal cells by engagement of Fas with Fas ligand but are unable to kill the *lpr* cells, which lack Fas.

by accident when immunologists were measuring the cytolytic ability of lymphocytes isolated from mice with tumors. They originally predicted that these lymphocytes would exhibit a specific ability to kill the tumor cells to which they had been exposed. To do their experiments they included multiple controls, comparing the activity of these lymphocytes of interest with those taken from mice that had no tumors, as well as mice that had unrelated tumors. Much to their surprise, the investigators discovered that even the control lymphocytes, which either had not been exposed to any tumor or had been exposed to a very different type of tumor, were able to kill the tumor cells of the first mouse. In fact, the killing they were seeing did not seem to be following the rules of specificity that are the hallmark of a lymphocyte response to conventional antigens and pathogens.

Further study of this nonspecific tumor-cell killing revealed that, in fact, neither T nor B lymphocytes were involved at all. Instead, a population of larger, more granular lymphocytes was responsible.

The cells, called "natural killer" cells for their nonspecific cytotoxicity, make up 5% to 10% of the circulating lymphocyte population. Despite the absence of specific antigen receptors (i.e., antibody, T-cell receptors), they play a major role in immune defenses against infected cells, stressed cells, and tumor cells. They can also contribute to autoimmunity when dysregulated. More versatile than originally anticipated, NK cells also play a regulatory role in both innate and adaptive immune responses to conventional antigens by secreting cytokines that alter the immune response. They have also recently been shown to be critically important for the development of a normal placenta, not via their cytotoxic ability, but via their ability to recognize the presence of a different (paternal) MHC and initiate the remodeling of blood vessels.

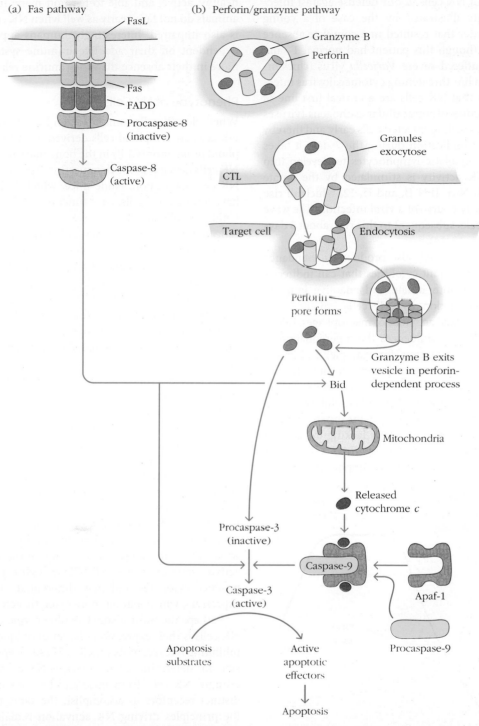

(a) Fas pathway

FasL

Fas

FADD

Procaspase-8 (inactive)

Caspase-8 (active)

(b) Perforin/granzyme pathway

Granzyme B

Perforin

Granules exocytose

CTL

Target cell   Endocytosis

Perforin pore forms

Granzyme B exits vesicle in perforin-dependent process

Bid

Mitochondria

Released cytochrome *c*

Procaspase-3 (inactive)

Caspase-9

Caspase-3 (active)

Apaf-1

Procaspase-9

Apoptosis substrates

Active apoptotic effectors

Apoptosis

FIGURE 13-13 Two pathways of target-cell apoptosis stimulated by CTLs. (a) The Fas pathway. Ligation of trimeric Fas units by CTL-borne Fas ligand leads to the association of Fas with the adapter molecule FADD, which in turn results in a series of reactions that activate a caspase cascade, leading to apoptosis of the target cell. (b) The perforin-granzyme pathway. Granule exocytosis releases granzymes and perforin from the CTL into the space between the CTL and the target cell. Granzyme B enters the target cell by endocytosis. Granzyme B is then released into the cytoplasm in a perforin-dependent process. Granzyme B can cleave and activate the pro-apoptotic bcl-2 family member bid, which stimulates mitochondria to release cytochrome c, which activates death pathways; granzyme B can also cleave and partially activate caspase-3. Release of cytochrome c and activation of caspase-3 are both required to initiate the caspase cascade that leads to cell apoptosis. *[Adapted from M. Barry and C. Bleackley, 2002, Cytotoxic T lymphocytes: All roads lead to death, Nature Reviews Immunology 2:401.]*

The importance of NK cells in our defense against infections is compellingly illustrated by the case of a young woman with a disorder that resulted in a complete absence of these cells. Even though this patient had normal T- and B-cell counts, she suffered severe *Varicella* virus (chickenpox) infections and a life-threatening cytomegalovirus infection. We now know that NK cells are a critical first line of defense against infection of intracellular pathogens (viruses and some bacteria) by killing infected cells early and thereby controlling pathogen replication during the 7 days it takes precursors of CD8$^+$ cytolytic lymphocytes to develop into functional CTLs. NK activity is stimulated by the innate immune cytokines IFN-α, IFN-β, and IL-12, which all rise rapidly during the early course of a viral infection. The wave of NK cell activity peaks subsequent to this rise, about 3 days after infection (Figure 13-14).

As indicated above, NK cells also produce an abundance of immunologically important cytokines that can indirectly but potently influence both innate and adaptive immune responses. IFN-γ production by NK cells enhances the phagocytic and microbicidal activities of macrophages. IFN-γ derived from NK cells also influences the differentiation of CD4$^+$ T helper subsets, inhibiting (for example) T$_H$2 proliferation and stimulating T$_H$1 development via induction of IL-12 by macrophages and dendritic cells (see Chapter 11).

NK cells are potent enough that, in conjunction with other protective mechanisms provided by the innate immune system, they can protect animals totally lacking in adaptive immunity. This is nicely illustrated by *RAG-1* KO mice which have no antigen-specific B or T lymphocytes yet are healthy, active, and able to fend off many infections. These animals do not fare nearly as well when NK cell development is also impaired. Interestingly, humans appear to be more dependent on their adaptive immune systems and suffer more in their absence than their murine relatives.

Phenotype of NK Cells

Where do NK cells come from, and what do they look like? NK cells are lymphoid cells derived from the *common lymphoid progenitor* (CLP) in the bone marrow. Although some NK cells develop in the thymus, this organ is not required for NK cell maturation. Nude mice, which lack a thymus and have few or no T cells, have functional NK cell populations. Unlike T cells and B cells, NK cells do not undergo rearrangement of receptor genes; NK cells still develop in mice in which the recombinase genes *RAG-1* or *RAG-2* have been knocked out. Consistent with this observation, NK cells are also found in *severe combined immunodeficiency* (SCID) mice, which have no T and B cells because of a defect in enzymatic activity required for receptor rearrangement.

NK cells are quite heterogeneous. Different subpopulations can be distinguished based on differences in expression and secretion of specific immunologically relevant molecules. Whether this heterogeneity reflects different stages in their activation or maturation or truly distinct subpopulations remains unclear.

Most murine NK cells express CD122 (the 75-kDa β subunit of the IL-2 receptor), NK1.1 (a member of the NKR-P1 family), and CD49b (an integrin). NK cells also typically express CD2 and FcγRIII. In fact, cell depletion with monoclonal anti-FcγRIII antibody removes almost all NK cells from circulation. Human NK cells also express IL-2 receptors and FcγRIII, but do not express NK1.1. Rather, they are distinguished from other lymphocytes by CD56 expression, which varies in intensity depending on the maturation and activity state of an NK cell (CD56 high expressers tend to produce cytokines and may differentiate into CD56 low expressers, which exhibit more cytolytic activity).

Perhaps the most distinctive phenotypic characteristic of NK cells is their expression of a set of unique activating and inhibiting NK receptors (NKRs). These receptors are responsible for determining which targets NK cells will kill. Interestingly, NK cells from mice and humans use surprisingly distinct receptors to accomplish the same thing; however, the principles driving NK activation remain the same. In addition, the number and type of activating and inhibitory receptors expressed by NK cells vary widely, and the balance of signals received through these receptors determines whether NK cells induce death of their targets.

How NK Cells Recognize Targets: The Missing Self Model

Because NK cells do not express antigen-specific receptors, the mechanism by which these cells recognize tumor or infected cells and distinguish them from normal body cells

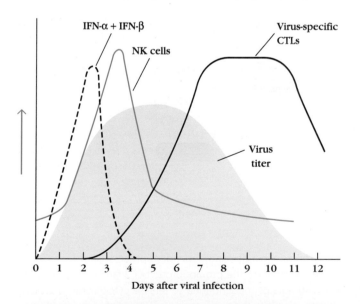

FIGURE 13-14 Time course of viral infection. IFN-α and IFN-β (dashed curve) are released from virus-infected cells soon after infection. These cytokines stimulate the NK cells, quickly leading to a rise in the NK-cell population (blue curve) from the basal level. NK cells help contain the infection during the period required for generation of CTLs (black curve). Once the CTL population reaches a peak, the virus titer (blue area) rapidly decreases.

baffled immunologists for years. Klas Karre advanced an interesting hypothesis that formed the foundations for our understanding of how NK cells distinguish self from non (or altered) self. He proposed that NK cells kill when they do not perceive the presence of normal self-proteins on a cell, the **missing self model**. The implied corollary to the proposal is that recognition of self inhibits the ability to kill.

Clues to the origins of such inhibitory signals came from early studies looking at which tumor targets NK cells killed best. Investigators examined multiple variables associated with NK preferences and discovered that killing correlated best with levels of MHC class I molecules expressed on the tumor cells. In one study, they examined the ability of CTLs and NK cells to kill B cells that were transformed into tumor cells by infection with *Epstein-Barr virus* (EBV). CTLs were unable to recognize and lyse these B cells. However, NK cells were very effective killers of these tumor cells. Ultimately, investigators realized that EBV infection down-regulated MHC class I, allowing the cells to evade CD8$^+$ T cell recognition. However, this absence of class I made them perfect targets for NK cells, which responded to the "missing self." The investigators "clinched" a role of MHC class I by transfecting the B-cell tumors with human MHC class I genes. NK cells were no longer able to kill the cells. The subsequent discovery of receptors on NK cells that produce inhibitory signals when they recognize MHC molecules on potential target cells provided direct support for this model. These inhibitory receptors were shown to prevent NK-cell killing, proliferation, and cytokine release.

Since many virus-infected and tumor cells decrease MHC expression, the missing self model (i.e., basing the decision to kill on whether a target expresses sufficient levels of MHC class I, a ubiquitous self-protein) makes good physiological sense. Although the fundamental paradigm has stood the test of time, not surprisingly the real situation is more complicated (Figure 13-15). As indicated above, it turns out that NK cells express two different categories of receptors: one that delivers signals that inhibit the cytotoxic activity of NK cells (those that recognize MHC class I products) and another that delivers signals that enhance cytotoxic activity. NK cells distinguish healthy cells from infected or cancerous ones by monitoring and integrating both sets of signals; whether a target is killed or not depends on the balance between activators and inhibitors. Inhibiting signals, in general, can trump activating signals; thus, cells expressing normal levels of unaltered MHC class I molecules tend to escape all forms of NK-cell-mediated killing.

NK Cell Receptors

NK receptors fall into two functional categories: inhibitory molecules that bind MHC class I and block the NK cell from killing, and activating molecules that trigger NK cell cytotoxicity if an inhibiting receptor is not occupied. Both of these groups also fall into two structural categories: members with lectin-like extracellular regions and members with immunoglobulin-like extracellular regions (Table 13-5).

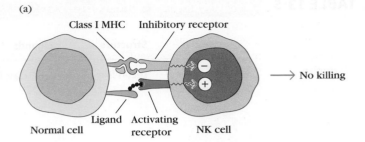

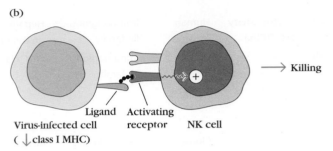

FIGURE 13-15 Model of how cytotoxic activity of NK cells is restricted to altered self cells. An activating receptor on NK cells interacts with its ligand on normal and altered self cells, inducing an activation signal that results in killing. However, engagement of inhibitory NK-cell receptors such as inhibitory KIRs and CD94-NKG2 by class I MHC molecules delivers an inhibition signal that counteracts the activation signal. Expression of class I molecules on normal cells (a) thus prevents their destruction by NK cells. Because class I expression is often decreased on altered self cells (b), the killing signal predominates, leading to their destruction.

Note that although lectins typically bind carbohydrates, most of the lectin-like NK-cell receptors bind proteins rather than carbohydrates.

The extracellular structure of the NK-cell receptor does not immediately identify it as inhibitory or activating. NK receptors with identical extracellular structures can have different intracellular domains and therefore different signaling properties. As with FcRs (described above) the intracellular sequences of the activating NK receptors generally contain ITAMs (see Chapter 3), and the intracellular sequences of inhibitory NK receptors contain ITIMs. (Interestingly, like NK receptors, Fc receptors also fall into two structural families: lectin-like and immunoglobulin-like.)

Inhibitory NK Receptors and Ligands Inhibitory receptors, which bind to MHC class I molecules, are the most important determinant of an NK cell's decision to kill. In humans, they are members of a diverse immunoglobulin-like family known as the *killer-cell immunoglobulin-like receptors* (**KIR**). Surprisingly, the KIR family appears to have evolved extremely rapidly. Functional KIR receptors found in primates do not exist in rodents. Mice use a distinct family of receptors, the lectin-like **Ly49** family, to achieve the same function as KIRs, namely inhibition of NK cell cytolytic activity through binding to MHC class I

TABLE 13-5   Features of NK cell receptors

	Species	Structure	Ligands	Activating or inhibiting	Examples
NKG2D	Mouse and human	Lectin-like	MHC class I	Most but not all activating	NKG2D (activating) binds MIC-A, MIC-B, and ULPB family members (human) or H60, Mult1, Rae-1 family members (mice) CD94-NKG2A (inhibiting, human) binds HLA-E
Natural cytotoxicity receptors (NCRs)	Human	Immunoglobulin family	Heparin sulfate on tumor cells	Most (all?) activating	NKp30, NKp44, and NKp46
KIR	Human	Immunoglobulin family	MHC class I (HLA-B and HLA-C)	Most inhibiting	KIR2DL1 (inhibiting) interacts with HLA-C KIR2DS4 (activating) interacts with HLA-Cw4 and other non-HLA proteins in tumors
Ly49	Mouse	Lectin-like	MHC class I and homologs	Most inhibiting	Ly49E (inhibiting) recognizes urokinase Ly49H (activating) recognizes a viral (MCMV) homolog of MHC class I

molecules on healthy cells. Functional Ly49 receptors do not exist in humans.

Both KIR and Ly49 family members are highly diverse and polymorphic. These inhibitory receptors are generally specific for polymorphic regions of MHC class I molecules ($H-2^k$ or $H-2^d$ in the mouse; HLA-A, HLA-B, or HLA-C in the human). Only a fraction of MHC class I and class I-like ligands for these diverse receptors have been identified. Given the genetic polymorphism of both KIR receptors and their MHC class I ligands, it is not surprising that individuals can inherit KIR-MHC class I combinations that are not ideal and could increase their susceptibilities to disease.

An exception to the rule of thumb that human NK inhibitory receptors are Ig-like molecules is the lectin-like inhibitory receptor CD94-NKG2A, a disulfide-bonded heterodimer made up of two glycoproteins: CD94 and a member of the NKG2 family. Whereas KIR receptors typically recognize polymorphisms of HLA-B or HLA-C, the CD94-NKG2A receptors recognize HLA-E on potential target cells. This exception also makes biological sense. Because HLA-E is not transported to the surface of a cell unless it has bound a peptide derived from the HLA-A, HLA-B, or HLA-C proteins, the amount of HLA-E on the surface serves as an indicator of the overall level of class I MHC biosynthesis in the cells. These inhibitory CD94-NKG2A receptors recognize surface HLA-E and send inhibitory signals to the NK cell, with the net result that killing of potential target cells is inhibited if they are expressing adequate levels of class I.

Unlike the antigen receptors expressed by B cells and T cells, NK receptor cells are not subject to allelic exclusion, and NK cells can express several different KIR or Ly49 receptors, each specific for a different MHC molecule or for a set

of closely related MHC molecules. Individual clones of human NK cells expressing a CD94-NKG2A receptor and as many as six different KIR receptors have been found. The ability of each NK cell to express multiple KIR or NKG2A receptors improves its chances of recognizing the polymorphic MHC class I variants expressed by an individual's cells, and therefore prevents NK from killing healthy cells simply because a KIR variant it expressed didn't happen to bind to the MHC class I variant expressed by that normal cell.

Activating NK Receptors and Ligands   Most activating receptors expressed by murine and human NK cells are structurally similar and many are C-lectin like—so named because they have calcium-dependent carbohydrate recognition domains. **NKG2D** has emerged as one of the most important activating receptors and acts through a signaling cascade similar to that initiated by CD28 in T cells. The ligands for NKG2D are nonpolymorphic MHC class I-like molecules that do not associate with β_2-microglobulin. They are often induced on cells undergoing stress, such as DNA damage or infection. Thus, they help to focus NK cell activity on cell and situations where their lytic activity would be most adaptive. One particularly interesting example of an activating lectin-like NK receptor is Ly49H, a mouse receptor that interacts with an MHC class I-like protein encoded not by the body but by a murine virus, MCMV. In the absence of this receptor, mice are poorly protected from this common virus.

Many molecules that are not exclusively expressed by NK cells also enhance NK activity. These include CD2 (the receptor for the adhesion molecule LFA-3), CD244 (the receptor for CD48), as well as receptors for inflammatory cytokines. The involvement of each of these proteins, again, makes biological sense, focusing NK cell attention on environments that are suffering inflammatory insults.

Finally, it is important to remember that most human and murine NK cells express the FcγIII receptor (CD16), which is responsible for most antibody-mediated recognition and killing of target cells by NK cells (ADCC). As described above, ADCC is a potent addition to our response to infection and malignancy. Cells infected with virus, for instance, often express viral envelope proteins on their surface. Antibodies produced by B cells that responded to these proteins will bind the cell surface and recruit NK cells, which will lyse the infected cell.

How NK Cells Induce Apoptosis of Their Targets

Regardless of which receptors are involved in regulating NK-cell lytic activity, NK cells kill targets by processes similar to those employed by CTLs. NK cells express and secrete FasL and readily induce death in Fas-bearing target cells. The cytoplasm of NK cells also has numerous granules containing perforin and granzymes. Like CTLs, NK cells develop an organized immunological synapse at the site of contact with a target cell, after which degranulation occurs, with release of perforin and granzymes at the junction between the interacting cells. Perforin and granzymes are thought to play the same roles in NK-mediated killing of target cells by apoptosis as they do in the CTL-mediated killing process.

NK Activity "Licensing" and Regulation

Even newly formed NK cells have large granules in their cytoplasm, and it was traditionally thought that NK cells were capable of killing from the moment they matured. It is now thought that most NK cells need to be licensed before they can use their cytotoxic machinery on any target (even with the proper activating and inhibiting receptor signaling). Licensing of an NK cell is thought to occur via a first engagement of their inhibitory, MHC class I binding, receptors. This event can be considered a way for the immune system to test an NK cell's ability to restrain its killing activity and an important strategy for maintaining tolerance to self in this cell subset. Specifically, licensing allows only those cells that have the capacity to be disarmed via inhibitory interactions to be armed.

Once licensed, NK cells are thought to continually browse tissues and target cells via their multiple inhibiting and activating receptors. Engagement of activating ligands on the surface of a tumor cell, virus-infected cell, or otherwise stressed cell signals NK cells to kill the target cell. If the NK cells' inhibitory receptors detect normal levels of MHC class I on potential target cells, these inhibitor signals override the activation signals. This would not only prevent the death of the target cell but would also abrogate NK-cell proliferation and cytokine (e.g., IFN-γ and TNF-α) production. The overall consequence of this strategy is to spare cells that express critical indicators of normal self, the MHC class I molecules, and to kill cells that lack indicators of self and do not express normal levels of class I MHC. A full under-

standing of the role of and requirement for licensing awaits further study.

NK Cell Memory

The line between NK cells and B and T lymphocytes continues to blur. B and T lymphocytes were once considered the only cells in the immune system that could generate memory responses. However, recent data indicate that some NK cells can also generate a memory response to antigen. The evidence for this remarkable property came from experiments showing that NK cells expressing a receptor that binds a viral protein could transfer memory of this antigen exposure to naïve animals that had not been previously exposed or infected (see Classic Experiment Box 13-2).

These observations show that at least some NK cells possess the developmental machinery to become memory cells, in other words to increase their longevity (although perhaps not as long as B and T lymphocytes) and improve their response times. We are left with a number of questions: Do all NK cells have the capacity to develop into memory cells? And given that NK cells do not express conventional antigen-specific receptors, against which antigens can they develop memory?

NKT Cells Bridge the Innate and Adaptive Immune Systems

The above discussions of cell-mediated immunity covered the CTL, an important component of adaptive immunity that expresses an antigen-specific TCR, and the NK cell, a cell with both innate and adaptive properties, bearing receptors that recognize self and "missing" self on cell surfaces. A third type of cytolytic lymphocyte has been identified with characteristics shared by both the CTL and the NK cell. This cell type, designated the **NKT cell** to reflect its hybrid quality, develops in the thymus, and, strictly speaking, is a member of the adaptive immune system. It undergoes antigen-receptor gene rearrangements and expresses a TCR complex on its surface. However, it also exhibits characteristics of cells in the innate immune system:

- The T-cell receptor on the human NKT cell is invariant, with the TCRα and TCRβ chains encoded by specific gene segments ($V_\alpha 24$-$J_\alpha 18$ and $V_\beta 11$, respectively) within the germ-line DNA; the cells expressing this αβ TCR combination are therefore sometimes referred to as **invariant NKT (iNKT) cells**.

- The TCR on NKT cells does not recognize MHC-bound peptides but rather a glycolipid presented by the non-polymorphic CD1d molecule (Figure 13-16).

- NKT cells can act as both helper cells (secreting cytokines that shape responses) and cytotoxic cells (killing target cells).

- NKT cells include both CD4$^+$ and CD4$^-$ subpopulations, which may also differ by cytokine production.

Rethinking Immunological Memory: NK Cells Join Lymphocytes as Memory-Capable Cells

Memory has a mystique in immunological circles, partly because it is the fundamental basis for the success of vaccination, which empowered societies by radically changing our relationship to disease, and partly because its molecular basis still remains elusive. As one of the hallmarks of the adaptive immune system, memory has always been considered an exclusive feature of B cells and T cells.

NK cells, the underappreciated lymphocyte cousins, have no antigen-specific receptors and until recently were assumed to have no capacity for memory. As recognition of their importance in responding to infection, killing tumor cells, and sculpting adaptive immune systems has grown, assumptions about their activities have been reexamined. To our surprise, the results of several clever experiments have revealed that NK cells do, indeed, share a capacity for antigen-specific memory with B and T cells.

O'Leary et al. showed in 2006 that memory of an antigen that causes hypersensitivity can be transferred by liver NK cells into a mouse that had never seen this antigen. Sun et al. showed in 2009 that memory of infection with murine cytomegalovirus (MCMV) could be transferred by NK cells. Below we discuss aspects of this latter study.

To see if NK cells had any memory capacity, the investigators wisely chose to examine the NK response to MCMV. NK cells are known to be an important component of protection against infection with MCMV and a large fraction of NK cells (up to 50%) express an activating receptor specific for MCMV (Ly49H). A schematic of their experimental approach and results is shown in Figure 1 and two key pieces of Sun et al.'s original data are shown in Figure 2.

The investigators devised a rigorously controlled system where they could track MCMV-specific NK cells. Specifically, they adoptively transferred (introduced a population of cells from one mouse to another) purified NK cells from wild-type mice into mice whose NK cells were not capable of expressing or signaling through Ly49H. (These mice are deficient in a signaling molecule, DAP12, that sends signals from Ly49 receptors.) The donor NK cells also differed from the host NK cells by expression of CD45.1, a CD45 allelic variant that can be easily identified by flow cytometry (see Chapter 20). This way, the investigators could control the number of Ly49H NK cells in the mouse and could specifically trace those they introduced.

They first found that the population of NK cells that were specific for MCMV (Ly49$^+$CD45.1$^+$) proliferated during the 7 days following infection then declined over the weeks after viral infection (see Figure 2b). This expansion and contraction

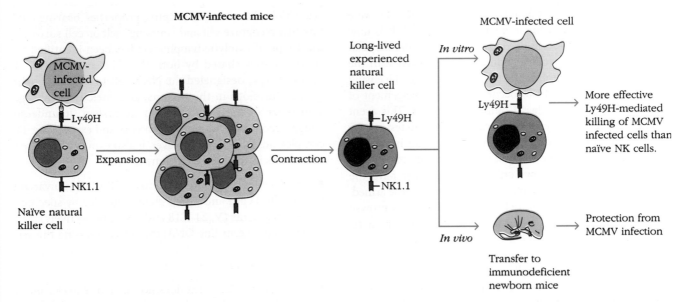

FIGURE 1

The investigator's experimental approach. To examine the possibility that NK cells exhibited hallmark characteristics of memory cells, investigators first looked to see if NK cells with the Ly49H activating receptor, which binds to the MCMV virus, would expand and persist. They next looked to see if the NK cells that persisted after infection could protect another animal from infection with MCMV. They adoptively transferred the NK cells that had been generated in the infected mouse into an uninfected mouse pup. Then they infected that mouse with MCMV and looked for signs of disease. Finally, they examined the behavior of these persistent NK cells in vitro, to see if they responded faster and better to signals through Ly49H. The results of these experiments are described in the text and some are shown in Figure 2. [Adapted from J. C. Sun, et al., 2009, Adaptive immune features of natural killer cells. Nature **457**:557–561.]

BOX 13-2

precisely paralleled the behavior of conventional antigen-specific T cells (specifically, CD4+ T cells) in a primary response.

They next asked a critical question. Did this antigen-specific NK-cell response result in memory? Specifically, did MCMV NK cells persist over time? And if they did, would they exhibit a more robust response if rechallenged with MCMV? The investigators show that even 70 days after infection, a population of Ly49H+ (MCMV specific), CD45.1+ (donor-derived cells) could be detected. When the activity of these cells was compared to cells from mice that hadn't been infected, the investigators found that they were, indeed, more active: twice as many NK cells from infected mice ("memory NK cells") could make IFN-γ than NK cells from uninfected mice.

However, to show definitively that this memory was physiologically relevant, the investigators asked if memory NK cells could rescue mice from infection. They isolated and transferred naïve and "memory" NK cells in varying numbers into neonatal mice, which are naturally immunodeficient and die if infected with MCMV. They asked whether the memory NK cells would enhance survival of infected neonatal mice. Figure 2c shows their results as a survival curve, which shows frequency of mice that remain alive versus time (in days) after infection. In the absence of antigen-specific NK cells (PBS control), 100% of mice die within 2 weeks of a primary infection with MCMV. However, when Ly49H+ NK cells are transferred, survival prospects increase. In fact, in the presence of Ly49H+ NK cells isolated from previously infected mice (Memory NK), 75% of the mice lived. Introduction of Ly49H+ cells from mice that had not been previously infected (Naïve NK) also improved survival, but it took at least 10-fold more of these naïve NK cells to provide equal protection.

These and other studies show that lymphocytes must share the immunological stage with at least some NK cells as actors with excellent memories. Whether such memory applies to other antigen-specific NK subsets—indeed to other immunological cells—and whether it has the same molecular features as lymphocyte memory remain open and worthy questions.

O'Leary, J. G., M. Goodarzi, D. L. Drayton, and U. H. von Andrian. 2006. T cell- and B cell-independent adaptive immunity mediated by natural killer cells. *Nature Immunology* **7**:507–516.

Sun, J. C., et al. 2009. Adaptive immune features of natural killer cells. *Nature* **457**:557–561.

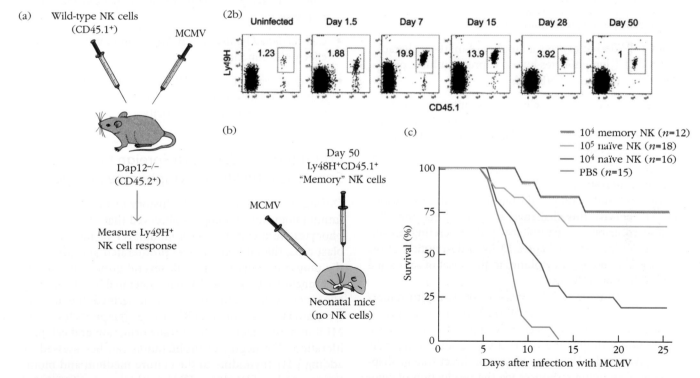

FIGURE 2
Experimental results. (a) The experimental design that allowed the investigators to ask how NK cells reactive to MCMV responded to infection. Investigators took advantage of strain differences in CD45 allotypes to trace responsive NK cells. Naïve NK cells from mice that expressed the CD45.1 allotype were introduced into mice that expressed the CD45.2 allotype. (CD45 is expressed by all blood cells.) These hosts were then infected with MCMV. Ly49H+ NK cells that expressed the CD45.1 marker were then identified by flow cytometry. (b) The flow cytometry profile of spleen cells isolated at varying times after infection and stained with fluorescent antibodies to Ly49H and CD45.1. The meaning of these results is discussed in the text. (c) The NK cells from part (b) that persisted through day 50 after MCMV were then isolated. Varying numbers of these "memory NK cells" and varying numbers of naïve, freshly isolated NK cells were transferred into animals that had not been exposed to virus. These animals were then infected with MCMV. Part (c) shows the percent survival of the mice in each group. The meaning of these results is discussed in the text. [Adapted from J. C. Sun, et al., 2009, Adaptive immune features of natural killer cells, Nature **457**:557–561.]

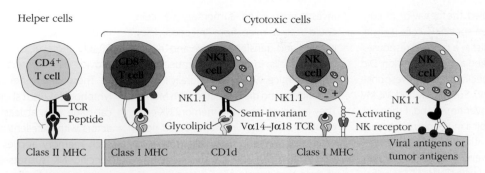

FIGURE 13-16 Comparison of ways effector cells of the immune system recognize their targets. CD4$^+$ and CD8$^+$ T cells express a diverse array of TCRαβ chains and interact with classical MHC class II-peptide and MHC class I-peptide complexes, respectively. NKT cells express an invariant TCRαβ chain and bind a unique MHC class I molecule (CD1d)-glycolipid complex. NK cells have two ways to recognize targets. Using their FcRs, they can attach themselves to antibodies that have bound viral or tumor antigens on the surface of a cell. They can also bind MHC class I with their inhibiting NK receptors (e.g., KIR). In the absence of this binding, they will kill a target cell. *[Adapted from J. U. Adams, 2005, Another kind of antigen,* The Scientist *19:15.]*

- NKT cell killing appears to depend predominantly on FasL-Fas interactions.

- NKT cells do not form memory cells.

- NKT cells do not express a number of markers characteristic of T lymphocytes but do express multiple proteins characteristic of NK cells.

The exact role of NKT cells in immunity remains to be defined. However, experiments show that mice lacking NKT cells mount a deficient response to certain low-dose infections by bacteria that express glycolipids that can be recognized by NKT cell receptors (e.g., *Sphingomonas* and *Ehrlichia*). Interestingly, high-dose *Sphingomonas* infection leads to sepsis and death in wild-type mice, but those lacking NKT cells survive this challenge, suggesting that the NKT cells also contribute to pathology by oversecreting inflammatory cytokines. (See Chapter 15 for a description of the role of proinflammatory cytokines in the onset of sepsis and the exacerbation of disease.)

Other data implicate NKT cells in immunity to tumors and suggest that NKT cells recognize lipid antigens specific to tumor cells. Finally, NKT cells also appear to contribute to viral immunity, despite the fact that viruses do not typically express glycolipids. They may play an indirect role in shaping the viral immune responses via the production of cytokines, including IFN-γ, IL-2, TNF-α, and IL-4.

Experimental Assessment of Cell-Mediated Cytotoxicity

Three experimental systems have been particularly useful for measuring the activation and effector phases of cell-mediated cytotoxic responses:

- The *mixed-lymphocyte reaction (MLR)* is an in vitro system for assaying T$_H$-cell proliferation in a cell-mediated response.

- *Cell-mediated lympholysis (CML)* is an in vitro assay of effector T$_C$ function.

- The *graft-versus-host (GVH) reaction* in experimental animals provides an in vivo system for studying cell-mediated (T$_H$ and T$_C$) cytotoxicity.

Co-Culturing T Cells with Foreign Cells Stimulates the Mixed-Lymphocyte Reaction

During the 1960s, early in the history of modern cellular immunology, immunologists observed that when rat lymphocytes were cultured on a monolayer of mouse fibroblast cells, the rat lymphocytes proliferated and destroyed the mouse fibroblasts. In 1970, several groups discovered that functional CTLs could also be generated by co-culturing allogeneic spleen cells in a system termed the **mixed-lymphocyte reaction (MLR)**. The T lymphocytes in an MLR undergo extensive blast transformation and cell proliferation. The degree of proliferation can be assessed by adding [^{3}H]-thymidine to the culture medium and monitoring uptake of label into DNA in the course of repeated cell divisions.

Both populations of allogeneic T lymphocytes proliferate in an MLR unless one population is rendered unresponsive by treatment with mitomycin C or lethal x-irradiation (Figure 13-17). When target cells are incapacitated the approach is referred to as a one-way MLR and the unresponsive population provides stimulation in the form of antigen presenting cells that express alloantigens foreign to the responder T cells. The responder cells include a mixture of T$_H$ and T$_C$ cells that

(a)

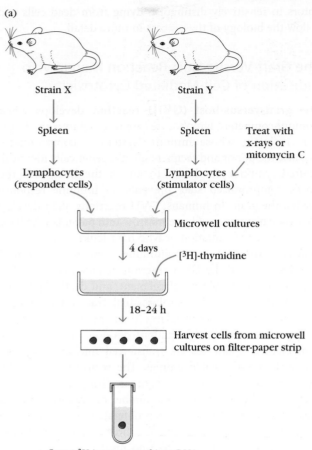

(b)

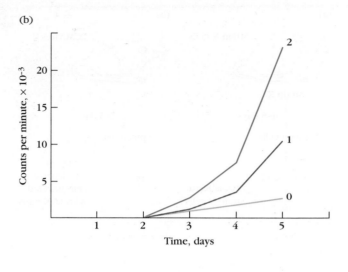

FIGURE 13-17 One-way mixed-lymphocyte reaction (MLR). (a) This assay measures the proliferation of lymphocytes from one strain (X, responder cells) in response to allogeneic cells that have been x-irradiated or treated with mitomycin C to prevent proliferation (Y, stimulator cells). The amount of [³H]-thymidine incorporated into the DNA is directly proportional to the extent of responder-cell prolifera-

tion. (b) The amount of [³H]-thymidine uptake in a one-way MLR depends on the degree of differences in class II MHC molecules between the stimulator and responder cells. Curve 0 = no class II MHC differences; curve 1 = one class II MHC difference; curve 2 = two class II MHC differences. These results demonstrate that the greater the class II MHC differences, the greater the proliferation of responder cells.

will divide when exposed to antigen presented by the stimulators. By 72 to 96 hours, an expanding population of functional CTLs (and/or helper T cells) is generated. With this experimental system, functional CTLs (and helper T cells) can be generated entirely in vitro, after which their activity can be assessed with various effector assays.

The significance of the role of T_H cells in the one-way MLR can be demonstrated by use of antibodies to the T_H-cell membrane marker CD4. In a one-way MLR, responder T_H cells recognize allogeneic class II MHC molecules on the stimulator cells and proliferate. Removal of the $CD4^+$ T_H cells from the responder population with anti-CD4 plus complement abolishes the MLR and prevents generation of CTLs. In addition to T_H cells, accessory cells such as dendritic cells are necessary for the MLR to proceed. When adherent cells are removed from the stimulator population, the proliferative response in the MLR is abolished and functional CTLs are no longer generated. The function of

these accessory cells is to activate the class II MHC-restricted T_H cells, whose proliferation is measured in the MLR. In the absence of T_H-cell activation, T_C proliferation is also halted.

CTL Activity Can Be Demonstrated by Cell-Mediated Lympholysis

Development of the **cell-mediated lympholysis (CML)** assay was a major experimental advance that contributed to understanding the mechanism of target-cell killing by CTLs. In this assay, suitable target cells are labeled intracellularly with chromium-51 (^{51}Cr) by incubating the target cells with $Na_2{}^{51}CrO_4$. After the ^{51}Cr diffuses into a cell, it binds to cytoplasmic proteins, reducing passive diffusion of the label out of the cell. When specifically activated CTLs are incubated for 1 to 4 hours with such labeled target cells, the targets lyse and ^{51}Cr is released. The amount of ^{51}Cr released correlates directly

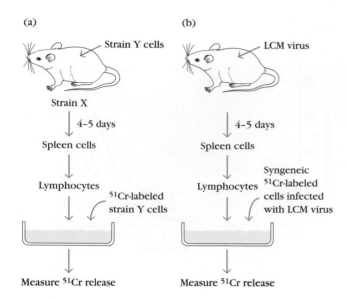

FIGURE 13-18 In vitro cell-mediated lympholysis (CML) assay. This assay has been used to measure the activity of cytotoxic T lymphocytes (CTLs) against (a) allogeneic cells or (b) virus-infected cells. Target cells are loaded with ^{51}Cr and cultured with CTLs. When lysed, targets release ^{51}Cr into the supernatant, which is isolated and assessed for radioactivity with a gamma counter.

with the number of target cells lysed by the CTLs. By means of this assay, the specificity of CTLs for allogeneic cells, tumor cells, virus-infected cells, and chemically modified cells has been demonstrated (Figure 13-18).

The T cells responsible for CML were identified by selectively depleting different T-cell subpopulations by adding complement to antibodies specific for each population (antibody-plus-complement lysis). In general, the activity of CTLs exhibits class I MHC restriction; that is, they can kill only target cells that present antigen associated with syngeneic class I MHC molecules. Occasionally, however, class II–restricted CD4$^+$ T cells have been shown to function as CTLs.

Currently, more and more investigators are using fluorescence-based variants of this assay to detect cytolysis. Instead of loading target with radioactive chromium, they load them with fluorescent molecules, including *carboxyfluorescein-succinimidyl-ester* (CFSE). Lysis can be measured by the loss of fluorescent cells in a mixed-cell population via flow cytometry. Variations in staining methods can allow investi-

gators to sensitively distinguish dying from dead cells and follow the biology of the process in more detail.

The Graft-Versus-Host Reaction Is an in Vivo Indication of Cell-Mediated Cytotoxicity

The **graft-versus-host (GVH) reaction** develops when immunocompetent lymphocytes are injected into an allogeneic recipient whose immune system is compromised. Because the donor and recipient are not genetically identical, grafted lymphocytes begin to attack the host, and the host's compromised state prevents an immune response against the graft. In humans, GVH reactions often develop after transplantation of bone marrow into patients who have been exposed to radiation or who have leukemia, immunodeficiency diseases, or autoimmune anemias. The clinical manifestations of the GVH reaction include diarrhea, skin lesions, jaundice, spleen enlargement, and death. Epithelial cells of the skin and gastrointestinal tract often become necrotic, causing the skin and intestinal lining to be sloughed.

Experimentally, GVH reactions develop when immunocompetent lymphocytes are transferred into an allogeneic neonatal or x-irradiated animal. The recipients, especially neonatal ones, often exhibit weight loss. The grafted lymphocytes generally are carried to a number of organs, including the spleen, where they begin to proliferate in response to the allogeneic MHC antigens of the host. This proliferation induces an influx of host cells and results in visible spleen enlargement, or splenomegaly. The intensity of a GVH reaction can be assessed by calculating the *spleen index* as follows:

$$\text{Spleen index} = \frac{\text{weight of experimental spleen/total body weight}}{\text{weight of control spleen/total body weight}}$$

A spleen index of 1.3 or greater is considered to be indicative of a positive GVH reaction. Spleen enlargement results from proliferation of both CD4$^+$ and CD8$^+$ T-cell populations. NK cells also have been shown to play a role in the GVH reaction, and these cells may contribute to some of the skin lesions and intestinal wall damage observed. Corticosteroids, which inhibit immune cell activity, are used to treat GVH, but can make a patient more vulnerable to infection. Most therapies focus on preventing GVH. Procuring a graft from a donor that is precisely HLA matched to the recipient is clearly the most effective approach.

S U M M A R Y

- The effector functions of the immune system have both humoral (antibody) and cellular arms. Both cooperate to clear the body of infection and, in some cases, abnormal cells (e.g., tumors).

- The humoral arm of the immune system refers to the activities of antibodies secreted by B lymphocytes. The

antigen specificity, isotype, and interactions with FcRs are all important features of antibody effector function.

- Antibodies inhibit and clear infection by blocking the ability of pathogens to infect (neutralization), by coating pathogens so that they are recognized and phagocytosed by innate immune cells (opsonization), by recruiting complement to the pathogen (complement fixation), and by directing other cells of the innate immune system to kill infected cells (antibody-dependent cell cytotoxicity).

- Each antibody isotype has different effector functions: IgM and some IgG antibodies fix complement; IgG antibodies mediate ADCC; IgE antibodies induce release of inflammatory molecules from granulocytes to kill parasites; and IgA antibodies are a major isotype in bodily secretions and block entry of bacteria and toxins to the bloodstream.

- Fc receptors are responsible for many of the effector functions of antibodies, including phagocytosis of antibody-antigen complexes by macrophages; transcytosis of antibodies through epithelial cell layers; lysis of antibody-bound infected cells by NK cells (ADCC); protecting serum antibodies from degradation; and regulating the activities of innate immune cells.

- FcRs are expressed by many cell types in the body and generate signals when bound to antibody-antigen complexes. FcR generated signals can either activate or inhibit the activity of cells. Whether an FcR is activating or inhibiting depends on whether it includes or associates with an ITAM (activating) or an ITIM (inhibiting) protein motif.

- The effector cells of the cell-mediated immune system include cells of the innate immune system (natural killer [NK] cells) and cells of the adaptive immune system: helper CD4$^+$ T cells (T$_H$ cells) and CD8$^+$ cytotoxic T lymphocytes (CTLs or T$_C$ cells). NKT cells are also participants and have features of both the innate and adaptive immune systems.

- Compared with naïve T$_H$ and T$_C$ (CTL) cells, effector cells are more easily activated, express higher levels of cell-adhesion molecules, exhibit different trafficking patterns, and produce both soluble and membrane effector molecules.

- In order to become functional CTLs, naïve T$_C$ precursors (CTL-Ps) must engage with APCs that have been previously activated (licensed). CD4$^+$ T-cell help is not necessary for this first activation step, but is required for optimal generation of memory and proliferation of CTLs.

- Antigen-specific effector T cell populations can be identified and tracked by labeling with MHC tetramers.

- CTLs induce cell death via two mechanisms: the perforin-granzyme pathway and the Fas-FasL pathway. Both trigger apoptosis in the target cells.

- The perforin-mediated cytotoxic response of CTLs involves several steps: TCR-MHC mediated recognition of target cells; formation of CTL/target-cell conjugates and immune synapse formation; repositioning of CTL cytoplasmic granules toward the target cell; granule release; formation of pores in the target-cell membrane; dissociation of the CTL from the target; and death of the target cell. Fas-mediated killing also involves conjugate and immune synapse formation where Fas-FasL interactions occur.

- Other nonantigen-specific cells of the innate immune system (NK cells and granulocytes) can also kill target cells via antibody-dependent cell-mediated cytotoxicity (ADCC), resulting in the release of lytic enzymes, perforin, or TNF that damage the target-cell membrane.

- NK cells induce apoptosis of tumor cells and virus-infected cells by mechanisms similar to CTLs (perforin-induced pore formation, as well as Fas-FasL interactions), but are regulated by distinct receptors.

- NK cells kill those cells that have lost or reduced their levels of MHC class I, a ubiquitously expressed self protein—a phenomenon described by the missing self model.

- NK cell killing is regulated by a balance between positive signals generated by the engagement of activating NK receptors and negative signals from inhibitory NK receptors.

- Many NK cell-activating and inhibitory receptors bind MHC class I products and their homologs. The expression of relatively high levels of class I MHC molecules on normal cells protects them against NK-cell-mediated killing by engaging inhibitory receptors.

- NK-cell receptors fall into two major structural groups based on their extracellular regions: the lectin-like receptors and the Ig-like receptors. Whether receptors are activating or inhibiting depends on their intracellular regions: ITAM expressing receptors are activating, ITIM expressing receptors are inhibiting.

- NKT cells have characteristics common to both T lymphocytes and NK cells; most express an invariant TCR and markers common to NK cells. They exhibit both helper and cytotoxic activity and kill cells predominantly via FasL-Fas interactions.

- Immune cell cytotoxicity can be measured in vitro by mixed lymphocyte reaction (MLR), or cell-mediated lympholysis (CML) via radioactivity or fluorescence. They can be assessed in vivo by graft versus host (GVH) responses and increases in spleen size (splenomegaly).

REFERENCES

Bevan, M. J. 2004. Helping the CD8(+) T-cell response. *Nature Reviews Immunology* **4**:595–602.

Bourgeois, C., and C. Tanchot. 2003. Mini-review CD4 T cells are required for CD8 T cell memory generation. *European Journal of Immunology* **33**:3225–3231.

Brooks, C. G. 2008. Ly49 receptors: Not always a class I act? *Blood* **112**:4789–4790.

Cruz-Muñoz, M. E., and A. Veillette. 2010. Do NK cells always need a license to kill? *Nature Immunology* **11**:279–280.

de Saint Basile, G., G. Ménasché, and A. Fischer. 2010. Molecular mechanisms of biogenesis and exocytosis of cytotoxic granules. *Nature Reviews Immunology* 10:568–579.

Diana, J., and A. Lehuen. 2009. NKT cells: Friend or foe during viral infections? *European Journal of Immunology* 39:3283–3291.

Gould, H. J., and B. J. Sutton. 2008. IgE in allergy and asthma today. *Nature Reviews Immunology* 8:205–217.

Jenkins, M. R., and G. M. Griffiths. 2010. The synapse and cytolytic machinery of cytotoxic T cells. *Current Opinion in Immunology* 22:308–313.

Klenerman, P., V. Cerundolo, and P. R. Dunbar. 2002. Tracking T cells with tetramers: New tales from new tools. *Nature Reviews Immunology* 2:263–272.

Mancardi, D.A., et al. 2008. FcgammaRIV is a mouse IgE receptor that resembles macrophage FcepsilonRI in humans and promotes IgE-induced lung inflammation. *Journal of Clinical Investigations* 118:3738–3750.

Mescher, M. F., et al. 2007. Molecular basis for checkpoints in the CD8 T cell response: Tolerance versus activation. *Seminars in Immunology* 19:153–161.

Mestas, J., and C. C. W Hughes. 2004. Of Mice and Not Men: Differences between Mouse and Human Immunology. *Journal of Immunology* 172:2731–2738.

Moretta, L. 2010. Dissecting CD56^{dim} human NK cells. *Blood.* 116:3689–3691.

Nimmerjahn, F., and J. V. Ravetch. 2008. Fc-gamma receptors as regulators of immune responses. *Nature Reviews Immunology* 8:34–47.

Nimmerjahn, F., and J. V. Ravetch. 2010. Antibody mediated modulation of immune responses. *Immunological Reviews.* 236:265–275.

Orr, M. T., and L. L. Lanier. 2010. Natural killer cell education and tolerance. *Cell* 142:847–856.

Parham, P. 2005. MHC class I molecules and KIRs in human history, health and survival. *Nature Reviews Immunology* 5:201–214.

Rajagopalan, S., and E. O. Long. 2005. Understanding how combinations of HLA and KIR genes influence disease. *Journal of Experimental Medicine* 201:1025–1029.

Roopenian, D. C., and S. Akilesh. 2007. The neonatal Fc receptor comes of age. *Nature Reviews Immunology* 7:715–725.

Salfield, J. G. 2007. Isotype selection in antibody engineering. *Nature Biotechnology.* 25:1369–1372.

Schneider-Merck, T., et al. 2010. Human IgG2 antibodies against epidermal growth factor receptor effectively trigger ADCC but, in contrast to IgG1, only by cells of myeloid lineage. *Journal of Immunology* 184:512–520.

Sun, J. C., J. N. Beilke, and L. L. Lanier. 2009. Adaptive immune features of natural killer cells. *Nature* 457:557–561.

Sun, J. C., and L. L. Lanier. 2009. Natural killer cells remember: An evolutionary bridge between innate and adaptive immunity? *European Journal of Immunology* 39:2059–2064.

Trambas, C. M., and G. M. Griffiths. 2003. Delivering the kiss of death. *Nature Immunology* 4:399–403.

Useful Web Sites

www.youtube.com/watch?v=84MlWh1XN0Q

www.youtube.com/watch?v=yRnuwTDR1og YouTube offers many animations of CTL and NK killing. Use your knowledge from this book to critique their accuracy. The preceding links provide two examples.

www.signaling-gateway.org The "molecule pages" are accessible and up-to-date descriptions of the characteristics of many signaling receptors (including the FcRs and NK receptors described in this chapter).

www.nature.com/nri/posters/nkcells/nri1012_nkcells_poster.pdf This is a poster of our current understanding of the diversity of NK receptors, offered for free by *Nature Reviews Immunology*.

http://en.wikipedia.org/wiki/Fc_receptor This Wikipedia site offers a credible and particularly well-organized description of the Fc receptors. Wikipedia is a very useful first source for information relevant to this and other chapters. However, it is important to be aware that not all information on the site is validated. It is always important to go to the primary sources to verify what you read.

 STUDY QUESTIONS

CLINICAL FOCUS QUESTIONS

1. One inherited combination of KIR and MHC genes leads to increased susceptibility to a form of arthritis. Would you expect this to be caused by increased or decreased activation of NK cells? Could an increase in susceptibility to diabetes be explained using the same logic? Explain.

2. Griscelli syndrome (type 2) is a rare and fatal disease caused by a loss of function of the Rab27A gene. It is also characterized by both loss of pigment (partial albinism) and immunodeficiency, which has been associated with a failure in cytotoxic T-cell activity. Look up the function of Rab27A online. Why might a deficiency in Rab27A lead to both these symptoms? What other cell types might you expect to be affected?

1. Investigators have developed antibodies that bind to FcγRIII and CD32 and potently block the interactions of these receptors with their ligands. Which of the following humoral immune responses would these antibodies block? Explain.

Neutralization

Opsonization

Complement fixation

ADCC

2. Indicate whether each of the following statements is true or false. If you believe a statement is false, explain why.

   a. Fc receptor engagement always results in internalization and destruction of the antibody-antigen complex.
   b. IgE mediates ADCC.
   c. Cytokines can regulate which branch of the immune system is activated.
   d. Both CTLs and NK cells release perforin after interacting with target cells.
   e. Antigen activation of naïve CTL-Ps requires a costimulatory signal delivered by interaction of CD28 and CD80/86.
   f. CTLs use a single mechanism to kill target cells.
   g. CD4$^+$ T cells are absolutely required for the activation of naïve CD8$^+$ T cells.
   h. The ability of an NK cell to kill a target is determined by signals received via both activating and inhibiting receptors.
   i. Human NK cells express functional Ly49 receptors.

3. You have a monoclonal antibody specific for LFA-1. You perform CML assays of a CTL clone, using target cells for which the clone is specific, in the presence and absence of this antibody. Predict the relative amounts of ^{51}Cr released in the two assays. Explain your answer.

4. You decide to co culture lymphocytes from the strains listed in the following table in order to observe the *mixed-lymphocyte reaction* (MLR). In each case, indicate which lymphocyte population(s) you would expect to proliferate.

Population 1	Population 2	Proliferation
C57BL/6 (H-2^b)	CBA (H-2^k)	
C57BL/6 (H-2^b)	CBA (H-2^k) Mitomycin C-treated	
C57BL/6 (H-2^b)	(CBAxC57BL/6)F1(H-2$^{k/b}$)	
C57BL/6 (H-2^b)	C57L (H-2^b)	

5. In the mixed-lymphocyte reaction (MLR), the uptake of [^{3}H]-thymidine is often used to assess cell proliferation.

   a. Which cell type proliferates in the MLR?
   b. How could you prove the identity of the proliferating cell?
   c. Explain why production of IL-2 also can be used to assess cell proliferation in the MLR.

6. Indicate whether each of the properties listed below is exhibited by NK cells, CTLs, NKT cells, all, or none.

a. _____ Can make IFN-γ
b. _____ Can make IL-2
c. _____ Is class I MHC restricted
d. _____ Expresses CD8
e. _____ Is required for B-cell activation
f. _____ Is cytotoxic for target cells
g. _____ Is the effector cell in a CML assay
h. _____ Expresses NK1.1
i. _____ Expresses CD4
j. _____ Expresses CD3
k. _____ Recognizes lipid
l. _____ Can express the IL-2 receptor
m. _____ Expresses the αβ T-cell receptor
n. _____ Expresses NKGD2 receptors
o. _____ Responds to soluble antigens alone
p. _____ Produces perforin
q. _____ Expresses FasL.

7. Mice from several different inbred strains were infected with LCM virus, and several days later their spleen cells were isolated. The ability of the primed spleen cells to lyse LCM-infected, ^{51}Cr-labeled target cells from various strains was determined. In the following table, indicate with a (+) or (−) whether the spleen cells listed in the left column would cause ^{51}Cr release from the target cells listed in the headings across the top of the table.

Source of primed spleen cells	^{51}Cr release from LCM-infected target cells			
	B10.D2 (H-2^d)	B10 (H-2^b)	B10.BR (H-2^k)	(BALB/ cxB10) F1 (H-2$^{b/d}$)
B10.D2 (H-2^d)				
B10 (H-2^b)				
B10.BR (H-2^k)				
(BALB/cxB10) F1 (H-2$^{b/d}$)				

8. A mouse is infected with influenza virus. How could you assess whether the mouse has T$_H$ and T$_C$ cells specific for influenza?

9. NK cells do not express TCR molecules, yet they bind to class I MHC molecules on potential target cells. Explain how NK cells lacking TCRs can specifically recognize infected cells.

10. Consider the following three mouse strains:

    H-2^d mice in which both perforins have been knocked out.

    H-2^d mice in which Fas ligand has been knocked out

    H-2^d mice in which both perforin and Fas ligand have been knocked out

    Each strain is immunized with LCM virus. One week after immunization, T cells from these mice are harvested and tested for cytotoxicity. Which of the following targets would T cells from each strain be able to lyse?

a. Target cells from normal LCMV-infected H-2^b mice
b. Target cells from normal H-2^d mice
c. Target cells from H-2^d mice in which both perforin and Fas have been knocked out
d. Target cells from LCMV-infected normal H-2^d mice
e. Target cells from H-2^d mice where both perforin and Fas have been knocked out

11. You wish to determine the frequency of class I–restricted T cells in an HIV-infected individual that are specific for a peptide generated from gp120, a component of the virus. Assume that you know the HLA type of the subject. What method would you use, and how would you perform the analysis? Be as specific as you can.

12. Indicate whether each of the following statements regarding Fas-mediated or perforin-mediated programmed cell death is true or false. If you believe a statement is false, explain why.

    a. Both mechanisms induce apoptosis of the target cell.
    b. Target cells must express Fas ligand to be killed via the Fas-mediated pathway.
    c. Only the perforin mediated pathway stimulates a caspase cascade.
    d. Both pathways require granzyme to induce apoptosis.
    e. Both pathways ultimately activate caspase-3.
    f. Granzyme is responsible for the assembly of membrane pores.

ANALYZE THE DATA E. Jäger and colleagues isolated an antigen from a tumor in a cancer patient who expressed HLA-A2. To characterize the cell-mediated immune response of the patient to this tumor antigen, the investigators generated a series of peptide fragments from the tumor antigen and pulsed antigen-presenting cells with these peptides. They then measured the patient CTL response against each of these targets. Answer the following questions based on the data in the table and what you have learned from multiple chapters in this book. ("Pulsing" refers to a technique in which antigen-presenting cells are exposed to high concentrations of peptides in solution. They exchange these peptides for those that were in the grooves of their surface MHC molecules. This is a convenient way to generate antigen-presenting cells that express specific MHC-peptide complexes of interest.)

    a. Which epitope(s) of the tumor antigen was/were recognized by T cells? Explain your answer.
    b. Immunologists have identified anchor residues on HLA-A2 molecules that are the most important for antigen presentation. What amino acids are most likely to bind HLA-A2 anchor residues if these amino acids must be conserved? Explain your answer.
    c. Some of the peptides inspected by these investigators are poorly immunogenic. One explanation is that less immunogenic peptides may bind anchor residues ineffectively.

Propose a different but reasonable hypothesis to explain why some of the peptides are poor immunogens.

    d. The investigators pulsed antigen-presenting cells with peptides and used CTLs from the patient's peripheral blood to perform CTL assays. What HLA-A molecule is expressed by these antigen-presenting cells? Explain your answer.
    e. In what way is peptide 2 an unusual epitope?

Peptide fragments derived from antigen isolated from tumor			
Number	**Peptide sequence**	**Position**	**Specific lysis***
1	SLAQDAPPLPV	108–118	0
2	SLLMWITQCFL	157–167	55
3	QLSISSCLQQL	146–156	1
4	QLQLSISSCL	144–153	1
5	LLMWITQCFL	158–167	15
6	RLTAADHRQL	136–145	5
7	FTVSGNILTI	126–135	1
8	ITQCFLPVFL	162–171	1
9	SLAQDAPPL	108–116	3
10	PLPVPGVLL	115–123	1
11	WITQCFLPV	161–169	5
12	SLLMWITQC	157–165	78
13	RLLEFYLAM	86–94	7
14	SISSCLQQL	148–156	6
15	LMWITQCFL	159–167	4
16	QLQLSISSC	144–152	1
17	CLQQLSLLM	152–160	1
18	QLSLLMWIT	155–163	84
19	NILTIRLTA	131–139	2
20	GVLLKEFTV	120–128	3
21	ILTIRLTAA	132–140	12
22	TVSGNILTI	127–135	4
23	GTGGSTGDA	7–15	9
24	ATPMEAELA	97–105	1
25	FTVSGNILT	126–134	1
26	LTAADHRQL	137–145	9

*Specific lysis is a relative measure of lytic activity and can range from 0 to 100%.

The Immune Response in Space and Time

Imagine trying to understand American or European football from snapshots taken of players on the field or from examinations of the behavior and phenotype of a few players taken off the field. Significant information about the dynamic games can certainly be gleaned, particularly by clever observers who note differences in the color of jerseys, the directions players are facing, implied motion of feet relative to the position of a ball, and variations in physiognomy. Imagine then having the capacity to pull one or two players off the field and examine them and their behavior, with and without a ball. This would certainly help. However, one is still likely to miss the relevance of each position on the team, the significance of substitutions, and the rationale behind a stop in play, and one is very likely to misinterpret the antics of players who have just sacked a quarterback or flopped in dramatic agony in front of the goal. There is little doubt that watching and listening to the entire game in real time would both simplify and enhance attempts to understand the rules and motions, and would reveal events that would not be predicted even by the most rigorous analysis of individual or small group interactions, anatomy, and behaviors.

Understanding the rules of the immune system is an even more daunting endeavor. Immune cells are the most dynamic cells in the body. They browse cells in every tissue, and organize responses in nearly every niche. Their movements are choreographed in multidimensional space by rules that have been gleaned by some of the most elegant cellular, molecular genetic, and biochemical experiments in biology. However, there is no substitute for the information that can be gained by directly visualizing the multiple cellular players.

Immunologists, cell biologists, physicists, and engineers have been perfecting techniques to watch cells in situ and in vivo for decades, and the technologies now available to watch cell movement in living organs have ushered in a new experimental era (see Advances Box 14-1). The data coming from these imaging studies are both satisfying and exciting. They have confirmed many of the indications and predictions from elegant

A T cell (blue) migrates along the fibroblastic reticular cell network in a lymph node. [Reprinted from Immunity, 25 /6, Bajénoff, M., et al., Stromal Cell Networks Regulate Lymphocyte Entry, migration, and Territoriality in Lymph Nodes, 1-13, Copyright 2006, with permission from Elsevier]

- Immune Cell Behavior before Antigen Is Introduced
- Immune Cell Behavior during the Innate Immune Response
- Immune Cell Behavior during the Adaptive Immune Response
- Immune Cell Behavior in Peripheral Tissues

molecular, biochemical, and cellular experiments in more static conditions, answered outstanding questions, challenged other interpretations, and raised new questions about the complex cellular and molecular interactions that must be integrated in multiple dimensions during an immune response.

In this chapter, featuring animations and dynamic images of immune cell behavior in vivo that should allow you to visualize much of the information that you have digested in the previous chapters, we will follow the immune response in space and time. Because the techniques that have allowed us to see cells in real time and in real tissues that are still developing, not all immune events have been directly witnessed or recorded. However, by exposing you to the beginning of what is bound to become a rich archive of video images, we hope also to provide lasting images that you can probe and query conceptually.

Dynamic Imaging Techniques

As described in Chapter 20, **fluorescence microscopy** allows one to visualize fluorochrome tagged (or autofluorescent) molecules and structures within a tissue and/or cell. Most conventional fluorescence microscopy uses mercury lamps as light sources. These lamps generate light of multiple, discrete wavelengths (each of which can excite a variety of fluorochromes) and shed this light on a relatively wide area of tissue. The light emitted by the fluorochromes is distinguished by filters that allow specific wavelengths to pass.

Confocal microscopy was an advance that markedly improved the focus and resolution of images by both narrowing the area of tissue that is excited by laser light and limiting the focal plane from which emitted light is collected (by limiting which emitted light is collected via a pinhole). The laser light can scan a plane, generating an "optical section." These sections can be layered atop one another digitally to generate three-dimensional images.

Two-photon microscopy (also known as 2P or *two-photon scanning laser microscopy*, TPSLM), also increases resolution but via a different technique: by directing low-energy (high-wavelength) laser light onto the field of interest. When, and only when, two low-energy photons simultaneously reach their target—and add their energies—they produce light of the energy (and wavelength) required to activate a fluorochrome of interest. Although both confocal and 2P microscopy can yield time-lapse animations of cells observed (by generating many optical sections over time), 2P microscopy has a key advantage: the longer wavelengths of light that can be used to excite the fluorochrome are significantly less damaging to cells and structures and can penetrate tissue much more deeply (100–200 μm). Therefore, cellular activities well within small intact, live, organs can be imaged for relatively long periods of time.

Investigators have also developed a suite of analytical tools that allow one to quantitate the cellular movements visualized by dynamic imaging. Fine measurements of motility rates and directions have enabled investigators, for instance, to determine if a cell population is moving by random walk or via a more directed, chemokine-gradient mediated fashion.

Intravital microscopy adds another feature to these fluorescent microscopy approaches and introduces technologies that allow one to expose and image living tissue (e.g., the popliteal or inguinal lymph of a living, anesthetized mouse).

HOW TO VIEW A DYNAMIC IMAGE

Videos of cell trafficking are now relatively common additions to published papers and typically appear in online supplemental material. Several are also included in this chapter. Some guidelines may be helpful when watching these videos so you can keep a critical eye on what they do and do not show. Awareness of the variables that influence the images you are watching can help explain discrepancies or inspire technological advances, novel interpretations, and new experiments.

- ***Which cells are labeled?*** Investigators can typically label only two or three cells or features at a time, and it is very important to recognize that this means you are seeing only a fraction of the cellular interactions that are occurring. Black areas are not empty but are dense with other cells and extracellular material. Curiosity about these unstained areas will profoundly enhance your appreciation for the limitations of the data, and will inspire new experimental questions. For example, it was first thought that T and B cells wended their way through lymph nodes simply by following soluble gradients of chemokines, but both static and dynamic imaging data

revealed networks of reticular cells and fibers on which the cells travel in random walks that are assisted, if not directed by, chemokines (see Videos 14-2, 14-3, and 14-4; and Figures 14-4, 14-5, and 14-6). In sum, always ask yourself what isn't there, not just what is.

- ***How are cells labeled?*** Cells can be fluorescently labeled with *green fluorescent protein* (GFP) or its brighter *yellow* and *red* derivatives (YFP and RFP). Some reporter mice express GFP downstream of a ubiquitously expressed promoter (e.g., actin), so that all cells isolated from this mouse will fluoresce intrinsically (Figure 1a). Often, mice are genetically manipulated so that expression of a fluorescent protein is driven by a tissue-specific gene. In this way, only a certain cell type is illuminated (Figure 1b). Cells can also be labeled in vitro with reagents that pass through the cell membrane and remain stable inside the cell (Figure 1c). The stability of the label is one important consideration. The disappearance of a fluorescent set of cells could mean the fading of a stain, or the reduction in expression of a fluorescent protein (e.g., after cell proliferation), not necessarily the loss of a cell population.

- ***When and how are cells introduced?*** Most current dynamic imaging systems involve the introduction of fluorescently tagged isolated cell populations into a recipient mouse. This is done for a variety of good reasons, including the need to label only specific subsets of cells as well as the desire to synchronize the arrival of cell populations into tissues or lymphoid organs. Manipulations on cells performed in vitro, however, can alter cell function. Antigen-presenting cells, for instance, can be nonspecifically activated in culture, which can change their trafficking patterns and cell interactions.

(a) Actin-GFP

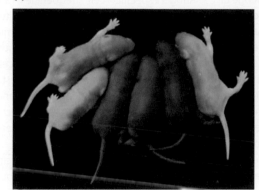

(b) CD11c-YFP dendritic cells imaged in mouse ear

(c) CFSE staining

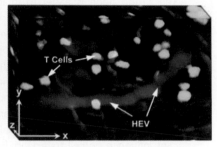

FIGURE 1

Fluorescent labeling of immune cells. (a) This transgenic mouse expresses GPF under an actin promoter (b) In this transgenic mouse, YFP is downstream of the CD11c promoter, which drives protein expression in most DCs. The Weninger lab used two-photon intravital microscopy to visualize the movements of DCs (yellow) in the skin after infection with the protozoa *Leishmania major*. DCs associate with extracellular matrix, which can be seen because they produce what are known as "second harmonic" signals and fluoresce. The migration of these cells can be visualized in this YouTube video: www.youtube.com/watch?v=XOeRJPIMpSs. (c) Cells (e.g., T cells) can be labeled with membrane permeable dyes that fluoresce green (CFSE) or red (CMTMR). Cells pick up the dye quickly in vitro and can be injected intravenously (adoptively transferred). C. Miller et al. were among the first to visualize CFSE-labeled T-cell behavior in a lymph node, a static example of which is shown here with HEV labeled red. *[(a) Hiroshi Kubota, Mary R. Avarbock and Ralph L. Brinster. Growth factors essential for self-renewal and expansion of mouse spermatogonial stem cells. PNAS Nov. 23, 2004, Vol. 101, No. 47, 16489–16494. © 2004 National Academy of Sciences, USA. Courtesy James Hayden, RBP, Hiroshi Kubota and Ralph Brinster: School of Veterinary Medicine, University of Pennsylvania. (b) Ng LG, Hsu A, Mandell MA, Roediger B, Hoeller C, et al. (2008) Migratory Dermal Dendritic Cells Act as Rapid Sensors of Protozoan Parasites. PLoS Pathog 4(11): e1000222. (c) Mark J. Miller, Sindy H. Wei, Michael, D. Cahalan and Ian Parker. Autonomous T cell trafficking examined in vivo with intravital two-photon microscopy. PNAS, 2003, Vol. 100, No. 5: 2604–2609. © 2003 National Academy of Sciences, USA.]*

Some isolation procedures can inhibit cell motility. The sequence of cell introduction is also an important variable. Some investigators introduce antigen-presenting cells prior to the introduction of T cells; others will introduce antigen into a mouse with yellow fluorescent DCs and then follow with T cells, and so on. Each investigator will have a rationale for his or her strategy, and it is useful to be aware and compare. Awareness of the potential artifacts introduced by in vitro manipulations will improve your ability to understand and critique the videos.

- **When and how is antigen introduced?** The ability to track cells during an immune response requires that antigen be introduced, of course. Antigen can be noninfectious—for example, foreign protein or peptide (e.g., ovalbumin) to which transgenic T and B cells are specific—or infectious (e.g., *Toxoplasma gondii*; lymphocytic choriomeningitis virus, LCMV). Antigens can be introduced intravenously, intradermally, subcutaneously, intranasally,

and so on, and, particularly in the case of noninfectious antigen, with some combination of adjuvants (e.g., alum and cytokines). The mode of introduction can dramatically influence the quality, quantity, timing, location, and outcome of the immune response. For example, respiratory system immunity is best inspired by introducing antigen intranasally. Antigen introduction by infection with living pathogens will inspire a natural, full-fledged innate immune response that is likely to have different, more complex effects on immune cells and microenvironments than noninfectious antigen or killed pathogens. These differences are all of interest to immunologists, and some are even now being addressed directly by dynamic imaging. However, when antigen delivery and quality are not the direct topics of interest, it is arguably even more important to be aware of the variables that can influence what you see.

- **How are investigators visualizing antigen-specific immune cells?** In

order to understand the behavior of antigen-specific lymphocytes, one has to be able to distinguish them from the millions of other lymphocytes that are not specific for the particular antigen of interest. Investigators will often introduce fluorescent lymphocytes whose antigen specificity is known (e.g., cells isolated from mice that express transgenic T-cell receptors or B-cell receptors). This approach allows one to manipulate cell frequencies and specifically trace the behavior of antigen-specific cells. However, transgenic systems are not available for every antigen of interest and can introduce their own artifacts (e.g., receptor levels tend to be high, and high cell frequencies do not mimic physiological responses). Investigators have found some clever ways around this by introducing an antigen against which receptor transgenics react (e.g., ovalbumin or ovalbumin peptide) into the genome of the pathogen. They have also found ways to infer antigen specificity from

(continued)

behavior after introducing T cells that are generally fluorescent.

- **What type of tissue is being examined, and under what conditions?** The lymph node is easier to image than the spleen, which has more structures that autofluoresce, with two-photon intravital microscopy techniques. The lymph node that drains the site of infection will provide different information from the lymph nodes that do not. Temperature and culture conditions will have a significant influence on cell trafficking. Isolated organs must be maintained in perfusion chambers that stabilize these conditions, and can be easier to manipulate and maintain in focus. However, cellular traffic patterns in explanted organs may not reflect all that is physiologic. Nodes that are surgically exposed in an anesthetized animal can be imaged for longer periods and may reveal more physiologically relevant patterns. However, surgery can inspire inflammation and stress that could also influence trafficking.

QUESTIONS DYNAMIC IMAGING ALLOWS US TO ASK AND ANSWER

Dynamic imaging often "simply" offers visually satisfying, indeed stunning, confirmation of implications and predictions offered by other indirect, albeit elegant approaches. However, the approach can also offer information about the sequencing of immune cell interactions within a microanatomical context that could not be gained by other approaches. Dynamic imaging has already revealed unexpectedly important sites of immune-cell interaction (e.g., the subcapsular sinus of a lymph node), unexpected participants (neutrophils) in lymph node activity, unique shape changes and motility pat-

terns (e.g., among cells in skin grafts, or within germinal centers) that challenge or enhance conventional immunological wisdom. It has allowed investigators to directly address questions that are raised by other data. Which cells do naïve T cells encounter first during an immune response? What types of interactions are associated with activation? death? tolerance? Where do CD8$^+$ T cells first encounter antigen? When and where do they encounter CD4$^+$ T-cell help? Where and when do cells leave their microenvironments? And ever more specific questions can be asked. For instance, do somatically mutating dark-zone germinal-center B cells travel often to the light zone to test their antigen specificity? Are antigen-specific T cells that enter a tumor capable of engaging in long-term interactions that result in cell death? What kind of Ca^{2+} signaling occurs with serial versus stable interactions? Do the functional fates of T cells that arrive in a lymph node before and after antigen differ? When do regulatory T cells suppress a response? Dynamic imaging still has much to offer, and is developing perhaps most rapidly in directions that allow us to combine cell trafficking information with intracellular signaling information. We are beginning to appreciate not just the timing and context of cellular interactions, but their relationship to molecular events, and there is much more to come.

Dynamic imaging cannot, of course, explain everything in the immune response. It is limited to observations of cells and structures that one chooses to (and is able to) label at one time. It is limited to one or two areas of a mouse at a time, and for only a limited time (20–60 minutes). It is dependent on experimental systems that are to an extent contrived—manipulated so that antigen and antigen

specificity can be traced fluorescently. Antigen-specific cell frequencies are higher than they are in physiological situations. And, as noted above, dynamic imaging techniques have only just begun to allow us to visualize intra- and extracellular interactions simultaneously. Investigators are remarkably innovative, and advancements and integrations of techniques that will give us a more complete picture are yet to come.

Dynamic imaging is not only an exciting technique to have at our disposal, but it also is now arguably a necessary part of the immunology community's experimental arsenal. We are likely going to become even more dependent on the information that dynamic imaging can offer about cellular movements deep in living tissues, about the influence of therapies on these movements, about the effect of distinct vaccine regimens on cell behavior, and about the coordination of cellular movements within intracellular and extracellular molecular events.

However, perhaps dynamic imaging will have its greatest influence on education. The immune system is a marvelous tangle of multidimensional complexity and organization that is difficult to describe in words. A simple video can clarify in minutes what our rich yet imperfect language of immunology cannot easily articulate. It is, of course, our responsibility as scientists, young and old, to bring a critical eye to what pictures and films show us and do not show us. Richard Avedon, a twentieth-century American photographer, perhaps said it best: "All photos are accurate. None of them is the truth." However, moving images of cells in multidimensional time and space may bring us closer to the truth. They certainly have the capacity to make immunology more accessible and to inspire the desire to see—and know—more.

We start the chapter with a general summary of the behavior of innate and adaptive cells before an immune response. We then follow the central themes of the previous chapters by first examining the behavior of innate immune cells after the introduction of an antigen

and then by examining the behavior of the main cell types that regulate the adaptive immune system. We focus primarily on the behavior of the immune response in lymph nodes, the secondary lymphoid organ that has been most accessible to dynamic imaging techniques.

Finally, we close with recent examples of immune cell behavior during the response to physiological antigen (pathogen and alloantigen) in vivo. We emphasize where and when cellular interactions occur, providing more context for the understanding of molecular and cellular participants that you have gained from previous chapters.

Two Advances features provide more detailed information relevant to our understanding of immune cell behavior. The first describes cutting-edge dynamic imaging approaches that have allowed us to see cells in real time. The description allows you to better understand the experimental basis for the images you see, and also provides the background you need to critique the images that you encounter in primary literature. The second describes the molecular players in cell trafficking and the molecular events that regulate the passage of immune cells from blood to tissue, and it can be used as a resource to understand more advanced literature.

Immune Cell Behavior before Antigen Is Introduced

Hematopoietic stem cells in the bone marrow give rise to all cells involved in an immune response (Chapter 2). Myeloid cells, part of the innate immune system, include antigen-presenting cells (dendritic cells and macrophages) and granulocytes (neutrophils, basophils, and eosinophils). Once mature, these cells exit the bone marrow and circulate throughout the body, some taking up residence in tissues and organs. Lymphoid cells, part of the adaptive immune system, also begin their development in the bone marrow, but complete it in distinct niches. B cells complete their maturation among the osteoblasts of the bone marrow. T-cell precursors leave the bone marrow to mature in a distinct organ, the thymus. During B and T lymphocyte development, the genes that encode the B-cell receptor and T-cell receptor undergo gene rearrangement so that each mature B or T cell expresses a unique antigen receptor (Chapter 7). Both types of lymphocytes undergo negative selection so that those that survive development will not be autoreactive. Immature T cells (thymocytes) also undergo positive selection for T-cell receptor specificities that can recognize self-MHC (*major histocompatibility complex*)/foreign peptide combinations with some affinity. Those lymphocytes that successfully survive selection events exit from the thymus and bone marrow as naïve T and naïve B lymphocytes, respectively (Chapters 9 and 10).

Naïve Lymphocytes Circulate Between Secondary and Tertiary Lymphoid Tissues

After a brief transit time (30 minutes) in the blood, nearly 45% of all mature, naïve lymphocytes produced by the thymus

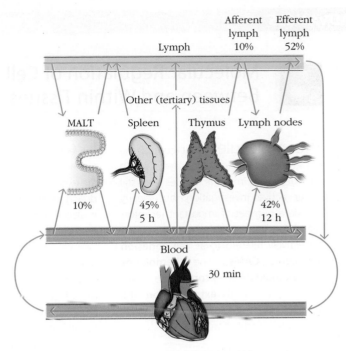

FIGURE 14-1 Lymphocyte recirculation routes. The percentage of the lymphocyte pool that circulates to various sites and the average transit times in the major sites are indicated. Lymphocytes migrate from the blood into lymph nodes through specialized areas in postcapillary venules called high-endothelial venules (HEVs).

and bone marrow travel by blood directly to the spleen, where they browse for approximately 5 hours (Figure 14-1). An almost equal proportion (42%) of lymphocytes enter various peripheral lymph nodes, where they remain, scanning stromal cell surfaces, for 12 to 18 hours. A smaller number of lymphocytes (10%) migrate to secondary lymphoid tissues in the skin and gastrointestinal, pulmonary, and genitourinary mucosa, which have the most direct contact with the external environment. Several protein interactions regulate lymphocyte homing to specific tissues, including chemokine receptor/chemokine, selectin/addressin and integrin/adhesion molecule engagements. We will consider some of these below, but describe them more fully in Advances Box 14-2.

Continual lymphocyte recirculation maximizes the numbers of naïve lymphocytes that encounter antigen. An individual lymphocyte may make a complete circuit from the blood to the tissues and lymph and back again once or twice per day. Since only about one in 10^5 lymphocytes can recognize a particular antigen, a large number of T or B cells must contact antigen on a given *antigen-presenting cell* (APC) within a short time to generate a specific immune response. The odds that such a small percentage of lymphocytes capable of interacting with an antigen actually makes contact with that antigen are improved by the extensive recirculation of lymphocytes and profoundly increased by factors that regulate, organize, and direct the circulation of lymphocytes

Molecular Regulation of Cell Migration Between and Within Tissues

The molecules that control the movement of the cellular players within and between immune tissues are still under active investigation, but clearly include a number of important receptor/ligand families: **selectins** and **integrins** (collectively referred to as **cell adhesion molecules [CAMs]**), and **chemokines** and **chemokine receptors**. These molecules regulate both **extravasation**, the transit of cells from blood to tissue, as well as trafficking within tissues.

CELL-ADHESION MOLECULES: SELECTINS, MUCINS, INTEGRINS, AND IG-SUPERFAMILY PROTEINS

Cell-adhesion molecules (CAMs) are versatile molecules that play a role in all cell-cell interactions, both among cells that are sessile within organs and also between the highly mobile immune cells and the tissues they travel to and from. In the immune system, CAMs play a role in helping leukocytes to adhere to the vascular endothelium prior to extravasation. They also increase the strength of the functional interactions between cells of the immune system, including T_H cells and APCs, T_H and B cells, and CTLs and their target cells.

A number of endothelial and leukocyte CAMs have been cloned and characterized, providing new details about the extravasation process. Most of these CAMs belong to four families of proteins: the *selectin* family, the *mucin-like* family, the *integrin* family, and the *immunoglobulin (Ig)* superfamily (Figure 1).

Selectins and Mucin-Like Proteins

The *selectin* family of membrane glycoproteins has an extracellular lectin-like domain that enables these molecules to bind to specific carbohydrate groups that decorate glycosylated *mucin-like proteins*. The selectin family includes three molecules, *L-, E-, and P-selectin*, also called *CD62L, CD62E,* and *CD62P*. Most circulat-

ing leukocytes express L-selectin (CD62L). On the other hand, E- selectin and P-selectin are expressed on vascular endothelial cells during an inflammatory response. P-selectin is stored within Weibel-Palade bodies, a type of granule within the endothelial cell. On activation of the endothelial cell, the granule fuses with the plasma membrane, resulting in the expression of P-selectin on the cell surface. Expression of E-selectin requires the synthesis of new proteins and occurs after the endothelium has been stimulated with pro-inflammatory cytokines. Selectin molecules are responsible for the initial stickiness of leukocytes to vascular endothelium.

Mucin-like proteins are a group of heavily glycosylated serine- and threonine-rich proteins that bind to selectins. Those relevant to the immune system include CD34 and GlyCAM-1, found on HEVs in peripheral lymph nodes, and MadCAM-1, found on endothelial cells in the intestine.

Selectins bind to sulfated carbohydrate moieties, including the sialyl-Lewis^x carbohydrate "cap" that decorate these molecules. L-selectin (CD62L) expressed on naïve lymphocytes specifically interacts with carbohydrate residues referred to generally as *peripheral node addressin* (*PNAd*), which is found associated with multiple molecules (e.g., CD34, GlyCAM-1, and MacCAM-1) on endothelial cells. On the other hand, E- and P-selectin, which are expressed on inflamed endothelium, interact with carbohydrate moieties on *P-selectin glycoprotein ligand-1* (*PSGL-1* or *CD162*), which is found on all white blood cells. Selectin interactions are critical for the first step in extravasation: leukocyte rolling.

Integrins

The integrins are heterodimeric proteins consisting of an α and a β chain that are noncovalently associated at the cell surface. Most integrins bind extracellular

matrix molecules and participate in cell-extracellular matrix interactions throughout the body. Some subfamilies bind cell-surface adhesion molecules and are involved in cell-cell interactions. Leukocytes express several integrins that are also expressed on other cell types, but also express a specific subfamily of integrins known as the $\beta 2$-integrins (or CD18-integrins), which bind to members of the Ig superfamily as well as proteins associated with the inflammatory response: *Ig superfamily cellular-adhesion molecules* (ICAMs). $\beta 2$ integrins interact with multiple alpha chains also known as CD11. *Lymphocyte function–associated antigen-1* (LFA-1 or CD11a/CD18) is one of the best-characterized integrins and regulates many immune-cell interactions, including naïve T-cell/APC encounters. It initially binds weakly to ICAM-1, but a signal from the T-cell receptor will alter its conformation and result in stronger binding (see Figure 1). Such inside-out signaling is an important feature of multiple integrins and allows cells to probe other cell surfaces before making a commitment to an interaction. $\beta 7$ integrins (e.g., CD103 [$\alpha E \beta 7$] and $\alpha 4 \beta 7$) are also expressed by leukocytes, including intraepithelial lymphocytes and regulatory T cells. These can interact with ICAMs and *cadherins*, which are expressed by multiple epithelial cells. CD103/E-cadherin (CD324) interactions may be important for localization of lymphocyte subpopulations in the gut (and other mucosal) tissues.

The combination of integrins expressed on a given cell type allows these cells to bind to different CAMs expressed on the surface of the vascular endothelium (and other epithelial cells). Clustering of integrins on the cell surface also increases the likelihood of effective binding and plays a role in leukocyte migration. The importance of integrin molecules in leukocyte extravasation is demonstrated by

BOX 14-2

(a) General structure of CAM families

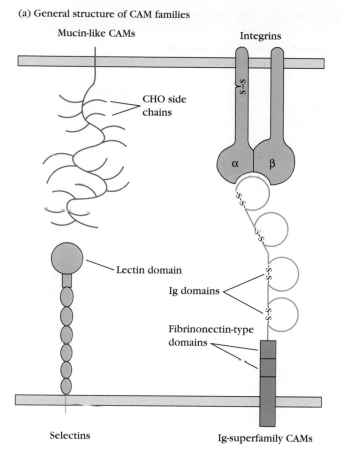

(b) Selected CAMs belonging to each family

Mucin-like CAMs:	Selectins:
GlyCAM-1	L-selectin
CD34	P-selectin
PSGL-1	E-selectin
MAdCAM-1	

Ig-superfamily CAMs:	Integrins:
ICAM-1, -2, -3	α4β1 (VLA-4, LPAM-2)
VCAM-1	α4β7 (LPAM-1)
LFA-2 (CD2)	α6β1 (VLA-6)
LFA-3 (CD58)	αLβ2 (LFA-1)
MAdCAM-1	αMβ2 (Mac-1)
	αXβ2 (CR4, p150/95)
	αEβ7 (CD103)

FIGURE 1

Schematic diagrams depicting the general structures of the four families of cell-adhesion molecules (a) and a list of representative molecules in each family (b). The lectin domain in selectins interacts primarily with carbohydrate (CHO) moieties on mucin-like molecules (collectively referred to as PNAd and PSGL-1). Both component chains in integrin molecules contribute to the binding site, which interacts with an Ig domain in CAMs belonging to the Ig superfamily. MAdCAM-1 contains both mucin-like and Ig-like domains and can bind to both selectins and integrins.

leukocyte adhesion deficiency (LAD) type 1, an autosomal recessive disease caused by a mutation in CD18 and characterized by recurrent bacterial infections and impaired healing of wounds. A similar deficiency is seen in individuals with impaired expression of selectins and has been coined LAD type 2.

Ig-superfamily CAMs

The **Ig-superfamily CAMs (ICAMs)**, which can act as ligands for β2 integrins, contain a variable number of immunoglobulin-like domains and thus are classified as members of the immunoglobulin superfamily. Included in this group are ICAM-1 (CD54), ICAM-2 (CD102), ICAM-3 (CD50), and VCAM-1 (CD106), which are expressed on vascular endothelial cells and bind to various integrin molecules.

ICAMs can also exhibit *homotypic binding*—the binding of an ICAM to 'itself'. For example, PECAM-1 molecules on one cell can bind to PECAM-1 molecules on

another cell. Junctional cell adhesion molecule-1 (JAM-1, CD321), which is also located within the endothelial junctional complex, can interact with another JAM-1 molecule (as well as CD11a/CD18 (LFA-1)) to play a role in transendothelial migration. Homotypic adhesion among Ig-superfamily members can also be found among other cell types, including neural cells.

An interesting cell-adhesion molecule called MAdCAM-1 has both Ig-like domains and mucin-like domains. This molecule is expressed on mucosal endothelium and directs lymphocyte entry into mucosa. It binds to integrin α4β7 (LPAM) via its immunoglobulin-like domain and to L-selectin (CD62L) via its mucin-like domain. *Platelet-endothelial cell adhesion molecule-1 (PECAM-1, CD31)* is found on the surface of leukocytes (neutrophils, monocytes, and a subset of T lymphocytes) and within the junctional complex of endothelial cells.

CHEMOKINES AND CHEMOKINE RECEPTORS

Chemokines, the other superfamily of proteins that play a major role in regulating immune cell trafficking, are small polypeptides, most of which contain 90 to 130 amino acid residues (see Appendix III for the complete list). They help localize circulating cells to sites of inflammation, to primary and secondary lymphoid tissue, and to specific microenvironments within that tissue. They do so not only by exerting a chemotactic influence, but also by controlling adhesion and activation of multiple types of leukocyte populations and subpopulations. Consequently, they are major regulators of leukocyte traffic. Some chemokines are primarily involved in inflammatory processes; others are constitutively expressed and play important homeostatic or developmental roles. *"Housekeeping"* chemokines are produced in lymphoid organs and tissues or in nonlymphoid sites such as skin, where they

(continued)

TABLE 1	Molecules involved in extravasation of leukocytes			
Leukocyte	Molecules involved in rolling	Chemokines involved in activation	Molecules involved in adhesion	Comments
Neutrophils	L-selectin and PSGL-1	IL-8 and macrophage inflammatory protein 1b (MIP-1b [aka CCL4])	LFA-1 and MAC-1	First to the site of inflammation: responds to C5a, bacterial peptides containing N-formyl peptides, and leukotrienes within inflamed tissues
(Inflammatory) Monocytes	L-selectin	Monocyte-chemoattractant protein-1 (MCP-1, CCL2)	VLA-4	Home to tissues after neutrophils; respond to bacterial peptide fragments and complement fragments within inflamed tissues
Naïve Lymphocytes	L-selectin, LFA-1, VLA-4 (in low-affinity forms)	CCL21, CCL19, and CXCL12 (for T cells) CXCL13 (for B cells)	LFA-1 and VLA-4	Travel across high-endothelial venules to enter the lymph node

direct normal trafficking of lymphocytes between primary and secondary lymphoid organs (e.g., from bone marrow to spleen, from thymus to lymph node, etc.). B-cell and T-cell lymphopoiesis are also dependent on appropriate chemokine expression, which regulates the microenvironment locale of developing cells.

Chemokine-mediated effects are not limited to the immune system (Table 1). Mice that lack either the chemokine CXCL12 (also called SDF-1) or its receptor show major defects in the development of the brain and the heart. Members of the chemokine family have also been shown to play regulatory roles in the development of blood vessels (angiogenesis) and wound healing.

The *inflammatory chemokines* are typically induced in response to infection. Contact with pathogens or the action of proinflammatory cytokines, such as TNF-α, up-regulates the expression of inflammatory cytokines at sites of developing inflammation. Chemokines cause leukocytes to move into various tissue sites by inducing the adherence of these cells to the vascular endothelium. After migrating into tissues, leukocytes travel in the direction of increasing localized concentrations of chemokines, resulting in the

targeted recruitment of phagocytes and effector lymphocyte populations to inflammatory sites. The assembly of leukocytes at sites of infection, orchestrated by chemokines, is an essential part of mounting an appropriately focused response to infection.

The chemokine family consists of over 40 members, with additional variation contributed by alternate RNA splicing pathways during transcription, post-translational processing, and isoforms (see Appendix III). The chemokines possess four conserved cysteine residues, and based on the position of two of the four invariant cysteine residues, almost all fall into one of two distinctive subgroups (see Figure 4-17):

- *CC subgroup* chemokines, in which the conserved cysteines are contiguous
- *CXC subgroup* chemokines, in which the conserved cysteines are separated by some other amino acid (X)

The exceptions are CX3CL1, which has three amino acids between the conserved cysteines, and XCL1 and XCL2, which lack two of the four conserved cysteines.

Chemokine action is mediated by receptors whose polypeptide chains traverse the membrane seven times. The

receptors are members of the G-protein-linked family of receptors and are grouped according to the type of chemokine(s) they bind. The *CC receptors* (CCRs) recognize CC chemokines, and *CXC receptors* (CXCRs) recognize CXC chemokines. As with cytokines, the interaction between chemokines and their receptors is both strong ($K_d > 10^{-9}$) and highly specific. However, as shown in Appendix III, most receptors bind more than one chemokine. For example, CXCR2 recognizes at least six different chemokines, and many chemokines can bind to more than one receptor.

When a receptor binds an appropriate chemokine, it activates heterotrimeric large G proteins, initiating a signal transduction process that generates such potent second messengers as cyclic AMP (cAMP), IP_3, Ca^{2+}, and activated small G proteins. Dramatic changes are effected by the chemokine-initiated activation of these signal transduction pathways. Within seconds, an appropriate chemokine can cause abrupt and extensive changes to the shape of a leukocyte. Chemokines can also promote greater adhesiveness to endothelial walls by activating leukocyte integrins (an event that is also dependent on the shear forces generated

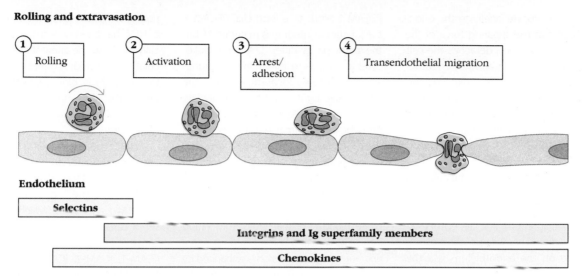

Rolling and extravasation

FIGURE 2
The sequential but overlapping steps in leukocyte extravasation. Tethering and rolling are mediated by binding of selectin molecules to sialylated carbohydrate moieties on mucin-like CAMs. Chemokines then bind to G-protein-linked receptors on the leukocyte, triggering an activating signal. This signal (in combination with the shear forces applied by blood flow) induces a conformational change in the integrin molecules, enabling them to adhere firmly to Ig superfamily molecules on the endothelium. Finally, leukocytes traverse the tight endothelial junction and subsequently migrate into the underlying tissue (transmigration).

by blood flow). Finally, chemokines induce the generation of microbicidal oxygen radicals in phagocytes, and the release of granular contents, including proteases, histamine, and cytotoxic proteins from granulocytes.

Differences in chemokine-receptor expression profiles mediate differences in leukocyte activity

Differences in the patterns of expression of chemokine receptors by leukocytes coupled with the production of distinctive profiles of chemokines by destination tissues and sites provide rich opportunities for the differential regulation of activities and movements of different leukocyte populations. Among the major populations of human leukocytes, neutrophils express CXCR1, CXCR2, and CXCR4; eosinophils express CCR1 and CCR3. Resting B cells express CCR5, and activated B cells will transiently up-regulate CCR7 (see below). Although resting naïve T cells display CCR7, CCR7 is not present on activated T cells, which may express CCR1, CCR4, CCR5, CCR8, CXCR3, and others. Indeed, T-cell effector subsets (T_H1, T_H2, T_H17, T_{REG}) can be distinguished not only by patterns of cytokine production but

also by patterns of chemokine receptor expression. T_H2 and T_{REG} cells express CCR4 and CCR8, and a number of other receptors not expressed by T_H1 cells. On the other hand, T_H1 cells express CCR5, and CXCR3, whereas most T_H2 cells do not. Many T_H17 cells express CCR6. A more comprehensive list of chemokines and their receptors is included in Appendix III.

EXTRAVASATION

Extravasation is a central feature of cell trafficking to both normal and inflamed tissue and is coordinated by both adhesion-molecule interactions and chemokine interactions. "The Inner Life of the Cell" (available for free at http://multimedia.mcb.harvard.edu/media.html) is a stunning artistic and scientifically relevant animation that depicts the surface and signaling interactions that govern the travels of a white blood cell from the bloodstream into the surrounding tissues. Its first few minutes, in particular, provide an excellent visual reference to the molecular events we describe below.

As an inflammatory response develops, various cytokines and other inflammatory mediators produced by innate immune cells act on the local blood ves-

sels, activating them to increase expression of cell-adhesion molecules, including selectins. Leukocytes passing by this site recognize the activated endothelium and adhere strongly enough so that they are not swept away by the flowing blood. The bound leukocytes then extravasate, migrating between endothelial cells to gain access to the inflamed, underlying tissue.

Leukocyte extravasation is often divided into four steps: (1) *rolling*, mediated by selectins; (2) *activation* by chemokines; (3) *arrest and adhesion*, mediated by integrins binding to Ig-family members; and (4) *transendothelial migration* (Figure 2).

Rolling

In the first step of extravasation, leukocytes attach loosely to the endothelium by a low-affinity selectin-carbohydrate interaction. In an inflammatory response, cytokines and other inflammatory mediators act on the local endothelium, inducing expression of adhesion molecules of the selectin family. These E- and P-selectin molecules bind to a variety of mucin-associated carbohydrate molecules (PLSG-1) on the leukocyte membrane (e.g., a neutrophil). These interactions

(continued)

tether the leukocyte briefly to the endothelial cell, but the shearing force of the circulating blood soon detaches the cell. Selectin molecules on another endothelial cell again tether the leukocyte; this process is repeated so that the cell tumbles end over end along the endothelium, a type of binding called *rolling*.

Chemokine activation

The process of rolling slows the cell long enough to allow interactions between chemokines presented on the surface of the endothelium and receptors on the leukocyte. The various chemokines expressed on the endothelium and the repertoire of chemokine receptors on the leukocyte provide a degree of specificity to the recruitment of white blood cells to the infection site. Signal transduction events caused by the binding of a chemokine to the chemokine receptor on the surface of the leukocyte result in change of conformation and clustering of *integrins* (e.g., LFA-1) on the leukocyte. Recent data suggest that this event is also dependent on the shear forces generated by blood flow.

Arrest and adhesion

The inside-out signaling induced by chemokine receptor engagement enhances the affinity of leukocyte integrins for their ICAM partners on the endothelium, lessening the likelihood that they will be carried away by the flowing blood.

Transmigration

The leukocyte then squeezes between two neighboring endothelial cells without disrupting the integrity of the endothelial barrier. It accomplishes this by *homotypic* (self-self) binding of *platelet-endothelial-cell adhesion molecule-1* (PECAM-1; CD31) expressed on both the leukocyte and the endothelial cell. PECAM-1 is normally found within the endothelial junction in a homotypic interaction. Thus, when the leukocyte

PECAM-1 binds to endothelial PECAM-1, the junctional integrity is maintained, but the participants differ. LFA-1 on leukocytes also binds to JAM-1, another adhesion molecule located within the endothelial tight junction, to mediate egress from the bloodstream. Other integrins binding to matrix proteins within the basement membrane and extracellular matrix allow the leukocytes to follow a gradient of chemoattractants to the site of infection.

This sequence of events applies to all white blood cells that cross from blood into tissues: rolling is mediated by selectins, and integrin adhesion is enhanced by chemokines. A comparison of the specific molecules associated with extravasation of several leukocyte subsets is shown in Table 1.

TRAFFICKING BETWEEN AND WITHIN TISSUES: LYMPHOCYTES—A CASE STUDY

Lymphocyte extravasation and high-endothelial venules

Naïve lymphocytes enter the lymph node via specialized regions of vascular endothelium called **high-endothelial venules (HEVs)** (see Figure 14-2). These postcapillary venules are composed of specialized cells with a plump, cuboidal ("high") shape that contrasts sharply in appearance with the flattened endothelial cells that line the rest of the capillary. Each of the secondary lymphoid organs, with the exception of the spleen, contains HEVs.

Naïve lymphocytes do not exhibit a preference for a particular type of secondary lymphoid tissue but instead circulate indiscriminately to secondary lymphoid tissue throughout the body. The development and maintenance of HEVs in lymphoid organs is influenced by cytokines produced in response to antigen capture. For example, HEVs fail to develop in animals raised in a germ-free environment. The role of antigenic activa-

tion of lymphocytes in the maintenance of HEVs has been demonstrated by surgically blocking the afferent lymphatic vasculature to a node, so that antigen entry to the node is blocked. Within a short period, the HEVs show impaired function and eventually revert to a more flattened morphology.

HEVs express a variety of CAMs. Like other vascular endothelial cells, HEVs express CAMs of the selectin family (E- and P-selectin); the mucin-like family (GlyCAM-1 and CD34, which express the set of carbohydrate ligands for L-selectin known as PNAd); and the Ig superfamily (ICAM-1, ICAM-2, ICAM-3, VCAM-1, and MAdCAM-1). Some adhesion molecules that are distributed in a tissue-specific manner have been called **vascular addressins (VAs)** because they serve to direct the extravasation of different populations of recirculating lymphocytes to particular lymphoid organs.

Lymphocyte migration within a lymph node

After cells migrate into tissues, they are directed to the site of infection by cues generated by the innate immune system. One good example of the molecular cues that guide cells within tissues comes from studies of lymphocyte trafficking in secondary lymphoid organs. Chemokines coordinate the compartmentalization of lymphocyte subpopulations into their respective zones within the lymph nodes and are organized by the reticular networks. Fibroblastic reticular cell networks in the T-cell zone are decorated with CCL21 and CCL19, which attract naïve, CCR7$^+$ T cells to the T-cell zone. The networks are also decorated with IL-7, which enhances the survival of T cells. Follicular DC networks are decorated with CXCL13, which attracts naïve, CXCR5$^+$ B cells to the follicles. Both networks also express ICAM-1 and VCAM-1, which enhance the adhesion of naïve T and B cells, which express ligands LFA-1 and VCAM-4.

(a)

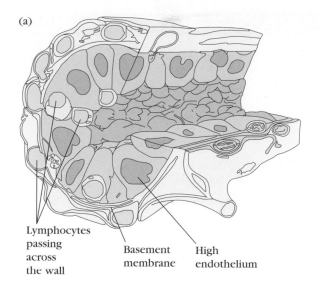

Lymphocytes passing across the wall

Basement membrane

High endothelium

(b)

(c)

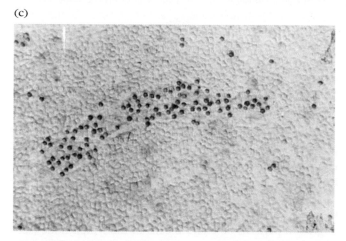

FIGURE 14-2 Lymphocyte migration through HEVs. (a) Schematic cross-sectional diagram of a lymph node postcapillary venule with high endothelium. Lymphocytes are shown in various stages of attachment to the HEV and in migration across the wall into the cortex of the node. (b) Transmission electron micrograph capturing lymphocytes (L) migrating through endothelial cell (E) layer. (c) Micrograph of frozen sections of lymphoid tissue. Some 85% of the lymphocytes (darkly stained) are bound to HEVs (in cross-section), and constitute only 1% to 2% of the total area of the tissue section. *[Part a is based on a drawing from Barbara Gould and adapted from A. O. Anderson and N. D. Anderson, 1981, in Cellular Functions in Immunity and Inflammation, J. J. Oppenheim et al., eds., Elsevier North-Holland; part b is from S. D. Rosen, 1989, Current Opinions in Cell Biology 1:913]*

and APCs within the secondary lymphoid organs. Early estimates using static imaging and in vitro data suggested that up to 500 T cells probe the surface of a single dendritic cell (DC) per hour. Dynamic imaging data now reveal that one DC can be surveyed by more than 5,000 T cells per hour—and suddenly it is much easier to imagine that 1 in 100,000 antigen-specific cells can find the MHC-peptide complex to which it can bind.

Lymphocytes exit the blood and enter the lymph node by extravasating at the *high-endothelial venules* (HEVs) present in the lymph node cortex (Figure 14-2). It has been estimated that as many as 1.4×10^4 to 3×10^4 lymphocytes extravasate through a single lymph node's HEVs every second.

Many, although not all, of the molecules that regulate extravasation are known. Briefly, HEVs express ligands (referred to as *addressins*) for L-selectin (CD62L), which is expressed on the surface of multiple cell types, including naïve T and B cells. Contact with these lectins slows the cells, allowing them to interact with other adhesion molecules and chemokines that stimulate cell movement between endothelial cells and into the lymph node tissue. Depending on the specific chemokine receptors that they express, lymphocytes will then home to specific microenvironments within a lymph node (see Chapter 2 and below).

Naïve Lymphocytes Sample Stromal Cells in the Lymph Nodes

After crossing the high-endothelial venules, mature, naïve lymphocytes enter the cortex of the lymph node (see Chapter 2). From here, B lymphocytes and T lymphocytes are guided by different chemokine interactions to distinct microenvironments. B cells enter the B-cell follicles, and T cells enter the paracortex (or T-cell zone)—a deeper part of the cortex (Overview Figure 14-3). Here they will scan for antigens, probing the surfaces of cells with their antigen receptors.

The movements of naïve lymphocytes that enter and scan the lymph node are remarkable to watch and, when first visualized, immediately (and perhaps even permanently) reverse impressions from fixed microscopic images that lymphocytes are rather dull and inert. These movements also revealed details about the microenvironment that had not been anticipated by static images of the tissues.

Naïve Lymphocytes Browse for Antigen along Reticular Networks in the Lymph Node

In 2002, the Michael Cahalan lab published one of the first videos of naïve lymphocytes trafficking within lymph nodes (Figure 14-4 and Accompanying Video 14-1). This group and others noted that the patterns of movement of the fluorescently tagged T cells within the "empty" (not fluorescent) black background of the image suggested that the cells were not moving freely; rather, their journeys appeared to be influenced by "invisible" structures in the lymph node tissue. These observations and others inspired the discovery of the

Cell Traffic in a Lymph Node

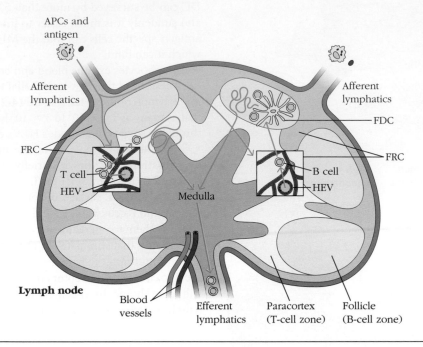

T and B lymphocytes travel into a lymph node from the blood, migrating across the wall of high-endothelial venules (HEVs) into the cortex (extravasation). T lymphocytes browse the surfaces of cells on the fibroblastic reticular cell (FRC) network, and if they do not bind to antigen within 16 to 20 hours, they leave the node via the efferent lymphatics in the medulla. B lymphocytes travel into the follicles and browse the surfaces of follicular dendritic cells. They, too, leave the node via the efferent lymphatics if they do not bind antigen. Dendritic cells and other antigen-presenting cells migrate into the lymph node via afferent lymphatics and join resident dendritic cells on the FRC network, where they are scanned by lymphocytes. Antigen typically arrives at the lymph node via the afferent lymphatics, entering the subcapsular sinuses either as processed peptide presented by antigen-presenting cells, or as unprocessed protein that can be coated with complement.

---

fibroblastic reticular network (see Figure 2-9b), a network of fibrous conduits and cells that provide pathways for traveling leukocytes. Indeed, when the Cahalan lab illuminated the network, investigators saw that naïve T cells interacted directly with the conduits, using them as guides for their otherwise random probing (see Figure 14-4).

The Germain lab visualized naïve lymphocyte cell probing in the lymph node in even more detail. One of that lab's three-dimensional images shows a T cell hugging a reticular fiber (Figure 14-5 and Accompanying Video 14-2). Their live video of naïve T lymphocytes diving in and out of the lymph-node parenchyma, following the tracks laid down by fibroblastic reticular cells, compellingly underscores the structural basis for the movements of T cells visualized before the network was definitively revealed (Figure 14-6 and Accompanying Video 14-3).

Naïve B cells that transit across HEVs into the lymph node cortex also begin to travel along fibroblastic reticular fibers. However, they soon change allegiance to the fibers established by the follicular DCs (Figure 14-7 and Accompanying Video 14-4).

Differences in chemokine receptor expression and chemokine distribution are responsible for the differences in migration patterns of T and B cells in the lymph node (see Chapter 2). Fibroblastic reticular cells in the T-cell zone are decorated with CCL21 and CCL19, which attract naïve, CCR7$^+$ T cells to the T-cell zone. In contrast, follicular DC networks are decorated with CXCL13, which attracts naïve, CXCR5$^+$ B cells to the follicles. (See Advances Box 14-2 for a more complete discussion of the contributions of chemokines to immune cell trafficking.)

The spaces probed by fluorescently labeled lymphocytes in these videos is occupied not only by the fibrous network, but by stromal cells that present antigen to T and B lymphocytes. The Nussenzweig lab revealed that DCs are present in every microenvironment within the lymph node (Figure 14-8). In the subcapsular sinus, actively migrating DCs travel among a bed of sessile macrophages, which are long-term

(a)

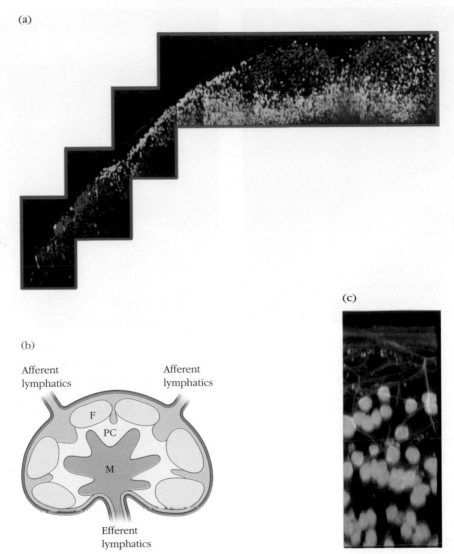

(b)

Afferent lymphatics

Afferent lymphatics

F

PC

M

Efferent lymphatics

(c)

FIGURE 14-4 AND ACCOMPANYING VIDEO 14-1
Two-photon imaging of live T and B lymphocytes within a mouse lymph node. Fluorescently labeled T lymphocytes (green) and B lymphocytes (red) were injected into a mouse and visualized by two-photon microscopy after they homed to the inguinal lymph node. (a, b) The images show that T and B cells localize to distinct regions of the lymph node: T cells in the paracortex (PC), B cells in the follicles (F). (c) A magnified image of T cells (fluorescing green) inter-

acting with the fibroblastic reticular cell network (stained red). The accompanying time-lapse video shows the movements of injected B (red) and T (green) lymphocytes, both of which continually extend and retract processes, probing the (unlabeled) follicular dendritic cell and fibroblastic reticular cell networks in their respective compartments. B cells move more slowly, on average, than T cells, and also do not change shape as dramatically. *[Miller et al. Science 7 June 2002: Vol. 296 no. 5574 pp. 1869–1873. Reprinted with permission from AAAS.]*

residents of this microenvironment (see Figure 14-8 and Accompanying Video 14-5). In the T-cell zone of the lymph node, however, DCs do not crawl as vigorously. Instead, they are part of the fibroreticular network, extending their long processes along the conduits and allowing naïve T cells to scan their surfaces during their travels (see Figure 14-8b). Follicular DCs are also an intrinsic part of the network in the B-cell zone (B-cell follicle), where their surfaces are also scanned by B lymphocytes (see Figure 14-8c).

Over the 12 to 18 hours that naïve T cells are in each lymph node, they probe DC surfaces in the paracortex, traveling in random directions on routes established by the

reticular network. Naïve B cells spend as much time scanning for pathogens as they, too, travel in random directions on follicular DC networks. If they do not find and bind antigen, naïve B and T cells up-regulate the S1P1 receptor, which allows them to leave the lymph node via efferent lymphatics in the medullary sinus (see Figure 14-3). Egress is dependent on interactions between S1P and its receptor, S1P1, which is ultimately up-regulated by naïve T and B cells that do not encounter antigen. These cells leave via portals in the medullary sinus (Figure 14-9 and Accompanying Video 14-6) and return to circulation, so that they may resume scanning for antigens in other lymph nodes.

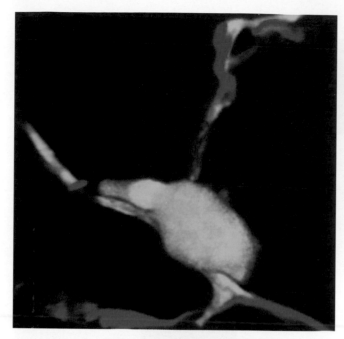

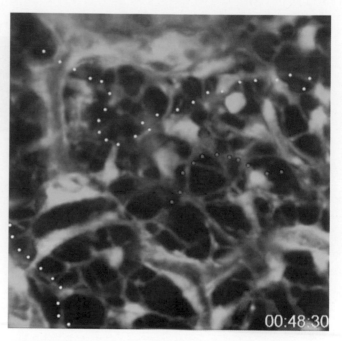

FIGURE 14-5 AND ACCOMPANYING VIDEO 14-2
T cell interacting with fibroblastic reticular network. Fluorescently labeled T cells (cyan) were injected into a mouse and visualized by confocal microscopy after homing to the lymph node. This image shows how closely a T cell interacts with the FRC (stained red and green), as it scans the surface of dendritic cells whose processes are also associated with the network (not shown). The video shows a three-dimensional depiction of this T cell-fibroblastic reticular network. *[Bajénoff, M., et al., Stromal Cell Networks Regulate Lymphocyte Entry, migration, and Territoriality in Lymph Nodes, Immunity **25**: 1–13. December 2006 Elsevier Inc.]*

FIGURE 14-6 AND ACCOMPANYING VIDEO 14-3
T lymphocytes migrate along the fibroblastic reticular cell network. T cells that were fluorescently labeled (red) were injected into a mouse whose fibroblastic reticular cells (and other nonblood cells) fluoresced green. Cell migrations in the paracortex of the popliteal lymph node of an anesthetized animal were imaged via intravital microscopy. The three dotted lines trace the movements of three different T cells, showing how they are guided by routes laid down by the reticular network. *[Reprinted from Bajénoff, M., et al., Stromal cell networks regulate lymphoctye entry, migration, and territoriality in lymph nodes. Immunity **25**: 989–1001, Copyright 2006, with permission of Elsevier.]*

Immune Cell Behavior during the Innate Immune Response

Infectious agents range in size and complexity from intracellular bacteria and viruses to extracellular bacteria, eukaryotic protozoa, and fungi, to large multicellular pathogens, such as worms. Although the immune response and its molecular and cellular players are tailored to each type of antigen, the broad similarities in immune cell behavior are illustrative and the best place to begin.

As you know from Chapter 5, the innate immune system is the first line of defense against infection, and has two roles: (1) to initiate the clearance or killing of pathogens and (2) to alert the adaptive immune system to the presence of pathogens. Myeloid cells (granulocytes, macrophages, and DCs) are the main players in this arm of immunity and respond to infection within hours, even minutes.

Both extracellular and intracellular pathogens gain entry to the body via breaks in the skin or mucosal surfaces, such as the lung, the gut, and the reproductive tract. Both types are recognized by *p*attern *r*ecognition *r*eceptors (PRRs) of the innate immune system that are on the surface of the

diverse cell participants in the innate immune response. For example, *Herpes Simplex Virus*, an intracellular pathogen, binds to both external (TLR2) and internal (TLR9) Toll-like receptors expressed by microglial cells, which are the myeloid-lineage APCs of the nervous system. *Staphylococcus aureus*, an extracellular bacterium that typically gains entry via the skin, binds to TLR2 on DCs in dermal (skin) tissue.

The binding of pathogen to PRRs on innate immune cells induces intracellular signals that enhance their ability to (1) coordinate the killing of pathogens and (2) alert the adaptive immune system to the infection. Granulocytes and APCs begin to produce cytokines and chemokines that attract more innate cells (e.g., neutrophils, NK cells) to the site of infection. Tissue macrophages, for instance, are induced to secrete chemokines such as IL-8 that effectively and quickly attract neutrophils. In turn, neutrophils make more IL-8 and attract even more neutrophils that swarm and kill pathogens at sites of infection and response. Such behavior is illustrated in Figure 14-10 and Accompanying Video 14-7, which show neutrophils (green) swarming around protozoal parasites (red) in a draining lymph node 3 hours after infection. These swarms are also likely to occur in tissues, at the sites of infection.

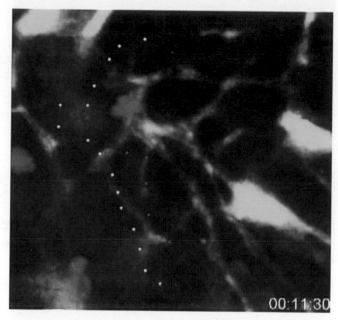

FIGURE 14-7 AND ACCOMPANYING VIDEO 14-4
B cells migrate along follicular dendritic cell networks.
B cells that were fluorescently labeled (red) were injected into a mouse whose follicular dendritic cells (and other nonblood cells) fluoresced green. Cell migrations in the follicles of the popliteal lymph node of an anesthetized animal were imaged via intravital microscopy. The dotted lines, shown in the figure and in the second part of the video, trace the movements of two individual B cells, showing that they are guided by routes established by the follicular network. *[Reprinted from Bajénoff, M., et al., Stromal cell networks regulate lymphoctye entry, migration, and territoriality in lymph nodes. Immunity **25**: 989–1001, Copyright 2006, with permission of Elsevier.]*

APCs, which reside in and can also be attracted to sites of infection, become even more effective phagocytes after they recognize pathogen. They also become more effective at processing pathogenic proteins and presenting them in class II MHC molecules via the exogenous pathway, and class I MHC via cross-presentation (see Chapter 8).

Antigen-Presenting Cells Travel to Lymph Nodes and Present Processed Antigen to T Cells

Once alerted to the presence of pathogen, APCs perform the second major function of the innate immune system by migrating to draining (closest) lymph nodes, and present processed antigen to naïve lymphocytes. Because they are the most efficient activators of naïve T lymphocytes, DCs play an especially important role. They enter the lymph node via afferent lymphatics (joining DCs that form part of the reticular network) and present MHC-peptide complexes to probing, naïve CD4$^+$ and CD8$^+$ T cells (Figure 14-11a).

The spleen is also a destination for some activated APCs. However, given the spleen's focus on blood-borne pathogens, it is likely that APCs that have more direct access to such pathogens (such as APCs associated with the marginal zone and/or splenic sinuses) play an even more important role in the immune response generated in this secondary immune organ.

Unprocessed Antigen also Gains Access to Lymph-Node B Cells

T cells need to encounter processed antigen on innate immune cells (APCs) to initiate the immune response. But B

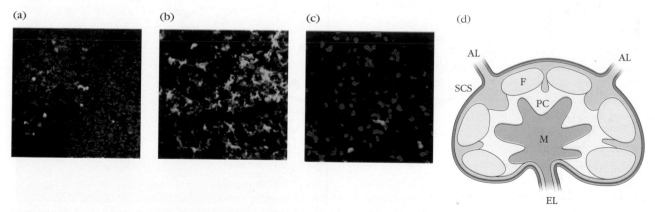

FIGURE 14-8 AND ACCOMPANYING VIDEO 14-5
Dendritic cells (DCs) are present in all lymph node microenvironments. Intravital microscopy of inguinal lymph nodes of anesthetized mice, all of whose dendritic cells fluoresce green. Several areas of the lymph node are visualized (a, b, c) and their relative locations are depicted (d), including the subcapsular sinus (SCS) (a), where the green dendritic cells can be seen surrounded by macrophages (which have taken up a red fluorescent dye); the paracortex (PC) (b), where dendritic cells mingle with T cells (red); and the follicle (F) (c), where dendritic cells are not as numerous but can be found among the B cells (blue). The video shows the movements of dendritic cells (green) in the subcapsular sinus, where they crawl among more sessile red fluorescent macrophages, probing their surface for antigen that they can take up, process, and ultimately present to T cells once they travel to the paracortex. *[Lindquist, Randall, Shakhar, Guy, Dudziak, Wardemann, Hedda, Eisenreich, Thomas, Distin, Michael L., and Nessenzweig, Michel C., Visualizing dendritic cell networks in vivo. Nature Immunology 5, 1243–1250 (2004) doi:10.1038/ni1139]*

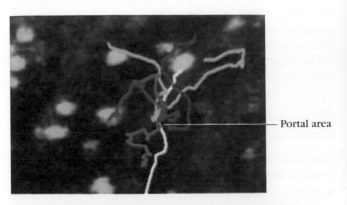

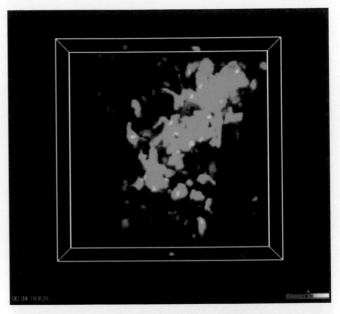

FIGURE 14-9 AND ACCOMPANYING VIDEO 14-6

Lymphocytes exit the lymph node through portals.
Fluorescently labeled T cells (green) were injected into a mouse, and their movements were visualized by two-photon microscopy in an isolated lymph node that was also stained with a red fluorescent dye that binds the walls of the medullary sinus. T cells transit from the lymph node tissue across an endothelial lining into the medullary sinus (red) at discrete areas called portals. The movements of several T cells that are passing through one particular portal area into the medullary sinus were traced. These tracings are shown as dotted lines in the video and are overlaid in the static figure (a), where they intersect in the portal area. *[Reprinted by permission from Macmillan Publishers Ltd: Wei, S.H., et al., Sphingosine 1-phosphate type 1 receptor agonism inhibits transendothelial migration of medullary T cells to lymphatic sinuses, Nature Immunology 6: 432–444, copyright 2005.]*

FIGURE 14-10 AND ACCOMPANYING VIDEO 14-7

Neutrophil swarming. Mice whose neutrophils fluoresced (green) were infected with *Toxoplasma gondii* that was fluorescently labeled (red). Three hours later, the behavior of neutrophils in isolated, draining lymph nodes was imaged by two-photon microscopy. This image shows a neutrophil swarm around protozoal parasites (red) in the subcapsular sinus of a lymph node. The video reveals the swarming behavior more dramatically and shows a large group of neutrophils (upper right) migrating en masse to join a smaller group that has initiated an interaction with the parasite (lower left). *[Reprinted from Chtanova T., et al., Dynamics of Neutrophil Migration in Lymph Nodes during Infection, Immunity 29: 487–496, supplemental data. Copyright 2008 with permission of Elsevier.]*

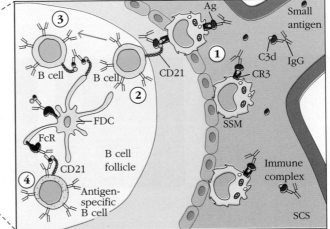

FIGURE 14-11 How antigen travels into a lymph node.

Antigen is delivered from tissues to the lymph node via the afferent lymphatics. (a) Dendritic cells that have processed antigen at the site of infection travel to the paracortex and are scanned by naïve T cells. (Soluble antigen can also be engulfed and processed by resident dendritic cells that are part of the dendritic cell network in the paracortex.) (b) (1) Opsonized soluble antigen (here shown coated with complement) will meet CD169⁺ macrophages in the subcapsular sinus (SSC). Subcapsular sinus macrophages (SSM) will, in turn, transfer the antigen to the interior of the lymph node, where it can be (2) picked up by cells with complement receptors, including the nonantigen-specific B cell shown here. (3) These nonantigen-specific B cells travel through the follicle and deliver the antigen to follicular dendritic cells. (4) Antigen-specific B cells encounter antigen as they probe the surface of follicular dendritic cells in the follicle. *[Adapted from T. Junt, E. Scandella, and B. Burkhard Ludewig (2008, October), Form follows function: Lymphoid tissue microarchitecture in antimicrobial immune defence, Nature Reviews Immunology 8:764–775.]*

cells, whose receptors recognize unprocessed antigen, also need to be activated by antigen if they are going participate. How do they encounter antigen? Is the innate immune system involved in this encounter?

Investigators used dynamic imaging approaches to follow the route of subcutaneously injected, whole antigen. They found that unprocessed antigen did, in fact, reach lymph nodes—within minutes (even seconds) of injection. Processed peptide on the surface of DCs reached the lymph node within 1 to 3 hours. This first wave of unprocessed antigen was responsible for activating antigen-specific B cells browsing the follicles of the lymph node; they responded within 30 to 60 minutes.

Studies now show that, depending on its size and solubility, an unprocessed antigen can gain access to a lymph node in several ways. If small and soluble enough, whole antigen can travel directly to the lymph node via the blood. Larger antigen particles and pathogens can be carried to the lymph node via the afferent lymphatics—on the "backs" of APCs that have bound opsonized pathogen. These unprocessed small and large antigens typically end up in the subcapsular sinus of a lymph node, which is outside the cortical follicles where B cells browse for antigen. How does this whole antigen ultimately gain access to B cells in the follicles?

Antigen that is opsonized by complement and/or antibody complexes (see Chapter 6) is trapped by a specialized group of macrophages (CD169$^+$ macrophages) that reside in the subcapsular sinus (Figure 14-11b). These CD169$^+$ macrophages transfer the antigen to cells within the lymph node, including nonantigen-specific B cells and other macrophages that express complement and Fc receptors. In turn, these B cells and macrophages transfer antigen to follicular DCs, which are scanned continually by naïve B lymphocytes.

The importance of complement receptors in the capture and transport of soluble antigen to the lymph node was directly demonstrated by a study that compared the abilities of normal versus complement-receptor-deficient B cells to capture antigens from subcapsular sinus macrophages (Figure 14-12 and Accompanying Video 14-8). Normal B cells can be seen sampling antigen present on the surface of macrophages (right panel). They grab and run away with chunks of the red protein. Complement-receptor-deficient B cells also probe the macrophages, but come away completely empty (left panel).

Immune Cell Behavior during the Adaptive Immune Response

In 2004, the Jenkins lab published two remarkable simulations of the activities of T-cell, B-cell, and dendritic-cell interactions in a lymph node before and after the introduction of antigen. Although not live images, their animations were based directly on published data about leukocyte behavior and trafficking. They remain a most illustrative model of what may happen within a lymph node after infection. In fact, they offer visual speculation of events that

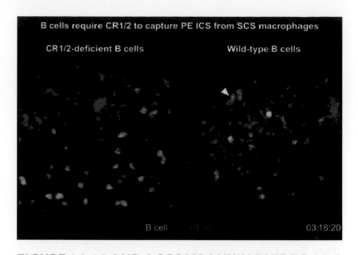

FIGURE 14-12 AND ACCOMPANYING VIDEO 14-8 B cells capture antigen from macrophages in the subcapsular sinus of the lymph node. The interaction between antigen-specific B cells (green) and macrophages that had captured this antigen (red) in the subcapsular sinus of a lymph node was imaged by two-photon microscopy. The image and video compare the behavior of wild-type and complement-receptor-deficient (Cr-1- and Cr-2-deficient) B cells in lymph nodes 1 hour after the antigen is introduced. In both situations, the macrophages in the subcapsular sinus capture the red fluorescent antigen (labeled "PE-IC" for phycoerythrin-immune complex) quickly and become coated with it. Wild-type B cells (right) interact with these red macrophages (yellow arrow) and grab the antigen, carrying it away (typically to follicles). However, although complement-receptor-deficient B cells (left) also probe macrophages for antigen (white arrows in the video), they come away empty. (The blue fluorescent structures are extracellular matrix fibers that help form the capsule and sinus.) *[Reprinted by permission from Macmillan Publishers Ltd: Phan, T. G., et al., Subcapsular encounter and complement-dependent transport of immune complexes by lymph node B cells, Nature Immunology 8:992–1000, supplementary movie 6, Copyright 2007.]*

clearly take place, but have not yet been fully captured in one tissue in real time: the functional response of the three main cell types (T cell, B cell, and APC) from antigen exposure through proliferation, complete with fluorescent indications of cell activation status.

Video 14-9 shows naïve T and B cells browsing antigen presenting cells in the lymph node in the absence of antigen. You should recognize the activities of these cells from what we have described above. Each lymphocyte type browses in their own niche: B cells (red) in the follicles and T cells (green) in the T-cell zone (the paracortex in lymph nodes). Dendritic cells (purple) are also moving through the lymph node—predominantly in the T-cell zone (paracortex).

After (soluble) antigen is introduced (Video 14-10), you will see marked changes in the behavior of each cell type. An activated DC (depicted in brighter purple) enters the lymph node from the site of infection. Shortly afterwards, it engages an antigen-specific T cell. This T cell becomes activated, turning more yellow, a change in color meant to depict T-cell receptor signaling. It then divides. During the same period antigen-specific B cells in the follicle also become activated (turning more yellow), after engaging soluble antigen. These activated B cells markedly change their patterns of movement and move toward the T-cell zone to 'flirt' with antigen-specific T cells. Finally, an activated T cell engages the activated B cell, resulting in B cell division.

Videos 14-9 and 14-10, particularly Video 14-10, provide an excellent reference for the more detailed description of immune cell behavior in lymph nodes that we will describe below. We will focus most closely on the interactions and travels of the three major lymphocyte subpopulations—CD4$^+$ T cells, B cells, and CD8$^+$ T cells—after they encounter antigen.

Naïve CD4$^+$ T Cells Arrest Their Movements after Engaging Antigens

As you have learned from Chapter 11, CD4$^+$ T cells are required for the initiation of optimal humoral and cellular adaptive immune responses. They are activated when they engage MHC-peptide on the surface of DCs that have been exposed to pathogen. Depending on the specific combination of signals they receive from DCs, activated CD4$^+$ T cells differentiate into distinct effector helper cells. Some will enhance the cellular immune response by contributing to the efforts of APCs to induce the differentiation of killer CD8$^+$ T cells. Others will enhance the humoral immune response by providing help to B cells that have been previously activated by antigen engagement.

The first studies of live CD4$^+$ T cells interacting with antigen presented by DCs in a lymph node came from work in the Cahalan laboratory, which showed that naïve T cells travel relatively quickly through the paracortex (10 μm/min or more) prior to engaging antigen. After binding MHC-peptide complexes, they slow down considerably (to 4–6 μm/min). These "arrested" T cells form both transient and stable engagements with DCs.

A compelling example of the effect of antigen on T-cell movements is shown in Figure 14-13 and Accompanying Video 14-11. T cells specific for a male antigen are labeled red, and DCs are green. Within minutes after antigen is injected, crawling T cells "arrest" and form stable contacts with DCs expressing the antigen.

From their observations, investigators initially proposed that there were three stages to this first encounter between a naïve T cell and a DC (Figure 14-14): (1) an initial period of 1 to 2 hours when cells made brief (5- to 15-minute) serial contacts with DCs, followed by (2) a 2- to 14-hour period of more stable (> 6-hour) engagements with DCs, which in turn was followed by (3) a period during which cells disengaged and returned to serial sampling, exhibiting a "swarming" behavior around DCs. These three stages preceded the first signs of proliferation, which began 24 to 36 hours after initial antigen encounter and continued into the fourth and fifth day after antigen introduction.

It is now known that T cells and DCs can form stable interactions quickly; transient interactions are not common to all

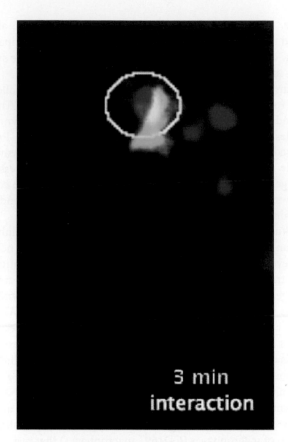

FIGURE 14-13 AND ACCOMPANYING VIDEO 14-11
T cells arrest after antigen encounter. In this experiment, cell activities were imaged via two-photon microscopy in the lymph node of a live, anesthetized mouse. Within minutes after antigen (male-specific peptide) is injected, a rapidly moving T cell (fluorescing red) "arrests" its motion as it forms a stable contact with the more slowly crawling dendritic cell (green) that has processed and presented the antigen. The static image indicates the stable interaction between T cell and dendritic cell 30 minutes after initiating contact. *[Celli, S., et al. (2007), Real-time manipulation of T cell-dendritic cell interactions in vivo reveals the importance of prolonged contacts for CD4 T cell activation.* Immunity *27:25–34, movie 10.]*

responses. Rather, it appears as if the kinetics of early encounters differs depending on the quality, quantity, and availability of antigen as well as the activation state of a DC. Nonetheless, in order to proliferate and differentiate optimally, CD4$^+$ T cells must ultimately become involved in committed, long-term relationships (8 hours or more) with a DC.

B Cells Seek Help from CD4$^+$ T Cells at the Border between the Follicle and Paracortex of the Lymph Node

B cells need two signals to be fully activated. They are partially activated after they find and bind antigen during their travels through the B-cell follicle. They then need help from an activated T cell, which releases cytokines that regulate differentiation and provides ligands that stimulate CD40 signaling. Several different effector helper cells—T$_H$1, T$_H$2, and T$_{FH}$

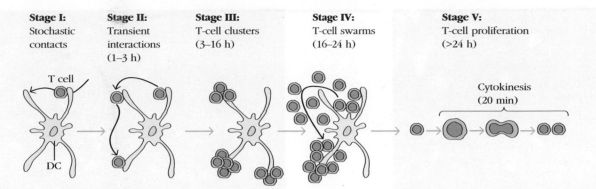

Stage I:
Stochastic contacts

Stage II:
Transient interactions
(1–3 h)

Stage III:
T-cell clusters
(3–16 h)

Stage IV:
T-cell swarms
(16–24 h)

Stage V:
T-cell proliferation
(>24 h)

Cytokinesis
(20 min)

T cell

DC

FIGURE 14-14 Stages observed during the encounter between a naïve T cell and a dendritic cell. Investigators observed that T cells made transient contact with dendritic cells both early and late in a response, prior to proliferation. They described several sequential stages, depicted here. These stages may not be required—other experiments show that CD4+ and CD8+ T cells can form stable interactions very quickly, bypassing the first step. However, most agree that the outcome of the interaction varies with its quality—and stable interactions appear required for optimal differentiation and proliferation of effector cells. Proliferation occurs after cells have detached from the dendritic cell, although they seem to continue to make transient contacts. *[Adapted from M. J. Miller, O. Safrina, I. Parker, and M. D. Cahalan (2004), Imaging the single cell dynamics of CD4+ T cell activation by dendritic cells in lymph nodes,* Journal of Experimental Medicine *200:847–856.]*

subpopulations, for example—can provide help that activated B cells need to proliferate and differentiate into producers of high-affinity antibodies and memory cells. Each will encourage the production of different antibody isotypes, tailoring the response to the type of pathogen that initiated the activity.

Activated B cells seek T-cell help at the border between the follicle and the paracortex of the lymph node. The Cyster lab pioneered dynamic imaging studies of antigen-specific B cells and generated now-classic images that revealed a B-cell/T cell choreography that was not fully anticipated by other approaches. They show B cells in the follicle moving purposefully toward the T-cell zone hours after binding antigen via the B-cell receptor (BCR) (Figure 14-15 and Accompanying Video 14-12). The investigators also showed that this migration was dependent on the chemokine receptor CCR7. When B cells bind antigen, they receive signals that up-regulate CCR7, which is not ordinarily expressed by naïve B cells. B cells can now follow chemotactic signals generated by the paracortex and migrate toward helper T cells. In other words, by inducing the up-regulation of CCR7, antigen binding redirects the B cell to the T-cell zone, where it can receive the appropriate help.

The investigators captured the encounter between an activated B cell and activated helper CD4+ T cell at the border between the follicle and T-cell zone (Figure 14-16 and Accompanying Video 14-13). Antigen-specific B-cell/T-cell pairs form stable connections within a day of antigen introduction and migrate together, with the B cell in full control of the antigen-specific T cell, which trails behind "helplessly." Such behavior was not evident between T and B cells that did not share antigen specificity.

The intimate and stable interactions established between T and B cells reflects the type of help that helper T cells must deliver to B cells. The cell pair forms an immunological synapse between them, with TCR, CD28, and adhesion mole-

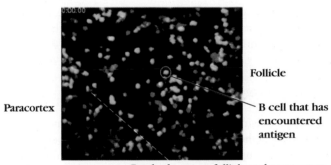

Follicle

Paracortex

B cell that has encountered antigen

Border between follicle and paracortex

FIGURE 14-15 AND ACCOMPANYING VIDEO 14-12
B cells travel to the border between the follicle and paracortex after being activated by antigen. Antigen-specific (BCR-transgenic) B cells (green) and nonantigen-specific (wild-type) B cells (red) were transferred together into recipient mice. Soluble antigen was injected 1 to 2 days later, and the behavior of the B cells in draining lymph nodes was imaged by two-photon microscopy. The dotted line in the static image roughly indicates the border between the B-cell zone (follicle) and T-cell zone (paracortex). An antigen-specific B cell that has been activated by antigen (1–3 hours after it was introduced) is circled, and its route can be followed in the video. It moves "purposefully" through other meandering B cells to the follicle border. (The time in hours:minutes:seconds is shown in the upper-left corner.) *[Okada T, Miller MJ, Parker I, Krummel MF, Neighbors M, et al. (2005) Antigen-Engaged B Cells Undergo Chemotaxis toward the T Zone and Form Motile Conjugates with Helper T Cells. PLoS Biology 3:e150. Video S5.]*

cules on the surface of T cells engaging class II MHC-peptide complexes, CD80/86 molecules, and other adhesion molecules on B cells. Into this synaptic space, the T cell delivers cytokines (e.g., IL-4) that stimulate B-cell differentiation. The synapse is also the site of stimulating interactions between CD40L on the T cell and CD40 on the B cell. CD40 signals are required for B-cell proliferation and differentiation.

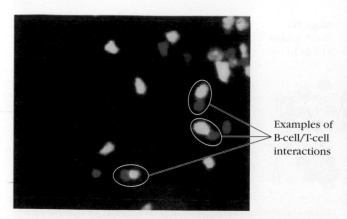

FIGURE 14-16 AND ACCOMPANYING VIDEO 14-13
Antigen-specific B and T cells interact at the border between the follicle and paracortex. Antigen-specific B cells (red) and antigen-specific T cells (green) were transferred into a mouse that had been immunized with a soluble antigen. Eight hours after antigen was introduced, the interactions between T and B cells were observed in an isolated lymph node via two-photon microscopy. The static image shows examples of stable contacts made between T cells and B cells in the paracortex, along the border of the follicle. The video shows how the more actively migrating B cells (red) drag the interacting T cells (green) behind them as they travel. *[Okada T, Miller MJ, Parker I, Krummel MF, Neighbors M, et al. (2005) Antigen-Engaged B Cells Undergo Chemotaxis toward the T Zone and Form Motile Conjugates with Helper T Cells. PLoS Biology 3: e150. Video S6.]*

After receiving T-cell help, activated B cells travel to the outer edges of the follicle, following cues that are dependent on a G-protein-coupled receptor (EBI2). Here B cells continue to proliferate and, in some cases, continue to receive T-cell help. Some of these activated B cells will differentiate into plasma cells and leave the lymph node; others will return to the interior of the follicle, seeding a germinal center, where they undergo more rounds of proliferation and somatic hypermutation.

Dynamic Imaging Approaches Have Been Used to Address a Controversy about B-Cell Behavior in Germinal Centers

The movements of cells within germinal centers have been visualized by several groups. Cyster and other labs revealed that germinal-center B cells were unusually motile and extended long processes—something more characteristic of DCs than lymphocytes (Figure 14-17 and Accompanying Video 14-14).

Immunologists studying B cells have also used dynamic imaging studies to address a controversy about the behavior of activated B cells in the germinal center. Recall that the germinal center is divided into two major subzones based, originally, on their appearance within classically stained sections under a light microscope (see Figure 12-11b). The dark zone contains proliferating B cells and is thought to be the main site of somatic hypermutation (see Chapter 12). The light zone is replete with follicular DCs and their reticular network, which present antigen-antibody complexes and provide a scaffold for further B-cell/T-helper cell interactions

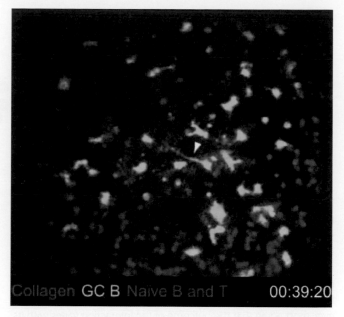

FIGURE 14-17 AND ACCOMPANYING VIDEO 14-14
Germinal-center B cells differ in their behavior from naïve B cells. In this experiment, fluorescently labeled antigen-specific B cells (green) were transferred into a mouse, which was immunized with soluble antigen 1 day later. Naïve (B and T) lymphocytes, fluorescently labeled red, were introduced 6 days later. The behaviors of antigen-activated B cells versus naïve lymphocytes were observed by two-photon imaging of isolated lymph nodes. Antigen-experienced B cells (green) occupy the germinal centers and differ dramatically in shape and activity from naïve lymphocytes (red). They extend processes that actively probe the microenvironment in ways more reminiscent of DCs. Most naïve B (and T cells) remain outside the germinal center, where they probe less actively and extend shorter processes. (Collagen fibers in the lymph node microenvironment fluoresce blue in this system.) *[Allen, C.D., Okada, T., Tang, H.L., and Cyster, J.G. (2007b). Imaging of germinal center selection events during affinity maturation. Science 315:528–531. Movie S1.]*

that regulate germinal center selection events. The light zone is thought to be the primary site of differentiation to plasma cells and memory B cells, as well as the site for positive selection of antigen-specific B cells that are somatically mutated in the dark zone. Positive selection is thought to occur when new B-cell specificities generated in the dark zone move to the light zone to sample antigens and test their affinities against antigens. Those that bind with high affinity received T-cell help and survived to return to the dark zone to generate even higher-affinity specificities. The highest-affinity clones arise after several cycles of somatic hypermutation in the dark zone, and antigen sampling in the light zone.

Based on this model of positive selection, investigators made a simple prediction: dynamic imaging experiments should show B cells trafficking back and forth between the zones (Figure 14-18). However, the experiments showed that germinal center B cells much prefer to travel *within* rather than *between* the light and dark zones! In fact, only a few percent were found to travel between the zones every hour. In addition,

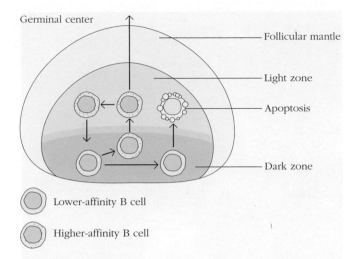

Germinal center

Follicular mantle

Light zone

Apoptosis

Dark zone

Lower-affinity B cell

Higher-affinity B cell

FIGURE 14-18 One proposed model for trafficking of germinal-center B cells between dark and light zones. Germinal-center B cells are thought to proliferate predominantly in the dark zone, where they also undergo somatic mutation. High-affinity clones are thought to be positively selected in the light zone, where both antigen and T-cell help are available. This proposal requires cells to move from dark to light zones, perhaps even more than once. Two-photon intravital microscopy experiments surprised many when they showed that (1) B cells preferred to traffic within rather than between zones, (2) germinal-center B cells can divide in the light zone, and (3) the B cells made surprisingly brief contact with T helper cells when they did encounter them in the light zone. This recirculation model is still favored, but is currently under modification based on these and other experiments. *[Adapted from A. E. Hauser, M. J. Shlomchik, and A. M. Huberman, (2007), In vivo imaging studies shed light on germinal-centre development,* Nature Reviews Immunology *7:499.]*

B-cell division was not simply confined to the dark zone but was also observed in the light zone. Finally, B cells made only brief contact with T-helper cells in the light zone, challenging another assumption that high-affinity B-cell clones would make stable connections with T cells that allow them to outcompete lower-affinity B-cell clones for T-cell help.

Although these observations inspired new models in which B cells behave and develop independently in light and dark zones, other analyses suggested that the frequency of travel between dark and light zones is still compatible with traditional models of positive selection for high-affinity clones (see Chapter 12). B cells may simply sample antigen less frequently than once assumed and may not establish long-term contacts with T helper cells. The brevity of germinal-center B-cell interactions with T helper cells suggest that only the highest-affinity B clones, which can process and present more antigen, engage T cells effectively, even if transiently.

In this case, dynamic images raised more questions than they resolved. However, they inspired novel speculation that may lead us to a more accurate understanding of the B-cell response.

CD8⁺ T Cells Are Activated in the Lymph Node via a Multicellular Interaction

Naïve CD8$^+$ T cells are the precursors of *cytotoxic T cells* (CTLs). Major participants in the cellular immune response, CTLs rove the body for infected cells, which they can kill very efficiently by inducing apoptosis. Like naïve CD4$^+$ T cells, naïve CD8$^+$ T cells are also activated by interacting with DCs in secondary lymphoid tissues; however, they recognize class I MHC-peptide rather than class II MHC-peptide combinations. Like B cells, they also require CD4$^+$ T-cell help to be optimally activated.

The behavior of naïve and activated CD8$^+$ T cells in the lymph node has been traced using dynamic imaging techniques. The movements of those cells after antigen introduction resemble those of CD4$^+$ T cells: CD8$^+$ T cells arrest when they contact antigen on DCs and engage in both transient and stable associations. Some investigations, however, suggest that CD8$^+$ T cells form stable contacts with DCs more quickly than CD4$^+$ T cells, even at low antigen concentrations (Figure 14-19 and Accompanying Video 14-15). Like CD4$^+$ T cells, CD8$^+$ T cells will divide up to 10 times over a

No peptide 10^{-9} M 10^{-7} M

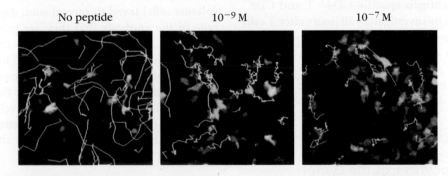

FIGURE 14-19 AND ACCOMPANYING VIDEO 14-15 CD8⁺ T cells arrest when they encounter antigen on DCs in the lymph node. This figure shows tracings (white) of the trajectories of antigen-specific CD8$^+$ T cells (red) interacting over a 5-minute period with DCs (green) in the lymph node of mice that had been immunized with two different concentrations of peptide antigen (10^{-9} M and 10^{-7} M) or no antigen at all. The trajectories of T cells interacting with antigen-exposed DCs are considerably shorter than those interacting with DCs from mice that were not immunized, indicating that the T cells arrest when they encounter antigen. *[Reprinted from Beuneu, H., Lemaitre, F., Deguine, J., Moreau, H.D., Bouvier, I., Garcia, Z., Albert, M.L., and Bousso, P., Visualizing the functional diversification of CD8$^+$ T cell responses in lymph nodes.* Immunity *33:412–23.]*

4- to 5-day period after antigen engagement. Mature CTLs ultimately leave the lymph node to travel to sites of infection.

Dynamic imaging techniques also allowed investigators to experimentally address two important questions about T-cell activation in vivo. First, when during these stimulatory interactions does a naïve CD8$^+$ T cell start to differentiate into a functional CTL? To answer this question, investigators engineered CD8$^+$ T cells that fluoresced whenever they expressed IFNγ, a cytokine that is produced by differentiated CTLs. Using dynamic imaging techniques, they generated time-lapse videos of the behavior of these cells in a lymph node. Their results were unexpected. Antigen-specific (TCR-transgenic) CD8$^+$ T cells could differentiate into IFNγ—producing cells quickly—prior to cell division (less than 24 hours after antigen encounter). The amount of IFNγ expressed by each activated CD8$^+$ T cell also varied widely. These findings suggest that functional differentiation of CD8$^+$ T cells begins as soon as T cells engage antigen on DCs. They also suggest that how well a cell generates IFNγ depends not simply on the affinity of the TCR for antigen (which was the same for each T cell). The quality of signals delivered by the DC and/or with the quality of CD4$^+$ T-cell help are likely to play major roles.

Second, how do CD4$^+$ T cells provide help, given that they cannot directly engage CD8$^+$ T cells? CD8$^+$ T cells, of course, interact with class I MHC-peptide complexes. CD4$^+$ T cells interact with class II MHC-peptide complexes, which are not expressed by CD8$^+$ cells. The simplest model for the delivery of CD4$^+$ T-cell help envisions an interaction among three cells: a single DC that presents peptides from an antigen protein in both class I and class II MHC to a CD8$^+$ T cell and a CD4$^+$ T cell, respectively (see Figure 14-11a for a schematic example of this *tricellular complex*).

The probability that three cells with the appropriate antigen specificities and complexes could find each other in a physiological context seemed low to immunologists. However, to their surprise and delight, dynamic imaging experiments confirmed that just such an interaction occurs in a lymph node. Cellular trios (consisting of an antigen-expressing DC, and antigen-specific CD4$^+$ T and CD8$^+$ T cells) were observed by investigators 20 hours after T cells were exposed to antigen (Figure 14-20 and Accompanying Video 14-16). Further data suggest these interactions may not be as improbable as supposed. In fact, they are facilitated by chemokine interactions: DCs activated by CD4$^+$ T cells produce chemokines (CCL3, CCL4, CCL5) that specifically attract CD8$^+$ T cells that express the chemokine receptors CCR4 and CCR5.

Activated Lymphocytes Exit the Lymph Node and Recirculate

Successful activation of all three lymphocyte subsets results in proliferation and differentiation into effector and memory lymphocytes. These mature cells must leave the lymph

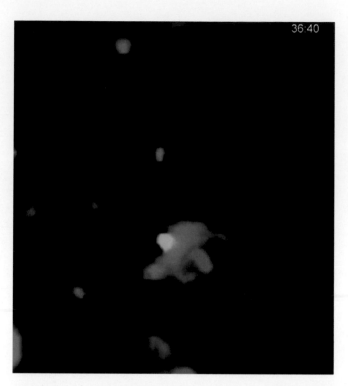

FIGURE 14-20 AND ACCOMPANYING VIDEO 14-16 The formation of a tricellular complex in the lymph node during CD8$^+$ T-cell activation. This dynamic image and video depict a lymph node of a live, anesthetized mouse that had first been injected subcutaneously with antigen-expressing dendritic cells (blue), and subsequently injected intravenously with antigen-specific CD4$^+$ T cells (red) and CD8$^+$ T cells (green). The static image shows a stable complex that includes all three fluorescently labeled cells. The video reveals the sequence of interactions that occurred, showing that a CD4$^+$ T cell (red) and antigen-expressing dendritic cell interacts first and is then joined by CD8$^+$ T cell (green). *[Reprinted with permission from Macmillan Publishers, Ltd: Castellino, F., Huang, A.Y., Altan-Bonnet, G., Stoll, S., Scheinecker, C., and Germain, R.N 2006. Chemokines enhance immunity by guiding naïve CD8$^+$ T cells to sites of CD4$^+$ T cell-dendritic cell interaction. Nature **440**:890–5.]*

node to perform their functions. Antibody-producing B cells (plasma cells) travel to several sites, depending on the isotype of the antibody that they are producing. Early IgM producers release antibodies from the medulla of the lymph node, many IgG producers go to the bone marrow, and IgA producers localize to mucosal-associated lymphoid tissue (MALT), particularly at the gut. Effector and memory CD4$^+$ and CD8$^+$ T cells travel to multiple organs and sites of infection, following chemokine and cytokine cues that have been generated by the innate immune response at those sites.

A Summary of Our Current Understanding

Biochemical and dynamic imaging studies have helped to generate an overview of the timing and organization of the fundamental T, B, and DC behaviors during a primary adaptive

A Summary of B-Cell and T-Cell Behavior in a Lymph Node after the Introduction of Antigen

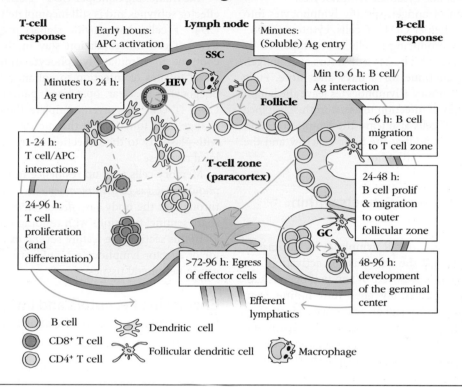

T-cell response

Early hours: APC activation

Lymph node

Minutes: (Soluble) Ag entry

B-cell response

SSC

Minutes to 24 h: Ag entry

HEV

Min to 6 h: B cell/ Ag interaction

Follicle

1-24 h: T cell/APC interactions

T-cell zone (paracortex)

~6 h: B cell migration to T cell zone

24-96 h: T cell proliferation (and differentiation)

24-48 h: B cell prolif & migration to outer follicular zone

GC

>72-96 h: Egress of effector cells

48-96 h: development of the germinal center

Efferent lymphatics

○ B cell
◉ CD8⁺ T cell
◎ CD4⁺ T cell

Dendritic cell

Follicular dendritic cell

Macrophage

(See text for full description.)

immune response in a lymph node (Overview Figure 14-21). Not every detail of Figure 14-21 is likely to be correct—and not every cellular player is represented. However, this graphic overview should provide a useful reference and reminder of the fundamental events that govern the development of the adaptive immune response in space and time.

Briefly, naïve lymphocytes—B cells, CD4⁺ T cells, and CD8⁺ T cells—continually circulate through secondary lymphoid tissue, browsing for antigens that they can bind. After entering a lymph node via HEVs, lymphocytes are guided by fibrous networks as they randomly explore their microenvironments. B cells scan the surface of follicular DCs in the follicle (B-cell zone) for unprocessed antigen that they might bind. T cells scan the processed peptide-MHC complexes on the surface of DCs in the paracortex or T-cell zone.

Antigen arrives in a lymph node in waves. Soluble, unprocessed, and opsonized antigen can arrive within minutes and is passed from cell to cell to the follicular dendritic cells that are scanned by naïve B cells. DCs that have processed antigen at the sites of infection arrive in the paracortex hours after infection.

When a probing lymphocyte engages an antigen, its movements slow down as it begins to commit to the interactions that induce their differentiation into effector cells and proliferation, processes that start early but continue over a 4- to 5-day period. A CD4⁺ T cell that engages class II MHC-peptide complexes on the surfaces of DCs may differentiate into one of several types of effector helper cells. Which effector cell type it will become depends on both quantitative (affinity) and qualitative (costimulatory molecule engagements and cytokine interactions) variables. Some CD4⁺ T cells gain the ability to help B cells to differentiate into antibody-producing cells. Some gain the ability to enhance cytotoxic cell differentiation and activity.

Within hours after successfully engaging antigen in follicles with their BCRs, B cells process and present antigen peptides in their own MHC molecules and migrate to the edges of a follicle, where they seek contact with activated CD4⁺ T cells that have differentiated into effectors that help B cells. Some differentiate directly into plasma cells; others reenter the follicle and establish a germinal center, where their BCR isotype and affinities are shaped by selection events and additional CD4$^+$ T-cell help. Within roughly the same period of time, CD8$^+$ T cells engage in multicellular complexes with DCs and helper CD4$^+$ T cells and start to differentiate into bona fide cytotoxic lymphocytes.

The differentiation of antigen-specific lymphocytes into effector and memory helper CD4$^+$ T cells, cytotoxic CD8$^+$ T cells, and antibody-producing B cells is accompanied by proliferation, which is robust by 24 to 48 hours after antigen encounter, and can continue for 4 days and more. Fully mature effector and memory T lymphocytes leave the lymph node and travel to sites of infection to assist in the clearance of pathogen. Effector B cells also leave the lymph node and travel to several sites, including the bone marrow and the mucosal tissues, to release antibodies.

The Immune Response Contracts within 10 to 14 Days

The proliferative activity of all lymphocytes ultimately declines, typically because the stimulus for activation and proliferation (pathogen) is removed. When pathogen is successfully cleared (or successfully goes into hiding from immune cells), inflammatory signals will cease. Most effector cells of the immune system have limited life spans and will ultimately die in the absence of stimulation. The life span of innate immune cells is particularly short. Neutrophils, for instance, live for little over 5 days.

The contraction of effector lymphocytes, most of which also have a finite lifespan, can be accelerated by self-limiting interactions. Engagements between T cells themselves can result in cell death. All T cells express the death receptor Fas, and activated T cells express FasL. Fas-FasL interactions induce apoptosis and may help to reduce the numbers of lymphocytes at the end of a response. Regulatory T cells may also help to quell normal immune responses by releasing inhibitory cytokines (Chapter 11). By approximately 10 days after initiation of the adaptive immune response, the lymph node has quieted and returned to its antigen naïve state and structure.

Immune Cell Behavior in Peripheral Tissues

Once the effector and memory cells that are generated after antigen encounter leave the lymphoid tissue, they begin to recirculate among peripheral tissues. Unlike naïve lymphocytes, effector lymphocyte populations do not reenter secondary lymphoid tissue but, instead, exhibit tissue-selective homing behavior.

Antibody-producing B cells (plasma cells) travel to several sites, including the medullary region of the lymph node and the bone marrow, where they release large quantities of antibodies. Pathogen-specific cytotoxic CD8$^+$ T cells follow chemokine cues to the sites of cellular infection, where they induce apoptosis of cells that express the class I MHC-peptide complexes that reveal their infection by intracellular pathogens.

The trafficking of helper CD4$^+$ T cells varies according to effector subtypes and is still incompletely understood. Some CD4$^+$ T cells that provide signals promoting B-cell and CD8$^+$ T-cell differentiation stay in the lymph node and continue to stimulate lymphocyte differentiation. Others travel to sites of infection to enhance the ability of tissue phagocytes to clear opsonized pathogen. Some effector CD4$^+$ T cells may even complete their differentiation or change their functions at the sites of infection, responding with plasticity to the collection of signals and needs of the local response.

Memory cells take up residence in both secondary lymphoid tissue (as *central memory cells*) and peripheral tissues throughout the body (as *effector memory cells*). There they share sentinel functions with residing APCs.

In this section, we feature several examples of the behavior of effector lymphocytes confronting physiological antigen in peripheral tissues.

Chemokine Receptors and Integrins Regulate Homing of Effector Lymphocytes to Peripheral Tissues

Not surprisingly, the trafficking of effector and memory lymphocytes to peripheral tissues is regulated by changes in the repertoire of chemokine receptors expressed on the memory or effector cell surface and the chemokines generated by innate immune cells at sites of infection. In addition, tissues display unique sets of adhesion molecules that help select for the effector subset.

In general, effector cells and effector memory cells avoid redirection back to secondary lymphoid tissues because they decrease expression of L-selectin (CD62L), which prevents them from entering via HEVs. Instead, effector cells express adhesion molecules and chemokine receptors that coordinate their homing to relevant tissues.

For example, a subset of memory and effector T cells that home to the intestinal mucosa expresses high levels of the integrins α4β7 (LPAM-1) and CD11a/CD18 (LFA-1, αLβ2), which bind to MAdCAM and various ICAMs on intestinal lamina propria venules (Figure 14-22). Cells homing to the gut mucosa also express the chemokine receptor CCR9, which binds to CCL25 in the small intestine. IgA-secreting B cells are also recruited into gut tissue via the chemokines CCL25 and CCL28.

Other memory/effector cells home preferentially to the skin because they express high levels of *cutaneous lymphocyte antigen* (CLA) and LFA-1, which bind to E-selectin and ICAMs on dermal venules of the skin (see Figure 14-22). The chemokines CCL17, CCL27, and CCL1 also play a role in the recruitment of skin-homing T cells.

(a)

Mucosal-homing
effector T cell

(b)

Skin-homing
effector T cell

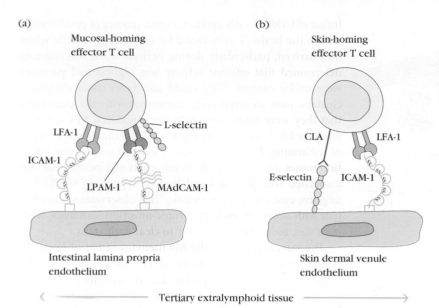

LFA-1 — L-selectin

CLA — LFA-1

ICAM-1

E-selectin — ICAM-1

LPAM-1 — MAdCAM-1

Intestinal lamina propria
endothelium

Skin dermal venule
endothelium

← Tertiary extralymphoid tissue →

FIGURE 14-22 Examples of homing receptors and vascular addressins involved in selective trafficking of naïve and effector T cells. Various subsets of effector T cells express high levels of particular homing receptors that allow them to home to endothelium in particular tertiary extralymphoid tissues. The initial interactions in homing of effector T cells to (a) mucosal and (b) skin sites are illustrated.

Effector Lymphocytes Respond to Antigen in Multiple Tissues

Dynamic imaging techniques have been used to visualize the immune response to specific pathogen, rather than noninfectious experimental antigen. Below we describe recent studies that specifically follow the behavior of T cells responding to physiologic insults in the periphery. We explore the first two in the most depth. One follows the behavior of T cells in response to infection with the protozoa that causes toxoplasmosis (*Toxoplasma gondii*), a pathogen that is transmitted from cats to humans (and other animals) via feces and produces an infection that is often asymptomatic. However, "Toxo" can cause harm to the fetus of a pregnant woman. The second study traces the behavior of T cells responding to—and rejecting—an allogeneic skin graft. The observations should enhance the development of therapeutic efforts to inhibit graft rejection.

We close with several short, tantalizing descriptions of interactions between immune cells and physiological antigens that have been revealed by recent dynamic imaging studies. As you will see, these studies provide a powerful reminder of the influence of inflammation, which is difficult to mimic in the absence of real infection. The studies also begin to reveal immune cell behavior outside secondary lymphoid tissue.

CD8⁺ T-Cell Response to Infection by *Toxoplasma gondii*

Investigators have recently used dynamic imaging methods to trace the behavior of the T cell specific for *Toxoplasma gondii* (Toxo), a protozoal parasite that infects many different tissues, including the brain. More than half of the world's population has been infected by Toxo, which is acquired by ingesting contaminated raw meat, soil, or litter exposed to feces from infected cats. Most of us never know that we have been invaded, and some of us harbor the parasite in cysts for

long periods of time. In those with weakened immune systems, however, Toxo infection can damage the brain and eyes. The pathogen also can cross the placenta and cause disease in fetuses, whose immune systems are underdeveloped. Understanding our response to this pathogen is an important step in controlling it.

In order to follow the immune response to this pathogen, investigators cleverly modified the parasite so that it expresses an antigen (an *ova*lbumin [OVA] peptide) that can be recognized by TCR transgenic (OT-1) CD8⁺ T cells, which, as you now know (see Advances Box 14-1), can be more easily traced and manipulated. The Hunter lab has imaged CD8⁺ T cells responding to this infection in both the lymph nodes and the brain. Their work shows that CD8⁺ T cells in the lymph node respond very rapidly (within 36 hours) after intravenous injection of *T. gondii*, forming stable contacts with DCs that peak between 36 and 48 hours. As predicted by work described above, crawling naïve antigen-specific T cells slow down considerably (from 6–8 μm/min to 4 μm/min) after encounter with DCs that have been exposed to antigen. Their movements pick up again after 48 hours, when they are actively proliferating and seem to rely on transient contacts, again as predicted by investigations with noninfectious systems.

Interestingly, responding CD8⁺ T cells and DCs appear to collect in the lymph node's subcapsular sinus, a localization that was also seen after viral infection (Figure 14-23). Investigators speculate that this may be an active site of inflammation in the lymph node that attracts effector cells. DCs also change in appearance, becoming vacuolated—not as a result of infection, but in response to general inflammation.

CD8⁺ T cells reactive to *T. gondii* antigens appeared in the brain by day 3, and peaked at day 22, an increase that was coincident with peak proliferation of CD8⁺ T cells in secondary lymphoid organs. The investigators provide evidence that the increase in T cells in the brain was due to the continual

(a)

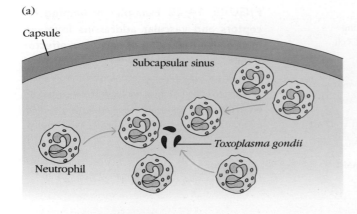

(b)

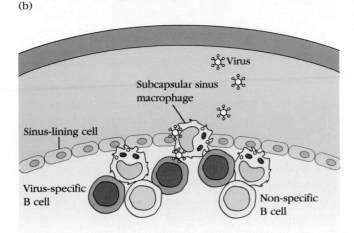

(c)

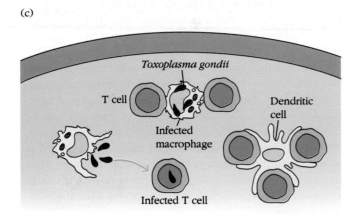

FIGURE 14-23 Examples of immune response to pathogens within the subcapsular sinus of a lymph node. (a) *Toxoplasma gondii*, a protozoal parasite, can gain direct access to the subcapsular sinus and attracts swarms of neutrophils. (b) After a virus gains access, it accumulates on CD169+ macrophages that line the sinus. Virus-specific B cells will then gather from the follicular side of the lining, sampling antigen from the macrophages. (c) Memory CD8+ T cells also have access to the subcapsular sinus and have been shown to cluster around cells infected with *T. gondii*. This makes them vulnerable to further infection. *[Adapted from J. L. Coombes, and E. A. Robey, E.A. 2010, Dynamic imaging of host-pathogen interactions in vivo, Nature Reviews. Immunology **10**:353–364, Figure 3.]*

influx of CD8+ T cells and not a consequence of proliferation within the brain. T cells could be surprisingly motile when they arrived, particularly during periods when investigators determined that effector activity was highest and parasites were under control. They could also form more stationary clusters, near infected cells, consistent with the possibility that they were interacting with brain-resident APCs, which were not labeled in this experiment. Unfortunately, the ability of infiltrating *T. gondii*-reactive T cells to control the infection wanes over time (> 40 days), a loss of function that was associated with the up-regulation of expression of PD-1, a negative costimulatory molecule. This observation is consistent with observations that chronic infection "exhausts" lymphocytes, reducing their ability to clear pathogens.

The infection altered the microenvironment of both the lymph node and the brain in several unanticipated ways. Inflammation (antigen specific and nonspecific) disrupted the reticular network of the lymph node and decreased the levels of CCL21. This disruption may have assisted the establishment of cell-cell contacts during infection, but also was likely to abrogate responses to new infections, by taking away the cues for naïve T-cell migration. In the brain, however, the infection appeared to result in the establishment of a new reticular network decorated with CCL21. This was a surprise and presumably helps organize T-cell entry and response in this tertiary tissue. T-cell interactions with this network after infection imply that this network plays a significant role in T-cell trafficking within the brain (Figure 14-24

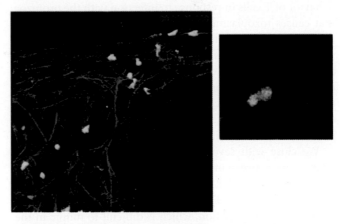

FIGURE 14-24 AND ACCOMPANYING VIDEO 14-17
A reticular network is established at the site of infection by *Toxoplasma gondii*. This dynamic image was taken of brain tissue in mice that had been infected for 4 weeks with the protozoan *Toxoplasma gondii*, and then injected intravenously with protozoan-specific T cells (green). The image shows a fibrous network fluorescing blue, that has been established specifically in response to the infection (it is not present in uninfected tissue). The videos show T cells using the routes established by the network to migrate through the tissue. The magnified image shows T cells extending processes that interact intimately with the fibers. *[Reprinted with permission from Macmillan Publishers, Ltd: Castellino, F., Huang, A.Y., Altan-Bonnet, G., Stoll, S., Scheinecker, C., and Germain, R.N, 2006. Chemokines enhance immunity by guiding naïve CD8+ T cells to sites of CD4+ T cell-dendritic cell interaction. Nature **440**:890–5.]*

and Accompanying Video 14-17). Other studies intriguingly suggest that such reticular networks are routinely established in other tertiary tissues after infection.

Host Immune Cell Response to a Skin Graft

The rejection of grafts that are MHC mismatched (allografts) requires both CD4$^+$ and CD8$^+$ T-cell activity. However, the sequence of interactions that lead to rejection is not fully understood, and the role of host (recipient) versus graft (donor) APCs in mediating the activation is still disputed. Specifically, in order to be activated, naïve CD4$^+$ and CD8$^+$ T cells must interact with APCs that express the graft's alloantigens. Are the activating APCs coming from the graft or from the host?

The Bousso lab performed an elegant set of dynamic imaging experiments to address this and other questions. MHC-mismatched skin was grafted onto a mouse ear and imaged by two-photon intravital microscopy. The mismatched skin was from a mouse whose DCs fluoresced yellow (a CD11c-YFP reporter mouse), so their motions could be traced. These fluorescent DCs exited the graft and were gone within 6 days. The investigators looked for these graft-derived DCs in the draining lymph nodes of the recipient mouse and found them as early as 3 days after engraftment, but they had the look of dying cells and never reappeared.

The investigators then asked which cells came into the skin graft from the host mouse. To do this, they put the graft onto mice whose cells all express green fluorescent protein (GFP). They found that CD11b$^+$ (myeloid) green fluorescent cells infiltrated the graft within 3 days. These cells were dominated by neutrophils at first, but over time they included more and more monocytes and DCs. Interestingly, such myeloid cells were also found in grafts from control, MHC-matched mice (isografts). The investigators took advantage of their system to look more closely at the locations of these cells and found that GFP$^+$ cells infiltrated the dermis of both allografts and isografts. However, only the allografts showed evidence of GFP$^+$ cells infiltrating all of the donated skin, including the outer layer (the epidermis), indicating that deep infiltration was antigen dependent.

To see if the infiltrating cells carried alloantigen back to draining lymph nodes, the investigators introduced a clever twist to their system (Figure 14-25a). First they put their grafts on MHC-mismatched hosts with GFP$^+$ cells and let the green, myeloid cells (neutrophils and monocytes) infiltrate the graft over a 6-day period. They then removed the graft, with its new green host cell infiltrates, and put it on another MHC-mismatched mouse, but one that did not have any GFP$^+$ cells. They could then trace the fate of the green infiltrates. Three days later, they found some of these GFP$^+$ cells in the T-cell zones of the draining lymph node, suggesting that infiltrating cells from the host could play a role in initiating the T-cell response to the graft. This is consistent with suggestions that host APCs *cross-present* the graft's alloantigens to CD8$^+$ T cells. (Recall from Chapter 8 that is the process by which APCs pick up extracellular antigens and present them as class I MHC-peptide complexes to CD8$^+$ T cells.)

They strengthened their case for the role of host APCs in cross-presenting alloantigen via a clever, albeit complex scheme. Essentially they grafted skin from a "strain A" class I MHC-deficient mouse onto an antigen-mismatched ("strain B") mouse that expressed normal class I MHC (Figure 14-25b). In this system, the skin graft's own APCs were completely unable to present antigen to class I MHC-restricted CD8$^+$ T cells; only the infiltrating cells from the recipient would be able to do so. They allowed strain B cells to infiltrate the graft, then removed it, transplanting it onto another strain B mouse, but one that was class I MHC deficient. If and only if the infiltrating (strain B) cells from the first host successfully cross-presented foreign antigen from the strain A graft would the CD8$^+$ T cells in this second recipient respond (and initiate rejection). Indeed, in vivo imaging revealed that antigen-specific CD8$^+$ T cells from this second recipient responded to the graft—confirming that host DCs that infiltrated the graft pick up and present foreign antigen to CTLs, which are responsible for graft rejection.

Finally, the investigators visualized the activity of the graft-rejecting cytotoxic CD8$^+$ T cells. Dividing T cells could be found in the draining lymph node as early as two days after graft transplant. However, such cells were not found in the graft itself until day 8, and seemed to enter from the graft edges. CD8$^+$ T cells were found at the border between the dermis (deep) and epidermis (more superficial) border of the allograft and were often in proximity to dying cells, which were likely their targets. CD8$^+$ T cells were slower moving in the allografts (versus isografts), which again is consistent with indications that antigen-specific cells arrest when they meet their antigen. By day 10, when the tissue was undergoing active rejection, CD8$^+$ T cells were found throughout the graft.

Dendritic Cell Contribution to *Listeria* Infection

As you know, DCs are important for initiating an adaptive immune response against infection; however, in some cases they are responsible for maintaining infection! *Listeria monocytogenes* is an intracellular bacterium that resides in soil and can cause food-associated illness and fatalities (Figure 14-26; also see www.dnatube.com/video/2506/Intracellular-Listeria-Infection for an illustrative video). In 2011, one of the worst food-borne disease outbreaks in the United States was traced to *Listeria* associated with cantaloupe melons.

Recall that the spleen is the site of response to blood-borne pathogens: all white blood cells circulate through sinuses in the red pulp, which also is a site where DCs and monocytes can sample antigen (see Figure 2-10). White blood cells enter and scan the spleen's version of the T-cell zone, the *periarteriolar lymphoid sheath* (PALS). B cells enter and scan follicles, which are very similar to those found in lymph nodes. The area that separates the red pulp from the white pulp, the *marginal zone* (MZ), is a relatively

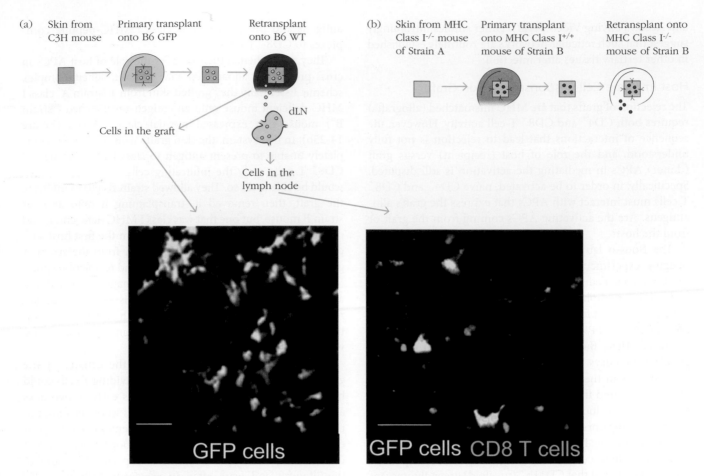

FIGURE 14-25 In vivo imaging of the immune response to an antigen-mismatched skin graft. This figure shows the approaches taken by investigators to see which antigen-presenting cells—host or graft (donor) APC?—activate the CD8$^+$ T cells that will reject a skin graft. (a) Skin from a mouse of one strain (C3H) was grafted onto the ear of a mouse of an MHC-mismatched strain (B6) in which all cells expressed GFP. The cells that infiltrated the skin from this B6-GFP mouse (pseudo colored yellow) were predominantly myeloid (neutrophils, monocytes, macrophages). After time was allowed for infiltration, the investigators removed the graft and placed it on a wild-type B6 mouse whose CD8$^+$ T cells (shown in green). This allowed the investigators to see where the infiltrating cells (yellow) would go. They found the cells in the draining lymph node (dLN) where they were interacting with CD8$^+$ T cells. These CD8$^+$ T cells ultimately distributed themselves through the graft, killing the MHC-mismatched cells. (b) Skin from a strain A mouse that expressed no class I MHC was transplanted on the ear of an antigen-mismatched (strain B) mouse that expressed class I MHC. In this situation, *none of the donor graft cells and only the infiltrating recipient cells could present antigen to CD8$^+$ T cells.* The investigators let this graft accumulate recipient cells, then retransplanted it onto an antigen-matched class I$^{-/-}$ mouse. The T cells from this mouse responded to the strain B cells that had infiltrated the primary graft and picked up the strain A antigen. *[Reprinted by permission from Macmillan Publishers Ltd: Celli, S., Albert, M.L., and Bousso, P., 2011. Visualizing the innate and adaptive immune responses underlying allograft rejection by two-photon microscopy, Nature Medicine 17: 744–749.]*

unique microenvironment that is the first line of defense against blood-borne pathogens. It is also the site of residence of DCs and monocytes as well as a unique group of B cells (MZ B cells) that do not circulate.

Dynamic imaging of fluorescently tagged *Listeria* bacteria showed that they arrive in the red pulp of the spleen within seconds of intravenous injection (see Figure 14-26). (During a physiologic infection, when bacteria are ingested, they probably take longer to arrive in the red pulp—either within white blood cells that they infected in the mucosa of the gut or directly from blood that they penetrated during infection.)

Within minutes, *Listeria* associates with and is presumably phagocytosed by the DCs in the red pulp sinuses, thereby infecting them. *Listeria*-specific CD8$^+$ T cells accumulate in these sinuses, forming stable contacts with the DCs that activate them. Investigators found that DCs in these sinuses recruit other innate immune cells, including monocytes, and unwittingly pass their *Listeria* infection to these cells (via their extensive processes, down which *Listeria* travel). These circulating cells, which are critical for controlling the pathogen, also become vehicles of infection that spreads throughout the body.

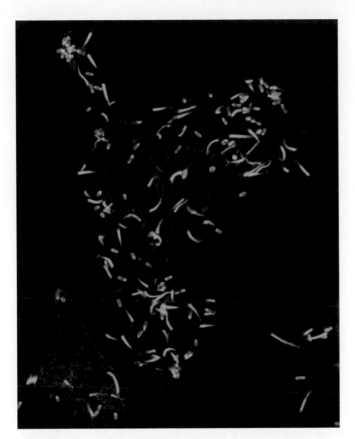

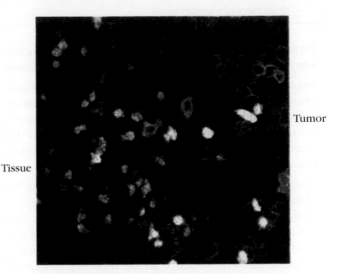

FIGURE 14-27 T cells infiltrating a tumor. Dynamic imaging of interactions of antigen-specific T cells (yellow) with tumor expressing that antigen (red) shows that some T cells infiltrate the solid tumor, but many gather in tissue borders that are not as tumor ridden. *[Deguine, J., et al., 2010. Intravital Imaging Reveals Distinct Dynamics for Natural Killer and CD8 T Cells during Tumor Regression, Immunity 3: 632–644. Copyright 2010 with permission from Elsevier.]*

FIGURE 14-26 Intracellular *Listeria* infection. *Listeria* (green) use the host cell actin (red) to propel themselves through the cell cytoplasm (cometlike tails of red actin can be seen trailing the green bacteria). *[Courtesy Alain Charbit.]*

T-Cell Response to Tumors

Dynamic imaging studies have also visualized an immune reaction to experimentally generated tumors. These show that antigen-specific, activated T cells can gain access to the tumor (Figure 14-27). When they do, they migrate vigorously and exhibit effective cytolytic behavior, associating for long periods of time (6 hours or more) with tumor cells and inducing apoptosis. However, the tumors failed to regress, and T-cell response seemed limited by (1) poor access to the tissue and (2) poor antigen presentation by tumor-associated cells, not by the availability of activated CTLs or their ability to mediate cytolysis. These studies suggest that strategies to improve antigen presentation and tumor accessibility may be more effective than strategies that simply increase antigen-specific T-cell number.

Regulatory T-Cell Responses

Dynamic imaging has also allowed us to see when and where regulatory T cells exert their suppressive activity, and the results satisfyingly confirm predictions made from "static" experiments, suggesting that regulatory T cells can suppress immune responses in more than one way (see Chapter 9). In

a model of Type I diabetes (the result of T-cell-mediated destruction of β cells of the pancreatic islet), investigators traced the activities of fluorescently labeled antigen-specific regulatory T cells introduced prior to disease onset. They exhibited two distinct behaviors. Some prevented diabetogenic effector cells from making productive clustering contacts with cells presenting auto-antigen in the pancreatic islets, consistent with proposals that regulatory T cells can inhibit effector T-cell function by quelling the activation potential of resident APCs. Other regulatory T cells formed stable connections with effector CD8$^+$ T cells and inhibited their ability to kill pancreatic target cells, an effect associated with TGF-β secretion by regulatory T cells.

Memory T-Cell Responses

Dynamic imaging of the trafficking of memory T and B cells is in its early stages. However, experiments have already revealed intriguing features. For instance, memory CD8$^+$ T cells appear to localize to and reside in the B-cell follicle after *Listeria* infection. A clever set of experiments revealed that T cells that are latecomers to the scene of an immune response in a lymph node tend to generate central memory T cells, perhaps because they do not have as much access to fully activating antigen interactions that would lead to effector cell generation.

The Analyze the Data study question at the end of this chapter also describes another interesting investigation that reveals differences in the trafficking of CD4$^+$ and CD8$^+$ memory T cells.

SUMMARY

- Dynamic imaging takes advantage of advances in microscopy to visualize the movements of cells deep in live tissue.

- Innate immune cells, including granulocytes and antigen-presenting cells (APCs), are the first to respond to pathogen. Innate immune cells bind pathogen via pattern recognition receptors (PRRs), which generate signals that result in the release of inflammatory signals, including chemokines and cytokines.

- Chemokines attract other innate immune cells to the site of infection. Neutrophils are generally the first cell type to move from the bloodstream into inflammatory sites; they can swarm around pathogens as they attack. They also produce more chemokines that attract APCs, including dendritic cells (DCs), which are the most efficient activators of naïve T cells.

- DCs at the site of infection engulf antigen, presenting processed peptide in class II MHC and cross-presenting it in class I MHC. They travel to draining lymph nodes via the afferent lymphatics to alert the adaptive immune system of the presence of pathogen.

- Antigen can enter the lymph node via multiple routes. Soluble, unprocessed antigen enters the tissue quickly and directly, or can be trapped by macrophages lining the sinuses and relayed to follicular DCs through the activity of noncognate B cells (and others).

- Processed antigen (in the form of MHC-peptide complexes) arrives within a few hours on the surface of DCs, which travel to the T-cell zone (paracortex). Dendritic cells extend long processes and can interact with up to 5000 T cells per hour.

- Lymphocytes undergo constant recirculation between the blood, lymph, lymphoid organs, and tertiary extralymphoid tissues, increasing the chances that the small number of lymphocytes specific for a given antigen (about 1 in 10^5 cells) will actually encounter that antigen.

- Cell-adhesion molecules (CAMs) and chemokine/chemokine receptor interactions regulate the migration of leukocytes both into and within lymphoid organs and inflamed tissues.

- Extravasation of all white blood cells involves four steps: rolling, activation, arrest/adhesion, and transendothelial migration.

- Naïve B and T lymphocytes express CD62L that helps them to home secondary lymphoid organs, where they extravasate across high-endothelial venules (HEVs).

- Naïve lymphocytes are very motile cells in the lymph node and crawl along reticular fibers in their respective lymph node zones. B lymphocytes browse for antigen along follicular DC networks in the follicle. T lymphocytes interact with fibroblastic reticular cell networks in the paracortex, where DCs that express antigen can also be found. Their

movements are regulated in part by chemokine interactions. Naïve lymphocytes that do not encounter antigen leave the lymph node after about 12 to 18 hours and reenter circulation to probe another lymph node.

- T lymphocytes that encounter antigen to which they bind slow down considerably and ultimately "arrest," forming stable contacts with APCs (in the CD4$^+$ and CD8$^+$ T cells). These contacts last between 1 and 24 hours, and if productive result in proliferation. Cells can experience up to 10 divisions over the following 4 to 5 days.

- B cells that encounter antigen up-regulate CCR7 and travel by chemotaxis from the center of the follicle to the T-cell zone, where they wait for T-cell help. Antigen-specific T-cell/B-cell interactions are strong and stable. The T/B cell pair moves actively at the follicular/T-cell border.

- CD8$^+$ T cells quickly form stable interactions with antigen-bearing DCs in the cortex.

- Stable tricellular complexes that include a CD8$^+$ T cell, a CD4$^+$ helper T cell, and a dendritic cell have been directly observed in dynamic images, showing that activated CD8$^+$ T cells may be specifically attracted by chemokines produced by DCs activated by CD4$^+$ T cells.

- Germinal-center B cells are unusually motile and distinct in morphology (they look more like DCs). Although they tend to circulate within, rather than between, the light zone and dark zone, a small percentage traffic from dark to light zones, presumably to sample antigen and receive T-cell help. Their contact with helper T cells is surprisingly brief, a behavior that may put high-affinity B cells, which express higher densities of MHC-peptide at a competitive advantage.

- Effector and memory lymphocytes leave the lymph node via portals in the medullary region of the lymph node, an event that depends on interactions between S1P1 receptors on white blood cells and S1P in their environment. They travel back into circulation via efferent lymphatics. Effector lymphocytes down-regulate CD62L, expressing a different profile of chemokine receptors that allow them to interact with inflamed vascular endothelium of tertiary tissues.

- Dynamic imaging of cells responding to a protozoal parasite that infects the brain reveals that inflammation can temporarily alter the microenvironment (reticular structure) of a lymph node and can remodel the tissue that is the site of infection by establishing reticular systems for lymphocyte trafficking.

- Dynamic images of cells responding to an MHC-mismatched skin graft show that the APCs of the recipient are capable of sampling and presenting alloantigen to host CD8$^+$ T cells, which ultimately distribute themselves throughout the graft and destroy it.

REFERENCES

Allen, C. D., T. Okada, and J. G. Cyster. (2007a). Germinal-center organization and cellular dynamics. *Immunity* **27**:190–202.

Allen, C. D., T. Okada, H. L. Tang, and J. G. Cyster. (2007b). Imaging of germinal center selection events during affinity maturation. *Science* **315**:528–531.

Bajénoff, M., et al. (2006). Stromal cell networks regulate lymphocyte entry, migration, and territoriality in lymph nodes. *Immunity* **25**:989–1001.

Bajénoff, M., et al. (2007). Highways, byways and breadcrumbs: Directing lymphocyte traffic in the lymph node. *Trends in Immunology* **28**:346–352.

Beuneu, H., et al. (2010). Visualizing the functional diversification of CD8$^+$ T cell responses in lymph nodes. *Immunity* **33**:412–423.

Bousso, P. (2008). T-cell activation by dendritic cells in the lymph node: Lessons from the movies. *Nature Reviews Immunology* **8**:675–684.

Bousso, P., and E. Robey. (2003). Dynamics of CD8$^+$ T cell priming by dendritic cells in intact lymph nodes. *Nature Immunology* **4**: 579–585.

Cahalan, M. D. (2011). Imaging transplant rejection: A new view. *Nature Medicine* **17**:662–663.

Cahalan, M. D., and I. Parker. (2005). Close encounters of the first and second kind: T-DC and T-B interactions in the lymph node. *Seminars in Immunology* **17**:442–451.

Cahalan, M. D., and I. Parker. (2008). Choreography of cell motility and interaction dynamics imaged by two-photon microscopy in lymphoid organs. *Annual Review of Immunology* **26**:585–626.

Castellino, F., and R. N. Germain. (2006). Cooperation between CD4$^+$ and CD8$^+$ T cells: When, where, and how. *Annual Review of Immunology* **24**:519–540.

Castellino, F., et al. (2006). Chemokines enhance immunity by guiding naïve CD8$^+$ T cells to sites of CD4$^+$ T cell-dendritic cell interaction. *Nature* **440**:890–895.

Catron, D. M., A. A. Itano, K. A. Pape, D. L. Mueller, and M. K. Jenkins. (2004). Visualizing the first 50 hr of the primary immune response to a soluble antigen. *Immunity* **21**:341–347.

Celli, S., M. L. Albert, and P. Bousso. (2011). Visualizing the innate and adaptive immune responses underlying allograft rejection by two-photon microscopy. *Nature Medicine* **17**:744–749.

Celli, S., F. Lemaître, and P. Bousso. (2007). Real-time manipulation of T cell-dendritic cell interactions in vivo reveals the importance of prolonged contacts for CD4$^+$ T cell activation. *Immunity* **27**:625–634.

Coombes, J. L., and E. A. Robey. (2010). Dynamic imaging of host-pathogen interactions in vivo. *Nature Reviews Immunology* **10**:353–364.

Cyster, J. G. (2010). Shining a light on germinal center B cells. *Cell* **143**:503–505.

Germain, R. N., M. J. Miller, M. L. Dustin, and M. C. Nussenzweig. (2006). Dynamic imaging of the immune system: progress, pitfalls and promise. *Nature Reviews Immunology* **6**:497–507.

Hauser, A. E., Shlomchik, M. J., and Haberman, A. M. (2007). In vivo imaging studies shed light on germinal-centre development. *Nature Reviews Immunology* **7**:499–504.

Jenkins, M. K. (2008). Imaging the immune system. *Immunological Review* **221**:5–6.

John, B., et al. (2009). Dynamic imaging of CD8($^+$) T cells and dendritic cells during infection with *Toxoplasma gondii*. *PLoS Pathogens* **5**:e1000505.

Junt, T., E. Scandella, and B. Ludewig. (2008). Form follows function: Lymphoid tissue microarchitecture in antimicrobial immune defence. *Nature Reviews Immunology* **8**:764–775.

Lindquist, R. L., et al. (2004). Visualizing dendritic cell networks in vivo. *Nature Immunology* **5**:1243–1250.

Miller, M. J., O. Safrina, I. Parker, and M. D. Cahalan. (2004). Imaging the single cell dynamics of CD4$^+$ T cell activation by dendritic cells in lymph nodes. *Journal of Experimental Medicine* **200**:847–856.

Miller, M. J., S. H. Wei, I. Parker, and M. D. Cahalan. (2002). Two-photon imaging of lymphocyte motility and antigen response in intact lymph node. *Science* **296**:1869–1873.

Mueller, S. N., and H. D. Hickman. (2010). In vivo imaging of the T cell response to infection. *Current Opinion in Immunology* **22**:293–298.

Okada, T., et al. (2005). Antigen-engaged B cells undergo chemotaxis toward the T zone and form motile conjugates with helper T cells. *PLoS Biology* **3**:e150.

Pereira, J. P., L. M. Kelly, and J. G. Cyster. (2010). Finding the right niche: B-cell migration in the early phases of T-dependent antibody responses. *International Immunology* **22**:413–419.

Phan, T. G., I. Grigorova, T. Okada, and J. G. Cyster. (2007). Subcapsular encounter and complement-dependent transport of immune complexes by lymph node B cells. *Nature Immunology* **8**:992–1000.

Schwickert, T. A. (2007). In vivo imaging of germinal centres reveals a dynamic open structure. *Nature* **446**:83–87.

Shwab, S. R., and J. G. Cyster. (2007). Finding a way out: Lymphocyte egress from lymphoid organs. *Nature Immunology* **8**:1295–1301.

Weber, C., L. Fraemohs, and E. Dejana. (2007). Adhesion molecules in vascular inflammation. *Nature Reviews Immunology* **7**:467–477.

Wei, S. H., et al. (2007). Ca2$^+$ signals in CD4$^+$ T cells during early contacts with antigen-bearing dendritic cells in lymph node. *Journal of Immunology* **179**:1586–1594.

Wilson, E. H., et al. (2009). Behavior of parasite-specific effector CD8$^+$ T cells in the brain and visualization of a kinesis-associated system of reticular fibers. *Immunity* **30**:300–311.

Useful Websites

http://artnscience.us/index.html Immunologist Art Anderson has developed a lovely Web site that appreciates the importance of integrating excellent immunological information with images.

www.youtube.com/watch?v=XOeRJPIMpSs This YouTube video shows fluorescently labeled dendritic cells migrating through skin.

http://multimedia.mcb.harvard.edu/media. html "The Inner Life of the Cell" is an artistic and informative animation of the intercellular and intracellular molecular events associated with leukocyte extravasation.

Video Links

Copy and paste these links into a web browser to view the videos. If a link does not work, try a different browser or player. Each video associated with figures in this chapter can also be found in the on-line ImmunologyPortal associated with Kuby Immunology (Chapter 14).

http://multimedia.mcb.harvard.edu/media. html Video 14-1 Two-photon imaging of live T and B lymphocytes within a mouse lymph node.

http://download.cell.com/immunity/mmcs/ journals/1074-7613/PIIS1074761306004833.mmc3. avi Video 14-2 T cell interacting with fibroblastic reticular network.

http://download.cell.com/immunity/mmcs/ journals/1074-7613/PIIS1074761306004833.mmc6. avi Video 14-3 T lymphocytes migrate along the fibroblastic reticular cell network.

http://download.cell.com/immunity/mmcs/ journals/1074-7613/PIIS1074761306004833. mmc14.avi Video 14-4 B cells migrate along follicular dendritic cell networks.

http://www.nature.com/ni/journal/v5/n12/extref/ ni1139-S8.mov Video 14-5 Dendritic cells (DCs) are present in all lymph node microenvironments.

http://www.nature.com/ni/journal/v6/n12/extref/ ni1269-S11.mov Video 14-6 Lymphocytes exit the lymph node through portals.

http://www.ncbi.nlm.nih.gov/pmc/articles/ PMC2569002/bin/NIHMS71239-supplement-07. mov Video 14-7 Neutrophil swarming.

http://www.nature.com/ni/journal/v8/n9/extref/ ni1494-S6.mpg Video 14-8 B cells capture antigen from macrophages in the subcapsular sinus of the lymph node.

http://download.cell.com/immunity/mmcs/ journals/1074-7613/PIIS1074761304002389.mmc1. mov Video 14-9 Movements of DC and naïve lymphocytes in the absence of antigen.

http://download.cell.com/immunity/mmcs/ journals/1074-7613/PIIS1074761304002389.mmc2. mov Video 14-10 Movements of DC and antigen-specific CD4$^+$ T cells and B cells after subcutaneous injection of a soluble antigen.

http://www.sciencedirect.com/science/Miami MultiMediaURL/1-s2.0-S1074761307004517/1-s2.0- S1074761307004517-mmc10.mov/272197/html/ S1074761307004517/c23e22d1cf621035a8f 3f278912e7171/mmc10.mov Video 14-11 T cells arrest after antigen encounter.

http://www.plosbiology.org/article/fetchSingle Representation.action?uri=info:doi/10.1371/journal. pbio.0030150.sv001 Video 14-12 B cells travel to the border between the follicle and paracortex after being activated by antigen.

http://www.plosbiology.org/article/fetchSingle Representation.action?uri=info:doi/10.1371/journal. pbio.0030150.sv006 Video 14-13 Antigen-specific B and T cells interact at the border between the follicle and paracortex.

http://www.sciencemag.org/content/suppl/ 2007/01/31/1136736.DC1/1136736s1.mpg Video 14-14 Germinal-center B cells differ in their behavior from naïve B cells.

http://www.sciencedirect.com/science/Miami MultiMediaURL/1-s2.0-S1074761310003213/1-s2.0- S1074761310003213-mmc2.mov/272197/html/S107 4761310003213/02ec1444124cb3ef651dc939 ef632e2b/mmc2.mov Video 14-15 CD8$^+$ T cells arrest when they encounter antigen on DCs in the lymph node.

http://www.nature.com/nature/journal/v440/ n7086/extref/nature04651-s9.mov Video 14-16 The formation of a tricellular complex in the lymph node during CD8$^+$ T cell activation.

http://www.sciencedirect.com/science/Miami MultiMediaURL/1-s2.0-S1074761309000600/1-s2.0- S1074761309000600-mmc12.mov/272197/html/ S1074761309000600/8623d017a55653b6c73fb81 e2da216e5/mmc12.mov Video 14-17 A reticular network is established at the site of infection by *Toxoplasma gondii*.

ANALYZE THE DATA A recent paper (T. Gebhardt et al., Different patterns of peripheral migration by memory CD4$^+$ and CD8$^+$ T cells, 2011, *Nature* **477**:216–219) presented the first intravital microscopy data directly comparing the behavior of memory CD4$^+$ and CD8$^+$ T cells in response to infection. Their data were surprising.

Gebhardt et al.'s experimental strategy: The investigators infected the skin of mice with Herpes Simplex Virus and adoptively transferred CD4$^+$ T cells (which fluoresce green in the images and video associated with the study) specific for the virus as well as CD8$^+$ T cells (which fluoresce red), both of which expressed T-cell receptors specific for the virus. They observed the movements of these cells in the infected skin during the effector phase of the immune response (8 days after infection) as well as 5 or more weeks later when the only cells in the tissue would be memory cells. Remember that skin has two layers: an outer layer (epidermis) and an inner layer (dermis).

Their results: During the effector phase, both CD4$^+$ and CD8$^+$ T cells initially moved similarly through the dermis. However, gradually, these two cell populations distributed themselves differently: CD4$^+$ T cells stayed in the dermis, CD8$^+$ T cells moved to the epidermis (!).

The investigators then looked at the memory cell populations in infected skin weeks later. What they saw is depicted in the following video (memory CD8$^+$ T cells [red], memory CD4$^+$ T cells [green]): www.nature.com/nature/journal/v477/n7363/extref/nature10339-s4.mov

Your assignment: Take a look at this time-lapse video and read its legend.

a. Describe what you see, identifying at least two specific differences between CD4$^+$ and CD8$^+$ T-cell behavior.

b. Propose a molecular difference that could explain one of these distinctions.

c. Advance one hypothesis about the adaptive value of the difference(s) you observed. That is, what advantage (if any) may such a difference in CD4$^+$ and CD8$^+$ T-cell behavior provide an animal responding to a skin infection?

EXPERIMENTAL DESIGN QUESTION You want to directly test claims that CCR5 is important for the localization of naïve B cells to the follicles of a lymph node. You have all the reagents that you need to perform a two-photon intravital microscopy, including (1) an anti-CCR5 antibody that you know will block the interactions between CCR5 and its ligand (and can be injected), and (2) a CCR5$^{-/-}$ mouse. Design an experiment that will definitively test these claims. Define what you will measure, and sketch one figure predicting your results.

CLINICAL FOCUS QUESTION *Leukocyte adhesion deficiency I* (LAD I) is a rare genetic disease that results from a defect or deficiency in CD18. Patients with this condition usually do not live past childhood because they cannot fight off bacterial infections. Why? Given what you have learned in this chapter,

advance a proposal. Be specific and concise. What approaches might be taken to treat this disease?

1. Which of the following are features of two-photon microscopy? (See Advances Box 14-1.)

   a. Excites fluorochromes with shorter wavelengths of light than confocal microscopy

   b. Achieves higher resolutions than confocal microscopy

   c. Can be used to generate three-dimensional images

   d. Can be used to generate time-lapse videos

   e. Can penetrate tissue to deeper depths than confocal microscopy

   f. Damages tissue more seriously than confocal microscopy

2. You want to track the behavior of T cells specific for the influenza virus in a mouse lymph node. You have a mouse whose cells all express *yellow fluorescent protein* (YFP). You decide to isolate T cells from this mouse and introduce them into a mouse that has been immunized with influenza, as well as into a control mouse that was given no antigen. You look at the lymph nodes of both mice, expecting to see a difference in the behavior of the cells. However, you do not see much of a difference. At first you wonder if all you read is true—perhaps T cells do not arrest when they encounter antigen! But then you realize that your experimental design was flawed. What was the problem?

3. Indicate whether each of the following statements is true or false. If you think a statement is false, explain why.

   a. Chemokines are chemoattractants for lymphocytes but not other leukocytes.

   b. T cells, but not B cells, express the chemokine receptor CCR7.

   c. Antigen can only come into the lymph node if it is associated with an antigen-presenting cell.

   d. Lymphocytes increase their motility after they engage dendritic cells (DCs) expressing an antigen to which they bind.

   e. T cells crawl along the follicular DC network as they scan DCs for antigen in the lymph node.

   f. Lymphocytes make use of reticular networks only in secondary lymphoid organs.

   g. Leukocyte extravasation follows this sequence: adhesion, chemokine activation, rolling, transmigration.

   h. Most secondary lymphoid organs contain high-endothelial venules (HEVs).

4. Provide an example of lymphocyte chemotaxis during an immune response.

5. Describe where CD8$^+$ T cells and B cells receive T-cell help within a secondary lymphoid organ.

6. How might the behavior of an antigen-specific CD8$^+$ T cell differ in the lymph node of a CCL3-deficient animal versus a wild-type animal?

7. Extravasation of neutrophils and of lymphocytes occurs by generally similar mechanisms, although some differences distinguish the two processes.

   a. List in order the basic steps in leukocyte extravasation.
   b. Which step requires chemokine activation and why?
   c. Naïve lymphocytes generally do not enter tissues other than the secondary lymphoid organs. What confines them to this system?

8. Naïve T and naïve B-cell subpopulations migrate preferentially into different parts of the lymph node. What is the basis for this compartmentalization? Identify both structural and molecular influences.

9. True or False? Germinal-center B cells differ in morphology and motility from other B cells in the follicle.

10. Place a check mark next to the molecules that interact with each other.

    a. _____ Chemokine and L-selectin
    b. _____ E-selectin and L-selectin
    c. _____ CCL19 and CCR7
    d. _____ ICAM and chemokine
    e. _____ Chemokine and G-protein-coupled receptor
    f. _____ BCR and MHC
    g. _____ TCR and MHC

11. Predict how a deficiency in each of the following would affect T-cell and B-cell trafficking in a lymph node during a response to antigen. How might they affect an animal's health?

    a. L-selectin
    b. CCR7
    c. CCR5
    d. S1P1

Allergy, Hypersensitivities, and Chronic Inflammation

The same immune reactions that protect us from infection can also inflict a great deal of damage, not simply on a pathogen, but on our own cells and tissues. As you have learned, the immune response uses multiple strategies to reduce damage to self by turning off responses when pathogen is cleared and avoiding reactions to self antigens. However, these checks and balances can break down, leading to immune-mediated reactions that are more detrimental than protective. Some immune-mediated disorders are caused by a failure of immune tolerance. These autoimmune disorders will be discussed in Chapter 16. Others are caused by an inappropriately vigorous innate and/or adaptive response to antigens that pose little or no threat. These disorders, called hypersensitivities, will be the main focus of this chapter. Finally, some disorders are caused by a failure to turn off innate or adaptive responses, resulting in a chronic inflammatory state. We will close this chapter with a discussion of the causes and consequences of chronic inflammation, a condition that is of interest to many because of its intriguing association with the current obesity epidemic.

Two French scientists, Paul Portier and Charles Richet, were the first to recognize and describe hypersensitivities. In the early twentieth century, as part of their studies of the responses of bathers in the Mediterranean to the stings of Portuguese man-o'-war jellyfish (*Physalia physalis*), they demonstrated that the toxic agent in the sting was a small protein. They reasoned that eliciting an antibody response that could neutralize the toxin may serve to protect the host. Therefore, they injected low doses of the toxin into dogs to elicit an immune response, and followed with a booster injection a few weeks later. However, instead of generating a protective antibody response, the unfortunate dogs responded immediately to the second injection with vomiting, diarrhea, asphyxia, and death. Richet coined the term "*anaphylaxis*," derived from the Greek and translated loosely as "against protection" to describe this overreaction of the immune system, the first description of a hypersensitivity reaction. Richet was

Young girl sneezing in response to flowers.
[Brand New Images/Getty Images]

- Allergy: A Type I Hypersensitivity Reaction
- Antibody-Mediated (Type II) Hypersensitivity Reactions
- Immune Complex-Mediated (Type III) Hypersensitivity
- Delayed-Type (Type IV) Hypersensitivity (DTH)
- Chronic Inflammation

subsequently awarded the Nobel Prize in Physiology or Medicine in 1913.

Since that time, immunologists have learned that there are multiple types of hypersensitivity reactions. **Immediate hypersensitivity** reactions result in symptoms that manifest themselves within very short time periods after the immune stimulus, like those described above. Other types of hypersensitivity reactions take hours or days to manifest themselves, and are referred to as *delayed-type hypersensitivity* (**DTH**) reactions. In general, immediate hypersensitivity reactions result from antibody-antigen reactions, whereas DTH is caused by T-cell reactions.

As it became clear that different immune mechanisms give rise to distinct hypersensitivity reactions, two immunologists, P. G. H. Gell and R. R. A. Coombs, proposed a classification scheme to discriminate among the various types of hypersensitivity (see Figure 15-1). **Type I hypersensitivity** reactions are mediated by IgE antibodies,

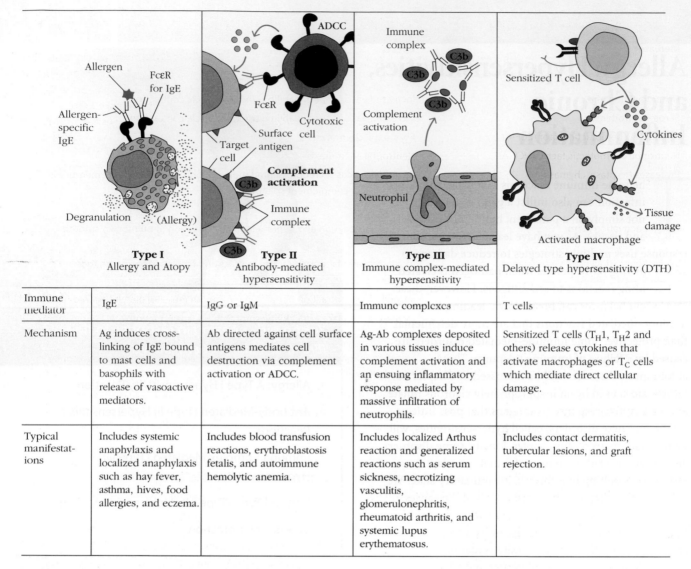

	Type I Allergy and Atopy	**Type II** Antibody-mediated hypersensitivity	**Type III** Immune complex-mediated hypersensitivity	**Type IV** Delayed type hypersensitivity (DTH)
Immune mediator	IgE	IgG or IgM	Immune complexes	T cells
Mechanism	Ag induces cross-linking of IgE bound to mast cells and basophils with release of vasoactive mediators.	Ab directed against cell surface antigens mediates cell destruction via complement activation or ADCC.	Ag-Ab complexes deposited in various tissues induce complement activation and an ensuing inflammatory response mediated by massive infiltration of neutrophils.	Sensitized T cells (T_H1, T_H2 and others) release cytokines that activate macrophages or T_C cells which mediate direct cellular damage.
Typical manifestations	Includes systemic anaphylaxis and localized anaphylaxis such as hay fever, asthma, hives, food allergies, and eczema.	Includes blood transfusion reactions, erythroblastosis fetalis, and autoimmune hemolytic anemia.	Includes localized Arthus reaction and generalized reactions such as serum sickness, necrotizing vasculitis, glomerulonephritis, rheumatoid arthritis, and systemic lupus erythematosus.	Includes contact dermatitis, tubercular lesions, and graft rejection.

FIGURE 15-1 The four types of hypersensitivity reactions.

and include many of the most common allergies to respiratory allergens, such as pollen and dust mites. **Type II hypersensitivity** reactions result from the binding of IgG or IgM to the surface of host cells, which are then destroyed by complement- or cell-mediated mechanisms. In **type III hypersensitivity** reactions, antigen-antibody complexes deposited on host cells induce complement fixation and an ensuing inflammatory response. **Type IV hypersensitivity** reactions result from inappropriate T-cell activation. It should be noted that, although this classification method has proven to be a useful analytical and descriptive tool, many clinical hypersensitivity disorders include molecular and cellular contributions from components belonging to more than one of these categories. The subdivisions are not as frequently evoked in real-world clinical settings as they once were.

The term **allergy** first appeared in the medical literature in 1906, when the pediatrician Clemens von

Pirquet noted that the response to some antigens resulted in damage to the host, rather than in a protective response. Although most familiar respiratory allergies result from the generation of IgE antibodies toward the eliciting agent, and therefore are type I hypersensitivity reactions, other common reactions that are associated with allergy, such as the response to poison ivy, result from T-cell-mediated, type IV responses.

Allergy: A Type I Hypersensitivity Reaction

More than half of the U.S. population (54.3%) suffers from type I hypersensitivity reactions, which encompass the most common allergic reactions, including hay fever, asthma, atopic dermatitis, and food allergies. The incidence of allergy continues to rise in the human population, and understanding immune mechanisms behind the response has already led to new therapies. Below we describe the molecular and cellular participants in the various type I hypersensitivities, as well as the rationale behind current treatments.

IgE Antibodies Are Responsible for Type I Hypersensitivity

Type I hypersensitivity reactions (allergies) are initiated by the interaction between an IgE antibody and a multivalent antigen. Classic Experiment Box 15-1 describes the brilliant series of experiments by K. Ishizaka and T. Ishizaka in the 1960s and 1970s that led to the identification of IgE as the class of antibody responsible for allergies. In normal individuals, the level of IgE in serum is the lowest of any of the immunoglobulin classes, making further physiochemical studies of this molecule particularly difficult. It was not until the discovery of an IgE-producing myeloma by Johansson and Bennich in 1967 that extensive analyses of IgE could be undertaken.

Many Allergens Can Elicit a Type I Response

Healthy individuals generate IgE antibodies only in response to parasitic infections. However, some people, referred to as **atopic**, are predisposed to generate IgE antibodies against common environmental antigens, such as those listed in Table 15-1. Chemical analysis revealed that most, if not all, allergens are either protein or glycoprotein in nature, with multiple antigenic sites, or epitopes, per molecule. For many years, scientists tried unsuccessfully to find any structural commonalities among molecules that induced distress in

susceptible individuals, but recently several features shared by many allergens have begun to provide clues to the biological basis of their activity.

First, many allergens have intrinsic enzymatic activity that affects the immune response. For example, allergen extracts from dust mites and cockroaches as well as from fungi and bacteria are relatively high in protease activity. Some of these proteases have been shown to be capable of disrupting the integrity of epithelial cell junctions, and allowing allergens to access the underlying cells and molecules of the innate and adaptive immune systems. Others, including a protease (*Der p 1*) produced by the dust mite, cleave and activate complement components at the mucosal surface. Still others cleave and stimulate protease-activated receptors on the surfaces of immune cells, enhancing inflammation. Thus, one factor that distinguishes allergenic from nonallergenic molecules may be the presence of enzymatic activity that affects the cells and molecules of the immune system.

Second, many allergens contain potential pathogen-associated molecular patterns, or PAMPS (see Chapter 5), capable of interacting with receptors of the innate immune system, and initiating a cascade of responses leading to an allergic response. However, it is unclear why this cascade is stimulated in only a subset of individuals.

Third, many allergens enter the host via mucosal tissues at very low concentrations, which tend to predispose the individual to generate T_H2 responses, leading to B-cell secretion of IgE.

IgE Antibodies Act by Cross-Linking Fcε Receptors on the Surfaces of Innate Immune Cells

IgE antibodies alone are not destructive. Instead, they cause hypersensitivity by binding to Fc receptors specific for their constant regions (FcεRs). These are expressed by a variety of innate immune cells, including mast cells, basophils, and eosinophils (Chapter 2). The binding of IgE antibodies to FcεRs activates these granulocytes, inducing a signaling cascade that causes cells to release the contents of intracellular granules into the blood, a process called degranulation (see Figure 15-2). The contents of granules vary from cell to cell, but typically include histamine, heparin, and proteases. Together with other mediators that are synthesized by activated granulocytes (leukotrienes, prostaglandins, chemokines, and cytokines), these mediators act on surrounding tissues and other immune cells, causing allergy symptoms.

Interestingly, the serum half-life of IgE is quite short (only 2 to 3 days). However, when bound to its receptor on an innate immune cell, IgE is stable for weeks. IgE actually binds two different receptors, the high-affinity FcεRI, which is responsible for most of the symptoms we associate with allergy, and the lower-affinity FcεRII, which regulates the production of IgE by B cells, as well as its transport across cells (Chapter 13). The role of FcεRI in type I hypersensitivities is confirmed by experiments conducted in mice that lack an FcεRI α chain; they are resistant to localized and systemic allergic responses, despite having normal numbers of mast cells.

TABLE 15-1	Common allergens associated with type I hypersensitivity	
Plant pollens	**Foods**	
Rye grass	Nuts	
Ragweed	Seafood	
Timothy grass	Eggs	
Birch trees	Peas, Beans	
Drugs	Milk	
Penicillin		
Sulfonamides	**Insect products**	
Local anesthetics	Bee venom	
Salicylates	Wasp venom	
	Ant venom	
Mold spores	Cockroach calyx	
Animal hair and dander	Dust mites	
Latex		
Foreign serum		
Vaccines		

The Discovery and Identification of IgE as the Carrier of Allergic Hypersensitivity

In a stunning series of papers published between 1966 and the mid-1970s, Kimishige Ishizaka and Teruko Ishizaka, working with a number of collaborators, identified a new class of immunoglobulins, which they called IgE antibodies, as being the major effector molecules in type 1 antibody-mediated hypersensitivity reactions.

The Ishizakas built on work performed in 1921 by K. Prausnitz and H. Kustner, who injected serum from an allergic person into the skin of a nonallergic individual. When the appropriate antigen was later injected into the same site, a **wheal and flare reaction** (swelling and reddening) was detected. This reaction, called the P-K reaction after its originators, was the basis for the first biological assay for IgE activity.

In their now classic experiments, the Ishizakas assayed for the presence of allergen-specific antibody using the wheal and flare reaction. They also employed **radioimmunodiffusion**, using the ability of radioactive allergen E derived from ragweed pollen to bind to and precipitate pollen-specific antibodies as an additional assay; the antibodies formed a radioactive precipitate on binding to the ragweed allergen. (Note that both the antigen and the immunoglobulin class are designated "E" in this series of experiments.)

The Ishizakas reasoned that the best starting material for purifying the protein

responsible for the P-K reaction would be the serum of an allergic individual who displayed hypersensitivity to ragweed pollen E. (Serum is that component of the blood remaining after the cells and clotting components have been removed.) To purify the serum protein responsible for the allergic reaction, they took whole human serum and subjected it to *ammonium sulfate precipitation* (different proteins precipitate out at varying concentrations of ammonium sulfate), and *ion exchange chromatography*, which separates proteins on the basis of their intrinsic charge.

Different fractions from the chromatography column were tested by radioimmunodiffusion for their ability to bind to radioactive antigen E from ragweed pollen. Portions of the different fractions were also injected at varying dilutions into volunteers, along with antigen, to test for a P-K reaction. Finally, each fraction was also tested semiquantitatively for the presence of IgG and IgA antibodies. The results of these experiments are shown for the serum from one of the three individuals they tested (see Table 1 below).

From the results in this table, it is clear that the ability of proteins in the various fractions to induce a P-K reaction did not correlate with the amounts of either IgG or IgA antibodies, but it did correlate with the amounts of antibodies that could be detected in an immunodiffusion reaction

with radioactive antigen E. Perhaps another antibody class was responsible for the line of immunoprecipitation on the immunodiffusion gel?

The fractions containing the highest concentration of protein able to bind to allergen E were further purified using gel chromatography, which separates proteins on the basis of size and molecular shape. Again, the presence of the protein was detected on the basis of its ability to bind to radioactive antigen E and to induce a P-K reaction in the skin of a test subject who had been injected with antigen E.

The resulting protein still contained small amounts of IgG and IgA antibodies, which were eliminated by mixing the fractions with antibodies directed toward those human antibody subclasses, and then removing the resultant immunoprecipitate. The Ishizakas' final protein product was 500 to 1000 times more potent than the original serum in its ability to generate a P-K reaction, and the most active preparation generated a positive skin reaction at a dilution of 1:8000. None of its reactivities correlated with the presence of any of the other known classes of antibody, and so a new class of antibody was named, IgE, based on its ability to bind to allergens and bring about a P-K reaction.

As described in Chapter 3, the level of IgE in the serum is the lowest of all the antibody classes, falling within the range

The High-Affinity IgE Receptor, FcεRI

Mast cells and basophils constitutively express high levels of the high-affinity IgE receptor, FcεRI, which binds IgE with an exceptionally high-affinity constant of $10^{10} M^{-1}$. This affinity helps overcome the difficulties associated with responding to an exceptionally low concentration of IgE in the serum (1.3×10^{-7} M). Eosinophils, Langerhans cells, monocytes, and platelets also express FcεRI, although at lower levels.

Most cells express a tetrameric form of FcεRI, which includes an α and β chain and two identical disulfide-linked γ chains (Figure 15-3a). Monocytes and platelets express an alternative form lacking the β chain. The α chains of the FcεRI, members of the immunoglobulin superfamily, directly

bind the IgE heavy chains, whereas the β and γ chains are responsible for signal transduction. They contain *immunoreceptor tyrosine-based activation motifs* (ITAMs) (see Chapter 3) that are phosphorylated in response to IgE cross-linking.

IgE-mediated signaling begins when an allergen cross-links IgE that is bound to the surface FcεRI receptor (Figure 15-2). Although the specific biochemical events that follow cross-linking of the FcεRI receptor vary among cells and modes of stimulation, the FcεRI signaling cascade generally resembles that initiated by antigen receptors and growth factor receptors (Chapter 3). Briefly, IgE cross-linking induces the aggregation and migration of receptors into membrane lipid rafts, followed by phosphorylation of ITAM motifs by

TABLE 1	Data from original paper identifying the immunoglobulins responsible for skin sensitization				
Serum donor	Fraction	Minimum dose for P-K reaction	Relative amount IgE	Relative amount IgG	Relative amount IgA
U	A	0.04	+	+	−
→	B	0.008	++	+	−
	C	0.26	+	−	+/−
	D	> 0.9	−	−	−
A	A	0.002	++	++	−
→	B	0.0006	+++	+	+/−
	C	0.0014	++	+	+/−
	D	0.005	++	+	+
	E	0.017	++	−	+
	F	0.13	+	−	+

Modified from original table entitled "Distribution of skin-sensitizing activity and of γG (IgG) and γA (IgA) globulin following diethylaminoethyl (DEAE) Sephadex column fractionation," in Ishizaka, K., and T. Ishizaka, (1967), Identification of γE-antibodies as a carrier of reaginic activity. *Journal of Immunology* **99**(6):1187–1198.

of 0.1 to 0.4 μg/ml, although atopic individuals can have as much as 10 times this concentration of IgE in their circulation. However, in 1967, Johansson and Bennich discovered an IgE-producing myeloma, which enabled a full biochemical analysis of the molecule. The structure of IgE is described in Chapter 3.

In Table 1, which is modified from the original data in this classic 1967 paper, serum protein fractions from two separate donors were evaluated for the relative amounts of IgA or IgG (referred to as γA and γG, respectively, in this publication),

using rabbit antisera against the two immunoglobulin subclasses, and for the presence of the putative "IgE" using radio-immunodiffusion (as described in this chapter's text). IgG is the most common class of immunoglobulin in serum, and IgA was included because early experiments had suggested that IgA may be the antibody responsible for the wheal and flare reaction. The "Minimum dose for P-K" column refers to the quantity of the fraction required to yield a measureable wheal and flare response. In this column, the lower the number, the higher the

amount of P-K responsive antibody in the fraction (i.e., fraction B had the highest amount of the allergenic antibody).

It can readily be seen that the fractions showing the strongest P-K responses (highlighted arrows: →) are also those in which the highest amounts of the so-called γE were measured by radio-immunodiffusion. P-K reactions did not correlate with either IgG or with IgA levels in the serum from this, or from two other donors.

Ishizaka, K., and T. Ishizaka. (1967). Identification of γE-antibodies as a carrier of reaginic activity. *Journal of Immunology* **99**(6):1187–1198.

associated tyrosine kinases. Adapter molecules then latch onto the phosphorylated tyrosine residues and initiate signaling cascades culminating in enzyme and/or transcription factor activation. Figure 15-4 identifies just a few of the signaling events specifically associated with mast cell activation.

FcεRI signaling leads to mast cell and basophil (1) degranulation of vesicles containing multiple inflammatory mediators (Figure 15-5a), (2) expression of inflammatory cytokines and (3) conversion of arachidonic acid into leukotrienes and prostaglandins, two important lipid mediators of inflammation. These mediators have multiple short-term and long-term effects on tissues that will be described in more detail below (Figure 15-5b).

The Low-Affinity IgE Receptor, FcεRII

The other IgE receptor, designated FcεRII, or CD23, has a much lower affinity for IgE (K_d of $1 \times 10^6 \, M^{-1}$) (Figure 15-3b). CD23 is structurally distinct from FcεRI (it is a lectin and a type II membrane protein) and exists in two isoforms that differ only slightly in the N-terminal cytoplasmic domain. CD23a is found on activated B cells, whereas CD23b is induced on various cell types by the cytokine IL-4. Both isoforms also exist as membrane-bound and soluble forms, the latter being generated by proteolysis of the surface molecule. Interestingly, CD23 binds not only to IgE, but also to the complement receptor CD21.

The outcome of CD23 ligation depends on which ligands it binds to (IgE or CD21) and whether it does so as a soluble

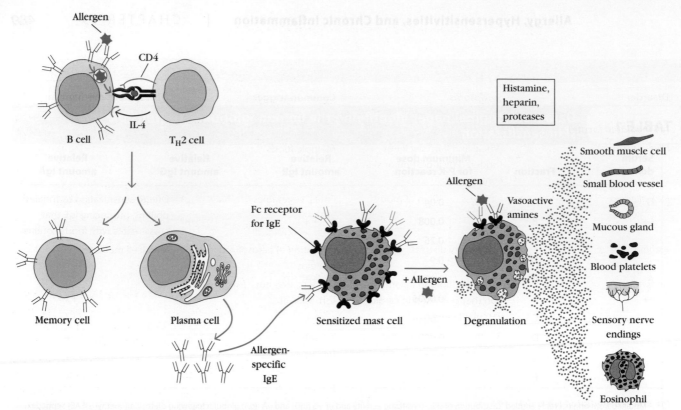

FIGURE 15-2 General mechanism underlying an immediate type I hypersensitivity reaction. Exposure to an allergen activates T$_H$2 cells that stimulate B cells to form IgE-secreting plasma cells. The secreted IgE molecules bind to IgE-specific Fc receptors (FcεRI) on mast cells and blood basophils. (Many molecules of IgE with various specificities can bind to the FcεRI.) Second exposure to the allergen leads to cross-linking of the bound IgE, triggering the release of pharmacologically active mediators (vasoactive amines) from mast cells and basophils. The mediators cause smooth muscle contraction, increased vascular permeability, and vasodilation.

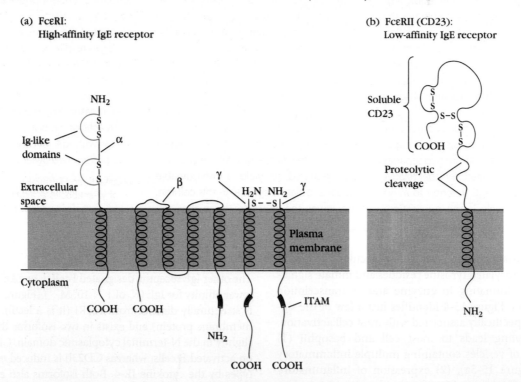

FIGURE 15-3 Schematic diagrams of the high-affinity FcεRI and low-affinity FcεRII receptors that bind the Fc region of IgE.
(a) FcεRI consists of a α chain that binds IgE, a β chain that participates in signaling, and two disulfide-linked γ chains that are the most important component in signal transduction. The β and γ chains contain a cytoplasmic ITAM, a motif also present in the Igα/Igβ (CD79α/β) heterodimer of the B-cell receptor and in the T-cell receptor complex. (b) The single-chain FcεRII is unusual because it is a type II transmembrane protein, oriented in the membrane with its NH$_2$-terminus directed toward the cell interior and its COOH-terminus directed toward the extracellular space.

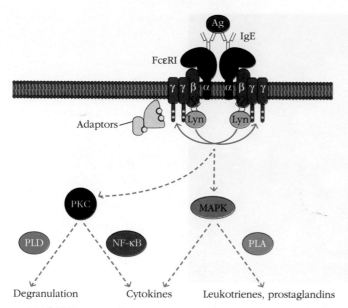

FIGURE 15-4 Signaling pathways initiated by IgE allergen cross-linking. By cross-linking Fcε receptors, IgE initiates signals that lead to mast cell degranulation, cytokine expression, and leukotriene and prostaglandin generation. The signaling cascades initiated by the FcεRI are similar to those initiated by antigen receptors (see Chapter 3). This figure illustrates only a few of the players in the complex signaling network. Briefly, cross-linking of FcεRI activates tyrosine kinases, including Lyn, which phosphorylate adaptor molecules, which organize signaling responses that lead to activation of multiple kinases, including protein kinase C (PKC) and various mitogen-activated protein kinases (MAPKs). These, in turn, activate transcription factors (e.g., NFκB) that regulate cytokine production. They also activate lipases, including phospholipase D (PLD), which regulates degranulation, and phospholipase A (PLA), which regulates the metabolism of the leukotriene and prostaglandin precursor arachidonic acid. [K. Roth, W. M. Chen, and T. J. Lin, 2008, Positive and negative regulatory mechanisms in high-affinity IgE receptor-mediated mast cell activation, Archivum immunologiae et therapiae experimentalis **56**:385–399.]

or membrane-bound molecule. For example, when soluble CD23 (sCD23) binds to CD21 on the surface of IgE-synthesizing B cells, IgE synthesis is increased. However, when membrane-bound CD23 binds to soluble IgE, further IgE synthesis is suppressed. Atopic individuals express relatively high levels of surface and soluble CD23.

IgE Receptor Signaling Is Tightly Regulated

Given the powerful nature of the molecular mediators released by mast cells, basophils, and eosinophils following FcεRI signaling, it should come as no surprise that the responses are subject to complex systems of regulation. Below we offer just a few examples of ways in which signaling through the FcεRI receptor can be inhibited.

Co-Clustering with Inhibitory Receptors

Recall from earlier chapters that the intracellular regions of some lymphocyte receptors, including FcγRIIB, bear immu-

noreceptor *tyrosine inhibitory motifs* (ITIMs as distinct from ITAMs) (see Figure 13-3). Cross-linking of these receptors leads to *inhibition*, rather than to *activation* of cellular responses. Interestingly, mast cells express both the activating FcεRI and inhibiting FcγRIIB. Therefore, if an allergen binds both IgG and IgE molecules, it will trigger signals through both Fc receptors. The inhibitory signal dominates. This phenomenon is part of the reason that eliciting IgG antibodies against common allergens through desensitization therapies can help atopic patients. The more anti-allergen IgG they have, the higher the probability that allergens will co-cluster FcεRI receptors with inhibitory FcγRIIB receptors.

Inhibition of Downstream Signaling Molecules

Because many of the reactions in the activation pathway downstream from the FcεRI pathways are phosphorylations, multiple phosphatases, such as SHPs, SHIPs, and PTEN, play an important role in dampening receptor signaling. In addition, the tyrosine kinase, Lyn, can play a negative as well as a positive role in FcεRI signaling by phosphorylating and activating the inhibitory FcγRIIB. Finally, FcεRI signaling through Lyn and Syk kinases also activates E3 ubiquitin ligases, including c-Cbl. Cbl ubiquitinylates Lyn and Syk, as well as FcεRI itself, triggering their degradation. Thus, FcεRI activity contributes to its own demise.

Innate Immune Cells Produce Molecules Responsible for Type I Hypersensitivity Symptoms

The molecules released in response to FcεRI cross-linking by mast cells, basophils, and eosinophils are responsible for the clinical manifestations of type I hypersensitivity. These inflammatory mediators act on local tissues as well as on populations of secondary effector cells, including other eosinophils, neutrophils, T lymphocytes, monocytes, and platelets.

When generated in response to parasitic infection, these mediators initiate beneficial defense processes, including vasodilation and increased vascular permeability, which brings an influx of plasma and inflammatory cells to attack the pathogen. They also inflict direct damage on the parasite. In contrast, mediator release induced by allergens results in unnecessary increases in vascular permeability and inflammation that lead to tissue damage with little benefit.

The molecular mediators can be classified as either primary or secondary (Table 15-2). Primary mediators are preformed and stored in granules prior to cell activation, whereas secondary mediators are either synthesized after target-cell activation or released by the breakdown of membrane phospholipids during the degranulation process. The most significant primary mediators are histamine, proteases, eosinophil chemotactic factor (ECF), neutrophil chemotactic factor (NCF), and heparin. Secondary mediators include platelet-activating factor (PAF), leukotrienes, prostaglandins, bradykinins, and various cytokines and chemokines.

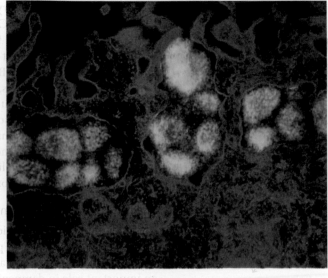

(a)

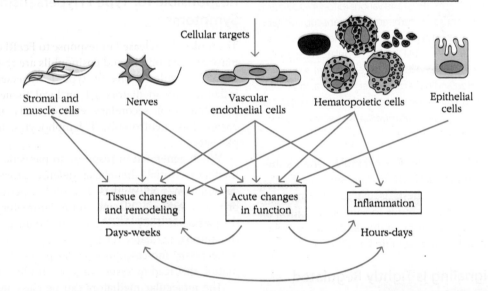

(b)

Cytoplasmic granules

Mast cell activation

Cytokines, chemokines, growth factors, lipid mediators, granule-associated mediators

Cellular targets

Stromal and muscle cells   Nerves   Vascular endothelial cells   Hematopoietic cells   Epithelial cells

Tissue changes and remodeling

Acute changes in function

Inflammation

Days-weeks   Hours-days

FIGURE 15-5 Mast cell activity. (a) A mast cell in the process of degranulating (colorized EM image of mast cell membrane). (b) Mast cell mediators and their effects. Different stimuli activate mast cells to secrete different amounts and/or types of products. Activated mast cells immediately release preformed, granule-associated inflammatory mediators (including histamine, proteases, and heparin) and are induced to generate lipid mediators (such as leukotrienes and prostaglandins), chemokines, cytokines, and growth factors (some of which can also be packaged in granules). These mediators act on different cell types, and have both acute and chronic effects. When produced over long periods of time, mast cell mediators have a significant influence on tissue structure by enhancing proliferation of fibroblasts and epithelial cells, increasing production and deposition of collagen and other connective tissue proteins, stimulating the generation of blood vessels, and more. [(a) http://t3.gstatic.com/images?q=tbn:ANd9GcQh3-PP1n1mbEilsiHmciOq-eniEL5D-iMw3HTWIbVZsQ-TW0QPbQ. (b) Figure 1 in Galli, S. J., and S. Nakae. (2003). Mast Cells to the Defense. Nature Immunology 4:1160–1162.]

TABLE 15-2 Principal mediators involved in type I hypersensitivity.

Mediator	Effects
	Primary
Histamine, heparin	Increased vascular permeability; smooth muscle contraction
Serotonin (rodents)	Increased vascular permeability; smooth muscle contraction
Eosinophil chemotactic factor (ECF-A)	Eosinophil chemotaxis
Neutrophil chemotactic factor (NCF-A)	Neutrophil chemotaxis
Proteases (tryptase, chymase)	Bronchial mucus secretion; degradation of blood vessel basement membrane; generation of complement split products
	Secondary
Platelet-activating factor	Platelet aggregation and degranulation; contraction of pulmonary smooth muscles
Leukotrienes (slow reactive substance of anaphylaxis, SRS-A)	Increased vascular permeability; contraction of pulmonary smooth muscles
Prostaglandins	Vasodilation; contraction of pulmonary smooth muscles; platelet aggregation
Bradykinin	Increased vascular permeability; smooth muscle contraction
Cytokines	
IL-1 and TNF-α	Systemic anaphylaxis; increased expression of adhesion molecules on venular endothelial cells
IL-4 and IL-13	Increased IgE production
IL-3, IL-5, IL-6, IL-10, TGF-β, and GM-CSF	Various effects (see text)

The varying manifestations of type I hypersensitivity in different tissues and species reflect variations in the primary and secondary mediators present. Below we briefly describe the main biological effects of several key mediators.

Histamine

Histamine, which is formed by decarboxylation of the amino acid histidine, is a major component of mast-cell granules, accounting for about 10% of granule weight. Its biological effects are observed within minutes of mast-cell activation. Once released from mast cells, histamine binds one of four different histamine receptors, designated H_1, H_2, H_3, and H_4. These receptors have different tissue distributions and mediate different effects on histamine binding. Serotonin is also present in the mast cells of rodents and has effects similar to histamine.

Most of the biologic effects of histamine in allergic reactions are mediated by the binding of histamine to H_1 receptors. This binding induces contraction of intestinal and bronchial smooth muscles, increased permeability of venules (small veins), and increased mucous secretion. Interaction of histamine with H_2 receptors increases vasopermeability (due to contraction of endothelial cells) and vasodilation (by relaxing the smooth muscle of blood vessels), stimulates exocrine glands, and increases the release of acid in the stomach. Binding of histamine to H_2 receptors on mast cells and basophils suppresses degranulation; thus, histamine ultimately exerts negative feedback on the further release of mediators.

Leukotrienes and Prostaglandins

As secondary mediators, the leukotrienes and prostaglandins are not formed until the mast cell undergoes degranulation and phospholipase signaling initiates the enzymatic breakdown of phospholipids in the plasma membrane. An ensuing enzymatic cascade generates the prostaglandins and the leukotrienes.

In a type I hypersensitivity-mediated asthmatic response, the initial contraction of human bronchial and tracheal smooth muscles is at first mediated by histamine; however, within 30 to 60 seconds, further contraction is signaled by leukotrienes and prostaglandins. Active at nanomole levels, the leukotrienes are approximately 1000 times more effective at mediating bronchoconstriction than is histamine, and they are also more potent stimulators of vascular permeability and mucus secretion. In humans, the leukotrienes are thought to contribute significantly to the prolonged bronchospasm and buildup of mucus seen in asthmatics.

Cytokines and Chemokines

Adding to the complexity of the type I reaction is the variety of cytokines released from mast cells and basophils. Mast cells, basophils, and eosinophils secrete several interleukins, including IL-4, IL-5, IL-8, IL-13, GM-CSF, and TNF-α. These cytokines alter the local microenvironment and lead to the recruitment of inflammatory cells such as neutrophils and eosinophils. For instance, IL-4 and IL-13 stimulate a T_H2 response and thus increase IgE production by B cells.

IL-5 is especially important in the recruitment and activation of eosinophils. The high concentrations of TNF-α secreted by mast cells may contribute to shock in systemic anaphylaxis. IL-8 acts as a chemotactic factor, and attracts further neutrophils, basophils, and various subsets of T cells to the site of the hypersensitivity response. GM-CSF stimulates the production and activation of myeloid cells, including granulocytes and macrophages.

Type I Hypersensitivities Are Characterized by Both Early and Late Responses

Type I hypersensitivity responses are divided into an immediate early response and one or more late phase responses (Figure 15-6). The early response occurs within minutes of allergen exposure and, as described above, results from the release of histamine, leukotrienes, and prostaglandins from local mast cells.

However, hours after the immediate phase of a type I hypersensitivity reaction begins to subside, mediators released during the course of the reaction induce localized inflammation, called the late-phase reaction. Cytokines released from mast cells, particularly TNF-α and IL-1, increase the expression of cell adhesion molecules on venular endothelial cells, thus facilitating the influx of neutrophils, eosinophils, and T_H2 cells that characterizes this phase of the response.

Eosinophils play a principal role in the late-phase reaction. Eosinophil chemotactic factor, released by mast cells during the initial reaction, attracts large numbers of eosinophils to the affected site. Cytokines released at the site, including IL-3, IL-5, and GM-CSF, contribute to the growth and differentiation of these cells, which are then activated by binding of antibody-coated antigen. This leads to degranulation and further release of inflammatory mediators that contribute to the extensive tissue damage typical of the late-phase reaction. Neutrophils, another major participant in late-phase reactions, are attracted to the site of an ongoing type I reaction by neutrophil chemotactic factor released from degranulating mast cells. Once activated, the neutrophils release their granule contents, including lytic enzymes, platelet-activating factor, and leukotrienes.

Recently, a third phase of type I hypersensitivity has been described in models of type I hypersensitivity reactions in the skin. This third phase starts around 3 days post antigen challenge and peaks at day 4. It is also characterized by massive eosinophil infiltration but, in contrast to the second phase, requires the presence of basophils. As shown in Figure 15-7, which illustrates an example of the late-phase responses in the mouse ear, cytokines and proteases released from basophils act on tissue-resident cells such as fibroblasts. These fibroblasts then secrete chemokines that are responsible for the recruitment of larger numbers of eosinophils and neutrophils to the skin lesion. Subsequent degranulation of the eosinophils and neutrophils adds to the considerable tissue damage at the site of the initial allergen contact. These experiments illustrate how multiple granulocyte subsets can cooperate in the induction of chronic allergic inflammation.

There Are Several Categories of Type I Hypersensitivity Reactions

The clinical manifestations of type I reactions can range from life-threatening conditions, such as systemic anaphylaxis and severe asthma, to localized reactions, such as hay fever and eczema. The nature of the clinical symptoms depends on the route by which the allergen enters the body, as well as on its concentration and on the prior allergen exposure of the host. In this section, we describe the pathology of the various type I hypersensitivity reactions.

Systemic Anaphylaxis

Systemic anaphylaxis is a shocklike and often fatal state that occurs within minutes of exposure to an allergen. It is usually initiated by an allergen introduced directly into the bloodstream or absorbed from the gut or skin. Symptoms include labored respiration, a precipitous drop in blood pressure leading to **anaphylactic shock**, followed by contraction of smooth muscles leading to defecation, urination, and bronchiolar constriction. This leads to asphyxiation, which can lead to death within 2 to 4 minutes of exposure to the allergen. These symptoms are all due to rapid antibody-mediated degranulation of mast cells and the systemic effects of their contents.

A wide range of antigens has been shown to trigger this reaction in susceptible humans, including the venom from bee, wasp, hornet, and ant stings; drugs such as penicillin, insulin, and antitoxins; and foods such as seafood and nuts. If not treated quickly, these reactions can be fatal. Epinephrine, the drug of choice for treating systemic anaphylactic reactions, counteracts the effects of mediators such as histamine and the leukotrienes, relaxing the smooth muscles of the airways and reducing vascular permeability. Epinephrine also improves cardiac output, which is necessary to prevent vascular collapse during an anaphylactic reaction.

Localized Hypersensitivity Reactions

In localized hypersensitivity reactions (atopy), the pathology is limited to a specific target tissue or organ, and often occurs at the epithelial surfaces first exposed to allergens. Atopic allergies include a wide range of IgE-mediated disorders, such as allergic rhinitis (hay fever), asthma, atopic dermatitis (eczema), and food allergies.

The most common atopic disorder, affecting almost 50% of the U.S. population, is *allergic rhinitis* or hay fever. Symptoms result from the inhalation of common airborne allergens (pollens, dust, viral antigens), which react with IgE molecules bound to sensitized mast cells in the conjunctivae and nasal mucosa. Cross-linking of IgE receptors induces the release of histamine and heparin from mast cells, which then cause vasodilation, increased capillary permeability, and production of exudates in the eyes and respiratory tract. Tearing, runny nose, sneezing, and coughing are the main symptoms.

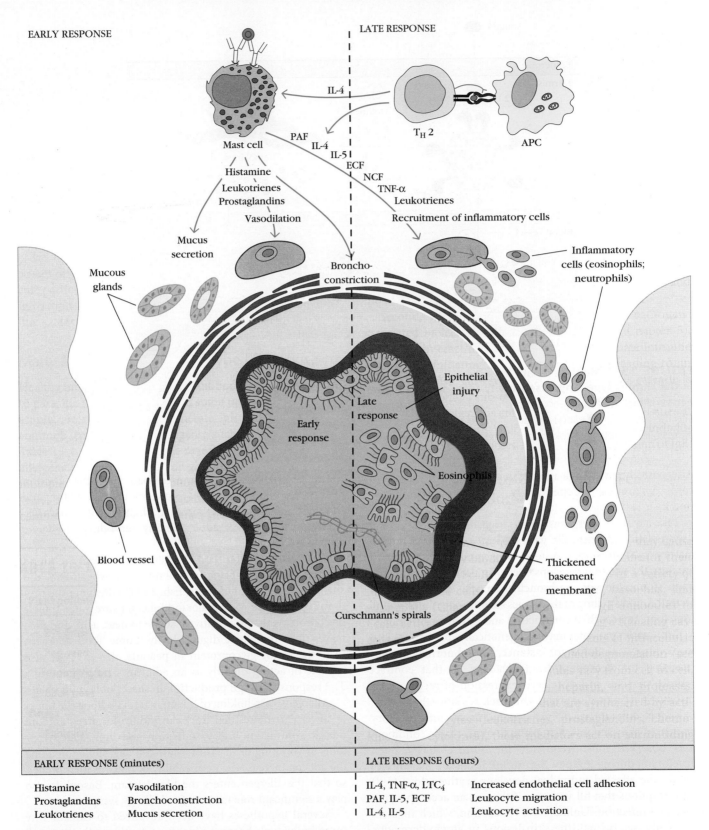

EARLY RESPONSE

LATE RESPONSE

IL-4

Mast cell

PAF
IL-4
IL-5
ECF
NCF
TNF-α
Leukotrienes

Histamine
Leukotrienes
Prostaglandins

Vasodilation

Mucus
secretion

Broncho-
constriction

$T_H 2$

APC

Recruitment of inflammatory cells

Inflammatory
cells (eosinophils;
neutrophils)

Mucous
glands

Epithelial
injury

Late
response

Early
response

Eosinophils

Blood vessel

Thickened
basement
membrane

Curschmann's spirals

EARLY RESPONSE (minutes)		LATE RESPONSE (hours)	
Histamine	Vasodilation	IL-4, TNF-α, LTC₄	Increased endothelial cell adhesion
Prostaglandins	Bronchoconstriction	PAF, IL-5, ECF	Leukocyte migration
Leukotrienes	Mucus secretion	IL-4, IL-5	Leukocyte activation

FIGURE 15-6 The early and late inflammatory responses in asthma. The immune cells involved in the early and late responses are represented at the top. The effects of various mediators on an airway, represented in cross-section, are illustrated in the center and also described in the text.

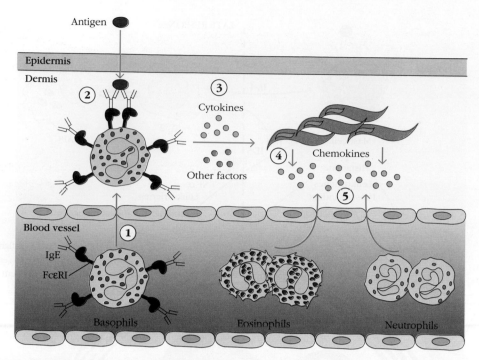

FIGURE 15-7 Multiple innate immune cells are involved in the chronic type 1 hypersensitivity response in a mouse ear-swelling model. Late in a type I response, basophils migrate into the dermis of the ear (1) and are activated by IgE/antigen complexes (2). They release cytokines and other inflammatory mediators (3) that stimulate stromal cells (e.g., fibroblasts) to release chemokines (4), which attract other granulocytes, including eosinophils and neutrophils to the tissue, contributing to a chronic inflammation and tissue damage. [*Adapted from H. Karasuyama et al., 2011, Nonredundant roles of basophils in immunity,* Annual Review of Immunology *29:45–69.*]

Alternatively, an asthma attack can be induced by exercise or cold, apparently independently of allergen stimulation (intrinsic asthma).

Like hay fever, allergic asthma is triggered by activation and degranulation of mast cells, with subsequent release of inflammatory mediators, but instead of occurring in the nasal mucosa, the reaction develops deeper in the lower respiratory tract. Contraction of the bronchial smooth muscles, mucus secretion, and swelling of the tissues surrounding the airway all contribute to bronchial constriction and airway obstruction.

Asthmatic patients may be genetically predisposed to atopic responses. Some, for instance, have abnormal levels of receptors for neuropeptides (substance P) that contract smooth muscles and decreased expression of receptors for vasoactive intestinal peptide, which relaxes smooth muscles.

Atopic dermatitis (*allergic eczema*) is an inflammatory disease of skin and another example of an atopic condition. It is observed most frequently in young children, often developing during infancy. Serum IgE levels are usually elevated. The affected individual develops erythematous (red) skin eruptions that fill with pus if there is an accompanying bacterial infection. Unlike a DTH reaction, which involves T_H1 cells (see below), the skin lesions in atopic dermatitis contain T_H2 cells and an increased number of eosinophils.

Food Allergies: A Common Type of Atopy on the Rise

Food allergies, whose incidence is on the rise, are another common type of atopy. In children, food allergies account for more anaphylactic responses than any other type of allergy. They are highest in frequency among infants and toddlers (6%–8%) and decrease slightly with maturity. Approximately 4% of adults display reproducible allergic reactions to food.

The most common food allergens for children are found in cow's milk, eggs, peanuts, tree nuts, soy, wheat, fish, and shellfish. Among adults, nuts, fish, and shellfish are the predominant culprits. Most major food allergens are water-soluble glycoproteins that are relatively stable to heat, acid, and proteases and, therefore, digest slowly. Some food allergens (e.g., the major glycoprotein in peanuts, *Ara h 1*) are also capable of acting directly as an adjuvant and promoting a T_H2 response and IgE production in susceptible individuals.

Allergen cross-linking of IgE on mast cells along the upper or lower gastrointestinal tract can induce localized smooth muscle contraction and vasodilation and thus such symptoms as vomiting or diarrhea. Mast-cell degranulation along the gut can increase the permeability of mucous membranes, so that the allergen enters the bloodstream. Basophils also play a significant role in acute food allergy symptoms.

Several hypotheses have been advanced to explain why some individuals become sensitive to antigens that are well tolerated by the rest of the population. One possibility is that a temporary viral or bacterial infection may lead to a short-lived increase in the permeability of the gut surface, allowing increased absorption of allergenic antigens and sensitization. Alternatively, sensitization may occur via the respiratory

TABLE 15-3 Immune basis for some food allergies

Disorder	Symptoms	Common trigger	Notes about mechanism
IgE mediated (acute)			
Hives (urticaria)	Wheal and flare swellings triggered by ingestion or skin contact	Multiple foods	
Oral allergy	Itchiness, swelling of mouth	Fruits, vegetables	Due to sensitization by inhaled pollens, producing IgE that cross-reacts with food proteins
Asthma, rhinitis	Respiratory distress	Inhalation of aerosolized food proteins	Mast-cell mediated
Anaphylaxis	Rapid, multiorgan inflammation that can result in cardiovascular failure	Peanuts, tree nuts, fish, shellfish, milk, etc.	
Exercise-induced anaphylaxis	As above, but occurs when one exercises after eating trigger foods	Wheat, shellfish, celery (may be due to changes in gut absorption associated with exercise)	
IgE and cell mediated (chronic)			
Atopic dermatitis	Rash (often in children)	Egg, milk, wheat, soy, etc.	May be skin T cell mediated
Gastrointestinal inflammation	Pain, weight loss, edema, and/or obstruction	Multiple foods	Eosinophil mediated
Cell mediated (chronic)			
Intestinal inflammation brought about by dietary protein (e.g., enterocolitis, proctitis)	Most often seen in infants: diarrhea, poor growth, and/or bloody stools	Cow's milk (directly or via breast milk), soy, grains	TNF-α mediated

Adapted from S. H. Sicherer and H. A. Sampson, 2009, Food allergy, *Annual Review of Medicine* **60**:261–277.

route or via absorption of allergens through the skin. This is thought to be the case for one type of allergic reaction to apple proteins. Exposure to birch pollen can induce respiratory type I hypersensitivity, and the IgE that is generated cross-reacts to a protein from apples, leading to a severe digestive allergic response. Finally, various dietary conditions may bias an individual's responses in the direction of T_H2 generation. These include reduced dietary antioxidants, altered fat consumption, and over- or under-provision of Vitamin D. Table 15-3 lists various immune mechanisms that play a role in food allergy. Note that although most are IgE mediated, some are mediated by T cells.

The efficiency of the gut barrier improves with maturity, and the food allergies of many infants resolve without treatment as they grow, even though allergen-specific IgE can still be detected in their blood. However, resolution is not always achieved, and in some cases the continuation of the allergic state reflects a reduced frequency of regulatory T cells in allergic versus nonallergic individuals.

Depending on where the allergen is deposited, patients with atopic dermatitis and food hypersensitivity can also exhibit other symptoms. For example, some individuals develop asthmatic attacks after ingesting certain foods. Others develop atopic urticaria, commonly known as *hives*, when a food allergen is carried to sensitized mast cells in the skin, causing swollen (edematous), erythematous eruptions.

There Is a Genetic Basis for Type I Hypersensitivity

The susceptibility of individuals to atopic responses has a strong genetic component that has been mapped to several possible loci by candidate gene association studies, genome-wide linkage analyses, and genome expression studies (see Clinical Focus Box 15-2). As might be expected from the pathogenesis of allergy and asthma, many of the associated gene loci encode proteins involved in the generation and regulation of immune responsiveness (e.g., innate immune receptors, cytokines and chemokines and their receptors, MHC proteins) as well as with airway remodeling (e.g., growth factors and proteolytic enzymes). Other proteins that have been implicated in the hereditary predisposition to allergy and asthma include transcription factors and proteins that regulate epigenetic modifications.

CLINICAL FOCUS

The Genetics of Asthma and Allergy

It has long been appreciated that a predisposition to asthma and allergic responses runs strongly in families, suggesting the presence of an hereditary component. In addition, twin studies in humans and mice have implicated both environmental and epigenetic, as well as genetic, factors in determining the susceptibility of an individual to hypersensitivity responses. With this degree of complexity, it is not surprising that the identification of the genes involved in controlling an individual's vulnerability to hypersensitivity responses has proven to be a difficult task. However, since the late 1980s, the geneticist's toolkit has markedly expanded with the availability of **genome wide sequence** information in addition to libraries of **single nucleotide polymorphisms (SNPs)**. These tools, along with more classical genetic approaches have been used to map hypersensitivity susceptibility genes.

One approach to determining which genes are associated with a particular pathological state is to use knowledge of the disease to develop and then genetically test an hypothesis regarding **potential candidate genes** (i.e. "educated guesses"). For example, we know that asthma is associated with high numbers of differentiated T_H2 cells, and high levels of IL-4 expression introduce a bias in the differentiation of activated CD4 T cells toward the T_H2 state. We can therefore hypothesize that asthma sufferers may

exhibit polymorphisms in structural or regulatory regions of the IL-4 gene, leading to unusually high levels of IL-4 production.

Using this theoretical framework, geneticists selected a region on human chromosome 5, 5q31-33, for detailed analysis. This region contains a cluster of cytokine genes, among which are the genes for IL-3, IL-4, IL-5, IL-9, and IL-13, as well as the gene encoding granulocyte-macrophage colony stimulating factor. Careful study of this region revealed a polymorphism associated with the predisposition to asthma that maps to the promoter region of IL-4—a confirmation of the hypothesis advanced above. In addition, two alleles of IL-9 associated with a predisposition to atopy were also identified.

A second approach starts with a statistically based search for genes associated with particular disease states and is referred to as a **genome wide association survey (GWAS)**. The genomes of individuals who do and those who do not have the disease in question are mapped with respect to the presence of SNPs. Statistically significant association of disease with a particular polymorphism then provides the motivation for detailed sequence analysis in the region of the SNP, and a search for likely candidate genes. Cloning of genes begins in the region identified by the candidate SNP and then proceeds by a sequential search of contiguous

sequences until a gene of interest is identified. This technique is referred to as **positional cloning**, and several genes important in asthma and atopy have been identified in this way.

Figure 1 illustrates some of the products of genes already identified as having polymorphisms relevant to the development of asthma or atopy. However, this investigation is far from complete. Sometimes the same SNP has been shown to have different effects in various racial or ethnic populations, illustrating the complexity of such disease-associated genetic studies.

Diagnostic Tests and Treatments Are Available for Type I Hypersensitivity Reactions

Type I hypersensitivity is commonly assessed by skin testing, an inexpensive and relatively safe diagnostic approach that allows screening of a wide range of antigens at once. Small amounts of potential allergens are introduced at specific skin sites (e.g., the forearm or back), either by intradermal injection or by dropping onto a site of a superficial scratch. Thirty minutes later, the sites are reexamined. Redness and swelling

(the result of local mast cell degranulation) indicate an allergic response (Figure 15-8). Less commonly, physicians may elect to measure the serum levels of either total or allergen-specific IgE using ELISA or Western blot technologies (see Chapter 20).

Treatment of type I hypersensitivity reactions always begins with measures to avoid the causative agents. However, no one can avoid contact with aeroallergens such as pollen, and a number of immunological and pharmaceutical interventions are now available.

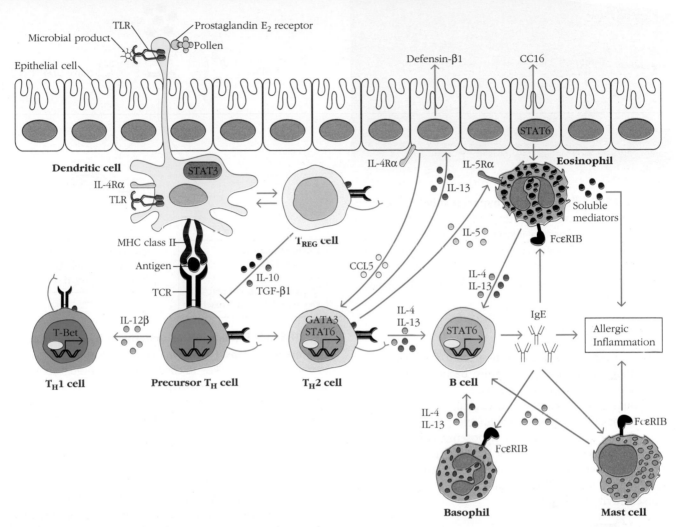

FIGURE 1
Products of genes that exhibit polymorphisms associated with predisposition to allergy. The genes coding for the products shown have been discovered by multiple genome wide survey techniques. They can be divided into three broad categories. One group of genes codes for proteins that trigger an immune response. These include pattern recognition receptors (TLR2, TLR4, TLR6), pollen receptors (Prostaglandin E2 receptor), and inhibitory cytokines (IL-10, TGFβ), as well as genes coding for proteins that regulate antigen presentation (MHC class II). Another group of genes codes for proteins that regulate T_H2 differentiation and innate immune cell responses. These include polarizing cytokines (IL-4, IL-12), signaling and transcriptional regulators (STAT6, T-bet GATA3), Fcε receptors, effector cytokines (IL-4, IL-5, and IL-13), and cytokine receptors (IL-4R, IL-5R, IL-13R). Another group of genes is expressed by epithelial cells and smooth muscle cells and codes for proteins that regulate the tissue response. These include chemokines (CCL6), defensins (Defensin β2), and other signaling molecules. [Vercelli, D. Discovering susceptibility genes for asthma and allergy. Nature Reviews. Immunology **8**:169–182.]

Hyposensitization

For many years, physicians have been treating allergic patients with repeated exposure (via ingestion or injection) to increasing doses of allergens, in a regimen termed *hyposensitization* or immunotherapy. This mode of treatment attacks the disease mechanism of the allergic individual at the source and, when it works, is by far the most effective way to manage allergies. Hyposensitization can reduce or even eliminate symptoms for months or years after the desensitization course is complete.

How does hyposensitization work? Two main mechanisms have been proposed (Figure 15-9). Repeated exposure to low doses of allergen may induce an increase in regulatory T cells producing the immunosuppressive cytokines TGF-β and/or IL-10 (a form of tolerance). It may also induce an increase in noninflammatory IgG (specifically IgG4) antibodies specific for the allergens (desensitization). These antibodies either competitively inhibit IgE binding or induce co-clustering of antigen with inhibitory Fc receptors as described above. Regardless of mechanism, hyposensitization

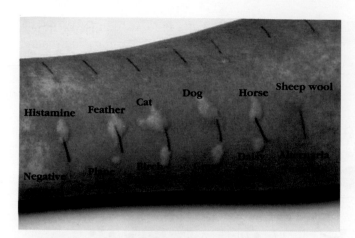

FIGURE 15-8 Skin testing for hypersensitivity. This photograph shows an example of a skin test for a variety of antigens. These were introduced by superficial injection and read after 30 minutes. The positive control for a response is histamine; the negative control is typically just saline. This individual is clearly atopic; the skin test reveals responses to multiple animal and plant allergens. [*Southern Illinois University/Getty Images*]

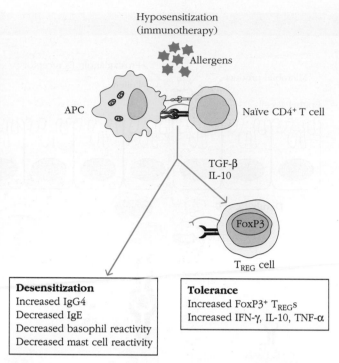

FIGURE 15-9 Mechanisms underlying hyposensitization treatment for type I allergy. This figure illustrates two major mechanisms that are likely to contribute to successful hyposensitization treatment (immunotherapy). Repeated injection or ingestion of low doses of antigen may lead to immune tolerance via the induction of regulatory T cells that quell the immune response to the allergen. Alternatively or in addition, it may induce the generation of IgG antibodies (specifically IgG4), which either compete with IgE for binding to antigen or induce co-clustering of FcεRI with inhibitory FcγRII receptors (see chapter text). This inhibits basophil and mast cell activity, reducing symptoms (desensitization). [*A. M. Scurlock, B. P. Vickery, J. O'B. Hourihane, and A. W. Burks, 2010, Pediatric food allergy and mucosal tolerance, Mucosal Immunology 3(4):345–354, www.nature.com/mi/journal/v3/n4/fig_tab/mi201021f1.html.*]

results in a reduction of allergen-specific T_H2 cells, and a concomitant decrease in eosinophils, basophils, mast cells, and neutrophils in the target organs.

Although often strikingly successful, hyposensitization does not work in every individual for every allergen. Patients whose disease is refractory to hyposensitization, or who choose not to use it, can try other therapeutic strategies that have taken advantage of our growing knowledge of mechanisms behind mast cell degranulation and the activity of hypersensitivity mediators.

Antihistamines, Leukotriene Antagonists, and Inhalation Corticosteroids

For many years now, antihistamines have been the most useful drugs for the treatment of allergic rhinitis. These drugs inhibit histamine activity by binding and blocking histamine receptors on target cells. The H_1 receptors are blocked by the first-generation antihistamines such as diphenhydramine and chlorpheniramine, which are quite effective in controlling the symptoms of allergic rhinitis. Unfortunately, since they are capable of crossing the blood-brain barrier, they also act on H_1 receptors in the nervous system and have multiple side effects. Because these first-generation drugs bind to muscarinic acetylcholine receptors, they can also induce dry mouth, urinary retention, constipation, slow heartbeat, and sedation. Second-generation antihistamines such as fexofenadine, loratidine, and desloratidine were developed in the early 1980s, and exhibit significantly less cross-reactivity with muscarinic receptors.

Leukotriene antagonists, specifically montelukast, have also been used to treat type I hypersensitivities and are com-

parable in effectiveness to antihistamines. Finally, inhalation therapy with low-dose corticosteroids reduces inflammation by inhibiting innate immune cell activity and has been used successfully to reduce frequency and severity of asthma attacks.

Immunotherapeutics

Therapeutic anti-IgE antibodies have been developed; one such antibody, omalizumab, has been approved by the FDA and is available as a pharmacological agent. Omalizumab binds the Fc region of IgE and interferes with IgE-FcεR interactions. This reagent has been used to treat both allergic rhinitis and allergic asthma. However, for the treatment of allergic rhinitis, omalizumab is no more effective than second-generation antihistamines and is rarely prescribed because of its higher cost. Other monoclonal reagents are also being evaluated for their clinical value.

Other Medications Used to Control Allergic Asthma

In addition to prescribing therapies that focus on inhibiting the molecular and cellular causes of type I hypersensitivities, physicians prescribe drugs that alleviate the symptoms. In particular, drugs that enhance production of the second messenger cAMP help to counter the bronchoconstriction of asthma and the degranulation of mast cells. Epinephrine, or epinephrine agonists (like albuterol), do this by binding to their G protein-coupled receptors, which generate signals that generate cAMP (see Chapter 3). Theophylline, another commonly used drug in the treatment of asthma, does this by antagonizing *phospho*diesterase (PDE), which normally breaks down cAMP.

The Hygiene Hypothesis Has Been Advanced to Explain Increases in Allergy Incidence

Asthma incidence has increased dramatically in the developed world over the past two decades. This observation supported suggestions that reduction in air quality associated with industrialization played a role in respiratory hypersensitivity. Indeed, the incidence and severity of asthma among those growing up in inner cities is significantly higher. However, it became clear that air quality, although very important, was not the only factor contributing to the increase in asthma incidence.

Another contributing cause was suggested by surprising studies from Europe, the United States, Australia, and New Zealand demonstrating that children exposed to a farm environment either prenatally or neonatally were significantly less likely to suffer from hay fever, atopic dermatitis, asthma, and wheezing compared with the control population. In addition, exposure of a pregnant mother or baby to barns and stables, farm animals, hay and grain products, and/or unprocessed cow's milk all resulted in a decreased tendency to develop type I hypersensitivity later in life. Exposure of children to day care situations and older siblings also correlated with a reduction in asthma incidence. What all these conditions have in common is early exposure to pathogens and potential allergens.

Intrigued by this commonality, investigators advanced the *hygiene hypothesis*, which proposed that exposure to some pathogens during infancy and youth benefits individuals by stimulating immune responses and establishing a healthy balance of T-cell subset activities so that no one response dominates. For instance, the immune system of newborn babies may be biased in the T_H2 direction by the uterine environment. T_H1-T_H2 balance may be restored by the occurrence of infections in the developing neonate. However, in the sanitary conditions promoted by Western medicine, the neonatal immune system may not have the exposure to infections that would otherwise reorient it to generate T_H1-type responses.

Evidence that pathogen exposure induces NK-mediated interferon γ secretion, which biases the responses of the subject away from the T_H2 direction and thus away from antibody production that contributes to asthma and other allergies, also supports this view. Other investigations, however, focus on the possibility that exposure to viral, bacterial, and parasitic pathogens helps establish the broad array of regulatory T cells that is critical for moderating normal immune responses and quelling autoimmune reactions.

The hygiene hypothesis has been advanced to explain increases in the incidence of all allergic responses (e.g., food allergies) as well as increases in the frequency of people suffering from autoimmune diseases (Chapter 16). Studies testing various predictions of the hypothesis are ongoing, and it continues to be modified in its particulars. Understanding the cellular and molecular players that contribute to allergy is clearly important in evaluating the data from these studies.

Antibody-Mediated (Type II) Hypersensitivity Reactions

Type II hypersensitivity reactions involve antibody-mediated destruction of cells by immunoglobulins of heavy chain classes other than IgE. Antibody bound to a cell-surface antigen can induce death of the antibody-bound cell by three distinct mechanisms (see Chapters 6 and 13). First, certain immunoglobulin subclasses can activate the complement system, creating pores in the membrane of a foreign cell. Secondly, antibodies can mediate cell destruction by *antibody-dependent cell-mediated cytotoxicity* (ADCC), in which cytotoxic cells bearing Fc receptors bind to the Fc region of antibodies on target cells and promote killing of the cells. Finally, antibody bound to a foreign cell also can serve as an opsonin, enabling phagocytic cells with Fc or C3b receptors to bind and phagocytose the antibody-coated cell.

In this section, we examine three examples of type II hypersensitivity reactions. Certain autoimmune diseases involve autoantibody-mediated cellular destruction by type II mechanisms and will be described in Chapter 16.

Transfusion Reactions Are an Example of Type II Hypersensitivity

Several proteins and glycoproteins on the membrane of red blood cells are encoded by genes with several allelic forms. An individual with a particular allele of a blood-group antigen can recognize other allelic forms in transfused blood as foreign, and mount an antibody response. Blood types are referred to as A, B, or O, and the antigens that are associated with the blood types are identified as A, B, and H, respectively.

Interestingly, the blood type antigens (ABH) are carbohydrates, rather than proteins. This was demonstrated by simple experiments in which the addition of high concentrations of particular simple sugars was shown to inhibit antibody binding to red blood cells bearing particular types of red

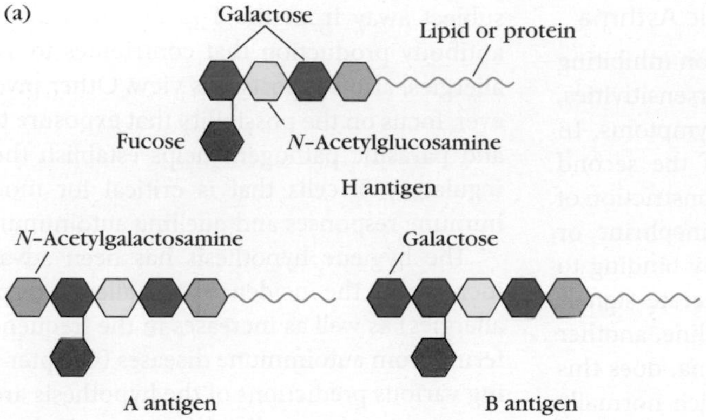

(a)

(b)

Genotype	Blood-group phenotype	Antigens on erythrocytes *(agglutinins)*	Serum antibodies *(isohemagglutinins)*
AA or AO	A	A	Anti–B
BB or BO	B	B	Anti–A
AB	AB	A and B	None
OO	O	H	Anti–A and anti–B

FIGURE 15-10 ABO (ABH) blood groups. (a) Structure of terminal sugars, which constitute the distinguishing epitopes of the A, B, and H blood antigens. All individuals express the H antigen, but not all individuals express the A or B antigens. The blood group of those who express neither A or B antigens (but, like all people express the H antigen) is referred to as O. (b) ABO genotypes, corresponding phenotypes, agglutinins (antigens), and isohemagglutinins (antibodies that react to nonhost antigens).

blood cell antigens. These inhibition reactions revealed that antibodies directed to group A antigens predominantly bound to N-acetyl glucosamine residues, those to group B antigens bound to galactose residues, and those directed toward the so-called H antigens bound to fucose residues (Figure 15-10a). Note that the H antigen is present in all blood types.

The A, B, and H antigens are synthesized by a series of enzymatic reactions catalyzed by glycosyltransferases. The final step of the biosynthesis of the A and B antigens is catalyzed by A and B transferases, encoded by alleles *A* and *B* at the *ABO* genetic locus. Although initially detected on the surface of red blood cells, antigens of the ABO blood type system also occur on the surface of other cells as well as in bodily secretions.

Antibodies directed toward ABH antigens are termed *isohemagglutinins*. Figure 15-10b shows the pattern of blood cell antigens and expressed isohemagglutinins normally found within the human population. Most adults possess IgM antibodies to those members of the ABH family they do not express. This is because common microorganisms express carbohydrate antigens very similar in structure to the carbohydrates of the ABH system and induce a B-cell response. B cells generating antibodies specific for the ABH antigens expressed by the host, however, undergo negative selection.

For example, an individual with blood type A recognizes B-like epitopes on microorganisms and produces isohem-

agglutinins to the B-like epitopes. This same individual does not respond to A-like epitopes on the same microorganisms because they have been tolerized to self-A epitopes. If a type A individual is transfused with blood containing type B cells, a *transfusion reaction* occurs in which the preexisting anti-B isohemagglutinins bind to the B blood cells and mediate their destruction by means of complement-mediated lysis. Individuals with blood type O express only the H antigen. Although they can donate blood to anyone, they have antibodies that will react to both A-type or B-type blood. Their anti-A- and anti-B-producing B cells were never exposed to A or B antigens and therefore were never deleted.

The clinical manifestations of transfusion reactions result from massive intravascular hemolysis of the transfused red blood cells by antibody plus complement. These manifestations may be either immediate or delayed. Reactions that begin immediately are most commonly associated with ABO blood-group incompatibilities, which lead to complement-mediated lysis triggered by the IgM isohemagglutinins. Within hours, free hemoglobin can be detected in the plasma; it is filtered through the kidneys, resulting in hemoglobinuria. As the hemoglobin is degraded, the porphyrin component is metabolized to bilirubin, which at high levels is toxic to the organism. Typical symptoms of bilirubinemia include fever, chills, nausea, clotting within blood vessels, pain in the lower back, and hemoglobin in the urine. Treatment involves prompt termination of the transfusion and maintenance of urine flow with a diuretic, because the accumulation of hemoglobin in the kidney can cause acute tubular necrosis.

Antibodies to other blood-group antigens such as Rh factor (see below) may result from repeated blood transfusions because minor allelic differences in these antigens can stimulate antibody production. These antibodies are usually of the IgG class. These incompatibilities typically result in delayed hemolytic transfusion reactions that develop between 2 and 6 days after transfusion. Because IgG is less effective than IgM in activating complement, complement-mediated lysis of the transfused red blood cells is incomplete. Free hemoglobin is usually not detected in the plasma or urine in these reactions. Rather, many of the transfused cells are destroyed at extravascular sites by agglutination, opsonization, and subsequent phagocytosis by macrophages. Symptoms include fever, increased bilirubin, mild jaundice, and anemia.

Hemolytic Disease of the Newborn Is Caused by Type II Reactions

Hemolytic disease of the newborn develops when maternal IgG antibodies specific for fetal blood-group antigens cross the placenta and destroy fetal red blood cells. The consequences of such transfer can be minor, serious, or lethal. Severe hemolytic disease of the newborn, called *erythroblastosis fetalis*, most commonly develops when the mother and fetus express different alleles of the *Rh*esus (Rh) antigen. Although there are actually five alleles of the Rh antigen, expression of the D allele elicits the strongest immune response. We therefore designate individuals bearing the D allele of the Rh antigen as Rh$^+$.

An Rh$^-$ mother fertilized by an Rh$^+$ father is in danger of developing a response to the Rh antigen and rejecting an Rh$^+$ fetus. During pregnancy, fetal red blood cells are separated from the mother's circulation by a layer of cells in the placenta called the trophoblast. During her first pregnancy with an Rh$^+$ fetus, an Rh$^-$ woman is usually not exposed to enough fetal red blood cells to activate her Rh-specific B cells. However, at the time of delivery, separation of the placenta from the uterine wall allows larger amounts of fetal umbilical cord blood to enter the mother's circulation. These fetal red blood cells stimulate Rh-specific B cells to mount an immune response, resulting in the production of Rh-specific plasma cells and memory B cells in the mother. The secreted IgM antibody clears the Rh$^+$ fetal red cells from the mother's circulation, but memory cells remain, a threat to any subsequent pregnancy with an Rh$^+$ fetus. Importantly, since IgM antibodies do not pass through the placenta, IgM anti-Rh antigens are no threat to the fetus.

Activation of IgG-secreting memory cells in a subsequent pregnancy results in the formation of IgG anti-Rh antibodies, which, however, can cross the placenta and damage the fetal red blood cells (Figure 15-11). Mild to severe anemia can develop in the fetus, sometimes with fatal consequences. In addition, conversion of hemoglobin to bilirubin can present an additional threat to the newborn because the lipid-soluble bilirubin may accumulate in the brain and cause brain damage. Because the blood-brain barrier is not complete until after birth, very young babies can suffer fatal brain damage from bilirubin. Fortunately, bilirubin is rapidly broken down on exposure of the skin to *ultraviolet* (UV) light, and babies who display the telltale jaundiced appearance that signifies high levels of blood bilirubin are treated by exposure to UV light in their cribs (Figure 15-12).

Hemolytic disease of the newborn caused by Rh incompatibility in a second or later pregnancy can be almost entirely prevented by administering antibodies against the Rh antigen to the mother at around 28 weeks of pregnancy and within 24 to 48 hours after the first delivery. Anti-Rh antibodies are also administered to pregnant women after amniocentesis. These antibodies, marketed as *Rhogam*, bind to any fetal red blood cells that may have entered the mother's circulation and facilitate their clearance before B-cell activation and ensuing memory-cell production can take place. In a subsequent pregnancy with an Rh$^+$ fetus, a mother who has been treated with Rhogam is unlikely to produce IgG anti-Rh antibodies; thus, the fetus is protected from the damage that would occur when these antibodies cross the placenta.

The development of hemolytic disease of the newborn caused by Rh incompatibility can be detected by testing maternal serum at intervals during pregnancy for antibodies to the Rh antigen. A rise in the titer of these antibodies as pregnancy progresses indicates that the mother has been exposed to Rh antigens and is producing increasing amounts of antibody. Treatment depends on the severity of the reaction. For a severe reaction, the fetus can be given an intrauterine blood-exchange transfusion to replace fetal Rh$^+$ red blood cells with Rh$^-$ cells. These transfusions are given every 10 to 21 days until delivery. In less severe cases, a blood-exchange transfusion is not given until after birth, primarily to remove bilirubin; the infant is also exposed to low levels of UV light to break down the bilirubin and prevent cerebral damage. The mother can also be treated during the pregnancy by *plasmapheresis*. In this procedure, a cell separation machine is used to separate the mother's blood into two fractions: cells and plasma. The plasma containing the anti-Rh antibody is discarded, and the cells are reinfused into the mother in an albumin or fresh plasma solution.

The majority of cases (65%) of hemolytic disease of the newborn, however, are caused by ABO blood-group incompatibility between the mother and fetus and are not severe. Type A or B fetuses carried by type O mothers most commonly develop these reactions. A type O mother can develop IgG antibodies to the A or B blood-group antigens through exposure to fetal blood-group A or B antigens in successive pregnancies. Usually, the fetal anemia resulting from this incompatibility is mild; the major clinical manifestation is a

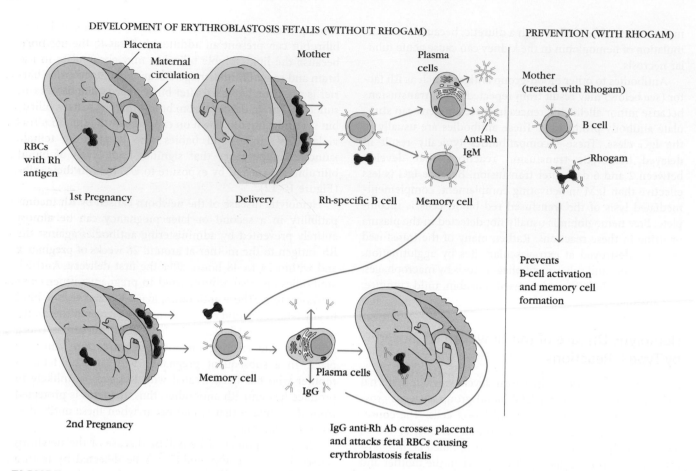

DEVELOPMENT OF ERYTHROBLASTOSIS FETALIS (WITHOUT RHOGAM)

PREVENTION (WITH RHOGAM)

1st Pregnancy

Delivery   Rh-specific B cell   Memory cell

2nd Pregnancy

IgG anti-Rh Ab crosses placenta
and attacks fetal RBCs causing
erythroblastosis fetalis

FIGURE 15-11 Destruction of Rh positive red blood cells during erythroblastosis fetalis of the newborn. Development of erythroblastosis fetalis (hemolytic disease of the newborn) is caused when an Rh⁻ mother carries an Rh⁺ fetus (*left*). The effect of treatment with anti-Rh antibody, or Rhogam, is shown on the *right*.

FIGURE 15-12 Ultraviolet light is used to treat bilirubinemia of the newborn. [*Stephanie Clarke/123RF*]

slight elevation of bilirubin, with jaundice. Exposure of the infant to low levels of UV light is often enough to break down the bilirubin and avoid cerebral damage. In severe cases, transfusion may be required.

Hemolytic Anemia Can Be Drug Induced

Certain antibiotics (e.g., penicillin, cephalosporins, and streptomycin), as well as other well-known drugs (including ibuprofen and naproxen), can adsorb nonspecifically to proteins on red blood cell membranes, forming a drug-protein complex. In some patients, such drug-protein complexes induce formation of antibodies. These antibodies then bind to the adsorbed drug on red blood cells, inducing complement-mediated lysis and thus progressive anemia. When the drug is withdrawn, the hemolytic anemia disappears. Penicillin is notable in that it can induce all four types of hypersensitivity with various clinical manifestations (Table 15-4).

TABLE 15-4	Penicillin-induced hypersensitive reactions	
Type of reaction	Antibody or lymphocytes induced	Clinical manifestations
I	IgE	Urticaria, systemic anaphylaxis
II	IgM, IgG	Hemolytic anemia
III	IgG	Serum sickness, glomerulonephritis
IV	T_H1 cells	Contact dermatitis

Immune Complex-Mediated (Type III) Hypersensitivity

The reaction of antibody with antigen generates immune complexes. Generally, these complexes facilitate the clearance of antigen by phagocytic cells and red blood cells. In some cases, however, the presence of large numbers and networks of immune complexes can lead to tissue-damaging type III hypersensitivity reactions. The magnitude of the reaction depends on the number and size of immune complexes, their distribution within the body, and the ability of the phagocyte system to clear the complexes and thus minimize the tissue damage. The deposition of these complexes initiates a reaction that results in the recruitment of complement components and neutrophils to the site, with resultant tissue injury.

Immune Complexes Can Damage Various Tissues

The formation of antigen-antibody complexes occurs as a normal part of an adaptive immune response. It is usually followed by Fc receptor-mediated recognition of the complexes by phagocytes, which engulf and destroy them and/or by complement activation resulting in the lysis of the cells on which the immune complexes are found. However, under certain conditions, immune complexes are inefficiently cleared and may be deposited in the blood vessels or tissues, setting the stage for a type III hypersensitivity response. These conditions include (1) the presence of antigens capable of generating particularly extensive antigen-antibody lattices, (2) a high intrinsic affinity of antigens for particular tissues, (3) the presence of highly charged antigens (which can affect immune complex engulfment) and (4) a compromised phagocytic system. All have been associated with the initiation of a type III response.

Immune complexes bind to mast cells, neutrophils, and macrophages via Fc receptors, triggering the release of vasoactive mediators and inflammatory cytokines, which interact with the capillary epithelium and increase the permeability of the blood vessel walls. Immune complexes then move through the capillary walls and into the tissues where they are deposited and set up a localized inflammatory response. Complement fixation results in the production of the anaphylatoxin chemokines C3a and C5a, which attract more neutrophils and macrophages. These in turn are further activated by immune complexes binding to their Fc receptors to secrete proinflammatory chemokines and cytokines, prostaglandins, and proteases. Proteases digest the basement membrane proteins collagen, elastin, and cartilage. Tissue damage is further mediated by oxygen free radicals released by the activated neutrophils. In addition, immune complexes interact with platelets and induce the formation of tiny clots. Complex deposition in the tissues can give rise to symptoms such as fever, urticaria (rashes), joint pain, lymph node enlargement, and protein in the urine. The resulting inflammatory lesion is referred to as vasculitis if it occurs in a blood vessel, glomerulonephritis if it occurs in the kidney, or arthritis if it occurs in the joints.

Immune Complex-Mediated Hypersensitivity Can Resolve Spontaneously

If immune complex-mediated disease is induced by a single large bolus of antigen that is then gradually cleared, it can resolve spontaneously. Spontaneous recovery is seen, for example, when glomerulonephritis is initiated following a streptococcal infection. Streptococcal antigen-antibody complexes bind to the basement membrane of the kidney and set up a type III response, which resolves as the bacterial load is eliminated. Similarly, patients being treated by injections of passive antibody can develop immune responses to the foreign antibody and generate large immune complexes. This was seen initially during the use of horse antibodies in the treatment of diphtheria in the early 1900s. On repeated injections with the horse antibodies, patients developed a syndrome known as serum sickness, which resolved as soon as the antibodies were withdrawn. Serum sickness is an example of a systemic form of immune complex disease, which resulted in arthritis, skin rash, and fever.

A more modern manifestation of the same problem occurs in patients who receive therapeutic mouse-derived monoclonal antibodies designed to treat previously intractable cancers. After several such treatments, some patients generate their own antibodies against the foreign monoclonals and develop serum sickness–like symptoms. We know now that the injection of the mouse antibodies caused a generalized type III reaction, and in many cases the therapeutic antibodies were actually cleared before they could reach their pathogenic target. To avoid this response, current therapeutic antibodies are genetically engineered to remove most of the foreign epitopes (they are *humanized*).

TABLE 15-5	Examples of diseases resulting from type III hypersensitivity reactions.
Autoimmune diseases	
Systemic lupus erythematosus	
Rheumatoid arthritis	
Multiple sclerosis	
Drug reactions	
Allergies to penicillin and sulfonamides	
Infectious diseases	
Poststreptococcal glomerulonephritis	
Meningitis	
Hepatitis	
Mononucleosis	
Malaria	
Trypanosomiasis	

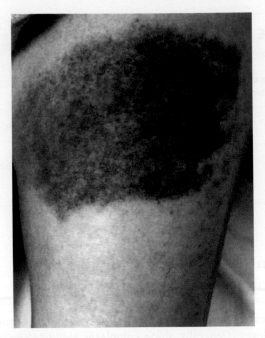

FIGURE 15-13 An Arthus reaction. This photograph shows an Arthus reaction on a thigh of a 72-year-old woman. This occurred at the site of injection of a chemotherapeutic drug, 3 to 4 hours after the patient received a second injection (15 days after the first). This response was accompanied by fever and significant discomfort. *[From P. Boura et al., 2006, Eosinophilic cellulitis (Wells' syndrome) as a cutaneous reaction to the administration of adalimumab, Annals of the Rheumatic Diseases 65:839–840. doi:10.1136/ard.2005.044685.]*

Autoantigens Can be Involved in Immune Complex-Mediated Reactions

If the antigen in the immune complex is an autoantigen, it cannot be eliminated and type III hypersensitivity reactions cannot be easily resolved. In such situations, chronic type III responses develop. For example, in systemic lupus erythematosus, persistent antibody responses to autoantigens are an identifying feature of the disease, and complexes are deposited in the joints, kidneys, and skin of patients. Examples of diseases resulting from type III hypersensitivity reactions are found in Table 15-5.

Arthus Reactions Are Localized Type III Hypersensitivity Reactions

One example of a localized type III hypersensitivity reaction has been used extensively as an experimental tool. If an animal or human subject is injected intradermally with an antigen to which large amounts of circulating antibodies exist (or have been recently introduced by intravenous injections), antigen will diffuse into the walls of local blood vessels and large immune complexes will precipitate close to the injection site. This initiates an inflammatory reaction that peaks approximately 4 to 10 hours post injection and is known as an Arthus reaction. Inflammation at the site of an Arthus reaction is characterized by swelling and localized bleeding, followed by fibrin deposition (Figure 15-13).

A sensitive individual may react to an insect bite with a rapid, localized type I reaction, which can be followed, some 4 to 10 hours later, by the development of a typical Arthus reaction, characterized by pronounced erythema and edema. Intrapulmonary Arthus-type reactions induced by bacterial

spores, fungi, or dried fecal proteins can also cause pneumonitis or alveolitis. These reactions are known by a variety of common names reflecting the source of the antigen. For example, farmer's lung develops after inhalation of actinomycetes from moldy hay, and pigeon fancier's disease results from inhalation of a serum protein in dust derived from dried pigeon feces.

Delayed-Type (Type IV) Hypersensitivity (DTH)

Type IV hypersensitivity, commonly referred to as *Delayed-Type Hypersensitivity* (DTH), is the only hypersensitivity category that is purely cell mediated rather than antibody mediated. In 1890, Robert Koch observed that individuals infected with *Mycobacterium tuberculosis* developed a localized inflammatory response when injected intradermally (in the skin) with a filtrate derived from a mycobacterial culture. He therefore named this localized skin reaction a *tuberculin reaction*. Later, as it became apparent that a variety of other antigens could induce this cellular response (Table 15-6), its name was changed to delayed-type, or type IV, hypersensitivity. The hallmarks of a type IV reaction are its initiation by T cells (as distinct from antibodies), the delay required for

TABLE 15-6	Intracellular pathogens and contact antigens that induce delayed-type (type IV) hypersensitivity

Intracellular bacteria	Intracellular viruses
Mycobacterium tuberculosis	Herpes simplex virus
Mycobacterium leprae	Variola (smallpox)
Brucella abortus	Measles virus
Listeria monocytogenes	
Intracellular fungi	**Contact antigens**
Pneumocystis carinii	Picrylchloride
Candida albicans	Hair dyes
Histoplasma capsulatum	Nickel salts
Cryptococcus neoformans	Poison ivy
	Poison oak

Intracellular parasites

Leishmania sp.

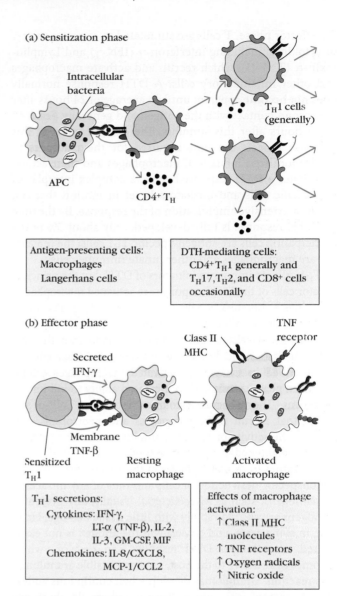

(a) Sensitization phase

Intracellular bacteria

APC

CD4⁺ T_H

T_H1 cells (generally)

Antigen-presenting cells: Macrophages Langerhans cells	DTH-mediating cells: CD4⁺T_H1 generally and T_H17, T_H2, and CD8⁺ cells occasionally

(b) Effector phase

Secreted IFN-γ

Membrane TNF-β

Class II MHC

TNF receptor

Sensitized T_H1   Resting macrophage   Activated macrophage

T_H1 secretions: Cytokines: IFN-γ, LT-α (TNF-β), IL-2, IL-3, GM-CSF, MIF Chemokines: IL-8/CXCL8, MCP-1/CCL2	Effects of macrophage activation: ↑ Class II MHC molecules ↑ TNF receptors ↑ Oxygen radicals ↑ Nitric oxide

FIGURE 15-14 The DTH response. (a) In the sensitization phase after initial contact with antigen (e.g., peptides derived from intracellular bacteria), T_H cells proliferate and differentiate into T_H1 cells. Cytokines secreted by these T cells are indicated by the black balls. (b) In the effector phase after subsequent exposure of sensitized T_H cells to antigen, T_H1 cells secrete a variety of cytokines and chemokines. These factors attract and activate macrophages and other non-specific inflammatory cells. Activated macrophages are more effective in presenting antigen, thus perpetuating the DTH response, and function as the primary effector cells in this reaction. Other helper T-cell subsets are now thought to participate in DTH (T_H2 and T_H17) and CD8⁺ T cells also contribute.

the reaction to develop, and the recruitment of macrophages (as opposed to neutrophils or eosinophils) as the primary cellular component of the infiltrate that surrounds the site of inflammation.

The most common type IV hypersensitivity is the contact dermatitis that occurs after exposure to *Toxicodendron* species, which include poison ivy, poison oak, and poison sumac. This is a significant public health problem. Approximately 50% to 70% of the U.S. adult population is clinically sensitive to exposure to *Toxicodendron*; only 10% to 15% of the population is tolerant. Some responses can be severe and require hospitalization.

The Initiation of a Type IV DTH Response Involves Sensitization by Antigen

A DTH response begins with an initial sensitization by antigen, followed by a period of at least 1 to 2 weeks during which antigen-specific T cells are activated and clonally expanded (Figure 15-14a). A variety of *antigen-presenting cells* (APCs) are involved in the induction of a DTH response, including Langerhans cells (dendritic cells found in the epidermis) and macrophages. These cells pick up antigen that enters through the skin and transport it to regional lymph nodes, where T cells are activated. In some species, including humans, the vascular endothelial cells express class II MHC molecules and can also function as APCs in the development of the DTH response. Generally, the T cells activated during the sensitization phase of a traditional DTH response are CD4⁺, primarily of the T_H1 subtypes. However, recent studies indicate that T_H17, T_H2, and CD8⁺ cells can also play a role.

The Effector Phase of a Classical DTH Response Is Induced by Second Exposure to a Sensitizing Antigen

A second exposure to the sensitizing antigen induces the effector phase of the DTH response (see Figure 15-14b). In the effector phase, T cells are stimulated to secrete a variety of cytokines, including interferon-γ (IFN-γ) and Lymphotoxin-α (TNF-β), which recruit and activate macrophages and other inflammatory cells. A DTH response normally does not become apparent until an average of 24 hours after the second contact with the antigen and generally peaks 48 to 72 hours after this stimulus. The delayed onset of this response reflects the time required for the cytokines to induce localized influxes of macrophages and their activation. Once a DTH response begins, a complex interplay of nonspecific cells and mediators is set in motion that can result in extensive amplification of the response. By the time the DTH response is fully developed, only about 5% of the participating cells are antigen-specific T_H1 cells; the remainder are macrophages and other innate immune cells.

T_H1 cells are important initiators of DTH, but the principal effector cells of the DTH response are activated macrophages. Cytokines elaborated by helper T cells, including IFN-γ and Lymphotoxin-α, induce blood monocytes to adhere to vascular endothelial cells, migrate from the blood into the surrounding tissues, and differentiate into activated macrophages. As described in Chapter 2, activated macrophages exhibit enhanced phagocytosis and an increased ability to kill microorganisms. They produce cytokines, including TNF-α and IL-1β, that recruit more monocytes and neutrophils, and enhance the activity of T_H1 cells, amplifying the response.

The heightened phagocytic activity and the buildup of lytic enzymes from macrophages in the area of infection lead to nonspecific destruction of cells and thus of any intracellular pathogens, such as *Mycobacteria*. Usually, any presented pathogens are cleared rapidly with little tissue damage. However, in some cases, and especially if the antigen is not easily cleared, a prolonged DTH response can develop, which becomes destructive to the host, causing a visible granulomatous reaction. Granulomas develop when continuous activation of macrophages induces them to adhere closely to one another. Under these conditions, macrophages assume an epithelioid shape and sometimes fuse to form multinucleated giant cells (Figure 15-15a). These giant cells displace the normal tissue cells, forming palpable nodules, and releasing high concentrations of lytic enzymes, which destroy surrounding tissue. The granulomatous response can damage blood vessels and lead to extensive tissue necrosis.

The response to *Mycobacterium tuberculosis* illustrates the double-edged nature of the DTH response. Immunity to this intracellular bacterium involves a DTH response in which activated macrophages wall off the organism in the lung and contain it within a granuloma-type lesion called a tubercle (see Figure 15-15b). Often, however, the release of concentrated lytic enzymes from the activated macrophages within the tubercles damages the very lung tissue that the immune response aims to preserve.

The DTH Reaction Can be Detected by a Skin Test

The presence of a DTH reaction can be measured experimentally by injecting antigen intradermally into an animal and

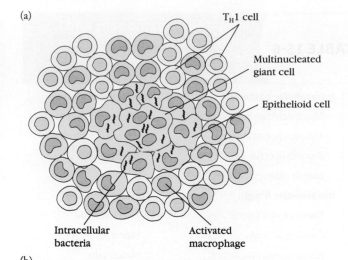

(a)

- T_H1 cell
- Multinucleated giant cell
- Epithelioid cell
- Intracellular bacteria
- Activated macrophage

(b)

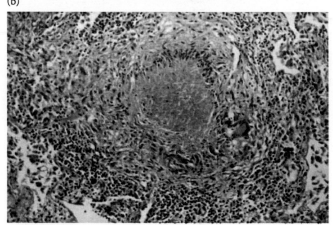

FIGURE 15-15 A prolonged DTH response can lead to formation of a granuloma, a nodule-like mass. (a) Lytic enzymes released from activated macrophages in a granuloma can cause extensive tissue damage. (b) Stained section of a granuloma associated with tuberculosis. *[Biophoto Associates/Getty Images]*

observing whether a characteristic skin lesion develops days later at the injection site. A positive skin-test reaction indicates that the individual has a population of sensitized T_H1 cells specific for the test antigen. For example, to determine whether an individual has been exposed to *M. tuberculosis*, PPD, a protein derived from the cell wall of this mycobacterium, is injected intradermally. Development of a red, slightly swollen, firm lesion at the site between 48 and 72 hours later indicates previous exposure. Note, however, that a positive test does not allow one to conclude whether the exposure was due to a pathogenic form of *M. tuberculosis* or to a vaccine antigen, which is used in some parts of the world.

Contact Dermatitis Is a Type IV Hypersensitivity Response

Contact dermatitis is one common manifestation of a type IV hypersensitivity. The simplest form of contact dermatitis occurs when a reactive chemical compound binds to skin proteins and these modified proteins are presented to T cells in the context of the appropriate MHC antigens. The reactive chemical may be a pharmaceutical, a component of a cosmetic or a hair dye, an industrial chemical such as formaldehyde or turpentine, an artificial hapten such as fluoro-dinitrobenzene, a metal ion such as nickel, or the active allergen from poison ivy.

Sometimes, however, the antigen is presented to T cells via a less familiar route. Some chemicals are first metabolized by the body to form a reactive metabolite, which, in turn, binds to proteins presented by the MHC. Some investigators propose that other molecules have the capacity to bind directly to the T-cell receptor and directly stimulate an effector or memory T cell. These ideas are new and still being evaluated. What seems clear, however, is that many small molecules and drugs can attach to cell-surface proteins of T cells as well as of APCs and give rise to the hypersensitivity reactions characteristic of contact dermatitis.

The classical DTH response to *Mycobacterium* antigens described above is mediated by CD4$^+$, T_H1 T cells. However, we now know that T-cell hypersensitivity reactions directed toward other antigens are mediated by a variety of T-cell subtypes and cytokines. The severe dermatitis associated with some type IV hypersensitivities to drugs, for example, is caused by CD8$^+$ T cells and NK cells. These cytotoxic cells induce death of keratinocytes and sloughing of the skin or mucus membrane. Diseases in which this mechanism is active include erythema multiforme, Stevens-Johnson syndrome, and toxic epidermal necrolysis; they can be fatal. The

allergen in these cases can be associated with drugs as common as the nonsteroidal anti-inflammatory medication ibuprofen. The incidence of such complications is higher in males than in females and usually occurs in young adults.

At present the best way to avoid a DTH response it to avoid the causative antigen. Once hypersensitivity has developed, topical or oral corticosteroids can be used to suppress the destructive immune response.

Chronic Inflammation

Regardless of whether an immune response is activated appropriately or inappropriately, a full-blown immune response makes one feel ill, largely because of the activity of inflammatory mediators released by innate immune cells. In most cases, the misery subsides when the insult (antigen, allergen, or toxin) is removed. However, in some circumstances, an inflammatory stimulus persists, generating a chronic inflammatory response that has systemic effects, which have been associated most directly with Type 2 diabetes. Recent studies also raise a possibility that chronic inflammation exacerbates heart disease, kidney disease, Alzheimer's, and cancer (Figure 15-16).

Infections Can Cause Chronic Inflammation

Chronic inflammatory conditions have a variety of causes, some of which are still being identified (Figure 15-16). Some are the result of infections that persist because

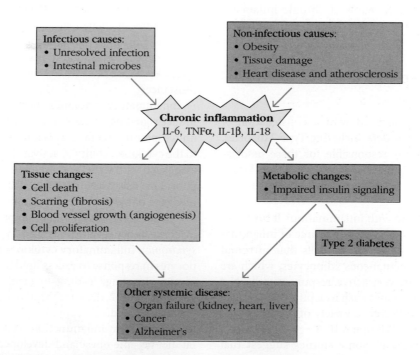

FIGURE 15-16 Causes and consequences of chronic inflammation. Chronic inflammation has infectious and noninfectious causes, including obesity. Chronic inflammatory conditions, regardless of cause, have common systemic consequences, some of which are related to the effects of inflammatory mediators on metabolism (Type 2 diabetes) and some of which are related to the effects of inflammatory mediators on tissue organization and cell proliferation (e.g., cancer). Other disorders have also been associated with chronic inflammation although the mechanisms behind the association may be indirect and are still being studied.

pathogen has continual access to the body. For instance, gum disease and unhealed wounds make a body vulnerable to continual microbe invasion and immune stimulation. Gut pathogens can also contribute. Although our commensal bacteria play an important role in dampening our reaction to microbes that we ingest, this protective mechanism can fail, and gut microbes can contribute to a chronic inflammatory state.

Some chronic inflammatory conditions are caused by pathogens that evade the immune system and remain active in the body, inspiring continual low-level inflammatory reactions. Fungi and mycobacteria are two examples of pathogens that are not always successfully cleared and have the ability to continually stimulate immune cells that release inflammatory molecules and cytokines.

There Are Noninfectious Causes of Chronic Inflammation

Interestingly, pathogens are not the only causes of chronic inflammation. Physical damage to tissue also releases molecules (*d*amage-*a*ssociated *m*olecular *p*atterns, or DAMPs) that induce the secretion of inflammatory cytokines. If tissue damage is not resolved because of continual mechanical interference, for instance, the inflammatory stimulus persists. Tumors, autoimmune disorders, atherosclerosis, and heart disease, each of which results in tissue damage that stimulates immune responses, are other examples of disorders that can cause chronic inflammation. The biomedical community was startled, however, to discover that one of the most common noninfectious causes of chronic inflammation today is obesity, a condition that did not, at first, suggest a relationship to inflammation.

Obesity Is Associated with Chronic Inflammation

Obesity has long been associated with a constellation of metabolic and systemic disorders, including Type 2 diabetes. The biological mechanisms responsible for these associations are still being investigated. However, recent work suggests that many of the systemic effects of obesity are mediated by inflammation.

What does fat have to do with inflammation? It turns out that the immune system is not the only source of inflammatory cytokines. Visceral adipocytes, fat cells that surround organs (as opposed to subcutaneous adipocytes, which are located under the skin), are very active, responsive cells. Not only do they generate hormones such as leptin that regulate metabolism, but they also secrete a variety of proinflammatory mediators, including TNF-α and IL-6.

What triggers this release? Some studies suggest that intracellular stress responses associated with excessive lipid buildup induce signals that enhance production of cytokines and inflammatory mediators. Free fatty acids, a common consequence of obesity, may also play a role.

They appear to have the capacity to bind toll-like receptors on adipocytes, initiating a signaling cascade analogous to that experienced by innate immune cells that recognize pathogen.

Obesity is now recognized as a major, if not the major, cause of chronic inflammation, which, as you will see below, has severe consequences. Interestingly, approximately 6% of individuals who are considered obese by weight do not generate inflammatory cytokines and show few signs of metabolic dysfunction. The basis for the ability of these individuals to tolerate excess fat is an area of active investigation.

Chronic Inflammation Can Cause Systemic Disease

The specific consequences of chronic inflammation vary with the tissue of origin as well as the sex, age, and health status of the individual. However, given that those who suffer from chronic inflammation all exhibit increased circulation of inflammatory mediators, including the classical proinflammatory cytokine trio (IL-1, IL-6, and TNF-α), it is not surprising that many suffer similar systemic disorders (Figure 15-16).

Chronic Inflammation and Insulin Resistance

Type 2 diabetes is one of the most common consequences of chronic inflammation. Diabetes results from a failure in insulin signaling, a failure that leads to general metabolic dysfunction. Type 1 diabetes (Chapter 16) is caused by the autoimmune-mediated destruction of pancreatic islet cells that make insulin. Type 2 diabetes, however, is caused by a failure of cells to respond to insulin, a state known as *insulin resistance*. What does inflammation have to do with insulin resistance?

Inflammatory cytokines, particularly TNF-α and IL-6, induce signaling cascades that inhibit the ability of the insulin receptor to function. This interference is in large part due to the cytokines' ability to activate JNK, a MAPK that is often associated with stress and inflammatory responses. JNK can phosphorylate and inactivate IRS-1, a key downstream mediator of insulin receptor signaling.

This observation also helps explain the long-recognized association between obesity and Type 2 diabetes (*metabolic syndrome*). Inflammatory cytokines released by visceral adipocytes in response to excess lipid induce signals that inhibit insulin signaling, leading to insulin resistance, a primary cause of Type 2 diabetes. This perspective is directly supported by studies in mouse models where obesity was uncoupled from inflammation. Wild-type mice fed a high-fat diet became obese and developed Type 2 diabetes. Mice that lack JNK also became obese on the same diet, but did not develop diabetes. (See Clinical Focus Box 15-3 for a more complete discussion of the association between obesity, inflammation, and diabetes.)

CLINICAL FOCUS

BOX 15-3

Type 2 Diabetes, Obesity, and Inflammation

As late as the 1960s, scientists and physicians searching for information on Type 2 diabetes, obesity, or inflammation would have looked in separate chapters of physiology or pathology books. Obesity was viewed as a problem of poor nutritional management or resources, psychology, or (in the case of rare hormonal disorders) endocrinology. Type 2 diabetes was known to be a result of the inability to effectively use insulin, resulting in high blood glucose levels (hyperglycemia) and was, therefore, seen solely as the province of endocrinologists. Inflammation had been understood for close to a thousand years to be a result of immune system activity. However, research conducted over the past few decades has resulted in a dramatic shift in our thinking regarding the centrality of inflammatory responses in many human pathologies, including Type 2 diabetes. Furthermore, knowledge about how inflammatory mediators work and the cells from which they are released has helped us to understand the linkages between obesity and Type 2 diabetes. Here we will explore the intersecting biological pathways that connect obesity, inflammation and Type 2 diabetes.

Obesity and Type 2 diabetes currently represent a public health problem of stunning proportions in the United States. As of summer 2011, 33.8% of adults in the United States are obese (defined as having a body mass index [BMI] of 30 or above), and childhood obesity rates are rising rapidly, with 17% of children and adolescents aged from 2 to 19 years falling into this category. Nor is the problem restricted to the United States. The World Health Organization reports that 300 million adults are obese and as many as 1 billion are reported to be overweight worldwide. But why should this be a problem, and what does it have to do with immunology?

In Type 1 diabetes, the insulin-producing β cells of the islets of Langerhans are subject to autoimmune attack and are destroyed. However, in Type 2 diabetes, patients experience a state of insulin resistance, in which the body still makes insulin, but the responses to it are dulled and the amount of insulin in the circulation is unable to do its job of driving dietary sugar out of the bloodstream and into the waiting cells.

The first indication that Type 2 diabetes may result from, or at least be exacerbated by, inflammatory signals came almost a hundred years ago, when it was discovered that patients receiving salicylate (aspirin) for inflammatory conditions showed an increase in insulin sensitivity. Studies published in the early years of the twenty-first century further demonstrated that patients suffering from a variety of infectious diseases, including hepatitis C and HIV, as well as those with autoimmune diseases such as rheumatoid arthritis, displayed insulin resistance. These diseases share the common feature of inducing an active inflammatory response. In each case, the insulin resistance was improved upon treatment with anti-inflammatory drugs.

What do we know about the mechanism by which inflammation results in insulin resistance? Insulin signals a cell to import glucose by binding to a cell surface receptor that is a member of the receptor tyrosine kinase (RTK) family. When insulin binds to the receptor on the external surface of the cell, a signal is transmitted to the intracellular part of the receptor, activating the intrinsic tyrosine kinase activity of the receptor (see Chapter 3). The two halves of the insulin-bound dimeric receptor phosphorylate one another on tyrosine residues, and these residues then act as docking sites for other proteins in the signaling cascade. A set of six proteins termed the *I*nsulin *R*eceptor *S*ubstrate (IRS) proteins are among the early, pivotally important substrates of the insulin receptor, and they bind to it through their phosphotyrosine binding domains. The IRS proteins are then phosphorylated by the RTKs and subsequently act as adapter molecules for transmitting the insulin signal to downstream molecules such as the kinases PI3 kinase and Fyn, and the docking proteins Grb2 and SHP2. However, if the IRS proteins are phosphorylated on serine residues by IRS serine/threonine kinases, their signaling capacity is inhibited (see Figure 1).

Inflammatory cytokines, such as IL-6, IL-1β and TNF-α, bind to cell-surface receptors and signal the activation of kinases, including JNK, which phosphorylate IRS-1 on serine residues, inhibiting its activity. Thus, inflammatory cytokines act to inhibit the insulin signal, leading to insulin resistance. However, what is the source of the excess inflammatory cytokines released in individuals with Type 2 diabetes?

Patients with Type 2 diabetes are frequently (although not always) obese, and so investigators began to explore the relationships between obesity and the generation of inflammatory cytokines. Mice fed high-fat diets, or those with a genetic predisposition to obesity, were found to develop chronically elevated levels of inflammatory mediators, such as TNF-α, IL-1β, and IL-6, and increased local concentrations of chemokines, such as CCL2, which draw immune cells, particularly macrophages, into adipose tissue. Investigators therefore advanced the hypothesis that the nutrients themselves might be activating signaling pathways leading to the release of these mediators. But with what receptors are the nutrients interacting?

One clue has come from genetically modified mice. Animals in which the gene encoding the TLR4 receptor has been eliminated are protected from the insulin resistance engendered by eating a high-fat diet. This suggests that the Toll receptors may be recognizing the excess nutrients and initiating the inflammatory

(continued)

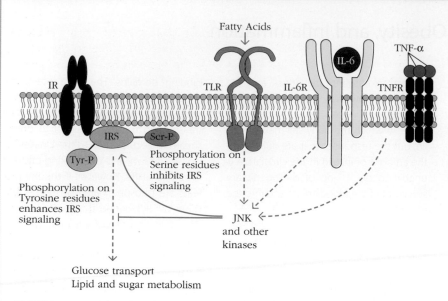

FIGURE 1

Signaling events that link obesity to insulin resistance. Several factors trigger signaling cascades that interfere with *insulin receptor* (IR) signaling cascades. Inflammatory cytokines, including TNF-α and IL-6, are produced by adipocytes themselves, as well as immune cells. These trigger signaling events in multiple cells that activate kinases, including JNK, which inactivate IRS by phosphorylating it at a serine residue. Thus, insulin signaling is impaired. JNK can also be activated (and insulin signaling inactivated) by the interaction between toll like receptors and free fatty acids, which are increased in obesity. *[Based on Gray, S., and J. K. Kim. (2011, October). New insights into insulin resistance in the diabetic heart.* Trends in Endocrinology and Metabolism **22**(10):394–403.]

response. Indeed, TLR4 and TLR2 have both been shown to be responsive to high levels of free fatty acids. A second observation which implicates the TLR4 receptor in the sensing of a nutrient-rich environment is that serum *lipopolysaccharide* (LPS) in mice is increased after feeding. Perhaps intestinal permeability is increased immediately after feeding, which allows the introduction of inflammatory molecules such as LPS directly into the circulation. In an animal fed a normal diet, this inflammation is short-lived. However, it is possible that in animals living in the nutrient-rich environment characteristic of obesity, intestinal permeability remains relatively high long after a meal is completed and thus the animal is chronically exposed to low levels of inflammatory signals.

If Toll receptors are signaling the release of inflammatory cytokines, what is the nature of the cell that carries these Toll receptors? A significant number of studies have demonstrated the presence of macrophages and mast cells in adipose (fat) tissue, suggesting that these two cell types are at least in part responsible for the detection of excess nutrients and the development of inflammatory signals in adipose tissues. The roles of macrophages and mast cells in the development of obesity-associated inflammation are supported by the finding that depletion of cells bearing the CD11c marker (macrophages, dendritic cells, and neutrophils) increases insulin sensitivity. In addition, it is possible that macrophages, newly recruited into expanding adipose tissue, may differentiate into a more potent proinflammatory phenotype than macrophages normally resident in lean adipose tissue. Improved insulin sensitivity in adipose tissue was also found upon depletion of mast cells, NKT cells, and CD8+ T cells, whereas increased levels of CD4+ and T_REG cells are associated with higher levels of insulin sensitivity. T_REG cells associated with adipose tissue secrete large amounts of

Chronic Inflammation and Susceptibility to Other Diseases

As part of the normal healing process, inflammatory cytokines also enhance blood vessel flow and blood vessel formation (angiogenesis), induce proliferation and activation of fibroblasts and immune cells, and regulate death of infected or damaged cells. Together, these events induce tissue remodeling that gives immune cells better access to pathogens and scarring that heals wounds. However, continual stimulation of this healing process has deleterious consequences. Overstimulation of fibroblasts leads to excessive tissue scarring (fibrosis), which can physically and severely impair organ function. Continual stimulation of cell proliferation enhances the probability of mutations

and may contribute to tumor formation or growth. Enhancement of blood vessel formation can also enhance survival of cells in solid tumors.

Although focusing on similarities between acute and chronic inflammation is helpful in understanding some of the consequences described, it is also important not to oversimplify the relationship. For instance, although neutrophil infiltration is a cardinal feature of acute inflammation, monocytes, macrophages, and lymphocytes accumulate during chronic inflammation. Fibroblasts associated with chronic inflammation secrete distinct cytokines and may represent distinct lineages. These differences and others underscore the importance of thoughtfully considering and customizing approaches to ameliorate inflammatory conditions.

IL-10, which acts, in part, to reduce the inflammatory response. Finally, recent work shows that some adipocytes (fat cells) themselves are capable of generating inflammatory cytokines, including TNF-α and IL-6. Adipocytes also express Toll-like receptors that trigger cytokine release. Excess internal lipid also appears to stimulate internal stress responses by adipocytes that enhance cytokine production. So it appears that macrophages, mast cells, NKT cells, CD8 T cells, and adipocytes all play a part in the release of proinflammatory mediators in adipose tissues.

Another stimulus for the secretion of proinflammatory cytokines from adipose tissue-associated cells may be hypoxia, which develops in rapidly expanding adipose tissue. If an animal fed a high-fat diet puts on weight fairly rapidly, the newly developing vasculature (blood supply) is not always able to keep pace and the tissue may be temporarily starved of oxygen. This is known as a state of ischemia, and ischemic tissues recruit macrophages. Once present in ischemic tissues, macrophages are induced to express proinflammatory proteins, as well as angiogenic proteins, which aid in signaling the development of an increased blood supply. Adipocytes may also contribute to this response directly.

Once an animal starts down the road to obesity, its problems can become self-perpetuating. Adipocytes that are full of fat will tend to leak free fatty acids into the circulation, and these in turn induce further inflammation. Indeed, free fatty acids and cellular stress have also been shown to be additional triggers for IRS protein kinases. In addition, high levels of proinflammatory cytokines block the formation of new adipocytes and reduce the secretion of adiponectin, an important regulator of adipocyte production. As the obese adipocyte expands, it approaches its mechanical limit. Cellular responses to mechanical stress, or death, lead to the release of cytokines and additional fatty acids into the circulation. Furthermore, the swollen adipocyte also undergoes endoplasmic reticulum (ER) stress, in which an unusually high number of unfolded proteins accumulates in the ER. This, in turn, provokes the release of inflammatory cytokines and chemokines.

However, the problems of the Type 2 diabetic are not confined just to the adipocytes. In the liver, normally the site of glucose homeostasis, increased levels of inflammatory cytokines also help to induce insulin resistance. Gluconeogenesis (the formation of new glucose) is normally inhibited by insulin, but under inflammatory conditions gluconeogenesis is no longer suppressed and the high blood glucose levels characteristic of the Type 2 diabetic are further increased. In the pancreas, the site of insulin produc-tion, the high blood glucose levels initially induce hyperproliferation of the pancreatic β cells, but eventually apoptosis of the insulin-producing cells occurs, further exacerbating the state of high blood glucose. In the brain of animals fed a high-fat diet, inflammatory pathways are also activated in the hypothalamus, leading to resistance to the effects of both insulin and leptin, a hormone that normally signals satiety, thus setting up a positive feedback loop: the fatter the animal, the more it needs to eat to achieve satiety.

In summary, current research suggests that obese animals exhibit a state of chronic inflammation resulting from the release of nutrient-stimulated inflammatory mediators by adipocytes, themselves, as well as macrophages and mast cells. These inflammatory mediators in turn act on adipocytes and other cells to reduce their sensitivity to insulin, leading to the syndrome we now know as Type 2 diabetes. Inflammation stimulated by nutrient excess is chronic and is associated with a reduced metabolic rate. Some have suggested that it be distinguished from the acute inflammatory state set up by infectious stimuli by using the term *metaflammation*. Clinical research has already demonstrated some success in using anti-inflammatory medications such as salicylate homologs and IL-1 receptor antagonists in the treatment of insulin resistance.

SUMMARY

- Hypersensitivities are immune disorders caused by an inappropriate response to antigens that are not pathogens.

- Hypersensitivities are classically divided into four categories (types I–IV) that differ by the immune molecules and cells that cause them and the way they induce damage.

- The hygiene hypothesis has been advanced to explain increases in asthma and allergy incidence in the developed world and proposes that early exposure to pathogens inhibits allergy by balancing the representation and polarization of regulatory and helper T cell subsets.

- Allergy is a type I hypersensitivity reaction that is mediated by IgE antibodies.

- IgE antibodies bind to antigen via their variable regions and to one of two types of Fc receptors via their constant regions.

- Mast cells, basophils, and to a lesser extent eosinophils express FcεRI, and are the main mediators of allergy symptoms.

- Cross-linking of FcεRI receptors by allergen/IgE complexes initiates multiple signaling cascades that resemble those initiated by antigen receptors.

- Mast cells, basophils, and eosinophils that are stimulated by Fcε cross-linking release their granular contents (including histamine, proteases) in a process called degranulation. They also generate and secrete inflammatory cytokines and lipid inflammatory molecules (leukotrienes and prostaglandins).

- Degranulation releases both nonprotein (histamine) and protein (protease) molecules that induce rapid inflammatory responses (e.g., vasodilation and edema, smooth muscle constriction).

- Mast cell activity can also be down-regulated by inhibitory signals, including phosphatase activity, inhibitory FcR signaling, and ubiquitinylation and degradation of signaling molecules.

- Allergy symptoms vary depending on where the IgE response occurs and whether it is local or systemic. Asthma, atopic dermatitis, and food allergies are examples of local allergic responses. Anaphylaxis refers to a systemic IgE response and can be caused by systemic (e.g., intravenous) introduction of the same allergen that induces local responses.

- Individuals predisposed to allergic responses are referred to as atopic. Both genetic and environmental factors contribute to allergy susceptibility.

- Why some antigens induce allergy and others do not is still not fully understood. Some antigens that cause allergy appear to have intrinsic protease activity.

- Skin tests are an effective way to diagnose allergies.

- Allergies can be treated by hyposensitization, which increases IgG responses to allergens, in turn inhibiting IgE activity. They are also treated with pharmacological inhibitors of inflammation, including antihistamines, leukotriene inhibitors, and corticosteroids.

- Type II hypersensitivities are caused by IgG and IgM antibodies binding to an antigen on red blood cells and inducing cell destruction by recruiting complement or by ADCC.

- Transfusion reactions are caused by antibodies that bind to A, B, or H carbohydrate antigens, which are expressed on the surface of red blood cells. Individuals with different blood types (A, B, or O) express different carbohydrate antigens. They are tolerant to their own antigens, but generate antibodies against the antigen (A or B) that they do not express. All individuals express antigen H, so no antibodies are generated to this carbohydrate.

- Hemolytic disease of the newborn is caused by maternal antibody reaction to the Rh antigen, which can happen if mother is Rh⁻ and father is Rh⁺.

- Drug-induced hemolytic anemia is caused by antibody responses to red blood cells that have bound drug molecules or metabolites.

- Immune complexes of antibody and antigen can cause type III hypersensitivities when they cannot be cleared by phagocytes. This may be due to peculiarities of the antigen itself, or disorders in phagocytic machinery.

- Uncleared immune complexes can induce degranulation of mast cells and inflammation, and can be deposited in tissues and capillary beds where they induce more innate immune activity, blood vessel inflammation (vasculitis), and tissue damage.

- Arthus reactions are examples of immune complex (type III) hypersensitivity reactions and can be induced by insect bites, as well as inhalation of fungal or animal protein. They are characterized by local and sometimes severe inflammation of blood vessels.

- Delayed-type hypersensitivity (type IV hypersensitivity) is cell mediated, not antibody mediated. Examples include contact dermatitis caused by poison ivy, as well as the tuberculin reaction. DTH responses are responsible for granulomas associated with tuberculosis.

- DTH requires T cells to be sensitized to antigen. Subsequent reexposure to antigen results in cytokine generation, inflammation, and the recruitment of macrophages, which produce DTH symptoms 2 to 4 days after reexposure.

- T_H1 cells are classically associated with DTH, but other helper cell subsets have also been implicated recently.

- Chronic inflammation is a pathological condition characterized by persistent, increased expression of inflammatory cytokines.

- Chronic inflammation has infectious and noninfectious origins. It can be caused by an infection that is not fully resolved as well as by chronic tissue damage brought about by wounds, tumors, autoimmune disease, or organ disease. Obesity is now recognized as one of the most common causes of chronic inflammation.

- Obesity can result in chronic inflammation in part because visceral fat cells (adipocytes) can be stimulated to produce inflammatory cytokines directly.

- Chronic inflammation, regardless of cause, increases the susceptibility of individuals to several systemic diseases, including Type 2 diabetes.

- Inflammatory cytokines associated with chronic inflammation contribute to insulin resistance (Type 2 diabetes) by interfering with the activity of enzymes (e.g., JNK) downstream of the insulin receptor.

- Cytokines produced during chronic inflammation can also induce tissue scarring (leading to organ dysfunction) as well as cell proliferation and angiogenesis (which may contribute to tumor development).

REFERENCES

Abramson, J., and I. Pecht. (2007). Regulation of the mast cell response to the type 1 Fc epsilon receptor. *Immunological Reviews* **217**: 231–254.

Acharya, M., et al. (2010). CD23/FcepsilonRII: molecular multi-tasking. *Clinical and Experimental Immunology* **162**:12–23.

Adam, J., W. J. Pichler, and D. Yerly. (2011). Delayed drug hypersensitivity: Models of T-cell stimulation. *British Journal of Clinical Pharmacology* **71**:701–707.

Boura-Halfon, S., and Y. Zick. (2009). Phosphorylation of IRS proteins, insulin action, and insulin resistance. *American Journal of Physiology—Endocrinology and Metabolism* **296**:E581–591.

Cavani, A., and A. De Luca. Allergic contact dermatitis: Novel mechanisms and therapeutic perspectives. *Current Drug Metabolism* **11**:228–233.

Donath, M. Y., and S. E. Shoelson. (2011). Type 2 diabetes as an inflammatory disease. *Nature Reviews Immunology* **11**:98–107.

Galli, S. J., S. Nakae, and M. Tsai. (2005). Mast cells in the development of adaptive immune responses. *Nature Immunology* **6**:135–142.

Gerull, R., M. Nelle, and T. Schaible. (2011). Toxic epidermal necrolysis and Stevens-Johnson syndrome: A review. *Critical Care Medicine* **39**:1521–1532.

Gilfillan, A. M., and J. Rivera. (2009). The tyrosine kinase network regulating mast cell activation. *Immunological Reviews* **228**:149–169.

Gladman, A. C. (2006). Toxicodendron dermatitis: Poison ivy, oak, and sumac. *Wilderness and Environmental Medicine* **17**:120–128.

Gregor, M. F., and G. S. Hotamisligil. (2011). Inflammatory mechanisms in obesity. *Annual Review of Immunology* **29**:415–445.

Holgate, S. T., R. Djukanovic, T. Casale, T., and J. Bousquet. (2005). Anti-immunoglobulin E treatment with omalizumab in allergic diseases: An update on anti-inflammatory activity and clinical efficacy. *Clinical & Experimental Allergy* **35**:408–416.

Hotamisligil, G. S. (2010). Endoplasmic reticulum stress and the inflammatory basis of metabolic disease. *Cell* **140**:900–917.

Karasuyama, H., K. Mukai, K. Obata, Y. Tsujimura, and T. Wada. (2011). Nonredundant roles of basophils in immunity. *Annual Review of Immunology* **29**:45–69.

Madore, A. M., and C. Laprise. (2010). Immunological and genetic aspects of asthma and allergy. *Journal of Asthma and Allergy* **3**:107–121.

Minai-Fleminger, Y., and F. Levi-Schaffer. (2009). Mast cells and eosinophils: The two key effector cells in allergic inflammation. *Inflammation Research* **58**:631–638.

Mullane, K. (2011). Asthma translational medicine: Report card. *Biochemical Pharmacology* **82**:567–585.

Nizet, V., and M. E. Rothenberg. (2008). Mitochondrial missile defense. *Nature Medicine* **14**:910–912.

Roediger, B., and W. Weninger. (2011). How nickel turns on innate immune cells. *Immunology and Cell Biology* **89**:1–2.

Rosenwasser, L. J. (2011). Mechanisms of IgE Inflammation. *Current Allergy and Asthma Reports* **11**:178–183.

Rothenberg, M. E., and S. P. Hogan. (2006). The eosinophil. *Annual Review of Immunology* **24**:147–174.

Sicherer, S. H., and H. A. Sampson. (2009). Food allergy: Recent advances in pathophysiology and treatment. *Annual Review of Medicine* **60**:261–277.

Vercelli, D. (2008). Discovering susceptibility genes for asthma and allergy. *Nature Reviews Immunology* **8**:169–182.

Wan, Y. I., et al. (2011). A genome-wide association study to identify genetic determinants of atopy in subjects from the United Kingdom. *Journal of Allergy and Clinical Immunology* **127**:223–231, 231 e1–3.

Yamamoto, F. (2004). Review: ABO blood group system—ABH oligosaccharide antigens, anti-A and anti-B, A and B glycosyl-transferases, and ABO genes. *Immunohematology* **20**:3–22.

Useful Web Sites

https//chriskresser.com/how-inflammation-makes-you-fat-and-diabetic-and-vice-versa This is an interesting and credible series of commentaries by Chris Kresser who did not go to medical school, but graduated from an alternative medicine program and is open about his interest in examining the assumptions that underlie medical practices. His online articles on obesity and inflammation are informed and clearly written.

www.youtube.com/watch?v=IGDXNHMwcVs An above-average and roughly accurate YouTube animation on the cells and molecules involved in type I hypersensitivity (allergy). www.youtube.com/watch?v=y3bOgdvV-_M&feature=related is another YouTube animation about type I hypersensitivity that is clear, if overly simplified.

faculty.ccbcmd.edu/courses/bio141/lecguide/unit5/hypersensitivity/type3/typeiii.html An animation from Dr. Gary Kaiser's Microbiology website (Community College of Baltimore County) that helps one visualize type III (immune-complex mediated) hypersensitivity and how it induces vasculitis.

www.youtube.com/watch?v=wfmsFfC8WIM A YouTube mini-lecture on delayed-type hypersensitivity (focusing on the immunology behind poison ivy). Includes some good pictures of poison ivy lesions.

1. You have in your possession a number of mouse strains, each of which lacks a specific gene (gene knockout animals). How might the type I hypersensitivity response of each knockout strain (a–e) differ from a wild-type mouse? How might the type II hypersensitivity response differ?

   Explain your answer.

   a. Mouse is unable to generate a ε heavy chain.
   b. Mouse is unable to generate a high-affinity FcεRI receptor.
   c. Mouse is unable to generate a low-affinity FcεRII receptor.
   d. Mouse is deficient in the ability to generate the complement attack complex.
   e. Mouse is unable to express CD21.

2. What is the difference between primary and secondary pharmacological mediators in the type I hypersensitivity response? Name two of each.

3. How does histamine suppress its own release?

4. Immunotherapy of type I hypersensitivity responses is aimed at raising the levels of IgG antibodies specific for allergens. Describe one mechanism by which allergen-specific IgG can dampen down the IgE response to an allergen.

5. A mother has an Rh$^+$ blood type and the father has an Rh$^-$ blood type. Under these circumstances, the family pediatrician is not worried about the possibility of a type II hypersensitivity reaction. However, if the converse is true, and the mother is Rh$^-$ and the father Rh$^+$, the pediatrician does worry and asks the obstetrician to inject the mother with antibodies toward the end of her first pregnancy. Explain his reasoning in both cases.

6. A mother has an Rh$^-$ and the father an Rh$^+$ blood type. The first baby born to the parents was Rh$^+$. However, the parents elect for the mother not to receive Rhogam. Are all future babies of this couple at risk for type II hypersensitivity reactions? Why?/Why not?

7. Define type III hypersensitivity, illustrating the initiating cells and molecules, the cells and molecules that bring about the pathological effects and indicating two triggers for this type of response.

8. Indicate which type(s) of hypersensitivity reaction (I–IV) apply to the following characteristics. Each characteristic can apply to one, or more than one, type.

   a. Is an important defense against intracellular pathogens.
   b. Can be induced by penicillin.
   c. Involves histamine as an important mediator.
   d. Can be induced by poison oak in sensitive individuals.
   e. Can lead to symptoms of asthma.
   f. Occurs as a result of mismatched blood transfusion.
   g. Systemic form of reaction is treated with epinephrine.
   h. Can be induced by pollens and certain foods in sensitive individuals.
   i. May involve cell destruction by antibody-dependent cell-mediated cytotoxicity.
   j. One form of clinical manifestation is prevented by Rhogam.
   k. Localized form characterized by wheal-and-flare reaction.

9. As described in the text, a small group of obese individuals (6% or so) do not suffer from the chronic inflammatory state characteristic of most forms of obesity.

   a. These individuals do not develop Type 2 diabetes. Why not? Provide a specific, molecular answer.
   b. Some have suggested that these individuals express genetic polymorphisms that make them less susceptible to obesity-generated inflammation. Describe one possible genetic polymorphism that could uncouple obesity from inflammation.

16

Tolerance, Autoimmunity, and Transplantation

Early in the twentieth century, Paul Ehrlich realized that the immune system could go awry. Instead of reacting only against foreign antigens, it could focus its attack on the host. This condition, which he termed *horror autotoxicus*, can result in a clinical syndrome generically referred to as **autoimmunity**. This inappropriate response of the immune system, directing humoral and/or T-cell-mediated immune activity against self components, is the cause of **autoimmune diseases** such as rheumatoid arthritis (RA), multiple sclerosis (MS), systemic lupus erythematosus (SLE, or lupus) and certain types of diabetes. Autoimmune reactions can cause serious damage to cells and organs, sometimes with fatal consequences. In some cases the damage to self cells or organs is caused by antibodies; in other cases, T cells are the culprit.

Simply stated, autoimmunity results from some failure of the host's immune system to distinguish self from nonself, causing destruction of self cells and organs. Although on the rise, autoimmunity is still rare, suggesting that mechanisms to protect an individual from this sort of anti-self immune attack must exist, and they do. This process and the mechanisms that control it are collectively termed **tolerance,** or **self-tolerance.** Establishing self-tolerance is complicated, involving both the elimination of immune cells that can react against self-antigens and active inhibition of immune responses against self proteins. Our understanding of the mechanisms that control self-tolerance have really blossomed in the last decade, giving rise to new ways of understanding and treating autoimmune disease.

When self-tolerance processes are working correctly, host tissues should remain undisturbed by the immune system and only foreign invaders should be attacked. As we know from Chapter 9 and will address further below, the mechanisms that maintain self-tolerance do so by establishing what is "us," making the "them" clearer. These elaborate mechanisms that maintain self-tolerance also cause rejection of any transplanted tissues or cells that carry new proteins, as occurs whenever the donor is not genetically identical to the recipient. *Transplantation* refers to the act of transferring cells, tissues, or organs

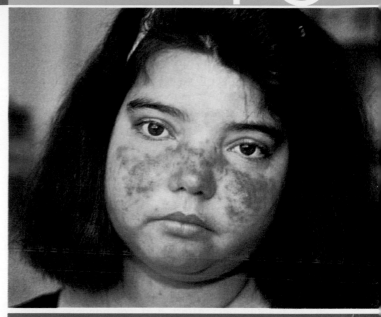

Characteristic butterfly rash in an SLE patient.
[From L. Steinman, 1993, Scientific American 269(3):80.]

- Establishment and Maintenance of Tolerance
- Autoimmunity
- Transplantation Immunology

from one site to another—typically from one individual to another. Transfers between two sites on the same individual (e.g., skin) or between identical twins, although not free of complication, are more likely to survive. Many diseases can be cured by implantation of a healthy organ, tissue, or cells (a graft). The development of surgical techniques that allow this has removed one barrier to successful transplantation, but others remain. A worldwide shortage of organs for transplantation leaves tens of thousands of individuals waiting for a transplant, sometimes for many years. And yet, the most formidable barrier to making transplantation a routine medical treatment is the immune system.

In this chapter, we first describe our current understanding of the general mechanisms that establish and maintain tolerance to self antigens. When these mechanisms fail, autoimmunity, the second major topic of this chapter, can result. Several common human autoimmune diseases resulting from failures of these mechanisms are described. These can be divided into two broad categories: organ specific and multiorgan (systemic) autoimmune diseases. Such diseases affect 5% to 8% of the human population; they are often chronic and debilitating, necessitating prolonged medical intervention. Several experimental animal models used to study autoimmunity, as well as therapies for treating autoimmune disease, are also discussed. In the final part of this chapter, we turn to the topic of transplantation, or situations in which self-tolerance can work against us. We begin by discussing the immunologic processes governing graft rejection, followed by current therapeutic modalities for suppressing these responses. We end with some of the most commonly transplanted tissues and the potential future role of xenotransplantation (cross-species grafts) in clinical medicine.

Establishment and Maintenance of Tolerance

The term *tolerance* applies to the many layers of protection imposed by the immune system to prevent the reaction of its cells and antibodies against host components. In other words, individuals should not typically respond aggressively against their own antigens, although they will respond to pathogens or even cells from another individual of the same species. Until fairly recently, this was thought to be mediated by the elimination of cells that can react against self antigens, yielding a state of unresponsiveness to self. Contemporary studies of tolerance provide evidence for a much more active role of immune cells in the selective inhibition of responses to self antigens. Rather than ignore self proteins, the immune system protects them. For instance, the discovery and characterization of regulatory T cells, which in fact recognize self proteins, have revolutionized the field of tolerance and autoimmunity, not to mention transplanation.

In the first step of this process, a phenomenon termed **central tolerance** deletes T- or B-cell clones before the cells are allowed to mature if they possess receptors that recognize self antigens with high affinity (see Chapter 9). Central tolerance occurs in the primary lymphoid organs: the bone marrow for B cells and the thymus for T cells (Figure 16-1a). Because central tolerance is not perfect and some self-reactive lymphocytes find their way into the periphery and secondary lymphoid tissues, additional safeguards limit their activity. These backup precautions include **peripheral tolerance,** which renders some self-reactive lymphocytes in secondary lymphoid tissues inactive and generates others that actively inhibit immune responses against self (Figure 16-1b). The possibility of damage from self-reactive lymphocytes is further limited by the life span of activated lymphocytes, which is regulated by programs that induce cell death (apoptosis) following receipt of specific signals.

The mechanisms mediating peripheral tolerance vary. Under normal circumstances, encounter of mature lymphocytes with an antigen leads to stimulation of the immune response. However, presenting the antigen in some alternative form, time, or location may instead lead to tolerance. Antigens that induce tolerance are called **tolerogens** rather than *immunogens*. Here, context is important; the same chemical compound can be both an immunogen and a tolerogen, depending on how and where it is presented to the immune system. For instance, an antigen presented to T cells without appropriate costimulation results in a form of tolerance known as **anergy** (unresponsiveness to antigenic stimulus), whereas the same antigen presented with costimulatory molecules can become a potent immunogen. When some antigens are introduced orally, tolerance can be the result, whereas the same antigen given as an intradermal or subcutaneous injection can be immunogenic. In other instances, mucosally administered antigens provide protective immunity, such as in the case of Sabin's oral polio vaccine. However, there is one general truth: tolerogens are antigen specific. The inactivation of an immune response does not result in general immune suppression, but rather is specific for the tolerogenic antigen.

In adults, most encounters with foreign antigen lead to an immune response aimed at eradication. This is not true in the fetus, where, due to the immature state of the immune system, exposure to antigens frequently results in tolerance. Other than fetal exposure, factors that promote tolerance rather than stimulation of the immune system by a given antigen include the following:

- High doses of antigen

- Long-term persistence of antigen in the host

- Intravenous or oral introduction

- Absence of adjuvants (compounds that enhance the immune response to antigen)

- Low levels of costimulation

- Presentation of antigen by immature or unactivated antigen-presenting cells (APCs)

In the 1960s, researchers believed that all self-reactive lymphocytes were eliminated during their development in the bone marrow and thymus. The conventional wisdom suggested that failure to eliminate these lymphocytes led to autoimmune consequences. More recent experimental evidence has countered that belief. Healthy individuals have been shown to possess mature, recirculating, self-reactive lymphocytes. Since the presence of these self-reactive lymphocytes in the periphery does not predict or inevitably result in autoimmune reactions, their activity must be regulated in healthy individuals through other mechanisms. Mechanisms that maintain tolerance include the induction of cell death or cell anergy in lymphocytes and limitations on the activity of self-reactive cells by regulatory processes (see Figure 16-1b).

Cell death plays an important role in establishing and maintaining both central and peripheral tolerance. This is evidenced by the development of systemic autoimmune diseases in mice with naturally occurring mutations of either the death receptor, Fas, or *Fas* ligand (FasL). As discussed in Chapter 9,

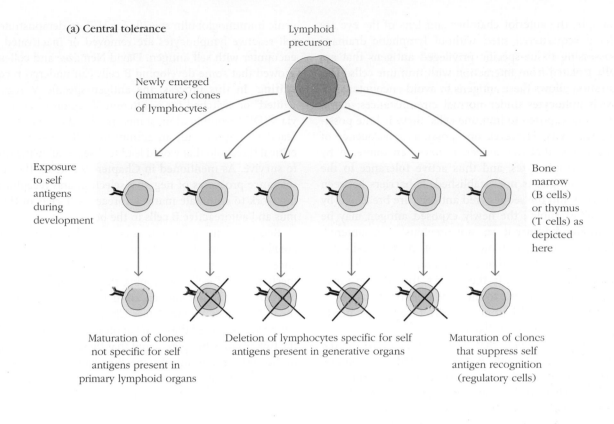

(a) Central tolerance

Lymphoid precursor

Newly emerged (immature) clones of lymphocytes

Exposure to self antigens during development

Bone marrow (B cells) or thymus (T cells) as depicted here

Maturation of clones not specific for self antigens present in primary lymphoid organs

Deletion of lymphocytes specific for self antigens present in generative organs

Maturation of clones that suppress self antigen recognition (regulatory cells)

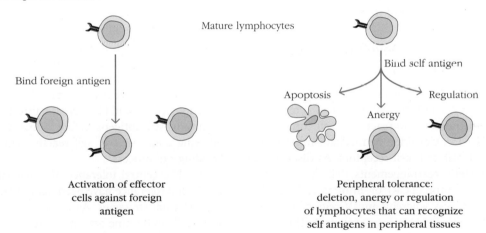

(b) Peripheral tolerance

Mature lymphocytes

Bind foreign antigen

Bind self antigen

Apoptosis

Anergy

Regulation

Activation of effector cells against foreign antigen

Peripheral tolerance: deletion, anergy or regulation of lymphocytes that can recognize self antigens in peripheral tissues

FIGURE 16-1 Central and peripheral tolerance. (a) Central tolerance is established by deletion of lymphocytes in primary lymphoid organs (thymus for T cells and bone marrow for B cells) if they possess receptors that can react with self antigens or by the emergence of regulatory T cells that can inhibit self-reactive cells. (b) Peripheral tolerance involves deleting, rendering anergic or actively suppressing escaped lymphocytes that possess receptors that react with self antigens. This process occurs in secondary lymphoid organs.

activated T cells express increased levels of Fas and FasL. In both B and T cells, engagement of Fas by FasL induces a rapid apoptotic death known as *activation-induced cell death* (AICD). Mice that carry inactivating mutations in Fas (*lpr/lpr*) or FasL (*gld/gld*) are not able to engage the AICD pathway and develop autoimmune disease early in life (see below).

Antigen Sequestration Is One Means to Protect Self Antigens from Attack

In addition to the various mechanisms of central and peripheral tolerance, an effective means to avoid self-reactivity is sequestration or compartment of antigens.

For example, the anterior chamber and lens of the eye are considered sequestered sites, without lymphatic drainage and possessing tissue-specific privileged antigens that are normally isolated from interaction with immune cells. This sequestration allows these antigens to avoid encounter with reactive lymphocytes under normal circumstances; if the antigen is not exposed to immune cells, there is little possibility of reactivity. However, one possible consequence of sequestration is that the antigen is never encountered by developing lymphocytes, and thus active tolerance to the sequestered antigen is not established. If barriers between immune cells and the sequestered antigens are breached (by trauma, for example), the newly exposed antigen may be seen as foreign because it was not previously encountered. Trauma to one eye that allows entry of immune cells can result in inflammation in that eye, leading to tissue destruction and impaired vision. In these cases, the other eye may also become inflamed due to the sudden entry of clones of these recently activated immune cells recognizing newly discovered tissue-specific antigens.

This is not to suggest that a lack of exposure to the immune system is the only factor that mediates immune privilege. A locally immunosuppressive microenvironment in tissues conventionally considered immune privileged, such as the eye and central nervous system (CNS), in addition to other active mechanisms, is believed to bias the immune response toward tolerance in these locations. We discuss further some of these pathways in the following sections, as well as in Chapters 9 and 10.

Central Tolerance Limits Development of Autoreactive T and B Cells

One mechanism strongly influencing central tolerance is the deletion during early stages of maturation of lymphocyte clones that have the potential to react with self components later (see Figure 16-1a). Consider the mechanisms that generate diversity in T-cell or B-cell receptors. As discussed in Chapter 7, the genetic rearrangements that give rise to a functional *T-cell receptor* (TCR) or *immunoglobulin* (Ig) occur through a process whereby any V-region gene segment can associate with any D or J gene segment. This means that the generation of variable regions that react with self antigens is almost inevitable. If this were allowed to occur frequently, such TCR or Ig receptors could produce mature functional T or B cells that recognize self antigens, and autoimmune disease would ensue.

Although our understanding of the precise molecular mechanisms mediating central tolerance in T and B cells is not complete, we do know that these cells undergo a developmentally regulated event called **negative selection**. This results in the induction of death in some, but not all, cells that carry potentially autoreactive TCR or Ig receptors. A classic experiment by C. C. Goodnow and colleagues, described in Chapter 10, mated transgenic mice expressing *hen egg white lysozyme* (HEL) with mice expressing a trans-

genic immunoglobulin specific for HEL to demonstrate that self-reactive lymphocytes are removed or inactivated after encounter with self antigen. David Nemazee and colleagues showed that some developing B cells can undergo **receptor editing**. In this process, the antigen-specific V region is "edited" or switched for a different V-region gene segment via V(D)J recombination, sometimes producing a less autoreactive receptor with an affinity for self antigens below a critical threshold that would lead to disease, allowing the cell to survive. As mentioned in Chapters 9 and 10, the central tolerance processes of negative selection and receptor editing work to eliminate many autoreactive T cells in the thymus and autoreactive B cells in the bone marrow.

Receptor editing, as well as clonal deletion or apoptosis, are now recognized as mechanisms that lead to central tolerance in developing B cells (see Chapter 10). By similar mechanisms, T cells developing in the thymus that have a high affinity for self antigen are deleted, primarily through the induction of apoptosis (see Chapter 9). More recently, it has been discovered that some of these self-reactive T cells in the thymus may be spared, and that these cells may function in the periphery as antigen-specific regulatory cells working to dampen immune responses to the antigens that they recognize (see below and Chapter 9).

Peripheral Tolerance Regulates Autoreactive Cells in the Circulation

Numerous studies have shown that central tolerance is not a foolproof process; all possible self-reactive lymphocytes are not deleted. In fact, it is now clear that lymphocytes with specificity for self antigens are not uncommon in the periphery. Two factors contribute to this: (1) not all self antigens are expressed in the central lymphoid organs where negative selection occurs, and (2) there is a threshold requirement for affinity to self antigens before clonal deletion is triggered, allowing some weakly self-reactive clones to survive the weeding-out process.

Just like central tolerance, the mechanisms that control peripheral tolerance have been demonstrated by a variety of experimental strategies. As we saw in Chapter 11, in order for T cells to become activated, the TCR must bind antigen presented by self-MHC (**major *h*istocompatibility *c*omplex molecules**) (signal 1), while at the same time the T cell must undergo costimulatory engagement (signal 2). Early experiments by Marc Jenkins and colleagues showed that when $CD4^+$ T-cell clones are stimulated in vitro through the TCR alone, without costimulation, they become anergic. Subsequent data showed that the interaction between CD28 on the T cell and CD80/86 (B7) on the APC provided the costimulatory signal required for T-cell activation. This led to a careful examination of costimulation, revealing the existence of other molecules that could bind to CD80/86 and the discovery of a related molecule, called CTLA-4. This molecule *inhibits* rather than stimulates T-cell activation upon binding CD80/86. We now appreciate that many such molecules deliver supplementary signals during T-cell activation, and the group of molecules that regulate T-cell behavior are now often referred to as *immunomodulatory*, to cover both costimulatory and inhibitory behavior. CTLA-4 expression is induced only after T cells are activated, providing a mechanism to dampen T-cell activity and regulate the immune response. Mice lacking CTLA-4 display massive proliferation of lymphocytes and widespread autoimmune disease, suggesting an essential role for this molecule in maintaining peripheral tolerance.

Peripheral tolerance in B cells appears to follow a similar set of rules. For instance, experiments with transgenic mice have demonstrated that when mature B cells encounter most soluble antigens in the absence of T-cell help, they become anergic and never migrate to germinal centers. In this way, maintenance of T-cell tolerance to self antigens enforces B-cell tolerance to the same antigens.

In T cells, a third mechanism for maintaining tolerance, in addition to T-cell anergy and apoptosis, is through the activity of **regulatory T cells** (T_{REG} cells). Acting in secondary lymphoid tissues and at sites of inflammation, T_{REG} cells recognize specific self antigens, and sometimes foreign antigens, via TCR interactions. However, they *down-regulate* immune processes when they engage with these antigens in the periphery. These cells can be generated both naturally, in the thymus (**nT_{REG} cells**, Figure 16-2), and after induction by antigen in the periphery (**iT_{REG} cells**; see below). In fact, many of the circulating T cells with specificity for self antigens may be such regulatory cells. Some scientists postulate a division of labor, with nT_{REG} cells specializing in regulating responses against self antigen to inhibit autoimmune disease and iT_{REG} cells controlling reactions against benign foreign antigens at mucosal surfaces, where the immune system comes in constant contact with the outside world (e.g., gut commensals or respiratory allergens). In a very recent study, specifically blocking iT_{REG} but not nT_{REG} cell development at the maternal-fetal interface correlated with a drop in the number of regulatory cells specific for paternal antigens and an increase in fetal death, suggesting that these cells may also be involved in regulating the immune response to fetal alloantigens and may influence pregnancy outcomes.

The existence of immune inhibitory T cells was first proposed in the early 1970s by scientists who identified this activity within the CD8$^+$ subset and called these cells CD8$^+$ suppressor T cells. However, a lack of available reagents and expertise for isolating, propagating, and characterizing these cells meant that many decades passed without sufficient supporting evidence of their existence. During this time, the idea of naturally occurring immunosuppressive cells fell out of favor, only to be resurrected 25 years later by S. Sakaguchi and colleagues with the characterization of a subset of CD4$^+$ T cells with immunosuppressive abilities. Regulatory cells are currently the focus of much research, where subsets of CD8$^+$ and CD4$^+$ T cells, as well as certain types of APCs, have been found to possess these immune-dampening capabilities.

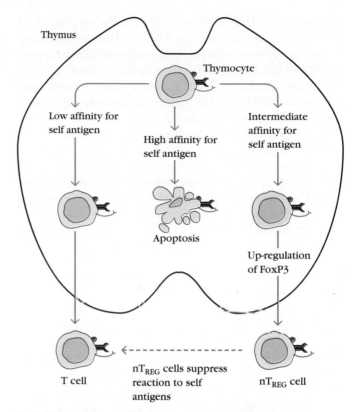

FIGURE 16-2 T regulatory (T_{REG}) cells generated from thymocytes during negative selection in the thymus can inhibit self-reactive T cells in the periphery. During T cell development in the thymus, thymocytes with high affinity for self antigens undergo apoptosis, while those with low affinity are positively selected and released. Thymocytes with intermediate affinity for antigens encountered in the thymus up-regulate the transcription factor FoxP3 and become natural nT_{REG} cells, which are released into the periphery and serve to keep self-reactive T-cell responses in check. [M. Kronenberg and A. Rudensky. 2005. *Regulation of immunity by self-reactive T cells.* Nature. **435**:598–604.]

Regulatory CD4$^+$ T Cells

As discussed in Chapter 9, T_{REG} cells were most recently characterized as a unique subset of CD4$^+$ T cells that express high levels of the IL-2R α chain (CD25), the low-affinity receptor for this cytokine. Naturally occurring nT_{REG} cells arise from a subset of T cells expressing receptors with intermediate affinity for self antigens in the thymus (see Figure 16-2). Certain of these cells up-regulate the transcription factor FoxP3 and then develop into cells that migrate out of the thymus and are capable of suppressing reactions to self antigens.

Evidence that CD4$^+$ T_{REG} cells can control the immune response to self antigens has now been demonstrated in many experimental settings. In experiments in *nonobese diabetic* (NOD) mice and *BioBreeding* (BB) rats, two strains prone to the development of autoimmune-based diabetes, the onset of diabetes was delayed when these animals were injected with normal CD4$^+$ T cells from histocompatible donors. Further characterization of the CD4$^+$ T cells of nondisease-prone mice revealed that a subset expressing high levels of CD25 was responsible for the suppression of diabetes. This population was further characterized using transgenic reporter mice expressing a *green fluorescent protein* (GFP) fused with the transcription factor FoxP3 (FoxP3-GFP mice). The GFP$^+$ T cells but not the GFP$^-$ T cells from these mice could be used to transfer the immune-suppressive activity, identifying this transcription factor as a major regulator controlling the development of these cells.

CD4$^+$ T$_{REG}$ cells have also been found to suppress responses to some *nonself* antigens. For example, these cells may control allergic responses against innocuous environmental substances and/or responses to the commensal microbes that make up the normal gut flora. In mice experimentally manipulated to lack CD4$^+$ T$_{REG}$ cells (approximately 5–10% of their peripheral CD4$^+$ T-cell population), inflammatory bowel disease is common. In strains of mice that are genetically resistant to the induction of the autoimmune disease *experimental autoimmune encephalomyelitis* (EAE; a murine model of MS), depletion of this CD4$^+$ T cell subset renders the mice susceptible to disease, suggesting that these regulatory cells play a role in suppressing autoimmunity.

The pathway by which some CD4$^+$ T cells develop regulatory functions in the thymus is still unclear, despite intense investigation. During development, engagement of the TCR with self antigens can result in either death of the developing thymocyte by negative selection or generation of this regulatory capacity (nT$_{REG}$s). The difference between these fates may be due to other intercellular interactions (i.e., the binding of CD28 with CD80/86 or CD40 with CD40L) or the presence of certain cytokines. The importance of FoxP3, which appears to be both essential and sufficient for the induction of immunosuppressive function, can be seen in humans inheriting a mutated form of this X-linked gene, which causes a multiorgan autoimmune disease (see Autoimmunity below and Chapter 18).

Naïve T cells that have escaped to the periphery also can be induced to express FoxP3 and acquire regulatory function (iT$_{REG}$ cells). Factors that favor the development of iT$_{REG}$ cells include the presence of certain cytokines during antigenic stimulation, chronic low-dose antigen exposure, and lack of costimulation or the presentation of antigen by immature *dendritic cells* (DCs). For example, there is significant evidence that iT$_{REG}$ cells and other regulators of immunity are present in the *gut-associated lymphoid tissue* (GALT). These cells are continuously exposed to gut microbes and food-borne antigens, which themselves may play a significant role in regulating immunity (see Clinical Focus Box 16-1). The microenvironment of the GALT is rich in lymphoid tissue-derived *transforming growth factor beta* (TGF-β), which is believed to encourage the development of iT$_{REG}$ cells. There is also some evidence that T cells can switch types, meaning that in certain circumstances T$_{REG}$ cells (immunosuppressive) can acquire effector function (immune activating), and vice versa. The cues that determine this switch are not known, although the amount of IL-2 and other key cytokines in the microenvironment may be a contributing factor.

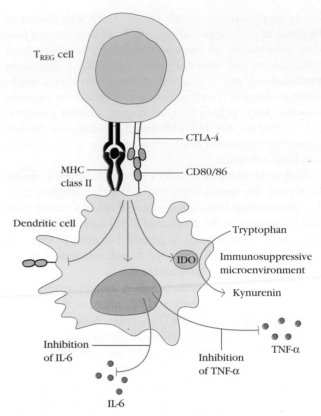

FIGURE 16-3 CTLA-4 mediated inhibition of APCs by T$_{REG}$ cells. One of the proposed mechanisms used by T$_{REG}$ cells to inhibit APCs involves signaling through the CD80/CD86 (B7) receptor. In the APC, this engagement results in decreased expression of CD80/86, activation of indoleamine-2,3-dioxygenase (IDO; an enzyme that converts tryptophan to kynurenin, creating an immunoinhibitory microenvironment), and changes in transcription leading to decreased expression of IL-6 and TNF-α. *[Adapted from Wing K, Sakaguchi S. 2010. Regulatory T cells exert checks and balances on self tolerance and autoimmunity.* Nature Immunology *11:7–13]*

In studying the mechanisms by which T$_{REG}$ cells inhibit immune responses, both contact-dependent and contact-independent processes have been observed. CD4$^+$ T$_{REG}$ cells have been shown to kill APCs or effector T cells directly, by means of granzyme and perforin. T$_{REG}$ cells may also modulate the function of other cells responding to antigen by surface receptor engagement. One prime example of this is the expression of CTLA-4. T$_{REG}$ cells express high levels of this immune inhibitory receptor. As shown in Figure 16-3, interaction of CTLA-4 on T$_{REG}$ cells with CD80/86 on an APC can lead to inhibition of APC function, including reduced expression of costimulatory molecules and proinflammatory cytokines, such as IL-6 and *tumor necrosis factor-α* (TNF-α). At the same time, these targeted APCs begin to express soluble factors that inhibit local immune cells, including *indoleamine-2,3-dioxygenase* (IDO). T$_{REG}$ cells themselves also secrete immune inhibitory cytokines, such as IL-10, TGF-β, and IL-35, suppressing the activity of other nearby T cells and APCs. Finally, because T$_{REG}$ cells express only the low-affinity IL-2R (CD25) but not the β or γ subunits, which are required for signal

Box 16-1

 It Takes Guts to Be Tolerant

In the past decade, a significant appreciation has developed for the gastrointestinal tract and the host's microflora in regulating the immune response. Specifically, the commensal organisms that comprise the gut microbiota appear to work actively at inducing tolerance to themselves. Maybe even more important, the gut and the organisms that reside there seem to play key roles in maintaining a level of systemic homeostasis that allows for the development of tolerance to self and that sets the stage for effective identification and elimination of pathogens (see also Clinical Focus Box 1-3).

Formerly thought to be "ignored" by immune cells, host microbiota are now known to participate in a two-way communication with the immune system. This cross-talk results in advantages for both the host and the microbe. For the microbe, this interaction drives immune tolerance to the bugs, allowing them to continue to thrive in their home. For the host, there appear to be multiple advantages to immune health, depending somewhat on the composition of the microbe(s) in question. For instance, germ-free mice have been found to harbor defects in both humoral and adaptive immunity, and in some strains there is an increased susceptibility to the development of autoimmunity. In one recent study looking at the early stages of human rheumatoid arthritis, significantly less of specific bacterial species were found in the intestinal flora of afflicted individuals. Collectively, these observations in humans and in animal models suggest that communication between commensal microbes and the host immune system may influence the induction or severity of some autoimmune diseases.

The antibody responses of germ-free mice to exogenous antigens are depressed, as are responses to Conconavalin A (Con A),

a strong T-cell mitogen. Germ-free animals were found to exhibit a T_H2 cytokine bias in response to antigenic challenge, which could be restored to balance by colonization with just *a single microbe*, specifically *Bacteroides fragilis*. Further, this restored immune balance was most strongly associated with microbial expression of one particular molecule: the surface-expressed polysaccharide A. This strongly suggests that single microorganisms, and in some cases maybe even single molecules expressed by these microbes, can have systemic effects on the balance of immunity in the host.

This increased awareness of gut-associated immune regulation has spawned several interesting hypotheses related to tolerance. One, called the microflora or altered microbiota hypothesis, posits that changes in gut microflora due to dietary modifications and/or increased antibiotic use have disrupted normal microbially mediated pathways important for regulating immune tolerance. There is already a clear link between intestinal microbiota and health of the gut, including a role for certain microbes in regulating inflammatory syndromes of the bowel. Further evidence for this comes from transplantation studies, where individuals treated with immune ablation therapy and high-dose antibiotics show overcolonization with certain, sometimes pathogenic, microbes. These individuals frequently manifest correlating immune defects, which can be reversed by directly manipulating the gut microflora.

But how do our commensal microbes influence the immune balance? Some posit that intestinal epithelial cells (IECs) and mucosal dendritic cells (DCs), which express several key innate receptors, including Toll-like receptors and NOD-like receptors (TLRs and NLRs), may be involved.

Mucosal DCs in the gut are known to sample the contents of the intestinal lumen. These cells could be another connection between commensal microbes and the maintenance of tolerance. In a study by R. Medzhitov and colleagues, mice engineered to lack the MyD88 gene, important for signaling through these innate receptors, were more susceptible to intestinal injury and autoimmunity. This suggests that signaling through these receptors can in some instances induce tolerance rather than an inflammatory response. This could be explained by the delivery of tolerogenic and homeostatic signals by as yet undefined antigens carried by the gut microbiota. Experiments to identify the specific ligands and receptors important for this cross-talk between microflora and the host immune system are ongoing.

In case you were thinking of taking this role of the gut in immune balance with a grain of salt, you might want to think again. Sodium chloride may be a new addition to the list of dietary contributors to immune pathway development. In vitro studies have shown that the addition of sodium chloride to cultures of $CD4^+$ T cells can drive development of T_H17 cells. Expansion of this cell type has been linked to the development of certain autoimmune diseases. It now appears that "We are what we eat" extends to the immune system, which may have a discriminating palate of its own.

Rakoff-Nahoum S, Paglino J, Eslami-Varzaneth F, Edberg S and Medzhitov R. 2004. Recognition of commensal microflora by toll-like receptors is required for intestinal homeostasis. Cell 118:229–41.

Round J, O'Connell R and Masmanian S. 2010. Coordination of tolerogenic immune responses by the commensal microbiota. Journal of Autoimmunity. 34:220–225.

transduction, they can act as a sponge, absorbing this growth- and survival-promoting cytokine and further discouraging expansion of local immunostimulatory effector T cells.

This pathway of inhibition is believed to be quite antigen specific; T_{REG} cells inhibit APCs presenting *their cognate*

antigen or effector T cells that *share their same antigen specificity* and not T cells with a different specificity. However, again taking advantage of FoxP3-GFP mice bred with TCR transgenic animals, it was shown that it is possible for FoxP3$^+$ T cells to inhibit T cells recognizing *other* antigens,

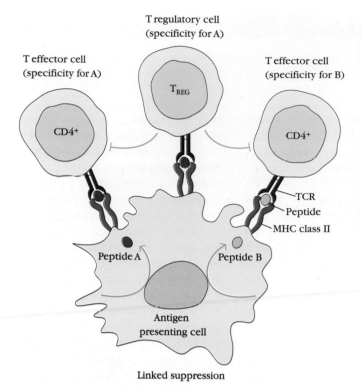

FIGURE 16-4 Linked suppression mediated by T_REG cells.
In cases where a single antigen-presenting cell (APC) engages simultaneously with T cells of different specificity, inhibitory signals meant for one can be transmitted to both and lead to a "spreading" of immune suppression to include other antigens. *[R. I.. Lechler, O. A. Garden, and L. A. Turka, 2003, The complementary roles of deletion and regulation in transplantation tolerance,* Nature Reviews Immunology *3:147–158]*

as occurs when both the T_REG cell and the "bystander" T cell recognizing another antigen interact with the same APC. The result is inhibition of the APC, via both contact-dependent and -independent pathways, as well as inhibition of the bystander T cell through soluble inhibitory factors and decommissioning of the APC (Figure 16-4). This simultaneous processing and presentation of different antigens might happen naturally in vivo when the antigens in question are parts of the same pathogen, although based on the findings in these experimental systems this was not required. This phenomenon, termed *linked suppression*, has now been seen in multiple experimental systems and may represent another way that T_REG cells support local self-tolerance in tissues lacking any pathogen-induced danger signals.

Regulatory CD8+ T Cells

Although an immunosuppressive role for CD8+ T cells was suggested as early as 1970, it took over 35 years to confirm and partially characterize this cell population. Although much work is yet to be done, the fact that CD8+ T_REG cells can play a role in inhibiting responses to self antigens is now fairly well established. For instance, adoptive transfer of a subset of CD8+ T cells can induce tolerance to heart transplants in recipient rats and can protect mice from the auto-

immune disease EAE. Although the specific phenotype/s of these cells and the mechanisms they use are still under debate, some central themes have begun to coalesce.

First, unlike in the case of CD4+ T_REG cells, the contribution to this population from thymic selection (nT_REG cells) is likely very small. In studies using TCR ovalbumin-specific transgenic mice where all the T cells are specific for this nonself antigen, no FoxP3-expressing CD8+ nT_REG cells were identified. In other experimental systems, only rare CD8+ T cells emigrating from the thymus with this natural regulatory phenotype could be detected. However, in the presence of antigen and TGF-β, CD8+ regulatory T cells expressing FoxP3 can be induced (iT_REG). In fact, the plasticity of this phenotype may be quite significant. Some hypothesize that almost any naïve T cell (CD4+ or CD8+) presented with antigen and the right cocktail of cytokines along with lack of costimulation will develop into an iT_REG cell, with the potential to revert back to immune stimulatory later!

The phenotype of these cells is also complicated. Several phenotypic markers have been associated with separate CD8+ T-cell populations that suppress immune responses. These include CD8+αα (as opposed to the more common α and β chains), both high- and low-affinity receptors for IL-2 (CD25 and CD122, respectively) and dendritic cell markers (CD11c), as well as others. In many cases, but not all, the master transcriptional regulator FoxP3 is also present.

As with their CD4+ counterparts, CD8+ T_REG cells likely use a range of mechanisms to inhibit other cells from responding to antigen. Whether this is mediated by separate populations of cells or by cells with the potential for many methods of inhibition is still unclear. Like with CD4+ T_REG cells, three main pathways seem to exist: APC lysis, inhibition of APC function, and regulation of effector T cells that share cognate antigen with the CD8+ T_REG cell.

The hypothesis that regulatory CD8+, as well as some CD4+, T cells work primarily by "decommissioning" specific APCs has received much attention of late. For example, CD8+ T cells can efficiently use conventional cytolysis to kill APCs presenting a self antigen, leading to a reduced frequency of presentation of this autoantigen and therefore fewer effector T cells (both CD4+ and CD8+) activated against this self molecule. Alternatively, regulatory cells may signal APCs to reduce their costimulatory potential, with a similar net effect. A reduction in CD80/86 expression on the APC may be even more efficient in favoring immunosuppression; it bars an APC from stimulating T cells and, at the same time, encourages further production of iT_REG cells upon interaction with the APC (see Figure 16-3). In fact, a reduction in the expression of various costimulatory molecules, but not MHC class I or class II, has been seen in several experimental systems designed to study T_REG cells.

Some mouse CD8+ T_REG cells, like their CD4+ counterparts, recognize antigen using conventional MHC molecules. However, the most well-characterized CD8+ T_REG cells are restricted to the nonclassical MHC class I molecules: Qa-1 in mice and HLA-E in humans. These nonclassical MHC molecules play key roles in presenting lipid antigens (see Chapter 8). Studies using mice engineered to lack Qa-1-restricted CD8$^+$ T$_{REG}$ cells showed that these animals develop aggressive autoimmune reactions against self antigens, suggesting that this population is involved in regulating CD4$^+$ T-cell responses to self antigens to maintain peripheral tolerance. In various in vitro experimental systems, CD8$^+$ iT$_{REG}$ cells have been found to express inhibitory cytokines, including IL-10 and TGF-β, although whether these cells are actually using any of these cytokines in vivo to suppress immune responses is still an open question.

Autoimmunity

Simply stated, autoimmune disease is caused by failure of the tolerance processes described above to protect the host from the action of self-reactive lymphocytes. These diseases result from the destruction of self proteins, cells, and organs by auto-antibodies or self-reactive T cells. For example, auto-antibodies are the major offender in Hashimoto's thyroiditis, in which antibodies reactive with thyroid-specific antigens cause severe tissue destruction. On the other hand, many autoimmune diseases are characterized by tissue destruction mediated directly by T cells. A well-known example is rheumatoid arthritis (RA), in which self-reactive T cells attack the tissue in joints, causing an inflammatory response that results in swelling and tissue destruction. Other examples of T-cell-mediated autoimmune disease include insulin-dependent or Type 1 diabetes mellitus (T1DM) and multiple sclerosis (MS). Table 16-1 lists several of the more common autoimmune disorders, as well as their primary immune mediators.

Autoimmune disease is estimated to affect between 3% and 8% of individuals in the industrialized world, making this a rising problem in terms of morbidity and mortality

---

TABLE 16-1   Some autoimmune diseases in humans

Disease	Self antigen/Target gene	Immune effector
ORGAN-SPECIFIC AUTOIMMUNE DISEASES		
Addison's disease	Adrenal cells	Auto-antibodies
Autoimmune hemolytic anemia	RBC membrane proteins	Auto-antibodies
Goodpasture's syndrome	Renal and lung basement membranes	Auto-antibodies
Graves' disease	Thyroid-stimulating hormone receptor	Auto-antibody (stimulating)
Hashimoto's thyroiditis	Thyroid proteins and cells	T$_H$1 cells, auto-antibodies
Idiopathic thrombocytopenia purpura	Platelet membrane proteins	Auto-antibodies
Type 1 diabetes mellitus	Pancreatic beta cells	T$_H$1 cells, auto-antibodies
Myasthenia gravis	Acetylcholine receptors	Auto-antibody (blocking)
Myocardial infarction	Heart	Auto-antibodies
Pernicious anemia	Gastric parietal cells; intrinsic factor	Auto-antibody
Poststreptococcal glomerulonephritis	Kidney	Antigen-antibody complexes
Spontaneous infertility	Sperm	Auto-antibodies
SYSTEMIC AUTOIMMUNE DISEASES		
Ankylosing spondylitis	Vertebrae	Immune complexes
Multiple sclerosis	Brain or white matter	T$_H$1 cells and T$_C$ cells, auto-antibodies
Rheumatoid arthritis	Connective tissue, IgG	Auto-antibodies, immune complexes
Scleroderma	Nuclei, heart, lungs, gastrointestinal tract, kidney	Auto-antibodies
Sjögren's syndrome	Salivary gland, liver, kidney, thyroid	Auto-antibodies
Systemic lupus erythematosus (SLE)	DNA, nuclear protein, RBC and platelet membranes	Auto-antibodies, immune complexes
Immune dysregulation polyendocrinopathy enteropathy X-linked (IPEX)	Multiorgan, loss of *FoxP3* gene	Missing regulatory T cells
Autoimmune polyendocrinopathy-candidiasis-ectodermal dystrophy (APECED)	Multiorgan, loss of *aire* gene	Defective central tolerance

TABLE 16-2 Experimental animal models of autoimmune diseases

Animal model	Possible human disease counterpart	Inducing antigen	Disease transferred by T cells
SPONTANEOUS AUTOIMMUNE DISEASES			
Nonobese diabetic (NOD) mouse	Insulin-dependent diabetes mellitus (1)	Unknown	Yes
(NZB X NZW) F₁ mouse	Systemic lupus erythematosus (SLE)	Unknown	Yes
Obese-strain chicken	Hashimoto's thyroiditis	Thyroglobulin	Yes
EXPERIMENTALLY INDUCED AUTOIMMUNE DISEASES*			
Experimental autoimmune myasthenia gravis (EAMG)	Myasthenia gravis	Acetylcholine receptor	Yes
Experimental autoimmune encephalomyelitis (EAE)	Multiple sclerosis (MS)	Myelin basic protein (MBP); proteolipid protein (PLP)	Yes
Autoimmune arthritis (AA)	Rheumatoid arthritis (RA)	M. tuberculosis (proteoglycans)	Yes
Experimental autoimmune thyroiditis (EAT)	Hashimoto's thyroiditis	Thyroglobulin	Yes

* These diseases can be induced by injecting appropriate animals with the indicated antigen in complete Freund's adjuvant. Except for autoimmune arthritis, the antigens used correspond to the self antigens associated with the human disease counterpart. Rheumatoid arthritis involves reaction to proteoglycans, which are self antigens associated with connective tissue.

around the globe. These diseases are often categorized as either organ-specific or systemic, depending on whether they affect a single organ or multiple systems in the body. Another method of grouping involves the immune component that does the bulk of the damage: T cells versus antibodies. In this section, we describe several examples of both organ-specific and systemic autoimmune disease. In each case, we discuss the antigenic target (when known), the causative process (either cellular or humoral), and the resulting symptoms. When available, examples of animal models used to study these disorders are also considered (Table 16-2). Finally, we touch on the factors believed to be involved in induction or control of autoimmunity, and the treatments for these conditions.

Some Autoimmune Diseases Target Specific Organs

Autoimmune diseases are caused by immune stimulatory lymphocytes or antibodies that recognize self components, resulting in cellular lysis and/or an inflammatory response in the affected organ. Gradually, the damaged cellular structure is replaced by connective tissue (fibrosis), and the function of the organ declines. In an organ-specific autoimmune disease, the immune response is usually directed to a target antigen unique to a single organ or gland, so that the manifestations are largely limited to that organ. The cells of the target organs may be damaged directly by humoral or cell-mediated effector mechanisms. Alternatively, anti-self anti-bodies may overstimulate or block the normal function of the target organ.

Hashimoto's Thyroiditis

In Hashimoto's thyroiditis, an individual produces auto-antibodies and sensitized T_H1 cells specific for thyroid antigens. This disease is much more common in women, often striking in middle-age (see Clinical Focus Box 16-2). Antibodies are formed to a number of thyroid proteins, including thyroglobulin and thyroid peroxidase, both of which are involved in the uptake of iodine. Binding of the auto-antibodies to these proteins interferes with iodine uptake, leading to decreased thyroid function and hypothyroidism (decreased production of thyroid hormones). The resulting delayed-type hypersensitivity (DTH) response is characterized by an intense infiltration of the thyroid gland by lymphocytes, macrophages, and plasma cells, which form lymphocytic follicles and germinal centers. (See Chapter 15 for a description of the DTH response.) The ensuing inflammatory response causes a goiter, or visible enlargement of the thyroid gland, a physiological response to hypothyroidism.

Type 1 Diabetes Mellitus

Type 1 diabetes mellitus (T1DM) or insulin-dependent diabetes, affects almost 2 in 1000 children in the U.S.; roughly double the incidence observed just 20 years ago. It is seen mostly in youth under the age of 14 and is less common than Type 2, or non-insulin dependent diabetes mellitus. T1DM is caused by an autoimmune attack against insulin-producing

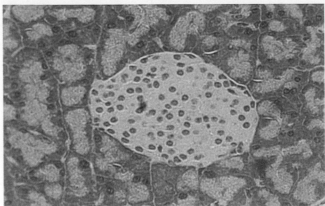

(b)

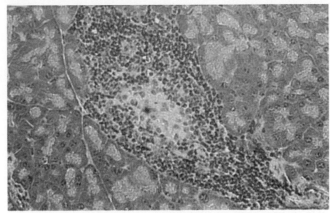

FIGURE 16-5 Photomicrographs of an islet of Langerhans in (a) pancreas from a normal mouse and (b) pancreas from a mouse with a disease resembling insulin-dependent diabetes mellitus. Note the lymphocyte infiltration into the islet (insulitis) in (b). *[From M. A. Atkinson and N. K. Maclaren, 1990,* Scientific American *263:1, 62.] [© 2007 W. H. Freeman and Company.]*

cells (beta cells) scattered throughout the pancreas, which results in decreased production of insulin and consequently increased levels of blood glucose. The attack begins with cytotoxic *T lymphocyte* (CTL) infiltration and activation of macrophages, frequently referred to as insulitis (Figure 16-5b), followed by cytokine release and the production of autoantibodies, which leads to a cell-mediated DTH response. The subsequent beta-cell destruction is thought to be mediated by cytokines released during the DTH response and by lytic enzymes released from the activated macrophages. Autoantibodies specific for beta cells may contribute to cell destruction by facilitating either antibody-mediated complement lysis or *antibody-dependent cell-mediated cytotoxicity* (ADCC).

The abnormalities in glucose metabolism associated with T1DM result in serious metabolic problems that include ketoacidosis (accumulation of ketone, a breakdown product from fat) and increased urine production. The late stages of the disease are often characterized by atherosclerotic vascular lesions (which cause gangrene of the extremities due to

impeded vascular flow), renal failure, and blindness. If untreated, death can result. The most common therapy for T1DM is daily administration of insulin. Although this is helpful, sporadic doses are not the same as metabolically regulated, continuous, and controlled release of the hormone. Unfortunately, diabetes can remain undetected for many years, allowing irreparable loss of pancreatic tissue to occur before treatment begins.

One of the best-studied animal models of this disease is the NOD mouse, which spontaneously develops a form of diabetes that resembles human T1DM. This disorder also involves lymphocytic infiltration of the pancreas and destruction of beta cells, and carries a strong association with certain MHC alleles. Disease is mediated by bone-marrow-derived cells; normal mice reconstituted with an injection of bone marrow cells from NOD mice will develop diabetes, and healthy NOD mice that have not yet developed disease can be spared by reconstitution with bone marrow cells from normal mice. NOD mice housed in germ-free environments show a higher incidence of diabetes compared to those in normal housing, suggesting that microbes may influence the development of autoimmune disease. In genome-wide scans, over 20 *insulin-dependent diabetes* (*Idd*) loci associated with disease susceptibility have been identified, including at least one member of the TNF receptor family.

Myasthenia Gravis

Myasthenia gravis is the classic example of an autoimmune disease mediated by blocking antibodies. A patient with this disease produces auto-antibodies that bind the *acetylcholine receptors* (AchRs) on the motor end plates of muscles, blocking the normal binding of acetylcholine and inducing complement-mediated lysis of the cells. The result is a progressive weakening of the skeletal muscles (Figure 16-6). Ultimately, the antibodies cause the destruction of the cells bearing the receptors. The early signs of this disease include drooping eyelids and inability to retract the corners of the mouth. Without treatment, progressive weakening of the muscles can lead to severe impairment of eating as well as problems with movement. However, with appropriate treatment, this disease can be managed quite well and afflicted individuals can lead a normal life. Treatments are aimed at increasing acetylcholine levels (e.g., using cholinesterase inhibitors), decreasing antibody production (using corticosteroids or other immunosuppressants), and/or removing antibodies (using plasmapheresis).

One of the first experimentally induced autoimmune disease animal models was discovered serendipitously in 1973, when rabbits immunized with AChRs purified from electric eels suddenly became paralyzed. (The original aim was to generate monoclonal antibodies for research use.) These rabbits developed antibodies against the foreign AChR that cross-reacted with their own AChRs. These autoantibodies then blocked muscle stimulation by Ach at the synapse and led to progressive muscle weakness. Within a

Why Are Women More Susceptible Than Men to Autoimmunity? Gender Differences in Autoimmune Disease

Of the nearly 50 million individuals in the United States believed to be living with autoimmune disease, approximately 78% are women. As shown in the following table, female-biased predisposition to autoimmunity is more apparent in some diseases than others. For example, the female-to-male ratio of individuals who suffer from diseases such as *multiple sclerosis* (MS) or *rheumatoid arthritis* (RA) is approximately two or three females to one male. There are nine women for every one man afflicted with systemic lupus erythematosus (SLE). However, these statistics do not tell the entire story. In some diseases, such as MS, the severity can be worse in men than in women. That women are more susceptible to autoimmune disease has been recognized for many years. The reasons are not entirely understood, although recent advances are helping to clarify this difference.

Although it may seem unlikely, considerable evidence suggests significant gender differences in immune responses. In general, females tend to mount more vigorous humoral and cellular immune responses. Immune cell activation, cytokine secretion after infection, numbers of circulating CD4$^+$ T cells and mitogenic

responses are all higher in women than men. Immunization studies conducted in mice and humans show that females produce a higher titer of antibodies than males; this is true during both primary and secondary responses. In transplantation, women also suffer from a higher rate of graft rejection. As one might guess, this enhanced immunity in females means that males, in general, are slightly more prone to infections.

The prevailing view is that sex hormone differences between men and women account for at least part of the observed gender difference in the rates of autoimmunity. Some of this evidence comes from observations made in SLE, where young women of child-bearing age are at greatest risk for the disease. Lupus flares during pregnancy (a high estrogen state) and increased rates of remission following menopause (a low estrogen state) also point to sex hormones as potential regulators of this autoimmune disease. The general consensus is that estrogens, the more female-specific hormones, are associated with enhanced immunity whereas androgens, or male-based hormones, are associated with its suppression.

In mice, whose gender differences are easier to study, a large body of literature documents gender differences in immune responses. Female mice are much more likely than male mice to develop T$_H$1 responses and, in infections for which pro-inflammatory T$_H$1 responses are beneficial, are more likely to be resistant to the infection. An excellent example is infection by viruses such as *vesicular stomatitis virus* (VSV), *herpes simplex virus* (HSV), and *Theiler's murine encephalomyelitis virus* (TMEV). Clearance of these viruses is enhanced by T$_H$1 responses. In other cases, however, a pro-inflammatory response can be deleterious. For example, a T$_H$1 response to *lymphocytic choriomeningitis virus* (LCMV) correlates with more severe disease and significant pathology. Thus, female mice are more likely to succumb to infection with LCMV. The importance of gender in LCMV infection is underscored by experiments demonstrating that castrated male mice behave immunologically like females and are more likely to experience autoimmune disease than uncastrated males.

Why this dichotomy between the sexes? One hypothesis posits that this increased risk of autoimmunity in women

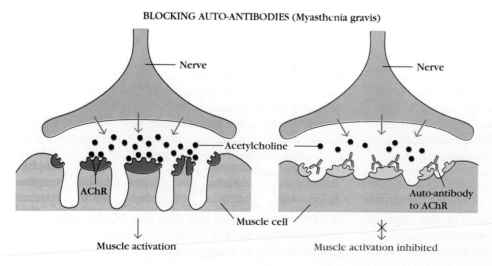

BLOCKING AUTO-ANTIBODIES (Myasthenia gravis)

FIGURE 16-6 In myasthenia gravis, binding of auto-antibodies to the acetylcholine receptor (AChR; *right*) blocks the normal binding of acetylcholine (burgundy dots) and subsequent muscle activation (*left*). In addition, the anti-AChR auto-antibody activates complement, which damages the muscle end plate; the number of acetylcholine receptors declines as the disease progresses. [© 2013 W. H. Freeman and Company.]

Box 16-2

is a by-product of the evolutionary role of women as bearers of children. Pregnancy may give us a clue as to how sex plays a role in regulating immune response. During pregnancy, it is critical that the mother tolerate the fetus, which is a type of foreign semi-allograft. This makes it very likely, maybe even crucial for successful implantation and fetal development, that the female immune system undergoes important modifications during pregnancy. Recall that females normally tend to mount more T_H1-like responses than T_H2 responses. During pregnancy, however, women mount more T_H2-like responses. It is thought that pregnancy-associated levels of sex steroids may promote an anti-inflammatory environment. In this regard, it is notable that diseases enhanced by T_H2-like responses, such as SLE, which has a strong antibody-mediated component, can be exacerbated during pregnancy, whereas diseases that involve inflammatory responses, such as RA and MS, sometimes are ameliorated in pregnant women.

Another effect of pregnancy is the presence of fetal cells in the maternal circulation, creating a state called microchimerism. Fetal cells can persist in the maternal circulation for decades. These long-lived fetal cells may play a role in the development of autoimmune disease. Furthermore, the exchange of cells during pregnancy is bidirectional (cells of the mother may also

TABLE 1	Gender prevalence ratios for selected autoimmune diseases[a]
Autoimmune disease	**Ratio (female/male)**
Hashimoto's thyroiditis/hypothyroidism	50:1
Systemic lupus erythematosus	9:1
Sjögren's syndrome	9:1–20:1
Graves' disease/hyperthyroidism	7:1
Rheumatoid arthritis	3:1–4:1
Scleroderma	3:1–4:1
Myasthenia gravis	2:1–3:1
Multiple sclerosis	2:1
Type 1 diabetes mellitus	1:1–2:1
Ulcerative colitis	1:1
Autoimmune myocarditis	1:1.2

[a]Modified from Gleicher and Barad, 2007.

appear in the fetal circulatory system), so that the presence of the mother's cells in the male circulation could also be a contributing factor in autoimmune disease. Although some studies have linked the presence of microchimerism with certain autoimmune syndromes, other studies have contradicted these findings, casting some doubt on this as a significant mode of induction of autoimmunity.

Zandman-Goddard G, Peeva E and Shoenfeld Y 2007. Gender and autoimmunity. Autoimmunity Reviews, 6(6): 366–72.

Gleicher N and Barad DH. 2007. Gender as risk factor for autoimmune diseases. Journal of Autoimmunity, 28 (1): 1–6.

year, this animal model, called *experimental autoimmune myasthenia gravis* (EAMG), led to the discovery that autoantibodies to the AChR were also the cause of myasthenia gravis in humans.

Some Autoimmune Diseases Are Systemic

In systemic autoimmune diseases, the immune response is directed toward a broad range of target antigens and involves a number of organs and tissues. These diseases reflect a general defect in immune regulation that results in hyperactive T cells and/or B cells. Tissue damage is typically widespread, both from cell-mediated immune responses and from direct cellular damage caused by auto-antibodies or by accumulation of immune complexes.

Systemic Lupus Erythematosus

One of the best examples of a systemic autoimmune disease is **systemic lupus erythematosus (SLE)**. Like several of the other autoimmune syndromes, this disease is more common in women, with approximately a 9:1 ratio (see Clinical Focus Box 16-2). Onset of symptoms typically appears between 20 and 40 years of age and is more frequent in African American and Hispanic women than in Caucasians, for unknown reasons. In identical twins where one suffers from SLE, the other has up to a 60% chance of developing SLE, suggesting a genetic component. However, although close relatives of an SLE patient are 25 times more likely to contract the disease, still only 2% of these individuals ever develop SLE.

Affected individuals may produce auto-antibodies to a vast array of tissue antigens, such as DNA, histones, RBCs,

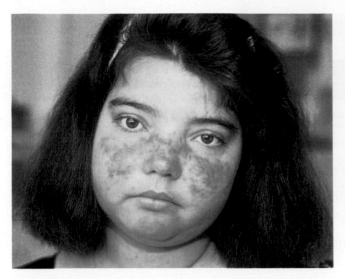

FIGURE 16-7 Characteristic "butterfly" rash over the cheeks of a woman with systemic lupus erythematosus. *[From L. Steinman, 1993, Scientific American 269(3):80.]*

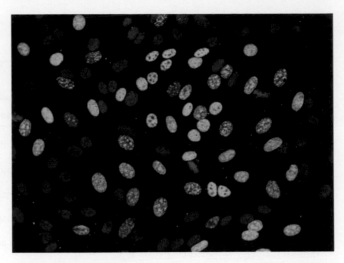

FIGURE 16-8 Diagnostic test for anti-nuclear antibodies using serum from an SLE patient. Serum dilutions from a patient are mixed with cells attached to a glass slide. Fluorescently labeled secondary antibodies directed against human antibodies are then added and reveal staining of the nucleus under a fluorescence microscope. *[Courtesy ORGENTEC Diagnostika GmbH]*

platelets, leukocytes, and clotting factors. Signs and symptoms include fever, weakness, arthritis, skin rashes (Figure 16-7), and kidney dysfunction. Auto-antibodies specific for RBCs and platelets can lead to complement-mediated lysis, resulting in hemolytic anemia and thrombocytopenia, respectively. When immune complexes of auto-antibodies with various nuclear antigens are deposited along the walls of small blood vessels, a type III hypersensitivity reaction develops. The complexes activate the complement system and generate membrane-attack complexes and complement fragments (C3a and C5a) that damage the wall of the blood vessel, resulting in vasculitis and glomerulonephritis. In severe cases, excessive complement activation produces elevated serum levels of certain complement fragments, leading to neutrophil aggregation and attachment to the vascular endothelium. Over time, the number of circulating neutrophils declines (neutropenia) and occlusions of various small blood vessels develop (vasculitis), which can lead to widespread tissue damage. Laboratory diagnosis of SLE involves detection of antinuclear antibodies directed against double-stranded or single-stranded DNA, nucleoprotein, histones, and nucleolar RNA. Indirect immunofluorescent staining with serum from SLE patients produces characteristic nuclear-staining patterns (Figure 16-8).

New Zealand Black (NZB) mice and F_1 hybrids of NZB x New Zealand White (NZW) mice spontaneously develop autoimmune diseases that closely resemble SLE. NZB mice develop autoimmune hemolytic anemia between 2 and 4 months of age, at which time various auto-antibodies can be detected, including antibodies to erythrocytes, nuclear proteins, DNA, and T lymphocytes. F_1 hybrids develop glomerulonephritis from immune-complex deposits in the kidney and die prematurely. As in human SLE, the incidence of autoimmunity in F_1 hybrids is greater in females.

Multiple Sclerosis

Multiple sclerosis (MS) is the most common cause of neurologic disability associated with disease in Western countries. MS occurs in women two to three times more frequently than men (see Clinical Focus Box 16-2) and, like SLE, frequently develops during childbearing years (approximately 20–40 years of age). Individuals with this disease produce autoreactive T cells that participate in the formation of inflammatory lesions along the myelin sheath of nerve fibers in the brain and spinal cord. Since myelin functions to insulate the nerve fibers, a breakdown in the myelin sheath leads to numerous neurologic dysfunctions, ranging from numbness in the limbs to paralysis or loss of vision.

Epidemiological studies indicate that MS is most common in the Northern Hemisphere and, interestingly, in the United States. Populations north of the 37th parallel have a markedly higher prevalence of MS than those south of this latitude. Individuals born south of the 37th parallel who move north before age 15 assume the higher relative risk. These provocative data suggest that an environmental component early in life affects the risk of contracting MS. Genetic influences are also important. The average person in the United States has about one chance in 1000 of developing MS; this increases to 1 in 50 to 100 for children or siblings of people with MS, and to 1 in 3 for an identical twin. The cause of MS is not well understood. Infection by certain viruses, such as Epstein-Barr virus (EBV), may predispose a person to MS. Some viruses can cause demyelinating diseases, but the data linking viruses to MS are not definitive.

EAE, one of the best-studied animal models of autoimmune disease, is mediated solely by T cells. It can be induced in a variety of species by immunization with *m*yelin *b*asic *p*rotein (MBP) or *p*roteo*l*ipid *p*rotein (PLP)—both components of the myelin sheaths surrounding neurons in the CNS—in complete Freund's adjuvant. Within 2 to 3 weeks, the animals develop cellular infiltration of the CNS, resulting in demyelination and paralysis. Most of the animals die, but others have milder symptoms, and some develop a chronic form of the disease that resembles relapsing and remitting MS in humans. Those that recover are resistant to the development of disease from a subsequent challenge with MBP and adjuvant. In these experiments, exposure of immature T cells to self antigens that normally are not present in the thymus presumably led to tolerance to these antigens. EAE was also prevented in susceptible rats by prior injection of MBP directly into the thymus. MBP is normally sequestered from the immune system by the blood-brain barrier, but in EAE the immune system is exposed to it under nonphysiologic conditions. EAE in small mammals does provide a system for testing treatments for human MS.

Rheumatoid Arthritis

Rheumatoid arthritis (RA) is a fairly common autoimmune disorder, most often diagnosed between the ages of 40 to 60 and more frequently seen in women. The major symptom is chronic inflammation of the joints (Figure 16-9), although the hematologic, cardiovascular, and respiratory systems are also frequently affected. Many individuals with RA produce a group of auto-antibodies called **rheumatoid factors** that are reactive with determinants in the Fc region of IgG—in other words, antibodies specific for antibodies! The classic rheumatoid factor is an IgM antibody that binds to normal circulating IgG, forming IgM-IgG complexes that are deposited in the joints. These immune complexes can activate the complement cascade, resulting in a type III hypersensitivity reaction, which leads to chronic inflammation of the joints. Treatments for RA include nonspecific drugs aimed at reducing inflammation, such as *n*onsteroidal *a*nti-*i*nflammatory *d*rugs (NSAIDs) and corticosteroids.

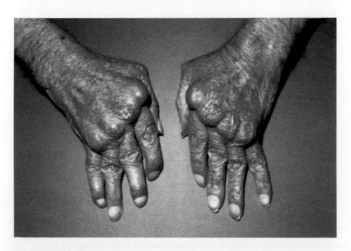

FIGURE 16-9 Swollen hand joints in a woman with rheumatoid arthritis. *[James Stevenson/Photo Researchers]*

More disease-specific immune modifiers have also been introduced, including antibodies that block TNF-α and IL-6 (see treatment below).

Both Intrinsic and Extrinsic Factors Can Favor Susceptibility to Autoimmune Disease

Overzealous immune activation can lead to autoimmunity, although suboptimal immune stimulation results in insufficiency that can allow microbes to take control. But what tips the balance toward a break in tolerance and the development of autoimmunity? Experiments with germ-free mouse models, discordance data in identical twins, and epidemiologic studies of geographic associations all suggest roles for both the environment and genes in susceptibility to the development of autoimmunity.

Environmental Factors Favoring the Development of Autoimmune Disease

As mentioned earlier, some autoimmune syndromes are more common in certain geographic locations or in particular climates. This suggests a link between environmental exposures (some of which may be microbial) or lifestyle factors, such as diet, in the development of autoimmune disease. For instance, we now know that cross-talk between gut microflora and the systemic immune system may help regulate peripheral tolerance, which could impact the development of autoimmune disease. Animals maintained in germ-free environments often show heightened susceptibility to autoimmune disease compared to their germ-laden counterparts. How commensal microbes influence tolerance and autoimmunity is an active area of research (see Clinical Focus Box 16-1).

Infections may also influence the induction of autoimmunity. For example, tissue pathology following infection may result in the release of sequestered self antigens that are presented in a way that fosters immune activation rather than tolerance induction. Likewise, the molecular structures of certain microbes may share chemical features with self components, resulting in the activation of immune cells with cross-reactive potential.

The Role of Genes in Susceptibility to Autoimmunity

Certain alleles within the MHC have been linked to several different autoimmune disorders. The strongest association between an HLA allele and autoimmunity is seen in ankylosing spondylitis, an inflammatory disease of vertebral joints. Individuals who express HLA-B27 are 90 times more likely to develop ankylosing spondylitis than individuals with a different HLA allele at this locus. This does not imply causation; not all individuals who express HLA-B27 experience this syndrome, suggesting that the relationship between MHC alleles and development of autoimmune disease is multifaceted. Interestingly, unlike many other autoimmune diseases, 90% of the cases of ankylosing spondylitis are seen in males.

Some non-MHC inherited genetic mutations can have causative effects on the development of autoimmunity. Not surprisingly, these tend to play prominent roles in immune regulation. Inactivating mutations in two immune-related genes, *aire* and *FoxP3*, result in forms of immune deficiency that impact central and peripheral tolerance, respectively (see above, as well as Chapters 9 and 18). A mutant form of *aire* has been shown to cause autoimmune *polyendocrinopathy-candidiasis-ectodermal dystrophy* (APECED), a systemic disease resulting from defective deletion or inactivation of autoreactive T cells in the thymus. The cause of another human autoimmune disorder—*immune dysregulation, polyendocrinopathy, enteropathy,* and *X-linked syndrome* (IPEX) has been mapped to mutations in the *FoxP3* gene. The product of this gene is required for the formation of many, but not all, regulatory T cells, suggesting that this disease is caused by an inability to generate the T_{REG} cells needed to maintain peripheral tolerance.

Many other genes with more subtle or even cumulative effects on susceptibility to autoimmunity have also been discovered. Not surprisingly, most of these play some role in the immune response. Genes for cytokines and their receptors, antigen processing and presentation, c-type lectin receptors, signaling pathways, adhesion molecules, and costimulatory or inhibitory receptors have all been linked to specific autoimmune diseases (Table 16-3). In many instances, multiple genes (sometimes with compounding environmental factors) may be required in order to predispose an individual to a particular autoimmune disease. In some cases, a single gene can heighten susceptibility to multiple different autoimmune disorders. For instance, a mutant form of PTPN22, a tyrosine phosphatase, results in reduced

TCR signaling capacity and has been linked to T1DM, RA, and SLE. It is believed that attenuation of TCR signaling during positive and negative selection may be what predisposes carriers of this allele to autoimmunity.

The Role of Certain T Helper Cell Types in Autoimmunity

In both organ-specific and systemic autoimmunity, CD4[+] rather than CD8[+] T cells have been linked to disease pathogenesis. However, the T helper (T_H) cell type or set of cytokines most closely associated with autoimmunity depends somewhat on the model system or human disease in question. As we know from Chapter 11, the antigen, the type of APC to make first encounter, and the surface receptors used during this engagement set the stage for the transition from innate to adaptive immunity. In this transition, the cytokine milieu will help determine which subsets of a T_H cell will predominate. The induction of autoimmunity is likewise a complex process, where even experimental models of the same human disease can be induced by different means, making outcomes in each case difficult to correlate. Nevertheless, a few themes have emerged from both human and animal studies of autoimmune disease.

Much of the initial data collected from various studies of autoimmune disease supported a role for autoreactive T_H1 cells and IFN-γ. For example, IFN-γ levels in the CNS of mice with EAE correlate with the severity of disease, and treatment with this cytokine exacerbates MS in humans. Likewise, adoptive transfer of IFN-γ-producing CD4[+] T_H cells from mice with EAE can induce disease in naïve hosts. On the flip side, elimination of IFN-γ using either neutralizing antibodies or removal of the gene does not protect animals from EAE; in fact, it worsens the symptoms.

TABLE 16-3   Examples of genetic associations with autoimmune disease

Disease	C-type lectin	Cytokines, their receptors and regulators	Innate immune response	Adhesion and costimulation	Antigen processing and presentation
Type 1 diabetes	CLEC16A	IL-2R		CTLA4	VNTR-Ins PTPN22
Rheumatoid arthritis (RA)	DCIR	STAT4	REL, C5-TRAFI	CD40	PTPN22, MHC2TA
Ankylosing spondylitis (AS)		IL-1A, IL-23R	KIR complex		ERAP1
Multiple sclerosis (MS)	CLEC16A	IL-2RA, IL-7R		CD40	
Systemic lupus erythematosus (SLE)		STAT4, IRF5	TNFAIP3	TNFSF4	PTPN22
Crohn's disease	CLEC16A	IL-23R	NOD2, NCF4	TNFSF15	PTPN2

Source: R. Thomas, 2010, The balancing act of autoimmunity: Central and peripheral tolerance versus infection control, *International Reviews of Immunology* **29**:211, Table 1 with modifications.

TABLE 16-4	Common pro-inflammatory environmental factors in autoimmune diseases	
Group	**Examples**	**Disease association examples**
Infection	Viral	Type I diabetes
	Bacterial	Reiter's syndrome
	Fungal	Mucocutaneous candidiasis (APECED)
Toxins	Smoking	Rheumatoid arthritis
	Fabric dyes	Scleroderma
	Iodine	Thyroiditis
Stress	Psychological	Multiple sclerosis, Systemic lupus erythematosus (SLE)
	Oxidative, metabolic	Rheumatoid arthritis
	Ultraviolet light	SLE
	Endoplasmic reticulum	Ulcerative colitis
Food	Gluten	Celiac disease
	Breastfeeding cessation	Type I diabetes
	Gastric bypass	Spondyloarthropathy

Source: R. Thomas, 2010. The balancing act of autoimmunity: Central and peripheral tolerance versus infection control. *International Reviews of Immunology* **29**:211, Table 2.]

These conflicting results led to the study of other cytokines or T cell types that may be involved in the induction of autoimmunity, especially those connected to IFN-γ. Recall from Chapter 11 that IL-12 and IL-23, which can be produced by APCs during activation, encourage the production of other cytokines, such as either IFN-γ or IL-17, favoring T-cell development along T_H1 or T_H17 pathways, respectively. Studies have shown that mice engineered to lack the gene for the p40 subunit of IL-12, which happens to be shared with IL-23, are protected from EAE. This protection is due to inhibition of IL-23, a cytokine required to sustain T_H17 cells. Mice lacking IL-17A are less susceptible to both EAE and collagen-induced arthritis, a model for human RA. In follow-up studies of patients with RA and psoriasis, elevated IL-17 expression has been found at the site of inflammation, and increased serum levels of IL-17 and IL-23 have been observed in patients with SLE. Collectively, these findings support the notion that T_H17 cells may be an important driver of some autoimmune diseases.

Several Possible Mechanisms Have Been Proposed for the Induction of Autoimmunity

In addition to genetic and environmental predisposing factors, autoimmunity likely develops from a number of different events. Disease may be induced by certain genetic mutations, the release of sequestered antigens, overstimulation of antigen-specific receptors, and stochastic events. In most cases, a combination of these is the cause. Another issue is the sex difference in autoimmune susceptibility, with diseases such as Hashimoto's thyroiditis, SLE, MS, and RA preferentially affecting women. Factors that may account for this, such as hormonal differences between the sexes and the potential effects of fetal cells in the maternal circulation following pregnancy, are discussed in the Clinical Focus Box 16-2.

As a result of random V(D)J recombination, over half of all antigen-specific receptors recognize self proteins. Not all of these are deleted during negative selection (see Chapter 9). Potentially self-reactive T and B cells found in the periphery are normally held in check by anergic or regulatory mechanisms, such as T_{REG} cells. However, exposure to carcinogens or infectious agents that favor DNA damage or polyclonal activation can potentially interfere with this regulation and/or lead to the expansion and survival of rare T- or B-cell clones with autoimmune potential (Table 16-4). Genes that, when mutated, could favor expansion include those encoding antigen receptors, signaling molecules, costimulatory or inhibitory molecules, apoptosis regulators, or growth factors (see Table 16-3). Gram-negative bacteria, cytomegalovirus, and EBV are all known polyclonal activators, inducing the proliferation of numerous clones of B cells that express IgM in the absence of T-cell help. If B cells reactive to self antigens are activated by this mechanism, autoantibodies can appear.

A role for particular microbial agents in autoimmunity was postulated for several reasons beyond their potential for DNA damage or polyclonal activation. As discussed earlier, some autoimmune syndromes are associated with certain geographic regions, and immigrants to an area can acquire enhanced susceptibility to the disorder associated with that region. This, coupled with the fact that a number of viruses and bacteria possess antigenic determinants that are similar or even identical to normal host-cell components, led to a hypothesis known as **molecular mimicry**. This proposes that some pathogens express protein epitopes resembling self components in either conformation or primary sequence. For instance, rheumatic fever, a disease caused by autoimmune destruction of heart muscle cells, can develop *after* a Group A *Streptococcus* infection. In this case, antibodies to streptococcal antigens have been shown to cross-react with the heart muscle proteins, resulting in immune complex deposition and complement activation, a type II hypersensitivity reaction (see Chapter 15). In one study, 600 different monoclonal antibodies specific for 11 different viruses were evaluated for their reactivity with normal tissue antigens. More than 3% of the virus-specific antibodies tested also bound to normal tissue, suggesting that molecular similarity between foreign and host antigens may be fairly common. In these cases, susceptibility may also be influenced by the MHC haplotype of the individual, since certain class I and class II MHC molecules may be more effective than others in presenting the homologous peptide for T-cell activation.

Release of sequestered antigens is another proposed mechanism of autoimmune initiation, one that may in some cases also be connected with environmental exposures. The induction of self-tolerance in T cells results from exposure of immature thymocytes to self antigens in the thymus, followed by clonal deletion or inactivation of any self-reactive cells (see Chapter 9). Tissue antigens that are not expressed in the thymus will not engage with developing T cells and will thus not induce self-tolerance. Trauma to tissues following either an accident or an infection can release these sequestered antigens into the circulation. For instance, the release of heart muscle antigens following myocardial infarction (heart attack) can lead to the formation of auto-antibodies that target healthy heart muscle cells. Studies involving injection of normally sequestered antigens directly into the thymus of susceptible animals support this proposed mechanism: injection of CNS myelin proteins or pancreatic beta cells can inhibit the development of EAE or diabetes, respectively. In these experiments, exposure of immature T cells to self antigens normally not present in the thymus presumably led to central and possibly also peripheral tolerance to these antigens.

It is worth reiterating that, although certain events may be associated with the development of autoimmunity, a complex combination of genotype and environmental factors likely influences the balance of self-tolerance versus development of autoimmune disease.

Autoimmune Diseases Can Be Treated by General or Pathway-Specific Immunosuppression

Ideally, treatment for autoimmune diseases should reduce only the autoimmune response, leaving the rest of the immune system intact. However, implementing this strategy has proven difficult. The current therapies to treat autoimmune disease fall into two categories: broad spectrum immunosuppressive treatments and more recent mechanism- or cell-type-specific strategies (Table 16-5).

Broad-Spectrum Therapies

Most first-generation therapies for autoimmune diseases are not cures but merely palliatives, reducing symptoms to provide the patient with an acceptable quality of life. For example, most general immunosuppressive treatments (e.g., corticosteroids, azathioprine, and cyclophosphamide) are strong anti-inflammatory drugs that suppress lymphocytes

by inhibiting their proliferation or by killing these rapidly dividing cells. Side effects of these drugs include general cytotoxicity, an increased risk of uncontrolled infection, and the development of cancer.

In some autoimmune diseases, removal of a specific organ or set of toxic compounds can alleviate symptoms. Patients with myasthenia gravis often have thymic abnormalities (e.g., thymic hyperplasia or thymomas), in which case thymectomy can increase the likelihood of remission. Plasmapheresis may also provide significant if short-term benefit for diseases involving antigen-antibody complexes (e.g., myasthenia gravis, SLE, and RA), where removal of a patient's plasma antibodies temporarily eliminates these complexes.

Strategies That Target Specific Cell Types

When antibodies and/or immune complexes are heavily involved in autoimmune pathology, strategies aimed at B cells can improve clinical symptoms. For example, a monoclonal antibody against the B-cell-specific antigen CD20 (Rituximab) depletes a subset of B cells and provides short-term benefit for RA. However, most cell-type specific agents used to treat autoimmune disorders target T cells or their products because these cells are either directly pathogenic or provide help to autoreactive B cells.

The first anti-T-cell antibodies used to treat autoimmune disease targeted the CD3 molecule and were designed to deplete T cells without signaling through this receptor. Although somewhat effective in the treatment of T1DM, this method still induced broad-spectrum immune suppression. Anti-CD4 antibodies successfully reversed MS and arthritis in animal models, although human trials of this treatment have shown no efficacy. A possible reason for this failure is that anti-CD4 may interfere with the activity of $CD4^+CD25^+$ regulatory T cells, a cell type we now know is key to the regulation of tolerance.

With this in mind and with the discovery of the T_H17 subset, scientists are beginning to target specific T helper cell pathways. In several mouse models of autoimmunity, including MS, T1DM, SLE, and IPEX, the transfer of T_{REG} cells can clearly inhibit disease pathogenesis. The greatest difficulty with translating this from mouse to human is in selecting a population of T_{REG} cells, as FoxP3 in humans does not correlate well with immunosuppressive activity. Therefore, most of the emphasis in the clinical applications of this approach is currently directed toward mimicking T_{REG}-like mechanisms of suppression (e.g., using IL-10) or inhibiting T_H17-mediated effects (e.g., using IL-17 or IL-23 blocking antibodies).

Therapies That Block Steps in the Inflammatory Process

Since chronic inflammation is a hallmark of debilitating autoimmune disease, steps in the inflammatory process are potential targets for intervention. Drugs that block TNF-α, one of the early mediators in the inflammatory process, are

| TABLE 16-5 | Drugs currently approved by the FDA or undergoing clinical trials to treat autoimmune disease or suppress the immune response, arranged according to mechanism of action |

Name	Brand name	Mechanism of action	Target disease
T- OR B-CELL DEPLETING AGENTS			
Lymphocyte immune globulin (horse), anti-thymocyte globulin (rabbit)	Atgam (horse), Thymoglobulin (rabbit)	Depleting horse/rabbit poly-clonal anti-thymocyte antibody	Renal transplant; aplastic anemia
Muronomab (OKT3)	Orthoclone OKT3	Mouse anti-human CD3ε mAb	Acute transplant rejection; graft-versus-host disease (GVHD)
Zanolimumab (HuMax-CD4)		Human anti-CD4 mAb, partially depleting	Rheumatoid arthritis (RA)
Rituximab (IDEC-C2B8)	Rituzan	Chimeric anti-CD20 mAb	RA
TARGETING TRAFFICKING/ADHESION			
Fingolimod (FTY720)		S1P receptor agonist	Relapsing/remitting multiple sclerosis (MS); renal transplant
TARGETING TCR SIGNALING			
Cyclosporine A	Gengraf, Neoral, Sandimmune	Calcineurin inhibitor	Transplant; severe active RA; severe plaque psoriasis
Tacrolimus (FK506)	Prograf (systemic), Protopic (topical)	Calcineurin inhibitor	Transplant; moderate-severe atopic dermatitis; ulcerative colitis (UC); RA; myasthenia gravis; GVHD
TARGETING COSTIMULATORY AND ACCESSORY MOLECULES			
Abatacept (BMS-188667)	Orencia	Fc fusion protein with extracellular domain of CTLA-4, blocks CD28-CD80/86 interaction	RA; lupus nephritis; inflammatory bowel disease (IBD); juvenile idiopathic arthritis (JIA)
Belatacept (BMS-224818, LEA29Y)		Same as Abatacept, higher affinity	Transplant
TARGETING CYTOKINES/CYTOKINE SIGNALING			
Sirolimus	Rapamune	mTOR inhibitor	Renal transplant, GVHD

Source: Scott M. Steward-Tharp, Yun-Jeong Song, Richard M. Siegel, John J. O'Shea, 2010, New insights into T cell biology and T cell-directed therapy for autoimmunity, inflammation, and immunosuppression. *Annals of the New York Academy of Sciences* **1183:**123, Table 1 with modifications.

widely used to treat RA, psoriasis, and Crohn's disease. An IL-1 receptor antagonist is approved for treatment of RA, as are antibodies directed against the IL-6 receptor and IL-15. Other anti-inflammatory, cytokine-based experimental treatments for autoimmunity include targeting the IL-2 receptor (CD25 and CD122), IL-1, and IFNs. More broadly, the class

of drugs designated as *statins*, used by millions to reduce cholesterol levels, have been found to lower serum levels of C-reactive protein (an acute-phase protein and indicator of inflammation), reduce levels of pro-inflammatory cytokines, and decrease expression of adhesion molecules on endothelial cells. Clinical trials of statins for treatment of RA and MS have shown encouraging results. The addition of such drugs with prior FDA approval and extensive safety testing is a tremendous advantage, considering that 95% of agents fail human trials due to safety concerns.

Compounds that block the chemokine or adhesion molecule signals controlling lymphocyte movement into sites of inflammation can also thwart autoimmune processes. The most well-characterized inhibitor of cell trafficking is FTY720. This compound is an analog of sphingosine 1-phosphate (S1P), which is involved in the migration of lymphocytes into the blood and lymph. By acting as a receptor antagonist, it inhibits egress of all subsets of T cells, resulting in a reduction of up to 85% in circulating blood lymphocytes. Importantly, FTY720, which has been effective in treating MS, is also reported to inhibit T_H1 and T_H17 cells, and to enhance T_{REG} cell activity.

Strategies That Interfere with Costimulation

T cells require both antigenic stimulation via the TCR (signal 1) and costimulation (signal 2) in order to become fully activated (see Chapter 11). Without costimulation, T cells undergo apoptosis, become anergic, or are induced as immune inhibitors. Therefore, one way to control T-cell activation would be to regulate costimulation. To this end, a fusion protein was generated consisting of the extracellular domain of CTLA-4 combined with the human IgG1 constant region. CTLA-4 binds to its CD80/86 partner with an affinity that is approximately 20 times greater than that of CD28. This therapeutic fusion protein, called abatacept (Orencia) and approved for the treatment of RA, has also been studied with limited success in patients with MS, T1DM, SLE, and inflammatory bowel disease. This drug blocks CD80/86 on APCs from engaging with CD28 on T cells.

Antigen-Specific Immunotherapy

The holy grail of immunotherapy to treat autoimmune disease is a strategy that specifically targets the autoreactive cells, sparing all other leukocytes. In an ideal world, clinical therapies that induce tolerance to the auto-antigen could reverse the course of autoimmune disease. However, even in cases when the auto-antigen is known, as with T1DM and MS, there is a risk of exacerbating the disease by introduction of the tolerogen, as was seen in some of the early trials. Nonetheless, glatiramer acetate (GA), a polymer of four basic amino acids found commonly in MBP and approved for treating MS, selectively increased the number of T_{REG} cells and modulated the function of APCs, suggesting that this strategy may also be effective for the treatment of other autoimmune disorders. This drug is the only FDA-approved antigen-specific treatment currently available for an autoimmune disease, although others are in the pipeline. BHT-3021, a DNA-based product aimed at tolerizing the immune system to proinsulin, if approved, would be the first antigen-based treatment for type 1 diabetes.

Transplantation Immunology

Alexis Carrel reported the first systematic study of transplantation in 1908; he interchanged both kidneys in a series of cats, some of which maintained urinary output for up to 25 days. Although all the cats eventually died, the experiment established that a transplanted organ could carry out its normal function in the recipient. The first human kidney transplant, attempted in 1935 by a Russian surgeon, failed because a mismatch of blood types between donor and recipient caused almost immediate rejection of the kidney. We now know that this rapid immune response, termed hyperacute rejection, is mediated by preformed antibodies (described below). Finally, in 1954 a team in Boston headed by Joseph Murray performed the first successful human kidney transplant between identical twins, followed 3 years later by the first transplant between nonidentical individuals. Today, the transfer of various organs and tissues between individuals is performed with ever-increasing frequency and rates of success, at least for their short-term survival.

Although a supply of organs is provided by accident victims and, in some cases, living donors, many more people are in need of transplants than can be accommodated with available organs. According to the U.S Department of Health and Human Services, as of December 2012 over 116,000 individuals in the United States are on the waiting list for an organ transplant (see http://optn.transplant.hrsa.gov for real-time data). The majority of those on the list (over 75%) require a kidney, for which the median waiting period ranges from 3 to 5 years.

Immunosuppressive agents can delay or prevent rejection of transplanted organs, but they have side effects. New treatments that promise longer transplant survival and more specific tolerance to the graft without suppressing other immune function are under development. This section describes the mechanisms underlying graft rejection, procedures that are presently used to prolong graft survival, and the current status of transplantation as a clinical tool.

Graft Rejection Occurs Based on Immunologic Principles

The degree and type of immune response to a transplant varies with the type and source of the grafted tissue. The following terms denote different types of transplants:

- **Autograft** is self tissue transferred from one body site to another in the same individual. Examples include transferring healthy skin to a burned area in burn patients and using healthy blood vessels to replace blocked coronary arteries.

- **Isograft** is tissue transferred between genetically identical individuals. This occurs in inbred strains of mice or identical human twins, when the donor and recipient are syngeneic.

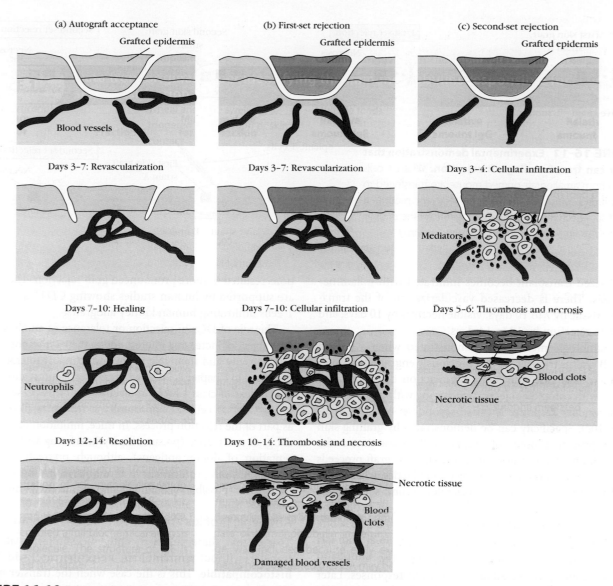

FIGURE 16-10 Schematic diagrams of the process of graft acceptance and rejection. (a) Acceptance of an autograft is completed within 12 to 14 days. (b) First-set rejection of an allograft begins 7 to 10 days after grafting, with full rejection occurring by 10 to 14 days. (c) Second-set rejection of an allograft begins within 3 to 4 days, with full rejection by 5 to 6 days. The cellular infiltrate that invades an allograft (b, c) contains lymphocytes, phagocytes, and other inflammatory cells. *[© 2013 W. H. Freeman and Company.]*

- **Allograft** is tissue transferred between genetically different members of the same species. In mice this means transferring tissue from one strain to another, and in humans this occurs in transplants in which the donor and recipient are not genetically identical (the majority of cases).

- **Xenograft** is tissue transferred between different species (e.g., the graft of a baboon heart into a human). Because of significant shortages of donated organs, raising animals for the specific purpose of serving as organ donors for humans is under serious consideration.

Autografts and isografts are usually accepted, owing to the genetic identity between donor and recipient (Figure 16-10a). Because an allograft is genetically dissimilar to the host and

therefore expresses unique antigens, it is often recognized as foreign by the immune system and is therefore rejected. Obviously, xenografts exhibit the greatest genetic and antigenic disparity, engendering a vigorous graft rejection response.

Specificity and Memory in Allograft Rejection

The rate of allograft rejection varies according to the tissue involved; skin grafts are generally rejected faster than other tissues, such as kidney or heart. Despite these time differences, the immune response culminating in graft rejection always displays the attributes of specificity and memory. If an inbred mouse of strain A is grafted with skin from strain B, primary graft rejection, known as first-set rejection, occurs (Figure 16-10b). The skin first becomes revascularized between days 3 and 7. As the reaction develops, the

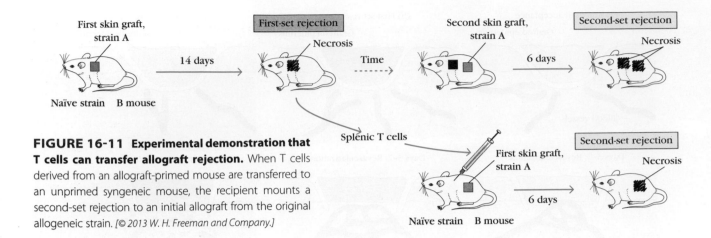

FIGURE 16-11 Experimental demonstration that T cells can transfer allograft rejection. When T cells derived from an allograft-primed mouse are transferred to an unprimed syngeneic mouse, the recipient mounts a second-set rejection to an initial allograft from the original allogeneic strain. [© 2013 W. H. Freeman and Company.]

vascularized transplant becomes infiltrated with inflammatory cells. There is decreased vascularization of the transplanted tissue by 7 to 10 days, visible necrosis by 10 days, and complete rejection by 12 to 14 days.

Immunologic memory is demonstrated when a second strain-B graft is transferred to a previously engrafted strain-A mouse. In this case, the anti-graft reaction develops more quickly, with complete rejection occurring within 5 to 6 days. This secondary response is called second-set rejection (Figure 16-10c). Specificity can be demonstrated by grafting skin from an unrelated mouse of strain C at the same time as the second strain-B graft. Rejection of the strain-C graft proceeds according to the slower, first-set rejection kinetics, whereas the strain-B graft is rejected in an accelerated second-set fashion.

Role of T Cells in Graft Rejection

Using adoptive transfer studies in the early 1950s, Avrion Mitchison showed that donor lymphocytes but not serum antibody could transfer allograft rejection responses. Later studies further defined these as T cells. For instance, nude mice, which lack a thymus and consequently lack functional T cells, were found to be incapable of allograft rejection; these mice even accept xenografts. In other studies, T cells derived from an allograft-primed mouse were shown to transfer second-set allograft rejection to an unprimed syngeneic recipient as long as that recipient was grafted with the same allogeneic tissue (Figure 16-11).

Analysis of the T-cell subpopulations involved in allograft rejection has implicated both CD4[+] and CD8[+] populations. In one study, mice were injected with monoclonal antibodies to deplete one or both types of T cells and then the rate of graft rejection was measured. As shown in Figure 16-12, removal of the CD8[+] population alone had no effect on graft survival, and the graft was rejected at the same rate as in control mice (15 days). Removal of the CD4[+] T-cell population alone prolonged graft survival from 15 days to 30 days. However, removal of both CD4[+] and CD8[+] T cells resulted in long-term survival (up to 60 days) of the allografts. This study indicated that both CD4[+] and CD8[+] T cells participated in rejection and that the collaboration of the two subpopula-

tions resulted in more pronounced graft rejection. These data are supported by human studies showing CD4[+] and CD8[+] T cells infiltrating human kidney allografts.

The role of DCs in rejection or tolerance of an allograft is the subject of increasing interest due to their immunostimulatory capacity and their role in the induction of tolerance. As discussed in Chapter 8, DCs can present exogenous antigens in the context of class I MHC molecules via cross-presentation, giving CD8[+] T cells the opportunity to recognize alloantigens as part of the rejection process. In mice, inhibition of DCs can aid graft acceptance (presumably by interfering with the presentation of donor antigens), although pretreatment with donor DCs can promote survival of both heart and pancreas transplants (possibly by inducing tolerance to donor antigens).

Antigenic Profiles and Transplantation Tolerance

Tissues that share sufficient antigenic similarity, allowing transfer without immunologic rejection, are said to be **histocompatible**. This is the case when the transfer occurs

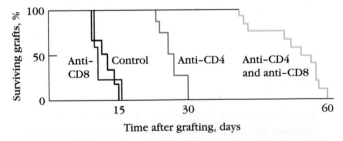

FIGURE 16-12 The role of CD4[+] and CD8[+] T cells in allograft rejection is demonstrated by the curves showing survival times of skin grafts between mice mismatched at the MHC. Animals in which the CD8[+] T cells were removed by treatment with an anti-CD8 monoclonal antibody (red) showed little difference from untreated control mice (black). Treatment with monoclonal anti-CD4 antibody (blue) improved graft survival significantly, and treatment with both anti-CD4 and anti-CD8 antibody prolonged graft survival most dramatically (green). [Cobbold SP, Martin G, Qin S, Waldmann H. 1986. Monoclonal antibodies to promote marrow engraftment and tissue graft tolerance. Nature **323**:164–166.]

between identical twins. Tissues that display significant antigenic differences are *histoincompatible* and typically induce an immune response that leads to tissue rejection. Of course, there are many shades of gray in the degree of histocompatibility between a donor and recipient. The antigens involved are encoded by more than 40 different loci, but the loci responsible for the most vigorous allograft rejection reactions are located within the MHC. The organization of the MHC—called the H-2 complex in mice and the HLA complex in humans—was described in detail in Chapter 8 (see Figures 8-7 and 8-8). Because the genes in the MHC locus are closely linked, they are usually inherited as a complete set from each parent, called a **haplotype**.

In inbred strains of mice, offspring inherit the same haplotype from each parent, meaning they are homozygous at the MHC locus. When mice from two different inbred strains are mated, the F_1 progeny each inherit one maternal and one paternal haplotype (see Figure 8-9). These heterozygous F_1 offspring express the MHC type from both parents (*b/k*, in Figure 8-10), which means they are tolerant to the alleles from both haplotypes and can accept grafts from either parent. However, neither of the parental strains can accept grafts from the F_1 offspring because each parent lacks one of the F_1 haplotypes and will therefore reject these MHC antigens.

In outbred populations, there is a high degree of heterozygosity at most loci, including the MHC. In matings between members of an outbred species, there is only a 25% chance that any two offspring will inherit identical MHC haplotypes unless the parents share one or more haplotypes. Therefore, for purposes of organ or bone marrow grafts, it can be assumed that there is a 25% chance of MHC identity between any two siblings. With parent-to-child grafts, the donor and recipient will always have one haplotype in common (50% match), which is why these grafts are so common. However, in this case, the donor and recipient are still nearly always mismatched for all or most alleles inherited from the other parent, providing a target for the immune system.

Role of Blood Group and MHC Antigens in Graft Tolerance

The most intense graft rejection reactions are due to differences between donor and recipient in ABO blood-group and MHC antigens. The blood-group antigens are expressed on RBCs, epithelial cells, and endothelial cells, requiring the donor and recipient to first be screened for ABO compatibility. If the recipient carries antibodies to any of these antigens, the transplanted tissue will induce rapid antibody-mediated lysis of the incompatible donor cells. For this reason, most transplants are conducted between individuals with a matching ABO blood group.

Next, the MHC compatibility between potential donors and a recipient is determined. The first choice is usually parents or first-order siblings with at least a partial MHC match, followed by other family members and even friends. Given our current success with immunosuppression and immune

tolerance induction protocols, solid organ transplants between individuals with even a total HLA mismatch can be successful. The most rigorous testing is conducted in bone marrow transplants, where at least partial HLA matching is crucial.

Several tests can be used to determine the HLA compatibility, and the choice is somewhat dependent on the organ or tissue in question. Molecular assays using sequence-specific primers to establish which HLA alleles are expressed by the recipient and potential donors (called *tissue typing*) has become more common in recent years, especially in bone marrow transplantation. Molecular assays provide greater specificity and higher resolution than assays that characterize MHC molecules serologically, using antigen-antibody interactions alone, which was standard practice in the past.

The presence of any preformed antibodies against potential donor HLA alloantigens must also be evaluated in the recipient. We generate antibodies against nonself HLA proteins for a number of reasons, but transplant recipients who have received prior allografts are especially likely to possess them. Testing for this is called *cross-matching*, and is the most important level of compatibility testing that occurs prior to solid organ transfer; a positive cross-match means that the recipient has antibodies against HLA proteins carried by the donor. The most common method used today is the Luminex assay, which employs fluorochrome-labeled microbeads impregnated with specific HLA proteins (Figure 16-13). Each HLA protein is associated with a fluorochrome of a different intensity. These HLA-impregnated beads are mixed with recipient serum, allowing clinicians to determine more precisely which donor specific anti-HLA antibodies are present

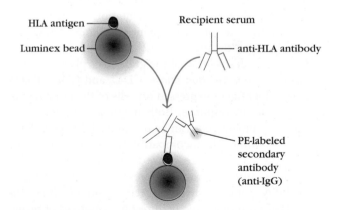

FIGURE 16-13 The Luminex cross-matching assay. Microbeads impregnated with fluorochromes of different intensity each carry a different HLA protein. Recipient serum is incubated with these beads, and any antibody binding is detected using *phycoerythrin* (PE)-labeled secondary anti-human immunoglobulin. Laser excitation and detection are used to determine the fluorochrome intensity of bound beads, and therefore the associated HLA molecule(s) with which the serum reacts. *[B. D. Tait, 2009, Luminex technology for HLA antibody detection in organ transplantation.* Nephrology *14:247–254, with modifications.]*

in the recipient prior to transplantation. The importance of careful cross-matching was shown in a seminal 1969 study, where up to 80% of kidney transplant patients with a positive cross-match experienced almost immediate transplant rejection, while only 5% of patients with a negative cross-match had this outcome.

The MHC makeup of donor and host is not the sole factor determining tissue acceptance. Even when MHC antigens are identical, the transplanted tissue can be rejected because of differences at various other loci, including the **minor histocompatibility locus**. As described in Chapter 8, the major histocompatibility antigens are recognized directly by T_H and T_C cells, a phenomenon termed *alloreactivity*. In contrast, minor histocompatibility antigens are recognized only when peptide fragments are presented in the context of self-MHC molecules. Rejection based on only minor histocompatibility differences is usually less vigorous but can still lead to graft rejection. Therefore, even in cases of HLA-identical matches, some degree of immune suppression is usually still required.

The Sensitization Stage of Graft Rejection

Graft rejection occurs in stages and can be caused by both humoral and cell-mediated immune responses to alloantigens (primarily, MHC molecules) expressed on cells of the graft. Antibody-mediated, DTH, and cell-mediated cytotoxicity reactions have all been implicated, but the latter are primarily credited with orchestrating this response. As we discuss later, immediate hyperacute rejection is caused primarily by preexisting anti-HLA antibodies. When graft rejection occurs in the absence of this preexisting immunity, it can be divided into two stages: (1) a sensitization phase, which occurs shortly after transplantation when antigen-reactive lymphocytes of the recipient proliferate in response to alloantigens on the graft, and (2) a later effector stage, in which immune destruction of the graft takes place.

During the *sensitization phase*, $CD4^+$ and $CD8^+$ T cells recognize alloantigens expressed on cells of the foreign graft and proliferate in response. Both major and minor histocompatibility alloantigens can be recognized. In general, the response to minor histocompatibility antigens is weak, although the combined response to several minor differences can be quite vigorous. The response to major histocompatibility antigens involves recognition of either the donor MHC molecule directly (*direct presentation*) or recognition of peptides from donor HLA in the cleft of the recipient's own MHC molecules (*indirect presentation*).

A host T_H cell becomes activated when it interacts with an APC that both expresses an appropriate antigenic-ligand/MHC-molecule complex and provides the requisite costimulatory signal. Because DCs are found in most tissues and constitutively express high levels of class II MHC molecules, activated allogeneic donor DCs can mediate direct presentation in grafts or in the draining lymph node, to which they can sometimes migrate. APCs of host origin can also migrate into a graft and endocytose the foreign alloantigens (both major and minor histocompatibility molecules), where they become activated and present alloantigens indirectly as processed peptides bound to self-MHC molecules. The cross-presentation ability of DCs (see Chapter 8) also allows them to present endocytic antigens in the context of class I MHC molecules to $CD8^+$ T cells, which can then participate in allograft rejection. In addition to DCs, other cell types have been implicated in alloantigen presentation and immune activation leading to graft rejection, including Langerhans cells and endothelial cells lining the blood vessels. Both of these cell types express class I and class II MHC antigens.

However, as we know from Chapter 11, T cells that respond to antigen via the TCR in the *absence* of costimulation or danger signals can become tolerant. This may help to explain a long-standing clinical observation: transfusion of donor blood into a graft recipient *prior* to transplantation can facilitate acceptance of a subsequent graft from that donor. This suggests that exposure to donor cells in this nonactivating context induced tolerance to donor alloantigens. Newer immunomodulation protocols based on this observation, as well as related experimental studies, are currently underway to design techniques to effectively induce donor-specific tolerance prior to engraftment.

Effector Stage of Graft Rejection

A variety of mechanisms participate in the effector stage of allograft rejection (Figure 16-14). The most common are cell-mediated reactions; less common mechanisms (except during hyperacute rejection) are antibody-mediated complement lysis and destruction by ADCC. The hallmark of graft rejection involving cell-mediated reactions is an influx of immune cells into the graft. Among these are T cells, especially $CD4^+$, and APCs, often macrophages. Histologically, the infiltration in many cases resembles that seen during a DTH response, in which cytokines produced by T cells promote inflammatory cell infiltration (see Figure 15-14). Although probably less important, recognition by host $CD8^+$ T cells of either foreign class I alloantigens on the graft or alloantigenic peptides cross-presented in the context of class I MHC by DCs can lead to CTL-mediated killing.

In each of these effector mechanisms, cytokines secreted by T_H cells play a central role. For example, IL-2 and IFN-γ produced by T_H1 cells have been shown to be important mediators of graft rejection. These two cytokines promote T-cell proliferation (including CTLs), DTH responses, and the synthesis of IgG by B cells, with resulting complement activation. A number of cytokines that encourage the expression of MHC class I and class II molecules (e.g., the interferons and TNFs) increase during graft rejection episodes, inducing a variety of cell types within the graft to increase surface expression of these proteins. Many of the cytokines most closely associated with T_H2 and T_H17 cells have also been implicated in graft rejection. Elevations in IL-4, -5, and -13, responsible for B-cell activation and eosinophil accumulation in allografts, and in IL-17, have all been linked to transplant rejection. Recent studies showing that

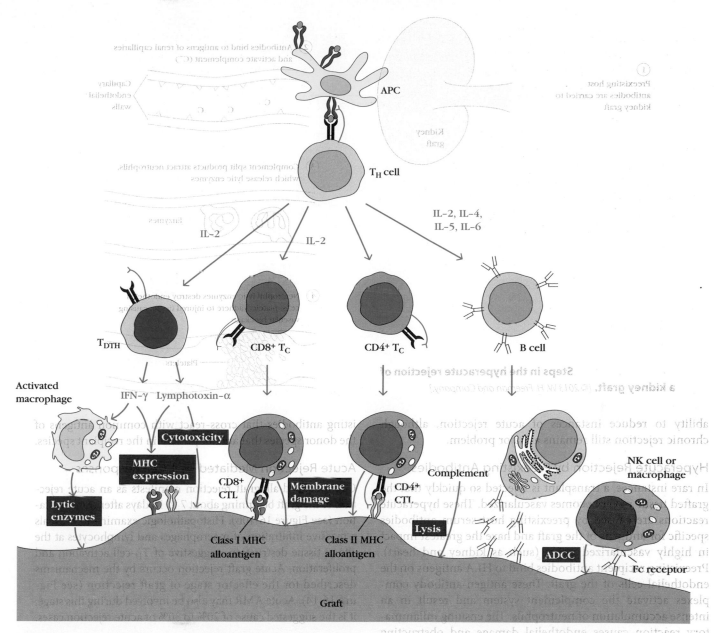

FIGURE 16-14 Effector mechanisms involved in allograft rejection. The generation or activity of various effector cells depends directly or indirectly on cytokines (blue) secreted by activated T_H cells. ADCC = *antibody-dependent cell-mediated cytotoxicity.* [© 2013 W. H. Freeman and Company.]

neutralization of IL-17 could extend the survival of cardiac allografts in the mouse have generated much interest in this cytokine and in the role of T_H17 cells in graft rejection.

Finally, *antibody-mediated rejection* (AMR), although less frequent, is still a major issue in clinical transplantation. These antibodies are often directed against donor HLA molecules or endothelial antigens. The hallmarks of this response, which is dependent on T-cell maintenance of these alloreactive B cells, are the activation of complement and deposition of C4d, especially among endothelial cells lining graft capillaries. AMR, although most associated with the earliest stages of rejection, can occur at any time during the clinical course of allograft rejection.

Graft Rejection Follows a Predictable Clinical Course

Graft rejection reactions, although somewhat variable in their time courses depending on the type of tissue transferred and the immune response involved, follow a fairly predictable course. Hyperacute rejection reactions typically occur within the first 24 hours after transplantation, acute rejection reactions usually begin in the first few weeks after transplantation, and chronic rejection reactions can occur from months to years after transplantation. Careful cross-matching can avoid most cases of hyperacute rejection. Current immunosuppressive agents have greatly advanced our

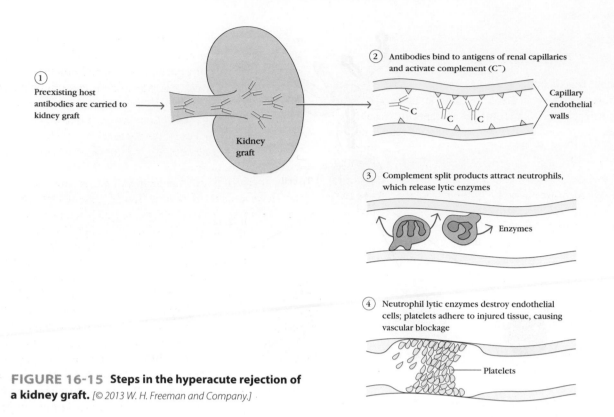

FIGURE 16-15 Steps in the hyperacute rejection of a kidney graft. *[© 2013 W. H. Freeman and Company.]*

ability to reduce instances of acute rejection, although chronic rejection still remains a major problem.

Hyperacute Rejection by Preexisting Antibodies

In rare instances, a transplant is rejected so quickly that the grafted tissue never becomes vascularized. These hyperacute reactions are caused by preexisting host serum antibodies specific for antigens of the graft and have the greatest impact in highly vascularized grafts (such as kidney and heart). Preexisting recipient antibodies bind to HLA antigens on the endothelial cells of the graft. These antigen-antibody complexes activate the complement system and result in an intense accumulation of neutrophils. The ensuing inflammatory reaction causes endothelial damage and obstructing blood clots within the capillaries, preventing vascularization of the graft (Figure 16-15).

Several mechanisms can account for the presence of preexisting antibodies specific for allogeneic MHC antigens. These include repeated blood transfusions that induced antibodies to MHC antigens expressed on allogeneic WBCs in the blood; repeated pregnancies, in which women develop antibodies against paternal alloantigens of the fetus; exposure to infectious agents, which can elicit MHC cross-reactive antibodies; or a previous transplant, which induced high levels of antibodies to the allogeneic MHC antigens present in that graft. In some cases, preexisting antibodies specific for blood-group antigens may also be present and can mediate hyperacute rejection. However, with careful cross-matching and ABO blood-group typing, many instances of hyperacute rejection can be avoided. Xenotransplants (see below) are also often rejected in a hyperacute manner because of preex-

isting antibodies that cross-react with common antigens of the donor species that are not present in the recipient species.

Acute Rejection Mediated by T-Cell Responses

Cell-mediated allograft rejection manifests as an acute rejection of the graft beginning about 7 to 10 days after transplantation (see Figure 16-10b). Histopathologic examination reveals a massive infiltration of macrophages and lymphocytes at the site of tissue destruction, suggestive of T_H-cell activation and proliferation. Acute graft rejection occurs by the mechanisms described for the effector stage of graft rejection (see Figure 16-14). Acute AMR may also be involved during this stage; it is the suggested cause of 20% to 30% of acute rejection cases.

Chronic Rejection Phase

Chronic rejection reactions develop months or years after acute rejection reactions have subsided. The mechanisms include both humoral and cell-mediated responses by the recipient. Although immunosuppressive drugs and advanced tissue-typing techniques have dramatically increased survival of allografts during the first years, little progress has been made in long-term survival. In data collected in the United States as of 2008, 1-year kidney graft survival rates approached 97%. However, even in cases of a living donor—the most ideal scenario—10-year survival rates were only 60% (based on procedures performed in 2000). Immunosuppressive drugs usually do little to manage chronic rejection, which not infrequently necessitates another transplant. In up to 60% of cases in which there is some form of chronic allograft dysfunction, antidonor antibodies can be found in the recipient, suggesting that in addition to cell-mediated responses, AMR may also be involved in chronic rejection.

Immunosuppressive Therapy Can Be Either General or Target-Specific

Allogeneic transplantation always requires some degree of immunosuppression if the transplant is to survive. Most immunosuppressive treatments are nonspecific, resulting in generalized suppression of responses to all antigens, not just those of the allograft. This places the recipient at increased risk of infection and cancer. In fact, infection is the most common cause of transplant-related death. Many immunosuppressive measures slow the proliferation of activated lymphocytes, thus affecting any rapidly dividing nonimmune cells (e.g., gut epithelial cells or bone marrow hematopoietic stem cells), and leading to serious or even life-threatening complications. Patients on long-term immunosuppressive therapy are also at increased risk of hypertension and metabolic bone disease. Fine-tuning of their immunosuppressive cocktail, and eventual weaning off these drugs, is an ongoing process for most, if not all, transplant recipients.

Total Lymphoid Irradiation to Eliminate Lymphocytes

Because lymphocytes are extremely sensitive to x-rays, x-irradiation can be used to eliminate them in the transplant recipient just before grafting. Although not a part of most immunosuppressive regimens, this is often used in bone marrow transplantation or to treat **graft-versus-host disease (GVHD)**, in which the graft rejects the host. In total lymphoid irradiation, the recipient receives multiple x-ray exposures to the thymus, spleen, and lymph nodes before the transplant, and the recipient is engrafted in this immunosuppressed state. Because the bone marrow is not x-irradiated, lymphoid stem cells proliferate and renew the population of recirculating lymphocytes. These newly formed lymphocytes appear to be more likely to become tolerant to the antigens of the graft.

Generalized Immunosuppressive Therapy

In 1959, Robert Schwartz and William Dameshek reported that treatment with 6-mercaptopurine suppressed immune responses in animal models. Joseph Murray and colleagues then screened a number of its chemical analogues for use in human transplantation. One, azathioprine, when used in combination with corticosteroids, dramatically increased survival of allografts. Murray received a Nobel prize in 1991 for this clinical advance, and the developers of the drug, Gertrude Elion and George Hitchings, received the Nobel prize in 1987.

Azathioprine (Imuran) is a potent mitotic inhibitor often given just before and after transplantation to diminish both B- and T-cell proliferation. Other mitotic inhibitors that are sometimes used in conjunction with immunosuppressive agents are cyclophosphamide and methotrexate. Cyclophosphamide is an alkylating agent that inserts into the DNA helix and becomes cross-linked, leading to disruption of the DNA chain. It is especially effective against rapidly dividing cells and is therefore sometimes given at the time of grafting to block T-cell proliferation. Methotrexate acts as a folic-acid antagonist

to block purine biosynthesis. Because mitotic inhibitors act on all rapidly dividing cells, they cause significant side effects, especially affecting the gut and liver, in addition to their target, bone marrow-derived cells. Most often, these mitotic inhibitors are combined with immunosuppressive drugs such as corticosteroids (e.g., prednisone and dexamethasone). These potent anti-inflammatory agents exert their effects at many levels of the immune response and therefore help prevent acute episodes of graft rejection.

More specific immune suppression became possible with the development of several fungal metabolites, including cyclosporin A (CsA), FK506 (tacrolimus), and rapamycin (also known as sirolimus). Although chemically unrelated, these exert similar effects, blocking the activation and proliferation of resting T cells. Some of these also prevent transcription of several genes encoding important T-cell activation molecules, such as IL-2 and the high-affinity IL-2 receptor (IL-2Rα). By inhibiting T_H-cell proliferation and cytokine expression, these drugs reduce the subsequent activation of various effector populations involved in graft rejection, making them a mainstay in heart, liver, kidney, and bone marrow transplantation. In one study of 209 kidney transplants from deceased donors, the 1-year survival rate was 80% among recipients receiving CsA and 64% among those receiving other immunosuppressive treatments. Similar results have been obtained with liver transplants (Figure 16-16). Despite these impressive results, CsA does have some side effects, most notably toxicity to the kidneys. FK506 and rapamycin are 10 to 100 times more potent immunosuppressants than CsA and therefore can be administered at lower doses and with fewer side effects.

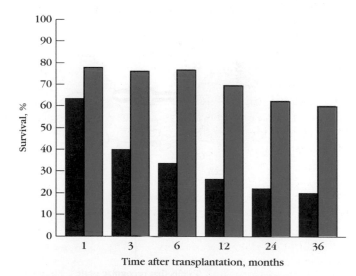

FIGURE 16-16 Comparison of the survival rates of liver transplants following azathioprine versus cyclosporin A treatment. Transplant survival rates are shown over a 3-year period for 84 liver transplant patients immunosuppressed using a combination of azathioprine plus corticosteroids (black) compared with another 55 patients treated with cyclosporin A plus corticosteroids (blue). [*Sabesin SM, Williams JW. 1987. Current status of liver transplantation.* Hospital Practice 22:75–86.]

Specific Immunosuppressive Therapy

The ideal immunosuppressant would be antigen-specific, inhibiting the immune response to the alloantigens present in the graft while preserving the recipient's ability to respond to other foreign antigens. Although this goal has not yet been achieved, several more targeted immunosuppressive agents have been developed. Most involve the use of monoclonal antibodies (mAbs) or soluble ligands that bind specific cell-surface molecules. One limitation of most first-generation mAbs came from their origin in animals. Recipients of these frequently developed an immune response to the nonhuman epitopes, rapidly clearing the mAbs from the body. This limitation has been overcome by the construction of humanized mAbs and mouse-human chimeric antibodies (see Chapter 20).

Many different mAbs have been tested in transplantation settings, and the majority work by either depleting the recipient of a particular cell population or by blocking a key step in immune signaling. Antithymocyte globulin (ATG), prepared from animals exposed to human lymphocytes, can be used to deplete lymphocytes in recipients prior to transplantation, but has significant side effects. A more subset-specific strategy uses a mAb to the CD3 molecule of the TCR, called OKT3, and rapidly depletes mature T cells from the circulation. This depletion appears to be caused by binding of antibody-coated T cells to Fc receptors on phagocytic cells, which then phagocytose and clear the T cells from the circulation. In a further refinement of this strategy, a cytotoxic agent such as diphtheria toxin is coupled with the mAb. Antibody-bound cells then internalize the toxin and die. Another technique uses mAbs specific for the high-affinity IL-2 receptor CD25 (Basiliximab). Since this receptor is expressed only on activated T cells, this treatment specifically blocks proliferation of T cells activated in

response to the alloantigens of the graft. However, since T_{REG} cells also express CD25 and may aid in alloantigen tolerance, this strategy may have drawbacks. More recently, a mAb against CD20 (Rituximab) has been used to deplete mature B cells and is aimed at suppressing AMR responses. Finally, in cases of bone marrow transplantation, mAbs against T-cell-specific markers have been used to pretreat the donor's bone marrow to destroy immunocompetent T cells that may react with the recipient tissues, causing GVHD (described below).

Because cytokines appear to play an important role in allograft rejection, these compounds can also be specifically targeted. Animal studies have explored the use of mAbs specific for the cytokines implicated in transplant rejection, particularly TNF-α, IFN-γ, and IL-2. In mice, anti-TNF-α mAbs prolong bone marrow transplants and reduce the incidence of GVHD. Antibodies to IFN-γ and to IL-2 have each been reported in some cases to prolong cardiac transplants in rats.

As described in Chapter 11, T_H-cell activation requires a costimulatory signal in addition to the signal mediated by the TCR. The interaction between CD80/86 on the membrane of APCs and the CD28 or CTLA-4 molecule on T cells provides one such signal (see Figure 11-3). Without this costimulatory signal, antigen-activated T cells become anergic (see Figure 11-4). CD28 is expressed on both resting and activated T cells, while CTLA-4 is expressed only on activated T cells and binds CD80/86 with a 20-fold-higher affinity. In mice, D. J. Lenschow, J. A. Bluestone, and colleagues demonstrated prolonged graft survival by blocking CD80/86 signaling with a soluble fusion protein consisting of the extracellular domain of CTLA-4 fused to human IgG1 heavy chain (called CTLA-4Ig). This new drug, belatacept, was shown to induce anergy in T cells directed against the grafted tissue and has been

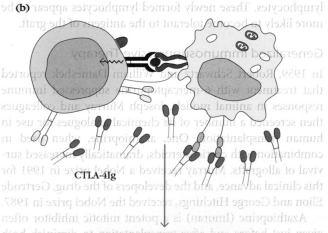

(a)

CD28   CD80/86

T cell   APC

T cells that recognize graft antigens become activated

Graft rejected

(b)

CTLA-4Ig

T cells that recognize graft antigens lack costimulation and become anergic

Graft survives

FIGURE 16-17 Blocking costimulatory signals at the time of transplantation can cause anergy instead of activation of the T cells reactive against the graft. T-cell activation requires both the interaction of the TCR with its ligand and the reaction of costimulatory receptors with their ligands (a). In (b), contact between one of the costimulatory receptors, CD28 on the T cell, and its ligand, CD80/86 on the APC, is blocked by reaction of CD80/86 with the soluble ligand CTLA-4Ig. The CTLA-4 is coupled to an Ig H chain, which slows its clearance from the circulation. This process specifically suppresses graft rejection without inhibiting the immune response to other antigens. [© 2013 W. H. Freeman and Company.]

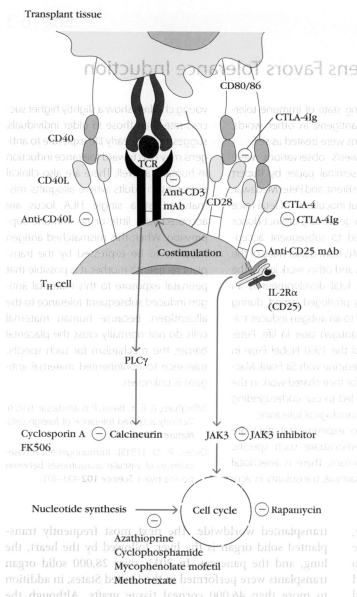

FIGURE 16-18 Sites of action for various agents used in clinical transplantation. [© 2013 W. H. Freeman and Company.]

approved by the FDA for prevention of organ rejection in adult kidney transplant patients (Figure 16-17).

Some of the treatments used to suppress transplant rejection in clinical settings are summarized in Figure 16-18, along with their sites of action.

Immune Tolerance to Allografts Is Favored in Certain Instances

Sometimes, an allograft may be accepted with little or no use of immunosuppressive drugs. Obviously, with tissues that lack alloantigens (e.g., cartilage or heart valves), no immunologic barrier to transplantation exists. Acceptance of an allograft can be favored in one of two situations: when cells or tissue are grafted to a so-called privileged site that is sequestered from immune system surveillance, or when a state of tolerance has been induced biologically, usually by previous exposure to the

antigens of the donor in a manner that causes immune tolerance rather than sensitization. One unique example of the latter, involving fetal exposure to foreign antigens, occurs in species where nonidentical twins share a placenta during fetal development (see Classic Experiments Box 16-3).

Cells and Cytokines Associated with Graft Tolerance

There is now significant evidence that FoxP3-expressing T_{REG} cells play a role in transplantation tolerance. In *clinical operational tolerance*, where the graft survives despite the removal of all immunosuppressive therapy, there is an increase in the number of T_{REG} cells in the circulation and in the graft. These cells are believed to inhibit alloreactive cells using a combination of direct contact and expression of immunosuppressive cytokines, such as TGF-β, IL-10, and IL-35. To date, difficulties identifying and isolating this population of T cells have limited their use as a treatment for inducing transplant tolerance. However, strategies that use existing or induced T_{REG} cells to limit graft rejection, especially in GVHD, are an active area of research.

Immunologically Privileged Sites

An allograft placed in an immunologically privileged site, or an area without significant immune cell access (e.g., the anterior chamber of the eye, cornea, uterus, testes, and brain), is less likely to experience rejection. Each of these sites is characterized by an absence of lymphatic vessels, and sometimes also blood vessels. Consequently, the alloantigens of the graft are not able to sensitize the recipient's lymphocytes, and the graft has an increased likelihood of acceptance even when HLA antigens are not matched. The privileged status of the cornea has allowed corneal transplants to be highly successful. Ironically, the successful transplantation of allogeneic pancreatic islet cells into the thymus in a rat model of diabetes suggests that the thymus may also be a unique type of immunologically privileged site.

Immunologically privileged sites fail to induce an immune response because they are effectively sequestered from the cells of the immune system. This suggests the possibility of physically sequestering grafted cells. In one study, pancreatic islet cells were encapsulated in semipermeable membranes and then transplanted into diabetic mice. The islet cells survived and produced insulin. The transplanted cells were not rejected, because the recipient's immune cells could not penetrate the membrane. This novel transplant method may have application for treatment of human diabetics.

Inducing Transplantation Tolerance

Methods for inducing tolerance to allow acceptance of allografts have been studied extensively in animal models, and some of the discoveries have now been applied to humans. The current favorite involves the induction of a state of mixed hematopoietic chimerism, where donor and recipient hematopoietic cells coexist in the host prior to transplantation. The seed for this strategy originated from

Early Life Exposure to Antigens Favors Tolerance Induction

In 1945, Ray Owen, an immunologist working at the California Institute of Technology, reported a novel observation in cattle. His discovery would advance our understanding of immune tolerance and provide grist for many transplant immunologists who followed in his footsteps. He noticed that nonidentical or dizygotic cattle twins retained the ability to accept cells or tissue from their genetically distinct sibling throughout their lives. This was not true for the nonidentical twins of other mammalian species that did not share a placenta in utero. In cattle, the shared placenta allowed free blood circulation from one twin to the other throughout the embryonic and fetal period. Although the twins may have inherited distinct paternal and maternal antigens, they did not recognize those of their placental partner as foreign and could therefore later accept grafts from them. He hypothesized that exposure to the alloantigens of their placental sibling during this early stage of life somehow

induced a lifelong state of immune tolerance to these antigens; in other words, these alloantigens were treated as self.

In 1953, Owen's observations were extended in a seminal paper by Rupert Billingham, Leslie Brent, and Peter Medawar. They showed that inoculation of fetal mice with cells from a genetically distinct donor mouse strain led to subsequent acceptance of skin grafts from donor mice of the same strain. This and other work led to the hypothesis that fetal development is an immunologically privileged period, during which exposure to an antigen induces tolerance to that antigen later in life. Peter Medawar shared the 1960 Nobel Prize in Physiology or Medicine with Sir Frank MacFarlane Burnet, for their shared work in the discoveries that led to our understanding of acquired immunological tolerance.

Although no experimental data are available to demonstrate such specific tolerance in humans, there is anecdotal evidence. For example, transplants in very

young children show a slightly higher success rate than those in older individuals, suggesting that early life exposure to antigens may bias toward tolerance induction in humans as well. There are also clinical examples in adults where allografts mismatched at a single HLA locus are accepted with little or no immune suppression. When this mismatched antigen happens to be expressed by the transplant recipient's mother, it is possible that perinatal exposure to this maternal antigen induced subsequent tolerance to the alloantigen. Because human maternal cells do not normally cross the placental barrier, the mechanism for such specific tolerance to noninherited maternal antigens is unknown.

Billingham, R. E., L. Brent, P. B. Medawar. (1953). 'Actively acquired tolerance' of foreign cells. *Nature* **172**:603–606.

Owen, R. D. (1945). Immunogenetic consequences of vascular anastomoses between bovine twins. *Science* **102**:400–401.

animal studies and observations in humans. For instance, transplant recipients who underwent total myeloablative therapy followed by donor bone marrow transfer *prior* to receiving a solid organ from the same donor displayed enhanced tolerance for the solid organ graft. A modified protocol involving a less intense non-myeloablative procedure followed by bone marrow transfer resulted in mixed chimerism that, even when quite transient, was still associated with improved graft outcomes. The mechanism for this induction of tolerance is still unclear; both central deletion of alloreactive T cells and an enhancement of immune suppression by T_{REG} cells are hypothesized.

Some Organs Are More Amenable to Clinical Transplantation Than Others

For a number of illnesses, a transplant is the only means of therapy. Figure 16-19 summarizes the major organ and cell transplants being performed today. Certain combinations of organs, such as heart and lung or kidney and pancreas, are being transplanted simultaneously with increasing frequency.

Since the first kidney transplant was performed in the 1950s, it is estimated that over 500,000 kidneys have been

transplanted worldwide. The next most frequently transplanted solid organ is the liver, followed by the heart, the lung, and the pancreas. In 2011, over 28,000 solid organ transplants were performed in the United States, in addition to more than 46,000 corneal tissue grafts. Although the clinical outcomes have improved considerably in the past few years, major obstacles still exist. Immunosuppressive drugs greatly increase the short-term survival of the transplant, but medical problems arise from their use and chronic rejection remains a lingering problem. The need for additional transplants after rejection exacerbates the shortage of organs that is a major obstacle to the widespread use of transplantation. Research on artificial organs continues, but there are no reports of universal long-term successes. This makes the idea of looking outside our species more compelling to some (see Clinical Focus Box 16-4).

The frequency with which a given organ or tissue is transplanted depends on a number of factors, including alternative treatment options, organ availability, and the level of difficulty of the procedure. Several factors contribute to the kidney being the most commonly transplanted organ. Many common diseases (e.g., diabetes) result in kidney failure that can be alleviated by transplantation. Because kidneys come in pairs and we can survive with only one, this organ is

Transplants Performed in 2011

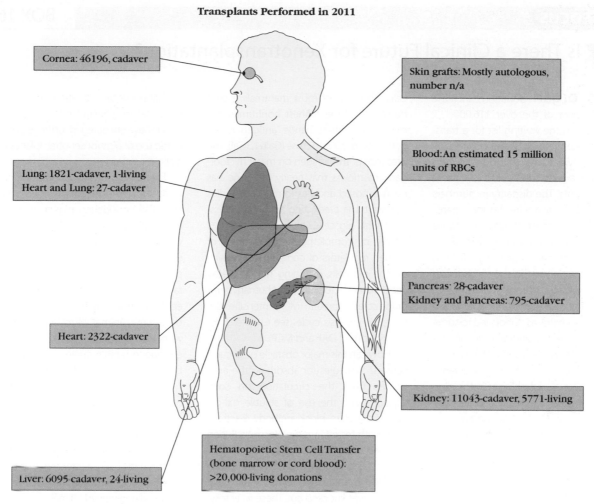

Cornea: 46196, cadaver

Skin grafts: Mostly autologous, number n/a

Lung: 1821-cadaver, 1-living
Heart and Lung: 27-cadaver

Blood: An estimated 15 million units of RBCs

Pancreas: 28-cadaver
Kidney and Pancreas: 795-cadaver

Heart: 2322-cadaver

Kidney: 11043-cadaver, 5771-living

Hematopoietic Stem Cell Transfer (bone marrow or cord blood): >20,000-living donations

Liver: 6095-cadaver, 24-living

FIGURE 16-19 Transplantation routinely conducted in clinical practice. For the solid organs, the number of transplants performed in the United States in 2011 is indicated. Estimates are included for other transplants if available. *[© 2013 W. H. Freeman and Company.]*

available from living as well as deceased donors. In most transplant situations, transfer from a living donor affords enhanced chances of graft survival. Surgical procedures for kidney transfer are also simpler than for the liver or heart. Because many transplants of this type have been conducted for many years, patient-care procedures and effective immunosuppressive regimens are well established. Matching of blood and histocompatibility groups presents no special problems, and transplants can even be conducted across significant mismatches. This contrasts with bone marrow transplants, which must be at least partially matched. The major problems faced by patients waiting for a kidney are the shortage of organs and the increasing number of sensitized recipients. The latter results from rejection of a first transplant, which leaves the recipient sensitized to the alloantigens in that graft. As with all nonidentical transplants, it is typically necessary to maintain kidney transplant patients on some form of immunosuppression for their entire lives. Unfortunately, this gives rise to complications, including risks of cancer and infection, as well as other side effects such as hypertension and metabolic bone disease.

After the kidney, bone marrow is the most frequent transplant. This procedure is increasingly used to treat hematologic diseases, including leukemia, lymphoma, and immunodeficiencies, especially *severe combined immunodeficiency* (SCID; see Chapter 18). Although the supply of bone marrow, which is a renewing resource, is less of a problem than is the supply of kidneys, finding a matched donor is a major obstacle. However, current tissue typing techniques can quickly identify donors with at least partial HLA matches. Bone marrow transplant recipients are typically immunologically suppressed before transfer, making graft rejection rare. However, the presence of foreign immunocompetent cells means that GVHD is a real risk, although the use of immunosuppressive drugs and pretreatment to deplete T cells from the graft have improved outcomes.

Perhaps the most dramatic forms of transplantation involve the transfer of heart, lung, or both: situations where the recipient must be kept alive via artificial means during surgery. The human heart can remain viable for a limited time in ice-cold buffer solutions, which delay tissue damage. However,

BOX 16-4

Is There a Clinical Future for Xenotransplantation?

Unless organ donations increase drastically, most of the over 116,000 U.S. patients still on the waiting list for a transplant at the end of 2012 will not receive one. In fact, fewer than 30,000 donor organs will become available in time to save these waiting patients. The disparity in numbers of individuals on a waiting list for a transplant and the number of available organs grows every year, making it increasingly unlikely that human organs can fill this need. One solution to this shortfall is to utilize animal organs, a process called xenotransplantation. Clinical attempts at using nonhuman primates as donors led to some short-term success, including function of a baboon liver for 70 days and a chimpanzee kidney for 9 months in human recipients. Although there are advantages in the phylogenetic similarity between primate species, the use of nonhuman primates as organ donors has several important disadvantages. This similarity carries with it elevated risk in terms of the transfer of pathogenic viruses, not to mention the impracticalities and ethical concerns that arise in the use of these close cousins.

The use of pigs to supply organs for humans has been under serious consideration for many years. Pigs breed rapidly, have large litters, can be housed in pathogen-free environments, and share considerable anatomic and physiologic similarity with humans. In fact, pigs have served as donors of cardiac valves for humans for years. However, balancing the advantages of pig donors are several serious difficulties. For example, if a pig kidney were implanted into a human by techniques standard for human transplants, it would likely fail in a rapid and dramatic fashion due to hyperacute rejection. This antibody-mediated rejection is due to the presence on the pig

cells (and cells of most mammals except humans and the highest nonhuman primates) of a disaccharide antigen called galactosyl-α-1,3-galactose (Galα1,3Gal). The presence of this antigen on many microorganisms means that nearly everyone has been exposed and has formed antibodies against it. The preexisting antibodies cross-react with pig cells, which are then lysed rapidly by complement. The absence of human regulators of complement activity on the pig cells, including human *decay accelerating factor* (DAF) and human *membrane cofactor protein* (MCP), intensifies the complement lysis cycle (see Chapter 6 for descriptions of DAF and MCP).

How can this major obstacle be circumvented? Strategies for absorbing the antibodies from the circulation on solid supports and the use of soluble gal-gal disaccharides to block antibody reactions were both tested. A more elegant solution involved genetically engineered pigs in which the gene for the enzyme responsible for the addition of Galα1,3Gal to pig proteins was knocked out. These *galactosyl transferase gene knockout* (GalT-KO) pigs have been used as heart or kidney donors for baboons in experimental systems. K. Kuwaki and colleagues transplanted GalT-KO pig hearts into baboons immunosuppressed with antithymocyte globulin and an anti-CD154 monoclonal antibody (the CD40 ligand found mostly on T cells) and then maintained with commonly used immunosuppressive drugs. The mean survival time was 92 days, and one GalT-KO pig heart transplant survived in a baboon for 179 days. K. Yamada and coworkers demonstrated kidney function in recipients of GalT-KO pig kidneys for up to 83 days using a regimen of simultaneous thymus transplant in an attempt to establish tolerance in

the baboon recipients. Although these studies were not conclusive, some promising results have encouraged further exploration of the use of pigs for xenotransplantation in a clinical setting.

Even if all issues of antigenic difference were resolved, additional concerns remain for those considering pigs as a source of transplanted tissue. Pig endogenous retroviruses introduced into humans as a result of xenotransplantation could cause significant disease. Opponents of xenotransplantation raise the specter of another HIV-type epidemic resulting from human infection by a new animal retrovirus. Continuing work on development of pigs free of endogenous pig retroviruses could reduce the possibility of this bleak outcome.

Will we see the use of pig kidneys in humans in the near future? The increasing demand for organs is driving the commercial development of colonies of pigs suitable for such purposes. Although kidneys are the most sought-after organ at present, other organs and cells from the specially bred and engineered animals will find use if they are proven to be safe and effective. A statement issued in 2000 from the American Society of Transplantation and the American Society of Transplant Surgeons endorses the use of xenotransplants if certain conditions are met, including the demonstration of feasibility in a nonhuman primate model, proven benefit to the patient, and lack of infectious disease risk. Although certain barriers remain to the clinical use of xenotransplants, serious efforts are in motion to overcome these difficulties.

Ekser B and Cooper D. 2010. Overcoming barriers to xenotransplantation: prospects for the future. *Expert Reviews in Clinical Immunology*, 6(2):219–230.

eventually lack of oxygen (ischemia) and the resulting deprivation of ATP lead to irreversible organ death. The surgical methods for implanting a heart have been available since the first heart transplant was carried out in 1964 in South Africa by Dr. Christian Barnard. Today, the 1-year survival rate for

heart transplants has climbed to greater than 80%. Brain-dead accident victims with an intact circulatory system and a functioning heart are the typical source of these organs. HLA matching is often not possible because of the limited supply of these organs and the urgency of the procedure.

The liver is important because it clears and detoxifies substances in the body. Malfunction of this organ can be caused by viral diseases (e.g., hepatitis) or exposure to harmful chemicals (e.g., chronic alcoholism), although most liver transplants are actually performed to correct congenital abnormalities. This organ has a complicated circulatory network, posing some unique technical challenges. However, its large size also presents opportunity; the liver from a single donor can often be split and given to at least two different recipients.

One of the more common diseases in the United States is diabetes mellitus. This disease is caused by malfunction of insulin-producing islet cells in the pancreas. Newer protocols that avoid whole-organ transfer involve harvesting donor islet cells and perfusing them into the recipient's liver, where they become permanently established in the liver sinusoids. Initial results indicate that 53% of recipients are insulin independent after such a transplant, some for up to 2 years. Several factors favor survival of functioning pancreatic cells, the most important being the condition of the islet cells used for implantation.

Most skin transplants are conducted with autologous tissue. However, after severe burns, foreign skin thawed from frozen deposits may also be used. These grafts generally act as biologic dressings because the cellular elements are no longer viable and the graft does not grow in the new host. True allogeneic skin grafting using fresh viable donor skin has been undertaken, but rejection is a significant issue that must be managed with aggressive immunosuppressive therapy, which unfortunately also increases the already high infection risk.

This list of commonly transplanted tissue is in no way comprehensive, and will surely expand with time. Improved procedures for inducing tolerance and controlling rejection, along with any advances in future organ availability, would add significantly to this list. For instance, the recent use of intracerebral neural cell grafts has restored function in victims of Parkinson's disease. In studies conducted thus far, the source of neural donor cells was human embryos; the possibility of using those from other animal species is being tested. Likewise, the transfer of composite tissues (e.g., whole digit, limb, and even facial transplants) is still relatively rare and extremely complicated, but advances are being made.

SUMMARY

- A major task of the immune system is to distinguish self from nonself. Failure to do so results in immune attacks against cells and organs of the host with the possible onset of autoimmune disease.

- Mechanisms to prevent self-reactivity (i.e., tolerance) operate at several levels. Central tolerance serves to delete self-reactive T or B lymphocytes; peripheral tolerance inactivates or regulates self-reactive lymphocytes that survive the initial screening process.

- Human autoimmune diseases can be divided into organ-specific and systemic diseases. The organ-specific diseases involve an autoimmune response directed primarily against a single organ or gland. The systemic diseases are directed against a broad spectrum of tissues.

- There are both spontaneous and experimental animal models for autoimmune diseases. Spontaneous autoimmune diseases result from genetic defects, whereas experimental animal models have been developed by immunizing animals with self antigens in the presence of adjuvant.

- There is evidence for genetic and environmental influences on autoimmunity. In particular, certain alleles of MHC have been strongly linked to autoimmunity. In addition, defects in many different genes involved in immunity can predispose individuals to autoimmune disease. However, environmental factors, including microflora and infection, can also have impacts on autoimmune susceptibility.

- CD4$^+$ rather than CD8$^+$ T cells are most associated with autoimmunity. There is evidence for both T$_H$1 and T$_H$17 cells in the development of autoimmunity, depending on the disease in question.

- A variety of mechanisms have been proposed for induction of autoimmunity, including release of sequestered antigens, molecular mimicry, and polyclonal stimulation of lymphocytes. Evidence exists for each of these mechanisms, reflecting the many different pathways leading to autoimmune reactions.

- Current therapies for autoimmune diseases include treatment with generally immunosuppressive drugs as well as treatments that inhibit specific cell types or pathways, such as B cells, T cells, adhesion molecules, costimulation, and T$_H$17 cells. Strategies aimed at enhancement of T$_{REG}$ cells, induction of tolerance, and antigen-specific targeting are also under development.

- Graft rejection is an immunologic response displaying the attributes of specificity, memory, and self-nonself recognition. There are three major types of rejection reactions:

  - Hyperacute rejection, mediated by preexisting host antibodies against graft antigens

  - Acute graft rejection, in which T$_H$ cells and/or CTLs mediate tissue damage

  - Chronic rejection, which involves both cellular and humoral immune components

- The immune response to tissue antigens encoded within the major histocompatibility complex is the strongest force in rejection.

- The match between a recipient and potential graft donors is assessed by typing blood-group antigens and MHC antigens, and evaluating existing anti-donor antibodies (cross-matching).

- The process of graft rejection can be divided into a sensitization stage, in which T cells are stimulated, and an effector stage, in which they attack the graft.

- In most clinical situations, graft rejection is suppressed by nonspecific immunosuppressive agents or by total lymphoid x-irradiation.

- Experimental approaches using monoclonal antibodies offer the possibility of more specific immunosuppression. These antibodies may act by:
  - Depleting certain populations of reactive cells
  - Blocking TCR engagement or interfering with costimulation
  - Inhibiting the trafficking of certain cell types
  - Interfering with specific cytokine signaling

- Certain sites in the body are immunologically privileged, including the cornea of the eye, brain, testes, and uterus, and transplants in these sites may not be rejected despite genetic mismatch between donor and recipient.

- Specific tolerance to alloantigens can be induced by exposure to these antigens in utero or as a neonate. In some cases, prior exposure of adults to alloantigens in the form of hematopoietic cells, creating a state of mixed chimerism, can favor later success of grafts expressing these same alloantigens.

- Of all the organs or cell types amenable to transplantation, kidney transplants are the most common, and this organ is in the greatest demand.

- A major complication in bone marrow transplantation is GVHD, mediated by the lymphocytes contained within the donor marrow that target the recipient's cells.

- The critical shortage of organs available for transplantation may be solved in the future by using organs from nonhuman species (xenotransplants).

REFERENCES

Abdelnoor, A. M., et al. 2009. Influence of HLA disparity, immunosuppressive regimen used, and type of kidney allograft on production of anti-HLA class-I antibodies after transplant and occurrence of rejection. *Immunopharmacology and Immunotoxicology* **31**(1):83–87.

Anderson, M. S., et al. 2005. The cellular mechanism of Aire control of T cell tolerance. *Immunity* **23**:227.

Costa, V. S., T. C. Mattana, and M. E. da Silva. 2010. Unregulated IL-23/IL-17 immune response in autoimmune diseases. *Diabetes Research and Clinical Practice* **88**(3):222–226.

Chinen, J., and R. H. Buckley. 2010. Transplantation immunology: Solid organ and bone marrow. *Journal of Allergy and Clinical Immunology* **125**(2 Suppl 2):S324--335.

Damsker, J. M., A. . Hansen, and R. R. 2010. Th1 and Th17 cells: Adversaries and collaborators. *Annals of the New York Academy of Sciences* **1183**:211–221.

Gorantla, V. S., et al. 2010. T regulatory cells and transplantation tolerance. *Transplantation Reviews (Orlando)* **24**(3):147–159.

Hafler, D. A., et al. 2005. Multiple sclerosis. *Immunological Reviews* **204**:208.

Hogquist, K. A., T. A. Baldwin, and S. C. Jameson. 2005. Central tolerance: Learning self-control in the thymus. *Nature Reviews Immunology* **5**:772.

Issa, F., A. Schiopu, and K. J. Wood. 2010. Role of T cells in graft rejection and transplantation tolerance. *Expert Review of Clinical Immunology* **6**(1):155–169.

Kapp, J. A., and R. P. Bucy. 2008. CD8$^+$ suppressor T cells resurrected. *Hum Immunol* **69**(11):715–720. Kunz, M., and S. M. Ibrahim. 2009. Cytokines and cytokine profiles in human autoimmune diseases and animal models of autoimmunity. *Mediators of Inflammation* **2009**:979258.

Lu, L., and H. Cantor. 2008. Generation and regulation of CD8($^+$) regulatory T cells. *Cellular and Molecular Immunology* **5**(6):401–406.

Pomié, C., I. Ménager-Marcq, and J. P. van Meerwijk. 2008. Murine CD8$^+$ regulatory T lymphocytes: The new era. *Human Immunology* **69**(11):708–714.

Ricordi, C., and T. B. Strom. 2004. Clinical islet transplantation: Advances and immunological challenges. *Nature Reviews. Immunology* **4**:259.

Rioux, J. D., and A. K. Abbas. 2005. Paths to understanding the genetic basis of autoimmune disease. *Nature* **435**:584.

Round, J. L., R. M. O'Connell, and S. K. Mazmanian. 2010. Coordination of tolerogenic immune responses by the commensal microbiota. *Journal of Autoimmunity* **34**(3):J220–J225.

Sakaguchi, S. 2004. Naturally arising CD4$^+$ regulatory T cells for immunologic self-tolerance and negative control of immune responses. *Annual Review of Immunology* **22**:531.

Sayegh, M. H., and C. B. Carpenter. 2004. Transplantation 50 years later—progress, challenges, and promises. *New England Journal of Medicine* **351**:26.

Steward-Tharp, S. M., Y. J. Song, R. M. Siegel, and J. J. O'Shea. 2010. New insights into T cell biology and T cell-directed therapy for autoimmunity, inflammation, and immunosuppression. *Annals of the New York Academy of Sciences* **1183**:123–148.

Tait, B. D. 2009. Solid phase assays for HLA antibody detection in clinical transplantation. *Current Opinion in Immunology* **21**(5):573–577.

Thomas, R. 2010. The balancing act of autoimmunity: Central and peripheral tolerance versus infection control. *International Reviews of Immunology* **29**(2):211–233.

Turka, L. A., and R. I. Lechler. 2009. Towards the identification of biomarkers of transplantation tolerance. *Nature Reviews Immunology* 9(7):521–526.

Turka, L. A., K. Wood, and J. A. Bluestone. 2010. Bringing transplantation tolerance into the clinic: Lessons from the ITN and RISET for the Establishment of Tolerance consortia. *Current Opinion in Organ Transplantation* 15(4):441–448.

Veldhoen, M. 2009. The role of T helper subsets in autoimmunity and allergy. *Current Opinion in Immunology* 21(6):606–611.

von Boehmer, H., and F. Melchers 2010. Checkpoints in lymphocyte development and autoimmune disease. *Nature Immunology* 11(1):14–20.

Waldmann, H., and S. Cobbold. 2004. Exploiting tolerance processes in transplantation. *Science* 305:209.

Wing, K., and S. Sakaguchi. 2010. Regulatory T cells exert checks and balances on self tolerance and autoimmunity. *Nature Immunology* 11(1):7–13.

Useful Websites

www.lupus.org/index.html The site for the Lupus Foundation of America contains valuable information for patients and family members as well as current information about research in this area.

www.niams.nih.gov Home page for the National Institute of Arthritis and Musculoskeletal and Skin Diseases. This site contains links to other arthritis sites.

www.niddk.nih.gov Home page for the National Institute of Diabetes and Digestive and Kidney Diseases. This site contains an exhaustive list of links to other diabetes health-related sites.

www.unos.org The United Network for Organ Sharing site has information concerning solid-organ transplantation for patients, families, doctors, and teachers, as well as up-to-date numbers on waiting patients.

www.marrow.org The National Marrow Donor Program website contains information about all aspects of bone marrow transplantation.

http://optn.transplant.hrsa.gov/data The Organ Procurement and Transplantation Network site is run by the U.S. Department of Health and Human Services. It maintains real-time numbers on waiting patients, as well as data on organ transplants in the United States.

www.srtr.org The Scientific Registry of Transplant Recipients is a national database of transplantation statistics.

www.who.int/transplantation/knowledgebase/en The World Health Organization runs this site as a clearinghouse of information relating to organ, tissue, and cell donation and transplantation worldwide.

www.immunetolerance.org This website, run by the U.S.-based Immune Tolerance Network, is aimed at translating basic research findings in tolerance induction into therapy for autoimmunity, allergy, and transplantation.

 STUDY QUESTIONS

CLINICAL FOCUS QUESTION What are some of the possible reasons why females are more susceptible to autoimmune diseases than males?

1. Explain why all self-reactive lymphocytes are not eliminated in the thymus or bone marrow. How are the surviving self reactors prevented from harming the host?

2. Why is tolerance critical to the normal functioning of the immune system?

3. What is the importance of receptor editing to B-cell tolerance?

4. For each of the following autoimmune diseases (a–j), select the most appropriate characteristic (1–10) listed below.

   Disease

   a. _____ Experimental autoimmune encephalitis (EAE)
   b. _____ Graves' disease
   c. _____ Systemic lupus erythematosus (SLE)
   d. _____ Type 1 diabetes mellitus (T1DM)
   e. _____ Rheumatoid arthritis (RA)
   f. _____ Hashimoto's thyroiditis
   g. _____ Experimental autoimmune myasthenia gravis (EAMG)
   h. _____ Myasthenia gravis
   i. _____ Multiple sclerosis (MS)
   j. _____ Autoimmune hemolytic anemia

   Characteristics

   (1) Auto-antibodies to acetylcholine receptor
   (2) T_H1-cell reaction to thyroid antigens
   (3) Auto-antibodies to RBC antigens
   (4) T-cell response to myelin
   (5) Induced by injection of *myelin basic protein* (MBP) plus complete Freund's adjuvant
   (6) Auto-antibody to IgG
   (7) Auto-antibodies to DNA and DNA-associated protein
   (8) Auto-antibodies to receptor for thyroid-stimulating hormone
   (9) Induced by injection of acetylcholine receptors
   (10) T_H1-cell response to pancreatic beta cells

5. Experimental autoimmune encephalitis (EAE) has proved to be a useful animal model of autoimmune disorders.

   a. Describe how this animal model is made.
   b. What is unusual about the animals that recover from EAE?
   c. How has this animal model indicated a role for T cells in the development of autoimmunity?

6. Molecular mimicry is one mechanism proposed to account for the development of autoimmunity. How has induction of EAE with myelin basic protein contributed to the understanding of molecular mimicry in autoimmune disease?

7. Describe at least three different mechanisms by which a localized viral infection might contribute to the development of an organ-specific autoimmune disease.

8. Monoclonal antibodies have been administered for therapy in various autoimmune animal models. Which monoclonal antibodies have been used, and what is the rationale for these approaches?

9. Indicate whether each of the following statements is true or false. If you think a statement is false, explain why.

   a. T_H1 cells have been associated with development of autoimmunity.
   b. Immunization of mice with IL-12 prevents induction of EAE by injection of MBP plus adjuvant.
   c. The presence of the *HLA B27* allele is diagnostic for ankylosing spondylitis, an autoimmune disease affecting the vertebrae.
   d. A defect in the gene encoding Fas can reduce programmed cell death by apoptosis.

10. For each of the following autoimmune disorders (a–d), indicate which of the following treatments (1–5) may be appropriate:

   Disease
   a. Hashimoto's thyroiditis
   b. Systemic lupus erythematosus
   c. Graves' disease
   d. Myasthenia gravis

   Treatment
   (1) Cyclosporin A
   (2) Thymectomy
   (3) Plasmapheresis
   (4) Kidney transplant
   (5) Thyroid hormones

11. Which of the following are examples of mechanisms for the development of autoimmunity? For each possibility, give an example.

   a. Polyclonal B-cell activation
   b. Tissue damage
   c. Viral infection
   d. Increased expression of TCR molecules
   e. Increased expression of class II MHC molecules

12. Indicate whether each of the following statements is true or false. If you think a statement is false, explain why.

   a. Acute rejection is mediated by preexisting host antibodies specific for antigens on the grafted tissue.
   b. Second-set rejection is a manifestation of immunologic memory.
   c. Host dendritic cells can migrate into grafted tissue and act as APCs.

   d. All allografts between individuals with identical HLA haplotypes will be accepted.
   e. Cytokines produced by host T_H cells activated in response to alloantigens play a major role in graft rejection.

13. Indicate whether a skin graft from each donor to each recipient listed in the following table would result in rejection (R) or acceptance (A). If you believe a rejection reaction would occur, indicate whether it would be a first-set rejection (FSR), occurring in 12 to 14 days, or a second-set rejection (SSR), occurring in 5 to 6 days. All the mouse strains listed have different H-2 haplotypes.

Donor	Recipient
BALB/c	C3H
BALB/c	Rat
BALB/c	Nude mouse
BALB/c	C3H, had previous BALB/c graft
BALB/c	C3H, had previous C57BL/6 graft
BALB/c	BALB/c
BALB/c	(BALB/c x C3H)F_1
BALB/c	(C3H x C57BL/6)F_1
(BALB/c x C3H)F_1	BALB/c
(BALB/c x C3H)F_1	BALB/c, had previous F_1 graft

14. Graft-*versus*-*host* *disease* (GVHD) frequently develops after certain types of transplantations.

   a. Briefly outline the mechanisms involved in GVHD.
   b. Under what conditions is GVHD likely to occur?
   c. Some researchers have found that GVHD can be diminished by prior treatment of the graft with monoclonal antibody plus complement or with monoclonal antibody conjugated with toxins. List at least two cell-surface antigens to which monoclonal antibodies could be prepared and used for this purpose, and give the rationale for your choices.

15. What is the biologic basis for attempting to use soluble CTLA-4Ig or anti-CD40L to block allograft rejection? Why might this be better than treating a graft recipient with CsA or FK506?

16. Immediately after transplantation, a patient is often given extra strong doses of anti-rejection drugs and then allowed to taper off as time passes. Describe the effects of the commonly used anti-rejection drugs azathioprine, cyclosporine A, FK506, and rapamycin. Why is it possible to decrease the use of some of these drugs at some point after transplantation?

CLINICAL FOCUS QUESTION What features would be desirable in an ideal animal donor for xenotransplantation? How would you test your model prior to doing clinical trials in humans?

Infectious Diseases and Vaccines

S urviving infectious disease outbreaks was one of the primary drivers for our earliest forays into the study of immunology. This led to the development and use of rudimentary vaccines even before we understood how a vaccine could induce protective immunity (see Chapter 1). Since those early vaccination trials of Edward Jenner and Louis Pasteur, vaccines have been developed for many diseases that were once major afflictions of humankind. For example, the incidence of diphtheria, measles, mumps, pertussis (whooping cough), rubella (German measles), poliomyelitis, and tetanus, which once collectively claimed the lives of millions, has declined dramatically as vaccination has become more common. Clearly, vaccination is a cost-effective weapon for disease prevention, and yet the need for safe and effective vaccines for many life-threatening infectious diseases remains.

These and other public health concerns led to the development of agencies to help organize the accumulating data concerning infectious disease, such as the World Health Organization (WHO) and the U.S.-based Centers for Disease Control and Prevention (CDC). These organizations monitor public health and disease, guide health care policy discussions, respond to sudden infectious disease outbreaks, and report regularly on their findings. Although the local and international expenditures on these practices are questioned at times, there is no doubt that these and present-day biomedical advances have led us to an age in which rapid and often effective response to sudden infectious disease outbreaks is commonplace. It has also allowed us to better appreciate the conditions and policies that can limit outbreaks of infectious disease.

Although vaccination or naturally acquired protective immunity can provide critical defense against many pathogens, infectious diseases still cause the death of millions each year. Although the number varies greatly by region, about 25% of deaths worldwide are associated with communicable diseases, which kill an estimated 11 million to 12 million people each year (Figure 17-1). Sanitation, antibiotics, and vaccination have reduced the impact of infectious disease, but infections still account

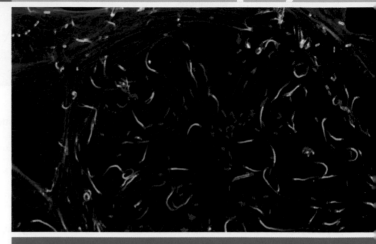

The bacteria *Listeria monocytogenes* polymerizing host cell actin into comet tails. *[Courtesy Matteo Bonazzi, PhD, Edith Gouin, and Pascale Cossart]*

- The Importance of Barriers to Infection and the Innate Response
- Viral Infections
- Bacterial Infections
- Parasitic Infections
- Fungal Infections
- Emerging and Re-emerging Infectious Diseases
- Vaccines

for almost half of the leading causes of death in the developing world, especially among the very young.

Adding to the endemic infectious disease burden most heavily borne by the developing world, new diseases are emerging and others are resurfacing. Influenza and *West Nile* virus (WNV) strains prevalent in birds have adapted to cause human infection. Previously rare infections by certain bacteria or fungi are increasing because of the rise in the numbers of individuals with impaired immunity, primarily due to the prevalence of *human immunodeficiency virus* (HIV)-induced *acquired immunodeficiency syndrome* (AIDS). Increasing antibiotic resistance in existing pathogens, such as *Staphylococcus aureus* and *Mycobacterium tuberculosis*, has some infections

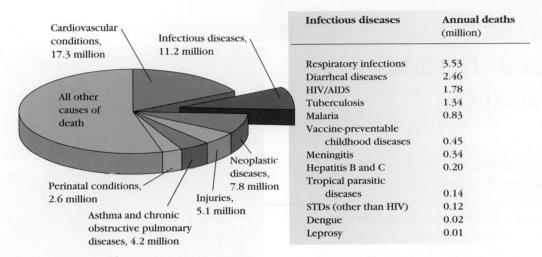

Infectious diseases	Annual deaths (million)
Respiratory infections	3.53
Diarrheal diseases	2.46
HIV/AIDS	1.78
Tuberculosis	1.34
Malaria	0.83
Vaccine-preventable childhood diseases	0.45
Meningitis	0.34
Hepatitis B and C	0.20
Tropical parasitic diseases	0.14
STDs (other than HIV)	0.12
Dengue	0.02
Leprosy	0.01

FIGURE 17-1 Infectious diseases are among the leading causes of death worldwide. The 11.2 million deaths attributable annually directly to infections are broken down by category in this table. *[based on WHO 2008 global burdens of disease estimates]*

spreading at an alarming rate in developing as well as in industrialized countries. In certain instances, a common infectious agent has become associated with a new disease. Such is the case for the recently identified disease necrotizing fasciitis, caused by the so-called flesh-eating strain of *Streptococcus pyogenes*, a bacterium most commonly associated with the now rare disease scarlet fever.

In this chapter, the concepts of immunity described throughout the text are applied to selected infectious diseases caused by the four main types of pathogens (viruses, bacteria, fungi, and parasites). We focus on particular infectious diseases that affect large numbers of people, that illustrate specific immune concepts, and that use novel strategies to subvert the immune response, as well as some diseases that have warranted recent headlines. The chapter concludes with a section on vaccines, divided by the type of vaccine design being applied and including examples of specific pathogens that have been successfully targeted using these strategies.

The Importance of Barriers to Infection and the Innate Response

First and foremost, in order for a pathogen to establish an infection in a susceptible host, it must breach physical and chemical barriers. One of the first and most important of these barriers consists of the epithelial surfaces of the skin and the lining of the gut. The difficulty of penetrating these surfaces ensures that most pathogens never gain productive entry into the host. In addition, the epithelia produce chemicals that are useful in preventing infection. The secretion of gastric enzymes by specialized epithelial cells lowers the pH of the stomach and upper gastrointestinal tract, and other

specialized cells in the gut produce antibacterial peptides. In addition, normal commensal flora present at mucosal surfaces (the gastrointestinal, genitourinary, and respiratory tracts) can competitively inhibit the binding of pathogens to host cells. When pathogen dose and virulence are minimal, these barriers can often block productive infection altogether.

Interventions that introduce barriers to infection in *intermediate* hosts can be used as an indirect strategy to disrupt the cycle of infectious disease in humans. For example, many pathogens make use of arthropod vectors, such as the mosquito, for parts of their life cycle and as vehicles for transmission to humans. Very recent studies in Dengue virus, transmitted by the bite of an infected mosquito and the cause of an often fatal hemorrhagic fever in humans, suggest that it may be possible to engineer mosquitoes that are resistant to infection with the virus. When these engineered mosquitoes were released into the wild, they began to supplant the wild-type, virus-susceptible mosquito population, suggesting that they may have the potential to break the cycle of transmission. This and other exciting new avenues of research that target animal disease vectors could advance infectious disease eradication without the requirement to intervene with the human immune response. Of course, this strategy is not a possibility with most infectious diseases, for which there is no animal vector.

When the basic human barriers to infection are breached, more directed innate immune responses come into play at or near the site of infection. These early responses are often tailored to the type of pathogen, using molecular pattern recognition receptors (see Chapter 5). Some bacteria produce endotoxins such as lipopolysaccharide (LPS), which stimulate macrophages or endothelial cells to produce cytokines, such as IL-1, IL-6, and TNF-α. These cytokines can activate nearby innate cells, encouraging phagocytosis of the bacteria. The cell walls of many gram-positive bacteria contain a peptidoglycan that activates the alternative complement pathway, leading to opsonization and phagocytosis or lysis (see Chapter 6).

TABLE 17-1 Mechanisms of humoral and cell-mediated immune responses to viruses

Response type	Effector molecule or cell	Activity
Humoral	Antibody (especially secretory IgA)	Blocks binding of virus to host cells, thus preventing infection or reinfection
	IgG, IgM, and IgA antibody	Blocks fusion of viral envelope with host cell's plasma membrane
	IgG and IgM antibody	Enhances phagocytosis of viral particles (opsonization)
	IgM antibody	Agglutinates viral particles
	Complement activated by IgG or IgM antibody	Mediates opsonization by C3b and lysis of enveloped viral particles by membrane-attack complex
Cell mediated	IFN-λ secreted by T_H or T_C cells	Has direct antiviral activity
	Cytotoxic T lymphocytes (CTLs)	Kill virus-infected self cells
	NK cells and macrophages	Kill virus-infected cells by antibody-dependent cell-mediated cytotoxicity (ADCC)

Viruses commonly induce the production of interferons, which can inhibit viral replication by inducing an antiviral response in neighboring cells. Viruses are also controlled by *n*atural *k*iller (NK) cells, which frequently form the first line of defense in these infections (see Chapter 5). In many cases, these innate responses can lead to the resolution of infection.

Cells responding via innate immunity at the infection site receive signals that help coordinate the subsequently more specific adaptive immune response. During this very pathogen-specific stage of the immune response, final eradication of the foreign invader often occurs, typically leaving a memory response capable of halting secondary infections. However, just as adaptive immunity in vertebrates has evolved over many millennia, pathogens have evolved a variety of strategies to escape destruction by the adaptive immune response. Some pathogens reduce their own antigenicity either by growing within host cells, where they are sequestered from immune attack, or by shedding their membrane antigens. Other pathogen strategies include camouflage (expressing molecules with amino acid sequences similar to those of host cell membrane molecules or acquiring a covering of host membrane molecules); suppressing the immune response selectively or directing it toward a pathway that is ineffective at fighting the infection; and continual variation in surface antigens. Examples of each of these strategies are included in the following sections.

Viral Infections

Viruses are small segments of nucleic acid with a protein or lipoprotein coat that require host resources for their replication. Typically, a virus enters a cell via a cell-surface receptor for which it has affinity and preempts cell biosynthetic machinery to replicate all components of itself, including its genome. This genome replication step is often error prone, generating numerous mutations. Because large numbers of new viral particles (virions) are produced in a replication cycle, many different mutants with individual survival advantages can be selected for the ability to propagate most effectively in the host.

A virus is more likely to thrive if it does not kill its host, as sustained coexistence in the host favors the survival and spread of the virus. However, the mutability of the viral genome sometimes gives rise to lethal variants that do not conform to this state of equilibrium with their host. If such mutants cause the early death of their host, survival of the virus requires that it spread to new hosts rapidly. Among the other survival strategies available to viruses is a long latency period before severe illness, during which time the host may pass the virus to others unknowingly, as in the case of HIV. One additional strategy used by viruses is facile transmission, such as with influenza and the smallpox virus, where infection is efficiently transferred during even a short acute illness. The life cycle of some viruses pathogenic for humans, such as WNV, may also include nonhuman hosts, providing them with additional reservoirs.

A number of specific immune effector mechanisms, together with nonspecific defense mechanisms, prevent or eliminate most viral infections (Table 17-1). Passage across the mucosa of the respiratory, genitourinary, or gastrointestinal tracts accounts for most instances of viral transmission. Entrance of the virus may also occur through broken skin, usually as a result of an insect bite or puncture wound. The outcome of this infection depends on how effectively the host's defensive mechanisms resist the offensive tactics of the virus.

The innate immune response to viral infection primarily begins with the recognition of *p*athogen *a*ssociated *m*olecular *p*atterns (PAMPs) and leads to the generation of antiviral effectors. For example, *d*ouble-*s*tranded *RNA* (dsRNA) molecules and other virus-specific structures are detected by one of several PAMP receptors, inducing the expression of type I interferons (IFN-α and IFN-β), the assembly of intracellular inflammasome complexes, and the activation of NK cells.

Type I interferons can induce an antiviral response or resistance to viral replication by binding to the IFN-α/-β receptor, thereby activating the JAK-STAT pathway and the production of new transcripts, one of which encodes an enzyme that leads to viral RNA degradation (see Figure 5-16). IFN-α/-β binding also induces dsRNA-dependent *protein kinase* (PKR), which leads to inactivation of protein synthesis, thus blocking viral replication in infected cells. The binding of type I interferon to NK cells induces lytic activity, making them very effective in killing virally infected cells. This activity is enhanced by IL-12, a cytokine that is produced by dendritic cells very early in the response to viral infection.

Many Viruses Are Neutralized by Antibodies

Antibodies specific for viral surface antigens are often crucial in containing the spread of a virus during acute infection and in protecting against reinfection. Antibodies are particularly effective if they are localized at the site of viral entry into the body and if they bind to key viral surface structures, interfering with their ability to attach to host cells. For example, influenza virus binds to sialic acid residues in cell membrane glycoproteins and glycolipids, rhinovirus binds to *intercellular adhesion molecules* (ICAMs), and *Epstein-Barr virus* (EBV) binds to type 2 complement receptors on B cells. The advantage of the attenuated oral polio vaccine, discussed later in this chapter, is that it induces production of secretory IgA, which effectively blocks attachment of poliovirus to epithelial cells lining the gastrointestinal tract.

Viral neutralization by antibody sometimes involves mechanisms that operate after viral attachment to host cells. For example, antibodies may block viral penetration by binding to epitopes that are necessary to mediate fusion of the viral envelope with the plasma membrane. If the induced antibody is of a complement-activating isotype, lysis of enveloped virions can ensue. Antibody or complement can also agglutinate viral particles and function as an opsonizing agent to facilitate Fc- or C3b-receptor-mediated phagocytosis of the free virions.

Cell-Mediated Immunity Is Important for Viral Control and Clearance

Although antibodies have an important role in containing the spread of a virus in the acute phases of infection, they cannot eliminate established infection once the viral genome is integrated into host chromosomal DNA. Once such an infection is established, cell-mediated immune mechanisms are most important in host defense. In general, both $CD8^+$ T_C cells and $CD4^+$ T_H1 cells are required components of the cell-mediated antiviral defense. Activated T_H1 cells produce a number of cytokines, including IL-2, IFN-γ, and *tumor necrosis factor*-α (TNF-α), which defend against viruses either directly or indirectly. IFN-γ acts directly by inducing an antiviral state in nearby cells. IL-2 acts indirectly by assisting the development of *cytotoxic T lymphocyte* (CTL) precursors into an effector population. Both IL-2 and IFN-γ acti-

vate NK cells, which play an important role in host defense and lysis of infected cells during the first days of many viral infections, until a specific CTL response develops.

In most viral infections, specific CTL activity arises within 3 to 4 days after infection, peaks by 7 to 10 days, and then declines. Within 7 to 10 days of primary infection, most virions have been eliminated, paralleling the development of CTLs. CTLs specific for the virus eliminate virus-infected self cells and thus eliminate potential sources of new virus. Virus-specific CTLs confer protection against that virus in nonimmune recipients following adoptive transfer. Transfer of a CTL clone specific for influenza virus strain X protects mice against strain X but not against influenza virus strain Y.

Viruses Employ Several Different Strategies to Evade Host Defense Mechanisms

Despite their restricted genome size, a number of viruses encode proteins that interfere with innate and adaptive levels of host defense. Presumably, the advantage of such proteins is that they enable viruses to replicate more effectively amid host antiviral defenses. As described above, the induction of type I interferon is a major innate defense against viral infection, but some viruses have developed strategies to evade the action of IFN-α/-β. These include hepatitis C virus, which has been shown to overcome the antiviral effect of the interferons by blocking or inhibiting the action of PKR (see Figure 5-16).

Another mechanism for evading host responses is inhibition of antigen presentation by infected host cells. *Herpes simplex virus* (HSV) produces an immediate-early protein (synthesized shortly after viral replication) that very effectively inhibits the human transporter molecule needed for antigen processing (TAP; see Figure 8-17). Inhibition of TAP blocks antigen delivery to class I MHC molecules in HSV-infected cells, thus preventing presentation of viral antigen to $CD8^+$ T cells. This results in the trapping of empty class I MHC molecules in the endoplasmic reticulum and effectively shuts down a $CD8^+$ T-cell response to HSV-infected cells. Likewise, adenoviruses and *cytomegalovirus* (CMV) use distinct molecular mechanisms to reduce the surface expression of class I MHC molecules, again inhibiting antigen presentation to $CD8^+$ T cells. Other viruses, such as measles virus and HIV, reduce levels of class II MHC molecules on the surface, thus blocking the function of antigen-specific antiviral helper T cells.

Complement activation is another of the antibody-mediated destruction pathways of viruses, resulting in opsonization and elimination of the virus by phagocytic cells. A number of viruses, such as vaccinia virus, evade complement-mediated destruction by secreting a protein that binds to the C4b complement component, inhibiting the classical complement pathway. HSV also makes a glycoprotein component that binds to the C3b complement component, inhibiting both the classical and alternative pathways.

A number of viruses escape immune attack by constantly changing their surface antigens. The influenza virus is a prime example, as discussed below. Antigenic variation among rhinoviruses, the causative agent of the common cold, is responsible for our inability to produce an effective vaccine for colds. Nowhere is antigenic variation greater than in HIV, the causative agent of AIDS, estimated to accumulate mutations 65 times faster than the influenza virus. A section of Chapter 18 is dedicated to HIV and AIDS.

Viruses such as EBV, CMV, and HIV cause generalized or specific immunosuppression. In some cases, immunosuppression is caused by direct viral infection of lymphocytes or macrophages. The virus can then either directly destroy the immune cells by cytolytic mechanisms or alter their function. In other cases, immunosuppression is the result of a cytokine imbalance or diversion of the immune responses toward pathways less effective at virus eradication. For instance, EBV, the cause of mononucleosis, produces a protein that is homologous to IL-10; like IL-10, this protein suppresses cytokine production by the T_H1 subset, resulting in an immunosuppressed state.

Influenza Has Been Responsible for Some of the Worst Pandemics in History

The influenza virus infects the upper respiratory tract and major central airways in humans, horses, birds, pigs, and even seals. Between 1918 and 1919, the largest influenza pandemic (worldwide epidemic) in recent history occurred, killing between 20 million and 50 million people. The sequence of this virus has recently been reconstructed, leading to much controversy about its publication (Clinical Focus Box 17-1). Two other less major pandemics occurred in the twentieth century, caused by influenza strains that were new or had not circulated in the recent past, leaving most people with little immunity to them.

CLINICAL FOCUS BOX 17-1

 ## The 1918 Pandemic Influenza Virus: Should It Publish or Perish?

The most virulent and devastating of the pandemic strains of influenza virus in recent history was seen in 1918 and 1919. Worldwide deaths from that so-called "Spanish flu" strain may have reached 50 million in less than 1 year, compared with the roughly 10,000 to 15,000 who die yearly from nonpandemic strains. Approximately 675,000 of the victims of Spanish flu were located the United States, with certain areas, such as Alaska and the Pacific Islands, losing more than half of their population during the outbreak. Mortality rates for the 1918 pandemic flu were surprisingly high, especially among young and healthy individuals, reaching 2.5% in infected individuals compared to less than 0.1% during other flu epidemics. Most of these deaths were the result of a virulent pneumonia, which felled some patients in as little as 5 days.

Thanks to present-day molecular techniques and chance, the recent reconstruction and sequencing of the virus that caused the 1918 pandemic became possible. After several failed attempts, a research team led by Jeffrey Tautenberger published the final genetic sequences of the deadly 1918 flu virus. Their results were made possible following isolation of viral RNA from Spanish flu victims using formalin-fixed lung autopsy samples and

tissue collected from an Inuit woman who was buried in permafrost in Alaska. Analysis of the sequence revealed that this highly virulent strain was derived from an avian virus and differed significantly from other human influenza A strains, making it the most "bird-like" of the influenza strains ever isolated from humans.

The reconstructed virus sequence became the object of intense study, as well as controversy. Thanks to a cDNA reconstruction of live virus, scientists were able to study the virulence factors at play in this deadly strain. In mouse studies, they found that the reconstructed virus spread rapidly in the respiratory tract and produced high numbers of progeny, causing pervasive damage in the lungs. Using recombinant virus strains, they found that three polymerase genes and the HA gene appeared to account for the high lethality of the virus; replacing the polymerase genes significantly reduced the virulence of the strain while a new HA gene completely blocked its ability to kill the host.

Word of the imminent publication of the sequence of the 1918 influenza caused a scientific and public controversy. On the one hand, many virologists, molecular biologists, and epidemiologists were eager to glimpse this highly virulent sequence for clues to what determinants might play

a role in lethality and what measures could be taken to avoid this in the future. On the other hand, some feared that this sequence might be used for evil rather than for good, leading to a potential do-it-yourself recombinant reconstructionist frenzy, culminating in a weaponized version of the influenza virus. In the end, the sequences were published and follow-up studies led to the conclusion that the 1918 virus is sensitive to the seasonal flu vaccine and even treatable with available anti-flu drugs; this may have calmed many fears.

Nevertheless, controversy persists. Recently, the National Science Advisory Board for Biosecurity (NSABB), a U.S.-based body that advises the community about research concerning agents deemed a national security threat, recommended against the release of data related to the current avian H5N1 strain. These data would reveal the mutations behind the virus's transmissibility to humans. After 8 months of deliberations, *Science* magazine published a special open-access issue in June 2012 describing this work and related policy issues, placing another one in the publish-rather-than-perish column.

T. M. Tumpey et al. 2005. The 1918 Flu Virus Is Resurrected. *Science* **310:**77–80; and P. Palese. 2012. Don't Censor Life-Saving Science. *Nature* **481:**115.

(a)

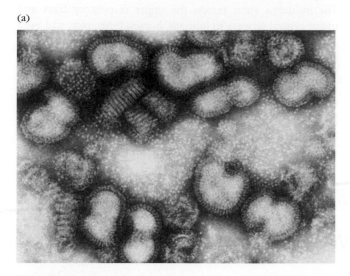

(b)

FIGURE 17-2 **Influenza virus.** (a) Electron micrograph of influenza virus reveals roughly spherical viral particles enclosed in a lipid bilayer with protruding hemagglutinin and neuraminidase glycoprotein spikes. (b) Schematic representation of influenza structure. The envelope is covered with neuraminidase and hemagglutinin spikes. Inside is an inner layer of matrix protein surrounding the nucleocapsid, which consists of eight ssRNA molecules associated with nucleoprotein. The eight RNA strands encode 10 proteins: PB1, PB2, PA, HA (hemagglutinin), NP (nucleoprotein), NA (neuraminidase), M1, M2, NS1, and NS2. *[Source: (a) Courtesy of G. Murti, Department of Virology, St. Jude Children's Research Hospital, Memphis, Tenn.]*

Properties of the Influenza Virus

Influenza virions are surrounded by an outer envelope, a lipid bilayer derived from the plasma membrane of the infected cell, plus various virus-specific proteins. Imbedded in this envelope are two key viral glycoproteins, **hemagglutinin (HA)** and **neuraminidase (NA)** (Figure 17-2a). HA trimers are responsible for the attachment of the virus to host cells, binding to the sialic acid groups on host-cell glycoproteins and glycolipids. NA is an enzyme that cleaves *N*-acetylneuraminic (sialic) acid from nascent viral glycoproteins and host-cell membrane glycoproteins, facilitating viral budding from the infected host cell. Thus these two structures are essential for viral attachment and for exit of new virus from infected cells—so important in fact that we track new strains of influenza based on their antigenic subtypes of HA and NA (e.g., H1N1 versus H5N1 virus). Within the envelope, an inner layer of matrix protein surrounds the nucleocapsid, which consists of eight different strands of single-stranded RNA (ssRNA) associated with protein and RNA polymerase (Figure 17-2b). Each RNA strand encodes one or more different influenza proteins.

There are three basic types of influenza (A, B, and C), each differing in the makeup of its nuclear and matrix proteins. Type A is the most common and is responsible for the major human pandemics. Influenza virus strains are tracked yearly by the Centers for Disease Control and Prevention (CDC) and the World Health Organization (WHO). According to WHO nomenclature, each virus strain is defined by its animal host of origin (specified if other than human), geographical origin, strain number, year of isolation, and the antigenic structures of HA and NA. For example, A/Sw/Iowa/15/30 (H1N1) designates strain-A isolate 15 that arose in swine in Iowa in 1930 and has antigenic subtypes 1 for both HA and NA (refer to Table 17-2, pandemic strains).

Variation in Epidemic Influenza Strains

To date, there are 13 different antigenic subtypes for HAs and 9 for NAs. Antigenic variation in HA and NA is generated by two different mechanisms: antigenic drift and antigenic shift. **Antigenic drift** involves a series of spontaneous point mutations that occur gradually, resulting in minor changes in HA and NA over time. **Antigenic shift** results in the sudden emergence of a new subtype of influenza, where the structures of HA and/or NA are considerably different from that of the virus present in a preceding year.

The immune response contributes to the emergence of these antigenically distinct influenza strains. In a typical year, the predominant virus strain undergoes antigenic

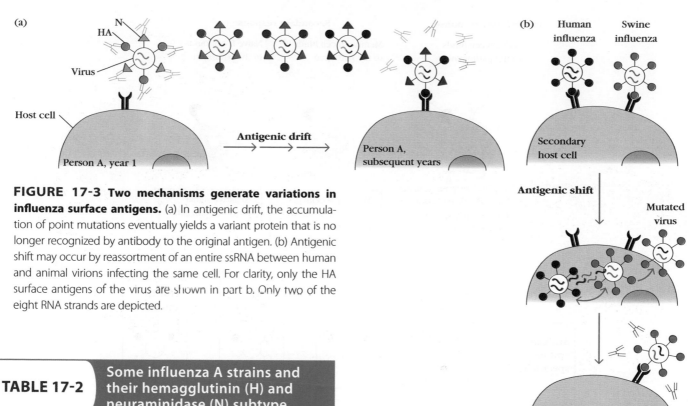

FIGURE 17-3 Two mechanisms generate variations in influenza surface antigens. (a) In antigenic drift, the accumulation of point mutations eventually yields a variant protein that is no longer recognized by antibody to the original antigen. (b) Antigenic shift may occur by reassortment of an entire ssRNA between human and animal virions infecting the same cell. For clarity, only the HA surface antigens of the virus are shown in part b. Only two of the eight RNA strands are depicted.

TABLE 17-2	Some influenza A strains and their hemagglutinin (H) and neuraminidase (N) subtype	

Species	Virus strain designation	Antigenic subtype
Human	A/Puerto Rico/8/34	H0N1
	A/Fort Monmouth/1/47	H1N1
	A/Singapore/1/57	H2N2
	A/Hong Kong/1/68	H3N2
	A/USSR/80/77	H1N1
	A/Brazil/11/78	H1N1
	A/Bangkok/1/79	H3N2
	A/Taiwan/1/86	H1N1
	A/Shanghai/16/89	H3N2
	A/Johannesburg/33/95	H3N2
	A/Wuhan/359/95	H3N2
	A/Texas/36/95	H1N1
	A/Hong Kong/156/97	H5N1
	A/California/04/2009	H1N1
Swine	A/Sw/Iowa/15/30	H1N1
	A/Sw/Taiwan/70	H3N2
Horse (equine)	A/Eq/Prague/1/56	H7N7
	A/Eq/Miami/1/63	H3N8
Bird	A/Fowl/Dutch/27	H7N7
	A/Tern/South America/61	H5N3
	A/Turkey/Ontario/68	H8N4
	A/Chicken/Hong Kong/258/97	H5N1

drift, generating minor antigenic variants. As individuals infected with influenza mount an effective immune response, they will eliminate that strain. However, the accumulation of point mutations sufficiently alters the antigenicity of some variants so that they are able to escape immune elimination (Figure 17-3a) and become a new variant of influenza that is transmitted to others, causing another local epidemic cycle. The role of antibody in such immunologic selection can be demonstrated in the laboratory by mixing an influenza strain with a monoclonal antibody specific for that strain and then culturing the virus in cells. The antibody neutralizes all unaltered viral particles, and only those viral particles with mutations resulting in altered antigenicity escape neutralization and are able to continue the infection. Within a short time in culture, a new influenza strain emerges, just as it does in nature. In this way, influenza evolves during a typical flu season, such that the dominant strains at the start and end of the season are antigenically distinct. This is why we are offered a new flu vaccine each year. The vaccine formulation is based on carefully constructed models tracking the dominant variant(s) from the end of the previous season. And our guesses are not always 100% accurate, making some years' influenza vaccinations more effective than others.

Episodes of antigenic shift are thought to occur through a different mechanism. The primary mechanism is genetic

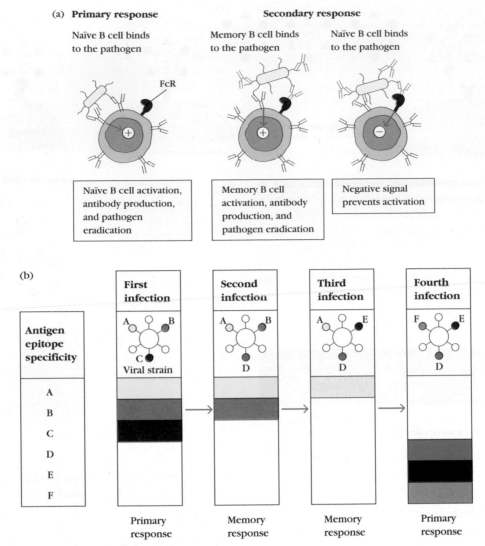

FIGURE 17-4 The presence of preformed antibody inhibits primary responses to a pathogen. (a) During a primary response, naïve B cells are activated and produce antibodies specific to epitopes on the pathogen. During a secondary response to a variant of that pathogen, memory B cells specific to epitopes encountered in the past will be reactivated and help to eradicate the pathogen. The Fc regions of antibodies bound to the surface of the pathogen will bind to the FcRs on naïve B cells and inhibit them from responding, even to new epitopes on the pathogen. (b) This inhibition of primary responses against unique epitopes on pathogens that elicit memory cell responses is called original antigenic sin. No immune response is mounted to each new epitope during subsequent exposures to the pathogen until the pathogen expresses a significant number of unique epitopes and memory cells can no longer eradicate the organism. In this case, a new primary response is mounted and disease symptoms are severe until a new adaptive response has been established, resetting original antigenic sin. *[Source: Adapted from P. Parham, 2009,* The Immune System, *New York: Garland Science. a: Fig. 10.23; b: Fig. 10.25.]*

reassortment between influenza virions from humans and those from various animals (Figure 17-3b). The fact that the influenza genome contains eight separate strands of ssRNA makes possible the reassortment of individual RNA strands of human and animal virions within a secondary (non-human) host cell infected with both viruses—in other words, shuffling of the DNA segments derived from the animal and human strains. Both pigs and birds can harbor human influenza A viruses, as well as their own and maybe those of other species, making them perfect conduits for in vivo genetic reassortment between human influenza A

viruses and the potential sources for strains that have been coined "swine flu" or "bird or avian flu." As one might imagine, the contribution of viral proteins from the viruses of these animals is frequently "new" to humans who will have little or no immunity, sparking pandemics.

Original Antigenic Sin and Susceptibility to Influenza

Immunologic memory is an amazing and beautiful thing. For pathogens that don't change much from one encounter to the next, it can provide us with lifelong protection. However, preformed immunity can come with caveats for pathogens that have evolved to vary their antigenic structure. During secondary encounters with a pathogen that bears a strong molecular resemblance to a pathogen seen in the past (i.e., we have developed adaptive immunity to some of the epitopes before), memory cells specific for previously encountered epitopes are engaged rapidly and efficiently. As long as these cells and their products, like antibodies, can dispatch the pathogen efficiently, there is no need to mount a *primary* response to any new epitopes carried by that pathogen. In fact, the presence of antibodies attached to a pathogen, either as residuals from a recent infection or produced by reactivation of memory B cells, will divert naïve B cells from responding (Figure 17-4a). This occurs when the Fc region of the pathogen-associated antibody binds with Fc receptors on naïve B cells, inducing anergy.

In other words, if there is a way to take care of an infection with memory, this will be the default pathway. This concept is referred to as **original antigenic sin**, or the tendency to focus an immune attack on those structures that were present during the original, or primary, encounter with a pathogen and for which we have established memory. This means that our immune systems effectively ignore the subtle changes occurring each year in pathogens that drift antigenically, like influenza virus (Figure 17-4b). Once the organism has drifted sufficiently that there are only "new" epitopes, or insufficient numbers of key epitopes to effectively dispatch with existing memory cells, a new primary response is mounted. In such a year, we experience a bad case of the flu. Since we all begin our journeys of original antigenic sin at different times and in response to different antigenic variants, we don't typically all get a bad case of the flu at the same time; the exceptions are pandemic influenza years (see Clinical Focus Box 17-1).

Bacterial Infections

Immunity to bacterial infections is achieved by means of antibody unless the bacterium is capable of intracellular growth, in which case *delayed-type hypersensitivity* (DTH) has an important role. Bacteria enter the body either through a number of natural routes (e.g., the respiratory, gastrointestinal, and genitourinary tracts) or through normally inaccessible routes opened up by breaks in mucous membranes or skin. Depending on the number of organisms entering and their virulence, different levels of host defense are enlisted. If the inoculum size and the virulence are both low, then localized tissue phagocytes may be able to eliminate the bacteria via nonspecific innate defenses. Larger inocula or organisms with greater virulence tend to induce antigen-specific adaptive immune responses.

In some bacterial infections, disease symptoms are caused not by the pathogen itself but by the immune response. As described in Chapter 4, pathogen-stimulated overproduction of cytokines leads to the symptoms associated with bacterial septic shock, food poisoning, and toxic shock syndrome. For instance, cell wall endotoxins of some gram-negative bacteria activate macrophages, resulting in release of high levels of IL-1 and TNF-α, which can cause septic shock. In staphylococcal food poisoning and toxic shock syndrome, exotoxins produced by the pathogens function as superantigens, which can activate all T cells that express T-cell receptors with a particular V_β domain (see Table 11-2). The resulting systemic production of cytokines by activated T_H cells is overwhelming, causing many of the symptoms of these diseases.

Immune Responses to Extracellular and Intracellular Bacteria Can Differ

Infection by extracellular bacteria induces production of antibodies, which are ordinarily secreted by plasma cells in regional lymph nodes and the submucosa of the respiratory and gastrointestinal tracts. The humoral immune response is the main protective response against extracellular bacteria. The antibodies act in several ways to protect the host from the invading organisms, including removal of the bacteria and inactivation of bacterial toxins (Figure 17-5). Extracellular bacteria can be pathogenic because they induce a localized inflammatory response or because they produce toxins. The toxins—endotoxin or exotoxin—can be cytotoxic but also may cause pathogenesis in other ways. An excellent example of this is the toxin produced by diphtheria, which blocks protein synthesis. Endotoxins, such as LPS, are generally components of bacterial cell walls, whereas exotoxins, such as diphtheria toxin, are secreted by the bacteria.

Antibody that binds to accessible antigens on the surface of a bacterium can, together with the C3b component of complement, act as an opsonin that increases phagocytosis and thus clearance of the bacterium. In the case of some bacteria—notably, the gram-negative organisms—complement activation can lead directly to lysis of the organism. Antibody-mediated activation of the complement system can also induce localized production of immune effector molecules that help to develop an amplified and more effective inflammatory response. For example, the complement fragments C3a and C5a act as anaphylatoxins, inducing local mast-cell degranulation and thus vasodilation and the extravasation of lymphocytes and neutrophils from the blood into tissue spaces (see Figure 17-5). Other complement split products serve as chemotactic factors for neutrophils and macrophages, thereby contributing to the buildup of phagocytic cells at the site of infection. Antibody to a bacterial toxin may bind to the toxin and neutralize it; the antibody-toxin complexes are then cleared by phagocytic cells in the same manner as any other antigen-antibody complex.

Although innate immunity is not very effective against intracellular bacterial pathogens, intracellular bacteria can activate NK cells, which in turn provide an early defense against these organisms. Intracellular bacterial infections tend

Antibody-Mediated Mechanisms for Combating Infection by Extracellular Bacteria

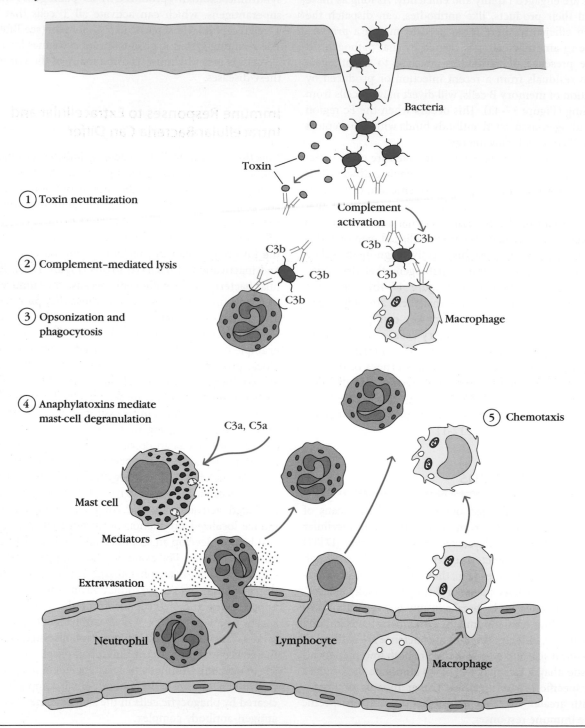

(1) Toxin neutralization

(2) Complement–mediated lysis

(3) Opsonization and phagocytosis

(4) Anaphylatoxins mediate mast-cell degranulation

(5) Chemotaxis

Bacteria

Toxin

Complement activation

C3b

Macrophage

C3a, C5a

Mast cell

Mediators

Extravasation

Neutrophil   Lymphocyte

Macrophage

(1) Antibody neutralizes bacterial toxins. (2) Complement activation on bacterial surfaces leads to complement-mediated lysis of bacteria. (3) Antibody and the complement split product C3b bind to bacteria, serving as opsonins to increase phagocytosis. (4) C3a and C5a, generated by antibody-initiated complement activation, induce local mast-cell degranulation, releasing substances that mediate vasodilation and extravasation of lymphocytes and neutrophils. (5) Other complement products are chemotactic for neutrophils and macrophages.

TABLE 17-3	Host immune responses to bacterial infection and bacterial evasion mechanisms	
Infection process	**Host defense**	**Bacterial evasion mechanisms**
Attachment to host cells	Blockage of attachment by secretory IgA antibodies	Secretion of proteases that cleave secretory IgA dimers (*Neisseria meningitidis, N. gonorrhoeae, Haemophilus influenzae*) Antigenic variation in attachment structures (pili of *N. gonorrhoeae*)
Proliferation	Phagocytosis (Ab- and C3b-mediated opsonization)	Production of surface structures (polysaccharide capsule, M protein, fibrin coat) that inhibit phagocytic cells Mechanisms for surviving within phagocytic cells Induction of apoptosis in macrophages (*Shigella flexneri*)
	Complement-mediated lysis and localized inflammatory response	Generalized resistance of gram-positive bacteria to complement-mediated lysis Insertion of membrane-attack complex prevented by long side chain in cell-wall LPS (some gram-negative bacteria)
Invasion of host tissues	Ab-mediated agglutination	Secretion of elastase that inactivates C3a and C5a (*Pseudomonas*)
Toxin-induced damage to host cells	Neutralization of toxin by antibody	Secretion of hyaluronidase, which enhances bacterial invasiveness

to induce a cell-mediated immune response, specifically DTH. In this response, cytokines secreted by CD4$^+$ T cells are important—most notably IFN-γ, which activates macrophages to kill ingested pathogens more effectively.

Bacteria Can Evade Host Defense Mechanisms at Several Different Stages

There are four primary steps in bacterial infection:

1. Attachment to host cells

2. Proliferation

3. Invasion of host tissue

4. Toxin-induced damage to host cells

Host-defense mechanisms act at each of these steps, and many bacteria have evolved ways to circumvent some of them (Table 17-3).

Some bacteria express molecules that enhance their ability to attach to host cells. A number of gram-negative bacteria, for example, have pili (long hairlike projections), which enable them to attach to the membrane of the intestinal or genitourinary tract (Figure 17-6). Other bacteria, such as *Bordetella pertussis*, the cause of whooping cough, secrete adhesion molecules that attach to both the bacterium and the ciliated epithelial cells of the upper respiratory tract.

Secretory IgA antibodies specific for such bacterial structures can block bacterial attachment to mucosal epithelial cells and are the main host defense against bacterial attachment. However, some bacteria, such as the species of *Neisseria* that cause gonorrhea and meningitis, evade the IgA response by secreting proteases that cleave secretory IgA at the hinge region; the resulting Fab and Fc fragments have a

shortened half-life in mucous secretions and are not able to agglutinate microorganisms.

Some bacteria evade the antibody responses of the host by changing their surface antigens. In *Neisseria gonorrhoeae*, for example, pilin (the protein component of the pili) has a highly variable structure, generated by gene rearrangements of its coding sequence. The pilin locus consists of 1 or 2 expressed genes and 10 to 20 silent genes. Each gene is arranged into six regions called *minicassettes*. Pilin variation is generated by a process of gene conversion, in which one or more minicassettes from the silent genes replace a minicassette of the expression gene. This process generates enormous antigenic

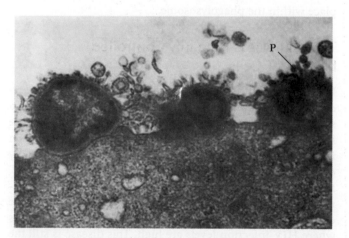

FIGURE 17-6 Electron micrograph of *Neisseria gonorrhoeae* attaching to urethral epithelial cells. Pili (P) extend from the gonococcal surface and mediate the attachment. *[Source: M. E. Ward and P. J. Watt, Adherence of* Neisseria gonorrhoeae *to urethral mucosal cells: an electron microscope study of human gonorrhea. 1972,* Journal of Infectious Disease **126**:601.]

variation, which may contribute to the pathogenicity of *N. gonorrhoeae* by increasing the likelihood that expressed pili will go undetected by antibody, allowing them to bind firmly to epithelial cells and avoid neutralization by IgA.

Bacteria may also possess surface structures that inhibit phagocytosis. A classic example is *Streptococcus pneumoniae*, whose polysaccharide capsule prevents phagocytosis very effectively. The 84 serotypes of *S. pneumoniae* differ from one another by distinct capsular polysaccharides, and during infection the host produces antibody against the infecting serotype. This antibody protects against reinfection with the same serotype but will not protect against infection by a different serotype. In this way, genetic variants of *S. pneumoniae* can cause disease many times in the same individual. On other bacteria, such as *Streptococcus pyogenes*, a surface protein projection called the M protein inhibits phagocytosis, a key step in bacterial removal. Some pathogenic staphylococci are able to assemble a protective coat from host blood proteins. These bacteria secrete a coagulase enzyme that precipitates a fibrin coat around them, shielding them from phagocytic cells.

Mechanisms for interfering with the complement system help other bacteria survive. In some gram-negative bacteria, for example, long side chains on the lipid A moiety of the cell wall core polysaccharide help to resist complement-mediated lysis. *Pseudomonas* secretes an enzyme, elastase, that inactivates both the C3a and C5a anaphylatoxins, thereby diminishing the localized inflammatory reaction.

A number of bacteria escape host-defense mechanisms through their ability to survive within phagocytic cells. Bacteria such as *Listeria monocytogenes* escape from the phagolysosome to the cytoplasm, a favorable environment for their growth. Other bacteria, such as members of the *Mycobacterium* genus, block lysosomal fusion with the phagolysosome or resist the oxidative attack that typically takes place within the phagolysosome.

Tuberculosis Is Primarily Controlled by CD4$^+$ T Cells

Until recently, tuberculosis, caused by *Mycobacterium tuberculosis*, was the leading cause of death in the world from a single infectious agent. Today, *M. tuberculosis* and HIV vie for the lead in deaths due to an infectious agent, with an increasing number of individuals infected by both. Roughly one-third of the world's population is infected with *M. tuberculosis*. Although tuberculosis was believed to be eliminated as a public health problem in the United States, the disease re-emerged in the early 1990s, particularly in areas where HIV-infection levels are high. This disease is still the leading killer of individuals with AIDS.

M. tuberculosis spreads easily, and pulmonary infection usually results from inhalation of small droplets of respiratory secretions containing a few bacilli. The inhaled bacilli are ingested by alveolar macrophages in the lung and are able to survive and multiply intracellularly by inhibiting forma-

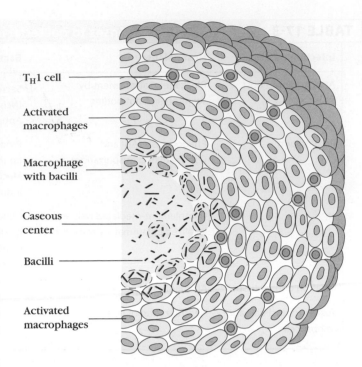

FIGURE 17-7 A tubercle formed in pulmonary tuberculosis.

T$_H$1 cell

Activated macrophages

Macrophage with bacilli

Caseous center

Bacilli

Activated macrophages

tion of phagolysosomes. When the infected macrophages lyse, large numbers of bacilli are released.

The most common clinical pattern of infection with *M. tuberculosis*, seen in 90% of infected individuals, is pulmonary tuberculosis. In this pattern, CD4$^+$ T cells are activated within 2 to 6 weeks after infection and secrete cytokines that induce the infiltration of large numbers of activated macrophages. These cells wall off the organism inside a **granuloma** called a tubercle (Figure 17-7), a cluster of small lymphocytes surrounding infected macrophages. The localized concentrations of lysosomal enzymes in these granulomas can cause extensive tissue necrosis. The massive activation of macrophages that occurs within tubercles often results in the concentrated release of lytic enzymes. These enzymes destroy nearby healthy cells, resulting in circular regions of necrotic tissue, which eventually form a lesion with a caseous (cheeselike) consistency. As these lesions heal, they become calcified and are readily visible on x-rays of the lungs as a defined shadow. Much of the tissue damage seen with *M. tuberculosis* is thus actually due to pathology associated with the cell-mediated immune response.

Because the activated macrophages suppress proliferation of the phagocytosed bacilli, infection is contained. Cytokines produced by CD4$^+$ T cells (T$_H$1 subset) play an important role in the response by activating macrophages so that they are able to kill the bacilli or inhibit their growth. The role of IFN-γ in the immune response to mycobacteria has been demonstrated with knockout mice lacking IFN-γ. These mice died when they were infected with an attenuated strain of mycobacteria, whereas IFN-γ$^+$ wild-type mice survived.

The CD4$^+$ T-cell-mediated immune response mounted by the majority of people exposed to *M. tuberculosis* controls the infection and later protects against reinfection. However, in about 10% of infected individuals, the disease progresses to chronic pulmonary tuberculosis or to extrapulmonary tuberculosis. This progression may occur years after the primary infection. In this clinical pattern, accumulation of large concentrations of mycobacterial antigens within tubercles leads to chronic CD4$^+$ T-cell activation and ensuing macrophage activation, with high concentrations of lytic enzymes causing the necrotic caseous lesions to liquefy into a rich medium that allows the tubercle bacilli to proliferate extracellularly. Eventually the lesions rupture, and the bacilli disseminate in the lung and/or are spread through the blood and lymphatic vessels.

Tuberculosis has traditionally been treated for long periods of time with several different antibiotics, sometimes in combination. The intracellular growth of *M. tuberculosis* makes it difficult for drugs to reach the bacilli, necessitating up to 9 months of daily treatment. Recent clinical trials showed a slightly more effective combination of antibiotics that can be taken less frequently and for a shorter period. Infected individuals with latent tuberculosis were more likely to complete this course of antibiotics, reducing the chances of disease spread.

At present, the only vaccine for *M. tuberculosis* is an attenuated strain of *M. bovis* called Bacille Calmette-Guérin (BCG). This vaccine is fairly effective against extrapulmonary tuberculosis but less so against the more common pulmonary tuberculosis; in some cases, BCG vaccination has even increased the risk of infection. Moreover, after BCG vaccination the tuberculin skin test cannot be used as an effective monitor of exposure to wild-type *M. tuberculosis*. Because of these drawbacks, this vaccine is not used in the United States but is used in several other countries. However, the alarming increase in multidrug-resistant strains has stimulated renewed efforts to develop a more effective tuberculosis vaccine.

Diphtheria Can Be Controlled by Immunization with Inactivated Toxoid

Diphtheria is the classic example of a bacterial disease caused by a secreted exotoxin. Immunity to *Corynebacterium diphtheriae*, the causative agent, can be induced by immunization with an inactivated form of the toxin, known as a **toxoid**. Natural infection with *C. diphtheriae* occurs only in humans, and is spread by respiratory droplets. The organism colonizes the nasopharyngeal tract and causes little tissue damage, with only a mild inflammatory reaction. Virulence is due to its potent exotoxin, which destroys the underlying tissue and results in heart, liver, and kidney damage, as well as to suffocation following formation of a tough fibrous membrane in the respiratory tract. Interestingly, the exotoxin is encoded not by the bacteria but by the *tox* gene carried by a bacterial virus (called phage β). The toxin inhib-

its protein synthesis and is extremely potent: a single molecule has been shown to kill a cell.

Immunization with the toxoid-based vaccine caused a rapid decrease in the number of cases of diphtheria, although sporadic outbreaks have occurred in areas where vaccination coverage is allowed to lapse. Diphtheria toxoid is administered in a vaccine combination with *Bordetella pertussis* (the cause of whooping cough) and tetanus toxoid (called DPT, for *d*iphtheria, *p*ertussis, and *t*etanus). DPT or DTaP, is given to children beginning at 6 to 8 weeks of age as part of the normal course of childhood immunizations (see below). Immunization with the toxoid induces the production of antibodies (antitoxin), which can bind to the toxin and neutralize its activity. Because antitoxin levels decline slowly over time, booster shots are recommended at 10-year intervals to maintain antitoxin levels within the protective range.

Parasitic Infections

The term *parasite* encompasses a vast number of protozoan and helminthic organisms (worms). The diversity of the parasitic universe makes it difficult to generalize, but a major difference between these types of parasites is that the protozoans are unicellular eukaryotes that usually live and multiply *within* host cells for at least part of their life cycle, whereas helminths are multicellular organisms that can be quite large and have the ability to live and reproduce *outside* their human host. Most clinically relevant protozoan parasites also require an intermediate host for a portion of their life cycle and for transmission to human hosts.

Parasites can evade the immune system, allowing them to chronically infect their human hosts and exact a lifelong toll. Malaria, African sleeping sickness, Chagas's disease, leishmaniasis, and toxoplasmosis are among the most common parasitic diseases. Experimental systems, especially mouse models of infection, have defined how immunity to certain parasites is achieved, but the diversity of parasites and the complexity of the infections they cause preclude easy generalizations.

Protozoan Parasites Account for Huge Worldwide Disease Burdens

Infections caused by parasites account for an enormous disease burden worldwide, especially in tropical or subtropical regions and developing countries where sanitation and living conditions are poor. The presence of sewage-tainted water promotes parasite spread and transmission to humans. Likewise, many of the protozoan parasites spend part of their time in arthropod hosts, such as mosquitoes, flies, or ticks, which serve as an essential microenvironment for completion of their life cycle and as a vector for transmission to humans. Targeting these blood-feeding vectors can therefore be quite effective in interrupting protozoan propagation and reducing transmission rates.

The type and effectiveness of immune response to protozoan infection depends in part on the location of the parasite within the host and the life cycle stage of the parasite. Many protozoans spend part of their time free within the bloodstream; humoral antibody is most effective during these stages. At other stages they may grow intracellularly, making cell-mediated immune reactions the most effective host defense. In the development of vaccines for protozoan diseases, the life cycle stages of these pathogens and the branch of the immune response that is most likely to confer protection must be carefully considered.

Malaria

Malaria is the number-one parasitic cause of death worldwide. Half of the world's population lives in a malaria endemic zone, and nearly 10% of the world population is infected by the causative agent of malaria: one of several species of the genus *Plasmodium*, of which *P. falciparum* is the most virulent. The alarming development of multiple-drug resistance in *Plasmodium* and the increased pesticide resistance of the *Anopheles* mosquito, the arthropod vector of *Plasmodium*, underscore the importance of developing new strategies to hinder the spread of malaria.

Plasmodium has an extremely complex life cycle. Female *Anopheles* mosquitoes serve as the vector and host for part of the parasite's life cycle. (Because male *Anopheles* mosquitoes feed on plant juices, they do not transmit *Plasmodium*.) Human infection begins when sporozoites, one of the life cycle stages of *Plasmodium*, enter the bloodstream as an infected female mosquito takes a blood meal (Figure 17-8). Sporozoites are long, slender cells that are covered by a 45-kDa protein called circumsporozoite (CS) antigen, which mediates their adhesion to hepatocytes. The binding site on the CS antigen is a conserved region that has a high degree of sequence homology with known human cell adhesion molecules. In hepatocytes, the parasites differentiate into merozoites, which infect red blood cells, initiating the major symptoms and pathology of malaria. Eventually some of the merozoites differentiate into male and female gametocytes, which may be ingested by a female *Anopheles* mosquito during a blood meal from an infected individual. Within the mosquito's gut, the male and female gametocytes differentiate into gametes that fuse to form a zygote, which multiplies and differentiates into sporozoites within the mosquito's salivary gland, initiating the cycle again.

The symptoms of malaria are recurrent chills, fever, and sweating that peak roughly every 48 hours, when successive generations of merozoites are released from infected red blood cells. An infected individual eventually becomes weak and anemic. The merozoites can block capillaries, causing intense headaches, renal failure, heart failure, or cerebral damage (called cerebral malaria), often with fatal consequences. Some malaria symptoms may be caused by excessive production of cytokines, a hypothesis stemming from the observation that cancer patients treated with recombinant TNF-α developed symptoms that mimicked malaria. The connection

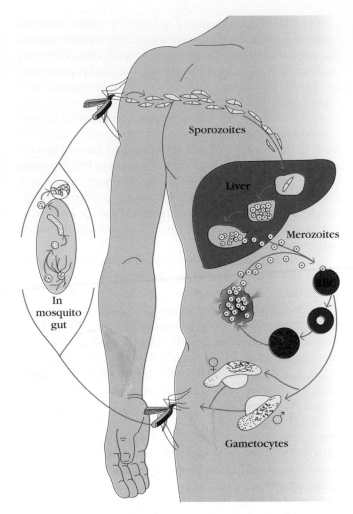

FIGURE 17-8 The life cycle of *Plasmodium*. Sporozoites enter the bloodstream when an infected mosquito takes a blood meal. The sporozoites migrate to the liver, where they multiply, transforming liver hepatocytes into giant multinucleate schizonts, which release thousands of merozoites into the bloodstream. The merozoites infect red blood cells, which eventually rupture, releasing more merozoites. Eventually some of the merozoites differentiate into male and female gametocytes, which are ingested by a mosquito and differentiate into gametes in the mosquito's gut. The gametes fuse to form a zygote that differentiates to the sporozoite stage within the salivary gland of the mosquito.

between TNF-α and malaria symptoms was studied by infecting mice with a mouse-specific strain of *Plasmodium*, which causes rapid death by cerebral malaria. Injection of these mice with antibodies to TNF-α prevented this rapid death.

In regions where malaria is endemic, the immune response to *Plasmodium* infection is poor. Children younger than 14 years old mount the weakest immune response and consequently are most likely to develop malaria. In some regions, the childhood mortality rate for malaria reaches 50%. Even in adults, the degree of immunity is far from complete. Most people living in endemic regions have lifelong low-level *Plasmodium* infections.

A number of factors may contribute to these low levels of immune response to *Plasmodium*. The maturational changes allow the organism to keep changing its surface molecules, resulting in continual changes in the antigens seen by the immune system. The intracellular phases of the life cycle reduce the degree of immune activation generated by the pathogen and allow the organism to multiply shielded from attack. The most accessible stage, the sporozoite, circulates in the blood for such a short time before infecting hepatocytes (approx. 30 minutes) that effective immune activation is unlikely to occur. Even when an antibody response does develop to sporozoites, *Plasmodium* overcomes that response by sloughing off the CS surface antigens, thus rendering the antibodies ineffective. The development of drug resistance by *Plasmodium* has complicated drug treatment choices for malaria, making the search for a vaccine very important.

African Sleeping Sickness

Two species of African trypanosomes cause African sleeping sickness, a chronic, debilitating disease transmitted to humans and cattle by the bite of the tsetse fly. In the bloodstream, the trypanosome, a flagellated protozoan, differentiates into a long, slender form that continues to divide every 4 to 6 hours. The disease progresses from an early, systemic stage in which trypanosomes multiply in the blood to a neurologic stage in which the parasite infects cells of the central nervous system, leading to meningoencephalitis and eventual loss of consciousness—thus the name.

The surface of the *Trypanosoma* parasite is covered with a variable surface glycoprotein (VSG). Several unusual genetic processes generate extensive variation in these surface structures, enabling the organism to escape immunologic clearance. An individual trypanosome carries a large repertoire of VSG genes, each encoding a different VSG primary sequence, but expresses only a single VSG gene at a time. *Trypanosoma brucei*, for example, carries more than 1000 VSG genes in its genome. Activation of a VSG gene results in duplication of the gene and its transposition to a transcriptionally active expression site (ES) at the telomeric end of specific chromosomes (Figure 17-9a). Activation of a new VSG gene displaces the previous gene from the telomeric ES, like placing a new DVD in the reader. Trypanosomes have multiple transcriptionally active ES sites, so that a number of VSG genes can potentially be expressed; unknown control mechanisms limit expression to a single VSG expression site at any time.

As parasite numbers increase after infection, an effective humoral response develops to the VSG covering the surface of the parasite. These antibodies eliminate most of the parasites from the bloodstream, both by complement-mediated lysis and by opsonization and subsequent phagocytosis. However, about 1% of the organisms bear an antigenically different VSG because of transposition of that VSG gene into the ES. These parasites escape the initial antibody response, begin to proliferate in the bloodstream, and go on to populate the next wave of parasitemia in the host. The successive waves of parasitemia reflect a unique mechanism of antigenic shift by which the trypanosomes evade the immune response to their surface antigens. Each new variant that arises in the course of a single infection escapes the humoral antibodies generated in response to the preceding variant, and so waves of parasitemia recur (Figure 17-9b). The new variants arise not by clonal outgrowth from a single escape variant cell, but from the expansion of multiple cells that have activated the same VSG gene in the current wave of parasitic growth. It is not known how this process is coordinated. This continual shifting of surface epitopes has made vaccine development extremely difficult.

Leishmaniasis

The protozoan parasite *Leishmania major* illustrates how powerfully different host responses can impact disease outcome, leading to either clearance of the parasite or death from the infection. *Leishmania* is a flagellated protozoan that lives in the phagosomes of macrophages and is transmitted by sandflies. It usually results in one of two syndromes: a localized cutaneous lesion that is generally painless and self-resolving, or a systemic form of the disease, called visceral **leishmaniasis**, which is nearly always fatal without treatment.

Resistance to leishmaniasis correlates well with the production of IFN-γ and the development of a T_H1 response. Strains of mice that are naturally resistant to *Leishmania* develop a T_H1 response and produce IFN-γ upon infection. If IFN-γ production or signaling is blocked in these strains, the mice become highly susceptible to *Leishmania*-induced fatality. However, a few strains of mice, such as BALB/c, are naturally susceptible to *Leishmania*-induced death. BALB/c animals mount a T_H2-type response to *Leishmania* infection, producing high levels of IL-4 and essentially no IFN-γ. Studies have shown that a small subset of CD4$^+$ T cells in the susceptible animals recognize a particular epitope on *L. major*, and produces high levels of IL-4 early in the response to the parasite, skewing the response towards a T_H2-dominated pathway. Understanding how different T-helper responses affect the outcome of infections could contribute to the rational design of effective treatments and vaccines against this and other pathogens.

A Variety of Diseases Are Caused by Parasitic Worms (Helminths)

Parasitic worms, or helminths, are responsible for a wide variety of diseases in humans and animals. The adult forms are large, multicellular organisms that can often be seen with the naked eye. Parasitic worms are frequently categorized based on their structure and site of infection: nematodes (roundworms), cestodes (tapeworms), and trematodes (flukes). Most enter their animal hosts through the intestinal tract; helminth eggs can contaminate food, water, feces, and soil. Although helminths are exclusively extracellular and therefore more accessible to the immune system than protozoans, most

Successive Waves of Parasitemia after Infection with *Trypanosoma* Result from Antigenic Shifts in the Parasite's Variant Surface Glycoprotein (VSG)

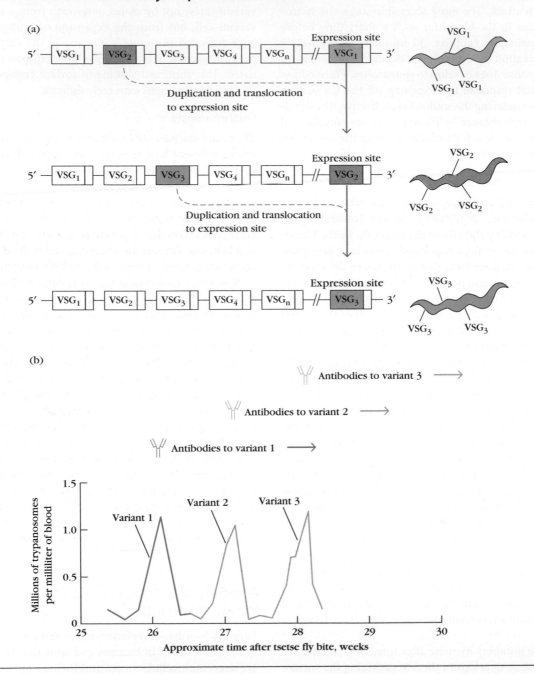

(a) Antigenic shifts in trypanosomes occur by the duplication of gene segments encoding variant VSG molecules and their translocation to an expression site located close to the telomere. (b) Antibodies develop against each variant as the numbers of these parasites rise, but each new variant that arises is unaffected by the humoral antibodies induced by the previous variant. [*Source: Part (b) adapted from J. Donelson, 1988, The Biology of Parasitism, New York: Alan R. Liss.*]

infected individuals carry few parasites. Unlike protozoan parasites, helminths do not multiply within their hosts. Thus, the immune response is not strongly engaged, and the level of immunity generated can be very poor.

More than 300 million people are infected with *Schistosoma*, which causes the chronic, debilitating, and sometimes-fatal disease **schistosomiasis**. Infection occurs through contact with free-swimming infectious larvae that are released from an infected snail and bore into the skin, frequently while individuals wade through contaminated water. As they mature, they migrate in the body, with the final site of infection varying by species. The females produce eggs, some of which are excreted and infect more snails. Most symptoms of schistosomiasis are initiated by the eggs, which invade tissues and cause hemorrhage. A chronic state can develop in which the unexcreted eggs induce cell-mediated DTH reactions, resulting in large granulomas that can obstruct the venous blood flow to the liver or bladder.

An immune response does develop to the schistosomes, but it is usually not sufficient to eliminate the adult worms. Instead, the worms survive for up to 20 years, causing prolonged morbidity. Adult schistosome worms have several unique mechanisms that protect them from immune defenses. These include decreasing the expression of antigens on their outer membrane and enclosing themselves in a glycolipid-and-glycoprotein coat derived from the host, masking the presence of their own antigens. Among the antigens observed on the adult worm are the host's own ABO blood-group and histocompatibility antigens! The immune response is, of course, diminished by this covering made of the host's self antigens, which must contribute to the lifelong persistence of these organisms.

The major contributors to protective immunity against schistosomiasis are controversial. The immune response to infection with *S. mansoni* is dominated by T_H-2-like mediators, with high titers of antischistosome IgE antibodies, localized increases in degranulating mast cells, and an influx of eosinophils (Figure 17-10, *top*). These cells can then bind the antibody-coated parasite using their Fc receptors for IgE or IgG, inducing degranulation and death to the parasite via antibody-dependent cell-mediated cytotoxicity (ADCC; see Figure 13-14). One eosinophil mediator, called basic protein, has been found to be particularly toxic to helminths. However, immunization studies in mice suggest that a T_H1 response, characterized by IFN-γ and macrophage accumulation, may actually be *more* effective for inducing protective immunity (Figure 17-10, *bottom*). In fact, inbred strains of mice with deficiencies in mast cells or IgE can still develop protective immunity to *S. mansoni* following vaccination. Based on these observations, it has been suggested that the ability to induce an *ineffective* T_H2-like response may have evolved in schistosomes as a clever defense mechanism to ensure that IL-10 and other T_H-1 inhibitors are induced in order to block initiation of a more *effective* T_H1-dominated pathway.

Fungal Infections

Fungi are a diverse and ubiquitous group of organisms that occupy many niches and also perform services for humans, including the fermentation of bread, cheese, wine, and beer, as well as the production of penicillin. As many as a million species of fungi are known to exist; only about 400 are potential agents of human disease. Infections may result from introduction of exogenous organisms due to injury or inhalation, or from endogenous organisms such as the commensals present in the gut and on the skin.

Fungal diseases, or **mycoses**, are classified based on the following criteria:

- Site of infection—superficial, cutaneous, subcutaneous, or deep and systemic
- Route of acquisition—exogenous or endogenous
- Virulence—primary or opportunistic

These categories (summarized in Table 17-4) are not mutually exclusive. For example, an infection such as coccidiomycosis may progress from a cutaneous lesion to a systemic infection of the lungs. Cutaneous infections include attacks on skin, hair, and nails; examples are ringworm, athlete's foot, and jock itch. Subcutaneous infections are normally introduced by trauma and accompanied by inflammation; if inflammation is chronic, extensive tissue damage may ensue. Deep mycoses involve the lungs, the central nervous system, bones, and the abdominal viscera. These infections can occur through ingestion, inhalation, or inoculation into the bloodstream. A very rare and deadly outbreak of fungal meningitis in 2012 was linked to *Exserohilum rostratum*, a fungal contaminant in a preparation of corticosteroids used for epidural injections.

Virulence can be divided into primary, indicating the rare agents with high pathogenicity, and opportunistic, denoting weakly virulent agents that primarily infect individuals with compromised immunity. Most fungal infections of healthy individuals are resolved rapidly, with few clinical signs. The most commonly encountered and best-studied human fungal pathogens are *Cryptococcus neoformans*, *Aspergillus fumigatus*, *Coccidioides immitis*, *Histoplasma capsulatum*, and *Blastomyces dermatitidis*. Diseases caused by these fungi are named for the agent; for example, *C. neoformans* causes cryptococcosis and *B. dermatitidis* causes blastomycosis. In each case, infection with these environmental agents is aided by predisposing conditions that include AIDS, immunosuppressive drug treatment, and malnutrition.

Innate Immunity controls Most Fungal Infections

The barriers of innate immunity control most fungi. Commensal organisms also help control the growth of potential pathogens, as demonstrated by long-term treatment with broad-spectrum antibiotics, which destroy normal mucosal bacterial flora and often lead to oral or vulvovaginal infection

Overview of the Immune Response Generated against *Schistosoma mansoni*

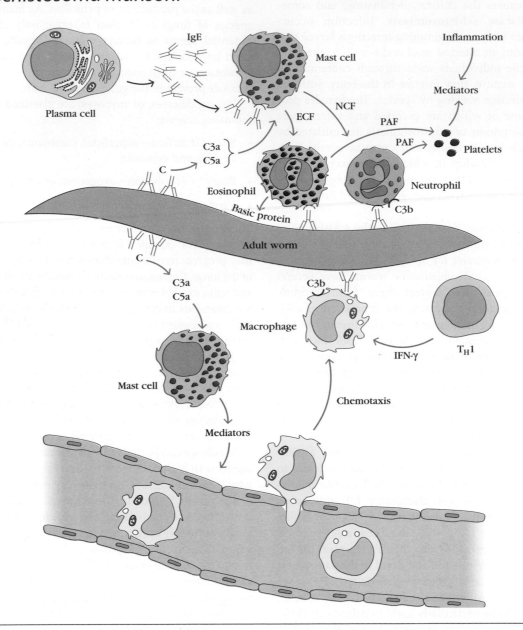

The response includes an IgE humoral component (top) and a cell-mediated component involving CD4$^+$ T cells (bottom). C = complement; ECF = eosinophil chemotactic factor; NCF = neutrophil chemotactic factor; PAF = platelet-activating factor.

with *Candida albicans*, an opportunistic agent. Phagocytosis by neutrophils is a strong defense against most fungi, and so people with neutropenia (low neutrophil count) are generally more susceptible to fungal disease.

Resolution of infection in normal, healthy individuals is often rapid and initiated by recognition of common fungal cell wall PAMPs. The three most medically relevant cell wall components include β-glucans (polymers of glucose), mannans (long chains of mannose), and chitin (a polymer of N-acetylglucosamine). The importance of certain *pattern recognition receptors* (PRRs) in resolving fungal infection has been demonstrated by the increased susceptibility to

TABLE 17-4	Classification of fungal diseases	
Site of infection	Superficial	Epidermis, no inflammation
	Cutaneous	Skin, hair, nails
	Subcutaneous	Wounds, usually inflammatory
	Deep or systemic	Lungs, abdominal viscera, bones, CNS
Route of acquisition	Exogenous	Environmental, airborne, cutaneous, or percutaneous
	Endogenous	Latent reactivation, commensal organism
Virulence	Primary	Inherently virulent, infects healthy host
	Opportunistic	Low virulence, infects immunocompromised host

mycoses seen in individuals with defects in these components. For instance, certain molecular variants of dectin 1, a C-type lectin receptor (see Chapter 5), are associated with chronic mucocutaneous candidiasis. Toll-like receptors 2, 4 and 9, as well as complement receptor 3 (CR3), are also involved in the innate response to fungi. In sum, recognition of these cell wall components leads to the activation of complement (via both alternative and lectin pathways) along with the induction of phagocytosis and destruction of fungal cells. The key role of CR3, which recognizes complement deposited on the β-glucans of fungal cells, was confirmed by the fact that mortality from experimental infections of mice with *Cryptococcus* increased after an antibody to CR3 was administered.

Like other microbes, fungi have evolved mechanisms to evade the innate immune response. These include production of a capsule, as in the case of *C. neoformans,* which blocks PRR binding. Another evasion strategy employed by this organism involves fungi-induced expulsion from macrophages that does not kill host cells and therefore avoids inflammation.

Immunity against Fungal Pathogens Can Be Acquired

A convincing demonstration of acquired immunity against fungal infection is the protection against subsequent attacks following an infection. This protection is not always obvious for fungal disease because primary infection often goes unnoticed. However, positive skin reactivity to fungal antigens is a good indicator of prior infection and the presence of memory responses. For instance, a granulomatous inflammation controls spread of *C. neoformans* and *H. capsulatum,* indicating the presence of acquired cell-mediated immunity. However, the organism may remain in a latent state within the granuloma, reactivating only if the host becomes immunosuppressed.

The presence of antibodies is another sign of prior exposure and lasting immunity, and antibodies against *C. neoformans* are commonly found in healthy subjects. However, probably the most convincing argument for preexisting immunity against fungal pathogens comes from the frequency of normally rare fungal diseases in patients with compromised immunity. AIDS patients suffer increased incidences of mucosal candidiasis, histoplasmosis, coccidiomycosis, and cryptococcosis. These observations in T-cell-compromised AIDS patients and data showing that B-cell–deficient mice have *no* increased susceptibility to fungal disease indicate that cell-mediated mechanisms of immunity likely control most fungal pathogens.

The study of immunity to fungal pathogens has become more pressing with the advent of AIDS and the increase in individuals receiving immunosuppressive drugs for other conditions. This has led to the observation that strong T_H1 responses and the production of IFN-γ, important for optimal macrophage activation, are most commonly associated with protection against fungi. Conversely, T_H2 and T_{REG} cell responses, or their products, are associated with susceptibility to mycoses. This is apparent in patients displaying distinct T helper responses to coccidioidomycosis, where T_H1 immune activity is associated with a mild, asymptomatic infection and T_H2 responses result in a severe and often relapsing form of the disease. Although the role for other cell types is less certain, recently a regulatory role for T_H17 cells in controlling adaptive immunity against fungi has been postulated, where these cells are hypothesized to help support T_H1- and discourage T_H2-cell activation.

Emerging and Re-emerging Infectious Diseases

At least yearly, it seems, we hear about a new virus or bacterium arising, accompanied by severe illness or death. Newly described human pathogens are referred to as *emerging pathogens.* Examples include HIV, SARS, WNV, the widely publicized Ebola virus, and *Legionella pneumophila,* the bacterial causative agent for Legionnaires' disease. These often appear to come from nowhere and, as far as we know, are caused by new human pathogens. On the other hand, *re-emerging pathogens* are those that were formerly rare or largely eradicated but suddenly begin to infect a widening number of individuals. The re-emergence of these diseases is

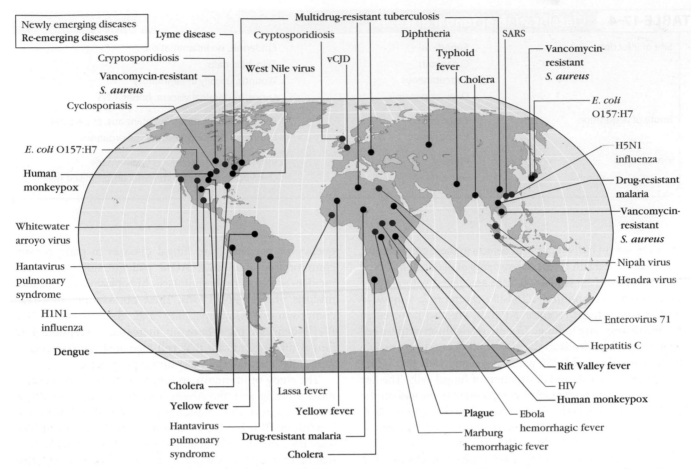

FIGURE 17-11 Examples of emerging and re-emerging diseases showing points of origin. Red represents newly emerging diseases, black re-emerging diseases. *[Source: Adapted from A. S. Fauci, 2001, Infectious diseases: Considerations for the 21st Century,* Clinical Infectious Diseases *32:675.]*

not surprising if we consider that some microbes can adapt to a range of environments, and that many microbes can exist in dormant states or can survive in intermediate hosts. Examples of many of the emerging and re-emerging infectious diseases are shown in Figure 17-11.

Some Noteworthy New Infectious Diseases Have Appeared Recently

Ebola was first recognized after an outbreak in Africa in 1976. The disease received a great deal of attention because of the severity and rapid progression to death after the onset of symptoms. By 1977, the causative virus had been isolated and classified as a filovirus, a type of RNA virus that includes the similarly deadly Marburg virus, a close relative of Ebola. The most pathogenic strain, Ebola-Zaire, causes a particularly severe hemorrhagic fever, killing roughly 50% to 90% of those infected, often within days of the onset of symptoms. Although the risk of death is very high after infection, it is fairly easy to control the spread of the virus by isolating infected individuals. This is an example of a pathogen that likely resides in some wild animals, which helps it to remain

in the environment for long periods of time and only infect humans when some form of contact is made. Although the animal reservoir for Ebola has not been found, fruit bats are the most likely candidate.

Legionnaires' disease is a virulent pneumonia first reported in attendees at an American Legion convention in Philadelphia in 1976. Of the 221 afflicted, 34 died. The organism causing the disease turned out to be a bacterium, later named *Legionella pneumophila*. It proliferates in cool, damp areas, including the condensing units of large commercial air-conditioning systems. Infection can spread when the air-conditioning system emits an aerosol containing the bacteria. Improved design of air-conditioning and plumbing systems has greatly reduced the danger from this disease.

In November 2002, an unexplained atypical pneumonia was seen in the Guandong province of China, proving resistant to any treatment. A physician who had cared for some of these patients traveled to Hong Kong and infected guests in his hotel, who then seeded a multinational outbreak that lasted until May 2003. By the time the disease, called *severe acute respiratory syndrome* (SARS), was contained, 8096 cases had been reported, with 774 deaths. A rapid response

(a)　　　　　　　　　　　　　　(b)

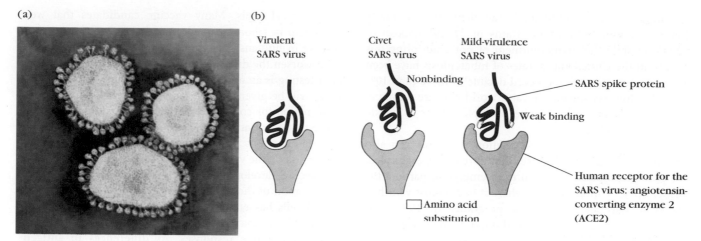

FIGURE 17-12 The coronavirus that caused the outbreak of severe acute respiratory syndrome, or SARS. (a) The virus is studded with spikes that in cross-section give it the appearance of a crown, hence the name coronavirus. (b) The human receptor for the SARS virus is angiotensin-converting enzyme 2 (ACE2). The spike protein from the highly virulent form of the virus binds with high affinity to this receptor; the variant of this virus found in the civet cat and in a version that causes mild human disease have two different amino acid substitutions in the spike protein. The civet cat SARS virus does not bind to the human receptor, and the version that causes mild disease in humans binds 1000 times more weakly than the most virulent form. *[Source: Part a from Dr. Linda Stannard, UCT/Photo Researchers; part b adapted from K. V. Holmes, 2005, Structural biology. Adaptation of SARS coronavirus to humans.* Science ***309:1822.]***

by the biomedical community identified the etiologic agent as a coronavirus, so named because the spike proteins emanating from these viruses give them a crownlike appearance (Figure 17-12a). This virus was soon traced to the civet cat after the earliest cases of SARS were found to occur in animal vendors. Human coronaviruses had been known for many years, primarily as the cause of a mild form of the common cold, but this newly emerged variant had not been seen previously. X-ray crystallography showed that a mutated civet SARS virus spike protein could bind to a human cell surface protein, human *angiotensin-converting enzyme type 2* (ACE2), 1000 times more tightly than did the original virus spike protein, answering the question of how this virus jumped from the civet cat to humans (Figure 17-12b). Animal models for SARS showed that antibodies to the outer spike protein could thwart replication of the virus, leading to the later development of an intranasal vaccine that could induce a strong local humoral response as well as cell-mediated immunity.

West Nile virus (WNV), first identified in Uganda in 1937, was not seen outside Africa or western Asia until 1999, when it showed up in the New York City area. By 2006, it had been reported in all but six states in the continental United States. WNV is a flavivirus that replicates very well in certain species of birds and is carried by mosquitoes from infected birds to so-called dead-end hosts such as horses and humans. Transmission between humans via mosquitos is inefficient because the titer of virus in human blood is low and the amount of blood transferred by the insect bite is small and does not contain sufficient virus to cause infection. WNV may, however, be transferred from human to human by blood transfusion and may be passed from infected pregnant mothers to their newborns. It is only a health hazard in individuals with compromised immune function, in whom it can cross the blood-brain barrier and cause life-threatening encephalitis or meningitis. The primary public health control measure is education of the public about mosquito control.

Why are these new diseases emerging and others re-emerging around the globe? One reason may be the crowding of the world's poorest populations into ever smaller areas within huge cities. Another factor is the great increase in international travel: an individual can become infected on one continent and then spread the disease on another continent all in the same day. The mass distribution and importation of food can also expose large populations to potentially contaminated food. Indiscriminate use of antibiotics in humans and in veterinary applications also fosters resistance in pathogens to the drugs commonly used against them. Finally, changes in climate and weather patterns, along with the extension of human populations into the previously unbroached domains of animals, can introduce new pathogens into human populations.

Diseases May Re-emerge for Various Reasons

Tuberculosis is a re-emerging disease now receiving considerable attention. Twenty years ago, public health officials were convinced that tuberculosis would soon disappear as a major health consideration in the United States. A series of events conspired to interrupt the trend, including the AIDS epidemic and other conditions that result in immunosuppressed individuals, allowing *Mycobacterium* strains to gain a foothold and evolve resistance to the conventional battery of antibiotics. These individuals can then pass on newly emerged, antibiotic-resistant strains of *M. tuberculosis* to others. Despite the disappointment of not eradicating this disease in the United States, rates of tuberculosis have been declining annually, with a record decline of more than 10% in 2009. However, incidence rates worldwide are falling at less than 1% annually, despite multiple efforts to tackle tuberculosis globally.

Laxity in adherence to vaccination programs can also lead to re-emergence of diseases that were nearly eradicated. For example, diphtheria began to re-emerge in parts of the former Soviet Union in 1994, where it had almost vanished thanks to European vaccination programs. By 1995, over 50,000 cases were reported and thousands died. The social upheaval and instability that came with the breakup of the Soviet Union was almost certainly a major factor in the re-emergence of this disease, due to lapses in vaccination and other public health measures. Even in the United States, an increasing trend in some regions to delay or opt out of childhood vaccination has led to sporadic local outbreaks in previously rare childhood diseases such as measles and whooping cough. The WHO and the CDC both actively monitor new infections, and work closely to detect and identify new infectious agents and key re-emerging pathogens. Their collective data help to provide up-to-date information for travelers to parts of the world where such emerging and re-emerging infectious agents may pose a risk.

Vaccines

Preventative vaccines have led to the control or elimination of many infectious diseases that once claimed millions of lives. Since October 1997, not a single naturally acquired smallpox case has been reported anywhere. On the heels of the global victory over smallpox, the program to eradicate polio went into overdrive. Led by the WHO and several large philanthropic donors, numbers of polio cases were reduced worldwide by over 95% by the year 2000. Alas, cultural and religious resistance to vaccination led to polio's reappearance in recent years, although many vaccination programs are now back on track. Worldwide vaccination campaigns can also be credited with the control of at least 10 other major infectious diseases (measles, mumps, rubella, typhoid, tetanus, diphtheria, pertussis, influenza, yellow fever, and rabies), many of which previously affected mostly babies and young children.

Still, a crying need remains for vaccines against many other diseases, including malaria, tuberculosis, and AIDS. More work is needed for existing vaccines as well: to improve safety and efficacy of some, or to lower the cost and simplify the delivery of existing vaccines so that they can reach those who need them most, especially in developing countries. Based on WHO data, millions of infants still die due to diseases that could be prevented by existing vaccines.

Development of effective new vaccines is a long, complicated, and costly process, rarely reaching the stage of years-long clinical trials. Many vaccine candidates that were successful in laboratory and animal studies fail to prevent disease in humans, have unacceptable side effects, or may even worsen the disease they were meant to prevent. Stringent testing is an absolute necessity, because approved vaccines will be given to large numbers of well people. Clear information for consumers about adverse side effects, even those that occur at very low frequency, must be made available and carefully balanced against the potential benefit of protection by the vaccine.

Vaccine development begins with basic research. An appreciation of the differences in epitopes recognized by T and B cells has enabled immunologists to design vaccine candidates to maximize activation of both cellular and humoral immune responses. As differences in antigen-processing pathways became evident, design strategies and additives can be employed to activate specific, desired immune pathways and to maximize antigen presentation. Targeting strategies to elicit protection at mucosal surfaces, the most common site of infection, are also underway. Finally, techniques like genetic engineering are being employed to develop vaccines that maximize the immune response to selected epitopes and that simplify delivery. Here we describe current vaccine strategies, some vaccines presently in use, and new strategies that may lead to future vaccines.

Protective Immunity Can Be Achieved by Active or Passive Immunization

Immunization is the process of eliciting a long-lived state of protective immunity against a disease-causing pathogen. Exposure to the live pathogen followed by recovery is one route to immunization. **Vaccination**, or intentional exposure to forms of a pathogen that do not cause disease (a **vaccine**), is another. In an ideal world, both engage antigen-specific lymphocytes and result in the generation of memory cells, providing long-lived protection. However, vaccination does not ensure immunity, and a state of immune protection can be achieved by means other than infection or vaccination. For example, the transfer of antibodies from mother to fetus or the injection of antiserum against a pathogen both provide immune protection. Without the development of memory B or T cells specific to the organism, however, this state of immunity is only temporary. Thus, vaccination is an event, whereas immunization (the development of a protective memory response) is a potential outcome of that event. Here we describe current use of passive and active immunization techniques.

Passive Immunization by Delivery of Preformed Antibody

Edward Jenner and Louis Pasteur are recognized as the pioneers of vaccination, or early attempts to induce active immunity, but similar recognition is due to Emil von Behring and Hidesaburo Kitasato for their contributions to passive

TABLE 17-5	Common agents used for passive immunization

Disease	Agent
Black widow spider bite	Horse antivenin
Botulism	Horse antitoxin
Cytomegalovirus	Human polyclonal Ab
Diphtheria	Horse antitoxin
Hepatitis A and B	Pooled human immunoglobulin
Measles	Pooled human immunoglobulin
Rabies	Human or horse polyclonal Ab
Respiratory disease	Monoclonal anti-RSV*
Snake bite	Horse antivenin
Tetanus	Pooled human immunoglobulin or horse antitoxin
Varicella zoster virus	Human polyclonal Ab

*Respiratory syncytial virus

Source: Adapted from A. Casadevall, 1999, Passive antibody therapies: progress and continuing challenges. Clinical Immunology **93:5**.

immunity. These investigators were the first to show that immunity elicited in one animal can be transferred to another by injecting serum from the first.

Passive immunization, in which preformed antibodies are transferred to a recipient, occurs naturally when maternal IgG crosses the placenta to the developing fetus. Maternal antibodies to diphtheria, tetanus, streptococci, rubeola, rubella, mumps, and poliovirus all afford passively acquired protection to the developing fetus. Later, maternal antibodies present in breast milk can also provide passive immunity to the infant in the form of maternally produced IgA.

Passive immunization can also be achieved by injecting a recipient with preformed antibodies, called antiserum, from immune individuals. Before vaccines and antibiotics became available, passive immunization was the only effective therapy for some otherwise fatal diseases, such as diphtheria, providing much needed humoral defense (see Chapter 1, Clinical Focus Box 1-2). Currently, several conditions still warrant the use of passive immunization, including the following:

- Immune deficiency, especially congenital or acquired B-cell defects

- Toxin or venom exposure with immediate threat to life

- Exposure to pathogens that can cause death faster than an effective immune response can develop

Babies born with congenital immune deficiencies are frequently treated with passive immunization, as are chil-

dren experiencing acute respiratory failure caused by respiratory syncytial virus (RSV). Passive immunity is used in unvaccinated individuals exposed to the organisms that cause botulism, tetanus, diphtheria, hepatitis, measles, and rabies (Table 17-5), or to protect travelers and health-care workers who expect exposure to potential pathogens for which they lack protective immunity. Antiserum provides protection against poisonous snake and insect bites. In all these instances, it is important to remember that *passive immunization does not activate the host's immune response*. Therefore, it generates no memory response and protection is transient.

Although passive immunization may be effective, it should be used with caution because certain risks are associated with the injection of preformed antibody. If the antibody was produced in another species, such as a horse (one of the most common animal sources), the recipient can mount a strong response to the isotypic determinants of the foreign antibody, or the parts of the antibody that are unique to the horse species. This anti-isotype response can cause serious complications. Some individuals will produce IgE antibody specific for horse-specific determinants. High levels of these IgE-horse antibody immune complexes can induce pervasive mast-cell degranulation, leading to systemic anaphylaxis (see Chapter 15). Other individuals produce IgG or IgM antibodies specific for the foreign antibody, which form complement-activating immune complexes. The deposition of these complexes in the tissues can lead to type III hypersensitivity reactions. Even when human antiserum, or gamma globulin, is used, the recipient can generate an anti-allotype response to the human immunoglobulin (recognition of within species antigenic differences), although its intensity is usually much less than that of an anti-isotype response.

Active Immunization to Induce Immunity and Memory

Whereas the aim of passive immunization is transient protection or alleviation of an existing condition, the goal of active immunization is to trigger the adaptive immune response in a way that will elicit protective immunity and immunologic memory. When active immunization is successful, a subsequent exposure to the pathogenic agent elicits a secondary immune response that successfully eliminates the pathogen or prevents disease mediated by its products. Active immunization can be achieved by natural infection with a microorganism, or it can be acquired artificially by administration of a vaccine. In active immunization, as the name implies, the immune system plays an active role—proliferation of antigen-reactive T and B cells is induced and results in the formation of protective memory cells. This is the primary goal of vaccination.

Active immunization with various types of vaccines has played an important role in the reduction of deaths from infectious diseases, especially among children. In the United States, vaccination of children begins at birth. The American

TABLE 17-6 Recommended childhood immunization schedule in the United States, 2012

Vaccine	Birth	1 mo	2 mo	4 mo	6 mo	9 mo	12 mo	15 mo	18 mo	19–23 mo	2–3 yr	4–6 yr
Hepatitis B	Hep B	← Hep B →			←		Hep B		→			
Rotavirus			RV	RV	RV							
Diphtheria, tetanus, pertussis			DTaP	DTaP	DTaP			← DTaP →				DTaP
Haemophilus influenzae type b			Hib	Hib	Hib		← Hib →					
Pneumococcal			PCV	PCV	PCV		← PCV →					← PPSV →
Inactivated poliovirus			IPV	IPV	←		IPV		→			IPV
Influenza					←			Influenza (yearly)				→
Measles, mumps, rubella							← MMR →					MMR
Varicella							← Varicella →					Varicella
Hepatitis A				(Two doses at least 6 months apart)			← Dose 1		→			← HepA series →
Meningococcal								←	MCV4			→

Recommendations in effect as of 12/23/11. Any dose not administered at the recommended age should be administered at a subsequent visit, when indicated and feasible. A combination vaccine is generally preferred over separate injections of equivalent component vaccines.

MCV4 and PPSV ranges are recommended for certain high risk groups. See website for further details.

Source: Modified version of 2012 American Academy of Pediatrics recommendations, found on CDC chart, www.cdc.gov/vaccines/schedules/index.html.

Academy of Pediatrics sets nationwide recommendations (updated in 2012) for childhood immunizations in this country, as outlined in Table 17-6. The program includes the following vaccines for children from birth to age 6:

- *Hep*atitis *B* vaccine (HepB)
- Diphtheria-*t*etanus-(*a*cellular) *p*ertussis (DTaP) combined vaccine
- *H*aemophilus *i*nfluenzae type *b* (Hib) vaccine to prevent bacterial meningitis and pneumonia
- *I*nactivated (Salk) *p*olio *v*accine (IPV)
- *M*easles-*m*umps-*r*ubella (MMR) combined vaccine
- Varicella zoster vaccine for chickenpox
- *M*eningococcal *v*accine (MCV4) against *Neisseria meningitidis*
- Pneumococcal *c*onjugate *v*accine (PCV) or *p*neumococcal *polysaccharide vaccine* (PPSV) against *Streptococcus pneumoniae*
- Influenza virus vaccine (seasonal flu)
- *Hep*atitis *A* vaccine (HepA)
- *R*otavirus (RV)

New recommendations as of February 2012 for young adults add males ages 11 to 12, to the individuals recommended to receive vaccination against sexually-transmitted *h*uman *pap*illoma*v*irus (HPV), the primary cause of cervical cancer in women. Since 2006 this vaccine has been recommended for females, starting at age 11 to 12 (see Clinical Focus Box 19-1).

As indicated in Table 17-6, children typically require boosters (repeated inoculations) at appropriately timed intervals to achieve protective immunity. In the first months of life, the reason for this may be persistence of circulating maternal antibodies in the young infant. For example, passively acquired maternal antibodies bind to epitopes on the DTaP vaccine and block adequate activation of the immune system; therefore, this vaccine must be given several times *after* the maternal antibody has been cleared from an infant's circulation in order to achieve adequate immunity. Passively acquired maternal antibody also interferes with the effectiveness of the measles vaccine; for this reason, the MMR vaccine is not given before 12 to 15 months of age. In developing countries, however, the measles vaccine is administered at 9 months, even though maternal antibodies are still present, because 30% to 50% of young children in these countries contract the disease before 15 months of age. Multiple immunizations with the polio vaccine are required to ensure that an adequate immune response is generated to each of the three strains of poliovirus that make up the vaccine.

The widespread use of vaccines for common, life-threatening diseases in the United States has led to a dramatic decrease in the incidence of these diseases. As long as these immunization programs are maintained, especially in young children, the incidence of these diseases typically remains low. However, the occurrence or even the suggestion of possible side effects to a vaccine, as well as general trends toward reduced vaccination in children, can cause a drop in vaccination rates that leads to re-emergence of that disease. For example, rare but significant side effects from the original pertussis attenuated bacterial vaccine included seizures, encephalitis, brain damage, and even death. Decreased usage of the vaccine led to an increase in the incidence of whooping cough, before the development of an acellular pertussis vaccine (the *aP* in DTaP), as effective as the older vaccine but with fewer side effects, became available in 1991. However, recent trends toward altered vaccination schedules or refusal to vaccinate children may be fueling outbreaks of this disease. There were 18,000 cases and several deaths in U.S. children in just the first part of 2012, putting that year on track to be the worst in the past five decades. Despite the safety record of this vaccine and the frightening rise in this potentially deadly childhood disease, some parents still elect not to vaccinate their children (see Chapter 1, Clinical Focus Box 1-1).

Recommendations for vaccination of adults depend on the risk group. Vaccines for meningitis, pneumonia, and influenza are often given to groups living in close quarters (e.g., military recruits and incoming college students) or to individuals with reduced immunity (e.g., the elderly). Depending on their destination, international travelers are also routinely immunized against endemic diseases such as cholera, yellow fever, plague, typhoid, hepatitis, meningitis,

typhus, and polio. Immunization against the deadly disease anthrax had been reserved for workers coming into close contact with infected animals or animal products. Concerns about the potential use of anthrax spores by terrorists or in biological warfare has widened use of the vaccine to military personnel and civilians in areas believed to be at risk.

Vaccination is not 100% effective. With any vaccine, a small percentage of recipients will respond poorly and therefore will not be adequately protected. This is not a serious problem if the majority of the population is immune to an infectious agent, significantly reducing the pathogen reservoir. In this case, the chance of a susceptible individual contacting an infected individual is very low. This phenomenon is known as **herd immunity**. The appearance of measles epidemics among college students and preschool-age children in the United States resulted partly from an overall decrease in vaccinations, which had lowered the herd immunity of the population (Figure 17-13). Among preschool-age children, 88% of those who developed measles were unvaccinated. Most of the college students who contracted measles had been vaccinated as children but only once. The failure of the single vaccination to protect them may have resulted from the lingering presence of passively acquired maternal antibodies (disappearing at 12 to 15 months of age), which reduced their overall response to the vaccine. This prompted the revised recommendation that children receive two MMR immunizations: one at 12 to 15 months of age and the second at 4 to 6 years.

The CDC has called attention to the decline in vaccination rates and herd immunity among U.S. children. For example, based on data collected by California health

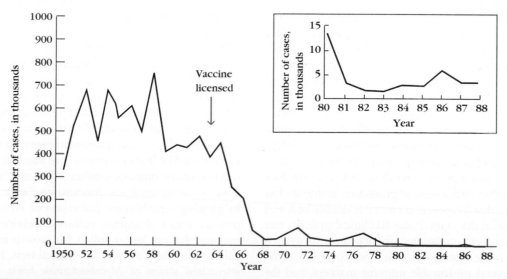

FIGURE 17-13 Introduction of the measles vaccine in 1962 led to a dramatic decrease in the annual incidence of this disease in the United States. Occasional outbreaks of measles in the 1980s (inset) occurred mainly among unvaccinated young children and among college students; most of the latter had been vaccinated but only once, when they were very young. [*Source: Data from Centers for Disease Control and Prevention.*]

officials, pockets of vaccine noncompliance, mostly due to fear of adverse consequences, have likely decreased local herd immunity, resulting in outbreaks. Such a decrease portends serious consequences, as illustrated by recent events in the newly independent states of the former Soviet Union. By the mid-1990s, a diphtheria epidemic was raging in many regions of these new countries, linked to a decrease in herd immunity resulting from decreased vaccination rates after the breakup of the Soviet Union. Mass immunization programs now control this infectious disease.

There Are Several Vaccine Strategies, Each with Unique Advantages and Challenges

Three key factors must be kept in mind in the development of a successful vaccine: the vaccine must be safe, it must be effective in preventing infection, and the strategy should be reasonably achievable given the population in question. Population considerations can include geographical locale, access to the target group (which may require several vaccinations), complicating co-infections or nutritional states, and, of course, cost. Critical for success is the branch of the immune system that is activated, and therefore vaccine designers must recognize the important differences between activation of the humoral and cell-mediated branches, or divergent pathways within these branches. Protection must also reach the relevant site of infection; mucosal surfaces are the prime candidates. An additional factor is the development of long-term immunologic memory. For example, a vaccine that induces a protective primary response may fail to induce the formation of memory cells, leaving the host unprotected after the primary response subsides.

Before vaccines can progress from the laboratory bench to the bedside, they must go through rigorous testing in animals and humans. Most vaccines in development never progress beyond animal testing. The type of testing depends on what animal model systems are available, but frequently involves rodents and/or nonhuman primates. When these animal studies prove fruitful, follow-up clinical trials in humans can be initiated. Phase I clinical trials assess human safety; a small number of volunteers are monitored closely for adverse side effects. Only once this hurdle has been successfully passed can a trial move on to phase II, where effectiveness against the pathogen in question is evaluated. In this case, development of a measurable immune response to the immunogen is tested. However, even a positive result at this juncture does not necessarily mean that a state of protective immunity has been achieved or that long-term memory is established, and many vaccines fail at this stage. Phase III clinical trials are run in expanded volunteer populations, where protection against "the real thing" is the desired outcome, and where safety, the evaluation of several measurable immune markers, and the incidence of wild-type infections with the relevant pathogen are all monitored carefully over time.

The relative significance of memory cells in immunity depends, in part, on the incubation period of the pathogen.

For influenza virus, which has a very short incubation period (1 or 2 days), disease symptoms are already underway by the time memory cells are reactivated. Effective protection against influenza therefore depends on maintaining high levels of neutralizing antibody by repeated immunizations; those at highest risk are immunized each year, usually at the start of the flu season. For pathogens with a longer incubation period, the presence of detectable neutralizing antibody at the time of infection is not necessary. The poliovirus, for example, requires more than 3 days to begin to infect the central nervous system. An incubation period of this length gives the memory B cells time to respond by producing high levels of serum antibody. Thus, the vaccine for polio is designed to induce high levels of immunologic memory. After immunization with the Salk vaccine (an inactivated form, see below), serum antibody levels peak within 2 weeks and then decline. However, the memory response continues to climb, reaching maximal levels at 6 months post vaccine and persisting for years. If an immunized individual is later exposed to the poliovirus, these memory cells will respond by differentiating into plasma cells that produce high levels of serum antibody, which defend the individual from the effects of the virus.

In the remainder of this section, various approaches to the design of vaccines—both currently used vaccines and experimental ones—are described, with an examination of the ability of the vaccines to induce humoral and cell-mediated immunity and the production of memory cells. As Table 17-7 indicates, the common vaccines currently in use consist of live but attenuated organisms, inactivated (killed) bacterial cells or viral particles, as well as protein or carbohydrate fragments (subunits) of the target organism. Several new types of vaccine candidate provide potential advantages in protection, production, or delivery to those in need. The primary characteristics and some advantages and disadvantages of the different types of vaccines are included in the following discussions.

Live, Attenuated Vaccines

In some cases, microorganisms can be attenuated or disabled so that they lose their ability to cause significant disease (pathogenicity) but retain their capacity for transient growth within an inoculated host. Some agents are naturally attenuated by virtue of their inability to cause disease in a given host, although they can immunize these individuals. The first vaccine used by Jenner is of this type: vaccinia virus (cowpox) inoculation of humans confers immunity to smallpox but does not cause smallpox. Attenuation can often be achieved by growing a pathogenic bacterium or virus for prolonged periods under abnormal culture conditions. This selects mutants that are better suited for growth in the abnormal culture conditions than in the natural host. For example, an attenuated strain of *Mycobacterium bovis* called **Bacillus Calmette-Guérin (BCG)** was developed by growing *M. bovis* on a medium containing increasing concentrations of bile. After 13 years, this strain had adapted to growth in strong bile and had become sufficiently attenuated that it was suitable as

| TABLE 17-7 | Classification of common vaccines for humans | | |

Vaccine type	Diseases	Advantages	Disadvantages
WHOLE ORGANISMS			
Live attenuated	Measles Mumps Polio (Sabin vaccine) Rotavirus Rubella Tuberculosis Varicella Yellow fever	Strong immune response; often lifelong immunity with few doses	Requires refrigerated storage; may mutate to virulent form
Inactivated or killed	Cholera Influenza Hepatitis A Plague Polio (Salk vaccine) Rabies	Stable; safer than live vaccines; refrigerated storage not required	Weaker immune response than live vaccines; booster shots usually required
PURIFIED MACROMOLECULES			
Toxoid (inactivated exotoxin)	Diphtheria Tetanus	Immune system becomes primed to recognize bacterial toxins	
Subunit (inactivated exotoxin)	Hepatitis B Pertussis Streptococcal pneumonia	Specific antigens lower the chance of adverse reactions	Difficult to develop
Conjugate	*Haemophilus influenzae* type B Streptococcal pneumonia	Primes infant immune systems to recognize certain bacteria	
OTHER			
DNA	In clinical testing	Strong humoral and cellular immune response; relatively inexpensive to manufacture	Not yet available
Recombinant vector	In clinical testing	Mimics natural infection, resulting in strong immune response	Not yet available

a vaccine for tuberculosis. Due to variable effectiveness and difficulties in follow-up monitoring, BCG is not used in the United States. The Sabin form of the polio vaccine and the measles vaccine both consist of attenuated viral strains.

Attenuated vaccines have advantages and disadvantages. Because of their capacity for transient growth, such vaccines provide prolonged immune system exposure to the individual epitopes on the attenuated organisms and more closely mimic the growth patterns of the "real" pathogen, resulting in increased immunogenicity and efficient production of memory cells. Thus, these vaccines often require only a single immunization, a major advantage in developing coun-

tries, where studies show that a significant number of individuals fail to return for boosters. The ability of many attenuated vaccines to replicate within host cells makes them particularly suitable for inducing cell-mediated responses.

The oral polio vaccine (OPV) designed by Albert Sabin, consisting of three attenuated strains of poliovirus, is administered orally to children. The attenuated viruses colonize the intestine and induce production of secretory IgA, an important defense against naturally acquired poliovirus. The vaccine also induces IgM and IgG classes of antibody and ultimately protective immunity to all three strains of virulent poliovirus. Unlike most other attenuated vaccines, OPV

Reported polio cases

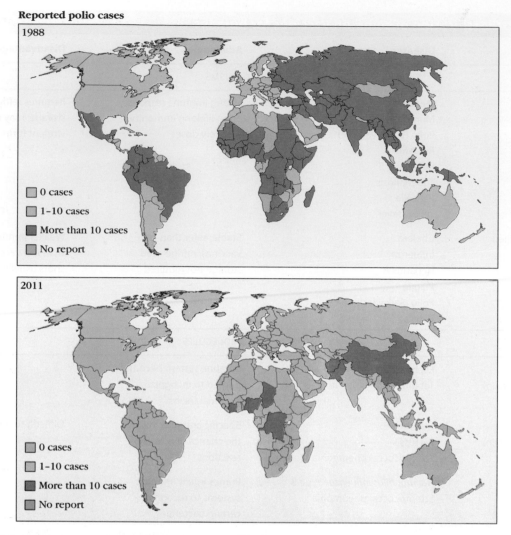

FIGURE 17-14 Progress toward the worldwide eradication of polio. Comparison of polio infection numbers for 1988 with those for 2011 show considerable progress in most parts of the world, although some areas in Africa and Asia have shown recent increases. Some experts question whether the use of live attenuated oral polio vaccine (OPV) will cause reversion to pathogenic forms at a rate sufficiently high to prevent total eradication of this once-prevalent crippling disease. [Source: Data from World Health Organization.]

requires boosters, because the three strains of attenuated poliovirus interfere with each other's replication in the intestine. With the first immunization, one strain will predominate in its growth, inducing immunity to that strain. With the second immunization, the immunity generated by the previous immunization will limit the growth of the previously predominant strain in the vaccine, enabling one of the two remaining strains to colonize the intestines and induce immunity. Finally, with the third immunization, immunity to all three strains is achieved.

A major disadvantage of attenuated vaccines is that these live forms may mutate and revert to virulent forms in vivo, resulting in paralytic disease in the vaccinated individual and serving as a source of pathogen transmission. The rate of reversion of the OPV is about 1 case in 2.4 million doses of vaccine. This reversion can also allow patho-

genic forms of the virus to find their way into the water supply, especially in areas where sanitation is not rigorous or wastewater must be recycled. This possibility has led to the exclusive use of the inactivated polio vaccine in this country (see Table 17-6). The projected eradication of paralytic polio (Figure 17-14) may be impossible as long as OPV is used anywhere in the world. The alternative inactivated polio vaccine created by Jonas Salk will likely be substituted as the number of cases decreases, although there are problems with delivery in developing countries. The ultimate goal is to achieve a polio-free world in which no vaccine is needed.

Attenuated vaccines also may be associated with complications similar to those seen in the natural disease. A small percentage of recipients of the measles vaccine, for example, develop postvaccine encephalitis or other complications, although the risk of vaccine-related complications is still much lower than risks from infection. An independent study showed that 75 million doses of measles vaccine were given between 1970 and 1993, with 48 cases of vaccine-related encephalopathy (approx. 1 per 1.5 million). This low incidence compared with the rate of encephalopathy associated with infection argues for the efficacy of the vaccine. An even more convincing argument for vaccination is the high death rate associated with measles infection, even in developed countries.

In addition to culturing methods, genetic engineering provides a way to attenuate a virus irreversibly, by selectively removing genes that are necessary for virulence or for growth in the vaccinee. This has been done with a herpesvirus vaccine for pigs, in which the thymidine kinase gene was removed. Because thymidine kinase is required for the virus to grow in certain types of cells (e.g., neurons), removal of this gene rendered the virus incapable of causing disease. A live, attenuated vaccine against influenza was developed recently under the name FluMist. The virus was grown at lower-than-normal temperatures until a cold-adapted strain resulted that is unable to grow at human body temperature of 37°C. This live attenuated virus is administered intranasally and causes a transient infection in the upper respiratory tract, sufficient to induce a strong immune response. The virus cannot spread beyond the upper respiratory tract because of its inability to grow at the elevated temperatures of the inner body. Because of the ease of administration and induction of good mucosal immunity, cold-adapted, nasally administered flu vaccines are on the rise.

Inactivated or "Killed" Vaccines

Another common means to make a pathogen safe for use in a vaccine is by treatment with heat or chemicals. This kills the pathogen, making it incapable of replication, but still allows it to induce an immune response to at least some of the antigens contained within the organism. It is critically important to maintain the structure of epitopes on surface antigens during inactivation. Heat inactivation is often unsatisfactory because it causes extensive denaturation of proteins; thus, any epitopes that depend on higher orders of protein structure are likely to be altered significantly. Chemical inactivation with formaldehyde or various alkylating agents has been successful. The Salk polio vaccine is produced by formaldehyde inactivation of the poliovirus.

Although live attenuated vaccines generally require only one dose to induce long-lasting immunity, killed vaccines often require repeated boosters to achieve a protective immune status. Because they do not replicate in the host, killed vaccines typically induce a predominantly humoral antibody response and are less effective than attenuated vaccines in inducing cell-mediated immunity or in eliciting a secretory IgA response, key components of an ideal protective and mucosally based response.

Even though the pathogens they contain are killed, inactivated whole-organism vaccines still carry certain risks. A serious complication with the first Salk vaccines arose when formaldehyde failed to kill all the virus in two vaccine lots, leading to paralytic polio in a high percentage of recipients. Risk is also encountered in the manufacture of the inactivated vaccines. Large quantities of the infectious agent must be handled prior to inactivation, and those exposed to the process are at risk of infection. However, in general, the safety of inactivated vaccines is greater than that of live attenuated vaccines. Inactivated vaccines are commonly used against both viral and bacterial diseases, including the classic yearly flu vaccine and vaccines for hepatitis A and cholera. In addition to their relative safety, their advantages include stability and ease of storage and transport.

Subunit Vaccines

Many of the risks associated with attenuated or killed whole-organism vaccines can be avoided with a strategy that uses only specific, purified macromolecules derived from the pathogen. The three most common applications of this strategy, referred to as a subunit vaccine, are inactivated exotoxins or toxoids, capsular polysaccharides or surface glycoproteins, and key recombinant protein antigens (see Table 17-7). One limitation of some subunit vaccines, especially polysaccharide vaccines, is their inability to activate T_H cells. Instead, they activate B cells in a thymus-independent type 2 (TI-2) manner, resulting in IgM production but little class switching, no affinity maturation, and little, if any, development of memory cells. However, vaccines that conjugate a polysaccharide antigen to a protein carrier can alleviate this problem by inducing T_H cell responses (see below).

Some bacterial pathogens, including those that cause diphtheria and tetanus, produce exotoxins that account for all or most of the disease symptoms resulting from infection. Diphtheria and tetanus vaccines have been made by purifying the bacterial exotoxin and then inactivating it with formaldehyde to form a toxoid. Vaccination with the toxoid induces antitoxoid antibodies, which are capable of binding to the toxin and neutralizing its effects. Conditions for the production of toxoid vaccines must be closely controlled and balanced to avoid excessive modification of the epitope structure while also accomplishing complete detoxification. As discussed previously, passive immunity can also be used to provide temporary protection in unvaccinated individuals exposed to organisms that produce these exotoxins, although no long-term protection is achieved in this case.

The virulence of some pathogenic bacteria depends primarily on the antiphagocytic properties of their hydrophilic polysaccharide capsule. Coating the capsule with antibodies and/or complement greatly increases the ability of macrophages and neutrophils to phagocytose such pathogens. These findings provide the rationale for vaccines consisting of purified capsular polysaccharides. The current vaccine for *Streptococcus pneumoniae* (the organism which causes pneumococcal pneumonia) consists of 13 antigenically distinct capsular polysaccharides (PCV13). The vaccine induces formation of opsonizing antibodies and is now on the list of vaccines recommended for all infants (see Table 17-6). The vaccine for *Neisseria meningitidis*, a common cause of bacterial meningitis, also consists of purified capsular polysaccharides. Some viruses carry surface glycoproteins (e.g., an envelope protein from HIV-1) that have been tested for use in antiviral vaccines, with little success. However, glycoprotein-D from HSV-2 has been shown to prevent genital herpes in clinical trials of some vaccines, suggesting that this may be a viable approach for some antiviral vaccines as well.

Theoretically, the gene encoding any immunogenic protein can be cloned and expressed in cultured cells using recombinant DNA technology, and this technique has been applied widely in the design of many types of subunit vaccines. For example, the safest way to produce sufficient quantities of the purified toxins that go into the generation of toxoid vaccines involves cloning the exotoxin genes from pathogenic organisms into easily cultured host cells. A number of genes encoding surface antigens from viral, bacterial, and protozoan pathogens have also been successfully cloned into cellular expression systems for use in vaccine development. The first such recombinant antigen vaccine approved for human use is the hepatitis B vaccine, developed by cloning the gene for the major *hepatitis B surface antigen* (HBsAg) and expressing it in yeast cells. The recombinant yeast cells are grown in large fermenters, allowing HBsAg to accumulate in the cells. The yeast cells are harvested and disrupted, releasing the recombinant HBsAg, which is then purified by conventional biochemical techniques. Recombinant hepatitis B vaccine induces the production of protective antibodies and holds much worldwide promise for protecting against this human pathogen.

Recombinant Vector Vaccines

Recall that live attenuated vaccines prolong antigen delivery and encourage cell-mediated responses, but have the disadvantage that they can sometimes revert to pathogenic forms. Recombinant vectors maintain the advantages of live attenuated vaccines while avoiding this major disadvantage. Individual genes that encode key antigens of especially virulent pathogens can be introduced into attenuated viruses or bacteria. The attenuated organism serves as a vector, replicating within the vaccinated host and expressing the gene product of the pathogen. However, since most of the genome of the pathogen is missing, reversion potential is virtually eliminated. Recombinant vector vaccines have been prepared utilizing existing licensed live, attenuated vaccines and adding to them genes encoding antigens present on newly emerging pathogens. Such chimeric virus vaccines can be more quickly tested and approved than an entirely new product. A very recent example of this is the yellow

fever vaccine that was engineered to express antigens of WNV. A number of organisms have been used as the vector in such preparations, including vaccinia virus, the canarypox virus, attenuated poliovirus, adenoviruses, attenuated strains of *Salmonella*, the BCG strain of *Mycobacterium bovis*, and certain strains of *Streptococcus* that normally exist in the oral cavity.

Vaccinia virus, the attenuated vaccine used to eradicate smallpox, has been widely employed as a vector for the design of new vaccines. This large, complex virus, with a genome of about 200 genes, can be engineered to carry several dozen foreign genes without impairing its capacity to infect host cells and replicate. The procedure for producing a vaccinia vector that carries a foreign gene from another pathogen is outlined in Figure 17-15. The genetically engineered vaccinia expresses high levels of the inserted gene product, which can then serve as a potent immunogen in an inoculated host. Like the smallpox vaccine, genetically engineered vaccinia vector vaccines can be administered simply by scratching the skin, causing a localized infection in host cells. If the foreign gene product expressed by the vaccinia vector is a viral envelope protein, it is inserted into the membrane of the infected host cell, inducing development of cell-mediated as well as antibody-mediated immunity.

Other attenuated vectors may prove to be safer than vaccinia in vaccine preparations. A relative of vaccinia, the canarypox virus, is also large and easily engineered to carry multiple genes. Unlike vaccinia, it does not appear to be virulent, even in individuals with severe immune suppression. Another possible vector is an attenuated strain of *Salmonella typhimurium*, which has been engineered with genes from the bacterium that causes cholera. The advantage of this vector for use in vaccines is that *Salmonella* infects cells of the mucosal lining of the gut and therefore will induce secretory IgA production. Similar strategies are underway for organisms that enter via oral or respiratory routes, targeting bacteria that are normal flora at these sites as vectors for the addition of pathogen-specific genes. Eliciting immunity at the mucosal surface could provide excellent protection at the portal of entry for many common infectious agents, such as cholera and gonorrhea.

DNA Vaccines

A more recent vaccination strategy, called a DNA vaccine, utilizes plasmid DNA encoding antigenic proteins that are injected directly into the muscle of the recipient. This strategy relies on the host cells to take up the DNA and produce the immunogenic protein in vivo, thus directing the antigen through endogenous MHC class I presentation pathways, helping to activate better CTL responses. The DNA appears either to integrate into the chromosomal DNA or to be maintained for long periods in an episomal form, and is often taken up by dendritic cells or muscle cells in the injection area. Since muscle cells express low levels

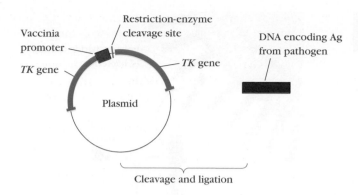

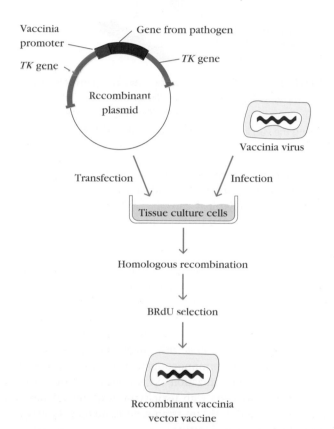

FIGURE 17-15 Production of vaccine using a recombinant vaccinia vector. (top) The gene that encodes the desired antigen (orange) is inserted into a plasmid vector adjacent to a vaccinia promoter (pink) and flanked on either side by the vaccinia thymidine kinase (*TK*) gene (green). (bottom) When tissue culture cells are incubated simultaneously with vaccinia virus and the recombinant plasmid, the antigen gene and promoter are inserted into the vaccinia virus genome by homologous recombination at the site of the nonessential TK gene, resulting in a TK⁻ recombinant virus. Cells containing the recombinant vaccinia virus are selected by addition of *bromo*deoxyuridine (BrdU), which kills TK⁺ cells.

of class I MHC molecules and do not express costimulatory molecules, delivery to local dendritic cells may be crucial to the development of antigenic responses to DNA vaccines. Tests in animal models have shown that DNA vaccines are able to induce protective immunity against a number of

pathogens, including influenza and rabies viruses. The addition of a follow-up booster shot with protein antigen (called a DNA prime and protein boost strategy), or inclusion of supplementary DNA sequences in the vector, may enhance the immune response. One sequence that has been added to some vaccines is the common CpG DNA motif found in some pathogens; recall that this sequence is the ligand for TLR9 (see Chapter 5).

DNA vaccines offer some potential advantages over many of the existing vaccine approaches. Since the encoded protein is expressed in the host in its natural form—there is no denaturation or modification—the immune response is directed to the antigen exactly as it is expressed by the pathogen, inducing both humoral and cell-mediated immunity. To stimulate both arms of the adaptive immune response with non-DNA vaccines normally requires immunization with a live attenuated preparation, which incurs additional risk. DNA vaccines also induce prolonged expression of the antigen, enhancing the induction of immunological memory. Finally, DNA vaccines present important practical advantages (see Table 17-7). No refrigeration of the plasmid DNA is required, eliminating long-term storage challenges. In addition, the same plasmid vector can be custom tailored to insert DNA encoding a variety of proteins, which allows the simultaneous manufacture of a variety of DNA vaccines for different pathogens, saving time and money. An improved method for administering DNA vaccines entails coating gold microscopic beads with the plasmid DNA and delivery of the coated particles through the skin into the underlying muscle with an air gun (called a gene gun). This allows rapid delivery of vaccine to large populations without the need for huge numbers of needles and syringes, improving both safety and cost.

Human trials are underway with several different DNA vaccines, including those for malaria, HIV, influenza, Ebola, and herpesvirus, along with several vaccines aimed at cancer therapy (Table 17-8). Although there are currently no licensed human DNA vaccines, three such vaccines have been licensed for veterinary use, including a WNV vaccine that is protective in horses. This vaccine has been tested in humans; after three doses, most volunteers demonstrated titers of neutralizing antibody similar to those seen in horses, as well as CD8⁺ and CD4⁺ T cell responses against the virus. Since the widespread development of DNA vaccines for use in humans is still in its early stages, the risks associated with the use of this strategy are still largely unknown.

Conjugate or Multivalent Vaccines Can Improve Immunogenicity and Outcome

One significant drawback to the techniques that do not utilize a live vaccine strategy is that they may induce weak or limited adaptive responses. To address this, schemes have been developed that employ the fusing of a highly immunogenic

TABLE 17-8	Diseases for which DNA vaccines have entered clinical trials	
Infectious diseases	**Cancer**	
Human immunodeficiency virus	B-cell lymphoma	
Influenza	Prostate cancer	
(Seasonal, Pandemic)	Melanoma	
	Breast cancer	
Malaria	Ovarian cancer	
Hepatitis B virus	Cervical cancer	
Severe acute respiratory syndrome		
Marburg hemorrhagic fever	Hepatocellular cancer	
Ebola virus	Bladder cancer	
Human papillomavirus	Lung cancer	
West Nile virus	Sarcoma	
Dengue fever	Renal cell cancer	
Herpes simplex virus		
Measles		
Malaria		

Source: M. Liu 2011. DNA vaccines: an historical perspective and view to the future. *Immunological Reviews,* **239:**62–84.

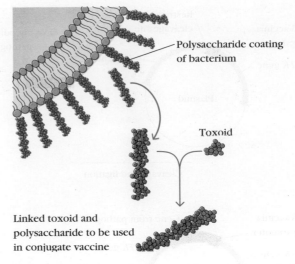

FIGURE 17-16 A conjugate vaccine protects against *Haemophilus influenzae* **type b (Hib).** The vaccine is prepared by conjugating the surface polysaccharide of Hib to a protein molecule, making the vaccine more immunogenic than either alone.

Labels: Polysaccharide coating of bacterium; Toxoid; Linked toxoid and polysaccharide to be used in conjugate vaccine

protein to a weak vaccine immunogen (a conjugate) or mixing in extraneous proteins (multivalent) to enhance or supplement immunity to the pathogen.

The vaccine against *H*aemophilus *influenzae* type *b* (Hib), a major cause of bacterial meningitis and infection-induced deafness in children, is a conjugate formulation included in the recommended childhood regimen (see Table 17-6). It consists of type b capsular polysaccharide covalently linked to a protein carrier, tetanus toxoid (Figure 17-16). Introduction of the conjugate Hib vaccines has resulted in a rapid decline in Hib cases in the United States and other countries that have introduced this vaccine. The polysaccharide-protein conjugate is considerably more immunogenic than the polysaccharide alone; and because it activates T_H cells, it enables class switching from IgM to IgG. Although this type of vaccine can induce memory B cells, it cannot induce memory T cells specific for the pathogen. In the case of the Hib vaccine, it appears that the memory B cells can be activated to some degree in the absence of a population of memory T_H cells, thus accounting for its efficacy. Like Hib, the MCV4 vaccine uses a similar strategy and is another

example of a conjugate vaccine that protects young children from meningitis (see Table 17-6). It is a multivalent vaccine consisting of individual capsular *Neisseria* polysaccharide antigens joined to the highly immunogenic diphtheria toxoid protein.

In a recent study, immunization with β-glucan isolated from brown alga and conjugated to diphtheria toxoid raised antibodies in mice and rats that protected against challenge with both *Aspergillus fumigatus* and *Candida albicans*. The protection was transferred by serum or vaginal fluid from the immunized animals, indicating that the immunity is antibody based. Infections with fungal pathogens are a serious problem for immunocompromised individuals. The availability of immunization or antibody treatment could circumvent problems with toxicity of antifungal drugs and the emergence of resistant strains, an issue that is especially important in hospital settings.

Since subunit polysaccharide or protein vaccines tend to induce humoral but not cell-mediated responses, a method is needed for constructing vaccines that contain both immunodominant B-cell and T-cell epitopes. Furthermore, if a CTL response is desired, the vaccine must be delivered intracellularly so that the peptides can be processed and presented via class I MHC molecules. One innovative means of producing a multivalent vaccine that can deliver many copies of the antigen into cells is to incorporate antigens into protein micelles, lipid vesicles (called liposomes), or immunostimulating complexes (Figure 17-17a). Mixing proteins in detergent and then removing the detergent forms micelles. The individual proteins orient themselves with their hydrophilic residues toward the aqueous environment and the hydrophobic

(a) Detergent-extracted membrane antigens or antigenic peptides

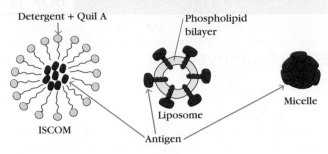

(b) ISCOM delivery of antigen into cell

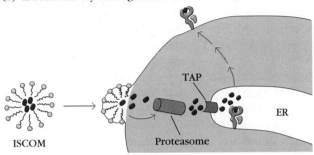

FIGURE 17-17 Multivalent subunit vaccines. (a) Protein micelles, liposomes, and *immunostimulating complexes* (ISCOMs) can all be prepared from detergent-extracted antigens or antigenic peptides. In micelles and liposomes, the hydrophilic residues of the antigen molecules are oriented outward. In ISCOMs, the long fatty-acid tails of the external detergent layer are adjacent to the hydrophobic residues of the centrally located antigen molecules. (b) ISCOMs and liposomes can deliver agents inside cells, so they mimic endogenous antigens. Subsequent processing by the endogenous pathway and presentation with class I MHC molecules induces a cell-mediated response. ER = endoplasmic reticulum; TAP = transporter associated with antigen processing.

residues at the center to exclude their interaction with the aqueous environment. Protein-containing liposomes are prepared by mixing the proteins with a suspension of phospholipids under conditions that form lipid bilayer vesicles; the proteins are incorporated into the bilayer with the hydrophilic residues exposed. *Immunostimulating complexes* (ISCOMs) are lipid carriers prepared by mixing protein with detergent and a glycoside called Quil A, an adjuvant. Membrane proteins from various pathogens, including influenza virus, measles virus, hepatitis B virus, and HIV, have been incorporated into micelles, liposomes, and ISCOMs and are being assessed as potential vaccines. In addition to their increased immunogenicity, liposomes and ISCOMs appear to fuse with the plasma membrane to deliver the antigen intracellularly, where it can be processed by the endogenous pathway, leading to CTL responses (Figure 17-17b).

Adjuvants Are Included to Enhance the Immune Response to a Vaccine

In an ideal case, vaccines mimic most of the key immunologic events that occur during a natural infection, eliciting strong and comprehensive immune responses but without the risks associated with live agents. A discussion of vaccines would be incomplete without mentioning the importance of *adjuvants*, substances that are added to vaccine preparations to enhance the immune response to the antigen or pathogen with which they are mixed. This is especially important when the vaccine preparation is a pathogen subunit or other nonliving form of the organism, where immunogenicity can be quite low. These additives mixed with the vaccines, such as those described above, can both enhance the immune response and help with delivery of the vaccine to the immune system.

For almost 80 years, the only adjuvant used in human vaccines was aluminum salts (called alum), a fairly good enhancer of T_H2 responses but a weaker stimulator of T_H1 pathways. Alum is mixed in an emulsion with the immunogen and is primarily thought to work by creating slow release delivery of the immunogen at the injection site, which helps in sustained stimulation of the immune response. It may also help to recruit APCs and encourage the formation of large antigen complexes that are more likely to be phagocytosed by these cells.

In recent years, two new adjuvants have been licensed for use in human vaccines. One, MF59, is an oil-in-water emulsion that, like alum, is believed to aid in slow antigen delivery. The other, AS04, contains alum plus a TLR4 agonist. TLR4 is a PRR for bacterial *lipopolysaccharides* (LPS), and signaling by this adjuvant ultimately encourages T_H1 responses. AS04 is currently used in vaccines against HPV and HSV-2. All of these adjuvants have been found to enhance the production of antibodies as compared to unadjuvanted vaccine preparations.

Some creative uses of existing adjuvants and next generation adjuvants or immune modulators are also under development. These include compounds designed to stimulate certain PRRs—specifically, several TLRs and at least one NOD-like receptor. Although adjuvants have improved humoral immunity, few if any actually enhance cellular immunity, especially CD8$^+$ T-cell responses. Recently, a combination of two already licensed adjuvant components was used in mice treated with an influenza virus peptide and was found to generate protective CD8$^+$ T-cell memory responses. This strategy is attractive, as it utilizes adjuvants with an already established track record of safety and efficacy in humans. One very recent and particularly novel strategy utilizes a subunit vaccine followed by a chemokine, aimed at eliciting protective immunity and then *moving* the immune cells to the relevant mucosal surface (see Advances Box 17-2).

BOX 17-2

A Prime and Pull Vaccine Strategy for Preventing Sexually Transmitted Diseases

Most pathogens breach the physical barriers of the body at mucosal surfaces, such as the gastrointestinal, respiratory, or genital tracts. Therefore, some form of protective immunity stationed at these portals would be the most effective means of interrupting infection by such pathogens. However, until recently the nuances of the specialized secondary lymphoid structures that serve these surfaces (collectively referred to as *mucosal associated lymphoid tissues*, or MALT) and the unique mechanisms of establishing and maintaining adaptive immune memory at these structures have not been fully appreciated. Traditional vaccination strategies, such as subcutaneous injection of antigen, often fail to elicit protective IgA responses at mucosal sites but instead drive the development of serum IgG, which is less protective against most mucosally encountered pathogens.

Positioning protective immunity at mucosal sites of entry is being studied in the design of vaccines to protect against *sexually transmitted infections* (STIs). According to the WHO, there are more than 30 different sexually acquired pathogens, including viral, bacterial, and parasitic organisms. It is estimated that up to 1 million people become infected with an STI every day! STIs are one of the top five reasons for individuals to seek health care in developing countries, and 30% to 40%

of female infertility is linked to postinfection damage of the fallopian tubes. The most common STIs are chlamydia, gonorrhea, syphilis, hepatitis B, and HIV. Worldwide, these STIs take their greatest toll on young women of childbearing age. Based on anatomy, women are more likely to become infected following unprotected sex with a carrier, whereas men are more likely to transmit these infections through semen. Although many of these infections are treatable, an absence of or delay in symptoms means that many unwitting carriers transmit infection to their partners. Even low levels of genital infection can increase the likelihood of transmission of new STIs, including HIV.

Establishing and maintaining memory cells that home to the genital tract and can protect against STIs is difficult. In experimental systems, intranasal or intravaginal immunogen delivery has been shown to elicit antigen-specific T- and B-cell responses that traffic to the female genital tract, where protective IgA can also be found. However, as in the case of HIV, the ideal vaccine would recruit antigen-specific CD8[+] T cells and IgA-secreting B cells to the genital tract, but not activated CD4[+] T cells. The latter are potential targets for new infections and could therefore inadvertently enhance transmission rates in vaccinees.

Akiko Iwasaki's group at Yale University has recently applied a new vaccine strat-

egy called "prime and pull" in a mouse model of genital herpes. They used a conventional subcutaneous injection of attenuated HSV-2 to prime systemic T cell responses in mice, followed by a topical application of the chemokine CXCL9 to the vaginal canal of the mice. This chemokine resulted in the specific recruitment of effector T cells with a memory phenotype to the mucosal tissues of the vagina: the "pull." Mice treated with this prime and pull strategy showed a significant increase in survival after challenge with live herpesvirus compared with mice that received only the priming vaccine injection.

If this scheme proves safe and effective in human trials, the pull arm of this approach could theoretically be appended to conventional vaccines already approved for use in humans, shortening the lengthy pipeline for most "new" vaccines. Likewise, the chemotactic agent could be modified to recruit different target cells, or applied to other mucosal tissues, reeling in the memory cells elicited from existing vaccines directly to the sites most in need of protection.

H. Shin and A. Iwasaki. 2012. A vaccine strategy that protects against genital herpes by establishing local memory T cells. *Nature* **491:** 463–467.

A. Iwasaki. 2010. Antiviral immune responses in the genital tract: Clues for vaccines. *Nature Reviews Immunology* **10:** 699–711.

SUMMARY

- Physical barriers, such as the skin and mucosal secretions, can block infection. Innate immune responses form the initial defense against pathogens that breach these barriers. These include the nonspecific production of complement components, phagocytic cells, and certain cytokines that are elicited in response to local infection by various pathogens.

- The immune response to viral infections involves both humoral and cell-mediated components. Some viruses mutate rapidly and and/or have acquired specific mechanisms to evade parts of the innate or adaptive immune response to them.

- The immune response to extracellular bacterial infections is generally mediated by antibody. Antibody can activate complement-mediated lysis of the bacterium, neutralize toxins, and serve as an opsonin to increase phagocytosis. Host defense against intracellular bacteria depends largely on CD4$^+$ T-cell-mediated responses.

■ Parasites are a very broad infectious disease category. Both humoral and cell-mediated immune responses have been implicated in immunity to protozoan infections. In general, humoral antibody is effective against blood-borne stages of the protozoan life cycle, but once protozoans have infected host cells, cell-mediated immunity is necessary. Protozoans escape the immune response through several evasion strategies.

■ Helminths are large parasites that normally do not multiply within human hosts. Only low levels of immunity are induced to these organisms, partly because so few organisms are carried by an affected individual or exposed to the immune system. Helminths can form chronic infections and generally are attacked by antibody-mediated defenses.

■ Fungal diseases, or mycoses, are rarely severe in normal, healthy individuals but pose a greater problem for those with immunodeficiency. Both innate immunity and adaptive immunity control infection by fungi.

■ Emerging and re-emerging pathogens include some newly described organisms and others previously thought to have been controlled by public health practices. Factors leading to the emergence of such pathogens include increased travel, poor sanitation, and intense crowding of some populations.

■ A state of immunity can be induced by passive or active immunization. Short-term passive immunization is induced by the transfer of preformed antibodies. Natural infection or vaccination can induce active immunization and lead to long-term immunity.

■ Five types of vaccines are currently used or under experimental consideration in humans: live, attenuated (avirulent) microorganisms; inactivated (killed) microorganisms; purified macromolecules (subunits); viral vectors carrying recombinant genes (live, recombinant); and DNA vaccines.

■ Realizing the optimum benefit of vaccines will require cheaper manufacture and improved delivery methods for existing vaccines.

■ Live vaccines (either attenuated or viral vectors) have the advantage of inducing both humoral and cell-mediated immunity, and can produce more effective overall protective immunity. However, live, attenuated vaccines carry the risk of reversion, which is not an issue with recombinant forms.

■ Isolated protein components of pathogens expressed in cell culture can be used to create effective vaccines, especially when the toxic effects of the pathogen are due to discrete protein products.

■ Polysaccharide and other less immunogenic vaccines may be conjugated to more immunogenic proteins to enhance or maximize the immune response.

■ Introduction of plasmid DNA encoding protein antigens from a pathogen can be used to induce both humoral and cell-mediated responses. Such DNA vaccines for several diseases are presently in human clinical trials.

■ Adjuvants are substances added to vaccine preparations that help aid in delivery of the vaccine to the immune system and that enhance responses. A new generation of adjuvants are being developed that target specific pattern recognition receptors (PRRs) or that modify responses in ways that may help to direct the immune response toward more protective pathways.

REFERENCES

Alcami, A., and U. H. Koszinowski. 2000. Viral mechanisms of immune evasion. *Trends in Microbiology* **8**:410–418.

Bachmann, M. F., and G. T. Jennings. 2010. Vaccine delivery: A matter of size, geometry, kinetics and molecular patterns. *Nature Reviews Immunology* **11**:787–796.

Bloom, B. R., ed. 1994. *Tuberculosis: Pathogenesis, Protection and Control.* Washington, DC: ASM Press.

Borst, P., et al. 1998. Control of VSG gene expression sites in *Trypanosoma brucei. Molecular and Biochemical Parasitology* **91**:67–76.

Coffman, R. L., A. Sher, and R. A Seder. 2010. Vaccine adjuvants: Putting innate immunity to work. *Immunity* **33**(4):492–503.

Iwasaki, A. 2010. Antiviral immune responses in the genital tract: Clues for vaccines. *Nature Reviews Immunology* **10**(10):699–711.

Kaufmann, S. H., A. Sher, and R. Ahmed, eds. 2002. *Immunology of Infectious Diseases.* Washington, DC: ASM Press.

Knodler, L. A., J. Celli, and B. B. Finlay. 2001. Pathogenic trickery: Deception of host cell processes. *Nature Reviews Molecular Cell Biology* **2**:578–588.

Li, F., et al. 2005. Structure of SARS coronavirus spike receptor-binding domain complexed with receptor. *Science* **309**:1864–1868.

Liu, M. A. 2011.DNA vaccines: An historical perspective and view to the future. *Immunology Reviews* **239**(1):62–84.

Lorenzo, M. E., H. L. Ploegh, and R. S. Tirabassi. 2001. Viral immune evasion strategies and the underlying cell biology. *Seminars in Immunology* **13**:1–9.

Merrell, D. S., and S. Falkow. 2004. Frontal and stealth attack strategies in microbial pathogenesis. *Nature* **430**:250–256.

Macleod, M. K., et al. 2011. Vaccine adjuvants aluminum and monophosphoryl lipid A provide distinct signals to generate protective cytotoxic memory CD8 T cells. *Proceedings of the National Academy of Sciences USA* **108**(19):7914–7919.

Morens, D. M., G. K. Folkers, and A. S. Fauci. 2008. Emerging infections: A perpetual challenge. *Lancet Infectious Diseases* **8**(11):710–719.

Romani, L. 2011. Immunity to fungal infections. *Nature Reviews Immunology* **11**(4):275–288.

Schofield, L., and G. E. Grau. 2005. Immunological processes in malaria pathogenesis. *Nature Reviews Immunology* **5**:722–735.

Skowronski, D. M., et al. 2005. Severe acute respiratory syndrome. *Annual Review of Medicine* **56**:357.

Torosantucci, A., et al. 2005. A novel glyco-conjugate vaccine against fungal pathogens. *Journal of Experimental Medicine* **202**:597–606.

Yang, Y.-Z., et al. 2004. A DNA vaccine induces SARS coronavirus neutralization and protective immunity in mice. *Nature* **248**:561.

Useful Websites

www.niaid.nih.gov   The National Institute of Allergy and Infectious Diseases is the institute within the National Institutes of Health that sponsors research in infectious diseases, and its Web site provides a number of links to other relevant sites.

www.who.int   This is the home page of the World Health Organization, the international organization that monitors infectious diseases worldwide.

www.cdc.gov   The Centers for Disease Control and Prevention (CDC) is a United States government agency that tracks infectious disease outbreaks and vaccine research in the United States.

www.upmc-biosecurity.org   The University of Pittsburgh Center for Biosecurity Web site provides information about select agents and emerging diseases that may pose a security threat.

www.gavialliance.org   The Global Alliance for Vaccines and Immunization (GAVI) is a source of information about vaccines in developing countries and worldwide efforts at disease eradication. This site contains links to major international vaccine information sites.

www.ecbt.org   Every Child by Two offers useful information on childhood vaccination, including recommended immunization schedules.

 STUDY QUESTIONS

INFECTIOUS DISEASES

CLINICAL FOCUS QUESTION

1. The effect of the MHC on the immune response to peptides of the influenza virus nucleoprotein was studied in H-2^b mice that had been previously immunized with live influenza virions. The CTL activity of primed lymphocytes was determined by in vitro CML assays using H-2^k fibroblasts as target cells. The target cells had been transfected with different H-2^b class I MHC genes and were infected either with live influenza or incubated with nucleoprotein synthetic peptides. The results of these assays are shown in the following table.

   a. Why was there no killing of the target cells in system A even though the target cells were infected with live influenza?

   b. Why was a CTL response generated to the nucleoprotein in system C, even though it is an internal viral protein?

   c. Why was there a good CTL response in system C to peptide 365–380, whereas there was no response in system D to peptide 50–63?

   d. If you were going to develop a synthetic peptide vaccine for influenza in humans, how would these results obtained in mice influence your design of a vaccine?

2. Describe the nonspecific defenses that operate when a disease-producing microorganism first enters the body.

3. Describe the various specific defense mechanisms that the immune system employs to combat various pathogens.

4. What is the role of the humoral response in immunity to influenza?

5. Describe the unique mechanisms each of the following pathogens has for escaping the immune response: (a) African trypanosomes, (b) *Plasmodium* species, and (c) influenza virus.

6. M. F. Good and co-workers analyzed the effect of MHC haplotype on the antibody response to a malarial circumsporozoite

Target cell (H-2^k fibroblast)	Test antigen	CTL activity of influenza-primed H-2^b lymphocytes (% lysis)
(A) Untransfected	Live influenza	0
(B) Transfected with class I D^b	Live influenza	60
(C) Transfected with class I D^b	Nucleoprotein peptide 365–380	50
(D) Transfected with class I D^b	Nucleoprotein peptide 50–63	2
(E) Transfected with class I K^b	Nucleoprotein peptide 365–380	0.5
(F) Transfected with class I K^b	Nucleoprotein peptide 50–63	1

(CS) peptide antigen in several recombinant congenic mouse strains. Their results are shown in the table below:

Strain	H-2 alleles					Antibody response to CS peptide
	K	IA	IE	S	D	
B10.BR	k	k	k	k	k	<1
B10.A (4R)	k	k	b	b	b	<1
B10.HTT	s	s	k	k	d	<1
B10.A (5R)	b	b	k	d	d	67
B10	b	b	b	b	b	73
B10.MBR	b	k	k	k	q	<1

Source: Adapted from M. F. Good et al., 1988, The T cell response to the malaria circumsporozoite protein: an immunological approach to vaccine development. *Annual Review of Immunology* **6:**663–688.

a. Based on the results of this study, which MHC molecule(s) serve(s) as restriction element(s) for this peptide antigen?

b. Since antigen recognition by B cells is not MHC restricted, why is the humoral antibody response influenced by the MHC haplotype?

7. Fill in the blanks in the following statements.

a. The current vaccine for tuberculosis consists of an attenuated strain of *M. bovis* called _____.

b. Variation in influenza surface proteins is generated by _____ and _____.

c. Variation in pilin, which is expressed by many gram-negative bacteria, is generated by the process of _____.

d. The mycobacteria causing tuberculosis are walled off in granulomatous lesions called _____, which contain a small number of _____ and many _____.

e. The diphtheria vaccine is a formaldehyde-treated preparation of the exotoxin, called a _____.

f. A major contribution to nonspecific host defense against viruses is provided by _____ and _____.

g. The primary host defense against viral and bacterial attachment to epithelial surfaces is _____.

h. Two cytokines of particular importance in the response to infection with *M. tuberculosis* are _____, which stimulates development of T_H1 cells, and _____, which promotes activation of macrophages.

8. Despite the fact that there are no licensed vaccines for them, life-threatening fungal infections are not a problem for the general population. Why? Who may be at risk for them?

9. Discuss the factors that contribute to the emergence of new pathogens or the re-emergence of pathogens previously thought to be controlled in human populations.

10. Which of the following are strategies used by pathogens to evade the immune system? For each correct choice, give a specific example.

a. Changing the antigens expressed on their surfaces

b. Going dormant in host cells

c. Secreting proteases to inactivate antibodies

d. Having low virulence

e. Developing resistance to complement-mediated lysis

f. Allowing point mutations in surface epitopes, resulting in antigenic drift

g. Increasing phagocytic activity of macrophages

11. Which of the following is a characteristic of the inflammatory response against extracellular bacterial infections?

a. Increased levels of IgE

b. Activation of self-reactive $CD8^+$ T cells

c. Activation of complement

d. Swelling caused by release of vasodilators

e. Degranulation of tissue mast cells

f. Phagocytosis by macrophages

12. Your mother may have scolded you for running around outside without shoes. This is sound advice because of the mode of transmission of the helminth *Schistosoma mansoni*, the causative agent of schistosomiasis.

a. If you disobeyed your mother and contracted this parasite, what cells of your immune system would fight the infection?

b. If your doctor administered a cytokine to drive the immune response, which would be a good choice, and how would this supplement alter maturation of plasma cells to produce a more helpful class of antibody?

13. Usually, the influenza virus changes its structure very slightly from one year to the next. However, although we are being exposed to these "modified" influenza strains every year, we do not always come down with the flu, even when the virus successfully breaches physical barriers. Sometimes, we do get a really bad case of the flu, despite the fact that we presumably have memory cells left from an earlier primary response to influenza. Aside from higher doses of virus and the possibility of a particularly pathogenic strain, why is it that some years we get very sick and other years we do not? For example, I might get a bad case of the flu while you experience no disease, and yet we are both being exposed to the exact same virus strain that year. What is happening here? Assume that you are not receiving the yearly influenza vaccine.

VACCINES

CLINICAL FOCUS QUESTION A connection between the new pneumococcus vaccine and a relatively rare form of arthritis has been reported. What data would you need to validate this report? How would you proceed to evaluate this possible connection?

1. Indicate whether each of the following statements is true or false. If you think a statement is false, explain why.

a. Transplacental transfer of maternal IgG antibodies against measles confers short-term immunity on the fetus.

b. Attenuated vaccines are more likely to induce cell-mediated immunity than killed vaccines are.

c. One disadvantage of DNA vaccines is that they don't generate significant immunologic memory.

d. Macromolecules generally contain a large number of potential epitopes.

e. A DNA vaccine only induces a response to a single epitope.

2. What are the advantages and disadvantages of using live attenuated organisms as vaccines?

3. A young girl who had never been immunized to tetanus stepped on a rusty nail and got a deep puncture wound. The doctor cleaned out the wound and gave the child an injection of tetanus antitoxin.

   a. Why was antitoxin given instead of a booster shot of tetanus toxoid?
   b. If the girl receives no further treatment and steps on a rusty nail again 3 years later, will she be immune to tetanus?

4. What are the advantages of the Sabin polio vaccine compared with the Salk vaccine? Why is the Sabin vaccine no longer recommended for use in the United States?

5. Why doesn't the live attenuated influenza vaccine (FluMist) cause respiratory infection?

6. In an attempt to develop a synthetic peptide vaccine, you have analyzed the amino acid sequence of a protein antigen for (a) hydrophobic peptides and (b) strongly hydrophilic peptides. How might peptides of each type be used as a vaccine to induce different immune responses?

7. Explain the phenomenon of herd immunity. How does it relate to the appearance of certain epidemics?

8. You have identified a bacterial protein antigen that confers protective immunity to a pathogenic bacterium and have cloned the gene that encodes it. The choices are either to express the protein in yeast and use this recombinant protein as a vaccine or to use the gene for the protein to prepare a DNA vaccine. Which approach would you take and why?

9. Explain the relationship between the incubation period of a pathogen and the approach needed to achieve effective active immunization.

10. List the three types of purified macromolecules that are currently used as vaccines.

11. Some parents choose not to vaccinate their infants. Reasons include religion, allergic reactions, fear that the infant will develop the disease the vaccine is raised against, and, recently, a fear, unsupported by research, that vaccines can cause autism. What would be the consequence if a significant proportion of the population was not vaccinated against childhood diseases such as measles or pertussis?

12. For each of the following diseases or conditions, indicate what type of vaccination is used:

a. Polio	1. Inactivated
b. Chickenpox	2. Attenuated
c. Tetanus	3. Inactivated exotoxin
d. Hepatitis B	4. Purified macromolecule
e. Cholera	
f. Measles	
g. Mumps	

13. While on a backpacking trip you are bitten by a poisonous snake. The medevac comes to airlift you to the nearest hospital, where you receive human immunoglobulin treatment (gammaglobulin or antiserum) against the poisonous snake venom. You recover from your snakebite and return home for some TLC. One year later during an environmental studies field trip, you are bitten once again by the same type of snake. Please answer the following questions:

   a. Since you fully recovered from the first snakebite, are you protected from the effects of the poison this second time (i.e., did you develop adaptive immunity)?
   b. Immunologically, what occurred the first time you were bitten and treated for the bite?
   c. Compared to the first snakebite, are you more sensitive, less sensitive, or equally sensitive to the venom from the second bite?

ANALYZE THE DATA T. W. Kim and coworkers (Enhancing DNA vaccine potency by combining a strategy to prolong dendritic cell life with intracellular targeting strategies. *Journal of Immunology* **171**:2970, 2003) investigated methods to enhance the immune response against human papillomavirus (HPV)16 E7 antigen. Groups of mice were vaccinated with the following antigens incorporated in DNA vaccine constructs:

- + HPV E7 antigen
- + E7 + heat-shock protein 70 (HSP70)
- + E7 + calreticulin
- + E7 + Sorting signal of lysosome-associated membrane protein 1 (Sig/LAMP-1)

A second array of mice received the same DNA vaccines and was coadministered an additional DNA construct incorporating the anti-apoptosis gene *Bcl-xL*. To test the efficacy of these DNA vaccine constructs in inducing a host response, spleen cells from vaccinated mice were harvested 7 days after injection, the cells were incubated overnight in vitro with MHC class I–restricted E7 peptide (aa 49–57), and then the cells were stained for both CD8 and IFN-γ (part (a) of the figure on the following page). In another experiment Kim and group determined how effective their vaccines were if mice lacked CD4$^+$ T cells (part (b), shown on the following page).

   a. Which DNA vaccine(s) is (are) the most effective in inducing an immune response against papillomavirus E7 antigen? Explain your answer.
   b. Propose a hypothesis to explain why expressing calreticulin in the vaccine construct was effective in inducing CD8$^+$ T cells.
   c. Propose a mechanism to explain the data in part (a) of the figure.
   d. If you were told that the +E7 +Sig/LAMP-l construct is the only one that targets antigen to the class II MHC processing pathway, propose a hypothesis to explain why antigen that would target MHC II molecules enhances a CD8$^+$ T-cell response. Why do you think a special signal was necessary to target antigen to MHC II?
   e. What four variables contribute to the E7-specific CD8$^+$ T lymphocyte response in vitro as measured in part (b) of the figure?

(a)

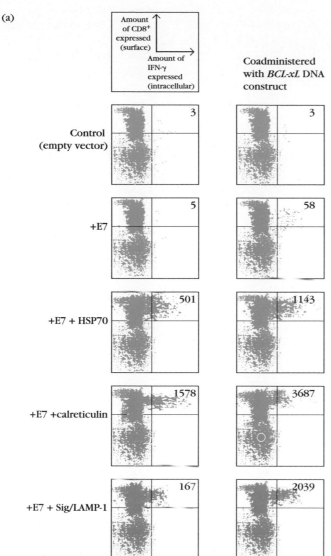

Amount of CD8+ expressed (surface) →

Amount of IFN-γ expressed (intracellular) →

Coadministered with *BCL-xL* DNA construct

Control (empty vector)	3	3
+E7	5	58
+E7 + HSP70	501	1143
+E7 +calreticulin	1578	3687
+E7 + Sig/LAMP-1	167	2039

Intracellular cytokine staining followed by flow cytometry analysis to determine the E7 -specific CD8+ T-cell response in mice vaccinated with DNA vaccines using intracellular targeting strategies. A DNA construct including the anti-apoptosis gene *Bcl-xL* was coadministered to the group on the right. The number at top right in each graph is the number of cells represented in the top right quadrant.

(b)

Number of E7-specific IFN-γ+ CD8+ T cells /3 × 10⁵ splenocytes

$$\text{Number of E7-specific IFN-}\gamma^{+}\text{ CD8}^{+}\text{ T cells / }3 \times 10^{5}\text{ splenocytes}$$

- With peptide
- Without peptide

+Bcl-xL

+Bcl-xL

Wild-type mice CD4 knockout mice

E7-specific CD8+ T lymphocyte response in CD4 knockout mice vaccinated with +E7 +Sig/LAMP-1 DNA construct, with or without +*Bcl-xL* DNA construct.

Immunodeficiency Disorders

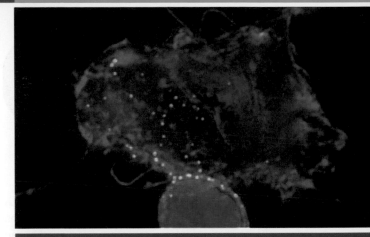

Interaction between dendritic cell and T cell indicating passage of HIV-1 (green dots) between the cells. *[Courtesy of Thomas J. Hope, Northwestern University.]*

- Primary Immunodeficiencies
- Secondary Immunodeficiencies

L
ike any complex multicomponent system, the immune system can be subject to failures of some or all of its parts. These failures can have dire consequences. When the system loses its sense of self and begins to attack the host's own cells, the result is **autoimmunity**, described in Chapter 16. When the system errs by failing to protect the host from disease-causing agents, the result is **immunodeficiency**, the subject of this chapter.

Immunodeficiency resulting from an inherited genetic or developmental defect in the immune system is called a **primary immunodeficiency**. In such a condition, the defect is present at birth, although it may not manifest until later in life. These diseases can be caused by defects in virtually any gene involved in immune development or function, innate or adaptive, humoral or cell mediated, plus genes not previously associated with immunity. As one can imagine, the nature of the component(s) that fail(s) determines the degree and type of the immune defect; some immunodeficiency disorders are relatively minor, requiring little or no treatment, although others can be life threatening and necessitate major intervention.

Secondary immunodeficiency, also known as acquired immunodeficiency, is the loss of immune function that results from exposure to an external agent, often an infection. Although several external factors can affect immune function, by far the most well-known secondary immunodeficiency is *acquired immunodeficiency syndrome* (AIDS), which results from infection with the **human immunodeficiency virus (HIV)**. A global summary of the AIDS epidemic conducted by the Joint United Nations Programme on HIV/AIDS (UNAIDS) shows that by the end of 2011 (the most recent data available) over 34 million people were living with HIV and 2.5 million new infections occurred in just that year (330,000 of them in children under age 15 years). In 2011, AIDS killed approximately 1.7 million people. The good news is that, largely thanks to increased access to antiretroviral drugs, this was roughly a 24% decrease in the rate of AIDS-related deaths as compared to just 6 years earlier. People with AIDS, like individuals with severe inherited immunodeficiency, are at risk of **opportunistic infections**, caused by microorganisms that healthy individuals can easily eradicate but that cause disease and even death in those with significantly impaired immune function.

The first part of this chapter describes the most common primary immunodeficiencies, examines progress in identifying new defects that can lead to these types of disorders, and considers approaches to their study and treatment. The rest of the chapter describes acquired immunodeficiency, with a focus on HIV infection and AIDS, along with the current status of therapeutic and prevention strategies for combating this often fatal disorder.

Primary Immunodeficiencies

To date, over 150 different types of primary, or inherited, immunodeficiency have been identified. Theoretically, any component important to immune function that is defective can lead to some form of immunodeficiency. Collectively, *primary immunodeficiency disorders* (PIDs) have helped immunologists to appreciate the importance of specific cellular events or proteins that are required for proper immune system function. Most of these disorders are monogenic, or

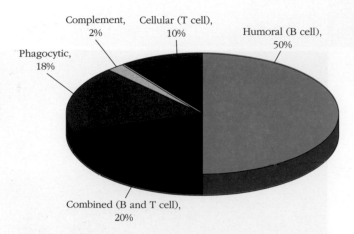

FIGURE 18-1 Distribution of primary immunodeficiencies by type. Primary immunodeficiency can involve either innate processes (phagocytosis, complement, or other defects) or the adaptive immune response (humoral, cellular, or both). Of these categories, adaptive immune disruptions are the most common, with antibody defects making up the largest portion of these. *[Song et al., 2011,* Clinical and Molecular Allergy *9:10. doi:10.1186/1476-7961-9-10]*

caused by defects in a single gene, and are extremely rare. Primary immunodeficiency diseases vary in severity from mild to nearly fatal. They can be loosely categorized as affecting either innate immunity or adaptive responses, and are often grouped by the specific components of the immune system most affected (Figure 18-1). The most common forms of primary immunodeficiency, and frequently the least severe, are those that impair one or more antibody isotype. However, due to the complex interconnections of the immune response, defects in one pathway can also manifest in other arms of the immune response, and different gene defects can produce the same phenotype, making strict categorization complicated.

The cellular consequences of a particular gene disruption depend on the specific immune system component involved and the severity of the disruption (Figure 18-2). Defects that interrupt early hematopoietic cell development affect everything downstream of this step, as is the case for reticular dysgenesis, a disease in which all hematopoietic cell survival is impaired. Defects in more highly differentiated downstream compartments of the immune system, such as in selective immunoglobulin deficiencies, have consequences that tend to be more specific and usually less severe. In some cases, the loss of a gene not specifically associated with immunity has been found to have undue influence on cells of the hematopoietic lineage, such as the lymphoid cell destruction seen in *a*denosine *d*eaminase *d*eficiency (ADA), which disables both B and T cells, leading to a form of severe combined immunodeficiency. Decreased production of phagocytes, such as neutrophils, or the inhibition of phagocytic processes typically manifest as increased susceptibility to bacterial or fungal infections, as seen in defects that affect various cells within the myeloid lineage (Figure 18-2,

orange). In general, defects in the T-cell components of the immune system tend to have a greater overall impact on the immune response than genetic mutations that affect only B cells or innate responses. This is due to the pivotal role of T cells in directing downstream immune events, and occurs because defects in this cell type often affect both humoral and cell-mediated responses.

Immunodeficiency disorders can also stem from developmental defects that alter a specific organ. This is most commonly seen in sufferers of DiGeorge syndrome, where T-cell development is hindered by a congenital defect that blocks growth of the thymus. Since many B-cell responses require T-cell help, most of the adaptive immune response is compromised in patients who suffer the complete form of the disease in which little or no thymic tissue is present, even though B cells are intact. Finally, a more recent category of immunodeficiency syndrome has come to light, illustrating the importance of immune regulation, or "the brakes" of the immune system. APECED and IPEX are both immunodeficiency disorders that result in *overactive* immune responses, or autoimmunity, due to the dysregulation of self-reactive T cells. Some of the most well-characterized primary immunodeficiency disorders with known genetic causes are listed in Table 18-1, along with the specific gene defect and resulting immune impairment.

The nature of the immune defect will determine which groups of pathogens are most challenging to individuals who inherit these immunodeficiency disorders (Table 18-2). Inherited defects that impair B cells, resulting in depressed expression of one or more of the antibody classes, are typically characterized by recurring bacterial infections. These symptoms are similar to those exhibited by some of the individuals who inherit mutations in genes that encode complement components. Phagocytes are so important for the removal of fungi and bacteria that individuals with disruptions of phagocytic function suffer from more of these types of infections. Finally, the pivotal role of the T cell in orchestrating the direction of the immune response means that disruptions to the performance of this cell type can have wide-ranging effects, including depressed antibody production, dysregulation of cytokine expression, and impaired cellular cytotoxicity. In some instances, such as when T- and B-cell responses to self are not properly regulated, autoimmunity can become the primary symptom.

The first part of this section on primary immunodeficiency diseases looks at defects within adaptive immunity, starting with the most extreme cases, characterized by defects to T cells, B cells, or both. This is followed by a discussion of disruptions to innate responses, including cells of the myeloid lineage, receptors important for innate immunity, and complement defects. The autoimmune consequences stemming from dysregulation of the immune system are also described. Finally, we look at the current treatment options available to affected individuals and the use of animal models of primary immunodeficiency in basic immunology research.

OVERVIEW FIGURE

18-2

Primary Immunodeficiencies Result from Congenital Defects in Specific Cell Types

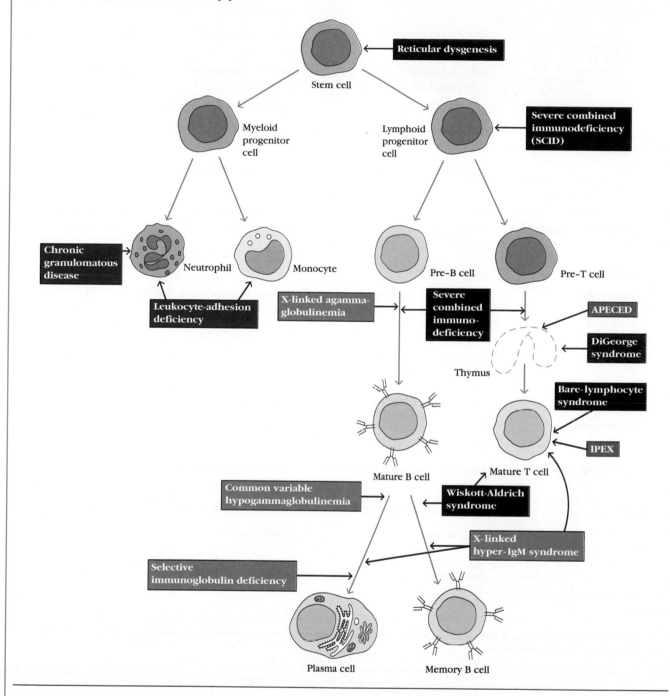

Orange = phagocytic deficiencies, green = humoral deficiencies, red = cell-mediated deficiencies, pink = regulatory cell deficiencies, and purple = combined immunodeficiencies, or defects that affect more than one cell lineage. APECED = *autoimmune polyendocrinopathy and ectodermal dystrophy*. IPEX = *immune dysregulation, polyendocrinopathy, enteropathy, X-linked syndrome*.

TABLE 18-1   Some primary human immunodeficiency diseases and underlying genetic defects

Immunodeficiency disease	Specific defect	Impaired function	Inheritance mode*
Severe combined immunodeficiency (SCID)	*RAG1/RAG2* deficiency	No TCR or Ig gene rearrangement	AR
	ADA deficiency	Toxic metabolite in T and B cells	AR
	PNP deficiency		AR
	JAK-3 deficiency	Defective signals from IL-2, -4, -7, -9, -15, -21	AR
	IL-2Rγ deficiency		XL
	ZAP-70 deficiency	Defective signal from TCR	AR
Bare-lymphocyte syndrome (BLS)	Defect in class II MHC gene promoter	No class II MHC molecules	AR
Wiskott-Aldrich syndrome (WAS)	Cytoskeletal protein (WASP)	Defective T-cells and platelets	XL
Mendelian susceptibility to mycobacterial diseases (MSMD)	IFN-γR IL-12/IL-12R STAT1	Impaired immunity to mycobacteria	AR or AD
DiGeorge syndrome	Thymic aplasia	T-cell development	AD
Gammaglobulinemias	X-linked agammaglobulinemia	Bruton's tyrosine kinase (Btk); no mature B cells	XL
	X-linked hyper-IgM syndrome	Defective CD40 ligand	XL
	Common variable immunodeficiency	Low IgG, IgA; variable IgM	Complex
	Selective IgA deficiency	Low or no IgA	Complex
Chronic granulomatous disease	gp91^phox	No oxidative burst for phagocytic killing	XL
	p67^phox, p47^phox, p22^phox		AR
Chediak-Higashi syndrome	Defective intracellular transport protein (LYST)	Inability to lyse bacteria	AR
Leukocyte adhesion defect	Defective integrin β2 (CD18)	Leukocyte extravasation	AR
Autoimmune polyendocrinopathy and ectodermal dystrophy (APECED)	AIRE defect	T-cell tolerance	AR
Immune dysregulation, polyendocrinopathy, enteropathy, X-linked (IPEX) syndrome	FoxP3 defect	Absence of T_{REG} cells	XL

* AR = autosomal recessive; AD = autosomal dominant; XL= X linked; "Complex" inheritance modes include conditions for which precise genetic data are not available and that may involve several interacting loci.

TABLE 18-2	Patterns of infection and illness associated with primary immunodeficiency diseases

	Disease	
Disorder	OPPORTUNISTIC INFECTIONS	OTHER SYMPTOMS
Antibody	Sinopulmonary (pyogenic bacteria) Gastrointestinal (enterovirus, giardia)	Autoimmune disease (autoantibodies, inflammatory bowel disease)
Cell-mediated immunity	Pneumonia (pyogenic bacteria, *Pneumocystis carinii*, viruses) Gastrointestinal (viruses), mycoses of skin and mucous membranes (fungi)	
Complement	Sepsis and other blood-borne infections (streptococci, pneumococci, neisseria)	Autoimmune disease (systemic lupus erythematosus, glomerulonephritis)
Phagocytosis	Skin abscesses, reticuloendothelial infections (staphylococci, enteric bacteria, fungi, mycobacteria)	
Regulatory T cells	N/A	Autoimmune disease

Source: Adapted from H. M. Lederman, 2000, The clinical presentation of primary immunodeficiency diseases, Clinical Focus on Primary Immune Deficiencies. Towson, MD: Immune Deficiency Foundation 2(1):1.

Combined Immunodeficiencies Disrupt Adaptive Immunity

Among the most severe forms of inherited immunodeficiency are a group of disorders termed **combined immunodeficiencies (CIDs)**: diseases resulting from an absence of T cells or significantly impaired T-cell function, combined with some disruption of antibody responses. Defects within the T-cell compartment generally also affect the humoral system because T_H cells are typically required for complete B-cell activation, antibody production, and isotype switching. Therefore, some depression in the level of one or more antibody isotypes and an associated increase in susceptibility to bacterial infection are common with CIDs. T-cell impairment can lead to a reduction in both delayed-type hypersensitivity responses and cell-mediated cytotoxicity, resulting in increased susceptibility to almost all types of infectious agents, but especially viruses, protozoa, and fungi. For instance, infections with species of *Mycobacteria* are common in CID patients, reflecting the importance of T cells in eliminating intracellular pathogens. Likewise, viruses that are otherwise rarely pathogenic (such as cytomegalovirus or even live, attenuated measles vaccine) may be life threatening for individuals with CIDs. The following section first discusses the most severe CIDs, such as when there is an absence of both T and B cells, followed by less severe forms of the disease, in which more minor disruptions to particular components of the T- and B-cell compartments are observed.

Severe Combined Immunodeficiency (SCID)

The most extreme forms of CID make up a family of disorders termed **severe combined immunodeficiency (SCID)**.

These stem from genetic defects that lead to a virtual or absolute lack of functional T cells in the periphery. As a general rule, these defects target steps that occur early in T-cell development or that affect the stem cells that feed the lymphoid lineage. The four general categories of events that have been found to result in SCID include the following:

1. Defective cytokine signaling in T-cell precursors, caused by mutations in certain cytokines, cytokine receptors, or regulatory molecules that control their expression

2. Premature death of the lymphoid lineage due to accumulation of toxic metabolites, caused by defects in the purine metabolism pathways

3. Defective V(D)J rearrangement in developing lymphocytes, caused by mutations in the genes for RAG1 and RAG2, or other proteins involved in the rearrangement process

4. Disruptions in pre-TCR or TCR signaling during development, caused by mutations in tyrosine kinases, adapter molecules, downstream messengers, or transcription factors involved in TCR signaling

Depending on the underlying genetic defect, an individual with SCID may have a loss of only T cells (T^-B^+) or both T and B cells (T^-B^-). In either case, both cellular and humoral immunity are either severely depressed or absent. Clinically, SCID is characterized by a very low number of circulating lymphocytes and a failure to mount immune responses mediated by T cells. In many cases, the thymus will not fully develop without a sufficient number of T cells, and the few circulating T cells present in some SCID patients often do not respond to stimulation by mitogens, indicating that they cannot proliferate in response to antigens. In many cases, myeloid and erythroid cells (red-blood-cell precursors) appear normal in number and function, indicating that only lymphoid cells are affected.

Infants born with SCID experience severe recurrent infections that, without early, aggressive treatment, can quickly prove fatal. Although both the T and B lineages may be affected, the initial manifestation in these infants is typically infection by fungi or viruses that are normally dealt with by cellular immune responses. This is because antibody deficits can be masked in the first few months of life by the presence of passive antibodies derived from transplacental circulation or breast milk. Infants with SCID often suffer from chronic diarrhea, recurrent respiratory infections, and a general failure to thrive. The life span of these children can be prolonged by preventing contact with all potentially harmful microorganisms—for example, by confinement in a sterile atmosphere. However, extraordinary effort is required to prevent contact with all opportunistic microorganisms; any object, including food, that comes in contact with the sequestered SCID patient must first be sterilized. Such isolation is feasible only as a temporary measure, pending replacement therapy treatments and/or bone marrow transplantation (more on these below).

The immune system is so compromised in SCID patients that common microbes and even live-attenuated vaccines can cause persistent infection and life-threatening disease. For this reason, it is important to diagnose SCID early, especially prior to the administration of live vaccines, such as the rotavirus vaccine, which is recommended at 2 months of age (see Chapter 17). A screening test for SCID has been developed that utilizes the standard blood samples collected from neonates via heel or finger pricks. This rapid *p*olymerase *c*hain *r*eaction (PCR)-based assay looks for evidence of gene recombination as in excised DNA from the TCR or BCR locus, called *T*-cell *r*eceptor *e*xcision *c*ircles (TRECs) and κ-*d*eleting *r*ecombination *e*xcision *c*ircles (KRECs). In 2010, recommendations to screen every newborn for SCID were approved. To date, approximately half of the babies born in the United States receive standard newborn screening for SCID, before live vaccines are administered and when the implementation of aggressive therapy is most beneficial.

Deficiency in cytokine signaling is at the root of the most common forms of SCID, and defects in the gene encoding the common gamma (γ) chain of the IL-2 receptor (*IL2RG*; see Figure 4-8) are the most frequent culprits. This particular form of immunodeficiency is often referred to as *X-linked SCID* (or SCIDX1) because the affected gene is located on the X chromosome, and the disorder is thus more common in males. Defects in this chain impede signaling not only through IL-2R but also through receptors for IL-4, -7, -9, -15, and -21, all of which use this chain in their structures. This leads to widespread defects in B-, T-, and NK-cell development. Although this chain was first identified as a part of the IL-2 receptor, impaired IL-7 signaling is likely the source

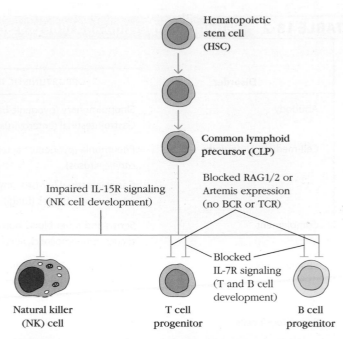

FIGURE 18-3 Defects in lymphocyte development and signaling can lead to severe combined immunodeficiency (SCID). SCID may result from defects in the recombination-activating genes (*RAG1* and *RAG2*) or the DNA excision-repair pathway (e.g., *Artemis*) required for synthesis of the functional immunoglobulins and T-cell receptors in developing lymphocytes. Likewise, defects in the common γ chain of receptors for IL-2, -4, -7, -9, and -15, required for the hematopoietic development of lymphocytes, or JAK-3, which transduces these signals (not shown), can also lead to SCID.

of both T-and B-cell developmental defects, while lack of IL-15 signaling is believed to account for the block to NK cells (Figure 18-3). Deficiency in the kinase JAK-3, which associates with the cytoplasmic region of the common gamma (γ) chain, can produce a phenotype similar to X-linked SCID, as this enzyme is required for the intracellular signaling cascade utilized by all of these cytokine receptors (see Chapter 4).

Defects in the pathways involved in the recombination events that produce immunoglobulin and T-cell receptors highlight the importance of early signaling through these receptors for lymphocyte survival. Mutations in the *r*ecombinase *a*ctivating *g*enes (*RAG1* and *RAG2*) and genes encoding proteins involved in the DNA excision-repair pathways employed during gene rearrangement (e.g., *Artemis*) can also lead to SCID (see Figure 18-3). In these cases, production of antigen-specific receptors is blocked at the pre-T- and pre-B-cell receptor stages of development, leading to a virtual absence of functioning T and B cells, while leaving the numbers and function of NK cells largely intact (see Clinical Focus Box 7-3).

Another relatively common defect resulting in SCID is **adenosine deaminase (ADA) deficiency.** Adenosine deaminase catalyzes conversion of adenosine or deoxyadenosine to inosine or deoxyinosine, respectively. Its deficiency results in the intracellular accumulation of toxic adenosine metabolites, which interferes with purine metabolism and DNA synthesis. This housekeeping enzyme is found in all cells, so these toxic compounds also produce neurologic and metabolic symptoms, including deafness, behavioral problems, and liver damage. Defects in T, B, and NK cells are due to toxic metabolite-induced apoptosis of lymphoid precursors in primary lymphoid organs. Deficiency in another purine salvage pathway enzyme, *purine nucleoside phosphorylase* (PNP), produces a similar phenotype via much the same mechanism.

In some instances, the genetic defects associated with SCID lead to perturbations in hematopoiesis. In **reticular dysgenesis (RD)**, the initial stages of hematopoietic cell development are blocked by defects in the *adenylate kinase 2 gene* (*AK2*), favoring apoptosis of myeloid and lymphoid precursors and resulting in severe reductions in circulating leukocytes (see Figure 18-2). The resulting general failure leads to impairment of both innate and adaptive immunity, resulting in susceptibility to infection by all types of microorganisms. Without aggressive treatment, babies with this very rare form of SCID usually die in early infancy from uncontrolled infection.

MHC Defects That Can Resemble SCID

A failure to express MHC molecules can lead to general failures of immunity that *resemble* SCID without directly impacting lymphocytes themselves. For example, without class II MHC molecules, positive selection of CD4$^+$ T cells in the thymus is impaired, and with it, peripheral T helper cell responses are impaired. This type of immunodeficiency is called **bare-lymphocyte syndrome** and is the topic of Clinical Focus Box 8-4. The important and ubiquitous role of class I MHC molecules is highlighted in patients with defective class I expression. This rare immunodeficiency disorder can be caused by mutations in the TAP genes, which are vital to antigen processing and presentation by class I MHC molecules (see Figure 8-17). This defect, which typically allows for some residual expression of class I molecules, results in impaired positive selection of CD8$^+$ T cells, depressed cell-mediated immunity, and heightened susceptibility to viral infection.

Developmental Defects of the Thymus

Some immunodeficiency syndromes affecting T cells are grounded in failure of the thymus to undergo normal development. These thymic malfunctions can have subtle to profound outcomes on T-cell function, depending on the nature of the defect. **DiGeorge syndrome (DGS)**, also called velocardiofacial syndrome, is one example. This disorder typically results from various deletions in a region on chromosome 22 containing up to 50 genes, with the *T-box* transcription factor (*TBX1*) thought to be most influential. This transcription factor is highly expressed during particular stages of embryonic development, when facial structures,

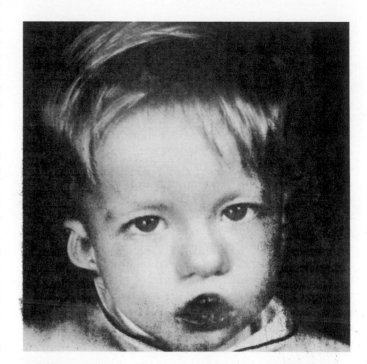

FIGURE 18-4 A child with DiGeorge syndrome showing characteristic dysplasia of ears and mouth and abnormally wide distance between the eyes. *[R. Kretschmer et al., 1968, New England Journal of Medicine **279**:1295; photograph courtesy of F. S. Rosen.]*

heart, thyroid, parathyroid, and thymus tissues are forming (Figure 18-4). For this reason, the syndrome is sometimes also called the third and fourth pharyngeal pouch syndrome. Not surprisingly, DGS patients present with symptoms of immunodeficiency, hypoparathyroidism, and congenital heart anomalies, where the latter are typically the most critical. Although most DGS sufferers show some degree of immunodeficiency, the degree varies widely. In very rare cases of complete DGS, where no thymic tissue develops, severe depression of T-cell numbers and poor antibody responses due to lack of T-cell help leave patients susceptible to all types of opportunistic pathogens. Thymic transplantation and passive antibody treatment can be of value to these individuals, although severe heart disease can limit long-term survival even when the immune defects are corrected. In the majority of DGS patients, in which some residual thymic tissue develops and functional T cells are found in the periphery, treatments to avoid bacterial infection, such as antibiotics, are often sufficient to compensate for the immune defects.

Wiskott-Aldrich Syndrome (WAS)

Although SCID is caused by genetic defects that result in the loss of T cells or major T-cell impairment, a number of other CIDs can result from less severe disruptions to T-cell function. The defect in patients suffering from **Wiskott-Aldrich syndrome (WAS)** occurs in an X-linked gene named for this disease (*WASP*), which encodes a cytoskeletal protein highly expressed in hematopoietic cells (see Table 18-1). The WAS protein (WASP) is required for assembly and reorganization of actin filaments in cells of the hematopoietic lineage, events critical to proper immune synapse formation and intracellular signaling. Clinical manifestations, which usually appear early in the first year of life, vary widely and severity depends on the specific mutation, but eczema and thrombocytopenia (low platelet counts and smaller than normal platelets, which can result in near fatal bleeding) are both common. Humoral defects, including lower than normal levels of IgM, as well as impaired cell-mediated immunity, are also common features. WAS patients often experience recurrent bacterial infections, especially by encapsulated strains such as *S. pneumoniae*, *H. influenzae* type b (Hib), and *S. aureus*. As the disease develops, autoimmunity and B-cell malignancy are not uncommon, suggesting that regulatory T-cell functions are also impaired. Mild forms of the disease can be treated by targeting the symptoms—transfusions for bleeding and passive antibodies or antibiotics for bacterial infections—but severe cases and long-term corrective measures require hematopoietic stem cell transfer.

Hyper IgM Syndrome

An inherited deficiency in CD40 ligand (CD40L or CD154) leads to impaired communication between T cells and antigen-presenting cells (APCs), highlighting the role of this surface molecule in this costimulatory process. In this X-linked disorder, T_H cells fail to express functional CD40L on their plasma membrane, which typically interacts with the CD40 molecule present on B cells and dendritic cells (DCs). This costimulatory engagement is required for APC activation, and its absence in B cells interferes with class switching, B-cell responses to T-dependent antigens, and the production of memory cells (Figure 18-5). The B-cell response to T-*independent* antigens, however, is unaffected, accounting for the presence of IgM antibodies in these patients, which range from normal to high levels and give the disorder its common name, **hyper IgM syndrome (HIM)**. Without class switching or hypermutation, patients make very low levels of all other antibody isotypes and fail to produce germinal centers during a humoral response, which highlights the role of the CD40-CD40L interaction in the

generation of these structures. Because CD40-CD40L interactions are also required for DC maturation and IL-12 secretion, defects in this pathway result in increased susceptibility to intracellular pathogens. Affected children therefore suffer from a range of recurrent infections, especially in the respiratory tract. Although this form of immunodeficiency results in alterations in antibody production and presents with symptoms similar to HIM variants seen in the next section on antibody deficiencies, it is classified as a CID. This is because the underlying deficiency is present in T cells, leading to a secondary defect in B-cell activation. Several other recessively inherited variants of HIM syndrome have been linked to downstream events, such as mutations in one of the enzymes involved in class switching, with the net result of depressed production of all antibody isotypes except IgM.

Hyper IgE Syndrome (Job Syndrome)

Another primary immunodeficiency is characterized by skin abscesses, recurrent pneumonia, eczema, and elevated levels of IgE, accompanied by facial abnormalities and bone fragility. This multisystem disorder, known as **hyper IgE syndrome (HIE)**, is most frequently caused by an autosomal dominant mutation in the *STAT3* gene. This gene is involved in the intracellular signaling cascade induced by IL-6 and TGF-β receptor ligation, and is important for T_H17 cell differentiation (see Figure 11-11). Its absence is thought to lead to dysregulation of T_H pathway development and may be the reason for overproduction of IgE. Patients with Job syndrome have lower-than-normal levels of circulating T_H17 cells, and naïve cells isolated from these individuals are not capable of producing IL-17 or IL-22 in response to antigenic stimulation. Depressed T_H17 responses, which are important for clearance of fungal and extracellular bacterial infections, explain the susceptibility of these patients to *C. albicans* and *S. aureus*. STAT3 defects also inhibit IL-10 signaling and the development of regulatory T cells, which is evident in the reduction of induced T_{REG} cells in these patients. Although STAT3 is involved in the signal transduction of many cytokines and therefore could play a role in the elevation of IgE in these patients, no clear mechanism for this has been defined.

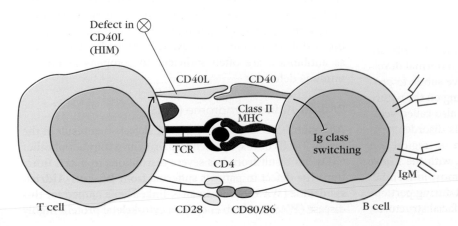

Defect in ⊗
CD40L
(HIM)

CD40L   CD40

Class II
MHC

TCR   Ig class
switching

CD4   IgM

T cell   CD28   CD80/86   B cell

FIGURE 18-5 Defects in components of APC-T cell interactions can give rise to primary immunodeficiency. Defects in CD40/CD40L costimulation between T cells and APCs lead to a block in APC maturation. In B cells, this manifests as a defect in class switching, leading to elevated levels of IgM and no other isotypes (called hyperIgM syndrome, or HIM). In DCs, this blocks maturation and the secretion of costimulatory cytokines, such as IL-12, which are important for T-cell differentiation.

B-Cell Immunodeficiencies Exhibit Depressed Production of One or More Antibody Isotypes

Immunodeficiency disorders caused by B-cell defects make up a diverse spectrum of diseases ranging from the complete absence of mature recirculating B cells, plasma cells, and immunoglobulin, to the selective absence of only certain classes of immunoglobulins. Patients with inherited B-cell defects are usually subject to recurrent bacterial infections but display normal immunity to most viral and fungal infections because the T-cell branch of the immune system is largely unaffected. In patients with these types of immunodeficiencies, the most common infections are caused by encapsulated bacteria such as staphylococci, streptococci, and pneumococci, because antibody is critical for the opsonization and clearance of these organisms. Although the underlying defects have been identified for some of these conditions, several of the more common deficiencies, such as common variable immunodeficiency and selective IgA deficiency, appear to involve multiple genes and a continuum of phenotypes.

X-Linked Agammaglobulinemia

*X-l*inked *a*gammaglobulinemia (X-LA), or Bruton's hypogammaglobulinemia, is characterized by extremely low IgG levels and by the absence of other immunoglobulin classes. Babies born with this disorder have virtually no peripheral B cells (< 1%) and suffer from recurrent bacterial infections. X-LA is caused by a defect in Bruton's tyrosine kinase (Btk), which is required for signal transduction through the BCR (see Figure 3-28 and Clinical Focus Box 3-2). Without functional Btk, B-cell development in the bone marrow is arrested at the pro-B- to pre-B-cell stage, and the B lymphocytes in these patients remain in the pre-B stage, with heavy chains rearranged but light chains in their germ-line configuration. Present-day use of antibiotics and replacement therapy in the form of passively administered antibodies can make this disease quite manageable.

Common Variable Immunodeficiency Disorders

The defects underlying the complex group of diseases belonging to this category are more different than they are similar. However, sufferers of *c*ommon *v*ariable *i*mmunodeficiency *d*isorders (CVIDs) do share recurrent infection resulting from immunodeficiency, marked by reduction in the levels of one or more antibody isotype and impaired B-cell responses to antigen, all with no other known cause. This condition can manifest in childhood or later in life, when it is sometimes called late-onset hypogammaglobulinemia or, incorrectly, acquired hypogammaglobulinemia. Respiratory tract infection by common bacterial strains is the most common symptom, and can be controlled by administration of immunoglobulin. Most cases of CVID have undefined genetic causes, and most patients have normal numbers of B cells, suggesting that B-cell development is not the underlying defect in most cases. Reflecting the diversity of this set of diseases, inheritance can follow autosomal recessive or autosomal dominant patterns, although most cases are sporadic. Several different proteins involving various steps of the B-cell activation cascade have been implicated in recent years.

Selective IgA Deficiency

A number of immunodeficiency states are characterized by significantly lowered amounts of specific immunoglobulin isotypes. Of these, IgA deficiency is by far the most common, affecting approximately 1 in 700. Individuals with selective IgA deficiency typically exhibit normal levels of other antibody isotypes and may enjoy a full life span, troubled only by a greater-than-normal susceptibility to infections of the respiratory and genitourinary tracts, the primary sites of IgA secretion. Family-association studies have shown that IgA deficiency sometimes occurs in the same families as CVID, suggesting some overlap in causation. The spectrum of clinical symptoms of IgA deficiency is broad; most of those affected are asymptomatic (up to 70%), whereas others may suffer from an assortment of serious complications. Problems such as intestinal malabsorption, allergic disease, and autoimmune disorders can be associated with low IgA levels. The reasons for this variability in the clinical profile are not clear but may relate to the ability of some, but not all, patients to substitute IgM for IgA as a mucosal antibody. The defect in IgA deficiency is related to the inability of IgA-expressing B cells to undergo normal differentiation to the plasma-cell stage. A gene or genes outside of the immunoglobulin gene complex is suspected of being responsible for this fairly common syndrome.

Disruptions to Innate Components May Also Impact Adaptive Responses

Most innate immune defects are caused by problems in the myeloid-cell lineage or in complement (see Figure 18-2). Most of these defects result in depressed numbers of phagocytic cells or defects in the phagocytic process that are manifested by recurrent microbial infection of greater or lesser severity. The phagocytic processes may be faulty at several stages, including cell motility, adherence to and phagocytosis of organisms, and intracellular killing by macrophages.

Leukocyte Adhesion Deficiency

As described in Chapter 3, cell-surface molecules belonging to the integrin family of proteins function as adhesion molecules and are required to facilitate cellular interaction. Three of these, LFA-1, Mac-1, and gp150/95 (CD11a, b, and c, respectively), have a common β chain (CD18) and are variably present on different monocytic cells; CD11a is also expressed on B cells. An immunodeficiency related to dysfunction of the adhesion molecules is rooted in a defect localized to the common β chain and affects expression of all three of the molecules that use this chain. This defect, called *leukocyte adhesion deficiency* (LAD), causes susceptibility to infection with both gram-positive and gram-negative bacteria as well as various fungi. Impairment of adhesion of leukocytes to vascular endothelium limits recruitment of cells to sites of inflammation. Viral immunity is somewhat impaired, as would be predicted from the defective T–B-cell cooperation arising from the adhesion defect. LAD varies in its severity; some affected individuals die within a few years, whereas others survive into their forties. The reason for the variable disease phenotype in this disorder is not known.

Chronic Granulomatous Disease

Chronic granulomatous disease (CGD) is the prototype of immunodeficiency that impacts phagocytic function and arises in at least two distinct forms: an X-linked form in about 70% of patients and an autosomal recessive form found in the rest. This group of disorders is rooted in a defect in the *nicotinamide adenine dinucleotide phosphate* (NADPH) oxidative pathway by which phagocytes generate superoxide radicals and other reactive compounds that kill phagocytosed pathogens. For this reason, CGD patients suffer from infection by bacterial and fungal pathogens, as well as excessive inflammatory responses that lead to the formation of granulomas (a small mass of inflamed tissue). Genetic causes have been mapped to several missing or defective *phagosome oxidase* (phox) proteins that participate in this pathway (see Table 18-1). Standard treatment includes the use of antibiotics and antifungal compounds to control infection. Of late, the addition of IFN-γ to this regimen has been shown to improve CGD symptoms in both humans and animal models. Although the mechanism for this is still debated, in vitro studies have shown that IFN-γ treatment induces TNF-α and the production of *nitric oxide* (NO, another oxidative mediator) and enhances the uptake of inflammation-inducing apoptotic cells, which could play a role in inhibiting the formation of granulomas during inflammation in these patients.

Chediak-Higashi Syndrome

This rare autosomal recessive disease is an example of a lysosomal storage and transport disorder. *Chediak-Higashi syndrome* (CHS) is characterized by recurrent bacterial infections as well as defects in blood clotting, pigmentation, and neurologic function. Immunodeficiency hallmarks include neutropenia (depressed numbers of neutrophils) as well as impairments in T cells, NK cells, and granulocytes. CHS is associated with oculocutaneous albinism, or light-colored skin, hair, and eyes, accompanied by photosensitivity. The underlying cause has been mapped to mutations in the *lysosomal trafficking regulator* (*LYST*) gene that cause defects in the LYST protein, which is important for transport of proteins into lysosomes as well as for controlling lysosome size, movement, and function. Disruption to this and related

organelles, such as the melanosomes of skin cells (melanocytes), results in enlarged organelles and defective transport functions. Affected phagocytes produce giant granules, a diagnostic hallmark, but are unable to kill engulfed pathogens, and melanocytes fail to transport melanin (responsible for pigmentation). Similar enlarged lysosome-like structures in platelets and nerve cells are also thought to interfere with blood clotting and neurologic function, respectively. Exocytosis pathways are likewise affected, which could account for the defects in killing seen in T_C cells and NK cells, as well as impaired chemotactic responses. Without early antimicrobial therapy followed by bone marrow transplant, patients often die due to opportunistic infection before reaching 10 years of age. However, no therapies are currently available to treat the defects in other cells, so even when immune function is restored, neurologic and other complications continue to progress.

Mendelian Susceptibility to Mycobacterial Diseases

Recently, a set of immunodeficiency disorders has been grouped into a mixed-cell category based on the shared characteristic of single gene (*Mendelian*) inheritance of *susceptibility to mycobacterial diseases* (MSMD). Discovery of the underlying defects in MSMD highlights the connections between innate and adaptive immunity, as well as the key role played by IFN-γ in fighting infection by mycobacteria, intracellular organisms that can cause tuberculosis and leprosy. During natural mycobacterial infection, macrophages in the lung or DCs in the draining lymph node recognize these bacteria through *pattern recognition receptors* (PRRs), such as TLR2 and TLR4, which trigger migration to lymph nodes followed by APC activation and differentiation. In the presence of strong costimulation, such as engagement of CD40 on the APC with CD40L on the T cell, these activated APCs produce significant amounts of IL-12 and IL-23, which can bind to their receptors on T_H cells and NK cells, respectively. This leads to production of cytokines such as IFN-γ, IL-17, and TNF-α. In a positive feedback loop, T_H cells in this environment differentiate into T_H1-type cells, further producers of IFN-γ. Upon binding to the IFN-γR on APCs, this cytokine induces a signaling cascade, involving Janus kinases and STAT1, which results in enhanced phagocytosis and optimal phago-lysosomal fusion, effectively killing engulfed bacteria.

This story of mycobacterial infection makes the defective genes or proteins now implicated in MSMD no great surprise (Figure 18-6). To date, six genes within the IFN-γ/IL-12/IL-23 pathways have been linked to MSMD, including those encoding IL-12, IL-12R, IFN-γR (both chains), STAT1, and a kinase downstream of IL-12 signaling (TYK2). Another gene linked to MSMD, called *NEMO*, controls the behavior of the signal transduction molecule NF-κB (see Figure 3-17), which can affect CD40-dependent induction of IL-12. However, most mutations in *NEMO* lead to more widespread immune defects and susceptibility patterns than are seen in typical MSMD patients. The specific gene and type of mutation

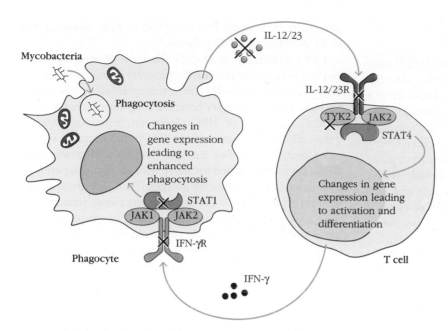

FIGURE 18-6 Genetic defects resulting in Mendelian susceptibility to mycobacterial diseases (MSMD). Many primary immunodeficiencies associated with increased susceptibility to mycobacterial infection are associated with defects in either the IFN-γ pathway (e.g., IFN-γR or the related STAT1 signaling molecule) or IL-12/23 signaling pathway (e.g., IL-12, IL-12R, and the associated TYK2 signaling molecule). These pathways are particularly important for clearing intravesicular infections. [Cottle, L. E., Mendelian susceptibility to mycobacterial disease, Clinical Genetics 2011:79, 17–22, with modifications.]

determine which other pathogens, if any, also pose a risk to these patients and influence prognosis as well as treatment options.

Complement Deficiencies Are Relatively Common

Immunodeficiency diseases resulting from defects in the complement system, which has innate as well as adaptive triggers, are described in Chapter 6. Depending on the specific component that is defective, these immunodeficiencies can manifest as a generalized failure to activate complement (e.g., C4 defects) or failures of discrete pathways or functions (e.g., alternative pathway activation). Most complement deficiencies are associated with increased susceptibility to bacterial infections and/or immune-complex diseases. For example, deficiency in properdin, which stabilizes the C3 convertase in the alternative complement pathway, is caused by a defect in a gene located on the X chromosome and is specifically associated with increased risk of infection with species of *Neisseria*. These types of bacterial infection are also more common in those with defects in the late components of complement, including C5–C9. Defects in *mannose-binding lectin* (MBL) result in increased susceptibility to a variety of infections by bacterial or fungal agents. Recall from Chapter 6 that MBL is a key initiator of the complement attack on many pathogens and is an important component of the innate immune response to many organisms.

Immunodeficiency That Disrupts Immune Regulation Can Manifest as Autoimmunity

In addition to recognizing and eliminating foreign antigens, the adaptive immune system must learn to recognize self MHC proteins and to be proactive in suppressing reactions to self antigens in the host. These processes are carried out by the induction of tolerance in the thymus and by the surveillance activities of regulatory T cells (T_{REG} cells; see Chapters 9 and 16). Disruptions to genes involved in these immune regulatory or homeostatic processes, although caused by inborn immunodeficiencies, actually manifest as immune overactivity, or autoimmunity (see below and Chapter 16).

Autoimmune Polyendocrinopathy and Ectodermal Dystrophy

Individuals with a defect in the autoimmune regulatory gene *AIRE*, discussed in detail in Chapter 9, suffer from a disease called **autoimmune polyendocrinopathy and ectodermal dystrophy (APECED)**. The AIRE protein is expressed in medullary epithelial cells of the thymus, where it acts as a transcription factor to control expression of a whole host of tissue-restricted antigens. Proper expression of these peripheral proteins in the thymus facilitates the negative selection of potentially autoreactive T cells before they can exit into the circulation. It appears that depressed expression of *AIRE* in these individuals results in reduced levels of tissue-specific antigens in thymic epithelial cells, allowing the escape of autoreactive T cells into the periphery, where they precipitate organ-specific autoimmunity. APECED patients experience inhibition of endocrine function, including hypoadrenalism, hypoparathyroidism, and hypothyroidism, along with chronic candidiasis. Autoimmune responses against antigens present in these endocrine organs, as well as the adrenal cortex, gonads, and pancreatic β-cells, are observed in these individuals. Although autoantibodies to these tissues are also observed, these may result from the tissue destruction mediated by pathogenic T cells.

Immune Dysregulation, Polyendocrinopathy, Enteropathy, X-linked (IPEX) Syndrome

Although many T cells with the ability to recognize self antigens are destroyed in the thymus during negative selection, one class of CD4$^+$ T cells with regulatory capabilities survive and actively inhibit reactions to these self antigens. The development and function of these T$_{REG}$ cells are controlled by a master regulator and transcription factor, called FoxP3 (see Chapters 9 and 16). Patients with **immune dysregulation, *polyendocrinopathy, e*nteropathy, X-linked (IPEX) syndrome** have inherited a mutated *FoxP3* gene and lack expression of this protein, leading to a near absence of T$_{REG}$ cells. Without these regulatory cells in the periphery, autoreactive T cells that have escaped central tolerance in the thymus go unchecked, leading to systemic autoimmune disease. Affected infants exhibit immune destruction of the bowel, pancreas, thyroid, and skin, and they often die in the first 2 years of life due to sepsis and failure to thrive.

Immunodeficiency Disorders Are Treated by Replacement Therapy

Although there are no sure-fire cures for immunodeficiency disorders, there are several treatment possibilities. In addition to the use of antimicrobial agents, and the drastic option of total isolation from exposure to any opportunistic pathogen, treatment options for immunodeficiencies include the following:

- Replacement of a missing protein
- Replacement of a missing cell type or lineage
- Replacement of a missing or defective gene

For disorders that impair antibody production, the classic course of treatment is administration of the missing protein immunoglobulin. Pooled human gammaglobulin given either intravenously or subcutaneously protects against recurrent infection in many types of immunodeficiency. Maintenance of reasonably high levels of serum immunoglobulin (5 mg/ml serum) will prevent most common infections in the agammaglobulinemic patient. Advances in the preparation of human monoclonal antibodies and in the ability to genetically engineer chimeric antibodies with mouse V regions and human-derived C regions make it possible to prepare antibodies specific for important pathogens (see Chapter 20).

Advances in molecular biology make it possible to clone the genes that encode other immunologically important proteins, such as cytokines, and to express these genes in vitro, using bacterial or eukaryotic expression systems. The availability of such proteins allows new modes of therapy in which immunologically important proteins may be replaced or their concentrations increased in the patient. For example, the delivery of recombinant IFN-γ has proven effective for patients with CGD, and recombinant adenosine deaminase has been successfully administered to ADA-deficient SCID patients.

Cell replacement as therapy for some immunodeficiencies has been made possible by progress in bone marrow, or *h*ematopoietic *s*tem *c*ell (HSC), transplantation (see Chapter 16) and is the primary potential long-term cure for SCID patients. Transfer of HSCs from an immunocompetent donor allows development of a functional immune system (see Clinical Focus Box 2-2). Success rates of greater than 90% have been reported for those who are fortunate enough to have an *h*uman *l*eukocyte *a*ntigen (HLA)-identical donor. In cases where there is partial HLA matching, treatment in the first few months of life has the best prognosis for relatively long-lasting results. These procedures can also be relatively successful with SCID infants when haploidentical (complete match of one HLA gene set or haplotype) donor marrow is used. In this case, T cells are depleted to avoid graft versus host disease, and CD34$^+$ stem cells are enriched before introducing the donor bone marrow into the recipient. A variation of bone marrow transplantation is the injection of parental CD34$^+$ cells in utero when the birth of an infant with SCID is expected.

If a single gene defect has been identified, as in adenosine deaminase or IL-2Rγ defects, replacement of the defective gene, or gene therapy, may be a treatment option. During the last 20 years, several clinical tests of gene therapy for these two types of SCID have been undertaken, with mixed results. In these trials, CD34$^+$ HSCs are first isolated from the bone marrow or umbilical cord blood of HLA-identical or haploidentical donors. These cells are transduced with the corrected gene and then introduced into the patient, in some cases after myeloablation conditioning, a pre-treatment that destroys existing leukocytes, "making space" for engraftment of the new cells. As of 2012, we are now more than two decades out from these initial trials. In general, gene therapy for the IL-2Rγ defect has been slightly more successful than have similar treatments for ADA deficiency, yielding immunodeficiency correction rates of approximately 85% versus 70%, respectively. This is likely due to the presence of ADA defects in cell types other than leukocytes. Although these therapies have allowed a majority of the treated individuals to recover significant immune function, there have been some setbacks. In 5 of the first 20 cases of gene therapy for X-linked SCID, insertion of the vector used to transfer the corrected gene led to mutagenesis and development of leukemia. Most of these cases were successfully resolved; however, one individual died from this cancer, effectively halting additional trials. Studies are currently underway to redesign the vectors used, to make them safer, and to specifically direct integration into inactive regions of the genome.

Animal Models of Immunodeficiency Have Been Used to Study Basic Immune Function

There is a good reason PIDs are sometimes called nature's experiments. Many of the molecular details of how the immune system works have come from discovering and studying the broken parts. More important, observations made in humans with PIDs and in animal models of immunodeficiency have taught us what questions to ask. Does the immune system play a role in surveillance for cancer? (The answer is yes, and it's not always a good one; see Chapter 19.) How can a defect in the IL-2R lead to a total lack of B cells as well as T cells? How can mutation of a single gene related to immunity cause immunologic attack of a whole range of different self proteins? (See above for answers to both.) And the list goes on.

Experimental animals with spontaneous or engineered primary immunodeficiencies have provided fertile ground for manipulating and studying basic immune processes. By comparing the phenotypes of animals with and without these blocks in certain components of the immune system, scientists have been able to tease out many details of normal immune processes. The two most widely used animal models of primary immunodeficiency are the athymic, or nude, mouse and the SCID mouse. However, development of other genetically altered animals in which a single target immune gene is knocked out or mutated has also yielded valuable information about the role of these genes in combating infection, and has highlighted some unexpected connections between the immune system and other systems in the body.

Nude (Athymic) Mice

A genetic trait designated *nu* (now called Foxn1*nu*), which is controlled by a recessive gene on chromosome 11, was discovered in 1962 by Norman Roy Grist. Mice homozygous for this trait *(nu/nu,* or nude mice)* are hairless and have a vestigial thymus (Figure 18-7). Heterozygotic *nu/wt* littermates have hair and a normal thymus. We now know that the mutated gene *FOXN1* encodes a transcription factor, mainly expressed in the thymus and skin epithelial cells, that plays a role in cell differentiation and survival, suggesting that the hair loss and immunodeficiency may be caused by the same defect. Like humans born with severe immunodeficiency, these mice do not survive for long without intervention, and 50% or more die within the first 2 weeks after birth from opportunistic infection if housed under standard conditions. When these animals are to be used for experimental purposes, precautions include the use of sterilized food, water, cages, and bedding. The cages are protected from dust by placing them in a laminar flow rack or by using cage-fitted air filters.

Nude mice have now been studied for many years and have been developed into a tool for biomedical research. For example, because these mice can permanently tolerate both allografts and xenografts (tissue from another species), they have a number of practical experimental uses in the study of transplantation and cancer. Hybridomas (immortalized B cells) or solid tumors from any origin can be grown in the nude mouse, allowing their propagation and the evaluation of new tumor imaging techniques or pharmacological treatments for cancer in these animals.

The SCID Mouse

In 1983, Melvin and Gayle Bosma and their colleagues described an autosomal recessive mutation in mice that gave rise to a severe deficiency in mature lymphocytes. They designated the trait SCID because of its similarity to human severe combined immunodeficiency. The SCID mouse was shown to have early B- and T-lineage cells but a virtual absence of lymphoid cells in the thymus, spleen, lymph nodes, and gut tissue, the usual locations of functional T and B cells. Precursor cells in the SCID mouse appeared to be unable to differentiate into mature functional B and T lymphocytes. Inbred mouse lines carrying this defect, which have now been propagated and studied in great detail, neither make antibody nor carry out delayed-type hypersensitivity or graft rejection reactions. Lacking much of their adaptive response, they succumb to infection early in life if not kept in extremely pathogen-free environments. Hematopoietic cells other than lymphocytes develop normally in the SCID mouse; red blood cells, monocytes, and granulocytes are present and functional. Like humans, SCID mice may be rendered immunologically competent by transplantation of stem cells from a matched donor.

The mutation was discovered in a gene called *protein kinase, DNA activated, catalytic polypeptide (PRKDC)*, which was later shown to participate in the double-stranded DNA break-repair pathway important for antigen-specific receptor gene recombination in B and T cells. This defect is a leaky mutation: a certain number of SCID mice do produce immunoglobulin, and about half of these mice can also reject skin allografts, suggesting components of both humoral and adaptive immunity are present. This leaky phenotype has somewhat limited the widespread use of these mice in research laboratories. However, their ability, like the nude mouse, to accept engrafted tissue from any species has

FIGURE 18-7 A nude mouse (Foxn1*nu*/ Foxn1*nu*). This defect leads to absence of a thymus or a vestigial thymus and cell-mediated immunodeficiency. *[Courtesy of the Jackson Laboratory, Bar Harbor, Maine.]*

led to the development of chimeric mice reconstituted with a *hu*manized immune system (called hu-SCID). These human cells can develop in a normal fashion and, as a result, hu-SCID mice contain T cells, B cells, and immunoglobulin of human origin. In one important application, these mice can be infected with HIV-1, a pathogen that does not infect mouse cells. This provides an animal model in which to test therapeutic or prophylactic strategies against HIV infection.

RAG Knockout Mice

The potential utility of a mouse model that lacks adaptive immunity, or certain components of adaptive responses, led to the engineering of mice with more targeted mutations. Arguably, the most widely used have been mice with deletions in one of the recombination-activating enzymes, RAG1 and RAG2, responsible for the rearrangement of immunoglobulin or T-cell receptor genes. Since both enzymes are required for recombination, the phenotypes of the two are almost identical, although absence of RAG2 blocks B-cell and T-cell development at an earlier stage and more completely. Unlike nude or SCID mice, RAG2 knockout mice exhibit "tight" defects in both B-cell and T-cell compartments; precursor cells cannot rearrange the genes for antigen-specific receptors or proceed along a normal developmental path, and thus both B and T cells are absent. With a SCID phenotype, RAG knockout mice can be used as an alternative to nude or conventional SCID mice. Their applications include experimental cancer and infectious disease research, as well as more targeted investigations of immune gene function. RAG knockout mice can be the background strain for the production of transgenic mice carrying specific rearranged T-cell or B-cell receptor genes. For example, since these loci have already rearranged, T-cell receptor transgenes will not require the RAG enzymes, and can develop "normally" in the thymus, allowing immunologists to study the events that occur during positive and negative selection while observing the behavior of a million or more T cells of the same clonotype. Although the degree to which this represents truly typical in vivo development of a T cell is questionable, this model has been widely used to ask and to answer many important questions related to MHC restriction and tolerance.

Secondary Immunodeficiencies

As described above, a variety of defects in the immune system can give rise to immunodeficiency. In addition to the inherited primary immunodeficiencies, there are also acquired (secondary) immunodeficiencies. Although AIDS resulting from HIV infection is the most well known of these, other factors, such as drug treatment, metabolic disease, or malnutrition, can also impact immune function and lead to secondary deficiencies. As in primary immunodeficiency, symptoms include heightened susceptibility to common infectious agents and sometimes opportunistic infections. The effect depends on the degree of immune suppression and inherent host susceptibility factors, but can range from no clinical symptoms to almost complete collapse of the immune system, as in HIV-induced AIDS. In most cases, withdrawal of the external condition causing the deficiency can result in restoration of immune function. The first part of this section will cover secondary immunodeficiency due to some non-HIV causes, and the remainder will deal with AIDS.

One secondary immunodeficiency that has been recognized for some time but has an unknown cause is acquired **hypogammaglobulinemia**. This condition is sometimes confused with CVID, a condition that shows genetic predisposition (see above). Symptoms include recurrent infection, and the condition typically manifests in young adults who have very low but detectable levels of total immunoglobulin with normal T-cell numbers and function. However, some cases do involve T-cell defects, which may grow more severe as the disease progresses. The disease is generally treated by immunoglobulin therapy, allowing patients to live a relatively normal life. Unlike the similar primary deficiencies described above, there is no evidence for genetic transmission of this disease. Mothers with acquired hypogammaglobulinemia deliver normal infants. However, at birth these infants will be deficient in circulating immunoglobulin due to the lack of IgG in maternal circulation that can be passively transferred to the infant.

Another form of secondary immunodeficiency, **agent-induced immunodeficiency**, results from exposure to any of a number of environmental agents that induce an immunosuppressed state. These could be immunosuppressive drugs used to combat autoimmune diseases such as rheumatoid arthritis or the corticosteroids commonly used during transplantation procedures to blunt the attack of the immune system on donor organs. The mechanism of action of these immunosuppressive agents varies, as do the defects in immune function, although T cells are a common target. As described in Chapter 16, recent efforts have been made to use more specific means of inducing tolerance to allografts to circumvent the unwanted side effects of general immunosuppression. In addition, cytotoxic drugs or radiation treatments given to treat various forms of cancer, as well as accidental radiation exposure, frequently damage rapidly dividing cells in the body, including those of the immune system, inducing a state of temporary immunodeficiency as an unwanted consequence. Patients undergoing such therapy must be monitored closely and treated with antibiotics or immunoglobulin if infection appears.

Extremes of age are also natural factors in immune function. The very young and elderly suffer from impairments to immune function not typically seen during the remainder of the life span. Neonates, and especially premature babies, can be very susceptible to infection, with degree of prematurity linked to the degree of immune dysfunction. Although all the basic immune components are in place in full-term, healthy newborns, the complete range of innate and adaptive immune functions take some time to mature. Along with presence of passive maternal antibody for about the first 6 months of life, this is part of the reason for a gradual vaccination program against the common childhood infectious diseases that peak around 1 year of age (see Chapter 17). In later life, individuals again experience an increasing risk of infection, especially by bacteria and viruses, as well as more malignancies. Cell-mediated immunity is generally depressed, and although there are increased numbers of memory B cells and circulating IgG, the diversity of the B-cell repertoire is diminished.

The single most common cause of acquired immunodeficiency, even dwarfing the number of individuals worldwide affected by AIDS, is severe malnutrition, affecting both innate and adaptive immunity. Sustained periods with very low protein-calorie diets (hypoproteinemia) are associated with depression in T-cell numbers and function, although deleterious B cell effects may take longer to appear. The reason for this is unclear, although some evidence suggests a bias toward anti-inflammatory immune pathways (e.g., IL-10 and T_{REG} cells) when protein is scarce. In addition to protein, an insufficiency in micronutrients, such as zinc and ascorbic acid, likely contributes to the general immunodeficiency and increased susceptibility to opportunistic infection that occurs with malnutrition. This can be further complicated by stress and infection, both of which may contribute to diarrhea, further reducing nutrient absorption in the gut. Deficiency in vitamin D, required for calcium uptake and bone health, has also been linked to an inhibition in the ability of macrophages to act against intracellular pathogens, such as *M. tuberculosis*, endemic in many regions of the world where people are at greatest risk of malnutrition. Severe malnourishment thus ranks as one of the most preventable causes of poor immune function in otherwise healthy individuals, and when combined with chronic infection (as with HIV/AIDS, tuberculosis, or cholera) can be all the more deadly.

HIV/AIDS Has Claimed Millions of Lives Worldwide

In recent years, all other forms of immunodeficiency have been overshadowed by an epidemic of severe immunodeficiency caused by the infectious agent called human immunodeficiency virus (HIV). HIV causes acquired immunodeficiency syndrome (AIDS) and was first recognized as opportunistic infections in a cluster of individuals on both coasts of the United States in June 1981. This group of patients displayed unusual infections, including the opportunistic fungal pathogen *Pneumocystis carinii*, which causes *P. carinii* pneumonia (PCP) in people with immunodeficiency. Previously, these infections were limited primarily to individuals taking immunosuppressive drugs. In addition to PCP, some of those early patients had Kaposi's sarcoma, an extremely rare skin tumor, as well as other, rarely encountered opportunistic infections. More complete evaluation showed that all patients had a common marked deficiency in

cell-mediated immune responses and a significant decrease in the subpopulation of T cells that carry the CD4 marker (T helper cells). When epidemiologists examined the background of the first patients with this new syndrome, they found that the majority were homosexual males. In those early days before we knew the cause or transmission route, and as the number of AIDS cases climbed throughout the world, people thought to be at highest risk for AIDS were homosexual males, promiscuous heterosexual individuals of either sex and their partners, intravenous drug users, people who received blood or blood products prior to 1985, and infants born to HIV-infected mothers. We now know that all these initial patients had intimate contact with an HIV-infected individual or exposure to HIV-tainted blood.

Since its discovery in the early 1980s, AIDS has increased to epidemic proportions throughout the world. As of December 2011, approximately 34 million people were living with HIV infection, 1.3 million in the United States. Although reporting of AIDS cases is mandatory in the United States, many states do not require reporting of cases of HIV infection that have not yet progressed to AIDS, making the count of HIV-infected individuals an estimate. The demographic profile of new HIV infections is evolving in the United States, where racial and ethnic minorities, especially men, are being disproportionately affected (Figure 18-8).

The toll of HIV/AIDS in the United States is dwarfed by figures for other parts of the world. The global distribution of those afflicted with HIV is shown in Figure 18-9. In sub-Saharan Africa, the region most affected, an estimated 23.5 million people were living with HIV at the end of 2010, and another 4 million were in South and Southeast Asia. Epidemiologic statistics estimate that more than 24 million people worldwide have died from AIDS since the beginning

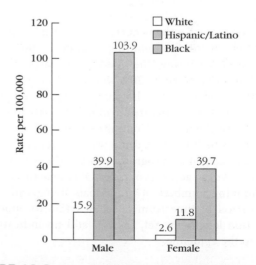

FIGURE 18-8 Rate of new HIV-1 infections in the United States in 2009, sorted by race/ethnicity and gender. Recent demographic data suggest a disproportionate and widening increase in the number of new HIV-1 infections among blacks and Hispanics as compared to whites, especially among men. *[Centers for Disease Control, www.cdc.gov/hiv/resources.]*

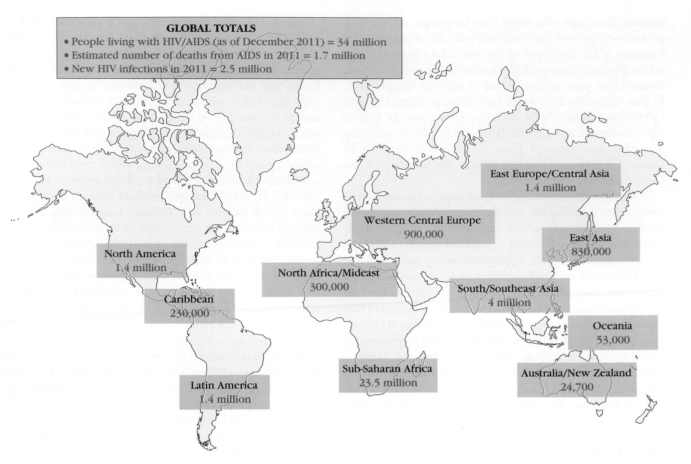

GLOBAL TOTALS
- People living with HIV/AIDS (as of December 2011) = 34 million
- Estimated number of deaths from AIDS in 2011 = 1.7 million
- New HIV infections in 2011 = 2.5 million

East Europe/Central Asia
1.4 million

Western Central Europe
900,000

East Asia
830,000

North America
1.4 million

North Africa/Mideast
300,000

South/Southeast Asia
4 million

Caribbean
230,000

Oceania
53,000

Sub-Saharan Africa
23.5 million

Australia/New Zealand
24,700

Latin America
1.4 million

FIGURE 18-9 The global AIDS epidemic. As of 2011, approximately 34 million people worldwide were living with HIV; most of them were in sub-Saharan Africa and Southeast Asia. Although the rate of new infections is decreasing, 2.5 million people are estimated to have contracted HIV in 2011. [*UNAIDS Report on the Global AIDS Epidemic (2012), www.unaids.org/globalreport/global_report.htm.*]

of the epidemic, leaving millions of children orphaned. Despite a better understanding of how HIV is transmitted, estimates indicate the occurrence of 2.5 million new HIV infections in 2011, amounting to almost 7,000 new infections each day! Now the good news: rates of new HIV infections decreased by 20% worldwide in 2011 compared to 2001, and access to lifesaving drugs has significantly expanded. These gains are attributable partly to the United Nations Declaration of Commitment on HIV/AIDS, signed in 2001, which has paved the way for stepped-up prevention and education programs around the world as well as expanded drug access programs. Of course this has also led to climbing numbers of individuals living with AIDS, as the period of time from onset of AIDS to opportunistic infection lengthens. Yet, there is still no indication of an end to the epidemic.

The Retrovirus HIV-1 Is the Causative Agent of AIDS

Within a few years after recognition of AIDS as an infectious disease, the causative agent, now known as HIV-1, was discovered and characterized in the laboratories of Luc Montagnier in Paris and Robert Gallo in Bethesda, Maryland (Figure 18-10). About 2 years later, the infectious agent was found to be a **retrovirus** of the lentivirus genus, which display long incubation periods (*lente* is Latin for "slow"). Retroviruses carry their genetic information in the form of RNA, and when the virus enters a cell this RNA is reverse-transcribed (RNA to DNA, rather than the other way around) by a virally encoded enzyme, *reverse transcriptase* (RT). This copy of DNA, which is called a **provirus**, is integrated into the cell genome and is replicated along with the cell DNA. When the provirus is expressed to form new virions (viral particles), the cell lyses. Alternatively, the provirus may remain latent in the cell until some regulatory signal starts the expression process. The discovery of a retrovirus as the cause of HIV was novel, since at the time only one other human retrovirus, *human T-cell lymphotropic virus I* (HTLV-I), had been identified. Although comparisons of their genomic sequences revealed that HIV-1 is not a close relative of HTLV-I, similarities in overall characteristics led to use of the name HTLV-III for the AIDS virus in early reports.

About 5 years after the discovery of HIV-1, a close retroviral cousin, HIV-2, was isolated from some AIDS sufferers in

Structure of HIV

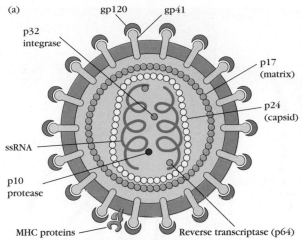

(a)

gp120   gp41

p32
integrase

p17
(matrix)

p24
(capsid)

ssRNA

p10
protease

MHC proteins   Reverse transcriptase (p64)

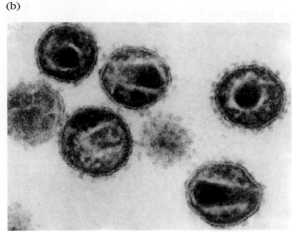

(b)

(a) Cross-sectional schematic diagram of HIV. Each virion carries 72 glycoprotein projections composed of gp120 and gp41: gp41 is a transmembrane molecule that crosses the lipid bilayer of the viral envelope, gp120 is associated with gp41, and together they interact with the target receptor (CD4) and coreceptor (CXCR4 or CCR5) on host cells. The viral envelope derives from the host cell and contains some host-cell membrane proteins, including class I and class II MHC molecules. Within the envelope is the viral matrix (p17) and the core, or nucleocapsid (p24). The HIV genome consists of two copies of single-stranded RNA (ssRNA), which are associated with two molecules of reverse transcriptase (p64) plus p10, a protease, and p32, an integrase. (b) Electron micrograph of HIV virions magnified 200,000 times. The glycoprotein projections are faintly visible as "knobs" extending from the periphery of each virion. *[Part a adapted from B. M. Peterlin and P. A. Luciw, 1988,* AIDS *2:S29; part b from a micrograph by Hans Geldenblom of the Robert Koch Institute (Berlin), in R. C. Gallo and L. Montagnier, 1988,* Scientific American ***259**(6):41.]*

Africa. Unlike HIV-1, its prevalence is mostly limited to areas of Western Africa, and disease progresses much more slowly, if at all. In Guinea-Bissau, where HIV-2 is most common, up to 8% of the population may be persistently infected and yet most of these individuals experience a nearly normal lifespan. There is some hope that scientists can gain a better understanding of HIV-1 from the study of the more benign cohabitation of HIV-2 and its human host.

Viruses related to HIV-1 have been found in nonhuman primates, and some of these are believed to be the original source of HIV-1 and -2 in humans. These viruses, variants of simian *i*mmunodeficiency *v*irus (SIV), can cause immunodeficiency disease in certain infected monkeys. Typically, SIV strains cause no disease in their natural hosts but produce immunodeficiency similar to AIDS when injected into another species. For example, the virus from African green monkeys (SIV_{agm}) is present in a high percentage of normal, healthy African green monkeys in the wild. However, when SIV_{agm} is injected into macaques, it causes a severe and often lethal immunodeficiency. HIV-1 is believed to have evolved from a strain of SIV that jumped the species barrier from African chimpanzees to humans, although HIV-2 is thought to have arisen from a separate but similar transfer from SIV-

infected sooty mangabeys. Both of these events are believed to have occurred some time during the twentieth century, making this a relatively new pathogen for the human population.

A number of other animal retroviruses more or less similar to HIV-1 have been reported. These include the *f*eline *i*mmunodeficiency *v*irus and *b*ovine *i*mmunodeficiency *v*irus (FIV and BIV, respectively) and the mouse leukemia virus. Study of these animal viruses has yielded information concerning the general nature of retrovirus action and pathways to the induction of immunodeficiency. Because HIV does not replicate in typical laboratory animals, model systems to study it are few. Only the chimpanzee supports infection with HIV-1 at a level sufficient to be useful in vaccine trials, but infected chimpanzees rarely develop AIDS, which limits the value of this model in the study of viral pathogenesis. In addition, the number of chimpanzees available for such studies is low, and both the expense and the ethical issues involved preclude widespread use of this infection model. The SCID mouse (see above) reconstituted with human lymphoid tissue for infection with HIV-1 has been useful for certain studies of HIV-1 infection, especially in the development of drugs to combat viral replication.

CLINICAL FOCUS

Prevention of Infant HIV Infection by Anti-Retroviral Treatment

It is estimated that almost 700,000 infants became infected with HIV through mother-to-child transmission in 2005 (just prior to the widespread implementation of standard prophylactic treatment for HIV-infected mothers). The majority of these infections resulted from transmission of virus from HIV-infected mothers during childbirth or by transfer of virus from milk during breast-feeding. The incidence of maternally acquired infection can be reduced by treatment of the infected mother with a course of zidovudine or azidothymidine (AZT, a *n*ucleoside analog *r*everse-*t*ranscriptase *i*nhibitor [NRTI]) for several months prior to delivery and treatment of her infant for 6 weeks after birth. This treatment regimen is standard practice in the United States. However, the majority of worldwide HIV infection in infants occurs in sub-Saharan Africa and other less-developed areas, where the cost and timing of the zidovudine regimen render it an impractical solution to the problem of maternal-infant HIV transmission.

A 1999 clinical trial of the anti-retroviral drug nevirapine (Viramune, an NRTI) brought hope for a practical way to combat HIV transmission at birth in less-than-ideal conditions of clinical care. The trial took place at Mulago Hospital in Kampala,

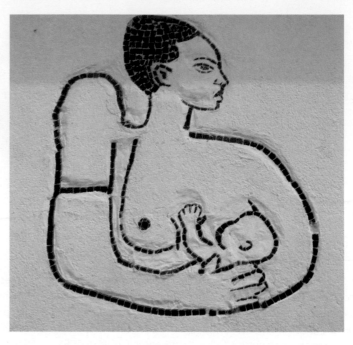

FIGURE 1

Mural showing mother and child on an outside wall of Mulago Hospital Complex in Kampala, Uganda, site of the clinical trial demonstrating that maternal-infant HIV-1 transmission at birth was greatly reduced by nevirapine. [*Courtesy of Thomas Quinn, Johns Hopkins University.*]

Uganda, and enrolled 645 mothers who tested positive for HIV infection (Figure 1). About half of the mothers were given a single dose of nevirapine at the onset of labor, and their infants were given a single dose one day after birth. The dose and timing were dictated by the customary rapid

discharge at the hospital. The control arm of the study involved a more extensive course of zidovudine, but local conditions did not allow the full course typically administered to infected mothers in the United States. The subjects were followed for at least 18 months. The infected mothers breast-fed

HIV-1 Is Spread by Intimate Contact with Infected Body Fluids

Although some of the details involving the mechanism by which HIV-1 infects an individual are still incomplete, the big picture of transmission routes has been resolved. Epidemiological data indicate that the most common means of transmission include vaginal and anal intercourse, receipt of infected blood or blood products, and passage from HIV-infected mothers to their infants. Before routine tests for HIV-1 were in place, patients who received blood transfusions and hemophiliacs who received blood products were at risk for HIV-1 infection. Exposure to infected blood accounts for the high incidence of AIDS among intravenous drug users, who often share hypodermic needles. Infants born to

mothers who are infected with HIV-1 are at high risk of infection; without prophylaxis, over 25% of these newborns may become infected with the virus. However, elective Cesarean section delivery and antiretroviral treatment programs for HIV⁺ pregnant women and their newborns are making a real dent in these numbers (see Clinical Focus Box 18-1).

In the worldwide epidemic, it is estimated that approximately 75% of the cases of HIV transmission are attributable to sexual contact. The presence of other *s*exually *t*ransmitted *d*iseases (STDs) increases the likelihood of transmission, and in situations where STDs flourish, such as unregulated prostitution, these infections probably represent a powerful cofactor for sexual transmission of HIV-1. The open lesions and activated inflammatory cells (some of which may express receptors for HIV) associated with STDs favor the

BOX 18-1

99% of the infants in this study, so the test measures a considerable interval during which the infants were exposed to risk of infection at birth and while nursing.

The infants were tested for HIV-1 infection at several times after birth up to 18 months of age using an RNA PCR test and at later times for HIV-1 antibody (after maternal antibody would no longer interfere with the test). The overall rate of infection for infants born to untreated mothers is estimated to be about 37%. In the United States, when the full course of zidovudine is used, the rate drops to 20%. The highly encouraging results of the Uganda study revealed infection in only 13.5% of the babies in the nevirapine group when tested at 16 weeks of age. Of those given a short course of zidovudine, 22.1% were infected at this age compared to 40.2% in a small placebo group. In the 18-month follow-up of the test group, 15.7% of those given nevirapine were infected, whereas 25.8% given zidovudine were positive for HIV-1.

These promising results led to World Health Organization (WHO) recommendations that this nevirapine regimen be used in all instances where mother-to-child transmission of HIV was a danger. But what about the risks associated with breast-feeding? In 2007, it was estimated that up to 200,000 infants become infected with HIV annually through breast milk. In resource-limited countries, poor early nutrition and susceptibility to disease are key factors in infant death rates; breast-feeding significantly reduces these risks. Therefore, a follow-up investigation called the *B*reastfeeding, *A*ntiretrovirals, and *N*utrition (BAN) study was conducted in Lilongwe, Malawi, in 2006 to 2008. This investigation employed a dose of nevirapine for the mother and child at birth, followed by 7 days of treatment of the pair with two NRTIs. The 2369 mother-infant pairs were then randomized to one of three post-delivery prophylaxis groups: maternal treatment, infant treatment, or placebo control. Mothers in the maternal regimen received a triple-drug cocktail for 28 weeks post delivery, including two NRTIs and either an NNRTI or a protease inhibitor. The babies in the infant treatment group received nevirapine daily for the same period of time, while initial participants in the control population received no additional treatment (in the latter part of the study, no new pairs were enrolled in this group). All mothers were encouraged to breast-feed for 6 months and wean before 7 months. Comparing the rate of HIV transmission that occurred during breast-feeding (from weeks 2–28) in the control population to the treatment groups, results demonstrated a 53% protective effect for the maternal regimen and a 74% protective effect in the infant treatment group.

Collectively, the conclusions from these efficacy and safety trials have led to new worldwide recommendations for protecting newborns from maternally transmitted HIV. In 2009, the WHO revised its recommendations to encourage breast-feeding for at least 12 months, based on studies in Africa showing increased infant mortality rates in babies weaned as early as 6 months. The hope is that with an extended nursing period that includes nevirapine prophylaxis, both the danger of death from disease or malnutrition and the risk of HIV transmission can be significantly reduced, even in severely resource-limited settings.

As mentioned above, these studies were designed to conform to the reality of maternal health care in less developed nations. Nevirapine has significant advantages in this regard, including stability of the drug at room temperature, reasonable cost, and relatively few side effects. The dose of nevirapine administered to the mother and infant at birth costs about 200 times less than the zidovudine regimen used in the United States to block HIV transmission at birth. In fact, the treatment is sufficiently inexpensive as to suggest that it may be cost-effective to treat all mothers and infants at the time of delivery in areas where the rates of infection are high and the nevirapine treatment costs are less than the tests used to determine HIV infection.

---

transfer and attachment of the virus during intercourse. Estimates of transmission rates per exposure vary widely and depend on many factors, such as the presence of STDs and number of virions. However, male-to-female transfer between discordant couples during vaginal intercourse is approximately twice as risky to the female as to the male, based solely on anatomical considerations, and receptive partners in anal intercourse are even more at risk. Data from studies in India and in Africa indicate that men who are circumcised are at significantly lower risk of acquiring HIV-1 via sexual contact, possibly because foreskin provides a source of cells that can become infected or harbor the virus. However, this did not work in reverse: circumcised males were equally likely to transmit HIV-1 to their sexual partners. No similarly protective effect of circumcision was seen for other STDs, including herpes simplex type 2, syphilis, or gonorrhea.

Identifying the initial events that take place during HIV transmission is logistically and ethically challenging, as we know that immediate antiviral treatment significantly diminishes the odds of infection. Nonetheless, hypotheses concerning the most likely sequence of events have been pieced together based on observations in humans and animals, including in vitro studies using explanted human tissue and in vivo studies in macaques, a nonhuman primate. Based on these observations, we believe that free virus and virus-infected cells, which can both be found in vaginal secretions and semen, contribute to infection. Direct infection of the many activated but resting memory CD4$^+$ T cells present within the vaginal mucosa is likely the primary initial source

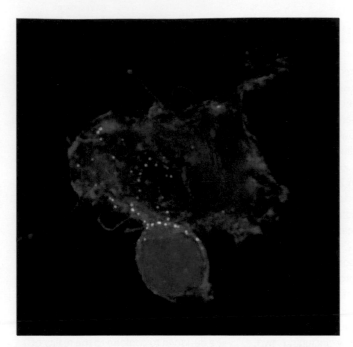

FIGURE 18-11 Interaction between dendritic cell and T cell, indicating passage of HIV-1 (green dots) between the cells. Note that particles cluster at the interface between the large dendritic cell and the smaller T cell. *[Courtesy of Thomas J. Hope, Northwestern University.]*

of infection in the female genital tract (the most studied location). In macaques, a foothold or attachment can be established in as little as 30 to 60 minutes, and high numbers of infected $CD4^+$ cells are seen within 1 day of exposure. Replication of HIV-1 in the vaginal mucosa was also shown to help activate local $CD4^+$ T cells, providing yet more targets for the virus and creating a vicious cycle. Likewise, the inflammation associated with STDs is thought to enhance the number of T_H cells and their susceptibility to infection. Viral transcytosis through epithelial cells, or endocytic transport from the luminal to the basal surface of the cell, is another possible route. *Langerhans cells* (LCs), a type of intraepithelial DC with long processes that reach close to the vaginal lumen, have also been shown to take up virus, although they may not become infected. These and other DCs may transport intact infectious virus, possibly for days, within endocytic compartments, and can transfer this virus to $CD4^+$ T cells (Figure 18-11). The role of macrophages in these early events, as transporters or targets of infection, is somewhat controversial but not suspected to be a major contributor. Finally, free virus can squeeze between epithelial cells or gain access through microabrasions, eventually encountering susceptible cells or afferent lymphatic vessels.

Whether free or cell-associated, the virus then migrates through the submucosa to the draining lymph node, where the adaptive immune response can be initiated. However, once there, further viral spread is facilitated, some through cell-to-cell hand-off via infectious synapses, as many more cells with the proper surface receptors are encountered.

Although the female genital tract is a relatively robust barrier to most infectious agents (with the adaptive immune response taking over from there), emerging evidence based on viral sequence analysis suggests that a single HIV-1 virion may be responsible for all or most of the systemic infection in many male-to-female transfers.

Because transmission of HIV-1 infection requires direct contact with infected blood, milk, semen, or vaginal fluid, preventive measures can be taken to block these events. Scientific researchers and medical professionals who take reasonable precautions, which include avoiding exposure of broken skin or mucosal membranes with fluids from their patients, significantly decrease their chances of becoming infected. When exposure does occur, rapid administration of anti-HIV treatment can often prevent systemic infection. The use of condoms when having sex with individuals of unknown infection status also significantly reduces chances of infection. One factor contributing to the spread of HIV is the long period after infection during which no clinical signs may appear but during which the infected individual may infect others. Thus, universal use of precautionary measures is important whenever infection status is uncertain.

In Vitro Studies Have Revealed the Structure and Life Cycle of HIV-1

The HIV-1 genome and encoded proteins have been fairly well characterized, and the functions of most of these proteins are known (Figure 18-12). HIV-1 carries three structural genes (*gag, pol,* and *env*) and six regulatory or accessory genes (*tat, rev, nef, vif, vpr,* and *vpu*). The structural genes and the proteins they encode were the first to be sequenced and meticulously characterized. The *gag* gene encodes several proteins, including the capsid and matrix, which enclose the viral genome and associated proteins. The *pol* gene codes for the three main enzymes (in addition to those supplied by the host cell) that are required for the viral life cycle: protease, integrase, and reverse transcriptase. In fact, the protease enzyme is required to process precursor proteins of many of the other viral peptides. As we will see shortly, these uniquely viral enzymes are some of the main targets for therapeutic intervention. The final structural gene, *env*, is the source of the surface proteins gp120 and gp41, involved in attachment of the virus to the $CD4^+$ viral receptor and its coreceptor, either CXCR4 or CCR5. The regulatory genes expressed by HIV-1, which took longer to characterize, encode functions such as modulating CD4 and class I MHC expression, inactivating host proteins that interfere with viral transcription, and facilitating intracellular viral transport.

Much has been learned about the life cycle of HIV-1 from in vitro studies, where cultured human T cells have been used to map out virus attachment and post-attachment intracellular events (Figure 18-13a). HIV-1 infects cells that carry the CD4 antigen on their surface; in addition to T_H cells, these can include monocytes and macrophages, as well as other cells expressing CD4. This preference for $CD4^+$ cells

(a)

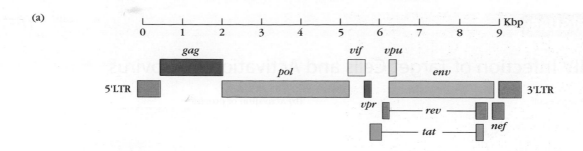

(b)

Gene	Protein product	Function of encoded proteins
gag	53-kDa precursor ↓	*Nucleocapsid proteins*
	p17	Forms outer core-protein layer (matrix)
	p24	Forms inner core-protein layer (capsid)
	p9	Is component of nucleoid core
	p7	Binds directly to genomic RNA
env	160-kDa precursor ↓	*Envelope glycoproteins*
	gp41	Is transmembrane protein associated with gp120 and required for fusion
	gp120	Protrudes from envelope and binds CD4
pol	Precursor ↓	*Enzymes*
	p64	Has reverse transcriptase and RNase activity
	p51	Has reverse transcriptase activity
	p10	Is protease that cleaves *gag* precursor
	p32	Is integrase
		Regulatory proteins
tat	p14	Strongly activates transcription of proviral DNA
rev	p19	Allows export of unspliced and singly spliced mRNAs from nucleus
		Auxiliary proteins
nef	p27	Down-regulates host-cell class I MHC and CD4
vpu	p16	Is required for efficient viral assembly and budding. Promotes extracellular release of viral particles, degrades CD4 in ER
vif	p23	Promotes maturation and infectivity of viral particle
vpr	p15	Promotes nuclear localization of preintegration complex, inhibits cell division

FIGURE 18-12 Genetic organization of HIV-1 (a) and functions of encoded proteins (b). The three major genes—*gag, pol,* and *env*—encode polypeptide precursors that are cleaved to yield the nucleocapsid core proteins, enzymes required for replication, and to envelop core proteins. Of the remaining six genes, three (*tat, rev,* and *nef*) encode regulatory proteins that play a major role in controlling expression, two (*vif* and *vpu*) encode proteins required for virion maturation, and one (*vpr*) encodes a weak transcriptional activator. The 5′ long terminal repeat (LTR) contains sequences to which various regulatory proteins bind. The organization of the HIV-2 and SIV genomes is very similar except that the *vpu* gene is replaced by *vpx* in both of them.

18-13

 HIV Infection of Target Cells and Activation of Provirus

(a) Infection of target cell

(b) Activation of provirus

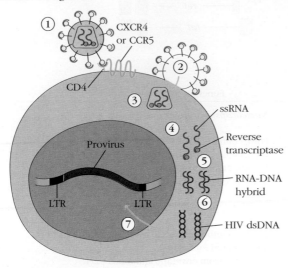

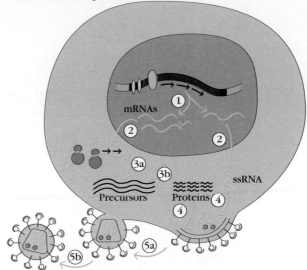

① HIV gp120 binds to CD4 on target cell.

② HIV gp41 binds to a chemokine receptor (CXCR4 or CCR5) and fuses with the target cell membrane.

③ Nucleocapsid containing viral genome and enzymes enters cells.

④ Viral genome and enzymes are released following removal of core proteins.

⑤ Viral reverse transcriptase catalyzes reverse transcription of ssRNA, forming RNA-DNA hybrids.

⑥ Original RNA template is partially degraded by ribonuclease H, followed by synthesis of second DNA strand to yield HIV dsDNA.

⑦ The viral dsDNA is then translocated to the nucleus and integrated into the host chromosomal DNA by the viral integrase enzyme.

① Transcription factors stimulate transcription of proviral DNA into genomic ssRNA and, after processing, several mRNAs.

② Viral RNA is exported to cytoplasm.

③a Host-cell ribosomes catalyze synthesis of viral precursor proteins.

③b Viral protease cleaves precursors into viral proteins.

④ HIV ssRNA and proteins assemble beneath the host-cell membrane, into which gp41 and gp120 are inserted.

⑤a The membrane buds out, forming the viral envelope.

⑤b Released viral particles complete maturation; incorporated precursor proteins are cleaved by viral protease present in viral particles.

(a) Following entry of HIV into cells and formation of dsDNA, integration of the viral DNA into the host-cell genome creates the provirus.
(b) The provirus remains latent until events in the infected cell trigger its activation, leading to formation and release of viral particles.

is due to a high-affinity interaction between gp120 and the CD4 molecule on the host cell. However, this interaction alone is not sufficient for viral entry and productive infection. Expression of another cell-surface molecule, called a coreceptor, is required for HIV-1 access to the cell. Each of the two known coreceptors for HIV-1, CCR5 and CXCR4, belongs to a separate class of molecule known as a chemokine receptor. The role of chemokine receptors in the body is to bind their natural ligands, chemokines, which are chemotactic messengers driving the movement of leukocytes (see Chapter 4). The infection of a T cell is assisted by the CXCR4 coreceptor, while the analogous CCR5 seems to be the pre-

ferred coreceptor for viral entry into monocytes and macrophages, and is now a target for antiviral intervention.

After HIV-1 has entered a cell, the RNA genome of the virus is reverse-transcribed and a cDNA copy integrates into the host genome. The integrated provirus is transcribed, and the various viral RNA messages are spliced and translated into proteins that, along with a complete new copy of the RNA genome, are used to form new viral particles (Figure 18-13b). These initial viral proteins are cleaved by the virally encoded protease into the forms that make up the nuclear capsid in a mature infectious viral particle. Virus expression leads to newly formed virions that bud from the surface of the infected cell, often

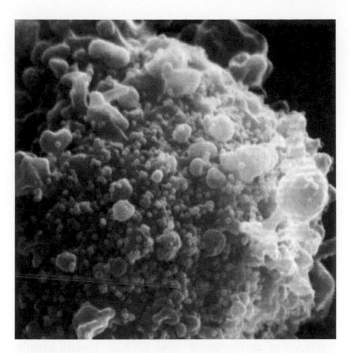

FIGURE 18-14 Once the HIV provirus has been activated, buds representing newly formed viral particles can be observed on the surface of an infected T cell. The extensive cell damage resulting from budding and release of virions leads to the death of infected cells. *[Courtesy of R. C. Gallo, 1988, HIV—The cause of AIDS: An overview on its biology, mechanisms of disease induction, and our attempts to control it. Journal of Acquired Immune Deficiency Syndromes 1:521.]*

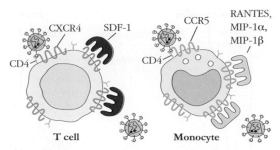

FIGURE 18-15 CXCR4 and CCR5 serve as coreceptors for HIV infection. Although CD4 binds to the envelope glycoprotein of HIV-1, a second coreceptor is necessary for entry and infection. T-cell–tropic strains of HIV-1 use the coreceptor CXCR4, whereas the macrophage-tropic strains use CCR5. Both are receptors for chemokines, and their normal ligands (SDF-1, RANTES, and MIP) can block HIV infection of the cell.

causing cell lysis (Figure 18-14). However, HIV-1 can also become latent, or remain unexpressed, for long periods of time in an infected cell. This period of dormancy makes the task of finding these latently infected cells especially difficult for the immune response to HIV-1; latent infection is believed to aid in the establishment of HIV reservoirs, or safe havens, where both drug therapy and antiviral immunity can have little impact.

Studies of the viral envelope protein gp120 identified a region called the V3 loop, which plays a role in the choice of receptors used by the virus. It is clear from these studies that a single amino acid difference in this region of gp120 may be sufficient to determine which receptor is used. Moreover, a mutation in the CCR5 gene that occurs with varying frequency, depending on ethnic background, imparts nearly total resistance to infection with the strains of HIV-1 that are most commonly encountered in sexual exposure. Individuals who are homozygous for this mutation express no CCR5 on the surface of their cells, making them impervious to viral strains that require this coreceptor but, remarkably, otherwise apparently unperturbed by loss of this chemokine receptor. This has led to some recent stories, and new hopes, of HIV elimination. For instance, one HIV-infected patient who received a set of bone marrow transplants (to treat leukemia) from a donor who *lacked* the CCR5 coreceptor protein may, or may not, be virus free. Confirmation and long-term follow-up of this finding, as well as development

of other techniques to exploit this coreceptor block as a method of virus elimination, are currently underway.

The discovery that CXCR4 and CCR5 serve as coreceptors for HIV-1 on T cells and macrophages, respectively, explained why some strains of HIV-1 preferentially infect T cells (T-tropic strains), whereas others prefer macrophages (M-tropic strains). T-tropic strains use the CXCR4 coreceptor, whereas M-tropic strains use CCR5 (Figure 18-15). This also helped to explain some observed roles of chemokines in virus replication. It was known from in vitro studies that certain chemokines, such as RANTES, had a negative effect on virus replication. CCR5 and CXCR4 cannot bind simultaneously to HIV-1 and to their natural chemokine ligands. Competition for the receptor between the virus and the natural chemokine ligand can thus block viral entry into the host cell. Early enthusiasm for the use of these chemokines as antiviral agents was dampened when significant RANTES expression was observed in some HIV-1 infected individuals who progress to disease, with no obvious antiviral effect. Despite this, an antagonist of CCR5 has recently been approved for use as a therapeutic inhibitor of HIV spread (see below).

Infection with HIV-1 Leads to Gradual Impairment of Immune Function

Isolation of HIV-1 and its growth in culture allowed purification of viral proteins and the development of tests for infection with the virus. The most commonly used test is an ELISA (see Chapter 20) to detect the presence of antibodies directed against proteins of HIV-1, especially the gag p24 protein (see Figures 18-10a and 18-12b), one of the most immunogenic of the HIV proteins. These antibodies generally appear in the serum of infected individuals within 6 to 12 weeks after exposure, but can take up to 6 months to appear. When antibodies appear in the blood, the individual is said to have seroconverted or to be seropositive for HIV-1. Positive p24 ELISA results are then confirmed using the more specific Western blot technique, which detects the presence of antibodies against several HIV-1 proteins.

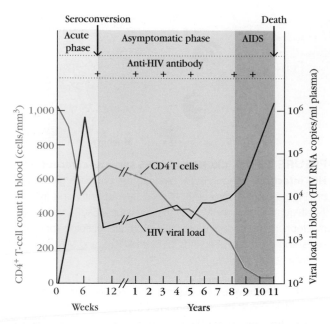

FIGURE 18-16 The typical course of an HIV infection.
Soon after infection, viral RNA is detectable in the serum. However, HIV infection is most commonly detected by the presence of anti-HIV antibodies after seroconversion, which normally occurs within a few months after infection. Clinical symptoms indicative of AIDS generally do not appear for up to 8 years after infection, but this interval is variable, especially when antiretroviral therapy is used. The onset of clinical AIDS is usually signaled by a decrease in T-cell numbers and a sharp increase in viral load. *[Adapted from A. Fauci et al., 1996, Immunopathogenic mechanisms of HIV infection. Annals of Internal Medicine 124:654.]*

Although the precise course of HIV-1 infection and disease onset varies considerably in different patients, a general scheme for the progression to AIDS can be outlined (Figure 18-16). First, there is the acute, or primary, stage of infection. This is the period immediately after infection, where there are often no detectable anti-HIV-1 antibodies. Estimates vary, but some reports find that more than half of the individuals undergoing primary infection experience flu-like symptoms, including fever, lymphadenopathy (swollen lymph nodes), and malaise approximately 2 to 4 weeks after exposure. During this acute phase, HIV-1 infection is spreading and the **viral load** (number of virions) in the blood as well as in other body fluids can be quite high, elevating the risk of transfer to others. For unknown reasons, seroconversion, or the appearance of antibody against HIV antigens, can take months to develop.

This stage is followed by an asymptomatic period during which there is a gradual decline in CD4+ T cells but usually no outward symptoms of disease. This is driven by an immune response involving both antibody and cytotoxic CD8+ T lymphocytes that keeps viral replication in check and drives down the viral load. The length of this asymptomatic window varies greatly and is likely due to a combination of host and viral factors. Although the infected individual

normally has no clinical signs of disease at this stage, viral replication continues, CD4+ cell levels gradually fall, and viral load in the circulation can be measured by PCR assays for viral RNA. These measurements of viral load have assumed a major role in the determination of the patient's status and prognosis. Even when the level of virus in the circulation is stable, large amounts of virus are produced in infected CD4+ T cells; as many as 10^9 virions are released every day and continually infect and destroy additional host T cells. Despite this high rate of replication, the virus is kept in check by the immune system throughout the asymptomatic phase of infection, and the level of virus in circulation from about 6 months after infection (the set point) can be a predictor of the course of disease. Low levels of virus in this period correlate with a longer asymptomatic period and opportunistic pathogen-free window.

Without treatment, most HIV-1 infected patients eventually progress to AIDS, where opportunistic infection is the hallmark. Diagnosis of AIDS occurs only once four criteria have been met: evidence of infection with HIV-1 (presence of antibodies or viral RNA in blood), greatly diminished numbers of CD4+ T cells (< 200 cells/µl of blood), impaired or absent delayed-type hypersensitivity reactions, and the occurrence of opportunistic infections (Table 18-3). The first overt indication of AIDS is often opportunistic infection with the fungus *Candida albicans*, which causes the appearance of sores in the mouth (thrush) and, in women, a vulvovaginal yeast infection that does not respond to treatment. A persistent hacking cough caused by *P. carinii* infection of the lungs is another early indicator. A rise in the level of circulating HIV-1 in the plasma (viremia) and a concomitant drop in the number of CD4+ T cells generally precedes this first appearance of symptoms. Late-stage AIDS patients generally succumb to tuberculosis, pneumonia, severe wasting diarrhea, or various malignancies. Without treatment, the time between acquisition of the virus and death from the immunodeficiency averages 9 to 11 years.

Active Research Investigates the Mechanism of Progression to AIDS

Of intense interest to immunologists are the events that take place between the initial encounter with HIV-1 and the takeover and collapse of the host immune system. Understanding how the immune system holds HIV-1 in check during the asymptomatic phase could aid in the design of effective therapeutic and preventive strategies. For this reason, the handful of HIV-1 infected individuals who remain asymptomatic for very long periods without treatment (long-term nonprogressors), estimated to be $< 2\%$ of HIV-1+ individuals in the United States, are the subject of intense study. Another group of heavily studied individuals consists of those in high-risk groups, such as many prostitutes. From the study of virus-immune system interactions in the handful of individuals in these high-risk groups who remain

TABLE 18-3 Stage definition for HIV infection among adults and adolescents

Stage*	CD4+ T-cell count		CD4+ T-cell percentage		Clinical evidence
1	≥ 500/μL	or	> 29%	and	No AIDS-defining condition
2	200–499/μL	or	14–28%	and	No AIDS-defining condition
3 (AIDS)	< 200/μL	or	< 14%	or	Presence of AIDS-defining condition

AIDS-DEFINING CONDITIONS:

- Candidiasis of bronchi, trachea, or lungs
- Candidiasis of esophagus
- Cervical cancer, invasive
- Coccidioidomycosis, disseminated or extrapulmonary
- Cryptococcosis, extrapulmonary
- Cryptosporidiosis, chronic intestinal (> 1 month duration)
- Cytomegalovirus disease (other than liver, spleen, or nodes)
- Cytomegalovirus retinitis (with loss of vision)
- Encephalopathy, HIV related
- Herpes simplex: chronic ulcers (> 1 month duration) or bronchitis, pneumonitis, or esophagitis
- Histoplasmosis, disseminated or extrapulmonary
- Isosporiasis, chronic intestinal (> 1 month duration)
- Kaposi's sarcoma
- Lymphoid interstitial pneumonia or pulmonary lymphoid hyperplasia complex
- Lymphoma, Burkitt (or equivalent term)
- Lymphoma, immunoblastic (or equivalent term)
- Lymphoma, primary, of brain
- *Mycobacterium avium* complex or *Mycobacterium kansasii*, disseminated or extrapulmonary
- *Mycobacterium tuberculosis* of any site, pulmonary, disseminated, or extrapulmonary
- *Mycobacterium*, other species or unidentified species, disseminated or extrapulmonary
- *Pneumocystis jirovecii* pneumonia
- Pneumonia, recurrent
- Progressive multifocal leukoencephalopathy
- *Salmonella* septicemia, recurrent
- Toxoplasmosis of brain
- Wasting syndrome attributed to HIV

* All require laboratory confirmation of HIV infection.

Source: AIDS-Defining Conditions, 2008, Centers for Disease Control, *www.cdc.gov/mmwr/preview/mmwrhtml/rr5710a2.htm.*

seronegative despite known and repeated exposure, we hope to gain clues to mechanisms of control or possibly even protection. In addition to the discovery of CCR5 deletion (see above), several interesting findings have emerged from studying high-risk populations, including the presence of strong CD8+ T-cell responses against HIV in many of these individuals, as well as HLA-associated influences on disease susceptibility.

Although the viral load in plasma remains fairly stable throughout the period of chronic HIV infection, examination of the lymph nodes and gastrointestinal (GI) tract tissue reveals a different picture. Fragments of nodes obtained by biopsy from infected subjects show high levels of infected cells at all stages of infection; in many cases, the structure of the lymph node is completely destroyed by virus long before plasma viral load increases above the

(a)   HIV⁻   (b)   HIV⁺

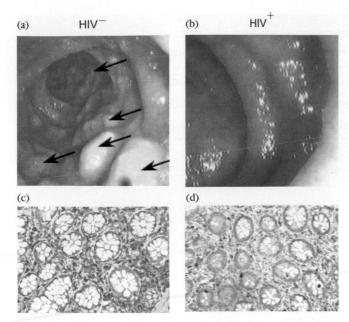

(c)   (d)

(e)

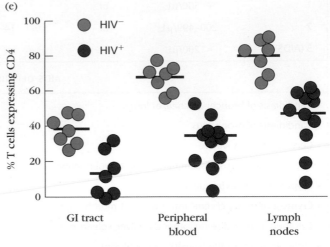

FIGURE 18-17 Endoscopic and histologic evidence for depletion of CD4⁺ T cells in the GI tract of AIDS patients. Panels (a) and (c) show the intestinal tract of a normal individual and a stained section from a biopsy of the same area (terminal ileum) with obvious large lymphoid aggregates (arrows, a) and CD4⁺ T cells stained with antibody (brown color, c). Similar analysis of samples from an HIV⁺ patient in the acute stage of infection in panels (b) and

(d) indicate absence of normal lymphoid tissue and sparse staining for CD4⁺ T cells. (e) Comparison of CD4⁺ T-cell numbers in samples from GI tract, peripheral blood (PB), and lymph nodes of AIDS-positive and -negative individuals. *[From J. M. Brenchley et al. 2004, CD4⁺ T cell depletion during all stages of HIV disease occurs predominantly in the gastrointestinal tract. Journal of Experimental Medicine **200:749**.]*

steady-state level. In fact, data from 2004 show that the gut may be the main site of HIV-1 replication and CD4⁺ T-cell depletion. Work from the laboratories of Ashley Haase and Daniel Douek indicates a dramatic depletion of lymphoid tissue and specifically CD4⁺ T cells from the GI tract during HIV infection, starting as early as the acute stages of infection (Figure 18-17). Subsequent investigations of the association between the GI tract and HIV have suggested that T_H17 cells, which express both the CCR5 and CXCR4 coreceptors, are the primary targets of infection and destruction. These T_H17 cells are thought to play an important role in homeostatic regulation of the innate and adaptive responses to microbial flora in the gut. Destruction of these cells and disruption of the integrity of the mucosal barrier in the GI tract may allow for the translocation of microbial products across the epithelial lining, explaining some of the rampant immune stimulation that is characteristic of HIV infection. In a deadly feedback loop, this immune stimulation generates yet more activated CD4⁺ cells, the favored targets for HIV infection and replication.

The severe decrease in CD4⁺ T cells is a clinical hallmark of AIDS, and several explanations have been advanced for it. In early studies, direct viral infection and destruction of CD4⁺ T cells was discounted as the primary cause, because the large numbers of circulating HIV-infected T cells predicted by the model were not found. More recent studies indicate that it is difficult to find the infected cells

because they are so rapidly killed by HIV (the half-life of an actively infected CD4⁺ T cell is less than 1.5 days) and because most of the infected cells may localize to the GI tract. Smaller numbers of CD4⁺ T cells become infected but do not actively replicate virus. These latently infected cells persist for long periods, and the integrated proviral DNA replicates in cell division along with cell DNA. Studies in which viral load is decreased by antiretroviral therapy show a concurrent increase in CD4⁺ T-cell numbers in the peripheral blood and eventually in the gut. Apoptosis due to nonspecific immune activation and bystander effects of free virus or infected cells acting on uninfected cells have also been postulated to play a role in HIV-induced lymphopenia.

Although depletion of CD4⁺ T cells is the primary focus of follow-up testing in HIV-infected individuals, other immunologic consequences involving both adaptive and innate immune functions can be observed during the progression to AIDS. These include a decrease or absence of delayed-type hypersensitivity to antigens for which the individual normally reacts, decreased serum immunoglobulins (especially IgG and IgA), and impaired cellular responses to antigens. Generally, the HIV-infected individual loses the ability to mount T-cell responses in a predictable sequence: responses to specific recall antigens (e.g., influenza virus) are lost first, then response to alloantigens declines, and finally mitogenic responses to stimuli disappear. Innate responses are also impacted, including NK and dendritic cell functions.

TABLE 18-4   Immunologic abnormalities associated with HIV infection

Stage of infection	Typical abnormalities observed
	LYMPH NODE STRUCTURE
Early	Infection and destruction of dendritic cells; some structural disruption, especially to gastrointestinal tract-associated lymphoid tissues
Late	Extensive damage and tissue necrosis; loss of follicular dendritic cells and germinal centers; inability to trap antigens or support activation of T and B cells
	T HELPER (T_H) CELLS
Early	Depletion of $CD4^+$ T cells, especially in the gut (T_H17 main targets); loss of in vitro proliferative response to specific antigen
Late	Further decrease in T_H-cell numbers and corresponding helper activities; no response to T-cell mitogens or alloantigens
	ANTIBODY PRODUCTION
Early	Enhanced nonspecific IgG and IgA production but reduced IgM synthesis
Late	No proliferation of B cells specific for HIV-1: no detectable anti-HIV antibodies in some patients; increased numbers of B cells with low CD21 expression and enhanced Ig secretion
	CYTOKINE PRODUCTION
Early	Increased levels of some cytokines
Late	Shift in cytokine production from T_H1 subset to T_H2 subset
	DELAYED-TYPE HYPERSENSITIVITY
Early	Highly significant reduction in proliferative capacity of T_H1 cells and reduction in skin-test reactivity
Late	Elimination of DTH response; complete absence of skin-test reactivity
	T CYTOTOXIC (T_C) CELLS
Early	Normal reactivity
Late	Reduction but not elimination of CTL activity due to impaired ability to generate CTLs from T_C cells

Table 18-4 lists some immune abnormalities common to HIV/AIDS.

Individuals infected with HIV-1 often display dysfunction of the central and peripheral nervous systems, especially in the later stages of infection. Viral sequences have been detected by HIV-1 probes in the brains of children and adults with AIDS, suggesting that viral replication occurs there. Quantitative comparison of specimens from brain, lymph node, spleen, and lung of AIDS patients with progressive encephalopathy indicated that the brain was heavily infected. A frequent complication in later stages of HIV infection is AIDS dementia, a neurological syndrome characterized by abnormalities in cognition, motor performance, and behavior. It remains unknown whether or not AIDS dementia and other pathological effects observed in the central nervous systems of infected individuals are a direct effect of HIV-1 on the brain, a consequence of immune responses to the virus, or a result of infection by opportunistic agents.

Therapeutic Agents Inhibit Retrovirus Replication

Development of a vaccine to prevent the spread of AIDS is the highest priority for immunologists. Meanwhile, drugs that can reverse the effects of HIV-1 have greatly improved the outlook for infected individuals. There are several strategies for the development of effective antiviral drugs that take advantage of the life cycle of HIV (Figure 18-18). The key to success for such therapies is that they must be

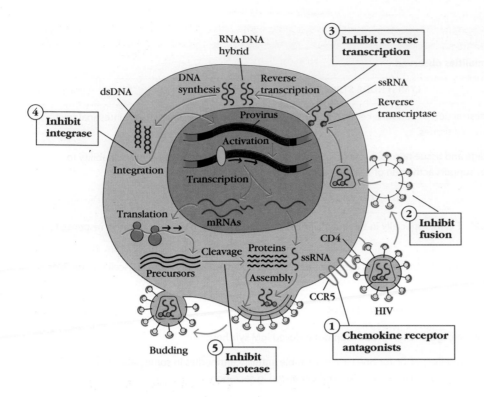

FIGURE 18-18 Stages in the viral replication cycle that provide targets for therapeutic antiretroviral drugs. The first licensed drugs with anti-HIV activity interfered with reverse transcription of viral RNA to cDNA (3), followed by drugs which blocked the viral protease that cleaves precursor proteins into the peptides needed to assemble new virions (5). In the past decade, newer drugs have come on the market that interfere with other steps in the viral life cycle, such as HIV coreceptor attachment (1) or fusion to the cell membrane (2), as well as the viral integrase necessary for insertion of proviral DNA into the host cell chromosome (4).

specific for HIV-1 and interfere minimally with normal cell processes. Thus far, antiviral agents targeting five separate steps in the viral life cycle have proven effective. The first success came with drugs that interfere with the reverse transcription of viral RNA to cDNA (③ in Figure 18-18); several drugs that use two possible mechanisms of action operate at this step. The second generation of drugs inhibits the viral protease (⑤ in Figure 18-18) required to cleave precursor proteins into the units needed for construction of new mature virions. This was followed by development of an inhibitor of the viral gp41 that blocks fusion of virus with the host cell membrane (② in Figure 18-18), inhibiting new infection of cells. The two newest antiviral agents on the market target attachment of the virus to the cell by competition for access to the CCR5 chemokine coreceptor used by the virus (① in Figure 18-18) or interfere with integrase (④ in Figure 18-18) required for insertion of the viral genome into host cell DNA. Table 18-5 lists the currently available categories of anti-HIV therapies, along with the year in which they were first approved for use. It should be stressed that the development of any drug to the point at which it can be used for patients is a long and arduous procedure. The drugs that pass the rigorous tests for safety and efficacy represent a small fraction of those that receive initial consideration.

There are two possible strategies for developing pharmaceutical agents that can interfere with reverse transcription: *nucleoside reverse transcriptase inhibitors* (NRTIs) and **non-nucleoside reverse transcriptase inhibitors** (NNRTIs). The prototype and earliest (approved in 1987) of the drugs that interferes with reverse transcription was zidovudine, or *azidothymidine* (AZT), in the NRTI class.

TABLE 18-5	Categories of HIV-1 drugs in clinical use	
Category	**FDA approval date***
Nucleoside/nucleotide analogues	1987
Nonnucleoside reverse transcriptase inhibitors	1996
Protease inhibitors	1995
Fusion/attachment inhibitors	2003
Chemokine coreceptor antagonists	2007
Integrase inhibitors	2007

*Year of first FDA approval for a drug to treat HIV-1 infection in that drug category.

Source: Antiretroviral Drug Profiles. (2012). HIV InSite, University of California, San Francisco. http://hivinsite.ucsf.edu/InSite?page=ar-drugs

The introduction of AZT, a nucleoside analogue and competitive inhibitor of the enzyme, into the growing cDNA chain of the retrovirus causes termination of the chain. AZT is effective in some but not all patients, and its efficacy is further limited because long-term use has adverse side effects and because resistant viral mutants develop in treated patients. The administered AZT is used not only by the HIV-1 *reverse transcriptase* (RT) but also by human DNA polymerase. The incorporation of AZT into the DNA of host cells kills them. Precursors of red blood cells and other rapidly dividing cells are especially sensitive to AZT, resulting in anemia and other side effects. NNRTI drugs inhibit the action of RT by binding to a different site on the enzyme. These noncompetitive inhibitors of RT have less of an adverse effect on host proteins, and therefore fewer side effects. However, they are still susceptible to the development of resistance as the virus mutates, and for this reason are typically only used in combination with other anti-HIV drugs that have different targets. Nevirapine, licensed in 1996, was the first such compound designed to treat HIV, but since then many more such drugs have come on the market.

A separate class of drugs, called **protease inhibitors**, target the HIV-1 protease that cleaves viral protein precursors into the peptides required for packaging into virions in the final stages of HIV replication. The first of the protease inhibitors, saquinavir, came to market in the mid-1990s, but many more have emerged since then. They are most commonly used today as a part of a multidrug cocktail designated as **highly active antiretroviral therapy (HAART)**. In most cases, this combines the use of three or more anti-HIV drugs from different classes. The combination strategy appears to overcome the ability of the virus to rapidly produce mutants that are drug resistant. In many cases, HAART has lowered plasma viral load to levels that are not detectable by current methods and has improved the health of AIDS patients to the point that they can again function at a normal level. The decrease in the number of AIDS deaths in the United States in recent years is attributed to this advance in therapy. Despite the optimism engendered by success with HAART, present drawbacks include the need for consistent adherence to this regimen, lest viral drug-resistant mutants be favored in the patient. In addition, some patient may experience serious side effects that become too severe to allow the use of HAART. This said, access to multiple drugs within most of the drug categories in the United States makes substitution possible.

The success of HAART has led researchers to wonder whether it might be possible to eradicate all virus from an infected individual and thus actually cure AIDS. Most AIDS experts are not convinced that this is feasible, mainly because of the persistence of latently infected CD4$^+$ T cells and macrophages, which can serve as a reservoir of infectious virus if and when the provirus becomes reactivated.

Even with a viral load beneath the level of detection by PCR assays, the immune system may not recover sufficiently to clear virus should it begin to replicate in response to some activation signal. In addition, virus may persist in sites, such as the brain, that are not readily penetrated by the antiretroviral drugs, even though the virus in circulation is undetectable. The use of immune modulators, such as recombinant IL-2 and IL-7 as adjuvants to HAART, is being examined as a strategy to help reconstitute the immune system and restore normal immune function, with mixed results.

Some forms of antiviral therapy may also work to *prevent* HIV infection in individuals at high risk. Based on studies completed in 2011 and 2012, when both partners in an HIV discordant couple (where one is infected and the other is not) were given a cocktail of two NRTIs, rates of sexual transmission of HIV dropped by 60% to 75%. Cost considerations aside, this *pre-exposure prophylaxis* (or PrEP) treatment was approved by the FDA in 2012 as a preventive measure for healthy, HIV-negative individuals at high risk of infection.

A Vaccine May Be the Only Way to Stop the HIV/AIDS Epidemic

The AIDS epidemic continues to rage despite the advances in therapeutic approaches outlined above. The expense of HAART (roughly $1000/month in the United States), the strict adherence required to avoid development of resistance, and the possibility of side effects preclude universal application. At present, it appears that the best option to stop the spread of AIDS is a safe, effective vaccine that prevents infection and/or progression to disease. The cause of AIDS was discovered over 25 years ago. Why don't we have an AIDS vaccine by now?

The best way to approach an answer to this question is to examine the specific challenges presented by HIV-1. HIV mutates rapidly, creating a moving target for both the immune response and any vaccine design. This means that there are many possible variants of the virus to contend with in any one geographical location, not even considering many different countries. We know from HIV seropositive individuals that the development of humoral immunity during a natural infection, even the presence of neutralizing antibodies, does not necessarily inhibit viral spread. This means that a strong humoral response alone is unlikely to block infection. Despite years of study, data from long-term nonprogressor and exposed seronegative individuals have shed little light on the immune correlates of protection from disease or infection. Without clear goalposts, it is difficult to know which types of cellular immune response will be most effective, or when we have reached enough of a response to protect against a natural infection. In addition, good animal models for testing these fledgling vaccines are limited and expensive.

TABLE 18-6   Major HIV-1 vaccine trials

Vaccine design	Study name	Status	Result
Purified protein (gp120)	VAX003, VAX004	Completed in 2003	No protection
Recombinant adenovirus vector (*gag/pol/nef*)	HVTN 502 STEP HVTN 503 Phambili	Terminated in 2007	No protection
Recombinant canarypox vector (*env/gag/protease*) + *env* gp120 protein boost	RV 144	Completed in 2009	30% reduction
DNA vaccine - 6-plasmids (*env/gag/pol/nef*) + Recombinant adenovirus vector boost (same genes)	HVTN 505	Began in 2009	N/A (in progress)

HVTN = HIV vaccine trials network

Source: Adapted from D. H. Barouch and B. Korber. 2010. HIV-1 vaccine development after STEP. *Annual Reviews in Medicine* **61;**153.

Still, valuable lessons have been learned from over a decade of HIV vaccine initiatives, both from studies in nonhuman primates and from clinical trials in humans (Table 18-6). The earliest of these trials in humans aimed at eliciting humoral immunity to neutralize incoming virus. This approach used purified envelope protein from HIV (specifically, gp120) as an immunogen. These studies, completed in the United States and Thailand in 2003, showed weak neutralizing antibody responses and no protection from infection. This was followed by a new wave of vaccine design aimed at eliciting cellular immunity to the virus using recombinant viral vectors that would better mimic natural infection. The initial vector chosen was *ade*novirus serotype 5 (Ad5), carrying recombinant DNA derived from the *gag, pol,* and *nef* genes of HIV-1. Ad5 is derived from a naturally occurring human virus to which between 30% and 80% of individuals (depending on geographic location) have previously been exposed, appearing to make this a safe vector choice. These trials included two stages of vaccination: an initial priming dose followed by a vaccine boost using the same vector. Initially conducted in the Americas, the Caribbean, and Australia, these human trials began in 2003 but were halted prematurely in 2007, midway through the trial period, because they did not demonstrate the desired reduction in viral load for those volunteers who became infected after receiving the vaccine. More important, the rate of infection appeared to be higher in the vaccinated group of the trial than in the placebo control population, especially among individuals who had pretrial immune responses to adenovirus. This was a resounding blow to the AIDS research community and sent many vaccine design teams back to the drawing board.

These disappointing results have led to new thinking in terms of targets for the next wave of HIV-1 vaccine design:

vaccines that elicit both humoral and cellular immunity. One trial using this strategy has recently been completed. It used a prime-boost combination with a recombinant DNA vector containing HIV-1 genes (the prime) followed by peptides derived from the virus (the boost). The vector chosen was canarypox, a virus that does not replicate in humans and that was engineered to carry DNA from the *env, gag,* and the portion of *pol* that encodes the protease of HIV-1. The protocol also included a boost with a companion vaccine consisting of an engineered version of HIV-1 gp120 protein. Results from this study were modest but promising, demonstrating an approximately 30% reduction in infection rates among the vaccine treated group compared to the matched placebo control population. Although statistically significant, these results are still far from the near 100% protection rate that is the aim for all vaccines against infectious disease. Nevertheless, this is one of the first really promising steps forward in HIV vaccine research. In 2009, trials were launched for a new DNA vaccine containing six plasmids representing multiple clades of *env* genes plus single clade *gag, pol,* and *nef* (prime), followed by an adenoviral vector containing the same genes (boost). Results from this trial, which expanded its scope in 2011, are still pending.

It is clear that development of an HIV vaccine is not a simple exercise in classic vaccinology. More research is needed to understand how this viral attack against the immune system can be thwarted. Although much has been written about the subject and large-scale initiatives are proposed, the path to an effective vaccine is not obvious. It is certain only that all data must be carefully analyzed and that all possible means of creating immunity must be tested. This is one of the greatest public health challenges of our time. An intense and cooperative effort will be required to devise, test, and deliver a safe and effective vaccine for AIDS.

SUMMARY

- Immunodeficiency results from the failure of one or more components of the immune system. Primary immunodeficiencies are based on genetic defects present at birth; secondary or acquired immunodeficiencies arise from a variety of external causes.

- Immunodeficiencies may be classified by the cell types involved and may affect either the lymphoid- or the myeloid-cell lineage, or both.

- Combined immunodeficiencies (CIDs) disrupt adaptive immunity by interfering with both T-cell and B-cell responses. Severe combined immunodeficiency, or SCID, is the most extreme form of CID and arises from a lack of functional T cells, which also manifests as no T-cell help for B-cell responses.

- Genetic failures of thymic development and major histocompatibility (MHC) expression can lead to CID because of the disruption of T-cell development.

- B-cell immunodeficiencies, which are associated with susceptibility to bacterial infection, can range from single isotype defects to total disruptions of humoral immunity.

- Selective immunoglobulin deficiencies are the less-severe form of B-cell deficiency and result from defects in more highly differentiated cell types.

- Myeloid immunodeficiencies cause impaired phagocytic function. Affected individuals suffer from increased susceptibility to bacterial infection.

- Complement deficiencies are relatively common and vary in their clinical impact, but are generally associated with increased susceptibility to bacterial infection.

- Immunodeficiencies that disrupt immune regulation can lead to overactive immune responses that manifest as autoimmune syndromes.

- Immunodeficiency disorders can be treated by replacement of defective or missing proteins, cells, or genes.

- Administration of human immunoglobulin is a common treatment, especially for those disorders that primarily disrupt antibody responses.

- Animal models for immunodeficiency include nude and SCID mice. Targeted gene knockout mice provide a means to study the role of specific genes in immune function.

- Secondary or acquired immunodeficiency can result from immunosuppressive drugs, infection, and malnutrition; the most well-known form of this is AIDS, caused by human immunodeficiency virus-1 (HIV-1), which is a retrovirus.

- HIV-1 infection is spread by contact with infected body fluids such as occurs during sex, intravenous drug use, and from mother to infant during childbirth or breast-feeding.

- The HIV-1 genome contains three structural and six regulatory genes that encode the proteins necessary for infection and propagation in host cells.

- HIV-1 uses the host CD4 molecule as well as a chemokine coreceptor (CCR5 or CXCR4) to attach to and fuse with host cells.

- Infection with HIV-1 results in gradual and severe impairment of immune function, marked by depletion of CD4$^+$ T cells, especially in the gut, and can result in death from opportunistic infection.

- Treatment of HIV infection with antiretroviral drugs that target specific steps in the viral life cycle, especially in combination, can lower the viral load and provide relief from some symptoms of infection. However, no cures for HIV are available.

- Efforts to develop a vaccine for HIV-1 have been only modestly successful. The millions of new infections occurring yearly emphasize the need for an effective vaccine.

REFERENCES

Belizário, J. E. 2009. Immunodeficient mouse models: An overview. *The Open Immunology Journal* 2:79–85.

Brenchley, J. M., and D. C. Douek. 2008. HIV infection and the gastrointestinal immune system. *Mucosal Immunity* 1(1):23.

Buckley, R. H. 2004. Molecular defects in human severe combined immunodeficiency and approaches to immune reconstitution. *Annual Review of Immunology* 22:625.

Calvazzano-Calvo, M., et al. 2005. Gene therapy for severe combined immunodeficiency. *Annual Review of Medicine* 56:585.

Chasela, C. S., et al. 2010. Maternal or infant antiretroviral drugs to reduce HIV-1 transmission. *The New England Journal of Medicine* 362(24):2271.

Chinen, J., and W. T. Shearer. 2009. Secondary immunodeficiencies, including HIV infection. *Journal of Allergy and Clinical Immunology* 125(2):S195.

Corey, L., et al. 2009. Post STEP modifications for research on HIV vaccines. *AIDS* 23(1):3.

Douek, D. C., et al. 2009. Emerging concepts in the immunopathogenesis of AIDS. *Annual Reviews in Medicine* 60:471.

Fischer, A. 2007. Human primary immunodeficiency diseases. *Immunity* 27:835.

Fischer, A., et al. 2010. 20 years of gene therapy for SCID. *Nature Immunology* 11(6):457.

Fried A. J., and F. A. Bonilla. 2009. Pathogenesis, diagnosis, and management of primary antibody deficiencies and infections. *Clinical Microbiology Reviews* 22(3):396.

Granich, R., et al., 2010. Highly active antiretroviral treatment for prevention of HIV transmission. *Journal of the International AIDS Society* 13:1.

Hladik, F., and T. J. Hope. 2009. HIV infection of the genital mucosa in women. *Current HIV/AIDS Reports* 6:20–28.

Hui-Qi, Q., S. P. Fisher-Hoch, and J. B. McCormick. 2011. Molecular immunity to mycobacteria: Knowledge from the mutation and phenotype spectrum analysis of Mendelian susceptibility to mycobacterial diseases. *International Journal of Infectious Diseases* 15(5):e305–e313.

Isgrò, A., et al. 2005. Bone marrow clonogenic capability, cytokine production, and thymic output in patients with common variable immunodeficiency. *Journal of Immunology* 174(8):5074–5081.

Jackson, J. B., et al. 2003. Intrapartum and neonatal single-dose nevirapine compared to zidovudine for prevention of mother-to-child transmission of HIV-1 in Kampala, Uganda: 18-month follow-up of the HIVNET 012 randomized trial. *Lancet* 362:859.

Moore, J. P., et al. 2004. The CCR5 and CXCR4 coreceptors—central to understanding the transmission and pathogenesis of human immunodeficiency virus type I infection. *AIDS Research and Human Retroviruses* 20:111.

Moraes-Vasconcelos, D., et al. 2008. Primary immune deficiency disorders presenting as autoimmune diseases: IPEX and APECED. *Journal of Clinical Immunology* 28 (Suppl 1): S11.

Notarangelo, L. D. 2010. Primary immunodeficiences. *Journal of Allergy and Clinical Immunology* 125(2):S182.

Ochs, H. D., et al. 2009. $T_H 17$ cells and regulatory T cells in primary immunodeficiency diseases. *Journal of Allergy and Clinical Immunology* 123(5):977.

Piacentini, L., et al. 2008. Genetic correlates of protection against HIV infection: The ally within. *Journal of Internal Medicine* 265:110.

Rerks-ngarm, S., et al., 2009. Vaccination with ALVAC and AIDSVAX to prevent HIV-1 infection in Thailand. *The New England Journal of Medicine* 361(23):2209.

Useful Websites

www.niaid.nih.gov/topics/immunedeficiency/ The National Institute of Allergy and Infectious Diseases (NIAID) maintains a Web site for learning more about primary immunodeficiency diseases, their treatment, and current research.

www.niaid.nih.gov/topics/hivaids/research/vaccines Another NIAID Web page that includes information about AIDS vaccines and links to documents about vaccines in general.

www.scid.net This Web site contains links to periodicals and databases with information about SCID.

http://hivinsite.ucsf.edu This HIV/AIDS information site is maintained by the University of California, San Francisco, one of many centers of research in this field.

www.unaids.org/en Information about the national and global AIDS epidemic can be accessed from this site.

http://hiv-web.lanl.gov This Web site, maintained by the Los Alamos National Laboratory, contains all available sequence data on HIV and SIV along with up-to-date reviews on topics of current interest to AIDS research.

www.aidsinfo.nih.gov This site, maintained by the National Institutes of Health, contains information and national guidelines on the treatment and prevention of AIDS.

http://clinicaltrials.gov This National Institutes of Health-sponsored Web site is a registry of all private and government funded clinical trials in the United States and worldwide.

 STUDY QUESTIONS

CLINICAL FOCUS QUESTION The spread of HIV/AIDS from infected mothers to infants can be reduced by single-dose regimens of the reverse transcriptase inhibitor nevirapine. What would you want to know before giving this drug to all mothers and infants (without checking infection status) at delivery in areas of high endemic infection?

1. Indicate whether each of the following statements is true or false. If you think a statement is false, explain why.

a. Complete DiGeorge syndrome is a congenital birth defect resulting in absence of the thymus.

b. X-linked agammaglobulinemia (XLA) is a combined B-cell and T-cell immunodeficiency disease.

c. The hallmark of a phagocytic deficiency is increased susceptibility to viral infections.

d. In chronic granulomatous disease, the underlying defect is in phagosome oxidase or an associated protein.

e. Injections of immunoglobulins are given to treat individuals with X-linked agammaglobulinemia.

f. Multiple defects have been identified in human SCID.

g. Mice with the SCID defect lack functional B and T lymphocytes.

h. Mice with SCID-like phenotype can be produced by knockout of *RAG* genes.

i. Children born with SCID often manifest with increased infection by encapsulated bacteria in the first months of life.

j. Failure to express class II MHC molecules in bare-lymphocyte syndrome affects cell-mediated immunity only.

2. For each of the following immunodeficiency disorders, indicate which treatment would be appropriate.

Immunodeficiency

a. Chronic granulomatous disease
b. ADA-deficient SCID
c. X-linked agammaglobulinemia
d. DiGeorge syndrome
e. IL-2R-deficient SCID
f. Common variable immunodeficiency

Treatment

1. Full bone marrow transplantation
2. Pooled human gamma globulin
3. Recombinant IFN-γ
4. Recombinant adenosine deaminase
5. Thymus transplant in an infant

3. Patients with X-linked hyper-IgM syndrome express normal genes for other antibody subtypes but fail to produce IgG, IgA, or IgE. Explain how the defect in this syndrome accounts for the lack of other antibody isotypes.

4. Patients with DiGeorge syndrome are born with either no thymus or a severely defective thymus. In the severe form, the patient cannot develop mature helper, cytotoxic, or regulatory T cells. If an adult suffers loss of the thymus due to accident or injury, little or no T-cell deficiency is seen. Explain this discrepancy.

5. Infants born with SCID experience severe recurrent infections. The initial manifestation in these infants is typically fungal or viral infections, and only rarely bacterial ones. Why are bacterial infections less of an issue in these newborns?

6. In B-cell immunodeficiencies, infections by bacteria are common ailments. However, not all types of bacteria prove equally problematic, and encapsulated bacteria, such as staphylococci, are often the most problematic. Why is this type of bacteria such a problem for individuals who inherit this type of immunodeficiency?

7. Granulocytes from patients with leukocyte adhesion deficiency (LAD) express greatly reduced amounts of CD11a, b, and c adhesion molecules.

a. What is the nature of the defect that results in decreased expression of these adhesion molecules in LAD patients?
b. What is the normal function of the integrin molecule LFA-1? Give specific examples.

8. How can an inherited defect in the IL-2 receptor cause the demise of developing B cells, as well as T cells, if B cells do not possess receptors for IL-2 signaling?

9. Primary immunodeficiency results from a defective component of the immune response. Typically, this manifests as failure to fight off one or more types of pathogen. Very occasionally, immunodeficiency results in autoimmune syndromes, where the immune response attacks self tissues. Explain how this might occur.

10. Immunologists have studied the defect in SCID mice in an effort to understand the molecular basis for severe combined immunodeficiency in humans. In both SCID mice and humans with this disorder, mature B and T cells fail to develop.

a. In what way do rearranged Ig heavy-chain genes in SCID mice differ from those in normal mice?
b. In SCID mice, rearrangement of κ light-chain DNA is not seen. Explain why.
c. If you introduced a rearranged, functional μ heavy-chain gene into progenitor B cells of SCID mice, would the κ light-chain DNA undergo a normal rearrangement? Explain your answer.

11. Indicate whether each of the following statements is true or false. If you think a statement is false, explain why.

a. HIV-1 and HIV-2 are both believed to have evolved following the species jump of SIV from chimpanzees to humans.
b. HIV-1 causes immune suppression in both humans and chimpanzees.
c. SIV is endemic in the African green monkey.
d. The anti-HIV drugs zidovudine and saquinavir both act on the same point in the viral replication cycle.
e. T-cell activation increases transcription of the HIV proviral genome.
f. HIV-infected patients meet the criteria for AIDS diagnosis as soon as their CD4$^+$ T-cell count drops below 500 cells/μl.
g. The polymerase chain reaction is a sensitive test used to detect antibodies to HIV.
h. If HAART is successful, viral load will typically decrease.

12. Various mechanisms have been proposed to account for the decrease in the numbers of CD4$^+$ T cells in HIV-infected individuals. What seems to be the most likely reason for depletion of CD4$^+$ T cells?

13. Would you expect the viral load in the blood of HIV-infected individuals in the early years of the asymptomatic phase of HIV-1 infection to vary significantly (assuming no drug treatment)? What about CD4$^+$ T-cell counts? Why?

14. If viral load begins to increase in the blood of an HIV-infected individual and the level of CD4$^+$ T cells decrease, what would this indicate about the infection?

15. Clinicians often monitor the level of skin-test reactivity, or delayed-type hypersensitivity, to commonly encountered infectious agents in HIV-infected individuals. Why do you think these immune reactions are monitored, and what change might you expect to see in skin-test reactivity with progression into AIDS?

16. Certain chemokines have been shown to suppress infection of cells by HIV, and proinflammatory cytokines enhance cell infection. What is the explanation for this?

17. Treatments with combinations of anti-HIV drugs (HAART) have reduced virus levels significantly in some treated patients and can delay the onset of AIDS. If an AIDS patient becomes free of opportunistic infection and has no detectable virus in the circulation, can that person be considered cured?

18. Suppose you are a physician who has two HIV-infected patients. Patient B. W. has a bacterial infection of the skin (*S. aureus*), and patient L. S. has a mycobacterial infection. The CD4$^+$ T-cell counts of both patients are about 250 cells/μl. Would you diagnose either patient or both of them as having AIDS?

19. **ANALYZE THE DATA.** Common variable immunodeficiency (CVID) causes low concentrations of serum Igs and leads to frequent bacterial infections in the respiratory and gastrointestinal tracts. People with CVID also have an increased prevalence of autoimmune disorders and cancers. Isgrò and colleagues (*Journal of Immunology*, 2005, **174**:5074) examined the bone marrow of several individu-

als with CVID. They looked at the T-cell phenotypes of CVID patients (as shown in the following table) as well as some of the cytokines made by these individuals (see accompanying graphs).

a. What is the impact of CVID on T helper cells?
b. Naïve CTLs require IL-2 production from T helper cells in order to become activated (see Chapter 14). How might CVID impact the generation of CTLs?
c. True or false? CVID inhibits cytokine production. Explain your answer. Speculate on the physiological impact of the cytokine pattern of CVID patients.
d. Would you predict an effect of CVID on the humoral immune response?

						Patients							Control
		1	**2**	**5**	**6**	**7**	**8**	**9**	**10**	**11**	**CVID**	**(*n* = 10)**
CD4$^+$	%	47	36	28	28	27	19	19	32	57	34	47.5
	cells/μl	296	234	278	1652	257	361	289	248	982	351	1024
Naïve CD4$^+$	%	4	6.8	25.8	4	11.6	7.9	14	12	2	10	52
	cells/μl	12	16	72	66	30	29	40	30	20	31	519
Activated CD4$^+$	%	2	5	22	3	9	8	12	9	1.5	8	37
	cells/μl	6	12	61	50	23	29	35	22	15	25	385
CD8$^+$	%	31	38	30	57	44	56	47	45	21	39	20
	cells/μl	195	247	298	3363	420	1065	714	348	362	414	404
Naïve CD8$^+$	%	25	30	43.9	7	16.9	12.9	20	13	15	22	58
	cells/μl	49	74	131	235	71	138	143	45	54	88	233

(a)

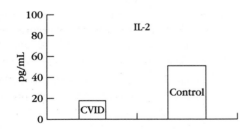

(b)

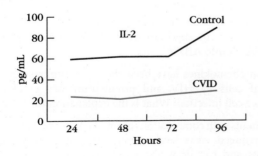
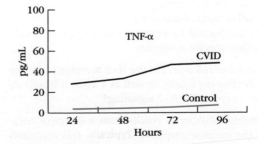

Cancer and the Immune System

A s the death toll from infectious disease has declined in the Western world, cancer has become the second leading cause of death, topped only by heart disease. Current estimates project that half of all men and one in three women in the United States will develop cancer at some point in their lifetimes, and that one in five will die from it. From an immunologic perspective, cancer cells can be viewed as altered self cells that have escaped normal growth-regulating mechanisms. This chapter examines the unique properties of cancer cells, giving particular attention to those properties that can be recognized by the immune system. We then describe the immune responses that develop against cancer cells, as well as the methods by which cancers manage to evade those responses. The final section surveys current clinical and experimental immunotherapies for cancer.

In most organs and tissues of a mature animal, a balance is maintained between cell renewal and cell death. The various types of mature cells in the body have a given life span; as these cells die, new cells are generated by the proliferation and differentiation of various types of stem cells. This cell growth and proliferation are essential for wound healing and homeostasis. Under normal circumstances in the adult, the production of new cells is regulated so that the number of any particular type of cell remains fairly constant. Occasionally, however, cells arise that no longer respond to normal growth control mechanisms; these cells proliferate in an unregulated manner, giving rise to cancer. In the following sections, we first cover common terminology related to cancer, and then discuss the pathways that can lead to this uncontrolled cell growth.

Terminology and Common Types of Cancer

Cells that give rise to clones of cells that can expand in an uncontrolled manner will produce a tumor or **neoplasm**. A tumor that is not capable of indefinite growth and does not invade the healthy surrounding tissue extensively is said to

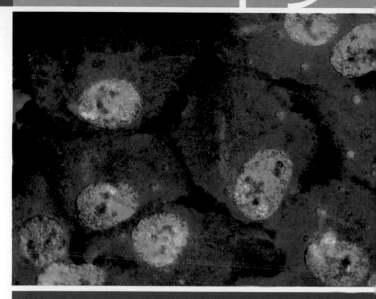

A cluster of prostate cancer cells, stained to visulize the nuclei (green), Golgi apparatus (pink), and actin filaments (purple). *[Nancy Kedersha/Getty Images]*

- Terminology and Common Types of Cancer
- Malignant Transformation of Cells
- Tumor Antigens
- The Immune Response to Cancer
- Cancer Immunotherapy

be **benign**. A tumor that continues to grow and becomes progressively more invasive is called **malignant**; the term *cancer* refers specifically to a malignant tumor. In addition to uncontrolled growth, malignant tumors exhibit **metastasis**, whereby small clusters of cancerous cells dislodge from the original tumor, invade the blood or lymphatic vessels, and are carried to other distant tissues, where they take up residence and continue to proliferate. In this way, a primary tumor at one site can give rise to a secondary tumor at another site (Figure 19-1).

Malignant tumors or cancers are classified according to the embryonic origin of the tissue from which the tumor is derived. Most (80–90%) are **carcinomas**, tumors that arise from epithelial origins such as skin, gut, or the epithelial lining of internal organs and glands. Skin cancers and the majority of cancers of the colon, breast, prostate, and lung are carcinomas. The **leukemias**, **lymphomas**, and **myelomas** are malignant tumors of hematopoietic cells derived from the bone marrow and account for about 9% of cancer incidence

Tumor Growth and Metastasis

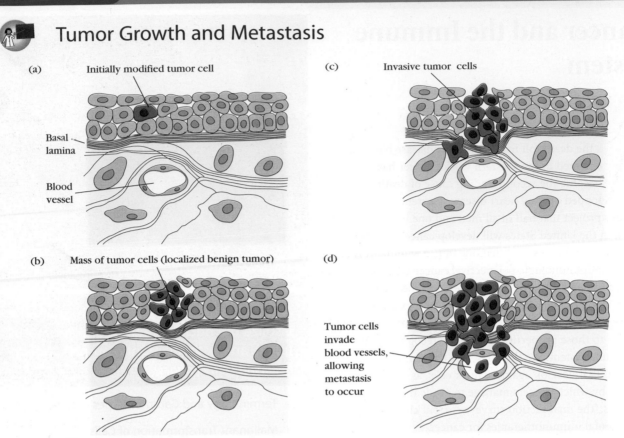

(a) A single cell develops altered growth properties at a tissue site. (b) The altered cell proliferates, forming a mass of localized tumor cells, or a benign tumor. (c) The tumor cells become progressively more invasive, spreading to the underlying basal lamina. The tumor is now classified as malignant. (d) The malignant tumor metastasizes by generating small clusters of cancer cells that dislodge from the tumor and are carried by the blood or lymph to other sites in the body. [Adapted from J. Darnell et al., 1990, Molecular Cell Biology, 2nd ed., Scientific American Books.]

in the United States. Leukemias proliferate as single detached cells, whereas lymphomas and myelomas tend to grow as tumor masses. **Sarcomas**, which arise less frequently (around 1% of the incidence of cancer in the United States), are derived from mesodermal connective tissues, such as bone, fat, and cartilage.

Historically, the leukemias were classified as acute or chronic according to the clinical progression of the disease. The acute leukemias appeared suddenly and progressed rapidly, whereas the chronic leukemias were much less aggressive and developed slowly as mild, barely symptomatic diseases. These clinical distinctions apply to untreated leukemias; with current treatments, the acute leukemias often have a good prognosis, and permanent remission is possible. Now the major distinction between acute and chronic leukemias is the maturity of the cell involved. Acute leukemias tend to arise in less mature cells, whereas chronic leukemias arise in mature cells, although each can arise from lymphoid

or myeloid lineages. The acute leukemias include **acute lymphocytic leukemia (ALL)** and **acute myelogenous leukemia (AML)**. These diseases can develop at any age and have a rapid onset. The chronic leukemias include **chronic lymphocytic leukemia (CLL)** and **chronic myelogenous leukemia (CML)**, which develop more slowly and are seen primarily in adults.

Malignant Transformation of Cells

Much has been learned about cancer from in vitro studies of primary cells. Treatment of normal cultured cells with specific chemical agents, irradiation, and certain viruses can alter their morphology and growth properties. In some cases, this process leads to unregulated growth and produces cells capable of growing as tumors when they are injected into animals. Such cells are said to have undergone **transformation**, or malignant transformation, and they often exhibit properties in vitro similar to those of cancer cells that form in vivo. For example, they have decreased requirements for the survival factors required by most cells (such as growth factors and serum), are no longer anchorage dependent, and grow in a density-independent fashion. Moreover, both cancer cells and transformed cells can be subcultured in vitro indefinitely; that is, for all practical purposes, they are immortal. Because of the similar properties of cancer cells and transformed cells, the process of malignant transformation has been studied extensively as a model of cancer induction.

DNA Alterations Can Induce Malignant Transformation

Transformation can be induced by various chemical substances (such as formaldehyde, DDT, and some pesticides), physical agents (e.g., asbestos), and ionizing radiation; all are linked to DNA mutations. For this reason, these agents are commonly referred to as **carcinogens**. The International Agency for Research on Cancer (IARC) tracks the most common causes of cancer in humans and maintains lists of potential cancer-causing agents (see Useful Web Sites). Infection with certain viruses, most of which share the property of integrating into the host cell genome and disrupting chromosomal DNA, can also lead to transformation. Table 19-1 lists the most common viruses associated with cancer. Although the activity of each of these agents is associated with cancer initiation, thanks to the variety of DNA repair mechanisms present in our cells, exposure to carcinogens does not always lead to cancer. Instead, a confluence of factors must occur before enough changes to normal cellular genes occurs to

TABLE 19-1	Common human infectious carcinogens		
Viral Agent		**Type**	**Cancer**
HTLV-1 Human T-cell leukemia virus -1		RNA	Adult T-cell leukemia or lymphoma
HHV-8 Human herpesvirus-8		DNA	Kaposi's sarcoma (especially in HIV+)
HPV Human papillomavirus		DNA	Cervical carcinoma
HBV and HCV Hepatitis B and C viruses		DNA	Liver carcinoma
EBV Epstein-Barr virus		DNA	Burkitt's lymphoma and nasopharyngeal carcinoma

Sources: The National Institute for Occupational Safety and Health (NIOSH) and the Centers for Disease Control and Prevention (CDC). [http://www.cdc.gov/niosh/topics/cancer/]

induce malignant transformation, as we discuss further in the following sections.

The Discovery of Oncogenes Paved the Way for Our Understanding of Cancer Induction

The discovery that disruption of normal cellular genes can lead to cancer came from studies done with cancer-causing viruses. In 1916, Peyton Rous, working at the Rockefeller Institute, showed that cells cultured with a particular retrovirus underwent malignant transformation. This virus was later named Rous sarcoma virus. Fifty years after his initial results, Rous was awarded the Nobel Prize in Physiology or Medicine for his discoveries, which led to the term **oncogene**, from the Greek word *ónkos* (which means "mass" or "tumor"). This term was originally used to describe the unique genetic material present in certain viruses that promotes malignant transformation of infected cells.

In 1971, Howard Temin suggested that oncogenes might not be unique to transforming viruses but might also be found in normal cells. Temin proposed that a virus might acquire a normal growth-promoting gene from the genome of a cell it infects. He called these normal cellular genes **proto-oncogenes** to distinguish them from the cancer-promoting sequences carried by some viruses (**viral oncogenes**, or v-*onc*). The following year, R. J. Huebner and G. J. Todaro proposed that mutations or genetic rearrangements of in situ proto-oncogenes by carcinogens or viruses might alter the normally regulated function of these genes, converting them into cancer inducers, called **cellular oncogenes** (c-*onc*; Figure 19-2). Considerable evidence supporting this hypothesis accumulated in subsequent years. For example, some malignantly transformed cells contain multiple copies of cellular oncogenes, leading to increases in oncogene products. Such amplification of cellular oncogenes has been observed in cells from various types of human cancers.

In the mid-1970s, J. Michael Bishop and Harold E. Varmus at the University of California, San Francisco, again used the Rous retrovirus to establish the origins of these viral oncogenes. Retroviruses have RNA genomes that must be reverse-transcribed into DNA during their life cycle, followed by integration of this DNA into the host-cell chromosome. Retroviruses therefore make intimate contact with the genomes of their host cell, especially those genes that happen to become their neighbors following chromosomal integration. In elegant studies that led to another Nobel Prize in 1989, Bishop and Varmus demonstrated that the oncogene carried by Rous sarcoma virus (later called v-*src*) was in fact just a version of a normal growth-promoting cellular gene (c-*src*) found throughout many species and probably acquired previously from a host cell during viral replication. By randomly incorporating this normal cellular gene into its own genome, the Rous sarcoma virus acquired the ability to induce malignant transformation in the cells it infected, giving it a selective advantage. The evidence that oncogenes alone could induce malignant transformation

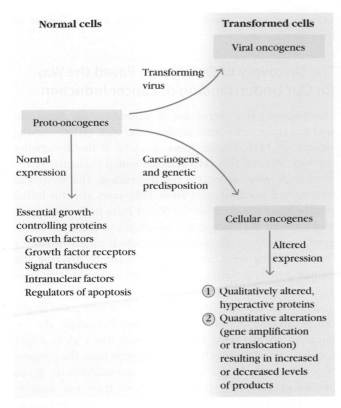

FIGURE 19-2 The relationship of proto-oncogenes to oncogenes. The process of malignant transformation can result from infection with transforming viruses, which carry oncogene sequences, or via carcinogen-induced conversion of a wild-type proto-oncogene into a cellular oncogene, sometimes compounded by genetic predisposition in the host cell. Cellular oncogenes arise due to DNA changes that alter expression of proto-oncogenes, including DNA mutations that result in the production of qualitatively different gene products (1), as well as DNA amplification or chromosomal translocation, leading to increased or decreased expression of these gene products (2).

came again from studies of the v-*src* oncogene; when this oncogene was cloned and transfected into normal cells in culture, the cells underwent malignant transformation.

Genes Associated with Cancer Control Cell Proliferation and Survival

Normal tissues maintain homeostasis through a tightly regulated process of cell proliferation balanced by cell death. An imbalance at either end of the scale encourages development of a cancerous state. The genes involved in these homeostatic processes work by producing proteins that either *encourage* or *discourage* cellular proliferation and survival. Not surprisingly, it is the disruption of these same growth-regulating genes that accounts for most, if not all, forms of cancer. The activities of these proteins can occur anywhere in the pathway, from signaling events at the surface of the cell, to intracellular signal transduction processes and nuclear events.

Normal cellular genes that are associated with the formation of cancer fall into three major categories based on their activities: oncogenes, tumor-suppressor genes, and genes involved in programmed cell death, or apoptosis (Table 19-2). As discussed earlier, oncogenes are involved in cell growth-promoting processes. On the flip side, **tumor-suppressor genes** play the opposite role in homeostasis, dampening cellular growth and proliferation. Unlike oncogenes, which become the villain when their activity is enhanced, tumor-suppressor genes, also known as **anti-oncogenes**, become involved in cancer induction when they *fail*. Finally, many of the genes involved in apoptosis have also been associated with cancer, as these genes either enforce or inhibit cell death signals.

Recently, several unifying hallmarks have been proposed that help to characterize the transition of non-neoplastic cells to a cancerous state. These include sustained proliferation, subversion of negative growth regulators, and resistance to cell death—controlled, respectively, by the activity of oncogenes, tumor-suppressor genes, and genes that regulate apoptosis. Other proposed hallmarks of cancer include replicative immortality, the growth of new blood vessels (also called angiogenesis), and the potential to invade other tissues (metastasis). Two underlying factors that help to *enable* the development of these hallmarks—genetic instability and inflammation—are also now recognized contributors to tumorigenesis. In the last several years, the significance of specific cells and chemical signals associated with the immune response in both suppressing and encouraging tumor development has been documented. Many of these will be discussed further in the section later in this chapter entitled "The Immune Response to Cancer."

Cancer-Promoting Activity of Oncogenes

The proteins encoded by a particular oncogene and its corresponding proto-oncogene have a very similar function. Sequence comparisons of viral and cellular oncogenes reveal that they are highly conserved in evolution. Although most cellular oncogenes consist of the typical series of exons and introns, their viral counterparts consist of uninterrupted coding sequences, suggesting that the virus acquired the oncogene through an intermediate RNA transcript from which the intron sequences were removed during RNA processing. The actual coding sequences of viral oncogenes and the corresponding proto-oncogenes exhibit a high degree of homology. In some cases, a single point mutation is all that distinguishes a viral oncogene from the corresponding proto-oncogene. In many cases, the conversion of a proto-oncogene into an oncogene accompanies a mere change in the *level* of expression of a normal growth-controlling protein.

One category of proto-oncogenes and their viral counterparts encodes growth factors or growth factor receptors. Included among these are *sis*, which encodes a form of platelet-derived growth factor, and *fms*, *erbB*, and *neu*, which encode growth factor receptors (see Table 19-2). In normal

TABLE 19-2	Functional classification of cancer-associated genes
Type/name	Nature of gene product
CATEGORY I: GENES THAT INDUCE CELLULAR PROLIFERATION	
Growth factors	
sis	A form of platelet-derived growth factor (PDGF)
Growth factor receptors	
fms	Receptor for colony-stimulating factor 1 (CSF-1)
erbB	Receptor for epidermal growth factor (EGF)
neu	Protein (HER2) related to EGF receptor
erbA	Receptor for thyroid hormone
Signal transducers	
src	Tyrosine kinase
abl	Tyrosine kinase
Ha-ras	GTP-binding protein with GTPase activity
N-ras	GTP-binding protein with GTPase activity
K-ras	GTP-binding protein with GTPase activity
Transcription factors	
jun	Component of transcription factor AP1
fos	Component of transcription factor AP1
myc	DNA-binding protein
CATEGORY II: TUMOR SUPRESSOR GENES, INHIBITORS OF CELLULAR PROLIFERATION*	
Rb	Suppressor of retinoblastoma
TP53	Nuclear phosphoprotein that inhibits formation of small-cell lung cancer and colon cancers
DCC	Suppressor of colon carcinoma
APC	Suppressor of adenomatous polyposis
NF1	Suppressor of neurofibromatosis
WT1	Suppressor of Wilm's tumor
CATEGORY III: GENES THAT REGULATE PROGRAMMED CELL DEATH	
bcl-2	Suppressor of apoptosis
Bcl-x_L	Suppressor of apoptosis
Bax	Inducer of apoptosis
Bim	Inducer of apoptosis
Puma	Inducer of apoptosis

*The activity of the normal products of the category II genes inhibits progression of the cell cycle. Loss of a gene or its inactivation by mutation in an indicated tumor-suppressor gene is associated with development of the indicated cancers.

cells, the expression of growth factors and their receptors is carefully regulated. Usually, one population of cells secretes a growth factor that acts on another population of cells carrying the receptor for the factor, thus stimulating proliferation of the second population. Inappropriate expression of either a growth factor or its receptor can result in uncontrolled proliferation. In breast cancer, increased synthesis of the growth factor receptor encoded by c-*neu* has been linked with a poor prognosis.

Other oncogenes in this category encode products that function in signal transduction pathways or as transcription factors. The *src* and *abl* oncogenes encode tyrosine kinases, and the *ras* oncogene encodes a GTP-binding protein (see Table 19-2). The products of these genes act as signal transducers. The *myc*, *jun*, and *fos* oncogenes all encode transcription factors. Overactivity of any of these oncogenes may result in unregulated proliferation.

Some transformations arise due to chromosomal translocations that bring proto-oncogenes under the control of other genomic regions. For instance, a number of B- and T-cell leukemias and lymphomas arise from translocations involving immunoglobulin or T-cell receptor loci, very transcriptionally active regions in these cells. One of the best characterized is the translocation of c-*myc* from its position on chromosome 8 to the immunoglobulin heavy-chain enhancer region on chromosome 14, accounting for 75% of Burkitt's lymphoma cases (Figure 19-3). In the remaining patients with Burkitt's lymphoma, c-*myc* remains on chromosome 8 but the κ or γ light-chain genes are translocated to c-*myc* on the tip of chromosome 8. As a result of these translocations, synthesis of the c-Myc protein, which functions as a transcription factor controlling the behavior of many genes involved in cell growth and proliferation,

increases and allows unregulated cellular growth. The consequences of enhanced and constitutive *myc* expression in lymphoid cells have been investigated in transgenic mice. In one study, mice were engineered to express a transgene consisting of all three c-*myc* exons and the immunoglobulin heavy-chain enhancer. Of 15 transgenic pups born, 13 developed lymphomas of the B-cell lineage within a few months of birth.

Cellular transformation has also been associated with mutations in proto-oncogenes. This may be a major mechanism by which chemical carcinogens or x-irradiation convert a proto-oncogene into a cancer-inducing oncogene. For instance, single-point mutations in c-*ras*, which encodes a GTPase, have been detected in carcinomas of the bladder, colon, and lung (see Table 19-2). Alterations that result in overactivity of the *ras* oncogene, part of the *epidermal growth factor* (EGF) receptor signaling pathway, are seen in up to 30% of all human cancers and nearly 90% of all pancreatic cancers.

Viral integration into the host-cell genome may in itself serve to convert a proto-oncogene into a transforming oncogene. For

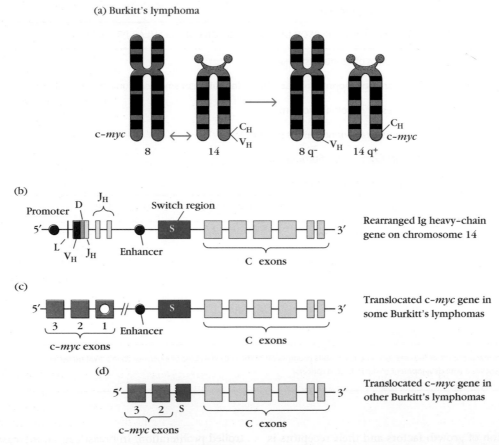

(a) Burkitt's lymphoma

FIGURE 19-3 Chromosomal translocations resulting in Burkitt's lymphoma. (a) Most cases of Burkitt's lymphoma arise following a chromosomal translocation that moves part of chromosome 8, containing the c-*myc* gene, to the Ig heavy-chain locus on chromosome 14, shown in more detail in (b). (c) In some cases, the entire c-*myc* gene is inserted near the Ig heavy-chain enhancer. (d) In other cases, only the coding exons (2 and 3) of c-*myc* are inserted at the μ switch site. The proto-oncogene *myc* encodes a transcription factor involved in the regulation of many genes with roles in cell growth, proliferation, and apoptosis.

example, *a*vian *l*eukosis *v*irus (ALV) is a retrovirus that does not carry any viral oncogenes, yet it is able to transform B cells into lymphomas. This particular retrovirus has been shown to integrate within the c-*myc* proto-oncogene, resulting in increased synthesis of c-Myc. Some DNA viruses have also been associated with malignant transformation, such as certain serotypes of the *h*uman *p*apilloma*v*irus (HPV), which are linked with cervical cancer (see Table 19-1).

Tumor-Suppressor Genes

A second category of cancer-associated genes—called tumor-suppressor genes, or anti-oncogenes—encodes proteins that *inhibit* excessive cell proliferation. In their normal state, tumor-suppressor genes prevent cells from progressing through the cell cycle inappropriately, functioning like brakes. A release of this inhibition is what can lead to cancer induction. The prototype of this category of oncogenes is *Rb*, the retinoblastoma gene (see Table 19-2). Hereditary retinoblastoma is a rare childhood cancer in which tumors develop from neural precursor cells in the immature retina. The affected child inherits a mutated *Rb* allele; later, somatic inactivation of the remaining *Rb* allele is what leads to tumor growth. Unlike oncogenes where a single allele alteration can lead to unregulated growth, tumor-suppressor genes require a "two-hit" disabling sequence, as the one functional allele is enough to suppress the development of cancer.

Probably the single most frequent genetic abnormality in human cancer, found in 60% of all tumors, is a mutation in the *TP53* gene. This tumor-suppressor gene encodes p53, a nuclear phosphoprotein with multiple cellular roles, including involvement in growth arrest, DNA repair, and apoptosis. Over 90% of small-cell lung cancers and over 50% of breast and colon cancers have been shown to be associated with mutations in *TP53*.

The Role of Apoptotic Genes

A third category of cancer-associated genes is sequences involved in programmed cell death, or apoptosis. These genes can encode proteins that either induce or block apoptosis. Pro-apoptotic genes act like tumor suppressors, normally inhibiting cell survival, whereas anti-apoptotic genes behave more like oncogenes, promoting cell survival. Thus, a failure of the former or overactivity of the latter can encourage neoplastic transformation of cells.

Included in this category are genes such as *bcl-2*, an anti-apoptosis gene (see Table 19-2). This oncogene was originally discovered because of its association with B-cell follicular lymphoma. Since its discovery, *bcl-2* has been shown to play an important role in regulating cell survival during hematopoiesis and in the survival of selected B cells and T cells during maturation. Interestingly, the Epstein-Barr virus (which can cause infectious mononucleosis) contains a gene that has sequence homology to *bcl-2* and may act in a similar manner to suppress apoptosis.

Malignant Transformation Involves Multiple Steps

The development from a normal cell to a cancerous cell is typically a multistep process of clonal evolution driven by a series of somatic mutations. These mutations progressively convert the cell from normal growth to a precancerous state and finally a cancerous state, where all barriers designed to contain cell growth have been surmounted. Induction of malignant transformation appears to involve at least two distinct phases: initiation and promotion. Initiation involves changes in the genome but does not, in itself, lead to malignant transformation. Malignant transformation can occur during the rampant cell division that follows the initiation phase, and results from the accumulation of new DNA alterations, typically affecting proto-oncogenes, tumor-suppressor genes or apoptotic genes, that lead to truly unregulated cellular growth.

The presence of myriad chromosomal abnormalities in precancerous and cancerous cells lends support to the role of multiple mutations in the development of cancer. This has been demonstrated in human colon cancer, which typically progresses in a series of well-defined morphologic stages (Figure 19-4). Colon cancer begins as small, benign tumors called adenomas in the colorectal epithelium. These precancerous tumors grow, gradually displaying increasing levels of intracellular disorganization until they acquire the malignant phenotype (Figure 19-4b). The morphologic stages of colon cancer have been correlated with a sequence of gene changes (Figure 19-4a) involving inactivation or loss of three tumor-suppressor genes (*APC*, *DCC*, and *TP53*) and activation of one cellular proliferation oncogene (*K-ras*).

Studies with transgenic mice also support the role of multiple steps in the induction of cancer. Transgenic mice expressing high levels of Bcl-2, a protein encoded by the anti-apoptotic gene *bcl-2*, develop a population of small resting B cells (derived from secondary lymphoid follicles) that have greatly extended life spans. Gradually, these transgenic mice develop lymphomas. Analysis of lymphomas from these mice has shown that approximately half have a c-*myc* translocation (a proto-oncogene) to the immunoglobulin H-chain locus. The synergism of Myc and Bcl-2 is highlighted in double-transgenic mice produced by mating the *bcl-2*$^+$ transgenic mice with *myc*$^+$ transgenic mice. These mice develop leukemia very rapidly.

The role of DNA mutations and the progressive genomic changes that can lead to the induction of cancer are clearly illustrated by diseases such as *x*eroderma *p*igmentosum (XP). This rare disorder is inherited in an autosomal recessive manner and is caused by defects in any one of a number of genes involved in normal DNA repair, collectively known as the nucleotide excision repair (NER) pathway. Individuals with this disease are unable to repair *u*ltraviolet (UV)-induced mutations, which occur naturally with even moderate exposure to sunlight and are typically repaired via the NER pathway. A buildup of unrepaired DNA in the

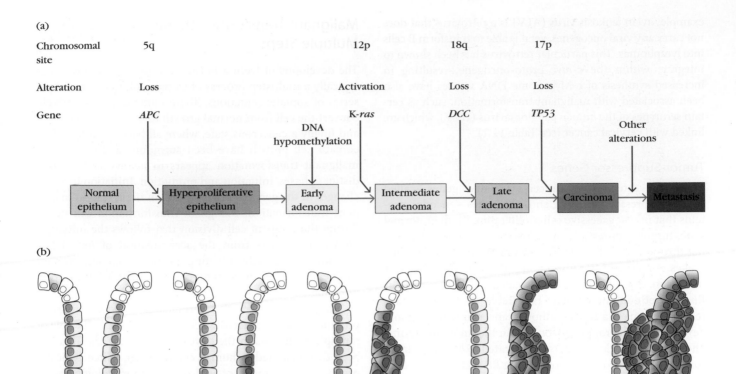

FIGURE 19-4 Model of sequential genetic alterations leading to metastatic colon cancer. Each of the stages indicated in (a) is morphologically distinct, as illustrated in (b), allowing researchers to determine the sequence of genetic alterations. In this sequence, benign colorectal polyps progress to carcinoma following mutations resulting in the inactivation or loss of three tumor-suppressor genes (*APC*, *DCC*, and *TP53*) and the activation of one oncogene linked to cellular proliferation (*K-ras*). [(a) Adapted from B. Vogelstein and K. W. Kinzler, 1993, The multistep nature of cancer, Trends in Genetics 9:138. (b) Adapted from P. Rizk and N. Barker, 2012, Gut stem cells in tissue renewal and disease: Methods, markers, and myths, Systems Biology and Medicine 4:5, 475–496.]

cutaneous cells of young children with XP leads to random genomic alterations, including to genes involved in regulating normal cell growth and division. This leads to unregulated growth of some cells, allowing further DNA mutations to occur and promoting the development of neoplasms, such as malignant melanoma or squamous cell carcinoma, the most common forms of skin cancer in XP patients. The mean age of skin cancer in children with XP is age 8, as compared to 60+ years of age in the general population. Figure 19-5 shows a child with the early skin manifestations common to XP.

Accumulating data from several recent studies suggest that within a growing tumor, not all cells have equal potential for unlimited growth. Clinical studies of at least three different types of tumors, including those originating in the gut, brain, and skin, all suggest that a subset of cells within a tumor may be the true engines of tumor growth. This subset (called cancer stem cells) displays true unlimited regenera-

tive potential, and is the major producer of the other cancer cells, populating the mass of the tumor by sharing this unrestrained ability to divide. If these new discoveries prove valid and can be applied more broadly, future therapies will undoubtedly require that we hone in on these rare cancer stem cells, cutting off the source of tumor cell expansion and, potentially, providing hope for cancer eradication.

Tumor Antigens

Neoplastic cells are self cells and thus most of the antigens associated with them are subject to the tolerance-inducing processes that maintain homeostasis and inhibit the development of autoimmunity. However, unique or inappropriately expressed antigens can be found in many tumors and are frequently detected by the immune system. Some of these antigens may be the products of oncogenes, where

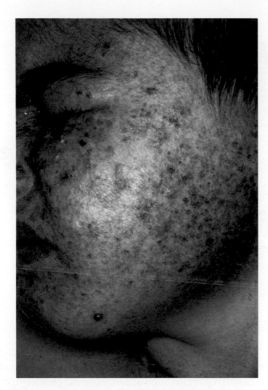

FIGURE 19-5 Xeroderma pigmentosum. This rare autosomal-recessive inherited disorder arises from mutations to one of several genes involved in DNA repair. This disorder is characterized by extreme skin sensitivity to ultraviolet light, abnormal skin pigmentation, and a high frequency of skin cancers, especially on sun-exposed skin of the face, neck and arms. *[CID/ISM/Phototake]*

TABLE 19-3	Examples of common tumor antigens	
Category	**Antigen/s**	**Associated cancer types**
Tumor-Specific Antigens (TSAs)		
Viral	HPV: L1, E6, E7	Cervical carcinoma
	HBV: HBsAg	Hepatocellular carcinoma
	SV40: Tag	Malignant pleural mesothelioma (cancer of the lung lining)
Tumor-Associated Antigens (TAAs)		
Overexpression	MUC1	Breast, ovarian
	MUC13/CA-125	Ovarian
	HER-2/neu	Breast, melanoma, ovarian, gastric, pancreatic
	MAGE	Melanoma
	PSMA	Prostate
	TPD52	Prostate, breast, ovarian
Differentiation Stage	CEA	Colon
	Gp100	Melanoma
	AFP	Hepatocellular carcinoma
	Tyrosinase	Melanoma
	PSA	Prostate
	PAP	Prostate

Abbreviations: SV40, simian virus 40; L, late gene; E, early gene; HBsAg, hepatitis B surface antigen; Tag, large tumor antigen; MUC, mucin; MAGE, melanoma-associated antigen; HER/neu, human epidermal receptor/neurological; PSMA, prostate-specific membrane antigen; TP, tumor protein; PSA, prostate-specific antigen; PAP, prostatic acid phosphatase.

Source: Adapted from Table 1 in J. F. Aldrich, et al., 2010. Vaccines and immunotherapeutics for the treatment of malignant disease, *Clinical and Developmental Immunology.*

there is no qualitative difference between the oncogene and proto-oncogene products; instead, it is merely increased levels of the oncogene product that can be recognized by the immune system.

Most tumor antigens give rise to peptides that are recognized by the immune system following presentation by self *m*ajor *h*istocompatibility *c*omplex (MHC) molecules. In fact, many of these antigens have been identified by their ability to induce the proliferation of antigen-specific *c*ytotoxic *T* *l*ymphocytes (CTLs) or helper T cells. To date, tumor antigens recognized by human T cells fall into one of four groups based on their source:

- Antigens encoded by genes exclusively expressed by tumors (e.g., viral genes)
- Antigens encoded by variant forms of normal genes that are altered by mutation
- Antigens normally expressed only at certain stages of development
- Antigens that are overexpressed in particular tumors

Table 19-3 lists several categories of common antigens associated with tumors. As one can imagine, many clinical research studies aim to utilize these antigens as diagnostic or prognostic indicators, as well as therapeutic targets for tumor elimination. There are two main types of tumor antigens, categorized by their uniqueness: **tumor-specific antigens (TSAs)** and **tumor-associated antigens (TAAs)**. Originally these were designated as *t*ransplantation antigens (TSTAs and TATAs), stemming from studies in which these antigens were discovered by transplanting them into recipient animals, inducing a rejection immune response.

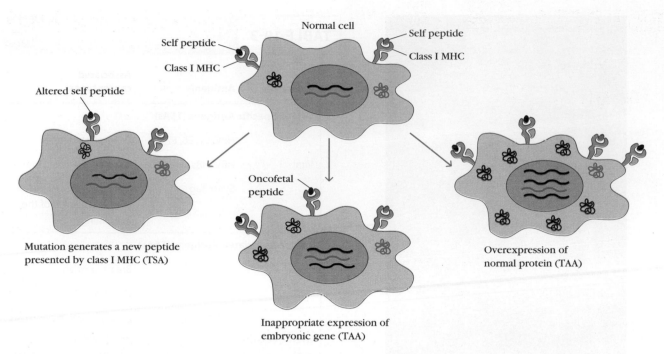

FIGURE 19-6 Different mechanisms generate tumor-specific antigens (TSAs) and tumor-associated antigens (TAAs). TSAs are unique to tumor cells, and can result from DNA mutations that lead to expression of altered self proteins or from expression of viral antigens (not shown) in transformed cells. TAAs are more common and represent normal cellular proteins displaying unique expression patterns, such as an embryonic protein expressed in the adult or overexpression of self proteins. Both types of tumor antigens can be detected by the immune system following presentation in class I MHC.

Tumor-Specific Antigens Are Unique to Tumor Cells

TSAs are unique proteins that may result from mutations in tumor cells that generate altered proteins and, therefore, new antigens. Cytosolic processing of these proteins then gives rise to novel peptides that are presented with class I MHC molecules (Figure 19-6), inducing a cell-mediated response by tumor-specific CTLs. TSAs have been identified on tumors induced with chemical or physical carcinogens, as well as on some virally induced tumors.

Demonstrating the presence of TSAs on spontaneously occurring or chemically induced tumors is particularly difficult. These antigens can be quite diverse and are only identified by their ability to induce T-cell-mediated rejection. The immune response to such tumors typically eliminates all of the tumor cells bearing sufficient numbers of these unique antigens, and thus selects for cells bearing few or none. Nonetheless, experimental methods have been developed to facilitate the characterization of TSAs, which have been shown to differ from normal cellular proteins by as little as a single amino acid. Further characterization of TSAs has demonstrated that many of these antigens are not cell membrane proteins; rather, they are short peptides derived from cytosolic proteins that have been processed and presented together with class I MHC molecules.

In contrast to chemically induced tumors, virally induced tumors express tumor antigens shared by all tumors induced by the same virus, making their characterization simpler. For example, when syngeneic mice are injected with killed cells from a particular polyoma virus-induced tumor, the recipients are protected against subsequent challenge with live cells from any polyoma-induced tumors. Likewise, when lymphocytes are transferred from mice with a virus-induced tumor into normal syngeneic recipients, the recipients reject subsequent transplants of all syngeneic tumors induced by the same virus, suggesting that lymphocytes recognize and kill cells expressing a virally derived TSA.

In some cases, the presence of virus-specific tumor antigens is an indicator of neoplastic transformation. In humans, Burkitt's lymphoma cells have been shown to express a nuclear antigen of the Epstein-Barr virus that may indeed be a tumor-specific antigen for this type of tumor. HPV E6 and E7 proteins are found in more than 80% of invasive cervical cancers, and they provide the clearest example of a virally encoded tumor antigen. In fact, the first clinically approved cancer vaccine is used to prevent infection with HPV and block the emergence of cervical cancer (see Clinical Focus Box 19-1).

Tumor-Associated Antigens Are Normal Cellular Proteins with Unique Expression Patterns

In contrast to TSAs, TAAs are not unique to the cancer. Instead, these represent normal cellular proteins typically

A Vaccine to Prevent Cervical Cancer, and More

Globally, cervical cancer is the third leading cause of death among women, and second only to breast cancer in terms of cancer deaths in women. Over 500,000 women each year develop cervical cancer (80% in developing countries), and approximately 275,000 women die from the disease annually. Periodic cervical examination (using the Papanicolaou test, or Pap smear) to detect abnormal cervical cells significantly reduces the risk for women. However, a health-care program that includes regular Pap smears is commonly beyond the means of the less advantaged and is largely unavailable in many developing countries.

Human papillomavirus (HPV), the most common sexually transmitted infection, is implicated in over 99% of cervical cancers. HPV is also associated with most cases of vaginal, vulval, anal, penile, and oropharyngeal (head and neck) cancers, as well as genital warts. Among the hundred-plus genotypes of HPV, approximately 40 are associated with genital or oral infections. Two of these, types 16 and 18, account for more than 70% of all instances of cervical cancer, whereas types 6 and 11 are most often involved in HPV-associated genital warts.

It is estimated that the majority of sexually active men and women become infected with HPV at some point in their lives. A study of female college students at the University of Washington published in 2003 showed that after 5 years, more than 60% of study participants (all of whom were HPV negative when enrolled in the study) became infected. Most infections are resolved without disease; it is persistent infection leading to cervical or anal intraepithelial neoplasia that is associated with high cancer risk.

Preventing cervical cancer, therefore, appears to be a matter of preventing HPV infection. Gardasil (manufactured by Merck), the first vaccine ever approved for the prevention of cancer, was licensed in 2006 for the prevention of infection with HPV and potential development of cervical cancer or genital warts. This quadrivalent formulation targets HPV types 6, 11, 16, and 18. Three years later, GlaxoSmithKline received a

license for Cervarix, a vaccine to prevent cervical cancer that targets only HPV types 16 and 18. These vaccines are between 95% and 99% effective in preventing infection by HPV. Conclusive evidence that this will translate into significantly reduced rates of cervical cancer in women, which can take many years to develop, will not be available until long-term follow-up studies have been completed.

In June 2006, the federal *Advisory Committee on Immunization Practices* (ACIP) recommended routine HPV vaccination for girls ages 11 to 12, and catch up immunizations for females ages 13 to 26 who have not already received the vaccine. Although the committee did not recommend routine immunization for boys at that time, it did suggest that Gardasil be made available to males ages 9 to 26. As of 2007, 25% of 13- to 17-year-old girls in the United States reported receiving at least one dose of this vaccine. In 2011, this number rose to 53% in girls, still far short of targeted numbers (~80%) and significantly lower than the rates of compliance for most other routine childhood vaccines (somewhere around 90%, depending on the age of the child).

In 2011, boys ages 11 to 13 were added to the ACIP list of recommendations for routine HPV vaccination. The hope is that this will curb the rising tide of anal and oropharyngeal cancers among men, but also cut back the infection cycle and impact rates of cervical cancer in women. However, HPV vaccination rates among young men remained at only 8% at the end of 2011. The idea was that, with Gardasil in particular, the ability to reduce the incidence of unsightly genital warts might provide added incentive for male vaccination.

With a safe and effective vaccine against a common and deadly cancer available for several years, why are the rates of immunization in young people still so low? The answer depends somewhat on the country in question, as well as social and economic factors. Especially in developing countries, cost and ease of use are major barriers. Development is currently underway for a second generation of HPV vaccines, which

are more cost effective, easier to administer and produce longer-lived immunity to a broader range of HPV genotypes.

Controversy based on social and ethical issues, as well as misinformation, are also high on the list of reasons why HPV vaccination rates are believed to remain so low. In U.S.-based studies of factors that influence decisions to vaccinate adolescents against HPV, mother's attitudes, physician recommendations, and misunderstandings in all groups were highlighted. For instance, since HPV is a sexually transmitted infection, most parents prefer to consider this an issue for "the future," assuming that their children are not sexually active and that there is plenty of time before a vaccine for a sexually transmitted disease should be considered. In fact, based on the Centers for Disease Control and Prevention (CDC) surveillance data from 2011, 47% of high school students in the U.S. have had sexual intercourse; greater than 6% beginning before the age of 13. The HPV vaccine regimen, which involves three intramuscular injections administered over a 6-month period, is most effective when completed *prior* to exposure, and produces the most robust immune response in 11- to 12-year-olds, the target population.

Physician recommendations are also key to making a dent in the rates of HPV vaccination. In a 2011 study, women 19 to 26 years old were asked about whether they had received an HPV vaccine. In the group that had received a provider recommendation, 85% were immunized, compared with only 5% among women who did not receive a physician recommendation. More public and professional information concerning the advantages of this vaccine before the onset of sexual activity, as well as the lifetime risk of disease caused by HPV, may help drive down the cycle of infection and worldwide deaths due to this sexually transmitted killer.

Winer RL, Lee SK, Hughes JP, Adam DE, Kiviat NB, Koutsky LA. Genital human papillomavirus infection: incidence and risk factors in a cohort of female university students. Am J Epidemiol. 2003 Feb 1;157(3):218–26.

CDC. Youth risk behavior surveillance—United States, 2011. MMWR 2012;61(SS-4).

expressed only during specific developmental stages, such as in the fetus, or at extremely low levels in normal conditions, but which are upregulated in tumor cells (see Figure 19-6). Those derived from mutation-induced reactivation of certain fetal or embryonic genes, called **oncofetal tumor antigens**, normally only appear early in embryonic development, before the immune system acquires immunocompetence. When transformation of cells causes them to appear at later stages of development on neoplastic cells of the adult, they are recognized as nonself and induce an immunologic response.

Two well-studied oncofetal antigens are **alpha-fetoprotein (AFP)** and **carcinoembryonic antigen (CEA)**. AFP, the most abundant fetal protein, drops from milligram levels in fetal serum to between 5 ng/ml and 50 ng/ml after birth. Elevated levels of this glycoprotein can also be found in women, especially during the early stages of pregnancy. Significantly elevated AFP levels in nonpregnant adults are not uncommon in ovarian, testicular, and liver cancers, where serum levels above 300 ng/ml can be indicative of small lesions even in asymptomatic individuals. Monitoring of these levels can help clinicians to make prognoses and to evaluate treatment efficacy, especially in liver cancer.

CEA is another oncofetal membrane glycoprotein found on gastrointestinal and liver cells of 2- to 6-month-old fetuses. Approximately 90% of patients with advanced colorectal cancer and 50% of patients with early colorectal cancer have increased levels of CEA in their serum; some patients with other types of cancer also exhibit increased CEA levels. However, because AFP and CEA can be found in trace amounts in some normal adults and in some noncancerous disease states, the presence of these oncofetal antigens is not necessarily diagnostic of tumors but can still be used to monitor tumor growth. If, for example, a patient has had surgery to remove a colorectal carcinoma, CEA levels are monitored after surgery; an increase in the CEA level is an indication that tumor growth has resumed.

In addition to embryonic antigens, the category of TAAs also includes the products of some oncogenes, such as several growth factors and growth factor receptors. These proteins, although transcribed in the adult, are normally tightly regulated and expressed only at low levels. For instance, a variety of tumor cells express the *epidermal growth factor* (EGF) receptor at levels 100 times greater than in normal cells. Another, melanotransferrin, designated p97, has fibroblast growth factor-like activities. Whereas normal cells express fewer than 8,000 molecules of p97 per cell, melanoma cells express 50,000 to 500,000 molecules per cell. The gene that encodes p97 has been cloned, and a recombinant vaccinia virus vaccine has been prepared that carries the cloned gene. When this vaccine was injected into mice, it induced both humoral and cell-mediated immune responses, which protected the mice against live melanoma cells expressing the p97 antigen. Targeting such oncogene products has yielded some significant clinical success, such as in the case of breast cancers overexpressing HER2 (discussed further in the final section of this chapter).

The Immune Response to Cancer

As mentioned earlier, cells have multiple built-in (or intrinsic) mechanisms to prevent cancer. One example is the NER pathway of DNA repair, which encourages cell senescence (permanent cell cycle arrest) or even apoptosis at the first signs of unregulated growth. However, in the event that this system fails, control mechanisms external to the cell can kick in. At their most basic, these cell-extrinsic mechanisms involve environmental signals that instruct a cell to activate internal pathways leading to growth arrest or apoptosis, to prevent cancer cell spread. For example, when epithelial-cell/extracellular-matrix associations are disrupted due to malignant transformation, death signals are triggered that block proliferation and spread of these contact dependent cells. Thus these extracellular attachments serve as inhibitors of cell death, which when broken set off a safety mechanism promoting apoptosis. If unregulated growth continues, identification and rejection of tumor cells by components of the immune system may help salvage homeostasis. Although several key immune cell types and effector molecules that participate in this response have been identified in recent years, much still remains to be learned about natural mechanisms of anti-tumor immunity and how best to induce these in clinical settings.

For over a century, the controversy over whether and how the immune system participates in cancer recognition and destruction has raged. However, data collected in the past couple of decades from both animal models and clinical studies has clearly defined a role for the immune response in tumor cell identification and eradication. To date, there are three proposed mechanisms by which the immune system is thought to control cancer:

- By destroying viruses that are known to transform cells

- By eliminating pathogens and reducing pro-tumor inflammation

- By actively identifying and eliminating cancerous cells

This final mechanism, involving tumor cell identification and eradication, is termed **immunosurveillance**. It posits that the immune system continually monitors for and destroys neoplastic cells. Evidence from animal models as well as immune deficiency disorders and induced immune suppression in humans supports this hypothesis. For instance, AIDS patients and transplant recipients on immunosuppressive drugs have a much higher incidence of several types of cancer than do individuals with fully competent immune systems.

However, recent data also point to potential pro-tumor influences of the immune response on cancer. For instance, chronic inflammation and immune-mediated selection for malignant cells may actually contribute to cancer cell spread and survival. Contemporary studies of immunity to cancer have now generated a more nuanced hypothesis of immune involvement in neoplastic regulation. This model is called

immunoediting; it incorporates observations of both tumor-inhibiting and tumor-enhancing processes mediated by the immune system. In the following section we describe this revised view of the role of the immune response in cancer.

Immunoediting Both Protects Against and Promotes Tumor Growth

In the mid-1990s, research in animal models of cancer suggested that natural immunity could eliminate tumors. Armed with this understanding, researchers identified some of the key cell types and effector molecules involved. Experimental studies showed that mice lacking intact T-cell compartments or interferon (IFN)-γ signaling pathways were more susceptible to chemically induced or transplanted tumors, respectively. Likewise, RAG2 knockout mice, which fail to generate T, B, or *natural killer T* (NKT) cells, were more likely to spontaneously develop cancer as they aged and were more susceptible to chemical carcinogens.

However, the real surprise came when scientists used the same chemical carcinogen to induce tumors in both wild-type and RAG2 knockout mice, then adoptively transferred these tumors into syngeneic wild-type recipients. (See Chapters 12 and 20 for further discussion of adoptive transfer.) All of the tumors coming from wild-type animals grew aggressively in their new hosts, whereas up to 40% of the tumors taken from immunodeficient mice were rejected by recipients. This suggested that tumors growing in immune-deficient environments are more immunogenic than those arising in an immunocompetent environment. These observations led to the idea that the immune system exerts a dynamic influence on cancer, inhibiting tumor cells but also *sculpting* them in a Darwinian process of selection: those that survive are better able to outwit the immune response and have a survival advantage.

The three currently proposed phases to the immunoediting hypothesis are elimination, equilibrium, and escape (Figure 19-7). The first phase, elimination, is the traditional view of the immune system as a major player in the identification and destruction of newly formed cancer cells. Equilibrium is the proposed second phase, characterized by a state of balance between destruction and survival of a small number of neoplastic cells. Ample clinical evidence now suggests that phase 2 can continue for up to decades after the emergence of a tumor. However, identifying residual transformed cells and targeting them during this window is challenging. Escape is the final phase of cancer progression, when the most aggressive and least immunogenic of the residual tumor cells begin to thrive and spread. Most basic research studies have focused on the role of the immune system in the elimination phase, where both innate and adaptive processes identify and target transformed cells for destruction, sometimes setting the stage for what will occur during the equilibrium and escape phases. The following sections discuss our current understanding of the role of the immune system in cancer eradication or survival during cancer immunoediting.

Key Immunologic Pathways Mediating Tumor Eradication Have Been Identified

As early as the 1860s, Rudolph Virchow observed that leukocytes infiltrate the site of a solid tumor, and he proposed a link between inflammation and cancer *induction*. However, as infiltrating leukocytes can be found in both cancers that progress and those that resolve, the role of inflammation in immunity to cancer was obscure. More recent scientific advances have allowed a more detailed analysis of the location and type of cells involved, leading to the identification of key indicators of cancer regression in certain types of tumors. With this newfound knowledge, advances are being made in cancer diagnostic tools, prognostic indicators, and targets for clinical intervention.

Much of the research aimed at probing the relationship between the immune system and cancer has employed mouse model systems. Many or even most mice rendered immune deficient via targeted gene knockouts or neutralizing antibodies display some increased incidence of cancer, be it spontaneous or carcinogen induced. Likewise, spontaneous or induced tumors in immune-competent mice have allowed scientists to model human cancer induction and test hypotheses concerning specific cell types and immune pathways in cancer eradication. Although many of the elements of both innate and adaptive immunity can be linked in some way to tumor-cell recognition and destruction, certain components appear to play key roles in immune-mediated cancer control.

Studies in mice and humans have led to the awareness that there are both "good" and "bad" forms of inflammation in the response to cancer (e.g. Figure 19-7, bottom right). On the good or anti-tumor side are innate responses dominated by immune-activating macrophages (called M1), cross-presenting *dendritic cells* (DCs; see Chapter 8), and NK cells. These cells and the cytokines they produce help elicit strong T_H1 and CTL responses, which are associated with a good prognosis and tumor regression. Conversely, the immune cell infiltrates found in tumors that are more likely to progress and metastasize include anti-inflammatory macrophages (M2) and *myeloid-derived suppressor cells* (MDSCs). Concomitantly, adaptive responses to cancer dominated by the T_H2 pathway (and in some cases, also T_H17 or T_{REG} cells) are associated with poorer clinical outcomes and reduced survival times. In the following sections, we discuss in greater detail our developing awareness of both the positive and negative relationships of some of these specific pathways to cancer immunity.

Innate Inhibitors of Cancer

Identified over 35 years ago, *natural killer* (NK) cells were among the first cell type to be recognized for their inherent ability to destroy tumor cells, from which their name derives. Mice rendered NK cell deficient, either by gene knockout or via neutralizing antibodies, show an increased

The Three Stages of Cancer Immunoediting

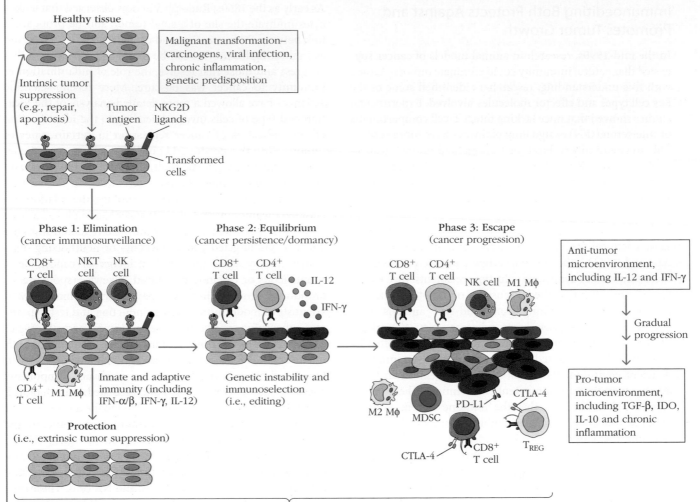

Recognition and targeting of tumor cells by the immune system is believed to occur in three phases. Phase I, Elimination: cancer cells are recognized by the immune system via their tumor antigens and targeted for destruction. In the process, some cells acquire mutations that allow them to resist immune destruction. Phase II, Equilibrium: low levels of abnormal cells persist, but their proliferation and spread are held in check by the adaptive immune response. Phase III, Escape: further mutation in the surviving tumor cells leads to the capacity for immortal growth and metastasis. Over time, inhibitory immune responses begin to dominate and immune activity shifts from anti- to pro-tumor. Tan cells are normal; pink to red cells represent progressive development of decreased immunogenicity in tumor cells. Abbreviations: MΦ, macrophage; MDSC, myeloid-derived suppressor cells; PD-L1, programmed death ligand 1. [Modified from M. D. Vesely, et al., 2011, Natural innate and adaptive immunity to cancer, Annual Review of Immunology 29:235–271.]

incidence of lymphomas and sarcomas. The importance of NK cells in tumor immunity is highlighted by the mutant mouse strain called beige and by **Chediak-Higashi syndrome** in humans. In both, a genetic defect causes marked impairment of NK cells, and in each case an associated increase is present in certain types of cancer.

NK cell recognition mechanisms use a series of surface receptors that respond to a balance of activating and inhibiting signals delivered by self cells (see Chapter 13). Since many transforming viruses can induce the downregulation of MHC expression, detecting "missing self" is likely at least one of the ways in which NK cells participate in tumor-cell identification and eradication. However, inducing signals in the form of molecular expressions of danger can also engage NK cell-activating receptors (e.g., NKG2D) delivering what is now referred to as an "altered" or "induced-self" signal (see Figure 19-7). Various forms of cellular stress, including viral infection, heat shock, UV radiation, and other agents that induce DNA damage can trigger expression of the ligands for these activating NK cell receptors. Using activating signals induced by DNA damage pathways, NK cells may thus be able to distinguish cancerous or precancerous cells from healthy neighboring cells. Once engaged, these cells use cytolytic granules that include such compounds as perforin to target their killing machinery at cells expressing these activating ligands. In fact, deficiency in perforin, a cytolytic compound used by both NK cells and CTLs to kill target cells, is linked to increased cancer susceptibility. Indirectly, NK cells may also participate in cancer eradication by secreting IFN-γ, a potent anticancer cytokine that encourages DCs to stimulate strong CTL responses in vitro (see *adaptive responses* below).

Numerous observations indicate that activated macrophages also play a significant role in the immune response to tumors. For example, macrophages are often observed to cluster around tumors, and the presence of proinflammatory macrophages, such as type M1, is correlated with tumor regression. Like NK cells, macrophages are not MHC restricted and express Fc receptors, enabling them to bind to antibody on tumor cells and to mediate *antibody-dependent cell-mediated cytotoxicity* (ADCC; discussed further below). The anti-tumor activity of activated macrophages is likely mediated by lytic enzymes, as well as reactive oxygen and nitrogen intermediates. In addition, activated macrophages secrete a cytokine called *tumor necrosis factor alpha* (TNF-α) that has potent anti-tumor activity.

More recently, a previously unsuspected role for the eosinophil in cancer immunity has also come to light. Mice engineered to lack eotaxin or CCL11, two chemoattractants for eosinophils, or IL-5, a stimulatory cytokine for this cell type, were all found to be more susceptible to carcinogen-induced cancers than wild-type mice. In addition, IL-5 transgenic animals, which display higher levels of circulating eosinophils, are more resistant to chemically induced sarcomas.

Adaptive Cell Types Involved in Cancer Eradication

In experimental animals, tumor antigens induce humoral and cell-mediated immune responses that lead to the destruction of transformed cells expressing these proteins. Animals that lack either αβ or γδ T cells are more susceptible to a number of induced and spontaneous tumors. Several tumors have been shown to induce CTLs that recognize tumor antigens presented by class I MHC on these neoplastic cells. In fact, strong anti-tumor CTL activity correlates significantly with tumor remission and is primarily credited with maintaining the equilibrium stage of cancer detection, or a state of immune-mediated neoplastic cell dormancy (see

Figure 19-7). Evidence for this comes from studies of mice treated with low-dose carcinogens. In the fraction of animals that do not develop cancer, rendering the animals immunodeficient can result in the sudden onset of cancer. Importantly, blocking NK-cell responses at this stage does not result in eruption of occult cancer, but blocking CD8$^+$ T-cell responses or IFN-γ does, highlighting the importance of adaptive immunity during this stage of cancer. The pressures of adaptive immunity on cancerous cells during this relatively long stage are believed to sculpt tumors (thus the *immunoediting* name), driving the selective survival of neoplastic cells that are less immunogenic and have accumulated mutations most favorable for immune evasion. One example of this is increased expression of PD-L1 on tumor cells, which binds to PD-1 on CTLs and inhibits engagement (see Figure 19-7).

In clinical studies of cancer, the frequency of *tumor-infiltrating lymphocytes* (TILs)—a combination of T cells, NKT cells, and NK cells—correlates with a prognosis of cancer regression. For instance, in one seminal study of ovarian cancer, 38% of women with high numbers of TILs compared with 4.5% of women with low numbers of TILs survived more than 5 years past diagnosis. However, beyond numbers, the type of infiltrating cells may be even more crucial and in some cases can have more prognostic power than clinical cancer staging. In general, a high frequency of CD8$^+$ T cells, and sometimes a high ratio of CTL to T$_{REG}$ cells is associated with enhanced survival. However, the story with T$_{REG}$ cells is complicated; several studies have observed a positive impact of this cell type on anticancer immune responses (discussed further in the following sections).

B cells respond to tumor-specific antigens by generating anti-tumor antibodies that can foster tumor-cell recognition and lysis. Using their Fc receptors, NK cells and macrophages again participate in this response, mediating ADCC (see Chapter 13). However, some anti-tumor antibodies serve a more detrimental role, blocking CTL access to tumor-specific antigens and enhancing survival of the cancerous cells. For this reason, a clear positive or negative role for B cells in cancer immunity is less obvious.

The Role of Cytokines in Cancer Immunity

Animal models in which cytokines or cytokine response pathways are eliminated have helped us to identify the role of specific cytokines in tumor-cell eradication. IFN-γ and the regulatory components of this pathway are clearly important in cancer elimination, as mice lacking these are more susceptible to a number of different tumors. This cytokine can exert direct anti-tumor effects on transformed cells, including enhanced class I MHC expression, making neoplastic cells better targets for CD8$^+$ T cell recognition and destruction (see Figure 19-7). Both type I (α/β) and type II (γ) interferons have immune cell-enhancing activities that can render these cells more efficient at tumor-cell removal.

The cytokine IL-12 has recently received much attention for its ability to enhance anti-tumor immunity. Administration of exogenous IL-12 protects mice from one type of chemically induced tumor. Mice genetically deficient for IL-12 also develop more papillomas (a type of epithelial cell cancer) than do wild-type animals. This may be due in part to IL-12 driving the development to T-cell pathways: this cytokine encourages DCs to activate strong T_H1 and CTL responses.

The cytokine TNF-α was named for its anticancer activity. When it was injected into tumor-bearing animals, it induced hemorrhage and necrosis of the tumor. However, this cytokine was later shown to have both tumor-inhibiting and tumor-promoting effects. Various carcinogen-treated $TNF^{-/-}$-mice have been found to display either more sarcomas or fewer skin carcinomas than wild-type mice, depending upon the mouse strain and the means of cancer induction, suggesting that TNF-α has a complex role in tumor immunity.

Some Inflammatory Responses Can Promote Cancer

As we know, the immune response involves a balance of activating and inhibiting pathways—both some gas and some brakes. Without the brakes, uncontrolled immune activity can lead to pathologic inflammation and even autoimmunity. In the immune response to cancer, inflammatory responses can serve a positive role, as we have seen above, but they also have the potential to promote cancer and create pro-tumor microenvironments, such as occurs during the escape phase of cancer immunoediting (see Figure 19-7, bottom right). In this section we discuss some specific immune response components that either support tumor growth or that direct immunity toward pathways of natural immunosuppression.

Chronic Inflammation

Chronic inflammation is believed to create a pro-tumor microenvironment via several mechanisms. First, inflammatory responses increase cellular stress signals and can lead to genotoxic stress, increasing mutation rates in cells and thus fostering tumorigenesis. Second, the growth factors and cytokines secreted by leukocytes often induce cellular proliferation, and during mutation events, nonimmune tumor cells can acquire the ability to respond to these growth stimulators. In this way, some immune cells and the factors they produce can help sustain and advance cancer growth. Finally, inflammation is pro-angiogenic, increasing blood vessel growth and allowing for greater tumor-cell invasion into surrounding tissues or transport via lymphatic vessels, one of the hallmarks of cancer.

Enhancing Anti-Tumor Antibodies

Antibodies can be produced against tumor-specific antigens, and these may be important flags for tumor-cell

eradication. Based on this discovery, attempts were made to protect animals against tumor growth by active immunization with tumor antigens or by passive immunization with anti-tumor antibodies. Much to the surprise of the researchers involved, these immunizations did not usually protect against the tumor; in some cases, they actually enhanced tumor growth.

The tumor-enhancing ability of immune sera subsequently was studied via *cell-mediated lympholysis* (CML) reactions, which measure in vitro lysis of target cells by CTLs. Serum taken from animals with progressive tumor growth blocked the CML reaction, whereas serum taken from animals with regressing tumors had little or no blocking activity. In 1969, the Hellstroms (Ingegerd and Karl) extended these findings by showing that children with progressive neuroblastoma had high levels of some kind of blocking factor in their serum, whereas children with regressive neuroblastoma did not have these serum factors. Since these first reports, serum blocking factors have been found to be associated with a number of human tumors.

In some cases, anti-tumor antibody itself acts as a serum blocking factor. Presumably the antibody binds to tumor-specific antigens and masks the antigens from cytotoxic T cells. However, in many cases the blocking factors are not antibodies alone, but rather antibodies complexed with free tumor antigens. Although these immune complexes have been shown to block CTL responses, the mechanism of this inhibition is not clear. These complexes also may inhibit ADCC by binding to Fc receptors on NK cells or macrophages and blocking their activity. Therefore, although some clinical studies today utilize antibodies to treat cancer (see Clinical Focus Box 13-1), most of these are not directed against tumor-specific antigens.

Active Immunosuppression in Tumor Microenvironments

Soluble factors secreted by tumor cells or the immune cells that infiltrate a tumor can encourage the development of a local immunosuppressive microenvironment. For instance, increased expression of TGF-β and *indole-amine-2,3-dioxygenase* (IDO), which inhibit T_H1 responses, has been found in various human cancers (see Figure 19-7, bottom right). Another immunosuppressive cytokine, IL-10, may play a duplicitous role in cancer immunity. IL-10 has tumor-promoting, tolerance-inducing properties in some situations but can also encourage anticancer innate immune responses in others. Likewise, TGF-β expressed during the latter stages of cancer encourages progression, possibly by blocking local DC activation and inhibiting T-cell function, although the presence of this cytokine during early tumor growth can be tumor inhibiting. These microenvironment effects appear to be quite localized to the primary tumor site. Studies in mice with a primary tumor have shown that additional cancer cells of the same type introduced to a new site can be rejected by the immune system, even while the primary tumor remains intact.

In both animal models and human cancer studies, much recent attention has focused on the role of various naturally immunosuppressive cell types in tumor responses. The general consensus from most animal studies is that an abundance of T_{REG} cells, myeloid-derived suppressor cells (MDSCs), M2 macrophages, and T_H2 cells confers a local state of immunosuppression or T_H2-directed immunity, and allows for tumor-cell evasion (see Figure 19-7, bottom right). The jury is still out, however, on how well this translates to humans, where the role of T_{REG} cells in clinical cancer immunity is increasingly controversial. In initial clinical studies, the presence of $CD4^+CD25^+FoxP3^+$ T_{REG} cells in pancreatic and ovarian cancer, or an elevated ratio of these cells to CTLs, was predictive of poor prognosis. However, similar investigations in colorectal cancer as well as head and neck cancers showed the opposite result: the presence of T_{REG} cells correlated with survival. These conflicting findings, combined with a lack of specific markers to target these cells, has stalled attempts to direct new cancer therapies toward the control of regulatory T-cell subsets that might favor tumor cell destruction.

Some Tumor Cells Evade Immune Recognition and Activation

Although the immune system clearly can respond to tumor cells, many of which express tumor antigens, the fact that so many individuals die each year from cancer suggests that the immune response to tumors is often ineffective. Immunoediting, or selective pressure applied by the anti-tumor immune response, often selects for escape mutants that can evade the immune response. Some of these escape strategies involve loss or gain of proteins by the tumor cells that helps them evade immune cell recognition and activation.

Reduced MHC Expression in Tumor Cells

Defects in antigen processing and presentation are common among the escape mutants arising in many tumors. These could include mutations that lead to reduced MHC expression, secretion of TSAs (rather than surface expression), defective transporter associated with antigen processing (TAP) or β2-microglobulin, and IFN-γ insensitivity. Each of these types of mutations results in decreased class I MHC presentation of tumor antigens and profound inhibition of $CD8^+$ T-cell recognition (Figure 19-8). NK cells should

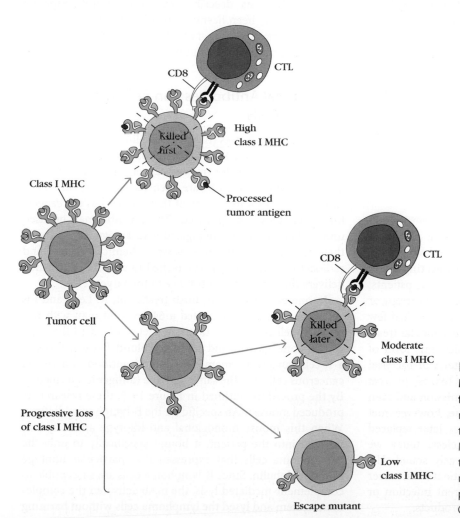

FIGURE 19-8 Down-regulation of class I MHC expression on tumor cells may allow for tumor escape mutants. The immune response itself may play a role in selection for tumor cells that express lower levels of class I MHC molecules, by preferentially eliminating those cells expressing high levels of class I molecules first. Malignant tumor cells that express fewer MHC molecules may thus escape CTL-mediated destruction.

recognize these cells lacking class I MHC. However, decreased expression of ligands that bind activating receptors on NK cells, also common among tumors, allows these cells to avoid NK cell-mediated killing.

Tumor Cell Subversion of Apoptosis Signals

The up-regulation of anti-apoptotic mediators and the expression of mutated or absent death receptors can lead to tumors that are resistant to programmed cell death signals. As mentioned earlier, faulty DNA repair mechanisms in transformed cells combined with immune-mediated pressures (e.g., selective destruction of cells expressing class I MHC) encourage the accumulation of tumor cells with these types of survival-enhancing mutations. In fact, the absence of MHC molecules on tumor cells is generally an indication of cancer progression and carries a poor prognosis.

Poor Costimulatory Signals Provided by Tumor Cells

As we know from Chapter 11, complete T-cell activation requires two signals: an activating signal, triggered by recognition of a peptide-MHC molecule complex by the T-cell receptor, and a costimulatory signal, triggered by the interaction of CD80/86 (B7) on antigen-presenting cells (APCs) with costimulatory molecules such as CD28 on T cells. Both signals are needed to induce IL-2 production and proliferation of T cells. By virtue of their status as self cells, tumors have fairly poor immunogenicity and tend to lack costimulatory molecules. Without sufficient numbers of APCs in the immediate vicinity of a tumor and with few stimulators to drive the activation of these cells, responding T cells may receive only a partial activating signal. This can lead to clonal anergy and immune tolerance. In fact, recently approved therapies for treating cancer specifically aim to enhance the costimulation provided to anti-tumor T cells.

Cancer Immunotherapy

Harnessing the immune system to fight cancer is not a new idea. In 1891, a bone carcinoma surgeon named William B. Coley first experimented with this approach, injecting bacteria directly into the inoperable tumor of one of his patients. This was an era prior to the development of chemotherapy or radiation treatment for cancer, and clinicians then had few choices. Based on success in this initial experimental treatment, Coley and other physicians continued this form of immunotherapy for many years, using bacteria or bacterial products that became known as "Coley's Toxins" to treat aggressive cancers. Published reports of remission and even elimination of tumors using this technique, however, met with much skepticism. This treatment was later replaced with chemo- and radiotherapy. Nonetheless, today we believe that Coley's observations were largely sound. His results were likely due to a boost to the patient's anticancer immune response as a result of a concurrent infection or presence of immunostimulatory bacterial products.

Present-day cancer treatment can take many forms. Beyond surgery and radiation treatment, which are most often employed in cases of larger, more discrete tumors, drug therapies can be used to target residual tumor cells and to attack dispersed cancers. Drug therapy for cancer falls loosely into four categories:

- Chemotherapies, aimed at blocking DNA synthesis and cell division
- Hormonal therapies, which interfere with tumor-cell growth
- Targeted therapies, such as small molecule inhibitors of cancer
- Immunotherapies, which induce or enhance the anti-tumor immune response

We will focus our attention here on immune-based therapies. These are designed to help eliminate a tumor by reviving, initiating, or supplementing the in vivo anti-tumor immune response or by neutralizing inhibitory pathways. Challenges in determining the efficacy of specific immunotherapies in clinical settings include the range of cancer cell types, tumor sizes, locations, and stages of disease, as well as questions of optimal dosing and schedule of treatment. The following sections describe several immunotherapeutic agents that have been licensed for use in humans, as well as some novel approaches that may yield clinically useful products to fight cancer in the future.

Monoclonal Antibodies Can Be Targeted to Tumor Cells

Monoclonal antibodies (mAbs) (see Clincal Focus Box 13-1 and Chapter 20) have long been used as experimental immunotherapeutic agents for treating cancer. At present, approximately 12 different mAbs are licensed for the treatment of cancer. Table 19-4 lists many of these, as well as the cancers for which they are approved. These mAbs may be used unmodified or can be conjugated with an agent to increase their efficacy. For instance, toxins, chemical agents, and radioactive particles can be attached to a mAb, which then delivers the conjugated substance to the target cell.

In one early success of mAb treatment, R. Levy and his colleagues successfully treated a 64-year-old man with terminal B-cell lymphoma that had metastasized to the liver, spleen, bone marrow, and peripheral blood. Because this was a B-cell cancer, the membrane-bound antibody on all the cancerous cells had the same idiotype (antigenic specificity). By the procedure outlined in Figure 19-9, these researchers produced mouse mAb specific for the B-lymphoma idiotype. When this mouse monoclonal anti-idiotype antibody was injected into the patient, it bound specifically to only the B-lymphoma cells that expressed that particular idiotype immunoglobulin. Since B-lymphoma cells are susceptible to complement-mediated lysis, the mAb activated the complement system and lysed the lymphoma cells without harming

| TABLE 19-4 | Monoclonal antibodies approved by the FDA and licensed for cancer treatment |

mAb name	Trade name	Target	Used to treat	Approved in:
Rituximab	Rituxan	CD20	Non-Hodgkin's lymphoma	1997
			Chronic lymphocytic leukemia (CLL)	2010
Trastuzumab	Herceptin	HER2	Breast cancer	1998
			Stomach cancer	2010
Gemtuzumab ozogamicin[2]	Mylotarg	CD33	Acute myelogenous leukemia (AML)	2000[1]
Alemtuzumab	Campath	CD52	CLL	2001
Ibritumomab tiuxetan[2]	Zevalin	CD20	Non-Hodgkin's lymphoma	2002
[131]I-Tositumomab[2]	Bexxar	CD20	Non-Hodgkin's lymphoma	2003
Cetuximab	Erbitux	EGFR	Colorectal cancer	2004
			Head and neck cancers	2006
Bevacizumab	Avastin	VEGF	Colorectal cancer	2004
			Non-small cell lung cancer	2006
			Breast cancer	2008
			Glioblastoma and kidney cancer	2009
Panitumumab	Vectibix	EGFR	Colorectal cancer	2006
Ofatumumab	Arzerra	CD20	CLL	2009
Denosumab	Xgeva	Rank ligand	Cancer spread to bone	2010
Ipilimumab	Yervoy	CTLA-4	Melanoma	2011
Brentuximab vedotin[2]	Adcetris	CD30	Hodgkin's lymphoma and one type of non-Hodgkin's lymphoma	2011

[1]General approval withdrawn in 2010 and now used only as a part of ongoing clinical trials

[2]Conjugated monoclonal antibodies

Source: American Cancer Society, www.cancer.org; and Table 2 from J. F. Aldrich et al., 2010, Vaccines and immunotherapeutics for the treatment of malignant disease, Clinical and Developmental Immunology, doi:10.1155/2010/697158.

other cells. After four injections with this anti-idiotype mAb, the tumors began to shrink and the patient entered an unusually long period of remission.

A custom approach targeting idiotypes like this is very costly and requires a specific reagent for each lymphoma patient. A more general mAb therapy for B-cell lymphoma is based on the fact that most B cells, whether normal or cancerous, bear lineage-distinctive antigens. For example, mAbs that target the B-cell marker CD20, such as Rituximab, are widely used to treat non-Hodgkin's lymphoma.

Some of the mAbs in clinical use can be coupled with radioactive isotopes, chemotherapy drugs, or potent toxins of biological origin. In such "guided missile" therapies, the toxic agents are delivered specifically to tumor cells. This ideally focuses the toxic effects on the tumor and spares normal tissues. Reagents known as **immunotoxins** have been constructed by coupling the inhibitor chain of a toxin (e.g., diphtheria toxin) to an antibody against a tumor-specific or tumor-associated antigen. In vitro studies have demonstrated that these "magic bullets" can kill tumor cells without harming normal cells, although none have yet been licensed for clinical use. Ibritumomab tiuxetan and ^{131}I-tositumomab are both examples of licensed radioisotope-conjugated mAbs for treating cancer. Each delivers a dose of radiation to cells bearing the CD20 cell-surface receptor, for which the mAb is specific, and can be used for the treatment of non-Hodgkin's lymphoma.

A variety of tumors express significantly increased levels of growth factors or their receptors, which are promising targets for anti-tumor mAbs. For example, in 25% to 30% of women with metastatic breast cancer, a genetic alteration of the tumor cells results in the increased expression of *h*uman

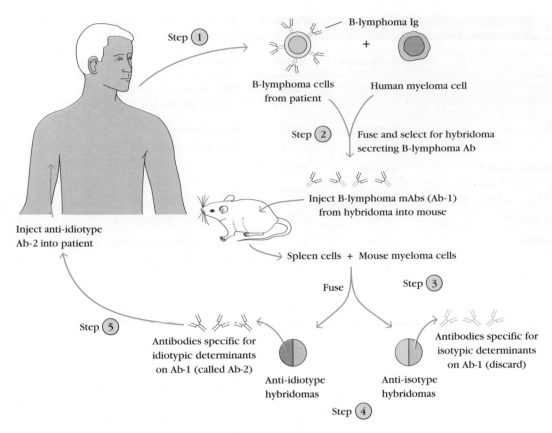

FIGURE 19-9 Development of a monoclonal antibody specific for idiotypic determinants on B-lymphoma cells. Because all the B-lymphoma cells in a patient are derived from a single transformed B cell, they all express the same membrane-bound antibody (Ab-1) with the same idiotype (i.e., the same antigenic specificity). In the procedure illustrated, a monoclonal anti-idiotypic antibody (Ab-2) against the B-lymphoma membrane-bound antibody is produced ex vivo (steps 1–4). This anti-idiotype antibody is then injected into the patient (step 5), where it binds selectively to the idiotypic determinants on the immunoglobulin of B-lymphoma cells, making these cells susceptible to complement-mediated lysis.

*e*pidermal-growth-factor–like *receptor 2* (HER2) encoded by the *neu* gene and expressed in only trace amounts in normal adults. Because of this difference in protein levels, a humanized mAb against HER2 has been successfully used to treat HER2-expressing breast cancers, selectively eliminating cancer cells without damaging normal cells. Several other mAbs that target specific growth factors or their receptors have also been approved for clinical use and are included in Table 19-4.

Cytokines Can Be Used to Augment the Immune Response to Tumors

The isolation and cloning of the various cytokine genes has facilitated the large-scale production of cytokines for use in clinical settings. Several of these have been used either singly or in combination to augment the immune response against cancer in clinical trials. Among these are all three interferons (IFN-α, -β, and -γ), the *tumor necrosis factors* (TNF-α and Lymphotoxin-α [TNF-β]), *granulocyte-macrophage colony stimulating factor* (GM-CSF) and several interleukins (IL-2, -4, -6, and -12). Trials with some of these, either used in vivo

or by treatment of cells ex vivo, have produced occasional encouraging results.

After some initially promising and long-lasting results in early trials, IL-2 was licensed for use in cancer therapy. Despite significant in vivo toxicity and the small fraction of patients impacted, many follow-up clinical trials incorporated IL-2 alone or in combination with other immunotherapies. Alas, many years of refinements with IL-2 regimens and combination drug testing have not shown this T-cell growth factor to be as effective as once hoped, and the mechanisms of action in those cases of durable response are still largely unknown. Today, IL-2 alone or in combination with IFN-α is still used for advanced kidney cancer and metastatic melanoma. One factor that may further complicate our understanding of the role of IL-2 in cancer is the impact of this cytokine on T_{REG} cells, a population that has been associated with both tumor-enhancing and tumor-inhibiting behaviors.

The major obstacles to in vivo cytokine therapy are the complexity of the cytokine network itself and systemic toxicity. This complexity makes it difficult to determine how a given recombinant cytokine will affect the production of other cytokines in the patient. Cytokines are also difficult to administer locally, and systemic administration of high levels of a given cytokine can lead to serious or even life-threatening consequences. Much of the systemic toxicity can be avoided when cytokines, such as GM-CSF or IL-2, are used only ex vivo to expand cultured leukocyte populations as a part of cell-based immunotherapy regimens (see below).

Tumor-Specific T Cells Can Be Expanded and Reintroduced into Patients

Early observations of lymphocyte infiltration into solid tumors suggested that these cells might be a source of tumor-specific immunity. Subsequent research has shown that in many cases, tumor-reactive T cells can be isolated from the peripheral blood, lymph nodes, and solid tumors of cancer patients. Despite the presence of these cells, tumors persist in these individuals, suggesting that these cells are not performing anti-tumor effector functions in vivo. A range of adoptive T-cell therapies aim to collect these cells from cancer patients, expand and activate them ex vivo, and re-infuse them into patients. These cells can come from solid tumors themselves (tumor infiltrating lymphocytes, or TILs), tumor-draining nodes, or peripheral blood. In one recent multi-center study, the TILs from patients with metastatic melanoma were collected and expanded with IL-2 in vitro to overcome their anergic state. Following lympho-depleting treatments designed to create a niche for these cells, patients were re-infused with large numbers of these activated, autologous TILs. Tumor regression responses reached up to 50%, with approximately 10% of the patients experiencing long-lived or complete remission.

New Therapeutic Vaccines May Enhance the Anti-Tumor Immune Response

Most vaccines, also called prophylactic vaccines, are designed to initiate an immune response before the onset of infection or disease. Therapeutic vaccines, on the other hand, aim to enhance or redirect an existing immune response. Immunogens that make successful prophylactic vaccines do not always work once infection has been established. For example, the HPV vaccine, which is up to 99% effective at preventing infection by the strains that most commonly cause cervical cancer, is not effective once a women is already infected with one of these strains. Therefore, vaccines for use in patients with existing infections or malignant cells must be designed to redirect an already engaged immune response.

Several cancer vaccine studies in mice have aimed at expanding or enhancing tumor-specific antigen presentation. Mouse DCs cultured in GM-CSF and incubated with tumor fragments, then re-infused into the mice, have been shown to activate both T_H cells and CTLs specific for the tumor antigens. When the mice were subsequently challenged with live tumor cells, they displayed tumor immunity. These experiments have led to a number of approaches to expand the population of APCs, so that these cells can activate T_H cells or CTLs specific for tumor antigens.

Employing a similar strategy, sipuleucel-T (Provenge) became the first approved therapeutic cancer vaccine. This vaccine is designed to fight metastatic prostate cancer by inducing an immune response to a common self prostate tumor antigen, prostatic acid phosphatase (PAP). This is an individual cell-based therapy, where DCs are first isolated from prostate cancer patients and then stimulated in vitro using a fusion protein consisting of PAP and GM-CSF (Figure 19-10). These antigen-stimulated and expanded autologous APCs are then re-infused into the patient, yielding survival increases of just over 4 months in men first treated with the drug. Exact mechanisms of action are still under study, although sipuleucel-T is believed to work by stimulating PAP-specific CTLs to kill prostate tumor cells. With as yet modest survival benefits and a cost of $93,000 for the recommended three infusions, the long-term potential for this new individual prostate cancer immunotherapy is still unclear.

Another approach that also utilizes the APC-activating cytokine GM-CSF is to transfect tumor cells with the gene for this chemical signal, creating a local source near the tumor cells. These engineered tumor cells, when reinfused into the patient, will secrete GM-CSF, enhancing the differentiation and activation of host APCs, especially DCs. As these DCs accumulate around the tumor cells, the GM-CSF secreted by the tumor cells will enhance the presentation of tumor antigens to T_H cells and CTLs by local DCs (see Figure 19-11).

Manipulation of Costimulatory Signals Can Improve Cancer Immunity

As discussed earlier, tumors cells frequently lack the costimulatory signals required for full T-cell activation. Several research groups have demonstrated that tumor immunity can be enhanced when these costimulatory signals are modified. For instance, when mouse CTL precursors (CTL-Ps) are incubated with melanoma cells in vitro, antigen recognition occurs. In the absence of a costimulatory signal, these CTL-Ps do not proliferate or differentiate into effector CTLs. However, when these melanoma cells are transfected with the gene that encodes CD80 (B7.1), the CTL-Ps differentiate into effector CTLs.

These findings offer the possibility that CD80/86-transfected tumor cells might be used to induce CTL responses in vivo. For instance, when P. Linsley, L. Chen, and colleagues injected melanoma-bearing mice with CD80/86+ melanoma cells, the melanomas completely regressed in more than 40% of the mice. S. Townsend and J. Allison used a similar approach to vaccinate mice against malignant melanoma. Normal mice were first immunized with irradiated, CD80

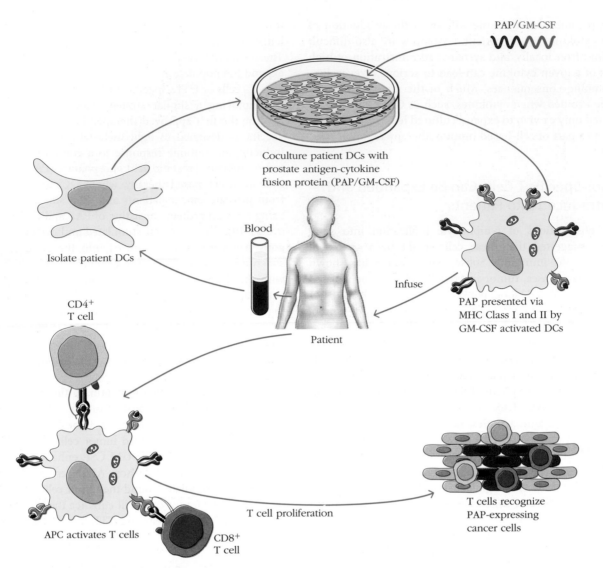

PAP/GM-CSF

Coculture patient DCs with
prostate antigen-cytokine
fusion protein (PAP/GM-CSF)

Isolate patient DCs

Blood

Infuse

PAP presented via
MHC Class I and II by
GM-CSF activated DCs

Patient

CD4+
T cell

T cell proliferation

APC activates T cells

CD8+
T cell

T cells recognize
PAP-expressing
cancer cells

FIGURE 19-10 Mechanism of action of sipuleucel-T, a prostate cancer vaccine. Autologous DCs are isolated from a patient's blood and cultured with a fusion protein consisting of the prostate cancer specific antigen PAP and the APC-activating cytokine GM-CSF (PAP/GM-CSF). DCs take up and process these antigens, after which they are reinfused into the patient in order to stimulate a T-cell response against PAP expressed on tumor cells in the prostate. Abbreviations: APC, antigen-presenting cell; DCs, dendritic cells; GM-CSF, granulocyte-macrophage colony-stimulating factor; PAP, prostatic acid phosphatase. [*Adapted from G. Di Lorenzo, C. Buonerba, and Philip W. Kantoff, 2011 September, Immunotherapy for the treatment of prostate cancer, Nature Reviews Clinical Oncology 8:551–561.*]

transfected melanoma cells and then challenged with unaltered malignant melanoma cells (Figure 19-12). The "vaccine" was found to protect a high percentage of the mice. It is hoped that a similar vaccine might prevent metastasis after surgical removal of a primary melanoma in human patients.

As we saw in Chapter 11, costimulatory molecules can also be involved in dampening the immune response. For example, engagement of CTLA-4 on T cells with the CD80/86 molecule on APCs results in T-cell inhibition. Studies in tumor-bearing mice have shown that mAbs against CTLA-4 can induce tumor rejection. Recently, two different mAbs capable of CTLA-4 blockade in humans have been developed. In clinical studies of patients with metastatic melanoma, up to 10% of patients treated with these humanized anti-CTLA-4 antibod-

ies experienced tumor regression, with some seeing long-term remission. Unfortunately, a significant number of patients experienced autoimmune side effects, highlighting the powerful role of CTLA-4 in the maintenance of self tolerance. Nonetheless, this angle is currently being pursued for other immune checkpoint regulatory molecules, such as PD-L1, where expression of this immuno-inhibiting molecule on cancer cells has been correlated with a poor clinical prognosis.

Combination Cancer Therapies Are Yielding Surprising Results

Standard chemotherapeutic treatments for cancer are cytotoxic to rapidly dividing cells, many of which are the targeted

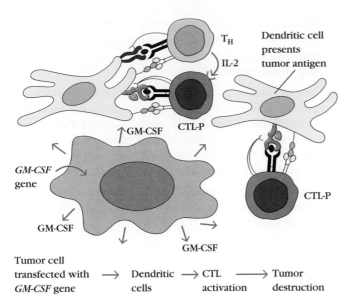

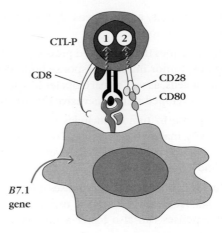

FIGURE 19-11 Use of GM-CSF-transfected tumor cells for cancer immunotherapy. Transfection of tumor cells with the gene encoding GM-CSF allows the tumor cells to secrete high levels of GM-CSF. This cytokine will activate DCs in the vicinity of the tumor, enabling the DCs to present tumor antigens to both T_H cells and cytotoxic T lymphocyte precursors (CTL-Ps).

FIGURE 19-12 Use of CD80 (B7.1)-transfected tumor cells for cancer immunotherapy. Tumor cells transfected with the *B7.1* gene express the costimulatory CD80 molecule, enabling them to provide both activating signal (1) and costimulatory signal (2) to CTL-Ps. As a result of the combined signals, the CTL-Ps differentiate into effector CTLs, which can mediate tumor destruction. In effect, the transfected tumor cell acts as an APC.

tumor. Common wisdom suggested that these treatments, which also kill other rapidly dividing cells, such as leukocytes (especially those in the myeloid lineage), were not a good combination with immunotherapeutic treatments, which aim to stimulate leukocytes. However, recent evidence suggests otherwise, and current theories postulate that some of the anticancer benefits of cytotoxic chemotherapeutic agents may in fact be due to their ability to induce immune activity against antigens released from killed tumor cells. With this in mind, early studies are underway to evaluate potential synergistic effects of combining chemo- or radiotherapy with immunotherapies for cancer. As with all new combination therapies, testing and outcome evaluation of the optimal timing, dose, and types of drugs is a complicated matrix. However, if the promise in initial studies showing synergistic effects holds true, the effective mixing of tumor-directed cytotoxicity with anti-tumor immune enhancement could help deliver the elusive one-two punch that cancer immunologists have been seeking.

SUMMARY

- Tumor cells differ from normal cells by changes in the regulation of growth, allowing them to proliferate indefinitely and eventually metastasize to other tissues.

- Normal cells can be transformed in vitro by chemical and physical carcinogens and by transforming viruses, all of which lead to DNA alterations.

- Transformed cells exhibit altered growth properties and are sometimes capable of inducing cancer when they are injected into animals.

- Proto-oncogenes encode proteins involved in normal cellular growth. The conversion of proto-oncogenes to oncogenes is a key step in the induction of most human cancer. This conversion may result from mutation in an oncogene, its translocation, or its amplification.

- Tumor-suppressor genes are normal cellular genes that dampen cell growth and proliferation. Disregulation of genes involved in pro- and anti-apoptotic pathways can also participate in cancer induction.

- A number of B- and T-cell leukemias and lymphomas are associated with translocated proto-oncogenes. In its new site, the translocated gene may come under the influence of enhancers or promoters that cause its transcription at higher levels than usual.

- Tumor cells can display tumor-specific antigens, which are novel proteins, and the more common tumor-associated antigens, which are modified normal proteins or proteins with changed patterns of expression.

- The tumor antigens recognized by T cells fall into one of four major categories: antigens encoded by genes with tumor-specific expression, antigens encoded by variant forms of normal genes that have been altered by mutation, antigens normally expressed only at certain stages of differentiation

- or differentiation lineages, and antigens that are overexpressed in particular tumors.

- Immunoediting describes the postulated process by which the developing anti-tumor immune response both identifies and kills tumor cells, as well as participates in selection of cells that have developed immunoevasive mutations.

- The immune response to tumors includes CTL-mediated lysis, NK-cell activity, macrophage-mediated tumor destruction, and destruction mediated by ADCC. Several cytokines, including IFN and TNF, help to mediate tumor-cell killing.

- Limited and directed inflammatory responses near tumors can be beneficial. However, prolonged or overly active inflammatory responses can also foster a microenvironment that promotes tumors.

- Tumors use several strategies to evade these immune responses, including accumulation of new mutations, poor costimulatory signaling, and active inhibition of local anti-tumor responses.

- Current cancer immunotherapy employs a variety of approaches. These include the use of mAbs, enhancement of costimulatory signaling to T cells, adoptive T cell transfer, cytokine augmentation, and therapeutic vaccines aimed at increasing the ability of APCs to activate adaptive cell-mediated tumor eradication.

- Combinations of cytotoxic chemotherapies with selected immunotherapies are yielding promising results against cancer.

REFERENCES

Aisenberg, A. C. 1993. Utility of gene rearrangements in lymphoid malignancies. *Annual Review of Medicine* **44**:75–84.

Aldrich, J. F., et al. 2010. Vaccines and immunotherapeutics for the treatment of malignant disease. *Clinical and Developmental Immunology*, doi10.1155/2010/697158.

Allison, J. P., A. A. Hurwitz, and D. R. Leach. 1995. Manipulation of costimulatory signals to enhance antitumor T-cell responses. *Current Opinion in Immunology* **7**:682–686.

Boon, T., P. G. Coulie, and B. Van den Eynde. 1997. Tumor antigens recognized by T cells. *Immunology Today* **18**:267–268.

Cho, H.-J., Y.-K. Oh, and Y. B. Kim. 2011. Advances in human papilloma virus vaccines: *A patent review*, Epub 2011, **21**(3):295–309.

Coulie, P. G., et al. 1994. A new gene coding for a differentiation antigen recognized by autologous cytolytic T lymphocytes on HLA-A2 melanomas. *Journal of Experimental Medicine* **180**:35–42.

Hanahan, D., and R. A. Weinberg. 2011. Hallmarks of cancer: *The next generation. Cell* **144**:646–674.

Houghton, A. N., J. S. Gold, and N. E. Blachere. 2001. Immunity against cancer: Lessons learned from melanoma. *Current Opinion in Immunology* **13**:134–140.

Hsu, F. J., et al. 1997. Tumor-specific idiotype vaccines in the treatment of patients with B-cell lymphoma. *Blood* **89**:3129–3135.

Lesterhuis, W. J., et. al. 2011. Cancer immunotherapy—revisited. *Nature Reviews in Drug Discovery* **10**:591–600.

Sahin, U., O. Tureci, and M. Pfreundschuh. 1997. Serological identification of human tumor antigens. *Current Opinion in Immunology* **9**:709–716.

Sharma, P., et. al. 2011. Novel cancer immunotherapy agents with survival benefit: Recent successes and next steps. *Nature Reviews in Cancer* **11**:805–812.

Srivastava, S. 2002. Roles of heat-shock proteins in innate and adaptive immunity. *Nature Reviews Immunology* **2**:185–194.

Townsend, S. E., F. W. Su, J. M. Atherton, and J. P. Allison. 1994. Specificity and longevity of antitumor responses induced by B7-transfected tumors. *Cancer Research*. **54** (24):6477-6483.

Vesely, M. D., et al. 2011. Natural innate and adaptive immunity to cancer. *Annual Review of Immunology* **29**:235–271.

Weinberg, R. A. 1996. How cancer arises. *Scientific American* **275**(3):62–70.

Useful Web Sites

http://seer.cancer.gov/statfacts/html/all.html

Surveillance Epidemiology and End Results (SEER), which provides U.S.-based cancer statistics, is a site maintained by the National Cancer Institute .

www.oncolink.org Oncolink offers comprehensive information about many types of cancer and is a good source of information about cancer research and advances in cancer therapy. The site is regularly updated and includes many useful links to other resources.

www.cancer.org/index The Web site of the American Cancer Society contains a great deal of information on the incidence, treatment, and prevention of cancer. The site also highlights significant achievements in cancer research.

www.cytopathnet.org A good resource for information on the cytological examination of tumors and on matters related to staining patterns that are typical of the cell populations found in a number of cancers.

www.iarc.fr The International Agency for Research on Cancer (IARC) is an extension of the World Health Organization that aims to identify the causes of cancer and promote international collaborations surrounding cancer research.

CLINICAL FOCUS QUESTION Why is cervical cancer a likely target for a vaccine that can prevent cancer? Can the approach being investigated for cervical cancer be applied to all types of cancer?

1. Indicate whether each of the following statements is true or false. If you think a statement is false, explain why.

   a. Hereditary retinoblastoma results from overexpression of a cellular oncogene.

   b. Translocation of the *c-myc* gene is found in many patients with Burkitt's lymphoma.

   c. Multiple copies of cellular oncogenes are sometimes observed in cancer cells.

   d. Viral integration into the cellular genome may convert a proto-oncogene into a transforming oncogene.

   e. All oncogenic retroviruses carry viral oncogenes.

   f. The immune response against a virus-induced tumor protects against another tumor induced by the same virus.

2. You are a clinical immunologist studying *acute lymphoblastic leukemia* (ALL). Leukemic cells from most patients with ALL have the morphology of lymphocytes but do not express cell-surface markers characteristic of mature B or T cells. You have isolated cells from ALL patients that do not express membrane Ig but do react with mAb against a normal pre-B-cell marker (B-200). You therefore suspect that these leukemic cells are pre-B cells. How would you use genetic analysis to confirm that the leukemic cells are committed to the B-cell lineage?

3. In a recent experiment, melanoma cells were isolated from patients with early or advanced stages of malignant melanoma. At the same time, T cells specific for tetanus toxoid antigen were isolated and cloned from each patient.

   a. When early-stage melanoma cells were cultured together with tetanus-toxoid antigen and the tetanus-toxoid-specific T-cell clones, the T-cell clones were observed to proliferate. This proliferation was blocked by addition of chloroquine, a drug that accumulates in lysosomes, or by addition of mAb to HLA-DR. Proliferation was not blocked by addition of mAb to HLA-A, -B, -DQ, or -DP. What might these findings indicate about the early-stage melanoma cells in this experimental system?

   b. When the same experiment was repeated with advanced-stage melanoma cells, the tetanus-toxoid T-cell clones failed to proliferate in response to the

tetanus-toxoid antigen. What might this indicate about advanced-stage melanoma cells?

   c. When early and advanced malignant melanoma cells were fixed with paraformaldehyde and incubated with processed tetanus toxoid, only the early-stage melanoma cells could induce proliferation of the tetanus-toxoid T-cell clones. What might this indicate about early-stage melanoma cells?

   d. How might you confirm the hypotheses in a, b, and c experimentally?

4. Describe three likely sources of tumor antigens.

5. Various cytokines have been evaluated for use in tumor immunotherapy. Describe four mechanisms by which cytokines mediate anti-tumor effects and the cytokines that induce each type of effect.

6. Infusion of transfected melanoma cells into cancer patients is a promising immunotherapy.

   a. Name two genes that have been transfected into melanoma cells for this purpose. What is the rationale behind the use of each of these genes?

   b. Why might use of such transfected melanoma cells also be effective in treating other types of cancers?

7. For each of the following descriptions, choose the most appropriate term:

Descriptions	*Terms*
a. A benign or malignant tumor	1. Sarcoma
b. A tumor that has arisen from endodermal tissue	2. Carcinoma
c. A tumor that has arisen from mesodermal connective tissue	3. Metastasis
d. A tumor that is invasive and continues to grow	4. Neoplasm
e. Tumor cells that have separated from the original tumor and grow in a different part of the body	5. Malignant
f. A tumor that is noninvasive	6. Leukemia
g. A tumor that has arisen from lymphoid cells	7. Transformation
h. A permanent change in the genome of a cell that results in abnormal growth	8. Lymphoma
i. Cancer cells that have arisen from hematopoietic cells that do not grow as a solid tumor	9. Benign

Experimental Systems and Methods

The days when an experimental methods chapter in an immunology textbook could neatly describe all the techniques used by practitioners of the subject are long gone. Immunologists use tools derived from the arsenals of structural biologists, biochemists, cell biologists, anatomists, microbiologists, and physiologists. In return, the science of immunology has donated an extensive toolbox of antibody-based and fluorescence-based techniques to the biological sciences.

In this chapter, we have attempted to provide students with the ability to understand the methodological choices made by professional immunologists. We hope that students will learn from it some of the advantages and limitations of many of the techniques they will encounter as they read in the immunological literature. Second, we have tried to provide students with the tools to understand the context in which particular methods or techniques are applicable, as they design their own original experiments. Fearlessness in following the interesting questions wherever they lead is one of the attributes of great scientists, and in this chapter we offer students a little insight into the broad array of technical possibilities available to them as they pursue their own questions. We have included some classical techniques, in order to support those reading in the earlier literature of the field, but we have also attempted to add some of the methods developed more recently. This chapter is designed to provide insight into the technical aspects of specific experiments described in previous chapters that advanced the field.

Space does not permit a detailed description of every technique, and this chapter is designed to provide an overall sense of the applicability of different immunological methods, rather than specific protocols for pursuing them. Students who wish to delve further into the details of any particular method can then locate specific protocols in a variety of sources, some of which are noted in the "Useful Web Sites" section at the end of the chapter. For purposes of concision, we have elected not to describe methods derived from molecular biology or biochemistry, as we believe that most students of

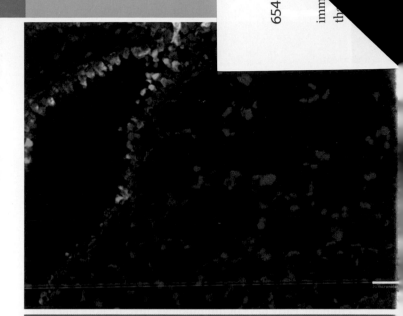

Mouse inflamed lung tissue stained for activated airway epithelial cells (green), infiltrating macrophages (red), and cell nuclei (blue). *[Image courtesy of Meera Nair, Laurel Monticelli, and David Artis, Perelman School of Medicine, University of Pennsylvania.]*

- Antibody Generation
- Immunoprecipitation- Based Techniques
- Agglutination Reactions
- Antibody Assays Based on Molecule Binding to Solid-Phase Supports
- Methods to Determine the Affinity of Antigen-Antibody Interactions
- Microscopic Visualization of Cells and Subcellular Structures
- Immunofluorescence-Based Imaging Techniques
- Flow Cytometry
- Magnetic Activated Cell Sorting
- Cell Cycle Analysis
- Assays of Cell Death
- Biochemical Approaches Used to Elucidate Signal Transduction Pathways
- Whole Animal Experimental Systems

...nology will already have been exposed to them, and
...t they are better handled in the relevant texts.

We begin by enumerating those methodologies that
are used to generate antibodies, and then describe some
of the many ways in which the interactions between
antibodies and antigens can be analyzed. We then briefly
describe some of the methods used to visualize cellular
and subcellular structures in the immune system, before
moving on to a discussion of various magnetic and
fluorescence-based techniques used for cell sorting and
cellular analysis at the population level. Assays that
analyze the cell cycle and measure cell death are then
addressed, and we complete the chapter with a brief
description of a number of commonly used whole animal
experimental systems.

Antibody Generation

From the early days of immunology, investigators and clinicians have made use of the ability of animals to respond to
immunization with the production of antibodies directed
toward injected antigens, such as viruses, bacteria, fungi, or
simple chemicals from the laboratory shelf. Antibodies harvested from the serum of immunized animals are the
secreted products of many clones of B cells and are thus
referred to as polyclonal antibodies. With subsequent immunizations using the same antigen, the average affinity of this
polyclonal antibody mixture for the antigen increases, as a
result of the process of affinity maturation, described in
Chapter 12.

Polyclonal Antibodies Are Secreted by Multiple Clones of Antigen-Specific B Cells

Polyclonal antibodies are generated by immunizing an experimental animal or a human subject with antigen one or more
times, bleeding the subject, and purifying the antibodies
from the subject's **serum**. Serum is what remains when both
the cellular components and the clotting factors have been
removed from the subject's blood. The addition of *adjuvants*
to the immunizing preparation elicits a stronger immune
response by deliberately engaging the innate immune system
to help in the activation of antigen-specific B and T cells.
Some current approaches to antibody generation employ Toll
receptor agonists such as poly I:C, an artificial nucleic acid
preparation, as an adjuvant and indeed, this approach is
being actively pursued in research into cancer immunotherapy. Traditionally, Freund's Complete Adjuvant, or alum was
used to maximize mouse antibody responses to antigens that
were mixed with the adjuvants prior to injection.

Because polyclonal antibodies are a mixture of antibodies
directed toward a variety of different epitopes of the immunizing antigen, they are particularly useful for techniques
such as agglutination or immunoprecipitation, which rely on
the ability of the antibody to form a large antigen-antibody
complex. The disadvantage of using a polyclonal preparation
is that some of the antibodies in the mixture may have ill-defined cross-reactivities with related antigens. Furthermore, since the antibody response matures with time post
immunization (Chapter 12), the range of cross-reactivities of
different preparations of polyclonal antibodies may vary
among different bleeds, even when they are derived from the
same donor animal.

A Monoclonal Antibody Is the Product of a Single Stimulated B Cell

The disadvantages of unforeseen cross-reactivities or variations in the fine specificity of polyclonal antibodies are
eliminated when using monoclonal antibodies (mAbs),
which are the product of a single, stimulated B cell. In 1975,
Georges Köhler and Cesar Milstein figured out how to generate large quantities of antibodies derived from a single
B-cell clone (Figure 20-1). By fusing a normal, activated,
antibody-producing B cell with a myeloma cell (a cancerous
plasma cell), they were able to generate a **hybridoma** that
possessed the immortal growth properties of the myeloma-cell parent and secreted the unique antibody produced by
the B-cell parent. Over time, myeloma-cell partners were
generated that had lost the ability to synthesize their own
immunoglobulin, thus ensuring that the only antibodies
secreted into the culture medium were those from the B-cell
fusion partner.

The original fusions used Sendai virus to disrupt the
plasma membrane of the cells; nowadays, chemical fusogens
such as polyethylene glycol are used instead. In general,
fusions between three or more cells are unstable, and the vast
majority of fused cells growing out of these cultures are the
products of the hybridization of two parent cells.

Hybrids formed by the fusion of two antibody-producing
B cells will not grow out of these cultures because B cells
have a relatively short half-life in vitro. However, hybrids
formed by the fusion of two or more cancer cells would
have the potential to grow out from the initial fusions, and
compete successfully for nutrients with the B cell-myeloma
hybrids. A method therefore had to be devised to eliminate
these tumor-tumor hybrids from the cultures of fused cells.
Köhler and Milstein solved this problem by using myeloma
cells lacking the enzyme *hypoxanthine guanine phosphoribosyl transferase* (HGPRT). HGPRT is necessary for one
of the two potential pathways of DNA synthesis, the *salvage*
pathway. The alternative pathway of DNA synthesis, the *de
novo* pathway, can be inhibited by the antibiotic aminopterin. Köhler and Milstein reasoned that, if they grew the
hybrid cultures in the presence of aminopterin, the mutant
tumor cells and tumor-tumor hybrids would be unable to
synthesize new DNA by either the salvage or the de novo
pathways and would eventually die. However, in the
hybridomas formed by fusion between B cells and tumor
cells, the B-cell parent would provide the HGPRT, and so

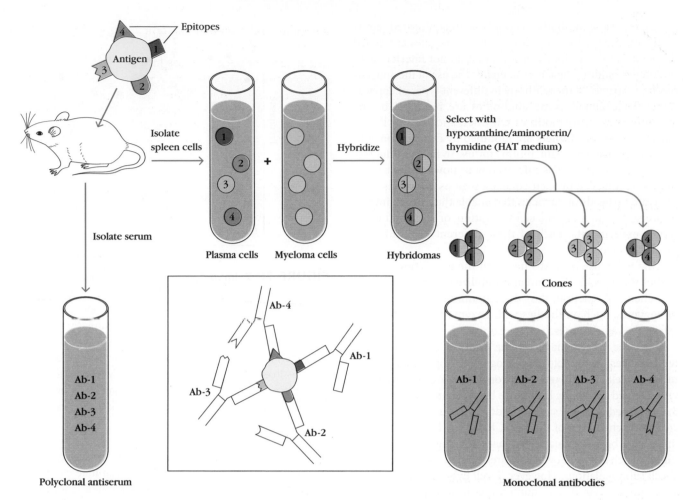

FIGURE 20-1 The conventional polyclonal antiserum produced in response to a complex antigen contains a mixture of monoclonal antibodies, each specific for one of the four epitopes shown on the antigen (inset). In contrast, a monoclonal antibody, which is derived from a single plasma cell, is specific for one epitope on a complex antigen. The outline of the basic method for obtaining a monoclonal antibody is illustrated here.

these hybrids would survive in the selection medium. Because the medium containing aminopterin is normally supplemented with hypoxanthine and thymidine to support nucleotide synthesis, the selective medium is known as "HAT medium."

The resulting clones of hybridoma cells randomly lose chromosomes over the first few days following fusion, but eventually they stabilize and can be cultured indefinitely, secreting large quantities of mAbs of predefined specificity and known cross-reactivity. The importance of hybridomas to the biological sciences was recognized when Köhler and Milstein were awarded the Nobel Prize in Physiology or Medicine in 1984.

The precise specificity, affinity, and cross-reactivity of mAbs are entirely stable with time, and they are particularly useful for diagnostic purposes. However, mAbs are less useful than polyclonal preparations in applications that require agglutination, since every antibody in the experiment has the identical binding site and, thus, fewer antibodies will be bound per antigen molecule.

Monoclonal Antibodies Can Be Modified for Use in the Laboratory or the Clinic

Monoclonal antibodies provide a reproducible binding site that will attach the antibody to its target cell or molecule, and so it is not surprising that the range of applications for mAbs has been limited only by the imagination of the investigators.

An mAb can be genetically modified so that only the binding site is retained, and the rest of the molecule is replaced by some other molecule. For example, because injection of large amounts of xenogeneic antibodies can induce inflammation, hybridomas secreting mAbs to be used in immunotherapy in the clinic are often subject to genetic manipulations, such that the binding sites of the original mouse mAb are cut and pasted onto the Fc regions of human antibodies. In addition, some antibodies are modified by conjugation to toxins designed to kill cells to which the antibody will bind. Many such chimeric and toxin-conjugated antibodies are now in regular clinical use.

A plethora of antibodies is available in conjugated (covalently attached) forms, in which other molecules are joined to regions of the antibody in ways that do not interfere with its antigen-binding capacity, but enable the use of the antigen-binding capacity of the antibody in different types of applications. For example, some antibodies are modified by the attachment of either biotin or enzymes, and used in ELISA assays (see below). Others are modified by conjugation with fluorescent dyes (or again, biotin) for use in immunofluorescent applications such as microscopy or flow cytometry. Yet others are attached by their constant regions to various types of synthetic beads or particles that enable their use in immunoprecipitation, magnetic cell separation, or electron microscopic experiments. Each of these different applications is discussed below.

One novel use of mAbs has been the generation of antibodies that specifically bind and stabilize the transition state of a chemical reaction, thus directly mimicking the activity of enzymes. Such antibodies with enzyme-like activities are referred to as **abzymes**.

Once the hybridoma technique had been established for B lymphocytes, immunologists recognized its potential usefulness to T-cell biology, and many long-term T-cell hybridomas with defined specificity have since been generated. In Chapter 3, we learned how one such hybridoma was used to characterize the biochemistry of the αβTCR. Other T-cell hybridoma lines have been invaluable in characterizing the conditions under which different cytokines are secreted and the nature of T-cell subpopulations. However, T-cell hybrids have not been as useful as their B-cell counterparts in terms of providing therapeutic, diagnostic, and general laboratory reagents, because they do not secrete a soluble receptor molecule and their binding specificity requires an MHC-based peptide.

Immunoprecipitation- Based Techniques

The multivalency of antibodies has allowed the development of techniques in which antibody-bound molecules can be precipitated from solution, or otherwise separated from nonbound molecules for further analysis. Some of these techniques are quite venerable; other applications are brand new. Still, all rely on the ability of antibodies to bind to more than one antigenic determinant on a single antigen, thus forming a large complex that will fall out of solution.

Immunoprecipitation Can Be Performed in Solution

When antibodies and soluble antigen are mixed in solution, the bi- or multivalent nature of immunoglobulins allows for a single antibody molecule to bind to more than one antigen (Figure 20-2). If the antigen is polyvalent (has more than one antibody binding site per antigen molecule), it may in its turn bind multiple different antibodies. Eventually, the

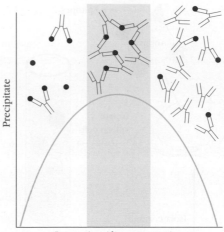

FIGURE 20-2 Immunoprecipitation in solution. When bi- or multivalent antibodies are mixed in solution with antigen, the antibodies can form cross-linkages with two or more antigen molecules, leading to the formation of a cross-linked precipitate (middle panel of graph). Precipitate formation requires that neither antigen (left panel in graph) nor antibody (right panel in graph) molecules are in excess. In either of these two cases, primarily monovalent binding takes place, as shown in the diagram. *[http://nfs.unipv.it/nfs/minf/ dispense/immunology/lectures/files/images/immunoprecipitation_electrophoresis.jpg]*

resulting cross-linked complex becomes so large that it falls out of solution as a precipitate. This precipitate can be spun out of the solution and the antigen separated from the precipitating antibodies by biochemical means.

Solution immunoprecipitation can be used to purify antigenic molecules from a heterogeneous mixture of soluble molecules, or to remove particular antigens from a solution. Immunoprecipitation occurs only when the antibody and antigen concentrations are essentially equivalent (see center panel in Figure 20-2). At either antigen (Figure 20-2, left panel) or antibody (Figure 20-2, right panel) excess, monovalent binding is favored that does not result in the formation of a precipitate. Recall from Chapter 3 that Kabat used immunoprecipitation with the ovalbumin antigen to remove anti-ovalbumin antibodies from solution, followed by electrophoresis; this experiment characterized antibodies as belonging to the γ globulin class of serum proteins.

Immunoprecipitation of Soluble Antigens Can Be Performed in Gel Matrices

Immune precipitates can form not only in solution but also in an agar matrix. When antigen and antibody diffuse toward one another in a gel matrix, a visible line of precipitation will form. As in a precipitation reaction in solution, visible precipitation occurs when the concentrations of antibody and antigen are equivalent to one another.

Immunodiffusion in gels is rapid, easy to perform and surprisingly accurate. In the Ouchterlony method, the most frequently employed variation of gel immunoprecipitation, both antigen and antibody diffuse radially from wells toward each other, thereby establishing a concentration gradient. At the relative antibody-antigen concentrations at which lattice formation is maximized, termed "equivalence," a visible line of precipitation, or "precipitin line," forms in the gel. More sophisticated analyses of Ouchterlony gels can offer information regarding the extent of cross-reactivity of antibody preparations with related antigens (see Figure 20-3).

Although various modifications of precipitation reactions were at one time the major types of assay used in immunology, other more sensitive methods are now available for antigen and antibody measurement and are described below. However, Ouchterlony assays are still occasionally used in the clinic. Table 20-1 presents a comparison of the *sensitivity*, or minimum amount of antibody detectable, of a number of immunoassays.

Immunoprecipitation Allows Characterization of Cell-Bound Molecules

A variant of immunoprecipitation is often used to isolate protein antigens from cell and tissue samples. A detergent extract of cells or tissues is mixed with antibodies to the protein of interest to form an antigen-antibody complex. To facilitate efficient retrieval of this complex, a secondary antibody, or other protein such as the bacterial Proteins A or G, can now be added. These secondary reagents all bind specifically to the Fc region of the first antibody and are normally preattached to a solid-phase support, such as a synthetic bead. Since the beads can be easily spun down in a

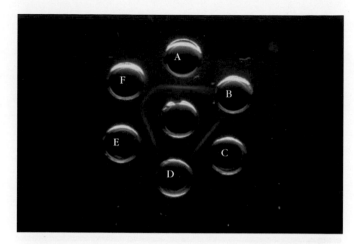

FIGURE 20-3 Immunodiffusion in agar gels can be used to assay for the presence of antibodies and determine cross-reactivity patterns between complex antigens and antibody samples. In this example, viral antigen was placed in the center well, and serum samples from different individuals were introduced into each of the surrounding wells. Note that the serum samples of individuals B and D are negative for antiviral antibodies, in that there is no line of precipitate between the serum sample on the outside and the viral sample in the center; all other serum samples are positive for antiviral antibodies. *[ASM MicrobeLibrary.org © Thomas Walton and Erica Suchman]*

centrifuge, the antibody-antigen-bead complex can be collected by centrifugation. Following centrifugation, the protein of interest can be separated from the precipitating antibodies by SDS gel electrophoresis. In a variant of this

TABLE 20-1	Sensitivity of various immunoassays
Assay	**Sensitivity*** **(μg antibody/ml)**
Precipitation reaction in fluids	20–200
Precipitation reaction in gels	
Ouchterlony double immunodiffusion	20–200
Agglutination reactions	
Direct	0.3
Agglutination inhibition	0.006–0.06
Radioimmunoassay (RIA)	0.0006–0.006
Enzyme-linked immunosorbent assay (ELISA)	~0.0001–0.01
ELISA using chemiluminescence	~0.00001–0.01†
Immunofluorescence	1.0
Flow cytometry	0.006–0.06

*The sensitivity depends on the affinity of the antibody used for the assay as well as the epitope density and distribution on the antigen.

† Note that the sensitivity of chemiluminescence-based ELISA assays can be made to match that of RIA.

Source: Updated and adapted from N. R. Rose et al., eds., 1997, *Manual of Clinical Laboratory Immunology*, 5th ed., Washington, DC: American Society for Microbiology.

technique, the beads attached to the secondary reagents may be magnetic, in which case the protein of interest is purified by passage over a magnetic column (see below).

Western blotting (see below) can then be used to ascertain the efficacy of the immunoprecipitation, to estimate the relative abundance of the bound protein in the tissue sample and to determine which other proteins co-immunoprecipitated and therefore are most likely associated with the target protein in its cellular location. Such co-immunoprecipitation studies were the first clue to the multimolecular natures of the TCR and BCR co-receptor complexes.

Agglutination Reactions

The cross-linking that occurs between di- or multivalent antibodies and multivalent, bacterial, or other cellular antigens can result in visible clumping of the complexes formed between cells bearing the antigens and the antibody molecules. This clumping reaction is called **agglutination**, and antibodies that produce such reactions are called **agglutinins**. Agglutination reactions are identical in principle to precipitation reactions; the only difference is that the cross-linked product is visible to the naked eye because of the larger size of the antigens.

Hemagglutination Reactions Can Be Used to Detect Any Antigen Conjugated to the Surface of Red Blood Cells

When antibodies bind antigens on the surface of *red blood cells* (RBCs), the resultant clumping reaction is referred to as **hemagglutination**. In the example shown in Figure 20-4, control buffer was added to well 10 of the microtiter tray. Antibodies to *sheep red blood cells* (SRBCs) were added to well one of this tray, and then this antiserum was serially diluted into wells 2 through 9, such that the concentration of antibodies to the SRBCs in well 2 was half that in well 1, and so on. The same number of SRBCs was then added to each well.

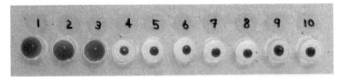

FIGURE 20-4 Demonstration of hemagglutination using antibodies against sheep red blood cells (SRBCs). The control tube (10) contains only SRBCs, which settle into a solid "button." The experimental tubes 1 to 9 contain a constant number of SRBCs plus serial twofold dilutions of anti-SRBC serum. The spread pattern in the experimental series indicates positive hemagglutination through tube 3. *[Louisiana State University Medical Center/MIP. Courtesy of Harriet C. W. Thompson.]*

In well 10, in the absence of any agglutinating antibody, the SRBCs settle into a tight button in the bottom of the well. This tight button represents a *negative* result in a hemagglutination assay. In well 1, the high concentration of anti-SRBC antibodies induced cross-linking of the SRBCs, so that they form a large clump and do not fall down to the bottom of the well. The diffuse shading of RBCs seen in well 1 represents a *positive* interaction between the antibodies and the SRBC surface antigen. The concentration of anti-SRBC antibodies in wells 2 and 3 remains high enough to allow hemagglutination, but once the antibodies have been diluted eightfold (well 4), there are too few antibodies to generate cross-links and the SRBCs can again settle into the bottom of the well. The responses in wells 1, 2, and 3 therefore represent a positive hemagglutination reaction.

Hemagglutination reactions are routinely performed to type RBCs. With tens of millions of blood-typing determinations run each year, this is one of the world's most frequently used immunoassays. In typing for the human ABO antigens, human RBCs are mixed with antisera to the A or B blood-group antigens. If the antigen is present on the cells, they agglutinate, forming a visible clump on the slide.

The ease and sensitivity of hemagglutination reactions and the fact that they do not require sensitive instrumentation for data analysis mean that hemagglutination assays can be adapted to measure antibodies directed against any antigen that can be attached to the RBC surface.

Hemagglutination Inhibition Reactions Are Used to Detect the Presence of Viruses and of Antiviral Antibodies

Hemagglutination inhibition reactions are also useful tools in the clinic and in the laboratory for the detection of viruses and of antiviral antibodies. Some viruses (most notably influenza) bear multivalent proteins or glycoproteins on their surfaces that interact with macromolecules on the RBC surface, and induce agglutination of the RBCs. For example, the influenza virus envelope bears a trimeric glycoprotein, hemagglutinin (HA). This HA molecule is subjected to mutation and selection, such that different strains of influenza bear different HA types, which in turn are bound by different antibodies. However, all HA molecules bind in a multivalent manner to the sialic acid residues on RBCs and agglutinate them.

To determine whether a patient has antibodies to a particular strain of influenza virus, a technician would perform a serial dilution of the patient's antiserum in a microtiter plate. The technician would then add the relevant virus and RBCs to each well, at concentrations known to induce hemagglutination. If the patient's antiserum has anti-HA antibodies that bind the particular influenza strain being tested, the antibodies will attach to the HA molecules on the surface of the virus and prevent those molecules from inducing hemagglutination. Therefore, the hemagglutination reaction will be inhibited at several antibody concentrations.

Bacterial Agglutination Can Be Used to Detect Antibodies to Bacteria

A bacterial infection often elicits the production of antibacterial antibodies specific for surface antigens on the bacterial cells; such antibodies can be detected by bacterial agglutination reactions. The principle of bacterial agglutination is identical to that for hemagglutination, but in this case the visible pellet is made up of bacteria, cross-linked by antibacterial antibodies.

Agglutination reactions can also provide quantitative information about the concentration of antibacterial antibodies in a patient's serum. The patients' sera are serially diluted, as described above. The last well in which agglutination is visible tells us the **agglutinin titer** of the patient, defined as *the reciprocal of the greatest serum dilution that elicits a positive agglutination reaction*. The agglutinin titer of an antiserum can be used to diagnose a bacterial infection. Patients with typhoid fever, for example, show a significant rise in the agglutination titer to *Salmonella typhi*. Agglutination reactions also provide a way to type bacteria. For instance, different species of the bacterium *Salmonella* can be distinguished by agglutination reactions with a panel of typing antisera.

Antibody Assays Based on Molecule Binding to Solid-Phase Supports

With increasing numbers of antibody-based assays in use in the clinic, and the consequent need for automation to ensure high throughput of patient diagnostic tests, many antibody-based assays now rely on antibodies or antigens that are bound to solid-phase supports, such as microtiter plates, microscope slides, or beads of different kinds.

Radioimmunoassays Are Used to Measure the Concentrations of Biologically Relevant Proteins and Hormones in Body Fluids

Although many *radio*immunoassays (RIAs) have now been replaced by enzyme-based immunoassays, some hormonal measurements are still made using this technology. Furthermore, the historical significance of RIAs necessitates a brief introduction here.

In 1960, two endocrinologists, S. A. Berson and Rosalyn Yalow, designed an exquisitely sensitive technique to determine levels of insulin/anti-insulin complexes in diabetic patients. Their technique soon proved its value for measuring hormones, serum proteins, drugs, and vitamins at levels that were orders of magnitude lower than had previously been detectable. Their accomplishment was recognized in 1977 (several years after Berson's death), by the award of a Nobel Prize to Yalow (Figure 20-5). The key to understanding the significance of this technological breakthrough is to realize that, prior to the development of

FIGURE 20-5 Rosalind Sussman Yalow. Creator of the radio-immunoassay technique and Nobel Laureate, 1977. *[Courtesy Department of Chemistry, Michigan State University.]*

the RIA, the concentrations of many biologically relevant proteins and hormones in body fluids were too low to detect by any known methods. Rather than representing merely an alteration in the sensitivity of established assays, the RIA therefore made it possible to measure substances that had hitherto been undetectable by *any* quantitative methodology.

Since the initial description of Yalow's assay, many technical variations have been developed to make the method more rapid and reliable. However, all variations depend on the availability of radioactively labeled antibody or antigen, and a method by which to separate antigen-antibody complexes from unbound reagents. We will describe one such assay here, but there are many ways to use this methodology to accomplish the desired experimental goal. Most RIAs still in use are based on the binding of antibody or antigen to a solid-phase support, such as the polystyrene or polyvinylchloride wells of a microtiter plate. The radioactive label that is most commonly used is ^{125}I, which binds to exposed tyrosine residues on proteins, with little effect on their overall structure.

The goal of the application we will describe is to determine the concentration of a particular cytokine in the blood

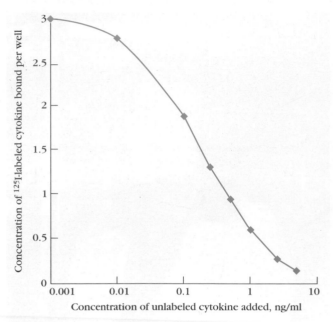

FIGURE 20-6   A competitive, solid-phase radioimmunoassay (RIA) to measure cytokine concentrations in serum. Anticytokine antibody is used to coat an RIA plate. A standard curve is obtained by adding increasing concentrations of unlabeled cytokine to a constant amount of ^{125}I-labeled cytokine. The experimental sample is then added to duplicate or triplicate wells containing the same amount of labeled cytokine as that used for the standard curve. As the unlabeled cytokine outcompetes the labeled form for binding to the plate, the amount of radioactivity per well drops in a predictable fashion, shown here. The amount of cytokine in the experimental samples can then be measured by interpolation from the standard curve.

of a patient. First, the wells of a microtiter plate are coated with a constant amount of antibody specific for the cytokine. The surface of the plastic binds tightly to proteins that stick, essentially irreversibly, to the plastic surface.

A known amount of radiolabeled cytokine is added to a set of control wells. A standard curve of unlabeled cytokines is then set up by adding increasing, known concentrations of cytokine to successive wells, along with the radiolabeled cytokine. As more unlabeled antigen competes with the labeled antigen, less and less radiolabeled cytokine will bind. After an incubation period, the amount of bound radiolabeled material is measured by washing off the unbound material and measuring the radioactivity in individual wells. An example of a standard curve generated in this way is shown in Figure 20-6.

The measurement of the amount of cytokine in the experimental samples is accomplished by treating the unknown samples in exactly the same way as the standard curve. The investigator then compares the amount of radioactivity bound to the plate in the experimental wells with the radioactive signal obtained in the standard curve wells containing known amounts of unlabeled cytokine.

One can measure either antigen or antibody concentrations using variations on this basic technique, which is extremely powerful, and capable of sensitivities in the picogram range. However, with the proliferation of RIAs came increasing levels of concern about the amount of radioactivity generated by research and clinical laboratories and the associated risks to the technical staff and to the environment. The next development in antibody-antigen solid-phase support assays was the *Enzyme Linked Immuno-Sorbent Assay*, or ELISA.

ELISA Assays Use Antibodies or Antigens Covalently Bound to Enzymes

ELISA assays are similar in principle to RIAs but, instead of using antibodies or antigens conjugated to radioisotopes, they use antibodies or antigens covalently bound to enzymes. The conjugated enzymes are selected on the basis of their ability to catalyze the conversion of a substrate into a colored, fluorescent, or chemiluminescent product. These assays match the sensitivity of RIAs and have the advantage of being safer and, often, less costly.

A number of variations of the basic ELISA assay have been developed (Figure 20-7). Each type of ELISA can be used qualitatively to detect the presence of antibody or antigen. Alternatively, a standard curve based on known concentrations of antibody or antigen can be prepared and used to determine the concentration of a sample.

Indirect ELISA

Antibody can be detected, or its concentration determined with an *indirect ELISA* assay (see Figure 20-7a). Serum or some other sample containing primary antibody (Ab_1) is added to an antigen-coated microtiter well and allowed to react with the antigen attached to the well. After any free Ab_1 is washed away, the antibody bound to the antigen is detected by adding an enzyme-conjugated secondary antibody (Ab_2) that binds to Ab_1. Any free Ab_2 is again washed away, and a substrate for the enzyme is added. The amount of colored, fluorescent, or luminescent reaction product that forms is measured using a specialized plate reader and compared with the amount of product generated when the same set of reactions is performed using a standard curve of known Ab_1 concentrations. (A *direct* ELISA assay would detect the amount of antigen on the plate using enzyme coupled antibodies, and is rarely used.)

This version of ELISA is the method of choice to detect the presence of serum antibodies against *human immunodeficiency virus* (HIV), the causative agent of AIDS. In this assay, recombinant envelope and core proteins of HIV are adsorbed as solid-phase antigens to microtiter wells. Individuals infected with HIV will produce serum antibodies to epitopes on these viral proteins. Generally, serum antibodies to HIV can be detected by ELISA within 6 weeks of infection.

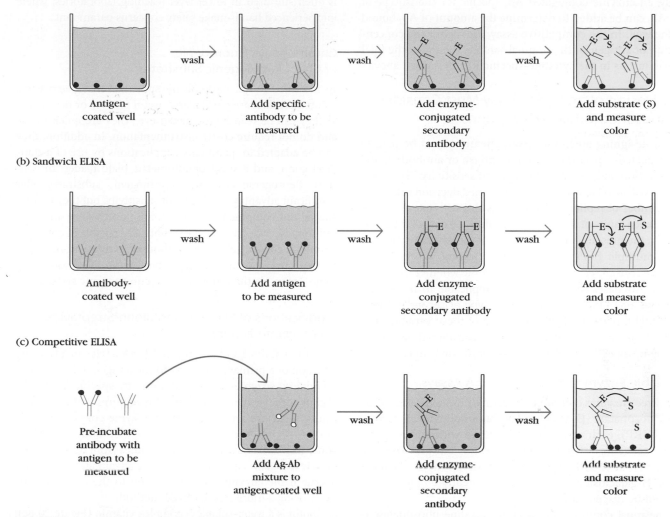

(a) Indirect ELISA

| Antigen-coated well | wash → | Add specific antibody to be measured | wash → | Add enzyme-conjugated secondary antibody | wash → | Add substrate (S) and measure color |

(b) Sandwich ELISA

Antibody-coated well — wash → Add antigen to be measured — wash → Add enzyme-conjugated secondary antibody — wash → Add substrate and measure color

(c) Competitive ELISA

Pre-incubate antibody with antigen to be measured — Add Ag-Ab mixture to antigen-coated well — wash → Add enzyme-conjugated secondary antibody — wash → Add substrate and measure color

FIGURE 20-7 Variations in enzyme-linked immunosorbent assay (ELISA) technique allow determination of antibody or antigen. Each assay can be used qualitatively or quantitatively by comparison with standard curves prepared with known concentrations of antibody or antigen. Antibody can be determined with an indirect ELISA (a), whereas antigen can be determined with a sandwich ELISA (b) or competitive ELISA (c). In the competitive ELISA, which is an inhibition-type assay that is identical in principle to the competition RIA described above, the concentration of antigen is inversely proportional to the color produced.

Sandwich ELISA

Antigen can be detected or measured by a sandwich ELISA (see Figure 20-7b). In this technique, the antibody (rather than the antigen) is immobilized on a microtiter well. A sample containing unknown amounts of antigen is allowed to react with the immobilized antibody. After the well is washed, a second enzyme-linked antibody specific for a different epitope on the antigen is added and allowed to react with the bound antigen. After any free second antibody is removed by washing, substrate is added, and the colored reaction product is measured.

A common variant on this assay uses a biotin-linked second antibody and then adds enzyme-linked avidin in an additional step (see below). Sandwich ELISAs have proven particularly useful for the measurement of soluble cytokine concentrations in tissue culture supernatants, as well as in serum and body fluids. Note that, for this assay to work, the two antibodies used for the antigen immobilization (capture) and detection phases respectively must bind to different determinants (epitopes) on the antigen. Sandwich ELISAs therefore routinely use a pair of monoclonal antibodies specific for different regions on the antigen.

Competitive ELISA

The competitive ELISA provides another extremely sensitive variation for measuring amounts of antigen (see Figure 20-7c). In this technique, antibody is first incubated in solution with a sample containing antigen. The antigen-antibody mixture is then added to an antigen-coated microtiter well. The more antigen present in the initial solution-phase sample, the less free antibody will be available to bind to the antigen-coated well. After washing off the unbound antibody, an enzyme-conjugated Ab_2 specific for the isotype of the Ab_1 can be added to determine the amount of Ab_1 bound to the well. In the competitive assay, the higher the concentration of antigen in the original sample, the lower the final signal, just as in the cytokine-specific RIA described above.

The Design of an ELISA Assay Must Consider Various Methodological Options

When designing an ELISA assay, attention must be paid to the expected concentration of the antigen or antibody in the test solution and hence to the required sensitivity. The investigator will also have to make an informed decision regarding the number of replicates required, as this will determine the confidence limits of the results and affect the cost of the assay. Finally, the present and future needs of the laboratory to perform multiple ELISAs for different antigens or antibodies must be considered, as this will determine whether the investigator purchases reagents that are applicable to more than one type of ELISA, or just a single type of secondary antibody. Below, we briefly discuss some of these variables and other technical factors that have to be considered by anyone setting up a particular ELISA assay for the first time.

Available Enzyme Systems for ELISA Assays

The three enzymes commonly used in ELISA assays are alkaline phosphatase, β-galactosidase, and *horseradish peroxidase* (HRP), each of which can be used with chromogenic substrates (substrates which are colorless but give rise to a product that absorbs light in the visible range). Of these three, β-galactosidase is the least frequently employed and will not be discussed further.

The most common substrate for alkaline phosphatase is *p-nitrophenyl phosphate* (pNPP), which is highly soluble and colorless. Alkaline phosphatase cleaves the phosphate group from this substrate to yield p-nitrophenol, which is yellow and absorbs light at 460 nm. Both substrate and product are nontoxic and inexpensive, and the substrate can be purchased in tablet form. However, the substrate will convert to product if left in solution at room temperature, which somewhat limits the range of the chromogenic ELISA assay. Chromogenic alkaline phosphatase assays are therefore extremely useful when a rapid, inexpensive, "yes or no" answer is called for, when the laboratory facilities are limited, and/or for multipurpose teaching laboratories. Recently, both fluorescent and luminescent assay kits for alkaline phosphatase have been made available by a variety of companies. The lower background level of light emission from such products dramatically enhances the sensitivity of the assays, allowing femtogram levels of antigens to be detected.

A similar array of substrate-product systems has been developed for the HRP-based ELISAs. Because fluorescent and luminescent substrates were available first for the HRP system, there are more HRP than alkaline phosphatase kits and variations currently available. The carcinogenic properties of some HRP substrates and products mean that alkaline phosphatase is often still used in lower-level teaching laboratories, where inexperienced hands make safety concerns paramount.

Chromogenic, Fluorogenic, or Chemiluminogenic Substrates

As indicated above, chromogenic substrates that give rise to a colored product are particularly useful for "yes or no" assays, as they can be read by the naked eye. They are quick to use and do not require costly instrumentation. In addition, they can be adapted to quantitative applications by use of a standard curve and a spectrophotometric plate reader. In contrast, fluorogenic and chemiluminogenic substrates offer significant advantages in terms of sensitivity, but they require specialized equipment. Indeed, some fluorescence and chemiluminescence-based assay systems are capable of detecting attomolar concentrations of antibodies or antigens. As might be expected, the substrates themselves are also somewhat more expensive than those used in chromogenic systems.

Modifications of ELISAs Using Biotin-Streptavidin Bonding Interactions

The original design of secondary ELISA assays involved the addition of a primary antibody to the antigen, followed by a secondary, enzyme-conjugated antibody specific for the Fc region of the primary antibody. This requires that each laboratory purchase a separate set of enzyme-conjugated antibodies specific for each class of primary antibody. Given that some ELISA kits use primary antibodies from different species, investigators quickly realized the advantages that would accrue if the enzyme could be bound to the primary antibody using a more standardized method.

Biotin is a water-soluble B complex vitamin (Figure 20-8a), which may have remained in chemical obscurity but for one significant property: It binds to the bacterial protein streptavidin with an affinity that is almost unparalleled in biology. Indeed, the K_d of the biotin-streptavidin interaction is of the order of 10^{-14}M, making it one of the strongest naturally occurring, noncovalent interactions in nature. Furthermore, this interaction is stable under a wide variety of conditions, including in the presence of both organic and nonorganic solvents, denaturants, and detergents, and in extremes of temperature. Streptavidin is a tetrameric protein capable of binding four molecules of biotin per molecule of streptavidin (Figure 20-8b).

Many chemical derivatives of biotin have been synthesized that enable covalent conjugation to antibodies, with minimal effect on the structure of the antibody or of biotin. Thus, one can use a variety of primary, biotin-conjugated antibodies with just one enzyme-conjugated stock of streptavidin. Biotin-streptavidin–based steps are now the norm in many immunological assays.

Students should note that one of the reasons for the sensitivity of ELISA assays lies in the opportunity for amplification at each step. Each molecule of primary antibody is capable of being conjugated by more than one molecule of biotin, and hence can

(a)

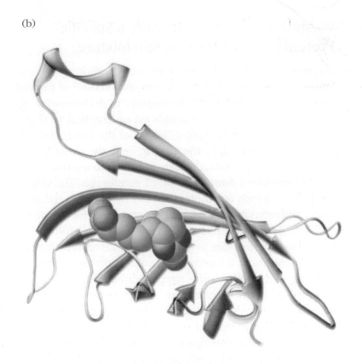

FIGURE 20-8 Biotin-streptavidin interactions are used in many immunologically based assay systems. (a) Biotin. (b) Monomeric streptavidin, in ribbon form, showing an individual bound biotin molecule. [(a), http://upload.wikimedia.org/wikipedia/commons/thumb/f/f9/Biotin_structure_JA.png/800px-Biotin_structure_JA.png; (b), http://upload.wikimedia.org/wikipedia/commons/5/5a/Streptavidin.png.]

bind more than one streptavidin molecule. Similarly, each enzyme molecule will process multiple substrates into product.

ELISPOT Assays Measure Molecules Secreted by Individual Cells

A modification of the ELISA, called the ELISPOT assay, allows the quantitative determination of the number of cells in a population that are producing a particular type of molecule. ELISPOT assays are frequently used to detect and quantify the number of cells in a population that are producing particular cytokines. They represent a variation on the sandwich ELISA described above.

In an ELISPOT assay for interferon γ (IFNγ), for example, assay plates are coated with anti-IFNγ antibody. This antibody is referred to as the "capture antibody" because its role is to capture interferon as it is secreted by individual cells and before it has had time to diffuse into the rest of the culture. A suspension of the cell population under investigation is then added to the coated plates and incubated with any appropriate stimulating agents. The cells settle onto the surface of the plate, and any interferon that is secreted by the stimulated cells is bound by the capture antibodies on the plate, creating a ring of IFNγ-antibody complexes around each interferon-producing cell. The plate is then washed to remove the cells, and an enzyme-linked antibody, the "detection antibody," specific for an antigenic determinant on IFNγ different from that bound by the capture antibody, is added and allowed to bind. After another washing step, an ELISPOT substrate is added. Substrates for ELISPOT assays are normally colorless but, when acted on by their cognate enzyme, they precipitate out of solution, leaving a clearly demarcated, colored "spot" wherever the enzyme-conjugated antibody bound. An example of data generated in an ELISPOT assay is shown in Figure 20-9.

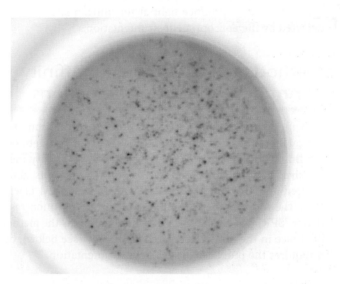

FIGURE 20-9 ELISPOT measurements of interferon γ secretion by NKT cells. A capture antibody to interferon γ was bound to a polyvinylidene difluoride–coated well of a microplate, and NKT cells were added at the appropriate concentration. After a 17-hour stimulation with phorbol myristoyl acetate and ionomycin, the cells were washed off and a biotin-conjugated, interferon-γ specific detection antibody was added and allowed to bind. Following removal of excess detecting antibody, streptavidin-conjugated alkaline phosphatase was added, followed (after washing out excess enzyme) by the substrate BCIP/NBT. BCIP/NBT is converted from a colorless solution to the brown-black substrate seen in the figure. The photograph shows the results of PMA/ionomycin stimulation of the NKT cells. Each brown-black spot represents the site of a cell that secreted interferon γ. These spots can be counted under a dissecting microscope at low power. [Nicole Cunningham and Jenni Punt, Haverford College.]

Western Blotting Can Identify a Specific Protein in a Complex Protein Mixture

Western blotting identifies and provides preliminary quantitation of a specific protein in a complex mixture of proteins. In Western blotting, a protein mixture is electrophoretically separated by SDS-polyacrylamide gel electrophoresis (Figure 20-10). In order to prevent diffusion of the bands that have been tightly focused by the electric field, the protein bands are then electrophoretically transferred on to a nitrocellulose or *polyvinylidene fluoride* (PVDF) membrane. The individual protein bands are then identified by flooding the membrane with enzyme-linked antibodies specific for the protein of interest. In an alternative version of the protocol that should now be familiar to the reader, the membrane may first be incubated with a biotin-conjugated antibody, followed by washing and addition of a streptavidin-conjugated enzyme. The protein-antibody complexes that form on the membrane are visualized by the addition of a chromogenic substrate that produces a highly colored and insoluble product at the site of the target protein. Even greater sensitivity can be achieved if a chemiluminescent compound with suitable enhancing agents is used to produce light at the antigen site, which is detected by the appropriate instrumentation.

Methods to Determine the Affinity of Antigen-Antibody Interactions

Two methods are commonly encountered in the immunological literature to determine the affinity of antigen-antibody binding. The older method, equilibrium dialysis, is easy, inexpensive, and illustrates several important concepts about antigen-antibody interactions. It will be described briefly, first. The more modern technique of *surface plasmon resonance* (SPR) has displaced equilibrium dialysis as the method of choice in the modern research laboratory (see below), but it requires the purchase of specific instrumentation.

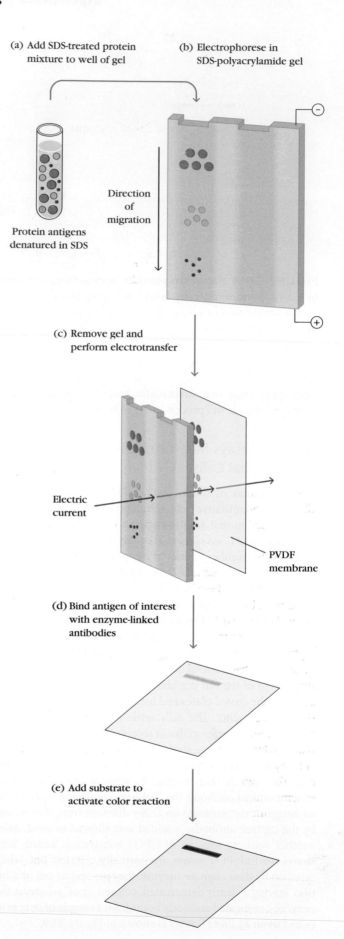

FIGURE 20-10 Western blotting uses antibodies to identify protein bands following gel electrophoresis. In Western blotting, a protein mixture is (a) treated with SDS, a strong denaturing detergent, (b) then separated by electrophoresis in an *SDS polyacrylamide gel* (SDS-PAGE), which separates the components according to their molecular weight; lower-molecular-weight components migrate farther than higher-molecular-weight components. (c) The gel is removed from the apparatus and applied to a protein-binding sheet of nitrocellulose or *polyvinylidene fluoride* (PVDF), and the proteins in the gel are transferred to the sheet by the passage of an electric current. (d) Addition of enzyme-linked antibodies detects the antigen of interest, and (e) the position of the antibodies is visualized by means of an ELISA reaction that generates a highly colored insoluble product that is deposited at the site of the reaction. Alternatively, a chemiluminescent ELISA can be used to generate light that is readily detected by exposure of the blot to a piece of photographic film.

Antibody affinity is a quantitative measure of binding strength between an antigen and an antibody. The combined strength of the noncovalent interactions between a *single* antigen-binding site on an antibody and a *single* epitope is the **affinity** of the antibody for that epitope and can be described by the association constant of the interaction (see Chapter 3).

$$K_a = \frac{[SL]}{[S][L]} \tag{1}$$

where
   [S] = the concentration of antibody binding sites,
   [L] = the concentration of free ligand and
   [SL] = the concentration of bound complexes.

Equilibrium Dialysis Can Be Used to Measure Antibody Affinity for Antigen

Equilibrium dialysis uses a chamber containing two compartments separated by a semipermeable membrane. Antibody is placed in one compartment, and a radioactively (or otherwise) labeled ligand, small enough to pass through the semipermeable membrane, is placed in the other compartment (Figure 20-11). In the absence of antibody, ligand added to compartment B will equilibrate on both sides of the membrane (see Figure 20-11a). In the presence of antibody however, some of the labeled ligand molecules will be bound to the antibody at equilibrium, trapping the ligand on the antibody side of the vessel, whereas unbound ligand will be equally distributed in both compartments. Thus, the total concentration of ligand will be greater in the compartment containing antibody (see Figure 20-11b). The difference in the ligand concentration in the two compartments represents the concentration of ligand bound to the antibody (i.e., the concentration of Ag-Ab complex). The higher the affinity of the antibody, the more ligand is bound.

Since the concentration of antibody placed into compartment A can be known, and the concentration of bound antigen (and therefore bound antibody) and free antigen can be deduced from the amounts of radioactivity in the antibody and nonantibody compartments respectively, the association constant can be calculated.

Equation 1 can be rewritten as

$$[SL] = K_a[S][L] = K_a \,([S]_t - [SL]) \times [L] \tag{2}$$

where $[S]_t$ = the total antibody binding site concentration, defined as $[S] + [SL]$.

(a)

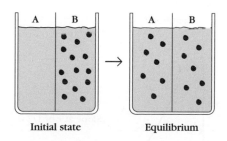

Control: No antibody present
(ligand equilibrates on both sides equally)

Initial state   Equilibrium

Experimental: Antibody in A
(at equilibrium more ligand in A due to Ab binding)

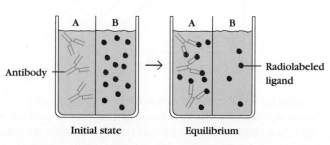

Antibody   Radiolabeled ligand

Initial state   Equilibrium

(b)

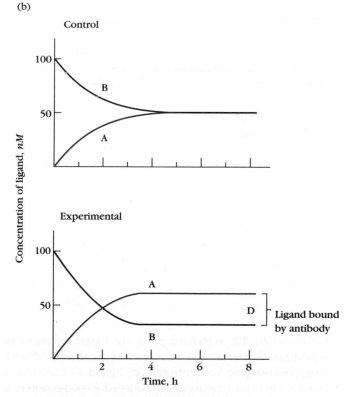

FIGURE 20-11 Determination of antibody affinity by equilibrium dialysis. (a) The dialysis chamber contains two compartments (A and B) separated by a semipermeable membrane. Antibody is added to one compartment and a radiolabeled ligand to another. At equilibrium, the concentration of radioactivity in both compartments is measured. (b) Plot of concentration of ligand in compartments A and B with time. At equilibrium, the difference in the concentration of radioactive ligand in the two compartments represents the concentration of ligand bound to antibody [SL] = the concentration of bound complexes.

If we now define:

[SL] = the concentration of bound ligand

and

[L] = concentration of free ligand]

then the ratio of bound to free ligand:

$$[SL]/[L] = K_a \, ([S_t] - [SL]) \qquad (3)$$

Equation (3) tells us that a plot of [SL]/[L] versus [SL] should yield a straight line with a slope of $-K_a$ and an intercept on the abscissa corresponding to the antibody-binding site concentration. (This is because when [SL]/[L] = 0, then $K_a[S_t] = K_a[SL]$ and therefore $S_t = [SL]$). This plot is referred to as a Scatchard plot. Note that this equation holds when the ligand is monovalent, which is the case for most small molecules whose affinity is measured in equilibrium dialysis experiments.

Sometimes, measurements of ligand binding are made at several different total antibody concentrations and, in this case, it is useful to be able to normalize our measurements to antibody concentration. If we divide equation (3) by the total concentration of antibody molecules, we obtain an equation that is familiar to many immunologists and biochemists:

$$\frac{r}{c} = K_a \, (n - r) \qquad (4)$$

where r is defined as the number of occupied sites per antibody molecule, n is defined as the total number of sites per antibody molecule, and c is the free ligand concentration = [L]. Thus,

$$r = \frac{[SL]}{[Ab]_t}$$

where $[Ab]_t$ = total molar antibody concentration and

$$n = \frac{[S]_t}{[Ab]_t}$$

From equation (4), we see that a plot of r/c versus r will yield a slope of $-K_a$ and an intercept on the abscissa of n (Figure 20-12). From this plot, one can therefore also gain information about the valency, n, of the antibody population under study.

If the plot of r/c versus r is a straight line, this is indicative of a single binding affinity (see Figure 20-12a), such as we would find in working with a mAb population. A heterogeneous mixture of antibodies would instead yield a curved line. With a curved-line plot (Figure 20-12b), it is still possible to determine the average affinity constant, by measuring the value of K_a when half the antigen-binding sites are filled. This can be done by using the appropriate software to determine the slope of the curve at the point where r = 1.

Although more sophisticated technologies now exist for the determination of antibody affinity (see below), they require expensive instrumentation, whereas equilibrium dialysis merely requires some dialysis tubing, antigen, antibody, and a way in which to distinguish bound from free

(a) Homogeneous antibody

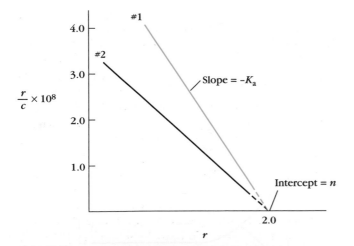

(b) Heterogeneous antibody

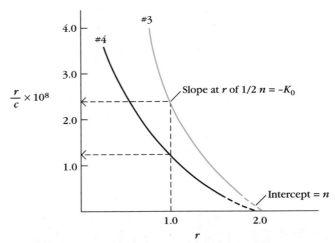

FIGURE 20-12 Scatchard plots are based on repeated equilibrium dialyses with a constant concentration of antibody and varying concentration of ligand. In these plots, r equals moles of bound ligand/mole antibody and c is the concentration of free ligand. From a Scatchard plot, both the equilibrium constant (K_a) and the number of binding sites per antibody molecule (n), or its valency, can be obtained. (a) If all antibodies have the same affinity, then a Scatchard plot yields a straight line with a slope of $-K_a$. The x intercept is n, the valency of the antibody, which is 2 for IgG and other divalent Igs. For IgM, which is pentameric, n = 10, and for dimeric IgA, n = 4. In this graph, antibody #1 has a higher affinity than antibody #2. (b) If the antibody preparation is polyclonal and has a range of affinities, a Scatchard plot yields a curved line whose slope is constantly changing. The average affinity constant K_0 can be calculated by determining the value of K_a when half of the binding sites are occupied (i.e., when r = 1 in this example). In this graph, antiserum #3 has a higher affinity ($K_0 = 2.4 \times 10^8$) than antiserum #4 ($K_0 = 1.25 \times 10^8$). Note that the curves shown in (a) and (b) are for divalent antibodies such as IgG.

antibody. Equilibrium dialysis is a general technique that measures the affinity of any large molecule for any small one capable of penetrating a dialysis membrane.

Surface Plasmon Resonance Is Now Commonly Used for Measurements of Antibody Affinity

Since the mid-1990s, equilibrium dialysis has been superceded as a method for affinity determination by **surface plas-** **mon resonance (SPR)**, which is more rapid and sensitive and can also provide information about antigen-antibody reaction rates (Figure 20-13). SPR relies on the occurrence of electromagnetic waves, called surface plasmons, that propagate at the interface of a metal and a solvent. The nature of the wave is sensitive to any alteration in this boundary, such as the adsorption of molecules to the metal surface. SPR works by detecting changes in the reflectance properties at the surface of an antigen-coated metal sensor when it binds antibody.

(a)

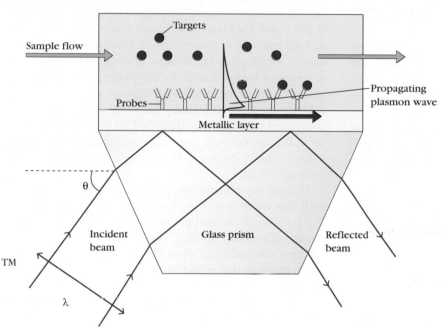

(b)

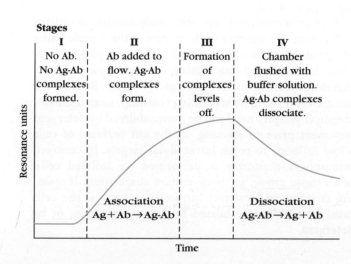

FIGURE 20-13 Surface plasmon resonance (SPR). (a) A buffer solution containing antibody is passed through a flow chamber, one wall of which contains a layer of immobilized antigen. As explained in the text, formation of antigen-antibody complexes on this layer causes a change in the resonant angle of a beam of polarized light against the back face of the layer. A sensitive detector records changes in the resonant angle as antigen-antibody complexes form. (b) Interpretation of a sensorgram. There are four stages in the plot of the detector response (expressed as resonance units, which represent a change of 0.0001 degree in the resonance angle) versus time. Stage I: Buffer is passed through the flow chamber. No Ag-Ab complexes are present, establishing a baseline. Stage II: Antibody is introduced into the flow and Ag-Ab complexes form. The ascending slope of this curve is proportional to the forward rate of the reaction. Stage III: The curve plateaus when all sites that can be bound at the prevailing antibody concentration are filled. The height of the plateau is directly proportional to the antibody concentration. Stage IV: The flow cell is flushed with buffer containing no antibody and the Ag-Ab complexes dissociate. The rate of dissociation is proportional to the slope of the dissociation curve. The ratio of the slopes, ascending over descending, equals $k_1/k_2 = k_a$.

Although the physics underlying SPR is rather sophisticated, the actual methodology is quite straightforward (see Figure 20-13a). A beam of polarized light is directed through a prism onto a chip coated with a thin gold film (coated with antigen on the opposite side), and reflected off the gold film toward a light-collecting sensor. At a unique angle, some incident light is absorbed by the gold layer, and its energy is transformed into surface plasmon waves. A sharp dip in the reflected light intensity can be measured at that angle, which is called the **resonant angle**. This angle depends on the color of the light, the thickness and conductivity of the metal film, and the optical properties of the material close to the gold layer's surfaces.

The SPR method takes advantage of the last of these factors, as the binding of antibodies to the antigen attached to the film produces a detectable change in the resonant angle, and the amount of the change is proportional to the number of bound antibodies. By measuring the rate at which the resonant angle changes during an antigen-antibody reaction, the rate of the antigen-antibody binding reaction can be determined (see Figure 20-13b).

Operationally, this is done by passing a solution of known concentration of antibody over the antigen-coated chip. A plot of the changes in the resonant angle versus time measured during an SPR experiment is called a *sensorgram*. In the course of an antigen-antibody reaction, the sensorgram plot rises until all of the sites capable of binding antibody (at a given concentration) have done so. Beyond that point, the sensorgram plateaus. The data from these measurements can be used to calculate k_1, the *association rate constant* for the antibody-antigen binding reaction.

Once the plateau has been reached on the sensorgram plot, solution containing no antibody can be passed through the chamber. Under these conditions, the antigen-antibody complexes dissociate, allowing calculation of the dissociation rate constant, k_2. Measurement of k_1 and k_2 allows determination of the affinity constant, K_a, since $K_a = k_1/k_2$.

Microscopic Visualization of Cells and Subcellular Structures

Imaging of cells and tissues can be accomplished using antibodies specific for antigens present in tissues that are conjugated to other molecules. If the molecule bound to the antibody is an enzyme, the presence of the antigens in the tissue sample can be visualized by washing away excess antibodies and adding substrates that are modified by the antibody-conjugated enzymes to create insoluble colored precipitates at the precise sites of antibody binding. Techniques that rely on this basic approach include immunocytochemistry and immunohistochemistry, and these methods use simple compound microscopes, coupled with computer-controlled systems to create images of fixed tissues. If the molecule conjugated to the antibody is a fluorescent dye, the antigen is visualized directly by fluorescence microscopy as described in the next section.

Immunocytochemistry and Immunohistochemistry Use Enzyme-Conjugated Antibodies to Create Images of Fixed Tissues

Immunocytochemistry and immunohistochemistry are both techniques that rely on the use of enzyme-conjugated antibodies to bind to proteins or other antigens in intact cells. A variety of enzyme-conjugated antibodies is available at this point, but classically the cell is treated in such a way that an antibody conjugated to an enzyme such as peroxidase is localized at the antigen binding site. This can be accomplished either directly, by using a peroxidase-conjugated antibody to bind to the cellular antigen, or indirectly, by using a biotin-conjugated primary antibody and streptavidin-conjugated peroxidase, as described above. Alternatively, as for ELISA, a secondary, peroxidase-conjugated antibody that binds to the Fc region of the primary, tissue-specific antibody can be used.

In these experiments, the peroxidase substrate is selected such that the enzyme's product forms a colored precipitate that is visible under the light microscope, and is deposited at the site of antibody binding. Using a range of enzymes with different substrates can allow the deposition of products of different colors that reflect the binding of different antibodies. Various chemical additives are available that can enhance the density and/or the color of the staining. The quality of the staining reaction depends on the ratio of specific to nonspecific staining, and so the investigator usually performs the staining reaction in the presence of relatively high concentrations of nonspecific proteins, such as nonfat dry milk (really!) in order to minimize nonspecific antibody binding. In expert hands, the images obtained from these techniques can be extraordinarily detailed and also aesthetically pleasing (Figure 20-14).

Immunocytochemistry and immunohistochemistry differ from one another in the nature of the sample being analyzed. In immunohistochemistry, the samples are prepared by sectioning intact tissue and the stained cells are therefore localized in their biological context. Because the sample is prepared by sectioning (cutting), immunohistochemical samples need not be permeabilized by detergent treatment prior to staining, as the cut surfaces of cells allow antibody to reach intracellular targets. In contrast, immunocytochemistry is performed on isolated cells, often those grown in tissue-culture suspension. If staining is designed to detect intracellular targets, the cells must first be permeabilized by organic fixatives, or by detergent.

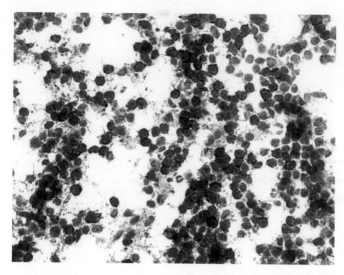

670

(a)

FIGURE 20-14 Immunohistochemical staining of lymph node. Human lymph node material was embedded in paraffin, fixed, and stained with antibody to CD4 (red stained cells). The lymph node was then counter-stained with hematoxylin (blue). *[Courtesy R&D Systems, Inc., Minneapolis, MN, USA.]*

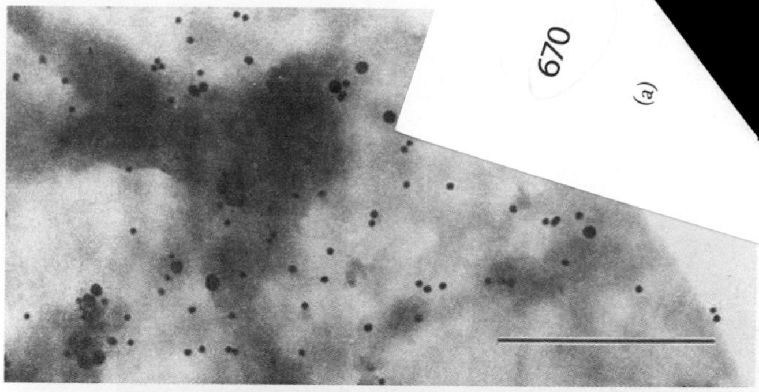

FIGURE 20-15 An immunoelectronmicrograph of the surface of a B-cell lymphoma was stained with two antibodies: one against class II MHC molecules labeled with 30-nm gold particles and another against MHC class I molecules labeled with 15-nm gold particles. The density of class I molecules exceeds that of class II on this cell. Bar = 500 nm. *[From A. Jenei et al., 1997, PNAS 94:7269–7274; courtesy of A. Jenei and S. Damjanovich, University Medical School of Debrecen, Hungary.]*

Immunoelectron Microscopy Uses Gold Beads to Visualize Antibody-Bound Antigens

In order to resolve fine structure details at high levels of magnification, scientists may need to use electron microscopy, rather than light microscopy. In such experiments, antibodies are coupled to electron dense particles such as colloidal gold particles. Wherever antibodies bind, gold deposits will be detected by the electron microscope. By using differently sized gold particles conjugated to particular antibodies, the subcellular relationship between two or more different antigenic epitopes can be analyzed (Figure 20-15).

Immunofluorescence-Based Imaging Techniques

The last decade has seen a veritable explosion in the manner in which fluorescence microscopy has been applied to studies of the immune system. Since students will be exposed to many of these techniques in the current primary literature, we present here a summary of the common variations of immunofluorescence. See also Box 14-1, which describes some important dynamic imaging techniques.

Fluorescence Can Be Used to Visualize Cells and Molecules

The phenomenon of fluorescence results from the property of some molecules to absorb light at one wavelength and

emit it at a longer wavelength. If the emitted light has a wavelength in the visible region of the spectrum, the fluorescent dye can be used to image any molecules bound by that dye. Antibody and streptavidin conjugates of a host of dyes have been used to provide spectacular images of cellular and subcellular structures, as well as to identify cells binding to those antibodies in flow cytometric experiments (see below). In addition, cells and animals can be engineered to express particular fluorescent proteins such as green *fluorescent protein* (GFP) or red *fluorescent protein* (RFP) under the control of cell- or tissue-specific promoters. The dyes can then be used to visualize the loci where proteins under the control of those same promoters are expressed. Other dyes specifically bind to particular macromolecules. For example, the blue dye 4′,6-*diamidino*-2-*phenylindole* (DAPI) specifically stains DNA. The protein phalloidin, which specifically binds filamentous actin, can also be conjugated to fluorescent probes (Figure 20-16a).

Immunofluorescence Microscopy Uses Antibodies Conjugated with Fluorescent Dyes

In standard immunofluorescence microscopy, the cell or tissue sample is stained with an antibody bound either directly or indirectly with a fluorescent dye, and then the slide holding the sample is placed on the microscope stage. Incident light excites the fluorescence in the sample, and filters ensure that only light of the required wavelength reaches the objective (Figure 20-16b). More modern instruments detect light at multiple different fluorescent wavelengths that is emitted after excitation of dyes conjugated to

(b)

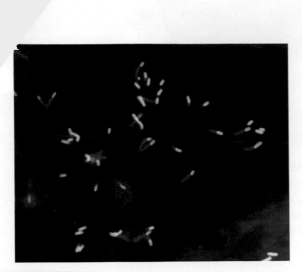

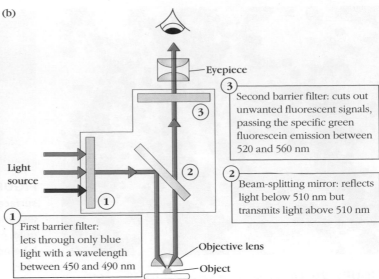

Eyepiece

3 Second barrier filter: cuts out unwanted fluorescent signals, passing the specific green fluorescein emission between 520 and 560 nm

2 Beam-splitting mirror: reflects light below 510 nm but transmits light above 510 nm

Light source

1 First barrier filter: lets through only blue light with a wavelength between 450 and 490 nm

Objective lens

Object

FIGURE 20-16 Fluorescently-labeled cells and the passage of light through a fluorescence microscope. (a) Fluorescence microscopic image of a human dendritic cell infected with engineered *Listeria monocytogenes*. The green fluorescence is generated by Alexa fluor 488-conjugated phalloidin (which binds actin filaments). Red fluorescence derives from red fluorescent protein expressed by the bacterium. DNA is stained with DAPI (blue). (b) Light from a source passes through a barrier filter that only allows passage of blue light of particular wavelengths. The light is then directed onto the sample by a dichroic mirror that reflects light of short wavelengths (below approximately 510 nm) but allows passage of higher wavelengths. When the blue light interacts with the sample, any fluorescent molecules excited by it emit fluorescence that then passes through the dichroic mirror, through a second barrier filter, and then is transmitted to the eyepiece. *[(a) Image courtesy of Dr. Keith Bahjat, Earle A. Chiles Research Institute.]*

various antibodies. The multiple color images can then be overlaid by the instrument's software to provide a single representation in which the locations of the antibody-bound molecules can be compared.

Confocal Fluorescence Microscopy Provides Three-Dimensional Images of Extraordinary Clarity

One of the limiting factors in obtaining clear images from fluorescence microscopy is the tendency of fluorescent molecules lying above and below the focal plane to contribute to the light that reaches the objective. In confocal microscopy, that artifact is eliminated by using an objective lens that focuses the light from the desired focal plane directly onto a pinhole aperture in front of the detector (Figure 20-17a). Light emitted from molecules located at other levels within the sample is stopped at the perimeter of the pinhole, and remarkably clear images of a single plane within the sample can thus be generated. In **laser scanning confocal microscopy**, the investigator uses lasers to provide the exciting light and computing power to move the focal plane in all three dimensions, thus enabling him/her to scan an x,y plane at different depths of focus, and reconstitute

powerful three-dimensional images. Figure 20-17b shows a spectacular image of blood vessels in mouse omentum, generated by this technique.

Multiphoton Fluorescence Microscopy Is a Variation of Confocal Microscopy

Two-photon and multi-photon microscopy are variations on confocal microscopy that offer even greater resolution in the development of three-dimensional images, following laser scanning of tissues (Box 14-1). In standard confocal microscopy, excitation of the fluorescent probes (dyes) occurs along the whole path of the laser beam through the tissue. This means that, although the emission beams are derived only from a single level within the sample, fluorescent probes are being excited throughout many levels of the tissue. Since fluorescent probes will eventually "bleach" (i.e., cease to emit light) after extensive excitation, this limits the useful life of the sample. It also means that additional light is emitted from the sample and must be filtered out of the final image.

In multiphoton fluorescence microscopy, long wavelength lasers are used that emit in the infrared region of the spectrum. These are relatively low energy, and so more than

(a) Principal of confocal microscopy

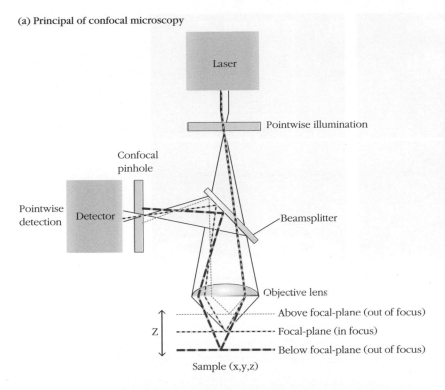

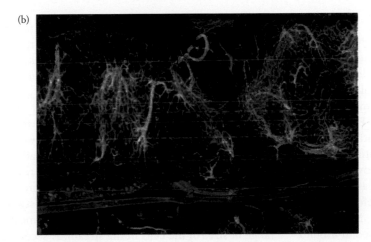

(b)

FIGURE 20-17 The principles of confocal microscopy. (a) The sample is illuminated by a laser beam that excites fluorescence from dyes in several different focal planes, represented here by the green, red, and purple lines. However, passage of light through the pinhole (shown on the right side of the image) filters out light emitted from all but a single focal plane, resulting in an extraordinarily clear image. Computer control of the exact plane from which light can be received through the pinhole allows the development of images from a number of focal planes and the creation of a composite three-dimensional representation such as that shown in Figure 20-1. (b) Confocal image of a mouse omentum. The omentum was stained with green dye-conjugated antibodies specific for CD31, which stains all endothelial vessels in the omentum. A second, red dye-conjugated antibody was then added specific for the molecule DARC, that is present only on the endothelial cells of the post-capillary end venules (the vessels from which extravasation of immune cells can occur). The red and green images were acquired separately and merged. Vessels staining both red and green appear yellow in this image. *[(b) Courtesy Aude Thiriot and Ulrich von Andrian, Harvard Medical School.]*

one photon must impinge on a fluorescent molecule in order to provide sufficient energy to excite the electrons. The low energy of these infrared lasers minimizes the extent of photobleaching and enhances the useful lifetime of the sample. Furthermore, the fact that at least two beams of light are required to bring about fluorescence excitation ensures that *excitation occurs only within the plane of intersection of the laser beams* (Figure 20-18). By moving the focal point of excitation within the x- and y- planes, information about a full optical section can be generated, and that whole process can then be repeated on additional z-levels, thus giving rise to a three-dimensional image. In Chapter 14, you have seen some of the powerful images developed with this technique.

Intravital Imaging Allows Observation of Immune Responses In Vivo

The topic of intravital imaging was addressed in Chapter 14, so it will be dealt with only briefly here. Intravital imaging takes advantage of the ability to maintain the circulation of lymph nodes and other immunological organs, after they have been gently lifted outside of an anaesthetized donor organism and onto a warmed microscope stage. There, using a multiphoton microscope, three-dimensional images of fluorescently labeled cells and structures can be generated and information gleaned about the behavior of immune cells essentially in vivo. Chapter 14 contains several links to movies taken in real time of stained immune system cells in situ using this powerful technology.

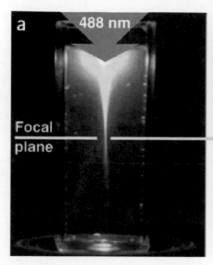 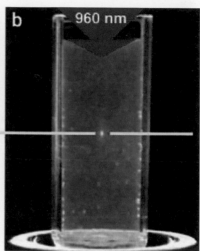

FIGURE 20-18 Difference in the depth of the region of fluorescence excitation affected by two-photon versus one-photon laser excitation. (a) A solution of the dye fluorescein is excited by a 488 nm laser. The excited molecules emit light in the green region of the spectrum (approximately 525 nm) that is clearly visible in this photograph as an extended cone. (b) The same solution is excited by a lower-energy 960 nm laser. The only molecules of fluorescein able to fluoresce are those that simultaneously absorb more than one photon of energy. This can occur only in the focal plane indicated by the yellow horizontal line. [*W. R. Zipfel, R. M. Williams, and W. W. Webb, 2002, Nonlinear magic: Multiphoton microscopy in the biosciences, Nature Biotechnology 21:1369–1377, Figure 2.*]

Flow Cytometry

Flow cytometry is the *sine qua non* (without which, nothing) of the modern immunologist's toolbox. It was developed by the Herzenbergs (Leonore and Leonard) and colleagues and found some of its earliest applications in analyzing cells from the blood, notably lymphocyte subpopulations. Flow cytometry per se is an analytical technique that quantifies the frequencies of cells binding to fluorescent antibodies and scattering light in characteristic ways (see below). When a **flow cytometer** is adapted to *sort* cell subpopulations on the basis of fluorescence and light scattering, it is referred to as a **Fluorescence Activated Cell Sorter (FACS)**. Monoclonal antibody and FACS technologies were developed at around the same time, and the two technological breakthroughs proved synergistic: the more antibodies that were available for cell typing and sorting, the more informative flow cytometric experiments became.

The fluorescent microscopy applications described so far are extremely valuable qualitative tools, but they do not provide quantitative data regarding the frequencies of cells within a population that stain with specific antibodies and that therefore belong to experimentally definable subpopulations. This shortcoming was addressed by the invention of the flow cytometer, which was designed to automate the analysis and separation of cells stained with different fluorescent antibodies.

The flow cytometer uses a laser beam and a series of detectors to identify both scattered light and fluorescent signals of particular wavelengths from single intact cells flowing in a focused stream past the laser (Figure 20-19a).

Cells in suspension are hydrodynamically focused into a narrow stream by being introduced inside a rapidly moving column of sheath fluid. Every time a cell passes in front of the laser beam, light is scattered, and this interruption of the laser signal is recorded.

One detector, called a photodiode, picks up light scattered in the forward direction (i.e., within a few degrees of the original path of the laser beam). High forward light scattering (within 5 to 10 degrees of the light path of the laser beam) provides an indication that the cells responsible for this light scattering are large. The more forward light scatter, the larger the cell, and so the amount of light scattered in the forward direction can be used as a rough measure of the range of sizes of the cells in the stream. A second light scattering detector is placed at approximately 90° to the path of the laser beam. The amount of side scattered light offers an indication of the extent of intracellular complexity of the scattering cells. The more side-angle light scatter, the more intracellular membranous structures, such as *endoplasmic reticulum* (ER) and mitochondria, are present in the scattering cell. Since much less light is detected by the side-angle scatter detectors than by the forward scatter photo-diode detector, side-scattered light is measured by sensitive photomultiplier tubes.

Fluorescently tagged antibodies bound to the surface antigens of particular cells can be excited by the laser. After excitation, they emit light that is recorded by a series of photomultiplier tubes located at a right angle to the laser beam. Each photomultiplier tube fluorescence detector is placed behind a series of dichroic filters and mirrors, so that it only receives and detects light within a particular range of

(a)

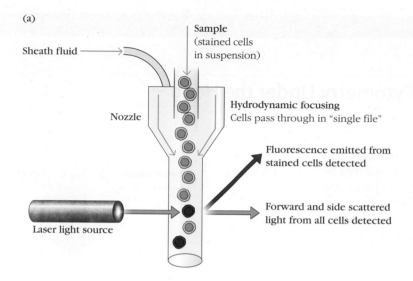

(b)

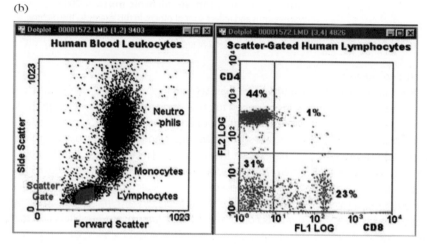

FIGURE 20-19 Principles of flow cytometry. (a) Cells introduced into the sample injection port are focused within a stream of sheath fluid and pass one by one in front of the laser beam. Forward-scattered light is detected by a photodiode. Side-scattered light and emitted fluorescence of various wavelengths is detected by photomultiplier tubes, after passage through a series of dichroic mirrors and light filters. All of the information obtained from individual cells is integrated by the software and can be expressed in a number of formats, such as that shown in (b). (b) On the left is a scatter plot of forward scatter (abscissa) versus side scatter (ordinate) of a sample of human white blood cells. Lymphocytes are gated and displayed in red. On the right is a plot of lymphocytes stained with anti-CD4 (ordinate) or anti-CD8 (abscissa) antibodies. [(a), www.sonyinsider.com/wp-content/uploads/2010/02/Flow-Cytometry-Diagram2.jpg; (b), Courtesy University of Massachusetts, Amherst, Department of Microbiology.]

wavelengths. Flow cytometers count every cell as it passes the laser beam and record the level of emitted fluorescence at a number of different wavelengths, as well as the amounts of forward- and side-scattered light for each cell; an attached computer stores all the data for each cell, which can then be called up by the analysis software as needed. This technology is advancing very quickly, and flow cytometers capable of detecting 8 to 12 fluorescence and light-scattering parameters are routinely used in clinical and research laboratories.

On the left of Figure 20-19b, we show an example of a typical forward and side scatter plot from human white blood cells. On the right of Figure 20-19b is a plot of the fluorescence intensity of cells derived from the region shown

gated in red in the scatter plot, which represents small, low-light-scattering cells, or lymphocytes. These cells have been stained with a fluorescein-conjugated antibody to CD8; therefore, CD8-bearing T cells will fluoresce in the green part of the spectrum. Green fluorescence is detected in the FL1, or green channel of the detector. Similarly, CD4[+] T cells have been stained with an antibody conjugated with a fluorochrome (e.g., *phycoerythrin*, or PE) that is detected in the FL2, or red channel. This plot therefore shows us that, of the cells falling into the lymphocyte gate, 44% were red-staining CD4-bearing cells and 23% were green-staining CD8-bearing cells. See Advances Box 20-1 for a more nuanced discussion of how a flow cytometer works.

Flow Cytometry Under the Hood

How does the flow cytometer actually measure fluorescence, and what data are you seeing when you view histograms, dot plots, and contour plots?

The flow cytometer is a conceptually simple instrument. It includes a *fluidics* system that sends cells in single file in front of one or more laser light sources that can excite fluorochromes. Light bounced or emitted from the cell is then guided by a network of mirrors and filters to light (photon) detectors, which record the signal.

The *filters* used to sort and guide the light coming from a cell fall into several broad categories. B*and pass (BP) filters* only allow light within a certain range of wavelengths to pass. For example, a BP filter might only allow passage of wavelengths from 520 nm to 550 nm. BP filters are routinely described by a pair of numbers, the first of which denotes the central wavelength of the BP filter and the second of which shows the range of wavelengths on either side of that central wavelength that will be allowed to pass through the filter (e.g., 530/30). L*ong pass (LP) filters* allow passage of any light of wavelengths longer than the filter specification but reflect (or stop) light of shorter wavelengths. A 640 LP filter, for example, would allow any wavelengths longer than 640 nm to pass through, but would reflect light of wavelengths shorter than 640 nm. *Short pass (SP) filters* operate in a converse manner, letting wavelengths lower than the filter specification pass, but reflecting higher wavelengths. *Dichroic mirrors* are very useful; they reflect light of specific wavelengths, allowing all other light to pass, and can also be referred to as SP or LP, depending on their characteristics. In the example shown in Figure 1, the dichroic mirror is allowing any light of wavelengths greater than 500 nm to pass through in the direction of the incident light, but it is reflecting light of wave-

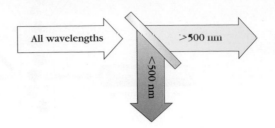

FIGURE 1
Flow cytometers contain dichroic mirrors. Dichroic mirrors allow passage of light of some wavelengths and reflect light of others. In this example, the dichroic mirror is enabling passage of light of wavelengths longer than 500 nm, but is reflecting light of shorter wavelengths in a direction perpendicular to the incident light. *[J. Punt]*

lengths lower than 500 nm in a direction perpendicular to the incident light. Dichroic mirrors are often placed at a 45° angle and split the light coming to them in this way, so that it can be directed to different photon detectors.

Because most of the light detected by a flow cytometer is scattered forward in the direction of the incident laser beam, this light is captured and detected by a photodiode. Photodiodes are not particularly sensitive, and therefore they are used to detect this *forward scattered* (FSC) light. In contrast, most light detectors in flow cytometers are the more sensitive detectors called *photomultiplier tubes (PMTs)*. In the most simplistic terms, these detectors convert photon energy into electrical energy. More specifically, they take advantage of the *photoelectric effect* and record (and amplify) electrons that are released when photons are absorbed. Each cell passing in front of a laser generates a pulse of electrons within PMTs that receive light signals. The pulse of electrons recorded over the time it takes for the cell to pass through the laser is called a voltage pulse. Each pulse has a *height* (H), *width* (W), and *area* (A), which is measured, digitized, and graphed on the plots that we have come to associate with flow cytometry data. In Figure 2, we illustrate the nature of the pulse gener-

ated as a cell passes vertically upward through the laser beam.

To better understand the information gained from a voltage pulse, consider the simple flow cytometer system represented in Figure 3. In this setup, one laser generates blue light (488 nm), and four PMT's positioned behind different sets of filters and mirrors only allow light of specific wavelengths to pass.

Now consider a group of lymphocytes from a mouse that have been stained with green fluorescent antibodies specific for CD4 (e.g., fluorescein *isothiocyanate*, or FITC anti-CD4) and red fluorescent antibodies specific for CD8 (e.g., *phycoerythrin*, or PE anti-CD8). These are passed, in single file, in front of the blue laser that is focused on the flow chamber, shown in purple in Figure 3.

A CD4[+] T cell passes by first. Some nonfluorescent 488 nm light from the laser is likely to be bounced to the side, perhaps by the nuclear material, mitochondria, or granules. This side-scattered (SSC) light will pass through the first dichroic mirror, which allows passage of light of wavelengths shorter than 560 nm. A beam splitter then sends some of this relatively low wavelength light to the SSC PMT, via a BP filter that is set to allow through light of the same wavelength as the original illuminating laser. This BP filter

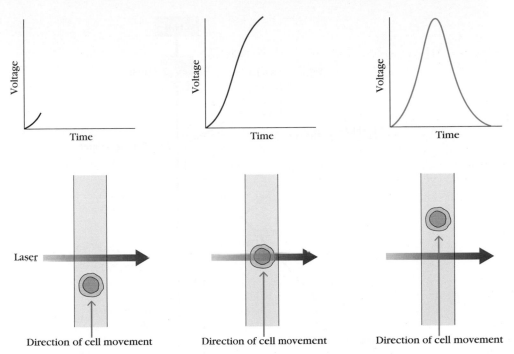

FIGURE 2
The voltage pulse generated by a cell passing in front of the laser beam. As cells pass in front of the laser beam, they emit or scatter light that is detected by a series of photomultiplier tubes (PMTs). The PMT receives the photons and converts them into a voltage pulse whose height, width and area is digitized and represented by the cytometer software. [J. Punt]

is labeled 488/10 in Box 20-1, Figure 3. A voltage pulse will be generated in the SSC PMT, as the cell passes by. The magnitude of the voltage pulse will be recorded and stored by the cytometer's software.

Similarly, light scattered in a forward direction will be detected by the photodiode placed to detect light scattered within 5–10 degrees of the direction of the laser, and the magnitude of the forward scatter signal will also be recorded and stored. Not shown in Box 20-1, Figure 3 is a filter block that is placed in the path of the laser beam that prevents any direct light from the laser from impinging on the FSC photodiode. Most flow cytometry applications will require a graph or "dot plot" of forward scatter versus side scatter of the cell population under study. (See Figure 20-19 for an example of a FSC versus SSC plot of a human blood cell population.) Each dot on these plots represents a single "event," which usually represents a single cell passing in front of the laser beam.

Recall that the cell that has scattered this light is a CD4-bearing cell. Because this CD4$^+$ cell is bound by FITC-labeled antibodies that will be activated by the 488 nm laser, it will also emit green light with a peak wavelength of approximately 525–530 nm. This green light, which is emitted in all directions, will be allowed to pass through the first dichroic mirror, and then through the BP filter in front of the green PMT, generating another voltage pulse in the green PMT. The shape of this pulse will be determined by the shape of the structure that is emitting the fluorescence. For instance, a fluorochrome that only labels the nuclei will generate a narrower voltage pulse than a fluorochrome that stains the entire cell (Figure 4; compare cells b versus a). Signals recorded from this green PMT are referred to as emanating from the *FL1* channel (Figure 3). The instrument software will integrate the information provided for each cell as it passes through the laser beam, so that all of the information pertaining to the detected fluorescence

and light-scattering properties are attributed to the correct cell.

When a CD8$^+$ T cell labeled with PE anti-CD8 passes in front of the laser, light scattered by that cell will also generate a voltage pulse in the SSC PMT and be detected by the FSC photodiode. In addition, the PE fluorochrome will be excited and emit orange light that will be reflected by the 560 nm SP dichroic mirror (because its wavelength is too long to pass through the filter) and also by the 640 nm LP dichroic mirror (because its wavelength, this time, is too short). These mirrors reflect the orange light directly through the 585 nm BP filter where it is detected by the orange PMT detector (FL2), generating a voltage pulse (see Figure 3). In theory, the CD8$^+$ T cell should not generate light that would find its way to the FL1 PMT; however, in reality, all cells emit some photons of a variety of wavelengths and will generate small voltage pulses in most PMTs. This inappropriate signal must be compensated for when conducting

(continued)

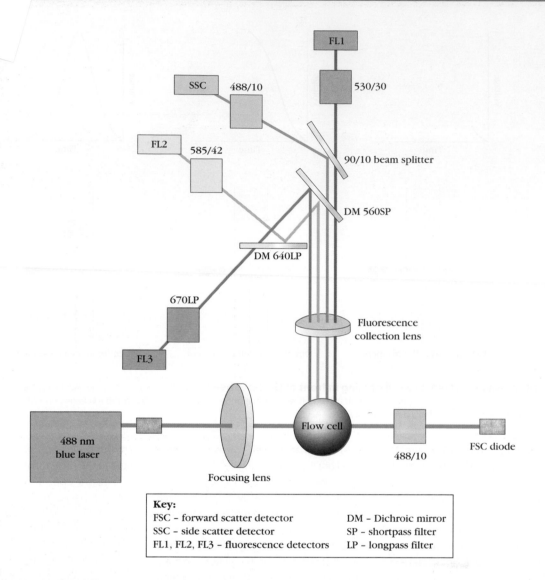

FIGURE 3

A simple flow cytometry setup. Cells passing through the flow cell are interrogated by the laser and scattered and fluorescent light is directed through the series of mirrors and filters to the appropriate PMTs. There, the induced voltages are digitized and represented by the software in graphical form. Since each parameter of light scatter or fluorescence is recorded for each cell detected, results can be displayed that include any combination of parameters for the cell population being studied. A variety of display styles are available depending on the graphing software used by the investigator. (See text for further details.) *[http://ars.els-cdn.com/content/image/1-s2.0-S0167779911001958-gr1.jpg]*

Flow cytometers can also detect fluorescence emissions from cells that have been permeabilized with detergent, which allows antibodies to enter the cells and stain intracellular components. This "intracellular staining" approach is used, for example, when analyzing cells for their ability to synthesize and release particular cytokines, as well as in cell cycle analysis and in detecting activities of enzymes for which fluorescent products are available (see below). In many medical centers, the flow cytometer is used to detect leukemias and classify them by cell type, which in turn influences decisions regarding future treatment. Likewise, the rapid measurement of T-cell subpopulations, an important prognostic indicator in AIDS, is routinely done by flow cytometric analysis. HIV infects and destroys CD4-bearing T cells. In order to monitor the clinical stage of the disease, labeled mAbs against CD4- and CD8-bearing T cells are used to determine their ratios in the patient's blood, as shown in Figure 20-19b. When the number of CD4 T cells falls below a certain level, the patient is at high risk for opportunistic infections.

(a)

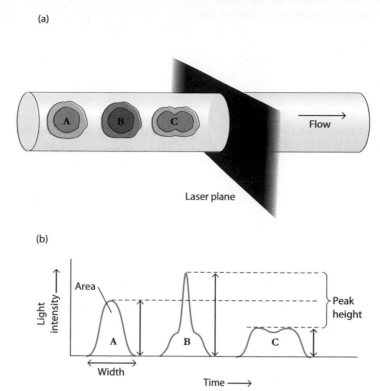

(b)

FIGURE 4

The nature of the voltage pulse is determined by the shape of the emitting structure. The size and shape of the voltage pulse is determined by the parameters of the cells and particles that are emitting the light signal. (See text for further details.) *[J. Punt]*

flow cytometric experiments; different instrument models perform this compensation in different ways.

Finally, consider a CD4$^+$ T cell and a CD8$^+$ T cell that are stuck together—that is, a doublet. When these pass through the laser, they will be considered one unit, and will scatter light, emit fluorescence, and generate voltage pulses in all three SSC, FL1, and FL2 PMTs as well as the FSC photodiode detector. This could lead an investigator to inaccurately assume that this is a single cell that expresses both CD4 and CD8. However, because the cell doublet is bigger, its voltage pulse will last longer and its width will be disproportionately large (Figure 4; consider cells A versus C). Fortunately, software programs allow investigators to remove from analysis cells that generate too wide a voltage pulse.

Each cell (or event) passing by a laser generates multiple voltage pulses at each PMT. In newer flow cytometers, the height, width, and area of each voltage pulse will be recorded and digitized. In older machines, only the height was routinely recorded. Investigators can choose to plot any of these measurements (e.g., FL1-H, SSC-W, FL2-A), although area (which provides information about the total light detected over the time the cell passes) is probably the most informative.

Magnetic Activated Cell Sorting

Fluorescence is not the only type of marker that immunologists can use to separate cells. A fine wool mesh made of ferromagnetic metal can be localized in a short column. Application of a magnetic field across the column will ensure that any magnetized material will stick to the mesh. By conjugating antibodies to magnetic beads or molecules, allowing the antibodies to bind to their antigens on the surface of particular cells, and then passing the cells through the column, those cells that bound to the magnetic beads can be held onto the mesh in the column, whereas those that did not bind the antibodies flow through. After washing off any nonspecifically bound cells with a stream of buffer, the magnetic bead-conjugated cells can then be released by removing the magnet from the outside of the column and passing buffer through. Magnetic cell separation is particularly useful for batch separation of large numbers of cells, whereas fluorescence-activated cell sorting, by sorting one cell at a time, makes fewer mistakes but is much slower. In the laboratory or the clinic, immunologists will often perform a batch sort using magnetic activated cell sorting, and then follow it up with a fluorescence-activated cell sort, in order to maximize the accuracy of cell separation.

Cell Cycle Analysis

Following activation, one of the first responses of lymphocytes is to divide. Immunologists have therefore been at the forefront of developing methodologies for cell cycle analysis. We will describe several methods (classical as well as more modern) that are commonly used by immunologists in analyzing the cell cycle status of populations of immune cells.

Tritiated (^{3}H) Thymidine Uptake Was One of the First Methods Used to Assess Cell Division

^{3}H thymidine uptake assays were the first to be used routinely to measure cell division in lymphocyte cultures. They rely on the fact that dividing cells synthesize DNA at a rapid pace, and radioactive thymidine in the culture fluid will therefore be quickly incorporated into high molecular weight DNA. In a ^{3}H thymidine uptake assay, cells subjected to proliferative signals are lysed at defined periods post-stimulation and their DNA is precipitated onto filters that bind to the high-molecular-weight nucleic acids, but allow unincorporated thymidine to wash straight through. The amount of radioactivity retained on the filters can then provide a measure of the amount of newly synthesized DNA and hence the number of cells undergoing division in the culture.

Colorimetric Assays for Cell Division Are Rapid and Eliminate the Use of Radioactive Isotopes

Motivated by reasons of safety and environmental responsibility to move away from the use of radioactivity-based measurements, scientists developed a number of different assays in which metabolically active cells cleave colorless substrates into colored, often insoluble products that can then be measured spectrophotometrically. In one such assay, the tetrazolium compound MTT (3-(4,5-Dimethylthiazol-2-yl)-2,5-diphenyltetrazolium bromide, a yellow tetrazole) is reduced by metabolically active cells to form insoluble, purple formazan dye crystals. The absorbance of each cell sample can then be read directly in the culture wells at 570 nm, the absorbance peak of formazan. The more metabolically active cells that are present in the culture, the more formazan will be generated, and so this assay provides a readout of the number of live cells as a function of time. In this way, the MTT assay can measure cell proliferation or cell death.

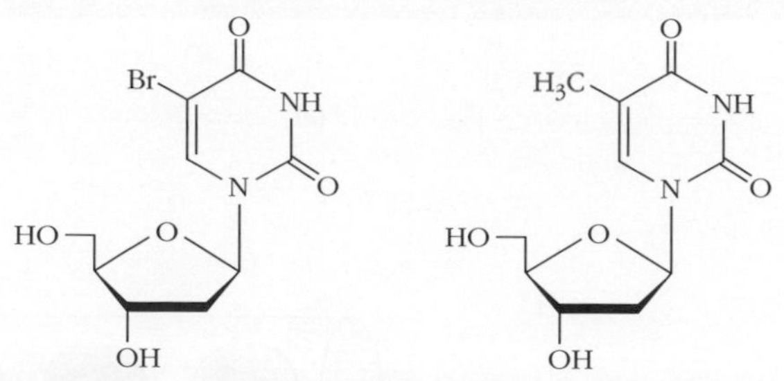

FIGURE 20-20 Bromodeoxyuridine incorporates into DNA in place of deoxythymidine during DNA synthesis. Bromodeoxyuridine (BrdU) is a thymidine analog, as the large bromine group serves to mimic the size and shape of the methyl group of thymidine. It is incorporated into DNA instead of thymidine and can be detected by anti-BrdU antibodies. *[http://openwetware.org/images/thumb/c/cc/BrdU_vs_dT.svg/250px-BrdU_vs_dT.svg.png.]*

Bromodeoxyuridine-Based Assays for Cell Division Use Antibodies to Detect Newly Synthesized DNA

When introduced into cells, *bromodeoxyuridine* (BrdU) is rapidly phosphorylated to bromodeoxyuridyl triphosphate (an analogue for deoxythymidine triphosphate) and is incorporated in its place into newly synthesized DNA (Figure 20-20). Cells that divide following BrdU incorporation can then be identified using antibodies to BrdU. In addition to serving as a label for newly divided cells, BrdU can also mark cells for light-induced cell death. If cells that have incorporated a high level of BrdU are exposed to light, they will photolyse; this has been used to selectively kill newly dividing cells. In recent years, other chemical analogues that mimic BrdU's functions as a marker of dividing cells have been generated, including EdU, which can be detected using specific reagents.

Propidium Iodide Enables Analysis of the Cell Cycle Status of Cell Populations

Propidium iodide is a fluorescent dye with a flat, planar structure that slides between the rungs of (or intercalates into) the DNA ladder in a quantitative manner (Figure 20-21a). By using the flow cytometer to measure the fluorescence from a population of propidium iodide-labeled cells, we can identify which cells within the populations are at each stage of the cell cycle. G1 cells will have half the DNA of G2 cells, or cells about to undergo mitosis, and cells that are currently replicating DNA and are therefore in S phase will have an intermediate value. Apoptotic cells and fragments that have begun to break down their DNA will appear as events with less than G1 amounts of DNA (Figure 20-21b). Doublet cells must first be excluded from analysis because they have the same amount of DNA per doublet as a cell in the G2 or M phase of the cell cycle; failure to exclude doublets will therefore result in an overestimate of the fraction of dividing cells.

(a)

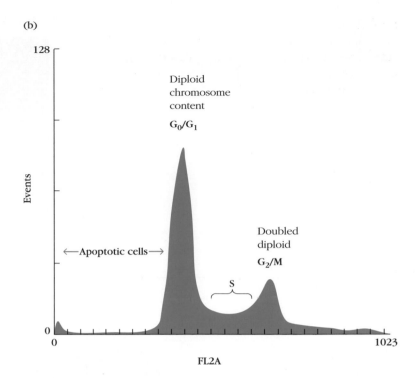

(b)

FIGURE 20-21 Propidium iodide intercalates into DNA and acts as a cell cycle and apoptosis indicator. (a) The flat planar ring structure of propidium iodide enables it to intercalate between the rungs of the DNA ladder. (b) A histogram of fluorescence measured in the FL2 channel shows cells bearing amounts of DNA characteristic of apoptotic cells, and cells in the G1, S, and G2/M phases of the cell cycle. *[(a), http://probes.invitrogen. com/media/structure/919.jpg; (b), www.meduniwien.ac.at/user/ johannes.schmid/PIstain2.jpg.]*

In order to make the most accurate measurements of the amount of dye taken up by cells, investigators routinely use the flow cytometer's ability to compute the area under the voltage pulse on the Fl2 channel (Advances Box 20-1). Other dyes that bind to DNA and allow for similar types of cell cycle analysis include DAPI, Hoechst 33342, and 7-*amino-actinomycin D* (AAD).

Carboxyfluorescein Succinimidyl Ester Can Be Used to Follow Cell Division

Carboxyfluorescein succinimidyl ester (CFSE) is more correctly named *carboxyfluorescein diacetate succinimidyl ester* (CFDASE). The diacetyl groups, seen at the top right and left of the molecule shown in Figure 20-22a, enable the CFDASE to enter the cell, and are then cleaved by intracellular esterases, so that the CFSE remains trapped within the cytoplasm. In the cytoplasm, molecules of CFSE are efficiently and covalently attached to intracytoplasmic proteins, with the succinimidyl ester acting as a leaving group. Somewhat surprisingly, attachment of CFSE to intracytoplasmic proteins occurs with essentially no deleterious effects on cellular metabolism or cell division.

The power of CFSE labeling lies in the fact that the amount of fluorescence emitted is cut in half each time the cell divides. This is illustrated in Figure 20-22b. The right-hand peak represents those cells that did not divide after CFSE incorporation. The peak to its immediate left represents cells that have divided once, the next one to the left represents cells that have divided twice, and so on. One can use the sorting capacity of the flow cytometer to physically separate those cells that have not divided, or have divided once, twice, or more times, and then analyze the separate cell populations for the expression of particular genes. In the example shown in Figure 20-22c, we see that the gene encoding the protein survivin is expressed hardly at all in nondividing T cells, but that it is expressed immediately on T-cell activation. By the time the cell has divided three times, the cells express two isoforms of the gene.

Assays of Cell Death

At the close of an immune response, most of the activated immune cells die. Furthermore, the outcome of many immune responses is the death of infected or affected cells, and so immunologists have developed a battery of methodologies to test for cell death.

The ^{51}Cr Release Assay Was the First Assay Used to Measure Cell Death

The ^{51}Cr release assay was the method of choice for measuring cytotoxic T cell- and natural killer cell-mediated killing for decades and was used in the experiments performed by Doherty and Zinkernagel that first described MHC restriction in T cell recognition. Target cells are first incubated in a solution of sodium ^{51}chromate, which is taken up into the cells. Excess chromium is washed out of the cell suspension and the radioactively labeled targets are mixed with the killer cell population at defined effector to target cell ratios. Death of the target cells is indicated by the release of ^{51}Cr into the supernatant of

(a)

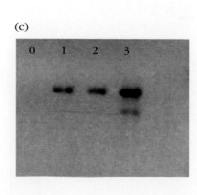

(c)

(b)

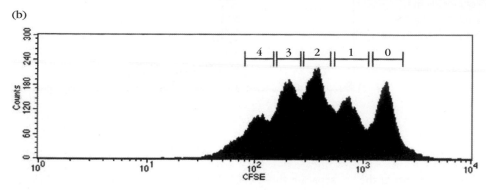

FIGURE 20-22 Staining with carboxyfluorescein succinimidyl ester allows assessment of the number of cell divisions undertaken after staining. (a) Structure of carboxyfluorescein diacetyl succinimidyl ester (CFSE). (b) Fluorescence histogram showing CFSE fluorescence from a population of dividing cells. The peak on the right represents those cells that have not divided since addition of the CFSE. Cells that have divided once have half the fluorescence of the undivided population and are shown in the peak immediately to the left of the undivided population. The other peaks have divided two, three, or four times respectively post stimulation. (c) Cells can be sorted according to how many times they have divided post addition of CFSE and tested for gene expression. In this example, the gene for the protein survivin is not expressed in nondividing T cells, but is expressed following stimulation with anti-CD3 and anti-CD28. After three cell divisions, the expression is up-regulated and two mRNA species (representing the RNA encoding two isoforms of the protein) are evident. *[(a) http://probes.invitrogen. com/media/structure/835.jpg; (b) and (c), Alexander Au, M.D., senior thesis, Haverford College.]*

the mixed cell culture, and is quantified by comparing the ^{51}Cr release from the test cells with that from detergent treated and control cells. ^{51}Cr is a γ-emitting radioactive isotope, and the radioactivity in the assay supernatants can therefore be readily measured in a γ counter. A modern alternative to this technique uses CFSE to label the cells and quantifies the release of fluorescent material into the supernatant with a fluorescent plate reader. These two assays measure all forms of cell death.

Fluorescently Labeled Annexin V Measures Phosphatidyl Serine in the Outer Lipid Envelope of Apoptotic Cells

In cells undergoing apoptosis (programmed cell death), but not other modes of cell death, the membrane phospholipid *phosphatidyl serine* flips from the interior to the exterior side of the plasma membrane phospholipid bilayer. Annexin V is a protein that binds to phosphatidyl serine in a calcium-dependent manner; fluorescently labeled Annexin V can therefore be used to tag apoptotic cells for detection using flow cytometry or a fluorescent plate reader.

The TUNEL Assay Measures Apoptotically Generated DNA Fragmentation

Another common method for the detection of apoptotic cell death uses the fact that apoptotic cells undergo a process of progressive DNA degradation, which results in the generation of short DNA fragments within the nucleus. The TUNEL assay relies on the use of the enzyme *terminal deoxyribonucleotidyl*

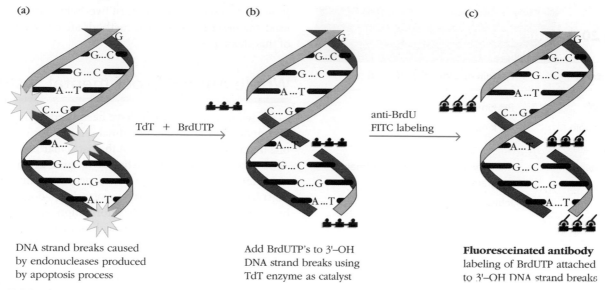

(a)

TdT + BrdUTP

DNA strand breaks caused
by endonucleases produced
by apoptosis process

(b)

anti-BrdU
FITC labeling

Add BrdUTP's to 3'–OH
DNA strand breaks using
TdT enzyme as catalyst

(c)

Fluoresceinated antibody
labeling of BrdUTP attached
to 3'–OH DNA strand breaks

FIGURE 20-23 Assessment of apoptosis using a TUNEL assay. (a) Apoptosis results in DNA fragmentation by intracellular nucleases. (b) BrdU nucleotide triphosphates are added onto the broken ends of fragmented DNA in fixed and permeabilized apoptotic cells using the enzyme TdT. (c). Fluoresceinated antibodies specific for BrdU can then be used to detect apoptotic cells. [www.phnxflow.com/images/DNA.gif.]

*t*ransferase (TdT) (see Chapter 7 for a detailed explanation of this enzyme's activity) to add bases onto the broken ends of DNA sequences in a nontemplated manner. The classic variation of the TUNEL method uses TdT to add BrdU to fixed and permeabilized cells. BrdU is incorporated into the newly synthesized DNA, and is then detected with fluorescently labeled anti-BrdU antibodies (Figure 20-23). More recent iterations of this method use the incorporation of short DNA segments prelabeled with a molecule that then binds to a fluorescent tag under very gentle conditions.

Caspase Assays Measure the Activity of Enzymes Involved in Apoptosis

The caspases are a family of *c*ysteine proteases that cleave proteins after *asp*artic acid residues. Different members of the caspase family are activated during apoptotic cascades, depending on their mode of initiation. For example, caspase 8 is activated upon engagement of the Fas receptor by Fas ligand. Several different types of caspase assays are now commercially available, including caspase detection kits that yield fluorescent products upon caspase-mediated cleavage, as well as kits that detect the cleaved and active forms of caspases using Western blot methodology.

Biochemical Approaches Used to Elucidate Signal Transduction Pathways

In order to identify the members of a signal transduction pathway and the nature of the interactions that occur between the various components, immunologists use a panoply of biochemical and genetic tools. Such an investigation usually begins when a scientist determines that interaction of a particular ligand and receptor has a certain outcome—for example, the activation of transcription of a particular gene. The question then becomes what are the intervening molecules that pass the signal from a receptor to the nucleus and result in the activation of transcription?

Biochemical Inhibitors Are Often Used to Identify Intermediates in Signaling Pathways

Certain families of enzymes (e.g., tyrosine kinases, serine/threonine kinases, caspases, and ubiquitinases) transduce molecular signals in a number of different pathways. Other pathways require the integrity of intracellular organelles, such as lysosomes. Commercially available chemical inhibitors have been developed for many enzyme families implicated in signal transduction, as well as for individual members of many of these families. Simultaneous application of an inhibitor with the signaling molecule can provide rapid information about whether the inhibited protein is implicated in the pathway under study, and inhibitor studies often prove a valuable way to start examining a new problem and gain some quick results. Chemical inhibitors that affect the functioning of intracellular compartments also exist, and their use allows an investigator to determine whether the pathway under investigation involves that compartment.

However, in using inhibitors to characterize a pathway, one must always be circumspect, as their specificity is not always fully characterized, and the signal transduction inhibition that is measured may not result from the inhibition of the protein that the manufacturer specified or the investigator believes is being studied. For example, inhibitors initially

TABLE 20-2	Some common inhibitors used in the dissection of signal transduction pathways
Inhibitor	**Protein or organelle affected**
Ly294002	PI3 kinase
Wortmannin	PI3 kinase
Rapamycin	mTOR
Chloroquine	Integrity of lysosomal compartments— pH gradient is collapsed
BX795	Inhibits IKKα
PepinhMyD	MyD88
PD98059	MAPKKK
Cyclosporine	Works with immunophilin to inhibit calcineurin
zVAD fmk	Members of the caspase family

developed for the Erk1 transcription factor are now known to also inhibit Erk5, a family member that was unknown when the first inhibitor was developed. Furthermore, the observed inhibition may result from interference with a related pathway. For example, the inhibited enzyme may prevent the production of one of the components of the pathway under study. Because of these caveats, inhibitor studies must always be supported with other means of analysis. Table 20-2 lists a few representative inhibitors of enzymes and transcription factors involved in signal transduction that students may encounter in their reading, along with their mode of action.

Many Methods Are Used to Identify Proteins That Interact with Molecules of Interest

As described earlier in the "Immunoprecipitation-Based Techniques" section, co-immunoprecipitation can be used to identify proteins that interact with a target molecule. For example, an antibody to an adapter protein can be used to immunoprecipitate that protein. After solubilizing the components of the precipitate and running them out on an SDS-PAGE gel, other bands may become visible which represent proteins that were interacting with the adapter protein in the intact cell. Sometimes, inspection of the molecular weight of the co-precipitating proteins, along with information about other proteins known to interact with that adapter, allows the investigator to develop a hypothesis about the identity of the co-precipitating molecule, which can then be confirmed with a Western blot. On other occasions, the investigator must subject a sample of the mystery band to microsequencing and identify it using the tools of bioinformatics. If Western blots fail to identify interacting

proteins, techniques such as yeast two-hybrid screens can be used, but description of those methods are beyond the scope of this chapter.

Whole Animal Experimental Systems

Many whole animal systems have been used in the study of immunology. The species of animal selected for study is the one that best meets the needs of the particular investigation. To test the effectiveness of particular vaccines against viruses or bacteria that affect only primates, primate animal models must be used. Studies of horses, goats, sheep, dogs, and rabbits have also yielded much information about immune responses and have provided us with numerous reagents for the study of human and mouse immunology. However, the species that has been used most frequently and with the greatest effectiveness in modeling the human immune system is the mouse. Mice are easy to handle, are genetically well characterized, and have a rapid breeding cycle. In this section, we will outline various types of murine animal models and clarify some of the more confusing nomenclature that students are likely to encounter as they read in the immunological literature. We will first briefly address the ethical questions that must and should arise in the minds of those who work with whole animals and the regulations that have been developed to protect nonhuman research subjects.

Animal Research Is Subject to Federal Guidelines That Protect Nonhuman Research Subjects

Our understanding of the immune response owes a great deal to animals, and huge advances have been made using a variety of animal models. Concerns about animal welfare, however, have accompanied these advances and have led many countries to adopt laws that regulate animal use on ethical grounds. In the United States, institutions and investigators that perform research on animals must comply with the Animal Welfare Act of 1966. Those researchers that receive federal funds must establish an Institutional Animal Care and Use Committee (IACUC) that oversees the practices of researchers and programs. All investigators must detail and justify their approaches before this committee and compliance is enforced by the U.S. Department of Agriculture (USDA).

Standards for the ethical treatment of animals are continually evaluated and updated as our awareness of animal biology, alternative technologies, and ethical concerns develop. Since it was signed into law, the Animal Welfare Act has been amended seven times in order to reflect these advances. In 2010, the European Union passed a law that updated and strengthened standards outlined in its 1986 animal welfare laws. In response to concerns that animal research was poorly described in the literature, the influential journal *Nature* established new policies in 2012 that require authors to include more detailed descriptions of approaches and standards associated with animal experimentation. Finally, a growing number of investigators have formally embraced the Three R's principles originally articulated in 1954 by Hume and described fully by Russell and Burch. These describe a commitment to *refine*, *reduce* and *replace* approaches that use animals in research. These principles were communicated to scientists first by a committee that included the Nobelist and immunologist Peter Medawar, and have been the focus of several prominent conferences in recent years.

Many individuals and groups feel that regulations still do not fully address or respect the needs of animals. Tension between ethical concerns and the desire to advance knowledge continues; although it inspires conflict at times, it also inspires continuing efforts to refine and improve policies.

Inbred Strains Can Reduce Experimental Variation

To control experimental variation caused by differences in the genetic backgrounds of experimental animals, immu-

nologists often work with inbred strains of mice produced by 20 or more generations of brother-sister mating. The rapid breeding cycle of mice makes them particularly well suited for the production of inbred strains in which the heterozygosity of alleles that is normally found in randomly outbred mice is replaced by homozygosity at all loci. Repeated inbreeding for 20 generations yields an inbred strain whose progeny are homozygous and identical (**syngeneic**) at more than 99% of all loci. Approximately 500 different inbred strains of mice are available, each designated by a series of letters and/or numbers (Table 20-3), and most of these strains are commercially available. Inbred strains have also been produced in rats, guinea pigs, hamsters, rabbits, and domestic fowl.

Recombinant inbred strains of mice are those in which two inbred strains (e.g., strains A and B) have been mated, resulting in a recombination within an interesting locus (e.g., within the MHC locus). Subsequent inbreeding of the animals with this recombination results in an inbred strain of mice bearing part of its MHC from strain A and the other

TABLE 20-3	Some common inbred mouse strains used by immunologists	
Strain	**Common substrains**	**Characteristics**
A	A/He	High incidence of mammary tumors in some substrains
	A/J	
	A/WySn	
AKR	AKR/J	High incidence of leukemia
	AKR/N	
	AKR/Cum	*Thy 1.2* allele in AKR/Cum, and *Thy 1.1* allele in other substrains (*Thy* gene encodes a T-cell surface protein)
BALB/c	BALB/cj	Sensitivity to radiation
	BALB/c AnN	Used in hybridoma technology
	BALB/cBy	Many myeloma cell lines were generated in these mice
CBA	CBA/J	Gene (*rd*) causing retinal degeneration in CBA/J
	CBA/H	
	CBA/N	Gene (*xid*) causing X-linked immunodeficiency in CBA/N
C3H	C3H/He	Gene (*rd*) causing retinal degeneration
	C3H/HeJ	High incidence of mammary tumors in many substrains (these carry a mammary-tumor virus that is passed via maternal milk to offspring)
	C3H/HeN	
C57BL/6	C57BL/6J	High incidence of hepatomas after irradiation
	C57BL/6By	High complement activity
	C57BL/6N	
C57BL/10	C57BL/10J	Very close relationship to C57BL/6 but differences in at least two loci
	C57BL/10ScSn	
	C57BL/10N	Frequent partner in preparation of congenic mice
C57BR	C57BR/cdj	High frequency of pituitary and liver tumors
		Very resistant to x-irradiation

[Adapted from P. Altman, 1979, Biological Handbooks, Vol. III: Inbred and Genetically Defined Strains of Laboratory Animals, Bethesda, MD: Federation of American Societies for Experimental Biology.]

from MHC B. Animals from this strain can then be used to determine which subregions of a locus contribute to which properties in the immune system of the animal. Such strains were used to delineate the functions of class 1 versus class 2 proteins encoded by the MHC locus.

Congenic Resistant Strains Are Used to Study the Effects of Particular Gene Loci on Immune Responses

Two strains of mice are **congenic** if they are genetically identical except at a single genetic locus or region. Any phenotypic differences that can be detected between congenic strains must therefore be encoded in the genetic region that differs between the two strains. Congenic strains that are identical with each other except at the MHC can be produced by taking F1 (first filial generation) mice derived from the mating of two strains (e.g., A and B), interbreeding them, and then selecting those F2s (second-generation mice) that reject a tumor from one of the strains (e.g., strain A). These mice must be homozygous for the strain B MHC, since if they were heterozygous for A and B at the MHC locus, they would accept the tumor and die. At this point, 50% of their overall genetic material derives from strain B. However, within that 50% is included both MHC alleles.

Next, these mice are backcrossed to strain A, and the process of intercrossing the progeny and testing for those animals which reject the strain A tumor is repeated. At each backcross generation, the fraction of the background (non-MHC) genes that are derived from strain B is cut in half, but only those mice homozygous for strain B MHC are retained. Between 15 and 20 rounds of these crosses lead to a new mouse strain that is strain A in all but its MHC, which is wholly derived from strain B. Such mice are termed A.B mice.

Adoptive Transfer Experiments Allow In Vivo Examination of Isolated Cell Populations

Lymphocyte subpopulations isolated from one animal can be injected into another animal of the same strain without eliciting a rejection reaction. This type of experimental system permitted immunologists to demonstrate for the first time that lymphocytes from an antigen-primed animal could transfer immunity to an unprimed syngeneic recipient. Adoptive transfer experiments using sorted lymphocyte subpopulations also proved the need for both T and B cells in the generation of an antibody response.

Adoptive transfer systems permit the in vivo examination of the functions of in vitro isolated cell populations. More sophisticated adoptive transfer models involve the transfer of cells between animals that differ in an **allotypic marker** that does not elicit a rejection reaction, but does enable the investigator to follow the fate of the injected cells using an antibody against that marker. Another commonly used adoptive transfer protocol involves the transfer of cells that have been fluorescently labeled into a recipient animal so that their fate can be followed using in vivo imaging technologies.

In some adoptive transfer protocols, it is important to eliminate the immune responsiveness of the host by exposing it to x-rays that kill host lymphocytes, prior to donor cell injection. If the host's hematopoietic cells might influence an adoptive transfer experiment, then high x-ray levels (900–1000 rads) are used to eliminate the entire hematopoietic system. Mice irradiated with such doses will die unless reconstituted with bone marrow from a syngeneic donor.

Transgenic Animals Carry Genes That Have Been Artificially Introduced

Development of techniques to introduce cloned foreign genes (**transgenes**) into mouse embryos has permitted immunologists to study the effects of many isolated genes on the immune response in vivo. If the introduced gene integrates stably into the germ-line cells, it will be transmitted to the offspring.

The first step in producing transgenic mice is the injection of foreign cloned DNA into a fertilized egg. In this technically demanding process, fertilized mouse eggs are held under suction at the end of a pipette and the transgene is microinjected into one of the pronuclei with a fine needle. In some fraction of the injected cells, the transgene integrates into the chromosomal DNA of the pronucleus and is passed on to the daughter cells of eggs that survive the process. The eggs, or early embryos, are then implanted in the oviduct of "pseudopregnant" females, and transgenic pups are born after 19 or 20 days of gestation (Figure 20-24).

With transgenic mice, immunologists have been able to study the expression patterns and functions of a large number of transgenes within the context of living animals. By constructing a transgene with a particular promoter, researchers can also artificially control the expression of the transgene. For example, the expression of genes controlled by the metallothionein promoter is activated by zinc, and transgenic mice carrying a transgene linked to a metallothionein promoter will express the transgene only when zinc is added to their water supply. Other promoters are functional only in certain tissues; the insulin promoter, for instance, promotes transcription only in pancreatic cells.

If a transgene is integrated into the chromosomal DNA within the one-cell mouse embryo, it will be integrated into both somatic cells and germ-line cells. The resulting transgenic mice thus can transmit the transgene to their offspring as a Mendelian trait. In this way, it has been possible to produce lines of transgenic mice in which every member of a line contains the same transgene. A variety of such transgenic lines are currently available commercially, or via collaborations with the producing labs.

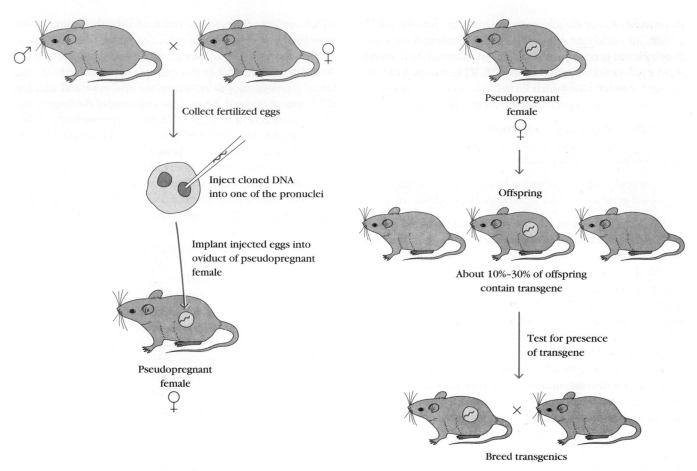

FIGURE 20-24 General procedure for producing transgenic mice. Fertilized eggs are collected from a pregnant female mouse. Cloned DNA (referred to as the transgene) is microinjected into one of the pronuclei of a fertilized egg. The eggs are then implanted into the oviduct of pseudopregnant foster mothers (obtained by mating normal females with a sterile male). The transgene will be incorporated into the chromosomal DNA of about 10% to 30% of the offspring and will be expressed in all of their somatic cells. If a tissue-specific promoter is linked to a transgene, then tissue-specific expression of the transgene will result.

Within the figure:

Collect fertilized eggs

Inject cloned DNA into one of the pronuclei

Implant injected eggs into oviduct of pseudopregnant female

Pseudopregnant female

Pseudopregnant female

Offspring

About 10%–30% of offspring contain transgene

Test for presence of transgene

Breed transgenics

Knock-in and Knockout Technologies Replace an Endogenous with a Nonfunctional or Engineered Gene Copy

One limitation of transgenic mice that are generated as described above is that the transgene is integrated randomly within the genome. This means that some transgenes insert in regions of DNA that are not transcriptionally active, and hence the genes are not expressed, whereas others may disrupt vital genes. To circumvent this limitation, researchers have developed the technology to target the desired gene to specific sites within the germ line of an animal, using homologous DNA recombination. This technique can be used to replace the endogenous gene with a truncated, mutated, or otherwise altered form of that gene, or alternatively to completely replace the endogenous gene with a DNA sequence of choice. For example, a nonfunctional form of a gene may be used to replace the normal allele, in order to determine the effects of losing the expression of the gene in the intact animal. Alternatively, knock-in technology may be used to

determine when and where the promoter for a particular gene is activated. In this latter case, a gene for a fluorescent protein, such as green fluorescent protein, can be specifically engineered into a site downstream from the promoter of the gene of interest. Every time that promoter is activated, the cells in which the promoter is turned on will glow green. How might this be accomplished?

We will describe one method for the generation of knockout mice using homologous DNA recombination. The same principles apply to the generation of knock-in mice, which would simply use different gene segments bounded by the homologous stretches of DNA. Production of gene-targeted knockout mice involves the following steps:

- Isolation and culture of **embryonic stem (ES) cells** from the inner cell mass of a mouse blastocyst

- Generation of the desired, altered form of the gene, bounded by sufficient DNA sequence from the native gene to facilitate homologous recombination

- Introduction of the desired gene into the cultured ES cells and selection of homologous recombinant cells in which the gene of interest has been incorporated. Sensitive *polymerase chain reaction* (PCR) techniques can be used to determine which ES cell colonies have incorporated the desired gene into the correct location.

- Injection of homologous recombinant ES cells into a recipient mouse blastocyst and surgical implantation of the blastocyst into a pseudopregnant mouse

- Mating of chimeric offspring heterozygous for the disrupted gene to produce homozygous knockout mice

The ES cells used in this procedure are obtained by culturing the inner cell mass of a mouse blastocyst in the presence of specific growth factors and on a feeder layer of fibroblasts. Under these conditions, the stem cells grow but remain pluripotent. One of the advantages of ES cells is the ease with which they can be genetically manipulated. Cloned DNA containing a desired gene can be introduced into ES cells in culture by various transfection techniques; the introduced DNA will be inserted by recombination into the chromosomal DNA of a small fraction of these.

In one model of generating knockout mice, the insertion constructs introduced into ES cells contain three genes: the target gene of interest and two selection genes, such as *neo*^{R},

which confers neomycin resistance, and the thymidine kinase gene from herpes simplex virus (*tk*^{HSV}), which confers sensitivity to gancyclovir, a cytotoxic nucleotide analogue (Figure 20-25a). The construct in this example is engineered with the target gene sequence disrupted by the *neo*^{R} gene and with the *tk*^{HSV} gene at one end, beyond the sequence of the target gene. Most constructs will insert at random by nonhomologous recombination rather than by gene-targeted insertion through homologous recombination. As shown in Figure 20-25a, those cells in which the construct inserts at random retain expression of the *tk*^{HSV} gene, whereas those cells in which the construct inserts by homologous recombination lose the *tk*^{HSV} gene and hence the sensitivity to gancyclovir. As illustrated in Figure 20-25b, a two-step selection scheme is used to obtain those ES cells that have undergone homologous recombination, whereby the disrupted gene replaces the target gene. The desired cells are resistant to both gancyclovir and neomycin. Other selection schemes exist, and individual scientists choose the one that best suits their needs, but this example protocol illustrates some of the general principles involved in generating and selecting those cells with the desired genetic alteration.

ES cells obtained by this procedure will be heterozygous for the knockout mutation in the target gene. These cells are clonally expanded in cell culture and injected into a mouse blastocyst, which is then implanted into a pseudopregnant female.

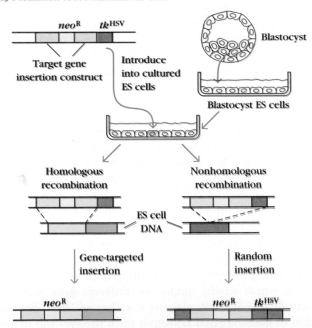

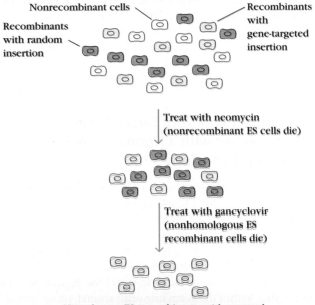

FIGURE 20-25 Formation and selection of mouse recombinant ES cells in which a particular target gene is disrupted. (a) In the engineered insertion construct, the target gene is disrupted with the *neo*^{R} gene, and the thymidine kinase *tk*^{HSV} gene is located outside the target gene. The construct is transfected into cultured ES cells. Recombination occurs in only about 1% of the cells, with nonhomologous recombination much more frequent than homologous recombination. (b) Selection with the neomycin-like drug G418 will kill any nonrecombinant ES cells because they lack the *neo*^{R} gene. Selection with gancyclovir will kill the nonhomologous recombinants carrying the *tk*^{HSV} gene, which confers sensitivity to gancyclovir. Only the homologous ES recombinants will survive this selection scheme. *[Adapted from H. Lodish et al., 1995, Molecular Cell Biology, 3rd ed., New York: Scientific American Books.]*

The transgenic offspring that develop are chimeric, composed of cells derived from the genetically altered ES cells and cells derived from normal cells of the host blastocyst. When the germ-line cells are derived from the genetically altered ES cells, the genetic alteration can be passed on to the offspring. If the recombinant ES cells are homozygous for black coat color (or another visible marker) and they are injected into a blastocyst homozygous for white coat color, then the chimeric progeny that carry the heterozygous knockout mutation in their germ line can be easily identified (Figure 20-26). When these are mated with each other, some of the offspring will be homozygous for the knockout mutation.

The Cre/*lox* System Enables Inducible Gene Deletion in Selected Tissues

In addition to the deletion of genes by gene targeting, experimental strategies have been developed that allow the specific deletion of a gene of interest only in selected tissues. This enables investigators to determine the effects of losing gene activity only in, for example, the tissues of the immune system, even if expression of those genes in other tissues is necessary for viability of the organism. These technologies rely on the use of site-specific recombinases from bacteria or yeast.

The most commonly used recombinase is Cre, isolated from bacteriophage P1. Cre recognizes a specific 34-bp site in DNA known as *loxP* and catalyzes a recombination event between two *loxP* sites such that the DNA between the two sites is deleted. Animals that ubiquitously express Cre recombinase will therefore delete all *loxP*-flanked sequences, whereas animals that express Cre only in certain tissues will only delete *loxP*-flanked sequences in those tissues. If the expression of the Cre recombinase gene is placed under the control of a tissue-specific promoter, then tissue-specific deletion of any DNA that is flanked by *loxP* sites will occur. For example, one could express Cre in B cells using the immunoglobulin promoter, and this would result in the targeted deletion of *loxP*-flanked DNA sequences only in B cells.

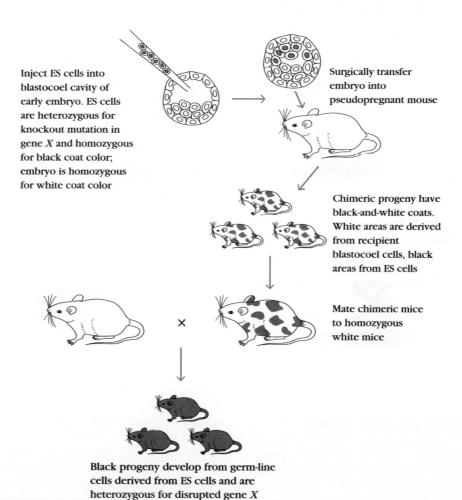

Inject ES cells into blastocoel cavity of early embryo. ES cells are heterozygous for knockout mutation in gene *X* and homozygous for black coat color; embryo is homozygous for white coat color

Surgically transfer embryo into pseudopregnant mouse

Chimeric progeny have black-and-white coats. White areas are derived from recipient blastocoel cells, black areas from ES cells

Mate chimeric mice to homozygous white mice

Black progeny develop from germ-line cells derived from ES cells and are heterozygous for disrupted gene *X*

FIGURE 20-26 General procedure for producing homozygous knockout mice. ES cells homozygous for a marker gene (e.g., black coat color) and heterozygous for a disrupted target gene are injected into an early embryo homozygous for an alternate marker (e.g., white coat color). The chimeric transgenic offspring, which have black-and-white coats, then are mated with homozygous white mice. The all-black progeny from this mating have ES-derived cells in their germ line, which are heterozygous for the disrupted target gene. Mating of these mice with each other produces animals homozygous for the disrupted target gene—that is, knockout mice. [*Adapted from M. R. Capecchi, 1989, Trends in Genetics 5:70.*]

This technology is particularly useful when the targeted deletion of a particular gene in the whole animal would have lethal consequences. For example, the DNA polymerase β gene is required for embryonic development, and deletion of this gene in the whole animal would therefore result in embryonic lethality. In experiments designed to delete the DNA polymerase β gene only in thymic tissues, scientists flanked the mouse DNA polymerase β gene with *loxP* and mated these mice with mice carrying a Cre transgene under the control of a T-cell promoter (Figure 20-27). The results of this mating are offspring that express the Cre recombinase specifically in T cells, allowing scientists to examine the effects of specifically deleting the enzyme DNA polymerase β in T cells.

The Cre/*lox* system can also be used to turn on gene expression in a particular tissue. Just as the lack of a particular gene may be lethal during embryonic development, the expression of a gene can be toxic. To examine tissue-specific expression of such a gene, it is possible to insert a translational stop sequence flanked by *loxP* into an intron at the beginning of the gene (Figure 20-27b). Using a tissue-specific promoter driving Cre expression, the stop sequence may be deleted in the tissue of choice and the expression of the potentially toxic gene examined in this tissue.

Some investigators have combined this technology with the use of an artificial inducer of Cre activity in order to control precisely when the gene is lost. Transgenic mice have been developed that express fusion proteins in which the Cre recombinase is linked to a second protein—for example, an altered estrogen receptor designed to respond to the drug tamoxifen (Novaldex). This Cre fusion protein is designed such that Cre is not active unless tamoxifen is present. Thus, one can place the Cre fusion protein expression under the control of a tissue-specific promoter, and precisely time the knockout of the gene in that tissue by the administration of tamoxifen. Other similar fusion proteins have been developed that allow Cre expression to come under the control of various antibiotics. These modifications of gene-targeting technology have been pivotal in the determination of the effects of particular genes in cells and tissues of the immune system.

(a)

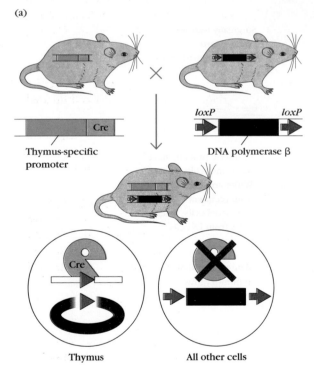

(b)

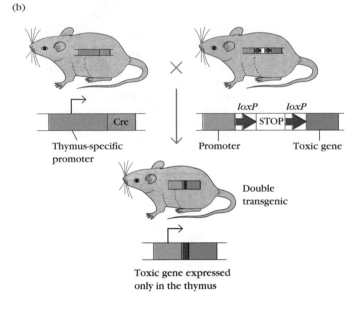

FIGURE 20-27 Gene targeting with Cre/loxP. (a) Conditional deletion by Cre recombinase. The targeted DNA polymerase β gene is modified by flanking the gene with *loxP* sites (for simplicity, only one allele is shown). Mice are generated from ES cells by standard procedures. Mating of the *loxP* modified mice with a Cre transgenic will generate double transgenic mice in which the *loxP*-flanked DNA polymerase β gene will be deleted in the tissue where Cre is expressed. In this example, Cre is expressed in thymus tissue, so that deletion of the *loxP*-flanked gene occurs only in the thymus of the double transgenic. Other tissues and organs still express the loxP-flanked gene. (b) Activation of gene expression using Cre/lox. A *loxP*-flanked translational STOP cassette is inserted between the promoter and the potentially toxic gene, and mice are generated from ES cells using standard procedures. These mice are mated to a transgenic line carrying the Cre gene driven by a tissue-specific promoter. In this example, Cre is expressed in the thymus, so that mating results in expression of the toxic gene (blue) solely in the thymus. Using this strategy, one can determine the effects of expression of the potentially toxic gene in a tissue-specific fashion. [*Adapted from B. Sauer, 1998, Methods **14**:381, with modifications to (a) from http://mammary.nih.gov/tools/molecular/ Wagner001/images/Cre-lox_3.GIF.*]

- Polyclonal antibodies are generated by immunizing an animal with an antigen (usually complexed with an adjuvant) one or more times, and then retrieving the antiserum.

- Monoclonal antibodies are generated by fusion of an antibody-producing B cell with a long-lived B-cell tumor.

- When bi- or multivalent antibodies are mixed with antigen, they can form a cross-linked matrix in solution, resulting in the formation of an immunoprecipitate. Immunoprecipitation can be used to purify proteins from a detergent extract of cells or tissues.

- Hemagglutination reactions measure the presence of antibodies to antigens located on red blood cells. Hemagglutination inhibition reactions measure the presence of antibodies to those viruses that induce hemagglutination, notably influenza.

- Bacterial agglutination reactions measure the presence of antibodies that bind specific bacterial strains.

- Radioimmunoassays provide sensitive means to measure antibody or antigen concentrations using radioactivity. Enzyme-linked immunosorbent assays (ELISA) use enzyme-conjugated antibodies to measure antigen and antibody concentrations without the need for radioisotopes. Different enzyme and substrate systems allow adaptation of ELISA technology to increase sensitivity. Substrates can be chromogenic, fluorogenic, or chemiluminogenic.

- Biotin-conjugated antibodies can be used with streptavidin-conjugated enzymes to increase the flexibility of ELISA and other assays.

- Western blotting uses antibodies to detect the presence of particular protein bands following SDS PAGE and transfer of bands onto a solid-phase support.

- Equilibrium dialysis is an inexpensive and relatively easy way to measure the affinity of antibody for antigens. Surface plasmon resonance measures forward and reverse rate constants of antibody binding, in addition to association constants.

- Immunocytochemistry and immunohistochemistry use antibodies covalently conjugated to enzymes to visualize cells and tissues. Once the antibodies are bound, substrates are added that are converted to products which form colored precipitates that are deposited at the site of antibody binding.

- Immunoelectron microscopy uses antibodies conjugated to gold beads in order to visualize antibody-bound structures at high resolution. Antibodies covalently conjugated to fluorescent probes provide vivid images of structures under the fluorescence microscope.

- By using a pinhole to only allow images from a particular depth of field, confocal microscopy enables the visualization of tissues at different focal planes. Software can then be used to re-create a three-dimensional picture of the tissue. Two- or multi-photon microscopy provides further resolution by requiring that two or more photons simultaneously impinge upon a fluorescent probe before emission is possible.

- Whole organs such as lymph nodes can be placed on a warmed microscope stage while maintaining lymphatic and blood circulation. By labeling particular cells in vivo with particular fluorescent probes, intravital images of ongoing immune responses can be captured.

- Flow cytometry allows quantitative measurements of the frequencies of cells binding to particular fluorescent antibodies or substrates and allows sorting of cells on the basis of their fluorescence and light-scattering properties.

- Magnetic-based cell sorting can be used to sort cells coated with antibodies coupled to magnetic beads or molecules.

- Dividing cells take up tritiated thymidine and incorporate it into high molecular weight DNA; the radioactivity of the DNA can be measured to provide an indicator of cell division. The number of actively metabolizing cells in a culture can be measured by the MTT assay.

- Bromodeoxyuridine (BrdU) is a thymidine analogue that is incorporated into actively replicating DNA. Antibodies to BrdU can be used to ascertain whether a cell divided after the addition of BrdU to the culture.

- Propidium iodide intercalates into the DNA helix in a quantitative manner. Flow cytometric profiles of propidium iodide allow analysis of the cell cycle status of a cell population.

- Carboxyfluorescein diacetate succinimidyl ester (CFSE) binds to the cytoplasmic proteins of cells in a quantitative manner; its intracellular concentration is cut in half with each cell division. CFSE fluorescence can therefore be used to measure the number of times a cell has divided since addition of the CFSE.

- ^{51}Cr is released from pre-labeled cells upon cell death. ^{51}Cr release can therefore be used to measure the extent of cell death within the labeled population.

- Fluorescently labeled Annexin V binds to phosphatidylserine exposed on the surface of cells undergoing apoptosis and can therefore be used as a measure of apoptosis in immunofluorescence or flow cytometric experiments.

- Apoptotic cells undergo DNA fragmentation. The TUNEL assay uses terminal deoxyribonucleotidyl transferase to add labeled nucleotides onto the DNA ends exposed on apoptotic fragmentation. The amount of added label is a measure of the extent of apoptosis.

- Caspases are activated in apoptotic cells, and fluorescence- or Western blot-based caspase assays therefore provide a measure of apoptosis.

- Inhibitors of enzymes known to be active in signaling pathways can be used to ascertain the components of newly described pathways.

- Immunoprecipitation and yeast two-hybrid screens are two common techniques to identify interacting pairs or sets of proteins.

- Animal research should be undertaken in a manner consistent with high ethical standards and is subject to federal guidelines.

- Different types of mouse strains have contributed immeasurably to immunological research. Inbred, congenic-resistant, and recombinant inbred strains of mice have each provided investigators with invaluable information.

- Adoptive transfer techniques introduce cells taken from one animal and transferred in various ways into a second animal.

- Transgenic animals carry genes that have been artificially introduced. Knock-in and knockout gene technologies enable the introduction of altered or inactive forms of genes into specific locations in the genome. The Cre/*lox* recombinase system allows investigators to target altered genes into specific locations in the genome in an inducible manner.

REFERENCES

Bonner, W. A., et al. 1972. Fluorescence activated cell sorting. *Review of Scientific Instruments* **43**:404–409.

Capecchi, M. R. 1989. Altering the genome by homologous gene targeting. *Science* **244**:1288–1292.

Capecchi, M. R. 1989. The new mouse genetics: Altering the genome by gene targeting. *Trends in Genetics* **5**:70–76.

Gossen, M., et al. 1995. Transcriptional activation by tetracyclines in mammalian cells. *Science* **268**:1766–1769.

Herzenberg, L. A., ed. 1996. *Weir's Handbook of Experimental Immunology*, 5th ed. Oxford University Press. Blackwell Scientific Press, Oxford, UK.

Köhler, G., and C. Milstein. 1975. Continuous cultures of fused cells secreting antibody of predefined specificity. *Nature* **256**:495–497.

Lerner, R. A., et al. 1991. At the crossroads of chemistry and immunology: Catalytic antibodies. *Science* **252**:659–667.

Malmqvist, M. 1993. Surface plasmon resonance for detection and measurement of antibody-antigen affinity and kinetics. *Current Opinion in Immunology* **5**:282.

Orban, P. C., et al., 1992. Tissue and site-specific DNA recombination in transgenic mice. *Proceedings of the National Academy of Sciences USA* **89**:6861–6865.

Rajewsky, K., et al. Conditional gene targeting. *Journal of Clinical Investigation* **98**:600–603.

Rich, R. L., and D. G. Myszka. 2003. Spying on HIV with SPR. *Trends in Microbiology* **11**:124–133.

Russell, W. M. S., and Burch, R. L. 1959. *The Principles of Humane Experimental Technique*. Methuen, London; reprinted by Universities Federation for Animal Welfare, Potters Bar, UK, 1992.

Weiss, A. J. 2012. Overview of membranes and membrane plates used in research and diagnostic ELISPOT assays. *Methods in Molecular Biology* **792**:243–256.

Zhu, Q., et al. 2010. Using 3 TLR ligands as a combination adjuvant induces qualitative changes in T cell responses needed for antiviral protection in mice. *Journal of Clinical Investigation* **120**:607–616.

Useful Web Sites

www.currentprotocols.com/WileyCDA/CurPro3Title/isbn-0471142735.html *Current Protocols in Immunology* is a frequently updated compendium of most techniques used by immunologists.

Many of the most useful sources of protocols and product information are those found on Web sites and product inserts from the manufacturers of relevant reagents. The following is a selection of the most useful Web sites, but students are encouraged to surf the Web and compare and contrast protocols from different manufacturers before finalizing their experimental designs.

www.miltenyibiotec.com/en/NN_628_Protocols.aspx Descriptions of protocols for use with Miltenyi magnetic beads are found here.

www.bdbiosciences.com/home.jsp The home page of BD Biosciences provides a wealth of information about flow cytometry.

www.jax.org The home page of Jackson Laboratories provides information about many inbred, transgenic, and other useful mouse strains.

http://cre.jax.org/introduction.html An introduction to Cre-lox technology provided by the Jackson Laboratories.

www.wikipedia.org

www.wikimedia.org Wikipedia and Wikimedia often provide useful and updated information and links to protocols.

www.youtube.com

www.jove.com YouTube and JoVE both provide video protocols of many types of immunological experiments.

www.immunoportal.com

bitesizebio.com ("Brain Food for Biologists") Both of these sites provide protocols and useful reviews of modern and competing technologies.

STUDY QUESTIONS

1. When might you elect to use a polyclonal rather than a monoclonal antibody preparation, and why?

2. You successfully used one batch of polyclonal sera to immunoprecipitate a protein of interest. Analysis of the precipitated protein by polyacrylamide gel electrophoresis showed a single band, indicating a pure protein. However, when you repeated the experiment using a second batch of the serum derived from the same animal, but at a later date, your protein precipitate was contaminated with other proteins that did not co-precipitate the first time around. Why?

3. For the following applications, would you elect to use a polyclonal antibody preparation, a monoclonal antibody, or more than one monoclonal antibody to detect your antigen? Explain your answer.
   a. Bacterial agglutination
   b. Immunoprecipitation
   c. Western blotting
   d. Detection of a cytokine using a solid-phase ELISA
   e. Diagnostic tissue typing

4. The following figure illustrates a hemagglutination inhibition assay. It shows the change with time in antibody titer of a newborn baby who survived the H1N1 influenza epidemic in 2009 in Thailand. The numbers along the bottom of the plate represent the dilution of the sera used in the experiment. The most concentrated serum is a 10-fold

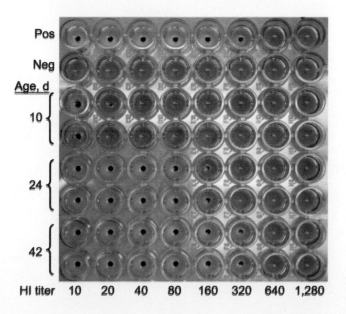

dilution from the patient's serum. In this experiment, sera obtained from different times and different sources have been serially diluted as indicated. Influenza virus and red blood cells have then been added to each well of the plate, and the ability of the various antisera to inhibit hemagglutination assessed.

The top row of this plate shows the hemagglutination inhibition that occurs when a positive-control anti-influenza antibody sample is added to the combination of virus and RBCs. It shows that the serum can be diluted out to 1 in 320 and still bind to the virus sufficiently to inhibit hemagglutination. The second row shows the hemagglutination that occurs when the virus is added in the absence of any neutralizing antibodies; it represents the negative control in this experiment.

The bottom six rows represent the hemagglutination inhibition capacity of the serum of the baby who is the subject of the experiment. The serum diluted in rows 3 and 4 (duplicate) was drawn from the baby when she was 10 days of age. The serum diluted in rows 5 and 6 was taken when the baby was 24 days old, and that in rows 7 and 8 when the baby was 42 days old.

   a. Did the baby's serum contain any influenza-specific antibodies when she was 10 days old?
   b. How old was the baby before her serum displayed the same hemagglutination capacity as the positive-control sample?
   c. Can we conclude from this experiment that the baby is making these antibodies herself?

5. Why might you elect to use an RIA rather than an ELISA with conventional chromogenic substrates? Would an ELISA with a chemiluminogenic substrate solve the problem?

6. The substrates used in an ELISPOT assay differ from those employed in a conventional ELISA assay, but may be similar to, or identical to, those used in Western blots. Why?

7. a. You have just started up your laboratory and are working on a shoestring budget until you hear about your first grant. You need to measure the affinity of a monoclonal antibody you're working on, and you have it available in radioactively labeled form. What method would you use, and why?
   b. Your experiment indicates that you now need to know the association rate of the antibody-antigen interaction. You receive good news about your grant. How do you proceed?

8. What advantages are offered by two- or multi-photon microscopy over more traditional confocal microscopy?

9. When might you elect to purify cell populations with magnetic-activated cell sorting, rather than fluorescence-activated cell sorting, and when might you choose fluorescence-based separation over magnetic-based methods?

10. Challenge question. This question should be answered with reference to Table 20-3 and material from Chapters 3 and 4.

    You are investigating the signaling pathway of a newly discovered cytokine that turns on the transcription of a well-defined set of genes. You discover that signal-induced transcription fails to occur in the presence of wortmannin, Ly294002, and BX795.

    a. What two enzymes does this implicate as part of your pathway?
    b. What transcription factor is implicated as mediating the transcription of the activated genes and why?
    c. What protein domains do you expect to find on intermediaries of the pathway downstream from PI3 kinase?

11. **Experimental Design Question.** You posit that under conditions of infection with a particularly virulent virus, B-1 B cells are entering inflamed regions of the lung. How might you test this hypothesis?

12. **Experimental Design Question.** You wish to knock out the expression of a particular gene, only in B cells, and you want to do the knockout only after you have exposed your B cells to antigen. What genetic constructs do you need to generate for this experiment?

The following table presents information about the nature, cellular distribution, and function of the CD antigens, which are the membrane molecules that have a variety of functions and serve as markers for particular cell types. Because many CD antigens are known by a variety of names, synonyms are indicated in addition to the official CD designations. The mass is listed (when known) for the CD antigens that are proteins but not for those that are carbohydrates or lipids. In summarizing the expression patterns for CD antigens, we have concentrated on cells of the immune system (leukocytes). However, many of these antigens are also expressed in other cell types, and we also give some examples of these. The description of the functions of each CD antigen also focuses on cells of the immune system, with an occasional example of functions in other cell types. The function is recorded as "not known" if there is no immunologically relevant function known for the CD marker. CD antigens are similarly named and share many properties in humans and mice, although there are some differences. The table presents information for human CD antigens.

Responsibility for naming and describing CD antigens rests with **Human Cell Differentiation Molecules (HCDM)** (www.hcdm.org/Home/tabid/36/Default.aspx), an organization that runs the **Human Leukocyte Differentiation Antigens (HLDA) Workshops**. The mission of HCDM is to characterize the structure, function, and distribution of leukocyte surface molecules and other molecules of the immune system. The Web site and international workshops are the product of a collaborative effort by numerous researchers and biomedical supply companies around the world and are updated regularly. The list of CD antigens was most recently updated at HLDA9 in March 2010; the new CD antigens are described on the HLDA Web site at www.hcdm.org/HLDA-9Workshop/tabid/60/Default.aspx.

An updated list of all CD protein molecules is available through the UniProt knowledge base at http://uniprot.org/docs/cdlist.

This searchable database, frequently updated, is sponsored and maintained by the Swiss Institute of Bioinformatics and provides information about the structure, sequence, cellular distribution, and functions of the proteins, as well as genetics, with links to relevant references.

The Web sites of several biosciences supply companies also include lists of CD antigens in both humans and mice, with links to antibody reagents the companies sell for detecting these markers. You can find the CD antigen lists of two of these companies at the following Web sites:

BD Biosciences: www.bdbiosciences.com/reagents/cdmarkers/index.jsp

BioLegend, Inc.: www.biolegend.com/support

An additional source of up-to-date information is PubMed, a Web site of the National Library of Medicine found at www.ncbi.nlm.nih.gov/pubmed.

A search engine at the PubMed site allows searches of published articles by key words as well as other variables, such as author name or year of publication. A key word search using the name of a particular CD antigen will provide a list of the most recent journal articles that address it. For a more comprehensive search on a particular antigen, the synonyms listed in this table can be included as alternative key words.

This table was updated based on information in the Web sites mentioned above, along with more recent literature references.

CD Antigen. MW. Synonyms and Properties.	Expression	Function
CD1a, -b, -c, -d, -e. 43–49 kDa. T6/Leu-6; MHC class I–like structures.	Dendritic cells, B cells, Langerhans cells, cortical thymocytes, monocytes, activated T cells, intestinal epithelial cells.	Antigen-presenting proteins that bind self and non-self lipid and glycolipid antigens and presents them to T-cell receptors on natural-killer T cells.
CD2. 45–58 kDa. LFA-2, T11, Leu-5, Tp50; CD58-binding adhesion molecule, sheep red blood cell (SRBC) receptor.	T cells, cortical thymocytes, NK cells, some B cells, some monocytes.	Interacts with LFA-3 and CD48/BCM1 to mediate adhesion between T cells and other cell types; contributes to T-cell activation.
CD3. composed of 3 polypeptide chains: γ, δ, ε. γ, 20-26 kDa; δ, 21 kDa; ε, 20 kDa. T3.	Thymocytes, T cells.	Signaling chains of the TCR. Essential roles in cell surface expression of the TCR and TCR signal transduction.
CD4. 55 kDa. T4, Leu3, L3T4, Ly4 (mouse), Ox38.	MHC class II–restricted T cells, some thymocytes, monocytes/macrophages.	Coreceptor for MHC class II–restricted T-cell activation; thymic differentiation marker for T cells; receptor for HIV.

(continued)

CD Antigen. MW. Synonyms and Properties.	Expression	Function
CD5. 67 kDa. Leu1, T1, Ly-1 (mouse), Ox19.	Mature T cells, cortical thymocytes; mature B-cell subset (B-1a B cells).	Positive or negative modulation of TCR and BCR signaling, depending on the type and developmental stage of cell displaying it.
CD6. 105-130 kDa. T12, Ox52, Tp120.	Most peripheral T cells, cortical thymocytes, B-cell subset, NK-cell subset.	Adhesion molecule, binds CD166; involved in costimulation, thymic selection, NK activation.
CD7. 40 kDa. gp40, Leu9, Tp41T-cell leukemia antigen.	Pluripotent hematopoietic cells, T cells, thymocytes, NK cells, pre-B cells.	Distinguishes primitive lymphoid progenitors from pluripotent stem cells. Plays a role in regulating peripheral T-cell and NK-cell cytokine production and sensitivity to LPS-induced shock. May have costimulatory activity for T cells.
CD8. Membrane-bound dimer of two chains, $\alpha\beta$ heterodimer or $\alpha\alpha$ homodimer. α, 32-34 kDa, β, 30-32 kDa. T8, Leu-2, Lyt-2 (mouse).	MHC class I–restricted T cells, some thymocytes, subset of dendritic cells, NK cells.	Coreceptor for MHC class I–restricted T-cell activation, thymic differentiation marker for T cells.
CD9. 24 kDa. MRP-1, p24, DRAP-27.	Platelets, pre-B cells, activated T cells, eosinophils, basophils, some epithelial and endothelial cells.	Modulation of cell adhesion and migration; triggers platelet activation and aggregation.
CD10. 100 kDa. Common acute lymphoblastic leukemia antigen (CALLA), EC 3.4.24.11 (neprilysin), enkephalinase, gp100, neutral endopeptidase (NEP), neprilysin, skin fibroblast elastase.	B-cell and T-cell precursors, neutrophils, bone marrow stromal cells, fibroblasts.	Membrane-bound neutral endopeptidase that cleaves a variety of inflammatory and vasoactive peptides.
CD11a. 180 kDa. Integrin αL chain; α chain of LFA-1 (leukocyte-function–associated molecule-1); forms LFA-1 by association with integrin β2 chain (CD18).	All leukocytes.	Subunit of LFA-1, a membrane glycoprotein that provides cell-cell adhesion by interaction with ICAMs-1-4 (intercellular adhesion molecules-1-4, CD54); functions in leukocyte-endothelial cell interaction, cytotoxic T-cell mediated killing, and antibody-dependent killing by granulocytes and monocytes.
CD11b. 170 kDa. Integrin αM chain, α chain of MAC-1 (CR3), forms MAC-1 by association with β2 integrin (CD18). C3biR, iC3b receptor, Ly40 (mouse), Ox42.	Granulocytes, monocytes, macrophages, NK cells, subsets of T and B cells, myeloid dendritic cells.	Implicated in various adhesive interactions of monocytes, macrophages, and granulocytes as well as in mediating the uptake of complement-coated particles. Identical to CR3, the receptor for the iC3b fragment of the third complement component. It probably recognizes the R-G-D peptide in C3b. Integrin alpha-M/beta-2 is also a receptor for fibrinogen, factor X, and ICAM1.
CD11c. 150 kDa. αX integrin chain, α chain of integrin αxβ2, forms CR4 by association with β2-integrin (CD18), leukocyte surface antigen, p150, 95.	Monocytes, macrophages, NK cells, granulocytes, subsets of T and B cells, dendritic cells.	Subunit of CR4 with CD18 that is similar to CD11b/CD18 complex, with which it acts cooperatively; the major form of CD11/CD18 on tissue macrophages, adhesion molecule; binds ICAMs 1,4, iC3b, and fibrinogen.
CD11d. 125 kDa. Integrin αD-chain, ITGAD, ADB2, associates with β2-integrin (CD18).	Monocytes, macrophages.	May play a role in the atherosclerotic process such as clearing lipoproteins from plaques and in phagocytosis of blood-borne pathogens, particulate matter, and senescent erythrocytes from the blood.
CDw12. 90–120 kDa.	Monocytes and granulocytes and their precursors, platelets, some epithelial and endothelial cells.	Not known.

CD Antigen. MW. Synonyms and Properties.	Expression	Function
CD13. 150-170 kDa. Aminopeptidase N (APN), EC 3.4.11.2, gp 150, Lap1.	Early progenitors of granulocytes and monocytes (CFU-GM), mature granulocytes and monocytes, bone marrow stromal cells, osteoclasts, a small number of large granular lymphocytes, T cells, some epithelial and endothelial cells.	Membrane-bound peptidase that catalyzes the removal of N-terminal amino acids from a variety of peptides; receptor mediating infection with human cytomegalovirus and coronavirus.
CD14. 53-55 kDa. LPS receptor (LPS-R).	Monocytes, macrophages, granulocytes (weak expression), Langerhans cells.	Receptor for endotoxin (lipopolysaccharide [LPS]) bound to LPS binding protein (LBP), activating innate immune responses; transfers the complex to TLR4. May also be involved in binding of peptidoglycans and lipoproteins to TLR2.
CD15. Terminal poly-*N*-acetyl lactosamine trisaccharide found on some glycolipids and glycoproteins. Lewis X, Lex, SSEA-1, 3-FAL.	Granulocytes.	Suggested to be the ligand for CD62E selectin, functions in cell-cell adhesion.
CD15s. Poly-*N*-acetyl lactosamine. Sialyl Lewis X (sLex).	Granulocytes, monocytes, macrophages, NK cells, activated T and B cells and memory helper T cells, high endothelial venules.	Strongest binding ligand for E-selectin functions in cell-cell adhesion.
CD15u. 3′-sulfated Lewis X.	Granulocytes, monocytes, macrophages, T and B cell subsets, NK cells, endothelial cells.	Ligand for P-selectin, functions in cell-cell adhesion.
CD15su. 6′-sulfated Lewis X.	Granulocytes, monocytes, macrophages, T and B cell subsets, NK cells, endothelial cells.	Ligand for L-selectin, functions in cell-cell adhesion.
CD16. 50–65 kDa. cD16a, FCγRIIIA.	NK cells, macrophages, subpopulation of T-cells, immature thymocytes and placental trophoblasts.	Low-affinity Fcγ receptor, binds IgG especially in complexes or aggregates; activates antibody-dependent processes such as phagocytosis and antibody-dependent cell-mediated cytotoxicity (ADCC).
CD16b. 48 dDa. FCγRIIIB.	Neutrophils, stimulated eosinophils.	Low-affinity Fcγ receptor, binds IgG, especially in complexes or aggregates. May serve as trap for immune complexes but does not activate phagocytosis or ADCC. May activate neutrophil transendothelial migration.
CD17. Lactosylceramide (LacCer); carbohydrate antigen.	Monocytes, granulocytes, platelets, subset of peripheral B cells (CD19⁺), T cells, dendritic cells.	May mediate adhesion; binds bacteria, may function in phagocytosis, trapping, motility, proliferation.
CD18. 95 kDa. β2 integrin chain that combines with CD11 α chains to form integrins.	All leukocytes.	Part of αβ integrins that bind ICAMs and function in adhesion and signaling of leukocytes; also part of receptors for complement and fibronectin (*see* entries for CD11a–d).
CD19. 95 kDa. B4, Leu-12.	B cells from earliest recognizable B-lineage cells during development to B-cell blasts but lost on maturation to plasma cells, follicular dendritic cells.	Part of B-cell coreceptor with CD21 and CD81; a critical signal transduction molecule that assembles with the BCR and regulates B-cell development, activation, and differentiation.
CD20. 33–37 kDa. B1, Bp35, Leu-16.	B cells, T cell subsets.	Ligation activates signaling pathways, may have a role in regulating B-cell activation, proliferation, and differentiation.
CD21. 145 kDa (membrane form); 110 kDa (soluble form). CR2, C3d receptor, Epstein-Barr virus (EBV) receptor. With CD19 and CD81, forms the B-cell coreceptor.	Mature B cells, subset of T cells, follicular dendritic cells, astrocytes.	Receptor for C3d, C3dg, and iC3b; with CD19 and CD81, part of the B-cell coreceptor complex that contributes to B-cell activation. Also serves as the receptor for EBV.

(continued)

CD Antigen. MW. Synonyms and Properties.	Expression	Function
CD22. 130 kDa. B-lymphocyte cell adhesion molecule (BL-CAM), Siglec-2, Leu-14m Lyb8 (mouse).	Surface of mature B cells, cytoplasm of late pro– and early pre–B cells.	Binds sialylated glycoproteins, including CD45. Promotes adhesion and may be involved in positive and negative signaling.
CD23. 45 kDa. FcεRII, B6, BLAST-2, Leu-20.	B cells (upregulated on activated cells), activated macrophages, follicular dendritic cells, eosinophils, platelets, intestinal epithelial cells.	Low-affinity IgE receptor. Regulates B-cell activation, growth, and IgE synthesis; triggers macrophages to release TNF, IL-1, IL-6, and GM-CSF; with food allergies, triggers transport in of IgE and IgE/allergen complexes across intestinal epithelium.
CD24. 35–45 kDa. BA-1, heat-stable antigen (HSA) in mouse.	B-cell lineage but lost at plasma-cell stage, T-cell subsets, monocytes, mature granulocytes, Langerhans cells, some epithelial cells.	Mucin-like adhesion molecule. Promotes antigen-activated B-cell proliferation; inhibits differentiation to plasma cells.
CD25. 55 kDa. IL-2Rα (IL-2 receptor α chain), Tac antigen, p55.	Activated B cells and T cells, regulatory T cells, immature thymocytes, activated monocytes, macrophages, NK cells, dendritic cell subset.	Low-affinity IL-2 receptor, associates with β and γ chains to form high-affinity IL-2R; activation marker; induces activation and proliferation of T cells, NK cells, B cells, and macrophages. Thymocyte differentiation marker; T_{reg} marker.
CD26. 110 kDa. dipeptidylpeptidase IV (DPP IV ectoenzyme); EC 3.4.14.5, adenosine deaminase–binding protein.	Activated T cells, mature thymocytes, B-cell subset, NK cells, macrophages, some epithelial cells, lymphatic endothelial cells.	Membrane-bound exopeptidase (cleaves certain dipeptides from protein N-termini); functions in T-cell costimulation; may be involved in lymphatic vessel adhesion.
CD27. 50–55 kDa. S152, T14; TNFRSF7 (tumor necrosis factor receptor superfamily member 7).	Mature thymocytes, T cells and B cell-subsets, NK cells.	Binds CD70. Costimulatory signal for T-and B-cell activation; role in murine T-cell development.
CD28. ~90 kDa (homodimeric form). T44, Tp44.	Mature thymocytes, most peripheral T cells, plasma cells, NK cells.	Costimulation of T-cell proliferation and cytokine production upon binding CD80 or CD86.
CD29A–D. 130 kDa. Integrin β1 chain, VLA-4β chain, platelet GPIIa.	Most leukocytes (weakly on granulocytes); isoforms B, C, D expressed on many non-hematopoietic cell types.	β subunit of VLA-1 integrin, binds VCAM, MadCAM-1, and fibronectin; involved in cell adhesion and recognition in a variety of processes, including embryogenesis, hemostasis, tissue repair, immune response, metastatic diffusion of tumor cells and development. Essential to the differentiation of hematopoietic stem cells with tumor progression and metastasis/invasion.
CD30. 105 kDa. Ber-H2, Ki-1; TNFRSF8.	Activated T, B, and NK cells, monocytes.	Binds CD30L (CD153). Costimulates lymphocyte proliferation and differentiation; may modulate cell survival/death.
CD31. 130–140 kDa. platelet endothelial cell adhesion molecule (PECAM-1), GPIIa, endocam.	Lymphocyte subsets, platelets, monocytes, granulocytes, endothelial cells.	Adhesion molecule; activates leukocyte transendothelial migration, especially under inflammatory conditions; may enhance phagocytosis of apoptotic cells and inhibit phagocytosis of viable cells.
CD32A–C. 40 kDa. FcγRII, FCRII, Ly-17 (mouse).	Isoforms variably expressed on B cells (B), NK cells (C), monocytes (A, B, C), macrophages (A, B, C), dendritic cells (B), Langerhans cells, neutrophils (A, C), eosinophils (A), and platelets (A, B), as well as on endothelial cells of the placenta (B).	Receptor for Fc portion of IgG in aggregates and complexes. Triggers IgG-mediated phagocytosis and activation of oxidative burst in neutrophils and monocytes and mediator release from granulocytes. FcγRIIB isoform does not trigger phagocytosis and on B cells is negative regulator of activation.

CD Antigen. MW. Synonyms and Properties.	Expression	Function
CD33. 67 kDa. gp67, p67, Siglec-3 (Sialic acid-binding Ig-like lectin 3).	Myeloid progenitors, monocytes, macrophages, dendritic cells, and granulocytes.	Binds sialic acid-containing oligosaccharides. Adhesion molecule; may inhibit proliferation of normal and leukemic myeloid cells. Serves as a marker for distinguishing between myeloid and lymphoid leukemias.
CD34. 105-120 kDa. gp105–120, mucosialin.	Hematopoietic stem and progenitor cells, small-vessel endothelial cells.	Bound by L-selectin. Cell-cell adhesion molecule with role in mediating hematopoietic stem cell adhesion to bone marrow stromal cells or extracellular matrix.
CD35. Many forms: 160–255 kDa. Complement receptor type one (CR1), C3b/C4bR.	B cells, some T-cell subsets, neutrophils, monocytes, eosinophils, follicular dendritic cells, and erythrocytes.	Receptor for C3b/C4b-coated particles, mediating their adherence and phagocytosis; facilitator of C3b and C4b cleavage, thus limiting complement activation.
CD36. 85 kDa Platelet glycoprotein IV (GPIV), GPIIIb, OKM5-antigen, PASIV, thrombospondin receptor.	Platelets, mature monocytes/macrophages, erythroid precursors, endothelial cells.	Multifunctional glycoprotein that acts as an adhesion molecule in platelet adhesion and aggregation and in platelet–monocyte or platelet–tumor cell interaction; role in recognition of apoptotic neutrophils and the phagocytic clearance of apoptotic cells. Scavenger receptor for oxidized LDL, plays role in cholesterol transport.
CD37. 40–52 kDa. gp 52–50. Tetraspanin-26.	B cells, low levels on T cells, monocytes, dendritic cells, granulocytes.	B-cell–expressed CD37 associates noncovalently with MHC class II, CD53, CD81, and CD82. Involved in the signal transduction pathway(s) that regulate cell development, activation, growth, and motility; may also be involved in T cell-B cell interactions.
CD38. 42 kDa. T10, ADP-ribosyl cyclase, cyclic ADP-ribose hydrolase.	Variable levels on the majority of hematopoietic precursor cells, lymphocytes, and some non-hematopoietic cells. High level of expression on B cells, activated T cells, and plasma cells.	Ectoenzyme that participates in nucleotide metabolism; synthesizes cyclic ADP-ribose, a second messenger for glucose-induced insulin secretion. Functions in signal transduction, positive and negative regulator of cell activation and proliferation, adhesion.
CD39. 78 kDa. Ectonucleoside triphosphate diphosphohydrolase 1 (ENTPD1), NTPDase-1.	Activated B cells, T-cell subsets, NK cells, macrophages, dendritic cells, Langerhans cells, microglia, some epithelial and endothelial cells, placenta.	ATP and ADP hydrolase. May modulate platelet activation, immune responses, and neurotransmission.
CD40. 45–48 kDa (monomer). Bp50; TNFRSF5.	Mature B cells but not plasma cells, activated monocytes, macrophages, dendritic cells, epithelial and endothelial cells, fibroblasts keratinocytes, CD34⁺ hematopoietic cell progenitors.	Binds to CD40 ligand (CD154). Provides essential costimulatory signals for B-cell activation, proliferation, differentiation, and isotype switching; apoptosis rescue signal for germinal center B cells. Stimulates cytokine production by macrophages and dendritic cells and up-regulates adhesion molecules on dendritic cells. Plays a critical role in the regulation of cell-mediated immunity as well as antibody-mediated immunity.

(continued)

CD Antigen. MW. Synonyms and Properties.	Expression	Function
CD41. αβ dimer: αIIb, 125 kDa; β3, 122 kDa. Integrin αIIb; Glycoprotein IIb (GP IIb), Human Platelet Antigen-3 (HPA-3) forms platelet fibrinogen receptor by association with GP IIIa.	Platelets, megakaryocytes.	Receptor for platelet fibrinogen; also binds fibronectin, plasminogen, prothrombin, thrombospondin, and vitronectin; mediates platelet aggregation; plays a central role in platelet activation, cohesion, coagulation, aggregation, and attachment.
CD42a–d. -a, 17-22 kDa; -b, 145 kDa; -c, 24 kDa; -d, 82 kDa. -a, GPIX; -b, GPIb-α; -c, GPIb-β; -d, GPV.	Platelets, megakaryocytes.	CD42a–d complex serves as receptor for von Willebrand factor and thrombin. Mediates adhesion of platelets to vascular endothelium especially after injury; amplification of platelet response to thrombin.
CD43. 115-135 kDa. leukocyte sialoglyco-protein, leukosialin, sialophorin, gpL115, Ly-48 (mouse).	All leukocytes except most resting B cells; major glycoprotein of thymocytes and T cells.	Possible role in regulating adhesion; may also regulate T-cell activation.
CD44. 85–250 kDa. Pgp-1 (phagocytic glyco-protein-1), ECMR III, HUTCH-1, Hermes, gp85; multiple isoforms.	Surface of most hematopoietic and non-hematopoietic cell types except platelets; CD44H is major isoform expressed on lymphocytes; CD44R is expressed on epithelial cells, monocytes, and activated leukocytes.	Receptor for hyaluronic acid (HA). Mediates cell-cell and cell-matrix interactions through its affinity for HA, and possibly also through its affinity for other ligands such as osteopontin, collagens, and matrix metalloproteinases (MMPs). Also involved in lymphocyte activation, recirculation and homing to lymphoid tissues and sites of inflammation, and in hematopoiesis. Adhesion with HA plays an important role in cell migration, tumor growth, and progression.
CD45. 180–240 kDa leukocyte common antigen (LCA), T200 on T cells, B220 on B cells, protein tyrosine phosphatase receptor type C (PTPRC); many different isoforms with different molecular weights are generated by alternative splicing of 3 exons that can be inserted immediately after an N-terminal sequence of 8 aa found on all isoforms. *See following isoforms.*	All hematopoietic cells except erythrocytes and platelets; especially high on lymphocytes (10% of their surface area comprising CD45); different isoforms characteristic of differentiated subsets of various hematopoietic cells.	Regulates activation of variety of cellular processes, including cell growth, differentiation, mitotic cycle, and oncogenic transformation. Essential role in T- and B-cell antigen-receptor-mediated activation; possible role in receptor-mediated activation in other leukocytes.
CD45RA. 205–220 kDa. Isoform of CD45 containing the A exon.	B cells, naïve T cells, monocytes, mature thymocytes.	Contributes to receptor-mediated signaling and cell activation.
CD45RB. 190–220 kDa. Isoform of CD45 containing the B exon.	B cells, T-cell subsets, NK cells, monocytes, macrophages, dendritic cells, granulocytes.	Contributes to receptor-mediated signaling and cell activation.
CD45RC. Isoform of CD45 containing the C exon.	B cells, CD8$^+$ T cells, subset of CD4$^+$ T cells, NK cells, mature thymocytes, monocytes, dendritic cells.	Contributes to receptor-mediated signaling and cell activation.
CD45RO. Isoform of CD45 containing none of the A, B, C exons.	Activated and memory T cells, B-cell subsets, immature thymocytes, activated monocytes, macrophages, dendritic cell subsets, granulocytes.	Contributes to receptor-mediated signaling and cell activation.

CD Antigen. MW. Synonyms and Properties.	Expression	Function
CD46. 64–68 kDa. Membrane cofactor protein (MCP), trophoblast leukocyte common antigen (TRA2.10).	Expressed by all cells except erythrocytes.	Inhibitory complement receptor: Acts as a cofactor for complement factor I, a serine protease that protects autologous cells against complement-mediated injury by cleaving C3b and C4b deposited on host tissue. May be involved in the fusion of the spermatozoa with the oocyte during fertilization. Also acts as a costimulatory factor for T-cells that induces the differentiation of CD4$^+$ into T-regulatory 1 cells. T-regulatory 1 cells suppress immune responses by secreting interleukin-10, and therefore are thought to prevent autoimmunity. A number of viral (e.g., measles) and bacterial (e.g., *S. pyogenes*) pathogens seem to exploit this property and directly induce an immunosuppressive phenotype in T-cells by binding to CD46.
CD47. 50–55 kDa Rh-associated protein, gp42, integrin-associated protein (IAP), neurophilin, MER6.	Most hematopoietic and non-hematopoietic cells; part of the Rh complex on erythrocytes, not expressed on Rh null erythrocytes.	Has roles in cell adhesion by acting as an adhesion receptor for THBS1 on platelets and in the modulation of integrins. Plays an important role in memory formation and synaptic plasticity in the hippocampus. Receptor for SIRPA, binding to which prevents phagocytosis by macrophages, maturation of immature dendritic cells, and inhibits cytokine production by mature dendritic cells. Interaction with SIRPG mediates cell-cell adhesion, enhances superantigen-dependent T-cell-mediated proliferation, and costimulates T-cell activation. May play a role in membrane transport and/or integrin-dependent signal transduction. May be involved in membrane permeability changes induced following virus infection.
CD48. 45 kDa. BLAST-1, BCM1, Sgp-60, SLAMF2.	Widely expressed on hematopoietic cells with the exception of some platelets and erythrocytes.	Adhesion molecule recognized by CD2, may participate in T-cell costimulation and adhesion; recently identified as a ligand of the leukocyte receptor CD244 (2B4), which on NK cells functions as an activating receptor.
CD49a. 200 kDa. Integrin α1 chain, very late antigen α1 chain (VLA-1 α chain).	Activated T cells, monocytes.	Integrin that associates with CD29, binds collagen, laminin-1; functions in adhesion.
CD49b. 160 kDa. Integrin α2 chain, VLA-2 α chain, GPIa.	B cells, activated T cells, NK-cell subsets, monocytes, platelets, megakaryocytes; epithelial and endothelial cells.	Integrin that associates with CD29, binds collagen, laminin, fibronectin, and E-cadherin; it is responsible for adhesion of platelets and other cells to collagens, modulation of collagen and collagenase gene expression, force generation and organization of newly synthesized extracellular matrix.

(continued)

CD Antigen. MW. Synonyms and Properties.	Expression	Function
CD49c. 150 kDa. Integrin α3 chain, VLA-3 α chain FRP-2.	Low levels on monocytes, B and T lymphocytes, various isoforms differentially expressed in brain, heart, muscle, and endothelial cells.	Integrin α3β1 is a receptor for fibronectin, laminin, collagen, epiligrin, thrombospondin and CSPG4. α3β1 may mediate LGALS3 stimulation by CSPG4 of endothelial cells migration.
CD49d. 150 kDa. Integrin α4 chain, VLA-4 α chain.	Many cell types, including T cells, immature thymocytes, B cells, NK cells, monocytes, eosinophils, basophils, mast cells, dendritic cells, erythroblastic precursor cells; not on normal red blood cells, platelets, or neutrophils.	Cell adhesion molecule that binds (depending on associated β chain) to cell surface ligands VCAM-1 and MAdCAM-1 and extracellular matrix proteins fibronectin and thrombospondin; contributes to leukocyte migration and homing; costimulatory molecule for T-cell activation.
CD49e. 135 kDa. Integrin α5 chain, VLA-5 α chain, Fibronectin receptor (FNR) α chain.	T cells, immature thymocytes, early and activated B cells, platelets, some epithelial and endothelial cells.	Integrin that associates with CD29 and mediates binding to fibronectin, providing a costimulatory signal to T cells; believed to be important for the maintenance of endothelial monolayer integrity along with CD49b; involved in adhesion, regulation of cell survival, and apoptosis.
CD49f. 125 kDa. Integrin α6 chain, VLA-6 α chain, platelet gpl.	Memory T cells, immature thymocytes, monocytes, platelets, megakaryocytes, some epithelial and endothelial cells, trophoblasts.	Integrin that associates with CD29 and CD104, binds laminin, participates in cell adhesion and migration, embryogenesis, and cell surface–mediated signaling.
CD50. 120–140 kDa. Intercellular adhesion molecule-3 (ICAM-3).	Most leukocytes, including immature thymocytes and Langerhans cells; endothelial cells.	Ligand for LFA-1 (CD11a/CD18) and is involved in integrin-dependent adhesion and cell migration. Recognized byCD209 (DC-SIGN) on dendritic cells. Acts as a costimulatory molecule for T-cell activation. Regulates leukocyte morphology.
CD51. 125 kDa. Integrin α-V chain, VNR-α chain, forms α-V integrin (vitronectin receptor) by association with CD61; known to form heterodimers with β1 CD29, β3 CD61, β5, β6, and β8 integrin subunits in various tissues.	Platelets, megakaryocytes, endothelial cells, certain activated leukocytes, NK cells, macrophages, and neutrophils; osteoclasts and smooth muscle cells.	Subunit of α-V integrin, which binds vitronectin, von Willebrand factor, fibronectin, fibrinogen, laminin, MMP-2, osteopontin, osteomodulin, prothrombin, thrombospondin; shown to mediate the binding of platelets to immobilized vitronectin without prior activation, also to interact with CD47; may bind apoptotic cells. Initiates bone resorption by mediating the adhesion of osteoclasts to osteopontin and may play a role in angiogenesis.
CD52. 25–29 kDa. CAMPATH-1, HE5 (human epididymis-specific protein 5).	Highly expressed on cortical thymocytes, lymphocytes, monocytes, macrophages, mast cells, epithelial cells lining male reproductive tract.	Expresses carbohydrate moieties that may be recognized; may play costimulatory role.
CD53. 32–42 kDa. OX44; tetraspanin 25.	B and T cells, mature thymocytes, NK cells, monocytes, macrophages, neutrophils, dendritic cells, osteoblasts, and osteoclasts.	Mediates signal transduction events involved in the regulation of cell development, activation, growth, and mortality; contributes to the transduction of CD2-generated signal in T and NK cells and may play a role in growth regulation. Cross-linking promotes activation of human B cells and rat macrophages.

CD Antigen. MW. Synonyms and Properties.	Expression	Function
CD54. 90, 75–115 kDa. Intercellular adhesion molecule-1 (ICAM-1).	Activated T and B cells, monocytes, activated endothelial cells.	Ligand for CD11a/CD18 or CD11b/CD18, shown to bind to fibrinogen and hyaluronan; receptor for rhinoviruses and for RBCs infected with malarial parasites. May play a role in development and promotion of adhesion; contributes to antigen-specific T-cell activation by antigen-presenting cells; contributes to the extravasation of leukocytes from blood vessels, particularly in areas of inflammation. Also, soluble form may inhibit the activation of CTL or NK cells by malignant cells.
CD55. 80 kDa (lymphocytes), 55 kDa (erythrocytes). Decay-accelerating factor (DAF).	Most cell types; also a soluble form in plasma and body fluids.	Member of the regulator of complement activation (RCA) family of proteins. Protective barrier against inappropriate complement activation and deposition on plasma membranes: interaction with cell-associated C4b and C3b polypeptides interferes with their ability to catalyze the conversion of C2 and factor B to enzymatically active C2a and Bb and thereby prevents the formation of C3 convertases that amplify the complement cascade. Also binds CD97. May contribute to lymphocyte activation. Also serves as a receptor for echovirus and coxsackie B virus.
CD56. 175–220 kDa. Leu-19, neural cell adhesion molecule (NCAM); multiple isoforms, predominant isoform on NK and T cells is 140 kDa membrane glycoprotein.	Human NK cells, subsets of CD4$^+$ and CD8$^+$ T cells; also neural tissue.	No clear immune function, although may be involved in tumor growth and spreading; cell adhesion molecule involved in neuron–neuron and neuron–muscle adhesion, neurite fasciculation, outgrowth of neurites, etc.
CD57. 110 kDa. HNK1, Leu-7; a carbohydrate antigen (oligosaccharide), component of many glycoproteins.	NK cells; subsets of T cells, B cells.	Recognized by L-selectin (CD62L) and P-selectin (CD62P) and functions in cell-cell adhesion.
CD58. 55–70 kDa. Lymphocyte function-associated antigen 3 (LFA-3).	Many leukocytes and other cell types; particularly high on memory T cells and dendritic cells.	Adhesion between CTL and target cells, antigen-presenting cells and T cells, and thymocytes and thymic epithelial cells; expressed on antigen-presenting cells; and enhances T-cell antigen recognition through binding to CD2, its only known ligand.
CD59. 18–25 kDa. IF-5Ag, HRF20 (homologous restriction factor 20), MACIF, MIRL, Protectin.	Most hematopoietic and nonhematopoietic cell types.	Potent species-specific inhibitor of the complement membrane attack complex (MAC)–mediated lysis. Acts by binding to the C8 and/or C9 complements of the assembling MAC, thereby preventing incorporation of the multiple copies of C9 required for complete formation of the osmolytic pore. Interacts with CD2 and Src kinases and may be involved in T-cell signal transduction and activation.

(continued)

CD Antigen. MW. Synonyms and Properties.	Expression	Function
CD60a. Carbohydrate found on the ganglioside GD3.	T-cell subsets, cortical thymocytes, granulocytes, platelets, some other cell types.	Involved in the regulation of T-cell apoptosis and induces mitochondrial permeability transition during apoptosis; costimulatory activity for T cells.
CD60b. 9-O-acetyl disialyl ganglioside GD3.	T-cell subsets, activated B cells.	May have costimulatory activity.
CD60c. 7-O-acetyl disialyl ganglioside GD3.	T-cell subsets.	May contribute to T-cell activation.
CD61. 110 kDa. Integrin β3 chain; platelet glycoprotein IIIa (GPIIIa).	Various isoforms associate with various integrin α chains on numerous cell types: platelets, megakaryocytes, macrophages, monocytes, mast cells, osteoclasts, endothelial cells, fibroblasts.	Integrin subunit, associates with CD41 (Integrin αIIb) or CD51 (integrin α-V chain); recognizes various soluble, membrane, and extracellular matrix proteins; participates in platelet aggregation and cell adhesion.
CD62E. 97–115 kDa. E-selectin, endothelial leukocyte adhesion molecule-1 (ELAM-1), LECAM-2.	Acutely activated vascular endothelium, chronic inflammatory lesions of skin and synovium.	C-type lectin endothelial adhesion molecule that mediates leukocyte (e.g., neutrophil) rolling on activated endothelium at inflammatory sites through interaction with PSGL1 through the sialyl Lewis X (CD15s) carbohydrate, CD43, and other leukocyte antigens; may also participate in angiogenesis and tumor-cell adhesion during metastasis via the blood.
CD62L. 74 kDa (lymphocytes), 95 kDa (neutrophils). L-selectin, leukocyte adhesion molecule-1 (LAM-1), LECAM-1, Leu-8, MEL-14, TQ-1.	Most peripheral blood B cells, T-cell subsets, NK-cell subsets, cortical thymocytes, monocytes, granulocytes.	C-type lectin adhesion molecule that mediates adherence (initial tethering and rolling, through interaction with PSGL1 through the sialyl Lewis X [CD15s] carbohydrate) of lymphocytes for homing to high endothelial venules of peripheral lymphoid tissue, and also leukocyte adhesion and rolling on activated endothelium at inflammatory sites.
CD62P. 140 kDa. P-selectin, granule membrane protein-140 (GMP-140), platelet activation–dependent granule-external membrane protein (PADGEM); C-type lectin, single-chain type 1 glycoprotein.	Activated platelets and endothelial cells.	C-type lectin adhesion molecule that binds to PSGL1 through the sialyl Lewis X (CD15s) carbohydrate on neutrophils and monocytes and mediates tethering and rolling of leukocytes on the surface of activated endothelial cells, the first step in leukocyte extravasation and migration toward sites of inflammation; mediates adherence to platelets; may contribute to inflammation-associated tissue destruction, atherogenesis, and thrombosis.
CD63. 40–60 kDa. Granulophysin, lysosomal membrane–associated glycoprotein 3 (LAMP-3), melanoma-associated antigen (ME491), neuroglandular antigen (NGA).	Some lymphocytes, platelets, degranulated neutrophils, monocytes, macrophages, fibroblasts, osteoclasts.	Mediates signal transduction events that play a role in the regulation of cell development, activation, growth, and motility; may play role in controlling protein and vesicle transport intracellularly and with the plasma membrane.
CD64. 72 kDa. FcγRI, FcγRIa, FCRI.	Monocytes, macrophages, blood and germinal center dendritic cells, granulocytes activated by IFN-γ or G-CSF, early myeloid-lineage cells.	High-affinity receptor for IgG. Activates phagocytosis, receptor-mediated endocytosis of IgG-antigen complexes, antigen capture for presentation to T cells; antibody-dependent cell-mediated cytotoxicity (ADCC); release of cytokines and reactive oxygen intermediates.

(continued)

CD Antigen. MW. Synonyms and Properties.	Expression	Function
CD65. Ceramide-dodecasaccharide, fuco-ganglioside Type II, VIM2.	Restricted to myeloid cells; with expression on most granulocytes and a proportion of monocytic cells; marker on myeloid leukemia cells.	Recognized by E-selectin (CD62E); may be involved in cell adhesion.
CD65s. Sialylated CD65, VIM2.	Granulocytes, monocytes; marker on myeloid leukemia cells.	Recognized by E-selectin (CD62E); may be involved in cell adhesion; possible involvement with phagocytosis.
CD66a. 140–180 kDa. NCA-160, carcino-embryonic antigen-related cell adhesion molecule 1 (CEAM1), biliary glycoprotein (BGP).	Granulocytes, epithelial cells.	Mediates cell-cell adhesion by homotypic and/ or heterotypic interactions with other CD66 molecules. Receptor for *Neisseria gonorrheae* and *N. meningitidis;* may trigger neutrophil activation; may also contribute to the interactions of activated granulocytes with each other or with endothelium or epithelium. May have tumor suppressor activity.
CD66b. 95–100 kDa. CEAM8, CD67, CGM6, NCA-95.	Granulocytes.	Similar to CD66a, receptor for *Neisseria gonorrheae* and *N. meningitides.* Adhesion molecule, trigger of neutrophil activation; enhances the respiratory burst activity of neutrophils, may also regulate the adhesion activity of CD11/CD18 in neutrophils.
CD66c. 90 kDa. CEACAM6, NCA-50/90.	Granulocytes, epithelial cells.	Similar to CD66a and CD66b; receptor for *Neisseria gonorrheae* and *N. meningitides.* Adhesion molecule, trigger of neutrophil activation, may regulate the adhesion activity of CD11/CD18 in neutrophils.
CD66d. 35 kDa. CEACAM3, CGM1.	Neutrophils.	Similar to CD66a–c; receptor for *Neisseria gonorrheae* and *N. meningitides;* regulates adhesion activity of CD11/CD18 in neutrophils.
CD66e. 180–200 kDa. CEACAM5.	Epithelial cells.	Homophilic and heterophilic adhesion. Also a receptor for *N. gonorrheae.* May play a role in the process of metastasis of cancer cells.
CD66f. 54–72 kDa. Pregnancy-specific β1 glycoprotein 1 (PSG-1).	Epithelial cells, fetal liver; produced in placenta and released.	Unclear; possible involvement in immune regulation and protection of fetus from maternal immune system; may be necessary for successful pregnancy as low levels in maternal blood predict spontaneous abortion.
CD67. Deleted (now called CD66b).	See CD66b	See CD66b
CD68. 110 kDa. gp110, macrosialin.	Highly expressed in monocytes and macrophages; also expressed on lymphocyte subsets, dendritic cells, granulocytes, myeloid progenitor cells, subset of CD34$^+$ hematopoietic progenitor cells.	Could play a role in phagocytic activities of tissue macrophages, both in intracellular lysosomal metabolism and extracellular cell-cell and cell-pathogen interactions. Binds to tissue- and organ-specific lectins or selectins, allowing homing of macrophage subsets to particular sites. Rapid recirculation of CD68 from endosomes and lysosomes to the plasma membrane may allow macrophages to crawl over selectin-bearing substrates or other cells.

(continued)

CD Antigen. MW. Synonyms and Properties.	Expression	Function
CD69. 60 kDa. Early activation antigen 1 (EA 1), Activation-inducer molecule (AIM), MLR3, very early activation (VEA); Leu-23.	Activated T, B, and NK lymphocytes; thymocyte subsets, monocytes, macrophages, granulocytes, Langerhans cells, platelets.	Involvement in early events of lymphocyte, monocyte, and platelet signaling and activation; may provide costimulatory signals in lymphocytes.
CD70. 75, 95, 170 kDa. CD27 ligand, Ki-24 antigen.	Activated T, B, and dendritic cells.	Member of TNF family; serves as ligand for CD27; role in costimulation of B and T cells and may augment the generation of cytotoxic T cells and cytokine production.
CD71. 95,190 kDa. T9, transferrin receptor; disulfide-linked homodimeric type 2 glycoprotein.	Proliferating cells, stem cell/precursor, and endothelial cells.	Iron uptake: binds ferrotransferrin at neutral pH and internalizes complex to acidic endosomal compartment, where iron is released.
CD72. 39–43 kDa. Ly-19.2, Ly-32.2, Lyb-2; C-type lectin, disulfide-linked homodimeric type 2 glycoprotein.	B cells (except plasma cells), some dendritic cells, and macrophages/monocytes.	Regulation of B-cell proliferation and differentiation
CD73. 69, 70, 72 kDa. Ecto-5′-nucleotidase; single-chain GPI-anchored glycoprotein.	Subpopulations of T cells (expression is confined to the CD28$^+$ subset) and B cells (about 75% of adult peripheral blood B cells), with expression increasing during development; follicular dendritic cells, epithelial cells, endothelial cells.	Possibly regulates the availability of adenosine for interaction with the cell surface adenosine receptor by converting AMP to adenosine; can mediate costimulatory signals for T-cell activation; may play a role in mediating the interaction between B cells and follicular dendritic cells.
CD74. 33, 35, 41 kDa. Class II–specific chaperone, Ii, invariant chain; homotrimers, single-chain type 2 glycoprotein.	Mostly found intracellularly in MHC class II–expressing cells, specifically, B cells, activated T cells, dendritic cells, macrophages, activated endothelial and epithelial cells.	Intracellular sorting of MHC class II molecules.
CD75. Lactosamines. Carbohydrate antigen.	B cells, subpopulation of peripheral-blood T cells and erythrocytes.	Cell adhesion; ligand for CD22.
CD75s. Formally known as CDw76. α2,6 Sialylated lactosamine Carbohydrate antigen.	Majority of B cells, subpopulation of T cells, subsets of endothelial and epithelial cells, possibly weakly expressed on erythrocytes.	Considered to be a binding partner for the B-cell-specific activation antigen CD22; cell differentiation and surface recognition; adhesion.
CD77. Pk blood-group antigen, Burkitt's lymphoma antigen (BLA), ceramide trihexoside (CTH), globotriaosylceramide (Gb3); glycosphingolipid antigen.	Germinal center B cells.	Critical cell surface molecule able to mediate an apoptotic signal; association with type 1 interferon receptor or with HIV coreceptor CXCR4 (CD184) may be essential for function; a receptor for lectins on the pili of a certain strain of *E. coli*. May be involved in the selection process within the germinal centers.
CD79a. 33–45 kDa. Ig-α, MB1; type 1 glycopeptide, disulfide-linked heterodimer with CD79b.	B cells.	Component of B-cell antigen receptor analogous to CD3; required for cell surface expression and signal transduction.
CD79b. 37 kDa. B29, Ig-β; type 1 glycopeptide, disulfide-linked heterodimer with CD79α.	B cells.	Also part of the B-cell antigen receptor, with CD79a.
CD80. 60 kDa. B7, B7.1, BB1, Ly-53; single-chain type 1 glycoprotein.	Activated B and T cells, macrophages, low levels on resting peripheral blood monocytes and dendritic cells.	Binds CD28 and CD152; costimulation of T-cell activation with CD86 when bound to CD28; inhibits T-cell activation when bound with CD152/CTLA-4.
CD81. 26 kDa. Target for antiproliferative antigen-1 (TAPA-1). Single-chain type 3,4-span protein, member of the 4-transmembrane-spanning protein superfamily (TM4SF).	Broadly expressed on hematopoietic cells; expressed by endothelial and epithelial cells; absent from erythrocytes, platelets, and neutrophils.	Member of CD19/CD21/Leu-13 signal transduction complex; mediates signal transduction events involved in the regulation of cell development, growth, and motility; participates in early T-cell development. Binds the E2 glycoprotein of hepatitis C virus.

CD Antigen. MW. Synonyms and Properties.	Expression	Function
CD82. 45–90 kDa. 4F9, C33, IA4, KAI1, R2; single-chain type 3-, 4-span glycoprotein, member of the 4-transmembrane-spanning protein superfamily (TM4SF).	Activated/differentiated hematopoietic cells, B and T cells, NK cells, monocytes, granulocytes, platelets, epithelial cells.	Signal transduction; may induce T-cell spreading and pseudopod formation, modulate T-cell proliferation, and provide costimulatory signals for cytokine production; possible role in activation of monocytes.
CD83. 43 kDa. HB15; single-chain type 1 glycoprotein.	Dendritic cells, B cells, Langerhans cells.	May play a role in antigen presentation and/or lymphocyte activation and regulation of the immune response.
CD84. 72-86 kDa. GR6.	Virtually all thymocytes, monocytes, platelets, circulating B cells, T-cell subsets.	Acts as an adhesion receptor by homophilic interactions to stimulate interferon gamma production by lymphocytes and activate platelet via a SH2D1A/SAP-dependent pathway.
CD85a. 110 kDa. LIR3.	Monocytes, macrophages, dendritic cells, granulocytes, a subpopulation of T lymphocytes.	Suppression of NK-cell-mediated cytotoxicity.
CD86. 80 kDa. B7.2, B70; single-chain type 1 glycoprotein.	Dendritic cells, memory B cells, germinal center B cells, monocytes, activated T cells, and endothelial cells.	Major T-cell costimulatory molecule, interacting with CD28 (stimulatory) and CD152/CTLA4 (inhibitory).
CD87. 32–56 kDa (monocytes). Urokinase plasminogen activator receptor (uPAR); single-chain GPI-anchored glycoprotein.	T cells, NK cells, monocytes, neutrophils; nonhematopoietic cells such as vascular endothelial cells, fibroblasts, smooth muscle cells, keratinocytes, placental trophoblasts, hepatocytes.	Receptor for uPA, which can convert plasminogen to plasmin; possible role in $\beta 2$ integrin-dependent adherence and chemotaxis; may play a role in the process of neoplastic and inflammatory cell invasion.
CD88. 43 kDa. C5a receptor, C5aR; type 3, 7-span glycoprotein, member of the 7-transmembrane-spanning protein superfamily (TM7SF).	Granulocytes, monocytes, dendritic cells, astrocytes, microglia, hepatocytes, alveolar macrophages, vascular endothelial cells.	C5a-mediated inflammation, activation of granulocytes, possible function in mucosal immunity.
CD89. 45–100 kDa. Fcα-receptor (Fcα-R), IgA Fc receptor, IgA receptor; single-chain type 1 glycoprotein.	Myeloid-lineage cells from promyelocytes to neutrophils and from promonocyte to monocytes; activated eosinophils, alveolar and splenic macrophages; subsets of T and B cells.	Induction of phagocytosis, degranulation, respiratory burst, killing of microorganisms.
CD90. 25–35 kDa. Thy-1; single-chain GPI-anchored glycoprotein.	Hematopoietic stem cells, neurons, connective tissue, thymocytes, peripheral T cells, human lymph node HEV endothelium.	Possible involvement in lymphocyte costimulation; possible inhibition of proliferation and differentiation of hematopoietic stem cells.
CD91. 515, 85 kDa. α-2-macroglobulin receptor (ALPHA2M-R), low-density lipoprotein receptor–related protein (LRP); single-chain type 1 glycoprotein.	Phagocytes, many nonhematopoietic cells.	Endocytosis-mediating receptor expressed in coated pits that appears to play a role in the regulation of proteolytic activity and lipoprotein metabolism.
CD92. 70 kDa. Formerly known as CDw92; CTL1, GR9.	Monocytes, granulocytes, peripheral blood lymphocytes (PBL), mast cells, endothelial cells, and epithelial cells.	Choline transporter.
CD93. 110 kDa. Formerly known as CDw93; GR11.	Monocytes, granulocytes, endothelial cells.	Cell adhesion and phagocytosis.
CD94. 70 kDa. Kp43; forms complex with NKG2 receptors, C-type lectin.	NK cells, subsets of CD8$^+$ αβ and γδ T cells.	Depending on NKG2 molecule associated, may activate or inhibit NK-cell cytotoxicity and cytokine release.
CD95. 45 kDa. APO-1, Fas antigen (Fas); TNF receptor superfamily, single-chain type 1 glycoprotein.	Activated T and B cells, monocytes, fibroblasts, neutrophils, NK cells.	Binds Fas ligand (FasL, CD178) and initiates apoptosis-inducing signals.

(continued)

CD Antigen. MW. Synonyms and Properties.	Expression	Function
CD96. 160 kDa. T-cell activation increased late expression (TACTILE); single-chain type 1 glycoprotein.	Activated T cells, NK cells.	Adhesion of activated T and NK cells during the late phase of immune response; also involved in antigen presentation and/or lymphocyte activation.
CD97. 75–85 kDa (PBMC, CD97a), 28 kDa (PBMC, CD97b). BL-KDD/F12; EGF-TM7 subfamily member, type 3-, 7-span glycoprotein, three isoforms.	Activated B and T cells, monocytes, granulocytes, and dendritic cells.	Binds to CD55, neutrophil migration.
CD98. 80 and 45 kDa. 4F2, FRP-1, RL-388 in mouse; disulfide-linked heterodimeric type 2 glycoprotein.	Not hematopoietic specific; activated and transformed cells; lower levels on quiescent cells; high levels on monocytes.	Role in regulation of cellular activation and aggregation.
CD99. 32 kDa. CD99R (epitope restricted to subset of CD99 molecules), E2, *MIC2* gene product; single-chain type 1 glycoprotein.	Leukocytes; highest on thymocytes.	Augments T-cell adhesion, induces apoptosis of double-positive thymocyte, participates in leukocyte migration, involved in T-cell activation and adhesion, binds to cyclophilin A.
CD100. 150 kDa (PHA blasts), 120 kDa (soluble). Disulfide-linked homodimeric type 1 glycoprotein.	Most hematopoietic cells except immature bone marrow cells, RBCs, and platelets; activated T cells, germinal center B cells.	Monocyte migration, T- and B-cell activation and T–B cell and T–dendritic cell interaction; shown to induce T-cell proliferation.
CD101. 120 kDa. P126, V7; disulfide-linked homodimeric type 1 glycoprotein.	Monocytes, granulocytes, dendritic cells, mucosal T cells, activated peripheral blood T cells; weak on resting T, B, and NK cells.	Possible costimulatory role in T-cell activation.
CD102. 55–65 kDa. Intercellular adhesion molecule-2 (ICAM-2); single-chain type 1 glycoprotein.	Vascular endothelial cells, monocytes, platelets, some populations of resting lymphocytes.	Like related proteins CD54 and CD50, binds CD11a/CD18 LFA-1; also reported to bind to CD11b/CD18 Mac-1; may play a role in lymphocyte recirculation. Shown to mediate adhesive interactions important for antigen-specific immune response, NK-cell-mediated clearance, lymphocyte recirculation and other cellular interactions important for immune response and surveillance; may also be involved in T-cell activation and adhesion.
CD103. 150 kDa, 25 kDa. HML-1, integrin αE chain; α-chain type 1 glycopeptide.	Intraepithelial lymphocytes (in tissues such as intestine, bronchi, inflammatory skin/breast/salivary glands), many lamina propria T cells, some lymphocytes in peripheral blood and peripheral lymphoid organs.	Binds to E-cadherin and integrin β7; role in the tissue-specific retention of lymphocytes at basolateral surface of intestinal epithelial cells; possible accessory molecule for activation of intraepithelial lymphocytes.
CD104. 220 kDa. β4 integrin chain, tumor-specific protein 180 antigen (TSP-180) in mouse; β-chain type 1 glycopeptide.	Immature thymocytes; neuronal, epithelial, and some endothelial cells; Schwann cells; trophoblasts.	Integrin that associates with CD49f, binds laminins and plectin, also interacts with keratin filaments intracellularly; involved in cell-cell, cell-matrix interactions, adhesion, and migration; important role in the adhesion of epithelia to basement membranes.
CD105. 90 kDa. Endoglin; TGFβ type III receptor; disulfide-linked homodimeric type 1 glycoprotein.	Endothelial cells of small and large vessels; activated monocytes and tissue macrophages; stromal cells of certain tissues, including bone marrow; pre–B cells in fetal marrow; erythroid precursors in fetal and adult bone marrow; syncytiotrophoblast throughout pregnancy and cytotrophoblasts transiently during first trimester.	Modulator of cellular responses to TGF-β1, may affect hematopoiesis and angiogenesis.

CD Antigen. MW. Synonyms and Properties.	Expression	Function
CD106. 100–110 kDa. INCAM-110, vascular-cell adhesion molecule-1 (VCAM-1); single-chain type 1 glycoprotein with multiple isoforms.	Endothelial cells, follicular and interfollicular dendritic cells, some macrophages, bone marrow stromal cells, nonvascular cell populations within joints, kidney, muscle, heart, placenta, and brain; can be induced on endothelia and other cell types in response to inflammatory cytokines.	Adhesion molecule that is ligand for VLA-4; involved in leukocyte adhesion, transmigration, and costimulation of T-cell proliferation; contributes to the extravasation of lymphocytes, monocytes, basophils, and eosinophils but not neutrophils from blood vessels.
CD107a. 100–120 kDa. Lysosome-associated membrane protein 1 (LAMP-1); single-chain type 1 glycoprotein.	Activated platelets, endothelial cells, tonsillar epithelium, granulocytes, T cells, macrophages, dendritic cells, lysosomal membrane, degranulated platelets, PHA-activated T cells, TNF-α-activated endothelium, FMLP-activated neutrophils.	Provides carbohydrate ligands to selectins, associated with enhanced metastatic potential of tumor cells.
CD107b. 100–120 kDa. Lysosome-associated membrane protein 2 (LAMP-2); single-chain type 1 glycoprotein, tissue-specific isoforms.	Granulocytes, lysosomal membrane, activated and degranulated platelets, TNF-α-activated endothelium, FMLP-activated neutrophils, tonsillar epithelium.	Protection, maintenance, and adhesion of lysosomes; associated with enhanced metastatic potential of tumor cells.
CD108. 80 kDa. Formerly known as CDw108. John-Milton-Hargen (JMH) human-blood-group antigen; GPI-anchored glycoprotein.	Erythrocytes, circulating lymphocytes, lymphoblasts.	May play a role in monocyte activation and in regulating immune cells.
CD109. 175 kDa. 8A3, E123 (7D1); GPI-anchored glycoprotein.	Activated T cells, activated platelets, human umbilical vein endothelial cells.	Negatively regulates TGFB1 pathway in keratinocytes, may affect hematopoiesis, cellular immunity, and hemostasis.
CD110. 85–92 kDa. Myeloproliferative leukemia virus oncogene (MPL), thrombopoietin receptor (TPO-R), C-MPI; cytokine receptor superfamily, single-chain type 1 glycoprotein.	Hematopoietic stem and progenitor cells, megakaryocyte progenitors, megakaryocytes, platelets.	Binds thrombopoietin; main regulator of megakaryocyte and platelet formation.
CD111. 75 kDa. Herpesvirus Ig-like receptor (HIgR), poliovirus receptor–related 1 (PRR1), poliovirus receptor–related 1(PVRL1), nectin 1, HveC; type 1 glycoprotein, member of the nectin family.	Expressed in multiple cell types intracellularly and in vesicle-like structures.	Ca^{2+}-independent immunoglobulin (Ig)–like cell-cell adhesion molecules; important involvement in the formation of many types of cell-cell junctions and cell-cell contacts, acts as receptors for human herpesvirus 1 (HHV-1), human herpesvirus 2 (HHV-2), and pseudorabies virus (PRV).
CD112. 72 kDa (long isoform), 64 kDa (short isoform). Herpesvirus entry protein (HVEB), poliovirus receptor–related 2 (PRR2), nectin 2; type 1 glycoprotein, member of the nectin family.	Multiple tissues and cell lineages.	May act as adhesion protein; involved in herpesvirus entry activity and may function as a coreceptor for HSV-1, HSV-2, and pseudorabies.
CD113. Poliovirus receptor–related 3 (PRR3), nectin 3; transmembrane protein.	Epithelial cells, testis, liver, and placenta.	Interacts with CD111 (nectin-1) and CD112 (nectin-2); may be involved in adhesion.
CD114. 130 kDa. CSF3R, HG-CSFR, granulocyte colony-stimulating factor receptor (G-CSFR).	All stages of granulocyte differentiation, monocytes, mature platelets, several non-hematopoietic cell types/tissues, including endothelial cells, placenta, trophoblastic cells.	Specific regulator of myeloid proliferation and differentiation.
CD115. 150 kDa. C-fms, colony-stimulating factor 1R (CSF-1R), macrophage colony-stimulating factor receptor (M-CSFR).	Monocytes, macrophages.	Acts as macrophage colony-stimulating factor (M-CSF) receptor to control the proliferation and differentiation of macrophages/monocytes.

(continued)

CD Antigen. MW. Synonyms and Properties.	Expression	Function
CD116. 80 kDa. GM-CSF receptor α chain; member of the cytokine receptor superfamily, α-chain type 1 glycoprotein.	Various myeloid cells, including macrophages, neutrophils, eosinophils; dendritic cells and their precursors; fibroblasts, endothelial cells.	Acts as primary binding subunit of the GM-CSF receptor to regulate growth and function of hematopoietic cells.
CD117. 145 kDa. c-KIT, stem-cell factor receptor (SCFR); single-chain type 1 glycoprotein.	Hematopoietic stem cells and progenitor cells, mast cells.	Stem-cell factor receptor, tyrosine kinase activity; early-acting hematopoietic growth factor receptor, necessary for the development of hematopoietic progenitors. Capable of inducing proliferation of mast cells and is a survival factor for primordial germ cells.
CD118. 190 kDa. LIF receptor, gp190; transmembrane protein belonging to the type I cytokine receptor superfamily, forms a heterodimer with gp130.	Adult and embryonic epithelial cells, monocytes, fibroblasts, embryonic stem cells, liver, placenta.	High-affinity receptor (in complex with gp130) for leukemia inhibitory factor (LIF); cell differentiation, signal transduction, and proliferation.
CD119. 90–100 kDa. IFN-γR, IFN-γRα; class 2 cytokine receptor, type 1 glycopeptide.	Monocytes, macrophages, T and B cells, NK cells, neutrophils, fibroblasts, epithelial cells, endothelium, a wide range of tumor cells.	Interferon-γ receptor; role in host defense and the initiation and effector phases of immune responses, including macrophage activation, B- and T-cell differentiation, activation of NK cells, up-regulation of the expression of MHC class I and II antigens.
CD120a. 50–60 kDa. TNFRI, p55; TNF receptor superfamily member, single-chain type 1 glycoprotein.	Constitutively on hematopoietic and non-hematopoietic cells; high on epithelial cells.	TNF and lymphotoxin-α receptor; mediates the signaling involved in proinflammatory cellular responses, programmed cell death, and antiviral activity.
CD120b. 75–85 kDa. TNFRII, p75; TNF receptor superfamily member, single-chain type 1 glycoprotein.	Constitutively on hematopoietic and non-hematopoietic cells; high on epithelial cells.	TNF and lymphotoxin-α receptor; mediates the signaling involved in proinflammatory cellular responses, programmed cell death, and antiviral activity.
CD121a. 80 kDa. IL-1R, IL-1R type 1, type 1 IL-1R; type 1 glycoprotein.	T cells, thymocytes, chondrocytes, synovial cells, hepatocytes, endothelial cells, keratinocytes; low levels on fibroblasts, lymphocytes, monocytes, macrophages, granulocytes, and dendritic, epithelial, and neural cells.	Type I interleukin-1 receptor; mediates thymocyte and T-cell activation, fibroblast proliferation, induction of acute phase proteins, and inflammatory reactions.
CD121b. 60–70 kDa. IL-1R type 2, type 2 IL-1R; single-chain type 1 glycoprotein.	B cells, macrophages, monocytes, neutrophils.	Type I interleukin-1 receptor, likely a decoy receptor.
CD122. 70–75 kDa. Interleukin-2 receptor β chain (IL-2Rβ); cytokine receptor superfamily, type 1 glycopeptide.	Activated T cells, B cells, NK cells, monocytes, macrophages, subset of resting T cells.	Critical component of IL-2- and IL-15-mediated signaling; role in T-cell-mediated immune response; promotes proliferation and activation of T cells, thymocytes, macrophages, B cells, and NK cells.
CD123. 70 kDa. IL-3 receptor α subunit (IL-3Rα).	Bone marrow stem cells, granulocytes, monocytes, megakaryocytes.	IL-3 receptor chain.
CD124. 140 kDa. 1L-4R and IL-13R α chain; cytokine receptor superfamily, a type 1 glycopeptide.	Mature B and T cells, hematopoietic precursors, fibroblasts, epithelial and endothelial cells, hematopoietic and nonhematopoietic cells.	Receptor subunit for IL-4 and IL-13, promotes T$_H$2 differentiation and regulates IgE production.
CD125. 60 kDa. IL-5R α chain; cytokine receptor superfamily, a type 1 glycopeptide; also exists as a soluble form.	Eosinophils, activated B cells, basophils, mast cells.	Low-affinity receptor for IL-5; α chain of IL-5 receptor. Soluble form antagonizes IL-5-induced eosinophil activation and proliferation.

CD Antigen. MW. Synonyms and Properties.	Expression	Function
CD126. 80 kDa. Interleukin-6 receptor (IL-6R); associates with CD130, cytokine receptor superfamily, a type 1 glycopeptide, also exists as soluble form.	T cells, monocytes, activated B cells, hepatocytes, some other nonhematopoietic cells.	Receptor for IL-6; soluble form is capable of binding to gp130 on cells and promoting IL-6-induced responses.
CD127. 65–90 kDa. IL-7 receptor (IL-7R), IL-7 receptor α (IL-7Rα) p90; cytokine receptor superfamily, a type 1 glycopeptide.	B-cell precursors, mature resting T cells, thymocytes.	IL-7 receptor α chain; may regulate immunoglobulin gene rearrangement.
CD128. Deleted; now CD181 and CD182.	*See* CD181, CD182.	*See* CD181, CD182.
CD129. 60–65 kDa. CDw129, IL9R α-chain; type 1 glycopeptide.	Activated T-cell lines, T and B cells, both erythroid and myeloid precursors.	IL-9R α chain; promotes the growth of activated T cells, generation of erythroid and myeloid precursors.
CD130. 130–140 kDa. IL-6Rβ, IL-11R, gp130; cytokine receptor superfamily, single-chain type 1 glycoprotein.	T cells, monocytes, endothelial cells; at high levels on activated and EBV-transformed B cells, plasma cells, and myelomas; lower levels on most leukocytes, epithelial cells, fibroblasts, hepatocytes, neural cells.	Binding to the CD126/IL-6R complex stabilizes it, resulting in the formation of a high-affinity receptor; required for signal transduction by interleukin 6, interleukin 11, leukemia inhibitory factor, ciliary neurotrophic factor, oncostatin M, cardiotrophin-1.
CD131. 120–140 kDa. GM-CSFR, IL-3R, IL-5R (β chain); common β subunit, cytokine receptor superfamily, β-chain type 1 glycopeptide.	Fibroblasts and endothelial cells; most myeloid cells, including early progenitors, and early B cells.	Receptor subunit required for signal transduction by IL-3, GM-CSF, and IL-5 receptors.
CD132. 64, 65–70 kDa. Common cytokine receptor γ chain, common γ chain, member of cytokine receptor superfamily, γ-chain type I glycopeptide.	T cells, B cells, NK cells, monocytes/macrophages, neutrophils.	Subunit of IL-2, IL-4, IL-7, IL-9, and IL-15 receptors.
CD133. 115–125 kDa. AC133, hematopoietic stem-cell antigen, prominin-like 1 (PROML1), prominin; pentaspan 5-transmembrane domain glycoprotein.	Hematopoietic stem cell, endothelial and epithelial cells.	Stem-cell marker, probably inhibits cell differentiation.
CD134. 47–51 kDa. OX40; TNF receptor superfamily, single-chain type 1 glycoprotein.	Activated and regulatory T cells.	Binding to OX40-ligand (CD252) results in the induction of B-cell proliferation, activation, Ig production; provides necessary costimulation for T-cell proliferation, activation, adhesion, differentiation; may inhibit apoptosis.
CD135. 155–160 kDa. FMS-like tyrosine kinase 3 (flt3), Flk-2 in mice, STK-1; tyrosine kinase receptor, type 3, single-chain type 1 glycoprotein.	Multipotential, myelomonocytic, and primitive B-cell progenitors.	Growth factor receptor for early hematopoietic progenitors, tyrosine kinase.
CD136. 150, 40 kDa. Macrophage-stimulating protein receptor (msp receptor), ron (p158-ron); tyrosine kinase receptor family, single-chain type 1 heterodimeric glycoprotein.	Macrophages; epithelial tissues, including skin, kidney, lung, liver, intestine, colon.	Induction of migration, morphological change, cytokine induction, phagocytosis, proliferation, and anti-apoptosis in different target cells; may play a role in inflammation, wound healing, the mechanisms of activation in invasive growth and movement of epithelial tumors.
CDw137. 39 kDa. 4-1BB, induced by lymphocyte activation (ILA), TNF receptor superfamily, single-chain type 1 glycoprotein.	T cells, B cells, monocytes, epithelial and hepatoma cells.	Costimulator of T-cell proliferation via binding to 4-1BBL.

(continued)

CD Antigen. MW. Synonyms and Properties.	Expression	Function
CD138. 92 kDa (immature B cells), 85 kDa (plasma cells). Heparin sulfate proteoglycan, syndecan-1; type 1 glycoprotein.	Pre–B cells, immature B cells and plasma cells, but not mature circulating B lymphocytes; basolateral surfaces of epithelial cells, embryonic mesenchymal cells, vascular smooth muscle cells, endothelium, neural cells, breast cancer cells.	Binds to many extracellular matrix proteins and mediates cell adhesion and growth.
CD139. 209–228 kDa. B-031.	B cells, monocytes, granulocytes, follicular dendritic cells, erythrocytes.	Not known.
CD140a. 180 kDa. PDGF receptor (PDGF-R), alpha platelet-derived growth factor receptor (PDGFRα); tyrosine kinase receptor, type 3 family.	Mesenchymal cells.	Receptor for platelet-derived growth factor (PDGF); involved in cell proliferation, differentiation, and survival and in signal transduction associated with PDGFR.
CD140b. 180 kDa. beta platelet-derived growth factor receptor (PDGFRβ); tyrosine kinase receptor, type 3 family.	Endothelial cells, subsets of stromal cells, on mesenchymal cells.	Receptor for platelet-derived growth factor (PDGF); involved in cell proliferation, differentiation, and survival and in signal transduction associated with PDGFR.
CD141. 105 kDa. Fetomodulin, thrombomodulin (TM); C-type lectin, single-chain type 1 glycoprotein.	Endothelial cells, megakaryocytes, platelets, monocytes, neutrophils.	Essential molecule for activation of protein C and initiation of the protein C anticoagulant pathway.
CD142. 45–47 kDa. Coagulation factor III, thromboplastin, tissue factor (TF); serine protease cofactor, single-chain type 1 glycoprotein.	High levels on epidermal keratinocytes, glomerular epithelial cells, and various other epithelia; inducible on monocytes and vascular endothelial cells by various inflammatory mediators.	Initiates coagulation protease cascade assembly and propagation, functions in normal hemostasis, and is a component of the cellular immune response; may play a role in tumor metastasis, breast cancer, hyperplasia, and angiogenesis.
CD143. 170, 90 kDa. EC 3.4.15.1, angiotensin-converting enzyme (ACE), kinase II, peptidyl dipeptidase A; type 1 glycoprotein.	Endothelial cells, activated macrophages, weakly on subsets of T cells, some dendritic-cell subsets.	An enzyme important in regulation of blood pressure; acts primarily as a peptidyl dipeptide hydrolase and is involved in the metabolism of two major vasoactive peptides, angiotensin II, and bradykinin.
CD144. 130 kDa. Cadherin-5 VE-cadherin; type 1 glycoprotein.	Endothelium, stem-cell subsets.	Control of endothelial cell-cell adhesion, permeability, and migration.
CDw145. 90 kDa, 110 kDa.	Highly on endothelial cells.	Not known.
CD146. 130 kDa. A32, MCAM, MUC18, mel-CAM, S-endo; type 1 glycoprotein.	Follicular dendritic cells, endothelium, melanoma, smooth muscle, intermediate trophoblast, a subpopulation of activated T cells.	Adhesion molecule.
CD147. 55–65 kDa. 5A11, basigin, CE9, HT7, M6, neurothelin, OX-47, extracellular-matrix metalloproteinase inducer (EMMPRIN), gp42 in mouse; type 1 glycoprotein.	All leukocytes, red blood cells, platelets, endothelial cells.	Potential cell-adhesion molecule and is involved in the regulation of T-cell function, spermatogenesis, embryo development, neural network formation, and tumor progression.
CD148. 200–260 kDa. HPTP-η, high-cell-density–enhanced PTP 1 (DEP-1), p260; single-chain type 1 glycoprotein belonging to protein tyrosine phosphatase (PTP) family.	Granulocytes, monocytes, weakly on resting T cells, and up-regulated following activation; high levels on memory T cells, dendritic cells, platelets, fibroblasts, nerve cells, Kupffer cells.	Regulates signaling pathways of a variety of cellular processes, including cell growth, differentiation, mitotic cycle and oncogenic transformation, contact inhibition of cell growth.
CD150. 75–95, 70 kDa. Formerly known as CDw150, IPO-3, signaling lymphocyte activation molecule (SLAM); single-chain type 1 glycoprotein.	Thymocytes, subpopulation of T cells, B cells, dendritic cells, endothelial cells.	B-cell and dendritic cell costimulation, T-cell activation; contributes to enhancement of immunostimulatory functions of dendritic cells.

CD Antigen. MW. Synonyms and Properties.	Expression	Function
CD151. 32 kDa. PETA-3, SFA-1. Tetraspanin family, single-chain type 3-, 4-span glycoprotein.	Platelets, megakaryocytes, immature hematopoietic cells, endothelial cells.	Adhesion molecule, may regulate integrin trafficking and/or function; enhances cell motility, invasion and metastasis of cancer cells.
CD152. ~33 kDa. Cytotoxic T lymphocyte–associated protein-4 (CTLA-4); disulfide-linked homodimeric type 1 glycoprotein.	Activated T cells and some activated B cells.	Negative regulator of T-cell activation; binds CD80 and CD86.
CD153. 40 kDa. CD30 ligand, CD30L; TNF superfamily, single-chain type 2 glycoprotein.	Activated T cells, activated macrophages, neutrophils, B cells.	Ligand for CD30; costimulates T cells.
CD154. ~33 kDa. CD40 ligand (CD40L), T-BAM, TNF-related activation protein (TRAP), gp39; TNF superfamily, homotrimeric type 2 glycoprotein.	Activated CD4$^+$ T cells, small subset of CD8$^+$ T cells and $\gamma\delta$ T cells; also activated basophils, platelets, monocytes, mast cells.	Ligand for CD40, inducer of B-cell proliferation and activation, antibody class switching and germinal center formation; costimulatory molecule and a regulator of T_H1 generation and function; role in negative selection and peripheral tolerance.
CD155. 80–90 kDa. Poliovirus receptor (PVR).	Monocytes, macrophages, thymocytes, CNS neurons.	Receptor for poliovirus; may affect cell migration and adhesion.
CD156a. 69 kDa. A disintegrin and metallo-proteinase domain 8 (ADAM8), MS2; single-chain type 1 glycoprotein.	Neutrophils, monocytes.	Involved in inflammation, cell adhesion; may play a role in muscle differentiation, signal transduction; possible involvement in extravasation of leukocytes.
CD156b. 100–120 kDa. A disintegrin and metalloproteinase domain 17 (ADAM17), TNF-α converting enzyme (TACE), snake venom–like protease (cSVP); processed and unprocessed form, type 1 glycoprotein.	T cells, neutrophils, endothelial cells, monocytes, dendritic cells, macrophages, polymorphonuclear leukocytes, myocytes.	Primary protease that cleaves transmembrane forms of TNF-α, and TGF-α to generate soluble forms.
CD157. 42–45, 50 kDa. BP-3/IF-7, BST-1, Mo5; single chain, GPI-anchored protein.	Granulocytes, monocytes, macrophages, some B-cell progenitors, some T-cell progenitors.	Support for growth of lymphocyte progenitors.
CD158a. 58, 50 kDa. KIR2DL1, EB6, MHC class I–specific NK receptors, p50.1, p58.1; member of the killer-cell immunoglobulin-like receptor (KIR) family, type 1 glycoprotein.	Most NK cells, some T-cell subsets.	Inhibition of NK-cell- and CTL-mediated cytolytic activity on interaction with the appropriate HLA-C alleles.
CD158b1. 58, 50 kDa. GL183, MHC class I–specific NK receptors, p50.2, p58.2; member of the killer-cell immunoglobulin-like receptor (KIR) family, type 1 glycoprotein.	Most NK cells, some T-cell subsets.	Inhibition of NK-cell- and CTL-mediated cytolytic activity on interaction with the appropriate HLA-C alleles.
CD159a. 43 kDa. NKG2A, Killer-cell lectin-like receptor subfamily C, member 1 (KLRC1); type 2 glycoprotein and a member of the NKG2 family, disulfide-linked heterodimers covalently bonded to CD94.	NK cell lines, CD8$^+$, $\gamma\delta$ T cells, on some T-cell subset clones and lines.	Potent negative regulator of NK cells and T-lymphocyte activation programs; implicated in activation and inhibition of NK-cell cytotoxicity and cytokine secretion.
CD160. 27 kDa. BY55, NK1, NK28; expressed as a disulfide-linked multimer at cell surface, type 2 GPI-anchored glycoprotein.	Peripheral blood NK cells and CD8$^+$ T cells, IELs.	Binds HLA-C, provides costimulatory signals in CD8$^+$ T lymphocytes.
CD161. ~40 kDa. NKR-P1A, Killer-cell lectin-like receptor subfamily B member 1 (KLRB1); C-type lectin, disulfide-bonded homodimeric type 2 glycoprotein.	Most NK cells, a subset of CD4$^+$ and CD8$^+$ T cells, thymocytes.	May play a role in NK-cell-mediated cytotoxicity function, induction of immature thymocyte proliferation.

(continued)

CD Antigen. MW. Synonyms and Properties.	Expression	Function
CD162. 110–120 kDa. PSGL-1; disulfide-linked homodimeric mucin-like type 1 glycoprotein.	Most peripheral-blood T cells, monocytes, granulocytes, B cells.	Major CD62P ligand on neutrophils and T lymphocytes; mediates adhesion and leukocyte rolling and tethering on endothelial cells.
CD162R. 140 kDa. Post-translational modification of PSGL-1(PEN5), PSGL1, selectin P ligand (SELPLG); poly-N-lactosamine carbohydrate.	NK cells.	Creates a unique binding site for L-selectin; may be a unique developmentally specific NK-cell marker.
CD163. 130 kDa. GHI/61, M130; single-chain type 1 glycoprotein.	Monocytes, macrophages, myeloid cells; low level on lymphocytes, bone marrow stromal cells, subset of erythroid progenitors.	Involved in hematopoietic progenitor cell-stromal cell interaction; endocytosis and clearance of hemoglobin/haptoglobin complexes.
CD164. 80–100 kDa. MUC-24, multiglycosylated core protein 24 (MGC-24); type 1 glycoprotein.	Lymphocytes, epithelial cells, monocytes, granulocytes.	Most exists intracellularly; facilitates adhesion of CD34$^+$ and plays a role in regulating hematopoietic cell proliferation.
CD165. 42 kDa. AD2, gp37; membrane glycoprotein.	Peripheral lymphocytes, immature thymocytes, monocytes, most platelets; low level on thymocytes and thymic epithelial cells.	Adhesive interactions, including adhesion between thymocytes and thymic epithelial cells; platelet formation.
CD166. 100–105 kDa. BEN, DM-GRASP, KG-CAM, neurolin, SC-1, activated-leukocyte cell adhesion molecule (ALCAM); single-chain type 1 glycoprotein.	Activated T cells, activated monocytes, epithelium, neurons, fibroblasts, cortical and medullary thymic epithelial cells.	Adhesion molecule that binds to CD6; involved in neurite extension by neurons via heterophilic and homophilic interactions; may have a role in T-cell development.
CD167a. 129 kDa. α subunit 54 kDa, β subunit 63 kDa. DDR1 (discoidin domain family member 1), receptor tyrosine kinase.	Mainly normal and transformed epithelial cells, dendritic cells.	Adhesion molecule and collagen receptor.
CD168. 80, 84, 88 kDa. RHAMM (receptor for hyaluronan-mediated motility); hyaluronan (HA)-binding receptor family.	Thymocytes, myelomonocytic lineage and dendritic cells; up-regulated on activated lymphocytes.	Expressed on a hyaluronan-binding receptor that participates in the hyaluronan-dependent motility of thymocytes, lymphocytes, hematopoietic progenitor cells, and malignant B lymphocytes; adhesion of early thymocyte progenitors to matrix.
CD169. 200 kDa. SIGLEC-1 (sialic acid binding Ig-like lectin-1); sialoadhesin, single-chain type 1 glycoprotein, soluble form results from alternate splicing.	Macrophages, dendritic cells.	Mediates cell-cell interactions by binding to sialylated ligands on neutrophils, monocytes, NK cells, B cells, and a subset of CD8$^+$ T cells.
CD170. 140 kDa. SIGLEC-5 (sialic acid–binding Ig-like lectin-5); type 1 glycoprotein.	Dendritic cells, macrophages, neutrophils.	Potential adhesion molecule; may function as a pattern of self–nonself recognition receptor and mediate negative signals.
CD171. 200–230 kDa. L1 cell adhesion molecule; single-chain type 1 glycoprotein.	Low to intermediate expression on human lymphoid and mylelomonocytic cells, including CD4$^+$ T cells, a subset of B cells, monocytes, monocyte-derived dendritic cells, many cells of the central and peripheral nervous system.	Cell-adhesion molecule that plays a role in the maintenance of lymph node architecture during an immune response; kidney morphogenesis; neural development.
CD172a. 110 kDa. Signal regulatory protein alpha (SIRPα), SIRPA; single-chain type 1 glycoprotein.	CD34$^+$ stem/progenitor cells, macrophages, monocytes, granulocytes, dendritic cells; also in CNS tissue.	Negative regulation of receptor tyrosine kinase–coupled signaling processes; binding to CD47 on another cell inhibits phagocytosis of that cell.
CD173. H2; blood group O antigen.	Among hematopoietic cells, present on erythrocytes and CD34$^+$ hematopoietic precursors.	Glycoprotein-borne oligosaccharide; function unknown.

CD Antigen. MW. Synonyms and Properties.	Expression	Function
CD174. Lewis Y blood-group antigen.	Among hematopoietic cells, present on epithelial cells and CD34$^+$ hematopoietic precursors.	New hematopoietic-progenitor-cell marker; glycoprotein-borne oligosaccharide. Function unknown but correlated with apoptosis; may be involved in hematopoietic-stem-cell homing.
CD175. Tn; carbohydrate, tumor-specific antigen.	Variety of leukemic cells, epithelial cells, hematopoietic bone marrow cells.	Not known.
CD175s. Sialyl-Tn; carbohydrate, tumor-specific antigen.	Endothelial and epithelial cells and erythroblasts.	Shown to bind to CD22, Siglec-3, -5, and -6; function unknown.
CD176. Thomsen-Friedrenreich (TF) antigen: histo-blood group–related carbohydrate antigen.	Endothelial cells and erythrocytes, different types of carcinomas; CD34$^+$ hematopoietic precursor cells of the bone marrow and on various hematopoietic cell lines.	May be involved in tumor metastasis, such as liver tumors and positive leukemia cells.
CD177. 58–64 kDa. NB1, human neutrophil antigen-2A (HNA-2a); single-chain GPI-anchored plasma membrane glycoprotein.	Surfaces and secondary granules of neutrophils, basophils, NK cells, T-cell subsets, monocytes and endothelial cells.	Neutrophil transmigration.
CD178. Monomer of cell surface form 40 kDa; soluble forms 26–30 kDa. Fas ligand, FasL; TNF superfamily; homotrimeric type 2 glycoprotein.	Constitutive or induced expression on many cell types, including T cells, NK cells, microglial cells, neutrophils, nonhematopoietic cells such as retinal and corneal parenchymal cells.	Ligand for the apoptosis-inducing receptor CD95 (Fas/APO-1); key effector of cytotoxicity and involved in Fas/FasL interaction, apoptosis, and regulation of immune responses. Proposed to transduce a costimulatory signal for CD8$^+$ and naïve CD4$^+$ T-cell activation.
CD179a. 16–18 kDa. VpreB; polypeptide.	Selectively expressed in pro-B and early pre-B cells.	Associates with CD179b to form surrogate light chain of the pre–B cell receptor; involved in early B-cell differentiation.
CD179b. 22 kDa. λ5; polypeptide.	Selectively expressed in pro-B and early pre-B cells.	Associates with CD179a to form surrogate light chain of the pre-B-cell receptor; involved in early B-cell differentiation.
CD180. 95–105 kDa. RP105, Ly64; leucine-rich repeat family (LRRF), type 1 glycoprotein.	Monocytes, dendritic cells, mantle zone B cells.	Induces activation that leads to up-regulation of costimulatory molecules, CD80 and CD86, and an increase in cell size; promotes B-cell susceptibility to BCR-induced cell death but not to CD95-induced apoptosis and may play a role in the transmission of a growth-promoting signal; controls LPS recognition and signaling in B cells.
CD181. 58–70 kDa. Formerly known as CDw128a, IL-8Rα, IL-8RA (interleukin-8 receptor A), CXCR1 (chemokine C-X-C motif receptor 1); G-protein-coupled receptor (GPCR), type 3-, 7-span glycoprotein.	Neutrophils, basophils, mast cells, eosinophils, a subset of T cells, monocytes, NK and endothelial cells, keratinocytes, melanoma cells.	One of two receptors for IL-8 (CXCL8) and other CXC chemokines; involved in adhesion and chemotaxis of various leukocyte populations.
CD182. 67–70 kDa. Formerly known as CDw128b, CXCR2, interleukin-8 receptor B (IL-8RB); G-protein-coupled receptor (GPCR), type 3-, 7-span glycoprotein.	Neutrophils, basophils, mast cells, eosinophils, a subset of T cells, monocytes, NK and endothelial cells, keratinocytes, melanoma cells.	One of two receptors for IL-8 (CXCL8) and other CXC chemokines; involved in adhesion and chemotaxis of various leukocyte populations.

(continued)

CD Antigen. MW. Synonyms and Properties.	Expression	Function
CD183. 40.6 kDa. CXCR3, IP10R, Mig-R; 7-transmembrane-spanning protein superfamily (TN7SF), type 3-, 7-span glycoprotein.	Activated T and T_H1 cells, some memory B cells and plasma cells, some NK cells, eosinophils, mast cells, and some dendritic cells.	Receptor for several CXC chemokines; induces chemotaxis of activated T cells, T_H1 cells, and other leukocytes; thought to be essential for T-cell recruitment to inflammatory sites.
CD184. ~40 kDa. CXCR4, fusin, LESTR (leukocyte-derived seven transmembrane domain receptor); transmembrane-spanning protein superfamily (TN7SF), type 3-, 7-span glycoprotein.	Variety of blood and tissue cells, including most leukocytes, lymphoid and myeloid precursor cells, endothelial and epithelial cells, astrocytes and neurons.	Receptor for CXCL12 chemokine. Mediates blood-cell migration and is involved in B-lymphopoiesis and myelopoiesis, cardiogenesis, blood vessel formation, and cerebellar development; costimulation of pre–B cell proliferation; induction of apoptosis. Coreceptor for HIV.
CD185. 42kDa. CXCR5, BLR1; G-protein-coupled receptor.	Mature B cells, follicular T cells, $CD8^+$ T cells, Burkitt's lymphoma cells, mature dendritic cells.	CXCL13 chemokine receptor; possible regulatory function in lymphomagenesis, B-cell differentiation and activation.
CD186. 39 kDa. CXCR6; G-protein-coupled receptor.	Subsets of T cells (T_H1, T_C, and memory T cells), NK cells, plasma cells.	CXCL16 receptor; attracts T-cell subsets. Coreceptor for some HIV strains.
CD191. 41 kDa. CCR1, MIP-1αR, RANTES-R; G-protein-coupled receptor.	Monocytes, macrophages, immature dendritic cells, mast cells, B cells, NK cells.	Receptor for several CC chemokines; recruits various leukocyte subsets.
CD192. 42 kDa. CCR2, MCP-1R; G-protein-coupled receptor.	Activated monocytes, macrophages, basophils, dendritic cell subsets, activated and memory T cells, B cells, and NK cells; endothelial cells.	Receptor for several CC chemokines; chemoattracts various leukocyte populations. HIV coreceptor.
CD193. 45 kDa. CCR3, CKR3; G-protein-coupled receptor.	Eosinophils, basophils, mast cells, immature dendritic cells, T_H2 cells.	Receptor for multiple CC chemokines. Selectively recruits T_H2 cells. HIV coreceptor.
CD194. 63KDa. CCR4, CKR-4; G-protein-coupled receptor.	Granulocytes, immature dendritic cells, some B and T cells (T_H2, skin, intestine).	Receptor for CCL17 and CCL22. Recruits T cells to skin and intestine.
CD195. 45 kDa. CCR5; 7-transmembrane domain G-protein-coupled receptor.	Macrophages, mast cells, basophils, immature dendritic cells, some B cells, activated T cells, T_H1 cells, T_{reg}, and NK cells.	Receptor for several CC chemokines. Regulates chemoattraction during inflammation, HIV coreceptor.
CD196. 40KDa. CCR6, DRY6; G-protein-coupled receptor.	Monocytes, neutrophils, eosinophils, dendritic cells, B, NK, activated and memory T cells.	CCL6 receptor. Recruits numerous leukocyte populations.
CD197. 45 kDa. CCR7.	Monocytes, most naïve T cells, a subset of memory T cells, T_{REG}s, B cells, mature dendritic cells, NK cells, CD4 or CD8 single-positive mature thymocytes.	Receptor for CCL19 and CCL21. Crucial roles in naïve T cells and antigen-loaded dendritic cells homing to secondary organs; involvement in T-lymphocyte adhesion and thymocyte migration.
CDw198. ~40 kDa. CCR8; G-protein-coupled receptor.	Memory T cells and skin-homing T cells, thymocytes, monocytes.	CCL1 receptor. Involved in T-cell homing to skin. HIV coreceptor.
CDw199. CCR9; G-protein-coupled receptor.	Intestine-homing T cells, IgA plasma cells, thymocytes, some dendritic cells.	Receptor for CCL25, involved in homing to intestine. Coreceptor for HIV-1.
CD200. 45–50 kDa. Ox2; single-chain type 1 glycoprotein.	Dendritic cells, thymocytes, B cells, vascular endothelium, trophoblasts, neurons, and some smooth muscle, activated T cells.	May play an immunoregulatory role.
CD201. 50 kDa. EPC-R.	Endothelial subset, HSC.	Signaling of activated protein C.

CD Antigen. MW. Synonyms and Properties.	Expression	Function
CD202b. 145 kDa. TEK, TIE2.	Endothelial and hematopoietic stem cells.	Crucial role in integrity of vessels during maturation, maintenance, and remodeling.
CD203c. 130, 150 kDa. E-NPP3, B10, PDNP3, E-NPP3, bovine intestinal phosphodiesterase; ecto-nucleotide pyrophosphatase/phosphodi-esterase (E-NPP) family; type 2 glycoprotein.	Basophils, mast cells, uterine tissue.	Involved in clearance of extracellular nucleotides.
CD204. 220 kDa. Macrophage scavenger receptor (MSR); trimeric integral membrane glycoprotein.	Alveolar macrophages, Kupffer cells of the liver, splenic red pulp macrophages, sinusoidal macrophages in lymph nodes, interstitial macrophages.	Role in the pathological deposition of cholesterol during atherogenesis via receptor-mediated uptake of low-density LDL; recognition and elimination of pathogenic microorganisms; endocytosis of macromolecules.
CD205. 205 kDa. DEC-205.	Dendritic cells, thymic epithelial cells.	Endocytosis.
CD206. 162–175 kDa. MMR (macrophage mannose receptor), MRC1 (mannose receptor, C type lectin); pattern recognition receptors, single-chain type 1 glycoprotein.	Subset of mononuclear phagocytes (but not circulating monocytes), immature dendritic cells, hepatic and lymphatic endothelial cells.	Pattern-recognition receptor involved in immune responses of macrophages and immature dendritic cells; facilitates endocytosis and phagocytosis; might be important in homeostasis.
CD207. 40 kDa. Langerin; C-type lectin, type 2 glycoprotein.	Subset of dendritic cells, Langerhans cells.	Endocytic receptor with a functional C-type lectin domain with mannose specificity that facilitates antigen recognition and uptake.
CD208. 70–90 kDa, DC-LAMP; lysosome-associated membrane protein (LAMP) family member.	Dendritic cells.	May participate in peptide loading onto MHC class II molecules.
CD209. 45.7 kDa. DC-SIGN (dendritic-cell-specific ICAM3-grabbing nonintegrin), HIV GP120-binding protein; C-type lectin.	Dendritic cells.	C-type lectin receptor; binds mannans. Activates phagocytosis; role in adhesion to endothelial cells; binds to HIV.
CD210. 90–110 kDa. IL-10Rα and β; member of class 2 cytokine receptor family, single-chain type 1 glycoprotein.	Mainly on hematopoietic cells, including T and B cells, NK cells, monocytes, and macrophages.	IL-10 receptor. Triggers anti-inflammatory responses, including inhibition of cytokine production and antigen-presenting cell functions.
CD212. 110, 85 kDa. IL-12Rβ; hematopoietin receptor family, single-chain type 1 receptor.	Activated CD4$^+$ and CD8$^+$ T cells, IL-2-activated CD56$^+$ NK cells, γδ T cells; PBL, cord blood lymphocytes, subsets of monocytes.	IL-12 receptor β chain; involved in cell signaling and immune regulation; pleiotropic effects on NK and T cells; induces IFN-γ production.
CD213a1. 65 kDa. IL13Rα1 (interleukin 13 receptor alpha 1) NR4 (in mouse); hemo-poietin receptor family, single-chain type 1 glycoprotein.	Most human tissues, hematopoietic and nonhematopoietic.	IL-13 and IL-4 receptor α chain; promote T$_H$2 phenotype and functions.
CD213a2. 60–70 kDa. IL13Rα2 (interleukin 13 receptor alpha 2); hemopoietin receptor family, single-chain type 1 glycoprotein.	B cells, monocytes, fibroblasts, immature dendritic cells.	Inhibits binding of IL-13 to the IL-13 cell-surface receptor.
CD217. 120 kDa. IL-17R; hemopoietin receptor family, single-chain type 1 glycoprotein.	B, T, NK cells, cord blood, PBL, thymocytes; fibroblasts, epithelial cells, monocytes, macrophages, granulocytes.	Receptor for IL-17, promotes inflammation.
CD218a. 70 kDa. IL-18Rα, IL-1Rrp; IL-1 receptor family.	T cells, NK cells, dendritic cells.	Binds to IL-18 and induces NF-κB activation.
CD218b. 70 kDa. IL-18Rβ, IL-18RAcP; IL-1 receptor family.	T cells, NK cells, dendritic cells.	Heterodimeric receptor that enhances IL-18 binding.

(continued)

CD Antigen. MW. Synonyms and Properties.	Expression	Function
CD220. 140, 70 kDa. Insulin-R.	Broad expression.	Receptor mediating insulin signaling.
CD221. 140, 70 kDa. IGF-1R.	Broad expression.	Binds IGF with high affinity; involved in signaling, cell proliferation, and differentiation.
CD222. 280–300 kDa. Man-6P receptor (mannose-6 phosphate receptor), IGF2R (insulin-like growth factor 2 receptor).	Fibroblasts, granulocytes, lymphocytes, myocytes.	Sorts newly synthesized lysosomal enzymes bearing M6P to lysosomes; activates latent TGF-β; affects cell adhesion and migration as well as angiogenesis.
CD223. 70 kDa. Lag-3 (lymphocyte activation gene 3); single-chain type 1 glycoprotein.	All subsets of activated T or NK cells.	May promote down-regulation of TCR signaling, leading to cell inactivation, a role in down-regulating an antigen-specific response; possibly helps activate $CD4^+$ and $CD8^+$ T cells to fully activate monocytes and dendritic cells to optimize MHC class I– and class II–mediated T-cell responses.
CD224. 100 kDa. GGT (gamma glutamyl transpeptidase).	Vascular endothelium, peripheral blood macrophages, subset of B cells; lymphocytes, monocytes, granulocytes, endothelial and stem cells.	Cellular detoxification, leukotriene biosynthesis, inhibition of apoptosis.
CD225. 17 kDa. Leu13, interferon-induced protein 17 (IFI17).	Leukocytes, endothelial cells, in multiple lineages.	Involved in lymphocyte activation and development; may play a role in controlling cell-to-cell interactions.
CD226. 65 kDa. DNAM-1, PTA-1 (platelet and T-cell activation antigen 1).	NK cells, platelets, monocytes, subset of T cells.	Involved in costimulation and adhesion.
CD227. MUC1 (mucin 1), episialin; very large glycosylated protein with small 25 kDa subunit and larger subunits of 300–700 kDa, type I transmembrane protein.	Epithelial cells, follicular dendritic cells, monocytes, subsets of lymphocytes, B and stem cells, some myelomas; some hematopoietic cell lineages.	Involved in adhesion and signaling.
CD228. 80–97 kDa. Melanotransferrin.	Stem cells, melanoma cells.	Cell adhesion, potential role in iron transport.
CD229. 100–120 kDa. T-lymphocyte surface antigen, Ly-9 (lymphocyte antigen 9); CD2-subset of immunoglobulin superfamily, single-chain type 1 glycoprotein.	Predominantly in the lymph nodes, spleen, thymus, peripheral blood leukocytes.	Involved in the activation of lymphocytes.
CD230. 30–40 kDa. Prion protein, prP27–30; GPI-anchored glycoprotein.	Broadly expressed on most cell types with the highest level on neurons and follicular dendritic cells; hematopoietic and non-hematopoietic cells.	Triggers prion diseases. Normal role uncertain; may inhibit apoptosis.
CD231. 28–45 kDa. T-cell acute lymphoblastic leukemia–associated antigen 1 (TALLA-1), A15, MXS1, TM4SF2, CCG-B7; tetraspan subfamily 1 member, type 3-, 4-span glycoprotein.	Strongly expressed on T-cell acute lymphoblastic leukemic cells (T-ALL), neuroblastoma cells, normal brain neurons.	May play a role in the regulation of cell development, activation, growth, and motility.
CD232. 200 kDa. Virus-encoded semaphorin protein receptor (VESPR), Plexin C1 (PLXNC1); plexin family molecules.	B and NK cells, monocytes, granulocytes.	May play a role in immune modulation; participates in differentiation and migration of monocytes, some classes of dendritic cells, neutrophils, B lymphocytes.
CD233. 95–110 kDa. Band 3, Diego blood group; typically a dimer.	Erythrocytes and basolateral membrane of some cells of the distal and collecting tubules of the kidney.	Maintains red-cell morphology via exchanging anion.
CD234. 35–45 kDa. FY-glycoprotein, Duffy blood group Ag/chemokine receptor (DARC), type 3-, 7-span acidic Fy-glycoprotein.	Erythrocytes, epithelial cells of the kidney collecting duct, lung alveoli, and thyroid; on neurons (Purkinje cells) in the cerebellum.	Chemokine decoy receptor; binds a number of chemokines, modulates the intensity of inflammatory reactions.

CD Antigen. MW. Synonyms and Properties.	Expression	Function
CD235a, -b, -ab. 35 kDa (a), 20 kDa (b). Glycophorin A, glycophorin B.	Erythrocytes.	May act as parasite receptor; probably affects cell aggregation.
CD236. 23 kDa, 32 kDa. Glycophorin C/D.	Variety of cells and tissues both erythroid and nonerythroid in nature.	Contains the Gerbich antigens; maintains mechanical stability in erythrocytes; parasite receptor.
CD236R. 32 kDa. Glycophorin C.	Variety of cells and tissues both erythroid and nonerythroid in nature.	Contains the Gerbich antigens; maintains mechanical stability in erythrocytes; parasite receptor.
CD238. 93 kDa. Kell.	Erythrocytes and subset of stem cells.	Blood group antigens. May be involved in erythrocyte differentiation and trafficking.
CD239. 78–85 kDa. B-CAM.	Erythrocytes and subset of stem cells.	Laminin alpha-5 receptor; potential role in intracellular signaling.
CD240CE. 30–32 kDa. Rh30CE.	Erythrocytes.	Rh factor. May help maintain erythrocyte mechanical properties.
CD240D. 30–32 kDa. Rh30D; type 1-, 2-transmembrane superfamily.	Erythrocytes.	Rh factor. May help maintain erythrocyte mechanical properties.
CD241. 50 kDa. RhAg, Rh50A; type 1-, 2-transmembrane superfamily.	Erythrocytes.	Rh factor. Forms a complex with CD47, LW, glycophorin B; may have membrane transport or channel function.
CD242. 42 kDa. ICAM-4; immunglobulin superfamily.	Erythrocytes.	Cell adhesion.
CD243. 170 kDa. Multidrug resistance protein 1 (MDR-1), P-glycoprotein (P-gp), ABC-B1, gP170; single-chain type 3-, 12-span glycoprotein.	Stem and progenitor cells.	Influences the uptake, tissue distribution, and elimination of P-gp-transported drugs and toxins; transporter in the blood-brain barrier.
CD244. 63–70 kDa. 2B4, NAIL (NK-cell activation-inducing ligand); single-chain type 1 glycoprotein.	NK cells, $\gamma\delta$ T cells, 50% of CD8$^+$ T cells, monocytes, basophils.	Activation receptor for NK cells and modulates NK-cell cytokine production, cytolytic function, and extravasation; involved in NK and T-cell interactions; may be important for the development of functional CD4$^+$ T cells and possibly serves to increase cell-cell adhesion
CD245. 220–240 kDa. p220/240.	All resting peripheral blood lymphocytes.	Signal transduction and costimulation of T and NK cells.
CD246. 80 kDa, 200 kDa. Anaplastic lymphoma kinase (ANK); single-chain type 1 glycoprotein.	Subset of T-cell lymphomas.	Receptor tyrosine kinase; may regulate cell growth and apoptosis.
CD247. 16 kDa. Zeta chain of CD3.	NK cells during thymopoiesis and mature T cells in the periphery.	T-cell activation.
CD248. 175 kDa. TEM1, endosialin; C-type lectin.	Endothelial tissues, stromal fibroblasts.	Tumor progression and angiogenesis.
CD249. 160 kDa. Aminopeptidase A; peptidase M1 family.	Epithelial and endothelial cells.	Renin-angiotensin system; potential role in the growth and differentiation of early B lineage cells.
CD252. 34 kDa. OX-40L, gp 34; TNF superfamily.	Activated B cells, cardiac myocytes.	CD40 ligand. T-cell costimulation.
CD253. 40KDa. TRAIL, Apo-2L, TL2, TNFSF10; TNF superfamily.	Activated T cells; broadly expressed in tissues.	Induces apoptosis.
CD254. 35 kDa. TRANCE, RANKL, OPGL; TNF superfamily.	Bone marrow stroma, activated T cells, lymph node.	Binds to OPG and RANK, differentiation of osteoclasts, augments dendritic cells to stimulate naïve T-cell proliferation.
CD256. 27 kDa. APRIL, TALL-2; TNF superfamily.	Monocytes and macrophages.	Binds TACI and BCMA; involved in B-cell proliferation.

(continued)

CD Antigen. MW. Synonyms and Properties.	Expression	Function
CD257. 34 KDa. BLys, BAFF, TALL-1; TNF superfamily.	Activated monocytes; exists as a soluble form.	B-cell growth and differentiation factor; costimulation of Ig production.
CD258. 29 kDa. LIGHT, HVEM-L; TNF superfamily.	Activated T cells, immature dendritic cells.	Binds LTβR, involved in T-cell proliferation; receptor for HVEM.
CD261. 56 kDa. TRAIL-R1, DR4; TNF receptor superfamily.	Activated T cells, peripheral blood leukocytes.	Activates FADD and caspase-8-mediated apoptosis.
CD262. 55 kDa. TRAIL-R2, DR5; TNF receptor superfamily.	Widely expressed; peripheral blood leukocytes.	Same as CD261.
CD263. 65 kDa. TRAIL-R3, DcR1, LIT; TNF receptor superfamily.	Peripheral blood leukocytes.	Receptor for TRAIL but lacks a death domain and is unable to induce apoptosis.
CD264. 35 kDa. TRAIL-R4, TRUNDO, DcR2; TNF receptor superfamily.	Peripheral blood leukocytes.	Receptor for TRAIL with a truncated death domain; inhibits apoptosis.
CD265. 97 kDa. RANK, TRANCE-R, ODFR; TNF receptor superfamily.	Broad.	Binds TRANCE; activates osteoclasts; T cell-dendritic cell interactions.
CD266. 14 kDa. TWEAK-R, FGF-inducible 14; TNF receptor superfamily.	Endothelial cell subset, placenta, kidney, heart.	Cell matrix interactions, endothelial growth and migration; receptor for TWEAK.
CD267. 32 KDa. TACI, TNFRSF13B.	B cells, activated T cells.	Binds BAFF and APRIL; regulates B cells.
CD268. 19 kDa. BAFFR, TNFRSF13C; TNF receptor superfamily.	B cells.	Binds BAFF; involved in B-cell survival.
CD269. 20 kDa. BCMA, TNFRSF17; TNF receptor superfamily.	Mature B cells.	Binds BAFF; involved in B-cell survival and proliferation.
CD271. 45 kDa. NGFR, p75 (NTR); TNF receptor superfamily.	Neurons, bone marrow mesenchymal cells.	Receptor for NGF, NT-3, NT-4; tumor suppressor, cell survival, and cell death.
CD272. 33 kDa. BTLA; glycoprotein.	T_H1 cells and activated B and T cells.	Inhibitory receptor on T lymphocytes with similarities to CTLA-4 (CD152) and PD-1 (CD279).
CD273. 25 kDa. B7DC, PDL2, programmed cell death 1 ligand 2; glycoprotein.	Dendritic cells, macrophages, monocytes, activated T cells.	Second ligand for PD-1 (CD279); this interaction, like that of PD-L1/PD-1, inhibits TCR-mediated proliferation and cytokine production.
CD274. 40 kDa. B7H1, PDL1, programmed cell death 1 ligand 1.	Macrophages, epithelial and dendritic cells, NK cells, activated T cells, and monocytes.	Putative ligand for PD-1 (CD279); shown to inhibit TCR-mediated proliferation and cytokine secretion; involved in costimulation and inhibition of lymphocytes.
CD275. 40 kDa, 60 kDa. B7H2, ICOSL; inducible T-cell costimulator ligand (ICOSL).	Macrophages, dendritic cells, weak T and B cells, activated monocytes.	Costimulation of T cells; reported to promote proliferation and cytokine production.
CD276. 40–45 kDa, 110 kDa. B7H3; B7 homolog 3.	Epithelial cells, activated monocytes and T cells, dendritic-cell subsets.	Stimulates T-cell proliferation and activation and IFN-γ production.
CD277. 56 kDa. BT3.1; B7 family: butyrophilin, subfamily 3, member A1.	B cells, T cells, NK cells, dendritic cells, monocytes, endothelial cells, stem-cell subsets.	May play a role in regulation of the immune response; acts as a regulator of T-cell activation and function.
CD278. 47–57 kDa. ICOS; inducible T-cell costimulator.	T_H2 cells, thymocyte subsets, activated T cells.	Plays a critical role in costimulating T-cell activation, development, proliferation, and cytokine production.
CD279. 50–55 kDa. Programmed cell death protein 1 (PD1 or PDCD1), hPD-1, SLEB2; single-chain type 1 glycoprotein.	Subset of thymocytes, activated T and B cells.	Inhibits TCR-mediated proliferation and cytokine production.
CD280. 180 kDa. ENDO180, uPARAP, mannose receptor C type 2, TEM22.	Myeloid progenitor cells, fibroblasts, chondrocytes, osteoclasts, osteocytes, subsets of endothelial and macrophage cells.	May affect cell motility and remodeling of the extracellular matrix.

CD Antigen. MW. Synonyms and Properties.	Expression	Function
CD281. 90 kDa. TLR1 (Toll-like receptor 1).	Many leukocytes; epithelial, endothelial, and some other cell types.	Binds mycobacterial and Gram-negative bacterial triacyl lipopeptides. Activates innate immune responses.
CD282. 90 kDa. TLR2 (Toll-like receptor 2).	Many leukocytes; epithelial, endothelial, and some other cell types.	Binds Gram-positive bacterial peptidoglycans, bacterial lipoproteins, trypanosome GPI-linked proteins, schistosome phosphatidylserine, and zymosan from yeast and other fungi. Activates innate immune responses.
CD283. 104 kDa. TLR3 (Toll-like receptor 3).	Many leukocytes; epithelial, endothelial, and some other cell types.	Binds viral double-stranded (ds) RNA and synthetic poly (I:C). Activates innate immune responses.
CD284. 85, 110, 130 kDa. TLR4 (Toll-like receptor 4).	Many leukocytes; epithelial, endothelial, and some other cell types.	Interacts with CD14 and MD-2. Binds Gram-negative bacterial lipopolysaccharide (LPS) with LBP-binding protein (LBP). Also binds F-protein from respiratory syncytial virus (RSV) and fungal mannans. Activates innate immune responses.
CD285. TLR5 (Toll-like receptor 5).	Many leukocytes; epithelial, endothelial, and some other cell types.	Binds bacterial flagellin. Activates innate immune responses.
CD286. TLR6 (Toll-like receptor 6).	Many leukocytes; epithelial, endothelial, and some other cell types.	Binds mycobacterial and Gram-positive bacterial diacyl lipopeptides and zymosan from yeast and other fungi. Activates innate immune responses.
CD287. TLR7 (Toll-like receptor 7).	Many leukocytes; epithelial, endothelial, and some other cell types.	Binds viral single-stranded (ss) RNA. Activates innate immune responses.
CD288. TLR8 (Toll-like receptor 8).	Many leukocytes; epithelial, endothelial, and some other cell types.	Binds viral single-stranded (ss) RNA. Activates innate immune responses.
CD289. 113 kDa. TLR9 (Toll-like receptor 9).	Many leukocytes; epithelial, endothelial, and some other cell types.	Binds bacterial unmethylated CpG DNA, malaria hemozoin, and herpesvirus. Activates innate immune responses.
CD290. 90 kDa. TLR10 (Toll-like receptor 10).	Many leukocytes; epithelial, endothelial, and some other cell types.	Microbial ligand unknown.
CD292. 50–55 kDa. 70-80 KDa. BMPR1A, ALK3; type 1 glycoprotein.	Bone progenitors.	BMP2 and -4 receptor, bone development; immune function unknown.
CDw293. 57 kDa. BMPR1B, ALK6; type 1 protein.	Bone progenitors.	BMP receptor, bone development; immune function unknown.
CD294. 55–70 kDa. CRT$_H$2, GPR44; G-protein-coupled receptor.	T$_H$2 cells, granulocytes.	Binds PGD2; provides stimulatory signals for T$_H$2 cells; involved in allergic inflammation.
CD295. 132 kDa. LeptinR (LEPR); type 1 cytokine receptor.	Broad.	Leptin receptor. Functions in adipocyte metabolism; normal lymphopoiesis.
CD296. 36 kDa. ART1, RT6, ART2; ADP-ribosyltransferase.	Peripheral T cells, NK-cell subset, heart and skeletal muscle.	Modifies integrins during differentiation; ADP ribosylation of target proteins.
CD297. 36 kDa. ART4, Dombrock blood group; ADP-ribosyltransferase.	Erythrocytes, activated monocytes.	ADP ribosylation of target proteins.
CD298. 32 kDa. Na$^+$/K$^+$ ATP-ase β-3 subunit.	Broad.	Ion transports.
CD299. 45 kDa. DC-SIGN-related, LSIGN, DC-SIGN2.	Endothelial subset.	Binds ICAM-3, HIV-1 gp120; coreceptor with DC-SIGN for HIV-1.
CD300a. 60 kDa. CMRF35H, IRC1, Irp60.	Monocytes, neutrophils, T- and B-cell subsets.	Inhibition.
CD300c. CMRF35A, LIR.	Monocytes, neutrophils, T- and B-cell subsets.	Unknown.

(continued)

CD Antigen. MW. Synonyms and Properties.	Expression	Function
CD300e. CMRF35L1.	Unknown.	May act as an activating receptor.
CD301. 40 kDa. MGL1, HML; C-type lectin superfamily.	Immature dendritic cells.	Binds Tn antigen; uptake of glycosylated antigens.
CD302. 19–28 kDa. DCL1, BIMLEC; type 1 transmembrane C-type lectin receptor.	Some myeloid and Hodgkin's cell lines.	Fusion protein in Hodgkin's lymphoma with DEC-205; unknown function.
CD303. 38 kDa. BDCA2, HECL; C-type lectin superfamily.	Plasmacytoid dendritic cells.	Inhibits IFN-α/β production.
CD304. 103 kDa. BDCA4, neuropilin 1; neuropilin family.	Neurons, subset of T cells, dendritic cells, endothelial cells, tumor cells.	Coreceptor with plexin, interacts with VEGF165 and semaphorins; axonal guidance, angiogenesis, cell survival, migration.
CD305. 31 kDa, 40 kDa. Leukocyte-associated Ig-like receptor-1 (LAIR), p40; type 1 glycoprotein.	Expression on B cells related to maturation stage; NK and T cells, monocyte-derived dendritic cells, thymocytes, thymic precursors.	Inhibitory receptor of cell function on NK cells, T and B cells; inhibits cellular activation and inflammation.
CD306. LAIR2.	T cells and monocytes.	Soluble form; involved in mucosal tolerance.
CD307. 55–106 kDa. IRTA; Fc receptor immunoglobulin superfamily.	B-cell subset, B-cell lymphomas.	B-cell development.
CD309. 230 kDa. VEGFR2, KDR; type I membrane tyrosine kinase.	Endothelial cells, angiogenic precursors, hemangioblast.	Binds VEGF; regulates cell adhesion and signaling; angiogenesis.
CD312. 90 kDa. EMR2.	Monocytes, macrophages, myeloid dendritic cells, low on granulocytes.	Cell adhesion and migration; involved in phagocytosis.
CD314. NKG2D, killer-cell lectin-like receptor subfamily K member 1; type II transmembrane protein with an extracellular C-type lectin-like domain.	NK cells primarily; also on activated macrophages and $\gamma\delta$ T cells and CD8$^+$ $\alpha\beta$ T cells.	Induces NK-cell activation and cytotoxicity.
CD315. 135 kDa. CD9P1, SMAP-6.	B-cell subsets, activated monocytes, endothelial and epithelial cells, hepatocytes, megakaryocytes.	May affect cell motility.
CD316. 62–78 kDa. IgSF8, KASP.	B cells, NK cells, T cells.	Involved in cell migration.
CD317. 29–33 kDa. BST2 (bone marrow stromal-cell antigen 2), PDCA-1 (plasmacytoid dendritic cell antigen-1), HM1.24; tetherin.	B and T cells, monocytes, NK and plasmacytoid dendritic cells, fibroblasts, plasma and stromal cells.	Antiviral activity due to virus tethering; may be involved in pre-B-cell growth.
CD318. 80/140 kDa. CDCP1, SIMA135.	Hematopoietic stem-cell subset, tumor cells.	Cell adhesion with extracellular matrix.
CD319. 37.4 kDa. CRACC, SLAM family member 7; multiple isoforms with possibly different functions.	NK cells, activated B-cells, NK-cell line but not in promyelocytic, B-, or T-cell lines.	Mediates NK-cell activation through an SAP-independent extracellular signal-regulated ERK-mediated pathway; may play a role in lymphocyte adhesion.
CD320. 29 kDa. 8D6A, 8D6. LDL receptor.	Follicular dendritic cells, germinal centers.	B-cell proliferation, tumor formation.
CD321. 35–40 kDa. JAM1, F11 receptor; immuglobulin superfamily, type 1.	Platelets, epithelial and endothelial cells.	Adhesions, tight junctions.
CD322. 33.2 kDa. Junctional adhesion molecule-2 (JAM-2), VE-JAM.	Endothelial cells, B cells, monocytes, T-cell subsets.	Cell-cell adhesion; central role in the regulation of transendothelial leukocyte migration to secondary organs.
CD324. 120 kDa. E-Cadherin; cadherin superfamily.	Epithelial and stem cells, erythroblasts, keratinocytes, trophoblasts, platelets.	Recently shown to act as an inhibitory receptor on γ/δ T cells; binds inhibitory receptors on CD8$^+$ T and NK cells; tumor suppression and cell growth and differentiation, cell adhesion.
CD325. 130 kDa. N-Cadherin (NCAD); CDw325; cadherin superfamily.	Brain; skeletal and cardiac muscle.	Cell adhesion, neuronal recognition.

CD Antigen. MW. Synonyms and Properties.	Expression	Function
CD326. 35–40 kDa. Ep-CAM, Ly74.	Most epithelial cell membranes.	May be involved in mucosal immunity.
CD327. 47 kDa (predicted). CD33L, CD33L1, OB-BP1, sialic acid–binding Ig-like lectin 6 (Siglec6); leptin-binding protein.	Leukocytes, B cells.	Mediates cell-cell recognition, binds leptin, possible modulator of leptin levels; likely inhibitory.
CD328. 65–75 kDa. AIRM1, QA79, p75, p75/AIRM1, sialic acid–binding Ig-like lectin 7 (Siglec7).	Variety of leukocytes in addition to T cells, including NK cells, monocytes, and granulocytes.	Cell-adhesion molecule; likely inhibitory function in NK- and T-cell activation.
CD329. 48 kDa (predicted). OBBP-like, sialic acid–binding Ig-like lectin 9 (Siglec9).	Variety of leukocytes in addition to T cells, including NK cells, monocytes, granulocytes.	Cell-adhesion molecule; likely inhibitory function in NK- and T-cell activation.
CD331. 60, 110–160 kDa. FGFR1, KAL2, N-SAM, Fms-like tyrosine kinase-2; transmembrane tyrosine kinase.	Fibroblasts, epithelial cells.	High-affinity receptor for fibroblast growth factors.
CD332. 135 kDa. FGFR2, BEK, KGFR; transmembrane tyrosine kinase.	Fibroblasts, epithelial cells.	High-affinity receptor for fibroblast growth factors.
CD333. 115–135 kDa. FGFR3, ACH, CEK2; transmembrane tyrosine kinase.	Fibroblasts, epithelial cells.	High-affinity receptor for fibroblast growth factors.
CD334. 88–110 kDa. FGFR4, JTK2, TKF; transmembrane tyrosine kinase.	Fibroblasts, epithelial cells.	High-affinity receptor for fibroblast growth factors.
CD335. 46 kDa. NKp46, NCR1, Ly94, natural cytotoxicity triggering receptor 1.	NK cells.	Cytotoxicity-activating receptor that may contribute to the increased efficiency with which activated NK cells mediate tumor cell lysis.
CD336. 44 kDa. NKp44, NCR2, Ly95, natural cytotoxicity triggering receptor 2.	NK cells.	Cytotoxicity-activating receptor that may contribute to the increased efficiency with which activated NK cells mediate tumor-cell lysis.
CD337. 30 kDa. NKp30, NCR3, Ly117.	NK cells.	Cytotoxicity-activating receptor that may contribute to the increased efficiency with which activated NK cells mediate tumor-cell lysis.
CD338. 72 kDa. ABCG2, BCRP, Bcrp1, MXR; G-protein-coupled receptor, 7-transmembrane.	Stem-cell subset.	Multidrug resistance transporter.
CD339. 134 kDa. Jagged-1, JAG1, JAGL1, hJ1.	Stromal cells, epithelial cells, myeloma cells.	Notch ligand; involved in hematopoiesis.
CD340. 185 kDa. HER2; NEU; ERBB2	Tumor tissues, epithelial cells.	Cell proliferation and differentiation, tumor cell metastasis.
CD344. 48–53 kDa. FZD4; EVR1; Fz-4; FzE4	Epithelial cells, endothelial cells, myeloid and neuronal progenitors.	A receptor for Wnt proteins; involved in cell proliferation and differentiation, embryonic development, and retinal vascularization.
CD349. 64 kDa. FZD9;FZD3	Mesenchymal stem cells, tissue progenitors.	A receptor for Wnt proteins; involved in B-cell development.
CD350. 65 kDa. FZD10; FzE7; FZ-10	Epithelial cells.	A receptor for Wnt proteins; involved in neural, lung, and limb development.

Data on the major biological activities and sources of many cytokines are presented in the following table. In most cases, mass is given (usually for human cytokines). In some instances, a given cytokine may have biological activities in addition to those listed here or may be produced by other sources as well as the ones cited here. This list of cytokines includes most cytokines of immunological interest. However, cytokines that are not closely identified with the immune system—for example, growth hormone—are not listed. Also, with the exception of IL-8, chemokines are not included in this compilation.

The following are the major references used for the information in this appendix:

PeproTech (www.peprotech.com)

Pubmed (www.ncbi.nlm.nih.gov/pubmed)

R&D Systems (www.rndsystems.com/browse_by_molecule.aspx)

Taub, D. T. (2004, September). Cytokine, growth factor, and chemokine ligand database. Current Protocols in *Immunology*. 6.29.1–6.29.89. (www.currentprotocols.com/WileyCDA/CPUnit/refId-im0629.html)

Uniprot (www.uniprot.org)

Cytokine. MW. Synonyms.	Sources	Activity
Interleukin 1 (IL-1). IL-1α 17.5 kDa, IL-1β 17.3 kDa. Lymphocyte-activating factor (LAF); mononuclear cell factor (MCF); endogenous pyrogen (EP).	Many cell types, including monocytes, macrophages, dendritic cells, NK cells, and non-immune system cells such as epithelial and endothelial cells, fibroblasts, adipocytes, astrocytes, and some smooth muscle cells.	Displays a wide variety of biological activities on many different cell types, including T cells, B cells, monocytes, eosinophils and dendritic cells, as well as fibroblasts, liver cells, vascular endothelial cells, and some cells of the nervous system. The in vivo effects of IL-1 include induction of local inflammation and systemic effects such as fever, the acute phase response, and stimulation of neutrophil production.
Interleukin 2 (IL-2). 15–20 kDa. T-cell growth factor (TCGF).	Activated T cells.	Stimulates proliferation and differentiation of T and B cells; activates NK cells.
Interleukin 3 (IL-3). 15.1 kDa (monomer), 30 kDa (dimer). Multipotential colony-stimulating factor (M-CSF); hematopoietic-cell growth factor (HCGF); mast-cell growth factor (MCGF).	Activated T cells, mast cells, basophils, and eosinophils.	Growth factor for hematopoietic cells; stimulates colony formation in neutrophil, eosinophil, basophil, mast cell, erythroid, megakaryocyte, and monocytic lineages but not of lymphoid cells.
Interleukin 4 (IL-4). 15–19 kDa. B-cell-stimulatory factor 1 (BSF-1).	T cells (particularly those of the T_H2 subset), mast cells, basophils, and bone marrow stromal cells.	Promote naïve T cell differentiation to T_H2 cells. Stimulates the growth and differentiation of B cells. Induces class switching to IgE. Promotes allergic responses.
Interleukin 5 (IL-5). 15 kDa. Eosinophil differentiation factor (EDF); eosinophil colony-stimulating factor (E-CSF).	T cells (particularly those of the T_H2 subset), mast cells, eosinophils.	Induces eosinophil formation and differentiation. Stimulates B cell growth and differentiation.
Interleukin 6 (IL-6). 26 kDa. B-cell stimulatory factor 2 (BSF-2); hybridoma/plasmacytoma growth factor (HPGF); hepatocyte-stimulating factor (HSF).	Some T cells and B cells, several non-lymphoid cells, including macrophages, bone marrow stromal cells, fibroblasts, endothelial and muscle cells, adipocytes, and astrocytes.	Regulates B- and T-cell functions; in vivo effects on hematopoiesis. Induces inflammation and the acute phase response.

(continued)

Cytokine. MW. Synonyms.	Sources	Activity
Interleukin 7 (IL-7). 20–28 kDa. Pre-B-cell growth factor; lymphopoietin-1 (LP-1).	Bone marrow and thymic stromal cells, intestinal epithelial cells.	Growth factor for T- and B-cell progenitors.
Interleukin 8 (IL-8). 6–8 kDa. Neutrophil-attractant/activating protein (NAP-1); neutrophil-activating factor (NAF); granulo-cyte chemotactic protein 1(GCP-1); CXCL8 chemokine.	Many cell types, including monocytes, macrophages, lymphocytes, granulo-cytes, and nonimmune system cells such as fibroblasts, endothelial and epi-thelial cells, and hepatocytes.	Chemokine that functions primarily as a chemoattractant and activator of neutro-phils; also attracts basophils and some subpopulations of lymphocytes; has angiogenic activity.
Interleukin 9 (IL-9). 32–40 kDa. P40; T-cell growth factor III.	Some activated T-helper-cell subsets.	Stimulates proliferation of T lymphocytes and hematopoietic precursors, may be involved in allergy and asthma.
Interleukin 10 (IL-10). 35–40 kDa. Cytokine synthesis inhibitory factor (CSIF).	Activated subsets of CD4$^+$ and CD8$^+$ T cells, macrophages, and dendritic cells.	Enhances proliferation of B cells, thymocytes, and mast cells; in cooperation with TGF-β, stimulates IgA synthesis and secretion by human B cells. Anti-inflammatory; antagonizes generation of the T$_H$1 subset of helper T cells.
Interleukin 11 (IL-11). 23 kDa.	Bone marrow stromal cells and IL-1–stimulated fibroblasts	Growth factor for plasmacytomas, mega-karyocytes, and macrophage progenitor cells.
Interleukin 12 (IL-12). Heterodimer con-taining a p35 subunit of 30–35 kDa, p40 subunit of 35–44 kDa. NK cell stimulatory factor (NKSF); cytotoxic lymphocyte matura-tion factor (CLMF).	Macrophages, B cells, and dendritic cells.	Important factor in inducing differentia-tion of T$_H$1 subset of helper T cells; induces IFN-γ production by T cells and NK cells and enhances NK and cytotoxic T cell activity.
Interleukin 13 (IL-13). 10 kDa.	Activated T cells (particularly those of the T$_H$2 subset), mast cells, and NK cells.	Role in T$_H$2 responses; up-regulates syn-thesis of IgE and suppresses inflammatory responses. Involved in pathology of asthma and some allergic conditions.
Interleukin 14 (IL-14). 60 kDa. High-molecular-weight B-cell growth factor (HMW-BCGF).	Activated T cells.	Enhances B-cell proliferation; inhibits antibody synthesis.
Interleukin 15 (IL-15). 14–15 kDa.	Many cell types but primarily dendritic cells and cells of the monocytic lineage.	Stimulates NK-cell and T-cell proliferation and development; helps to activate NK cells.
Interleukin 16 (IL-16). Homotetramer 60 kDa; monomer≈17 kDa. Lymphocyte chemoattractant factor (LCF).	Activated T cells and some other cell types.	Chemoattractant for CD4$^+$ T cells, mono-cytes, and eosinophils. Binding of IL-16 by CD4 inhibits HIV infection of CD4$^+$ cells.
Interleukin 17 (IL-17). 28–31 kDa. CTLA-8 (cytotoxic T lymphocyte–associated antigen 8). Family members IL-17A-F (see Table 4-5).	CD4$^+$ T cells (particularly those of the T$_H$17 subset), CD8$^+$, γδ T cells, NK cells, intraepithelial lymphocytes, and some other cells.	Promotes inflammation by increasing production by epithelial, endothelial, and fibroblast cells of proinflammatory cytokines such as IL-1, IL-6, TNF-α, G-CSF, GM-CSF, and chemokines that attract monocytes and neutrophils.
Interleukin 18 (IL-18). 18.2 kDa. Interferon gamma-inducing factor (IGIF).	Cells of the monocytic lineage and dendritic cells.	IL-1 family member. Promotes differen-tiation of T$_H$1 subset of helper T cells. Induces IFN-γ production by T cells and enhances NK-cell cytotoxicity.

Cytokine. MW. Synonyms.	Sources	Activity
Interleukin 19 (IL-19). Homotetramer 35–40 kDa.	LPS-stimulated monocytes and B cells.	A member of the IL-10 family of cytokines, induces reactive oxygen species and proinflammatory cytokines, which may promote apoptosis. Shown to promote T_H2 differentiation by inhibiting IFN-γ and augmenting IL-4 and IL-13 production.
Interleukin 20 (IL-20). 18 kDa.	Monocytes and keratinocytes.	A member of the IL-10 family of cytokines; has effects in epidermal tissues and psoriasis. Like IL-19, shown to promote T_H2 differentiation.
Interleukin 21 (IL-21). 15 kDa.	Activated CD4$^+$ T cells.	Enhances cytotoxic activity and IFN-γ production by activated NK and CD8$^+$ T cells. Contributes to B cell and follicular helper T cell activation in germinal centers.
Interleukin 22 (IL-22). Homodimer. 25 kDa. IL-10-related T-cell-derived inducible factor (IL-TIF).	CD4$^+$ T cells (particularly those of the T_H17 subset).	A member of the IL-10 family with roles in skin homeostasis and pathogenesis. Has both pro- and anti-inflammatory effects.
Interleukin 23 (IL-23). Heterodimer of p40 subunit of IL-12 (35–40 kDa) and p19 (18.7 kDa).	Activated dendritic cells and macrophages.	Induces T_H17 differentiation.
Interleukin 24 (IL-24). 35-40 kDa. IL-10B; MDA7 (melanoma differentiation associated protein 7).	Melanocytes, NK cells, B cells, subsets of T cells, fibroblasts, and melanoma cells.	Member of the IL-10 family. Induces TNF-α and IFN-γ and low levels of IL-1β, IL-12, and GM-CSF in human PBMC. Induces selective anticancer properties in melanoma cells by inhibiting proliferation and in breast carcinoma cells by promoting apoptosis.
Interleukin 25 (IL-25). 20 kDa. IL-17E; stroma-derived growth factor (SF20).	T_H2 subset of helper T cells, mast cells, basophils, eosinophils, intraepithelial lymphocytes, lung epithelial cells and macrophages, cells of GI tract and uterus.	Member of the IL-17 family. Induces production of T_H2 cytokines and suppresses T_H17 cytokines and eotaxin. May contribute to airway disease, through cytokine production, tissue reorganization, mucus secretion, and airway hyper reactivity. Proinflammatory.
Interleukin 26 (IL-26). 36 kDa homodimer. AK155.	Subset of T and NK cells.	Member of the IL-10 family. May have similar functions to IL-20.
Interleukin 27 (IL-27). Heterodimer composed of EBI3 (IL-27 β) and p28 (IL-27 α).	Produced by dendritic cells, macrophages, endothelial cells, and plasma cells.	Shown to induce clonal expansion of naïve CD4$^+$ T cells, to synergize with IL-12 to promote IFN-γ production from CD4$^+$ T cells, and to induce CD8 T-cell–mediated antitumor activity.
Interleukin 28 A/B (IL-28A/B). 22.3/21.7 kDa. Interferon-λ 2/3 (IFN-L2/3).	Monocyte-derived dendritic cells.	Co-expressed with IFN-β, participates in the antiviral immune response, and shown to induce increased level of both MHC class I and II; induces regulatory T-cell proliferation.

(continued)

Cytokine. MW. Synonyms.	Sources	Activity
Interleukin 29 (IL-29). Interferon-λ1 (IFN-L1).	Monocyte-derived dendritic cells.	Functions similarly to IL-28A/B.
Interleukin 30 (IL-30). IL-27p28.	Antigen-presenting cells.	Subunit of IL-27 heterodimer; functions same as IL-27.
Interleukin 31 (IL-31).	Mainly activated T_H2 T cells; can be induced in activated monocytes.	May be involved in recruitment of polymorphonuclear cells, monocytes, and T cells to a site of skin inflammation.
Interleukin 32 (IL-32). NK4.	Activated NK cells and PBMCs.	Member of the IL-1 family. Proinflammatory cytokine, has mitogenic properties, induces TNF-α.
Interleukin 33 (IL-33). Nuclear factor in high endothelial venules (NF-HEV).	High endothelial venule and smooth muscle cells.	IL-1 family member. Induces T_H2 cytokine production by T cells, mast cells, eosinophils, and basophils.
Interleukin 34 (IL-34).	Many cell types.	Promotes growth and development of myeloid cells.
Interleukin 35 (IL-35).	Regulatory T cells.	IL-12 family member. Induces and activates regulatory T cells. Suppresses inflammatory responses.
Interleukin 36 α, β, γ. (IL-36α, β, γ).	Dendritic cells, monocytes, T cells, keratinocytes, and epithelial cells.	IL-1 family members. Induce dendritic cells to produce proinflammatory cytokines and to express MHC class II, CD80, and CD86. Induce T cells to produce IFN-γ, IL-4, and IL-17.
Interleukin 37 (IL-37).	Monocytes, macrophages, dendritic cells, epithelial cells.	IL-1 family member. Inhibits innate immunity and inflammatory responses.
APRIL (a proliferation-inducing cytokine).	T cells, monocytes, macrophages, and dendritic cells.	Promotes B- and T-cell proliferation. Induces class switch recombination to IgA.
BAFF (human B-cell–activating factor). 18 kDa. TALL-1 (TNF and apoptosis ligand-related leukocyte-expressed ligand 1); BLyS (B-lymphocyte stimulator).	T cells, cells of the monocytic lineage, and dendritic cells.	Member of the TNF family, occurs in membrane-bound and soluble form. Supports proliferation of antigen-receptor–stimulated B cells. Differentiation and survival factor for immature B cells.
Cardiotrophin-1 (CT-1). 21.5 kDa.	Many cell types, including heart and skeletal muscle.	A member of the IL-6 family shown to stimulate hepatic expression of the acute phase proteins; induces cardiac myocyte hypertrophy; increases monocyte adhesion; involved in the metabolic syndrome.
Ciliary neurotrophic factor (CNTF). 24 kDa. Membrane-associated neurotransmitter stimulating factor (MANS).	Schwann cells and astrocytes.	A member of the IL-6 family that induces the expression of acute phase proteins in the liver and has been shown to function as an endogenous pyrogen. Also shown to function in ontogenesis and promote survival and regeneration of nerves.
Granulocyte colony-stimulating factor (G-CSF). 22 kDa.	Bone marrow stromal cells and macrophages.	Essential for growth and differentiation of neutrophils.
Granulocyte-macrophage colony-stimulating factor (GM-CSF). 22 kDa.	T cells, macrophages, fibroblasts, and endothelial cells.	Growth factor for hematopoietic progenitor cells and differentiation factor for granulocytic and monocytic cell lineages.

Cytokine. MW. Synonyms.	Sources	Activity
Macrophage colony-stimulating factor (M-CSF). Disulfide-linked homodimer of 45–90 kDa. Colony-stimulating factor 1 (CSF-1).	Many cell types, including lymphocytes, monocytes, fibroblasts, epithelial cells, and others.	Growth, differentiation, and survival factor for macrophage progenitors, macrophages, and granulocytes.
Interferon alpha (IFN-α). 16–27 kDa. Type 1 interferon; leukocyte interferon; lympho-blast interferon.	Cells activated by viral and other microbial components: macrophages, dendritic cells, and lymphocytes, virus-infected cells.	Induces resistance to virus infection. Inhibits cell proliferation. Increases expression of class I MHC molecules on nucleated cells.
Interferon beta (IFN-β). 22 kDa. Type 1 interferon; fibroblast interferon.	Cells activated by viral and other micro-bial components: fibroblasts, dendritic cells, and some epithelial cells, virus-infected cells.	Induces resistance to virus infection. Inhibits cell proliferation. Increases expression of class I MHC molecules.
Interferon gamma (IFN-γ). Monomer 17.1 kDa; dimer 40 kDa. Type 2 interferon; immune interferon; macrophage-activating factor (MAF); T-cell interferon.	T_H1 cells and some $CD8^+$ T cells and NK cells.	Supports T_H1 differentiation and is the key T_H1 cytokine. Induces class switching to IgG subclasses. Activates macro-phages and induces MHC class II expression. Weak antiviral and anti-proliferative activities.
Interferon lambda (IFN-λ). Same as IL-28 and IL-29.		
Leukemia inhibitory factor (LIF). 45 kDa. Differentiation-inhibiting activity (DIA); differentiation-retarding factor (DRF).	Many cell types, including T cells, cells of the monocytic lineage, fibroblasts, liver, and heart.	A member of the IL-6 family. Major experi-mental application: keeps cultures of ES cells in undifferentiated state to maintain their proliferation. In vivo, in combination with other cytokines, promotes hemato poiesis, stimulates acute phase response of liver cells, affects bone resorption, enhances glucose transport and insulin resistance, alters airway contractility, and causes loss of body fat.
Lymphotoxin alpha (LT-α). 25 kDa. Tumor necrosis factor beta (TNF-β); cytotoxin (CTX); differentiation-inducing factor (DIF); TNF ligand superfamily member 1 (TNFSF1).	Activated T cells, B cells, NK cells, macrophages, virus-infected hepatocytes.	Cytotoxic for some tumor and other cells. Required for development of lymph nodes and Peyer's patches and for formation of splenic B and T cell zones and germinal centers. Induces inflammation. Activates vascular endothe-lial cells and induces lymphangiogenesis. Required for NK cell differentiation.
Macrophage migration inhibitory factor (MIF). 12 kDa monomer, forms biologically active multimers.	Small amounts by many cell types; major producers are activated T cells, hepatocytes, monocytes, macrophages, and epithelial cells.	Activates macrophages and inhibits their migration.
Oncostatin M (OSM). 28–32 kDa. Onco M; ONC.	Activated T cells, monocytes, and adherent macrophages.	Many functions, including inhibition of growth of tumor cell lines; regulation of the growth and differentiation of cells during hematopoiesis, neurogenesis, and osteogenesis. Shown to enhance LDL uptake and also stimulates synthesis of acute phase proteins in the liver.

(continued)

Cytokine. MW. Synonyms.	Sources	Activity
Stem-cell factor (SCF). 36 kDa. Kit ligand (kitL) or steel factor (SLF).	Bone marrow stromal cells, cells of other organs such as brain, kidney, lung, and placenta.	Roles in development of hematopoietic, gonadal, and pigmental lineages; active in both membrane-bound and secreted forms.
Thrombopoietin (THPO). 60-70 kDa. Mega-karyocyte colony-stimulating factor; throm-bopoiesis-stimulating factor (TSF).	Liver, kidney, and skeletal muscle.	Megakaryocyte lineage-specific growth and differentiation factor that regulates platelet production.
Thymic stromal lymphopoietin (TSLP): 140 Da.	Epithelial cells, keratinocytes, basophils.	Acts on dendritic cells and CD4$^+$ T cells to induce T_H2 commitment and proliferation; supports B-cell proliferation and differentiation.
Transforming growth factor beta (TGF-β). ~25 kDa. Differentiation-inhibiting factor.	Some T cells (especially T_{REG}s), macrophages, platelets, and many other cell types.	Inhibits growth, differentiation, and function of a number of cell types, including T and B cells and monocytes/macrophages. Inhibits inflammation and enhances wound healing. Induces class switching to IgA.
Tumor necrosis factor alpha (TNF-α). 52 kDa. Cachectin, TNF ligand superfamily member 2 (TNFSF2).	Monocytes, macrophages, and other cell types, including activated T cells, NK cells, neutrophils, and fibroblasts.	Strong mediator of inflammatory and immune functions. Regulates growth and differentiation of a wide variety of cell types. Cytotoxic for many types of transformed and some normal cells. Promotes angiogenesis, bone resorption, and thrombotic processes. Suppresses lipogenic metabolism.
Tumor necrosis factor beta (TNF-β). Same as Lymphotoxin-α.		

Appendix III: Chemokines and Chemokine Receptors

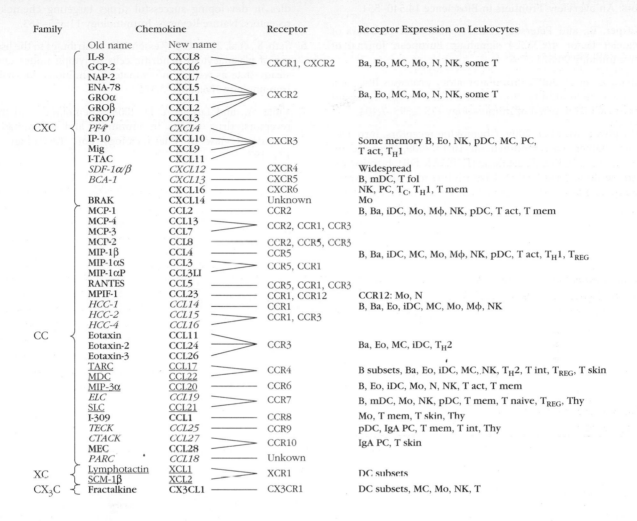

FIGURE 1 The chemokine system: an overview. Chemokines (family, old and new nomenclature), their receptors, and predominant receptor repertoires in different leukocyte populations are listed. Names in bold identify inflammatory chemokines, names in italics identify homeostatic chemokines, and underlined names refer to molecules belonging to both realms. Chemokine acronyms are as follows: BCA, B-cell activating chemokine; BRAK, breast and kidney chemokine; CTACK, cutaneous T-cell attracting chemokine; ELC, Epstein-Barr virus-induced receptor ligand chemokine; ENA-78, epithelial cell-derived neutrophils-activating factor (78 amino acids); GCP, granulocyte chemoattractant protein; GRO, growth-related oncogene; HCC, hemofiltrate CC chemokine; IP, IFN-inducible protein; I-TAC, IFN-inducible T-cell a chemoattractant; MCP, monocyte chemoattractant protein; MDC, macrophage derived chemokine; Mig, monokine induced by gamma interferon; MIP, macrophage inflammatory protein; MPIF, myeloid progenitor inhibitory factor; NAP, neutrophil-activating protein; PARC, pulmonary and activation-regulated chemokine; RANTES, regulated upon activation normal T cell-expressed and secreted; SCM, single C motif; SDF, stromal cell-derived factor; SLC, secondary lymphoid tissue chemokine; TARC, thymus and activation-related chemokine; TECK, thymus-expressed chemokine. Leukocyte acronyms are as follows: Ba, basophils; Eo, eosinophils; iDCs, immature dendritic cells; mDCs, mature DCs; MC, mast cells; Mo, monocytes; Mɸ, macrophages; N, neutrophils; PC, plasma cells; T naïve, naïve T cells; T act, activated T cells; T fol, T cells in follicles; T skin, skin-homing T cells; T mem, memory T cells; T int, intestine homing T cells; T_{REG}, regulatory T cells; Thy, thymocytes.

* Not a chemoattractant; signaling affects proliferation and various other functions.

REFERENCES

1. Bonecchi, R., et al. 2009. Chemokines and chemokine receptors: An overview. Frontiers in Bioscience 14:540–551.

2. Kasper, B., and Petersen, F. 2011. Molecular pathways of platelet factor 4/CXCL4 signaling. European Journal of Immunology 90:521–526.

3. Miao, Z., et al. 2007. Proinflammatory proteases liberate a discrete high-affinity functional FPRL1 (CCR12) ligand from CCL23. Journal of Immunology 178:7,395–7,404.

4. Murphy, P. M. et al. 2009. Chemokine receptors, introductory chapter. Last modified on 10/13/2009. Accessed on 09/13/2012. IUPHAR database (IUPHAR-DB), http://www.iuphar-db.org/DATABASE/FamilyIntroductionForward?familyId=14.

5. Schall, T. J., and Proudfoot, A. E. I. 2011. Overcoming hurdles in developing successful drugs targeting chemokine receptors. Nature Reviews. Immunology 11:355–363.

6. Seth, S., et al. 2011. CCR7 essentially contributes to the homing of plasmacytoid dendritic cells to lymph nodes under steady-state as well as inflammatory conditions. Journal of Immunology 186:3,364–3,372.

7. Viola, A., and Luster, A. D. 2008. Chemokines and their receptors: drug targets in immunity and inflammation. Annual Review of Pharmacology and Toxicology 48:171–197.

ABO blood-group antigen Antigenic determinants of the blood-group system defined by the agglutination of red blood cells exposed to anti-A and anti-B antibodies.

Abzyme A monoclonal antibody that has catalytic activity.

Acquired immunity See **adaptive immunity**.

Acquired immunodeficiency syndrome (AIDS) A disease caused by **human immunodeficiency virus (HIV)** that is marked by significant depletion of CD4$^+$ T cells and that results in increased susceptibility to a variety of opportunistic infections and cancers.

Activation-induced cytidine deaminase (AID) An enzyme that removes an amino group from deoxycytidine, forming deoxyuridine. This is the first step in the processes of both **somatic hypermutation** and **class switch recombination**.

Active immunity Adaptive immunity that is induced by natural exposure to a pathogen or by **vaccination**.

Acute lymphocytic leukemia (ALL) A form of cancer in which there is uncontrolled proliferation of a cell of the lymphoid lineage. The proliferating cells usually are present in the blood.

Acute myelogenous leukemia (AML) A form of cancer in which there is uncontrolled proliferation of a cell of the myeloid lineage. The proliferating cells usually are present in the blood.

Acute phase protein One of a group of serum proteins that increase in concentration in response to inflammation. Some **complement** components and **interferons** are acute phase proteins.

Acute phase response (APR) The production of certain proteins that appear in the blood shortly after many infections, often induced by proinflammatory cytokines generated at the site of infection. It is part of the host's early innate response to infection.

Acute phase response proteins Proteins synthesized in the liver in response to inflammation; serum concentrations of these proteins increase in inflammation.

Adapter Proteins Proteins that connect to other effector proteins in a signaling pathway and create a signaling scaffold.

Adaptive immunity Host defenses that are mediated by B cells and T cells following exposure to antigen and that exhibit specificity, diversity, memory, and self-nonself discrimination. See also **innate immunity**.

Adenosine deaminase (ADA) deficiency An immune deficiency disorder that is characterized by defects in adaptive immunity and is caused by the intracellular accumulation of toxic adenosine metabolites, especially in hematopoietic cells, which interferes with purine metabolism and DNA synthesis.

Adjuvants Factors that are added to a vaccine mixture to enhance the immune response to antigen by activating innate immune cells. Dead mycobacterium were among the original adjuvants, but more refined preparation include alum, cytokines, and/or lipids.

Adoptive transfer The transfer of the ability to make or participate in an immune response by the transplantation of cells of the immune system.

Affinity constant The ratio of the forward (k_1) to the reverse (k_{-1}) rate constant in an antibody-antigen reaction. Equivalent to the association constant in biochemical terms ($K_a = k_1/k_{-1}$).

Affinity Hypothesis A proposal stating that the fate of a developing T cell depends on the affinity of the interaction between its T cell receptor (TCR) and MHC-peptide ligand(s) it encounters in the thymus. High affinity interactions result in death by negative selection, lower affinity interactions in positive selection and maturation, and very low or no affinity interactions result in death by neglect (see Figure 9-7).

Affinity maturation The increase in average antibody affinity for an antigen that occurs during the course of an immune response or in subsequent exposures to the same antigen.

Affinity The strength with which a monovalent ligand interacts with a binding site. It is represented quantitatively by the affinity constant K_a.

Agent-induced immunodeficiency A state of immune deficiency induced by exposure to an environmental agent/s.

Agglutination inhibition The reduction of antibody-mediated clumping of particles by the addition of the soluble forms of the epitope recognized by the agglutinating antibody.

Agglutination The aggregation or clumping of particles (e.g., latex beads) or cells (e.g., red blood cells).

Agglutinin A substance capable of mediating the clumping of cells or particles; in particular, a hemagglutinin (HA) causes clumping of red blood cells.

Agglutinin titer The reciprocal of the greatest serum dilution that elicits a positive agglutination reaction.

AIRE A protein that regulates expression of tissue specific antigens in the thymus. It is expressed by a subset of medullary epithelial cells and regulates transcription.

Alleles Two or more alternative forms of a gene at a particular **locus** that confer alternative characters. The presence of multiple alleles results in **polymorphism**.

Allelic exclusion A process that permits expression of only one of the allelic forms of a gene. For example, a B cell expresses only one allele for an antibody heavy chain and one allele for a light chain (see Figure 7-11).

Allergy A **hypersensitivity** reaction that can include hay fever, asthma, **serum sickness**, systemic **anaphylaxis**, or contact dermatitis.

Allogeneic Denoting members of the same species that differ genetically.

Allograft A tissue transplant between **allogeneic** individuals.

Allotypes A set of **allotypic determinants** characteristic of some but not all members of a species.

Allotypic determinant An antigenic determinant that varies among members of a species or between different inbred strains of animals. The constant regions of antibodies possess allotypic determinants.

Allotypic marker A genetic marker that defines the presence of an allele on one strain of mouse that is not shared by other strains. Normally refers to allelic variants of immunoglobulin heavy chains.

Alpha-feto protein (AFP) See **oncofetal tumor antigen**.

Altered peptide model A proposal stating that developing T cells encounter different sets of peptides in the cortical region versus the medullary region of the thymus. Advanced to help explain differences in the subsets of cells that undergo positive versus negative selection.

Alternative pathway of complement activation A pathway of complement activation that is initiated by spontaneous hydrolysis of the C3 component of complement, resulting in the formation of a fluid-phase C3 convertase enzyme. This spontaneous initiation distinguishes the alternative pathway from the classical and lectin-mediated pathways that are both initiated by specific antigen binding by either antibodies or lectins respectively. However, one recently-discovered branch of the alternative pathway may begin with Properdin binding to the surface of bacteria from the *Neisseria* genus (see Figure 6-2).

Alternative tickover pathway   The alternative pathway of complement activation that is initiated by spontaneous hydrolysis of C3 molecule in the serum.

Alveolar macrophage   A macrophage found in the alveoli of the lung.

Anaphylactic shock   An acute, life threatening (Type I) whole-body allergic response to an antigen (e.g. drugs, insect venom). See also **Anaphylaxis**.

Anaphylatoxins   The **complement** split products C3a and C5a, which mediate **degranulation** of mast cells and basophils, resulting in release of mediators that induce contraction of smooth muscle and increased vascular permeability.

Anaphylaxis   An immediate type I hypersensitivity reaction, which is triggered by IgE-mediated mast cell **degranulation**. Systemic anaphylaxis leads to shock and is often fatal. Localized anaphylaxis involves various types of **atopic** reactions.

Anchor residues   The amino acid residues at key locations in a peptide sequence that fit into pockets of an make close molecular associations with complementary amino acids in the groove of an MHC molecule and which help to determine the peptide-binding specificity of particular MHC molecules.

Anergic, anergy   Unresponsive to antigenic stimulus.

Antagonist, antagonize   A molecule that inhibits the effect of another molecule.

Anti-allotype antibodies   Antibodies directed towards **allotypic determinants**.

Anti-Fab antibodies   Antibodies directed towards the **Fab regions** of other antibodies.

Anti-Fc antibodies   Antibodies specific for the **Fc regions** of other antibodies.

Anti-idiotypic antibodies   Antibodies directed towards antigenic determinants located in the antigen binding site of other antibodies.

Anti-isotype antibodies   Antibodies directed towards antigenic determinants located in the constant regions of antibodies, that are shared among all members of a species.

Anti-oncogenes   Another name for tumor suppressor genes.

Antibodies   Immunoglobulin proteins consisting of two identical heavy chains and two identical light chains, that recognize a particular **epitope** on an antigen and facilitates clearance of that antigen. Membrane-bound antibody is expressed by B cells that have not encountered antigen; secreted antibody is produced by **plasma cells**. Some antibodies are multiples of the basic four-chain structure.

Antibody molecule   See **antibodies**.

Antibody-dependent cell-mediated cytotoxicity (ADCC)   A cell-mediated reaction in which nonspecific cytotoxic cells that express **Fc receptors** (e.g., NK cells, neutrophils, macrophages) recognize bound antibody on a target cell and subsequently cause lysis of the target cell.

Antigen   Any substance (usually foreign) that binds specifically to an antibody or a T-cell receptor; often is used as a synonym for **immunogen**.

Antigen presentation   See **antigen processing**.

Antigen processing   Degradation of antigens by one of two pathways yielding antigenic peptides that are displayed bound to MHC molecules on the surface of antigen-presenting cells or altered self cells.

Antigen-presenting cell (APC)   Any cell that can process and present antigenic peptides in association with **class II MHC molecules** and deliver a **costimulatory signal** necessary for T-cell activation. Macrophages, dendritic cells, and B cells constitute the professional APCs. Nonprofessional APCs, which function in antigen presentation only for short periods include thymic epithelial cells and vascular endothelial cells.

Antigenic determinant   The site on an antigen that is recognized and bound by a particular antibody, TCR/MHC-peptide complex, or TCR-ligand-CD1 complex; also called **epitope**.

Antigenic drift   A series of spontaneous point mutations that generate minor antigenic variations in pathogens and lead to strain differences. See also **Antigenic shift**.

Antigenic peptide   In general, a peptide capable of raising an immune response, for example, in a peptide that forms a complex with MHC that can be recognized by a T-cell receptor.

Antigenic shift   Sudden emergence of a new pathogen subtype, frequently arising due to genetic reassortment that has led to substantial antigenic differences. See also **Antigenic drift**.

Antigenic specificity   See **specificity, antigenic**.

Antigenically committed   The state of a mature B cell displaying surface antibody specific for a single immunogen.

Antigenicity   The capacity to combine specifically with antibodies or T-cell receptor/MHC.

Antimicrobial peptides   Peptides/small proteins, such as defensins, less than 100-amino acids long that are produced constitutively or after activation by pathogens.

Antiserum   Serum from animals immunized with antigen that contains antibodies to that antigen.

Apoptosis   A process, often referred to as **programmed cell death**, where cells initiate a signaling pathway that results in their own demise. Apoptosis requires ATP and is typically dependent on the activation of internal caspases. In contrast to **necrosis**, it does not result in damage to surrounding cells.

Apoptosome   A wheel-like assemblage of molecules that regulate cell death initiated via the mitochondrial (intrinsic) pathway. Includes cytochrome-c, ATP, Apaf-1, and caspase-9.

APRIL   A member of the Tumor Necrosis Factor family of cytokines, important in B cell development and homeostasis.

Artemis   An enzyme that is a member of the Non-homologous End Joining (NHEJ) DNA repair pathway. During V(D)J recombination, Artemis opens the hairpin loops formed after RAG1/2-mediate cleavage of the immunoglobulin genes.

Association constant (Ka)   See **affinity constant**.

Atopic   Pertaining to clinical manifestations of type I (IgE-mediated) hypersensitivity, including allergic rhinitis (hay fever), eczema, asthma, and various food allergies.

Attenuate   To decrease the virulence of a pathogen and render it incapable of causing disease. Many vaccines are composed of attenuated bacteria or viruses that induce protective immunity without causing harmful infection.

Autocrine   A type of cell signaling in which the cell acted on by a cytokine is the source of the cytokine.

Autograft   Tissue grafted from one part of the body to another in the same individual.

Autoimmune diseases   A group of disorders caused by the action of ones own antibodies or T cells reactive against self proteins.

Autoimmune polyendrocrinopathy and ectodermal dystrophy (APECD)   An immune deficiency disorder in which depressed expression of *Aire* results in reduced levels of tissue-specific antigens in thymic epithelial cells, allowing the escape of autoreactive T cells into the periphery, where they precipitate organ-specific autoimmunity.

Autoimmunity   An abnormal immune response against self antigens.

Autologous   Denoting transplanted cells, tissues, or organs derived from the same individual.

Avidity   The strength of antigen-antibody binding when multiple epitopes on an antigen interact with multiple binding sites of an antibody. See also **affinity**.

B cell   See **B lymphocytes**.

B lymphocytes (B cells)   Lymphocytes that mature in the bone marrow and express membrane-bound antibodies. After interacting with antigen, they differentiate into antibody-secreting plasma cells and memory cells.

B-1 B cells   A subclass of B cells that predominates in the peritoneal and pleural cavity. B-1 B cells in general secrete low affinity IgM antibodies and do not undergo class switch recombination or somatic hypermutation. They thus occupy a niche between the innate and adaptive immune responses. Most, but not all, B-1 B cells express CD5 on their surface.

B-1b B cells   A subclass of B-1 cells that does not express the antigen CD5 on its cell surface, like most B-1 B cells.

B-2 B cells   The predominant class of B cells that are stimulated by antigens with T cell help the generate antibodies of multiple heavy chain classes whose genes undergo somatic hypermutation.

B-cell coreceptor   A complex of three proteins (CR2 (CD21), **CD19**, and TAPA-1) associated with the B-cell receptor. It is thought to amplify the activating signal induced by cross-linkage of the receptor.

B-cell receptor (BCR)   Complex comprising a membrane-bound immunoglobulin molecule and two associated signal-transducing Igα/Igβ molecules.

B-cell-specific activator protein (BSAP)   A transcription factor encoded by the gene Pax-5 that plays an essential role in early and later stages of B-cell development.

B-lymphocyte-induced maturation protein 1 (BLIMP-1)   Transcription factor vital to differentiation of B cells into plasma cells.

Bacillus Calmette-Guérin (BCG)   An attenuated form of *Mycobacterium bovis* used as a vaccine against another member of the genus, *M. tuberculosis*, the cause of tuberculosis. BCG can also be found as an adjuvant component in other vaccines.

Bacteremia   An infection in which viable bacteria are found in the blood.

BAFF   B-cell survival factor; a membrane-bound homolog of tumor necrosis factor, to which mature B cells bind though the **TACI** receptor. This interaction activates important transcription factors that promote B-cell survival, maturation, and antibody secretion.

BAFF receptor (BAFF-R)   Receptor for **BAFF**, a cytokine belonging to the tumor necrosis factor family that is important in B cell development and homeostasis

Bare-lymphocyte syndrome (BLS)   An immunodeficiency syndrome in which, without class II MHC molecules, positive selection of CD4$^+$ T cells in the thymus is impaired and, with it, peripheral T helper cell responses.

Basophil   A nonphagocytic granulocyte that expresses **Fc receptors** for IgE (see Figure 2-2b). Antigen-mediated cross-linkage of bound IgE induces **degranulation** of basophils.

BCG   See **Bacillus Calmette-Guérin**.

Bence-Jones proteins   Monoclonal light chains secreted by plasmacytoma tumors. Found in high concentrations in the urine of patients with multiple myeloma.

Benign   Pertaining to a nonmalignant form of a neoplasm or a mild form of an illness.

β-selection   The process during the **DN3** stage of T cell development where the functionality of thymocytes' rearranged TCRβ chains is tested. Only those thymocytes that have successfully rearranged a TCRβ chain and expressed it as a protein that can interact with pre-TCRα will deliver signals that ensure its survival, maturation to the CD4$^+$CD8$^+$ (DP) stage, and induce its proliferation.

β$_2$-microglobulin   Invariant subunit that associates with the polymorphic α chain to form **class I MHC molecules**; it is not encoded by MHC genes.

Bispecific antibody   Hybrid antibody made either by chemically cross-linking two different antibodies or by fusing hybridomas that produce different monoclonal antibodies.

Bone marrow   The living tissue found within the hard exterior of bone.

Booster   Inoculation given to stimulate and strengthen an immunologic memory response.

Bradykinin   An endogenously produced peptide that produces an **inflammatory response**.

Bronchus-associated lymphoid tissue (BALT)   Secondary lymphoid microenvironments in the lung mucosa system that support the development of the T and B lymphocyte response to antigens that enter the lower respiratory tract. Part of the mucosa associated lymphoid tissue system (MALT).

C (constant) gene segment   The 3′ coding of a rearranged immunoglobulin or T-cell receptor gene. There are multiple C gene segments in germ-line DNA, but as a result of gene rearrangement and, in some cases, RNA processing, only one segment is expressed in a given protein.

c-Kit (CD117)   Receptor for **stem cell factor (SCF)**.

C-reactive protein (CRP)   An acute phase protein that binds to phosphocholine in bacterial membranes and functions in opsonization; an increased level of serum CRP is an indicator of inflammation.

C-type lectin receptor (CLR)   A family of pattern-recognition receptors that contains C-type lectin carbohydrate-binding domains.

C3 convertase   Enzyme that breaks down the C3 component of complement into C3a and C3b.

C5 convertase   Enzyme that breaks down the C5 component of complement into C5a and C5b.

Calnexin   A protein resident of the ER that serves, along with **calreticulin**, as a molecular chaperone to assist in class I MHC molecule assembly.

Calreticulin   A protein resident of the ER that serves, along with **calnexin**, as a molecular chaperone to assist in class I MHC molecule assembly.

Cancer stem cells   A subset of cells within a tumor that has the stem-cell-like ability to give rise to all cells within that tumor and the ability to self-renew indefinitely. They are thought to be responsible for tumor growth.

Carcinoembryonic antigen (CEA)   An oncofetal antigen (found not only on cancerous cells but also on normal cells) that can be a tumor-associated antigen.

Carcinogen   Any chemical substances, physical agents or types of radiation that can induce DNA mutations and lead to the development of cancer.

Carcinoma   Tumor arising from endodermal or ectodermal tissues (e.g., skin or epithelium). Most cancers (>80%) are carcinomas.

Carrier   An immunogenic molecule containing antigenic determinants recognized by T cells. Conjugation of a carrier to a nonimmunogenic **hapten** renders the hapten immunogenic.

Carrier effect   A **secondary immune response** to a hapten depends on use of both the **hapten** and the **carrier** used in the initial immunization.

Cascade induction   The property of cytokines that pertains to their ability to induce one cell to release cytokines that then act upon another to induce the release of other cytokines and growth factors.

Caspase   A family of cysteine proteases that cleave after an aspartate residue. The term *caspase* incorporates these elements (*c*ysteine, *asp*artate, prote*ase*), which play important roles in the chain of reactions that leads to **apoptosis**.

Caspase recruitment domains (CARD)   Protein domain that binds caspase proteases.

CC subgroup   A subgroup of chemokines in which a disulfide bond links adjacent cysteines.

CD19   A quintessential B-cell marker, often used as such in flow cytometry experiments.

CD21   The B cell co-receptor molecule that also serves as a co-receptor for the complement components C3d and C3dg. Also known as CR2.

CD25   The high affinity IL-2 receptor chain (IL-2α) expressed on the surface of multiple immune cells, including some developing T cells, activated T cells, and many FoxP3$^+$ T cells.

CD3   A polypeptide complex containing three dimers: a γε heterodimer, a εδ heterodimer, and either a ξξ homodimer or a ξη heterodimer (see Figure 3-30). It is associated with the T-cell receptor and functions in signal transduction.

CD4   A glycoprotein that serves as a co-receptor on class II MHC–restricted T cells. Most helper T cells are CD4$^+$.

CD40   Member of the tumor necrosis factor receptor family. Signaling through **CD40L** on T cells to CD40 on B cells is necessary for **germinal center** formation, **somatic hypermutation** and **class switch recombination**.

CD40L   Ligand for CD40. CD40L is a member of the Tumor Necrosis Factor family of molecules and CD40 a member of the TNF receptor family. CD40:CD40L interactions are indispensable during T cell mediated B cell differentiation. B cells bear CD40 and T cell, CD40L.

CD44   Surface protein involved in cell-cell adhesion that is expressed by multiple immune cells, including some developing T cells, and some activated T cells. Differences in CD44 and CD25 expression distinguish very early stages of T cell development. CD44 is also associated with immune cell activation.

CD5 antigen   An antigen found on most B-1 B cells, (B-1a B cells), as well as on many T cells.

CD8   A dimeric protein that serves as a co-receptor on class I MHC–restricted T cells. Most cytotoxic T cells are CD8$^+$.

CDR3   The third complementarity-determining region, (or hypervariable region) of the immunoglobulin or TCR molecules (see Figure 3-18).

Cell adhesion molecules (CAMs)   A group of cell surface molecules that mediate intercellular adhesion. Most belong to one of four protein families: the **integrins**, **selectins**, mucin-like proteins, and **immunoglobulin superfamily** (see Box 14-2).

Cell line   A population of cultured tumor cells or normal cells that have been subjected to chemical or viral **transformation**. Cell lines can be propagated indefinitely in culture.

Cell-mediated immune response   Host defenses that are mediated by antigen-specific T cells. It protects against intracellular bacteria, viruses, and cancer and is responsible for graft rejection. Transfer of primed T cells confers this type of immunity on the recipient. See also **humoral immune response**.

Cell-mediated immunity   See **cell-mediated immune response**.

Cell-mediated lympholysis (CML)   In vitro lysis of allogeneic cells or virus-infected syngeneic cells by T cells (see Figure 13-18); can be used as an assay for CTL activity or class I MHC activity.

Cellular oncogene   See **proto-oncogene**.

Central memory T cells (T$_{CM}$)   A memory T cell subset that localizes to and resides in secondary lymphoid tissue. It participates in the secondary response to antigen and can give rise to new effector T cells. T$_{CM}$ may arise from effector T cells and/or from T cells that have been stimulated towards the end of an immune response.

Central tolerance   Elimination of self-reactive lymphocytes in primary generative organs such as the bone marrow and the thymus (see also **peripheral tolerance**).

Chediak-Higashi syndrome   An autosomal recessive immune deficiency disorder caused by a defect in lysosomal granules that impairs killing by NK cells.

Chemical barriers   Tissue layer that provides innate immune protection against infection by chemical means, such as low pH and presence of degradative enzymes.

Chemoattractant   A substance that attracts cells. Some chemoattractants also cause significant changes in the physiology of cells that bear receptors for them.

Chemokine receptors   Surface proteins expressed by immune cells that guide their migration among tissues and localization within tissues. They generate signals that regulate motility and adhesion when bound to chemokines secreted by a variety of immune and stromal cells.

Chemokines   Any of several secreted low-molecular-weight cytokines that mediate **chemotaxis** in particular leukocytes via receptor engagement and that can regulate the expression and/or adhesiveness of leukocyte **integrins** (see Appendix III).

Chemotactic factor   An agent that can cause leukocytes to move up its concentration gradient.

Chemotaxis   The induction of cell movement by the secretion of factors that either attract or repel the cell through the mediation of receptors for those factors.

Chimera   An animal or tissue composed of elements derived from genetically distinct individuals. The **SCID-human mouse** is a chimera. Also, a chimeric antibody that contains the amino acid sequence of one species in one region and the sequence of a different species in another (for example, an antibody with a human constant region and a mouse variable region).

Chimeric antibody   See **chimera**.

Chromogenic substrate   A colorless substance that is transformed into colored products by an enzymatic reaction.

Chronic granulomatous disease   Immunodeficiency caused by a defect in the enzyme NADPH (phagosome) oxidase resulting in failure to generate reactive oxygen species in neutrophils.

Chronic lymphocytic leukemia (CLL)   A type of leukemia in which cancerous lymphocytes are continually produced.

Chronic myelogenous leukemia (CML)   A type of leukemia in which cancerous lymphocytes of the myeloid lineage are continually produced.

Cilia   Hairlike projections on cells, including epithelial cells in the respiratory and gastrointestinal tracts; cilia function to propel mucus with trapped microbes out of the tract.

Class (isotype) switching   The change in the antibody class that a B cell produces.

Class I MHC genes   The set of genes that encode class I MHC molecules, which are glycoproteins found on nearly all nucleated cells.

Class I MHC molecules   Heterodimeric membrane proteins that consist of an α chain encoded in the MHC, associated noncovalently with β$_2$-microglobulin (see Figures 8-1 and 8-2). They are expressed by nearly all nucleated cells and function to present antigen to CD8$^+$ T cells. The classical class I molecules are H-2 K, D, and L in mice and HLA-A, -B, and -C in humans.

Class II MHC genes   The set of genes that encode class II MHC molecules, which are glycoproteins expressed by only professional antigen presenting cells.

Class II MHC molecules   Heterodimeric membrane proteins that consist of a noncovalently associated α and β chain, both encoded in the MHC (see Figures 8-1 and 8-2). They are expressed by **antigen-presenting cells** and function to present antigen to CD4$^+$ T cells. The classical class II molecules are H-2 IA and IE in mice and HLA-DP, -DQ, and -DR in humans.

Class III MHC genes   The set of genes that encode several different proteins, some with immune function, including components of the complement system and several inflammatory molecules.

Class III MHC molecules   Various proteins encoded in the **MHC** but distinct from class I and class II MHC molecules. Among others, they include some complement components and TNF-α and Lymphotoxin-α.

Class switch recombination (CSR)   The generation of antibody genes for heavy chain isotypes other than μ or δ by DNA recombination.

Class   The property of an antibody that is defined by the nature of its heavy chain (μ, δ, γ, α, or ε).

Classical pathway of complement activation   That pathway of **complement** activation that is initiated by antibody binding to antigen (see Figure 6-2).

CLIP   A protein that binds to the groove of MHC class II as it is assembled and carried to the cell surface. It prevents other peptides from associating with MHC class II until it encounters endocytosed proteins, when CLIP is digested and removed from the groove.

Clonal anergy   A physiological state in which cells are unable to be activated by antigen.

Clonal deletion   The induced death of members of a clone of lymphocytes with inappropriate receptors (e.g., those that strongly react with self during development).

Clonal selection hypothesis   This hypothesis states that antigen interacting with a receptor on a lymphocyte induces division and differentiation of that lymphocyte to form a clone of identical daughter cells. All daughter cells will bear the same receptor as the stimulated cell, and antibodies produced by B cells stimulated in this way will share the antigen-binding site with the membrane receptor of the stimulated cell. Following antigen elimination, representatives of the stimulated clone remain in the host as a source of immunological memory. Those clones of B cells that meet antigen at an immature stage of development will be eliminated from the repertoire.

Clonal selection   The antigen-mediated activation and proliferation of members of a clone of B cells that have receptors for the antigen (or for complexes of MHC and peptides derived from the antigen, in the case of T cells).

Clone   Cells arising from a single progenitor cell.

Clot   Coagulated mass; usually refers to coagulated blood, in which conversion of fibrinogen in the plasma to fibrin has produced a jelly-like substance containing entrapped blood cells.

Cluster of differentiation (CD)   A collection of monoclonal antibodies that all recognize an antigen found on a particular differentiated cell type or types. Each of the antigens recognized by such a collection of antibodies is called a CD marker and is assigned a unique identifying number.

Coding joints   The nucleotide sequences at the point of union of coding sequences during V(D)J rearrangement to form rearranged antibody or T-cell receptor genes.

Codominant   The expression of both the maternal and the paternal copy of a gene in a heterozygote.

Collectins   Family of calcium-dependent carbohydrate-binding proteins containing collagen-like domains.

Combined immunodeficiencies (CID)   Any of a number of immune deficiency disorders resulting from an absence of T cells or significantly impaired T-cell function, combined with some disruption of antibody responses.

Common lymphoid progenitor (CLP)   An immature blood cell that develops from the hematopoietic stem cell and gives rise to lymphocytes, including B and T cells and NK cells.

Common myeloid-erythroid progenitor (CMP)   An immature blood cell that develops from the hematopoietic stem cell and gives rise to all red blood cells and myeloid cells, including monocytes, macrophages, and granulocytes.

Complement   A group of serum and cell membrane proteins that interact with one another and with other molecules of innate and adaptive immunity to carry out key effector functions leading to pathogen recognition and elimination.

Complement system   See **complement**.

Complementarity-determining region (CDR)   Portions of the variable regions of antibody molecules that contain the antigen-binding residues.

Confocal microscopy   A type of fluorescence microscopy that, like two-photon microscopy, allows one to image fluorescent signals within one focal plane within a relatively thick tissue sample.

Conformational determinants   Epitopes of a protein that are composed of amino acids that are close together in the three-dimensional structure of the protein but may not be near each other in the amino acid sequence.

Congenic   Denoting individuals that differ genetically at a single genetic locus or region; also called coisogenic.

Constant (C) region   The nearly invariant portion of the immunoglobulin molecule that does not contain antigen-binding domains. The sequence of amino acids in the constant region determines the isotype (α, γ, δ, ε, and μ) of heavy chains and the type (κ and λ) of light chains.

Constant (C_L)   That part of the light chain that is not variable in sequence.

Cortex   The outer or peripheral layer of an organ.

Costimulatory receptors   Receptors expressed on the surface of T cells that deliver one of two signals required for T cell activation (Signal 2). They are activated when engaged by ligands, which are typically expressed by professional APC. The most common costimulator receptor is CD28.

Costimulatory signal   Additional signal that is required to induce proliferation of antigen-primed T cells and is generated by interaction of CD28 on T cells with CD80/86 on antigen-presenting cells. In B-cell activation, an analogous signal is provided by interaction of CD40 on B cells with CD40L on activated T_H cells.

CR1   Complement Receptor 1. Expressed on both erythrocytes and leukocytes and binds to C3b, C4b and their breakdown products. CR1 expression on erythrocytes is important in the clearance of immune complexes in the liver.

Cross-presentation   A protein processing and presentation pathway that occurs in some **pAPCs** where antigen acquired by endocytosis is redirected from the exogenous to the endogenous pathway, such that peptides associate with class I MHC molecules for presentation to CD8$^+$ T cells.

Cross-priming   The activation of CTL responses to antigens processed and presented via cross-presentation.

Cross-reactivity   Ability of a particular antibody or T-cell receptor to react with two or more antigens that possess a common **epitope**.

Cross-tolerance   The induction of CD8$^+$ T cell tolerance to an antigen processed and presented via cross-presentation.

CTL precursors (CTL-Ps)   Naïve CD8$^+$ T cells that have not yet been activated by antigen recognition. They do not yet express the cytotoxic machinery associated with fully mature killer T cells.

CXC subgroup   A family of chemokines that contain a disulfide bridge between cysteines separated by a different amino acid residue (X).

Cyclooxygenase 2 (COX2)   Enzyme responsible for the formation from arachidonic acid of prostaglandins and other pro-inflammatory mediators; target of non-steroidal anti-inflammatory drugs.

Cyclosporin A   A fungal product used as a drug to suppress allograft rejection. The compound blocks T-cell activation by interfering with transcription factors and preventing gene activation.

Cytokine storms   The pathological secretion of extremely high levels of cytokines induced by massive infection with particular pathogens. Typical symptoms include increased capillary permeability with resultant loss of blood pressure and shock, sometimes leading to death.

Cytokine-binding Homology Region (CHR)   A protein motif common to the cytokine binding receptors of several families.

Cytokines   Any of numerous secreted, low-molecular-weight proteins that regulate the intensity and duration of the immune response by exerting a variety of effects on lymphocytes and other immune cells that express the appropriate receptor (see Appendix II).

Cytosolic pathway   See **Endogenous pathway**.

Cytotoxic T lymphocytes (CTLs, or T_c cells)   An effector T cell (usually CD8$^+$) that can mediate the lysis of target cells bearing antigenic peptides complexed with a class I MHC molecule.

Damage-associated molecular patterns (DAMPs)   Components of dead/dying cells and damaged tissues that are recognized by pattern-recognition receptors.

Dark zone   A portion of the **germinal center** that is the site of rapid cell division by forms of B cells called centroblasts.

Death by neglect   Apoptosis of developing T cells (typically CD4$^+$CD8$^+$ thymocytes) that results when they do not receive TCR signals of adequate affinity. Most (90% or more) developing T cells undergo death by neglect.

Death domains   Protein motifs found in the cytoplasmic region of **Fas** and other proapoptotic signaling molecules. They engage the domains on other signaling molecules and initiate the formation of the **Death-Inducing Signaling Complex (DISC)** (see Figure 4-13).

Death-Inducing Signaling Complex (DISC)   An intracellular signaling aggregate formed in response to engagement of death receptors, including Fas. It includes the cytoplasmic tail of **Fas**, FADD, and procaspase-8, and initiates **apoptosis**.

Degranulation   Discharge of the contents of cytoplasmic granules by **basophils** and **mast cells** following cross-linkage (usually by antigen) of bound IgE. It is characteristic of **type I hypersensitivity**.

Delayed-type hypersensitivity (DTH)   A type IV hypersensitive response mediated by sensitized T$_H$ cells, which release various **cytokines** and **chemokines** (see Figure 15-14). The response generally occurs 2 to 3 days after T$_H$ cells interact with antigen. It is an important part of host defense against intracellular parasites and bacteria.

Dendritic cells (DCs)   Bone-marrow-derived cells that descend through the myeloid and lymphoid lineages and are specialized for antigen presentation to helper T cells.

Dermis   Layer of skin under the epidermis that contains blood and lymph vessels, hair follicles, nerves, and nerve endings.

Determinant-selection model   A hypothesis proposed to explain the variability in immune responsiveness to different MHC haplotypes. This model states that each MHC molecule binds a unique array of antigenic peptides, and some peptides are more successful in eliciting an effective immune response than others. See also **Holes-in-the-repertoire model**.

Diacylglyerol (DAG)   A lipid molecule generated upon cleavage of **phosphatidyl inositol bisphosphate** that is important in cell signaling.

Differentiation antigen   A cell surface marker that is expressed only during a particular developmental stage or by a particular cell lineage.

DiGeorge syndrome (DGS)   Congenital thymic aplasia (partial or total absence of the thymus) caused by deletion of a sequence on chromosome 22 during embryonic life. Consequences include immunodeficiency, facial abnormalities, and congenital heart disease.

Direct staining   A variation of fluorescent antibody staining in which the primary antibody is directly conjugated to the fluorescent label.

Dissociation constant K_d, the reciprocal of the **association constant** ($1/K_a$).

Diversity (D) segment   One of the gene segments encoding the immunoglobulin heavy chain or the TCR β or δ chains or its protein product.

DN1   The first in the four stages in the development of the most immature (CD4$^-$CD8$^-$ or double negative) thymocytes. DN1 cells express CD44 but not CD25 and are the progenitors that come from the bone marrow and have the potential to give rise to multiple lymphoid and myeloid cell lineages.

DN2   The second in the four stages in the development of the most immature (CD4$^-$CD8$^-$ or double negative) thymocytes. Commitment to the T cell lineage and rearrangement of the first TCR receptor genes occur among DN2 cells, which express both CD44 and CD25.

DN3   The third in the four stages in the development of the most immature (CD4$^-$CD8$^-$ or double negative) thymocytes. Only those DN3 cells that express a functional TCRβ chain continue to mature to the CD4$^+$CD8$^+$ stage and proliferate (**β-selection**). DN3 cells express CD25, but not CD44.

DN4   The last of the four stages in the development of the most immature (CD4$^-$CD8$^-$ or double negative) thymocytes. DN4 cells express neither CD44 nor CD25 and are in transition to the CD4$^+$CD8$^+$ (double positive or DP) stage of development.

Double immunodiffusion   A type of precipitation in gel analysis in which both antigen and antibody diffuse radially from wells toward each other, thereby establishing a concentration gradient. As equivalence is reached, a visible line of precipitation, a precipitin line, forms.

Double-negative (DN) cells   A subset of developing T cells (thymocytes) that do not express CD4 or CD8. At this early stage of T-cell development, DN cells do not express the TCR.

Double-positive (DP) cells   A subset of developing T cells (thymocytes) that express both CD4 and CD8. DP cells are an intermediate stage of developing thymocytes that express TCRs.

Downstream   (1) Towards the 3′ end of a gene; (2) Further away from the receptor in a signaling cascade.

E2A   A transcription factor that is required for the expression of the **recombination-activating genes (RAG)** as well as the expression of the **λ5** (lambda 5) component of the **pre-B-cell receptor** during B-cell development. It is essential for B-cell development.

Early B-cell factor (EBF)   A transcription factor that is essential for early B-cell development. It is necessary for the expression of RAG.

Early lymphoid progenitor cell (ELP)   A progenitor cell capable of dividing to give rise to either T or B lymphocyte progenitors.

Early pre-B-cell phase   The stage in B cell development at which the BCR heavy chain first appears on the cell surface in combination with the surrogate light chain, made up of **VpreB** and **λ5**.

Edema   Abnormal accumulation of fluid in intercellular spaces, often resulting from a failure of the lymphatic system to drain off normal leakage from the capillaries.

Effector caspases   The subset of caspase enzymes directly responsible for the cell apoptosis. Their cleavage activity results both in the breakdown of structural molecules (e.g. actin) or the activation of destructive molecules (e.g. endonucleases). Caspase-3 and caspase-7 are two well-characterized effector caspases.

Effector cell   Any cell capable of mediating an immune function (e.g., activated T$_H$ cells, CTLs, and plasma cells).

Effector memory T cells (T$_{EM}$)   A memory T cell subset that circulates among or resides in peripheral, non-lymphoid tissue. It is generated during the primary response and participates in the secondary response to antigen, exhibiting effector functions and proliferating more quickly than immune cells.

Effector response   Immune cell action that contributes to the clearance of infection. It includes responses mediated by helper T cells, which secrete cytokines that enhance the activity of several other immune cell subsets, by cytotoxic cells, including CD8$^+$ T cells and NK cells, and by antibody, which recruits soluble proteins (complement) and cells that can kill and clear pathogen. Also called effector function.

ELISA   See **enzyme-linked immunosorbent assay**.

Embryonic stem (ES) cell   Stem cell isolated from early embryo and grown in culture. Mouse ES cells give rise to a variety of cell types and are used to develop transgenic or knockout mouse strains.

Endocrine   Referring to regulatory secretions such as hormones or cytokines that pass from producer cell to target cell by the bloodstream.

Endocytosis   Process by which cells ingest extracellular macromolecules by enclosing them in a small portion of the plasma membrane, which invaginates and is pinched off to form an intracellular vesicle containing the ingested material.

Endogenous pathway   Intracellular route taken by antigen that is processed for presentation by MHC class I, typically associated with proteins generated in the cytosol.

Endosteal niche   Microenvironment in the bone marrow that fosters the development of hematopoietic stem cells and is postulated to associate specifically with self-renewing, long-term hematopoietic stem cells.

Endotoxins   Certain **lipopolysaccharide (LPS)** components of the cell wall of gram-negative bacteria that are responsible for many of the pathogenic effects associated with these organisms. Some function as **superantigens**.

Enzyme-linked immunosorbent assay (ELISA)   An assay for quantitating either antibody or antigen by use of an enzyme-linked antibody and a substrate that forms a colored reaction product (see Figure 20-7).

Eosinophils   Motile, somewhat phagocytic granulocytes that can migrate from blood to tissue spaces. They have large numbers of IgE receptors and are highly granular. They are thought to play a role in the defense against parasitic organisms such as roundworms (see Figure 2-2).

Epidermis   The outermost layer of the skin.

Epitope mapping   Localization of sites (epitopes) on an antigen molecule that are reactive with different antibodies or T-cell receptors.

Epitope   The portion of an antigen that is recognized and bound by an antibody or TCR-MHC combination; also called **antigenic determinant**.

Equilibrium dialysis   An experimental technique that can be used to determine the affinity of an antibody for antigen and its **valency** (see Figure 20-11).

ERAP   Endoplasmic reticulum aminopeptidase. An enzyme responsible for trimming amino acids from peptides in the ER in order to reach an optimal length for binding to class I MHC molecules.

Erythroblastosis fetalis   A type II hypersensitivity reaction in which maternal antibodies against fetal **Rh antigens** cause hemolysis of the erythrocytes of a newborn; also called *hemolytic disease of the newborn*.

Erythrocytes   Red blood cells.

Exocytosis   Process by which cells release molecules (e.g., cytokines, lytic enzymes, degradation products) contained within a membrane-bound vesicle by fusion of the vesicle with the plasma membrane.

Exogenous pathway   Intracellular route taken by antigen that is processed for presentation by MHC class II, typically associate with proteins that are endocytosed.

Exotoxins   Toxic proteins secreted by gram-positive and gram-negative bacteria; some function as **superantigens**. They cause food poisoning, toxic shock syndrome, and other disease states. See also **immunotoxin**.

Extravasation   Movement of blood cells through an unruptured blood vessel wall into the surrounding tissue, particularly at sites of inflammation.

F (ab′)₂ fragment   Two Fab units linked by disulfide bridges between fragments of the heavy chain. They are obtained by digestion of antibody with pepsin.

Fab (fragment antigen binding) region   Region at the N-terminus of the antibody molecule that interact with antigen. This antibody fragment, consisting of one light chain and part of one heavy chain, linked by an interchain disulfide bond, is obtained by brief papain digestion.

Fas (CD95)   A member of the Tumor Necrosis Factor Receptor family. On binding to its ligand, **FasL**, the Fas-bearing cell will often be induced to commit to an apoptotic program. Occasionally, however, Fas ligation leads to cell proliferation.

Fas ligand (FasL)   FasL is a member of the Tumor Necrosis Factor family of molecules and interacts with the Fas receptor, which is a member of the Tumor Necrosis Factor Receptor family. Signals delivered from FasL to Fas usually result in the death by **apoptosis** of the Fas-bearing cell.

Fc (fragment crystallizable) region   Region at the C terminus of the antibody molecule that interacts with Fc receptors on other cells and with components of the complement system. This crystallizable antibody fragment consists of the carboxyl-terminal portions of both heavy chains and is obtained by brief papain digestion.

Fc receptor (FcR)   Cell-surface receptor specific for the Fc portion of certain classes of immunoglobulin. It is present on lymphocytes, mast cells, macrophages, and other accessory cells.

FcγRIIb   A receptor that binds to the **Fc region** of antibodies engaged in antigen:antibody complexes. Signals through this receptor down-regulate B cell division and differentiation.

Fibrin   A filamentous protein produced by the action of thrombin on fibrinogen; fibrin is the main element in blood clotting.

Fibrinopeptide   One of two small peptides of about 20 amino acids released from fibrinogen by thrombin cleavage in the conversion to **fibrin**.

Fibroblast reticular cells (FRCs)   Stromal cells in secondary lymphoid tissue (and at some immune response sites in the periphery) that extend processes which provide the surface networks on which dendritic cells position themselves and T and B lymphocytes migrate as they probe for antigen. Associated with chemokines and cytokines that help guide cell movements.

Fibrosis   A process responsible for the development of a type of scar tissue at the site of chronic inflammation.

Ficolin   Member of a family of carbohydrate-binding proteins that contain a fibrinogen-like domain and a collagen-like domain.

Flow cytometer   An instrument that users lasers along with sophisticated optics to measure multiple fluorescent and light scattering parameters from thousands of cells as flow rapidly, one-by-one in front of the laser beam.

Fluorescence Activated Cell Sorter (FACS)   A flow cytometer equipped with the ability to sort cells sharing particular fluorescence and light scattering properties into different containers.

Fluorescence microscopy   A microscopic technique that allows the visualization of fluorescent signals generated from cells tagged with fluorescent antibodies or proteins.

Fluorescent antibody   An antibody with a **fluorochrome** conjugated to its Fc region that is used to stain cell surface molecules or tissues; the technique is called **immunofluorescence**.

Fluorochrome   A molecule that fluoresces when excited with appropriate wavelengths of light. See **immunofluorescence**.

fms-related tyrosine kinase 3 receptor (flt-3)   Binds to the membrane-bound flt-3 ligand on bone marrow stromal cells and signals the progenitor cell to begin synthesizing the **IL-7 receptor**.

Follicles   Microenvironments that specifically support the development of the B lymphocyte response in lymph nodes, spleen, and other secondary lymphoid tissue. They also become the site of development of the germinal center when a B cell is successfully activated.

Follicular dendritic cell (FDC)   A cell with extensive dendritic extensions that is found in the follicles of lymph nodes. Although they do not express class II MHC molecules, they are richly endowed with receptors for complement and Fc receptors for antibody. They are of a lineage that is distinct from class II MHC–bearing dendritic cells.

Follicular mantle zone   Zone of naïve, IgD-bearing B cells that surrounds the central region of a follicle engaged in a germinal center reaction. The non-antigen-specific, IgD-bearing cells are slowly pushed to the outside of the follicle as they are displaced by dividing cells in the germinal center.

Fragmentin   Enzymes present in the granules of cytotoxic lymphocytes that induce DNA fragmentation.

Framework region (FR)   A relatively conserved sequence of amino acids located on either side of the hypervariable regions in the variable domains of immunoglobulin heavy and light chains.

Freund's complete adjuvant (CFA)   A water-in-oil emulsion to which heat-killed mycobacteria have been added; antigens are administered in CFA to enhance their immunogenicity.

Freund's incomplete adjuvant   Freund's adjuvant lacking heat-killed mycobacteria.

G-Protein–Coupled Receptors (GPCRs)   Ligand receptors that interact with G proteins on the cytoplasmic side of the plasma membrane. G proteins are signal-transducing molecules that are activated when the receptor binds to its ligand. Receptor:ligand binding induces a conformational change in the G protein that induces it to exchange the GDP (which is in its binding site in the resting state), for GTP. The activated G protein:GTP complex then transduces the signal. GPCRs have a shared structure in which the proteins passes through the member a total of seven times.

γ (gamma)-globulin fraction   The electrophoretic fraction of serum that contains most of the immunoglobulin classes.

GATA-2 gene   A gene encoding a transcription factor that is essential for the development of several hematopoietic cell lineages, including the lymphoid, erythroid, and myeloid lineages.

Gene conversion   Process in which portions of one gene (the recipient) are changed to those of another gene (the donor). Homologous gene conversion is a diversification mechanism used for immunoglobulin V≈genes in some species.

Gene segments   Germ-line gene sequences that are combined with others to make a complete coding sequence; Ig and TCR genes are products of V, D, J gene segments.

Gene therapy   General term for any measure aimed at correction of a genetic defect by introduction of a normal gene or genes.

Generation of diversity   The generation of a diverse repertoire of antigen-binding receptors on B or T lymphocytes that occurs in the bone marrow or thymus, respectively.

Genome wide sequence   The sequence of all DNA (the entire genome) present in a cell.

Genotype   The combined genetic material inherited from both parents; also, the **alleles** present at one or more specific loci.

Germ-line theories   Classical theories that attempted to explain antibody diversity by postulating that all antibodies are encoded in the host chromosomes.

Germinal centers (GCs)   A region within lymph nodes and the spleen where T-dependent B-cell activation, proliferation, and differentiation occur (see Figure 12-4). Germinal centers are sites of intense B-cell somatic mutation and selection.

Graft-versus-host (GVH) reaction   A pathologic response to tissue transplantation in which immune cells in the transplanted tissue (graft) react against and damage host cells.

Graft-versus-host disease (GVHD)   A reaction that develops when a graft contains immunocompetent T cells that recognize and attack the recipient's cells.

Granulocytes   Any **leukocyte** that contains cytoplasmic granules, particularly the basophil, eosinophil, and neutrophil (see Figure 2-2).

Granuloma   A tumor-like mass or nodule that arises because of a chronic **inflammatory response** and contains many activated macrophages, T_H cells, and multinucleated giant cells formed by the fusion of macrophages.

Granzyme (fragmentin)   One of a set of enzymes found in the granules of T_C cells that can help to initiate apoptosis in target cells.

Grave's disease   An autoimmune disease in which the individual produces auto-antibodies to the receptor for thyroid-stimulating hormone TSH (see Table 16-1).

GTP-binding proteins   Proteins that bind Guanosine Tri-phosphate.

GTPase Activating Proteins (GAPs)   G proteins have an intrinsic GTPase activity that serves to limit the time during which G proteins can actively transduce a signal. GAPs enhance this GTPase activity, and thus further limit the signal through a GPCR.

Guanine-nucleotide Exchange Factors (GEFs)   Small proteins that catalyze the exchange of GTP for GDP in the guanine nucleotide binding sites of small and trimeric G proteins.

Gut-associated lymphoid tissue (GALT)   Secondary lymphoid microenvironments in the intestinal (gut) system that support the development of the T and B lymphocyte response to antigens that enter gut mucosa. Part of the **mucosa associated lymphoid tissue system** (MALT).

H-2 complex   Term for the **MHC** in the mouse.

HAART   See **Highly active antiretroviral therapy**.

Haplotype   The set of **alleles** of linked genes present on one parental chromosome; commonly used in reference to the **MHC** genes.

Hapten   A low-molecular-weight molecule that can be made immunogenic by conjugation to a suitable carrier.

Hapten-carrier conjugate   A covalent combination of a small molecule (**hapten**) with a large carrier molecule or structure.

Heavy (H) chain   The larger polypeptide of an antibody molecule; it is composed of one variable domain V_H and three or four constant domains (C_H1, C_H2, etc.) There are five major classes of heavy chains in humans (α, γ, δ, ε, and μ), which determine the **isotype** of an antibody (see Table 3-2).

Heavy-chain Joining segment (J_H)   One of the gene segments encoding the immunoglobulin heavy chain or its protein product (see Figure 7-3).

Heavy-chain Variable region   That part of the immunoglobulin heavy chain protein that varies from antibody to antibody and is encoded by the V, D, and J gene segments.

Heavy-chain Variable segment (V_H)   (1) One of the gene segments encoding the immunoglobulin heavy chain gene, or its protein product.

Helper T (T_H) cells   See **T helper (T_H) cells**.

Hemagglutination   The process of sticking together red blood cells using multivalent cells, viruses or molecules that bind to molecules on the red blood cell surface. Viruses such as influenza or antibodies are routinely measured by hemagglutination assays.

Hemagglutinin (HA)   Any substance that causes red blood cells to clump, or agglutinate. Most commonly the virally-derived glycoprotein found on the surface of influenza virus that binds to sialic acid residues on host cells causing them to agglutinate. See also **agglutinin**.

Hematopoiesis   The formation and differentiation of blood cells (see Figure 2-1).

Hematopoietic stem cell (HSC)   The cell type from which all lineages of blood cells arise.

Hematopoietin (Class I) cytokine family   The largest of the cytokine families, typified by Interleukin 2 (IL-2).

Hemolysis   Alteration or destruction of red blood cells, which liberates hemoglobin.

Heptamer   A conserved set of 7 nucleotides contiguous to each of the V, D, and J gene segments of all immunoglobulin and TCR gene segments. It serves as the recognition signal and binding site of the RAG1/2 protein complex.

Herd immunity   When the majority of the population is immune to an infectious agent, thus significantly reducing the pathogen reservoir due to the low chance of a susceptible individual contacting an infected individual.

Heteroconjugates   Hybrids of two different antibody molecules.

Heterotypic   An interaction between two molecules where the interacting domains have different structures from one another.

High-endothelial venule (HEV)   An area of a capillary venule composed of specialized cells with a plump, cuboidal ("high") shape through which lymphocytes migrate to enter various lymphoid organs (see Figure 14-2).

Highly active antiretroviral therapy (HAART)   A form of drug therapy used to treat infection with HIV that utilizes a combination of three or more anti-HIV drugs from different classes to inhibit viral replication and avoid selection of drug-resistant mutants.

Hinge   The flexible region of an immunoglobulin heavy chain between the C_H1 and C_H2 domains that allows the two binding sites to move independently of one another.

Histiocyte   An immobilized (sometimes called "tissue fixed") macrophage found in loose connective tissue.

Histocompatibility antigens   Family of proteins that determines the ability of one individual to accept tissue or cell grafts from another. The major histocompatibility antigens, which are encoded by the **MHC**, function in antigen presentation.

Histocompatible   Denoting individuals whose major histocompatibility antigens are identical. Grafts between such individuals are generally accepted.

HLA (human leukocyte antigen) complex   Term for the **MHC** in humans.

Holes-in-the-repertoire model   The concept that immune tolerance results from the absence of receptors specific for self antigens.

Homeostatic   Pertaining to processes that contribute to the maintenance and stability of a system, in this case, the immune system, under normal conditions.

Homing receptor   A receptor that directs various populations of lymphocytes to particular lymphoid and inflammatory tissues.

Homing   The differential migration of lymphocytes or other leukocytes to particular tissues or organs.

Homotypic   An interaction between two molecules where the interacting domains have identical or very similar structures to one another.

Human immunodeficiency virus (HIV)   The retrovirus that causes acquired immune deficiency syndrome (AIDS).

Human leukocyte antigen (HLA) complex   See **HLA complex**.

Humanized antibody   An antibody that contains the antigen-binding amino acid sequences of another species within the framework of a human immunoglobulin sequence.

Humoral immune response   Host defenses that are mediated by antibody present in the plasma, lymph, and tissue fluids. It protects against extracellular bacteria and foreign macromolecules. Transfer of antibodies confers this type of immunity on the recipient. See also **cell-mediated immune response**.

Humoral immunity   See **humoral immune response**.

Humoral   Pertaining to extracellular fluid, including the plasma, lymph, and tissue fluids.

Hybridoma   A **clone** of hybrid cells formed by fusion of normal lymphocytes with **myeloma cells**; it retains the properties of the normal cell to produce antibodies or T-cell receptors but exhibits the immortal growth characteristic of myeloma cells. Hybridomas are used to produce **monoclonal antibody**.

Hyper IgE syndrome (HIE)   An immune deficiency syndrome characterized by over expression of IgE and most frequently caused by mutations in the gene encoding STAT3. Also known as **Job syndrome**.

Hyper IgM syndrome (HIM)   An immune deficiency disorder that arises from inherited deficiencies in **CD40L**, resulting in impaired T cell-APC communication and a lack of isotype switching, manifesting as elevated levels of IgM but an absence of other antibody isotypes.

Hypersensitivity   Exaggerated immune response that causes damage to the individual. **Immediate hypersensitivity** (types I, II, and III) is mediated by antibody or immune complexes, and delayed-type hypersensitivity (type IV) is mediated by T_H cells.

Hypervariable   Those parts of the variable regions of the BCR and TCR that exhibit the most sequence variability and interact with the antigen. Otherwise known as the **complementarity determining regions**.

Hypogammaglobulinemia   Any immune deficiency disorder, either inherited or acquired, characterized by low levels of gammaglobulin (IgG).

Iccosomes   Immune-complex-coated cell fragments often found coating the spines of **follicular dendritic cells**.

Idiotope   A single **antigenic determinant** in the variable domains of an antibody or T-cell receptor; also called idiotypic determinant. Idiotopes are generated by the unique amino acid sequence specific for each antigen.

Idiotype   The set of antigenic determinants (**idiotopes**) characterizing a unique antibody or T-cell receptor.

IgD   Immunoglobulin D. An antibody class that serves importantly as a receptor on naïve B cells.

IgM   Immunoglobulin M. An antibody class that serves as a receptor on naïve B cells. IgM is also the first class of antibody to be secreted during the course of an immune response. Secreted IgM exists primarily in pentameric form.

Ikaros   A transcription factor required for the development of all lymphoid cell lineages.

IL-1 Receptor Activated Kinase (IRAK)   A family of kinases that participates in the signaling pathway from IL-1. IRAKs are also important in **TLR** signaling.

IL-10   A member of the interferon family of cytokines, that usually mediates an immuno-suppressive effect.

IL-17 family   A family of cytokines implicated in the early stages of the immune response. Most members of this family are pro-inflammatory in action.

IL-7 receptor   Receptor for the cytokine Interleukin 7, which is important for lymphocyte development.

Immature B cell   Immature B cells express a fully-formed IgM receptor on their cell surface. Contact with antigen at this stage of B cell development results in tolerance induction rather than activation. Immature B cells express lower levels of IgD and higher levels of IgM than do mature B cells. They also have lower levels of anti-apoptotic molecules and higher levels of Fas than mature B cells, reflective of their short-half lives.

Immediate hypersensitivity   An exaggerated immune response mediated by antibody (type I and II) or antigen-antibody complexes (type III) that manifests within minutes to hours after exposure of a sensitized individual to antigen (see Table 15-1).

Immune dysregulation, polyendocrinopathy, enteropathy, X-linked (IPEX) syndrome   An inherited immune deficiency disorder that manifests as an autoimmune syndrome caused by a lack of FoxP3 expression and near absence of regulatory T cells (T_{REG} cells).

Immunity   A state of protection from a particular infectious disease.

Immunization   The process of producing a state of immunity in a subject. See also **active immunity** and **passive immunity**.

Immunocompetent   Denoting a mature lymphocyte that is capable of recognizing a specific antigen and mediating an immune response; also an individual without any immune deficiency.

Immunodeficiency   Any deficiency in the immune response, whether inherited or acquired. It can result from defects in phagocytosis, humoral immunity, cell-mediated responses, or some combination of these (see Figure 18-2).

Immunodominant   Referring to epitopes that produce a more pronounced immune response than others under the same conditions.

Immunoediting   A recently formulated theory concerning the role of the immune system in responding to cancer. It includes three phases (elimination, equilibrium, and escape) and incorporates both positive (anti-tumor) and negative (pro-tumor) processes mediated by the immune system in responding to malignancy.

Immunoelectron microscopy   A technique in which antibodies used to stain a cell or tissue are labeled with an electron-dense material and visualized with an electron microscope.

Immunoelectrophoresis   A technique in which an antigen mixture is first separated into its component parts by electrophoresis and then tested by **double immunodiffusion**.

Immunofluorescence   Technique of staining cells or tissues with **fluorescent antibody** and visualizing them under a fluorescent microscope.

Immunogen   A substance capable of eliciting an immune response. All immunogens are **antigens**, but some antigens (e.g., haptens) are not immunogens.

Immunogenicity   The capacity of a substance to induce an immune response under a given set of conditions.

Immunoglobulin (Ig)   Protein consisting of two identical heavy chains and two identical light chains, that recognize a particular **epitope** on an antigen and facilitates clearance of that antigen. There are 5 types: IgA, IgD, IgE, IgG, and IgM. Also called **antibody**.

Immunoglobulin domains   Three dimensional structures characteristic of immunoglobulin and related proteins including T cell receptors, MHC proteins and adhesion molecules. Consists of a domain of 100 – 110 amino acids folded into two β-pleated sheets, each containing three of four antiparallel β strands and stabilized by an intrachain disulfide bond (see Figure 3-19).

Immunoglobulin fold   Characteristic structure in immunoglobulins that consists of a domain of 100 to 110 amino acids folded into two β-pleated sheets, each containing three or four antiparallel β strands and stabilized by an intrachain disulfide bond (see Figure 3-18).

Immunoglobulin superfamily   Group of proteins that contain **immunoglobulin-fold** domains, or structurally related domains; it includes immunoglobulins, T-cell receptors, MHC molecules, and numerous other membrane molecules.

Immunologic memory   The ability of the immune system to respond much more swiftly and with greater efficiency during a second, or later exposure to the same pathogen.

Immunoproteasome   A variant of the standard 20S proteasome, found in pAPCs and infected cells, that has unique catalytic subunits specialized to produce peptides that bind efficiently to class I MHC proteins.

Immunoreceptor tyrosine-based inhibitory motif (ITIM)   An amino acid sequence containing tyrosine residues in conserved sequence relationships with one another that serves as a docking site for downstream signaling molecules that will send an inhibitory signal to the cell. More formally known as immuno-receptor tyrosine-based inhibitory motif.

Immunoreceptor tyrosine-based activation motif (ITAM)   Amino acid sequence in the intracellular portion of signal-transducing cell surface molecules that interacts with and activates intracellular kinases after ligand binding by the surface molecule.

Immunosurveillance   A theory concerning anti-cancer responses that is now part of the immunoediting hypothesis (elimination phase) which posits that cells of the immune system continually survey the body in order to recognize and eliminate tumor cells.

Immunotoxin   Highly cytotoxic agents sometimes used in cancer treatment that are produced by conjugating an antibody (for instance, specific for tumor cells) with a highly toxic agent, such as a bacterial toxin.

Incomplete antibody   Antibody that binds antigen but does not induce agglutination.

Indirect staining   A method of immunofluorescent staining in which the primary antibody is unlabeled and is detected with an additional fluorochrome-labeled reagent.

Induced T_{REG} (iT_{REG}) cells   CD4$^+$ T cell subset that negatively regulates immune responses and is induced to develop by specific cytokine interactions in secondary lymphoid tissue that upregulate the master regulator FoxP3.

Inducible nitric oxide synthetase (iNOS)   An inducible form of NOS that generates the antimicrobial compound nitric oxide from arginine.

Inflamed   Manifesting redness, pain, heat, and swelling. See also **Inflammation**.

Inflammasomes   Multiprotein complex that promotes inflammation by processing inactive precursor forms of pro-inflammatory cytokines such as IL-1 and IL-18.

Inflammation   Tissue response to infection or damage which serves to eliminate or wall-off the infection or damage; classic signs of acute inflammation are heat (calor), pain (dolor), redness (rubor), swelling (tumor), and loss of function.

Inflammatory response   A localized tissue response to injury or other trauma characterized by pain, heat, redness, and swelling. The response includes both localized and systemic effects, consisting of altered patterns of blood flow, an influx of phagocytic and other immune cells, removal of foreign antigens, and healing of the damaged tissue.

Inhibitor of NF-κB (IκB)   A small protein that binds to the transcription factor **NF-κB** that inhibits its action, in part by retaining it in the cytoplasm.

Initiator caspases   The subset of caspase enzymes that initiate the process leading to cell apoptosis. Initiator caspases typically cleave and activate effector caspases, although they can also cleave other molecules that indirectly activate effector caspases (e.g. Bid). Caspase-8 is a well-characterized initiator caspase that is associated with the death receptor, **Fas**, and is cleaved and activated when Fas is engaged.

Innate immunity   Non-antigen specific host defenses that exist prior to exposure to an antigen and involve anatomic, physiologic, endocytic and phagocytic, anti-microbial, and inflammatory mechanisms, and which exhibit no adaptation or memory characteristics. See also **adaptive immunity**.

Inositol trisphosphate (IP$_3$)   A phosphorylated six-carbon sugar that binds to receptors in the membranes of the endoplasmic reticulum, leading to the release of Ca^{2+} ions into the cytoplasm.

Instructive model   A model advanced to explain the molecular basis for **lineage commitment**, the choice of a CD4$^+$CD8$^+$ thymocyte to become a CD4$^+$ versus a CD8$^+$ T cell. This model proposes that DP thymocytes that interact with MHC class II receive a distinct signal from DP thymoctyes that interact with MHC class I. These distinct signals induce differentiation into the helper CD4$^+$ or the cytotoxic CD8$^+$ lineage, respectively. This model is no longer accepted. See also **Kinectic signaling model**.

Insulin-dependent diabetes mellitus (IDDM)   An autoimmune disease caused by T cell attack on the insulin-producing beta cells of the pancreas, necessitating daily insulin injections.

Integrins   A group of heterodimeric cell adhesion molecules (e.g., LFA-1, VLA-4, and Mac-1) present on various leukocytes that bind to **Ig-superfamily CAMs** (e.g., ICAMs, VCAM-1) on endothelium (see Box 14-2).

Intercellular adhesion molecules (ICAMs)   Cellular adhesion molecules that bind to integrins. ICAMs are members of the immunoglobulin superfamily.

Interferon regulatory factor 4 (IRF-4)   Transcription factor important to the initiation of plasma cell differentiation.

Interferon regulatory factors (IRFs)   Transcription factors induced by signaling downstream of pattern-recognition, interferon, and other receptors that activate interferon genes.

Interferons (IFNs)   Several glycoprotein **cytokines** produced and secreted by certain cells that induce an antiviral state in other cells and also help to regulate the immune response (see Table 4-1).

Interleukin-1 (IL-1) family   Interleukin-1 was the first cytokine to be discovered. Members of this family interact with dimeric receptors to induce responses that are typically pro-inflammatory.

Interleukins (ILs)   A group of **cytokines** secreted by leukocytes that primarily affect the growth and differentiation of various hematopoietic and immune system cells (see Appendix II).

Interstitial fluid   Fluid found in the spaces between cells of an organ or tissue.

Intraepidermal lymphocytes   T cells found in epidermal layers.

Intraepithelial lymphocytes (IELs)   T cells found in the epithelial layer of organs and the gastrointestinal tract.

Intravital microscopy   A type of microscopy that allows one to image cell activity within living tissue and live organisms.

Invariant (Ii) chain   Component of the class II MHC protein that shows no genetic polymorphism. The Ii chain stabilizes the class II molecule before it has acquired an antigenic peptide.

Invariant NKT (iNKT) cells   A cytotoxic T cell subset that develops in the thymus and expresses very limited TCRαβ receptor diversity (one specific TCRα paired with only a few TCRβ chains) and recognize lipids associated with CD1, an MHC-like molecule.

Isograft   Graft between genetically identical individuals.

Isotype   (1) An antibody class that is determined by the constant-region sequence of the heavy chain. The five human isotypes, designated IgA, IgD, IgE, IgG, and IgM, exhibit structural and functional differences (see Table 4-4). Also refers to the set of **isotypic determinants** that is carried by all members of a species. (2) One of the five major kinds of heavy chains in antibody molecules (α, γ, δ, ε, and μ).

Isotype switching   Conversion of one antibody class (**isotype**) to another resulting from the genetic rearrangement of heavy-chain constant-region genes in B cells; also called class switching.

Isotypic determinant   An **antigenic determinant** within the immunoglobulin constant regions that is characteristic of a species.

iT$_{REG}$ cells   A type of T cell that, following antigen exposure in the periphery, is induced to express FoxP3 and acquire regulatory functions, suppressing immune activity against specific antigen. See also **nT$_{REG}$ cells**.

IκB kinase (IKK)   The enzyme that phosphorylates the inhibitory subunit of the transcription factor **NF-κB**. Phosphorylation of **IκB** results in its release from the transcription factor and movement of the transcription factor into the nucleus.

J (joining) chain   A polypeptide that links the heavy chains of monomeric units of polymeric IgM and di- or trimeric IgA. The linkage is by disulfide bonds between the J chain and the carboxyl-terminal cysteines of IgM or IgA heavy chains.

J (joining) gene segment   The part of a rearranged immunoglobulin or T-cell receptor gene that joins the variable region to the constant region and encodes part of the **hypervariable region**. There are multiple J gene segments in germ-line DNA, but gene rearrangement leaves only one in each functional rearranged gene.

JAK   See **Janus Activated Kinase**.

Janus Activated Kinase (JAK)   Kinase that typically transduce a signal from a Type 1 or Type 2 cytokine receptor to a cytoplasmically-located transcription factor belonging to the STAT (**Signal Transducer and Activator of Transcription**) family. On cytokine binding to the receptor, the JAK kinases are activated and phosphorylate the receptor molecule. This provides docking sites for a pair of STAT molecules which are phosphorylated, dimerize, and translocate to the nucleus to effect their transcriptional programs.

Job syndrome   See **Hyper IgE syndrome**.

Joining (J) segment   One of the gene segments encoding the immunoglobulin heavy or light chain or any of the four TCR chains or its protein product

Junctional flexibility   The diversity in antibody and T-cell receptor genes created by the imprecise joining of coding sequences during the assembly of the rearranged genes.

Kappa (κ) light chain   One of the two types of immunoglobulin **light chains** that join with heavy chains to form the B cell receptor and antibody heterodimer. Lambda (λ) is the other type.

Kinetic signaling model   A model advanced to explain the molecular basis for **lineage commitment**, the choice of a CD4$^+$CD8$^+$ thymocyte to become a CD4$^+$ versus a CD8$^+$ T cell. This model proposes that all DP thymocytes receiving T cell receptor signals decrease expression of CD8. Those thymocytes whose TCR binds MHC class II will continue to receive a signal stabilized by CD4-MHC class II interactions and will progress to the CD4$^+$ lineage. However those thymocytes whose TCR binds MHC class I will have this signal disrupted by the reduction in stabilizing CD8-MHC class I interactions. These cells require rescuing by cytokines (IL-7 or IL-15) which promote their development to the CD8$^+$ lineage.

KIR   Immunoglobulin-like receptors expressed by human natural killer cells that bind MHC class I molecules and inhibit cytotoxicity.

Knock-in genetics   A genetic manipulation that results in the insertion of a desired mutant form of a gene or a marker gene in a pre-selected site in the genome.

Knockout genetics   A genetic manipulation that results in the elimination of a selected gene from the genome.

Kupffer cell   A type of tissue-fixed macrophage found in liver.

λ5   A polypeptide that associates with **Vpre-B** to form the **surrogate light chain** of the **pre-B-cell receptor**.

Lambda (λ) chain   One of the two types of immunoglobulin **light chains** that join with heavy chains to form the B cell receptor and antibody heterodimer. Kappa (κ) is the other type.

Lamina propia   Layer of loose connective tissue under the intestinal epithelium where immune cells are organized. The site of the GALT and part of the mucosal immune system.

Laser scanning confocal microscopy   Microscopy that uses lasers to focus on a single plane within the sample.

Late pre-B-cell stage   At the late pre-B-cell stage, the pre-B cell receptor is lost from the B cell surface and light chain recombination begins in the genome.

Lck   A tyrosine kinase that operates early in the TCR signaling cascade. Associates non-covalently with the T cell co-receptor.

Leader (L) peptide   A short hydrophobic sequence of amino acids at the N-terminus of newly synthesized immunoglobulins; it inserts into the lipid bilayer of the vesicles that transport Ig to the cell surface. The leader is removed from the ends of mature antibody molecules by proteolysis.

Lectin pathway   Pathway of complement activation initiated by binding of serum protein MBL to the mannose-containing component of microbial cell walls (see Figure 6-2).

Lectins   Proteins that bind carbohydrates.

Leishmaniasis   The disease caused by the protozoan parasite *Leishmania major*.

Lepromatous leprosy   A disease caused by the intracellular bacteria *Mycobacterium leprae* and *Mycobacterium lepromatosis*, where skin, nerves, and upper respiratory tract are severely and chronically damaged by infection that is not successfully regulated by the immune response. This form of disease is associated with the production of a T$_H$2 rather than T$_H$1 response.

Leucine-rich repeats (LRRs)   Protein structural domains that contain many repeats of the leucine-containing 25-amino acid repeat sequence xLxxLxLxx.; the repeats stack up on each other. Found in TLR pattern-recognition receptors and in VLRs of jawless fish.

Leukemia   Cancer originating in any class of hematopoietic cell that tends to proliferate as single cells within the lymph or blood.

Leukocyte   A white blood cell. The category includes lymphocytes, granulocytes, platelets, monocytes, and macrophages.

Leukocyte adhesion deficiency (LAD)   An inherited immune deficiency disease in which the leukocytes are unable to undergo adhesion-dependent migration into sites of inflammation. Recurrent bacterial infections and impaired healing of wounds are characteristic of this disease.

Leukocytosis   An abnormally large number of leukocytes, usually associated with acute infection. Counts greater than 10,000/mm³ may be considered leukocytosis.

Leukotrienes   Several lipid mediators of inflammation and type I hypersensitivity, also called slow reactive substance of anaphylaxis (SRS-A). They are metabolic products of arachidonic acid.

Ligand   A molecule that binds to a receptor.

Light (L) chains   Immunoglobulin polypeptides of the lambda or kappa type that join with heavy-chain polypeptides to form the antibody heterodimer.

Light zone   A region of the germinal center that contains numerous **follicular dendritic cells**.

Lineage commitment   The development of a cell that can give rise to multiple cell types (multipotent) into one of those cell types. (1) In T cell development, the choice a CD4⁺CD8⁺ thymocyte makes to become a helper CD4⁺ versus cytotoxic CD8⁺ T cell. (2) In hematopoiesis, the choice a pluripotent stem cell makes to become either myeloid or lymphoid, as well as the subsequent choices to become specific immune cell subtypes.

Lipid rafts   Parts of the membrane characterized by highly ordered, detergent-insoluble, sphingolipid- and cholesterol-rich regions.

Lipopolysaccharide (LPS)   An oligomer of lipid and carbohydrate that constitutes the endotoxin of gram-negative bacteria. LPS acts as a polyclonal activator of murine B cells, inducing their division and differentiation into antibody-producing plasma cells.

Locus   The specific chromosomal location of a gene.

LPS tolerance   State of reduced responsiveness to LPS following an initial exposure to low/sublethal dose of LPS.

LRRs   See **leucine-rich repeats**.

Ly49   Receptors in the C-lectin protein family expressed by murine natural killer cells that bind MHC class I and typically inhibit cytotoxicity.

Lymph   Interstitial fluid derived from blood plasma that contains a variety of small and large molecules, lymphocytes, and some other cells. It circulates through the lymphatic vessels.

Lymph node   A small **secondary lymphoid organ** that contains lymphocytes, macrophages, and dendritic cells and serves as a site for filtration of foreign antigen and for activation and proliferation of lymphocytes (see Figure 2-8). See also **germinal center**.

Lymphatic system   A network of vessels and nodes that conveys lymph. It returns plasma-derived interstitial fluids to the bloodstream and plays an important role in the integration of the immune system.

Lymphatic vessels   Thinly walled vessels through which the fluid and cells of the lymphatic system move through the lymph nodes and ultimately into the thoracic duct, where it joins the bloodstream.

Lymphoblast   A proliferating lymphocyte.

Lymphocyte   A mononuclear leukocyte that mediates humoral or cell-mediated immunity. See also **B cell** and **T cell**.

Lymphoid progenitor cell   A cell committed to the lymphoid lineage from which all lymphocytes arise. Also known as **common lymphoid progenitor** (CLP).

Lymphoid-primed, multipotential progenitors (LMPPs)   Progenitor hematopoietic cells with the capacity to differentiate along either the lymphoid or the myeloid pathways.

Lymphoma   A cancer of lymphoid cells that tends to proliferate as a solid tumor.

Lymphotoxin-α (LT-α)   Also known as TNF-β, this cytokine is a member of the Tumor Necrosis Family. It is produced by activated lymphocytes and delivers a variety of signals to its target cells, including the induction of increased levels of MHC class II expression.

Lyn   A tyrosine kinase important in lymphocyte signaling.

Lysosome   A small cytoplasmic vesicle found in many types of cells that contains hydrolytic enzymes, which play an important role in the degradation of material ingested by **phagocytosis** and **endocytosis**.

Lysozyme   An enzyme present in tears, saliva, and mucous secretions that digests mucopeptides in bacterial cell walls and thus functions as a nonspecific antibacterial agent. Lysozyme from hen egg white (HEL) has frequently been used as an experimental antigen in immunological studies.

M cells   Specialized cells of the intestinal mucosa and other sites, such as the urogenital tract, that deliver antigen from the apical face of the cell to lymphocytes clustered in the pocket of its basolateral face.

Macrophages   Mononuclear phagocytic leukocytes that play roles in adaptive and innate immunity. There are many types of macrophages; some are migratory, whereas others are fixed in tissues.

Major histocompatibility complex (MHC) molecules   Proteins encoded by the major histocompatibility complex and classified as class I, class II, and class III MHC molecules. See also **MHC**.

Malignancy, malignant   Refers to cancerous cells capable of uncontrolled growth.

MALT (mucosal-associated lymphoid tissue)   Lymphoid cells and tissues organized below the epithelial layer of the body's mucosal surfaces. MALT is found in the gastrointestinal tract (GALT), bronchial tissues (BALT) and nasal tissue (NALT).

Mannose-binding lectin (MBL)   A serum protein that binds to mannose in microbial cell walls and initiates the lectin pathway of complement activation.

MAP kinase cascade   Mitogen activated protein kinase cascade. A series of reactions initiated by a cellular signal that results in successive phosphorylations of intra-cellular kinases and culminates in activation of transcription factors within the nucleus and often the initiation of cellular locomotion.

Marginal zone   A diffuse region of the spleen, situated on the periphery of the **periarteriolar lymphoid sheath (PALS)** between the **red pulp** and **white pulp**, that is rich in B cells (see Figure 2-10 and 12-18).

Mast cell   A bone-marrow-derived cell present in a variety of tissues that resembles peripheral blood basophils, bears **Fc receptors** for IgE, and undergoes IgE-mediated **degranulation**.

Medulla   The innermost or central region of an organ.

Megakaryocytes   Hematopoietic cells in the myeloid system that give rise to platelets.

Membrane-attack complex (MAC)   The complex of complement components C5–C9, which is formed in the terminal steps of the classical, lectin, and alternative complement pathways and mediates cell lysis by creating a membrane pore in the target cell.

Membrane-bound immunoglobulin (mIg)   A form of antibody that is bound to a cell as a transmembrane protein. It acts as the antigen-specific receptor of B cells.

Memory B cell   An antigen-committed, persistent B cell. B-cell differentiation results in formation of **plasma cells**, which secrete antibody, and **memory cells**, which are involved in the **secondary responses**.

Memory cells   Lymphocytes generated following encounters with antigen that are characteristically long lived; they are more readily stimulated than naïve lymphocytes and mediate a **secondary response** to subsequent encounters with the antigen.

Memory response   See **memory, immunologic**.

Memory T cells   T cells generated during a primary immune response that become long lived and more easily stimulated by the antigen to which they are specific. They include two main subsets, central and effector memory T cells, and are among the first participants in the faster more robust secondary immune response to antigen.

Memory, immunologic The attribute of the immune system mediated by **memory cells** whereby a second encounter with an antigen induces a heightened state of immune reactivity.

Metastasis The movement and colonization by tumor cells to sites distant from the primary site.

MHC (major histocompatibility complex) A group of genes encoding cell-surface molecules that are required for antigen presentation to T cells and for rapid graft rejection. It is called the H-2 complex in the mouse and the HLA complex in humans.

MHC restriction The characteristic of T cells that permits them to recognize antigen only after it is processed and the resulting antigenic peptide is displayed in association with either a **class I** or a **class II MHC molecule**.

MHC tetramers A soluble cluster of four MHC-peptide complexes used as a research tool to identify and trace antigen specific T cells in vitro and in vivo.

Microglial cell A type of macrophage found in the central nervous system.

Minor histocompatibility loci Genes outside of the MHC that encode antigens contributing to graft rejection.

Minor lymphocyte-stimulating (Mls) determinants Antigenic determinants encoded by endogenous retroviruses of the murine mammary tumor virus family that are displayed on the surface of certain cells.

Missing self model A model proposing that NK cell cytotoxicity is inhibited as long as they engage MHC class I with their receptors. However, when tumor cells and some virally infected cells reduce MHC class I expression, in other words, when they are "missing self" they are no longer protected from NK cytotoxicity.

Mitogen Activated Protein Kinase (MAPK) The first kinase in a MAP kinase cascade.

Mitogens Any substance that nonspecifically induces DNA synthesis and cell division. Common lymphocyte mitogens are concanavalin A, phytohemagglutinin, **lipopolysaccharide (LPS)**, pokeweed mitogen, and various **superantigens**.

Mixed-lymphocyte reaction (MLR) In vitro T-cell proliferation in response to cells expressing allogeneic MHC molecules; can be used as an assay for class II MHC activity.

Molecular mimicry One hypothesis used explain the induction of some autoimmune diseases, positing that some pathogens express antigenic determinants resembling host self components which can induce anti-self reactivity.

Monoclonal antibody Homogeneous preparation of antibody molecules, produced by a single clone of B lineage cells, often a hybridoma, all of which have the same antigenic specificity.

Monoclonal Deriving from a single clone of dividing cells.

Monocytes A mononuclear phagocytic leukocyte that circulates briefly in the bloodstream before migrating into the tissues where it becomes a **macrophage**.

Mucin A group of serine- and threonine-rich proteins that are heavily glycosylated. They are ligands for **selectins**.

Mucosal-associated lymphoid tissue (MALT) Lymphoid tissue situated along the mucous membranes that line the digestive, respiratory, and urogenital tracts.

Multiple myeloma A plasma-cell cancer.

Multiple sclerosis (MS) An autoimmune disease caused by auto-reactive T cells specific for components of the myelin sheaths which surround and insulate nerve fibers in the central nervous system. In Western countries, it is the most common cause of neurologic disability caused by disease.

Multipotential Can divide to form daughter cells that are more differentiated than the parent cell and that can develop along distinct blood-cell lineages.

Multipotential progenitor cells (MPPs) The stage of lymphoid differentiate that immediately precedes the **LMPP** stage. MPPs have lost the capacity for self-renewal that characterizes the true stem cell, but they retain the capacity to differentiate along many different hematopoietic lineages, including lymphoid, myeloid, erythroid, and megakaryocytic.

Multivalent Having more than one ligand binding site.

Mutational hot spots DNA sequences that are particularly susceptible to somatic hypermutation. Found in the variable regions of the immunoglobulin heavy and light chains.

Myasthenia gravis An autoimmune disease mediated by antibodies that block the acetylcholine receptors on the motor end plates of muscles, resulting in progressive weakening of the skeletal muscles.

Mycoses Any disease caused by fungal infection.

MyD88 Myeloid differentiation factor 88; adaptor protein that binds to all the IL-1 receptors and all TLRs except TLR3 and activates downstream signaling.

Myeloid progenitor cell A cell that gives rise to cells of the myeloid lineage. Also known as a **common myeloid progenitor** (CMP).

Myeloma A malignant tumor arising from cells of the bone marrow, specifically B cells.

Myeloma cell A cancerous plasma cell.

N-nucleotides, N-region nucleotides See **Non-templated (N) nucleotides**.

NADPH oxidase enzyme complex Phagosome oxidase, activated by pathogen binding to pattern-recognition receptors on phagocytic cells, that generates reactive oxygen species from oxygen.

Naïve Denoting mature B and T cells that have not encountered antigen; synonymous with *unprimed* and *virgin*.

Nasal-associated lymphoid tissue (NALT) Secondary lymphoid microenvironments in the nose that support the development of the T and B lymphocyte response to antigens that enter nasal passages. Part of the **mucosa associated lymphoid tissue system** (MALT).

Natural killer (NK) cells A class of large, granular, cytotoxic lymphocytes that do not have T- or B-cell receptors. They are antibody-independent killers of tumor cells and also can participate in **antibody-dependent cell-mediated cytotoxicity**.

Necrosis Morphologic changes that accompany death of individual cells or groups of cells and that release large amounts of intracellular components to the environment, leading to disruption and atrophy of tissue. See also **apoptosis**.

Negative costimulatory receptors Receptors expressed on the surface of some T cells that send signals that inhibit T cell activation. CTLA-4 is a common negative costimulatory molecule that is expressed on some activated T cells and helps to downregulate immune responses when antigen is cleared.

Negative selection The induction of death in lymphocytes bearing receptors that react too strongly with self antigens.

Neonatal Fc receptor (FcR$_N$) An class I MHC–like molecule that controls IgG and albumin half-life and transports IgG across the placenta.

Neoplasm Any new and abnormal growth; a benign or malignant tumor.

Neuraminidase (NA) An enzyme that cleaves N-acetylneuraminic acid (sialic acid) from glycoproteins. Most commonly a virally-expressed enzyme found on the surface of influenza virus that facilitates viral attachment to and budding from host cells.

Neutralize The ability of an antibody to prevent pathogens from infecting cells.

Neutralizing antibodies Antibodies that bind to pathogen and prevent it from infecting cells. Our most powerful vaccines induce neutralizing antibodies.

Neutrophil   A circulating, phagocytic granulocyte involved early in the **inflammatory response** (see Figure 2-2). It expresses **Fc receptors** and can participate in **antibody-dependent cell-mediated cytotoxicity**. Neutrophils are the most numerous white blood cells in the circulation.

NF-κB   An important transcription factor, most often associated with pro-inflammatory responses.

NKG2D   An activating receptor on NK cells that sends signals that enhance NK cytotoxic activity.

NKT cell   A subset of cytotoxic T cells that have features of both lymphocytes and innate immune cells. Like NK cells, they express NK1.1 and CD16, and like T cells they express TCRs, but these bind to glycolipids associated with an MHC class-I like molecule called CD1. iNKT cells are a subset of NKT cells that express very limited TCRαβ diversity.

Nod-like receptors (NLR)   Family of cytosolic pattern-recognition receptors with nuclear-binding/oligomerization and leucine-rich repeat domains; they have a nuclear-binding domain.

Non-nucleoside reverse transcriptase inhibitors (NNRTIs)   An antiretroviral drug inhibits the viral reverse transcriptase enzyme in a non-competitive fashion, by binding outside the substrate binding site and inducing a conformation change that inhibits DNA polymerization.

Non-templated (N) nucleotides   Nucleotides added to the V-D and D-J junctions of immunoglobulin and TCR genes by the enzyme **TdT**.

Nonamer   A sequence of nine nucleotides that is partially conserved and serves, with the absolutely conserved **heptamer**, as a binding site for the RAG1/2 recombination enzymes. The heptamer:nonamer sequences are found down- and up-stream of each of the V, D, and J segments encoding the immunoglobulin and T cell receptors.

Nonproductive rearrangement   Rearrangement in which gene segments are joined out of phase so that the triplet-reading frame for translation is not preserved.

Northern blotting   Common technique for detecting specific mRNAs, in which denatured mRNAs are separated electrophoretically and then transferred to a polymer sheet, which is incubated with a radiolabeled DNA probe specific for the mRNA of interest.

Notch   A surface receptor that when bound is cleaved to release a transcriptional regulator that regulates cell fate decisions. Notch activation is required for T cell development and determines whether a lymphocyte precursor becomes a B versus T cell.

nT$_{REG}$ cells   A type of T cell that is induced to express FoxP3 and acquire regulatory function during development in the thymus, and is responsible for suppressing immune activity against specific antigen. See also **iT$_{REG}$ cells**.

Nuclear Factor of Activated T cells (NFAT)   A transcription factor activated by TCR ligation. NFAT is held in the cytoplasm by the presence of a bound phosphate group. Activation through the TCR results in the activation of a phosphatase that releases the bound phosphate and facilitates entry of NFAT into the nucleus.

Nucleoside reverse transcriptase inhibitors (NRTIs)   An antiretroviral drug composed of nucleoside analogues which compete with native cellular nucleosides for binding to the viral reverse transcriptase enzyme but which lead to DNA chain termination and therefore halt viral DNA synthesis.

Nucleotide oligomerization domain/leucine-rich repeat-containing receptors (NLR)   see **Nod-like receptors (NLR)**.

Nude mouse   Homozygous genetic defect (*nu/nu*) carried by an inbred mouse strain that results in the absence of the thymus and consequently a marked deficiency of T cells and cell-mediated immunity. The mice are hairless (hence the name) and can accept grafts from other species.

Oncofetal tumor antigen   An antigen that is present during fetal development but generally is not expressed in tissues except following transformation. Alpha-feto protein (AFP) and carcinoembryonic antigen (CEA) are two examples that have been associated with various cancers (see Table 19-3).

Oncogene   A gene that encodes a protein capable of inducing cellular **transformation**. Oncogenes derived from viruses are written v-onc; their counterparts (**proto-oncogenes**) in normal cells are written c-onc.

One-turn recombination signal sequences   Immunoglobulin gene-recombination signal sequences separated by an intervening sequence of 12 base pairs.

Opportunistic infections   Infections caused by ubiquitous microorganisms that cause no harm to immune competent individuals but that pose a problem in cases of **immunodeficiency**.

Opsonin   A substance (e.g., an antibody or C3b) that promotes the phagocytosis of antigens by binding to them.

Opsonization, opsonize   Deposition of opsonins on an antigen, thereby promoting a stable adhesive contact with an appropriate phagocytic cell.

Original antigenic sin   The concept that a secondary response relies on the activity of memory cells rather than the activation of naïve cells, focusing on those structures that were present during the original, or primary, encounter with a pathogen and ignoring any new antigenic determinants.

Osteoclast   A bone macrophage.

P-addition   See **P-nucleotide addition**.

P-K reaction   Prausnitz-Kustner reaction, a local skin reaction to an allergen by a normal subject at the site of injected IgE from an allergic individual. (No longer used because of risk of transmitting hepatitis or AIDS.)

P-nucleotide addition   Addition of nucleotides from cleaved hairpin loops formed by the junction of V-D or D-J gene segments during Ig or TCR gene rearrangements.

P-region nucleotides   See **Palindromic (P) nucleotides**.

Palindromic (P) nucleotides   Nucleotides formed at the V-D and D-J junctions by asymmetric clipping of the hairpin junction formed by RAG1/2-mediated DNA cleavage (see Figure 7-8, Step 8).

PALS   See **Periarteriolar lymphoid sheath**.

Papain   A proteolytic enzyme derived from papayas and pineapples. Papain cleavage of immunoglobulins releases two **Fab** and one **Fc** fragment per IgG.

Paracortex   An area of the lymph node beneath the cortex that is populated mostly by T cells and interdigitating dendritic cells.

Paracrine   A type of regulatory secretion, such as a cytokine, that arrives by diffusion from a nearby cellular source.

Passive immunity   Temporary adaptive immunity conferred by the transfer of immune products, such as antibody (antiserum), from an immune individual to a nonimmune one. See also **active immunity**.

Passive immunotherapy   Treatment of an infectious disease by administration of previously generated antibodies specific for the infectious pathogen.

Pathogen   A disease-causing infectious agent.

Pathogen-associated molecular patterns (PAMPs)   Molecular patterns common to pathogens but not occurring in mammals. PAMPs are recognized by various **pattern-recognition receptors** of the innate immune system.

Pathogenesis   The means by which disease-causing organisms attack a host.

Pattern recognition receptors (PRRs)   Receptors of the innate immune system that recognize molecular patterns or motifs present on pathogens but absent in the host.

Pattern recognition   The ability of a receptor or ligand to interact with a class of similar molecules, such as mannose-containing oligosaccharides.

PAX5 transcription factor   A quintessential B cell transcription factor that controls the expression of many B cell specific genes.

Pentraxins   A family of serum proteins consisting of five identical globular subunits; CRP is a pentraxin.

Perforin   Cytolytic product of CTLs that, in the presence of Ca^{2+}, polymerizes to form transmembrane pores in target cells (see Figure 13-11).

Periarteriolar lymphoid sheath (PALS)   A collar of lymphocytes encasing small arterioles of the spleen.

Peripheral tolerance   Process by which self-reactive lymphocytes in the circulation are eliminated, rendered **anergic**, or otherwise inhibited from inducing an immune response.

Peyer's patches   Lymphoid follicles situated along the wall of the small intestine that trap antigens from the gastrointestinal tract and provide sites where B and T cells can interact with antigen.

PH domain   See **Pleckstrin homology (PH) domain**.

Phage display library   Collection of bacteriophages engineered to express specific V$_H$ and V$_L$ domains on their surface.

Phagocytes   Cells with the capacity to internalize and degrade microbes or particulate antigens; neutrophils and monocytes are the main phagocytes.

Phagocytosis   The cellular uptake of particulate materials by engulfment; a form of endocytosis.

Phagolysosome   An intracellular body formed by the fusion of a phagosome with a lysosome.

Phagosome   Intracellular vacuole containing ingested particulate materials; formed by the fusion of pseudopodia around a particle undergoing **phagocytosis**.

Phagosome oxidase (phox)   See **NADPH oxidase enzyme complex**.

Phosphatidyl Inositol bis-Phosphate (PIP$_2$)   A phospholipid found on the inner leaflet of cell membranes that is cleaved on cell signaling into the sugar **inositol trisphosphate** and **diacylglycerol**.

Phosphatidyl Inositol tris-Phosphate (PIP$_3$)   The product of **PIP$_2$** phosphorylation. PIP$_3$ is bound by signaling proteins bearing **PH domains**.

Phosphatidyl Inositol-3-kinase (PI3 kinase)   A family of enzymes that phosphorylate the inositol ring of phosphatidyl inositols at the 3 position. On immune signaling, the usual substrate is **PIP$_2$**.

Phosphatidyl serine   Membrane phospholipid. Normally located on the inner membrane leaflet, but flips to the outside leaflet in apoptotic cells.

Phospholipases Cγ (PLCγ)   A family of enzymes that cleave **phosphatidyl inositol bis-phosphate** into the sugar **inositol tris-phosphate** and the lipid **diacylglycerol**.

Physical barriers   Tissue layers, especially epithelia, whose physical integrity blocks infection.

PKCΘ   A serine/threonine protein kinase activated by **diacylglycerol** and important in TCR signaling.

Plasma cell   The antibody-secreting effector cell of the B lineage.

Plasma   The cell-free, fluid portion of blood, which contains all the clotting factors.

Plasmablasts   Cells that have begun the pathway of differentiation to plasma cells, but have not yet reached the stage of terminal differentiation and are therefore still capable of cell division.

Plasmacytomas   A plasma-cell cancer.

Plasmapheresis   A procedure that involves the separation of blood into two components, plasma and cells. The plasma is removed and cells are returned to the individual. This procedure is done during pregnancy when the mother makes anti-Rh antibodies that react with the blood cells of the fetus.

Plasmin   A serine protease formed by cleavage of plasminogen. Its major function is the hydrolysis of fibrin.

Platelets   Cells in the myeloid system that arise from megakaryocytes and regulate blood clotting.

Pleckstrin homology (PH) domain   A protein domain that binds specifically to **phosphatidyl inositol tris-phosphate** or **phosphatidyl inositol bis-phosphate**. These domain interactions are important in a wide variety of signaling cascades, including those initiated by ligand binding at the TCR and BCR.

Pleiotropic   Having more than one effect. For example, a cytokine that induces both proliferation and differentiation.

Poly-Ig receptor   A receptor for polymeric Ig molecules (IgA or IgM) that is expressed on the basolateral surface of most mucosal epithelial cells. It transports polymeric Ig across epithelia.

Polyclonal antibody   A mixture of antibodies produced by a variety of B-cell clones that have recognized the same antigen. Although all of the antibodies react with the immunizing antigen, they differ from each other in amino acid sequence.

Polygenic   The presence of multiple genes within the genome that encode proteins with the same function but slightly different structures.

Polymorphic, polymorphism   The presence of multiple allelic forms of a gene (**alleles**) within a population, as occurs within the major histocompatibility complex.

Positional cloning   A technique to identify a gene of interest that involves using known genetic markers to identifying a specific region in the genome associated with a characteristic of interest (e.g. disease susceptibility). This region is then cloned and sequenced to discover specific gene or genes that may be directly involved.

Positive selection   A process that permits the survival of only those T cells whose T-cell receptors recognize self MHC.

Potential candidate genes   List of genes that could be involved in the condition of interest, identified as a result research and educated guesses.

Pre-B-cell checkpoint   Developing B cells are tested at the pre-B cell stage to determine whether they can express a functional BCR heavy chain protein, in combination with the **VPreB** and **λ5** proteins, to form the pre-B-cell receptor. Those B cells that fail to form a functional pre-B-cell receptor are eliminated by apoptosis and are referred to as having failed to pass through the pre-B-cell checkpoint.

Pre-B-cell receptor   A complex of the Igα,Igβ heterodimer with membrane-bound Ig consisting of the μ heavy chain bound to the **surrogate light chain** Vpre-B/λ5.

Pre-T-cell receptor (pre-TCR)   A complex of the CD3 group with a structure consisting of the T-cell receptor β chain complexed with a 33-kDa glycoprotein called the pre-Tα chain.

Pre–B cell (precursor B cell)   The stage of B-cell development that follows the pro-B-cell stage. Pre-B cells produce cytoplasmic μ heavy chains and most display the **pre-B-cell receptor**.

pre–Tα chain   An invariant protein homolog of the TCRα chain that is expressed in early T cell development and pairs with newly rearranged TCRβ to form the pre-T cell receptor. This receptor delivers a signal that induces proliferation and differentiation to the CD4$^+$CD8$^+$ stage, when the TCRα chain is rearranged and if successfully translated into a protein takes the place of the pre-Tα.

Precipitin   An antibody that aggregates a soluble antigen, forming a macromolecular complex that yields a visible precipitate.

Primary foci   Clusters of antigen-stimulated dividing B cells found at the borders of the T and B cell areas of the lymph nodes and spleen that secrete IgM quickly after antigen stimulation.

Primary follicle   A lymphoid follicle, prior to stimulation with antigen, that contains a network of **follicular dendritic cells** and small resting B cells.

Primary immunodeficiency   An inherited genetic or developmental defect in some component/s of the immune system.

Primary lymphoid organs   Organs in which lymphocyte precursors mature into antigenically committed, immunocompetent cells. In mammals, the bone marrow and thymus are the primary lymphoid organs in which B-cell and T-cell maturation occur, respectively.

Primary response Immune response following initial exposure to antigen; this response is characterized by short duration and low magnitude compared to the response following subsequent exposures to the same antigen (**secondary response**).

Pro–B cell (progenitor B cell) The earliest distinct cell of the B-cell lineage.

Productive rearrangement The joining of V(D)J gene segments in phase to produce a VJ or V(D)J unit that can be translated in its entirety.

Professional antigen-presenting cell (pAPC) A myeloid cell with the capacity to activate T lymphocytes. Dendritic cells, macrophages, and B cells are all considered professional APCs because they can present antigenic peptide in both MHC class I and class II and express costimulatory ligands necessary for T cell stimulation.

Progenitor cell A cell that has lost the capacity for self renewal and is committed to the generation of a particular cell lineage.

Programmed cell death An induced and ordered process in which the cell actively participates in bringing about its own death (see also **Apoptosis**).

Proinflammatory Tending to cause inflammation; TNF-α, IL-6 and IL-1 are examples of proinflammatory cytokines.

Properdin Component of the alternative pathway of complement activation that stabilizes the alternative pathway **C3 convertase** C3bBb.

Prostaglandins A group of biologically active lipid derivatives of arachidonic acid. They mediate the **inflammatory response** and type I hypersensitivity reaction by inhibiting platelet aggregation, increasing vascular permeability, and inducing smooth-muscle contraction.

Protease inhibitors The common name for a class of antiviral medications that inhibit virus-specific proteases and therefore interfere with viral replication. Most often associated with anti-HIV drug therapy.

Proteasome A large multifunctional protease complex responsible for degradation of intracellular proteins.

Protectin Regulatory protein that binds to and inhibits the membrane attack complex of complement.

Protein A An F_C-binding protein present on the membrane of *Staphylococcus aureus* bacteria. It is used in immunology for the detection of antigen-antibody reactions and for the purification of antibodies.

Protein A/G A genetically engineered F_C-binding protein that is a hybrid of protein A and protein G. It is used in immunology for the detection of antigen-antibody reactions and for the purification of antibodies.

Protein G An F_C-binding protein present on the membrane of Streptococcus bacteria. It is used in immunology for the detection of antigen-antibody reactions and for the purification of antibodies.

Protein kinase C (PKC) A family of serine/threonine protein kinases activated by **diacylglycerol**.

Protein scaffold A set of proteins that binds together in a defined way to bring together molecules that would otherwise not come into contact. Protein scaffolds usually include a number of **adapter proteins** that then serve to bring enzyme and substrate proteins into apposition with the resultant activation or inhibition of the substrate protein.

Proto-oncogene A cancer-associated gene that encodes a factor that normally regulates cell proliferation, survival, or death; these genes are required for normal cellular functions. When mutated or produced in inappropriate amounts, a proto-oncogene becomes an oncogene, which can cause transformation of the cell (see Figure 19-2).

Provirus Viral DNA that is integrated into a host-cell genome in a latent state and must undergo activation before it is transcribed, leading to the formation of viral particles.

Pseudogene Nucleotide sequence that is a stable component of the genome but is incapable of being expressed. Pseudogenes are thought to have been derived by mutation of ancestral active genes.

Pseudopodia Membrane protrusions that extend from motile and phagocytosing cells.

Purine box factor 1 (PU.1) A transcription factor important in B cell development and survival.

Radioimmunoassay (RIA) A highly sensitive technique for measuring antigen or antibody that involves competitive binding of radiolabeled antigen or antibody (see Figure 20-6).

RAG1/2 (recombination-activating genes 1 and 2) The protein complex of RAG1 and RAG2 that catalyzes V(D)J recombination of B- and T-cell receptor genes. These proteins operate in association with a number of other enzymes to bring about the process of recombination, but RAG1/2, along with **TdT**, represent the lymphoid-specific components of the overall enzyme complex.

RAG1 (Recombination Activating Gene 1) See **RAG1/2**.

RAG2 (Recombination Activating Gene 2) See **RAG1/2**.

Ras Small, monomeric G protein important in signaling cascades.

Reactive nitrogen species Highly cytotoxic antimicrobial compounds formed by the combination of nitric oxide and superoxide anion within phagocytes such as neutrophils and macrophages.

Reactive oxygen species (ROS) Highly reactive compounds such as superoxide anion $\cdot O_2^-$, hydroxyl radicals $(OH\cdot)(OH^-)$, hydrogen peroxide (H_2O_2), and hypochlorous acid $(HClO)$ that are formed from oxygen under many conditions in cells and tissues, including microbe-activated innate responses of phagocytic cells; have anti-microbial activity.

Recent thymic emigrants (RTEs) Newly developed single positive $(CD4^+$ or $CD8^+)$ thymocytes that been allowed to leave the thymus. These new T cells still bear features of immature T cells and undergo further maturation in the periphery before joining the circulating fully mature, naïve T cell pool.

Receptor A molecule that specifically binds a **ligand**.

Receptor editing Process by which the T- or B-cell receptor sequence is altered after the initial recombination event, in order to reduce affinity for self antigens.

Recombinant inbred strains Mouse strains created by the mating of two inbred strains, with the formation of a recombination in an interesting locus such as the MHC. The mice bearing the recombination are then inbred to create a new inbred strain.

Recombination signal sequences (RSSs) Highly conserved heptamer and nonamer nucleotide sequences that serve as signals for the gene rearrangement process and flank each germ-line V, D, and J segment (see Figure 7-5).

Recombination-activating genes See **RAG1/2**.

Red pulp Portion of the spleen consisting of a network of sinusoids populated by macrophages and erythrocytes (see Figure 2-10). It is the site where old and defective red blood cells are destroyed.

Redundant Having the same effect as another signal.

Regulatory T cell (T_{REG}) A type of $CD4^+$ T cell that that negatively regulates immune responses. It is defined by expression of the master regulator FoxP3 and comes in two versions, the induced T_{REG}, which develops from mature T cells in the periphery, and the natural T_{REG}, which develops from immature T cells in the thymus.

Relative risk Probability that an individual with a given trait (usually, but not exclusively, a genetic trait) will acquire a disease compared with those in the same population group who lack that trait.

Resonant angle A property measured when using surface plasmon resonance to assess the affinity of the interaction between two molecules (see Figure 20-13).

Respiratory burst A metabolic process in activated phagocytes in which the rapid uptake of oxygen is used to produce reactive oxygen species that are toxic to ingested microorganisms.

Reticular dysgenesis (RD) A type of **severe combined immunodeficiency** (SCID) in which the initial stages of hematopoietic cell development are blocked by defects in the adenylate kinase 2 gene (*AK2*), favoring apoptosis of myeloid and lymphoid precursors and resulting in severe reductions in circulating leukocytes.

Retinoic acid-inducible gene-I-like receptors (RLRs) A family of cytosolic pattern-recognition receptors named for one member, RIG-I; known members are RNA helicases with CARD domains.

Retrovirus A type of RNA virus that uses a reverse transcriptase to produce a DNA copy of its RNA genome. HIV, which causes AIDS, and HTLV, which causes adult T-cell leukemia, are both retroviruses.

Rh antigen Any of a large number of antigens present on the surface of blood cells that constitute the Rh blood group. See also **erythroblastosis fetalis**.

Rheumatoid arthritis A common autoimmune disorder, primarily diagnosed in women 40 to 60 years old, caused by self-reactive antibodies called **rheumatoid factors**, which mediate chronic inflammation of the joints.

Rheumatoid factors Auto-antibodies found in the serum of individuals with rheumatoid arthritis and other connective-tissue diseases.

Rhogam Antibody against **Rh antigen** that is used to prevent **erythroblastosis fetalis**.

Rous sarcoma virus (RSV) A retrovirus that induces tumors in avian species.

Sarcoma Tumor of supporting or connective tissue.

Schistosomiasis A disease caused by the parasitic worm *Schistosoma*.

SCID See **Severe combined immunodeficiency**.

SCID-human mouse Immunodeficient mouse into which elements of a human immune system, such as bone marrow and thymic fragments, have been grafted. Such mice support the differentiation of pluripotent human hematopoietic stem cells into mature immunocytes and so are valuable for studies on lymphocyte development. See also **Severe combined immunodeficiency**.

SDS-polyacrylamide gel electrophoresis (SDS-PAGE) An electrophoretic method for the separation of proteins. It employs SDS to denature proteins and give them negative charges; when SDS denatured proteins are electrophoresed through polymerized-acrylamide gels, they separate according to their molecular weights.

Secondary follicle A **primary follicle** after antigenic stimulation; it develops into a ring of concentrically packed B cells surrounding a **germinal center**.

Secondary immune response The immune response to an antigen that has been previously introduced and recognized by adaptive immune cells. It is mediated primarily by memory T and B lymphocytes that have differentiated to respond more quickly and robustly to antigenic stimulation than the **primary response**.

Secondary immunodeficiency Loss of immune function that results from exposure to an external agent, often an infection.

Secondary lymphoid organs (SLOs) Organs and tissues in which mature, immunocompetent lymphocytes encounter trapped antigens and are activated into **effector cells**. In mammals, the **lymph nodes**, **spleen**, and **mucosal-associated lymphoid tissue (MALT)** constitute the secondary lymphoid organs.

Secondary response See **Secondary immune response**.

Secreted immunoglobulin (sIg) The form of antibody that is secreted by cells of the B lineage, especially plasma cells. This form of Ig lacks a transmembrane domain. See also **Membrane immunoglobulin (mIg)**.

Secretory component A fragment of the **poly-Ig receptor** that remains bound to Ig after transcytosis across an epithelium and cleavage.

Secretory IgA J chain–linked dimers or higher polymers of IgA that have transited epithelia and retain a bound remnant of the **poly-Ig receptor**.

Selectin One of a group of monomeric **cell adhesion molecules** present on leukocytes (L-selectin) and endothelium (E- and P-selectin) that bind to mucin-like CAMs (e.g., GlyCAM, PSGL-1) (see Box 14-2).

Self-MHC restriction The property of recognizing antigenic peptides only in the context of self MHC molecules.

Self-renewing Can divide to create identical copies of the parent cell.

Self-tolerance Unresponsiveness to self antigens.

Sepsis Infection of the bloodstream, frequently fatal.

Septic shock Shock induced by septicemia.

Septicemia Blood poisoning due to the presence of bacteria and/or their toxins in the blood.

Serum Fluid portion of the blood, which is free of cells and clotting factors.

Serum sickness A type III hypersensitivity reaction that develops when antigen is administered intravenously, resulting in the formation of large amounts of antigen-antibody complexes and their deposition in tissues. It often develops when individuals are immunized with antiserum derived from other species.

Severe combined immunodeficiency (SCID) A genetic defect in which adaptive immune responses do not occur, due to a lack of T cells and possibly B and NK cells.

SH2 domain A protein domain that binds phosphorylated tyrosine residues.

SH3 domain A protein domain that binds proline-rich peptides.

Signal A molecule that elicits a respond in a cell bearing a **receptor** for that signal. The response may be cell movement, cell division, activation of cell metabolism, or even cell death.

Signal joints In V(D)J gene rearrangement, the nucleotide sequences formed by the union of recombination signal sequences.

Signal peptide A small sequence of amino acids, also called the leader sequence, that guides the heavy or light chain through the endoplasmic reticulum and is cleaved from the nascent chains before assembly of the finished immunoglobulin molecule.

Signal Transducer and Activator of Transcription (STAT) Transcription factors that normal reside in the cytoplasm. Phosphorylation of cytokine receptors by **Janus Activated Kinases** results in the generation of binding sites on those receptors for the STATs, which relocate to the cytoplasmic regions of the receptor and are themselves phosphorylated by JAKs. Phosphorylated STATs them dimerize. Phosphorylation and dimerization expose nuclear localization signals and the STATs move to the nucleus where they act as transcription factors.

Signal transduction pathway A sequence of intra-cellular events brought about by receipt of a **signal** by a specific receptor.

Signaling Intracellular communication initiated by receptor-ligand interaction.

Single nucleotide polymorphisms (SNPs) A single base-pair variation in a DNA sequence among individuals in a population (i.e. an allelic variation). SNPs are often found in non-coding regions of the genome and can be used as genetic markers to locate genes that may underlie differences in individual responses to disease.

Single positive (SP) Either $CD4^+CD8^-$ or $CD8^+CD4^-$ mature T lymphocytes.

Slow-reacting substance of anaphylaxis (SRS-A) The collective term applied to leukotrienes that mediate inflammation.

Somatic hypermutation (SHM) The induced increase of mutation, 10^3- to 10^6-fold over the background rate, in the regions in and around rearranged immunoglobulin genes. In animals such as humans and mice, somatic hypermutation occurs in **germinal centers**.

Specificity, antigenic Capacity of antibody and T-cell receptor to recognize and interact with a single, unique **antigenic determinant** or **epitope**.

Spleen Secondary lymphoid organ where old erythrocytes are destroyed and blood-borne antigens are trapped and presented to lymphocytes in the **PALS** and **marginal zone** (see Figure 2-10).

Splenectomy   Surgical removal of the spleen.

Splenic vein   Vein that drains blood from the spleen and the site of egress of many splenic white blood cells.

Spliceosome   The complex of enzymes that mediates splicing of RNA.

Src-family kinases   A family of tyrosine kinase critically important in the early stages of signaling pathways in many cell types, including lymphocytes.

STAT   See **Signal Transducer and Activator of Transcription**.

Stem cell   A cell from which differentiated cells derive. Stem cells are classified as totipotent, pluripotent, multipotent, or unipotent depending on the range of cell types that they can generate. See Classic Experiment Box 2-1 and Clinical Focus Box 2-2.

Stem cell associated antigen-1 (Sca-1)   An antigen present on hematopoietic stem cells.

Stem cell factor (SCF)   A cytokine that exists in both membrane-bound and soluble forms. The SCF/c-Kit interaction is critical for the development, in adult animals, of **multipotential progenitor cells (MPPs)**.

Stem cell niches   Cellular microenvironments in the bone marrow (and some other tissues) that support the development of hematopoietic stem cells. Two are recognized: the **endosteal niche**, in proximity to bone cells (osteoblasts) and the **vascular niche**, in proximity to cells that line the blood vessels (endothelial cells).

Stochastic model   A model advanced to explain the molecular basis for **lineage commitment**, the choice of a $CD4^+CD8^+$ thymocyte to become a $CD4^+$ versus a $CD8^+$ T cell. This model proposes that all DP thymocytes receiving T cell receptor signals randomly decrease expression of either co-receptor CD4 or CD8. Only those thymocytes expressing the co-receptor that stabilizes the TCR-MHC interaction they experience will mature successfully. For example, a DP thymocyte that expresses a TCR with an affinity for MHC class I but randomly down-regulates CD8 will not receive adequate stimulation and will not mature. However, if that thymocyte randomly down-regulates CD4, it will maintain a TCR signal and mature to the CD8 lineage.

Streptavidin   A bacterial protein that binds to biotin with very high affinity. It is used in immunological assays to detect antibodies that have been labeled with biotin.

Stromal cell   A nonhematopoietic cell that supports the growth and differentiation of hematopoietic cells.

Sub-isotypes   A particular antibody subclass, e.g., IgG1 or IgA2.

Subcapsular sinus (SCS) macrophages   Macrophages that line the subcapsular sinus of the lymph nodes.

Subclasses   Variant sequences of the constant regions of antibodies of the IgG and IgA classes. There are four common variants of IgG and two of IgA in both mice and humans.

Superantigens   Any substance that binds to the V_β domain of the T-cell receptor and residues in the chain of class II MHC molecules. It induces activation of all T cells that express T-cell receptors with a particular V_β domain. It functions as a potent T-cell **mitogen** and may cause food poisoning and other disorders.

Surface plasmon resonance (SPR)   An instrumental technique for measurement of the affinity of molecular interactions based on changes of reflectance properties of a sensor coated with an interactive molecule (see Figure 20-13).

Surrogate light chain   The polypeptides **Vpre-B** and **λ5** that associate with μ heavy chains during the pre-B-cell stage of B-cell development to form the **pre-B-cell receptor**.

Switch (S) regions   In class switching, DNA sequences located upstream of each C_H segment (except C_δ).

Synergy   The property of two separate signals having an effect that is greater than the sum of the two signals.

Syngeneic   Denoting genetically identical individuals.

Systemic lupus erythematosus (SLE)   A multi-system autoimmune disease characterized by auto-antibodies to a vast array of tissue antigens such as DNA, histones, RBCs, platelets, leukocytes, and clotting factors.

T cell   See **T lymphocyte**.

T cytotoxic (T_C) cells   See **cytotoxic T lymphocytes**.

T follicular helper (T_{FH}) cells   Helper $CD4^+$ T cell subset that supports the development of B lymphocytes in the follicle and germinal center and expresses the master transcriptional regulator Bcl-6.

T helper (T_H) cells   T cells that are stimulated by antigen to provide signals that promote immune responses.

T helper type 1 (T_H1) cells   Helper $CD4^+$ T cell subset that enhances the cytotoxic immune response against intracellular pathogens and expresses the master transcriptional regulator T-Bet (see Table 11-3).

T helper type 17 (T_H17) cells   Helper $CD4^+$ T cell subset that enhances an inflammatory immune response against some fungi and bacteria and expresses the master transcriptional regulator ROR-γ.

T helper type 2 (T_H2) cells   Helper $CD4^+$ T cell subset that enhances B cell production of IgE and the immune response to pathogenic worms. It expresses the master transcriptional regulator GATA-3 (see Table 11-3).

T lymphocyte   A lymphocyte that matures in the thymus and expresses a T-cell receptor, CD3, and CD4 or CD8.

T-cell hybridoma   An artificially generated T cell tumor formed by fusing a single T cell with a naturally-occurring thymoma cell.

T-cell receptor (TCR)   Antigen-binding molecule expressed on the surface of T cells and associated with the **CD3** molecule. TCRs are heterodimeric, consisting of either an α and β chain or a γ and δ chain (see Figures 3-29 and 3-30).

T-dependent (TD) response   An antibody response elicited by a T-dependent antigen.

T-independent (TI) response   An antibody response elicited by a T-dependent antigen.

TACI receptor   Transmembrane activator and CAML interactor.

TAK1   See **Transforming growth factor-beta-kinase-1**.

TAP (transporter associated with antigen processing)   Heterodimeric protein present in the membrane of the rough endoplasmic reticulum (RER) that transports peptides into the lumen of the RER, where they bind to class I MHC molecules.

Tapasin   TAP-associated protein. Found in the ER that brings the TAP transporter into proximity with the class I MHC molecule, allowing it to associate with antigenic peptide.

TATA box   A Thymidine and Adenine-rich sequence upstream of many genes that serves as part of the RNA polymerase recognition and binding sequence.

TD antigen   See **Thymus-dependent (TD) antigen**.

Terminal deoxynucleotidyl transferase (TdT)   Enzyme that adds untemplated nucleotides at the V-D and D-J junctions of B and T cell receptor genes.

Tertiary lymphoid tissue   Aggregates of organized immune cells in organs that were the original site of infection.

T_H1 subset   See **T helper type 1 (T_H1) cells**.

T_H2 subset   See **T helper type 2 (T_H2) cells**.

Thoracic duct   The largest of the lymphatic vessels. It returns lymph to the circulation by emptying into the left subclavian vein near the heart.

Thromboxane   Lipid inflammatory mediator derived from arachidonic acid.

Thymocytes   Developing T cells present in the thymus.

Thymus   A primary lymphoid organ, in the thoracic cavity, where T-cell maturation takes place.

Thymus-dependent (TD) antigen Antigen that is unable to stimulate B cells to antibody production in the absence of help from T cells; response to such antigens involves isotype switching, **affinity maturation**, and memory-cell production.

TI-1 antigens T-Independent antigens, Type 1. Antigens that are capable of eliciting an antibody response in the absence of T cell help. TI-1 antigens are capable of being mitogenic.

TI-2 antigens T-Independent antigens, Type 2. Antigens that are capable of eliciting an antibody response in the absence of T cell help. TI-2 antigens are not capable of being mitogenic.

TIR domain Toll-IL-1R domain. Cytoplasmic domain of the IL-1 receptor and TLRs. Also present in the adaptors **MyD88** and **TRIF**, which interact with the receptors via TIR/TIR domain interactions.

Titer A measure of the relative strength of an antiserum. The titer is the reciprocal of the last dilution of an antiserum capable of mediating some measurable effect such as precipitation or agglutination.

TNF-α Tumor Necrosis Factor α. A pro-inflammatory cytokine.

Tolerance A state of immunologic unresponsiveness to particular antigens or sets of antigens. Typically, an organism is unresponsive or tolerant to self antigens.

Tolerogens Antigens that induce tolerance rather than immune reactivity. See also **immunogen**.

Toll-like receptors (TLRs) A family of cell-surface receptors found in invertebrates and vertebrates that recognize conserved molecules from many pathogens.

Toxoid A toxin that has been altered to eliminate its ability to cause disease but that still can function as an **immunogen** in a vaccine preparation.

Trabeculae Extensions of connective tissue (in this case, cartilage and bone) that provide a structural site and surface for the development of blood cells. Found at the ends of long bones (femur, tibia, humerus) as well as the edges of other bones.

TRAF6 TNF receptor-associated factor 6. Key signaling intermediate in IL-1R and TLR pathways. Activated by IRAK kinases and essential for activating downstream components.

Trafficking The differential migration of lymphoid cells to and from different tissues.

Transcytosis The movement of antibody molecules (polymeric IgA or IgM) across epithelial layers mediated by the poly-Ig receptor.

Transduce To pass on a signal from one part of the cell or one signaling molecule to the next in a signaling cascade.

Transformation Change that a normal cell undergoes as it becomes malignant, normally mediated by DNA alternations; also a permanent, heritable alteration in a cell resulting from the uptake and incorporation of foreign DNA into the genome.

Transforming growth factor-beta-kinase-1 (TAK1) Protein kinase downstream of IL-1R, TLRs, and other pattern-recognition receptors; part of complex with TAB1, TAB2 that is activated by TRAF6 and phosphorylates the IKK complex.

Transfusion reaction Type II hypersensitivity reaction to proteins or glycoproteins on the membrane of transfused red blood cells.

Transgene A cloned foreign gene present in an animal or plant.

Transitional B cells (T1, T2) Immature B cells that express the BCR but are not yet fully immunologically competent.

TRIF TIR-domain-containing adaptor protein. Adaptor recruited by TLR3 and TLR4 that mediates signaling pathway activating IRFs and interferon production.

Tuberculoid leprosy A disease caused by the intracellular bacteria, *Mycobacterium leprae* and *Mycobacterium lepromatosis*, where the immune response successfully encapsulates the bacteria and prevents widespread tissue damage. This more controlled form of disease is associated with the production of a T_H1 rather than T_H2 response.

Tumor-associated antigens (TSAs) Antigens that are expressed on particular tumors or types of tumors that are not unique to tumor cells. They are generally absent from or expressed at low levels on most normal adult cells. Formerly referred to as tumor-associated transplantation antigens (TATAs).

Tumor-specific antigens (TSAs) Antigens that are unique to tumor cells. Originally referred to as tumor-specific transplantation antigens (TSTAs).

Tumor-suppressor genes Genes that encode products that inhibit excessive cell proliferation or survival. Mutations in these genes are associated with the induction of malignancy.

Two-photon microscopy A type of microscopy that, like confocal microscopy, allows one to visualize fluorescent signals in one focal plane of a relatively thick tissue sample. It causes less damage than confocal microscopy and can be coupled with intravital imaging techniques.

Two-turn recombination signal sequences Immunoglobulin gene recombination signal sequences separated by an intervening sequence of 23 base pairs.

Tyk A kinase that belongs to the **JAK** family of kinases.

Type I hypersensitivity A pathologic immune reaction to non-infectious antigens mediated by IgE. It is the basis for allergy and atopy.

Type I interferons A group of cytokines belonging to the Interferon family of cytokines that mediates anti-viral effects. Type I interferons are released by many different cell types and are considered part of the innate immune system.

Type II hypersensitivity A pathologic immune reaction to non-infectious antigens mediated by IgG and IgM, which recruit complement or cytotoxic cells. It underlies blood transfusion reactions, Rh factor responses, and some hemolytic anemias.

Type II interferon A cytokine belonging to the interferon family that is normally secreted by activated T cells. Also known as Interferon γ.

Type III hypersensitivity A pathologic immune reaction to non-infectious antigens mediated by antibody-antigen immune complexes. It underlies damage associated with several disorders, including rheumatoid arthritis and systemic lupus erythematosus.

Type IV hypersensitivity A pathologic immune reaction to non-infectious antigens mediated by T cells. It underlies the response to poison ivy.

Ubiquitin A small signaling peptide that can either tag a protein for destruction by the proteasome, or, under some circumstances, activate that protein.

Unproductive An unproductive BCR or TCR gene rearrangement is one in which DNA recombination leads to a sequence containing a stop codon, that cannot therefore be transcribed and translated to form a functional protein.

Upstream (1) Towards the 5′ end of a gene; (2) Closer to the receptor in a signaling cascade.

V (variable) gene segment The 5′ coding portion of rearranged immunoglobulin and T-cell receptor genes. There are multiple V gene segments in germ-line DNA, but gene rearrangement leaves only one segment in each functional gene.

V (D)J recombinase The set of enzymatic activities that collectively bring about the joining of gene segments into a rearranged V(D)J unit.

Vaccination Intentional administration of a harmless or less harmful form of a pathogen in order to induce a specific adaptive immune response that protects the individual against later exposure to the pathogen.

Vaccine A preparation of immunogenic material used to induce immunity against pathogenic organisms.

Valence, valency Numerical measure of combining capacity, generally equal to the number of binding sites. Antibody molecules are bivalent or multivalent, whereas T-cell receptors are univalent.

Variability   (Antibody) variability is defined by the number of different amino acids at a given position divided by the frequency of the most common amino acid at that position (see Figure 3-18).

Variable (V) region   Amino-terminal portions of immunoglobulin and T-cell receptor chains that are highly variable and responsible for the **antigenic specificity** of these molecules.

Variable (V_L)   The variable region of an antibody **light chain**.

Vascular addressins   Tissue-specific adhesion molecules that direct the extravasation of different populations of circulating lymphocytes into particular lymphoid organs.

Vascular niche   Microenvironment in the bone marrow that fosters the development of hematopoietic stem cells and is postulated to associate specifically with hematopoietic stem cells that have begun to differentiate into mature blood cells.

Viral load   Concentration of virus in blood plasma; usually reported as copies of viral genome per unit volume of plasma.

Viral oncogene   Any cancer-promoting sequence carried by a virus that can induce transformation in infected host cells.

Vpre-B   A polypeptide chain that together with λ5 forms the **surrogate light chain** of the **pre-B-cell receptor**.

Western blotting   A common technique for detecting a protein in a mixture; the proteins are separated electrophoretically and then transferred to a polymer sheet, which is flooded with radiolabeled or enzyme-conjugated antibody specific for the protein of interest.

Wheal and flare reaction   A skin reaction to an injection of antigen that indicates an allergic response.

White pulp   Portion of the spleen that surrounds the arteries, forming a periarteriolar lymphoid sheath (PALS) populated mainly by T cells (see Figure 2-10).

Wiskott-Aldrich syndrome   An X-linked immune deficiency disorder caused by inheritance of a mutated *WASP* gene, which encodes a cytoskeletal protein highly expressed in hematopoietic cells and essential for proper immune synapse formation and intracellular signaling.

X-linked Hyper-IgM syndrome   An immunodeficiency disorder in which T_H cells fail to express CD40L. Patients with this disorder produce IgM but not other isotypes, they do not develop **germinal centers** or display somatic hypermutation, and they fail to generate **memory B cells**.

X-linked severe combined immunodeficiency (XSCID)   Immunodeficiency resulting from inherited mutations in the common γ chain of the receptor for IL-2, -4, -7, -9, and -15 that impair its ability to transmit signals from the receptor to intracellular proteins.

Xenograft   A graft or tissue transplanted from one species to another.

Yolk sac   Sac attached to the early embryo that provides nutrition for the embryo

ANSWERS TO STUDY QUESTIONS

Chapter 1

1. Jenner's method of using cowpox infection to confer immunity to smallpox was superior to earlier methods because it carried a significantly reduced risk of serious disease. The earlier method of using material from the lesions of smallpox victims conferred immunity but at the risk of acquiring the potentially lethal disease.

2. The Pasteur method for treating rabies consists of a series of inoculations with attenuated rabies virus. This process actively immunizes the recipient, who then mounts an immune response against the virus to stop the progress of infection. A simple test for active immunity would be to look for antibodies specific to the rabies virus in the recipient's blood at a time after completion of treatment, when all antibodies from a passive treatment would have cleared from circulation. Alternatively, one could challenge the recipient with attenuated rabies to see whether a secondary response occurred (this treatment may be precluded by ethical ramifications).

3. The immunized mothers would confer passive immunity upon their offspring because the anti-streptococcal antibodies, but not the B cells, cross the placental barrier and are present in the babies at birth. In addition, colostrum and milk from the mother would contain antibodies to protect the nursing infant from infection.

4. (a) H. (b) CM. (c) B. (d) H. (e) CM.

5. The four immunologic attributes are specificity, diversity, memory, and self/nonself recognition. Specificity refers to the ability of certain membrane-bound molecules on a mature lymphocyte to recognize only a single antigen (or small number of closely related antigens). Rearrangement of the immunoglobulin genes during lymphocyte maturation gives rise to antigenic specificity; it also generates a vast array of different specificities, or diversity, among mature lymphocytes. The ability of the immune system to respond to nonself-molecules, but (generally) not to self-molecules, results from the elimination during lymphocyte maturation of immature cells that recognize self-antigens. After exposure to a particular antigen, mature lymphocytes reactive with that antigen proliferate, differentiate, and adapt, generating a larger and more effective population of memory cells with the same specificity; this expanded population can respond more rapidly and intensely after a subsequent exposure to the same antigen, thus displaying immunologic memory.

6. The secondary immune response is faster (because it starts with an expanded population of antigen-specific cells), more effective (because the memory cells have learned and adapted during the primary response), and reaches higher levels of magnitude than the primary

response (again, because we begin with many more cells that have already honed their strategy).

7. Consequences of mild forms of immune dysfunction include sneezing, hives, and skin rashes caused by allergies. Asthma and anaphylactic reactions are more severe consequences of allergy and can result in death. Consequences of severe immune dysfunction include susceptibility to infection by a variety of microbial pathogens if the dysfunction involves immunodeficiency, or chronic debilitating diseases, such as rheumatoid arthritis, if the dysfunction involves autoimmunity. The most common cause of immunodeficiency is infection with the retrovirus HIV-1, which leads to AIDS.

8. (a) True. (b) True. (c) False. Most pathogens enter the body through mucous membranes, such as the gut or respiratory tract. (d) True. (e) False. Both are involved in each case. Innate immunity is deployed *first* during the primary response, and adaptive immunity begins later during that first encounter. During the secondary response, innate and adaptive immunity are again both involved. While innate responses are equally efficient, the second time around adaptive immunity uses memory cells to pick up where it left off at the end of the primary response and is therefore quicker and more effective in pathogen eradication during a secondary response. (f) False. These are two different types of disorder: autoimmunity occurs when the immune system attacks self, and immunodeficiency is when the immune system fails to attack nonself. The one caveat occurs in cases of immune deficiency involving immune *regulatory* components. Just like broken brakes, these can result in an overzealous immune attack on *self* structures that thus presents as autoimmunity. (g) False. The intravenous immunoglobulin provides protection for as long as it remains in the body (up to a few weeks), but this individual has not mounted his or her own immune response to the antigen and will therefore not possess any memory cells. (h) True. (i) False. The genes for encoding a T-cell receptor are rearranged and edited during T-cell development in the thymus so that each mature T cell carries a different T-cell receptor gene sequence. (j) False. The innate immune response does not generate memory cells (as the adaptive immune response does), so it is equally efficient during each infection.

9. Pasteur had inadvertently immunized his chickens during the first inoculation, using an old, attenuated bacterial strain. The old strain was no longer virulent enough to be fatal, but it was still able to elicit an adaptive immune response that protected the chickens from subsequent infections with fresh, virulent bacteria of the same type.

10. Viruses live inside host cells and require the host cell's machinery to replicate. Fungi are extracellular and are often kept in check by the immune system, but they can

be a problem for people with immune deficiencies. Fungi are the most homogenous in form. Parasites are the most varied in form and can range in size from single-celled, intracellular microorganisms to large macroscopic intestinal worms. Some parasites also go through several life-cycle stages in their human host, altering their antigenic structures and location so significantly between stages that they require completely different immune eradication strategies. Bacteria can cause intracellular or extracellular infections, which require different immune targeting and elimination methods. Many bacteria express cell surface molecular markers (PAMPs) that are recognized by receptors that are part of our innate immune system (PRRs).

11. Herd immunity is what occurs when enough members of a population have protective immunity to a pathogen, either through vaccination or prior infection, that they act as a buffer to spread and help protect those without immunity. The efficacy of herd immunity depends on pathogen characteristics such as how the pathogen is transmitted (airborne, fecal/oral, etc.) and whether it can survive for long outside the host. Herd characteristics include how we interact with one another as a "herd" (e.g., crowding indoors versus sparse populations and outdoor encounters) and the primary target population (e.g., young children versus adults). *(Other answers are also acceptable.)*

12. The instructional theory stated that the antibody structure is not determined until it is molded by antigen binding. In contrast, the selection theory states that the antigen receptor is first made (randomly) and then selected by antigen binding from a diverse group of receptors with predetermined structures.

13. (a) I. (b) A. (c) A. (d) A. (e) I. (f) A. (g) A. (h) A. (i) I. (j) A. (k) A.

14. Tolerance means that our immune system can discern between ourselves and foreign antigens, and does not attack self-antigens. Lymphocytes learn tolerance by being exposed to self-antigens during development, when most potentially self-reactive cells are either destroyed or inhibited from responding.

15. An antigen is anything that elicits an adaptive immune response—most commonly a part of some foreign protein or pathogen. An antibody is a soluble antigen-specific receptor molecule released by B cells that binds to an antigen and labels it for destruction. One interesting twist: antibodies can be antigens if they come from another species and are recognized as foreign by the host. This occurs in some instances of passive immunization, when antibodies from an animal such as a horse are given to humans, who then mount an adaptive immune response against horse-specific chemical patterns in the antibodies.

16. Pattern recognition receptors (PRRs) are germline-encoded receptors expressed in a variety of immune cells. They are designed by evolution to recognize molecules found on common pathogens, and they initiate the innate immune response when they bind these molecules. In contrast, B- and T-cell receptors are

expressed in lymphocytes, and the genes that encode these are produced by DNA rearrangement and editing, so that the receptor locus in each B or T cell's genome has a different sequence and encodes a different receptor. B- and T-cell receptors are diverse and have the ability to recognize a much greater assortment of antigens than PRRs, including those never before encountered by the immune system. B- and T-cell receptors are part of the adaptive immune response.

17. Cytokines are soluble molecular messengers that allow immune cells to communicate with one another. Chemokines are a subset of cytokines that are chemotactic, or have the ability to recruit cells to the site of infection.

Chapter 2

Research Focus Answer This is an open-ended question that describes real observations; several answers could be considered correct. The model you advance "simply" has to be logical and consistent with all the data presented. Perhaps the most straightforward possibility is the one that turns out to be true: Notch regulates the decision of a progenitor (the common lymphoid progenitor [CLP], in fact) to become either a B cell or a T cell. If Notch is active, CLPs become T cells (even in the bone marrow). If Notch is turned off, CLPs become B cells (even in the thymus). Other possibilities? Notch could induce apoptosis of B cells. However, if this were the case, why would there be more T cells in the bone marrow? Alternatively, Notch could influence the microenvironments that the cells develop in and, when active, make the bone marrow niches behave like thymic niches. When off, it could allow the thymic niche to "revert" to a bone marrow–like environment. This would work (even if it is a more complex scenario) and deserves full credit as an answer.

Clinical Focus Answer

(a) This outcome is unlikely; since the donor hematopoietic cells differentiate into T and B cells in an environment that contains antigens characteristic of both the host and donor, there is tolerance to cells and tissues of both. (b) This outcome is unlikely. T cells arising from the donor HSCs develop in the presence of the host's cells and tissues and are therefore immunologically tolerant of them. (c) This outcome is unlikely for the reasons cited in (a). (d) This outcome is a likely one because of the reasons cited in (a).

1. (a) Although T cells complete their maturation in the thymus, not the bone marrow, mature $CD4^+$ and $CD8^+$ T cells will recirculate back to the bone marrow. (b) Pluripotent hematopoietic stem cells are rare, representing less about 0.05% of cells in the bone marrow. (c) Hematopoietic stem cells can be mobilized from the bone marrow and circulate in the blood, which is now used as a source of stem cells for transplantation. (d) Macrophages will increase both class I and class II MHC expression after activation; however, T_H cells are $CD4^+$ and recognize antigenic peptide bound to class II MHC. (e) Immature, not mature B cells, are associated with osteoblasts, which help them to develop.

(f) Lymphoid follicles are found in all secondary lymphoid tissues, including that associated with mucosal tissues (MALT). (g) The follicular dendritic cell (FDC) network guides B cells within follicles. The follicular reticular cell (FRC) system guides T cells within the T cell zone (although in some cases it may also help B cells to get to follicles, so aspects of this statement are true). (h) Infection and associated inflammation stimulates the release of cytokines and chemokines that enhance blood cell development (particularly to the myeloid lineage). (i) FDC present soluble antigen on their surfaces to B cells, not T cells. (j) Dendritic cells can arise from both myeloid and lymphoid precursors. (k) B and T lymphocytes have antigen specific receptors on their surface, but NK cells, which are also lymphocytes do not. (l) B cells are generated outside the marrow in birds and ruminants. (m) This was true for all jawed vertebrates, but now may even be true for jawless vertebrates; in other words, this is now true, not false! (n) Recent data suggest that at least one jawless vertebrate, the lamprey, has T- and B-like cells.

2. (a) Myeloid progenitor. (b) Granulocyte-monocyte progenitor. (c) Hematopoietic stem cell. (d) Lymphoid progenitor.

3. The primary lymphoid organs are the bone marrow (bursa of Fabricius in birds) and the thymus. These organs function as sites for B-cell and T-cell maturation, respectively. The secondary lymphoid organs are the spleen, lymph nodes, and mucosal-associated lymphoid tissue (MALT) in various locations. MALT includes the tonsils, appendix, and Peyer's patches, as well as loose collections of lymphoid cells associated with the mucous membranes lining the respiratory, digestive, and urogenital tracts. All these organs trap antigen and provide sites where lymphocytes can interact with antigen and subsequently undergo clonal expansion.

4. Stem cells are capable of self-renewal and can give rise to more than one cell type, whereas progenitor cells have lost the capacity for self-renewal and are committed to a single cell lineage. Commitment of progenitor cells depends on their acquisition of responsiveness to particular growth factors.

5. The two primary roles of the thymus are the generation and the selection of a repertoire of T cells that will protect the body from infection.

6. In humans, the thymus reaches maximal size during puberty. During the adult years, the thymus gradually atrophies.

7. The SCID mouse has no T or B lymphocytes because they cannot rearrange their antigen specific receptors. Injection of normal stem cells will result in restoration of these cell types—an easily measured outcome. As HSCs are successively enriched in a preparation, the total number of cells that must be injected to restore these cell populations decreases.

8. Monocytes are the blood-borne precursors of macrophages. Monocytes have a characteristic kidney-bean-shaped nucleus and limited phagocytic and microbial killing capacity compared to macrophages.

Macrophages are much larger than monocytes and undergo changes in phenotype to increase phagocytosis, antimicrobial mechanisms (oxygen dependent and oxygen independent), and secretion of cytokines and other immune system modulators. Tissue-specific functions are also found in tissue macrophages.

9. The bursa of Fabricius in birds is the primary site where B lymphocytes develop. Bursectomy would result in a lack of circulating B cells and humoral immunity, and it would probably be fatal.

10. (a) The spleen has been classically considered an organ that 'filters' antigens from blood. The lymph node receives antigens principally from the afferent lymphatics. (b) Both the paracortex of the lymph node and the periarteriolar lymphoid sheath of the spleen are rich in T cells. B cells are found primarily in the follicles. (c) Germinal centers are found wherever there are follicles, which are present in all secondary lymphoid tissue, including lymph node, spleen, and the wide variety of mucosal associated lymphoid tissues. (d) True. (e) Afferent lymphatics are associated with lymph nodes, not spleen. (Efferent lymphatics can be found in both.) (f) Ikaros is required for lymphoid development which occurs primarily in the bone marrow (and thymus). However, lymphocytes populate the spleen and mount an immune response there, so the spleen would clearly be compromised in function in the absence of Ikaros and lymphocytes.

Descriptions

1. (d), 2. (l), 3. (c), 4. (m, n), 5. (g), 6. (i), 7. (n), 8. (b), 9. (o), 10. (h), 11. (c, f, g), 12. (c, f), 13. (c, e, f), 14. (j, k).

Chapter 3

1. (a) Cytoplasm (b) Nucleus (c) By de-phosphorylation with the enzyme calcineurin phosphatase, activated by binding to the calcium-calmodulin complex. Activation of the cell results in an increase in intracytoplasmic calcium ion concentration. (d) There are many possible answers to this question. For example, there may be multiple forms of NFAT, and so these drugs may just interact with particular forms of the enzyme. Other cells may have alternative pathways that can bypass the need for calcineurin, whereas T cells may not. In fact, cyclosporin binds to the protein immunophilin, which is specifically expressed in activated T cells, and it is the complex of cyclosporin and immunophilin that inhibits the activity of calcineurin.

2. Immunoglobulin proteins and the T-cell receptor share a common motif: the immunoglobulin fold, which may be the binding target for these antibodies.

3. (a) False. Receptors and ligands bind through noncovalent interactions such as hydrogen bonds, ionic bonds, hydrophobic interactions, and van der Waals interactions. (b) True. Receptor-ligand binding can activate the receptor and result in phosphorylation and receptor cross-linking.

4. (a) Reduction enabled the breakage of disulfide bonds between the heavy and light chains and the pairs of heavy chains in the IgG molecule. Alkylation ensured that the

bonds between the separated chains would not reform. Scientists could then separate the chains and figure out their molecular weight and how many chains belonged to each molecule. (b) Papain digested the molecule in the hinge region, releasing two antigen-binding fragments (Fabs) and one non–antigen-binding fragment, which spontaneously crystallized (Fc). This told investigators that there were two antigen-binding sites per molecule and suggested that part of the molecule was not variable in sequence (since it was regular enough in structure to crystallize). Pepsin isolated the divalent antigen-binding part of the molecule from the rest. (c) Antibodies were generated that specifically recognized the Fab and Fc fragments. Antibodies to Fab were found to bind to both heavy and light chains, thus indicating that the antigen-binding sites had components of both chains. Antibodies to Fc fragments bound only to the heavy chain.

5. An ITAM is an *Immunoreceptor Tyrosine-based Activation Motif*. It is a motif with a particular sequence containing phosphorylatable tyrosines in the cytoplasmic regions of several immunologically important proteins. Igα and Igβ are part of the signaling complex in B cells and are phosphorylated by the Src-family kinase Lyn, when the BCR moves into the lipid raft regions of the membrane upon activation.

6. An adapter protein bears more than one binding site for other proteins and serves to bring other proteins into contact with one another without, itself, having any enzymatic activity. Activation of the TCR results in activation of the Src-family kinase, Lck. Lck phosphorylates the tyrosine residues on the ITAMs of the CD3 complex associated with the TCR. ZAP-70 then binds to the pY residues via its SH2 regions, and is then itself phosphorylated and activated.

7. I would expect IgM to be able to bind more molecules of antigen than IgG, but perhaps not five times more. Steric hindrance/conformational constraints may prevent all 10 antigen-binding sites of IgM from being able to simultaneously bind to 10 antigenic sites.

8. Because you know that, in at least one case, activation of a cell can result in an alteration in the phenotype of the cytokine receptor, resulting in a dramatic increase in its affinity for the cytokine. For example, the IL2R exists in two forms: a moderate affinity form consisting of a βγ dimer and a high-affinity form, synthesized only following cell activation that consists of an αβγ trimer.

9. Src-family kinases have two tyrosine sites on which they can be phosphorylated: an inhibitory and an activatory site. In the resting state, the inhibitory site is phosphorylated by the kinase Csk, and the kinase folds up on itself, forming a bond between an internal SH2 group and the inhibitory phosphate, shielding the active site of the enzyme. Cleavage of the phosphate group from the inhibitory tyrosine allows the enzyme to open its structure and reveal the active site. Further activation of the enzyme then occurs when a second tyrosine is phosphorylated and stabilizes the activated state.

10. Since Lck lies at the beginning of the signaling pathway from the T-cell receptor, a T cell with an inactive form of lck will not be able to undergo activation to secrete IL2.

11. The signaling kinase Bruton's tyrosine kinase (Btk) is defective in 85% of cases of X-linked agammaglobulinemia. It is encoded on the X chromosome, and hence most cases of this disease occur in boys. Phosphorylation by Btk activates PLCγ2, which cleaves PIP_2 into inositol trisphosphate (IP_3) and diacylglycerol (DAG) following activation of pre-B cells through the pre-B-cell receptor, or of B cells through the immunoglobulin receptor. This leads eventually to activation of the NFAT and MAP kinase transcription factor pathways, culminating in B-cell differentiation and proliferation, which cannot occur in the absence of a functioning Btk protein.

12. Both the BCR and the TCR are noncovalently associated on their respective cell membranes with signal transduction complexes that function to transduce the signal initiated by antigen binding to the receptor into the interior of the cell. In the case of the B-cell receptor, the signal transduction complex is composed of Igα/Igβ. ITAMs on Igα/Igβ are phosphorylated by tyrosine kinases that are brought into close proximity with the B-cell receptor complex upon the oligomerization of the receptor and its movement into the lipid raft regions of the membrane, which occur upon antigen binding. The phosphorylated tyrosines of Igα/Igβ then serve as docking points for downstream components of the signal transduction pathway. In the case of the TCR, the signal transduction complex is the CD3 set of proteins, which contains six chains and serves a similar function to Igα/Igβ. ITAMs on CD3 are phosphorylated by Src family kinases, particularly Lck, and then serve as docking points for downstream components of the TCR signaling pathway. Lck in turn is associated with the TCR co-receptors CD8 and CD4.

13. (There are multiple possibilities for this answer, and we offer a few here).

    (a) The antibody is a Y-shaped molecule with the antigen-binding regions located at the two tips of the Y. At the junction of the three sections is a flexible hinge region that allows the two tips to move with respect to one another and hence to bind to antigenic determinants arranged at varying distances from one another on a multivalent antigen. (b) Within the variable region domains of the heavy and light chains, a common β-pleated sheet scaffold includes multiple anti-parallel β strands in a conserved conformation. However, at the turns between the β strands, there are varying numbers and sequences of amino acids, corresponding to hypervariable regions in the immunoglobulin variable regions sequences. These enable the creation of many different antigen-binding sites. (c) The constant regions of antibody molecules form the bridge between the antigen-binding region and receptors on phagocytic cells that will engulf antigen-antibody complex, or components of the complement system that will bind to the Fc regions of the antibody and aid in the disposal of the antigen. Different constant region structures bind to different Fc receptors and complement components.

14. Src-family kinases such as Lck are located at the beginning of many signal transduction pathways, and their activities are subjected to rigorous control mechanisms. Lck is maintained in an inactive state by being phosphorylated on an inhibitory tyrosine residue.

This phosphorylated tyrosine is then bound by an internal SH2 domain that holds the Lck in a closed, inert conformation. If the DNA encoding this tyrosine residue has been mutated or eliminated, such that this phosphorylation cannot occur, then there will be constitutive level of Lck activity. Since the enzyme that phosphorylates this inhibitory tyrosine is Csk, a reduction in Csk activity would have the same effect.

Chapter 4

1. All four are chemical signaling molecules, which are often, but not always, proteins. They deliver a message to a second cell, causing physiological changes in the second, target cell, which must bear receptors that are specific to the signaling molecule. Growth factors are often secreted constitutively, whereas cytokines and hormones are secreted in response to particular stimuli and their secretion is short lived. Most hormones are produced by specialized glands, and their action affects a small number of target cells. In contrast, cytokines are often produced by, and bind to, a variety of cells. Chemokines are specialized cytokines that provide chemotactic signals which induce target cells to move in a particular direction toward the cell secreting the chemokine.

2. Cells undergoing an immune response often move very close to one another and are held together by adhesive interactions. In this case, if, for example, the cytokine-releasing cell is a dendritic cell and the target cell is a T cell, the two cells will be held in close apposition to one another for several hours. In addition, the secretory apparatus of the dendritic cell will be oriented such that the cytokines will be secreted into the tiny space between the dendritic cell and the T cell. The actual concentration of cytokines experienced by the T-cell receptors at the point of contact with the dendritic cell will therefore be orders of magnitude higher than that in the surrounding circulation.

3. Activation of Type I and Type II cytokine receptors results in their dimerization and subsequent phosphorylation by *Janus Activated Kinases* (JAKs). These JAKs then phosphorylate inactive, cytoplasmic transcription factors, called *Signal Transducers and Activators of Transcription*, or STATs. Once phosphorylated, the STATs dimerize, with SH2 domains in each STAT binding to a phosphorylated tyrosine residue in its partner. This phosphorylation and dimerization allow the STATs to assume a conformation that allows them to dissociate from the cytokine receptor complex, enter the nucleus, and stimulate transcription from selected genes.

4. *Pleiotropy* is the capacity to bring about different end results in different cells.

   Synergy is the ability of two or more cytokines affecting a cell to bring about a response that is greater than the sum of each of the cytokines.

   Redundancy is the property that describes the fact that more than one cytokine can bring about the same effect.

   Antagonism is the tendency for two cytokines binding to the same cell to bring about opposite effects, or to reduce/eliminate the response to the other.

   Cascade induction is the ability of a cytokine to bind to one cell and to induce that cell to secrete additional cytokines.

5. A cytokine may induce the expression on the cell surface of new chemokine receptors and/or new adhesion molecules that would cause the cell to move to a new location and, once present, to be retained there.

6. Some cytokines in this class have distinct cytokine-binding subunits, but share a signaling subunit. For example, the dimeric receptors for IL-3 and GM-CSF have different binding, but the same signaling, subunits. If the cell is responding to IL-3, and all the signaling subunits are occupied in this response, binding of GM-CSF to the binding subunit alone would fail to transduce a signal. Thus, although the two receptors are not competing for binding to their respective ligands, they do compete for signaling efficacy.

7. When a Type I interferon binds to its receptor on a virally infected cell, the interferon signal results in the activation of a ribonuclease that breaks down cytoplasmic RNA. It is particularly effective against double-stranded RNA.

8. See Figure 4-14.

9. In the case of IL-1, cells secrete a protein called IL-1Ra, a soluble form of the IL-1 receptor, which is capable of binding to IL-1, but not of transducing an IL-1 signal. The balance of IL-1Ra, vs the signaling receptor will determine the amount of IL-1 able to productively interact with a cell-surface receptor. IL-18BP serves a similar purpose in the IL-18 system. Sometimes cells will cleave the interleukin-binding portion of the receptor from the surface of the cell. This has an effect similar to that described above, in that the cleaved receptor can bind to the cytokine, blocking its access to the productive receptor. This occurs in the case of the IL-2R, where the IL-2-binding portion of the IL-2Rα chain is secreted, and is able to bind the soluble IL-2 protein.

10. (a) The IL-2R is made up of three components: α, β, and γ. Quiescent (unactivated) T cells bear only the βγ dimer, which binds IL-2, but at intermediate affinity insufficient to result in a signal under physiological conditions. Upon antigen stimulation, the T cell synthesizes the α subunit, which binds to the βγ dimer and converts it into a high-affinity receptor capable of mediating the IL-2 signal at physiological cytokine concentrations.

   (b) See the following figure.

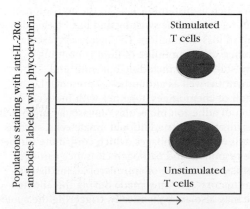

Populations staining with anti-IL-2Rβ and γ antibodies labeled with fluorescein

Chapter 5

Clinical Focus Answer Children with genetic defects in MyD88 and IRAK4 are particularly susceptible to infections with *Streptococcus pneumoniae*, *Staphylococcus aureus*, and *Pseudomonas aeruginosa*, and children with genetic variants of TLR4 are susceptible to Gram-negative urinary tract infections. Children with defects in the pathways activating the production or antiviral effects of IFNs–α, β are susceptible to herpes simplex virus encephalitis. These individuals are thought not to be susceptible to a broader array of infectious diseases because other innate and adaptive immune responses provide adequate protection. The supporting evidence for the adaptive immune response protecting against infections is that this limited array of susceptibilities is seen primarily in children, who become less susceptible as they get older, presumably as they develop adaptive immunological memory to these pathogens.

1. (a) defensins, lysozyme, psoriasin (b) phagocytosis, C-reactive protein, ficolins, MBL, surfactant proteins A, D; C-reactive protein, MBL (c) NADPH phagosome oxidase, O_2, ROS; NO, iNOS, arginine, RNS (d) Interferons-α, β, NK cells (e) Acute phase response, proinflammatory cytokines, IL-1, TNF-α (f) PRRs, PAMPs, antibodies, T-cell receptors (g) TLR2, TLR4 (h) TLR3, TLR7, TLR9 (i) TLR4, MyD88, TRIF (j) RLRs, NLRs (k) NF-κB, IRFs (l) IL-1, PRRs, caspase-1, inflammasomes, NLR (m) PAMPs, PRRs, dendritic cells (n) PRRs, defensins

2. B. Beutler showed that *lpr* mice were resistant to endotoxin (LPS) and that the genetic difference in these mice was lack of a functional TLR4 because of a single mutation in the TLR4 gene. R. Medzhitov and C. Janeway demonstrated that a protein with homology to *Drosophila* Toll (which turned out to be TLR4) activated the expression of innate immunity genes when expressed in a human cell line.

3. Inflammation is characterized by redness, heat, swelling, pain, and sometimes loss of local function. Cytokines made by PRR-activated resident innate cells act on the vascular endothelium, causing vascular dilation (producing redness and heat) and increasing permeability, resulting in influx of fluid and swelling (producing edema). Prostaglandins generated following the induced expression of COX2, together with mediators such as histamine, lead to the activation of local pain receptors. The swelling and local tissue damage can result in loss of function. The increased vascular permeability allows an influx of fluid containing protective substances, including opsonins and complement (as well as antibodies, if present). Local production of chemokines, together with induced expression of adhesion molecules on vascular endothelial cells, recruits to the site additional innate cells, such as neutrophils and macrophages, which contribute further to innate responses and pathogen clearance through phagocytosis and release of antimicrobial mediators. Proinflammatory cytokines made during this innate response may also act systemically, triggering the acute-phase response.

4. The innate immune system plays key roles in activating and regulating adaptive immune responses. Dendritic cells are key mediators of these roles. They bind pathogens at epithelial layers and deliver them to secondary lymphoid organs such as lymph nodes. After activation/maturation induced by TLR signaling, they present peptides from processed antigen to activate naïve $CD4^+$ and $CD8^+$ T cells. Depending on the pathogen and the PRRs to which it binds, dendritic cells are activated to produce certain cytokines that differentially induce naïve $CD4^+$ cells to differentiate into T_H subsets with different functions, usually appropriate for the particular pathogen. Also, PAMP binding to TLRs expressed on B and T lymphocytes can contribute to their activation by specific antigens to generate adaptive responses. Example of an adaptive response enhancing innate immune responses: The cytokine IFN-γ, produced by activated T_H1 cells, is a potent macrophage activator, including activating them to kill intracellular bacteria such as *M. tuberculosis*.

5. A major potential disadvantage of the adaptive immune system, with the *de novo* generation of diverse antigen receptors in each individual's B and T cells, is the possibility of autoimmunity, which may result in disease. Adaptive responses are also slow. Conserved PRRs that have evolved to recognize PAMPs are less likely to generate destructive responses to self components, and innate responses are activated rapidly. The disadvantages of potential autoreactivity and slow response are overwhelmed by the advantages of having adaptive as well as innate immunity. While the innate response is rapid and helps initiate and regulate the adaptive response, innate immunity cannot respond to new pathogens that may have evolved to lack PAMPs recognized by PRRs. Also, in general (except for NK cells) there is no immunological memory, so the innate response cannot be primed by initial exposure to pathogens. Given their complementary advantages and disadvantages, both systems are essential to maintaining the health of vertebrate animals.

Analyze the Data

a. Sepsis is a potentially dangerous systemic response to infection (usually from septicemia—blood bacterial infections) that includes fever, elevated heartbeat and breathing rate, low blood pressure, and compromised organ function due to circulatory defects. In its extreme form sepsis can lead to circulatory and respiratory failure, resulting in septic shock and death in many cases. Disseminated infections activate blood cells including monocytes and neutrophils, as well as resident macrophages and other cells in the spleen, liver, and other tissues, to release proinflammatory mediators including IL-1, TNF-α, and IL-6; these, in turn, systemically activate vascular endothelial cells, inducing their expression of cytokines, chemokines, adhesion molecules, and clotting factors that amplify the inflammatory response. Enzymes, ROS, and RNS released by activated neutrophils and other cells damage the vasculature; this damage, together with TNF-α-induced vasodilation and increased vascular permeability, results in fluid loss into the tissues that lowers blood pressure. TNF-α also stimulates release of clotting factors by vascular endothelial cells, resulting in intravascular blood clotting in capillaries. These effects on the blood vessels are particularly damaging to the kidneys and lungs, which are highly vascularized. High circulating TNF-α and IL-1 levels also adversely affect the heart.

b. ES-62 inhibits production of all of the tested proinflammatory mediators following activation with either the TLR2 or TLR4 ligands. As ES-62 binds to TLR4 but not TLR2, the results imply that ES-62 binding inhibits pathways downstream of both TLR4 and TLR2 that activate of synthesis and secretion of these mediators, rather than just inhibiting TLR4 itself (such as by blocking LPS binding).

This is an example of positive feedback regulation: activation of macrophages and neutrophils by bacterial infection stimulates increased levels of TLR4 and TLR2 expression, presumably making the cells better able to respond to the bacteria with protective innate and inflammatory responses. However, increased TLR expression might exacerbate the response in sepsis patients and increase the chances of getting septic shock.

ES-62 treatment completely inhibits expression of TLR4, so that the staining is the same as that of the isotype control antibody (i.e., there is no specific staining). In contrast, ES-62 has no effect on TLR2 expression on cells from sepsis patients.

The Western blot shows that incubation with ES-62 leads to a loss of MyD88 from cells, suggesting that ES-62 induces MyD88 degradation.

As ES-62 induces the loss of MyD88 and of TLR4 (but not TLR2) from the cell surface, ES-62 probably induces the degradation of both proteins. Clearly the loss of cell surface TLR4 would eliminate activation by the potent TLR ligand LPS. In addition, the absence of MyD88 would prevent the activation of signaling pathways downstream of all of the other TLRs except TLR3 (which signals through the TRIF adaptor) and therefore would explain ES-62's inhibition of activation of proinflammatory mediators by both TLR4 and TLR2 ligands shown in panel (a). Other experiments reported in the article by Puneet et al. (2011) confirm that ES-62 does induce the intracellular degradation of TLR4 and MyD88; ES-62 targets them to endosomes, which fuse with lysosomes.

Given the many complex effects of excessive levels of proinflammatory mediators on numerous organs that are associated with septic shock, it is unlikely that ES-62 injection would cure septic shock once it is underway. However, injections of ES-62 in sepsis patients might reduce the total levels of proinflammatory mediators to prevent septic shock from happening. This was confirmed in mouse sepsis studies described in the article by Puneet et al. (2011). Mice that had their intestines punctured surgically developed lethal septic shock from systemic infection with mixed intestinal bacteria. However, treatment with ES-62 beginning 1 hour after puncture provided complete protection against septic shock, and combined treatment with both ES-62 and antibiotics (commonly used in sepsis patients) when started as late as 6 hours after puncture also gave complete protection. As Puneet et al. (2011) conclude in their article, treatment with ES-62 (or related compounds) combined with antibiotics is worth exploring as an potentially effective treatment of sepsis patients.

Chapter 6

1. (a) True (b) True (c) True (d) True (e) False. The outer membrane of a virus is derived from the outer membrane of the host cell and is therefore susceptible to complement-mediated lysis. However, some viruses have developed mechanisms that enable them to evade complement-mediated lysis. (f) True

2. IgM undergoes a conformational change upon binding to antigen, which enables it to be bound by the first component of complement C1q. In the absence of antigen binding, the C1q binding site in the Fc region of IgM is inaccessible.

3. (a) The initiation of the classical pathway is mediated by the first component of complement, and C3 does not participate until after the formation of the active C3 convertase C4aC2b. The initiation of the alternate pathway begins with spontaneous cleavage of the C3 component, and therefore no part of the alternate pathway can operate in the absence of C3. (b) Clearance of immune complexes occurs only following opsonization of complexes by C3b binding, followed by phagocytosis or binding to the surface of erythrocytes via CR1 binding. Therefore, immune complex clearance is inhibited in the absence of C3. (c) Phagocytosis would be diminished in the absence of C3b-mediated opsonization. However, if antibodies specific for the bacteria are present, some phagocytosis would still occur.

4. Opsonizes pathogens, thus facilitating binding of immune system cells via complement receptors. Membrane attack complex (MAC) can induce lysis of pathogens. C3a, C4a, C5a act as chemoattractants to bring leukocytes to the site of infection. Increasing inflammatory response at the site of infection. Binding of C3 fragments by CD21 enhances B-cell activation by complement-coated antigen.

5. (a) The classical pathway is initiated by immune complexes involving IgM or IgG; the alternate pathway is generally initiated by binding of C3b to bacterial cell-wall components, and the lectin pathway is initiated by binding of lectins (e.g., MBL to microbial cell-wall carbohydrates). (b) The terminal reaction sequence after the C5 convertase generation is the same for all three pathways. The differences in the first steps are described in part (a) above. For the classical and the lectin pathways, the second step involves the binding of the serine protease complexes C4b2a (classical) and MASP1, MASP2 (lectin). These each act as a C3 convertase in each respective pathway, and the two pathways are identical from that point on. The alternative pathway uses a different C3 convertase, C3bBb. The formation of

Bb requires factor D, and the C3 convertase is stabilized on the cell surface by properdin. The alternative pathway C5 convertase is C3bBbC3b. (c) Healthy cells contain a variety of complement inhibitor proteins that prevent inadvertent activation of the alternative pathway. Some of these proteins are expressed only on the surface of host, and not on microbial cells; others are in solution, but are specifically bound by receptors on host cells. Mechanisms of action are explained in the answer to question 7.

6. (a) Induces dissociation and inhibition of the C1 proteases C1r and C1s from C1q. Serine protease inhibitor. (b) Accelerates dissociation of the C4b2aC3,

C3 convertase. Acts as a cofactor for factor I in C4b degradation. (c) Accelerates dissociation of C4b2a and C3bBb C3 convertases. (d) Acts as cofactor for factor I in degradation of C3b and C4b. (e) Binds C5b678 on host cells, blocking binding of C9 and the formation of the MAC. (f) Cleaves and inactivates the anaphylatoxins.

7. Immune complexes are cleared from the body following opsonization by C3b. In the absence of the early components of complement, C3b convertases are not formed, and hence no C3b will be available.

8. See table below.

	Complement component knocked out						
	C1q	C4	C3	C5	C9	Factor B	MASP-2
Formation of classical pathway C3 convertase	A	A	NE	NE	NE	NE	NE
Formation of alternative pathway C3 convertase	NE	NE	A	NE	NE	A	NE
Formation of classical pathway C5 convertase	A	A	A	NE	NE	NE	NE
Formation of lectin pathway C3 convertase	NE	A	NE	NE	NE	NE	A
C3b-mediated opsonization	D	D	A	NE	NE	D	D
Neutrophil chemotaxis and inflammation	D	D	D	D	NE	D	D
Cell lysis	D	D	A	A	A	D	D

Clinical Focus Answer (a) The two complement regulatory proteins DAF and Protectin are both attached to plasma membrane surfaces by glycosylphosphatidyl inositol linkages. In paroxysmal nocturnal hemoglobinuria (PNH), a defect in the enzyme PIG-A, which synthesizes these linkages, causes a decreased surface expression of both of these proteins. (b) Defects in PIG-A tend to be expressed somatically in cells early in the hematopoietic development. A given individual may express red blood cells that are wholly deficient, partially deficient, or wholly competent in *pig-A* expression. (c) PNH patients are essentially unable to express CD16 or CD66abce, indicating that those antigens are also most probably attached to the membranes using GPI linkages. Similarly, PNH patients are unable to express CD14, so it is also probably a GPI-linked protein. In contrast, PNH patients as well as normal control individuals are both able to express CD64, which is therefore most likely to be a transmembrane protein.

Analyze the Data (a) The arrow shows the population of cells that is undergoing apoptosis. CD46 is rapidly lost from the surface of cells undergoing apoptosis. C1q binds to the DNA that appears on the surface of apoptotic cells, and in the absence of the regulatory component CD46, the cell is susceptible to C3b deposition and C3b-mediated opsonization.

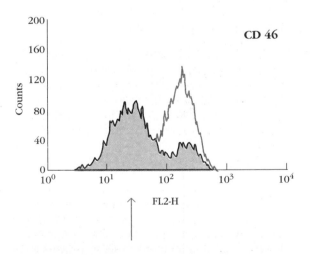

(b) Each of the circles shown represents cell populations with the indicated surface expression of C1q and CD46. T cells undergoing apoptosis begin to express DNA on their cell surfaces, which is bound by the first component of the classical pathway, C1q. Healthy T cells, on the other hand, express relatively high levels of the regulatory complement component, CD46, which is a cofactor for factor I.

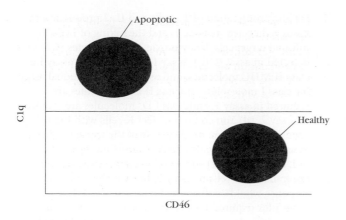

Chapter 7

1. (a) False: V_κ gene segments and C_λ are located on separate chromosomes and cannot be brought together during gene rearrangement. (b) True. (c) False. Naïve B cells produce a long primary transcript that carries the variable region and the mRNA for both the μ and δ constant regions. The switch in expression from the μ to the δ heavy chain occurs by mRNA splicing, not by DNA rearrangement. Switching to all other heavy-chain classes is mediated by DNA rearrangements. (d) True. (e) False. The variable regions of the β and δ TCR genes are encoded in three segments, analogous to the V, D, and J segments of the Ig heavy-chain variable region. The V_α and V_γ regions are each encoded in two segments.

2. V_H and J_H gene segments cannot join because both are flanked by recombination signal sequences (RSSs) containing a 23-bp (2-turn) spacer (see Figure 7-5b). According to the one-turn/two-turn joining rule, signal sequences having a two-turn spacer can join only with signal sequences having a one-turn (12-bp) spacer.

3. (a) 1, 2, 3. (b) 5. (c) 2, 3, 4. (d) 5. (e) 1, 3, 4, 5.

4. (a) P: Productive rearrangement of heavy-chain allele 1 must have occurred since the cell line expresses heavy chains encoded by this allele. (b) G: Allelic exclusion forbids the second heavy-chain allele from undergoing either productive or nonproductive rearrangement. (c) NP: In mice, the κ genes rearrange before the λ genes. Since the cell line expresses λ light chains, both κ alleles must have undergone nonproductive rearrangement, thus permitting λ-gene rearrangement to occur. (d) NP: Same reason as given in (c) above. (e) P: Productive rearrangement of the first λ-chain allele must have occurred since the cell line expresses λ light chains encoded by this allele. (f) G: Allelic exclusion forbids λ-chain allele 2 from undergoing either productive or nonproductive rearrangement (see Figure 7-11).

5. The κ-chain DNA must have the germ-line configuration because a productive heavy-chain rearrangement must occur before the light-chain (κ) DNA can begin to rearrange.

6. Random addition of N nucleotides at the D-J and V-DJ junctions contributes to the diversity within the CDR3 regions of heavy chains, but this process can result in a nonproductive rearrangement if the triplet reading frame is not preserved.

7. Whereas N-region addition occurs at the joints of Ig heavy but not light-chain variable regions, all TCR variable region joints may include N-region nucleotides. Somatic mutation adds diversity to the BCR, following antigen stimulation, but does not contribute to TCR diversity.

8. They used the fact that the receptor is a *membrane-bound protein* to isolate membrane-bound polysomes and used the RNA associated with the polysomes to generate cDNA probes specific for membrane receptor genes. They used the fact that the TCR is expressed in T cells but not B cells to remove all cDNAs that were expressed in both B and T cells. They hypothesized that the gene for the TCR would be encoded in recombining segments and that the pattern of DNA fragments encoding the receptor genes would be differentially arranged in different T cell clones.

9. (a) It must be a heavy chain because light chains do not have D regions. (b) The RSS. The heptamer of the heptamer-spacer-nonamer sequence directly abuts the end of the V region (c)(1). P-region nucleotides are formed by asymmetric cleavage of the hairpin at the coding joint prior to DNA ligation. The italicized GA residues on the coding strand, and the CT residues on the noncoding strand, could have been generated by that mechanism. (c)(2). We cannot know for certain that they were formed by P-nucleotide addition, as they could just as easily have been randomly inserted by TdT. (c)(3). The residues shown in bold have no place of origin in the original sequence, and so must have been added by N-nucleotide addition. (c)(4). Yes, we can, if no corresponding nucleotides can be found in the germ-line sequence, and they could not be accounted for by asymmetric hairpin joining.

10. (a) The restriction endonuclease must cut at a site within the constant region, as well as at sites upstream and downstream from it. (b) Recombination has occurred at only one of the two alleles. The germ-line bands complementary to both the constant and the variable region probes most probably derive from the other allele. (c) Given that the additional bands have remained in the same positions in the gel in both the germ-line and the myeloma DNA, it seems likely that there was a successful rearrangement at the first allele. (d) I would clone and sequence the DNA upstream from the C_κ region from both alleles and prove that one displayed a successful arrangement, although the other was still in the germ-line configuration.

Chapter 8

Clinical Focus Answer The TAP deficiency results in lack of class I molecules on the cell surface or a type I bare-lymphocyte syndrome. This leads to partial immunodeficiency in that antigen presentation is compromised, but there are NK cells and $\gamma\delta$ T cells to limit viral infection. Autoimmunity results from the lack of class I molecules that give negative signals through the killer-cell inhibitory receptor (KIR) molecules; interactions between KIR and class I molecules prevent the NK cells from lysing target cells. In their absence, self-cells are targets of autoimmune attack on skin cells, resulting in the lesions seen in TAP-deficient patients. The use of gene therapy to cure those affected with TAP deficiency is complicated by the fact that class I genes are expressed in nearly all nucleated cells. Because the class I product is cell bound, each deficient cell must be repaired to offset the effects of this problem. Therefore, although the replacement of the defective gene may be theoretically possible, ascertaining which cells can be repaired by transfection of the functional gene and reinfused into the host remains an obstacle.

1. (a) True (b) True (c) False: Class III MHC molecules are soluble proteins that do not function in antigen presentation. They include several complement components, TNF-α, and Lymphotoxin α. (d) False: The offspring of heterozygous parents inherit one MHC haplotype from each parent and thus will express some molecules that differ from those of each parent; for this reason, parents and offspring are histoincompatible. In contrast, siblings have a one in four chance of being histocompatible (see Figure 8-10c). (e) True (f) False: Most nucleated cells express class I MHC molecules, but neurons, placental cells, and sperm cells at certain stages of differentiation appear to lack class I molecules. (g) True

2. (a) Liver cells: Class I K^d, K^k, D^d, D^k, L^d, and L^k.
   (b) Macrophages: Class I K^d, K^k, D^d, D^k, L^d, and L^k. Class II $IA\alpha^k\beta^k$, $IA\alpha^d\beta^d$, $IA\alpha^k\beta^d$, $IA\alpha^d\beta^k$, $IE\alpha^k\beta^k$, $IE\alpha^d\beta^d$, $IE\alpha^d\beta^k$, $IE\alpha^k\beta^d$.

3.

	MHC molecules expressed on the membrane of the transfected L cells					
Transfected gene	D^k	D^b	K^k	K^b	IA^k	IA^b
None	+	–	+	–	–	–
K^b	+	–	+	+	–	–
$IA\alpha^b$	+	–	+	–	–	–
$IA\beta^b$	+	–	+	–	–	–
$IA\alpha^b$ and $IA\beta^b$	+	–	+	–	–	+

4. (a) SJL macrophages express the following MHC molecules: K^s, D^s, L^s, and IA^s. Because of the deletion of the IEα locus, IE^s is not expressed by these cells. (b) The transfected cells would express one heterologous IE molecule, $IE\alpha^k\beta^s$, and one homologous IE molecule, $IE\alpha^k\beta^k$, in addition to the molecules listed in (a).

5. See Figure 8-1, Figure 3-20, and Figure 3-19.

6. (a) The polymorphic residues are clustered in short stretches primarily within the membrane-distal domains of the class I and class II MHC molecules (see Figure 8-12). These regions form the peptide-binding groove of MHC molecules. (b) MHC polymorphism is thought to arise by gene conversion of short, nearly homologous DNA sequences within unexpressed pseudogenes in the MHC to functional class I or class II genes.

7. (a) The proliferation of T_H cells and IL-2 production by them is detected in assay 1, and the killing of LCMV-infected target cells by cytotoxic T lymphocytes (CTLs) is detected in assay 2. (b) Assay 1 is a functional assay for class II MHC molecules, and assay 2 is a functional assay for class I molecules. (c) Class II IA^k molecules are required in assay 1, and class I D^d molecules are required in assay 2. (d) You could transfect K cells with the IA^k gene and determine the response of the transfected cells in assay 1. Similarly, you could transfect a separate sample of L cells with the D^d gene and determine the response of the transfected cells in assay 2. In each case, a positive response would confirm the identity of the MHC molecules required for LCMV-specific activity of the spleen cells. As a control in each case, L cells should be transfected with a different class I or class II MHC gene and assayed in the appropriate assay. (e) The immunized spleen cells express both IA^k and D^d molecules. Of the listed strains, only A.TL and (BALB/c × B10.A) F_1 express both of these MHC molecules, and thus these are the only strains from which the spleen cells could have been isolated.

8. It is not possible to predict. Since the peptide-binding cleft is identical, both MHC molecules should bind the same peptide. However, the amino acid differences outside the cleft might prevent recognition of the second MHC molecules by the T-cell receptor on the T_C cells.

9. If RBCs expressed MHC molecules, then extensive tissue typing would be required before a blood transfusion, and only a few individuals would be acceptable donors for a given individual.

10. (a) No. Although those with the A99/B276 haplotype are at significantly increased relative risk, there is no absolute correlation between these alleles and the disease. (b) Nearly all of those with the disease will have the A99/B276 haplotype, but depending on the exact gene or genes responsible, this may not be a requirement for development of the disease. If the gene responsible for the disease lies between the A and B loci, then weaker associations to A99 and B276 may be observed. If the gene is located outside of the A and B regions and is linked to the haplotype only by association in a founder, then associations with other MHC genes may occur. (c) It is not possible to know how frequently the combination will occur relative to the frequency of the two individual alleles; linkage disequilibrium is difficult to predict. However, based on the data given, it may be speculated that the linkage to a disease that is fatal in individuals who have not reached reproductive years will have a negative effect on the frequency of the founder haplotype. An educated guess would be that the A99/B276 combination would be rarer than predicted on the basis of the frequency of the A99 and B276 alleles.

11. By convention, antigen-presenting cells are defined as those cells that can display antigenic peptides associated with class II MHC molecules and can deliver a costimulatory signal to $CD4^+$ T_H cells. A target cell is any cell that displays peptides associated with class I MHC molecules to $CD8^+$ T_C cells.

12. (a) Self-MHC restriction is the attribute of T cells that limits their response to antigen associated with self-MHC molecules on the membrane of antigen-presenting cells or target cells. In general, $CD4^+$ T_H cells are class II MHC restricted, and $CD8^+$ T_C cells are class I MHC restricted, although a few exceptions to this pattern occur. (b) Antigen processing involves the intracellular degradation of protein antigens into peptides that associate with class I or class II MHC molecules. (c) Endogenous antigens are synthesized within altered self-cells (e.g., virus-infected cells or tumor cells), are processed in the endogenous pathway, and are presented by class I MHC molecules to $CD8^+$ T_C cells. (d) Exogenous antigens are internalized by antigen-presenting cells, processed in the exogenous pathway, and presented by class II MHC molecules to $CD4^+$ T_H cells. (e) Anchor residues are the key locations (typically, positions 2/3 and 9) within an 8- to 10-amino-acid-long antigenic peptide that make direct contact with the antigen-binding cleft of MHC class I. The specific residues found at these locations distinguish the peptide fragments that can bind each allelic variant of class I. (f) An immunoproteasome is a variant of the classical proteasome, found in all cells, and is expressed in antigen-presenting cells and in infected target cells. The presence of this variant increases the production of antigenic fragments optimized for binding to MHC class I molecules.

13. (a) EN: Class I molecules associate with antigenic peptides and display them on the surface of target cells to $CD8^+$ T_C cells. (b) EX: Class II molecules associate with exogenous antigenic peptides and display them on the surface of APCs to $CD4^+$ T_H cells. (c) EX: The invariant chain interacts with the peptide-binding cleft of class II MHC molecules in the rough endoplasmic reticulum (RER), thereby preventing binding of peptides from endogenous sources. It also assists in folding of the class II α and β chains and in movement of class II molecules from the RER to endocytic compartments. (d) EX: Lysosomal hydrolases degrade exogenous antigens into peptides; these enzymes also degrade the invariant chain associated with class II molecules, so that the peptides and MHC molecules can associate. (e) EN: TAP, a transmembrane protein located in the RER membrane, mediates transport of antigenic peptides produced in the cytosolic pathway into the RER lumen where they can associate with class I MHC molecules. (f) B: In the endogenous pathway, vesicles containing peptide-class I MHC complexes move from the RER to the Golgi complex and then on to the cell surface. In the exogenous pathway, vesicles containing the invariant chain associated with class II MIIC molecules move from the RER to the Golgi and on to endocytic compartments. (g) EN: Proteasomes are large protein complexes with peptidase activity that degrade intracellular proteins within the cytosol. When associated with LMP2 and LMP7, which are encoded in the MHC region, and LMP10, which is not MHC encoded, proteasomes preferentially generate peptides that associate with class I MHC molecules. (h) B: Antigen-presenting cells internalize exogenous (external) antigens by phagocytosis or endocytosis. (i) EN: Calnexin is a protein within the RER membrane that acts as a molecular chaperone, assisting in the folding and association of newly formed class I α chains and β_2-microglobulin into heterodimers. (j) EX: After degradation of the invariant chain associated with class II MHC molecules, a small fragment called CLIP remains bound to the peptide-binding cleft, presumably preventing premature peptide loading of the MHC molecule. Eventually, CLIP is displaced by an antigenic peptide. (k) EN: Tapasin (TAP-associated protein) brings the transporter TAP into proximity with the class I MHC molecule and allows the MHC molecule to acquire an antigenic peptide (see Figure 8-18).

14. (a) Chloroquine inhibits the exogenous processing pathway, so that the APCs cannot display peptides derived from native lysozyme. The synthetic lysozyme peptide will exchange with other peptides associated with class II molecules on the APC membrane, so that it will be displayed to the T_H cells and induce their activation. (b) Delay of chloroquine addition provides time for native lysozyme to be degraded in the endocytic pathway.

15. (a) Dendritic cells: constitutively express both class II MHC molecules and costimulatory signals. B cells: constitutively express class II molecules but must be activated before expressing the CD80/86 costimulatory signal. Macrophages: must be activated before expressing either class II molecules or the CD80/86 costimulatory signal. (b) See Table 8-4. Many nonprofessional APCs function only during sustained inflammatory responses.

16. (a) R. (b) R. (c) NR. (d) R. (e) NR. (f) R.

17. (a) Intracellular bacteria, such as members of the *Mycobacterium* family, are a major source of nonpeptide antigens; the antigens observed in combination with CD1 are lipid and glycolipid components of the bacterial cell wall. (b) Members of the CD1 family associate with β_2-microglobulin and have structural similarity to class I MHC molecules. They are not true MHC molecules because they are not encoded within the MHC and are on a different chromosome. (c) The pathway for antigen processing taken by the CD1 molecules differs from that taken by class I MHC molecules. A major difference is that CD1 antigen processing is not inhibited in cells that are deficient in TAP, whereas class I MHC molecules cannot present antigen in TAP-deficient cells.

18. (b) The TAP1-TAP2 complex is located in the endoplasmic reticulum.

19. The offspring must have inherited HLA-A 3, HLA-B 59, and HLA-C 8 from the mother. Potential father 1 cannot be the biological father because although he shares HLA determinants with the offspring, the determinants are the same genotype inherited from the mother. Potential father 2 could be the biological father because he expressed the HLA genes expressed by the offspring that are not inherited from the mother (HLA-A 43, HLA-B 54, HLA-C 5). Potential father 3 cannot be the biological father because although he shares HLA determinants with the offspring the determinants are the same genes inherited from the mother.

20. Considering HLA-A, HLA-B and HLA-C only, a maximum of 6 different class I molecules are expressed in individuals who inherit unique maternal and paternal alleles at each loci. In the case of class II, considering only HLA-DP, DQ, and DR molecules where any α and β chain of each gene can pair to produce new maternal/paternal combinations, a maximum of 12 different class II molecules can be expressed (4 DP, 4 DQ, and 4 DR). Since humans can inherit as many as 3 functional DRβ genes, each of which is polymorphic, in practice fully heterozygous individuals have the ability to express more than 4 HLA-DR proteins.

21. Polygeny is defined as the presence of multiple genes in the genome with the same or similar function. In humans, MHC class I A, B, and C or class II DP, DQ, and DR are both examples of this (see Figure 8-7). Polymorphism is defined as the presence of multiple alleles for a given gene locus within the population. HLA-A1 versus HLA-A2 (for example, see Table 8-3) are examples of polymorphic alleles at the class I locus. Codominant expression is defined as the ability of an individual to simultaneously express both the maternal and the paternal alleles of a gene in the same cell. This process is what allows a heterozygous individual to express, for instance, both HLA-Cw2 and Cw4 alleles (see Figure 8-11). Polygeny ensures that even MHC homozygous individuals express a minimum of three different class I and class II proteins, each with a slightly different antigen-binding profile, expanding their repertoire of antigens that can be presented. MHC polymorphism and codominant expression in outbred populations help facilitate the inheritance and expression of different alleles at each locus, further increasing the number of different antigens that one individual can present. Codominant expression at the class II loci carries an added bonus: since these proteins are generated from two separate genes/chains, new combinations of α and β chains can arise, further enhancing the diversity of class II protein isoforms, or the number of unique MHC class II antigen-binding clefts.

22. The invariant chain is involved in MHC class II folding and peptide binding. Cells without this protein primarily retain misfolded class II proteins in the RER and are therefore unable to express MHC class II molecules on the cell surface. Since APCs are the primary cell types that express class II, cells with this mutant phenotype would be incapable of presenting exogenously processed antigens to naïve $CD4^+$ T cells.

23. Cross-presentation is the process by which some APCs can divert antigens collected from extracellular sources (exogenous pathway) to processing and presentation via MHC class I proteins (typically the realm of the endogenous pathway). This process is important for activation of naïve $CD8^+$ T cells to generate CTLs capable of detecting and lysing virally infected target cells. Dendritic cells, or a subset of this cell type, are thought to be the major players in this process, although "licensing" by antigen-specific $CD4^+$ T_H cells may first be required in order for DCs to engage in cross-presentation.

Analyze the Data (a) Yes. Comparing the relative amounts of L^d and L^q molecules without peptides, there are about half as many open L^q molecules as L^d molecules. The data suggest that L^q molecules form less stable peptide complexes than L^d molecules. (b) Part (a) in the figure shows that 4% of the L^d molecules don't bind MCMV peptide compared to 11% of the L^d after a W to R mutation. Thus, there appears to be a small decrease in peptide binding to L^d. It is interesting to note that nonspecific peptide binding increases severalfold after mutagenesis, based on the low amount of open-form L^d W97R (mutated L^d) versus native L^d. (c) Part (b) in the figure shows that 71% of the L^d molecules don't bind tum^- P91A14-22 peptide after a W to R mutation, compared to 2% for native L^d molecules. Thus, there is very poor binding of tum^- P91A14-22 peptide after a W to R mutation. (d) You would inject a mouse that expressed L^d because only 2% of the L^d molecules were open forms after the addition of tum^- P91A14-22 peptide compared to 77% free forms when L^q were pulsed with peptide. Therefore, L^d would present peptide better and probably activate T cells better than L^q. (e) Conserved anchor residues at the ends of the peptide bind to the MHC, allowing variability at other residues to influence which T-cell receptor engages the class I MHC-antigen complex.

Chapter 9

Clinical Focus Answers

1. FoxP3 (involved in development of T_{REG} cells), any TCR signaling molecules (which regulate TCR signal strength and therefore the outcome of thymic selection), MHC (which presents self-peptides), anything that prevents T cells from getting to the medulla (e.g., CCR7), and, finally, Nur77 and Bim, which are signaling molecules that regulate negative selection.

2. (a) The authors appear to be saying that the MHC variant, which would be expressed by medullary epithelial cells and dendritic cells in the thymus, may not be able to present (bind to) certain brain-specific self-peptides (or does so inefficiently). Therefore, some autoreactive $CD4^+$ T cells in the thymus may not be deleted. (Note that HLA-DR is an MHC Class II molecule.) (b) There is no one right question. Here are two possibilities. (1) If this were the case, wouldn't this also mean that peripheral dendritic cells would not be able to present the peptide, and therefore wouldn't activate the autoreactive T-cell escapees? (2) If you let an autoreactive helper T cell escape, don't you still need an autoreactive cytotoxic T cell to escape also? (This question depends on your knowing that autoimmune diseases like multiple sclerosis are caused in part by $CD8^+$ T-cell mediated damage.) (c) Given that immunologists are still pondering the issue, this is a challenging question. However, here are a couple of possibilities. (1) One accepts the possibility that this is a problem with negative selection. However, to understand how this leads to autoimmune disease, one can propose, for instance, that there is a difference between antigen presentation in the thymus and antigen presentation in the

periphery. Presentation of brain self-peptides by this MHC variant may be inefficient and allow autoreactive CD4$^+$ T cells to escape. But in the periphery this inefficiency can be overcome by enhancements in costimulatory molecule expression, levels of MHC expression, and so on among antigen-presenting cells that have been stimulated by pathogen or cell damage. (2) In addition (or alternatively), one can propose that this MHC variant compromises the development of regulatory T cells, not just the deletion of autoreactive CD4$^+$ T cells. Specifically, if the MHC Class II variant is unable to present the self-peptides, then it will neither mediate deletion of autoreactive CD4$^+$ T cells *nor* select for suppressive, autoreactive regulatory T cells. Note that we need to assume, in all cases, the existence of autoreactive CD8$^+$ T-cell escapees. This is not a radical assumption. As you now know, negative selection in the thymus is never perfect, and our ability to maintain tolerance depends to a significant extent on peripheral mechanisms.

3. (a) Fas-mediated signaling is important for the death-mediated regulation of lymphocyte populations in lymphocyte homeostasis. This regulation is essential because the immune response generates a sudden and often large increase in populations of responding lymphocytes. Inhibition of Fas-mediated cell death could result in progressively increasing and eventually unsustainable lymphocyte levels. The Canale-Smith syndrome demonstrates that without the culling of lymphocytes by apoptosis, severe and life-threatening disease can result. (b) Analysis of the patient T cells reveals a large number of double-negative (CD4$^-$, CD8$^-$) T cells and almost equal numbers of CD4$^+$ and CD8$^+$ cells. A normal subject would have very few ~5% double-negative cells and more CD4$^+$ than CD8$^+$ cells. See the Clinical Focus figure on page 322.

Study Question Answers

1. Knockout mice lacking class II MHC molecules fail to produce CD4$^+$ mature thymocytes, or those clacking class I MHC molecules fail to produce CD8$^+$ mature thymocytes, because at some level lineage commitment requires engagement between the MHC and the appropriate CD4/8 receptor.

   β-selection initiates maturation to the DN4 stage, proliferation, allelic exclusion, maturation to the DP stage, and TCRα locus rearrangement.

   Negative selection to tissue-specific antigens occurs in the medulla of the thymus, by mTECs and DCs.

   Most thymocytes die by neglect in the thymus because they either did not produce viable TCR, or because they do not bind to self-MHC.

   The extrinsic pathway of apoptosis can activate the intrinsic pathway through truncation of the Bcl-2 family protein Bid.

   Cytochrome *c* is an important downstream molecule in the intrinsic apoptotic pathway.

   Bcl-2 resides in the mitochondrial membrane and inhibits apoptosis by preventing cytochrome *c* release.

2. γδ T cells do not have a requirement that antigen be presented by MHC. Thus, they are not limited to recognition

of protein antigens. The recognition process seems to be closer to that of pattern recognition receptors found on innate immune system cells.

3.

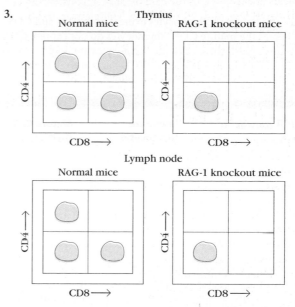

4. (a) There would be no CD4 SP, but all other stages would be present. The absence of MHC Class II would prevent positive selection and lineage commitment of CD4$^+$ T cells. (b) All stages would be present, but some of the mature cells would be reactive to tissue-specific antigens. (This would only be revealed by functional experiments.) AIRE regulates the expression of self-tissue-specific antigens by medullary epithelial cells. (c) All DN and DP cells would be present (β-selection would proceed unhindered). However, none of the DP cells would express normal TCRαβ dimers and could not be positively selected. (For the advanced, TCRγδ cell development would proceed normally—many of these are DN in phenotype, but a few are CD4 and CD8 SP.)

5. The first are CD3$^-$TCRβ$^-$ thymocytes and could simply be immature DN thymocytes. The second group are CD3$^+$TCRβ$^-$ and could be TCRγδ T cells!

6. (a) Flow cytometry. (b) Higher. (c) Lower.

   Positive selection would occur in H-2^k but not H-2^d background (MHC Class II). Most cells in the TCR transgenic would have the receptor specific for this MHC haplotype, so would get more than normal numbers of CD4 SP cells when positive selection occurs.

   (d) No mature single positive cells; may have reduced number of DP cells (negative selection is going to occur).

   Must speculate with this question—no absolutely clear answer. The cortical epithelium may not be able to mediate clonal deletion because it doesn't express the right costimulatory molecules. Investigators who have done experiments like this find evidence for negative selection of a sort, however. SP T cells develop, but they appear not to be easily activated.

7. (a) Thymocytes in A developed in a thymus whose cortical epithelial cells expressed H-2^d MHC molecules and

therefore became restricted to that MHC (via positive selection). They are not restricted to H-2^d, so ignore targets that express this MHC. (b) Same reasoning as above.

8. (a) The immature thymocytes express both CD4 and CD8, whereas the mature CD8$^+$ thymocytes do not express CD4. To distinguish these cells, the thymocytes are double-stained with fluorochrome-labeled anti-CD4 and anti-CD8 and analyzed in a FACS. (b) See the following table.

H-Y TCR transgenic mouse	Immature thymocytes	Mature CD8$^+$ thymocytes
H-2^k female	+	+
H-2^k male	+	−
H-2^d female	+	−
H-2^d male	+	−

(c) Because the gene encoding the H-Y antigen is on the Y chromosome, this antigen is not present in females. Thymocytes bearing the transgenic T-cell receptor, which is H-2^k restricted, would undergo positive selection in both male and female H-2^k transgenics. However, subsequent negative selection would eliminate thymocytes bearing the transgenic receptor, which is specific for H-Y antigen, in the male H-2^k transgenics. (d) Because the H-2^d transgenics would not express the appropriate MHC molecules, T cells bearing the transgenic T-cell receptor would not undergo positive selection.

9. (a) Class I K, D, and L molecules and class II IA molecules. (b) Class I molecules only. (c) The normal H-2^b mice should have both CD4$^+$ and CD8$^+$ T cells because both class I and class II MHC molecules would be present on thymic stromal cells during positive selection. H-2^b mice with knockout of the IA gene would express no class II molecules; thus, these mice would have only CD8$^+$ cells.

10. (a) Because the pre-T-cell receptor, which does not bind antigen, is associated with CD3, cells expressing the pre-TCR as well as the antigen-binding T-cell receptor would stain with anti-CD3. It is impossible to determine from this result how many of the CD3-staining cells are expressing complete T-cell receptors. The remaining cells are even more immature thymocytes that do not express CD3. (b) No. Because some of the CD3-staining cells express the pre-TCR or the αβ TCR instead of the complete γδ TCR, you cannot calculate the number of T$_C$ cells by simple subtraction. To determine the number of T$_C$ cells, you need fluorescent anti-CD8 antibody, which will stain only the CD8$^+$ T$_C$ cells.

Chapter 10

1. Fetal liver cells; B-1 B-cell progenitors are highly enriched in the fetal liver and are the first B cells to populate the periphery. The peritoneal and pleural cavities, as well as the spleen.

2. T2 cells have intermediate levels of IgD, whereas T1 cells have no to low amounts of IgD. T2 cells bear both CD21 and CD23, whereas neither antigen is expressed on T1 cells. T2 cells have higher levels of the receptor for the B-cell survival factor BAFF than do T1 cells. Interaction of antigen with T1 cells results in apoptosis; interaction of antigen with T2 cells sends survival and maturation signals.

3. Cell division at this stage allows the repertoire to maximize its use of B cells in which a heavy chain has been productively rearranged. Each daughter cell can then rearrange a different set of light-chain gene segments, giving rise to multiple B-cell clones bearing the same heavy chain, but different light-chain genes.

4. They can undergo apoptosis, in a process called negative selection. This occurs for B cells in the bone marrow and for T cells in the thymus.

They can become anergic—refractory to further stimulation—and eventually die. This occurs for both T and B cells.

Their receptors can undergo receptor editing. This occurs quite frequently in B cells. In T cells, the extent to which receptor editing occurs varies according to the nature of the animal (the nature of the transgene used to study editing), and therefore whether it is a meaningful mechanism in T-cell receptor development and selection is so far unclear.

5. First, knock out the ability of the animal to express that transcription factor, and then analyze the bone marrow for the occurrence of B-cell progenitors at each stage of development using flow cytometry. One would not expect to see any progenitors after the pre-pro-B-cell stage if the transcription factor is expressed then *and* is necessary for further B-cell development.

Second, make a fusion protein in which the promoter of the transcription factor is fused to a fluorescent protein, such as green or yellow fluorescent protein. Correlate the expression of the fluorescent proteins with the cell-surface markers. One would expect to see the fluorescent protein show up first in cells bearing markers characteristic of the pre-pro-B-cell stage.

Rearrangement has begun, with D to J$_H$ rearrangement occurring on the heavy chain. Test it using PCR, with primers complementary to sequences upstream of the D regions and downstream of the J regions, followed by sequencing, if necessary.

6. Create an animal in which the CXCL12 promoter is fused to a marker fluorescent protein, such as GFP. Make slides of bone marrow, taking care that the conditions did not break up cell attachments, and label the cells with markers characteristic of the target stage of development. Look for cell pairings between the CXCL12-labeled cells and progenitor cells labeled with the markers characteristic of the target stage of development.

7. Rearrangement starts first at the heavy-chain locus, beginning with D to J$_H$ and proceeding with V$_H$ to D. If the rearrangements at the first allele are not productive,

then rearrangement starts again on the second heavy-chain locus and proceeds in the same order. Successful rearrangement at a heavy-chain locus results in the expression of a heavy chain at the surface of the B-cell progenitor in combination with the surrogate light chain, to form the pre-B-cell receptor. This occurs at the beginning of the large pre-B-cell stage. Expression of the heavy chain at the cell-surface signals the cessation of further heavy-chain rearrangement.

At the light-chain locus in mice, rearrangement begins at one of the κ loci, and again, if it is not productive, it starts again on the other κ locus. If this is also not productive, the process repeats at the λ loci. In humans, the process is similar, but rearrangement may start at either the κ or the λ loci. Light-chain rearrangement is completed by the end of the small pre-B-cell stage, and the expression of the complete Ig receptor on the surface of the cell signals the beginning of the immature B-cell stage.

In T cells, rearrangement begins on successive β chain loci. In possession of V, D, and J segments, the β chain locus is analogous to the heavy-chain Ig locus. Successful rearrangement of the β chain results in expression of a pre-TCR on the cell surface, just as for the pre-BCR on B cells, coupled with the cessation of further β-chain gene segment recombination. Rearrangement at the α chain locus follows. One major difference between the processes of rearrangement in T and B cells is that allelic exclusion at the T-cell α chain locus is incomplete.

Analyze the Data (a) In the spleen, the *Dicer* knockout animal shows no mature B cells, indicated by the loss of the B220^{hi}CD19$^+$ population.
In the top bone marrow population, in the absence of *Dicer*, there are no sIgM-bearing B220^{hi} cells.

The second bone marrow panel shows retention of the progenitor cell marker, c-kit, in the *Dicer* knockout.

The third panel shows that in the absence of *Dicer* there is a loss of CD25, a marker characteristic of the point in development at which the pre-BCR is expressed on the cell surface. This suggests that the cells cannot get past this checkpoint.

(b) Since no IgM is expressed on the cell surface of the Dicer knockout animals, miRNAs must be necessary for progression to the pre-B-cell stage, at which IgM is first expressed on the cell surface. The presence of c-Kit and of low amounts of CD25 suggests that miRNAs may be acting at the preBCR checkpoint.

Analyze the Data (a) No. The fraction of cells labeled with Annexin V and therefore in the pre-apoptotic state is identical in the control and *Dicer* knockout populations.
(b) Yes. The fraction of cells labeled with Annexin V increases from 9.2% in the control to 65% in the *Dicer* knockout (c) Pulling together the data from the last two questions, I would hypothesize that miRNAs aid in controlling the expression of the preBCR on the cell surface, thereby allowing the cells to pass through the first checkpoint in development. Cells that cannot express preBCR die by apoptosis, and that is the process described in the second set of slides.

Chapter 11

Clinical and Experimental Focus Answer The data show that LIF inhibits T$_H$17 polarization (in its presence, the frequency of IL-17$^+$ cells is reduced by 50% after cells are exposed to T$_H$17 polarization conditions). T$_H$1 differentiation appears unaffected (the same frequency of IFNγ$^+$ cells are present after treatment with T$_H$1 polarization conditions). This suggests that LIF has a specific effect on the pathways that induce T$_H$17 differentiation or on those responsible for production of IL-17. It could interfere with any of the steps involved including (from outside to inside,) (1) signaling induced by TGF-β or by IL-6 polarizing cytokines, (2) RORγ expression itself, or (3) IL-17 expression itself. A reduction in T$_H$17 cells could result in less inflammation and the amelioration in disease that is seen in this model.

(It turns out that LIF acts in opposition to IL-6 and blocks its downstream signaler, STAT3. This abrogates the inhibitory effect that IL-6 has on FoxP3 expression, shifting the balance to T$_{REG}$ rather than T$_H$17 lineage commitment. So, disease amelioration is not just a consequence of fewer activated T cells, but a result of the increase in cells that quell T-cell responses.)

1. (a) Anergy. Signal 1 (if TCR is engaged) without costimulatory Signal 2 because CTLA-4 Ig will block the ability of CD28 to bind CD80/86. (b) No anergy. Signal 1 and Signal 2 are both generated. (c) Anergy. Signal 1 without Signal 2. (d) No anergy, but no activation either. Neither Signal 1 nor Signal 2 is generated.

2. (a) Very likely. Any activated professional APC, like a dendritic cell, up-regulates MHC molecules and costimulatory ligands, making them ideal activators of T cells. (b) Very unlikely. Activated dendritic cells travel to the draining lymph nodes (or spleen) and encounter naïve T cells there, not in peripheral tissues. Naïve T cells travel among secondary lymphoid organs, not peripheral tissues. However, effector T cells and some memory T cells, do travel to peripheral tissues and can be activated by dendritic cells there. (c) Very likely. TCR stimulation rapidly induces Ca^{2+} mobilization. (d) Very unlikely. The virus induced dendritic cells to make IL-12, one of the central polarizing cytokines for the T$_H$1 lineage. (e) Very unlikely. Central memory cells were certainly generated, too.

3. A mouse without GATA-3, the master regulator for T$_H$2 lineage commitment, will be unable to generate T$_H$2 cells, which are instrumental in mounting the immune response to worm infections. T$_H$2 cells help B cells to produce IgE, which has potent anti-parasite activity.

4. You will need to supply Signal 1 (anti-TCR), Signal 2 (anti-CD28), and Signal 3 (IL-12). CTLA-4 Ig and anti-CD80 both bind to the ligands for the costimulatory receptors and would not engage your T cells.

5. (a) Dendritic cells are best at activating naïve T cells—they express a high density of costimulatory ligands and MHC molecules. (b) ICOS is a positive costimulatory receptor (it is expressed on some effector T cells, including T$_{FH}$ cells). (c) Most cells do not express costimulatory ligands. Professional APCs (and thymic

epithelial cells) are among the only cells that do.
(d) ICOS and CTLA-4 also bind B7 family members (CD80 and CD86). PD1 also binds a B7 like molecule, PD-L1.
(e) Signal 3 is provided by cytokines, which include the polarizing cytokines that induce helper T-cell lineage differentiation. (f) It is a disease caused by T-cell response to superantigens (bacterial and/or viral), not autoantigens. (g) They mimic some TCR-MHC class II interactions. (h) They do not have any receptor for MHC class I and do not interact directly with $CD8^+$ T cells via their TCRs, which bind to MHC class II.
(i) Naïve T cells do not produce any effector cytokines.
(j) They are master transcriptional regulators of T helper cell lineage differentiation. (k) APCs can make some polarizing cytokines, but many of these cytokines originate from other cells, including other T cells, B cells, mast cells, and NK cells. (l) T-bet is a master transcriptional regulator of T_{FH} lineage differentiation.
(m) T_{FH} and T_H2 are classically the major sources of B-cell help, although all helper subsets can interact with B cells and influence Ig class switching. (n) They inhibit T-cell activation. (o) Effector cytokines have many different cellular targets, including B cells, endothelial cells, stromal cells in tissues, innate immune cells, and so on, as well as other T cells. (p) Central memory cells tend to reside in secondary lymphoid organs. (q) CCR7 attracts cells to secondary lymphoid tissue, and effector cells tend to rove the periphery. They typically down-regulate CCR7.

6. (a) True. It stimulates production of both FoxP3 and RORγ. (b) False. IL-6 in combination with TGF-β polarizes cells to the T_H17 lineage, an event that requires RORγ. IL-6 acts in part by inhibiting expression of FoxP3.

7. Your discussion should include a recognition of the three properties that distinguish T helper lineages: a unique set of polarizing cytokines, a unique master gene regulator, and a unique set of effector cytokines. Knowing which cytokines polarize a naïve T cell to this lineage as well as the identity of a unique master regulator that induces T_H9 lineage differentiation would strengthen the case for placing this in a unique T lineage category. Knowing how stable the phenotype is would also be useful. Does it differentiate into other subsets, or is it stable in effector function and phenotype? This property would not preclude T_H9 from being considered a distinct lineage (e.g., recall that T_H17 and T_{REG} cells can give rise to other lineages) but would add complexity to your assessment.

Chapter 12

1. TI-1 antigens are mitogenic and induce activation through both the BCR and innate immune receptors. TI-2 antigens bind tightly to the complement components C3d and C3dg, and so are bound by both the BCR and the complement receptor CD21 (CR2).

2.
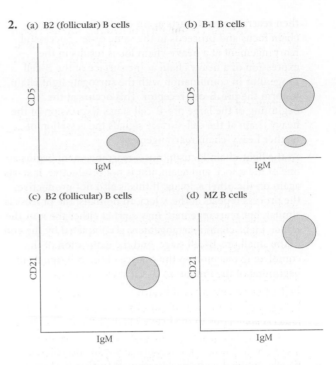

(a) and (b) B2 (follicular B cells) bear relatively high levels of IgM but do not express CD5. The majority of B-1 B cells, known as the B-1a fraction, does express CD5. However, there is a minority fraction of B-1 B cells, the B-1b fraction, that does not express CD5. (c) and (d) B2 B cells express normal levels of CD21, the complement receptor, whereas marginal zone B cells express particularly high levels of CD21.

3. No to both. Both class switch recombination and somatic hypermutation require the ability of T cells and B cells to interact with each other through the binding of B-cell CD40 molecules by the CD40L molecule on T cells.

4. Lymphotoxin-α is required for the generation of germinal centers, so my knockout mouse will have no germinal centers.

5. Since AID is required for both class switch recombination and for somatic hypermutation, I would expect my knockout mouse to be unable to express any classes of antibody other than IgM. Furthermore, I would expect the average affinity of the antibodies produced by the knockout mouse to be unchanged between primary and secondary stimulation, since, in the absence of AID, the antibodies genes will not be subject to somatic hypermutation.

6.
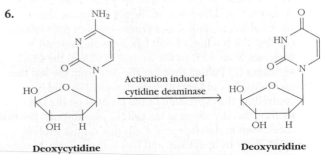

During SHM, the deamination of cytidine on one strand of the DNA encoding antibody variable regions leads to the formation of a mismatched G-U pair. The mismatch can then be recognized by a number of DNA repair mechanisms in the cell and resolved in one of several different ways. The simplest mechanism is the interpretation of the deoxyuridine as a deoxythymidine by the DNA replication apparatus. In this case, one of the daughter cells would have an A-T pair instead of the original G-C pair found in the parent cell. Alternatively, the mismatched uridine could be excised by a DNA uridine glycosylase enzyme. Error-prone polymerases would then fill the gap as part of the cell's short-patch base excision repair mechanism. Third, mismatch repair mechanisms could be induced that result in the excision of a longer stretch of DNA surrounding the mismatch. The excised strand could then be repaired by error-prone DNA polymerases, leading to a series of mutations in the region of the original mismatch.

In the case of class switch recombination (CSR), AID deaminates several cytidine residues in the switch (S) regions upstream of the two heavy-chain constant regions between which the class switch will occur (the donor and acceptor S regions). The resulting uridine residues are excised by uridine glycosylases, and the abasic sites are then nicked by endonucleases that create single-strand breaks at the abasic sites. These single-strand breaks are converted to double-strand breaks suitable for end joining by mismatch repair mechanisms. A constellation of enzymes then faithfully reconnects the two S regions, with the excision of the intervening DNA.

7. In order to survive, B cells need to receive signals from T cells. Since there are many more antigen-specific B cells than T cells within the germinal centers, B cells must compete with one another for T-cell binding. Since T cells are specific for peptide antigen displayed in the groove of class II MHC molecules, B cells that have internalized and displayed more antigen will have a selective advantage in attracting T-cell attention. B cells with higher-affinity receptors will bind, internalize, and display more antigen than B cells with lower-affinity receptors, and therefore compete successfully for T-cell help and survival signals. B cells with higher-affinity receptors have even been demonstrated to strip antigen from lower-affinity B cells.

8. The presence of circulating immune complexes serves as an indicator that the host organism has made a high concentration of antigen-specific antibodies and has succeeded in neutralizing the antigen. Therefore, no more antibody production is needed, and the host should not expend further energy in generating antibodies of this specificity. IgG-containing immune complexes are recognized by the Fc receptor FcγRIIb (CD32), and co-ligation of the immune complex by FcγRIIb and by the BCR results in phosphorylation of the ITIM on the cytoplasmic tail of FcγRIIb. Docking of the SHIP phosphatase at this receptor molecule allows it to dephosphorylate PIP3 to PIP2. This interferes with transmission of antigen signals at the B-cell receptor, resulting in the down-regulation of B-cell activation.

9. B-10 B cells have recently been shown to secrete the immunosuppressive cytokine IL-10 upon antigen stimulation.

10. (a) Small, soluble antigens can be directly acquired from the lymphatic circulation by follicular B cells, without the intervention of any other cells. These antigens enter the lymph node via the afferent lymph and pass into the subcapsular sinus (SCS) region. Some small antigens may diffuse between the SCS macrophages that line the sinus to reach the B cells in the follicles. (b) Other small antigens leave the sinus through a conduit network. Follicle B cells can access antigen through gaps in the layer of cells that form the walls of the conduits. (c) Larger antigens are bound by complement receptors on the surfaces of SCS macrophages. Antigen-specific B cells within the follicles can acquire the antigens directly from the macrophages and become activated.

Chapter 13

Clinical Focus Answers

1. Arthritis is characterized by inflammation of a joint leading to damaged tissue, swelling, and pain. Psoriatic arthritis is accompanied by skin lesions caused by immune attack (psoriasis). Association of KIR-MHC combinations with susceptibility to arthritic disease would likely stem from a deficiency of inhibitory signals (e.g., absence of MHC alleles that produce inhibitory ligands for specific KIR molecules), leading to damage of host cells and tissues. Diabetes is another autoimmune disease; exhibiting destruction of the host pancreatic islet cells that produce insulin, the same mechanisms predicted for arthritis could operate in diabetes. In both cases, absence of inhibitory NK signals could lead to damage inflicted by NK cells, directly or through their recruitment of other effector cells.

2. Rab27A is a GTPase that regulates the transport of intracellular vesicles (granules) to the cell membrane. Transport is required for the release of vesicular contents into the extracellular space. Many cells depend on this ability to function, including cytotoxic T cells, which release perforin and granzyme from internal vesicles, and melanocytes, which release pigment from vesicles (melanosomes). Without this capacity, an individual will be unable to kill infected cells and will exhibit a form of albinism. Many other cells could be affected, including granulocytes (eosinophils, basophils, and mast cells), although it is important to recognize that some express other Rab variants that compensate for the loss of Rab27 function.

1. FcγRIII and CD23 are both Fc receptors (see Table 13-3). Neutralization and complement fixation are antibody functions that do not rely on Fc receptors (although FcR can help mediate the clearance of neutralized antibody-pathogen complexes). Opsonization and ADCC are mediated by cells that express activating FcRs, including FcγRIII. CD23, however, is an inhibitory FcR that regulates (inhibits) the activity of other activating FcRs.

So, in short, antibodies to FcγRIII would block opsonization and ADCC, but not neutralization and complement fixation. Antibodies to CD23 would not inhibit any process. (Because they block inhibitory signals they could theoretically enhance opsonization and ADCC; it is not clear, however, that this would occur in all contexts.)

2. (a) False. Some FcRs, such as FcRγII (CD23), are inhibitory receptors. Others are expressed on cells that are not phagocytes. These can mediate transcytosis, regulate cell activation (e.g., antibody production of B cells), etc. (b) False. IgE mediates degranulation and activation of mast cells, basophils, and eosinophils. (c) True. (d) True. (e) True. (f) False. There are two pathways by which cytotoxic T cells kill target cells. One pathway is perforin dependent, and the other uses FAS ligand displayed by the CTL to induce death in FAS-expressing target cells. (g) False. Naïve CD8$^+$ T cells can be activated in the absence of CD4$^+$ T-cell help. T-cell help, however, is required for optimal proliferation and memory generation. (h) True. (i) False. Ly49 receptors are found on murine cells. Human NK cells express KIR receptors

3. The monoclonal antibody to LFA-1 should block formation of the CTL-target-cell conjugate. This should inhibit killing of the target cell and, therefore, should result in diminished ^{51}Cr release in the CML assay.

4. See table below.

Population 1	Population 2	Proliferation
C57BL/6 (H-2^b)	CBA (H-2^k)	1 and 2
C57BL/6 (H-2^b)	CBA (H-2^k) Mitomycin C-treated	1
C57BL/6 (H-2^b)	(CBAxC57BL/6)F1(H-2$^{k/b}$)	1
C57BL/6 (H-2^b)	C57L (H-2^b)	Neither

(*Brief explanation*): T cells from one inbred strain of mice will proliferate in response to alloantigens (MHC-peptide combinations on the surface of cells from another strain that differs by MHC haplotype) on the other strain. In the first row, cells are isolated from two strains that do differ by MHC haplotype (H-2^b and H-2^k). Cells from both strains will respond to each other. (T cells will recognize MHC-peptides on cells from the other strain (B cells and/ or some macrophages/monocytes that came along for the ride.) In the second row, however, cells from the CBA have been treated with a reagent that blocks proliferation. Although cells from both strains will "want" to respond, only those from the C57BL/6 isolates will be able to proliferate. In the third row, the C57BL/6 T cells will respond to cells from the F1 strain, but not vice versa. Why? T cells from the F1 strain are tolerant to both MHC haplotypes. In the fourth row, the strains do not differ by MHC haplotype. All cells are tolerant to H-2^b, and neither cell group will respond.

5. (a) T cells (CD4$^+$ and CD8$^+$) as long as both MHC class I and class II differ. (b) To demonstrate the identity of the proliferating cells, you could incubate them with distinct fluorochrome conjugated antibodies (e.g., fluorescein-labeled anti-CD4 monoclonal antibody and phycoerythrin-labeled anti-CD8 monoclonal antibody). The proliferating cells will be stained only with the anti-CD4 reagent. (c) As T cells recognize allogeneic MHC molecules on the stimulator cells, they are activated and begin to secrete IL-2, which then autostimulates T-cell proliferation. Thus, the extent of proliferation is directly related to the level of IL-2 produced.

6. (a) All. (b) All. (c) CTL. (d) CTL. (e) None. (f) All. (g) CTL. (h) Some NK and NKT. (i) Some NKT. (j) CTL and NKT. (k) NKT. (l) All. (m) CTL and NKT. (n) NK. (o) None. (p) All. (q) Some CTLs.

7. See table below.

Source of primed spleen cells	^{51}Cr release from LCM-infected target cells			
	B10.D2 (H-2^d)	B10 (H-2^b)	B10.BR (H-2^k)	(BALB/ cxB10) F1 (H-2$^{b/d}$)
B10.D2 (H-2^d)	+	−	−	+
B10 (H-2^b)	−	+	−	+
B10.BR (H-2^k)	−	−	+	−
(BALB/cxB10) F1 (H-2$^{b/d}$)	+	+	−	+

T cells will lyse targets expressing peptides from the antigen to which they were primed (LCMV) and expressing the MHC to which they are restricted (syngeneic MHC). These requirements are met in all cases where there is a positive symbol. The very observant student might also recognize that T cells of one strain will also react to alloantigens (cells from another strain expressing a distinct MHC haplotype)—the focus of question 4. In fact, this is true, and there will be some background death in all cases of MHC "mismatch" (in other words, you would also notice some "background" killing in conditions that are labeled with a "−"). However, because the T cells have been primed by immunization to LCMV, the LCMV-specific MHC restricted response would be a secondary response and would dominate the primary alloresponse.

8. To determine T$_C$ activity specific for influenza, perform a CML reaction by incubating spleen cells from the infected mouse with influenza-infected syngeneic target cells. To determine T$_H$ activity, incubate the spleen cells from the infected mouse with syngeneic APCs presenting influenza peptides; measure IL-2 production.

9. The "missing self" model has been used to explain how NK cells detect infected or tumor cells. If a potential target cell expresses normal levels of class I MHC molecules, inhibitory receptors on the NK cell (KIR,

CD94, NKG2) induce a signal transduction cascade that abrogates NK lytic activity. These negative signals override pro-killing signals generated via ligands binding to activating receptors on the NK cell (NKR-P1 and others). Some tumor cells and virally infected cells, however, reduce their expression of class I MHC and no longer stimulate NK inhibitory receptors. In this case, the activating (pro-killing) receptor signal dominates.

10. There are two pathways by which cytotoxic T cells kill target cells: one that is perforin dependent and one that uses FasL to induce death in Fas-expressing target cells. (And, as always, T cells will only lyse cells expressing the MHC-peptide combinations to which they are specific and restricted.) T cells from immunized perforin knockout H-2^d mice will be able to lyse (d). (These T cells will depend on FasL-Fas interactions to kill. Target cells don't need to express perforin to be susceptible, but they do need to express Fas.) T cells from immunized Fas ligand knockout H-2^d mice will be able to lyse (d), but they will also be able to lyse (e). These cells will depend on perforin-mediated pathways. Targets do not need to express perforin or Fas to be susceptible.) T cells from H-2^d mice in which both perforin and Fas ligand have been knocked out will not be able to lyse any of the cell types.

11. If the HLA (human MHC) type is known, MHC tetramers bound to the peptide generated from gp120 and labeled with a fluorescent tag can be used to specifically label all of the CD8$^+$ T cells in a sample that has T-cell receptors capable of recognizing this complex of HLA and peptide.

12. (a) True. (b) False. They need to express Fas, which transmits the pro-apoptotic signal. (c) False. Both mechanisms induce caspase activation. (d) False. Only the perforin-mediated pathway depends on granzyme activity. (e) True. (f) False. Perforin is responsible for the development of surface membrane and endocytic membrane pores.

Analyze the Data (a) Epitopes 2, 12, and 18 generated high CTL activity, and epitopes 5 and 21 generated medium activity. (b) It is possible that different peptides use distinct anchor residues, which would make this prediction more difficult. However, if we make the assumption that the same amino acids would be bound by HLA-A2, it appears that a leucine (L) on the amino terminus side separated by four amino acids from a threonine (T) is the only common motif for the five most immunogenic peptides (2, 12, 18, 5, and 21). All of the peptides that generate high CTL activity have two consecutive leucines on the amino terminal side as well. The problem with the threonines serving as anchors is that peptide 2 has four amino acids at the carboxyl side of the T, which seems to result in the end of the peptide extending out of the binding pocket. This would be a very unusual configuration. Peptide 12 may have a less dramatic but similar problem. Therefore, a leucine in pocket 2 of HLA-A2 would be consistent with the data generated by M. Matsumura et al. in 1992 (Emerging principles for the recognition of peptide antigens by MCH

class molecules, *Science* **257**:927). Thus, the main anchor may be a leucine residue at the amino end of the binding pocket, with possible contribution by T at the carboxyl end under some circumstances. (c) It is possible that there are no T cells specific for those peptides, even if they are presented in complex with MHC. Therefore, you would not see a CTL response. (d) CTLs only recognize antigen in the context of self-MHC molecules. Therefore, in order to assess CTL activity, the T2 cells also had to express HLA-A2. (e) Class I MHC molecules typically bind peptides containing 8 to 10 residues (see Figure 8-6). Peptide 2 is 11 residues long, suggesting that it bulges in the middle when bound. Since it appears to be a major epitope for CTL killing, bulging does not seem to interfere with CTL interaction and may contribute.

Chapter 14

Analyze the Data (a) This should be clear from the video. (b) Differences in expression of homing receptors and/or chemokine receptors would be reasonable hypotheses. Make sure to state specific possibilities (based on the information in Appendix III and examples in the text). (c) This requires creative (and rigorous) speculation on your part—we do not really know. Both these subpopulations are effector memory cells—consider what each will do if reactivated. Will they react with different kinetics? Will they stay in the same area of the tissue? Will they serve the same cell populations? Read the authors' discussion for their view of the possibilities.

Experimental Design Answer The best experimental designs will include your question, your prediction, and your experimental design. The design must include controls (both positive and negative, if possible—and more than one at times). You must also identify what you will measure and how you will interpret that measurement.

> *Question:* Do naïve B cells require CCR5 to localize to B-cell follicles? *Prediction:* Yes, they are absolutely dependent on this chemokine receptor; or no, they can use other chemokine receptors, although perhaps less efficiently.

> *Experimental Design:* Fundamentally, you must compare the in vivo activities of B cells that can use CCR5 with those that cannot. *One possibility:* compare the behavior of labeled, wild-type versus labeled, CCR5$^{-/-}$ B cells in two groups of mice. Alternatively or in addition, track the behavior of labeled, wild-type B cells in the presence or absence of the CCR5 blocking antibody.

Once you decide on your design, you must develop a protocol using intravital two-photon microscopy (dynamic imaging). You must be able to trace naïve B-cell movements (and, ideally, identify the B-cell follicle, too). Therefore, you need to fluorescently label those B cells: in vitro CFSE staining is probably the best approach in this case because you will be examining the behavior of B cells from different mice. It is not the only approach, however. (*Note:* To define the follicular area, you could also co-inject T cells that are labeled in a different color (few if any should be in the follicles) or come up with another more clever, original idea. Some investigators (as you may have noticed) do not directly label the follicle, but infer its location from the behavior of the cells).

Isolate, label, and inject the cells. Wait a specified time (based on previous studies in the literature or on your own experiments), anesthetize the mice, and record cell behavior in an exposed lymph node over time.

What will you measure? Go back to your question. Identifying the number of B cells that end up on a follicle over a period of time would answer your question directly. The figure you sketch could be a bar graph comparing these numbers in each experimental condition. Direction and speed of the cells may be two other useful parameters that could be generated from an analysis of trajectories—and you can describe how they will contribute to your understanding of the question.

Clinical Focus Answers There are many different possibilities, in theory. CD18 is part of the LFA-1 complex, which regulates extravasation of multiple subsets of leukocytes (see text and Advances Box 14-2). A CD18 deficiency could, therefore, inhibit the ability of innate immune cells to travel to the site of infection, naïve lymphocytes to enter secondary lymphoid organs, effector cells to recirculate effectively, and so on. All of these problems would severely decrease the ability of a child to fight off infection. Treatment could include genetic modification of bone marrow stem cells (reintroducing the *CD18* gene into hematopoietic stem cells), but should also include judicious antibiotic use. Check online resources for the what is possible. (Wikipedia.com and any government or university-based clinical site will likely provide good information.)

1. (a) True. (b) Not necessarily. Both provide higher resolution than conventional fluorescent microscopy. (c) True. (d) True. (e) True. (f) False.

2. You have not designed an experiment that allows you to focus on antigen-specific T cells—which will represent only a fraction of the whole population. You will need to find a better way to track these. (What are the possibilities? Use TCR transgenics specific for influenza [most, if not all, cells will be antigen specific], or modify the virus so that it expresses another antigen [e.g., OVA] that can be seen by TCR-transgenic cells. Or, try to isolate influenza-specific T cells [by tetramer staining? Difficult, but possible in theory]).

3. (a) False. All leukocytes respond to chemokines—they are one of the central regulators of immune cell migration. (b) False. Naïve B cells do not express CCR7 (which helps to send cells to the paracortex), but when activated by antigen binding, they up-regulate it so that they travel to the paracortex to find T-cell help. (c) False. Small antigens (typically opsonized [by complement, for instance]) arrive on their own via the afferent lymphatics. (d) False. They "arrest" their migrating behavior. (e) False. They crawl along the fibroblastic reticular cell network. (f) False. It appears from recent data that these networks can be established at sites of infection. (g) False. Strong adhesion requires chemokine activation. Rolling is the first event (mediated by selectins). (h) True.

4. The movement of an antigen-activated B cell from the follicle to the border between the follicle and paracortex is a classic example. However, there are many others, including the response of innate cells to signals generated by inflammation at the site of infection (e.g., neutrophils are attracted to IL-8 produced by other innate cells, including other neutrophils).

5. Both receive help in the lymph node cortex. B cells travel to the interface between follicle and paracortex to receive T-cell help and remain loosely associated with the follicle during their interaction with the T cell. The CD8$^+$ T cells receive help in the paracortex, where they interact with APCs and CD4$^+$ T cells.

6. Recall that CCL3 is produced by antigen-presenting cells that have been activated by CD4$^+$ helper T cells in the lymph node. CCL3 attracts CD8$^+$ T cells to form a tricellular complex so that they can receive optimal help. Without this cytokine, CD8$^+$ T cells may not find their way to antigen-presenting cell/CD4$^+$ T cell pairs and may not be optimally activated. On the other hand, other chemokines (e.g., CCL4) may be able to compensate.

7. (a) Rolling, chemokine interactions, adhesion, transmigration. (b) Adhesion. Adhesion molecules such as LFA-1 and VLA-4 are converted to their high-affinity forms by chemokine receptor signals (via an inside-out process). (c) The homing and chemokine receptors expressed by naïve lymphocytes attract them to secondary lymphoid tissues. For example, they express L-selectin (CD62L), which interacts with ligands on specialized endothelial structures (high-endothelial venules) located in the cortex of the lymph node.

8. Both the pattern of expression of chemokine receptors and chemokines regulate the compartmentalization of T and B cells in the lymph node. For example, naïve T cells express CCR7, which interacts with chemokines that decorate the fibroblastic reticular cell network in the paracortex. Naïve B cells express CCR5, which is expressed by cells in the follicle and by the follicular DC network. T-cell and B-cell movement is guided by the routes laid down by these networks.

9. True. Germinal-center B cells are more motile and extend unexpectedly long processes within the germinal center.

10. (a) (b) x (c) x (d) (e) x (f) (g) x

11. (a) Naïve T and B cells would not home properly to the HEV. (The individual would be significantly immunocompromised, although possibly able to compensate with some innate immune activity and splenic T-cell and B-cell activity.) (b) Naïve T cells would not home properly to the paracortex. Activated B cells would not home properly to the follicle-paracortex border. Would not be able to develop optimal adaptive immune responses unless compensated by other chemokines. (The individual would be immunocompromised.) (c) Naïve B cells would not home properly to the follicle. Would not be able to develop T-cell-dependent antibody responses unless compensated by other chemokines. (The individual

would be partially, but still significantly, immunocompromised.) (d) Naïve T and B cells, effector T and B cells, and effector memory T and B cells would not be able to leave the lymph node (and other tissues). (The individual would be immunocompromised unless compensated by other egress regulators and extra-lymphoid immune cell activity.)

Chapter 15

1. (a) No type I (which is mediated by IgE/FcεRI interactions), normal type II (which is mediated by IgG or IgM). (b) Same as above. (c) May have some type I reaction, but given that this receptor is important in regulating (both enhancing and suppressing) B-cell production of IgE, the response may be abnormal. (d) Type II responses would be most impaired because IgG and IgM exert their effects, in part, by recruiting complement, as well as by inducing ADCC. (e) Type I responses are likely to be suppressed. When bound by soluble versions of FcεRII on B cells, CD21 enhances IgE production. In its absence, the animal may not be able to generate as much IgE antibody as a wild-type mouse.

   (*Note:* All these answers assume that there are no other similar or redundant genes and proteins that could compensate for the absence of the gene in question.)

2. Primary mediators are found in mast cell and basophil granules and are released as soon as a mast cell is activated. These include histamine, proteases, serotonin, and so on (see chapter text). Secondary mediators are generated by mast cells and basophils in response to activation and are released later in the response. These include cytokines, leukotrienes, prostaglandins, and so on (see chapter text).

3. Histamine binds to at least four different histamine receptors. Binding to H2 receptors inhibits mast cell degranulation and therefore inhibits its own release.

4. IgG can bind to inhibitory Fc receptors (FcγRII), which are coexpressed by multiple cells that express FcεRIs. If an antigen engages both IgG and IgE, it can co-engage FcεRI and FcγRII, which results in inhibition of FcεRI signaling (see chapter text).

5. Rh mismatched moms and dads can generate both Rh$^+$ and Rh$^-$ fetuses. An Rh$^+$ mother will be tolerant to the Rh antigen and will not produce antibodies that could harm either an Rh$^-$ or Rh$^+$ fetus. However, an Rh$^-$ mother has the potential to generate an antibody response against an Rh$^+$ fetus, and could generate a harmful secondary response to a second, Rh$^+$ fetus. Rhogam (anti-Rh antibodies) will clear B cells (and antibodies) generated during the first pregnancy, preventing such a secondary response.

6. See above. Rh$^-$ babies are not at risk, but Rh$^+$ fetuses are.

7. Type III hypersensitivities are disorders brought about by immune complexes that cannot be cleared. They can activate innate immune cells that express Fc receptors and can activate complement, both of which induce inflammation. Such immune complex-mediated inflammation occurs in blood vessels, resulting in

vasculitis, as well as in tissue where the complexes are deposited when they pass through inflamed, vasodilated capillaries. Multiple insults can cause these hypersensitivities, including insect bites and inhalation of fungal spores or animal protein. (See chapter text.)

8. (a) IV (b) I, II, III, IV (c) I, mainly, but III also results in mast cell activation and histamine release. (d) IV (e) I (f) II (g) I, mainly (and others can benefit, too) (h) I (i) II (j) II (k) I

9. (a) In the absence of inflammation, insulin signaling will not be impaired by cytokine stimulation. Cytokine signals activate kinases, including JNK, that phosphorylate and inactivate IRS, a key downstream mediator of insulin receptor signaling. (Interestingly, this observation suggests that free fatty acids, alone, may not be enough to induce insulin resistance.) (b) This question requires speculation—there is no right answer. Some possibilities include (1) differences in IRS that make it less able to be phosphorylated at the serine residue, (2) differences in JNK that make it less likely to bind IRS, (3) differences in gene regions that regulate cytokine production by adipocytes, and so on. See the chapter text and use your imagination!

Chapter 16

Clinical Focus Answer The observations that women mount more robust immune responses and more T$_H$1-pathway-directed responses than men, as well as the effects of female sex hormones on the immune response, may in part explain gender differences in susceptibility to autoimmunity. Since the T$_H$1 type of response is proinflammatory, the development of autoimmunity may be enhanced.

1. The process called central tolerance eliminates lymphocytes with receptors displaying affinity for self antigens in the thymus or in the bone marrow. A self-reactive lymphocyte may escape elimination in these primary lymphoid organs if the self antigen is not encountered there or if the affinity for the self antigen is below what is needed to trigger the induction of apoptotic death. Self-reactive lymphocytes escaping central tolerance elimination are kept from harming the host by peripheral tolerance, which involves three major strategies: induction of cell death or apoptosis, induction of anergy (a state of nonresponsiveness), or induction of an antigen-specific population of regulatory T cells that keeps the self-reactive cells in check.

2. Tolerance is necessary to remove or regulate all self-reactive B and T lymphocytes. Without tolerance, which can be defined as unresponsiveness to an antigen, massive autoimmunity or self-reactivity would result.

3. Receptor editing is a process by which B cells (but not T cells) exchange the potentially autoreactive V region of the immunoglobulin with another V gene, thus changing antigen specificity and avoiding self-reactivity.

4. (a) 5, (b) 8, (c) 7, (d) 10, (e) 6, (f) 2, (g) 9, (h) 1, (i) 4, (j) 3

5. (a) EAE is induced by injecting mice or rats with myelin basic protein in complete Freund's adjuvant. (b) The

animals that recover from EAE are now resistant to EAE. If they are given a second injection of MBP in complete Freund's adjuvant, they no longer develop EAE. (c) If T cells from mice with EAE are transferred to normal syngeneic mice, the mice will develop EAE.

6. A number of viruses have been shown to possess proteins that share sequences with *myelin basic protein* (MBP). Since the encephalitogenic peptides of MBP are known, it is possible to test these peptides to see whether they bear sequence homology to known viral protein sequences. Computer analysis has revealed a number of viral peptides that bear sequence homology to encephalitogenic peptides of MBP. By immunizing rabbits with these viral sequences, it was possible to induce EAE. The studies on the encephalitogenic peptides of MBP also showed that different peptides induced EAE in different strains. Thus, the MHC haplotype will determine which cross-reacting viral peptides will be presented and, therefore, will influence the development of EAE.

7. (1) A virus might express an antigenic determinant that cross-reacts with a self-component. (2) A viral infection might induce localized expression of IFN-γ. IFN-γ might then induce inappropriate expression of class II MHC molecules on non-antigen-presenting cells, enabling self peptides presented together with the class II MHC molecules on these cells to activate T_H cells. (3) A virus might damage an organ, resulting in release of antigens that are normally sequestered from the immune system.

8. Anti-CD3 monoclonal antibodies have been used to block T-cell activity in T1DM. Rituximab, a monoclonal antibody against the B-cell-specific antigen CD20, depletes a subset of B cells and has been used to treat patients with rheumatoid arthritis (RA). Monoclonal antibodies against CD4, which depletes T_H cells, and one against IL-6 that blocks this pro-inflammatory cytokine, have also been used to treat RA. For psoriasis, a monoclonal antibody that recognizes the p40 subunit shared by IL-12 and IL-23 blocks this signaling pathway. Likewise, the fusion protein CTLA-4Ig blocks interactions between CD28 on T cells and CD80/86 on APCs, as treatment for RA, lupus, and inflammatory bowel disease. (See Table 16-5.)

9. (a) True. (b) False: IL-12, which promotes the development of T_H1 cells, increases the autoimmune response to MBP plus adjuvant. (c) False: The presence of *HLA B27* is strongly associated with susceptibility to ankylosing spondylitis but is not the only factor required for development of the disease. (d) True.

10. a. (5), b. (1) (3) (4), c. (1) (3), d. (1) (2) (3)

11. (a) Polyclonal B-cell activation can occur as a result of infection with gram-negative bacteria, cytomegalovirus, or Epstein-Barr virus (EBV), which induce nonspecific proliferation of B cells; some self-reactive B cells can be stimulated in this process. (b) If normally sequestered antigens are exposed, self-reactive T cells may be stimulated. (c) The immune response against a virus may cross-react with normal cellular antigens, as in the case of molecular mimicry. (d) Increased expression of TCR molecules should not lead to autoimmunity; however, if the expression is not regulated in the thymus, self-reactive cells could be produced. (e) Increased expression of class II MHC molecules has been seen in type 1 diabetes mellitus (T1DM) and Graves' disease, suggesting that inappropriate antigen presentation may stimulate self-reactive T cells.

12. (a) False: Acute rejection is cell mediated and probably involves the first-set rejection mechanism (see Figures 16-10b and 16-15). (b) True. (c) False: Passenger leukocytes are donor dendritic cells that express class I MHC molecules and high levels of class II MHC molecules. They migrate from the grafted tissue to regional lymph nodes of the recipient, where host immune cells respond to alloantigens on them. (d) False: A graft that is matched for the major histocompatibility antigens, encoded in the HLA, may be rejected because of differences in the minor histocompatibility antigens encoded at other loci. (e) True.

13. See table below.

Donor	Recipient	Response	Type of rejection
BALB/c	C3H	R	FSR
BALB/c	Rat	R	FSR
BALB/c	Nude mouse	A	
BALB/c	C3H, had previous BALB/c graft	R	SSR
BALB/c	C3H, had previous C57BL/6 graft	R	FSR
BALB/c	BALB/c	A	
BALB/c	(BALB/c X C3H) F$_1$	A	
BALB/c	(C3H X C57BL/6) F$_1$	R	FSR
(BALB/c X C3H) F$_1$	BALB/c	R	FSR
(BALB/c X C3H) F$_1$	BALB/c, had previous F$_1$ graft	R	SSR

14. (a) GVHD develops as donor T cells recognize alloantigens on cells of an immune-suppressed host. The response develops as donor T_H cells are activated in response to recipient MHC-peptide complexes displayed on APCs. Cytokines elaborated by these T_H cells activate a variety of effector cells, including NK cells, CTLs, and macrophages, which damage the host tissue. In addition, cytokines such as TNF may mediate direct cytolytic damage to the host cells. (b) GVHD develops when the donated organ or tissue contains immunocompetent lymphocytes and when the host is immune suppressed. (c) The donated organ or tissue could be treated with monoclonal antibodies to CD3, CD4, or the high-affinity IL-2 receptor (IL-2R) to deplete donor T_H cells. The rationale behind this approach is to diminish T_H-cell activation in response to the alloantigens of the host. The use of anti-CD3 will deplete all T cells; the use of anti-CD4 will deplete all T_H cells; the use of anti-IL-2R will deplete only the activated T_H cells.

15. The use of soluble CTLA4 or anti-CD40 ligand to promote acceptance of allografts is based on the requirement of a T cell for a costimulatory signal when its receptor is bound. Even if the recipient T cell recognizes the graft as foreign, the presence of CTLA4 or anti-CD40L will prevent the T cell from becoming activated because it does not receive a second signal through the CD40 or CD28 receptor (see Figure 16-17). Instead of becoming activated, T cells stimulated in the presence of these blocking molecules become anergic. The advantage of using soluble CTLA4 or anti-CD40L is that these molecules affect only those T cells involved in the reaction against the allograft. These allograft-specific T cells will become anergic, but the general population of T cells will remain normal. More general immunosuppressive measures, such as the use of CsA or FK506, cause immunodeficiency and subsequent susceptibility to infection.

16. Azathioprine is a mitotic inhibitor used to block proliferation of graft-specific T cells. Cyclosporin A, FK506 (tacrolimus), and rapamycin (sirolimus) are fungal metabolites that block activation and proliferation of resting T cells. Ideally, if early rejection is inhibited by preventing a response by specific T cells, these cells may be rendered tolerant of the graft over time. Lowering the dosage of the drugs is desirable because of decreased side effects in the long term.

Clinical Focus Answer The ideal animal for producing organs for xenotransplantation would have body size roughly equivalent to that of humans and could be genetically altered to eliminate any antigens that cause acute rejection. It should be free of any disease that can be passed to humans. The test of the organs must include transplantation to nonhuman primates and observation periods that are sufficiently long to ascertain that the organ remains fully functional in the new host and that no disease is transmitted.

Chapter 17

Infectious Diseases

Clinical Focus Answer

1. (a) Because the infected target cells expressed H-2^k MHC molecules but the primed T cells were H-2^k restricted.

(b) Because the influenza nucleoprotein is processed by the endogenous processing pathway and the resulting peptides are presented by class I MHC molecules. (c) Probably because the transfected class I D^b molecule is only able to present peptide 365–380 and not peptide 50–63. Alternatively, peptide 50–63 may not be a T-cell epitope. (d) These results suggest that a cocktail of several immunogenic peptides would be more likely to be presented by different MHC haplotypes in humans and would provide the best vaccines for humans.

2. Nonspecific host defenses include ciliated epithelial cells, bactericidal substances in mucous secretions, complement split products activated by the alternative pathway that serve both as opsonins and as chemotactic factors, and phagocytic cells.

3. Specific host defenses include humoral immunity which targets the destruction of extracellular infections (bacterial, fungal, or parasitic) or neutralization of all types of pathogens during extracellular stages, CTLs that identify and eliminate virally infected host cells, and T helper cells that secrete cytokines to assist other leukocytes in the elimination of both intracellular and extracellular pathogens.

4. Humoral antibody peaks within a few days of infection and binds to the influenza HA glycoprotein, blocking viral infection of host epithelial cells. Because the antibody is strain specific, its major role is in protecting against re-infection with the same strain of influenza.

5. (a) African trypanosomes are capable of antigenic shifts in the *variant surface glycoprotein* (VSG). The antigenic shifts are accomplished as gene segments encoding parts of the VSG are duplicated and translocated to transcriptionally active expression sites. (b) *Plasmodium* evades the immune system by continually undergoing maturational changes from sporozoite to merozoite to gametocyte, allowing the organism to continually change its surface molecules. In addition, the intracellular phases of its life cycle reduce the level of immune activation. Finally, the organism is able to slough off its circumsporozoite coat after antibody binds to it. (c) Influenza is able to evade the immune response through frequent antigenic changes in its hemagglutinin and neuraminidase glycoproteins. The antigenic changes are accomplished by the accumulation of small point mutations (antigenic drift) or through genetic reassortment of RNA between influenza virions from humans and animals (antigenic shift).

6. (a) IA^b. (b) Because antigen-specific, MHC-restricted T_H cells participate in B-cell activation.

7. (a) BCG (Bacille Calmette-Guérin). (b) Antigenic shift; antigenic drift. (c) Gene conversion. (d) Tubercles; T_H cells; activated macrophages. (e) Toxoid. (f) Interferon-α; interferon-γ. (g) Secretory IgA. (h) IL-12; IFN-γ.

8. Most fungal infections prevalent in the general population do not lead to severe disease and are dealt with by innate immune mechanisms and lead to protective adaptive responses. Problematic fungal infections are more commonly seen in those with some form of

immunodeficiency, such as patients with HIV/AIDS or those with immunosuppression caused by therapeutic measures.

9. One possible reason for the emergence of new pathogens is the crowding of the world's poorest populations into very small places within huge cities, because population density increases the spread of disease. Another factor is the increase in international travel. Other features of modern life that may contribute include mass distribution of food, which exposes large populations to potentially contaminated food, and unhygienic food preparation.

10. (a) Influenza virus changes surface expression of neuraminidase and hemagglutinin. (b) Herpesviruses remains dormant in nerve cells. Later emergence can cause outbreaks of cold sores or shingles (chickenpox virus). (c) *Neisseria* secretes proteases that cleave IgA. (d) False. (e) Several gram-positive bacteria resist complement-mediated lysis. (f) Influenza virus accumulates mutations from year to year. (g) False.

11. (a) No, IgE is raised against allergens and some parasites. (b) No, autoreactive T cells are activated only by intracellular infections. The statements in (c) through (f) are correct.

12. (a) Large, granular cells such as mast cells and eosinophils. Neutrophils and macrophages will also be involved. (b) Therapeutic cytokines such as IL-4 would help encourage IgE and the T_H2 response which is already present. However, cytokines that drive a T_H1 response, such as IL-12 or IFN-γ, may be more beneficial to longer-term immunity.

13. The answer comes from the concept of original antigenic sin, which posits that we only mount a primary response once we have exhausted the potential to use memory cells to eradicate the infection. Since most of our first encounters with influenza will vary, the years in which "all" of the key influenza epitopes are significantly "new" to each of us will also vary (see Figure 17-4b). It is only in these years that we experience a new primary response to influenza virus, and therefore symptoms of the flu are most severe.

Vaccines

Clinical Focus Answers Any connection between vaccination and a subsequent adverse reaction must be evaluated by valid clinical trials involving sufficient numbers of subjects in the control group (those given a placebo) and experimental group (those receiving the vaccine). This is needed to give a statistically correct assessment of the effects of the vaccine versus other possible causes for the adverse event. Such clinical studies must be carried out in a double-blind manner; that is, neither the subject nor the caregiver should know who received the vaccine and who received the placebo until the end of the observation period. In the example cited, it is possible that the adverse event (increased incidence of arthritis) was caused by an infection occurring near the time when the new vaccine was administered. Determining the precise cause of this side effect may not be possible, but ascertaining whether it is likely to be caused by this vaccine is feasible by appropriate studies of the vaccinated and control populations.

1. (a) True. (b) True. (c) False: Because DNA vaccines allow prolonged exposure to antigen, they are likely to generate immunologic memory. (d) True. (e) False: A DNA vaccine contains the gene encoding an entire protein antigen, which most likely contains multiple epitopes.

2. Because attenuated organisms are capable of limited growth within host cells, they are processed by the cytosolic pathway and presented on the membrane of infected host cells together with class I MHC molecules. These vaccines, therefore, usually can induce a cell-mediated immune response. The limited growth of attenuated organisms within the host often eliminates the need for booster doses of the vaccine. Also, if the attenuated organism is able to grow along mucous membranes, then the vaccine will be able to induce the production of secretory IgA. The major disadvantage of attenuated whole-organism vaccines is that they may revert to a virulent form. They also are more unstable than other types of vaccines, requiring refrigeration to maintain their activity.

3. (a) The antitoxin was given to inactivate any toxin that might be produced if *Clostridium tetani* infected the wound. The antitoxin was necessary because the girl had not been previously immunized and, therefore, did not have circulating antibody to tetanus toxin or memory B cells specific for tetanus toxin. (b) Because of the treatment with antitoxin, the girl would not develop immunity to tetanus as a result of the first injury. Therefore, after the second injury 3 years later, she will require another dose of antitoxin. To develop long-term immunity, she should be vaccinated with tetanus toxoid.

4. The Sabin polio vaccine is live and attenuated, whereas the Salk vaccine is heat killed and inactivated. The Sabin vaccine thus has the usual advantages of an attenuated vaccine compared with an inactivated one (see answer 2). Moreover, since the Sabin vaccine is capable of limited growth along the gastrointestinal tract, it induces production of secretory IgA. The attenuated Sabin vaccine can cause life-threatening infection in individuals, such as children with AIDS, whose immune systems are severely suppressed. Now that polio is rarely if ever seen in the United States, continuing use of a vaccine with the potential to revert to a more virulent form introduces an unwarranted element of risk to both the vaccinee and others who might contract the disease from them.

5. The virus strains used for the nasally administered vaccines are temperature-sensitive mutants that cannot grow at human body temperature (37 °C). The live attenuated virus can grow only in the upper respiratory tract, which is cooler, inducing protective immunity. These mutant viruses cannot grow in the warmer environment of the lower respiratory tract, where they could replicate and mutate into a disseminated influenza infection.

6. T-cell epitopes generally are internal peptides, which commonly contain a high proportion of hydrophobic residues. In contrast, B-cell epitopes are located on an antigen's surface, where they are accessible to antibody, and contain a high proportion of hydrophilic residues. Thus, synthetic hydrophobic peptides are most likely to

represent T-cell epitopes and induce a cell-mediated response, whereas synthetic hydrophilic peptides are most likely to represent accessible B-cell epitopes and induce an antibody response.

7. When the majority of a population is immune to a particular pathogen—that is, there is herd immunity—then the probability of the few susceptible members of the population contacting an infected individual is very low. Thus, susceptible individuals are not likely to become infected with the pathogen. If the number of immunized individuals decreases sufficiently, most commonly because of reduction in vaccination rates, then herd immunity no longer operates to protect susceptible individuals and infection may spread rapidly in a population, leading to an epidemic.

8. In this hypothetical situation, the gene can be cloned into an expression system and the protein expressed and purified in order to test it as a recombinant protein vaccine. Alternatively, the gene can be cloned into a plasmid vector that can be injected directly and tested as a DNA vaccine. Use of the cloned gene as a DNA vaccine is more efficient, because it eliminates the steps required for preparation of the protein and its purification. However, the plasmid containing the gene for the protective antigen must be suitably purified for use in human trials. DNA vaccines have a greater ability to stimulate both the humoral and cellular arms of the immune system than protein vaccines do and thus may confer more complete immunity. The choice must also take into consideration the fact that recombinant protein vaccines are in widespread use whereas DNA vaccines for human use are still in early test phases.

9. Pathogens with a short incubation period (e.g., influenza virus) cause disease symptoms before a memory-cell response can be induced. Protection against such pathogens is achieved by repeated reimmunizations to maintain high levels of neutralizing antibody. For pathogens with a longer incubation period (e.g., polio virus), the memory-cell response is rapid enough to prevent development of symptoms, and high levels of neutralizing antibody at the time of infection are unnecessary.

10. Bacterial capsular polysaccharides, inactivated bacterial exotoxins (toxoids), and surface protein antigens. The latter two commonly are produced by recombinant DNA technology. In addition, the use of DNA molecules to direct synthesis of antigens on immunization is being evaluated.

11. A possible loss of herd immunity in the population. Even in a vaccinated population of children, a small percentage may have diminished immunity to the diseased target due to differences among MHC molecule expression in a population, providing a reservoir for the disease. In addition, most vaccinated individuals, if exposed to the disease, will develop mild illness. Exposure of unvaccinated individuals to either source of disease would put them at risk for serious illness. Epidemics within adult populations would have more serious consequences, and infant mortality due to these diseases would increase.

12. (a) 1 or 2. (b) 2. (c) 3. (d) 4. (e) 1. (f) 2. (g) 2.

13. (a) No. The antisera you received 1 year ago protected you temporarily but those antibodies are now gone and you have no memory B cells to produce new antibodies during this second exposure. (b) Antibodies in the antiserum bound to the snake venom and neutralized its ability to cause damage. Phagocytes then engulfed and destroyed this antibody-coated venom. Because the snake venom was coated with antibodies, naïve B cells were not activated during this first exposure and therefore no adaptive immune response was mounted (see Figure 17-4a). (c) Equally sensitive. There are no residual cells or antibodies that were involved in the original encounter with this snake venom and, therefore, no recall response.

Analyze the Data (a) The pSG5DNA-Bcl-xL targeting calreticulin (CRT) and LAMP-1 are the most effective vaccines at inducing CD8$^+$ T cells to make IFN-γ. The pSG5DNA-Bcl-xL targeting HSP70 also activated CD8$^+$ T cells. However, the pSG5 construct without the anti-apoptosis gene targeting CRT also induced a good CD8$^+$ T-cell response. (b) Calreticulin is a chaperone protein associated with partially folded class I MHC molecules in the endoplasmic reticulum. Associating E7 antigen with the chaperone may enhance loading of class I MHC molecules with E7, making the antigen more available to T cells once it is expressed on cells. (c) The DNA vaccines co-injected with pSG5DNA-Bcl-xL were effective in inducing CD8$^+$ T cells, possibly because the expression of anti-apoptotic genes in dendritic cells allowed those cells to survive longer and present antigen to T cells for a longer time. The longer they presented antigen, the longer the host would respond to produce antigen-specific T cells. (d) The data in part (b) of the figure indicate that in the absence of CD4$^+$ helper T cells (in CD4 knockout mice), there is ineffective activation of CD8$^+$ T cells. Therefore, T-cell help is necessary to activate the CD8 response; by targeting antigen to MHC II, you more efficiently activate helper T cells. The Sig/E7/LAMP-1 construct was necessary because most antigens presented by MHC II molecules are processed by the endocytic pathway, and the Sig/E7/LAMP-1 construct targets antigen to the Golgi, where the E7 peptides can be exchanged with CLIP and inserted into MHC II. (e) Helper T cells (poor response in the CD4 knockout mice), long-lived dendritic cells (immunization with the pSG5DNA-Bcl-xL improves the response), antigen (the absence of peptide failed to induce a response), and the targeting of antigen to MHC II (immunizing with the Sig/E7/LAMP-1 construct is the only one that induces a CD8$^+$ T-cell response).

Chapter 18

Clinical Focus Answers Before nevirapine can be universally administered to all mothers at delivery, it is necessary to know that the benefit of this treatment outweighs the risk. The benefit of the drug, as learned from studies already conducted, is that it reduces significantly the transmission of HIV to infants born of infected mothers. A study of the risk of nevirapine administration to normal mothers and their infants must be carried out, and only if there are minimal or no side effects can universal treatment be recommended.

1. (a) True. (b) False: X-linked agammaglobulinemia is characterized by a reduction in B cells and an absence of

immunoglobulins. (c) False: Phagocytic defects result in recurrent bacterial and fungal infections. (d) True. (e) True. (f) True. (g) True. (h) True. (i) False: These children are usually able to eliminate common encapsulated bacteria with antibody plus complement but are susceptible to viral, protozoan, fungal, and intracellular bacterial pathogens, which are eliminated by the cell-mediated branch of the immune system. (j) False: Humoral immunity also is affected because class II-restricted T_H cells must be activated for an antibody response to occur.

2. (a) 3. (b) 4. (c) 2. (d) 5. (e) 1. (f) 2.

3. The defect in X-linked hyper-IgM syndrome is in CD40L expressed on B cells. CD40L mediates binding of B cells to T cells and sends costimulatory signals to the B cell for class switching. Without CD40L, class switching does not occur, and the B cells do not express other antibody isotypes.

4. As discussed in Chapter 9, the thymus is the location of differentiation and maturation of helper and cytotoxic T cells. Positive and negative selection occur in this organ as well. Thus, the thymocytes produced in the bone marrow of patients with DiGeorge syndrome do not have the ability to mature into effector cell types. In Chapter 2, we noted that the thymus decreases in size and function with age. In the adult, effector-cell populations have already been produced (peak thymus size occurs during puberty); therefore, a defect after this stage would cause less severe T-cell deficiency.

5. As long as the mother is not immunocompromised, maternal antibodies in the mother's serum will be passively transferred to the baby in utero. After birth, these IgG molecules will supply the newborn with passive immune protection from many common bacterial infections, which can then be quickly dispatched by antibody-mediated mechanisms. In immune competent babies, these maternal antibodies will eventually be replaced by the child's own immune response to the infectious agents they encounter. In children with SCID, this does not occur, and they gradually become more susceptible to bacterial infections as their maternally derived antibodies disappear.

6. Encapsulated bacteria, such as staphylococci, streptococci, and pneumococci, all require antibody-mediated opsonization for their clearance. These types of bacteria are therefore a particular problem for individuals who lack humoral immunity.

7. (a) Leukocyte adhesion deficiency results from biosynthesis of a defective β chain in LFA-1, CR3, and CR4, which all contain the same β chain. (b) LFA-1 plays a role in cell adhesion by binding to ICAM-1 expressed on various types of cells. Binding of LFA-1 to ICAM-1 is involved in the interactions between T_H cells and B cells, between CTLs and target cells, and between circulating leukocytes and vascular endothelial cells.

8. The high affinity IL-2 receptor is composed of two chains: the α chain and the common γ chain. Later, the γ chain was discovered to be a component of the receptors for five other cytokines: IL-4, -7, -9, -15, and -21. During

hematopoiesis, developing lymphocytes require IL-7 signaling, and therefore the complete IL-7R, in order to develop properly. Without the common γ chain, this does not occur and the development of lymphocytes is blocked.

9. Some components of the immune system are responsible for regulating or suppressing the activity of leukocytes (e.g., T_{REG} cells). When these pathways are defective, overactive immune responses can occur, leading to breaks in self tolerance that lead to attacks on self molecules, or autoimmune syndromes. One example is a disorder called immune dysregulation, polyendocrinopathy, enteropathy, X-linked (IPEX) syndrome, in which the gene encoding the transcription factor that controls development of T_{REG} cells, *Foxp3*, is defective. Another example is autoimmune polyendocrinopathy and ectodermal dystrophy (APECED), a disorder that arises from defects in the *AIRE* gene, encoding a transcription factor found in the thymus. The AIRE protein is responsible for the expression of tissue-restricted antigens. Without this protein, the negative selection of T cells in the thymus that recognize self antigen is disrupted and autoreactive T cells emerge and instigate organ-specific autoimmune attacks.

10. (a) The rearranged heavy-chain genes in SCID mice lack the D and/or J gene segments. (b) According to the model of allelic exclusion discussed in Chapter 5, a productive heavy-chain gene rearrangement must occur before κ-chain genes are rearranged. Since SCID mice lack productive heavy-chain rearrangement, they do not attempt κ light-chain rearrangement. (c) Yes: The rearranged heavy-chain gene would be transcribed to yield a functional μ heavy chain. The presence of the heavy chain then would induce rearrangement of the κ-chain gene (see Figure 7-11).

11. (a) False: HIV-1 is believed to have evolved from a strain of SIV that jumped the species barrier from African chimpanzees to humans, although HIV-2 is thought to have arisen from a separate but similar transfer from SIV-infected sooty mangabeys. (b) False: HIV-1 infects chimpanzees but does not cause immune suppression. (c) True. (d) False: Zidovudine (or azidothymidine, AZT) acts at the level of reverse transcription of the viral genome, whereas saquinavir is an inhibitor of the viral protease. (e) True. (f) False: A diagnosis of AIDS is based on both a low CD4$^+$ T-cell count (below 200 cells/μl, or 14%) and the presence of certain AIDS-defining conditions (see Table 18-3). (g) False: The PCR detects HIV proviral DNA in latently infected cells. (h) True.

12. The most likely reason for T-cell depletion in AIDS is the cytopathic effects of HIV infection. If the amount of virus in circulation is decreased by the use of antiviral agents, the number of T cells will increase.

13. No. For most of the asymptomatic phase of HIV infection, viral replication and the CD4$^+$ T-cell number are in a dynamic equilibrium; the level of virus and the percentage of CD4$^+$ T cells remains relatively constant.

14. An increase in the viral load and a decrease in CD4$^+$ T-cell levels indicates that HIV infection is progressing

from the asymptomatic phase into AIDS. Often, infection by an opportunistic agent occurs when the viral load increases and the CD4$^+$ T-cell level drops. The disease AIDS in an HIV-1 infected individual is defined by a CD4$^+$ T-cell count of less than 200 cells/μl and/or the presence of certain opportunistic infections (see Table 18-3).

15. Skin-test reactivity is monitored to indicate the functional activity of T$_H$ cells. As AIDS progresses and CD4$^+$ T cells decline, there is a decline in skin-test reactivity to common antigens.

16. Receptors for certain chemokines, such as CXCR4 and CCR5, also function as coreceptors for HIV-1. The chemokine that is the normal ligand for the receptor competes with the virus for binding to the receptor and can thus inhibit cell infection by blocking attachment of the virus (see Figure 18-15). Cytokines that activate T cells stimulate infection because they increase expression of receptors used by the virus.

17. No. There are latently infected cells that reside in lymph nodes or in other sites. These cells can be activated and begin producing virus, thus causing a relapse of the disease. In a true cure for AIDS, the patient must be free of all cells containing HIV DNA.

18. Patient L. S. fits the definition of AIDS, whereas patient B. W. does not. The clinical diagnosis of AIDS among HIV-infected individuals depends on both the T-cell count, or percentage, and the presence of various indicator diseases. See Table 18-3.

19. (a) CVID lowers the number and percentage of T helper cells. (b) Table 1 indicates that there are fewer naïve CD8$^+$ T cells in CVID patients. This could mean chronic activation in CVID has made these T cells less dependent on IL-2. Alternatively, the figure accompanying this question shows that bone marrow cells from CVID patients make less IL-2. If this is also true of T$_H$ cells, or there are fewer T$_H$ cells (as seen in the table accompanying this question), then the generation of CTLs may be impaired. This example of the complexity of the immune system demonstrates that specific responses are not always easily predictable. (c) False. The data in the figure accompanying this question shows that the kinetics and overall production of TNF-α is greater in CVID than in normal patients, but IL-2 is lower. Thus, there is no consistent impact. Based on what we know about TNF-α activity, it might be responsible for increased pathology or cell death, perhaps explaining the loss of both CD4$^+$ and CD8$^+$ cells in CVID patients. (d) According to information in Chapter 18, CVID lowers IgG, IgA, and IgM. However, based on the data presented in the table accompanying this question, one might predict that in the absence of T-cell help, there will be a significant impact on class switching.

Chapter 19

Clinical Focus Answers Because cervical cancer is linked to HPV infection, a preventive cancer vaccine may be possible; if HPV infection is prevented, then cervical cancer should also be prevented. Other cancers that may be targets for such

prevention include adult T-cell leukemia/lymphoma and the liver cancer that is linked to hepatitis B infection. Most cancers have not been clearly linked to an agent of infection, and therefore a preventive vaccine is not an option.

1. (a) False: Hereditary retinoblastoma results from inactivation of both alleles of *Rb*, a tumor-suppressor gene. (b) True. (c) True. (d) True. (e) False: Some oncogenic retroviruses do not have viral oncogenes. (f) True.

2. Cells of the pre-B-cell lineage have rearranged the heavy-chain genes and express the μ heavy chain in their cytoplasm (see Figure 10-6). You could perform Southern-blot analysis with a Cμ probe to see whether the heavy-chain genes have rearranged. You could also perform fluorescent staining with antibody specific for the cytoplasmic μ heavy chain.

3. (a) Early-stage melanoma cells appear to be functioning as APCs, processing the antigen by the exogenous pathway and presenting the tetanus toxoid antigen together with the class II MHC DR molecule. (b) Advanced-stage melanoma cells might have a reduction in the expression of class II MHC molecules, or they might not be able to internalize and process the antigen by the exogenous route. (c) Since the paraformaldehyde-fixed early melanoma cells could present processed tetanus toxoid, they must express class II MHC molecules on their surface. (d) Stain the early and advanced melanoma cells with fluorescent mAb specific for class II MHC molecules.

4. Tumor antigens may be encoded by genes expressed only by tumors, may be products of genes overexpressed by the tumor or genes normally expressed only at certain stages of differentiation, or may be products of normal genes that are altered by mutation. In some cases, viral products may be tumor antigens.

5. IFN-α, IFN-β, and IFN-γ enhance the expression of class I MHC molecules on tumor cells, thereby increasing the CTL response to tumors. IFN-γ also increases the activity of CTLs, macrophages, and NK cells, each of which plays a role in the immune response to tumors. TNF-α and Lymphotoxin-α have direct anti-tumor activity, inducing hemorrhagic necrosis and tumor regression. IL-2 activates TIL cells, which have anti-tumor activity.

6. (a) Melanoma cells transfected with the CD80/86 gene are able to deliver the costimulatory signal necessary for activation of CTL precursors to effector CTLs, which then can destroy tumor cells (Figure 19-12). Melanoma cells transfected with the GM-CSF gene secrete this cytokine, which stimulates the activation and differentiation of APCs, especially DCs, in the vicinity. The DCs then can present tumor antigens to T$_H$ cells and CTL precursors, enhancing the generation of effector CTLs (see Figure 19-10). (b) Because some of the tumor antigens on human melanomas are expressed by other types of cancers, transfected melanoma cells might be effective against other tumors carrying identical antigens.

7. (a) 4. (b) 2. (c) 1. (d) 5. (e) 3. (f) 9. (g) 8. (h) 7. (i) 6.

Chapter 20

1. If I wished to precipitate my antigen, I might elect to use a polyclonal preparation, as it would contain antibodies toward multiple different determinants on the antigen and therefore many antibody molecules could bind per antigen molecule, maximizing the chances that at least some of the antibodies could bind more than one antigen and facilitate precipitation.

2. With time, the population of B-cell clones that respond to an antigen in an individual will change. Some B-cell clones will die, and different clones will predominate. Overall, the affinity of the antibodies in the serum will increase according to the methods described in Chapter 12. However, this means that the proteins with which individual antibodies will cross-react will change as the range of binding sites modulates, and this is what has happened in your experiment.

3. (a) Ideally, polyclonal. I will have the most effective agglutination if antibodies can cross-link multiple sites on the bacterial surface. However, a monoclonal antibody would also work, as most antigens are repeated many times on the bacterial surface. (b) Ideally, polyclonal. To form a precipitate, I will need to cross-link multiple proteins. If each only has a single site at which an antibody can bind, a bivalent antibody can cross-link only two proteins, and that would be insufficient to create a precipitate. If I am precipitating a protein with multiple copies of the same site, then monoclonal antibodies could still work. (c) Here, either would work. The antibody binds to the band, localizing an enzymic reaction at the band and causing substrate conversion to product. A polyclonal antibody mixture would have the advantage that different antibodies could bind at different antigenic determinants on the target protein and therefore could give rise to a stronger signal. However, different bleeds of polyclonal sera would have different levels of cross-reactivity with other, structurally similar determinants on other proteins. Monoclonal antibodies will bind to predictable determinants, and, although they might still cross-react with structurally similar determinants on other proteins, those cross-reactivities are predictable and will not change from batch to batch. (d) Here, one would use two monoclonal antibodies directed toward different determinants on the cytokine. If the capture antibody were to be polyclonal, such that all the binding sites on the cytokine were bound, this would compete with a polyclonal detection antibody. (e) Monoclonal. Here, reproducibility of reactivity and cross-reactivity is demanded for clinical safety.

4. (a) Yes. There is some evidence of hemagglutination in the first well, which represents an antiserum dilution of 1 in 10. (b) By 42 days of age, the baby's serum has the same capacity for hemagglutination inhibition as the positive-control sample. (c) No. These could be maternal antibodies absorbed through breast milk or colostrum.

5. RIAs are orders of magnitude more sensitive than are ELISAs with chromogenic substrates. And yes, chemiluminogenic substrates allow more amplification and greater sensitivity, resulting in an assay that matches the RIA in its ability to measure low concentrations of antibodies or antigens.

6. ELISPOT assays measure the number of cells capable of secreting particular molecules, such as antibodies or cytokines. These assays therefore require that when substrate is converted to product, the product remains at the location where the substrate-to-product reaction occurred. Products in ELISPOT reactions are therefore insoluble. Similarly, Western blot assays use antibodies to determine the location of particular bands on a gel. Again, one requires that the product of the enzymic assay remains localized precisely where the enzyme is bound.

7. (a) Since cost is an issue, and I already have the antibody labeled, I would probably use equilibrium dialysis. (b) Surface plasmon resonance experiments will enable me to measure the association, as well as the dissociation rate constant, of the binding reaction between my antigen and antibody.

8. By using longer-wavelength, lower-energy light, two-photon microscopy induces less photobleaching of the tissue preparation. Further, since no fluorescence is emitted unless two exciting photons simultaneously impinge on the sample, the focal plane of the observed images can be more tightly defined.

9. Magnetic-activated cell sorting is most useful for batch separations of cell populations. It is faster than fluorescence-based methods, but not quite as precise. I would choose fluorescence-based sorting in situations when I need to be sure that there are no contaminating cells, because in FACS, cells are quite literally separated one at a time.

10. (a) PI3 kinase and IKKα (b) NFκB. NFκB activity is normally inhibited by the binding of the inhibitor of NFκB, IκB. When cells are activated by particular signals, the kinase IKK phosphorylates IκB, targeting it for destruction and enabling the passage of NFκB into the nucleus, thus activating transcription. (c) I expect one or more of the intermediate proteins to bear pleckstrin homology (PH) domains, as these domains bind to PIP_3, which is the product of the PI3 kinase catalyzed reaction.

11. There are many ways to test this idea, but we will offer just one here, which makes use of adoptive transfer and fluorescent labeling.

    Generate B-1 B cells in culture specific for the virus in question, and load them with CFSE. Inject the cells into the tail of a mouse infected with the virus, allow the cells to home for 12 to 24 hours, and then sacrifice the mouse and search for CFSE-labeled cells in the lung using tissue sections and immunofluorescence.

12. I will use a tamoxifen/Cre fusion protein whose expression is under the control of a B-cell-specific promoter, such as that controlling the expression of CD19. Therefore, Cre will be active only in B cells and only if I add the tamoxifen and I can therefore control exactly when the gene targeting will occur in reference to antigen immunization.

    I also need to generate an inactive, truncated form of the gene in question that is flanked by *loxP* sites. Ideally, the gene will have a selectable marker, such as neomycin resistance, and also will bear a sequence external to the *loxP* sites that will control for insertion of the gene into the correct location (see Figure 20-25a, left).

INDEX

Page numbers followed by f indicate figures; those followed by t indicate tables.